W9-BNA-946

PERMUTATIONS

The number of permutations of n elements taken r at a time, where $r \leq n$, is

$$_nP_r = \frac{n!}{(n-r)!}.$$

COMBINATIONS

The number of combinations of n elements taken r at a time, where $r \leq n$, is

$$\binom{n}{r} = {}_nC_r = \frac{n!}{(n-r)!r!}.$$

BINOMIAL PROBABILITY

If p is the probability of success in a single trial of a binomial experiment, the probability of x successes and $n - x$ failures in n independent repeated trials of the experiment is

$$\binom{n}{x} p^x(1-p)^{n-x}.$$

MEAN

The mean of the n numbers, $x_1, x_2, x_3, \ldots, x_n$, is

$$\bar{x} = \frac{x_1 + x_2 + \cdots + x_n}{n} = \frac{\Sigma(x)}{n}.$$

SAMPLE STANDARD DEVIATION

The standard deviation of a sample of n numbers, $x_1, x_2, x_3, \ldots, x_n$, with mean $\bar{x}$, is

$$s = \sqrt{\frac{\Sigma(x-\bar{x})^2}{n-1}}.$$

BINOMIAL DISTRIBUTION

Suppose an experiment is a series of n independent repeated trials, where the probability of a success in a single trial is always p. Let x be the number of successes in the n trials. Then the probability that exactly x successes will occur in n trials is given by

$$\binom{n}{x} p^x(1-p)^{n-x}.$$

The mean μ and variance σ^2 of a binomial distribution are, respectively,

$$\mu = np \quad \text{and} \quad \sigma^2 = np(1-p).$$

The standard deviation is

$$\sigma = \sqrt{np(1-p)}.$$

8th EDITION

Mathematics with Applications

In the Management, Natural, and Social Sciences

8th EDITION

Mathematics with Applications

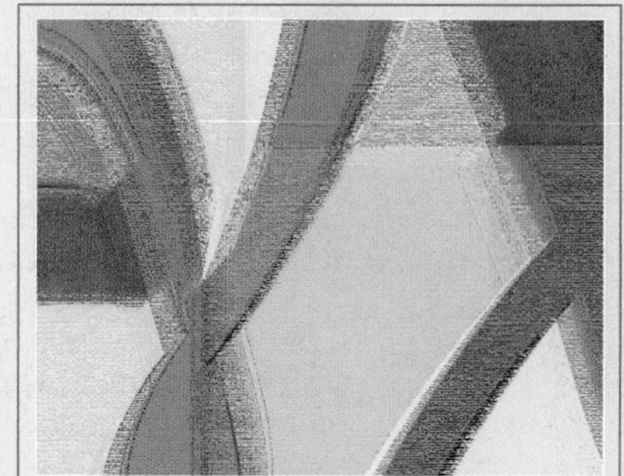

In the Management, Natural, and Social Sciences

MARGARET L. LIAL

AMERICAN RIVER COLLEGE

THOMAS W. HUNGERFORD

CLEVELAND STATE UNIVERSITY

Boston San Francisco New York
London Toronto Sydney Tokyo Singapore Madrid
Mexico City Munich Paris Cape Town Hong Kong Montreal

Publisher: Greg Tobin
Sponsoring Editor: William Hoffman
Executive Project Manager: Christine O'Brien
Assistant Editor: Stefanie Borge
Project Manager: Sandi Goldstein
Senior Production Supervisor: Karen Wernholm
Production Coordination: Elm Street Publishing Services, Inc.
Marketing Manager: Michael Boezi
Marketing Coordinator: Weslie Lewis
Media Producer: Ruth Berry
Software Editor/InterAct Math: Malcolm Litowitz
Software Editor/TestGen-EQ: David Malone
Manufacturing Buyer: Evelyn Beaton
Senior Designer: Barbara T. Atkinson
Text Design: Silvers Design
Cover Design: Leslie Haimes
Compositor: Nesbitt Graphics, Inc.
Illustrations: Techsetters, Inc.
Cover Photo: ©2002 Stone/Lonnie Duka

Photo Credits:
page 10 ©Chuck Savage/corbisstockmarket.com; **page 39** ©David Stoecklein/corbisstockmarket.com; **page 58** ©2002 PhotoDisc; **page 64** ©Mug Shots/corbisstockmarket.com; **page 95** ©Fritz Hoffmann/The Image Works; **page 99** AP/Wide World Photos; **page 107** ©Kathy Ferguson-Johnson/Photo Edit; **page 128** ©James L. Amos/CORBIS; **page 150** AP/Wide World Photos; **page 186** ©Reuters NewMedia Inc./CORBIS; **page 198** ©2002 Photo Disc; **page 215** ©Graham Neden/CORBIS; **page 225** Courtesy of NASA; **page 229** ©Brandon D. Cole/CORBIS; **page 266** ©Mug Shots/corbisstockmarket.com; **page 271** ©AP/Wide World Photos; **page 272** ©PhotoDisc; **page 309** ©Richard Hamilton Smith/CORBIS; **page 324** ©Owen Franken; **page 288** ©Lee Snider/CORBIS; **page 341** ©D. Boone/CORBIS; **page 359** ©Corbis Corporation/Tom Stewart, 2001; **page 384** ©Bob Rowan; Progressive Image/CORBIS; **page 409** ©PhotoDisc; **page 413** ©2002 PhotoDisc; **page 422** ©Gail Mooney/CORBIS; **page 438** ©Mary Boucher; **page 462** ©Corbis Corporation/Ariel Skelley, 2001; **page 479** ©Billy E. Barnes/PhotoEdit; **page 502** ©Mary Steinbacher/PhotoEdit; **page 507** ©Mug Shots/corbisstockmarket.com; **page 523** ©Kevin R. Morris/CORBIS; **page 543** ©Bob Daemmrich/The Image Works; **page 553** ©2002 PhotoDisc; **page 582** ©V.C.L./FPG; **page 593** AP/Wide World Photos; **page 594** ©David Frazier/Stone; **page 618** ©Sandy Felsenthal/CORBIS; **page 620** ©Myrleen Ferguson Cate/PhotoEdit; **page 657** ©2002 PhotoDisc; **page 692** ©NOVASTOCK/PhotoEdit; **page 710** ©SuperStock; **page 721** ©Alan Schein/corbisstockmarket.com; **page 735** ©2002 PhotoDisc; **page 749** ©Eyewire; **page 760** ©Macduff Everton/CORBIS; **page 792** ©David Young-Wolff/PhotoEdit; **page 817** ©Roy King/SuperStock; **page 819** ©2002 PhotoDisc; **page 831** ©PhotoDisc; **page 842** ©Grafton Marshall Smith/The Stock Market; **page 852** ©Tom Rosenthal/SuperStock; **page 853** ©Van Ravenswaay; Detley/CORBIS.

Library of Congress Cataloging-in-Publication Data

Lial, Margaret L.
Mathematics with applications.—8th ed. / Margaret L. Lial, Thomas W. Hungerford.
p. cm.
Includes bibliographical references and index.
ISBN 0-201-75529-7
1. Mathematics. I. Hungerford, Thomas W. II. Title.
QA37.3.L55 2002
510—dc21 2001056645

Copyright © 2003 Pearson Education, Inc. All rights reserved. No part of this publication may be reproduced, stored in a retrieval system, or transmitted, in any form or by any means, electronic, mechanical, photocopying, recording, or otherwise, without the prior written permission of the publisher. Printed in the United States of America.

3 4 5 6 7 8 9 10—QWT—050403

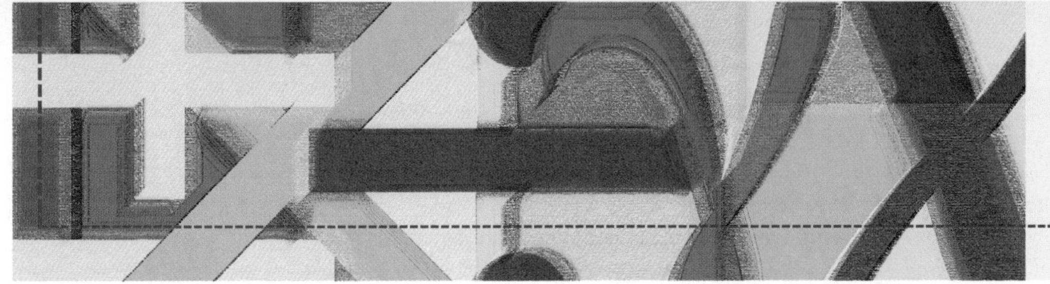

Contents

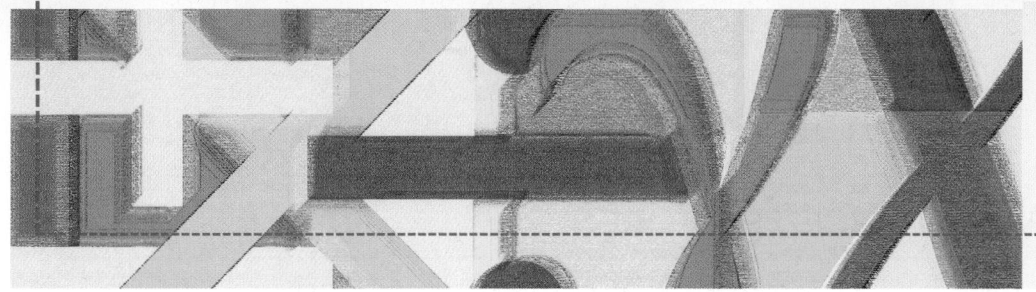

Preface

Like its predecessors, this edition of *Mathematics with Applications* is an applications-oriented text for students in business, management, social sciences, and health sciences. The only prerequisite is a course in algebra. Chapter 1 provides a thorough review of basic algebra for those students who need it.

The book is written at a level appropriate for its intended audience. We have done our best to present sound mathematics in an understandable manner, proceeding from the familiar to new material and from concrete examples to general rules and formulas. There is an ongoing focus on real-world problem solving, and almost every section includes relevant contemporary applications.

KEY ORGANIZATIONAL AND CONTENT CHANGES

The table of contents for this edition is substantially the same as that of the previous edition. Organizational changes are:

- Chapters 3 and 4 (Functions and Graphs, Polynomial and Rational Functions) of the previous edition were combined into a single chapter (Chapter 3, Functions and Graphs).
- The order of topics in Chapters 1, 2, and 9 is slightly changed.
- Many minor changes have been made throughout the book to improve clarity and coverage.

The major content changes are:

- A much more complete discussion of linear models appears in Section 2.3, including (optional) material on linear regression. Other regression models are mentioned at appropriate points later in the text.
- Section 4.4 has been substantially rewritten to give students more guidance in handling a variety of different types of logarithmic and exponential equations.
- Section 9.5 has been rewritten to clarify the properties of regular Markov chains.
- A new table summarizing common geometric area and volume formulas has been added to Appendix B.

NEW AND ENHANCED FEATURES

BALANCED APPROACH Multiple representations of a topic (symbolic, numerical, graphical, verbal) are given when appropriate. However, we do not believe that all representations are useful for all topics. So we are pedagogically selective in presenting alternative methods. Effective alternatives are discussed only when they are likely to increase student understanding.

TECHNOLOGY It is assumed that all students have a calculator that will handle exponential and logarithmic functions. Beyond that, however,

the use of technology in this text is optional.

Examples and exercises that definitely require some sort of technology (graphing calculators, spreadsheets, or other computer programs) are marked with the icon , so instructors who want to omit these discussions and exercises can easily do so. They should note, however, that some examples that do *not* require technology may include some mention of it or a graphing calculator screen (for instance, one showing the matrix associated with a system of linear equations).

Instructors who routinely use technology in their courses—and expect their students to do the same—will find more than enough material here to satisfy their needs. Here are some of the features they may want to incorporate into their courses.

- *Examples and Exercises marked with* A number of examples show students how various features of graphing calculators and spreadsheets can be applied to the topics in this book. These include, but are not limited to, graphing functions, solving equations, row-reducing matrices, finding linear regression models for data, statistical analysis, and approximations of definite integrals. There are also many exercises marked for technology use.
- *Technology Tips* These are placed at appropriate points in the text to inform students of various features of their graphing calculator or spreadsheet programs and to guide them in using these tools.
- *Graphing Calculator Appendix* Part 1 is a brief introduction to the features of graphing calculators that are relevant to topics in the text. Part 2 contains eleven programs for carrying out various procedures discussed in the text (including financial functions, a RREF program for TI-82 and Casio, and a new two-stage method program for linear programming problems with mixed constraints).

  An outline of the Graphing Calculator Appendix (including the complete list of programs) is on page 855. In order to facilitate easy updating of material, the full Appendix is on the text's Web site: www.aw.com/MWA8, where it can be downloaded and printed. Printed copies of the Appendix can be made available; contact your local Addison-Wesley sales representative.

REAL-DATA EXAMPLES AND APPLICATIONS Real-data exercises have long been a popular feature of this book. For this edition, almost all of these exercises have been updated and many new ones added. A significant number of real-data examples have also been introduced into the text.

CASE STUDIES A number of new end-of-chapter case studies have been added and others updated. The case studies are contemporary real-world applications of some of the mathematics presented in the chapter. They provide an (at least partial) answer to that old question, "What's this stuff good for?"

FULL-COLOR GRAPHICS AND PHOTOGRAPHY The new multicolor graphics in this edition not only make the book more attractive but are effectively used to enhance the mathematical exposition and to clarify many graphical presentations. Photographs now illustrate many of the applied exercises.

CONTINUING PEDAGOGICAL FEATURES

Popular features from earlier editions of *Mathematics with Applications* have been retained.

- Margin problems, keyed to the discussion in the text
- Extensive examples
- Boxed rules, definitions, and summaries
- "Cautions" that highlight common student difficulties and warn against errors
- An *Index of Applications* showing the wide variety of real-data applications in the text and allowing direct reference to particular topics
- Exercises keyed to the text
- Writing exercises (marked ✎)
- Connection exercises that involve material from earlier sections (marked )
- Exercises from CPA exams to give interested students valuable exposure to such exams
- End-of-chapter summaries and review exercises

COURSE FLEXIBILITY

This book can be used for a variety of courses, including the following.

Finite Mathematics and Calculus (one year or less). Use the entire book; cover topics from Chapters 1–4 as needed before proceeding to further topics.

Finite Mathematics (one semester or two quarters). Use as much of Chapters 1–4 as needed and then go into Chapters 5–10 as time permits and local needs require.

Calculus (one semester or quarter). Cover the precalculus topics in Chapters 1–4 as necessary and then use Chapters 11–14.

College Algebra with Applications (one semester or quarter). Use Chapters 1–8, with Chapters 7 and 8 being optional.

Chapter interdependence is as follows.

	Chapter	*Prerequisite*
1	Algebra and Equations	None
2	Graphs, Lines, and Inequalities	Chapter 1
3	Functions and Graphs	Chapters 1 and 2
4	Exponential and Logarithmic Functions	Chapter 3
5	Mathematics of Finance	Chapter 4
6	Systems of Linear Equations and Matrices	Chapters 1 and 2
7	Linear Programming	Chapters 3 and 6
8	Sets and Probability	None
9	Counting, Probability Distributions, and Further Topics in Probability	Chapter 8
10	Introduction to Statistics	Chapter 8
11	Differential Calculus	Chapters 1–4
12	Applications of the Derivative	Chapter 11
13	Integral Calculus	Chapters 11 and 12
14	Multivariate Calculus	Chapters 11–13

Contact your local Addison-Wesley sales representative to order a customized version of this text.

SUPPLEMENTS

For the Instructor

Instructor's Solutions Manual (ISBN 0-201-77009-1)
This manual contains detailed solutions to all text problems, suggested course outlines, and a chapter interdependence chart.

TestGen-EQ with QuizMaster (ISBN 0-201-79422-5)
Available on dual-platform Windows/Macintosh CD-ROM, TestGen enables instructors to build, edit, print, and administer tests using a computerized test bank of questions. The program also offers many options for organizing and displaying test banks and tests according to text chapter content. Tests can be printed or saved for online testing via a network or the Web, and the software can generate a variety of grading reports for tests and quizzes.

MyMathLab
MyMathLab is a complete online course available with this text and is ideal for helping students succeed in their course. This site offers a wide variety of resources, from dynamic multimedia—video clips, animations, audio explanations, and more—to an online grade book that allows students to track their performance on homework assignments and tests. With MyMathLab students can increase comprehension and get help with concepts they don't understand.

- The entire textbook is available online and is supplemented by multimedia content to help explain concepts.
- MyMathLab allows students to do practice work and to complete instructor-assigned tests and homework online. Based on these results, MyMathLab automatically builds an individual study plan that students can use to improve their skills.

- MyMathLab offers a full array of supplements such as work sheets, quizzes, and solution manuals and allows students to print them out and use them off-line.
- With MyMathLab, students can learn and practice with tutorial exercises tied directly to the exercises in the textbook.

MyMathLab requires a student access code. For more information, go to www.mymathlab.com.

For the Student

Student's Solutions Manual (ISBN 0-201-77010-5)
This manual contains detailed, carefully worked-out solutions to all odd-numbered section exercises and all Chapter Review and Case Study exercises.

InterAct Tutorial Web Site
www.interactmath.com
This interactive tutorial Web site provides algorithmically generated practice exercises that correlate directly to the exercises in the text. A detailed worked-out example and guided solution accompany each practice exercise. The Web site recognizes student errors and provides feedback.

InterAct MathXL®
www.mathxl.com
The MathXL Web site provides diagnostic testing and tutorial help, all online, using InterAct Math tutorial software and TestGen-EQ testing software. Students can take chapter tests correlated to this textbook, receive individualized study plans based on those test results, work practice problems and receive tutorial instruction for areas in which they need improvement, and take further tests to gauge their progress. Instructors can customize tests and track all student test results, study plans, and practice work. The site is free when an access code is bundled with a new text.

Addison-Wesley Math Tutor Center
The Addison-Wesley Math Tutor Center is staffed by qualified mathematics instructors who provide students with tutoring on examples and exercises answered at the back of the textbook. Tutoring is available via toll-free telephone, fax, e-mail, or whiteboard technology—which allows tutors and students to actually see the problems worked while they "talk" in real time over the Internet. This service is available five days a week, seven hours a day. For more information, go to www.aw.com/tutorcenter. An access card is required.

ACKNOWLEDGMENTS

We thank the following instructors, who reviewed the manuscript and made many helpful suggestions for improvement.

Erol Barbut, *University of Idaho*
Jean Davis, *Southwest Texas State University*
Frank Fitzgerald, *Boston University*
Kim Gregor, *Delaware Technical and Community College*

Joseph Guthrie, *University of Texas at El Paso*
Carol M. Hurwitz, *Manhattan College*
Michael J. Kallaher, *Washington State University*
Jeff Lee, *Texas Tech University*
James Mazzarella, *Holy Family College*
Shannon Michaux, *University of Colorado at Colorado Springs*
Kathy O'Dell, *University of Alabama at Huntsville*
Charles Odion, *Houston Community College*
Steven A. Schonefeld, *Tri-State University*
Surinder Sehgal, *Ohio State University*
Calvin Shipley, *Henderson State University*
Bhushan Wadhwa, *Cleveland State University*
Thelma West, *University of Louisiana at Lafayette*
Hing-Sing Yu, *University of Texas at San Antonio*
Cathy Zucco-Teveloff, *Trinity College*

We also wish to thank our accuracy checkers, who did an excellent job of checking both text and exercise answers: Elka Block, Tim Mogill, Frank Purcell, Deana Richmond, Laurie Semarne, and Tom Wegleitner. A second (and very helpful) accuracy check of the Exercises was done by Bill Fellers, Cindy Trimble, and the LaurelTech staff who prepared the *Solutions Manual* for the text. Special thanks go to Elka Block and Frank Purcell, whose careful proofreading has helped eliminate errors from the text; to Paul Van Erden, who created an accurate and complete index for us; and to Becky Troutman, who carefully compiled the Index of Applications.

We want to thank the staff of Addison-Wesley for their assistance with and contributions to this book, particularly Greg Tobin, Christine O'Brien, Stefanie Borge, Karen Wernholm, Barbara Atkinson, Michael Boezi, Weslie Lewis, and Ruth Berry. Finally, we wish to express our appreciation to freelance project editor Sandi Goldstein and to Phyllis Crittenden of Elm Street Publishing Services, Inc.

MARGARET L. LIAL
THOMAS W. HUNGERFORD

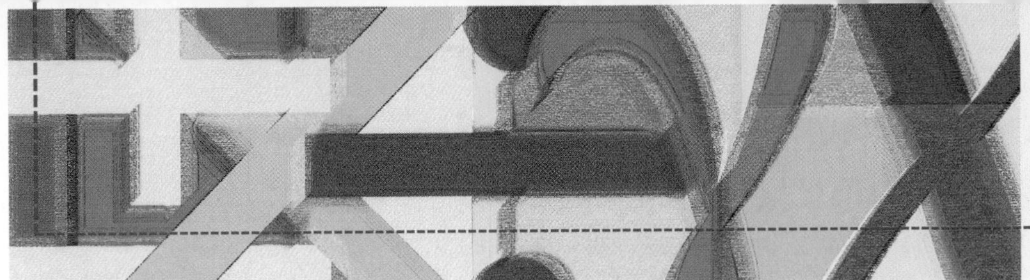

To the Student

Several features of the text are designed to assist you in understanding the concepts and learning the mathematical procedures involved. To help you learn new concepts and reinforce your understanding, there are numerous *side problems* in the margin. They are referred to in the text by numbers in colored circles, such as (2). When you see that symbol, you should work the indicated problem in the margin before going on.

Technology (such as graphing calculators, spreadsheets, or other computer programs) is not required to use this book. However, there are a number of optional examples and exercises that use graphing calculators; they are marked with the icon . There is also a *Graphing Calculator Appendix* that covers the basics of calculator use and provides some helpful programs for dealing with some of the topics in this book. The appendix can be found on the text's Web site: www.aw.com/MWA8. In addition, there are *Technology Tips* throughout the text that describe the proper menu or keys to be used in order to carry out a particular procedure on a graphing calculator.

The key to succeeding in this course is to remember that

mathematics is not a spectator sport.

You can't expect to learn mathematics without *doing* mathematics any more than you could learn to swim without getting wet. You have to take an active role, making use of all the resources at your disposal: your instructor, your fellow students, and this book.

There is no way that your instructor can possibly cover every aspect of a topic during class time. You simply won't develop the level of understanding you need to succeed unless you read the text carefully. In particular, you should read the text *before* starting the exercises. However, you can't read a math book the way you read a novel. You should have pencil, paper, and calculator handy to do the side problems, work out the statements you don't understand, and make notes on things to ask your fellow students and/or your instructor.

Finally, remember the words of the great Hillel: "The bashful do not learn." There is no such thing as a "dumb question" (assuming, of course, that you have read the book and your class notes and attempted the homework). Your instructor will welcome questions that arise from a serious effort on your part. So get your money's worth: Ask questions.

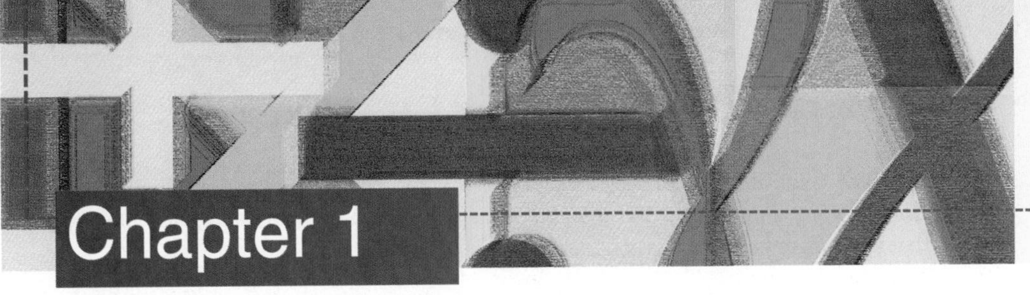

Algebra and Equations

Mathematics is widely used in business and the social and biological sciences, from determining the most efficient production schedule for a factory to mapping the human genome. Because algebra and equations are the basic mathematical tools for many such applications, we begin by reviewing the fundamental ideas of algebra. Your success in this course will depend on knowing the algebraic skills presented here.

1.1 THE REAL NUMBERS

Only real numbers will be used in this book.* The names of the most common types of real numbers are as follows.

The Real Numbers

Natural (counting) numbers	1, 2, 3, 4, . . .
Whole numbers	0, 1, 2, 3, 4, . . .
Integers	. . . , −3, −2, −1, 0, 1, 2, 3, . . .
Rational numbers	All numbers that can be written in the form p/q, where p and q are integers, with $q \neq 0$
Irrational numbers	Real numbers that are not rational

*Not all numbers are real numbers. An example of a number that is not a real number is $\sqrt{-1}$.

The relationships among these types of numbers are shown in Figure 1.1. Notice, for example, that the integers are also rational numbers and real numbers, but the integers are not irrational numbers.

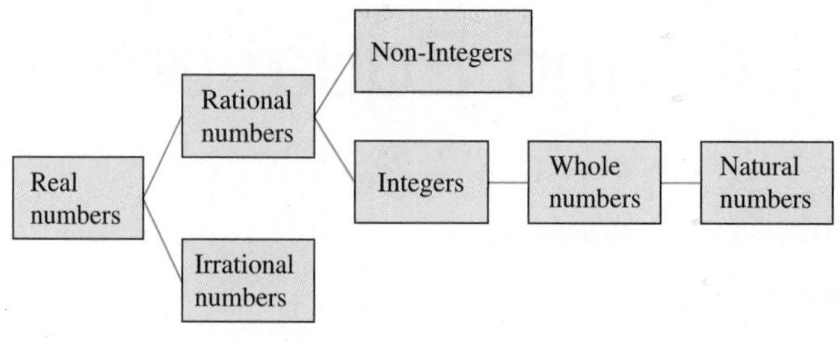

FIGURE 1.1

One example of an irrational number is π, the ratio of the circumference of a circle to its diameter. The number π can be approximated as $\pi \approx 3.14159$ ($\approx$ means "is approximately equal to"), but there is no rational number that is exactly equal to π.

1

Name all the types of numbers that apply to the following.

(a) -2

(b) $-5/8$

(c) $\pi/5$

Answers:

(a) Integer, rational, real

(b) Rational, real

(c) Irrational, real

EXAMPLE 1 What kind of number is each of the following?

(a) 6

The number 6 is a natural number, whole number, integer, rational number, and real number.

(b) 3/4

This number is rational and real.

(c) 3π

Because π is not a rational number, 3π is irrational and real.  *

All real numbers can be written in decimal form. A rational number, when written in decimal form, is either a terminating decimal, such as .5 or .128, or a repeating decimal in which some block of digits eventually repeats forever, such as 1.3333. . . or 4.7234234234. . . .† Irrational numbers are decimals that neither terminate nor repeat.

The only real numbers that can be entered exactly into a calculator are those rational numbers that are terminating decimals of no more than 10 or 12 digits (depending on the calculator). Similarly, the answers produced by a calculator are often 10–12 digit decimal *approximations*—accurate enough for most applications. As a general rule, *you should not round off any numbers during a long calculator computation,* so that your final answer will be as accurate as possible. For convenience, however, we usually round off the final answer.

The important basic properties of the real numbers are as follows.

*The use of margin problems is explained in the "To the Student" section preceding this chapter.

†Some graphing calculators have a FRAC key that automatically converts some repeating decimals to fraction form. FRAC programs for other graphing calculators are in the Program Appendix.

Properties of the Real Numbers

For all real numbers, a, b, and c, the following properties hold true.

Commutative properties $\quad a + b = b + a \quad ab = ba$

Associative properties $\quad (a + b) + c = a + (b + c) \quad (ab)c = a(bc)$

Identity properties There exists a unique real number 0, called the **additive identity,** such that

$$a + 0 = a \quad \text{and} \quad 0 + a = a.$$

There exists a unique real number 1, called the **multiplicative identity,** such that

$$a \cdot 1 = a \quad \text{and} \quad 1 \cdot a = a.$$

Inverse properties For each real number a, there exists a unique real number $-a$, called the **additive inverse** of a, such that

$$a + (-a) = 0 \quad \text{and} \quad (-a) + a = 0.$$

If $a \neq 0$, there exists a unique real number $1/a$, called the **multiplicative inverse** of a, such that

$$a \cdot \frac{1}{a} = 1 \quad \text{and} \quad \frac{1}{a} \cdot a = 1.$$

Distributive property $\quad a(b + c) = ab + ac$

2

Name the property illustrated in each of the following examples.

(a) $(2 + 3) + 9 = (3 + 2) + 9$

(b) $(2 + 3) + 9 = 2 + (3 + 9)$

(c) $(2 + 3) + 9 = 9 + (2 + 3)$

(d) $(4 \cdot 6)p = (6 \cdot 4)p$

(e) $4(6p) = (4 \cdot 6)p$

Answers:

(a) Commutative property

(b) Associative property

(c) Commutative property

(d) Commutative property

(e) Associative property

The next five examples illustrate the properties listed in the preceding box.

EXAMPLE 2 The commutative property says that the order in which you add or multiply two quantities doesn't matter.

(a) $(6 + x) + 9 = 9 + (x + 6)$

(b) $5 \cdot (9 \cdot 8) = (9 \cdot 8) \cdot 5$ ◀

EXAMPLE 3 When the associative property is used, the order of the numbers does not change, but the placement of parentheses does change.

(a) $4 + (9 + 8) = (4 + 9) + 8$

(b) $3(9x) = (3 \cdot 9)x$ ◀ 2

EXAMPLE 4 By the identity properties,

(a) $-8 + 0 = -8$ **(b)** $(-9) \cdot 1 = -9.$ ◀

3

Name the property illustrated in each of the following examples.

(a) $2 + 0 = 2$

(b) $-\frac{1}{4} \cdot (-4) = 1$

(c) $-\frac{1}{4} + \frac{1}{4} = 0$

(d) $1 \cdot \frac{2}{3} = \frac{2}{3}$

Answers:

(a) Identity property

(b) Inverse property

(c) Inverse property

(d) Identity property

Use the distributive property to complete each of the following.

(a) $4(-2 + 5)$

(b) $2(a + b)$

(c) $-3(p + 1)$

(d) $(8 - k)m$

(e) $5x + 3x$

Answers:

(a) $4(-2) + 4(5) = 12$

(b) $2a + 2b$

(c) $-3p - 3$

(d) $8m - km$

(e) $(5 + 3)x = 8x$

TECHNOLOGY TIP To enter -8 on a calculator, use the negation key (labeled $(-)$ or $+/-$), *not* the subtraction key. On most one-line scientific calculators, key in 8 $+/-$. On graphing calculators or two-line scientific calculators, key in either $(-)$ 8 or $+/-$ 8.

EXAMPLE 5 By the inverse properties, the statements in parts (a) through (d) are true.

(a) $9 + (-9) = 0$

(b) $-15 + 15 = 0$

(c) $-8 \cdot \left(\frac{1}{-8}\right) = 1$

(d) $\frac{1}{\sqrt{5}} \cdot \sqrt{5} = 1$ ◀

NOTE There is no real number x such that $0 \cdot x = 1$, so 0 has no multiplicative inverse. 3

EXAMPLE 6 By the distributive property,

(a) $9(6 + 4) = 9 \cdot 6 + 9 \cdot 4$

(b) $3(x + y) = 3x + 3y$

(c) $-8(m + 2) = (-8)(m) + (-8)(2) = -8m - 16$

(d) $(5 + x)y = 5y + xy$. ◀

NOTE As shown in Example 6(d), by the commutative property, the distributive property can also be written as $(a + b)c = ac + bc$. 4

Order of Operations

We avoid possible ambiguity when working problems with real numbers by using the following **order of operations,** which has been agreed on as the most useful. This order of operations is used by computers and calculators.

Order of Operations

If parentheses or square brackets are present:

1. Work separately above and below any fraction bar.
2. Use the rules below within each set of parentheses or square brackets. Start with the innermost set and work outward.

If no parentheses are present:

1. Find all powers and roots, working from left to right.
2. Do any multiplications or divisions in the order in which they occur, working from left to right.
3. Do any additions or subtractions in the order in which they occur, working from left to right.

EXAMPLE 7 Use the order of operations to evaluate each expression if $x = -2$, $y = 5$, and $z = -3$.

(a) $-4x^2 - 7y + 4z$

Use parentheses when replacing letters with numbers.

$$-4x^2 - 7y + 4z = -4(-2)^2 - 7(5) + 4(-3)$$
$$= -4(4) - 7(5) + 4(-3) = -16 - 35 - 12 = -63$$

(b) $$\frac{2(x-5)^2 + 4y}{z+4} = \frac{2(-2-5)^2 + 4(5)}{-3+4}$$
$$= \frac{2(-7)^2 + 20}{1}$$
$$= 2(49) + 20 = 118$$ ◀ 5

5

Evaluate the following if $m = -5$ and $n = 8$.

(a) $-2mn - 2m^2$

(b) $\dfrac{4(n-5)^2 - m}{m+n}$

Answers:

(a) 30

(b) $\dfrac{41}{3}$

Use a calculator to evaluate the following.

(a) $4^2 \div 8 + 3^2 \div 3$

(b) $[-7 + (-9)](-4) - 8(3)$

(c) $\dfrac{-11 - (-12) - 4 \cdot 5}{4(-2) - (-6)(-5)}$

(d) $\dfrac{36 \div 4 \cdot 3 \div 9 + 1}{9 \div (-6) \cdot 8 - 4}$

Answers:

(a) 5

(b) 40

(c) $\dfrac{19}{38} = \dfrac{1}{2} = .5$

(d) $-\dfrac{1}{4} = -.25$

EXAMPLE 8 Use a calculator to evaluate

$$\frac{-9(-3) + (-5)}{3(-4) - 5(2)}$$

Use extra parentheses (shown here in blue) around the numerator and denominator when you enter the number in your calculator and be careful to distinguish the negation key from the subtraction key.

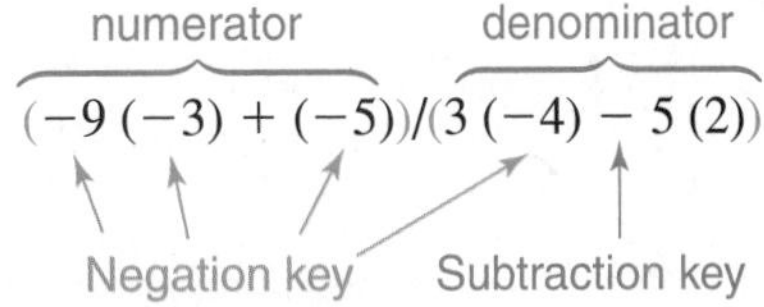

If you don't get -1 as the answer, then you are entering something incorrectly.

Square Roots

There are two numbers whose square is 16, namely, 4 and -4. The positive one, 4, is called the **square root** of 16. Similarly, the square root of a nonnegative number d is defined to be the *nonnegative* number whose square is d; it is denoted $\sqrt{d}$. For instance,

$$\sqrt{36} = 6 \text{ because } 6^2 = 36 \quad \text{and} \quad \sqrt{1.44} = 1.2 \text{ because } (1.2)^2 = 1.44.$$

No negative number has a square root in the real numbers. For instance, there is no real number whose square is -4, so -4 has no square root.

Every nonnegative real number has a square root. Unless an integer is a perfect square (such as $64 = 8^2$), its square root is an irrational number. A calculator can be used to obtain a rational approximation of these square roots.

TECHNOLOGY TIP On one-line scientific calculators, $\sqrt{40}$ is entered as 40 $\sqrt{\ }$. On graphing calculators and two-line scientific calculators, key in

$\sqrt{\ }$ 40 ENTER (or = or EXE).

7

Estimate each of the following.

(a) $\sqrt{73}$

(b) $\sqrt{22} + 3$

(c) Confirm your estimates in parts (a) and (b) with a calculator.

Answers:

(a) Between 8 and 9

(b) Between 7 and 8

(c) 8.5440; 7.6904

EXAMPLE 9 Estimate each of the following quantities. Verify your estimates with a calculator.

(a) $\sqrt{40}$

Since $6^2 = 36$ and $7^2 = 49$, $\sqrt{40}$ must be a number between 6 and 7. A typical calculator shows that $\sqrt{40} \approx 6.32455532$.

(b) $5\sqrt{7}$

$\sqrt{7}$ is between 2 and 3 because $2^2 = 4$ and $3^2 = 9$, so $5\sqrt{7}$ must be a number between $5 \cdot 2 = 10$ and $5 \cdot 3 = 15$. A calculator shows that $5\sqrt{7} \approx 13.22875656$. ◀ 7

CAUTION If c and d are positive real numbers, then $\sqrt{c + d}$ is *not* equal to $\sqrt{c} + \sqrt{d}$. For example, $\sqrt{9 + 16} = \sqrt{25} = 5$, but $\sqrt{9} + \sqrt{16} = 3 + 4 = 7$.

Draw a number line and graph the numbers −4, −1, 0, 1, 2.5, and 13/4 on it.

Answer:

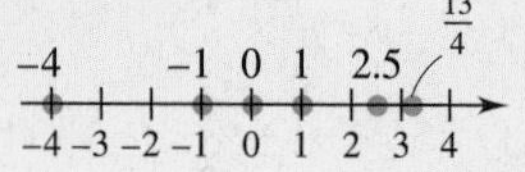

The Number Line

The real numbers can be illustrated geometrically with a diagram called a **number line.** Each real number corresponds to exactly one point on the line and vice-versa. A number line with several sample numbers located (or **graphed**) on it is shown in Figure 1.2. ◀ 8

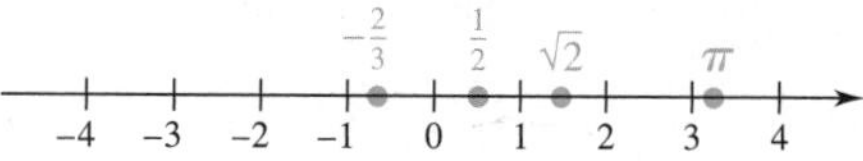

FIGURE 1.2

When comparing the sizes of two real numbers, the following symbols are used.

Symbol	Read	Meaning
$a < b$	a is less than b.	a lies to the *left* of b on the number line.
$b > a$	b is greater than a.	b lies to the *right* of a on the number line.

Note that $a < b$ means the same thing as $b > a$. The inequality symbols are sometimes joined with the equal sign, as follows.

Symbol	Read	Meaning
$a \le b$	a is less than or equal to b.	either $a < b$ or $a = b$
$b \ge a$	b is greater than or equal to a.	either $b > a$ or $b = a$

Only one part of an "either ... or" statement needs to be true for the entire statement to be considered true. So the statement $3 \le 7$ is true because $3 < 7$, and the statement $3 \le 3$ is true because $3 = 3$.

9

Write *true* or *false* for the following.

(a) $-9 \leq -2$

(b) $8 > -3$

(c) $-14 \leq -20$

Answers:

(a) True

(b) True

(c) False

EXAMPLE 10 Write *true* or *false* for each of the following.

(a) $8 < 12$

This statement says that 8 is less than 12, which is true.

(b) $-6 > -3$

The graph in Figure 1.3 shows that -6 is to the *left* of -3. Thus, $-6 < -3$, and the given statement is false.

FIGURE 1.3

(c) $-2 \leq -2$

Because $-2 = -2$, this statement is true. 

TECHNOLOGY TIP If your graphing calculator has inequality symbols (usually located on the TEST menu), you can key in statements such as "$5 < 12$" or "$-2 \geq 3$." When you press ENTER, the calculator will display 1 if the statement is true and 0 if it is false.

A number line can be used to draw the graph of a set of numbers, as shown in the next few examples.

10

Graph all real numbers x such that

(a) $-5 < x < 1$

(b) $4 < x < 7$.

Answers:

(a) [graph: -5 to 1]

(b) [graph: 4 to 7]

EXAMPLE 11 Graph all real numbers x such that $1 < x < 5$.

The graph includes all the real numbers between 1 and 5 and not just the integers. Graph these numbers by drawing a heavy line from 1 to 5 on the number line, as in Figure 1.4. Parentheses at 1 and 5 show that neither of these points belongs to the graph. ◀ 10

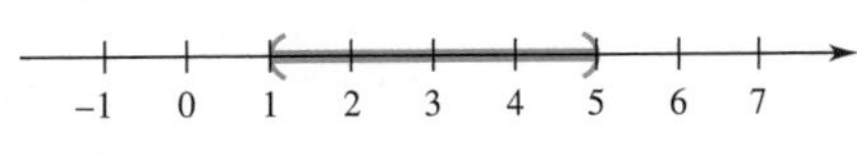

FIGURE 1.4

A set that consists of all the real numbers between two points, such as $1 < x < 5$ in Example 11, is called an **interval.** A special notation called **interval notation** is used to indicate an interval on the number line. For example, the interval including all numbers x, where $-2 < x < 3$, is written as $(-2, 3)$. The parentheses indicate that the numbers -2 and 3 are *not* included. If -2 and 3 are to be included in the interval, square brackets are used, as in $[-2, 3]$. The chart below shows several typical intervals, where $a < b$.

Inequality	*Interval Notation*	*Explanation*
$a \leq x \leq b$	$[a, b]$	Both a and b are included.
$a \leq x < b$	$[a, b)$	a is included; b is not.
$a < x \leq b$	$(a, b]$	b is included; a is not.
$a < x < b$	(a, b)	Neither a nor b is included.

(11)

Graph all real numbers x in the given interval.

(a) $(-\infty, 4]$

(b) $[-2, 1]$

Answers:

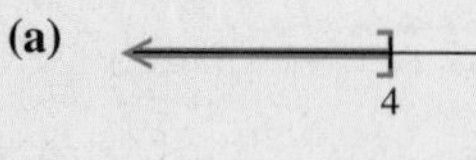

Interval notation is also used to describe sets such as the set of all numbers x, with $x \geq -2$. This interval is written $[-2, \infty)$. The set of all real numbers is written $(-\infty, \infty)$ in interval notation.

EXAMPLE 12 Graph the interval $[-2, \infty)$.

Start at -2 and draw a heavy line to the right, as in Figure 1.5. Use a square bracket at -2 to show that -2 itself is part of the graph. The symbol, ∞, read "infinity," *does not* represent a number. This notation simply indicates that *all* numbers greater than -2 are in the interval. Similarly, the notation $(-\infty, 2)$ indicates the set of all numbers x with $x < 2$. (11)

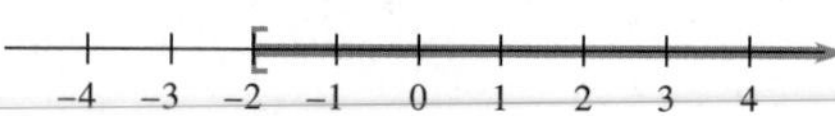

FIGURE 1.5

Absolute Value

Distance is always a nonnegative number. For example, the distance from 0 to -2 on a number line is 2, the same as the distance from 0 to 2. The **absolute value** of a number a is the distance on the number line from a to 0. Thus, the absolute value of both 2 and -2 is 2. We write the absolute value of the real number a as $|a|$. For example, the distance on the number line from 9 to 0 is 9, as is the distance from -9 to 0. (See Figure 1.6.) By definition, $|9| = 9$ and $|-9| = 9$.

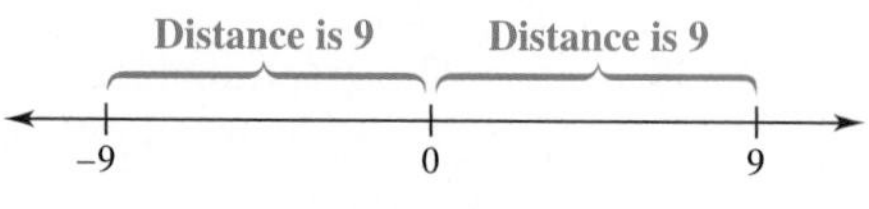

FIGURE 1.6

The facts that $|9| = 9$ and that $|-9| = 9 = -(-9)$ suggest the following algebraic definition of absolute value.

(12)

Find the following.

(a) $|-6|$

(b) $-|7|$

(c) $-|-2|$

(d) $|-3-4|$

(e) $|2-7|$

Answers:

(a) 6

(b) -7

(c) -2

(d) 7

(e) 5

Absolute Value

For any real number a,

$$|a| = a \quad \text{if } a \geq 0$$
$$|a| = -a \quad \text{if } a < 0.$$

Note the second part of the definition: for a negative number, say, -5, the negative of -5 is the positive number $-(-5) = 5$. Similarly, if a is any negative number, then $-a$ is a *positive* number. Thus, *for every real number a, $|a|$ is nonnegative.*

EXAMPLE 13 To evaluate $|8-9|$, you should first simplify the expression within the absolute value bars:

$$|8-9| = |-1| = 1.$$

Similarly, $-|-5-8| = -|-13| = -13$. (12)

1.1 Exercises

Label each of the following as true or false.

1. Every integer is a rational number.
2. No whole numbers are rational numbers.

Identify the properties that are illustrated in each of the following. Some may require more than one property. Assume all variables represent real numbers. (See Examples 2–6.)

3. $-7 + 0 = -7$
4. $3 + (-3) = (-3) + 3$
5. $0 + (-7) = -7 + 0$
6. $8 + (12 + 6) = (8 + 12) + 6$
7. $[5(-8)](-3) = 5[(-8)(-3)]$
8. $8(m + 4) = 8m + 8 \cdot 4$
9. How is the additive inverse property related to the additive identity property? the multiplicative inverse property to the multiplicative identity property?
10. Explain the distinction between the commutative and associative properties.

Evaluate each of the following if $p = -2$, $q = 4$, and $r = -5$. (See Examples 7 and 8.)

11. $-3(p + 5q)$
12. $2(q - r)$
13. $\dfrac{q + r}{q + p}$
14. $\dfrac{3q}{3p - 2r}$

Business *Lenders are required to state the* annual percentage rate *(APR) for every loan, using the formula APR = 12r, where r is the monthly interest rate. Find the APR when*

15. $r = 1.5$
16. $r = 1.67$

Find the monthly interest rate when

17. $APR = 9$
18. $APR = 19.5$

Evaluate each expression, using the order of operations given in the text. (See Examples 7 and 8.)

19. $8 - 4^2 - (-12)$
20. $8 - (-4)^2 - (-12)$
21. $-(3 - 5) - [2 - (3^2 - 13)]$
22. $\dfrac{2(3 - 7) + 4(8)}{4(-3) + (-3)(-2)}$
23. $\dfrac{2(-3) + 3/(-2) - 2/(-\sqrt{16})}{\sqrt{64} - 1}$
24. $\dfrac{6^2 - 3\sqrt{25}}{\sqrt{6^2 + 13}}$

State whether each of the following numbers is rational *or* irrational. *If the number is irrational, approximate it to four decimal places.*

25. 3π
26. $2/\pi$
27. $\sqrt{3}$
28. $\sqrt{4^2 - 3^2}$

Express each of the following statements in symbols, using $<$, $>$, $\leq$, or $\geq$.

29. 5 is less than 7.
30. -4 is greater than -9.
31. y is less than or equal to 8.3.
32. z is greater than or equal to -3.
33. t is positive.
34. c is at most 14.

Graph the following intervals on a number line. (See Examples 11 and 12.)

35. $(-8, -1)$
36. $[-1, 10]$
37. $[-2, 2)$
38. $(3, 7]$
39. $(-2, 3]$
40. $(-2, \infty)$
41. $(3, \infty)$
42. $(-\infty, 5)$
43. $(-\infty, -2]$
44. $[-4, \infty)$

Business *The* Consumer Price Index *(CPI) tracks the cost of a typical sample of consumer goods. The table shows the percentage increase for each year in a ten-year period.**

Year	1991	1992	1993	1994	1995
% change in CPI	3.1	2.9	2.7	2.7	2.5

Year	1996	1997	1998	1999	2000
% change in CPI	3.3	1.7	1.6	2.7	3.4

Let r denote the yearly percentage increase in the CPI. Find the number of years during this period that r satisfied the given inequality.

45. $r > 2.7$
46. $r < 2.7$
47. $r \leq 2.7$
48. $r \geq 3$
49. $r > 4$
50. $r \leq 1.7$

*U.S. Bureau of Labor Statistics.

Health *The Body Mass Index (BMI) is a number B that measures the relationship between a person's height H (in inches) and weight W (in pounds). It is given by this formula:*

$$B = \frac{.455W}{(.0254H)^2}.\text{*}$$

Federal guidelines suggest that the desirable range for B is $19 \le B \le 25$.

(a) *Find the BMI of the following athletes.*

(b) *Determine whether or not each athlete's BMI falls in the desirable range.*

51. Steffi Graf (119 pounds; 5 ft 9 in)

52. Jackie Joyner-Kersee (153 pounds; 5 ft 10 in)

53. Shaquille O'Neal (300 pounds; 7 ft 1 in)

54. Deion Sanders (195 pounds; 6 ft 1 in)

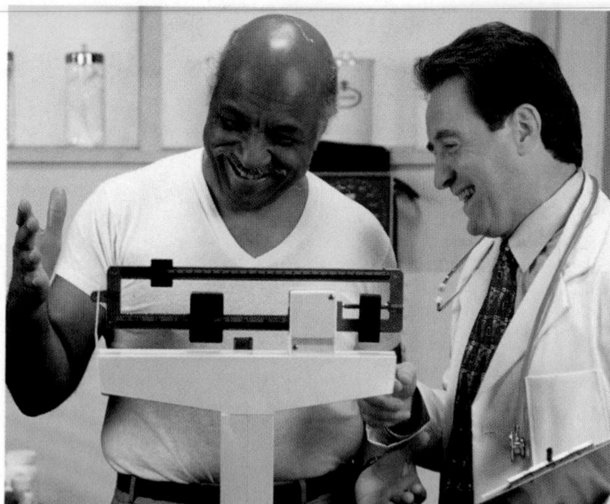

Physical Science *The wind-chill factor is a measure of the cooling effect that the wind has on a person's skin. It calculates the equivalent cooling temperature if there were no wind.*†

Temperature (°F)	Wind (mph)								
Calm	5	10	15	20	25	30	35	40	
40	36	34	32	30	29	28	28	27	
30	25	21	19	17	16	15	14	13	
20	13	9	6	4	3	1	0	−1	
10	1	−4	−7	−9	−11	−12	−14	−15	
0	−11	−16	−19	−22	−24	−26	−27	−29	
−10	−22	−28	−32	−35	−37	−39	−41	−43	
−20	−34	−41	−45	−48	−51	−53	−55	−57	
−30	−46	−53	−58	−61	−64	−67	−69	−71	
−40	−57	−66	−71	−74	−78	−80	−82	−84	

Suppose that we wish to determine the difference between two of these entries and we are interested only in the magnitude, or absolute value, of this difference. Then we subtract the two entries and find the absolute value. For example, the difference in wind-chill factors for wind at 20 miles per hour with a 20° temperature and wind at 30 miles per hour with a 10° temperature is $|-12° - 4°| = 16°$*, or equivalently,* $|4° - (-12°)| = 16°$.

Find the absolute value of the difference of the two indicated wind-chill factors.

55. Wind at 15 miles per hour with a 30° temperature and wind at 10 miles per hour with a −10° temperature

56. Wind at 20 miles per hour with a −20° temperature and wind at 5 miles per hour with a 30° temperature

57. Wind at 30 miles per hour with a −30° temperature and wind at 15 miles per hour with a −20° temperature

58. Wind at 40 miles per hour with a 40° temperature and wind at 25 miles per hour with a −30° temperature

Evaluate each of the following. (See Example 13.)

59. $|8| - |-4|$ **60.** $|-9| - |-12|$

61. $-|-4| - |-1 - 14|$ **62.** $-|6| - |-12 - 4|$

In each of the following problems, fill in the blank with either =, <, or >, so that the resulting statement is true.

63. $|5|$ _____ $|-5|$

64. $-|-4|$ _____ $|4|$

65. $|10 - 3|$ _____ $|3 - 10|$

66. $|6 - (-4)|$ _____ $|-4 - 6|$

67. $|-2 + 8|$ _____ $|2 - 8|$

68. $|3| \cdot |-5|$ _____ $|3(-5)|$

69. $|3 - 2|$ _____ $|3| - |2|$

70. $|5 - 1|$ _____ $|5| - |1|$

Write the expression without using absolute value.

71. $|a - 7|$ if $a < 7$

72. $|b - c|$ if $b \ge c$

73. In general, if a and b are any real numbers having the same sign (both negative or both positive), is it always true that $|a + b| = |a| + |b|$? Explain your answer.

74. If a and b are any two real numbers, is it always true that $|a - b| = |b - a|$? Explain your answer.

75. If a and b are any real numbers, is it always true that $|a + b| = |a| + |b|$? Explain your answer.

76. For which real numbers b does $|2 - b| = |2 + b|$? Explain your answer.

**Washington Post.*

†Table from the Joint Action Group for Temperature Indices, 2001.

Social Science *Use inequality symbols to rewrite each of the following statements, which are each based on an article in* The Sacramento Bee *newspaper.* Using x as the variable, describe what x represents in each exercise and then write an inequality. Example: At least 4000 foreign students attend the University of Southern California (USC). Let x represent the number of foreign students attending USC. Then $x \geq 4000$.*

77. Foreign students contribute more than \$1 billion annually to the California economy.

78. More than 60% of the international students come from Asian countries.

79. Less than 7.5% of foreign students now in the United States come from Middle Eastern countries.

80. No more than 10% of the foreign students in the United States come from Japan.

81. California has more than 13% of all foreign students in the United States.

82. Foreign students must prove they have at least \$22,000 in cash to spend here each year.

*"State colleges drawing foreigners despite cuts" by Lisa Lapin from *The Sacramento Bee,* December 2, 1992. Copyright, The Sacramento Bee, 1994. Reprinted by permission.

1.2 POLYNOMIALS

Polynomials are the fundamental tools of algebra and will play a central role in this course. In order to do polynomial arithmetic, you must first understand exponents. So we begin with them. You are familiar with the usual notation for squares and cubes:

$$5^2 = 5 \cdot 5 \quad \text{and} \quad 6^3 = 6 \cdot 6 \cdot 6.$$

We now extend this convenient notation to other cases.

> If n is a natural number and a is any real number, then
>
> a^n denotes the product $a \cdot a \cdot a \cdots a$ (n factors).
>
> The number a is the **base** and n is the **exponent.**

EXAMPLE 1 4^6, which is read "four to the sixth," is the number

$$4 \cdot 4 \cdot 4 \cdot 4 \cdot 4 \cdot 4 = 4096.$$

Similarly, $(-5)^3 = (-5)(-5)(-5) = -125$ and

$$\left(\frac{3}{2}\right)^4 = \frac{3}{2} \cdot \frac{3}{2} \cdot \frac{3}{2} \cdot \frac{3}{2} = \frac{81}{16}.$$

> Evaluate the following.
>
> (a) 6^3
>
> (b) 5^{12}
>
> (c) 1^9
>
> (d) $\left(\frac{7}{5}\right)^8$
>
> Answers:
>
> (a) 216
>
> (b) 244,140,625
>
> (c) 1
>
> (d) 14.75789056

EXAMPLE 2 Use a calculator to approximate the following.

(a) $(1.2)^8$

Key in 1.2 and then use the x^y key (labeled ∧ on some calculators); finally, key in the exponent 8. The calculator displays the (exact) answer 4.29981696.

(b) $\left(\frac{12}{7}\right)^{23}$

Don't compute 12/7 separately. Use parentheses and key in (12/7), followed by the x^y key and the exponent 23 to obtain the approximate answer 242,054.822. ①

Evaluate the following.

(a) $3 \cdot 6^2$

(b) $5 \cdot 4^3$

(c) -3^6

(d) $(-3)^6$

(e) $-2 \cdot (-3)^5$

Answers:

(a) 108

(b) 320

(c) −729

(d) 729

(e) 486

CAUTION A common error in using exponents occurs with expressions such as $4 \cdot 3^2$. The exponent of 2 applies only to the base 3, so that

$$4 \cdot 3^2 = 4 \cdot 3 \cdot 3 = 36.$$

On the other hand,

$$(4 \cdot 3)^2 = (4 \cdot 3)(4 \cdot 3) = 12 \cdot 12 = 144,$$

and so

$$4 \cdot 3^2 \neq (4 \cdot 3)^2.$$

Be careful to distinguish between expressions like -2^4 and $(-2)^4$.

$$-2^4 = -(2^4) = -(2 \cdot 2 \cdot 2 \cdot 2) = -16$$
$$(-2)^4 = (-2)(-2)(-2)(-2) = 16,$$

and so

$$-2^4 \neq (-2)^4.$$ (2)

By the definition of an exponent,

$$3^4 \cdot 3^2 = (3 \cdot 3 \cdot 3 \cdot 3)(3 \cdot 3) = 3^6.$$

This suggests the following property for the product of two powers of a number.

(3)

Simplify the following.

(a) $5^3 \cdot 5^6$

(b) $(-3)^4 \cdot (-3)^{10}$

(c) $(5p)^2 \cdot (5p)^8$

Answers:

(a) 5^9

(b) $(-3)^{14}$

(c) $(5p)^{10}$

If m and n are natural numbers and a is a real number, then

$$a^m \cdot a^n = a^{m+n}.$$

EXAMPLE 3 Simplify the following.

(a) $7^4 \cdot 7^6 = 7^{4+6} = 7^{10}$

(b) $(-2)^3 \cdot (-2)^5 = (-2)^{3+5} = (-2)^8$

(c) $(3k)^2 \cdot (3k)^3 = (3k)^5$

(d) $(m + n)^2 \cdot (m + n)^5 = (m + n)^7$ (3)

Polynomials

A polynomial is an algebraic expression like

$$5x^4 + 2x^3 + 6x, \quad 8m^3 + 9m^2 - 6m + 3, \quad 10p, \quad \text{or} \quad -9.$$

The letter used is called a *variable.* In formal terms, we define a polynomial as follows.

A **polynomial** in one variable x is an expression of the form

$$a_n x^n + a_{n-1} x^{n-1} + \cdots + a_1 x + a_0,$$

where n is a whole number, x is a variable, and $a_0, a_1, \ldots, a_n$ are real numbers (called the **coefficients** of the polynomial).

For example, the polynomial

$$8x^3 + 9x^2 - 6x + 3$$

is of the form

$$a_n x^n + a_{n-1}x^{n-1} + \cdots + a_1 x + a_0,$$

with $n = 3$, $a_n = a_3 = 8$, $a_{n-1} = a_2 = 9$, $a_1 = -6$, and $a_0 = 3$. Each of the expressions $8x^3$, $9x^2$, $-6x$, and 3 is called a **term** of the polynomial $8x^3 + 9x^2 - 6x + 3$. The coefficient a_0 of a polynomial (for instance, 3 in $8x^3 + 9x^2 - 6x + 3$) is called the **constant term.** Letters other than x may be used for the variable of a polynomial.

Only expressions that can be put in the form shown in the preceding box are polynomials. Consequently, the following expressions are *not* polynomials:

$$8x^3 + \frac{6}{x}, \quad \frac{9 + x}{2 - x}, \quad \text{and} \quad \frac{-p^2 + 5p + 3}{2p - 1}.$$

The **degree of a nonzero term** with only one variable is the exponent on the variable. For example, the term $9p^4$ has degree 4. The **degree of a polynomial** is the highest degree of any of its nonzero terms. Thus, the degree of $-p^2 + 5p + 3$ is 2. The **zero polynomial** consists of the constant term 0 and no other terms. No degree is assigned to the zero polynomial because it has no nonzero terms. A polynomial with two terms, such as $5x + 2$ or $x^3 + 7$, is called a **binomial,** and a polynomial with three terms, such as $3x^2 - 4x + 7$, is called a **trinomial.**

Addition and Subtraction

Two terms having the same variable with the same exponent are called **like terms;** other terms are called **unlike terms.** Polynomials can be added or subtracted by using the distributive property to combine like terms. Only like terms can be combined. For example,

$$12y^4 + 6y^4 = (12 + 6)y^4 = 18y^4$$

and

$$-2m^2 + 8m^2 = (-2 + 8)m^2 = 6m^2.$$

The polynomial $8y^4 + 2y^5$ has unlike terms, so it cannot be further simplified. The next example shows how to add and subtract polynomials by combining terms.

EXAMPLE 4 Add or subtract as indicated.

(a) $(8x^3 - 4x^2 + 6x) + (3x^3 + 5x^2 - 9x + 8)$

Combine like terms.

$$\begin{aligned}(8x^3 - 4x^2 + 6x) &+ (3x^3 + 5x^2 - 9x + 8)\\ &= (8x^3 + 3x^3) + (-4x^2 + 5x^2) + (6x - 9x) + 8 \quad \text{Commutative property and associative property}\\ &= 11x^3 + x^2 - 3x + 8 \quad \text{Distributive property}\end{aligned}$$

(b) $(-4x^4 + 6x^3 - 9x^2 - 12) + (-3x^3 + 8x^2 - 11x + 7)$

$$= -4x^4 + 3x^3 - x^2 - 11x - 5$$

4

Add or subtract.

(a) $(-2x^2 + 7x + 9) + (3x^2 + 2x - 7)$

(b) $(4x + 6) - (13x - 9)$

(c) $(9x^3 - 8x^2 + 2x) - (9x^3 - 2x^2 - 10)$

Answers:

(a) $x^2 + 9x + 2$

(b) $-9x + 15$

(c) $-6x^2 + 2x + 10$

(c) $(2x^2 - 11x + 8) - (7x^2 - 6x + 2)$

Use the definition of subtraction: $a - b = a + (-b)$. Here a and b are polynomials; a is $2x^2 - 11x + 8$ and $-b$ is

$$-(7x^2 - 6x + 2) = -7x^2 + 6x - 2.$$

Now perform the subtraction.

$$\begin{aligned}(2x^2 - 11x + 8) &- (7x^2 - 6x + 2)\\ &= (2x^2 - 11x + 8) + (-7x^2 + 6x - 2)\\ &= -5x^2 - 5x + 6\end{aligned}$$ ◀ 4

Multiplication

The distributive property is also used to multiply polynomials. For example, the product of $8x$ and $6x - 4$ is found as follows.

$$\begin{aligned}8x(6x - 4) &= 8x(6x) - 8x(4) && \text{Distributive property}\\ &= 48x^2 - 32x && x \cdot x = x^2\end{aligned}$$

5

Find the following products.

(a) $-6r(2r - 5)$

(b) $(8m + 3)(m^4 - 2m^2 + 6m)$

Answers:

(a) $-12r^2 + 30r$

(b) $8m^5 + 3m^4 - 16m^3 + 42m^2 + 18m$

EXAMPLE 5 Find each product.

(a) $$\begin{aligned}2p^3(3p^2 - 2p + 5) &= 2p^3(3p^2) + 2p^3(-2p) + 2p^3(5)\\ &= 6p^5 - 4p^4 + 10p^3\end{aligned}$$

(b) $$\begin{aligned}(3k - 2)(k^2 + 5k - 4) &= 3k(k^2 + 5k - 4) - 2(k^2 + 5k - 4)\\ &= 3k^3 + 15k^2 - 12k - 2k^2 - 10k + 8\\ &= 3k^3 + 13k^2 - 22k + 8\end{aligned}$$ ◀ 5

EXAMPLE 6 The product $(2x - 5)(3x + 4)$ can be found by using the distributive property twice.

$$\begin{aligned}(2x - 5)(3x + 4) &= 2x(3x + 4) - 5(3x + 4)\\ &= 2x \cdot 3x + 2x \cdot 4 + (-5) \cdot 3x + (-5) \cdot 4\\ &= 6x^2 + \underbrace{8x - 15x} - 20\\ &= 6x^2 - \quad 7x \quad - 20\end{aligned}$$ ◀

Observe the pattern in the second line of Example 6 and its relationship to the terms being multiplied.

$$(2x - 5)(3x + 4) = 2x \cdot 3x + 2x \cdot 4 + (-5) \cdot 3x + (-5) \cdot 4$$

$(2x - 5)(3x + 4)$ First terms

$(2x - 5)(3x + 4)$ Outside terms

$(2x - 5)(3x + 4)$ Inside terms

$(2x - 5)(3x + 4)$ Last terms

This pattern is easy to remember by using the acronym **FOIL** (**F**irst, **O**utside, **I**nside, **L**ast). The FOIL method makes it easy to find products such as this one mentally, without the necessity of writing out the intermediate steps.

6

Use FOIL to find these products.

(a) $(5k - 1)(2k + 3)$

(b) $(7z - 3)(2z + 5)$

Answers:

(a) $10k^2 + 13k - 3$

(b) $14z^2 + 29z - 15$

EXAMPLE 7

$$(3x + 2)(x + 5) = \underset{\text{First}}{\underset{\uparrow}{3x^2}} + \underset{\text{Outside}}{\underset{\uparrow}{15x}} + \underset{\text{Inside}}{\underset{\uparrow}{2x}} + \underset{\text{Last}}{\underset{\uparrow}{10}} = 3x^2 + 17x + 10$$

◀ 6

In business, the *revenue* from the sales of an item is given by

Revenue = (price per item) × (number of items sold).

The *cost* to manufacture and sell these items is given by

Cost = Fixed Costs + Variable Costs,

where the fixed costs include such things as buildings and machinery (which don't depend on how many items are made) and variable costs include such things as labor and materials (which vary, depending on how many items are made). Then

Profit = Revenue − Cost.

EXAMPLE 8

Hot Rocks Music sells CDs for \$8 each (wholesale) and can produce a maximum of 200,000 CDs. The variable cost of producing x thousand CDs is $3550x - 8x^2$ dollars, and the fixed costs for the manufacturing operation are \$215,000. If x thousand CDs are manufactured and sold, find expressions for the revenue, cost, and profit.

If x thousand CDs are sold at \$8 each, then

$$\text{Revenue} = (\text{price per item}) \times (\text{number of items sold})$$
$$R = 8 \times 1000x = 8000x,$$

where $x \le 200$ (because only 200,000 CDs can be made). The variable cost of making x thousand CDs is $3550x - 8x^2$, so that

$$\text{Cost} = \text{Fixed Costs} + \text{Variable Costs}$$
$$C = 215{,}000 + (3550x - 8x^2) \qquad (x \le 200)$$

Therefore, the profit is given by

$$P = R - C = 8000x - (215{,}000 + 3550x - 8x^2)$$
$$= 8000x - 215{,}000 - 3550x + 8x^2$$
$$P = 8x^2 + 4450x - 215{,}000 \qquad (x \le 200).$$

◀ 7

7

Suppose revenue is given by $7x^2 - 3x$, fixed costs are \$500, and variable costs are given by $3x^2 + 5x - 25$. Write an expression for

(a) Cost

(b) Profit.

Answers:

(a) $C = 3x^2 + 5x + 475$

(b) $P = 4x^2 - 8x - 475$

1.2 Exercises

Use a calculator to approximate these numbers. (See Examples 1 and 2.)

1. 12.1^7
2. $(-6.54)^{11}$
3. $(-18/7)^6$
4. $(7/9)^8$
5. Explain how the value of -3^2 differs from $(-3)^2$. Do -3^3 and $(-3)^3$ differ in the same way? Why or why not?
6. Describe the steps used to multiply 4^3 and 4^5. Is the product of 4^3 and 3^4 found in the same way? Explain.

Simplify each of the following. Leave answers with exponents. (See Example 3.)

7. $2^4 \cdot 2^3$
8. $3^8 \cdot 3^3$
9. $(-5)^2 \cdot (-5)^5$
10. $(-4)^4 \cdot (-4)^6$
11. $(2z)^5 \cdot (2z)^6$
12. $(6y)^3 \cdot (6y)^5$

Add or subtract as indicated. (See Example 4.)

13. $(3x^3 + 2x^2 - 5x) + (-4x^3 - x^2 + 8x)$
14. $(-2p^3 - 5p + 7) + (-4p^2 + 8p + 2)$
15. $(-4y^2 - 3y + 8) - (2y^2 - 6y - 2)$
16. $(7b^2 + 2b - 5) - (3b^2 + 2b - 6)$
17. $(2x^3 - 2x^2 + 4x - 3) - (2x^3 + 8x^2 - 1)$
18. $(3y^3 + 9y^2 - 11y + 8) - (-4y^2 + 10y - 6)$

Find each of the following. (See Examples 5–7.)

19. $-9m(2m^2 + 3m - 1)$
20. $2a(4a^2 - 6a + 3)$
21. $(3z + 5)(4z^2 - 2z + 1)$
22. $(2k + 3)(4k^3 - 3k^2 + k)$
23. $(6k - 1)(2k - 3)$
24. $(8r + 3)(r - 1)$
25. $(3y + 5)(2y - 1)$
26. $(5r - 3s)(5r + 4s)$
27. $(9k + q)(2k - q)$
28. $(.012x - .17)(.3x + .54)$
29. $(6.2m - 3.4)(.7m + 1.3)$
30. $2p - 3[4p - (3p + 1)]$
31. $5k - [k + (-3 + 5k)]$
32. $(3x - 1)(x + 2) - (2x + 5)^2$

Business *Find expressions for the Revenue, Cost, and Profit from selling x thousand items. (See Example 8.)*

	Item Price	*Fixed Costs*	*Variable Costs*
33.	\$5.00	\$150,000	$2250x$
34.	\$7.95	\$220,000	$4300x$
35.	\$7.50	\$300,000	$-3x^2 + 3480x - 325$
36.	\$12.75	\$450,000	$-4x^2 + 2880x - 295$

Business *The bar graph below shows the number (in millions) of U.S. households that are on-line, as reported by Jupiter Communications. The polynomial*

$$-.264x^2 + 6.84x + 8.9$$

gives a good approximation of the number of on-line households in year x, where x = 0 corresponds to 1995, x = 1 to 1996, and so on. For each year

(a) *use the bar graph to determine the number of on-line households;*
(b) *use the polynomial to determine the number of on-line households.*

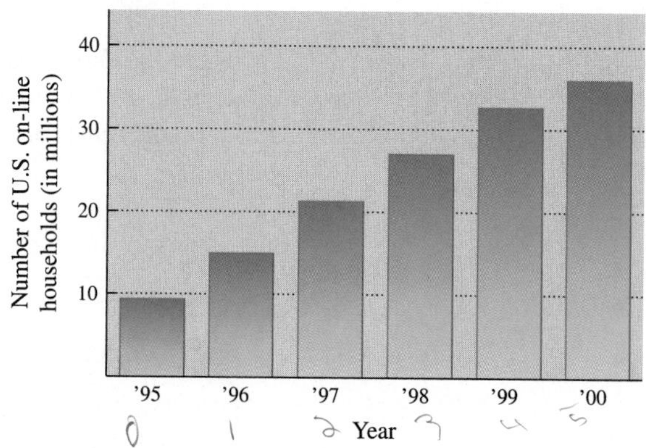

37. 1995
38. 1997
39. 1999
40. 2000

Assuming the polynomial approximation remains accurate, estimate the number of on-line households in each of these years.

41. 2002
42. 2003
43. 2004
44. 2005
45. **Physical Science** One of the most amazing formulas in all of ancient mathematics is the formula discovered by the Egyptians to find the volume of the frustum of a square pyramid, as shown in the figure. Its volume is given by $(1/3)h(a^2 + ab + b^2)$, where b is the length of the base, a is the length of the top, and h is the height.*

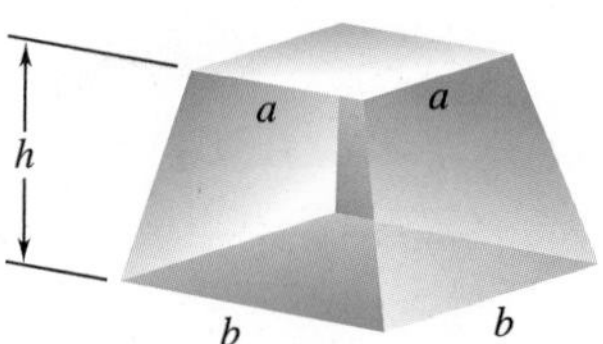

*H. A. Freebury, *A History of Mathematics*. (New York: MacMillan Company, 1968).

(a) When the Great Pyramid in Egypt was partially completed to a height h of 200 feet, b was 756 feet and a was 314 feet. Calculate its volume at this stage of construction.

(b) Try to visualize the figure if $a = b$. What is the resulting shape? Find its volume.

(c) Let $a = b$ in the Egyptian formula and simplify. Are the results the same?

46. Physical Science Refer to the formula and the discussion in Exercise 45.

(a) Use the expression $(1/3)h(a^2 + ab + b^2)$ to determine a formula for the volume of a pyramid with a square base b and height h by letting $a = 0$.

(b) The Great Pyramid in Egypt had a square base of 756 feet and a height of 481 feet. Find the volume of the Great Pyramid. Compare it with the 273-foot-tall Louisiana Superdome, which has an approximate volume of 125 million cubic feet.*

(c) The Superdome covers an area of 13 acres. How many acres does the Great Pyramid cover? (*Hint:* 1 acre = 43,560 ft^2.)

47. Suppose one polynomial has degree 3 and another also has degree 3. Find all possible values for the degree of their

(a) sum

(b) difference

(c) product.

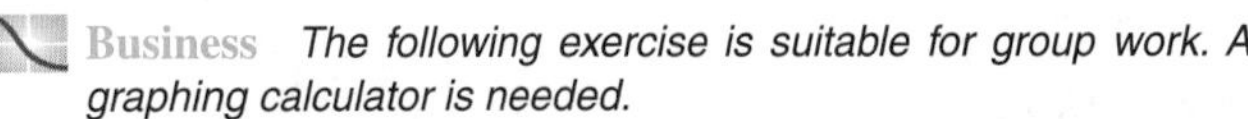
Business *The following exercise is suitable for group work. A graphing calculator is needed.*

48. (a) Use the table feature to make a table of values for the profit function in Example 8, using $x = 0, 5, 10, \ldots, 225$. Use the table to answer the following questions.

(b) What is the profit or loss (= negative profit) when 20,000 CDs are sold? When 40,000 are sold? Use data given in Example 8 to explain these answers.

(c) Approximately how many CDs must be sold in order for the company to make a profit?

(d) What is the profit from selling 100,000 CDs? From 150,000 CDs? From 200,000 CDs?

(e) Explain why the profit amounts shown in the table for 205,000 CDs and beyond are meaningless.

*Louisiana Superdome (www.superdome.com).

1.3 FACTORING

The number 18 can be written as a product in several ways: $9 \cdot 2$, $(-3)(-6)$, $1 \cdot 18$, etc. The numbers in each product (9, 2, −3, etc.) are called **factors** and the process of writing 18 as a product of factors is called **factoring.** Thus, factoring is the reverse of multiplication.

Factoring of polynomials provides a means to simplify many expressions and to solve certain types of equations. As is the usual custom, factoring of polynomials in this book will be restricted to finding factors with *integer* coefficients (otherwise there is an infinite number of possible factors).

Greatest Common Factor

The algebraic expression $15m + 45$ is made up of two terms, $15m$ and 45. Each of these terms has 15 as a factor. In fact, $15m = 15 \cdot m$ and $45 = 15 \cdot 3$. By the distributive property,

$$15m + 45 = 15 \cdot m + 15 \cdot 3 = 15(m + 3).$$

Both 15 and $m + 3$ are factors of $15m + 45$. Since 15 divides into all terms of $15m + 45$ and is the largest number that will do so, it is called the **greatest common factor** for the polynomial $15m + 45$. The process of writing $15m + 45$ as $15(m + 3)$ is called **factoring out** the greatest common factor.

EXAMPLE 1 Factor out the greatest common factor.

(a) $12p - 18q$

Both $12p$ and $18q$ are divisible by 6, and

$$\begin{aligned} 12p - 18q &= 6 \cdot 2p - 6 \cdot 3q \\ &= 6(2p - 3q). \end{aligned}$$

(b) $8x^3 - 9x^2 + 15x$

Each of these terms is divisible by x.

$$\begin{aligned} 8x^3 - 9x^2 + 15x &= (8x^2) \cdot x - (9x) \cdot x + 15 \cdot x \\ &= x(8x^2 - 9x + 15) \end{aligned}$$

(c) $5(4x - 3)^3 + 2(4x - 3)^2$

The quantity $(4x - 3)^2$ is a common factor. Factoring it out gives

$$\begin{aligned} 5(4x - 3)^3 + 2(4x - 3)^2 &= (4x - 3)^2[5(4x - 3) + 2] \\ &= (4x - 3)^2(20x - 15 + 2) \\ &= (4x - 3)^2(20x - 13). \end{aligned}$$

1

1

Factor out the greatest common factor.

(a) $12r + 9k$

(b) $75m^2 + 100n^2$

(c) $6m^4 - 9m^3 + 12m^2$

(d) $3(2k + 1)^3 + 4(2k + 1)^4$

Answers:

(a) $3(4r + 3k)$

(b) $25(3m^2 + 4n^2)$

(c) $3m^2(2m^2 - 3m + 4)$

(d) $(2k + 1)^3(7 + 8k)$

Factoring Trinomials

Factoring is the opposite of multiplication. Because the product of two binomials is usually a trinomial, we can expect factorable trinomials (that have terms with no common factor) to have two binomial factors. Thus, factoring trinomials requires using FOIL backward.

EXAMPLE 2 Factor $x^2 + 9x + 18$.

We must find integers b and d such that

$$\begin{aligned} x^2 + 9x + 18 &= (x + b)(x + d) \\ &= x^2 + dx + bx + bd \\ &= x^2 + (b + d)x + bd. \end{aligned}$$

Since the constant coefficients on each side of the equation must be equal, we must have $bd = 18$; that is, b and d are factors of 18. Similarly, the coefficients of x must be the same, so that $b + d = 9$. The possibilities are summarized in this table.

Factors b, d of 18	Sum $b + d$
$18 \cdot 1$	$18 + 1 = 19$
$9 \cdot 2$	$9 + 2 = 11$
$6 \cdot 3$	$6 + 3 = 9$

There's no need to list negative factors, such as $(-3)(-6)$, because their sum is negative. The table suggests that 6 and 3 will work. Verify that

$$(x + 6)(x + 3) = x^2 + 9x + 18.$$

2

2

Factor the following.

(a) $r^2 + 7r + 10$

(b) $x^2 + 4x + 3$

(c) $y^2 + 6y + 8$

Answers:

(a) $(r + 2)(r + 5)$

(b) $(x + 3)(x + 1)$

(c) $(y + 2)(y + 4)$

3

Factor the following.

(a) $x^2 - 4x + 3$

(b) $2y^2 - 5y + 2$

(c) $6z^2 - 13z + 6$

Answers:

(a) $(x - 3)(x - 1)$

(b) $(2y - 1)(y - 2)$

(c) $(3z - 2)(2z - 3)$

EXAMPLE 3 Factor $4y^2 - 11y + 6$.

We must find integers a, b, c, and d such that

$$\begin{aligned} 4y^2 - 11y + 6 &= (ay + b)(cy + d) \\ &= acy^2 + ady + bcy + bd \\ &= acy^2 + (ad + bc)y + bd. \end{aligned}$$

Since the coefficients of y^2 must be the same on both sides, we see that $ac = 4$. Similarly, the constant terms show that $bd = 6$. The positive factors of 4 are 4 and 1 or 2 and 2. Since the middle term is negative, we consider only negative factors of 6. The possibilities are -2 and -3 or -1 and -6. Now we try various arrangements of these factors until we find one that gives the correct coefficient of y.

$(2y - 1)(2y - 6) = 4y^2 - 14y + 6$ Incorrect

$(2y - 2)(2y - 3) = 4y^2 - 10y + 6$ Incorrect

$(y - 2)(4y - 3) = 4y^2 - 11y + 6$ Correct

The last trial gives the correct factorization.  3

4

Factor the following.

(a) $r^2 - 5r - 14$

(b) $3m^2 + 5m - 2$

(c) $6p^2 + 13pq - 5q^2$

Answers:

(a) $(r - 7)(r + 2)$

(b) $(3m - 1)(m + 2)$

(c) $(2p + 5q)(3p - q)$

EXAMPLE 4 Factor $6p^2 - 7pq - 5q^2$.

Again, we try various possibilities. The positive factors of 6 could be 2 and 3 or 1 and 6. As factors of -5 we have only -1 and 5 or -5 and 1. Try different combinations of these factors until the correct one is found.

$(2p - 5q)(3p + q) = 6p^2 - 13pq - 5q^2$ Incorrect

$(3p - 5q)(2p + q) = 6p^2 - 7pq - 5q^2$ Correct

Finally, $6p^2 - 7pq - 5q^2$ factors as $(3p - 5q)(2p + q)$. 4

NOTE In Examples 2–4, we chose positive factors of the positive first term. Of course, we could have used two negative factors, but the work is easier if positive factors are used.

The method shown above can be used to factor a **perfect square trinomial,** one that is the square of a binomial. The binomial can be predicted by observing the following patterns that always apply to a perfect square trinomial.

$$\left.\begin{aligned} x^2 + 2xy + y^2 &= (x + y)^2 \\ x^2 - 2xy + y^2 &= (x - y)^2 \end{aligned}\right\} \text{Perfect square trinomials}$$

EXAMPLE 5 Factor each trinomial.

(a) $16p^2 - 40pq + 25q^2$

Because $16p^2 = (4p)^2$ and $25q^2 = (5q)^2$, use the second pattern shown above with $4p$ replacing x and $5q$ replacing y to get

$$\begin{aligned} 16p^2 - 40pq + 25q^2 &= (4p)^2 - 2(4p)(5q) + (5q)^2 \\ &= (4p - 5q)^2. \end{aligned}$$

Factor each trinomial.

(a) $4m^2 + 4m + 1$

(b) $25z^2 - 80zt + 64t^2$

Answers:

(a) $(2m + 1)^2$

(b) $(5z - 8t)^2$

Make sure that the middle term of the trinomial being factored, $-40pq$ here, is twice the product of the two terms in the binomial $4p - 5q$.

$$-40pq = 2(4p)(-5q)$$

(b) $169x^2 + 104xy^2 + 16y^4 = (13x + 4y^2)^2$, since $2(13x)(4y^2) = 104xy^2$. 

Factoring Binomials

Three special factoring patterns are listed below. Each can be verified by multiplying on the right side of the equation. These formulas should be memorized.

$x^2 - y^2 = (x + y)(x - y)$	Difference of two squares
$x^3 - y^3 = (x - y)(x^2 + xy + y^2)$	Difference of two cubes
$x^3 + y^3 = (x + y)(x^2 - xy + y^2)$	Sum of two cubes

EXAMPLE 6 Factor each of the following.

(a) $4m^2 - 9$

Notice that $4m^2 - 9$ is the difference of two squares, since $4m^2 = (2m)^2$ and $9 = 3^2$. Use the pattern for the difference of two squares, letting $2m$ replace x and 3 replace y. Then the pattern $x^2 - y^2 = (x + y)(x - y)$ becomes

$$\begin{aligned} 4m^2 - 9 &= (2m)^2 - 3^2 \\ &= (2m + 3)(2m - 3). \end{aligned}$$

(b) $128p^2 - 98q^2$

First factor out the common factor of 2.

$$\begin{aligned} 128p^2 - 98q^2 &= 2(64p^2 - 49q^2) \\ &= 2[(8p)^2 - (7q)^2] \\ &= 2(8p + 7q)(8p - 7q) \end{aligned}$$

(c) $x^2 + 36$

The *sum* of two squares usually cannot be factored. To see this, check some possibilities.

$$(x + 6)(x + 6) = (x + 6)^2 = x^2 + 12x + 36$$
$$(x + 4)(x + 9) = x^2 + 13x + 36$$

Any product of two binomials will always have a middle term unless it is the *difference* of two squares.

(d) $4z^2 + 12z + 9 - w^2$

Notice that the first three terms can be factored as a perfect square.

$$4z^2 + 12z + 9 - w^2 = (2z + 3)^2 - w^2$$

Written in this form, the expression is the difference of squares, which can be factored as

$$(2z + 3)^2 - w^2 = [(2z + 3) + w][(2z + 3) - w]$$
$$= (2z + 3 + w)(2z + 3 - w).$$

(e) $256k^4 - 625m^4$

Use the difference of two squares pattern twice, as follows:

$$256k^4 - 625m^4 = (16k^2)^2 - (25m^2)^2$$
$$= (16k^2 + 25m^2)(16k^2 - 25m^2)$$
$$= (16k^2 + 25m^2)(4k + 5m)(4k - 5m).$$ ◀ 6

6

Factor the following.

(a) $9p^2 - 49$

(b) $y^2 + 100$

(c) $9r^2 + 12r + 4 - t^2$

(d) $81x^4 - 16y^4$

Answers:

(a) $(3p + 7)(3p - 7)$

(b) Cannot be factored

(c) $(3r + 2 + t)(3r + 2 - t)$

(d) $(9x^2 + 4y^2)(3x + 2y) \cdot (3x - 2y)$

EXAMPLE 7 Factor each of the following.

(a) $k^3 - 8$

Use the pattern for the difference of two cubes, since $8 = 2^3$, to get

$$k^3 - 8 = k^3 - 2^3 = (k - 2)(k^2 + 2k + 4).$$

(b) $m^3 + 125 = m^3 + 5^3 = (m + 5)(m^2 - 5m + 25)$

(c) $8k^3 - 27z^3 = (2k)^3 - (3z)^3 = (2k - 3z)(4k^2 + 6kz + 9z^2)$ ◀ 7

Factor the following.

(a) $a^3 + 1000$

(b) $z^3 - 64$

(c) $1000m^3 - 27z^3$

Answers:

(a) $(a + 10)(a^2 - 10a + 100)$

(b) $(z - 4)(z^2 + 4z + 16)$

(c) $(10m - 3z) \cdot (100m^2 + 30mz + 9z^2)$

EXAMPLE 8 Factor each of the following.

(a) $12x^2 - 26x - 10$

Look first for a common factor. Here there is a common factor of 2: $12x^2 - 26x - 10 = 2(6x^2 - 13x - 5)$. Now try to factor $6x^2 - 13x - 5$. Possible factors of 6 are 3 and 2 or 6 and 1. The only factors of -5 are -5 and 1 or 5 and -1. Try various combinations. You should find the trinomial factors as $(3x + 1)(2x - 5)$. Thus,

$$12x^2 - 26x - 10 = 2(3x + 1)(2x - 5).$$

(b) $16a^2 - 100 - 48ac + 36c^2$

Factor out the common factor of 4 first.

$$16a^2 - 100 - 48ac + 36c^2 = 4[4a^2 - 25 - 12ac + 9c^2]$$
$$= 4[(4a^2 - 12ac + 9c^2) - 25]$$ Rearrange terms and group.
$$= 4[(2a - 3c)^2 - 25]$$ Factor the trinomial.
$$= 4(2a - 3c + 5)(2a - 3c - 5)$$ Factor the difference of squares.

CAUTION Remember always to look first for a common factor. ◀ 

Factor.

(a) $6x^2 - 27x - 15$

(b) $18 - 8xy - 2y^2 - 8x^2$

Answers:

(a) $3(2x + 1)(x - 5)$

(b) $2(3 - 2x - y)(3 + 2x + y)$

1.3 Exercises

Factor out the greatest common factor in each of the following. (See Example 1.)

1. $12x^2 - 24x$
2. $5y - 25xy$
3. $r^3 - 5r^2 + r$
4. $t^3 + 3t^2 + 8t$
5. $6z^3 - 12z^2 + 18z$
6. $5x^3 + 35x^2 + 10x$
7. $3(2y - 1)^2 + 5(2y - 1)^3$
8. $(3x + 7)^5 - 2(3x + 7)^3$
9. $3(x + 5)^4 + (x + 5)^6$
10. $3(x + 6)^2 + 2(x + 6)^4$

Factor the polynomial. (See Example 2.)

11. $x^2 + 5x + 4$
12. $u^2 + 7u + 6$
13. $x^2 + 7x + 12$
14. $y^2 + 8y + 12$
15. $z^2 + 10z + 24$
16. $r^2 + 16r + 60$

Factor the polynomial. (See Examples 3 and 4.)

17. $2x^2 - 9x + 4$
18. $3w^2 - 8w + 4$
19. $15p^2 - 23p + 4$
20. $8x^2 - 14x + 3$
21. $4z^2 - 16z + 15$
22. $12y^2 - 29y + 15$
23. $6x^2 - 5x - 4$
24. $12z^2 + z - 1$
25. $10y^2 + 21y - 10$
26. $15u^2 + 4u - 4$
27. $6x^2 + 5x - 4$
28. $12y^2 + 7y - 10$

Factor each polynomial completely. Factor out the greatest common factor as necessary. (See Examples 2–6 and 8.)

29. $2a^2 + 3a - 5$
30. $6a^2 - 48a - 120$
31. $x^2 - 64$
32. $x^2 + 17xy + 72y^2$
33. $9p^2 - 24p + 16$
34. $3r^2 - r - 2$
35. $r^2 - 3rt - 10t^2$
36. $2a^2 + ab - 6b^2$
37. $m^2 - 6mn + 9n^2$
38. $8k^2 - 16k - 10$
39. $4p^2 - 9$
40. $8r^2 + r + 6$
41. $3x^2 - 24xz + 48z^2$
42. $9m^2 - 25$
43. $a^2 + 4ab + 5b^2$
44. $6y^2 - 11y - 7$
45. $-x^2 + 7x - 12$
46. $4y^2 + y - 3$
47. $3a^2 - 13a - 30$
48. $3k^2 + 2k - 8$
49. $21m^2 + 13mn + 2n^2$
50. $81y^2 - 100$
51. $20y^2 + 39yx - 11x^2$
52. $12s^2 + 11st - 5t^2$
53. $y^2 - 4yz - 21z^2$
54. $49a^2 + 9$
55. $121x^2 - 64$
56. $4z^2 + 56zy + 196y^2$
57. $5m^3(m^3 - 1)^2 - 3m^5(m^3 - 1)^3$
58. $9(x - 4)^5 - (x - 4)^3$

Factor each binomial. (See Example 7.)

59. $a^3 - 216$
60. $b^3 + 125$
61. $8r^3 - 27s^3$
62. $1000p^3 + 27q^3$
63. $64m^3 + 125$
64. $216y^3 - 343$
65. $1000y^3 - z^3$
66. $125p^3 + 8q^3$
67. When asked to factor $6x^4 - 3x^2 - 3$ completely, a student gave the following result:
$$6x^4 - 3x^2 - 3 = (2x^2 + 1)(3x^2 - 3).$$
Is this answer correct? Explain why.
68. When can the sum of two squares be factored? Give examples.
69. Explain why $(x + 2)^3$ is not the correct factorization of $x^3 + 8$ and give the correct factorization.
70. Describe how factoring and multiplication are related. Give examples.

1.4 RATIONAL EXPRESSIONS

We now consider **rational expressions,** such as

$$\frac{8}{x - 1}, \quad \frac{3x^2 + 4x}{5x - 6}, \quad \text{and} \quad \frac{2 + \frac{1}{y}}{y}.$$

Because rational expressions involve quotients, it is important to keep in mind values of the variables that make denominators 0. For example, 1 cannot be used as a replacement for x in the first rational expression above, and 6/5 cannot be used in the second one, since these values make the respective denominators equal 0. (1)

Operations with Rational Expressions

The rules for operations with rational expressions are the usual rules for fractions.

Operations with Rational Expressions

For all mathematical expressions P, $Q \neq 0$, R, and $S \neq 0$,

(a) $\dfrac{P}{Q} = \dfrac{PS}{QS}$ Fundamental property

(b) $\dfrac{P}{Q} \cdot \dfrac{R}{S} = \dfrac{PR}{QS}$ Multiplication

(c) $\dfrac{P}{Q} + \dfrac{R}{Q} = \dfrac{P+R}{Q}$ Addition

(d) $\dfrac{P}{Q} - \dfrac{R}{Q} = \dfrac{P-R}{Q}$ Subtraction

(e) $\dfrac{P}{Q} \div \dfrac{R}{S} = \dfrac{P}{Q} \cdot \dfrac{S}{R}$, $R \neq 0$. Division

The following examples illustrate these operations.

EXAMPLE 1 Write each of the following rational expressions in lowest terms (so that the numerator and denominator have no common factor with integer coefficients except 1 or -1).

(a) $\dfrac{12m}{-18}$

Both $12m$ and -18 are divisible by 6. By operation (a) above,

$$\frac{12m}{-18} = \frac{2m \cdot 6}{-3 \cdot 6} = \frac{2m}{-3} = -\frac{2m}{3}.$$

(b) $\dfrac{8x+16}{4} = \dfrac{8(x+2)}{4} = \dfrac{4 \cdot 2(x+2)}{4} = \dfrac{2(x+2)}{1} = 2(x+2)$

The numerator, $8x + 16$, was factored so that the common factor could be identified. The answer could also be written as $2x + 4$, if desired.

(c) $\dfrac{k^2+7k+12}{k^2+2k-3} = \dfrac{(k+4)(k+3)}{(k-1)(k+3)} = \dfrac{k+4}{k-1}$ (2)

(1)

What values of the variable make each denominator equal 0?

(a) $\dfrac{5}{x-3}$

(b) $\dfrac{2x-3}{4x-1}$

(c) $\dfrac{x+2}{x}$

(d) Why do we need to determine these values?

Answers:

(a) 3

(b) 1/4

(c) 0

(d) Because division by 0 is undefined.

(2)

Write each of the following in lowest terms.

(a) $\dfrac{12k+36}{18}$

(b) $\dfrac{15m+30m^2}{5m}$

(c) $\dfrac{2p^2+3p+1}{p^2+3p+2}$

Answers:

(a) $\dfrac{2(k+3)}{3}$ or $\dfrac{2k+6}{3}$

(b) $3(1+2m)$ or $3+6m$

(c) $\dfrac{2p+1}{p+2}$

The values of k in Example 1(c) are restricted to $k \neq 1$ and $k \neq -3$. From now on, such restrictions will be assumed when working with rational expressions.

EXAMPLE 2

(a) Multiply $\frac{2}{3} \cdot \frac{y}{5}$.

Use operation (b). Multiply the numerators and then the denominators.

$$\frac{2}{3} \cdot \frac{y}{5} = \frac{2 \cdot y}{3 \cdot 5} = \frac{2y}{15}$$

The result, $2y/15$, is in lowest terms.

(b) $\frac{3y+9}{6} \cdot \frac{18}{5y+15}$

Factor where possible.

$$\frac{3y+9}{6} \cdot \frac{18}{5y+15} = \frac{3(y+3)}{6} \cdot \frac{18}{5(y+3)}$$

$$= \frac{3 \cdot 18(y+3)}{6 \cdot 5(y+3)} \quad \text{Multiply numerators and denominators.}$$

$$= \frac{3 \cdot 6 \cdot 3(y+3)}{6 \cdot 5(y+3)} \quad 18 = 6 \cdot 3$$

$$= \frac{3 \cdot 3}{5} \quad \text{Write in lowest terms.}$$

$$= \frac{9}{5}$$

(c) $\frac{m^2+5m+6}{m+3} \cdot \frac{m^2+m-6}{m^2+3m+2}$

$$= \frac{(m+2)(m+3)}{m+3} \cdot \frac{(m-2)(m+3)}{(m+2)(m+1)} \quad \text{Factor.}$$

$$= \frac{(m+2)(m+3)(m-2)(m+3)}{(m+3)(m+2)(m+1)} \quad \text{Multiply.}$$

$$= \frac{(m-2)(m+3)}{m+1} \quad \text{Lowest terms}$$

$$= \frac{m^2+m-6}{m+1}$$ ◀ 3

3

Multiply.

(a) $\frac{3r^2}{5} \cdot \frac{20}{9r}$

(b) $\frac{y-4}{y^2-2y-8} \cdot \frac{y^2-4}{3y}$

Answers:

(a) $\frac{4r}{3}$

(b) $\frac{y-2}{3y}$

EXAMPLE 3

(a) Divide $\frac{8x}{5} \div \frac{11x^2}{20}$.

4

Divide.

(a) $\dfrac{5m}{16} \div \dfrac{m^2}{10}$

(b) $\dfrac{2y-8}{6} \div \dfrac{5y-20}{3}$

(c) $\dfrac{m^2-2m-3}{m(m+1)} \div \dfrac{m+4}{5m}$

Answers:

(a) $\dfrac{25}{8m}$

(b) $\dfrac{1}{5}$

(c) $\dfrac{5(m-3)}{m+4}$

Invert the second expression and multiply [operation (e)].

$$\frac{8x}{5} \div \frac{11x^2}{20} = \frac{8x}{5} \cdot \frac{20}{11x^2} \quad \text{Invert and multiply.}$$

$$= \frac{8x \cdot 20}{5 \cdot 11x^2} \quad \text{Multiply.}$$

$$= \frac{32}{11x} \quad \text{Lowest terms}$$

(b) $\dfrac{9p-36}{12} \div \dfrac{5(p-4)}{18}$

$$= \frac{9p-36}{12} \cdot \frac{18}{5(p-4)} \quad \text{Invert and multiply.}$$

$$= \frac{9(p-4)}{12} \cdot \frac{18}{5(p-4)} \quad \text{Factor.}$$

$$= \frac{27}{10} \quad \text{Multiply and write in lowest terms.}$$ 4

EXAMPLE 4 Add or subtract as indicated.

(a) $\dfrac{4}{5k} - \dfrac{11}{5k}$

When two rational expressions have the same denominators, subtract by subtracting the numerators and keeping the common denominator. [See operation (d).]

$$\frac{4}{5k} - \frac{11}{5k} = \frac{4-11}{5k} = -\frac{7}{5k}$$

(b) $\dfrac{7}{p} + \dfrac{9}{2p} + \dfrac{1}{3p}$.

These three denominators are different; addition requires the same denominators. Find a common denominator, one which can be divided by p, $2p$, and $3p$. A common denominator here is $6p$. Rewrite each rational expression, using operation (a), with a denominator of $6p$. Then, using operation (c), add the numerators and keep the common denominator.

$$\frac{7}{p} + \frac{9}{2p} + \frac{1}{3p} = \frac{6 \cdot 7}{6 \cdot p} + \frac{3 \cdot 9}{3 \cdot 2p} + \frac{2 \cdot 1}{2 \cdot 3p} \quad \text{Operation (a)}$$

$$= \frac{42}{6p} + \frac{27}{6p} + \frac{2}{6p}$$

$$= \frac{42+27+2}{6p} \quad \text{Operation (c)}$$

$$= \frac{71}{6p}$$

(c) $\frac{k^2}{k^2-1} - \frac{2k^2-k-3}{k^2+3k+2}$

Factor the denominators to find a common denominator.

$$\frac{k^2}{k^2-1} - \frac{2k^2-k-3}{k^2+3k+2} = \frac{k^2}{(k+1)(k-1)} - \frac{2k^2-k-3}{(k+1)(k+2)}$$

The common denominator is $(k+1)(k-1)(k+2)$. Write each fraction with the common denominator.

$$\begin{aligned}
&\frac{k^2}{(k+1)(k-1)} - \frac{2k^2-k-3}{(k+1)(k+2)} \\
&= \frac{k^2(k+2)}{(k+1)(k-1)(k+2)} - \frac{(2k^2-k-3)(k-1)}{(k+1)(k-1)(k+2)} \\
&= \frac{k^3+2k^2-(2k^2-k-3)(k-1)}{(k+1)(k-1)(k+2)} && \text{Subtract fractions.} \\
&= \frac{k^3+2k^2-(2k^3-3k^2-2k+3)}{(k+1)(k-1)(k+2)} && \text{Multiply } (2k^2-k-3)(k-1). \\
&= \frac{k^3+2k^2-2k^3+3k^2+2k-3}{(k+1)(k-1)(k+2)} && \text{Polynomial subtraction} \\
&= \frac{-k^3+5k^2+2k-3}{(k+1)(k-1)(k+2)} && \text{Combine terms.}
\end{aligned}$$

5

5

Add or subtract.

(a) $\frac{3}{4r} + \frac{8}{3r}$

(b) $\frac{1}{m-2} - \frac{3}{2(m-2)}$

(c) $\frac{p+1}{p^2-p} - \frac{p^2-1}{p^2+p-2}$

Answers:

(a) $\frac{41}{12r}$

(b) $\frac{-1}{2(m-2)}$

(c) $\frac{-p^3+p^2+4p+2}{p(p-1)(p+2)}$

Complex Fractions

Any quotient of two rational expressions is called a **complex fraction.** Complex fractions are simplified as follows.

EXAMPLE 5 Simplify each complex fraction.

(a) $\dfrac{6-\frac{5}{k}}{1+\frac{5}{k}}$

Multiply both numerator and denominator by the common denominator k.

$$\begin{aligned}
\frac{6-\frac{5}{k}}{1+\frac{5}{k}} &= \frac{k\left(6-\frac{5}{k}\right)}{k\left(1+\frac{5}{k}\right)} && \text{Multiply by } \frac{k}{k}. \\
&= \frac{6k-k\left(\frac{5}{k}\right)}{k+k\left(\frac{5}{k}\right)} && \text{Distributive property} \\
&= \frac{6k-5}{k+5} && \text{Simplify.}
\end{aligned}$$

TEST **(b)** $\dfrac{\dfrac{a}{a+1}+\dfrac{1}{a}}{\dfrac{1}{a}+\dfrac{1}{a+1}}$

Multiply both numerator and denominator by the common denominator of all the fractions, in this case $a(a + 1)$. Doing so gives

$$\frac{\frac{a}{a+1}+\frac{1}{a}}{\frac{1}{a}+\frac{1}{a+1}} = \frac{\left(\frac{a}{a+1}+\frac{1}{a}\right)a(a+1)}{\left(\frac{1}{a}+\frac{1}{a+1}\right)a(a+1)}$$

$$= \frac{a^2+(a+1)}{(a+1)+a} = \frac{a^2+a+1}{2a+1}.$$

As an alternative method of solution, first perform the indicated additions in the numerator and denominator and then divide.

$$\frac{\frac{a}{a+1}+\frac{1}{a}}{\frac{1}{a}+\frac{1}{a+1}} = \frac{\frac{a^2+1(a+1)}{a(a+1)}}{\frac{1(a+1)+1(a)}{a(a+1)}} = \frac{\frac{a^2+a+1}{a(a+1)}}{\frac{2a+1}{a(a+1)}}$$

$$= \frac{a^2+a+1}{a(a+1)} \cdot \frac{a(a+1)}{2a+1} = \frac{a^2+a+1}{2a+1}$$ ◀ 6

6

Simplify each complex fraction.

(a) $\dfrac{t-\dfrac{1}{t}}{2t+\dfrac{3}{t}}$

(b) $\dfrac{\dfrac{m}{m+2}+\dfrac{1}{m}}{\dfrac{1}{m}-\dfrac{1}{m+2}}$

Answers:

(a) $\dfrac{t^2-1}{2t^2+3}$

(b) $\dfrac{m^2+m+2}{2}$

1.4 Exercises

Write each of the following in lowest terms. Factor as necessary. (See Example 1.)

1. $\dfrac{8x^2}{40x}$

2. $\dfrac{27m}{81m^3}$

3. $\dfrac{20p^2}{35p^3}$

4. $\dfrac{18y^4}{27y^2}$

5. $\dfrac{5m+15}{4m+12}$

6. $\dfrac{10z+5}{20z+10}$

7. $\dfrac{4(w-3)}{(w-3)(w+3)}$

8. $\dfrac{-6(x+2)}{(x-4)(x+2)}$

9. $\dfrac{3y^2-12y}{9y^3}$

10. $\dfrac{15k^2+45k}{9k^2}$

11. $\dfrac{m^2-4m+4}{m^2+m-6}$

12. $\dfrac{r^2-r-6}{r^2+r-12}$

13. $\dfrac{x^2+3x-4}{x^2-1}$

14. $\dfrac{z^2-5z+6}{z^2-4}$

Multiply or divide as indicated in each of the following. Write all answers in lowest terms. (See Examples 2 and 3.)

15. $\dfrac{4p^3}{49}\cdot\dfrac{7}{2p^2}$

16. $\dfrac{24n^4}{6n^2}\cdot\dfrac{18n^2}{9n}$

17. $\dfrac{21a^5}{14a^3}\div\dfrac{8a}{12a^2}$

18. $\dfrac{2x^3}{6x^2}\div\dfrac{10x^2}{15x}$

19. $\dfrac{2a+b}{2c}\cdot\dfrac{15}{4(2a+b)}$

20. $\dfrac{4(x+2)}{w}\cdot\dfrac{3w}{8(x+2)}$

21. $\dfrac{15p-3}{6}\div\dfrac{10p-2}{3}$

22. $\dfrac{2k+8}{6}\div\dfrac{3k+12}{2}$

23. $\dfrac{9y-18}{6y+12}\cdot\dfrac{3y+6}{15y-30}$

24. $\frac{12r + 24}{36r - 36} \div \frac{6r + 12}{8r - 8}$

25. $\frac{4a + 12}{2a - 10} \div \frac{a^2 - 9}{a^2 - a - 20}$

26. $\frac{6r - 18}{9r^2 + 6r - 24} \cdot \frac{12r - 16}{4r - 12}$

27. $\frac{k^2 - k - 6}{k^2 + k - 12} \cdot \frac{k^2 + 3k - 4}{k^2 + 2k - 3}$

28. $\frac{n^2 - n - 6}{n^2 - 2n - 8} \div \frac{n^2 - 9}{n^2 + 7n + 12}$

29. In your own words, explain how to find the least common denominator for two fractions.

30. Describe the steps required to add three rational expressions. You may use an example to illustrate.

Add or subtract as indicated in each of the following. Write all answers in lowest terms. (See Example 4.)

31. $\frac{3}{5z} - \frac{2}{3z}$

32. $\frac{7}{4z} - \frac{5}{3z}$

33. $\frac{r + 2}{3} - \frac{r - 2}{3}$

34. $\frac{3y - 1}{8} - \frac{3y + 1}{8}$

35. $\frac{4}{x} + \frac{1}{3}$

36. $\frac{6}{r} - \frac{3}{4}$

37. $\frac{1}{m - 1} + \frac{2}{m}$

38. $\frac{8}{y + 2} - \frac{3}{y}$

39. $\frac{8}{3(a - 1)} + \frac{2}{a - 1}$

40. $\frac{5}{2(k + 3)} + \frac{2}{k + 3}$

41. $\frac{2}{5(k - 2)} + \frac{3}{4(k - 2)}$

42. $\frac{11}{3(p + 4)} - \frac{5}{6(p + 4)}$

43. $\frac{2}{x^2 - 2x - 3} + \frac{5}{x^2 - x - 6}$

44. $\frac{3}{m^2 - 3m - 10} + \frac{5}{m^2 - m - 20}$

45. $\frac{2y}{y^2 + 7y + 12} - \frac{y}{y^2 + 5y + 6}$

46. $\frac{-r}{r^2 - 10r + 16} - \frac{3r}{r^2 + 2r - 8}$

47. $\frac{3k}{2k^2 + 3k - 2} - \frac{2k}{2k^2 - 7k + 3}$

48. $\frac{4m}{3m^2 + 7m - 6} - \frac{m}{3m^2 - 14m + 8}$

In each of the following exercises, simplify the complex fraction. (See Example 5.)

49. $\frac{1 + \frac{1}{x}}{1 - \frac{1}{x}}$

50. $\frac{2 - \frac{2}{y}}{2 + \frac{2}{y}}$

51. $\frac{\frac{1}{x + h} - \frac{1}{x}}{h}$

52. $\frac{\frac{1}{(x + h)^2} - \frac{1}{x^2}}{h}$

53. **Social Sciences** When an average of ten cars per minute are arriving at the entrance gate of an amusement park and the average number being admitted per minute is x (with $x > 10$), then the average waiting time in minutes for each car is given by

$$\frac{x^2 - 10x + 25}{x^3 - 15x^2 + 50x}.*$$

(a) Reduce this fraction to lowest terms.
(b) Use the reduced fraction to determine the average waiting time *in seconds,* when $x = 11, 15, 20$.

54. **Business** In Example 8 of Section 1.2, we saw that the cost C of producing x thousand CDs is given by

$$C = -8x^2 + 3550x + 215{,}000 \qquad (x \leq 200)$$

(a) Write a rational expression in lowest terms that gives the average cost per CD when x thousand are produced. (*Hint:* The average cost is the total cost C divided by the number of CDs produced.)
(b) Find the average cost per CD for each of these production levels: 20,000, 50,000, and 150,000.

*L. Haefner, *Introduction to Transportation Systems,* New York: Holt, Rinehart and Winston, 1986.

1.5 EXPONENTS AND RADICALS

Exponents were introduced in Section 1.2. In this section the definition of exponents will be extended to include negative exponents and rational number exponents, such as 1/2 and 7/3.

Integer Exponents

In Section 1.2 we defined positive integer exponents and noted that $a^m \cdot a^n = a^{m+n}$ for positive integer values of m and n. Now we develop an analogous property for quotients. By definition,

$$\frac{6^5}{6^2} = \frac{6 \cdot 6 \cdot 6 \cdot \cancel{6} \cdot \cancel{6}}{\cancel{6} \cdot \cancel{6}} = 6 \cdot 6 \cdot 6 = 6^3.$$

Because there are 5 factors of 6 in the numerator and 2 factors of 6 in the denominator, the quotient has $5 - 2 = 3$ factors of 6. In general, we can make the following statement.

> If a is a nonzero real number and m, n are positive integers with $m > n$, then
>
> $$\frac{a^m}{a^n} = a^{m-n}.$$

Next, we want to give a meaning to expressions such as 3^0. If the quotient property in the preceding box is to continue to be valid, we must define 3^0 in such a way that

$$\frac{3^5}{3^5} = 3^{5-5} = 3^0.$$

Since $3^5/3^5 = 1$, it is reasonable to define $3^0 = 1$, and similarly in the general case. The symbol 0^0 is undefined.

> **Zero Exponent**
>
> If a is any nonzero real number, then
>
> $$a^0 = 1.$$

EXAMPLE 1 Evaluate the following.

(a) $6^0 = 1$

(b) $(-9)^0 = 1$

(c) $-(4)^0 = -(1) = -1$ ◀ 1

1

Evaluate the following.

(a) 17^0

(b) 30^0

(c) $(-10)^0$

(d) $-(12)^0$

Answers:

(a) 1

(b) 1

(c) 1

(d) -1

The next step is to define negative integer exponents. If they are to be defined in such a way that the quotient rule given above remains valid, then we must have, for example,

$$\frac{3^2}{3^4} = 3^{2-4} = 3^{-2}.$$

However,

$$\frac{3^2}{3^4} = \frac{\cancel{3} \cdot \cancel{3}}{\cancel{3} \cdot \cancel{3} \cdot 3 \cdot 3} = \frac{1}{3^2},$$

which suggests that 3^{-2} should be defined to be $1/3^2$. Thus, we have the following definition of a negative exponent.

Negative Exponent

If n is a natural number, and if $a \neq 0$, then

$$a^{-n} = \frac{1}{a^n}.$$

2

Evaluate the following.

(a) 6^{-2}

(b) -6^{-3}

(c) -3^{-4}

(d) $\left(\frac{5}{8}\right)^{-1}$

(e) $\left(\frac{1}{2}\right)^{-4}$

(f) $\left(\frac{7}{3}\right)^{-2}$

Answers:

(a) 1/36

(b) $-1/216$

(c) $-1/81$

(d) 8/5

(e) 16

(f) 9/49

EXAMPLE 2 Evaluate the following.

(a) $3^{-2} = \frac{1}{3^2} = \frac{1}{9}$

(b) $5^{-4} = \frac{1}{5^4} = \frac{1}{625}$

(c) $9^{-1} = \frac{1}{9^1} = \frac{1}{9}$

(d) $-4^{-2} = -\frac{1}{4^2} = -\frac{1}{16}$

(e) $\left(\frac{3}{4}\right)^{-1} = \frac{1}{\left(\frac{3}{4}\right)^1} = \frac{1}{\frac{3}{4}} = \frac{4}{3}$

(f) $\left(\frac{2}{3}\right)^{-3} = \frac{1}{\left(\frac{2}{3}\right)^3} = \frac{1}{\left(\frac{2^3}{3^3}\right)} = 1 \cdot \frac{3^3}{2^3} = \frac{3^3}{2^3} = \frac{27}{8}$ ◀

Parts (e) and (f) of Example 2 involve work with fractions that can lead to error. For a useful shortcut with such fractions, use the properties of division of rational numbers and the definition of a negative exponent to get

$$\left(\frac{a}{b}\right)^{-n} = \frac{1}{\left(\frac{a}{b}\right)^n} = \frac{1}{\left(\frac{a^n}{b^n}\right)} = 1 \cdot \frac{b^n}{a^n} = \frac{b^n}{a^n} = \left(\frac{b}{a}\right)^n.$$ (2)

When keying in negative exponents on a calculator, be sure to use the negation key (labeled (−) or +/−), not the subtraction key. Calculators normally display answers as decimals, as shown in Figure 1.7 for parts (a), (b), and (f) of Example 2. Some graphing calculators have a FRAC key that converts these decimals to fractions as shown in Figure 1.8.

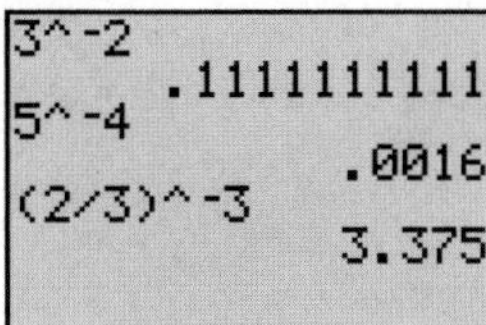

FIGURE 1.7

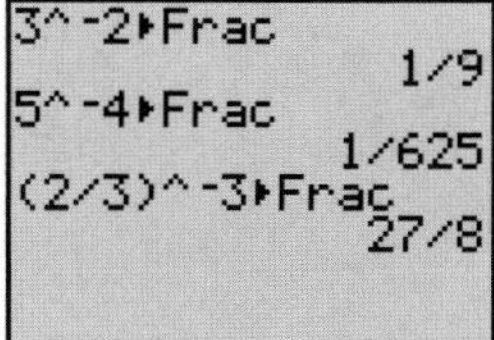

FIGURE 1.8

TECHNOLOGY TIP The FRAC key is in the MATH menu of TI graphing calculators. A FRAC program for other graphing calculators is in the Program Appendix. Fractions can be displayed on some graphing calculators by changing the number display format (in the MODES menu) to "fraction" or "exact."

Roots and Rational Exponents

The definition of a^n will now be extended to include rational values of n, such as 1/2 and 5/4. In order to do this, we need some terminology.

There are two numbers whose square is 16, 4 and -4. As we saw in Section 1.1, the positive one, 4, is called the *square root* (or second root) of 16. Similarly, there are two numbers whose fourth power is 16, 2 and -2. We call 2 the **fourth root** of 16. This suggests the following generalization.

If n is even, the ***n*th root of *a*** is the positive real number whose nth power is a.

All nonnegative numbers have nth roots for every natural number n, but *no negative number has a real, even nth root.* For example, there is no real number whose square is -16, so -16 has no square root.

We say that the **cube root** (or third root) of 8 is 2 because $2^3 = 8$. Similarly, since $(-2)^3 = -8$, we say -2 is the cube root of -8. Again, we can generalize.

If n is odd, the ***n*th root of *a*** is the real number whose nth power is a.

Every real number has an nth root for every *odd* natural number n.

We can now define rational exponents. If they are to have the same properties as integer exponents, we want $a^{1/2}$ to be a number such that

$$(a^{1/2})^2 = a^{1/2} \cdot a^{1/2} = a^{1/2 + 1/2} = a^1 = a.$$

Thus, $a^{1/2}$ should be a number whose square is a and it is reasonable to *define* $a^{1/2}$ to be the square root of a (if it exists). Similarly, $a^{1/3}$ is defined to be the cube root of a and we have the following definition.

Evaluate the following.

(a) $16^{1/2}$
(b) $16^{1/4}$
(c) $-256^{1/2}$
(d) $(-256)^{1/2}$
(e) $-8^{1/3}$
(f) $243^{1/5}$

Answers:

(a) 4
(b) 2
(c) -16
(d) Not a real number
(e) -2
(f) 3

If a is a real number and n is a positive integer, then

$a^{1/n}$ is defined to be the nth root of a (if it exists).

EXAMPLE 3 Evaluate the following roots.

(a) $36^{1/2} = 6$ because $6^2 = 36$.
(b) $100^{1/2} = 10$ because $10^2 = 100$.
(c) $-(225^{1/2}) = -15$ because $15^2 = 225$.
(d) $625^{1/4} = 5$ because $5^4 = 625$.
(e) $(-1296)^{1/4}$ is not a real number, although $-1296^{1/4} = -6$ because $6^4 = 1296$.
(f) $(-27)^{1/3} = -3$ because $(-3)^3 = -27$.
(g) $-32^{1/5} = -2$ because $2^5 = 32$. ◀ ③

A calculator can be used to evaluate expressions with fractional exponents. Whenever it's easy to do so, enter the fractional exponents in their equivalent decimal form. For instance, to find $625^{1/4}$, enter $625^{.25}$ on the calculator. When the decimal equivalent of a fraction is an infinitely repeating decimal, however, it is best to enter the fractional exponent directly using parentheses, as in Figure 1.9. If you omit the parentheses or use a shortened decimal approximation (such as .333 for 1/3), you will not get the correct answers because the calculator will not interpret these exponents as you intended, as shown in Figure 1.10.

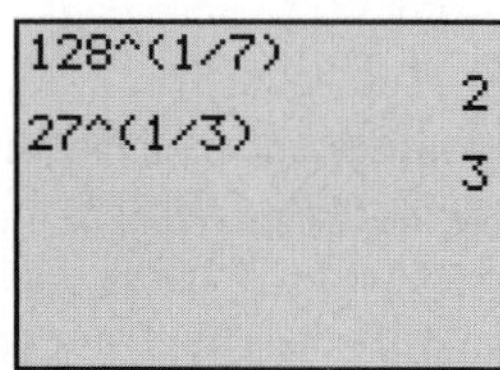

FIGURE 1.9

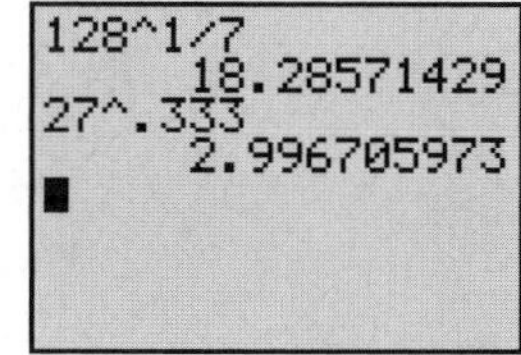

FIGURE 1.10

For more general rational exponents, the symbol $a^{m/n}$ should be defined so that the properties for exponents still hold. For example, by the product property, we want

$$(a^{1/3})^2 = a^{1/3} \cdot a^{1/3} = a^{1/3+1/3} = a^{2/3}$$

This suggests the following definition.

For all integers m and all positive integers n, and for all real numbers a for which $a^{1/n}$ is a real number,

$$a^{m/n} = (a^{1/n})^m.$$

Evaluate the following.

(a) $16^{3/4}$

(b) $25^{5/2}$

(c) $32^{7/5}$

(d) $100^{3/2}$

Answers:

(a) 8

(b) 3125

(c) 128

(d) 1000

EXAMPLE 4 Evaluate the following.

(a) $27^{2/3} = (27^{1/3})^2 = 3^2 = 9$

(b) $32^{2/5} = (32^{1/5})^2 = 2^2 = 4$

(c) $64^{4/3} = (64^{1/3})^4 = 4^4 = 256$

(d) $25^{3/2} = (25^{1/2})^3 = 5^3 = 125$

CAUTION When the base is negative, as in $(-8)^{2/3}$, many calculators produce an error message. On such calculators, you should first compute $(-8)^{1/3}$ and then square the result; that is, compute $[(-8)^{1/3}]^2$.

Rational exponents were defined so that one of the familiar properties of exponents remains valid. In fact, all of the rules for exponents developed earlier for integer exponents are valid for rational exponents, as summarized below and illustrated in Examples 5 and 6.

Properties of Exponents

For any rational numbers m and n, and any real numbers a and b for which the following exist,

(a) $a^m \cdot a^n = a^{m+n}$ Product property

(b) $\frac{a^m}{a^n} = a^{m-n}$ Quotient property

(c) $(a^m)^n = a^{mn}$

(d) $(ab)^m = a^m \cdot b^m$ Power properties

(e) $\left(\frac{a}{b}\right)^m = \frac{a^m}{b^m}$

(f) $a^0 = 1$

(g) $a^{-n} = \frac{1}{a^n}$

(h) $\left(\frac{a}{b}\right)^{-n} = \left(\frac{b}{a}\right)^n.$

EXAMPLE 5 Use the properties of exponents to simplify each of the following. Write answers with positive exponents.

(a) $7^{-4} \cdot 7^6 = 7^{-4+6} = 7^2$ Product property

(b) $\frac{9^{14}}{9^{-6}} = 9^{14-(-6)} = 9^{20}$ Quotient property

(c) $(2^{-3})^{-4} = 2^{(-3)(-4)} = 2^{12}$ Power property

5

Simplify the following.

(a) $9^6 \cdot 9^{-4}$

(b) $\dfrac{8^7}{8^{-3}}$

(c) $(13^4)^{-3}$

(d) $6^{2/5} \cdot 6^{3/5}$

(e) $\dfrac{8^{2/3} \cdot 8^{-4/3}}{8^2}$

Answers:

(a) 9^2

(b) 8^{10}

(c) $1/13^{12}$

(d) 6

(e) $1/8^{8/3}$ or $1/2^8$

(d) $\dfrac{27^{1/3} \cdot 27^{5/3}}{27^3} = \dfrac{27^{1/3+5/3}}{27^3}$ Product property

$= \dfrac{27^2}{27^3} = 27^{2-3}$ Quotient property

$= 27^{-1} = \dfrac{1}{27}$ Definition of negative exponent ◀ 5

You can use a calculator to check computations, such as those in Example 5, by computing the left and right sides separately and confirming that the answers are the same in each case. Figure 1.11 shows this for part (a) of Example 5.

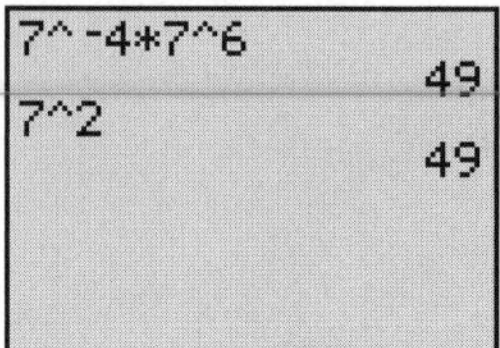

FIGURE 1.11

Simplify the following. Give answers with only positive exponents. Assume all variables represent positive real numbers.

(a) $\dfrac{(t^{-1})^2}{t^{-5}}$

(b) $\dfrac{(3z)^{-1}z^4}{z^2}$

(c) $3x^{1/4} \cdot 5x^{5/4}$

(d) $\left(\dfrac{2k^{1/3}}{p^{5/4}}\right)^2 \cdot \left(\dfrac{4k^{-2}}{p^5}\right)^{3/2}$

(e) $a^{5/8}(2a^{3/8} + a^{-1/8})$

Answers:

(a) t^3

(b) $z/3$

(c) $15x^{3/2}$

(d) $32/(p^{10}k^{7/3})$

(e) $2a + a^{1/2}$

▶ **EXAMPLE 6** Simplify each expression. Give answers with only positive exponents. Assume all variables represent positive real numbers.

(a) $\dfrac{(m^3)^{-2}}{m^4} = \dfrac{m^{-6}}{m^4} = m^{-6-4} = m^{-10} = \dfrac{1}{m^{10}}$

(b) $6y^{2/3} \cdot 2y^{-1/2} = 12y^{2/3-1/2} = 12y^{1/6}$

(c) $\left(\dfrac{3m^{5/6}}{y^{3/4}}\right)^2 = \dfrac{3^2(m^{5/6})^2}{(y^{3/4})^2} = \dfrac{3^2m^{5/3}}{y^{3/2}} = \dfrac{9m^{5/3}}{y^{3/2}}$

(d) $m^{2/3}(m^{7/3} + 2m^{1/3}) = (m^{2/3+7/3} + 2m^{2/3+1/3}) = m^3 + 2m$ ◀ 6

Radicals

The nth root of a was denoted above as $a^{1/n}$. An alternative notation for nth roots uses the **radical** symbol $\sqrt[n]{}$.

> If n is an even natural number and $a \geq 0$, or if n is an odd natural number,
> $$\sqrt[n]{a} = a^{1/n}.$$

In the radical expression $\sqrt[n]{a}$, a is called the **radicand** and n is called the **index.** When $n = 2$, the familiar square root symbol $\sqrt{a}$ is used instead of $\sqrt[2]{a}$.

7

Simplify.

(a) $\sqrt[3]{27}$

(b) $\sqrt[4]{625}$

(c) $\sqrt[6]{64}$

(d) $\sqrt[3]{\dfrac{64}{125}}$

Answers:

(a) 3

(b) 5

(c) 2

(d) 4/5

EXAMPLE 7 Simplify the following.

(a) $\sqrt[4]{16} = 16^{1/4} = 2$

(b) $\sqrt[5]{-32} = -2$

(c) $\sqrt[3]{1000} = 10$

(d) $\sqrt[6]{\dfrac{64}{729}} = \left(\dfrac{64}{729}\right)^{1/6} = \dfrac{64^{1/6}}{729^{1/6}} = \dfrac{2}{3}$ ◀ 7

The symbol $a^{m/n}$ also can be written in an alternative notation using radicals.

For all rational numbers m/n and all real numbers a for which $\sqrt[n]{a}$ exists,

$$a^{m/n} = (\sqrt[n]{a})^m \quad \text{or} \quad a^{m/n} = \sqrt[n]{a^m}.$$

Notice that $\sqrt[n]{x^n}$ cannot be written simply as x when n is even. For example, if $x = -5$,

$$\sqrt{x^2} = \sqrt{(-5)^2} = \sqrt{25} = 5 \neq x.$$

However, $|-5| = 5$, so that $\sqrt{x^2} = |x|$ when x is -5. This is true in general.

For any real number a and any natural number n,

$$\sqrt[n]{a^n} = |a| \text{ if } n \text{ is even}$$

and

$$\sqrt[n]{a^n} = a \text{ if } n \text{ is odd.}$$

To avoid this difficulty that $\sqrt[n]{a^n}$ is not necessarily equal to a, we shall assume that all variables in radicands represent only nonnegative numbers, as they usually do in applications.

The properties of exponents can be written with radicals as shown below.

For all real numbers a and b and positive integers n for which all indicated roots exist,

(a) $\sqrt[n]{a} \cdot \sqrt[n]{b} = \sqrt[n]{ab}$ **(b)** $\dfrac{\sqrt[n]{a}}{\sqrt[n]{b}} = \sqrt[n]{\dfrac{a}{b}}$ $(b \neq 0)$.

8

Simplify.

(a) $\sqrt{3} \cdot \sqrt{27}$

(b) $\sqrt{\dfrac{3}{49}}$

(c) $\sqrt{25 - 4}$

(d) $\sqrt{25} - \sqrt{4}$

Answers:

(a) 9

(b) $\dfrac{\sqrt{3}}{7}$

(c) $\sqrt{21}$

(d) 3

EXAMPLE 8 Simplify the following.

(a) $\sqrt{6} \cdot \sqrt{54} = \sqrt{6 \cdot 54} = \sqrt{324} = 18$

Alternatively, simplify $\sqrt{54}$ first.

$$\sqrt{6} \cdot \sqrt{54} = \sqrt{6} \cdot \sqrt{9 \cdot 6}$$
$$= \sqrt{6} \cdot 3\sqrt{6} = 3 \cdot 6 = 18$$

(b) $\sqrt{\dfrac{7}{64}} = \dfrac{\sqrt{7}}{\sqrt{64}} = \dfrac{\sqrt{7}}{8}$

(c) $\sqrt{9 + 16} = \sqrt{25} = 5$

When using a calculator to evaluate radical expressions, be sure to use parentheses. To compute $\sqrt{9 + 16}$ on a calculator, for example, you must enter it as $\sqrt{\ }$ (9 + 16) in order to get the correct answer 5. If you enter $\sqrt{\ }$ 9 + 16, most calculators will interpret it as 3 + 16 = 19, as shown in Figure 1.12. 8

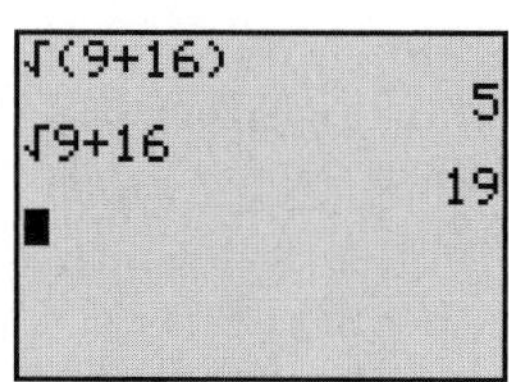

FIGURE 1.12

Multiplying radical expressions is much like multiplying polynomials.

9

Multiply.

(a) $(\sqrt{5} - \sqrt{2})(3 + \sqrt{2})$

(b) $(\sqrt{3} + \sqrt{7})(\sqrt{3} - \sqrt{7})$

Answers:

(a) $3\sqrt{5} + \sqrt{10} - 3\sqrt{2} - 2$

(b) -4

EXAMPLE 9 Multiply the following.

(a) $(\sqrt{2} + 3)(\sqrt{8} - 5) = \sqrt{2}(\sqrt{8}) - \sqrt{2}(5) + 3\sqrt{8} - 3(5)$ FOIL

$$= \sqrt{16} - 5\sqrt{2} + 3(2\sqrt{2}) - 15$$
$$= 4 - 5\sqrt{2} + 6\sqrt{2} - 15$$
$$= -11 + \sqrt{2}$$

(b) $(\sqrt{7} - \sqrt{10})(\sqrt{7} + \sqrt{10}) = (\sqrt{7})^2 - (\sqrt{10})^2$

$$= 7 - 10 = -3$$ 8

Rationalizing Denominators and Numerators

Before the invention of calculators, it was customary to **rationalize the denominators** of fractions (that is, write equivalent fractions with no radicals in the denominator) because this made many computations easier. Although there is no longer a computational reason, rationalizing denominators (and sometimes numerators) is still used today to simplify expressions and to derive useful formulas.

EXAMPLE 10 Rationalize each denominator.

(a) $\dfrac{4}{\sqrt{3}}$

The key is to multiply by 1, with 1 written as a radical fraction:

$$\frac{4}{\sqrt{3}} = \frac{4}{\sqrt{3}} \cdot 1 = \frac{4}{\sqrt{3}} \cdot \frac{\sqrt{3}}{\sqrt{3}} = \frac{4\sqrt{3}}{3}.$$

(b) $\dfrac{1}{3 - \sqrt{2}}$

The same technique works here, using $1 = \dfrac{3 + \sqrt{2}}{3 + \sqrt{2}}$:

$$\frac{1}{3-\sqrt{2}} = \frac{1}{3-\sqrt{2}} \cdot 1 = \frac{1}{3-\sqrt{2}} \cdot \frac{3+\sqrt{2}}{3+\sqrt{2}} = \frac{3+\sqrt{2}}{(3-\sqrt{2})(3+\sqrt{2})}$$

$$= \frac{3+\sqrt{2}}{9-2} = \frac{3+\sqrt{2}}{7}.$$

10

Rationalize the denominator.

(a) $\dfrac{2}{\sqrt{5}}$

(b) $\dfrac{1}{2 + \sqrt{3}}$

Answers:

(a) $\dfrac{2\sqrt{5}}{5}$

(b) $2 - \sqrt{3}$

EXAMPLE 11 Rationalize the numerator of $\dfrac{2 + \sqrt{5}}{1 + \sqrt{3}}$.

Multiply the fraction by $1 = \dfrac{2 - \sqrt{5}}{2 - \sqrt{5}}$:

$$\frac{2+\sqrt{5}}{1+\sqrt{3}} = \frac{2+\sqrt{5}}{1+\sqrt{3}} \cdot \frac{2-\sqrt{5}}{2-\sqrt{5}} = \frac{4-5}{2-\sqrt{5}+2\sqrt{3}-\sqrt{3}\sqrt{5}}$$

$$= \frac{-1}{2-\sqrt{5}+2\sqrt{3}-\sqrt{15}}.$$

1.5 Exercises

Evaluate each expression. Write all answers without exponents. (See Examples 1 and 2.)

1. 5^0 **2.** 8^0 **3.** 6^{-1}

4. 10^{-3} **5.** 2^{-5} **6.** -4^{-3}

7. -7^{-4} **8.** $\left(\frac{1}{3}\right)^{-2}$ **9.** $\left(\frac{1}{6}\right)^{-3}$

10. $\left(\frac{2}{5}\right)^{-4}$ **11.** $\left(\frac{4}{3}\right)^{-2}$

12. Explain why $-2^{-4} = -1/16$, but $(-2)^{-4} = 1/16$.

Evaluate each expression. Write all answers without exponents. Write decimal answers to the nearest tenth. (See Examples 3 and 4.)

13. $49^{1/2}$ **14.** $8^{1/3}$ **15.** $(7.51)^{1/4}$

16. $(68.93)^{1/5}$ **17.** $27^{2/3}$ **18.** $243^{2/2}$

19. $-64^{2/3}$ **20.** $-64^{3/2}$ **21.** $(8/27)^{-4/3}$

22. $(27/64)^{-1/3}$

Simplify each expression. Write all answers using only positive exponents. (See Example 5.)

23. $\frac{4^{-2}}{4^3}$ **24.** $\frac{9^{-4}}{9^{-3}}$

25. $4^{-3} \cdot 4^6$ **26.** $5^{-9} \cdot 5^{10}$

27. $8^{2/3} \cdot 8^{-1/3}$ **28.** $12^{-3/4} \cdot 12^{1/4}$

29. $\frac{8^9 \cdot 8^{-7}}{8^{-3}}$ **30.** $\frac{5^{-4} \cdot 5^6}{5^{-1}}$

31. $\frac{9^{-5/3}}{9^{2/3} \cdot 9^{-1/5}}$ **32.** $\frac{3^{5/3} \cdot 3^{-3/4}}{3^{-1/4}}$

Simplify each expression. Assume all variables represent positive real numbers. Write answers with only positive exponents. (See Example 6.)

33. $\frac{z^5 \cdot z^2}{z^4}$ **34.** $\frac{k^6 \cdot k^9}{k^{12}}$

35. $\frac{2^{-1}(p^{-1})^3}{2p^{-4}}$ **36.** $\frac{(5x^3)^{-2}}{x^4}$

37. $(q^{-5}r^2)^{-1}$ **38.** $(2y^2z^{-2})^{-3}$

39. $(2p^{-1})^3 \cdot (5p^2)^{-2}$ **40.** $(5^{-1}m^2)^{-3} \cdot (3m^{-2})^4$

41. $(2p)^{1/2} \cdot (2p^3)^{1/3}$ **42.** $(5k^2)^{3/2} \cdot (5k^{1/3})^{3/4}$

43. $p^{2/3}(2p^{1/3} + 5p)$ **44.** $2z^{1/2}(3z^{-1/2} + z^{1/2})$

Match the rational exponent expression in Column I with the equivalent radical expression in Column II. Assume that x is not zero.

I	II
45. $(-3x)^{1/3}$	**(a)** $\frac{3}{\sqrt[3]{x}}$
46. $-3x^{1/3}$	**(b)** $-3\sqrt[3]{x}$
47. $(-3x)^{-1/3}$	**(c)** $\frac{1}{\sqrt[3]{3x}}$
48. $-3x^{-1/3}$	**(d)** $\frac{-3}{\sqrt[3]{x}}$
49. $(3x)^{1/3}$	**(e)** $3\sqrt[3]{x}$
50. $3x^{-1/3}$	**(f)** $\sqrt[3]{-3x}$
51. $(3x)^{-1/3}$	**(g)** $\sqrt[3]{3x}$
52. $3x^{1/3}$	**(h)** $\frac{1}{\sqrt[3]{-3x}}$

Simplify each of the following. (See Examples 7–9.)

53. $\sqrt[3]{64}$ **54.** $\sqrt[6]{64}$

55. $\sqrt[4]{625}$ **56.** $\sqrt[5]{-243}$

57. $\sqrt[7]{-128}$ **58.** $\sqrt{44} \cdot \sqrt{11}$

59. $\sqrt[3]{81} \cdot \sqrt[3]{9}$ **60.** $\sqrt{49 - 16}$

61. $\sqrt{81 - 4}$ **62.** $\sqrt{49} - \sqrt{16}$

63. $\sqrt{81} - \sqrt{4}$

64. $(\sqrt{2} + 3)(\sqrt{2} - 3)$

65. $(\sqrt{5} + \sqrt{2})(\sqrt{5} - \sqrt{2})$

Rationalize the denominator of each of the following. (See Example 10.)

66. $\frac{3}{1 - \sqrt{2}}$ **67.** $\frac{2}{1 + \sqrt{5}}$

68. $\frac{4 - \sqrt{2}}{2 - \sqrt{2}}$ **69.** $\frac{\sqrt{3} - 1}{\sqrt{3} - 2}$

Rationalize the numerator of each of the following. (See Example 11.)

70. $\frac{1 + \sqrt{7}}{2 - \sqrt{3}}$ **71.** $\frac{2 - \sqrt{3}}{2 + \sqrt{3}}$

72. What is wrong with the statement $\sqrt[3]{4} \cdot \sqrt[3]{4} = 4$?

The following exercises are applications of exponentiation and radicals.

73. Business The theory of economic lot size shows that, under certain conditions, the number of units to order to minimize total cost is

$$x = \sqrt{\frac{kM}{f}}.$$

Here k is the cost to store one unit for one year, f is the (constant) setup cost to manufacture the product, and M is the total number of units produced annually. (See Section 12.3.) Find x for the following values of f, k, and M.

(a) $k = \$1$, $f = \$500$, $M = 100{,}000$
(b) $k = \$3$, $f = \$7$, $M = 16{,}700$
(c) $k = \$1$, $f = \$5$, $M = 16{,}800$

74. Social Science The threshold weight T for a person is the weight above which the risk of death increases greatly. One researcher found that the threshold weight in pounds for men aged 40–49 is related to height in inches by the equation $h = 12.3T^{1/3}$. What height corresponds to a threshold of 216 pounds for a man in this age group?

Natural Science *The following exercises refer to the table of wind-chill factors that precedes Exercises 55–58 of Section 1.1.*

75. The equation for calculating the wind-chill C was revised in 2001. Before that time, the following equation could be used to approximate the wind-chill:

$$C = 91.4 - (91.4 - T)(.478 + .301\sqrt{v} - .02v),$$

where T is the temperature and v is the wind velocity. Use this equation to find C when

(a) $T = -10$ and $v = 30$
(b) $T = -40$ and $v = 5$.
(c) How do the old wind-chill factors in parts (a) and (b) compare with the new ones given by the table?

76. Compare the wind-chill factors given by the old equation in Exercise 75 and the current table for a temperature of 0° and wind velocities of 10, 20, 30, and 40 mph. Under these conditions, how do the two formulas compare?

Business *The Consumer Price Index (CPI) in U.S. cities stood at* 100 *in* 1982–1984. *In recent years it can be approximated by*

$$161.44x^{.0442} \qquad (1 \le x \le 10),$$

where $x = 1$ *corresponds to* 1998, $x = 2$ *to* 1999, *and so on.* Approximate the CPI in the following years.*

77. First half of 1999 ($x = 2$)

78. Last half of 2000 ($x = 3.5$)

79. Last half of 2003

80. First half of 2004

Health *According to data from the National Organ Procurement and Transplantation Network, the number of kidney transplants (in thousands) in year x can be approximated by*

$$8.064(1.032^{x}) \qquad (3 \le x \le 18),$$

where $x = 3$ *corresponds to* 1988, $x = 4$ *to* 1989, *and so on. Find the approximate number of kidney transplants for each year.*

81. 1990

82. 1996

83. 1999

84. 2001

85. 2003

86. According to the United Network for Organ Sharing, approximately 46,800 people were waiting for a kidney transplant in 2000. About how many of these people did *not* get a transplant that year?

*U.S. Bureau of Labor.

87. Business Based on data from *Ward's Communications*, the number of sport utility vehicles (SUVs) sold each year (in millions) can be approximated by

$$\frac{12.8}{1 + 14.97(.838^{x})} \qquad (1 \le x \le 14),$$

where $x = 1$ corresponds to 1991, $x = 2$ to 1992, and so on.

(a) Approximately how many SUVs were sold in 1996? In 1999? In 2002?

(b) Assume that this formula remains valid over the next forty years (probably unlikely). Use the table feature of a graphing calculator to determine when SUV sales would reach 12,000,000. According to this approximation, would sales ever reach 13,000,000?

1.6 FIRST-DEGREE EQUATIONS

An **equation** is a statement that two mathematical expressions are equal; for example,

$$5x - 3 = 13, \qquad 8y = 4, \qquad -3p + 5 = 4p - 8.$$

The letter in each equation is called the variable. This section concentrates on **first-degree equations,** which are equations that involve only constants and the first power of the variable. All of the equations displayed above are first-degree equations, but neither of the following equations is first-degree: $2x^2 = 5x + 6$ (variable has exponent greater than 1) and $\sqrt{x + 2} = 4$ (because the variable is under the radical).

Is -4 a solution of the following equations?

(a) $3x + 5 = -7$

(b) $2x - 3 = 5$

(c) Is there more than one solution of the equation in part (a)?

Answers:

(a) Yes

(b) No

(c) No

A **solution** of an equation is a number that can be substituted for the variable in the equation to produce a true statement. For example, substituting the number 9 for x in the equation $2x + 1 = 19$ gives

$$2x + 1 = 19$$

$$2(9) + 1 \stackrel{?}{=} 19 \qquad \text{Let } x = 9.$$

$$18 + 1 = 19. \qquad \text{True}$$

This true statement indicates that 9 is a solution of $2x + 1 = 19$. (1)

The following properties are used to solve equations.

Properties of Equality

1. The same number may be added to or subtracted from both sides of an equation:

$$\text{If } a = b, \text{ then } a + c = b + c \quad \text{and} \quad a - c = b - c.$$

2. Both sides of an equation may be multiplied or divided by the same nonzero number:

$$\text{If } a = b \text{ and } c \neq 0, \text{ then } ac = bc \quad \text{and} \quad \frac{a}{c} = \frac{b}{c}.$$

(2)

Solve the following.

(a) $3p - 5 = 19$

(b) $4y + 3 = -5$

(c) $-2k + 6 = 2$

Answers:

(a) 8

(b) -2

(c) 2

EXAMPLE 1 Solve the equation $5x - 3 = 12$.

Using the first property of equality, add 3 to both sides. This isolates the term containing the variable on one side of the equation.

$$5x - 3 = 12$$

$$5x - 3 + 3 = 12 + 3 \qquad \text{Add 3 to both sides.}$$

$$5x = 15$$

Now arrange for the coefficient of x to be 1 by using the second property of equality.

$$5x = 15$$

$$\frac{5x}{5} = \frac{15}{5} \qquad \text{Divide both sides by 5.}$$

$$x = 3$$

The solution of the original equation, $5x - 3 = 12$, is 3. Check the solution by substituting 3 for x in the original equation. (2)

EXAMPLE 2 Solve $2k + 3(k - 4) = 2(k - 3)$.

First, simplify the equation by using the distributive property on the left-side term $3(k - 4)$ and right-side term $2(k - 3)$:

$$2k + 3(k - 4) = 2(k - 3)$$

$$2k + 3k - 12 = 2(k - 3) \qquad \text{Distributive property}$$

$$2k + 3k - 12 = 2k - 6 \qquad \text{Distributive property}$$

$$5k - 12 = 2k - 6 \qquad \text{Collect like terms on left side.}$$

One way to proceed is to add $-2k$ to both sides.

$$5k - 12 + (-2k) = 2k - 6 + (-2k) \qquad \text{Add } -2k \text{ to both sides.}$$

$$3k - 12 = -6$$

$$3k - 12 + 12 = -6 + 12 \qquad \text{Add 12 to both sides.}$$

$$3k = 6$$

$$\frac{1}{3}(3k) = \frac{1}{3}(6) \qquad \text{Multiply both sides by } \frac{1}{3}.$$

$$k = 2$$

The solution is 2. Check this result by substituting 2 for k in the original equation. ◀ (3)

(3)

Solve the following.

(a) $3(m - 6) + 2(m + 4) = 4m - 2$

(b) $-2(y + 3) + 4y = 3(y + 1) - 6$

Answers:

(a) 8

(b) -3

EXAMPLE 3 Use a calculator to solve $42.19x + 121.34 = 16.83x + 19.15$.

To avoid round-off errors in the intermediate steps, do all the algebra first, without using the calculator.

$$42.19x = 16.83x + 19.15 - 121.34 \qquad \text{Subtract 121.34 from both sides.}$$

$$42.19x - 16.83x = 19.15 - 121.34 \qquad \text{Subtract } 16.83x \text{ from both sides.}$$

$$(42.19 - 16.83)x = 19.15 - 121.34 \qquad \text{Distributive property}$$

$$x = \frac{19.15 - 121.34}{42.19 - 16.83} \qquad \text{Divide both sides by } (42.19 - 16.83).$$

Now use a calculator, being careful to put parentheses around numerator and denominator, and determine that $x \approx -4.0296$, as shown in Figure 1.13. Because this answer is approximate, it may not check exactly when substituted in the original equation. ◀

FIGURE 1.13

The next three examples show how to simplify the solution of first-degree equations involving fractions. We solve these equations by multiplying both sides of the equation by a *common denominator*. This step will eliminate the fractions.

EXAMPLE 4 Solve $\frac{r}{10} - \frac{2}{15} = \frac{3r}{20} - \frac{1}{5}$.

Here the denominators are 10, 15, 20, and 5. Each of these numbers can be divided into 60; therefore, 60 is a common denominator. Multiply both sides of the equation by 60.

$$60\left(\frac{r}{10} - \frac{2}{15}\right) = 60\left(\frac{3r}{20} - \frac{1}{5}\right)$$

$$60\left(\frac{r}{10}\right) - 60\left(\frac{2}{15}\right) = 60\left(\frac{3r}{20}\right) - 60\left(\frac{1}{5}\right) \qquad \text{Distributive property}$$

$$6r - 8 = 9r - 12$$

$$6r - 8 + (-6r) + 12 = 9r - 12 + (-6r) + 12 \qquad \text{Add } -6r \text{ and 12 to both sides.}$$

$$4 = 3r$$

$$r = \frac{4}{3} \qquad \text{Multiply both sides by 1/3.}$$

Check this solution in the original equation. ◀ 4

Solve the following.

(a) $\frac{x}{2} - \frac{x}{4} = 6$

(b) $\frac{2x}{3} + \frac{1}{2} = \frac{x}{4} - \frac{9}{2}$

Answers:

(a) 24

(b) -12

CAUTION Multiplying *both* sides of the *equation* by a number to eliminate fractions is valid. But multiplying a single fraction to simplify it is not valid. For instance, multiplying $\frac{3x}{8}$ by 8 *changes* it to $3x$, which is *not equal* to $\frac{3x}{8}$.

EXAMPLE 5 Solve $\frac{4}{3(k+2)} - \frac{k}{3(k+2)} = \frac{5}{3}$.

Multiply both sides of the equation by the common denominator $3(k + 2)$. Here $k \neq -2$, since $k = -2$ would give a 0 denominator, making the fraction undefined.

$$3(k+2) \cdot \frac{4}{3(k+2)} - 3(k+2) \cdot \frac{k}{3(k+2)} = 3(k+2) \cdot \frac{5}{3}$$

Simplify each side and solve for k.

$$4 - k = 5(k + 2)$$

$$4 - k = 5k + 10 \qquad \text{Distributive property}$$

$$4 - k + k = 5k + 10 + k \qquad \text{Add } k \text{ to both sides.}$$

$$4 = 6k + 10$$

$$4 + (-10) = 6k + 10 + (-10) \qquad \text{Add } -10 \text{ to both sides.}$$

$$-6 = 6k$$

$$-1 = k \qquad \text{Multiply both sides by } \frac{1}{6}.$$

5

Solve the equation.

$$\frac{5p+1}{3(p+1)} = \frac{3p-3}{3(p+1)} + \frac{9p-3}{3(p+1)}$$

Answer:

1

The solution is -1. Substitute -1 for k as a check.

$$\frac{4}{3(-1+2)} - \frac{-1}{3(-1+2)} \stackrel{?}{=} \frac{5}{3}$$

$$\frac{4}{3} - \frac{-1}{3} \stackrel{?}{=} \frac{5}{3}$$

$$\frac{5}{3} = \frac{5}{3}$$

The check shows that -1 is the solution. ◀ 5

CAUTION Whenever you multiply an equation by an expression involving the variable, as in Example 5, it is *essential* to check the solution, as the next example illustrates.

Solve each equation.

(a) $\frac{3p}{p+1} = 1 - \frac{3}{p+1}$

(b) $\frac{8y}{y-4} = \frac{32}{y-4} - 3.$

Answer:

Neither equation has a solution.

EXAMPLE 6 Solve $\frac{3x-4}{x-2} = \frac{x}{x-2}$.

Multiplying both sides by $x - 2$ produces

$$3x - 4 = x$$

$2x - 4 = 0$ Subtract x from both sides.

$2x = 4$ Add 4 to both sides.

$x = 2$ Divide both sides by 2.

Substituting 2 for x in the original equation produces fractions with 0 denominators. Since division by 0 is not defined, the original equation has no solution. ◀ 6

Sometimes an equation with several variables must be solved for one of the variables. This process is called **solving for a specified variable.**

7

Solve for x.

(a) $2x - 7y = 3xk$

(b) $8(4 - x) + 6p = -5k - 11yx$

Answers:

(a) $x = \frac{7y}{2-3k}$

(b) $x = \frac{5k+32+6p}{8-11y}$

EXAMPLE 7 Solve for x: $3(ax - 5a) + 4b = 4x - 2$.

Use the distributive property to get

$$3ax - 15a + 4b = 4x - 2.$$

Treat x as the variable, the other letters as constants. Get all terms with x on one side of the equation and all terms without x on the other side.

$3ax - 4x = 15a - 4b - 2$ Isolate terms with x on the left.

$(3a - 4)x = 15a - 4b - 2$ Distributive property

$x = \frac{15a-4b-2}{3a-4}$ Multiply by $\frac{1}{3a-4}$.

The final equation is solved for x, as required. 

Absolute Value Equations

Recall from Section 1.1 that the absolute value of a number a is either a or $-a$, whichever one is positive. For instance, $|4| = 4$ and $|-7| = -(-7) = 7$.

EXAMPLE 8 Solve $|x| = 3$.

Since $|x|$ is either x or $-x$, the equation says that

$$x = 3 \quad \text{or} \quad -x = 3$$
$$x = -3.$$

The solutions of $|x| = 3$ are 3 and -3.

(8)

Solve each equation.

(a) $|y| = 9$

(b) $|r + 3| = 1$

(c) $|2k - 3| = 7$

Answers:

(a) 9, −9

(b) −2, −4

(c) 5, −2

EXAMPLE 9 Solve $|p - 4| = 2$.

Since $|p - 4|$ is either $p - 4$ or $-(p - 4)$, we have

$$p - 4 = 2 \quad \text{or} \quad -(p - 4) = 2$$
$$p = 6 \qquad -p + 4 = 2$$
$$-p = -2$$
$$p = 2,$$

so that 6 and 2 are possible solutions. Checking them in the original equation shows that both are solutions. (8)

(9)

Solve each equation.

(a) $|r + 6| = |2r + 1|$

(b) $|5k - 7| = |10k - 2|$

Answers:

(a) 5, −7/3

(b) −1, 3/5

EXAMPLE 10 Solve $|4m - 3| = |m + 6|$.

The quantities in absolute value bars must either be equal or be negatives of one another to satisfy the equation. That is,

$$4m - 3 = m + 6 \quad \text{or} \quad 4m - 3 = -(m + 6)$$
$$3m = 9 \qquad 4m - 3 = -m - 6$$
$$m = 3 \qquad 5m = -3$$
$$m = -\frac{3}{5}.$$

Check that the solutions for the original equation are 3 and $-3/5$.  (9)

Applications

One of the main reasons for learning mathematics is to be able to use it to solve practical problems. There are no hard and fast rules for dealing with real-world applications, except perhaps to use common sense. However, you will find it much easier to deal with such problems if you don't try to do everything at once. After reading the problem carefully, attack it in stages, as suggested in the following guidelines.

Solving Applied Problems

Step 1 Decide on the unknown. Name it with some variable that you *write down*. Many students try to skip this step. They are eager to get on with the writing of the equation. But this is an important step. If you don't know what the variable represents, how can you write a meaningful equation or interpret a result?

Step 2 Decide on a variable expression to represent any other unknowns in the problem. For example, if x represents the width of a rectangle, and you know that the length is one more than twice the width, then *write down* that the length is $1 + 2x$.

Step 3 Draw a sketch or make a chart, if appropriate, showing the information given in the problem.

Step 4 Using the results of Steps 1–3, write an equation that expresses a condition that must be satisfied.

Step 5 Solve the equation.

Step 6 Check the solution in the words of the *original problem,* not just in the equation you have written.

The following examples illustrate this approach.

EXAMPLE 11 A financial manager has $14,000 to invest for her company. She plans to invest part of the money in tax-free bonds at 6% interest and the remainder in taxable bonds at 9%. She wants to earn $1005 per year in interest from the investments. Find the amount she should invest at each rate.

Step 1 Let x represent the amount to be invested at 6%.

Step 2 After x dollars are invested, the remaining amount is $14{,}000 - x$ dollars, which is to be invested at 9%.

Step 3 Interest for one year is given by rate × amount invested. For instance, 6% of x dollars is $.06x$. The given information is summarized in this chart.

Investment	*Amount Invested*	*Interest Rate*	*Interest Earned in 1 Year*
Tax-free bonds	x	6% = .06	$.06x$
Taxable bonds	$14{,}000 - x$	9% = .09	$.09(14{,}000 - x)$
Totals	14,000		1005

Step 4 Because the total interest is to be $1005, the last column of the tables shows that

$$.06x + .09(14{,}000 - x) = 1005.$$

An investor owns two pieces of property. One, worth twice as much as the other, returns 6% in annual interest, while the other returns 4%. Find the value of each piece of property if the total annual interest earned is $8000.

Answers:

6% return: $100,000; 4% return: $50,000

Step 5 Solve this equation.

$$\begin{aligned} .06x + .09(14{,}000 - x) &= 1005 \\ .06x + .09(14{,}000) - .09x &= 1005 \\ .06x + 1260 - .09x &= 1005 \\ -.03x &= -255 \\ x &= 8500. \end{aligned}$$

The manager should invest $8500 at 6% and $14,000 − $8500 = $5500 at 9%.

Step 6 Check these results in the original problem. If $8500 is invested at 6%, the interest is .06(8500) = $510. If $5500 is invested at 9%, the interest is .09(5500) = $495. So the total interest is $510 + $495 = $1005 as required. ◀ (10)

▶ EXAMPLE 12 Chuck travels 80 kilometers in the same time that Mary travels 180 kilometers. Mary travels 50 kilometers per hour faster than Chuck. Find the speed of each person.

Step 1 Use x to represent Chuck's speed and $x + 50$ to represent Mary's speed, which is 50 kilometers per hour faster than Chuck's.

Steps 2 and 3 Constant rate problems of this kind require the distance formula

$$d = rt,$$

where d is the distance traveled in t hours at a constant rate of speed r. The distance traveled by each person is given, along with the fact that the time traveled by each person is the same. Solve the formula $d = rt$ for t.

$$\begin{aligned} d &= rt \\ \frac{1}{r} \cdot d &= \frac{1}{r} \cdot rt \\ \frac{d}{r} &= t \end{aligned}$$

For Chuck, $d = 80$ and $r = x$, giving $t = 80/x$. For Mary, $d = 180$, $r = x + 50$, and $t = 180/(x + 50)$. Use these facts to complete a chart, which organizes the information given in the problem.

	d	r	t
Chuck	80	x	$\frac{80}{x}$
Mary	180	$x + 50$	$\frac{180}{x + 50}$

Step 4 Because both people traveled for the *same time*, the equation is

$$\frac{80}{x} = \frac{180}{x + 50}.$$

11

(**a**) Tom and Dick are in a run for charity. Tom runs at 7 mph and Dick runs at 5 mph. If they start at the same time, how long will it be until they are 1/2 mile apart?
(**b**) In part (a), suppose the run has a staggered start. If Dick starts first, and Tom starts 10 minutes later, how long will it be until they are neck and neck?

Answers:

(a) 15 minutes (1/4 hour)

(b) Tom has run 25 minutes.

Step 5 Multiply both sides of the equation by $x(x + 50)$.

$$x(x+50)\frac{80}{x} = x(x+50)\frac{180}{x+50}$$

$$80(x + 50) = 180x$$
$$80x + 4000 = 180x$$
$$4000 = 100x$$
$$40 = x$$

Step 6 Since x represents Chuck's speed, Chuck went 40 kilometers per hour. Mary's speed is $x + 50$, or $40 + 50 = 90$ kilometers per hour. Check these results in the words of the original problem. ◀ 11

EXAMPLE 13 An oil company distributor needs to fill orders for 89 octane gas but only has 87 and 93 octane gas on hand. How much of each type should be mixed together to produce 54,000 gallons of 89 octane gas?

The octane rating of a gasoline indicates that it has the same antiknock qualities as a standard fuel made of heptane and isooctane. The octane rating is the percentage of isooctane in the standard fuel.* So we assume that standard fuels are being mixed. Let x be the amount of 87 octane gas. Then $54{,}000 - x$ is the amount of 93 octane gas. We can summarize the relevant information in a chart.

Type of Gas	*Quantity*	*% Isooctane*	*Amount of Isooctane*
87 octane	x	87%	$.87x$
93 octane	$54{,}000 - x$	93%	$.93(54{,}000 - x)$
Mixture	54,000	89%	$.89(54{,}000)$

The amount of isooctane satisfies this equation:

$$\text{(Amount in 87 octane)} + \text{(Amount in 93 octane)} = \text{Amount in mixture}$$
$$.87x + .93(54{,}000 - x) = .89(54{,}000)$$

12

How much 89 octane gas and how much 94 octane gas are needed to produce 1000 gallons of 91 octane gas?

Answer:

600 gallons of 89 octane; 400 gallons of 94 octane

Use the distributive property on the left side and multiply on the right side:

$$.87x + 50{,}220 - .93x = 48{,}060$$
$$-.06x = -2160$$
$$x = \frac{-2160}{-.06} = 36{,}000$$

So the distributor should mix 36,000 gallons of 87 octane gas with $54{,}000 - 36{,}000 = 18{,}000$ gallons of 93 octane gas to obtain 54,000 gallons of 89 octane gas ◀ 12

*This is one of several possible ways of determining octane ratings.

1.6 Exercises

Solve each equation. (See Examples 1–6.)

1. $3x + 5 = 20$
2. $4 - 5y = 9$
3. $.6k - .3 = .5k + .4$
4. $2.5 + 5.04m = 8.5 - .06m$
5. $2a - 1 = 3(a + 1) + 7a + 5$
6. $3(k - 2) - 6 = 4k - (3k - 1)$
7. $2[x - (3 + 2x) + 9] = 2x + 4$
8. $-2[4(k + 2) - 3(k + 1)] = 14 + 2k$
9. $\frac{3x}{5} - \frac{4}{5}(x + 1) = 2 - \frac{3}{10}(3x - 4)$
10. $\frac{4}{3}(x - 2) - \frac{1}{2} = 2\left(\frac{3}{4}x - 1\right)$
11. $\frac{5y}{6} - 8 = 5 - \frac{2y}{3}$
12. $\frac{x}{2} - 3 = \frac{3x}{5} + 1$
13. $\frac{m}{2} - \frac{1}{m} = \frac{6m + 5}{12}$
14. $-\frac{3k}{2} + \frac{9k - 5}{6} = \frac{11k + 8}{k}$
15. $\frac{4}{x - 3} - \frac{8}{2x + 5} + \frac{3}{x - 3} = 0$
16. $\frac{5}{2p + 3} - \frac{3}{p - 2} = \frac{4}{2p + 3}$
17. $\frac{3}{2m + 4} = \frac{1}{m + 2} - 2$
18. $\frac{8}{3k - 9} - \frac{5}{k - 3} = 4$

Use a calculator to solve each of the following equations. Round your answer to the nearest hundredth. (See Example 3.)

19. $9.06x + 3.59(8x - 5) = 12.07x + .5612$
20. $-5.74(3.1 - 2.7p) = 1.09p + 5.2588$
21. $\frac{2.63r - 8.99}{1.25} - \frac{3.90r - 1.77}{2.45} = r$
22. $\frac{8.19m + 2.55}{4.34} - \frac{8.17m - 9.94}{1.04} = 4m$

Solve each equation for x. (See Example 7).

23. $4(a - x) = b - a + 2x$
24. $(3a + b) - bx = a(x - 2) \quad (a \neq -b)$
25. $5(b - x) = 2b + ax \quad (a \neq -5)$
26. $bx - 2b = 2a - ax$

Solve each equation for the specified variable. Assume all denominators are nonzero. (See Example 7.)

27. $PV = k$ for V
28. $i = prt$ for p
29. $V = V_0 + gt$ for g
30. $S = S_0 + gt^2 + k$ for g
31. $A = \frac{1}{2}(B + b)h$ for B
32. $C = \frac{5}{9}(F - 32)$ for F

Solve each equation. (See Examples 8–10.)

33. $|2h + 1| = 5$
34. $|4m - 3| = 12$
35. $|6 - 2p| = 10$
36. $|-5x + 7| = 15$
37. $\left|\frac{5}{r - 3}\right| = 10$
38. $\left|\frac{3}{2h - 1}\right| = 4$
39. $|3y - 2| = |4y + 5|$
40. $|1 - 3z| = |z + 2|$

Solve the following applied problems.

Business *The approximate annual interest rate of a loan paid off with monthly payments is given by*

$$A = \frac{24f}{b(p + 1)},$$

where f is the finance charge on the loan, p is the total number of payments, and b is the original balance of the loan. Use the formula to find the requested value in the following. Round A to the nearest percent and round other variables to the nearest whole numbers. (This formula is not accurate enough for the requirements of federal law.)

41. $f = \$800$, $b = \$4000$, $p = 36$; find A
42. $A = 5\%$, $b = \$1500$, $p = 24$; find f
43. $A = 6\%$, $f = \$370$, $p = 36$; find b
44. $A = 10\%$, $f = \$490$, $p = 48$; find b

Social Science *The population P of the United States (in millions) can be approximated by the equation*

$$P = 2.275x + 203.45,$$

where x is the number of years since 1970. Determine the year in which the United States has the given population.

45. 214,825,000 **46.** 248,950,000

47. 260,325,000 **48.** 283,075,000

Health *According to data from the U.S. Department of Health and Human Services, the total health care expenditures E in the United States (in billions of dollars) can be approximated by*

$$E = 55.99x + 702.76,$$

where x is the number of years since 1990. Determine the year in which health care expenditures were at the given level.

49. $1038.7 billion **50.** $1150.7 billion

51. $1318.7 billion **52.** $1374.6 billion

Business *When a loan is paid off early, a portion of the finance charge must be returned to the borrower. By one method of calculating the finance charge (called the* rule of 78*), the amount of unearned interest (finance charge to be returned) is given by*

$$u = f \cdot \frac{n(n+1)}{q(q+1)},$$

where u represents unearned interest, f is the original finance charge, n is the number of payments remaining when the loan is paid off, and q is the original number of payments. Find the amount of the unearned interest in each of the following.

53. Original finance charge = $800, loan scheduled to run 36 months, paid off with 18 payments remaining

54. Original finance charge = $1400, loan scheduled to run 48 months, paid off with 12 payments remaining

55. Natural Science The excess lifetime cancer risk R is a measure of the likelihood that an individual will develop cancer from a particular pollutant. For example, if $R = .01$, then a person has a 1% increased chance of developing cancer during a lifetime. The value of R for formaldehyde can be calculated using the equation $R = kd$, where k is a constant and d is the daily dose in parts per million. The constant k for formaldehyde can be calculated using the formula $k = .132(\frac{B}{W})$, where B is the total number of cubic meters of air a person breathes in one day and W is a person's weight in kilograms.*

(a) Find k for a person that breathes in 20 cubic meters of air per day and weighs 75 kg.

(b) Mobile homes in Minnesota were found to have a mean daily dose d of .42 parts per million. Calculate R.†

(c) For every 5000 people, how many cases of cancer could be expected each year from these levels of formaldehyde? Assume an average life expectancy of 72 years.

Business *Solve the following investment problems. (See Example 11.)*

56. Joe Gonzalvez received $52,000 profit from the sale of some land. He invested part at 5% interest and the rest at 4% interest. He earned a total of $2290 interest per year. How much did he invest at 5%?

57. Weijen Luan invests $20,000 received from an insurance settlement in two ways: some at 6% and some at 4%. Altogether, she makes $1040 per year interest. How much is invested at 4%?

58. Maria Martinelli bought two plots of land for a total of $120,000. On the first plot, she made a profit of 15%. On the second, she lost 10%. Her total profit was $5500. How much did she pay for each piece of land?

59. Suppose $20,000 is invested at 5%. How much additional money must be invested at 4% to produce a yield of 4.8% on the entire amount invested?

Solve the following applied problems. (See Example 12.)

60. A plane flies nonstop from New York to London, cities which are about 3500 miles apart. After one hour and six minutes in the air, the plane passes over Halifax, Nova Scotia, which is 600 miles from New York. Estimate the flying time from New York to London.

61. On vacation, Le Hong averaged 50 mph traveling from Denver to Minneapolis. Returning by a different route that covered the same number of miles, he averaged 55 mph. What is the distance between the two cities if his total traveling time was 32 hours?

62. Russ and Janet are running in the Apple Hill Fun Run. Russ runs at 7 mph, Janet at 5 mph. If they start at the same time, how long will it be before they are 2/3 mile apart?

Natural Science *Using the same assumptions about octane ratings as in Example 13, solve the following problems.*

63. How many liters of 94 octane gasoline should be mixed with 200 liters of 99 octane gasoline to get a mixture that is 97 octane?

64. A service station has 92 octane and 98 octane gasoline. How many liters of each gasoline should be mixed to provide 12 liters of 96 octane gasoline for a chemistry experiment?

Solve the following applied problems.

65. Business A major car rental firm charges $45.56 a day with unlimited mileage. A discount firm offers a similar car for $20 a day plus 18 cents per mile. How far must you drive in a day in order for the cost to be the same at both firms?

* A. Hines, T. Ghosh, S. Layalka, and R.Warder, *Indoor Air Quality and Control* Prentice Hall, 1993 (TD 883.1.I476 1993).

†I. Ritchie, and R. Lehnen, "An Analysis of Formaldehyde Concentration in Mobile and Conventional Homes" *Journal of Environmental Health* 47:300–305.

66. Natural Science A car radiator contains 8 quarts of fluid, 40% of which is antifreeze. How much fluid should be drained and replaced with pure (100%) antifreeze in order that the new mixture is 60% antifreeze?

67. Transportation In Massachusetts speeding fines are determined by the equation

$$y = 10(x - 65) + 50, \qquad x \geq 65,$$

where y is the amount of the fine (in dollars) for driving x miles per hour. If Paul was fined \$100 for speeding, how fast was he driving?

68. Health Based on data from the U.S. Office of Management and Budget, Medicare costs C (in billions of dollars) in year x can be approximated by $C = 18x - 35{,}753$. In what year were Medicare costs approximately 283 billion dollars?

69. Finance Jack borrowed his father's luxury car and promised to return it with a full tank of premium gas, which costs \$1.49 per gallon. From experience, he knows that he needs 15.5 gallons. But he has only \$22 (and no credit card), which isn't enough. He decides to get as much premium as possible and fill the remainder of the tank with regular gas, which costs \$1.29 per gallon. To the nearest thousandth of a gallon, how much of each type of gas should he get?

Geometry *Solve each of these geometry problems. (Hint: In each case, draw an appropriate figure and label its parts.)*

70 The length of a rectangular label is 3 centimeters less than twice the width. The perimeter is 54 centimeters. Find the width.

71. A puzzle piece in the shape of a triangle has a perimeter of 30 centimeters. Two sides of the triangle are each twice as long as the shortest side. Find the length of the shortest side.

72. A triangle has a perimeter of 27 centimeters. One side is twice as long as the shortest side. The third side is seven centimeters longer than the shortest side. Find the length of the shortest side.

73. A closed recycling bin is in the shape of a rectangular box. Find the height of the bin if its length is 18 feet, its width is 8 feet, and its surface area is 496 square feet.

1.7 QUADRATIC EQUATIONS

An equation that can be written in the form

$$ax^2 + bx + c = 0,$$

where a, b, and c are real numbers with $a \neq 0$, is called a **quadratic equation.** For example, each of

$$2x^2 + 3x + 4 = 0, \qquad x^2 = 6x - 9, \qquad 3x^2 + x = 6, \qquad \text{and} \qquad x^2 = 5$$

is a quadratic equation. A solution of an equation that is a real number is said to be a **real solution** of the equation.

One method of solving quadratic equations is based on the following property of real numbers.

Zero-Factor Property

If a and b are real numbers, with $ab = 0$, then $a = 0$ or $b = 0$ or both.

EXAMPLE 1 Solve the equation $(x - 4)(3x + 7) = 0$.

By the zero-factor property, the product $(x - 4)(3x + 7)$ can equal 0 only if at least one of the factors equals 0. That is, the product equals zero only if $x - 4 = 0$

1

Solve the following equations.

(a) $(y - 6)(y + 2) = 0$

(b) $(5k - 3)(k + 5) = 0$

(c) $(2r - 9)(3r + 5)(r + 3) = 0$

Answers:

(a) 6, −2

(b) 3/5, −5

(c) 9/2, −5/3, −3

Solve each equation by factoring.

(a) $y^2 + 3y = 10$

(b) $2r^2 + 9r = 5$

(c) $4k^2 = 9k$

Answers:

(a) 2, −5

(b) 1/2, −5

(c) 9/4, 0

or $3x + 7 = 0$. Solving each of these equations separately will give the solutions of the original equation.

$$x - 4 = 0 \quad \text{or} \quad 3x + 7 = 0$$
$$x = 4 \quad \text{or} \quad 3x = -7$$
$$x = -\frac{7}{3}$$

The solutions of the equation $(x - 4)(3x + 7) = 0$ are 4 and $-7/3$. Check these solutions by substitution in the original equation. ◀ 1

EXAMPLE 2 Solve $6r^2 + 7r = 3$.

Rewrite the equation as

$$6r^2 + 7r - 3 = 0.$$

Now factor $6r^2 + 7r - 3$ to get

$$(3r - 1)(2r + 3) = 0.$$

By the zero-factor property, the product $(3r - 1)(2r + 3)$ can equal 0 only if

$$3r - 1 = 0 \quad \text{or} \quad 2r + 3 = 0.$$

Solving each of these equations separately gives the solutions of the original equation.

$$3r = 1 \quad \text{or} \quad 2r = -3$$
$$r = \frac{1}{3} \qquad r = -\frac{3}{2}$$

Verify that both 1/3 and $-3/2$ are solutions by substituting them in the original equation. ◀ 2

An equation such as $x^2 = 5$ has two solutions, $\sqrt{5}$ and $-\sqrt{5}$. The same idea is true in general.

Square-Root Property

If $b > 0$, then the solutions of $x^2 = b$ are $\sqrt{b}$ and $-\sqrt{b}$.

The two solutions are sometimes abbreviated $\pm\sqrt{b}$.

EXAMPLE 3 Solve each equation.

(a) $m^2 = 17$

By the square-root property, the solutions are $\sqrt{17}$ and $-\sqrt{17}$, abbreviated $\pm\sqrt{17}$.

Solve each equation by using the square-root property.

(a) $p^2 = 21$

(b) $(m + 7)^2 = 15$

(c) $(2k - 3)^2 = 5$

Answers:

(a) $\pm\sqrt{21}$

(b) $-7 \pm \sqrt{15}$

(c) $(3 \pm \sqrt{5})/2$

(b) $(y - 4)^2 = 11$

Use a generalization of the square-root property, working as follows:

$$(y-4)^2 = 11$$

$$y - 4 = \sqrt{11} \quad \text{or} \quad y - 4 = -\sqrt{11}$$

$$y = 4 + \sqrt{11} \qquad\qquad y = 4 - \sqrt{11}.$$

Abbreviate the solutions as $4 \pm \sqrt{11}$. ◀ (3)

As suggested by Example 3(b), it is easy to solve any quadratic equation that can be written in the form $(x + d)^2 = k$. Fortunately, there is a simple algebraic technique that enables us to write any quadratic equation in this form.

EXAMPLE 4 Solve $4x^2 - 24x + 19 = 0$.

First, rearrange the equation so that the x terms are on the left and the constants on the right:

$$4x^2 - 24x = -19.$$

Now divide both sides by 4 so that 1 is the coefficient of x^2:

$$x^2 - 6x = -\frac{19}{4}.$$

Now take half the coefficient of x (namely, $-6/2 = -3$) and square it, obtaining 9. (The reason for this is explained below). Add 9 to both sides and factor the left side:

$$x^2 - 6x + 9 = -\frac{19}{4} + 9$$

$$(x - 3)^2 = \frac{17}{4}.$$

Use the square-root property to complete the solution.

$$x - 3 = \pm\sqrt{\frac{17}{4}} = \pm\frac{\sqrt{17}}{2}$$

$$x = 3 \pm \frac{\sqrt{17}}{2} = \frac{6 \pm \sqrt{17}}{2}$$

The two solutions are $\dfrac{6 + \sqrt{17}}{2}$ and $\dfrac{6 - \sqrt{17}}{2}$. ◀

To see why the technique used in Example 4 always works, look at this multiplication:

$$\left(x + \frac{b}{2}\right)^2 = x^2 + 2\left(\frac{b}{2}\right)x + \left(\frac{b}{2}\right)^2 = x^2 + bx + \left(\frac{b}{2}\right)^2.$$

If you read it from right to left, it shows that the expression $x^2 + bx$ can be made into a perfect square by adding the square of half of the coefficient of x, namely,

$\left(\frac{b}{2}\right)^2$. That's exactly what was done in Example 4 (where $b = -6$). This technique is called **completing the square.**

The method of completing the square can be used on the general quadratic equation,

$$ax^2 + bx + c = 0 \qquad (a \neq 0),$$

to convert it to one whose solutions can be found by the square-root property. This will give a general formula for solving any quadratic equation. Going through the necessary algebra produces the following important result, which you should memorize.

Quadratic Formula

The solutions of the quadratic equation $ax^2 + bx + c = 0$, where $a \neq 0$, are given by

$$x = \frac{-b \pm \sqrt{b^2 - 4ac}}{2a}.$$

CAUTION When using the quadratic formula, remember that the equation must be in the form $ax^2 + bx + c = 0$. Also, notice that the fraction bar in the quadratic formula extends under *both* terms in the numerator. Be sure to add $-b$ to $\pm\sqrt{b^2 - 4ac}$ *before* dividing by $2a$.

EXAMPLE 5 Solve $x^2 + 1 = 4x$.

First add $-4x$ to both sides to get 0 alone on the right side.

$$x^2 - 4x + 1 = 0$$

Now identify the letters a, b, and c. Here $a = 1$, $b = -4$, and $c = 1$. Substitute these numbers into the quadratic formula

$$x = \frac{-(-4) \pm \sqrt{(-4)^2 - 4(1)(1)}}{2(1)}$$

$$= \frac{4 \pm \sqrt{16 - 4}}{2}$$

$$= \frac{4 \pm \sqrt{12}}{2}$$

$$= \frac{4 \pm 2\sqrt{3}}{2} \qquad \sqrt{12} = \sqrt{4 \cdot 3} = \sqrt{4} \cdot \sqrt{3} = 2\sqrt{3}$$

$$= \frac{2(2 \pm \sqrt{3})}{2} \qquad \text{Factor } 4 \pm 2\sqrt{3}.$$

$$x = 2 \pm \sqrt{3}.$$

The $\pm$ sign represents the two solutions of the equation. First use + and then use − to find each of the solutions: $2 + \sqrt{3}$ and $2 - \sqrt{3}$. ◀ 4

4

Use the quadratic formula to solve each equation.

(a) $x^2 - 2x = 2$

(b) $u^2 - 6u + 4 = 0$

Answers:

(a) $x = 1 + \sqrt{3}$ or $1 - \sqrt{3}$

(b) $u = 3 + \sqrt{5}$ or $3 - \sqrt{5}$

Example 5 shows that the quadratic formula produces exact solutions. In many real-world applications, however, you must use a calculator to find decimal approximations of the solutions. The approximate solutions in Example 5 are

$$x = 2 + \sqrt{3} \approx 3.732050808 \quad \text{and} \quad x = 2 - \sqrt{3} \approx .2679491924,$$

as shown in Figure 1.14.

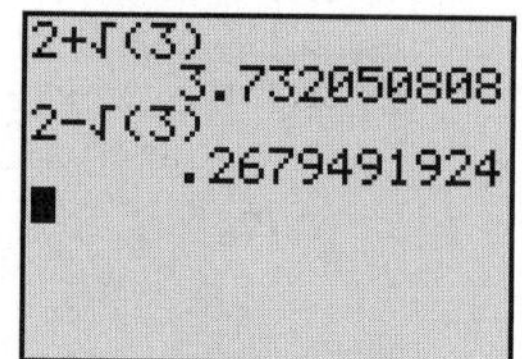

FIGURE 1.14

EXAMPLE 6 Use the quadratic formula and a calculator to solve

$$3.2x^2 + 15.93x - 7.1 = 0.$$

Compute $\sqrt{b^2 - 4ac} = \sqrt{15.93^2 - 4(3.2)(-7.1)} \approx 18.56461419$ and store the result in memory. This number may be inserted in any computation by using the memory recall key, which we shall denote here by M (it may be labeled differently on your calculator). Figure 1.15 shows that the approximate solutions are

$$x = \frac{-15.93 + M}{2 \cdot 3.2} \approx .4116584671 \quad \text{and} \quad x = \frac{-15.93 - M}{2 \cdot 3.2} \approx -5.389783467.$$

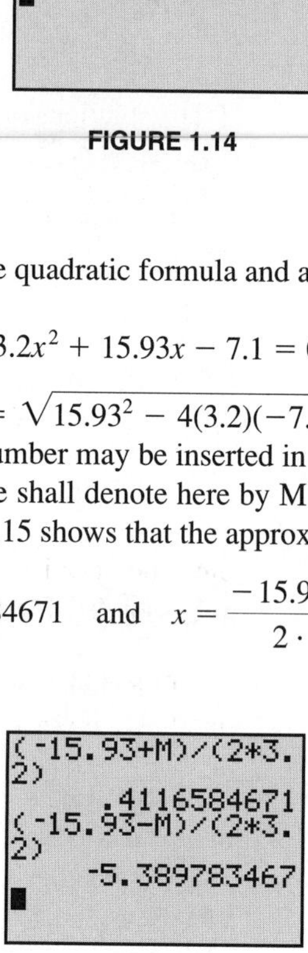

```
(-15.93+M)/(2*3.
2)
         .4116584671
(-15.93-M)/(2*3.
2)
        -5.389783467
```

FIGURE 1.15

5

Find approximate solutions for $5.1x^2 - 3.3x - 240.624 = 0$.

Answer:

$x = 7.2$ or $x \approx -6.5529$

Note the use of parentheses in Figure 1.15—omitting the parentheses leads to wrong answers. Also remember that these answers are *approximations*, so they may not check exactly when substituted in the original equation. 5

TECHNOLOGY TIP You can approximate the solutions of quadratic equations on a graphing calculator by using a quadratic formula program (see the Program Appendix) or using a built-in quadratic equation solver if your calculator has one. Then you need only enter the coefficients *a*, *b*, *c* to obtain the approximate solutions. See your instruction manual for details.

EXAMPLE 7 Solve $9x^2 - 30x + 25 = 0$.

Applying the quadratic formula with $a = 9$, $b = -30$, and $c = 25$, we have

$$x = \frac{-(-30) \pm \sqrt{(-30)^2 - 4(9)(25)}}{2(9)}$$

$$= \frac{30 \pm \sqrt{900 - 900}}{18} = \frac{30 \pm 0}{18} = \frac{30}{18} = \frac{5}{3}.$$

Therefore, the given equation has only one real solution. The fact that the solution is a rational number indicates that this equation could have been solved by factoring.

EXAMPLE 8 Solve $x^2 - 6x + 10 = 0$.

Apply the quadratic formula with $a = 1$, $b = -6$, and $c = 10$.

$$x = \frac{-(-6) \pm \sqrt{(-6)^2 - 4(1)(10)}}{2(1)}$$

$$= \frac{6 \pm \sqrt{36 - 40}}{2}$$

$$= \frac{6 \pm \sqrt{-4}}{2}$$

Since no negative number has a square root in the real number system, $\sqrt{-4}$ is not a real number. Hence the equation has no real solutions. (6)

6

Solve each equation.

(a) $9k^2 - 6k + 1 = 0$

(b) $4m^2 + 28m + 49 = 0$

(c) $2x^2 - 5x + 5 = 0$

Answers:

(a) 1/3

(b) $-7/2$

(c) No real solutions

Examples 5–8 show that the number of solutions of the quadratic equation $ax^2 + bx + c = 0$ is determined by $b^2 - 4ac$, the quantity under the radical, which is called the **discriminant** of the equation. (7)

Use the discriminant to determine the number of real solutions of each equation.

(a) $x^2 + 8x + 3 = 0$

(b) $2x^2 + x + 3 = 0$

(c) $x^2 - 194x + 9409 = 0$

Answers:

(a) 2

(b) 0

(c) 1

The Discriminant

The equation $\mathbf{ax^2 + bx + c = 0}$ has either two, or one, or no real solutions.

If $\mathbf{b^2 - 4ac > 0}$, there are two real solutions. (*Examples 5 and 6*)

If $\mathbf{b^2 - 4ac = 0}$, there is one real solution. (*Example 7*)

If $\mathbf{b^2 - 4ac < 0}$, there are no real solutions. (*Example 8*)

Applications

Quadratic equations arise in a variety of settings, as illustrated in the following examples. Example 9 depends on the following useful fact from geometry.

The Pythagorean Theorem

In a right triangle with legs of lengths a and b and hypotenuse of length c,

$$a^2 + b^2 = c^2.$$

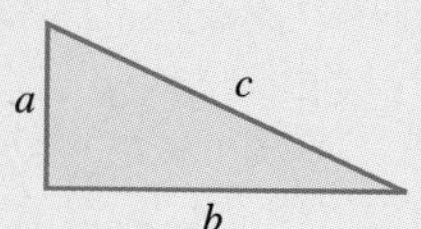

EXAMPLE 9 The size of a computer monitor is the diagonal measurement of its screen. The height of the screen is approximately 3/4 of its width. Kathy claims that John's 14-inch monitor has less than half the viewing area of her 21-inch monitor. John says, "No way! 14/21 is 2/3, so my monitor has 2/3 of the viewing area of yours." Who is right?

First, find the area of Kathy's screen. Let x be its width. Then its height is 3/4 of x, that is, $.75x$, as shown in Figure 1.16.

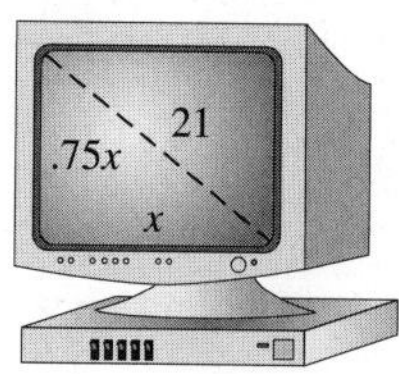

FIGURE 1.16

By the Pythagorean theorem,

$$\begin{aligned} x^2 + (.75x)^2 &= 21^2 \\ x^2 + .5625x^2 &= 441 && \text{Expand } (.75x)^2 \\ (1 + .5625)x^2 &= 441 && \text{Distributive property} \\ 1.5625x^2 &= 441 \\ x^2 &= 282.24 && \text{Divide both sides by 1.5625} \\ x &= \pm\sqrt{282.24} && \text{Square root property} \\ x &= \pm 16.8 \end{aligned}$$

We can ignore the negative solution since x is a width. Thus, the width is 16.8 inches and the height is $.75x = .75(16.8) = 12.6$ inches, so the area is

$$\text{Area} = \text{width} \times \text{height} = 16.8 \times 12.6 = 211.68 \text{ square inches.}$$

Next, find the area of John's screen by working Problem (8) at the side, which shows that John's screen has an area of 94.08 square inches. So, who is right?

(8)

Let x be the width of John's screen in Example 9.

(a) Write an equation that expresses the relationship between the width, height, and diagonal of the screen.

(b) Find the dimensions of the screen.

(c) Find its area.

Answer:

(a) $x^2 + (.75x)^2 = 14^2$

(b) 11.2 by 8.4 inches

(c) 94.08 square inches

EXAMPLE 10 A landscape architect wants to make an exposed gravel border of uniform width around a small shed behind a company plant. The shed is 10 feet by 6 feet. He has enough gravel to cover 36 square feet. How wide should the border be?

A sketch of the shed with border is given in Figure 1.17. Let x represent the width of the border. Then the width of the large rectangle is $6 + 2x$ and its length is $10 + 2x$.

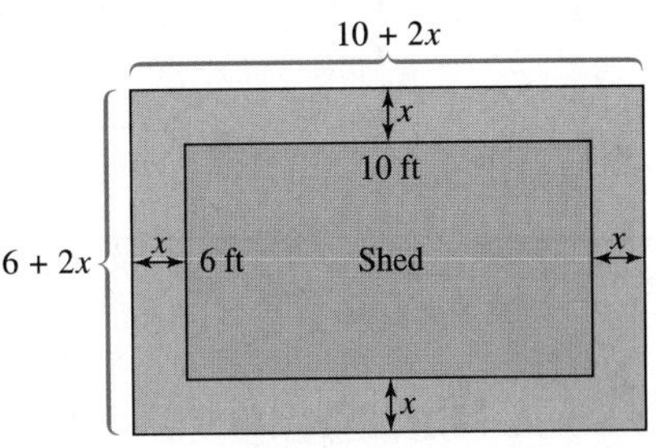

FIGURE 1.17

We must write an equation relating the given areas and dimensions. The area of the large rectangle is $(6 + 2x)(10 + 2x)$. The area occupied by the shed is $6 \cdot 10 = 60$. The area of the border is found by subtracting the area of the shed from the area of the large rectangle. This difference should be 36 square feet, giving the equation

$$(6 + 2x)(10 + 2x) - 60 = 36.$$

Solve this equation with the following sequence of steps.

$$60 + 32x + 4x^2 - 60 = 36$$
$$4x^2 + 32x - 36 = 0$$
$$x^2 + 8x - 9 = 0$$
$$(x + 9)(x - 1) = 0$$

The solutions are -9 and 1. The number -9 cannot be the width of the border, so the solution is to make the border 1 foot wide. ◀ (9)

(9)

> The length of a picture is 2 inches more than the width. It is mounted on a mat that extends 2 inches beyond the picture on all sides. What are the dimensions of the picture if the area of the mat is 99 square inches?
>
> Answer:
>
> 5 inches by 7 inches

In some applications it may be necessary to solve an equation in several variables for a specific variable.

(10)

> Solve each of the following equations for the indicated variable. Assume all variables are positive.
>
> **(a)** $k = mp^2 - bp$ for p
>
> **(b)** $r = \dfrac{APk^2}{3}$ for k
>
> Answers:
>
> **(a)** $p = \dfrac{b \pm \sqrt{b^2 + 4mk}}{2m}$
>
> **(b)** $k = \pm\sqrt{\dfrac{3r}{AP}}$ or $\dfrac{\pm\sqrt{3rAP}}{AP}$

EXAMPLE 11 Solve $v = mx^2 + x$ for x. (Assume m and v are positive.)

The equation is quadratic in x because of the x^2 term. Use the quadratic formula, first writing the equation in standard form.

$$v = mx^2 + x$$
$$0 = mx^2 + x - v$$

Let $a = m$, $b = 1$, and $c = -v$. Then the quadratic formula gives

$$x = \frac{-1 \pm \sqrt{1^2 - 4(m)(-v)}}{2m}$$

$$x = \frac{-1 \pm \sqrt{1 + 4mv}}{2m}.$$ ◀ (10)

1.7 Exercises

Use factoring to solve each equation. (See Examples 1 and 2.)

1. $(x + 3)(x - 12) = 0$

2. $(p - 16)(p - 5) = 0$

3. $x(x + 5) = 0$

4. $x^2 - 2x = 0$

5. $3z^2 = 6z$

6. $x^2 - 81 = 0$

7. $y^2 + 15y + 56 = 0$

8. $k^2 - 4k - 5 = 0$

9. $2x^2 = 5x - 3$

10. $2 = 12z^2 + 5z$

11. $6r^2 + r = 1$

12. $3y^2 = 16y - 5$

13. $2m^2 + 20 = 13m$

14. $10a^2 + 17a + 3 = 0$

15. $m(m - 7) = -10$

16. $z(2z + 7) = 4$

17. $9x^2 - 16 = 0$

18. $25y^2 - 64 = 0$

19. $16x^2 - 16x = 0$

20. $12y^2 - 48y = 0$

Solve each equation by using the square-root property. (See Example 3.)

21. $(r - 2)^2 = 7$

22. $(b + 5)^2 = 8$

23. $(4x - 1)^2 = 20$

24. $(3t + 5)^2 = 11$

Use the quadratic formula to solve each equation. If the solutions involve square roots, give both the exact and approximate solutions. (See Examples 5–8.)

25. $2x^2 + 5x + 1 = 0$

26. $3x^2 - x - 7 = 0$

27. $4k^2 + 2k = 1$

28. $r^2 = 3r + 5$

29. $5y^2 + 6y = 2$

30. $2z^2 + 3 = 8z$

31. $6x^2 + 6x + 5 = 0$

32. $3a^2 - 2a + 2 = 0$

33. $2r^2 - 7r + 5 = 0$

34. $8x^2 = 8x - 3$

35. $2x^2 - 7x + 30 = 0$

36. $3k^2 + k = 6$

37. $1 + \dfrac{7}{2a} = \dfrac{15}{2a^2}$

38. $5 - \dfrac{4}{k} - \dfrac{1}{k^2} = 0$

Use the discriminant to determine the number of real solutions of each equation. You need not solve the equations.

39. $25t^2 + 49 = 70t$

40. $9z^2 - 12z = 1$

41. $13x^2 + 24x - 6 = 0$

42. $22x^2 + 19x + 5 = 0$

Use a calculator and the quadratic formula to find approximate solutions of the equation. (See Example 6.)

43. $4.42x^2 - 10.14x + 3.79 = 0$

44. $3x^2 - 82.74x + 570.4923 = 0$

45. $7.63x^2 + 2.79x = 5.32$

46. $8.06x^2 + 25.8726x = 25.047256$

Solve the following problems. (See Example 3.)

47. Physical Science When wind resistance is ignored, a falling object travels a distance of d feet in t seconds, where $d = 16t^2$.

(a) How long does it take for the object to fall 36 feet?

(b) How long does it take for the object to fall five times farther than in part (a)?

48. Transportation According to the Federal Aviation Administration, the maximum recommended taxiing speed x (in miles per hour) for a plane on a curved runway exit is given by $R = .5x^2$, where R is the radius of the curve (in feet). Find the maximum taxiing speed for planes on such exits when the radius of the exit is

(a) 450 ft **(b)** 615 ft **(c)** 970 ft.

Solve the following problems. (See Example 6.)

49. Social Science The number of drivers who die in automobile accidents is related to the ages of the drivers, with teenagers and the eldery having the worst records. According to data from the National Highway Traffic Safety Administration, the driver fatality rate D per 1000 licensed drivers every 100 million miles can be approximated by the equation $D = .0031x^2 - .291x + 7.1$, where x is the age of the driver.

(a) For what ages is the driver fatality rate about 1 death per 1000?

(b) For what ages is the rate three times greater than in part (a)?

50. Social Science According to data from the U.S. Census Bureau, the population P of Cleveland, Ohio (in thousands), in year x can be approximated by $P = -.172x^2 + 16.82x + 487.62$, where $x = 0$ corresponds to 1900. In what year after 1940 was the population 877,620?

51. Business According to data from the International Civil Aviation Organization, the number N of airline passengers (in millions) in year x can be approximated by the equation $N = .68x^2 + 3.8x + 24$, where $x = 0$ corresponds to 1950. In what year did the number of passengers first reach

(a) one billion? **(b)** two billion?

(Recall that one billion is 1000 million.)

52. Social Science According to data from the U.S. Department of Health and Human Services, the cumulative number N of reported cases of AIDS in the United States in year x can be approximated by the equation $N = 3362.1x^2 - 17{,}270.3x + 24{,}043$, where $x = 0$ corresponds to 1980. In what year did the total reach 550,000?

Solve the following problems. (See Examples 9 and 10.)

53. Transportation Two trains leave the same city at the same time, one going north and the other east. The eastbound train travels 20 mph faster than the northbound one. After 5 hours the trains are 300 miles apart. Determine the speed of each train, using these steps.

(a) Let x denote the speed of the northbound train. Express the speed of the eastbound train in terms of x.

(b) Write expressions that give the distance traveled by each train after 5 hours.

(c) Use part (b) and the fact that the trains are 300 miles apart after 5 hours to write an equation. (A diagram of the situation may help.)

(d) Solve the equation and determine the speeds of the trains.

54. Chris and Josh have received walkie-talkies for Christmas. If they leave from the same point at the same time, Chris walking north at 2.5 mph and Josh walking east at 3 mph, how long will they be able to talk to each other if the range of the walkie-talkies is 4 miles? Round your answer to the nearest minute.

55. An ecology center wants to set up an experimental garden. It has 300 meters of fencing to enclose a rectangular area of 5000 square meters. Find the length and width of the rectangle as follows.

(a) Let $x =$ the length and write an expression for the width.

(b) Write an equation relating the length, width, and area, using the result of part (a).

(c) Solve the problem.

56. A landscape architect has included a rectangular flower bed measuring 9 feet by 5 feet in her plans for a new building. She wants to use two colors of flowers in the bed, one in the center and the other for a border of the same width on all four sides. If she can get just enough plants to cover 24 square feet for the border, how wide can the border be?

57. Joan wants to buy a rug for a room that is 12 feet by 15 feet. She wants to leave a uniform strip of floor around the rug. She can afford 108 square feet of carpeting. What dimensions should the rug have?

58. In 1991 Rick Mears won the (500 mile) Indianapolis 500 race. His speed (rate) was 100 mph (to the nearest mph) faster than that of the 1911 winner, Ray Harroun. Mears completed the race in 3.74 hours less time than Harroun. Find Mears's rate to the nearest whole number.

59. Physical Science If a ball is thrown upward with an initial velocity of 64 feet per second, then its height after t seconds is $h = 64t - 16t^2$. In how many seconds will the ball reach

(a) 64 ft? **(b)** 28 ft?

(c) Why are two answers possible?

60. Physical Science A particle moves horizontally with its distance (in centimeters) from a starting point at t seconds given by $d = 11t^2 - 10t$.

(a) How long will it take before the particle returns to the starting point?

(b) When will the particle be 100 centimeters from the starting point?

Solve each of the following equations for the indicated variable. Assume all denominators are nonzero and that all variables represent positive real numbers. (See Example 11.)

61. $S = \frac{1}{2}gt^2$ for t

62. $a = \pi r^2$ for r

63. $L = \frac{d^4k}{h^2}$ for h

64. $F = \frac{kMv^2}{r}$ for v

65. $P = \frac{E^2R}{(r + R)^2}$ for R

66. $S = 2\pi rh + 2\pi r^2$ for r

67. Solve the equation $z^4 - 2z^2 = 15$ as follows.

(a) Let $x = z^2$ and write the equation in terms of x.

(b) Solve the new equation for x.

(c) Set z^2 equal to each positive answer in part (b) and solve the resulting equation.

Solve each of the following equations. (See Exercise 67.)

68. $6p^4 = p^2 + 2$

69. $2q^4 + 3q^2 - 9 = 0$

70. $4a^4 = 2 - 7a^2$

71. $z^4 - 3z^2 - 1 = 0$

72. $2r^4 - r^2 - 5 = 0$

CHAPTER 1 SUMMARY

Key Terms and Symbols

1.1 $\approx$ is approximately equal to
π pi
$|a|$ absolute value of a
real number
natural (counting) number
whole number
integer
rational number
irrational number
properties of real numbers
additive inverse
multiplicative inverse
order of operations
square roots
number line
interval
interval notation
absolute value

1.2 a^n a to the power n
exponent or power
base
polynomial
variable
coefficient
term
constant term
degree of a nonzero term
degree of a polynomial
zero polynomial
binomial
trinomial
like terms
FOIL
revenue
fixed cost
variable cost
profit

1.3 factor
factoring
greatest common factor
perfect square trinomial
difference of two squares
sum and difference of two cubes

1.4 rational expression
operations with rational expressions
complex fraction

1.5 $a^{1/n}$ nth root of a
$\sqrt{a}$ square root of a
$\sqrt[n]{a}$ nth root of a
zero exponent
properties of exponents
radical
radicand
index
rationalizing the denominator
rationalizing the numerator

1.6 first-degree equation
solution of an equation
properties of equality
solving for a specified variable

1.7 quadratic equation
real solution
zero-factor property
square-root property
completing the square
quadratic formula
discriminant
Pythagorean theorem

Key Concepts

Factoring

$x^2 + 2xy + y^2 = (x + y)^2$ $\qquad x^3 - y^3 = (x - y)(x^2 + xy + y^2)$

$x^2 - 2xy + y^2 = (x - y)^2$ $\qquad x^3 + y^3 = (x + y)(x^2 - xy + y^2)$

$x^2 - y^2 = (x + y)(x - y)$

Rules for Radicals

Let a and b be real numbers, n be a positive integer, and m be any integer for which the following exist.

$a^{m/n} = \sqrt[n]{a^m} = (\sqrt[n]{a})^m$ $\qquad \sqrt[n]{a^n} = |a|$ if n is even $\qquad \sqrt[n]{a^n} = a$ if n is odd

$\sqrt[n]{a} \cdot \sqrt[n]{b} = \sqrt[n]{ab}$ $\qquad \dfrac{\sqrt[n]{a}}{\sqrt[n]{b}} = \sqrt[n]{\dfrac{a}{b}} \quad (b \neq 0)$

Rules for Exponents

Let a, b, r, and s be any real numbers for which the following exist.

$$a^{-r} = \frac{1}{a^r} \qquad a^0 = 1 \qquad \left(\frac{a}{b}\right)^r = \frac{a^r}{b^r}$$

$$a^r \cdot a^s = a^{r+s} \qquad (a^r)^s = a^{rs} \qquad a^{1/r} = \sqrt[r]{a}$$

$$\frac{a^r}{a^s} = a^{r-s} \qquad (ab)^r = a^r b^r \qquad \left(\frac{a}{b}\right)^{-r} = \left(\frac{b}{a}\right)^r$$

Absolute Value

Assume a and b are real numbers with $b > 0$.

The solutions of $|a| = b$ or $|a| = |b|$ are $a = b$ or $a = -b$.

Quadratic Equations

Facts needed to solve quadratic equations (in which a, b, and c are real numbers):

Factoring: If $ab = 0$, then $a = 0$ or $b = 0$ or both.

Square-Root Property: If $b > 0$, then the solutions of $x^2 = b$ are $\sqrt{b}$ and $-\sqrt{b}$.

Quadratic Formula: The solutions of $ax^2 + bx + c$ $(a \neq 0)$ are

$$x = \frac{-b \pm \sqrt{b^2 - 4ac}}{2a}.$$

Discriminant: There are two real solutions if $b^2 - 4ac > 0$, one real solution if $b^2 - 4ac = 0$, and no real solutions if $b^2 - 4ac < 0$.

Chapter 1 Review Exercises

Name the numbers from the list $-12, -6, -9/10, -\sqrt{7}, -\sqrt{4}, 0, 1/8, \pi/4, 6, \sqrt{11}$ that are

1. whole numbers

2. integers

3. rational numbers

4. irrational numbers

Identify the properties of real numbers that are illustrated in each of the following.

5. $7[(-3)4] = 7[4(-3)]$

6. $9(4 + 5) = (4 + 5)9$

7. $6(x + y - 3) = 6x + 6y + 6(-3)$

8. $11 + (8 + 3) = (11 + 8) + 3$

Express each statement in symbols.

9. x is at least 6.

10. x is negative.

Write the following numbers in numerical order from smallest to largest.

11. $-7, -3, 8, \pi, -2, 0$

12. $\frac{5}{6}, \frac{1}{2}, -\frac{2}{3}, -\frac{5}{4}, -\frac{3}{8}$

13. $|6 - 4|, -|-2|, |8 + 1|, -|3 - (-2)|$

14. $\sqrt{7}, -\sqrt{8}, -|\sqrt{16}|, |-\sqrt{12}|$

Write without absolute value bars.

15. $-|-6| + |3|$

16. $|-5| + |-9|$

17. $7 - |-8|$

18. $|-2| - |-7 + 3|$

Graph each of the following on a number line.

19. $x \geq -3$

20. $-4 < x \leq 6$

21. $x < -2$

22. $x \leq 1$

Use the order of operations to simplify.

23. $(-6 + 2 \cdot 5)(-2)$

24. $-4(-7 - 9 \div 3)$

25. $\dfrac{-8 + (-6)(-3) \div 9}{6 - (-2)}$

26. $\dfrac{20 \div 4 \cdot 2 \div 5 - 1}{-9 - (-3) - 12 \div 3}$

Perform each of the indicated operations.

27. $(3x^4 - x^2 + 5x) - (-x^4 + 3x^2 - 8x)$

28. $(-8y^3 + 8y^2 - 3y) - (2y^3 - 4y^2 - 10)$

29. $-2(q^4 - 3q^3 + 4q^2) + 4(q^4 + 2q^3 + q^2)$

30. $5(3y^4 - 4y^5 + y^6) - 3(2y^4 + y^5 - 3y^6)$

31. $(5z + 2)(3z - 2)$ **32.** $(8p - 4)(5p + 3)$

33. $(4k - 3h)(4k + 3h)$ **34.** $(2r - 5y)(2r + 5y)$

35. $(6x + 3y)^2$ **36.** $(2a - 5b)^2$

Factor as completely as possible.

37. $2kh^2 - 4kh + 5k$ **38.** $2m^2n^2 + 6mn^2 + 16n^2$

39. $3a^4 + 13a^3 + 4a^2$ **40.** $24x^3 + 4x^2 - 4x$

41. $10y^2 - 11y + 3$ **42.** $8q^2 + 3m + 4qm + 6q$

43. $4a^2 - 20a + 25$ **44.** $36p^2 + 12p + 1$

45. $144p^2 - 169q^2$ **46.** $81z^2 - 25x^2$

47. $8y^3 - 1$ **48.** $125a^3 + 216$

Perform each operation.

49. $\frac{4x}{5} \cdot \frac{35x}{12}$

50. $\frac{5k^2}{24} - \frac{75k}{36}$

51. $\frac{c^2 - 3c + 2}{2c(c - 1)} \div \frac{c - 2}{8c}$

52. $\frac{p^3 - 2p^2 - 8p}{3p(p^2 - 16)} \div \frac{p^2 + 4p + 4}{9p^2}$

53. $\frac{2y - 10}{5y} \cdot \frac{20y - 25}{12}$

54. $\frac{m^2 - 2m}{15m^3} \cdot \frac{5}{m^2 - 4}$

55. $\frac{2m^2 - 4m + 2}{m^2 - 1} \div \frac{6m + 18}{m^2 + 2m - 3}$

56. $\frac{x^2 + 6x + 5}{4(x^2 + 1)} \cdot \frac{2x(x + 1)}{x^2 - 25}$

57. $\frac{6}{15z} + \frac{2}{3z} - \frac{9}{10z}$

58. $\frac{5}{y - 2} - \frac{4}{y}$

59. $\frac{2}{5q} + \frac{10}{7q}$

60. Give two ways to evaluate $125^{2/3}$ and then compare them. Which do you prefer? Why?

Simplify each of the following. Write all answers without negative exponents. Assume all variables represent positive real numbers.

61. 5^{-3} **62.** 10^{-2}

63. -7^0 **64.** -3^{-1}

65. $\left(-\frac{6}{5}\right)^{-2}$ **66.** $\left(\frac{2}{3}\right)^{-3}$

67. $4^6 \cdot 4^{-3}$ **68.** $7^{-5} \cdot 7^{-1}$

69. $\frac{9^{-4}}{9^{-3}}$ **70.** $\frac{6^{-2}}{6^3}$

71. $\frac{9^4 \cdot 9^{-5}}{(9^{-2})^2}$ **72.** $\frac{k^4 \cdot k^{-3}}{(k^{-2})^{-3}}$

73. $4^{-1} + 2^{-1}$ **74.** $3^{-2} + 3^{-1}$

75. $125^{2/3}$ **76.** $128^{3/7}$

77. $9^{-5/2}$ **78.** $\left(\frac{144}{49}\right)^{-1/2}$

79. $\frac{5^{1/3} \cdot 5^{1/2}}{5^{3/2}}$ **80.** $\frac{2^{3/4} \cdot 2^{-1/2}}{2^{1/4}}$

81. $(3a^2)^{1/2} \cdot (3^2a)^{3/2}$ **82.** $(4p)^{2/3} \cdot (2p^3)^{3/2}$

Simplify each of the following.

83. $\sqrt[3]{27}$ **84.** $\sqrt[6]{-64}$

85. $\sqrt{24}$ **86.** $\sqrt{63}$

87. $\sqrt[3]{54p^3q^5}$ **88.** $\sqrt[4]{64a^5b^3}$

89. $\sqrt{\frac{5n^2}{6m}}$ **90.** $\sqrt{\frac{3x^3}{2z}}$

91. $2\sqrt{3} - 5\sqrt{12}$

92. $8\sqrt{7} + 2\sqrt{28}$

93. $(\sqrt{5} - 1)(\sqrt{5} + 1)$

94. $(\sqrt{7} - \sqrt{3})(\sqrt{7} + \sqrt{3})$

Rationalize each denominator.

95. $\frac{\sqrt{2}}{1 + \sqrt{3}}$

96. $\frac{4 + \sqrt{2}}{4 - \sqrt{5}}$

Social Science *In our system of government, the president is elected by the electoral college and not by individual voters. Because of this, smaller states have a greater voice in the selection of a president than they otherwise would have. Two political scientists have studied the problems of campaigning for president under the current system and have concluded that candidates should allot their money according to the formula*

$$\text{Amount for large state} = \left(\frac{E_{large}}{E_{small}}\right)^{3/2} \times \text{amount for small state.}$$

Here E_{large} represents the electoral vote of the large state, and E_{small} represents the electoral vote of the small state. Find the amount that should be spent in each of the following larger states if \$1,000,000 is spent in the small state and the following statements are true.

97. The large state has 48 electoral votes, and the small state has 3.

98. The large state has 36 electoral votes, and the small state has 4.

99. 6 votes in a small state; 28 in a large

100. 9 votes in a small state; 32 in a large

Solve each equation.

101. $2x - 5(x - 4) = 3x + 2$

102. $5y + 6 = -2(1 - 3y) + 4$

103. $\frac{2z}{5} - \frac{4z - 3}{10} = \frac{-z + 1}{10}$

104. $\frac{p}{p + 2} - \frac{3}{4} = \frac{2}{p + 2}$

105. $\frac{2m}{m - 3} = \frac{6}{m - 3} + 4$

106. $\frac{15}{k + 5} = 4 - \frac{3k}{k + 5}$

Solve for x.

107. $5ax - 1 = x$

108. $6x - 3y = 4bx$

109. $\frac{2x}{3 - c} = ax + 1$

110. $b^2x - 2x = 4b^2$

Solve each equation.

111. $|m - 3| = 9$

112. $|6 - x| = 12$

113. $\left|\frac{2 - y}{5}\right| = 8$

114. $|4k + 1| = |6k - 3|$

115. Business The number of private-label items in grocery stores is steadily growing. The number P of private-label items available (in thousands) in year x can be approximated by $P = 2.4x + 26.4$, where $x = 0$ corresponds to 1990.* Find the year in which the number of private-label items was
(a) 48,000 **(b)** 40,800.

116. Business A laser printer is on sale for 15% off. The sale price is \$306. What was the original price?

117. Health The percentage M of people who retire early and receive medical coverage from their employer in year x can be approximated by $M = -1.9x + 50.8$, where $x = 0$ corresponds to 1990.† Find the year in which the percentage of early retirees receiving employer sponsored medical coverage was
(a) 44.8% **(b)** 30.8%.

118. Finance Ellen borrowed \$500 from a credit union at 12% annual interest and got \$250 in cash with her credit card at 18% annual interest. What single rate of interest on \$750 results in the same total amount of interest?

119. Finance A real estate firm invests \$100,000 proceeds from a sale in two ways. The first portion is invested in a shopping center that provides an annual return of 8%. The rest is invested in a small apartment building with an annual return of 5%. The firm wants annual income of \$6800 from these investments. How much should be put into each investment?

120. Business To make a special mix for Valentine's Day, the owner of a candy store wants to combine chocolate hearts which sell for \$5 per pound with candy kisses which sell for \$3.50 per pound. How many pounds of each kind should be used to get 30 pounds of a mix that can be sold for \$4.50 per pound?

Determine the number of real solutions of each quadratic equation.

121. $x^2 - 6x = 4$

122. $-3x^2 + 5x + 2 = 0$

123. $4x^2 - 12x + 9 = 0$

124. $5x^2 + 2x + 1 = 0$

125. $x^2 + 3x + 5 = 0$

Find all real solutions of each equation.

126. $(b + 7)^2 = 5$

127. $(2p + 1)^2 = 7$

128. $2p^2 + 3p = 2$

129. $2y^2 = 15 + y$

130. $x^2 - 2x = 2$

131. $r^2 + 4r = 1$

132. $2m^2 - 12m = 11$

133. $9k^2 + 6k = 2$

134. $2a^2 + a - 15 = 0$

135. $12x^2 = 8x - 1$

136. $2q^2 - 11q = 21$

137. $3x^2 + 2x = 16$

138. $6k^4 + k^2 = 1$

139. $21p^4 = 2 + p^2$

140. $2x^4 = 7x^2 + 15$

141. $3m^4 + 20m^2 = 7$

142. $3 = \frac{13}{z} + \frac{10}{z^2}$

Solve each equation for the specified variable.

143. $p = \frac{E^2R}{(r + R)^2}$ for r

144. $p = \frac{E^2R}{(r + R)^2}$ for E

145. $K = s(s - a)$ for s

146. $kz^2 - hz - t = 0$ for z

*Datamonitor.

†William M. Mercer, Inc.

147. Energy The demand D for electricity in California (in thousands of gigawatt-hours) in year x can be approximated by $D = .74x^2 - 2.24x + 242.4$ $(0 \leq x \leq 10)$, where $x = 0$ corresponds to 1990.*

(a) In what year was demand about 294,000 gigawatt-hours?

(b) Assuming this model remains valid for the current decade, in what year will demand reach 396,000 gigawatt-hours?

148. Business A landscaper wants to put a cement walk of uniform width around a rectangular garden that measures 24 by 40 feet. She has enough cement to cover 740 square feet. To the nearest tenth of a foot, how wide should the walk be in order to use up all the cement?

149. Business A recreation director wants to fence off a rectangular playground beside an apartment building. The building forms one boundary, so she needs to fence only the other three sides. The area of the playground is to be 11,250 square meters. She has enough material to build 325 meters of fence. Find the length and width of the playground.

150. Transportation Two cars leave an intersection at the same time. One travels north and the other heads west traveling 10 mph faster. After 1 hour they are 50 miles apart. What were their speeds?

*Cambridge Energy Research Associates.

Case 1

Consumers Often Defy Common Sense*

Imagine two refrigerators in the appliance section of a department store. One sells for $700 and uses $85 worth of electricity a year. The other is $100 more expensive but costs only $25 a year to run. Given that either refrigerator should last at least 10 years without repair, consumers would overwhelmingly buy the second model, right?

Well, not exactly. Many studies by economists have shown that in a wide range of decisions about money—from paying taxes to buying major appliances—consumers consistently make decisions that defy common sense.

In some cases—as in the refrigerator example—this means that people are generally unwilling to pay a little more money up front to save a lot of money in the long run. At times, psychological studies have shown, consumers appear to assign entirely whimsical values to money, values that change depending on time and circumstances.

In recent years, these apparently irrational patterns of human behavior have become a subject of intense interest among economists and psychologists, both for what they say about the way the human mind works and because of their implications for public policy.

How, for example, can the United States move toward a more efficient use of electricity if so many consumers refuse to buy energy-efficient appliances even when such a move is in their own best interest?

At the heart of research into the economic behavior of consumers is a concept known as the discount rate. It is a measure of how consumers compare the value of a dollar received today with one received tomorrow.

Consider, for example, if you won $1000 in a lottery. How much more money would officials have to give you before you would agree to postpone cashing the check for a year?

Some people might insist on at least another $100, or 10 percent, since that is roughly how much it would take to make up for the combined effects of a year's worth of inflation and lost interest.

But the studies show that someone who wants immediate gratification might not be willing to postpone receiving the $1000 for 20 percent or 30 percent or even 40 percent more money.

In the language of economists, this type of person has a high discount rate: He or she discounts the value of $1000 so

*"Consumers often defy common sense" by Malcolm Gladwell from *The Washington Post*. Copyright © 1990. The Washington Post. Reprinted with permission.

much over a year that it would take hundreds of extra dollars to make waiting as attractive as getting the money immediately.

Of the two alternatives, waiting a year for more money is clearly more rational than taking the check now. Why would people turn down $1400 dollars next year in favor of $1000 today? Even if they needed the $1000 immediately, they would be better off borrowing it from a bank, even at 20 percent or even 30 percent interest. Then, a year later, they could pay off the loan—including the interest—with the $1400 and pocket the difference.

The fact is, however, that economists find numerous examples of such high discount rates implicit in consumer behavior.

While consumers were very much aware of savings to be made at the point of purchase, they so heavily discounted the value of monthly electrical costs that they would pay over the lifetime of their dryer or freezer that they were oblivious to the potential for greater savings.

Gas water heaters, for example, were found to carry an implicit discount rate of 100 percent. This means that in deciding which model was cheapest over the long run, consumers acted as if they valued a $100 gas bill for the first year as if it were really $50. Then, in the second year, they would value the next $100 gas bill as if it were really worth $25, and so on through the life of the appliance.

Few consumers actually make this formal calculation, of course. But there are clearly bizarre behavioral patterns in evidence.

Some experiments, for example, have shown that the way in which consumers make decisions about money depends a great deal on how much is at stake. Few people are willing to give up $10 now for $15 next year. But they are if the choice is between $100 now and $150 next year, a fact that would explain why consumers appear to care less about many small electricity bills—even if they add up to a lot—than one big initial outlay.

Exercises

1. Suppose a refrigerator that sells for $700 costs $85 a year for electricity. Write an expression for the cost to buy and run the refrigerator for x years.
2. Suppose another refrigerator costs $1000 and $25 a year for electricity. Write an expression for the total cost for this refrigerator over x years.
3. Over 10 years which refrigerator costs the most? By how much?
4. In how many years will the the total costs for the two refrigerators be equal?

Chapter 2

Graphs, Lines, and Inequalities

Graphical representations of data are commonly used in business, health sciences, and social sciences. Such data is often used to construct mathematical models useful for analyzing the current situation and predicting future performance. Lines, equations, and inequalities play an important role in developing such models. This chapter presents both algebraic and graphical methods for dealing with these topics.

2.1 GRAPHS

Just as the number line associates the points on a line with real numbers, a similar construction in two dimensions associates points in the plane with *ordered pairs* of real numbers. A **Cartesian coordinate system,** as shown in Figure 2.1, consists of a

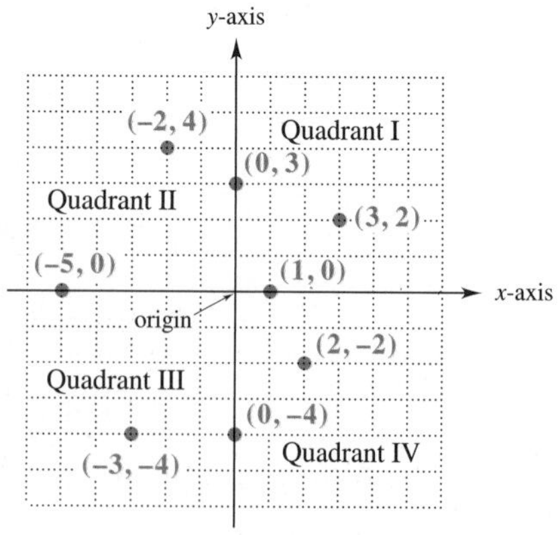

FIGURE 2.1

1

Locate $(-1, 6)$, $(-3, -5)$, $(4, -3)$, $(0, 2)$, and $(-5, 0)$ on a coordinate system.

Answer:

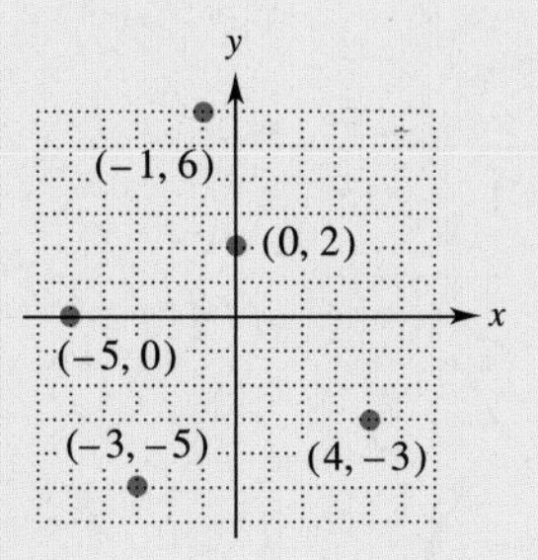

horizontal number line (usually called the ***x*-axis**) and a vertical number line (usually called the ***y*-axis**). The point where the number lines meet is called the **origin.** Each point in a Cartesian coordinate system is labeled with an **ordered pair** of real numbers, such as $(-2, 4)$ or $(3, 2)$. Several points and their corresponding ordered pairs are shown in Figure 2.1.

For the point labeled $(-2, 4)$, for example, -2 is the ***x*-coordinate** and 4 is the ***y*-coordinate.** You can think of these coordinates as directions telling you how to move to this point from the origin: you go 2 horizontal units to the left (x-coordinate) and 4 vertical units upward (y-coordinate). From now on, instead of referring to "the point labeled by the ordered pair $(-2, 4)$," we will say "the point $(-2, 4)$." 1

The x-axis and the y-axis divide the plane into four parts, or **quadrants,** which are numbered as shown in Figure 2.1. The points on the coordinate axes belong to no quadrant.

Equations and Graphs

A **solution of an equation** in two variables, such as

$$y = -2x + 3$$

or

$$y = x^2 + 7x - 2,$$

is an ordered pair of numbers such that the substitution of the first number for x and the second number for y produces a true statement.

2

Which of the following are solutions of $y = x^2 + 7x - 2$?

(a) $(1, 6)$

(b) $(-2, -20)$

(c) $(-1, -8)$

Answer:

(a) and (c)

EXAMPLE 1 Which of the following are solutions of $y = -2x + 3$?

(a) $(2, -1)$

This is a solution of $y = -2x + 3$ because "$-1 = -2 \cdot 2 + 3$" is a true statement.

(b) $(4, 7)$

Since $-2 \cdot 4 + 3 = -5$ and not 7, the ordered pair $(4, 7)$ is not a solution of $y = -2x + 3$. ◀ 2

Equations in two variables, such as $y = -2x + 3$, typically have an infinite number of solutions. To find one, choose a number for x and then compute the value of y that produces a solution. For instance, if $x = 5$, then $y = -2 \cdot 5 + 3 = -7$, so that the pair $(5, -7)$ is a solution of $y = -2x + 3$. Similarly, if $x = 0$, then $y = -2 \cdot 0 + 3 = 3$, so that $(0, 3)$ is also a solution.

The **graph** of an equation in two variables is the set of points in the plane whose coordinates (ordered pairs) are solutions of the equation. Thus, the graph of an equation is a picture of its solutions. Since a typical equation has infinitely many solutions, its graph has infinitely many points.

3

Graph $x = 5y$.

Answer:

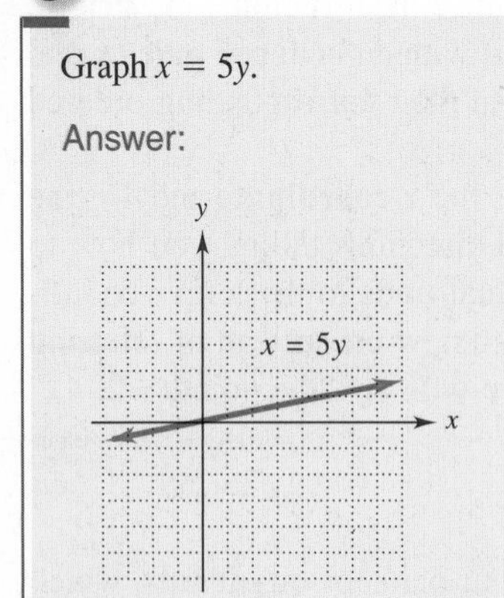

EXAMPLE 2 Sketch the graph of $y = -2x + 5$.

Since we cannot plot infinitely many points, we construct a table of y-values for a reasonable number of x-values, plot the corresponding points, and make an "educated guess" about the rest. The table of values and points in Figure 2.2 suggests that the graph is a straight line, as shown in Figure 2.3. ◀ 3

x	-1	0	2	4	5
$-2x + 5$	7	5	1	-3	-5

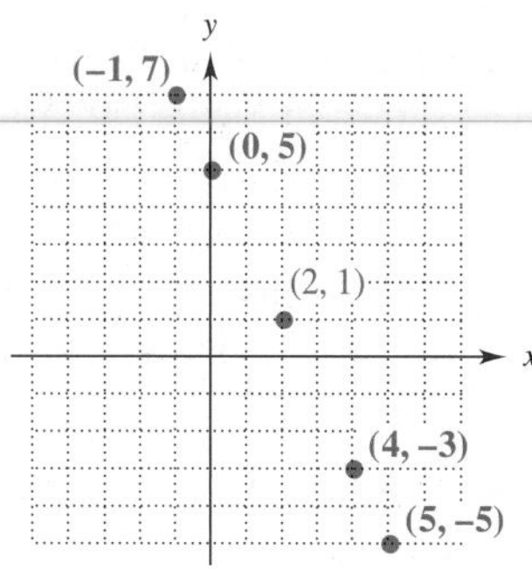

FIGURE 2.2

y-intercept
y = −2x + 5
x-intercept

FIGURE 2.3

An ***x*-intercept** of a graph is the x-coordinate of a point where the graph intersects the x-axis (the y-coordinate of this point is 0 since it's on the axis). Consequently, to find the x-intercepts of the graph of an equation, set $y = 0$ and solve for x. For instance, in Example 2, the x-intercept of the graph of $y = -2x + 5$ (see Figure 2.3) is found by setting $y = 0$ and solving for x.

$$0 = -2x + 5$$
$$2x = 5$$
$$x = \frac{5}{2}$$

Similarly, a ***y*-intercept** of a graph is the y-coordinate of a point where the graph intersects the y-axis (the x-coordinate of this point is 0—why?) The y-intercepts are found by setting $x = 0$ and solving for y. For example, the graph of $y = -2x + 5$ in Figure 2.3 has y-intercept 5. 4

4

Find the x- and y-intercepts of the graphs of these equations.

(a) $3x + 4y = 12$

(b) $5x - 2y = 8$

Answers:

(a) x-intercept 4, y-intercept 3

(b) x-intercept 8/5, y-intercept -4

EXAMPLE 3 Find the x- and y-intercepts of the graph of $y = 4 - x^2$ and sketch the graph.

Setting $x = 0$ in $y = 4 - x^2$, we see that the y-intercept is $y = 4$. Similarly, setting $y = 0$ gives $x^2 = 4$. Thus the x-intercepts are $x = 2$ and $x = -2$. Now make a table, being sure to take both positive and negative values for x, and plot the corresponding points, as in Figure 2.4. These points suggest that the entire graph looks like Figure 2.5. ◀

x	$4 - x^2$
−3	−5
−2	0
−1	3
0	4
1	3
2	0
3	−5

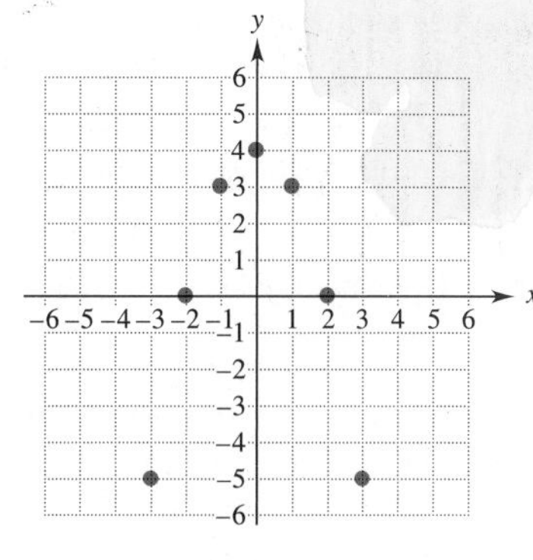

FIGURE 2.4

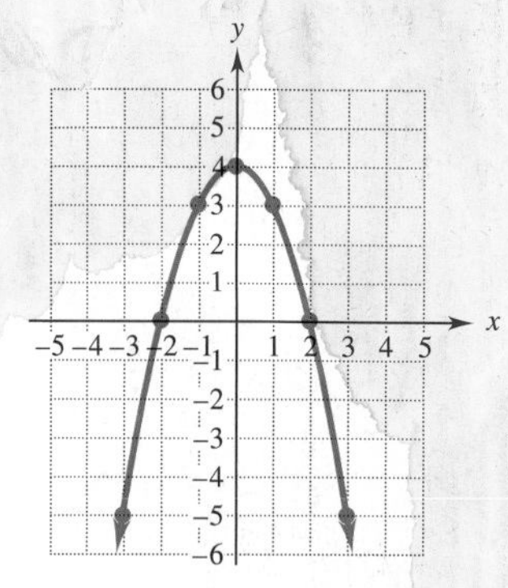

FIGURE 2.5

EXAMPLE 4 Sketch the graph of $y = \sqrt{x}$.

Notice that x can take only nonnegative values (since the square root of a negative number is not defined) and that the corresponding value of y is also nonnegative. Hence all the points on the graph will lie in the first quadrant. Computing some typical values, we obtain the graph in Figure 2.6. ◀

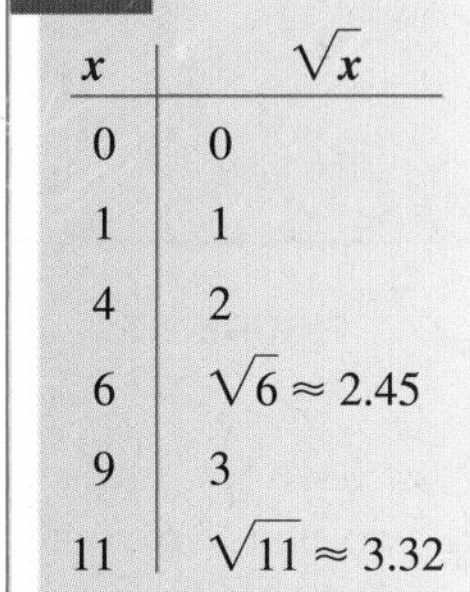

x	$\sqrt{x}$
0	0
1	1
4	2
6	$\sqrt{6} \approx 2.45$
9	3
11	$\sqrt{11} \approx 3.32$

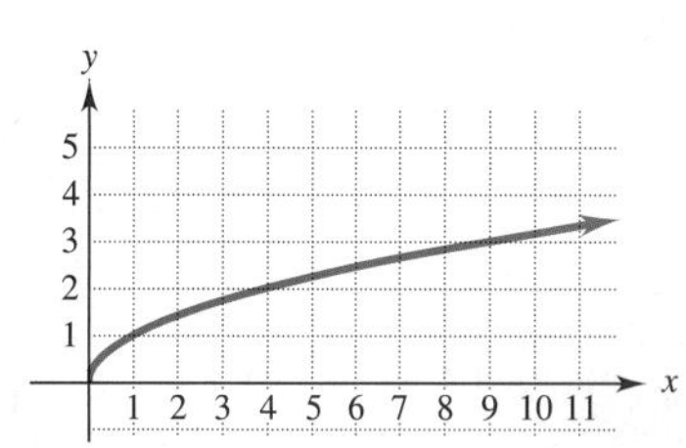

FIGURE 2.6

Example 2 shows that the x-intercept of the graph of $y = -2x + 5$ is the solution of $-2x + 5 = 0$. Example 3 shows that the x-intercepts of the graph of $y = 4 - x^2$ are the solutions of $4 - x^2 = 0$. Similar facts hold in the general case.

Intercepts and Equations

The real solutions of a one-variable equation of the form

expression in x = 0

are the x-intercepts of the graph of the two-variable equation

y = expression in x.

Technology and Graphs

A graphing calculator or computer graphing program follows essentially the same procedure used when graphing by hand: the calculator selects a large number of x-values (95 or more), equally spaced along the x-axis, and plots the corresponding points, simultaneously connecting them with line segments. Calculator-generated graphs are generally quite accurate, although they may not appear as smooth as hand-drawn ones. The next example illustrates the basics of graphing on a graphing calculator (computer graphing software operates similarly).

EXAMPLE 5 Use a graphing calculator to sketch the graph of the equation $2x^3 - 2y - 10x + 2 = 0$.

First, set the **viewing window**—the portion of the coordinate plane that will appear on the screen. Press the WINDOW key (labeled RANGE or PLOT-SETUP on some calculators) and enter the appropriate numbers, as in Figure 2.7 (which shows a T1-83; other calculators are similar). Then the calculator will display the portion of the plane shown in Figure 2.8, that is, the points (x, y) with $-9 \le x \le 9$ and $-6 \le y \le 6$.

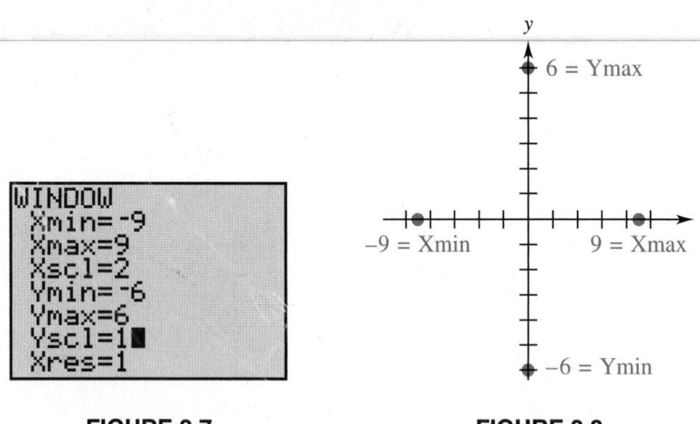

FIGURE 2.7 FIGURE 2.8

In Figure 2.7 we have set Xscl = 2 and Yscl = 1, which means the **tick marks** on the x-axis are two units apart and the tick marks on the y-axis are one unit apart (Figure 2.8).

Second, enter the equation to be graphed in the equation memory. To do this, you must first solve the equation for y (because a calculator only accepts equations of the form $y =$ expression in x):

$$2y = 2x^3 - 10x + 2$$
$$y = x^3 - 5x + 1.$$

Now press the Y= key (labeled SYMB on some calculators) and enter the equation, using the "variable key" for x, (This key is labeled X,T,θ,n or X,θ,T or x-VAR, depending on the calculator.) Figure 2.9 shows the equation entered on a TI-83; other calculators are similar. Now press GRAPH (or PLOT or DRW on some calculators) and obtain Figure 2.10.

FIGURE 2.9

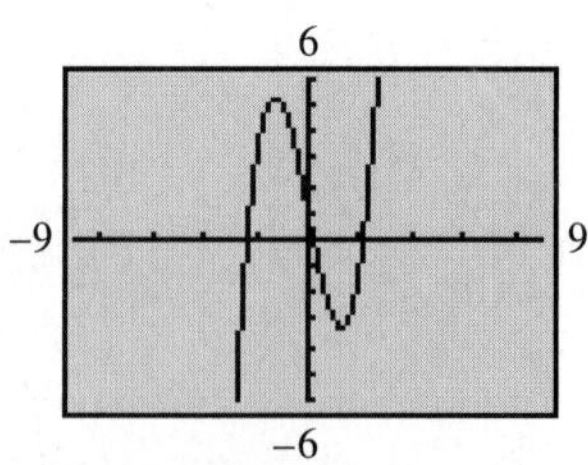

FIGURE 2.10

⑤ 

Use a graphing calculator to graph $y = 18x - 3x^3$ in the following viewing windows.

(a) $-10 \le x \le 10$ and $-10 \le y \le 10$ with Xscl = 1, Yscl = 1

(b) $-5 \le x \le 5$ and $-20 \le y \le 20$ with Xscl = 1, Yscl = 5

Answers:

(a)

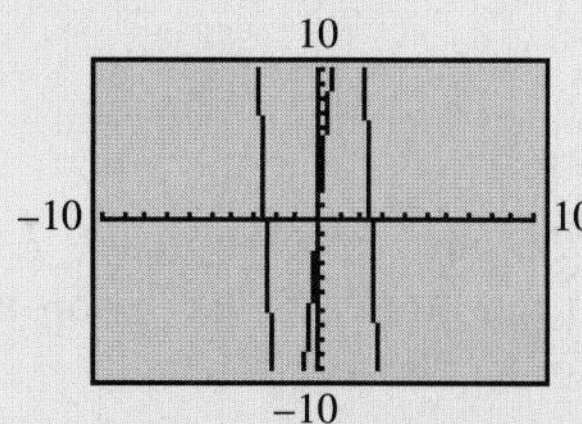

(b)

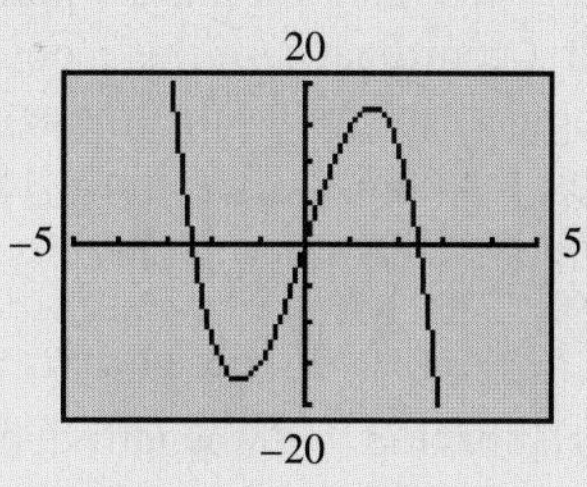

Finally, if necessary, change the viewing window to obtain a more readable graph. It is difficult to see the y-intercept in Figure 2.10, so press WINDOW and change the viewing window (Figure 2.11); then press GRAPH to obtain Figure 2.12 in which the y-intercept at $y = 1$ is clearly shown. (It isn't necessary to reenter the equation.) ◀ ⑤

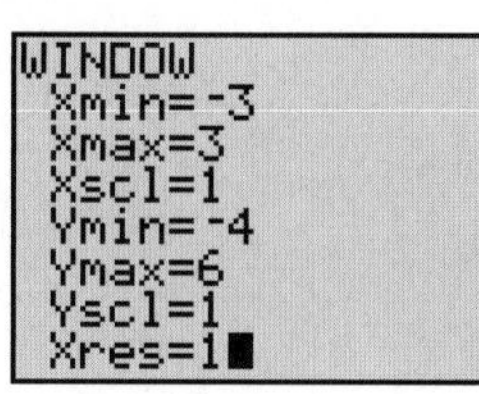

FIGURE 2.11

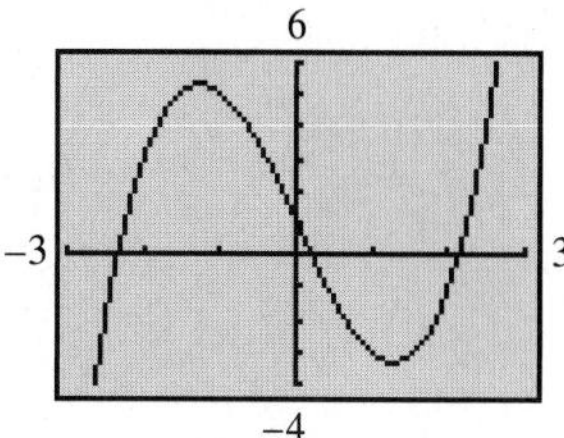

FIGURE 2.12

EXAMPLE 6 Use a graphing calculator to solve the equation $x^3 - 5x + 1 = 0$. The box on page 69 shows that the solutions of the equation are the x-intercepts of the graph of $y = x^3 - 5x + 1$ (Figure 2.12). To determine these x-intercepts as accurately as possible, use the **graphical root finder** on a graphing calculator. Look in the CALC or GRAPH MATH or PLOT FCN or G-SOLVE menu for "root," "zero," or "x-intercept." Check your instruction manual for the proper syntax. A typical root finder (Figure 2.13) shows that two of the solutions (x-intercepts) are $x \approx .2016$ and $x \approx 2.1284$. For the third solution, see Problem ⑥ at the side. ◀

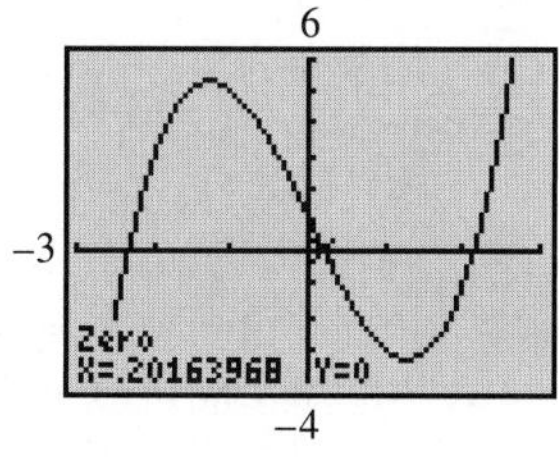

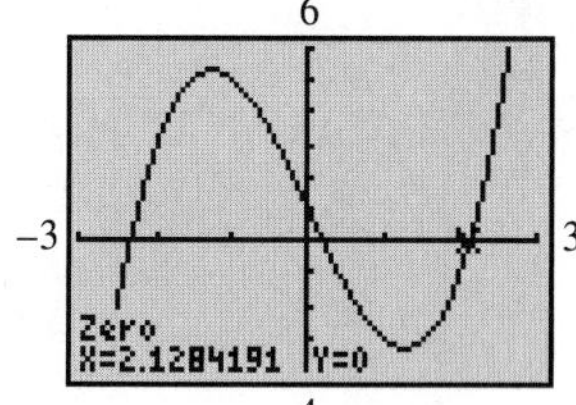

FIGURE 2.13

Use graphical root finder to approximate the third solution of the equation in Example 6.

Answer:

$x \approx -2.330059$

Graph Reading

Information is often given in graphical form, so you must be able to read and interpret graphs, that is, to translate graphical information into statements in English.

EXAMPLE 7 At various locations, the National Weather Service continuously records the temperature in graphical form. The results for March 20, 2001, in Cleveland, Ohio, are displayed in Figure 2.14 on the next page. The first coordinate of each point on the graph represents the time (measured in hours after midnight) and the second coordinate, the temperature at that time.

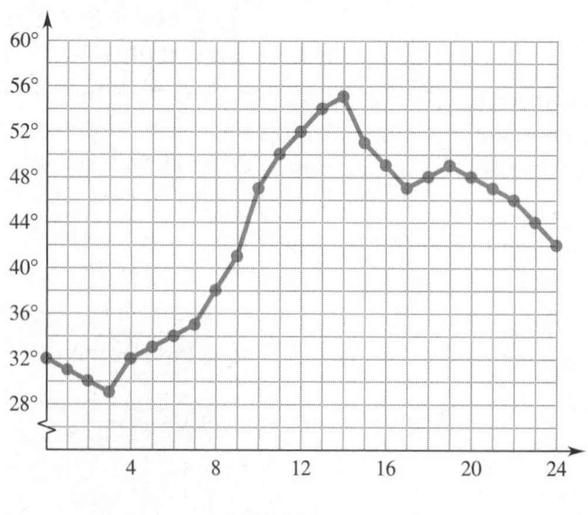

FIGURE 2.14

In Example 7, what are the highest and lowest temperatures during the day? When did they occur?

Answer:

Highest was 55° at 2 P.M.
Lowest was 29° at 3 A.M.

(a) What was the temperature at 9 A.M. and 6 P.M.?

The point (9, 41) is on the graph, which means that the temperature at 9 A.M. was 41°. Now 6 P.M. is 18 hours after midnight and the point (18, 48) is on the graph. So the temperature at 6 P.M. was 48°.

(b) At what times during the day was the temperature below 50°?

Look for the points whose second coordinates are less than 50, that is, points that lie below the horizontal line through 50°. The first coordinates of these points are the times when the temperature was below 50°. Figure 2.14 shows that these are the points with first coordinates less than 11 or greater than about 15.5. Since 15.5 hours corresponds to 3:30 P.M., the temperature was below 50° from midnight to 11 A.M. and from approximately 3:30 P.M. to midnight. ◀ 7

The next example deals with the basic business relationship that was introduced in Section 1.2:

$$\text{Profit} = \text{Revenue} - \text{Cost}.$$

EXAMPLE 8 Monthly revenue and costs for the Fenton Cell Phone Company are determined by the number t of phones produced and sold, as shown in Figure 2.15.

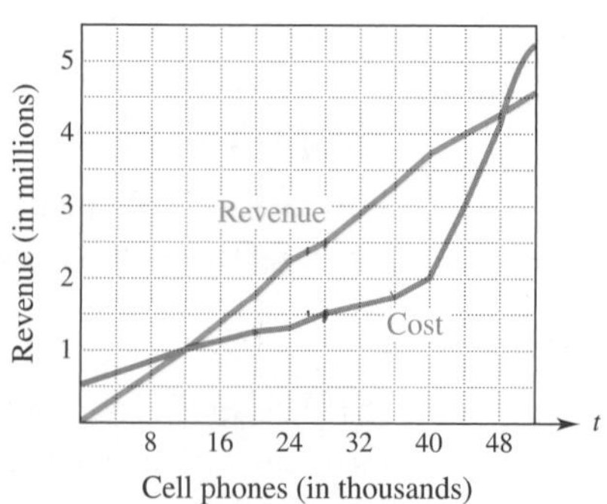

FIGURE 2.15

(a) How many phones should be produced each month if the company is to make a profit (assuming all phones produced are sold)?

Profit is Revenue − Cost, so the company makes a profit whenever revenue is greater than cost, that is, when the revenue graph is above the cost graph. Figure 2.15 shows that this occurs between $t = 12$ and $t = 48$, that is, when 12,000 to 48,000 phones are produced. If the company makes fewer than 12,000 phones, it will lose money (costs will be greater than revenue). They also lose money by making more than 48,000 phones (one reason might be that high production levels require large amounts of overtime pay, which drives costs up too much).

(b) Is it more profitable to make 40,000 or 44,000 phones?

On the revenue graph, the point with first coordinate 40 has second coordinate approximately 3.7, meaning that the revenue from 40,000 phones is about 3.7 million dollars. The point with first coordinate 40 on the cost graph is (40, 2), meaning that the cost of producing 40,000 phones is 2 million dollars. Therefore, the profit on 40,000 phones is about $3.7 - 2 = 1.7$ million dollars. For 44,000 phones, we have the approximate points (44, 4) on the revenue graph and (44, 3) on the cost graph. So the profit on 44,000 phones is $4 - 3 = 1$ million dollars. Consequently, it is more profitable to make 40,000 phones. ◀ (8)

(8)

In Example 8, find the profit from making

(a) 32,000 phones;

(b) 4000 phones.

Answers:

(a) About $1,200,000 (rounded)

(b) About −$500,000 (that is, a loss of $500,000)

2.1 Exercises

Determine whether the given ordered pair is a solution of the given equation. (See Example 1.)

1. $(1, -2)$; $3x - y - 5 = 0$
2. $(2, -1)$; $x^2 + y^2 - 6x + 8y = -15$
3. $(3, 4)$; $(x - 2)^2 + (y + 5)^2 = 4$
4. $(1, -1)$; $\frac{x^2}{2} + \frac{y^2}{3} = 1$

Sketch the graph of each of these equations. (See Example 2.)

5. $4y + 3x = 12$
6. $2x + 7y = 14$
7. $8x + 3y = 12$
8. $9y - 4x = 12$
9. $x = 2y + 3$
10. $x - 3y = 0$

List the x-intercepts and y-intercepts of each graph.

11.

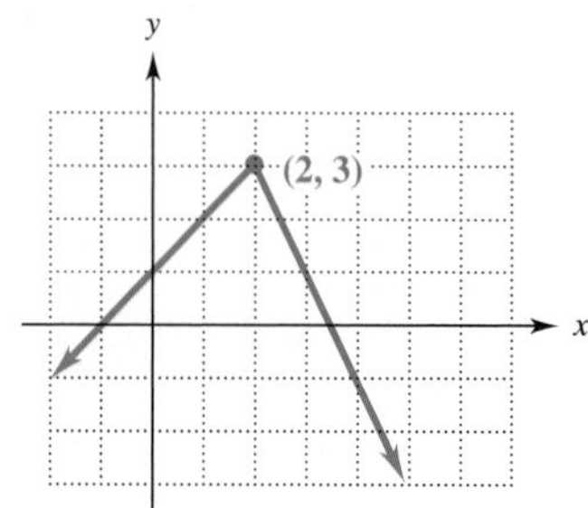

12.

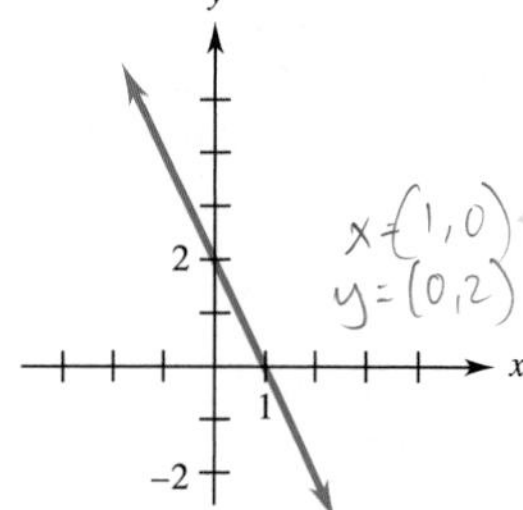

13.

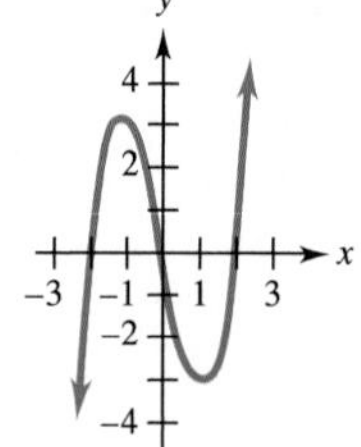

14.

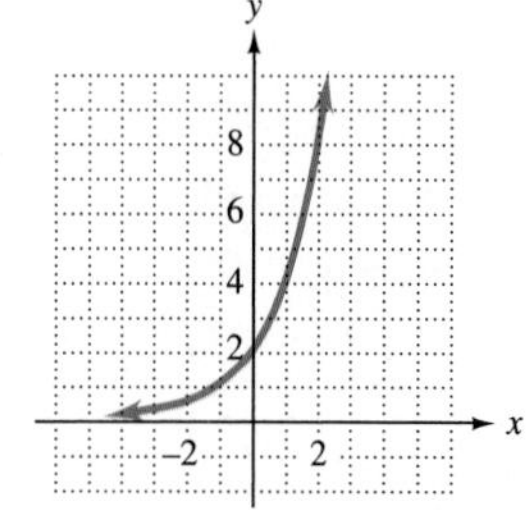

Find the x-intercepts and y-intercepts of the graph of each equation. You need not sketch the graph. (See Example 3.)

15. $3x + 4y = 24$ **16.** $x - 2y = 3$

17. $2x - 3y = 6$ **18.** $3x + y = 4$

19. $y = x^2 - 9$ **20.** $y = x^2 + 4$

Sketch the graph of the equation. (See Examples 2–4.)

21. $y = x^2$ **22.** $y = x^2 + 2$

23. $y = x^2 - 3$ **24.** $y = 2x^2$

25. $y = x^3$ **26.** $y = x^3 - 3$

27. $y = x^3 + 1$ **28.** $y = x^3/2$

29. $y = \sqrt{x + 2}$ **30.** $y = \sqrt{x - 2}$

31. $y = \sqrt{4 - x^2}$ **32.** $y = \sqrt{9 - x^2}$

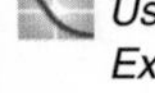 *Use a graphing calculator to find the graph of the equation. (See Example 5.)*

33. $y = x^2 + x + 1$ **34.** $y = 2 - x - x^2$

35. $y = (x - 3)^3$ **36.** $y = x^3 + 2x^2 + 2$

37. $y = x^3 - 3x^2 + x - 1$ **38.** $y = x^4 - 5x^2 - 2$

 *Use a graphing calculator for Exercises 39–40.*

39. Graph $y = x^4 - 2x^3 + 2x$ in a window with $-3 \le x \le 3$. Is the "flat" part of the graph near $x = 1$ really a horizontal line segment? (*Hint:* Use trace to move along the "flat" part and watch the y-coordinates. Do they remain the same [as they should on a horizontal segment]?)

40. **(a)** Graph $y = x^4 - 2x^3 + 2x$ in the **standard window** (the one with $-10 \le x \le 10$ and $-10 \le y \le 10$). Use the trace feature to approximate the coordinates of the lowest point on the graph.

(b) Use a minimum finder to obtain an accurate approximation of the lowest point. How does this compare with your answer in part (a)?

 *Use a graphing calculator to approximate all real solutions of the equation. (See Example 6.)*

41. $x^3 - 3x^2 + 5 = 0$

42. $x^3 + x - 1 = 0$

43. $2x^3 - 4x^2 + x - 3 = 0$

44. $6x^3 - 5x^2 + 3x - 2 = 0$

45. $x^5 - 6x + 6 = 0$

46. $x^3 - 3x^2 + x - 1 = 0$

 *Use a graphing calculator to work Exercises 47–50.*

47. **Physical Science** The surface area S of the right circular cone in the figure is given by $S = \pi r\sqrt{r^2 + h^2}$. What radius should be used to produce a cone of height 5 inches and surface area 100 square inches?

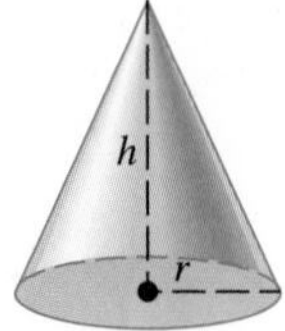

48. **Physical Science** The surface area of the right square pyramid in the figure is given by $S = b\sqrt{b^2 + 4h^2}$. If the pyramid has height 10 feet and surface area 100 square feet, what is the length of a side b of its base?

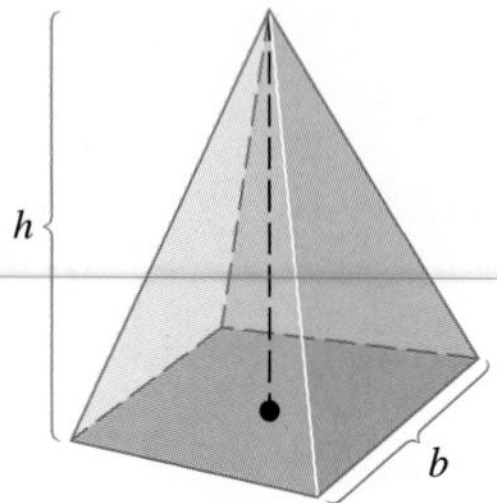

49. **Health** According to data from the National Cancer Institute, the rate of lung cancer cases per 100,000 women in the year x can be approximated by the equation $y = -.00075x^3 - .0083x^2 + 1.565x + 21.23$, where $x = 0$ corresponds to 1975. In what year did the female lung cancer rate first reach 40 per 100,000?

50. **Natural Science** Carbon monoxide combines with the hemoglobin of the blood to form carboxyhemoglobin (COHb), which reduces transport of oxygen to tissue. A 4% to 6% COHb level in the blood (typical for smokers) can cause symptoms such as blood flow alterations, visual impairment, and poorer vigilance ability. When $50 \le x \le 100$, the equation $T = .00787x^2 - 1.528x + 75.89$ approximates the exposure time T (in hours) necessary to reach this 4% to 6% level, where x is the amount of carbon monoxide present in the air in parts per million (ppm).*

(a) A kerosene heater or a room full of smokers is capable of producing 50 ppm of carbon monoxide. How long would it take for a nonsmoking person to start feeling the above symptoms?

(b) What carbon monoxide concentration will cause a person to reach a 4% to 6% COHb level in three hours?

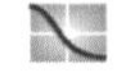 *Use the maximum/minimum finder on a graphing calculator to do Exercises 51 and 52.*

51. **Social Science** The number N of boys in the United States involved in high school sports (in millions) in year x can be approximated by $N = -.00015x^3 + .01x^2 - .16x + 4.1$, where $x = 0$ corresponds to 1975. Between 1975

*U.S. Department of Energy

and 2000, when were the fewest boys involved in sports? How many were involved then?

52. **Health** The number y of AIDS cases (in thousands) diagnosed in year x can be approximated by

$$y = .21x^3 - 4.6x^2 + 21.9x + 50 \qquad (0 \le x \le 9),$$

where $x = 0$ corresponds to 1990. When were the most cases diagnosed?

Physical Science *The temperature graphs for Fargo and Seattle on the same day are shown in the figure. Use them to do the following exercises. (See Example 7.)*

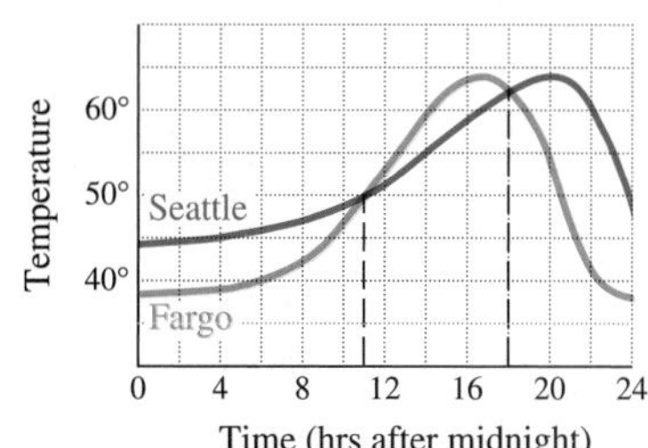

53. Approximately when did the temperature first reach 60° in Fargo? In Seattle?

54. At what times during the day did the two cities have the same temperature?

55. At what times during the day was it warmer in Fargo than in Seattle?

56. Was there any time when it was at least 10° warmer in Seattle than in Fargo?

Business *Use the revenue and cost graphs for the Fenton Cell Phone Company in Example 8 to do Exercises 57–60.*

57. Find the approximate cost of manufacturing the given number of phones.
(a) 20,000 **(b)** 36,000
(c) 48,000

58. Find the approximate revenue from selling the given number of phones.
(a) 12,000 **(b)** 24,000
(c) 36,000

59. Find the approximate profit from manufacturing the given number of phones.
(a) 20,000 **(b)** 28,000
(c) 36,000

60. The company must replace its aging machinery with better, but much more expensive, machines. In addition, raw material prices increase, so that monthly costs go up by \$250,000. Owing to competitive pressure, phone prices cannot be increased, so revenue remains the same. Under these new circumstances, find the approximate profit from manufacturing the given number of phones.
(a) 20,000 **(b)** 36,000
(c) 40,000

Physical Science *According to an article in the December 1994 issue of* Scientific American, *the coast-down time for a typical 1993 car as it drops 10 miles per hour from an initial speed depends on variations from the standard condition (automobile in neutral; average drag and tire pressure). The accompanying graph illustrates some of these conditions with coast-down time in seconds and initial speed in miles per hour.*

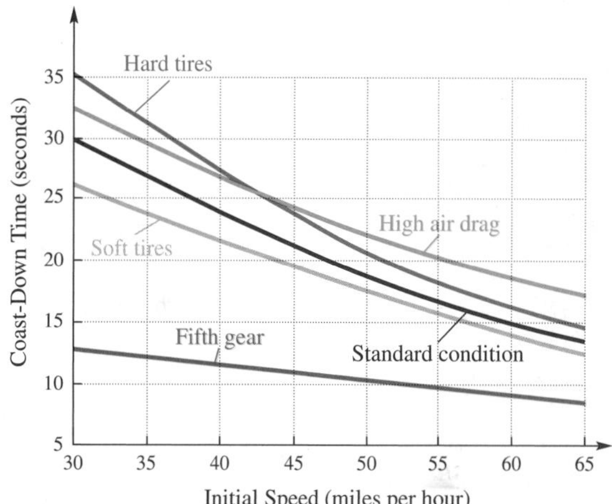

Use the graph to answer the following questions.

61. What is the approximate coast-down time in fifth gear if the initial speed is 40 miles per hour?

62. For what speed is the coast-down time the same for the conditions of high air drag and hard tires?

63. **Business** Sandi Goldstein takes out a 30-year mortgage on which her monthly payment is \$850. During the early years of the mortgage, most of each payment is for interest and the rather small remainder for principal. As time goes on the portion of each payment that goes for interest decreases, while the portion for principal increases, as shown in the accompanying graph.

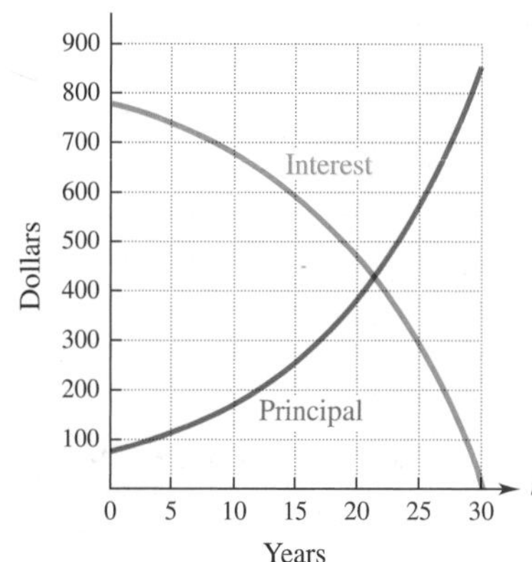

(a) Approximately how much of the \$850 monthly payment goes for interest in year 5? In year 15? In year 25?
(b) In what year will the monthly payment be equally divided between interest and principal?

Business *The accompanying graph shows non-agricultural employment (in thousands) in the United States during recent years.* Use it to answer the following questions.*

64. How many people were employed at the beginning of 1995? At the beginning of 2000?

65. During what period was employment below 112,000,000?

66. Approximately when was employment the lowest?

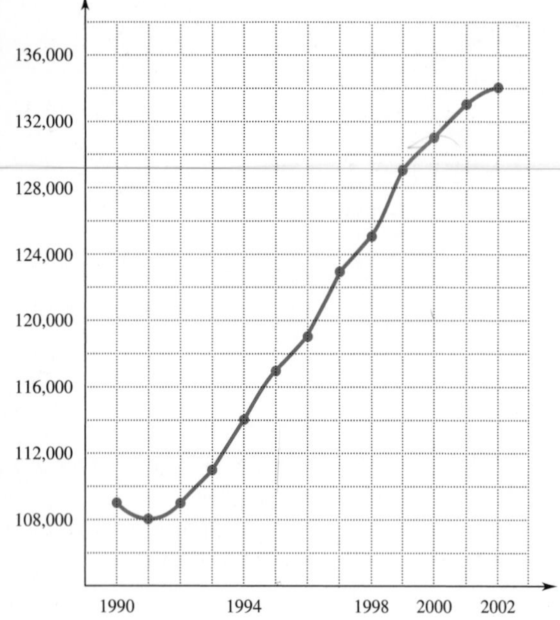

Business *The Consumer Confidence Index reflects people's feelings about their employment opportunities and income prospects. The accompanying graph from the* Seattle Post-Intelligencer *shows the index over a five-year period. Use it to answer the following questions.*

67. How did the Consumer Confidence Index vary during 1997 and 1998?

68. What was the lowest level of consumer confidence during 1998 and 1999? The highest level in these years?

69. During what periods was the index below 120?

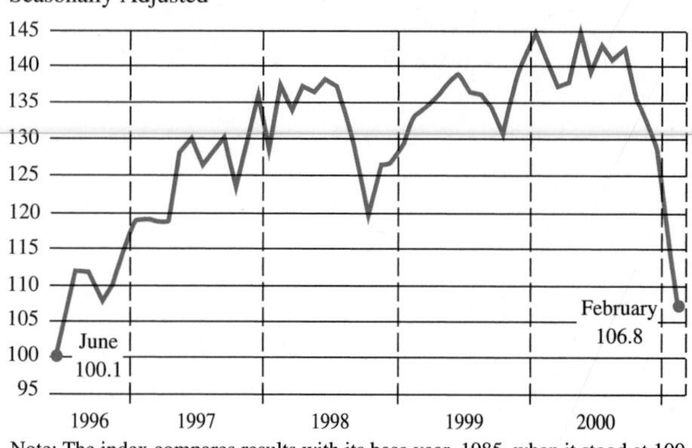

Note: The index compares results with its base year, 1985, when it stood at 100.
Source: The Conference Board THE ASSOCIATED PRESS

*U.S. Bureau of Labor Statistics; data for 2000–2002 are projections.

2.2 EQUATIONS OF LINES

Straight lines, which are the simplest graphs, play an important role in a wide variety of applications. They are considered here from both a geometric and an algebraic point of view.

The key geometric feature of a nonvertical straight line is how steeply it rises or falls as you move from left to right. The "steepness" of a line can be represented numerically by a number called the *slope* of the line.

To see how slope is defined, start with Figure 2.16, which shows a line passing through the two different points $(x_1, y_1) = (-3, 5)$ and $(x_2, y_2) = (2, -4)$. The difference in the two x-values,

$$x_2 - x_1 = 2 - (-3) = 5,$$

in this example is called the **change in x.** The Greek letter Δ (delta) is used to denote change. The symbol Δx (read "delta x") represents the change in x. In the same way, Δy represents the **change in y.** In this example,

$$\Delta y = y_2 - y_1 = -4 - 5 = -9.$$

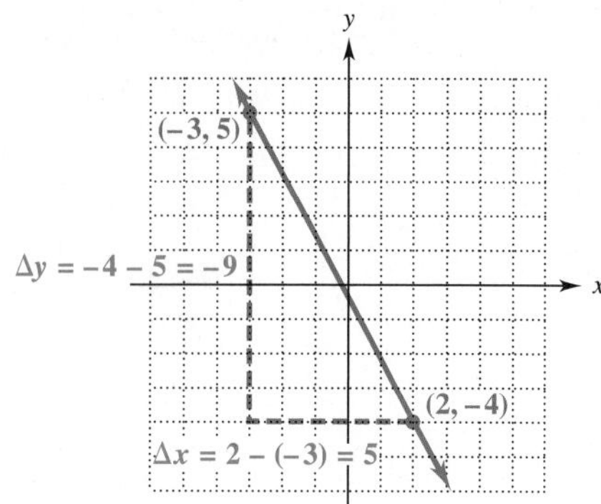

FIGURE 2.16

The **slope** of the line through the two points (x_1, y_1) and (x_2, y_2), where $x_1 \neq x_2$, is defined as the quotient of the change in y and the change in x:

$$\textbf{slope} = \frac{\textbf{change in } y}{\textbf{change in } x} = \frac{\Delta y}{\Delta x} = \frac{y_2 - y_1}{x_2 - x_1}.$$

The slope of the line in Figure 2.16 is

$$\text{slope} = \frac{\Delta y}{\Delta x} = \frac{-4 - 5}{2 - (-3)} = -\frac{9}{5}.$$

Using similar triangles from geometry, it can be shown that the slope is independent of the choice of points on the line. That is, the same value of the slope will be obtained for *any* choice of two different points on the line.

EXAMPLE 1 Find the slope of the line through the points $(-7, 6)$ and $(4, 5)$.

Let $(x_1, y_1) = (-7, 6)$ and $(x_2, y_2) = (4, 5)$. Use the definition of slope.

$$\text{slope} = \frac{\Delta y}{\Delta x} = \frac{y_2 - y_1}{x_2 - x_1} = \frac{5 - 6}{4 - (-7)} = \frac{-1}{11} = -\frac{1}{11}$$

The slope can also be found by letting $(x_1, y_1) = (4, 5)$ and $(x_2, y_2) = (-7, 6)$. In that case,

$$\text{slope} = \frac{6 - 5}{-7 - 4} = \frac{1}{-11} = -\frac{1}{11},$$

the same answer. ◀ (1)

(1)

Find the slope of the line through

(a) $(6, 11), (-4, -3)$;

(b) $(-3, 5), (-2, 8)$.

Answers:

(a) 7/5

(b) 3

CAUTION When finding the slope of a line, be careful to subtract the x-values and the y-values in the same order. For example, with the points (4, 3) and (2, 9), if you use $9 - 3$ for the numerator, you must use $2 - 4$ (*not* $4 - 2$) for the denominator.

EXAMPLE 2 Find the slope of the horizontal line in Figure 2.17.

Every point on the line has the same y-coordinate, -5. Choose any two of them to compute the slope, say, $(x_1, y_1) = (-3, -5)$ and $(x_2, y_2) = (2, -5)$:

$$\begin{aligned} \text{slope} &= \frac{-5 - (-5)}{2 - (-3)} \\ &= \frac{0}{5} \\ &= 0. \end{aligned}$$

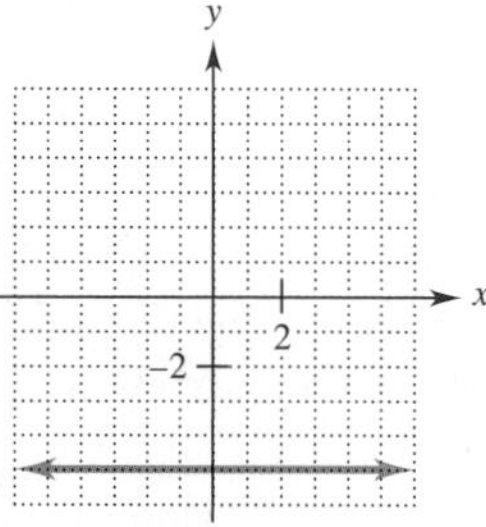

FIGURE 2.17

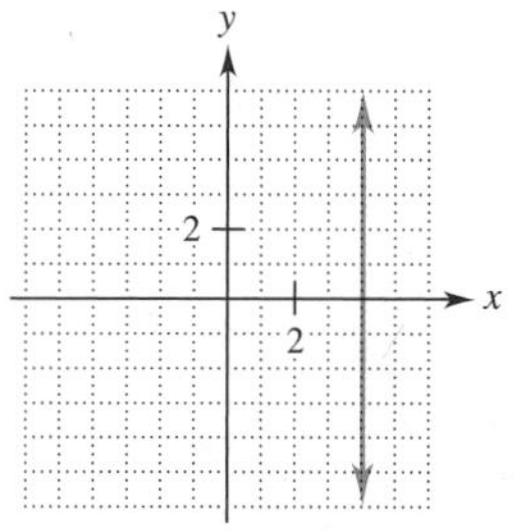

FIGURE 2.18

EXAMPLE 3 What is the slope of the vertical line in Figure 2.18?

Every point on the line has the same x-coordinate, 4. If you attempt to compute the slope with two of these points, say $(x_1, y_1) = (4, -2)$ and $(x_2, y_2) = (4, 1)$, you obtain

$$\begin{aligned} \text{slope} &= \frac{1 - (-2)}{4 - 4} \\ &= \frac{3}{0}. \end{aligned}$$

Division by 0 is not defined, so the slope of this line is undefined.

The arguments used in Examples 2 and 3 work in the general case and lead to the following conclusion.

> The slope of every horizontal line is 0.
>
> The slope of every vertical line is undefined.

Slope-Intercept Form

Slope can be used to develop an algebraic description of nonvertical straight lines. Assume that a line with slope m has y-intercept b, so that it goes through the point

$(0, b)$ (see Figure 2.19). Let (x, y) be any point on the line other than $(0, b)$. Using the definition of slope with the points $(0, b)$ and (x, y) gives

$$m = \frac{y - b}{x - 0}$$

$$m = \frac{y - b}{x}$$

$$mx = y - b \qquad \text{Multiply both sides by } x.$$

$$y = mx + b \qquad \text{Add } b \text{ to both sides. Reverse the equation.}$$

In other words, the coordinates of any point on the line satisfy the equation $y = mx + b$.

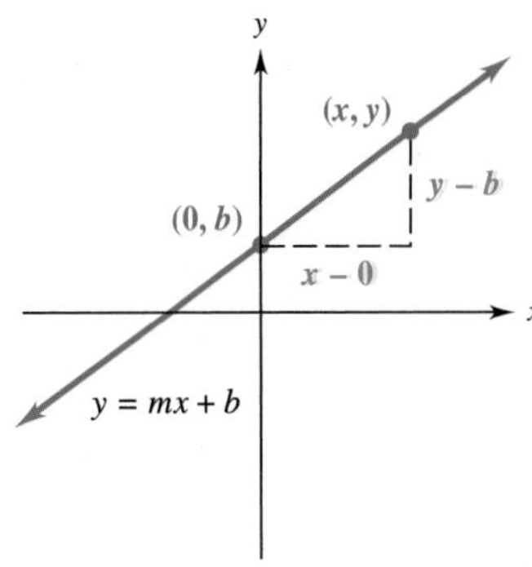

FIGURE 2.19

2

Find an equation for the line with

(a) y-intercept -3 and slope $2/3$;

(b) y-intercept $1/4$ and slope $-3/2$.

Answers:

(a) $y = \frac{2}{3}x - 3$

(b) $y = -\frac{3}{2}x + \frac{1}{4}$

Slope-Intercept Form

If a line has slope m and y-intercept b, then it is the graph of the equation

$$\boldsymbol{y = mx + b.}$$

This equation is called the **slope-intercept form** of the equation of the line.

EXAMPLE 4 Find an equation for the line with y-intercept $7/2$ and slope $-5/2$.

Use the slope-intercept form with $b = 7/2$ and $m = -5/2$.

$$y = mx + b$$

$$y = -\frac{5}{2}x + \frac{7}{2}$$ ◀ 2

EXAMPLE 5 Find the equation of the horizontal line with y-intercept 3.

The slope of the line is 0 (why?) and its y-intercept is 3, so its equation is

$$y = mx + b$$

$$y = 0x + 3$$

$$y = 3$$ ◀

The argument in Example 5 also works in the general case.

Find the slope and y-intercept for

(a) $x + 4y = 6$;

(b) $3x - 2y = 1$.

Answers:

(a) Slope $-1/4$; y-intercept $3/2$

(b) Slope $3/2$; y-intercept $-1/2$

If k is a constant, then the graph of the equation $y = k$ is the horizontal line with y-intercept k.

EXAMPLE 6 Find the slope and y-intercept for each of the following lines.

(a) $5x - 3y = 1$
Solve for y.

$$5x - 3y = 1$$
$$-3y = -5x + 1$$
$$y = \frac{5}{3}x - \frac{1}{3}$$

This equation is in the form $y = mx + b$, with $m = 5/3$ and $b = -1/3$. So the slope is $5/3$ and the y-intercept is $-1/3$.

(b) $-9x + 6y = 2$
Solve for y.

$$-9x + 6y = 2$$
$$6y = 9x + 2$$
$$y = \frac{3}{2}x + \frac{1}{3}$$

The slope is $3/2$ (coefficient of x) and the y-intercept is $1/3$. ◀ (3)

4

(a) List the slopes of the following lines:

E: $y = -.3x$, F: $y = -x$,

G: $y = -2x$, H: $y = -5x$.

(b) Graph all four lines on the same set of axes.

(c) How are the slopes of the lines related to their steepness?

Answers:

(a) Slope E $= -.3$; slope F $= -1$; slope G $= -2$; slope H $= -5$.

(b)

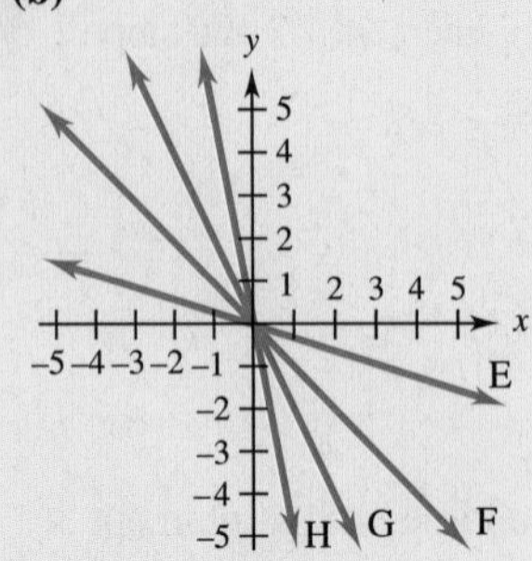

(c) The larger the slope in absolute value, the more steeply the line falls from left to right.

The slope-intercept form can be used to show how the slope measures the steepness of a line. Consider the straight lines A, B, C, and D given by the following equations; each has y-intercept 0 and slope as indicated.

A: $y = .5x$,	B: $y = x$,	C: $y = 3x$,	D: $y = 7x$
Slope .5	Slope 1	Slope 3	Slope 7

For these lines, Figure 2.20 shows that the bigger the slope, the more steeply the line rises from left to right. (4)

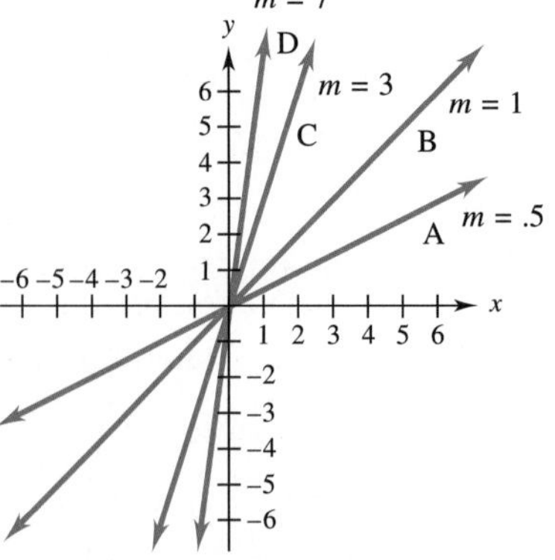

FIGURE 2.20

5

Graph the lines and label the intercepts.

(a) $3x + 4y = 12$

(b) $5x - 2y = 8$

Answers:

(a)

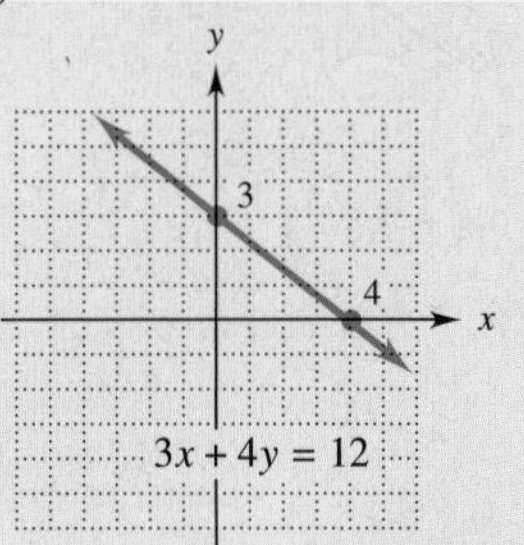

(b)

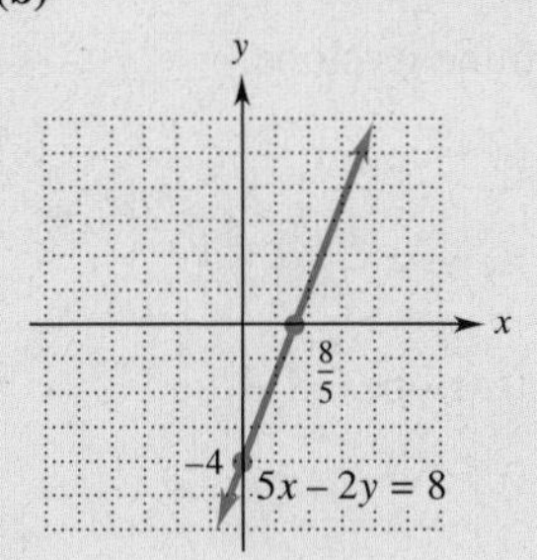

The preceding discussion and Problem 4 in the margin may be summarized as follows.

Direction of Line (moving from left to right)	*Slope*
Upward	**Positive** (larger for steeper lines)
Horizontal	**0**
Downward	**Negative** (larger in absolute value for steeper lines)
Vertical	**Undefined**

EXAMPLE 7 Sketch the graph of $x + 2y = 5$ and label the intercepts.

Find the x-intercept by setting $y = 0$ and solving for x.

$$x + 2 \cdot 0 = 5$$
$$x = 5$$

The x-intercept is 5 and (5, 0) is on the graph. The y-intercept is found similarly, by setting $x = 0$ and solving for y.

$$0 + 2y = 5$$
$$y = 5/2$$

The y-intercept is 5/2 and (0, 5/2) is on the graph. The points (5, 0) and (0, 5/2) can be used to sketch the graph (Figure 2.21). ◀ 5

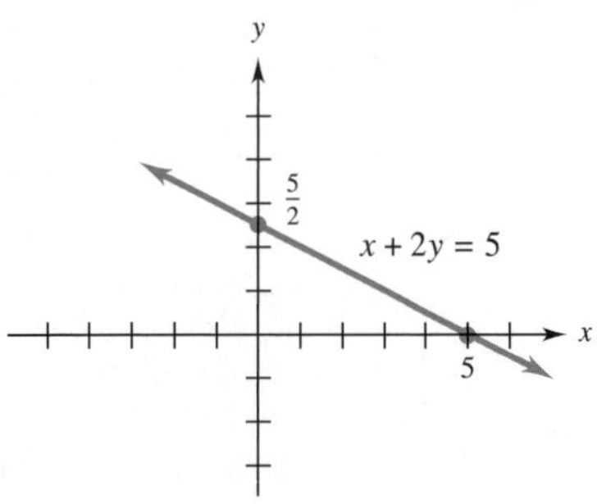

FIGURE 2.21

TECHNOLOGY TIP To graph a linear equation on a graphing calculator, you must first put the equation in slope-intercept form $y = mx + b$ so that it can be entered in the equation memory (called the Y= list on some calculators). Vertical lines cannot be graphed on most calculators.

Slopes of Parallel and Perpendicular Lines

We shall assume the following facts without proof. The first one is a consequence of the fact that slope measures steepness and parallel lines have the same steepness.

Two nonvertical lines are parallel whenever they have the same slope.

Two nonvertical lines are perpendicular whenever the product of their slopes is -1.

EXAMPLE 8 Determine whether each of the following pairs of lines are *parallel, perpendicular,* or *neither.*

(a) $2x + 3y = 5$ and $4x + 5 = -6y$

Put each equation in slope-intercept form by solving for y.

$$3y = -2x + 5 \qquad -6y = 4x + 5$$

$$y = -\frac{2}{3}x + \frac{5}{3} \qquad y = -\frac{2}{3}x - \frac{5}{6}$$

In each case the slope (coefficient of x) is $-2/3$, so the lines are parallel.

(b) $3x = y + 7$ and $x + 3y = 4$

Put each equation in slope-intercept form to determine its slope.

$$3x = y + 7 \qquad 3y = -x + 4$$

$$y = 3x - 7 \qquad y = -\frac{1}{3}x + \frac{4}{3}$$

slope 3   slope $-1/3$

Since $3(-1/3) = -1$, these lines are perpendicular.

(c) $x + y = 4$ and $x - 2y = 3$

Verify that the slope of the first line is -1 and the slope of the second is $1/2$. The slopes are not equal and their product is not -1, so the lines are neither parallel nor perpendicular. ◀ 6

6

Tell if the lines in each of the following pairs are *parallel, perpendicular,* or *neither.*

(a) $x - 2y = 6$ and $2x + y = 5$

(b) $3x + 4y = 8$ and $x + 3y = 2$

(c) $2x - y = 7$ and $2y = 4x - 5$

Answers:

(a) Perpendicular

(b) Neither

(c) Parallel

TECHNOLOGY TIP Perpendicular lines may not appear perpendicular on a graphing calculator, unless you use a *square window*—one in which a one-unit segment on the *y*-axis is the same length as a one-unit segment on the *x*-axis. To obtain such a window on most calculators, use a viewing window in which the *y*-axis is about 2/3 as long as the *x*-axis. The SQUARE (or ZSQUARE) key in the ZOOM menu will change the current window to a square window by automatically adjusting the length of one of the axes.

Point-Slope Form

The slope-intercept form of the equation of a line is usually the most convenient for graphing and for understanding how slopes and lines are related. However, it is not always the best way to *find* the equation of a line. In many situations (particularly in calculus), the slope and a point on the line are known and you must find the equation of the line. In such cases, the best method is to use the *point-slope form,* which we now explain.

Suppose a line has slope m and that (x_1, y_1) is a point on the line. Let (x, y) represent any other point on the line. Since m is the slope, then, by the definition of slope,

$$\frac{y - y_1}{x - x_1} = m.$$

Multiplying both sides by $x - x_1$ shows that

$$y - y_1 = m(x - x_1).$$

Point-Slope Form

If a line has slope m and passes through the point (x_1, y_1), then

$$\mathbf{y - y_1 = m(x - x_1)}$$

is the **point-slope form** of the equation of the line.

EXAMPLE 9 Find the equation of the line satisfying the given conditions.

Find both the point-slope and the slope-intercept forms of the equation of the line having the given slope and passing through the given point.

(a) $m = -3/5$, $(5, -2)$

(b) $m = 1/3$, $(6, 8)$

Answers:

(a) $y + 2 = -\frac{3}{5}(x - 5)$; $y = -\frac{3}{5}x + 1$.

(b) $y - 8 = \frac{1}{3}(x - 6)$; $y = \frac{1}{3}x + 6$.

(a) Slope 2; the point (5, 3) is on the line.

Use the point-slope form with $m = 2$ and $(x_1, y_1) = (5, 3)$. Substitute $x_1 = 5$, $y_1 = 3$, and $m = 2$ into the point-slope form of the equation.

$$y - y_1 = m(x - x_1)$$
$$y - 3 = 2(x - 5).$$

For some purposes, this form of the equation is fine; in other cases, you may want to rewrite it in the slope-intercept form.

(b) Slope -3; the point $(-4, 1)$ is on the line.

Use the point-slope form with $m = -3$ and $(x_1, y_1) = (-4, 1)$.

$$y - y_1 = m(x - x_1)$$
$$y - 1 = -3[x - (-4)] \quad \text{Point-slope form}$$

Using algebra we obtain the slope-intercept form of this equation.

$$y - 1 = -3(x + 4)$$
$$y - 1 = -3x - 12 \quad \text{Distributive property}$$
$$y = -3x - 11 \quad \text{Slope-intercept form}$$

◄ 7

The point-slope form can also be used to find an equation of a line given two different points on the line. The procedure is shown in the next example.

EXAMPLE 10 Find an equation of the line through (5, 4) and $(-10, -2)$.

Begin by using the definition of slope to find the slope of the line that passes through the two points.

$$\text{slope} = m = \frac{-2 - 4}{-10 - 5} = \frac{-6}{-15} = \frac{2}{5}$$

Find an equation of the line through

(a) (2, 3) and (−4, 6);

(b) (−8, 2) and (3, −6).

Answers:

(a) $2y = -x + 8$

(b) $11y = -8x - 42$

Use $m = 2/5$ and either of the given points in the point-slope form. If $(x_1, y_1) = (5, 4)$, then

$$y - y_1 = m(x - x_1)$$

$$y - 4 = \frac{2}{5}(x - 5) \qquad \text{Let } y_1 = 4,\ m = \frac{2}{5},\ x_1 = 5.$$

$$5(y - 4) = 2(x - 5) \qquad \text{Multiply both sides by 5.}$$

$$5y - 20 = 2x - 10 \qquad \text{Distributive property}$$

$$5y = 2x + 10.$$

Check that the result is the same when $(x_1, y_1) = (-10, -2)$. ◀ 8

Vertical Lines

The equation forms developed above do not apply to vertical lines because slope is not defined for such lines. However, vertical lines can easily be described as graphs of equations.

EXAMPLE 11 Find the equation whose graph is the vertical line in Figure 2.22.

Every point on the line has x-coordinate -1 and hence has the form $(-1, y)$. Thus every point is a solution of the equation $x + 0y = -1$, which is usually written simply as $x = -1$. Note that -1 is the x-intercept of the line. ◀

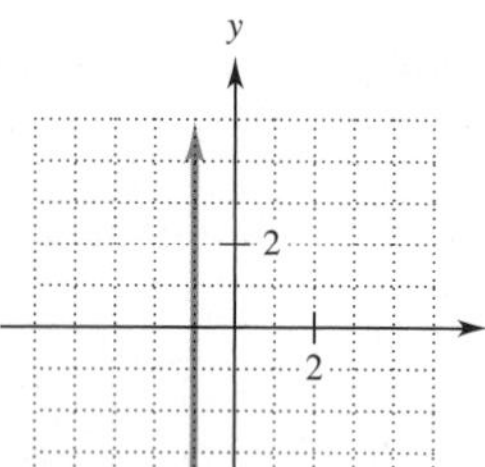

FIGURE 2.22

The argument in Example 11 also works in the general case.

If k is a constant, then the graph of the equation $x = k$ is the vertical line with x-intercept k.

Linear Equations

An equation in two variables whose graph is a straight line is called a **linear equation.** Linear equations have a variety of forms, as summarized in the following table. The last form listed is called the **general form** of a linear equation, and every linear equation can be written in this form. For example, $y = 4x - 5$ can be written in general forms as $4x - y = 5$ and $x = 6$ can be written in general form as $x + 0y = 6$.

Equation	*Description*
$x = k$	**Vertical line,** x-intercept k, no y-intercept, undefined slope
$y = k$	**Horizontal line,** y-intercept k, no x-intercept, slope 0
$y = mx + b$	**Slope-intercept form,** slope m, y-intercept b
$y - y_1 = m(x - x_1)$	**Point-slope form,** slope m, the line passes through (x_1, y_1).
$ax + by = c$	**General form.** If $a \neq 0$ and $b \neq 0$, the line has x-intercept c/a and y-intercept c/b.

Applications

Many relationships are linear, or almost linear, so that they can be approximated by linear equations.

EXAMPLE 12 In an experiment testing a person's reaction time y (in seconds) after undergoing x hours of stressful activity, the linear equation $y = .1957x + .1243$ was found to be a good approximation of the relationship between stress and reaction time during the first five hours.

(a) Assuming that reaction times continue to follow this pattern, what would be the approximate reaction time after eight and a half hours?

Substitute $x = 8.5$ in the equation and use a calculator to compute y.

$$y = .1957x + .1243$$
$$y = .1957(8.5) + .1243 = 1.78775$$

So the reaction time is approximately 1.8 seconds.

(b) When the reaction time is 1.5 seconds, how long has the subject been undergoing stressful activity?

Substitute $y = 1.5$ in the equation and solve for x.

$$1.5 = .1957x + .1243$$
$$1.5 - .1243 = .1957x$$
$$x = \frac{1.5 - .1243}{.1957} \approx 7.0296$$

Hence, the stressful activity has been going on for slightly more than 7 hours. 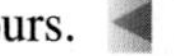

EXAMPLE 13 According to data from the College Board, the average cost of tuition and fees in public four-year colleges was \$2966 in the fall of 1996 and grew in an approximately linear fashion to \$3356 in the fall of 1999.

(a) Find a linear equation for this data.

Measure time along the x-axis and cost along the y-axis. Then the x-coordinate of each point is a year and the y-coordinate is the average cost of tuition and fees in that year. For convenience, let $x = 0$ correspond to 1990, so that $x = 6$ is 1996 and $x = 9$ is 1999. Then the given data points are (6, 2966) and (9, 3356). The slope of the line joining these points is

$$\frac{3356 - 2966}{9 - 6} = \frac{390}{3} = 130.$$

Using the point (6, 2966), we obtain the equation of this line

$$y - 2966 = 130(x - 6)$$
$$y - 2966 = 130x - 780$$
$$y = 130x + 2186$$

(b) Use this equation to estimate the average cost of tuition and fees in 1998.

Letting $x = 8$, we see that tuition and fees were

$$y = 130(8) + 2186 = \$3226.$$

(c) Assuming the equation remains valid in later years, estimate the average cost of tuition and fees in 2002.

When $x = 12$, then $y = 130(12) + 2186 = \$3746$.  9

9

The average cost of tuition and fees in private four-year colleges was \$12,824 in 1996 and \$15,380 in 1999.

(a) Let $x = 0$ correspond to 1990 and find a linear equation for this data.

(b) Estimate the average cost in 2001.

Answers:

(a) $y = 852x + 7712$

(b) \$17,084

2.2 Exercises

Find the slope of the line, if it is defined. (See Examples 1–3.)

1. Through $(2, 5)$ and $(0, 6)$
2. Through $(9, 0)$ and $(12, 15)$
3. Through $(-4, 7)$ and $(3, 0)$
4. Through $(-5, -2)$ and $(-4, 11)$
5. Through the origin and $(-4, 6)$
6. Through the origin and $(8, -2)$
7. Through $(-1, 4)$ and $(-1, 8)$
8. Through $(-3, 5)$ and $(2, 5)$

Find an equation of the line with the given y-intercept and slope m. (See Examples 4 and 5.)

9. $5, m = 3$
10. $-3, m = -7$
11. $1.5, m = -2.3$
12. $-4.5, m = 1.5$
13. $4, m = -3/4$
14. $-3, m = 2/3$

Find the slope and the y-intercept of the line whose equation is given. (See Example 6.)

15. $2x - y = 7$
16. $x + 2y = 7$
17. $6x = 2y + 4$
18. $4x + 3y = 12$
19. $6x - 9y = 14$
20. $4x + 2y = 0$
21. $2x - 3y = 0$
22. $y = 5$
23. $x = y - 5$
24. On one graph, sketch six straight lines that meet at a single point and satisfy this condition: one line has slope 0, two lines have positive slope, two lines have negative slope, and one line has undefined slope.
25. For which of the line segments in the figure is the slope
    (a) largest? (b) smallest?
    (c) largest in absolute value? (d) closest to 0?

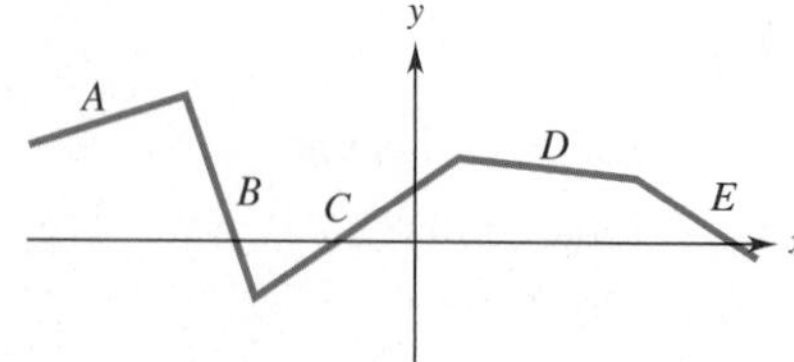

26. Match each equation with the line at the top of the next column that most closely resembles its graph. (*Hint:* Consider the signs of m and b in the slope-intercept form.)
    (a) $y = 3x + 2$ (b) $y = -3x + 2$
    (c) $y = 3x - 2$ (d) $y = -3x - 2$

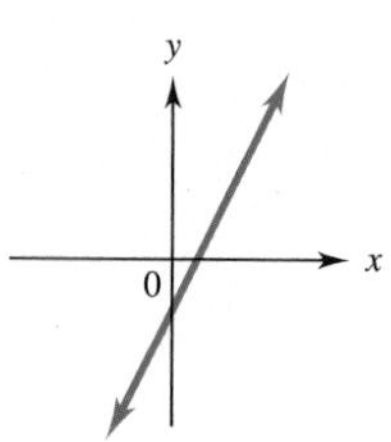

A

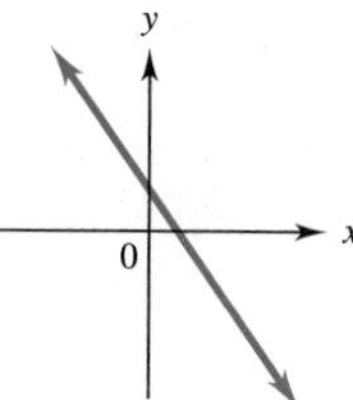

B

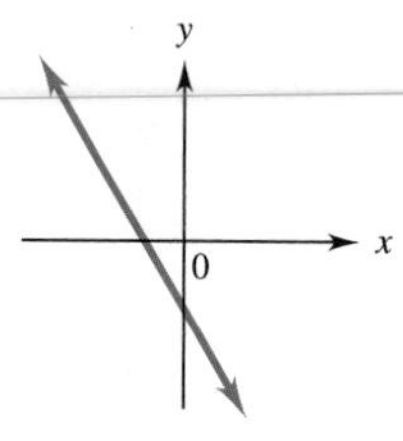

C

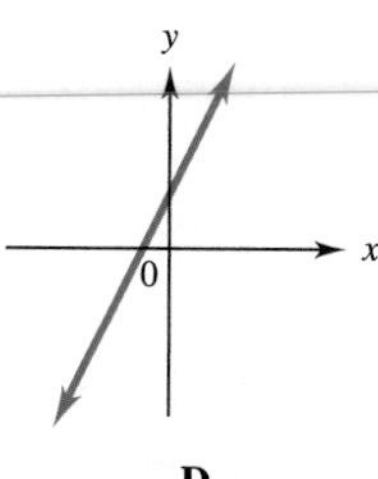

D

Sketch the graph of the equation and label its intercepts. (See Example 7.)

27. $2x - y = -2$
28. $2y + x = 4$
29. $2x + 3y = 4$
30. $-5x + 4y = 3$
31. $4x - 5y = 2$
32. $3x + 2y = 8$

Determine whether each pair of lines is parallel, perpendicular, *or* neither. *(See Example 8.)*

33. $4x - 3y = 6$ and $3x + 4y = 8$
34. $2x - 5y = 7$ and $15y - 5 = 6x$
35. $3x + 2y = 8$ and $6y = 5 - 9x$
36. $x - 3y = 4$ and $y = 1 - 3x$
37. $4x = 2y + 3$ and $2y = 2x + 3$
38. $2x - y = 6$ and $x - 2y = 4$
39. (a) Find the slope of each side of the triangle with vertices $(9, 6)$, $(-1, 2)$, and $(1, -3)$.
    (b) Is this triangle a right triangle? (*Hint:* Are two sides perpendicular?)
40. (a) Find the slope of each side of the quadrilateral with vertices $(-5, -2)$, $(-3, 1)$, $(3, 0)$, and $(1, -3)$.
    (b) Is this quadrilateral a parallelogram (opposite sides parallel)?

Find an equation of the line that passes through the given point and has the given slope. (See Examples 9 and 11.)

41. $(-1, 2), m = -2/3$
42. $(-4, -3), m = 5/4$
43. $(-2, -2), m = 2$

44. $(-2, 3), m = -1/2$

45. $(8, 2), m = 0$

46. $(2, -4), m = 0$

47. $(6, -5)$, undefined slope

48. $(-8, 9)$, undefined slope

Find an equation of the line that passes through the given points. (See Example 10.)

49. $(-1, 1)$ and $(2, 5)$

50. $(2, 5)$ and $(0, 6)$

51. $(1, 2)$ and $(3, 7)$

52. $(-1, -2)$ and $(2, -1)$

Find an equation of the line satisfying the given conditions.

53. Through the origin with slope 7

54. Through the origin and horizontal

55. Through $(5, 8)$ and vertical

56. Through $(7, 11)$ and parallel to $y = 6$

57. Through $(3, 4)$ and parallel to $4x - 2y = 5$

58. Through $(6, 8)$ and perpendicular to $y = 2x - 3$

59. x-intercept 5; y-intercept -5

60. Through $(-5, 2)$ and parallel to the line through $(1, 2)$ and $(4, 3)$

61. Through $(-1, 3)$ and perpendicular to the line through $(0, 1)$ and $(2, 3)$

62. y-intercept 3 and perpendicular to $2x - y + 6 = 0$

Business *The lost value of equipment over a period of time is called* depreciation. *The simplest method for calculating depreciation is* straight-line depreciation. *The annual straight-line depreciation of an item that cost x dollars with a useful life of n years is $D = (1/n)x$. Find the depreciation for items with the following characteristics.*

63. Cost: \$12,482; life 10 yr

64. Cost: \$39,700; life 12 yr

65. Cost: \$145,000; life 28 yr

66. Business Ral Corp. has an incentive compensation plan under which a branch manager receives 10% of the branch's income after deduction of the bonus but before deduction of income tax.* Branch income for 1988 before the bonus and income tax was \$165,000. The tax rate was 30%. The 1988 bonus amounted to

(a) \$12,600 **(b)** \$15,000

(c) \$16,500 **(d)** \$18,000.

Work the following applied problems. (See Examples 12 and 13.)

67. Transportation According to data from the U.S. Department of Transportation, the total number of passenger miles (in millions) traveled on all forms of transportation (air, train, bus, etc.) in year x can be approximated by the equation

$$y = .12x + 3.29$$

where $x = 0$ corresponds to 1990. Find the approximate number of passenger miles in each of these years:

(a) 1997 **(b)** 2001 **(c)** 2003.

(d) In what year will the total passenger miles reach 5 million?

68. Natural Sciences Deer ticks cause concern because they can carry Lyme disease. One study found a relationship between the density of acorns produced in the fall and the density of deer tick larvae the following spring.* The relationship can be approximated by the linear equation

$$y = 34x + 230,$$

where x is the number of acorns per square meter in the fall and y is the number of deer tick larvae per 400 square meters the following spring. According to this formula, approximately how many acorns per square meter would result in 1000 deer tick larvae per 400 square meters?

69. Business According to data from the U.S. Department of Commerce, movie-theater box-office receipts (in millions of dollars) in year x can be approximated by

$$y = 494.65x + 2971.03,$$

where $x = 0$ corresponds to 1995. Assume this equation remains valid until 2008.

(a) Estimate the box-office receipts in 2003.

(b) In what year will the box-office receipts reach \$8,000,000,000?

70. Business The Consumer Price Index (CPI) measures the cost of a common bundle of goods and services bought by the average consumer on a regular basis. It is scaled so that the CPI for 1982 is 100. For recent years, the CPI can be approximated by $y = 3.913x + 132.25$, where $x = 0$ corresponds to 1990.†

(a) To the nearest whole number, what was the CPI in 1999?

(b) If this trend continues, when will the CPI be 187.032?

71. Business An article in the *Economist* (March 31, 2001) states that worldwide automobile production was about 54 million vehicles in 2000 and is expected to be 63 million vehicles in 2008.

(a) Let the x-axis denote time and the y-axis the number of vehicles produced (in millions). Let $x = 0$ corre-

*Uniform CPA Examination, May 1989, American Institute of Certified Public Accountants.

**Science*, vol. 281, no. 5375, July 17, 1998, pp. 350–351.

†U.S. Bureau of Labor Statistics.

spond to the year 2000. Fill in the blanks: the given data is represented by the points (______, 54) and (8, ______).

(b) Find the equation determined by the two points in part (a).

(c) Use the linear equation in part (b) to estimate the number of vehicles produced in 2002.

(d) In what year will vehicle production first reach 60 million?

72. **Social Sciences** In 1974, a total of 86,821 people from other countries immigrated to the state of California. In 1996, the number of immigrants was 199,483.*

(a) If the change in foreign immigration to California is considered to be linear, write an equation expressing the number y of immigrants in terms of the number x of years after 1974.

(b) Use your result in part (a) to predict the foreign immigration to California in the year 2010.

73. **Social Sciences** The number of unmarried couples in the United States who are living together has been rising at a roughly linear rate in recent years. The number of cohabitating adults was 1.1 million in 1977 and 4.9 million in 1997.†

(a) Write an equation expressing the number y of cohabitating adults (in millions) in terms of the number x of years after 1977.

(b) Use your result in part (a) to predict the number of cohabitating adults in the year 2010.

74. **Social Sciences** The percentage of college students who are age 35 and older has been increasing at roughly a linear rate. In 1972 the percentage was 9%, and in 1998 it was 17%.‡

(a) Find an equation giving the percentage of college students age 35 and older in terms of time t, where t represents the number of years since 1970.

(b) If this linear trend continues, what percentage of college students will be 35 and over in 2010?

(c) If this linear trend continues, in what year will the percentage of college students 35 and over reach 31%?

75. **Physical Sciences** In 2001, the Intergovernmental Panel on Climate Change predicted that the average temperature on the earth would rise about .3°C per decade in the coming century.§ Let t measure the time in years since 1970, when the average global temperature was 15°C.

(a) Find a linear equation giving the average global temperature in degrees Celsius in terms of t, the number of years since 1970.

(b) Scientists have estimated that the sea level will rise by 65 cm if the average global temperature rises to 19°C. According to your answer to part (a), when would this occur?

76. **Health** For you to achieve the maximum benefit for the heart when exercising, your heart rate (in pulse beats per minute) should be in the target heart rate zone. The lower limit of this zone is found by taking 65% of the difference between 220 and your age. The upper limit is found by using 85%.*

(a) Find linear equations for the upper and lower limits (U and L) in terms of age x.

(b) What is the target heart rate zone for a 20 year old?

(c) What is the target heart rate zone for a 40 year old?

(d) Two women in an aerobics class stop to take their pulses and are surprised to find that they have the same pulse. One woman is 48 years older than the other and is working at the upper limit of her target heart rate zone. The younger woman is working at the lower limit of her target heart rate zone. What are the ages of the two women, and what is their pulse?

(e) Run for 10 minutes, take your pulse, and see if it in your target heart rate zone. (After all, this is listed as an exercise!)

77. **Physical Science** The graph shows the winning times (in minutes) at the Olympic Games for the 5000-meter run, together with a linear approximation of the data.†

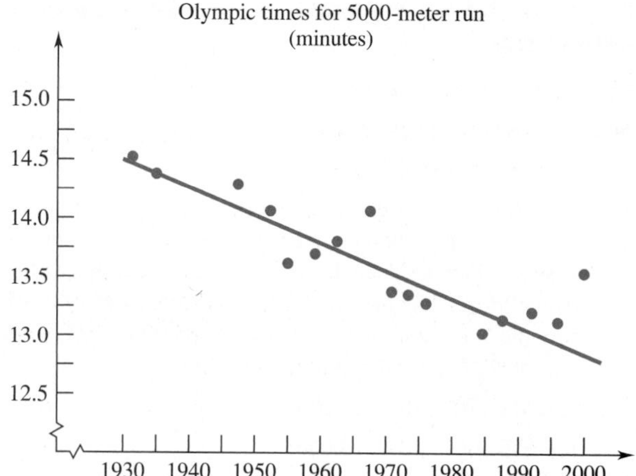

Legal Immigration to California in Federal Fiscal Year 1996, State of California Demographic Research Unit, June 1999.

†*New York Times*, February 15, 2000, p. F8.

‡*New York Times*, August 17, 1994, p. B9, and January 9, 2000, p. 4A.

§IPCC, *Climatic Change 2001: the Scientific Basis*.

*www.drkoop.com

†United States Olympic Committee.

(a) The equation for the linear approximation is $y = -.0221x + 57.14$. What does the slope of this line represent? Why is the slope negative?

(b) Can you think of any reason why there are no data points for the years 1940 and 1944?

78. **Business** The graph at the right shows the number of U.S. radio stations on the air along with a linear approximation of the data.*

(a) Use the data points (1950, 2773) and (1999, 12913) to estimate the slope of the line shown. Interpret this number.

(b) Discuss the accuracy of the linear approximation.

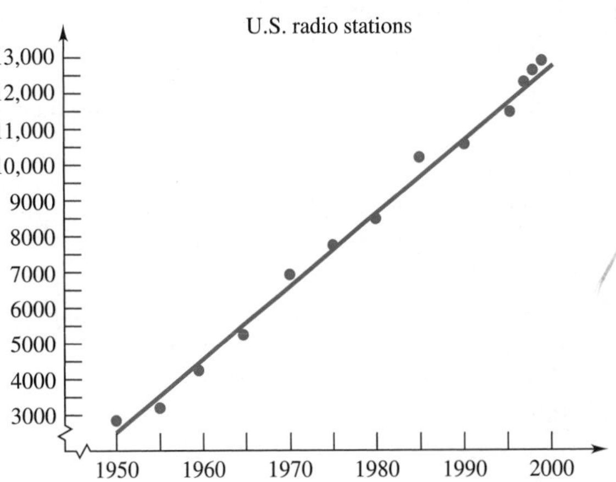

*National Association of Broadcasters.

2.3 LINEAR MODELS

In business and science, it is often necessary to make judgments based on data from the past. For instance, a stock analyst might use a company's profits in previous years to estimate next year's profits. A life insurance company might look at life expectancies of people born in various years to predict how much money it should expect to pay out next year.

In such situations, the available data is used to construct a mathematical model, such as an equation or a graph, which is used to approximate the likely outcome in cases where complete data is not available. In this section, we consider applications in which the data can be modeled by a linear equation.

The simplest way to construct a linear model is to use the line determined by two of the data points, as illustrated in the following example.

EXAMPLE 1 The profits of the General Electric Company (GE) (in billions of dollars) over a five-year period are shown in the table.*

Year	1996	1997	1998	1999	2000
Profit	7	8	9	11	13

(a) Let $x = 1$ correspond to 1996 and plot the points (x, y), where x is the year and y the profit.

*GE Annual Report 2000, March 2001. The profit figures (net earnings) have been rounded to the nearest billion.

The data points are (1, 7), (2, 8), (3, 9), (4, 11), and (5, 13), as shown in Figure 2.23.

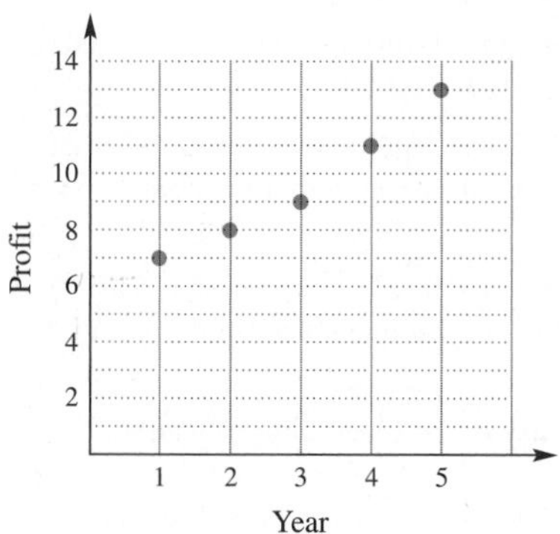

FIGURE 2.23

The data points are approximately linear, so a line should be a reasonable model.

(b) Use the data points (1, 7) and (3, 9) to find a line that models the data.

The slope of the line through (1, 7) and (3, 9) is $\frac{9-7}{3-1} = \frac{2}{2} = 1$. Using the point (1, 7), we find that the point-slope equation of this line is

$$y - 7 = 1(x - 1)$$
$$y = x + 6.$$

This line and the data points are shown in Figure 2.24. Although the line fits the first three data points perfectly, it underestimates profits in later years.

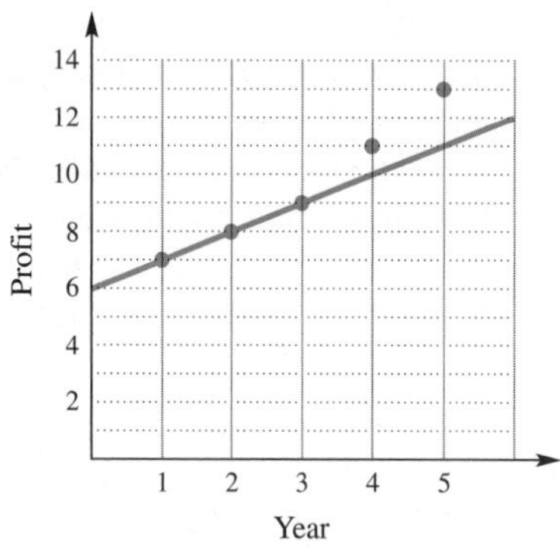

FIGURE 2.24

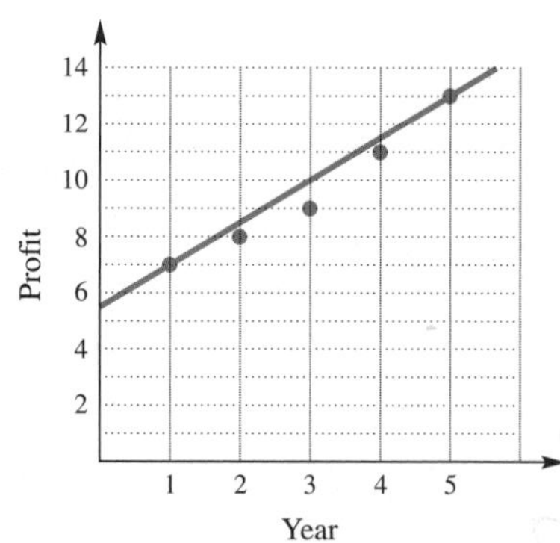

FIGURE 2.25

(c) Use the points (1, 7) and (5, 13) to find another line that models this data.

The slope is $\frac{13-7}{5-1} = \frac{6}{4} = \frac{3}{2} = 1.5$ and the equation is

$$y - 7 = 1.5(x - 1)$$
$$y = 1.5x + 5.5.$$

This line and the data points are graphed in Figure 2.25. Although this line passes through only two of the data points, it appears to fit the data reasonably well overall.

1

Use the points (2, 8) and (4, 11) to find another linear model for the data in Example 1.

Answer:

$y = 1.5x + 5$

(d) Use the model in part (c) to estimate the profits in 2001.
Substitute $x = 6$ (corresponding to 2001) in the equation:

$$y = 1.5x + 5.5$$
$$y = 1.5(6) + 5.5 = 14.5 \text{ billion.}$$ ◀ 1

Unlike the situation in Example 1, it isn't always possible to determine visually which of several lines best fits the data. For a given linear model, however, we can compute the difference between the actual amount p and the amount y given by the line. If the data point is (x, p) and the corresponding point on the line is (x, y), then the difference $p - y$ measures the error in the model for that particular value of x. The number $p - y$ is called a **residual.** As shown in Figure 2.26, the residual $p - y$ is the vertical distance from the data point to the line (positive when the data point is above the line, negative when it is below the line, and 0 when it is on the line).

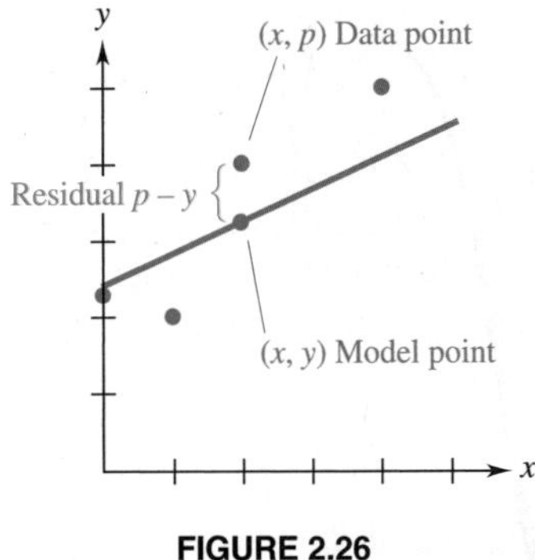

FIGURE 2.26

One way to determine how well a line fits the data points might be to compute the sum of its residuals, that is, the sum of the individual errors. Unfortunately, however, two different lines might each have 0 as the sum of their residuals, which doesn't enable us to decide which one is the better fit. (See Exercise 11 at the end of this section for an example of this possibility.) To avoid this difficulty, the sum of the *squares* of the residuals is used to measure how well a line fits the data points. When the sum of the squares is used, a smaller sum means a smaller error (and, hence, a better fit). The error is 0 only when *all* the data points lie on the line (a perfect fit).

EXAMPLE 2 In Example 1, two linear models for GE's profits were constructed: $y = x + 6$ and $y = 1.5x + 5.5$. For each model, determine the five residuals, the square of each residual, and the sum of the squared residuals.

The information for each model is summarized in the following tables.

$y = x + 6$

Data Point (x, p)	*Model Point* (x, y)	*Residual* $p - y$	*Squared Residual* $(p - y)^2$
(1, 7)	(1, 7)	0	0
(2, 8)	(2, 8)	0	0
(3, 9)	(3, 9)	0	0
(4, 11)	(4, 10)	1	1
(5, 13)	(5, 11)	2	4
			sum: 5

$y = 1.5x + 5.5$

Data Point (x, p)	*Model Point* (x, y)	*Residual* $p - y$	*Squared Residual* $(p - y)^2$
(1, 7)	(1, 7)	0	0
(2, 8)	(2, 8.5)	−.5	.25
(3, 9)	(3, 10)	−1	1
(4, 11)	(4, 11.5)	−.5	.25
(5, 13)	(5, 13)	0	0
			sum: 1.5

In side problem 1, you showed that $y = 1.5x + 5$ is also a linear model for GE's profits. Use this line and the data in Example 1 to find

(a) the residuals;

(b) the sum of the squares of the residuals

(c) Does this line fit the data better than the two lines in Example 2?

Answers:

(a) .5, 0, −.5, 0, .5

(b) .75

(c) Yes, because the sum of the squares of its residuals is smaller.

These results confirm what we saw visually in Example 1. The line $y = 1.5x + 5.5$ is a better fit for the data because the sum of the squares of its residuals is smaller than the sum for $y = x + 6$. 

Linear Regression (Optional)*

Mathematical techniques that are introduced in Chapter 14 can be used to prove the following result.

For any set of data points, there is one and only one line for which the sum of the squares of the residuals is as small as possible.

This *line of best fit* is called the **least squares regression line,** and the computational process for finding its equation is called **linear regression.** The linear regression formulas are quite complicated and require a large amount of computation. Fortunately, most graphing calculators and spreadsheet programs can do linear regression quickly and easily.

EXAMPLE 3 Recall that profits of GE (in billions of dollars) were as follows.

Year	1996	1997	1998	1999	2000
Profit	7	8	9	11	13

Use a graphing calculator to do the following.

(a) Plot the data points, with $x = 1$ corresponding to 1996.†

The data points are (1, 7), (2, 8), (3, 9), (4, 11), and (5, 13). Press STAT EDIT to bring up the statistics editor. Enter the x-coordinates as a list L_1 and the corresponding y-coordinates as a list L_2, as shown in Figure 2.27. To plot the data points, go to the STAT PLOT menu, choose a plot (here it's Plot 1), choose ON, and enter the lists L_1 and L_2, as shown in Figure 2.28. Then set the viewing window as usual and press GRAPH to produce Figure 2.29.

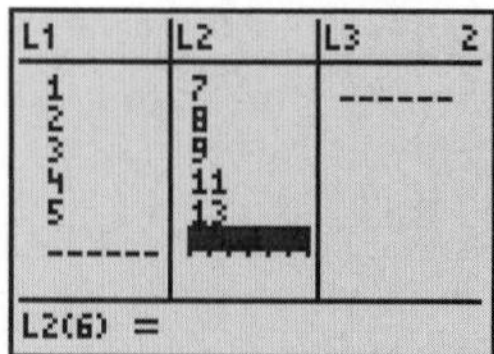

FIGURE 2.27

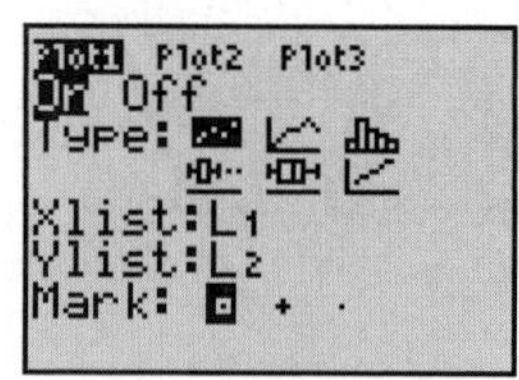

FIGURE 2.28

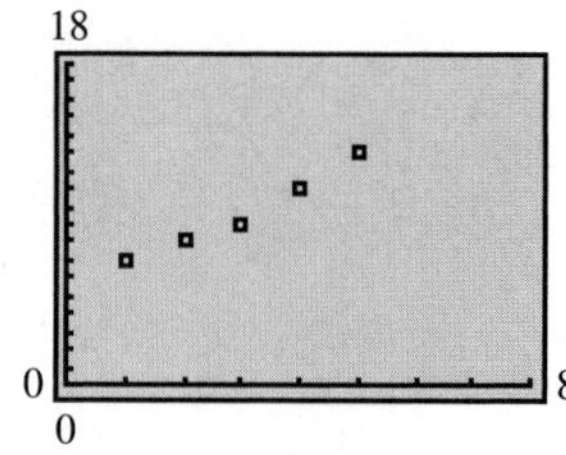

FIGURE 2.29

*Examples 3–6 require either a graphing calculator or a spreadsheet program.

†The process outlined here works for most TI graphing calculators. Other graphing calculators and spreadsheet programs operate similarly, but check your instruction manual, or see the Graphing Calculator Appendix, which is available at www.aw.com/MWA8.

Use the least squares regression line $y = 1.5x + 5.1$ and the data points in Example 3 to find

(a) the residuals;

(b) the sum of the squares of the residuals.

(c) How does this line compare to those in Example 1 and side problem 2?

Answers:

(a) .4, −.1, −.6, −.1, .4

(b) .7

(c) It fits the data best because the sum of the squares of its residuals is smallest.

(b) Find the least squares regression line for this data.

Go to the STAT CALC menu and choose LIN REG, which returns you to the home screen. As shown in Figure 2.30, enter the list names and the place where the equation of the regression line should be stored (here Y_1 is chosen; it's on the VARS Y-VARS FUNCTION menu); then press ENTER. Figure 2.31 shows that the equation of the regression line is

$$y = 1.5x + 5.1.$$

(The number r in Figure 2.31 will be discussed below.)

FIGURE 2.30

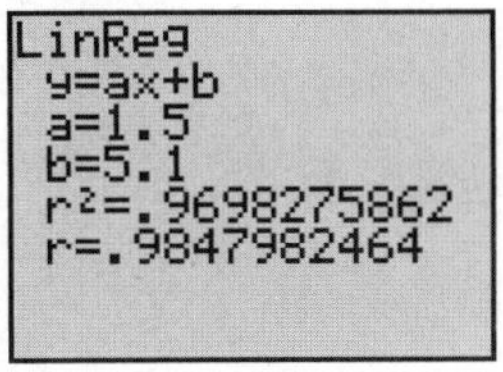

LinReg
y=ax+b
a=1.5
b=5.1
r²=.9698275862
r=.9847982464

FIGURE 2.31

(c) Graph the data points and the regression line on the same screen.

Press GRAPH to see the line plotted with the data points (Figure 2.32). 

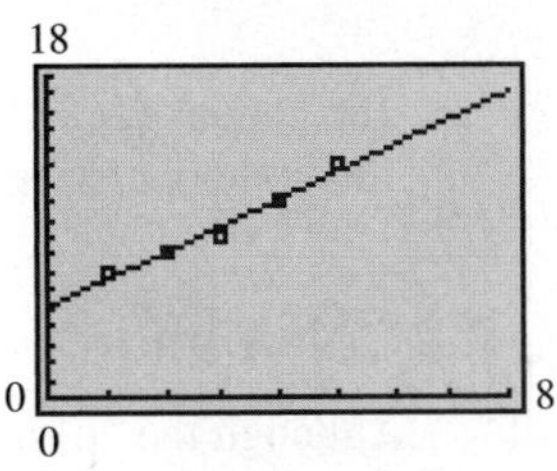

FIGURE 2.32

NOTE The mathematical techniques used to find the least squares regression line are discussed and illustrated with a simple example in the case at the end of Chapter 14. These techniques require multivariable calculus.

EXAMPLE 4 The table shows the number of accidental deaths per 100,000 population in the United States over the past century.*

Year	*Death Rate*	*Year*	*Death Rate*
1910	84.4	1960	52.1
1920	71.2	1970	56.2
1930	80.5	1980	46.5
1940	73.4	1990	36.9
1950	60.3	1998	34.1

*U.S. National Center for Health Statistics.

(a) Let $x = 0$ correspond to 1900. Use a graphing calculator or spreadsheet program to find the least squares regression line that models this data.

The data points are (10, 84.4), (20, 71.2), . . . , (98, 34.1). Enter the x-coordinates as list L_1 and the corresponding y-coordinates as list L_2 in a graphing calculator. Then find the regression line, as in Figure 2.33. Rounding the coefficients, the equation of the line is $y = -.565x + 90.525$.

FIGURE 2.33

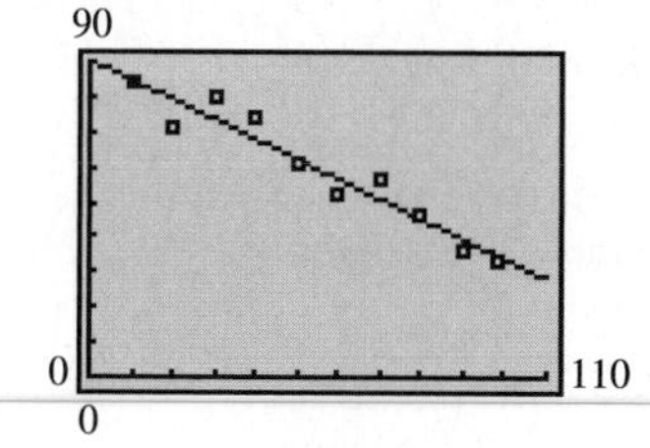

90

0 110

0

FIGURE 2.34

4

Using only the data from 1960 and later in Example 4, find the equation of the least squares regression line. Round the coefficients to three decimal places.

Answers:

$y = -.575x + 90.933$

(b) Plot the data points and the regression line on the same screen.
See Figure 2.34, which shows that the line is a reasonable model for the data.

(c) Use the regression line to predict the number of accidental deaths in 2004.
2004 corresponds to $x = 104$. Substitute $x = 104$ in the regression line equation:

$$y = -.565(104) + 90.525 \approx 31.8.$$

The model predicts that there will be approximately 31.8 accidental deaths per 100,000 population in 2004. ◀ 4

Correlation

Although the "best fit" line can always be found by linear regression, it may not be a good model. For instance, if the data points are widely scattered, no straight line will model the data accurately. When the linear regression line was computed in Examples 3 and 4, the calculator also displayed a number r and its square (Figures 2.31 and 2.33). This number is called the **coefficient of correlation.** It measures how closely the data points fit the regression line and thus indicates how good the regression line is for predictive purposes.

The coefficient of correlation r is always between -1 and 1. When $r = \pm 1$, the data points all lie on the regression line (a perfect fit). When the absolute value of r is close to 1, the line fits the data quite well, and when r is close to 0, the line is a poor fit for the data (but some other curve might be a good fit). Figure 2.35 shows how the value of r varies, depending on the pattern of the data points.

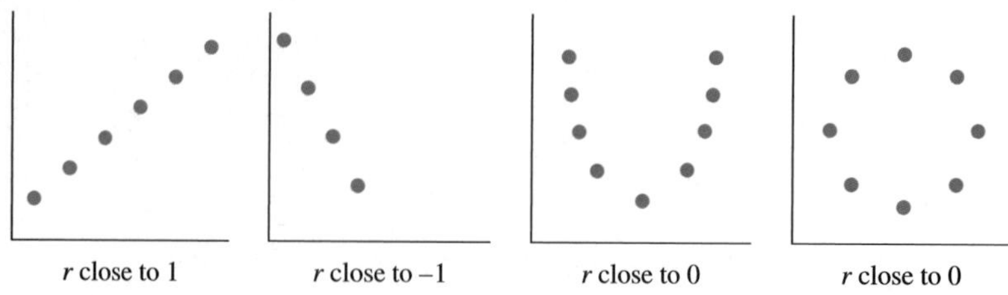

FIGURE 2.35

EXAMPLE 5 The number of unemployed people in the U.S. labor force (in millions) in recent years is shown in the table.*

Year	*Unemployed*	*Year*	*Unemployed*	*Year*	*Unemployed*
1987	7.425	1992	9.613	1997	6.739
1988	6.701	1993	8.940	1998	6.210
1989	6.528	1994	7.996	1999	5.880
1990	7.047	1995	7.404		
1991	8.628	1996	7.236		

Determine whether or not a linear equation is a good model for this data.

Let $x = 0$ correspond to 1980 and plot the data points (7, 7.425), (8, 6.701), etc., either by hand or using a graphing calculator, as in Figure 2.36. They do not form a linear pattern (unemployment tends to rise and fall). Alternatively, you could compute the regression equation for the data, as in Figure 2.37. The coefficient of correlation is $r \approx -.278$, a number quite close to 0, which indicates that the regression line is a poor fit for the data. Therefore, a linear equation is not a good model for this data. ◀ 5

5

Using only the data from 1992 and later in Example 5, find

(a) the equation of the least squares regression line;

(b) the coefficient of correlation.

(c) How well does this line fit the data?

Answers:

(a) $y \approx -.52x + 15.57$
(b) $r \approx -.98$
(c) It fits very well because $|r|$ is very close to 1.

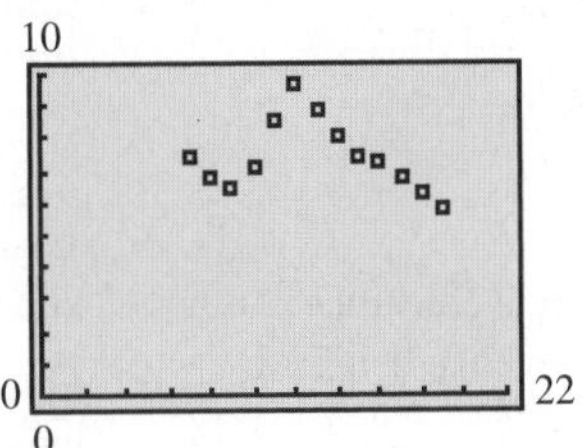

FIGURE 2.36

FIGURE 2.37

*U.S. Department of Labor, Bureau of Labor Statistics.

EXAMPLE 6 Enrollment projections (in millions) for all U.S. colleges and universities (including two-year institutions) in selected years are shown in the table.*

Year	1995	1998	1999	2000	2001	2002	2003
Enrollment	14.262	14.608	14.881	15.072	15.158	15.168	15.262

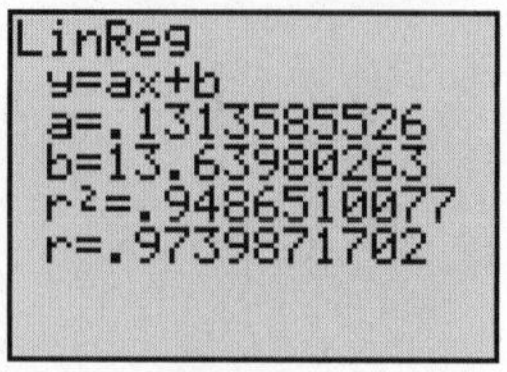

```
LinReg
 y=ax+b
 a=.1313585526
 b=13.63980263
 r²=.9486510077
 r=.9739871702
```

FIGURE 2.38

(a) Let $x = 0$ correspond to 1990. Use a graphing calculator or spreadsheet program to find a linear model for the data and determine how well it fits the data points.

The least squares regression line (with coefficients rounded) is

$$y = .1314x + 13.6398,$$

as shown in Figure 2.38. The coefficient of correlation is $r \approx .97$, which is very close to 1, so this line fits the data very well.

(b) Predict the enrollment in 2005.

Let $x = 15$ (corresponding to 2005) in the regression equation:

$$y = .1314(15) + 13.6398 = 15.6108.$$

Therefore, the enrollment in 2005 will be approximately 15,610,800 students.

(c) In what year will enrollment reach 16 million?

Let $y = 16$ and solve the regression equation for x.

$$16 = .1314x + 13.6398$$
$$2.3602 = .1314x$$
$$x \approx 17.96$$

Since these enrollment figures change once a year, use the nearest integer value for x, namely, 18. So enrollment will reach 16 million in 2008.

*As of fall of each year. *Statistical Abstract of the United States*, 2000.

2.3 Exercises

1. Physical Science The table shows equivalent Fahrenheit and Celsius temperatures.

Degrees Fahrenheit	32	68	104	140	176	212
Degrees Celsius	0	20	40	60	80	100

(a) Choose any two data points and use them to construct a linear equation that models the data, with x being Fahrenheit and y Celsius.

(b) Use the model in part (a) to find the Celsius temperature corresponding to

50° Fahrenheit and 75° Fahrenheit.

Physical Science *Use the linear equation derived in Exercise 1 to work the following problems.*

2. Convert each temperature.
(a) 58°F to Celsius **(b)** 50°C to Fahrenheit
(c) −10°C to Fahrenheit **(d)** −20°F to Celsius

3. According to the *World Almanac and Book of Facts*, 2001, Venus is the hottest planet, with a surface temperature of 864° Fahrenheit. What is this temperature in Celsius?

4. Find the temperature at which Celsius and Fahrenheit temperatures are numerically equal.

In each of the following problems, assume the data can be modeled by a straight line. Use two data points to find such a line and then answer the question. (See Example 1.)

5. **Business** The Consumer Price Index (CPI), which measures the cost of a typical package of consumer goods, stood at 130.7 in 1990 and 170.7 in 2000. Let $x = 0$ correspond to 1990 and estimate the CPI in 1995 and 2003.

6. **Social Science** The average monthly family payment for Temporary Assistance to Needy Families was \$473 in 1997 and \$589 in 1999.* Let $x = 0$ correspond to 1990 and estimate the average monthly payment in 2004.

7. **Business** The United States is China's largest export market. Imports from China grew from about 46 billion dollars in 1995 to 82 billion dollars in 1999. Let $x = 0$ correspond to 1990 and estimate the amount of imports in 2002.

8. **Physical Science** Suppose a baseball is thrown at 85 miles per hour. The ball will travel 320 feet when hit by a bat swung at 50 miles per hour and will travel 440 feet when hit by a bat swung at 80 miles per hour. Let y be the number of feet traveled by the ball when hit by a bat swung at x miles per hour. (*Note:* This is valid for $50 \leq x \leq 90$, where the bat is 35 inches long, weighs 32 ounces, and strikes a waist-high pitch so the place of the swing lies at 10° from the diagonal.)† How much farther will a ball travel for each mile per hour increase in the speed of the bat?

9. **Natural Science** The amount of tropical rain forests in Central America decreased from 130,000 square miles to about 80,000 square miles from 1969 to 1985. Let y be the amount (in ten thousands of square miles) x years after 1965. How large were the rain forests in the year 1997?

10. **Physical Science** Ski resorts require large amounts of water in order to make snow. Snowmass Ski Area in Colorado plans to pump at least 1120 gallons of water per minute for at least 12 hours a day from Snowmass Creek between mid-October and late December.‡ Environmentalists are concerned about the effects on the ecosystem. Find the minimum amount of water pumped in 30 days. (*Hint:* Let y be the total number of gallons pumped x days after pumping begins. Note that (0, 0) is on the graph of the equation.)

In each of the following problems, two linear models are given for the data. For each model,

(a) Find the residuals and their sum.
(b) Find the sum of the squares of the residuals.
(c) Decide which model is the better fit. (See Example 2.)

11. **Business** The weekly amount spent on advertising and the weekly sales revenue of a small store over a five-week period are shown in the table.

Advertising Expenditure x (in hundreds of dollars)	1	2	3	4	5
Sales Revenue y (in thousands of dollars)	2	2	3	3	5

Two equations that model the data are $y = .5x + 1.5$ and $y = x$.

12. **Business** The table shows the total outstanding consumer debt (excluding home mortgages) in billions of dollars in selected years.*

Year	1980	1985	1990	1995	1999
Consumer debt	349	585	789	1096	1395

Let $x = 0$ correspond to 1980. Two equations that model the data are $y = 54x + 314$ and $y = 55x + 350$.

In each of the following problems, determine whether or not a straight line is a good model for the data. You may do this either visually, by plotting the data points, or analytically, by finding the coefficient of correlation for the least squares regression line. (See Examples 5 and 6.)

13. **Natural Science** The table shows the average monthly temperature (in degrees Fahrenheit) in Cleveland, Ohio, based on data from 1961 to 1990.† Let $x = 2$ correspond to February, $x = 4$ to April, etc.

Month	Feb	April	June	Aug	Oct	Dec
Temperature	27.3	47.5	67.5	70.3	52.7	30.9

14. **Health** The table shows the number of deaths per 100,000 people from heart disease in selected years. Let $x = 0$ correspond to 1960.

Year	1960	1970	1980	1990	1999
Deaths	522	496	436	368	348

*Office of Planning, Research, and Evaluation, Administration for Children and Families, U.S. Department of Health and Human Services.

†Robert K. Adair, *The Physics of Baseball* (HarperCollins, 1990).

‡York Snow, Inc.

**Federal Reserve Bulletin.*

†National Climatic Data Center.

In the following exercises, find the required linear model as follows. If you do not have a graphing calculator or spreadsheet program, use the first and last data points to determine a line. If you do have a graphing calculator or spreadsheet program, find the least squares regression line. (See Examples 1, 3, 4, and 6.)

15. Health Use the data on death from heart disease in Exercise 14.

(a) Find a linear model for the data, with $x = 0$ corresponding to 1960.

(b) Estimate the number of deaths from heart disease in 2005.

16. Social Science The table shows how poverty level income cutoffs (in dollars) for a family of four have changed over time (in large part because of inflation).*

Year	1970	1975	1980	1985	1990	1995
Income	3968	5500	8414	10,989	13,359	15,569

(a) Find a linear model for the data, with $x = 0$ corresponding to 1970 and income measured in thousands of dollars (so that \$3968 is 3.968, etc.).

(b) What was the approximate poverty level in 2000? In 2003?

17. Physical Science While shopping for an air conditioner, Adam Bryer consulted the following table giving a machine's BTUs and the square footage (ft^2) that it would cool.†

ft^2 (x)	BTUs (y)
150	5000
175	5500
215	6000
250	6500
280	7000
310	7500
350	8000
370	8500
420	9000
450	9500

(a) Find a linear model for the data.

(b) To check the fit of the data to the line, use the results from part (a) to find the BTUs required to cool a room of 150 ft^2, 280 ft^2, and 420 ft^2. How well do the actual data agree with the predicted values?

(c) Suppose Adam's room measures 235 ft^2. Use the results from part (a) to decide how many BTUs it requires. If air conditioners are available only with the BTU choices in the table, which should Adam choose?

18. Business The number of new businesses started each year (in thousands) during the late 1990s is shown in the table.*

Year	1995	1996	1997	1998	1999
Starts	168	170	167	155	151

(a) Find a linear model for the data, with $x = 0$ corresponding to 1990.

(b) Estimate the number of business starts in 2003 and 2005.

The following exercises require a graphing calculator or spreadsheet program. (See Examples 3, 4, and 6.)

19. Business The graph shows Intel's net revenues (in billions of dollars) from 1991 to 2000.†

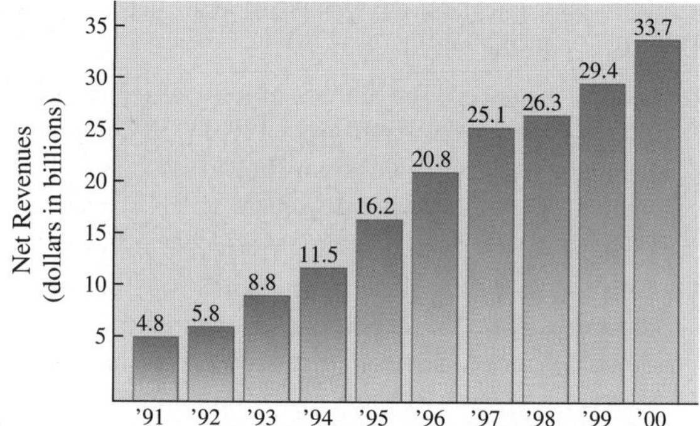

(a) List the data points, with $x = 0$ corresponding to 1990.

(b) Find the least squares regression line that models the data.

(c) If this trend continues, what will Intel's net revenues be in 2003 and 2005?

20. Natural Science Biologists have observed a linear relationship between the temperature and the frequency with which a cricket chirps. The following data were measured for the striped ground cricket.‡

*U.S. Bureau of Census, *Current Population Reports.*

†Morris Carey and James Carey, "On the House," *Sacramento Bee,* July 29, 2000.

**Statistical Abstracts of the United States 2000;* figures rounded to nearest thousand.

†Reprinted by permission of Intel Corporation, Copyright Intel Corporation 2002.

‡Reprinted by permission of the publishers from *The Songs of Insects* by George W. Pierce. Cambridge, Mass.: Harvard University Press, Copyright © 1948 by the President and Fellows of Harvard College.

Temperature °F (x)	Chirps per second (y)
88.6	20.0
71.6	16.0
93.3	19.8
84.3	18.4
80.6	17.1
75.2	15.5
69.7	14.7
82.0	17.1
69.4	15.4
83.3	16.2
79.6	15.0
82.6	17.2
80.6	16.0
83.5	17.0
76.3	14.4

(a) Find the least squares regression line that models the data.

(b) Use the results of part (a) to determine how many chirps per second you would expect to hear from the striped ground cricket if the temperature were 73°F.

(c) Use the results of part (a) to determine what the temperature is when the striped ground crickets are chirping at a rate of 18 times per second.

(d) Find the coefficient of correlation.

21. Social Science Over a fifteen-year period beginning in 1980, the average major league baseball game got longer and longer, as shown in the table.*

Year	Average Completion Time
1981	2:33
1982	2:34
1983	2:36
1984	2:35
1985	2:40
1986	2:44
1987	2:48
1988	2:45
1989	2:46
1990	2:48
1991	2:49
1992	2:49
1993	2:48
1994	2:54
1995	2:57

*New York Times, May 30, 1995, p. B9.

(a) Find the least squares regression line that models the data, with $x = 0$ corresponding to 1980 and y being the number of minutes beyond 2 hours (for instance, 2:36 corresponds to $y = 36$).

(b) If the trend in the data continues, when will the average completion time be 3 hours and 15 minutes?

22. Natural Science The following table shows the men's and women's world records (in seconds) in the 800-meter run.*

Year	Men's Record	Women's Record
1905	113.4	—
1915	111.9	—
1925	111.9	144.0
1935	109.7	135.6
1945	106.6	132.0
1955	105.7	125.0
1965	104.3	118.0
1975	103.7	117.48
1985	101.73	113.28
1995	101.11	113.28

(a) Find the equation for the least squares line for the men's record (y) in terms of the year (x). Use 5 for 1905, 15 for 1915, and so on.

(b) Find the equation for the least squares line for the women's record.

(c) Suppose the men's and women's records continue to improve as predicted by the equations found in parts (a) and (b). In what year will the women's record catch up with the men's record? Do you believe that will happen? Why or why not?

(d) Calculate the coefficient of correlation for both the men's and the women's record. What do these numbers tell you?

23. Health The table shows U.S. Health Care Financing Administration projections of personal health care expenditures (in billions of dollars).

Year	1990	1997	1999	2000	2001	2002
Expenditures	614.7	969.0	1078.3	1150.9	1227.3	1310.0

(a) Find the least squares regression line that models the data, with $x = 0$ corresponding to 1990.

(b) Does this model fit the data well? Why?

(c) Estimate health care expenditures in 2005.

*Brian J. Whipp and Susan A. Ward, "Will Women Soon Outrun Men?" *Nature*, vol. 355 January 2, 1992), p. 25. The data are from Peter Mathews, *Track and Field Athletics: The Records* (Guinness, 1986) pp. 11, 44, and from Robert W. Schutz and Yuanlong Liu in *Statistics in Sport*, edited by Jay Bennett (Arnold, 1998), p. 189.

(d) If this trend continues, when will expenditures reach 2000 billion dollars?

24. Health Employees of large and medium-sized businesses who want to be covered under company-sponsored health insurance plans are often required to pay part of the premium. A decade-long survey showed that participation rates declined as the employee contribution increased, as summarized in the following table.*

Average Monthly Employee Contribution ($)	62	69.50	76	84.50
Percentage of Participating Employees	83	82	77	76

(a) Find the least squares regression line that models the data, with x being the employee contribution and y the percentage of participating employees.
(b) Does this model fit the data well? Why?
(c) Estimate the participation rate when the monthly contribution is $100.
(d) At what contribution rate will participation fall below 50%?

*U.S. Bureau of Labor Statistics.

25. Business Total corporate profits before taxes (in billions of dollars) of U.S. companies during the 1990s are shown in the table.*

Year	1990	1994	1995	1996	1997	1998	1999
Profit	402	573	669	726	796	782	849

(a) Find the least squares regression line that models the data, with $x = 0$ corresponding to 1990.
(b) Does this model fit the data well? Why?
(c) If this trend continues, what will corporate profits be in 2004?
(d) If this trend continues, when will corporate profits reach $1475 billion?

*U.S. Bureau of Economic Analysis.

2.4 LINEAR INEQUALITIES

An **inequality** is a statement that one mathematical expression is greater than (or less than) another. Inequalities are very important in applications. For example, a company wants revenue to be *greater than* costs and must use *no more than* the total amount of capital or labor available.

Inequalities may be solved using algebraic or geometric methods. In this section we shall concentrate on algebraic methods for solving **linear inequalities,** such as

$$4 - 3x \leq 7 + 2x \quad \text{and} \quad -2 < 5 + 3m < 20,$$

and absolute value inequalities, such as $|x - 2| < 5$. The following properties are the basic algebraic tools for working with inequalities.

Properties of Inequality

For real numbers a, b, and c,

(a) if $a < b$, then $a + c < b + c$

(b) if $a < b$, and if $c > 0$, then $ac < bc$

(c) if $a < b$, and if $c < 0$, then $ac > bc$.

Throughout this section, definitions are given only for $<$, but they are equally valid for $>$, $\leq$, or $\geq$.

1

(a) First multiply both sides of $-6 < -1$ by 4, and then multiply both sides of $-6 < -1$ by -7.

(b) First multiply both sides of $9 \geq -4$ by 2 and then multiply both sides of $9 \geq -4$ by -5.

(c) First add 4 to both sides of $-3 < -1$, and then add -6 to both sides of $-3 < -1$.

Answers:

(a) $-24 < -4$; $42 > 7$

(b) $18 \geq -8$; $-45 \leq 20$

(c) $1 < 3$; $-9 < -7$

2

Solve these inequalities. Graph each solution.

(a) $5z - 11 < 14$

(b) $-3k \leq -12$

(c) $-8y \geq 32$

Answers:

(a) $z < 5$

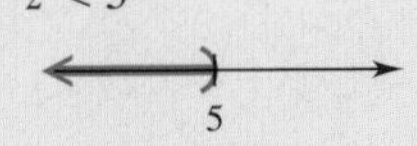

(b) $k \geq 4$

4

(c) $y \leq -4$

−4

CAUTION Pay careful attention to part (c): if both sides of an inequality are multiplied by a negative number, the direction of the inequality symbol must be reversed. For example, starting with the true statement $-3 < 5$ and multiplying both sides by the positive number 2 gives

$$-3 \cdot 2 < 5 \cdot 2,$$

or

$$-6 < 10,$$

still a true statement. On the other hand, starting with $-3 < 5$ and multiplying both sides by the negative number -2 gives a true result only if the direction of the inequality symbol is reversed:

$$-3(-2) > 5(-2)$$
$$6 > -10.$$ 1

EXAMPLE 1 Solve $3x + 5 > 11$. Graph the solution.

First, add -5 to both sides.

$$3x + 5 + (-5) > 11 + (-5)$$
$$3x > 6$$

Now multiply both sides by 1/3.

$$\frac{1}{3}(3x) > \frac{1}{3}(6)$$
$$x > 2$$

(Why was the direction of the inequality symbol not changed?) In interval notation (introduced in Section 1.1), the solution is the interval $(2, \infty)$, which is graphed on the number line in Figure 2.39. The parenthesis at 2 shows that 2 is not included in the solution.

FIGURE 2.39

As a partial check, note that 0, which is not part of the solution, makes the inequality false, while 3, which is part of the solution, makes it true.

$$3(0) + 5 \overset{?}{>} 11 \qquad 3(3) + 5 \overset{?}{>} 11$$
$$5 > 11 \quad \text{False} \qquad 14 > 11 \quad \text{True}$$ 2

EXAMPLE 2 Solve $4 - 3x \leq 7 + 2x$.

Add -4 to both sides.

$$4 - 3x + (-4) \leq 7 + 2x + (-4)$$
$$-3x \leq 3 + 2x$$

3

Solve these inequalities. Graph each solution.

(a) $8 - 6t \geq 2t + 24$

(b) $-4r + 3(r + 1) < 2r$

Answers:

(a) $t \leq -2$

(b) $r > 1$

Add $-2x$ to both sides. (Remember that *adding* to both sides never changes the direction of the inequality symbol.)

$$-3x + (-2x) \leq 3 + 2x + (-2x)$$
$$-5x \leq 3$$

Multiply both sides by $-1/5$. Since $-1/5$ is negative, change the direction of the inequality symbol.

$$-\frac{1}{5}(-5x) \geq -\frac{1}{5}(3)$$
$$x \geq -\frac{3}{5}$$

Figure 2.40 shows a graph of the solution $[-3/5, \infty)$. The bracket in Figure 2.40 shows that $-3/5$ is included in the solution. ◀ 3

FIGURE 2.40

4

Solve each of the following. Graph each solution.

(a) $9 < k + 5 < 13$

(b) $-6 \leq 2z + 4 \leq 12$

Answers:

(a) $4 < k < 8$

(b) $-5 \leq z \leq 4$

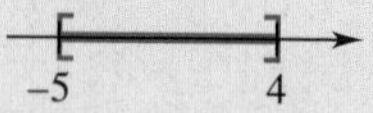

EXAMPLE 3 Solve $-2 < 5 + 3m < 20$. Graph the solution.

The inequality $-2 < 5 + 3m < 20$ says that $5 + 3m$ is *between* -2 and 20. We can solve this inequality with an extension of the properties given above. Work as follows, first adding -5 to each part.

$$-2 + (-5) < 5 + 3m + (-5) < 20 + (-5)$$
$$-7 < 3m < 15$$

Now multiply each part by 1/3.

$$-\frac{7}{3} < m < 5$$

A graph of the solution, $(-7/3, 5)$, is given in Figure 2.41. ◀ 4

FIGURE 2.41

EXAMPLE 4 The formula for converting from Celsius to Fahrenheit temperature is

$$F = \frac{9}{5}C + 32.$$

What Celsius temperature range corresponds to the range from 32°F to 77°F?

The Fahrenheit temperature range is $32 < F < 77$. Since $F = (9/5)C + 32$,

$$32 < \frac{9}{5}C + 32 < 77.$$

Solve the inequality for C.

$$32 < \frac{9}{5}C + 32 < 77$$

$$0 < \frac{9}{5}C < 45$$

$$\frac{5}{9} \cdot 0 < \frac{5}{9} \cdot \frac{9}{5}C < \frac{5}{9} \cdot 45$$

$$0 < C < 25$$

The corresponding Celsius temperature range is 0°C to 25°C. ◀ (5)

(5) In Example 4, what Celsius temperatures correspond to 5°F to 95°F?

Answer:

−15°C to 35°C

A product will break even, or produce a profit, only if the revenue R from selling the product at least equals the cost C of producing it, that is, if $R \geq C$.

EXAMPLE 5 A company analyst has determined that the cost to produce and sell x units of a certain product is $C = 20x + 1000$. The revenue for that product is $R = 70x$. Find the values of x for which the company will break even or make a profit on the product.

Solve the inequality $R \geq C$.

$$R \geq C$$

$$70x \geq 20x + 1000 \qquad \text{Let } R = 70x,\ C = 20x + 1000.$$

$$50x \geq 1000$$

$$x \geq 20$$

The company must produce and sell 20 items to break even and more than 20 to make a profit. ◀

Absolute Value Inequalities

You may wish to review the definition of absolute value in Section 1.1 before reading the following examples, which show how to solve inequalities involving absolute values.

EXAMPLE 6 Solve each inequality.

(a) $|x| < 5$

Because absolute value gives the distance from a number to 0, the inequality $|x| < 5$ is true for all real numbers whose distance from 0 is less than 5. This includes all numbers between −5 and 5, or numbers in the interval (−5, 5). A graph of the solution is shown in Figure 2.42.

FIGURE 2.42

(6)

Solve each inequality. Graph each solution.

(a) $|x| \le 1$

(b) $|y| \ge 3$

Answers:

(a) $[-1, 1]$

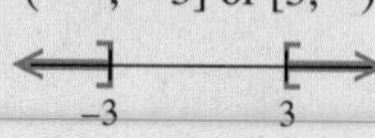

(b) All numbers in

$(-\infty, -3]$ or $[3, \infty)$

(b) $|x| > 5$

In a similar way, the solution of $|x| > 5$ is given by all those numbers whose distance from 0 is *greater* than 5. This includes the numbers satisfying $x < -5$ or $x > 5$. A graph of the solution, all numbers in

$$(-\infty, -5) \quad \text{or} \quad (5, \infty),$$

is shown in Figure 2.43. ◀ (6)

FIGURE 2.43

The examples above suggest the following generalizations.

> Assume a and b are real numbers with b positive.
>
> **1.** Solve $|a| = b$ by solving $a = b$ or $a = -b$.
>
> **2.** Solve $|a| < b$ by solving $-b < a < b$.
>
> **3.** Solve $|a| > b$ by solving $a < -b$ or $a > b$.

(7)

Solve each inequality. Graph each solution.

(a) $|p + 3| < 4$

(b) $|2k - 1| \le 7$

Answers:

(a) $(-7, 1)$

(b) $[-3, 4]$

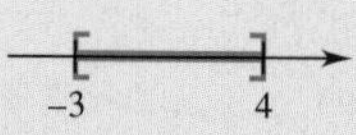

EXAMPLE 7 Solve $|x - 2| < 5$.

Replace a with $x - 2$ and b with 5 in property (2) in the box above. Now solve $|x - 2| < 5$ by solving the inequality

$$-5 < x - 2 < 5.$$

Add 2 to each part, getting the solution

$$-3 < x < 7,$$

which is graphed in Figure 2.44. ◀ (7)

FIGURE 2.44

EXAMPLE 8 Solve $|2 - 7m| - 1 > 4$.

First add 1 to both sides.

$$|2 - 7m| > 5$$

Now use property (3) from the box above to solve $|2 - 7m| > 5$ by solving the inequality

$$2 - 7m < -5 \quad \text{or} \quad 2 - 7m > 5.$$

Solve each part separately.

$$-7m < -7 \quad \text{or} \quad -7m > 3$$

$$m > 1 \quad \text{or} \quad m < -\frac{3}{7}$$

The solution, all numbers in $\left(-\infty, -\dfrac{3}{7}\right)$ or $(1, \infty)$, is graphed in Figure 2.45. ◀ (8)

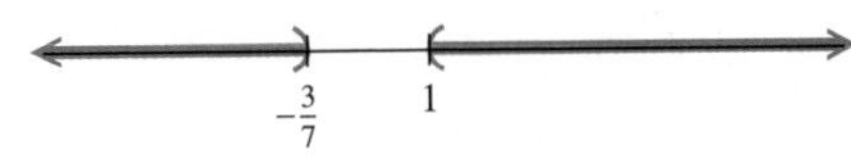

FIGURE 2.45

EXAMPLE 9 Solve $|2 - 5x| \geq -4$.

The absolute value of a number is always nonnegative. Therefore, $|2 - 5x| \geq -4$ is always true; so the solution is the set of all real numbers. Note that the inequality $|2 - 5x| \leq -4$ has no solution, because the absolute value of a quantity can never be less than a negative number. ◀ (9)

Absolute value inequalities can be used to indicate how far a number may be from a given number. The next example illustrates this use of absolute value.

EXAMPLE 10 Write each statement using absolute value.

(a) k is at least 4 units from 1.

If k is at least 4 units from 1, then the distance from k to 1 is greater than or equal to 4. See Figure 2.46(a). Write this statement using absolute value as follows:

$$|k - 1| \geq 4.$$

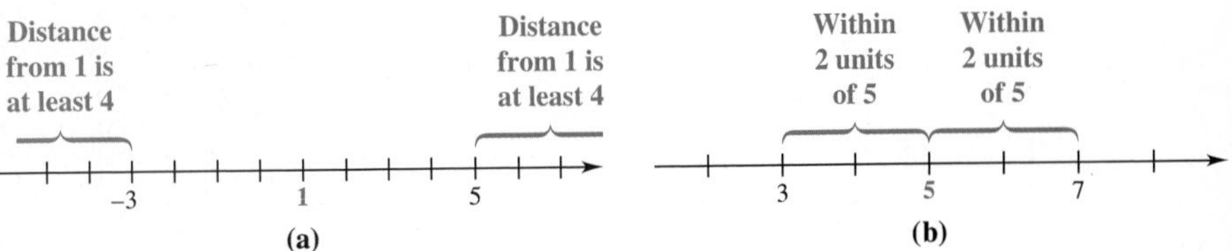

FIGURE 2.46

(b) p is within 2 units of 5.

This statement means that the distance between p and 5 must be less than or equal to 2. See Figure 2.46(b). Using absolute value notation, the statement is written as

$$|p - 5| \leq 2.$$ ◀ (10)

(8)

Solve each inequality. Graph each solution.

(a) $|y - 2| > 5$

(b) $|3k - 1| \geq 2$

(c) $|2 + 5r| - 4 \geq 1$

Answers:

(a) All numbers in $(-\infty, -3)$ or $(7, \infty)$

-3 7

(b) All numbers in $\left(-\infty, -\dfrac{1}{3}\right]$ or $[1, \infty)$

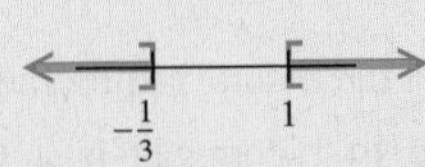

(c) All numbers in $\left(-\infty, -\dfrac{7}{5}\right]$ or $\left[\dfrac{3}{5}, \infty\right)$

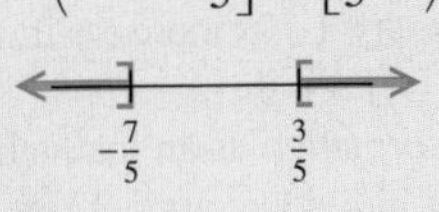

Solve each inequality.

(a) $|5m - 2| > -1$

(b) $|2 + 3a| < -3$

(c) $|6 + r| > 0$

Answers:

(a) All real numbers

(b) No solution

(c) All real numbers except -6

Write each statement using absolute value.

(a) m is at least 3 units from 5.

(b) t is within .01 of 4.

Answers:

(a) $|m - 5| \geq 3$

(b) $|t - 4| \leq .01$

2.4 Exercises

1. Explain how to determine whether a parenthesis or a bracket is used when graphing the solution of a linear inequality.

2. The three-part inequality $p < x < q$ means "p is less than x and x is less than q." Which one of the following inequalities is not satisfied by any real number x? Explain why.
   (a) $-3 < x < 5$ **(b)** $0 < x < 4$
   (c) $-7 < x < -10$ **(d)** $-3 < x < -2$

Solve each inequality and graph each solution. (See Examples 1–3.)

3. $-8k \le 32$
4. $-6a \le 36$
5. $-2b > 0$
6. $6 - 6z < 0$
7. $3x + 4 \le 12$
8. $2y - 5 < 9$
9. $-4 - p \ge 3$
10. $5 - 3r \le -4$
11. $7m - 5 < 2m + 10$
12. $6x - 2 > 4x - 8$
13. $m - (4 + 2m) + 3 < 2m + 2$
14. $2p - (3 - p) \le -7p - 2$
15. $-2(3y - 8) \ge 5(4y - 2)$
16. $5r - (r + 2) \ge 3(r - 1) + 5$
17. $3p - 1 < 6p + 2(p - 1)$
18. $x + 5(x + 1) > 4(2 - x) + x$
19. $-7 < y - 2 < 4$
20. $-3 < m + 6 < 2$
21. $8 \le 3r + 1 \le 13$
22. $-6 < 2p - 3 \le 5$
23. $-4 \le \dfrac{2k - 1}{3} \le 2$
24. $-1 \le \dfrac{5y + 2}{3} \le 4$
25. $\dfrac{3}{5}(2p + 3) \ge \dfrac{1}{10}(5p + 1)$
26. $\dfrac{8}{3}(z - 4) \le \dfrac{2}{9}(3z + 2)$

In the following exercises, write a linear inequality that describes the given graph.

27. 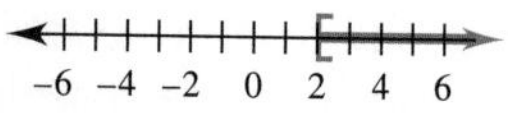

28. 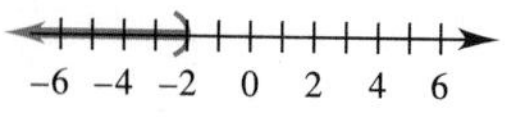

29. 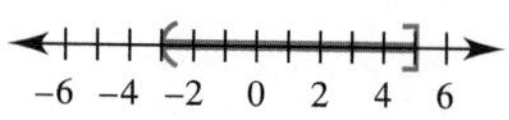

30. 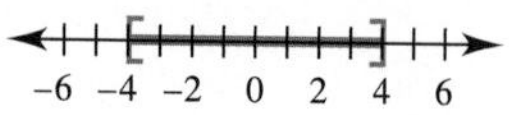

31. **Natural Science** Federal guidelines require drinking water to have fewer than .050 milligram per liter of lead. A test using 21 samples of water in a midwestern city found that the average amount of lead in the samples was .040 milligram per liter. All samples had lead content within 5% of the average.
   (a) Select a variable and write down what it represents.
   (b) Write a three-part inequality to express the sample results.
   (c) Did all the samples meet the federal requirement?

32. **Natural Science** Exposure to radon gas is a known lung cancer risk. According to the Environmental Protection Agency (EPA), the individual lifetime excess cancer risk R for radon exposure is between .0015 and .006, where $R = .01$ represents a 1% increase in risk of developing cancer.*
   (a) Write the preceding information as an inequality.
   (b) Determine the range of individual annual risk by dividing R by an average life expectancy of 75 years.

33. **Finance** The table below shows the federal income tax for a single person in 2002.

If Taxable Income Is: Over	*But Not Over*	*The Tax Is:*	*Of the Amount Over*
\$0	\$6,000	— 10%	\$0
6,000	27,950	\$600.00 + 15%	6,000
27,950	67,700	3,892.50 + 27%	27,950
67,700	141,250	14,625.00 + 30%	67,700
141,250	307,050	36,690.00 + 35%	141, 250
307,050	—	94,720.00 + 38.6%	307,050

   (a) Let x denote the taxable income. Write each of the six income ranges in the table as an inequality.
   (b) Let T denote the income tax. Write an inequality that gives the tax range in dollars for each of the six income ranges in the table.

Indoor-Air-Assessment: A Review of Indoor Air Quality Risk Characterization Studies, Report No. EPA/600/8-90/044, Environmental Protection Agency, 1991.

Solve each inequality. Graph each solution. (See Examples 6–9.)

34. $|p| > 7$

35. $|m| < 1$

36. $|r| \le 4$

37. $|a| < -2$

38. $|b| > -5$

39. $|2x + 5| < 3$

40. $\left|x - \frac{1}{2}\right| < 2$

41. $|3z + 1| \ge 7$

42. $|8b + 5| \ge 7$

43. $\left|5x + \frac{1}{2}\right| - 2 < 5$

44. $\left|x + \frac{2}{3}\right| + 1 < 4$

Physical Science *The given inequality describes the monthly average temperature T in degrees Fahrenheit in the given location. What range of temperatures corresponds to the inequality?*

45. $|T - 49| \le 20$; Santa Fe, New Mexico

46. $|T - 43| \le 24$; Marquette, Michigan

47. $|T - 62| \le 19$; Memphis, Tennessee

48. $|T - 10| \le 36$; Chesterfield Inlet, Canada

49. **Natural Science** Human beings emit carbon dioxide when they breathe. In one study, the emission rates of carbon dioxide by college students were measured both during lectures and exams. The average individual rate R_L (in grams per hour) during a lecture class satisfied the inequality $|R_L - 26.75| \le 1.42$, whereas during an exam the rate R_E satisfied the inequality $|R_E - 38.75| \le 2.17$.*

(a) Find the range of values for R_L and R_E.

(b) The class had 225 students. If T_L and T_E represent the total amounts of carbon dioxide (in grams) emitted during a one-hour lecture and one-hour exam respectively, write inequalities that describe the ranges for T_L and T_E.

50. **Social Science** When administering a standard intelligence test, we expect about 1/3 of the scores to be more than 12 units above 100 or more than 12 units below 100. Describe this situation by writing an absolute value inequality.

Work the following problems. (See Example 4.)

51. **Business** Projections from the U.S. Department of Commerce show that the number of Internet users (in millions) is approximately $11.2x - 49.5$, where $x = 0$ corresponds to 1990. During what years was the number of Internet users between 52 and 97 million?

52. **Business** Based on data from the Federal Communications Commission, the number of international long-distance calls (in millions) placed in the United States in year x is approximately $452.1x + 884.9$, where $x = 0$ corresponds to 1990. During what years was the number of such calls between 4,050,000,000 and 6,300,000,000?

53. **Health** At age x, with $0 \le x \le 80$, the average remaining life expectancy of a male is approximately $-.8x + 70.5$. For what age range is life expectancy between 54.5 and 42.5 years?

54. **Health** According to data from the U.S. Census Bureau the percentage of persons not covered by health insurance in year x is approximately $.3x + 13.9$, where $x = 0$ corresponds to 1990. During what years was the percentage of uninsured individuals between 15.4 and 16.9?

55. **Transportation** The cost of a taxi in New York City is \$2 plus 30 cents for each 1/5 of a mile. How far could you travel for at least \$5 but no more than \$11?

56. **Transportation** Anne Kelly went to a conference in Montreal for a week. She decided to rent a car and checked with two rental firms. Avery wanted \$56 per day, with no mileage fee. Hart wanted \$216 per week and \$.28 per mile (or part of a mile). How many miles must she drive before the Avery car is the better deal?

Business *In Exercises 57–62, find all values of x where the following products will at least break even. (See Example 5.)*

57. The cost to produce x units of wire is $C = 50x + 5000$, while the revenue is $R = 60x$.

58. The cost to produce x units of squash is $C = 100x + 6000$, while the revenue is $R = 500x$.

59. $C = 85x + 900$; $R = 105x$

60. $C = 70x + 500$; $R = 60x$

61. $C = 1000x + 5000$; $R = 900x$

62. $C = 2500x + 10{,}000$; $R = 102{,}500x$

Write each of the following statements using absolute value. (See Example 10.)

63. x is within 4 units of 2.

64. m is no more than 8 units from 9.

65. z is no less than 2 units from 12.

66. p is at least 5 units from 9.

*T. C. Wang, *ASHRAE Transactions* 81 (Part 1), 32 (1975).

2.5 POLYNOMIAL AND RATIONAL INEQUALITIES

This section deals with the solution of polynomial and rational inequalities, such as

$$r^2 + 3r - 4 \geq 0, \qquad x^3 - x \leq 0, \qquad \text{and} \qquad \frac{2x - 1}{3x + 4} < 5.$$

We shall concentrate on algebraic solution methods, but to understand why these methods work, we must first look at such inequalities from a graphical point of view.

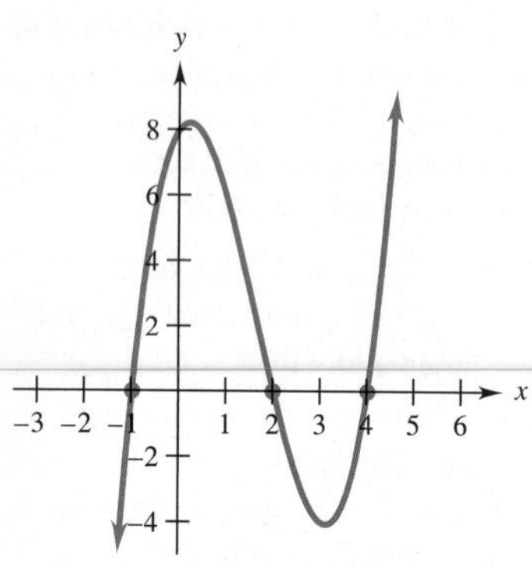

FIGURE 2.47

EXAMPLE 1 Use the graph of $y = x^3 - 5x^2 + 2x + 8$ in Figure 2.47 to solve each of the following inequalities.

(a) $x^3 - 5x^2 + 2x + 8 > 0$

Each point on the graph has coordinates of the form $(x, x^3 - 5x^2 + 2x + 8)$. The number x is a solution of the inequality exactly when the second coordinate of this point is positive, that is, when the point lies *above* the x-axis. So to solve the inequality, we need only find the first coordinates of points on the graph that are above the x-axis. This information can be read from Figure 2.47. The graph is above the x-axis when $-1 < x < 2$ and also when $x > 4$. Therefore, the solutions of the inequality are all numbers x in the intervals $(-1, 2)$ or $(4, \infty)$.

(b) $x^3 - 5x^2 + 2x + 8 < 0$

The number x is a solution of the inequality exactly when the second coordinate of the point $(x, x^3 - 5x^2 + 2x + 8)$ on the graph is negative, that is, when the point lies *below* the x-axis. Figure 2.47 shows that the graph is below the x-axis when $x < -1$ and also when $2 < x < 4$. Hence, the solutions are all numbers x in the intervals $(-\infty, -1)$ or $(2, 4)$. ◀

The solution process in Example 1 depends only on knowing the graph and its x-intercepts (that is, the points where the graph intersects the x-axis). This information can often be obtained algebraically, without doing any graphing, as illustrated in the next example.

EXAMPLE 2 Solve each of the following quadratic inequalities.

(a) $x^2 - x < 12$

First, rewrite the inequality as $x^2 - x - 12 < 0$. Now we don't know what the graph of $y = x^2 - x - 12$ looks like, but we can still find its x-intercepts by solving the equation

$$x^2 - x - 12 = 0$$
$$(x + 3)(x - 4) = 0.$$
$$x + 3 = 0 \quad \text{or} \quad x - 4 = 0$$
$$x = -3 \qquad\qquad x = 4$$

These numbers divide the x-axis (number line) into three regions, as indicated in Figure 2.48.

Region A	Region B	Region C
$x < -3$	$-3 < x < 4$	$x > 4$

−3 4

FIGURE 2.48

In each region, the graph of $y = x^2 - x - 12$ is an unbroken curve, so it will be entirely above or entirely below the axis. It can only pass from above to below the x-axis at the x-intercepts. To see whether the graph is above or below the x-axis when x is in region A, choose a value of x in region A, say $x = -5$, and substitute this in the equation.

$$y = x^2 - x - 12 = (-5)^2 - (-5) - 12 = 18$$

Therefore, the point $(-5, 18)$ is on the graph. Since its y-coordinate 18 is positive, this point lies above the x-axis, and hence the entire graph lies above the x-axis in region A.

Similarly, we can choose a value of x in region B, say $x = 0$. Then

$$y = x^2 - x - 12 = 0^2 - 0 - 12 = -12,$$

so that $(0, -12)$ is on the graph. Since this point lies below the x-axis (why?), the entire graph in region B must be below the x-axis. Finally, in region C, let $x = 5$. Then $y = 5^2 - 5 - 12 = 8$, so that $(5, 8)$ is on the graph and the entire graph in region C lies above the x-axis. We can summarize the results as follows.

Interval	$x < -3$	$-3 < x < 4$	$x > 4$
Test value in interval	-5	0	5
Value of $x^2 - x - 12$	18	-12	8
Graph	above x-axis	below x-axis	above x-axis
Conclusion	$x^2 - x - 12 > 0$	$x^2 - x - 12 < 0$	$x^2 - x - 12 > 0$

Solve each inequality. Graph the solution on the number line.

(a) $x^2 + 2x - 3 < 0$

(b) $2p^2 + 3p - 2 < 0$

Answers:

(a) $(-3, 1)$

−3 1

(b) $(-2, 1/2)$

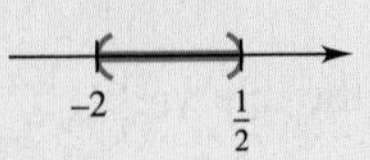

The last row shows that the only region where $x^2 - x - 12 < 0$ is region B, so the solutions of the inequality are all numbers x with $-3 < x < 4$, that is, the interval $(-3, 4)$, as shown in the number line graph in Figure 2.49.

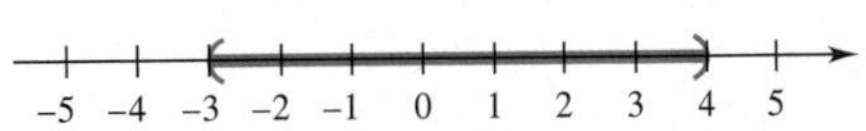

FIGURE 2.49

(b) $x^2 - x - 12 > 0$

Use the chart in part (a). The last row shows that $x^2 - x - 12 > 0$ only when x is in region A or region C. Hence, the solutions of the inequality are all numbers x with $x < -3$ or $x > 4$, that is, all numbers in the intervals $(-\infty, -3)$ or $(4, \infty)$. ◀ 1

EXAMPLE 3 Solve the quadratic inequality $r^2 + 3r \geq 4$.

First rewrite the inequality so that one side is 0.

$$r^2 + 3r \geq 4$$
$$r^2 + 3r - 4 \geq 0 \quad \text{Add } -4 \text{ to both sides.}$$

Now solve the corresponding equation.

$$r^2 + 3r - 4 = 0$$
$$(r - 1)(r + 4) = 0$$
$$r = 1 \quad \text{or} \quad r = -4$$

These numbers separate the number line into three regions, as shown in Figure 2.50. Test a number from each region.

Let $x = -5$ from region **A:** $(-5)^2 + 3(-5) - 4 = 6 > 0$.

Let $x = 0$ from region **B:** $(0)^2 + 3(0) - 4 = -4 < 0$.

Let $x = 2$ from region **C:** $(2)^2 + 3(2) - 4 = 6 > 0$.

We want the inequality to be positive or 0. The solution includes numbers in region A and in region C, as well as -4 and 1, the endpoints. The solution, which includes all numbers in the intervals $(-\infty, -4]$ or $[1, \infty)$, is graphed in Figure 2.50. ◀ (2)

(2)

Solve each inequality. Graph each solution.

(a) $k^2 + 2k - 15 \geq 0$

(b) $3m^2 + 7m \geq 6$

Answers:

(a) All numbers in $(-\infty, -5]$ or $[3, \infty)$

(b) All numbers in $(-\infty, -3]$ or $[2/3, \infty)$

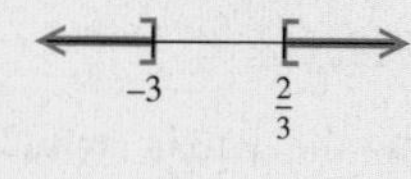

Region A $r < -4$ | Region B $-4 < r < 1$ | Region C $r > 1$

-6 -5 -4 -3 -2 -1 0 1 2 3 4 5

FIGURE 2.50

EXAMPLE 4 Solve $q^3 - 4q > 0$.

Solve the corresponding equation by factoring.

$$q^3 - 4q = 0$$
$$q(q^2 - 4) = 0$$
$$q(q + 2)(q - 2) = 0$$
$$q = 0 \quad \text{or} \quad q + 2 = 0 \quad \text{or} \quad q - 2 = 0$$
$$q = 0 \qquad q = -2 \qquad q = 2$$

These three numbers separate the number line into the four regions shown in Figure 2.51. Test a number from each region.

A: If $q = -3$, $(-3)^3 - 4(-3) = -15 < 0$.

B: If $q = -1$, $(-1)^3 - 4(-1) = 3 > 0$.

C: If $q = 1$, $(1)^3 - 4(1) = -3 < 0$.

D: If $q = 3$, $(3)^3 - 4(3) = 15 > 0$.

The numbers that make the polynomial positive are in the intervals

$$(-2, 0) \quad \text{or} \quad (2, \infty),$$

as graphed in Figure 2.51. ◀ (3)

(3)

Solve each inequality. Graph each solution.

(a) $m^3 - 9m > 0$

(b) $2k^3 - 50k \leq 0$

Answers:

(a) All numbers in $(-3, 0)$ or $(3, \infty)$

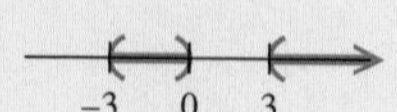

(b) All numbers in $(-\infty, -5]$ or $[0, 5]$

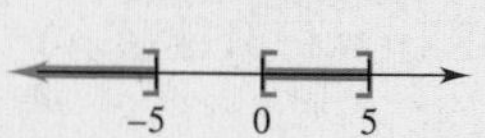

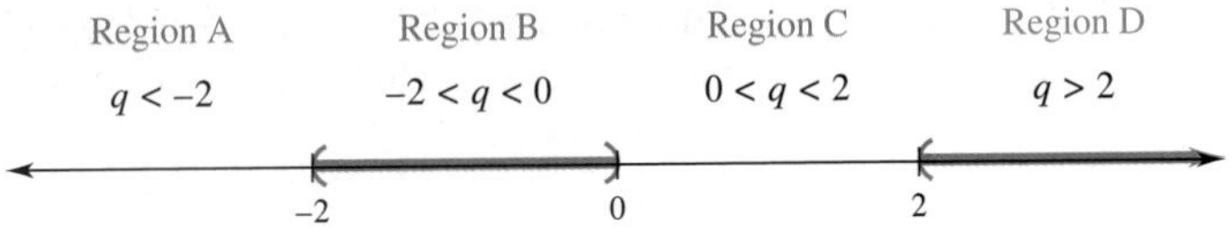

Region A	Region B	Region C	Region D
$q < -2$	$-2 < q < 0$	$0 < q < 2$	$q > 2$

-2 $\quad$ 0 $\quad$ 2

FIGURE 2.51

A graphing calculator can be used to solve inequalities, without the need to evaluate at a test number in each interval. It is also useful for finding approximate solutions when the x-intercepts of the graph cannot be found algebraically.

EXAMPLE 5 Use a graphing calculator to solve $x^3 - 5x^2 + x + 6 > 0$.

Begin by graphing $y = x^3 - 5x^2 + x + 6$ (Figure 2.52). Find the x-intercepts by solving $x^3 - 5x^2 + x + 6 = 0$. Since this cannot be readily done algebraically, use the graphical root finder to determine that the solutions (x-intercepts) are approximately $-.9254$, 1.4481, and 4.4774.

The graph is above the x-axis when $-.9254 < x < 1.4481$ and also when $x > 4.4774$. Therefore, the approximate solutions of the inequality are all numbers in the intervals $(-.9254, 1.4481)$ or $(4.4774, \infty)$. ◀ (4)

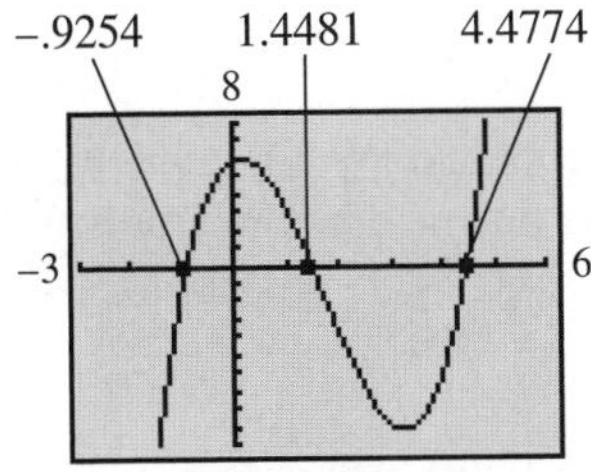

FIGURE 2.52

(4)

Use graphical methods to find approximate solutions of these inequalities.

(a) $x^2 - 6x + 2 > 0$

(b) $x^2 - 6x + 2 < 0$

Answers:

(a) All numbers in $(-\infty, .3542)$ or $(5.6458, \infty)$

(b) All numbers in $(.3542, 5.6458)$

Rational Inequalities

Inequalities with quotients of algebraic expressions are called **rational inequalities.** These inequalities can be solved in much the same way as polynomial inequalities.

EXAMPLE 6 Solve the rational inequality $\dfrac{5}{x+4} \geq 1$.

Write an equivalent inequality with one side equal to 0.

$$\frac{5}{x+4} \geq 1$$

$$\frac{5}{x+4} - 1 \geq 0$$

Write the left side as a single fraction.

$$\frac{5}{x+4} - \frac{x+4}{x+4} \geq 0 \qquad \text{Get a common denominator.}$$

$$\frac{5-(x+4)}{x+4} \geq 0 \qquad \text{Subtract fractions.}$$

$$\frac{5-x-4}{x+4} \geq 0 \qquad \text{Distributive property}$$

$$\frac{1-x}{x+4} \geq 0$$

The quotient can change sign only at places where the denominator is 0 or the numerator is 0. (In graphical terms, these are the only places where the graph of $y = \frac{1-x}{x+4}$ can change from above the x-axis to below.) This happens when

$$1 - x = 0 \quad \text{or} \quad x + 4 = 0$$
$$x = 1 \quad \text{or} \quad x = -4.$$

As in the earlier examples, the numbers -4 and 1 divide the x-axis into three regions. Test a number from each of these regions:

$$x < -4 \qquad \text{Let } x = -5\text{:} \quad \frac{1-(-5)}{-5+4} = -6 < 0.$$

$$-4 < x < 1 \qquad \text{Let } x = 0\text{:} \quad \frac{1-0}{0+4} = \frac{1}{4} > 0.$$

$$x > 1 \qquad \text{Let } x = 2\text{:} \quad \frac{1-2}{2+4} = -\frac{1}{6} < 0.$$

The test shows that numbers in $(-4, 1)$ satisfy the inequality. With a quotient, the endpoints must be considered individually to make sure that no denominator is 0. In this inequality, -4 makes the denominator 0, while 1 satisfies the given inequality. Write the solution as $(-4, 1]$.

CAUTION As suggested by Example 6, be very careful with the endpoints of the intervals in the solution of rational inequalities. (5)

(5)

Solve each inequality.

(a) $\frac{3}{x-2} \geq 4$

(b) $\frac{p}{1-p} < 3$

(c) Why is 2 excluded from the solution in part (a)?

Answers:

(a) $(2, 11/4]$

(b) All numbers in $(-\infty, 3/4)$ or $(1, \infty)$

(c) When $x = 2$, the fraction is undefined.

EXAMPLE 7 Solve $\frac{2x-1}{3x+4} < 5$.

Write an equivalent inequality with 0 on one side. Begin by subtracting 5 on both sides and combining the terms on the left into a single fraction.

$$\frac{2x-1}{3x+4} < 5$$

$$\frac{2x-1}{3x+4} - 5 < 0 \qquad \text{Get 0 on one side.}$$

$$\frac{2x-1-5(3x+4)}{3x+4} < 0 \qquad \text{Subtract.}$$

$$\frac{-13x-21}{3x+4} < 0 \qquad \text{Combine terms.}$$

Set the numerator and denominator each equal to 0 and solve the two equations.

$$-13x - 21 = 0 \quad \text{or} \quad 3x + 4 = 0$$
$$x = -\frac{21}{13} \quad \text{or} \quad x = -\frac{4}{3}$$

Use the values $-21/13$ and $-4/3$ to divide the number line into three intervals. Test a number from each interval in the inequality. The quotient is negative for numbers in $(-\infty, -21/13)$ or $(-4/3, \infty)$. Neither endpoint satisfies the given inequality.

Solve each rational inequality.

(a) $\frac{3y - 2}{2y + 5} < 1$

(b) $\frac{3c - 4}{2 - c} \geq -5$

Answers:

(a) $(-5/2, 7)$

(b) All numbers in $(-\infty, 2)$ or $[3, \infty)$

CAUTION In problems like Examples 6 and 7, we cannot begin by simply multiplying both sides by the denominator to simplify the inequality, because we do not know whether the variable denominator is positive or negative. (6)

TECHNOLOGY TIP Rational inequalities can also be solved graphically. In Example 7, for instance, after writing the inequality with a single fraction on the left and determining the values that make the numerator and denominator 0, graph the left side. The graph of $y = \frac{-13x - 21}{3x + 4}$ in Figure 2.53 contains an erroneous vertical line segment at $x = -4/3$, but you can still read the necessary information: The graph is below the x-axis to the left of $-21/13$ and to the right of $-4/3$.

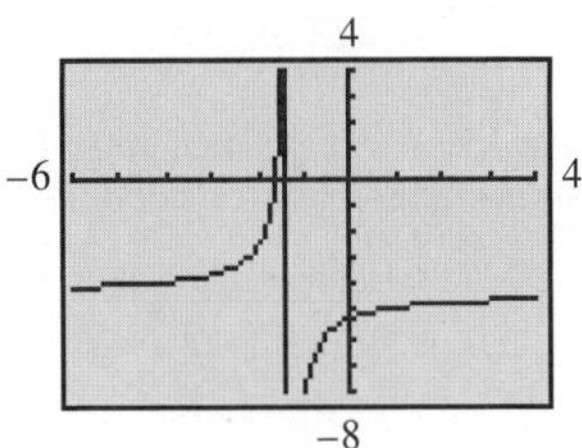

FIGURE 2.53

2.5 Exercises

Solve each of these quadratic inequalities. Graph the solutions on the number line. (See Examples 2 and 3.)

1. $(x + 5)(2x - 3) \leq 0$
2. $(5y - 1)(y + 4) > 0$
3. $r^2 + 4r > -3$
4. $z^2 + 6z < -8$
5. $4m^2 + 7m - 2 \leq 0$
6. $6p^2 - 11p + 3 \geq 0$
7. $4x^2 + 3x - 1 > 0$
8. $3x^2 - 5x > 2$
9. $x^2 \leq 25$
10. $y^2 \geq 4$
11. $p^2 - 16p > 0$
12. $r^2 - 9r < 0$

Solve these inequalities. (See Example 4.)

13. $x^3 - 9x \geq 0$
14. $p^3 - 25p \leq 0$
15. $(x + 6)(x + 1)(x - 4) \geq 0$
16. $(2x + 5)(x^2 - 1) \leq 0$
17. $(x + 4)(x^2 - 2x - 3) < 0$
18. $x^3 - 2x^2 - 3x \geq 0$
19. $6k^3 - 5k^2 < 4k$
20. $2m^3 + 7m^2 > 4m$
21. A student solved the inequality $p^2 < 16$ by taking the square root of both sides to get $p < 4$. She wrote the solution as $(-\infty, 4)$. Is her solution correct?

Use a graphing calculator to solve these inequalities. (See Example 5.)

22. $3x + 7 < 2x^2$
23. $.5x^2 - 1.2x < .1$
24. $3.1x^2 - 7.4x + 3.2 > 0$
25. $x^3 - 2x^2 - 5x + 7 \geq 2x + 1$
26. $x^4 - 6x^3 + 2x^2 < 5x - 2$
27. $2x^4 + 3x^3 < 2x^2 + 4x - 2$

28. $x^5 + 5x^4 > 4x^3 - 3x^2 - 2$

Solve the following rational inequalities. (See Examples 6 and 7.)

29. $\dfrac{r-3}{r-1} \geq 0$

30. $\dfrac{z+6}{z+3} > 1$

31. $\dfrac{a-2}{a-5} < -1$

32. $\dfrac{1}{3k-5} < \dfrac{1}{3}$

33. $\dfrac{1}{p-2} < \dfrac{1}{3}$

34. $\dfrac{7}{k+2} \geq \dfrac{1}{k+2}$

35. $\dfrac{5}{p+1} > \dfrac{12}{p+1}$

36. $\dfrac{x^2-4}{x} > 0$

37. $\dfrac{x^2-x-6}{x} < 0$

38. $\dfrac{x^2+x-2}{x^2-2x-3} < 0$

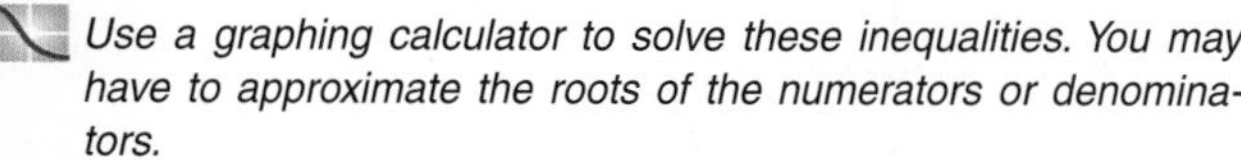

Use a graphing calculator to solve these inequalities. You may have to approximate the roots of the numerators or denominators.

39. $\dfrac{2x^2+x-1}{x^2-4x+4} \leq 0$

40. $\dfrac{x^3-3x^2+5x-29}{x^2-7} > 3$

Work these problems.

41. **Business** An analyst has found that his company's profits, in hundreds of thousands of dollars, are given by $P = 3x^2 - 35x + 50$, where x is the amount, in hundreds of dollars, spent on advertising. For what values of x does the company make a profit?

42. **Business** The commodities market is very unstable; money can be made or lost quickly on investments in soybeans, wheat, and so on. Suppose that an investor kept track of her total profit, P, at time t, in months, after she began investing, and she found that $P = 4t^2 - 29t + 30$. Find the time intervals where she has been ahead.

43. **Business** The manager of a large apartment complex has found that the profit is given by $P = -x^2 + 250x - 15{,}000$, where x is the number of apartments rented. For what values of x does the complex produce a profit?

44. **Physical Science** A physicist has found that the velocity of a moving particle is given by $2t^2 - 5t - 12$, where t is time in seconds since he began his observations. (Here t can be positive or negative; think of t seconds before his observations began.) Find the time intervals in which the velocity has been negative.

45. **Physical Science** A projectile is fired from ground level. After t seconds its height above the ground is $220t - 16t^2$ feet. For what time period is the projectile at least 624 feet above the ground?

46. **Social Science** According to data from the FBI Uniform Crime Report, the number of violent crimes per 100,000 inhabitants between 1984 and 1999 was approximately $-3.8x^2 + 87.4x + 233.2$, where $x = 0$ corresponds to 1980. During what years was the rate 712 or greater?

CHAPTER 2 SUMMARY

Key Terms and Symbols*

2.1 Cartesian coordinate system
x-axis
y-axis
origin
ordered pair
x-coordinate
y-coordinate
quadrant
solution of an equation
graph
x-intercept
y-intercept
[viewing window]
[graphical root finder]
[standard window]
graph reading

2.2 Δx change in x
Δy change in y
slope
slope-intercept form
point-slope form
linear equations
general form

2.3 linear models
residual
[least squares regression line]
[linear regression]
[coefficient of correlation]

2.4 linear inequality
properties of inequality
absolute value inequality

2.5 polynomial inequality
algebraic solution methods
[graphical solution methods]
rational inequality

*Terms in brackets deal with material in which a graphing calculator or other technology is used.

Key Concepts

The **slope** of the line through the points (x_1, y_1) and (x_2, y_2), where $x_1 \neq x_2$ is $m = \dfrac{y_2 - y_1}{x_2 - x_1}$.

The line with equation $\boldsymbol{y = mx + b}$ has slope m and y-intercept b.

The line with equation $\boldsymbol{y - y_1 = m(x - x_1)}$ has slope m and goes through (x_1, y_1).

The line with equation $\boldsymbol{ax + by = c}$ (with $a \neq 0$, $b \neq 0$) has x-intercept c/a and y-intercept c/b.

The line with equation $\boldsymbol{x = k}$ is vertical with x-intercept k, no y-intercept, and undefined slope.

The line with equation $\boldsymbol{y = k}$ is horizontal, with y-intercept k, no x-intercept, and slope 0.

Nonvertical **parallel lines** have the same slope, and **perpendicular lines,** if neither is vertical, have slopes with a product of -1.

Chapter 2 Review Exercises

Which of the ordered pairs $(-2, 3)$, $(0, -5)$, $(2, -3)$, $(3, -2)$, $(4, 3)$, $(7, 2)$ *are solutions of the given equation?*

1. $y = x^2 - 2x - 5$ **2.** $x - y = 5$

Sketch the graph of each equation.

3. $5x - 3y = 15$ **4.** $2x + 7y - 21 = 0$

5. $y + 3 = 0$ **6.** $y - 2x = 0$

7. $y = .25x^2 + 1$ **8.** $y = \sqrt{x + 4}$

9. The following temperature graph was recorded in Bratenahl, Ohio.
(a) At what times during the day was the temperature over 55°?
(b) When was the temperature below 40°?

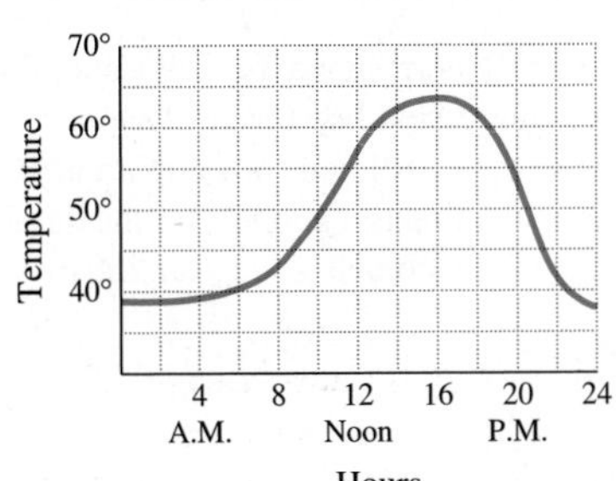

10. Greenville, South Carolina, is 500 miles south of Bratenahl, Ohio, and its temperature is 7° higher all day long (see the graph in Exercise 9). At what time was the temperature in Greenville the same as the temperature at noon in Bratenahl?

11. In your own words, define the slope of a line.

In Exercises 12–21, find the slope of the line.

12. Through $(-1, 4)$ and $(2, 3)$

13. Through $(5, -3)$ and $(-1, 2)$

14. Through $(7, -2)$ and the origin

15. Through $(8, 5)$ and $(0, 3)$

16. $2x + 3y = 30$

17. $4x - y = 7$

18. $x + 5 = 0$

19. $y = 3$

20. Parallel to $3x + 8y = 0$

21. Perpendicular to $x = 3y$

22. Graph the line through $(0, 5)$ with slope $m = -2/3$.

23. Graph the line through $(-4, 1)$ with $m = 3$.

24. What information is needed to determine the equation of a line?

Find an equation for each of the following lines.

25. Through $(5, -1)$, slope $2/3$

26. Through $(8, 0)$, slope $-1/4$

27. Through $(5, -2)$ and $(1, 3)$

28. Through $(2, -3)$ and $(-3, 4)$

29. Undefined slope, through $(-1, 4)$

30. Slope 0, through $(-2, 5)$

31. x-intercept -3, y-intercept 5

32. x-intercept $-2/3$, y-intercept $1/2$

33. Health According to the National Center for Health Statistics, 25.3% of adults were smokers in 1990. By 1998, only 24% smoked.

(a) Assuming the decline in smoking is linear, write an equation that gives the percentage of smokers in year x, with $x = 0$ corresponding to 1990.

(b) Is the slope of the line positive or negative? Why?

(c) Estimate the percentage of smokers in 2006.

34. Social Science In 1980 the percent of children living with two parents was 77 and in 1998 it was 68.*

(a) Assuming the decrease was linear, write an equation that relates the percent y and year x, with 1980 corresponding to $x = 0$.

(b) Graph the equation for the years 1990 to 2005.

(c) Estimate the number of children living in two-parent families in 2005.

35. Business The table shows the average hourly earnings of U.S. production workers.†

Year	1991	1993	1995	1997	1999
Earnings	10.32	10.83	11.43	12.28	13.24

(a) Use the first and last data points to find a linear model for the data, with $x = 0$ corresponding to 1990.

(b) If you have access to appropriate technology, find the least squares regression line for the data.

(c) Use the models from part (a) and/or (b) to estimate the hourly earnings in 1996. The actual average in 1996 was $11.81. How far off is the model?

(d) Use the model from part (a) and/or (b) to estimate the hourly earnings in 2004.

36. Finance The table shows the number of Automatic Teller Machine (ATM) transactions (in billions) in selected years.‡

Year	1985	1990	1994	1995
Transactions	3.5656	5.751	8.454	9.689

Year	1996	1997	1998	1999
Transactions	10.684	10.980	11.160	10.889

(a) Use the first and last data points to find a linear model for the data, with $x = 0$ corresponding to 1980.

(b) If you have access to appropriate technology, find the least squares regression line for the data.

(c) Use the model from part (a) and/or (b) to estimate the number of ATM transactions in 2003.

37. Transportation The table, based on data from the American Automobile Manufacturers Association, shows the average expenditure for a new car in various years.

Year	1992	1993	1994	1995
Expenditure	16,336	16,871	17,903	17,959

Year	1996	1997	1998	1999
Expenditure	18,777	19,551	20,849	21,022

(a) Let $x = 0$ correspond to 1990 and find the least squares regression line for the data.

(b) What is the coefficient of correlation? Does it indicate that the line is a good fit for the data?

(c) Estimate the expenditure on a new car in 2004.

38. Business The following table, from the U.S. Federal Deposit Insurance Corporation, shows the net after-tax income (in billions of dollars) of insured commercial banks.

Year	1990	1994	1995	1996
Income	16	44.6	48.7	52.4

Year	1997	1998	1999
Income	59.2	61.8	71.7

(a) Let $x = 0$ correspond to 1990 and find the least squares regression line for the data.

(b) What is the coefficient of correlation? Does it indicate that the line is a good fit for the data?

(c) Estimate the income for the year 2001.

Solve each inequality.

39. $-6x + 3 < 2x$

40. $12z \geq 5z - 7$

41. $2(3 - 2m) \geq 8m + 3$

42. $6p - 5 > -(2p + 3)$

43. $-3 \leq 4x - 1 \leq 7$

44. $0 \leq 3 - 2a \leq 15$

*U.S. Census Bureau.

†U.S. Bureau of Labor Statistics.

‡*Statistical Abstract of the United States*, 2000.

45. $|b| \leq 8$

46. $|a| > 7$

47. $|2x - 7| \geq 3$

48. $|4m + 9| \leq 16$

49. $|5k + 2| - 3 \leq 4$

50. $|3z - 5| + 2 \geq 10$

51. **Natural Science** Dr. Ryan has found that, over the years, 95% of the babies he has delivered have weighed y pounds, where $|y - 7.5| \leq 2$. What range of weights corresponds to this inequality?

52. **Natural Science** The number of milligrams of a certain substance per liter in drinking water samples all tested within .05 of 40 milligrams per liter. Write this information as an inequality, using absolute value.

53. **Business** Total energy production in the Asia-Pacific region was 415 million metric tons of oil equivalent (Mtoe) in 1990 and 645 Mtoe in 2000 and is increasing linearly.*
   (a) Find the linear equation that gives the energy production y in year x, with $x = 0$ corresponding to 1990.
   (b) When is energy production less than 599 Mtoe?
   (c) When is energy production more than 714 Mtoe?

54. **Business** One car rental firm charges $75 for a weekend rental (Friday afternoon through Monday morning) with unlimited mileage. A second firm charges $50 plus 5 cents per mile. For what range of miles driven is the second firm cheaper?

Solve each inequality.

55. $r^2 + r - 6 < 0$

56. $y^2 + 4y - 5 \geq 0$

57. $2z^2 + 7z \geq 15$

58. $3k^2 \leq k + 14$

59. $(x - 3)(x^2 + 7x + 10) \leq 0$

60. $(x + 4)(x^2 - 1) \geq 0$

61. $\frac{m + 2}{m} \leq 0$

62. $\frac{q - 4}{q + 3} > 0$

63. $\frac{5}{p + 1} > 2$

64. $\frac{6}{a - 2} \leq -3$

65. $\frac{2}{r + 5} \leq \frac{3}{r - 2}$

66. $\frac{1}{z - 1} > \frac{2}{z + 1}$

67. **Social Science** During the last century, the population of Cleveland, Ohio (in thousands), in year x was approximately $.06x^2 - 12.5x + 926$, where $x = 0$ corresponds to 1950. During what years since 1950 was the population at least 500,000?

*International Energy Agency, *Global Energy: The Changing Outlook,* 1992.

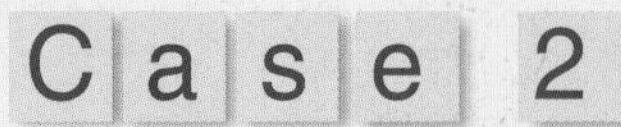

Using Extrapolation to Predict Life Expectancy

One reason for developing a mathematical model is to make predictions. If your model is a least squares line, you can predict the y value corresponding to some new x by substituting this x into an equation of the form $Y = mx + b$. (We use a capital Y to remind us that we're getting a predicted value rather than an actual data value.) Data analysts distinguish two very different kinds of prediction, *interpolation* and *extrapolation*. An interpolation uses a new x inside the x range of your original data. For example, if you have inflation data at five-year intervals from 1950 to 2000, estimating the rate of inflation in 1957 is an interpolation problem. But if you use the same data to estimate what the inflation rate was in 1920, or what it will be in 2020, you are extrapolating.

In general, interpolation is much safer than extrapolation, because data that are approximately linear over a short interval may be nonlinear over a larger interval. One way to detect nonlinearity is to look at *residuals,* which are the differences between the actual data values and the values predicted by the line of best fit. Here is a simple example:

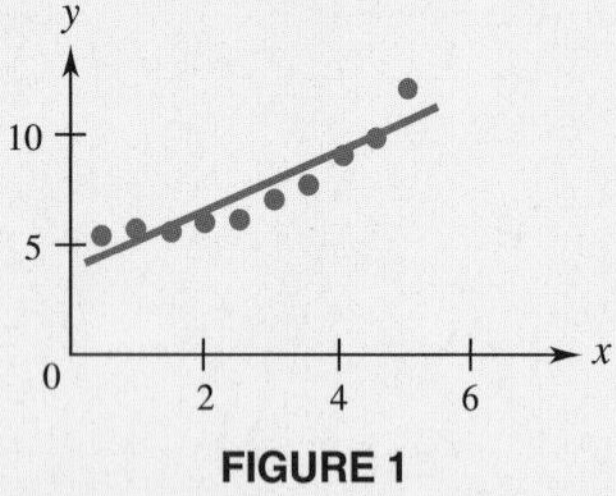

FIGURE 1

The regression equation for the best-fit line in Figure 1 is $Y = 3.431 + 1.334x$. Since the r value for this regression line is 0.93, our linear model fits the data very well. But we might notice that the predictions are a bit low at the ends and high in the middle. We can get a better look at this pattern by plotting the residuals. To find them, we put each value of the independent variable into the regression equation, calculate the predicted value Y, and subtract it from the actual y value. The residual plot in Figure 2 has the vertical axis rescaled to exaggerate the pattern. The residuals indicate that our data have a nonlinear U-shaped component that isn't captured by the linear fit. Extrapolating from this data set is probably not a good idea; our linear prediction for the value of y when x is 10 may be much too low.

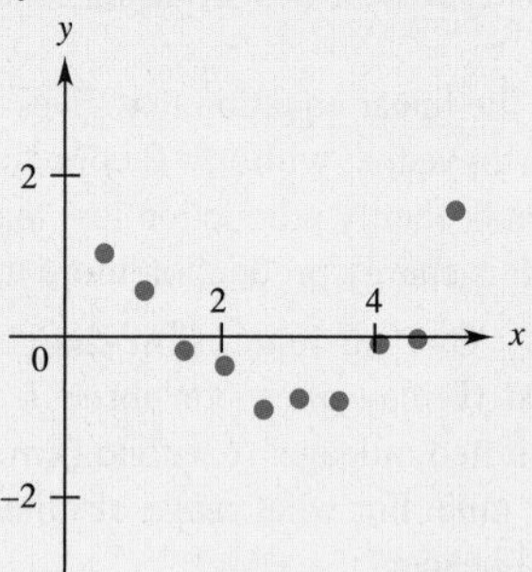

FIGURE 2

Exercises

*The following table gives the life expectancy at birth of females born in the United States in various years from 1950 to 1995.**

Year of Birth	*Life Expectancy in Years*
1950	71.3
1960	73.1
1970	74.7
1980	77.4
1985	78.2
1990	78.8
1995	78.9

1. If you have appropriate technology, verify that the least squares regression line that models this data (with coeffi-

**Health, United States, 1998,* Centers for Disease Control.

cients rounded) is $Y = .1827x + 62.2251$, where Y is the life expectancy of a woman born in year x and $x = 0$ corresponds to 1900.

2. Use the regression equation in Exercise 1 to guess a value for the life expectancy of females born in 1900.
3. Compare your answer with the actual life expectancy for females born in 1900, which was 48.3 years. Are you surprised?
4. Find the life expectancy (rounded to one decimal place) predicted by the regression equation for each year in the table and subtract it from the actual value in the second column. This gives you a table of residuals. Plot your residuals as points on a graph.
5. Now look at the residuals as a fresh data set and see if you can sketch the graph of a smooth function that fits the residuals well. How easy do you think it will be to predict the life expectancy at birth of females born in 2010?
6. What will happen if you try linear regression on the *residuals?* If you're not sure, use technology to find the regression equation for the residuals. Why does this result make sense?
7. Since most of the females born in 1985 are still alive, how did the Public Health Service come up with a life expectancy of 78.2 years for these women?

Chapter 3

Functions and Graphs

Functions are an extremely useful way of describing many real-world situations in which the value of one quantity varies with, depends on, or determines the value of another. In this chapter you will be introduced to functions, learn how to use functional notation, develop skill in constructing and interpreting the graphs of functions, and, finally, learn to apply this knowledge in a variety of situations.

3.1 FUNCTIONS

To understand the origin of the concept of function, we consider some "real life" situations in which one numerical quantity depends on, corresponds to, or determines another.

EXAMPLE 1 The amount of income tax you pay depends on the amount of your income. The way in which the income determines the tax is given by the tax law.

EXAMPLE 2 The weather bureau records the temperature over a 24-hour period in the form of a graph (see Figure 3.1). The graph shows the temperature that corresponds to each given time.

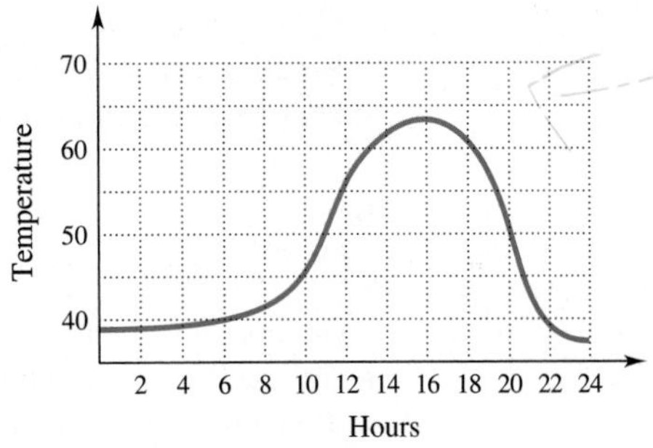

FIGURE 3.1

EXAMPLE 3 Suppose a rock is dropped straight down from a high point. From physics we know that the distance traveled by the rock in t seconds is $16t^2$ feet. So the distance depends on the time.

The first common feature shared by these examples is that each involves two sets of numbers, which we can think of as a set of inputs and a set of outputs.

	Set of Inputs	*Set of Outputs*
Example 1	All incomes	All tax amounts
Example 2	Hours since midnight	Temperatures during the day
Example 3	Seconds elapsed after dropping the rock	Distances the rock travels

The second common feature is that in each example there is a definite *rule* by which each input determines an output. In Example 1 the rule is given by the tax law, which specifies how each income (input) determines a tax amount (output). Similarly, the rule is given by the time/temperature graph in Example 2 and by the formula (distance $= 16t^2$) in Example 3.

Each of these examples could be represented by an idealized calculator that has a single operation key and can receive or display any real number. When a number is entered (*input*), and the "rule key" is pressed, an answer is displayed (*output*). (See Figure 3.2.) The formal definition of function has these same common features (input/rule/output), with a slight change of terminology.

FIGURE 3.2

A **function** consists of a set of inputs called the **domain,** a set of outputs called the **range,** and a rule by which each input determines *exactly one* output.

The domain in Example 1 consists of all possible income amounts; the rule is given by the tax law, and the range consists of all possible tax amounts. In Example 2 the domain is the set of hours in the day (that is, all real numbers from 0 to 24); the rule is given by the time/temperature graph, which shows the temperature at each time. The graph also shows that the range (the temperatures that actually occur during the day) includes all the numbers from 38 to 63.

Be sure that you understand the phrase "exactly one output" in the definition of the rule of a function. In Example 2, for instance, each time of day (input) determines exactly one temperature (output)—you can't have two different temperatures at the same time. However, it is quite possible to have the same temperature (output) at two different times (inputs). In other words:

In a function, each input produces a single output, but different inputs may produce the same output.

EXAMPLE 4 Which of the following rules describe functions?

(a) Use the optical reader at the checkout counter of the supermarket to convert codes to prices.

For each code, the reader produces exactly one price, so this is a function.

(b) Enter a number in a calculator and press the x^2 key.

This is a function because the calculator produces just one number x^2 for each number x that is entered.

(c) Assign to each number x the number y given by this table.

x	1	1	2	2	3	3
y	3	−3	5	−5	8	−8

Since at least one x-value corresponds to more than one y-value, this table does not define a function.

(d) Assign to each number x the number y given by this equation: $y = 3x - 5$.

Because the equation determines a unique value of y for each value of x, it defines a function. ◀ 1

Do the following define functions?

(a) The correspondence defined by the rule $y = x^2 + 5$, where x is the input and y is the output.

(b) Enter a nonzero number in a calculator and press the $1/x$ key.

(c) The correspondence between a computer, x, and several users of the computer, y

Answers:

(a) Yes

(b) Yes

(c) No

The equation $y = 3x - 5$ in part (d) of Example 4 defines a function, with x as input and y as output, because each value of x determines a *unique* value of y. In such a case, the equation is said to *define y as a function of x*.

2

Do the following define y as a function of x?

(a) $y = -6x + 1$

(b) $y = x^2$

(c) $x = y^2 - 1$

(d) $y < x + 2$

Answers:

(a) Yes

(b) Yes

(c) No

(d) No

EXAMPLE 5 Decide whether each of the following equations defines y as a function of x.

(a) $y = -4x + 11$

For a given value of x, calculating $-4x + 11$ produces exactly one value of y. (For example, if $x = -7$, then $y = -4(-7) + 11 = 39$.) Because one value of the input variable leads to exactly one value of the output variable, $y = -4x + 11$ defines y as a function of x.

(b) $y^2 = x$

Suppose $x = 36$. Then $y^2 = x$ becomes $y^2 = 36$, from which $y = 6$ or $y = -6$. Since one value of x can lead to two values of y, $y^2 = x$ does *not* define y as a function of x. ◀ 2

Almost all the functions in this book are defined by formulas or equations, as in part (a) of Example 5. The domain of such a function is determined by the following *agreement on domains.*

Unless otherwise stated, assume that the domain of any function defined by a formula or an equation is the largest set of real numbers that are meaningful replacements for the input variable.

For example, suppose

$$y = \frac{-4x}{2x - 3}.$$

Any real number can be used for x except $x = 3/2$, which makes the denominator equal 0. By the agreement on domains, the domain of this function is the set of all real numbers except 3/2.

EXAMPLE 6 Each of the following equations defines y as a function of x. Find the domain of each function.

(a) $y = x^4$

Any number can be raised to the fourth power, so the domain is the set of all real numbers, which is sometimes written as $(-\infty, \infty)$.

(b) $y = \sqrt{6 - x}$

For y to be a real number, $6 - x$ must be nonnegative. This happens only when $6 - x \geq 0$, or $6 \geq x$, making the domain the interval $(-\infty, 6]$.

(c) $y = \dfrac{1}{x + 3}$

Because the denominator cannot be 0, $x \neq -3$ and the domain consists of all numbers in the intervals

$$(-\infty, -3) \quad \text{or} \quad (-3, \infty).$$ ◀ 3

3

Give the domain.

(a) $y = 3x + 1$

(b) $y = x^2$

(c) $y = \sqrt{-x}$

Answers:

(a) $(-\infty, \infty)$

(b) $(-\infty, \infty)$

(c) $(-\infty, 0]$

CAUTION Notice the difference between the rule $y = \sqrt{6 - x}$ in Example 6(b) and the rule $y^2 = 6 - x$. For a particular x-value, say 2, the radical expression in the first rule represents a single positive number. The second rule, however, when $x = 2$, produces *two* numbers, $y = 2$ or $y = -2$.

Functional Notation

In actual practice, functions are seldom presented in the style of domain, rule, range, as they have been here. Functions are usually denoted by a letter (f is frequently used). If x is an input, then $f(x)$ denotes the output number that the function f produces from the input x. The symbol $f(x)$ is read "f of x." The rule is usually given by a formula, such as $f(x) = \sqrt{x^2 + 1}$. This formula can be thought of as a set of directions.

Name of function   Input number

$$f(x) = \sqrt{x^2 + 1}$$

Output number   Directions that tell you what to do with input x in order to produce the corresponding output $f(x)$; namely, "square it, add 1, and take the square root of the result."

For example, to find $f(3)$ (the output number produced by the input 3), simply replace x by 3 in the formula:

$$\begin{aligned} f(3) &= \sqrt{3^2 + 1} \\ &= \sqrt{10}. \end{aligned}$$

Similarly, replacing x by -5 and 0 shows that

$$\begin{aligned} f(-5) &= \sqrt{(-5)^2 + 1} \\ &= \sqrt{26} \end{aligned} \qquad \text{and} \qquad \begin{aligned} f(0) &= \sqrt{0^2 + 1} \\ &= 1. \end{aligned}$$

These directions can be applied to any quantities, such as $a + b$ or c^4 (where a, b, c are real numbers). Thus, to compute $f(a + b)$, the output corresponding to input $a + b$, we square the input [obtaining $(a + b)^2$], add 1 [obtaining $(a + b)^2 + 1$], and take the square root of the result:

$$\begin{aligned} f(a + b) &= \sqrt{(a + b)^2 + 1} \\ &= \sqrt{a^2 + 2ab + b^2 + 1}. \end{aligned}$$

Similarly, the output $f(c^4)$ corresponding to the input c^4 is computed by squaring the input $[(c^4)^2]$, adding 1 $[(c^4)^2 + 1]$, and taking the square root of the result:

$$\begin{aligned} f(c^4) &= \sqrt{(c^4)^2 + 1} \\ &= \sqrt{c^8 + 1}. \end{aligned}$$

EXAMPLE 7 Let $g(x) = -x^2 + 4x - 5$. Find each of the following.

(a) $g(-2)$
Replace x with -2.

$$\begin{aligned} g(-2) &= -(-2)^2 + 4(-2) - 5 \\ &= -4 - 8 - 5 \\ &= -17 \end{aligned}$$

(b) $g(x + h)$
Replace x by the quantity $x + h$ in the rule of g.

$$\begin{aligned} g(x + h) &= -(x + h)^2 + 4(x + h) - 5 \\ &= -(x^2 + 2xh + h^2) + (4x + 4h) - 5 \\ &= -x^2 - 2xh - h^2 + 4x + 4h - 5 \end{aligned}$$

(c) $g(x + h) - g(x)$
Use the result from part (b) and the rule for $g(x)$.

$$\begin{aligned} g(x + h) - g(x) &= (-x^2 - 2xh - h^2 + 4x + 4h - 5) - (-x^2 + 4x - 5) \\ &= -2xh - h^2 + 4h \end{aligned}$$

(d) $\dfrac{g(x + h) - g(x)}{h}$ (assuming $h \neq 0$)
The numerator was found in part (c). Divide it by h as follows.

$$\begin{aligned} \frac{g(x + h) - g(x)}{h} &= \frac{-2xh - h^2 + 4h}{h} \\ &= \frac{h(-2x - h + 4)}{h} \\ &= -2x - h + 4 \end{aligned}$$

The quotient found in Example 7(d)

$$\frac{g(x + h) - g(x)}{h}$$

is important in calculus, and we will see it again in Chapter 11. (4)

4

Let $f(x) = 5x^2 - 2x + 1$. Find the following.

(a) $f(1)$
(b) $f(3)$
(c) $f(1 + 3)$
(d) $f(1) + f(3)$
(e) $f(m)$
(f) $f(x + h) - f(x)$

Answers:

(a) 4
(b) 40
(c) 73
(d) 44
(e) $5m^2 - 2m + 1$
(f) $10xh + 5h^2 - 2h$

CAUTION Functional notation is *not* the same as ordinary algebraic notation. You cannot simplify an expression such as $f(x + h)$ by writing $f(x) + f(h)$. To see why, consider the answers to problems 4(c) and (d) at the side, which show that

$$f(1 + 3) \neq f(1) + f(3).$$

EXAMPLE 8 Suppose the projected sales (in thousands of dollars) of a small company over the next ten years are approximated by the function

$$S(x) = .08x^4 - .04x^3 + x^2 + 9x + 54.$$

A developer estimates that the total cost of building x large apartment complexes in a year is approximated by

$$A(x) = x^2 + 80x + 60,$$

where $A(x)$ represents the cost in hundred thousands of dollars. Find the cost of building

(a) 4 complexes;

(b) 10 complexes.

Answers:

(a) $39,600,000

(b) $96,000,000

(a) What are the projected sales this year?

The current year corresponds to $x = 0$ and the sales this year are given by $S(0)$. Substituting 0 for x in the rule of S, we see that $S(0) = 54$. So the current projected sales are $54,000.

(b) What will sales be in three years?

The sales three years from now are given by $S(3)$, which can be computed by hand or with a calculator:

$$S(x) = .08x^4 - .04x^3 + x^2 + 9x + 54$$
$$S(3) = .08(3)^4 - .04(3)^3 + (3)^2 + 9(3) + 54 \qquad \text{Let } x = 3.$$
$$= 95.4,$$

which means that sales are projected to be $95,400. 

If you have a graphing calculator, you should learn to use its table feature to evaluate functions.

EXAMPLE 9 Use the table feature of a graphing calculator to find the projected sales of the company in Example 8 for years 4 through 10.

Enter the sales equation $y = .08x^4 - .04x^3 + x^2 + 9x + 54$ in the equation memory of the calculator (often called the Y= list). Check your instruction manual for how to set the table to start at $x = 4$ and go through $x = 10$. Then display the table, as in Figure 3.3. It shows that sales are projected to rise from $123,920 in year 4 to $1,004,000 in year 10.

X	Y1
4	123.92
5	169
6	239.04
7	344.36
8	497.2
9	711.72
10	1004

X=4

FIGURE 3.3

3.1 Exercises

Which of the following rules define y as a function of x? (See Examples 1–5.)

1.

x	3	2	1	0	−1	−2	−3
y	9	4	1	0	1	4	9

2.

x	9	4	1	0	1	4	9
y	3	2	1	0	−1	−2	−3

3. $y = x^3$

4. $y = \sqrt{x - 1}$

5. $x = |y + 2|$

6. $x = y^2 + 3$

7. $y = \dfrac{-1}{x - 1}$

8. $y = \dfrac{4}{2x + 3}$

State the domain of each function. (See Example 6.)

9. $f(x) = 4x - 1$ **10.** $f(x) = 2x + 7$

11. $f(x) = x^4 - 1$ **12.** $f(x) = (2x + 5)^2$

13. $f(x) = \sqrt{-x + 3}$ **14.** $f(x) = \sqrt{4 - x}$

15. $f(x) = \frac{1}{x - 1}$ **16.** $f(x) = \frac{1}{x - 3}$

17. $f(x) = |5 - 4x|$ **18.** $f(x) = |-x - 6|$

For each of the following functions, find
(a) $f(4)$ **(b)** $f(-3)$ **(c)** $f(2.7)$ **(d)** $f(-4.9)$.
(See Example 7.)

19. $f(x) = 6$ **20.** $f(x) = 0$

21. $f(x) = 2x^2 + 4x$ **22.** $f(x) = x^2 - 2x$

23. $f(x) = \sqrt{x + 3}$ **24.** $f(x) = \sqrt{5 - x}$

25. $f(x) = x^3 - 6.2x^2 + 4.5x - 1$

26. $f(x) = x^4 + 5.5x^2 - .3x$

27. $f(x) = |x^2 - 6x - 4|$

28. $f(x) = |x^3 - x^2 + x - 1|$

29. $f(x) = \frac{\sqrt{x - 1}}{x^2 - 1}$

30. $f(x) = \sqrt{-x} + \frac{2}{x + 1}$

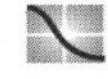
If you have a graphing calculator with table-making ability, display a table showing the (approximate) values of the given function at 3.5, 3.9, 4.3, 4.7, 5.1, and 5.5. (See Example 9.)

31. $g(x) = 3x^4 - x^3 + 2x$

32. $f(x) = \sqrt{x^2 - 2.4x + 8}$

For each of the following functions, find
(a) $f(p)$ **(b)** $f(-r)$ **(c)** $f(m + 3)$.
(See Example 7.)

33. $f(x) = 5 - x$ **34.** $f(x) = 3x + 7$

35. $f(x) = \sqrt{4 - x}$ **36.** $f(x) = \sqrt{-2x}$

37. $f(x) = x^3 + 1$ **38.** $f(x) = 2 - x^3$

39. $f(x) = \frac{3}{x - 1}$ **40.** $f(x) = \frac{-1}{2 + x}$

For each of the following functions find

$$\frac{f(x + h) - f(x)}{h}.$$

(See Example 7.)

41. $f(x) = 2x - 4$ **42.** $f(x) = 2 - 3x$

43. $f(x) = x^2 + 1$ **44.** $f(x) = x^2 - x$

Use a calculator to work these exercises. (See Example 8.)

45. Health The table contains incidence ratios by age for death from coronary heart disease (CHD) and lung cancer (LC) when comparing smokers (21–39 cigarettes per day) to nonsmokers.*

Age	*CHD*	*LC*
55–64	1.9	10
65–74	1.7	9

The incidence ratio of 10 means that smokers are 10 times more likely than nonsmokers to die of lung cancer between the ages of 55 and 64. If the incidence ratio is x, then the percent P (expressed as a decimal) of deaths caused by smoking is given by the function $P(x) = \frac{x - 1}{x}$.

(a) What is the percent of lung cancer deaths that can be attributed to smoking between the ages of 65 and 74?

(b) What is the percent of coronary heart disease deaths that can be attributed to smoking between the ages of 55 and 64?

46. Transportation The number of fliers on commuter airlines (10- to 30-seat planes) between 1975 and 2010 is approximated by $g(x) = .0138x^2 - .172x + 1.4$ (where $x = 0$ corresponds to 1975 and $g(x)$ is in millions).†

(a) How many fliers were there in 1975 and 1994?

(b) How many fliers are projected for 2006?

47. Social Science The number of Americans (in thousands) that are or are expected to be over 100 years old in year x can be approximated by the function $h(x) = .4018x^2 + 2.039x + 50$ (where $x = 0$ corresponds to 1994).‡

(a) How many Americans were over 100 in 1994? In 1996?

(b) Predict the number of Americans that will be over 100 years old in the year 2008.

48. Natural Science High concentrations of zinc ions in the water are lethal to rainbow trout. The function

$$f(x) = \left(\frac{x}{1960}\right)^{-.833}$$

gives the approximate average survival time (in minutes) for trout exposed to x milligrams per liter (mg/L) of zinc ions.§ Find the survival time (to the nearest minute) for the following concentrations of zinc ions.

(a) 110 **(b)** 525 **(c)** 1960 **(d)** 4500

*Walker, A., *Observations and Inference: An Introduction to the Methods of Epidemiology,* (Newton Lower Falls, MA: Epidemiology Resources, Inc., 1991).

†Based on data from the Federal Aviation Administration (in *USA Today,* March 27, 1995).

‡Based on data from the U.S. Census Bureau.

§C. Mason, *Biology of Freshwater Pollution.*

Use the table feature of a graphing calculator to do these exercises. (See Example 9.)

49. Finance Based on data from the U.S. Department of Education, federal student financial assistance (Pell grants, Perkins loans, work-study funds, etc.) is modeled by the function

$$g(x) = .104x^3 - 2.34x^2 + 18.23x - 9.6 \quad (4 \leq x \leq 10),$$

where $x = 0$ corresponds to 1990 and $g(x)$ is in billions of dollars.

(a) List the values of $g(x)$ for $x = 4, 5, \ldots, 10$.

(b) Explain why this model is not likely to be accurate after 2001. (*Hint:* find $g(15)$ and $g(18)$; do these numbers seem plausible?)

50. Natural Science The average monthly temperature in Austin, Texas, can be approximated by the function

$$f(x) = .03x^4 - .716x^3 + 4.38x^2 - 1.07x + 49 \quad (0 \leq x < 12),$$

where $x = 0$ corresponds to January 1, $x = 2$ to February 1, etc. List the average temperature on the first day of each month from April through October.

Work these problems.

51. Transportation The distance from Chicago to Seattle is approximately 2000 miles. A plane flying directly to Seattle passes over Chicago at noon. If the plane travels at 475 mph, find the rule of the function $f(t)$ that gives the distance of the plane from Seattle at time t hours (with $t = 0$ corresponding to noon).

52. Business A pretzel factory has daily fixed costs of \$1800. In addition, it costs 50 cents to produce each bag of pretzels. A bag of pretzels sells for \$1.20.

(a) Find the rule of the cost function $c(x)$ that gives the total daily cost of producing x bags of pretzels.

(b) Find the rule of the revenue function $r(x)$ that gives the daily revenue from selling x bags of pretzels.

(c) Find the rule of the profit function $p(x)$ that gives the daily profit from x bags of pretzels.

53. Business A bicycle factory has weekly fixed costs of \$36,000. The material and labor costs for each bike are \$125.

(a) Express the total weekly cost $C(x)$ as a function of the number x of bicycles that are made.

(b) What is the cost *per bicycle* of producing 600 bikes in a week?

54. Geometry Find an equation that expresses the area y of a square as a function of its

(a) side x (b) diagonal d

3.2 GRAPHS OF FUNCTIONS

The **graph** of a function $f(x)$ is defined to be the graph of the *equation* $y = f(x)$. For example, the graph of $f(x) = x^3 + x + 5$ is the graph of the equation $y = x^3 + x + 5$. Thus, the graph consists of all points in the coordinate plane whose coordinates are of the form $(x, x^3 + x + 5)$, that is, all points $(x, f(x))$. The same thing is true in the general case: Each point on the graph of a function f is an ordered pair whose first coordinate is an input number from the domain of f and whose second coordinate is the corresponding output number.

We begin with the simplest function graphs. In Section 2.2 we saw that the graph of an equation of the form $y = ax + b$ is a straight line. This equation also defines y as a function of x, which leads to the following terminology.

A **linear function** is a function whose rule can be written in the form

$$f(x) = ax + b$$

for some constants a and b.

EXAMPLE 1 The graph of the linear function $g(x) = .5x - 3$ is the graph of the equation $y = .5x - 3$. So the graph is a straight line with slope .5 and y-intercept -3, as shown in Figure 3.4.

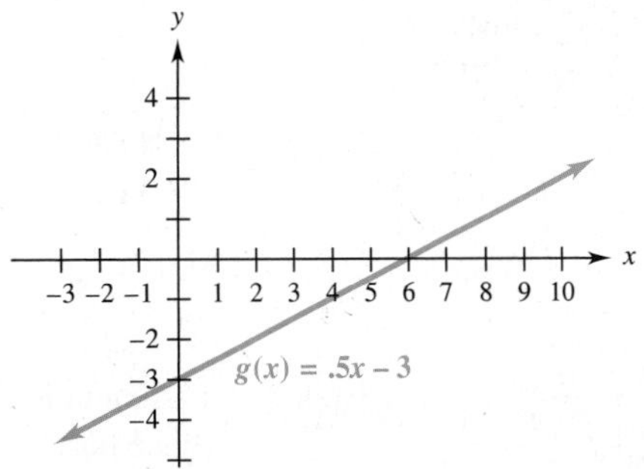

FIGURE 3.4

Piecewise Linear Functions

We now consider functions whose graphs consist of straight line segments. Such functions are called **piecewise linear functions** and are typically defined with different equations for different parts of the domain.

EXAMPLE 2 Graph the following function.

$$f(x) = \begin{cases} x + 1 & \text{if } x \leq 2 \\ -2x + 7 & \text{if } x > 2 \end{cases}$$

When $x \leq 2$, the graph consists of the part of the line $y = x + 1$ that lies to the left of $x = 2$. When $x > 2$, the graph consists of the part of the line $y = -2x + 7$ that lies to the right of $x = 2$. These line segments can be graphed by plotting the points determined by the following tables.

$x \leq 2$

x	-2	0	2
$y = x + 1$	-1	1	3

$x > 2$

x	2	3	4
$y = -2x + 7$	3	1	-1

Note that even though 2 is not in the interval $x > 2$, we found the ordered pair for that endpoint because the graph will extend right up to that point. Because this endpoint (2, 3) agrees with the endpoint for the interval $x \leq 2$, the two parts of the graph are joined at that point as shown in Figure 3.5. ◀

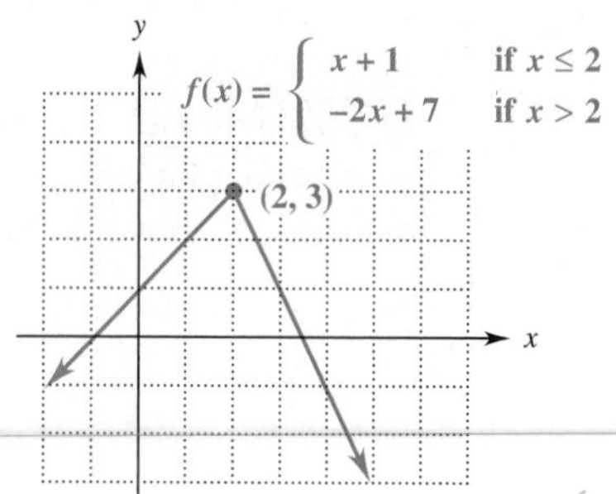

FIGURE 3.5

Graph each function.

(a) $f(x) = 2 - |x|$

(b) $f(x) = |5x - 7|$

Answers:

(a)

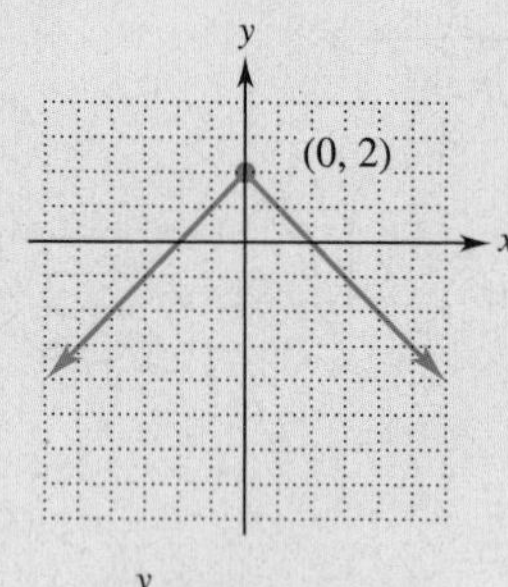

(b)

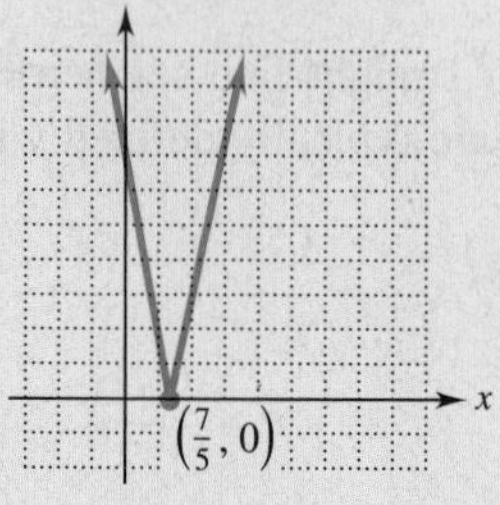

CAUTION In Example 2, notice that we did not graph the entire lines but only those portions with domain as given. Graphs of these functions should *not* be two intersecting lines.

EXAMPLE 3 Graph the **absolute value function,** whose rule is $f(x) = |x|$.

The function f is a piecewise linear function because from the definition of $|x|$,

$$f(x) = \begin{cases} x & \text{if } x \geq 0 \\ -x & \text{if } x < 0. \end{cases}$$

So the right half of the graph (that is, where $x \geq 0$) will consist of a portion of the line $y = x$. It can be graphed by plotting two points, say, (0, 0) and (1, 1). The left half of the graph (where $x < 0$) will consist of a portion of the line $y = -x$, which can be graphed by plotting $(-2, 2)$ and $(-1, 1)$, as shown in Figure 3.6. ◀ 1

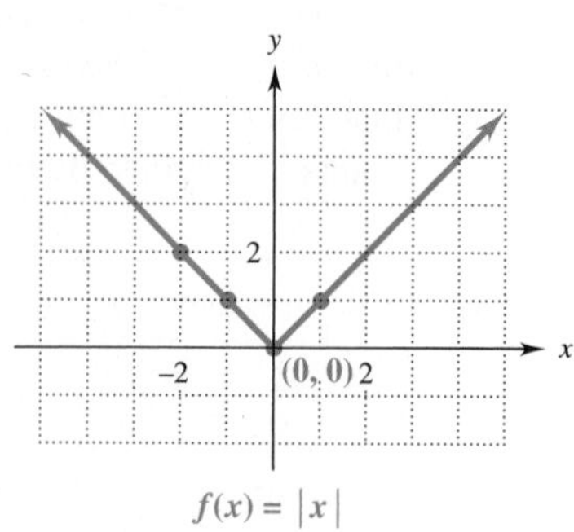

$f(x) = |x|$

FIGURE 3.6

2

Graph $f(x)$, where

$$f(x) = \begin{cases} -2x + 5 & \text{if } x < 2 \\ x - 4 & \text{if } x \geq 2. \end{cases}$$

Answer:

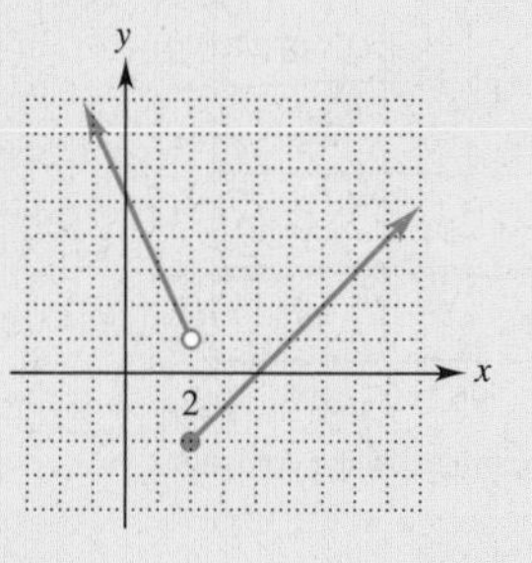

EXAMPLE 4 Graph the function

$$f(x) = \begin{cases} x - 2 & \text{if } x \leq 3 \\ -x + 8 & \text{if } x > 3. \end{cases}$$

The ordered pairs $(-2, -4)$, $(0, -2)$, and $(3, 1)$ satisfy $y = x - 2$, and the pairs $(4, 4)$ and $(6, 2)$ satisfy $y = -x + 8$. Use them to graph the two straight line segments that make up the graph of the function. The point $(3, 5)$, where the right half of the graph begins, is an endpoint, but is *not* part of the graph, as indicated by the open circle in Figure 3.7. ◀ 2

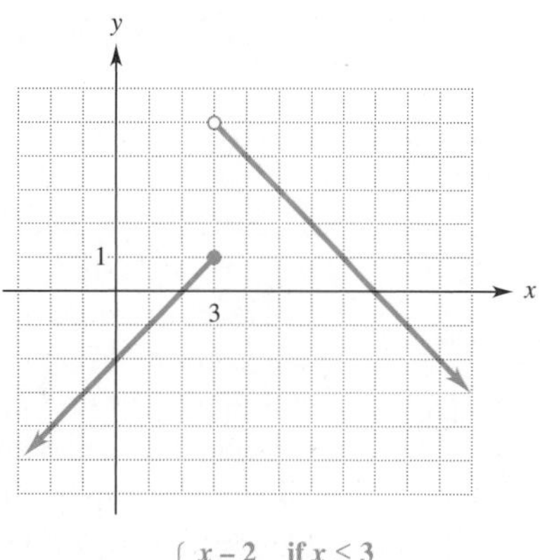

$$f(x) = \begin{cases} x - 2 & \text{if } x \leq 3 \\ -x + 8 & \text{if } x > 3 \end{cases}$$

FIGURE 3.7

TECHNOLOGY TIP To graph most piecewise linear functions on a graphing calculator, you must use a special syntax. For example, on TI and HP-39 calculators, the best way to obtain the graph in Example 4 is to graph two separate equations on the same screen:

$$y_1 = (x - 2)/(x \leq 3) \quad \text{and} \quad y_2 = (-x + 8)/(x > 3);$$

the inequality symbols are in the TEST (or CHAR) menu. However, most calculators will graph absolute value functions directly. To graph $f(x) = |x + 2|$, for instance, graph the equation $y = \text{abs}(x + 2)$. "Abs" (for absolute value) is on the keyboard or in the MATH menu.

Graphs used in business and the social sciences are often graphs of piecewise linear functions. It is just as important to know how to *read* such graphs as it is to construct them.

EXAMPLE 5 Figure 3.8 shows the graphs of three functions. Each function is the fatal crash rate per 10,000 drivers for a group of Ohio drivers.*

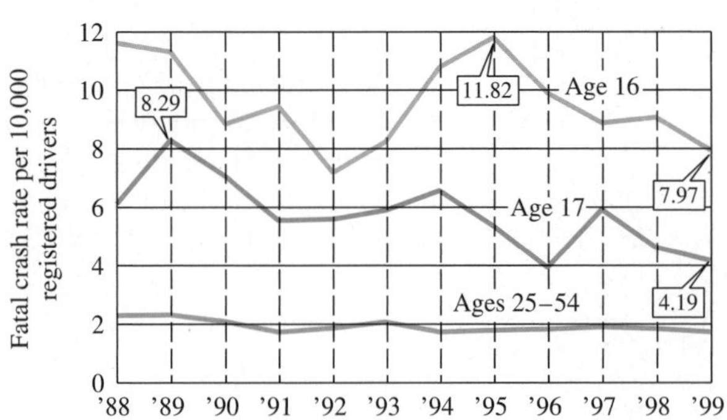

Source: © 2001 The Plain Dealer. All rights reserved. Reprinted with permission.

FIGURE 3.8

In Example 5, describe the fatal crash record of 17-year-old drivers. How does it compare to that of the other two groups?

Answer:

The crash rate shows a general downward trend since 1989 (with a very slight rise from 1991 to 1994 and a notable rise from 1996 to 1997). Their fatality rate is significantly lower than the 16-year-old rate, but two to three times that of older drivers.

(a) Describe the fatal crash record for 16-year-old drivers and for drivers at ages 25–54. How do they compare?

The graph suggests that fatal car crashes of 16-year-old drivers have followed a cyclic pattern, dropping significantly from 1988 to 1992, rising from 1992 to 1995, and dropping again in recent years (with a slight rise in 1998), with the rate generally being between 8 and 12. The graph for drivers ages 25–54 is almost a horizontal line, indicating a virtually constant fatality rate of about 2 per 10,000. Thus, the 16-year-old rate is about 4 to 6 times higher than the rate for older drivers, depending on the year.

(b) What were the worst and best years for 16- and 17-year-old drivers?

Sixteen-year-olds had their worst year in 1995, where their graph peaks, and their best year in 1992. The worst year for 17 year olds was 1989, and their best during this period was 1996. ◀ 3

4

Graph $y = [\frac{1}{2}x + 1]$.

Answer:

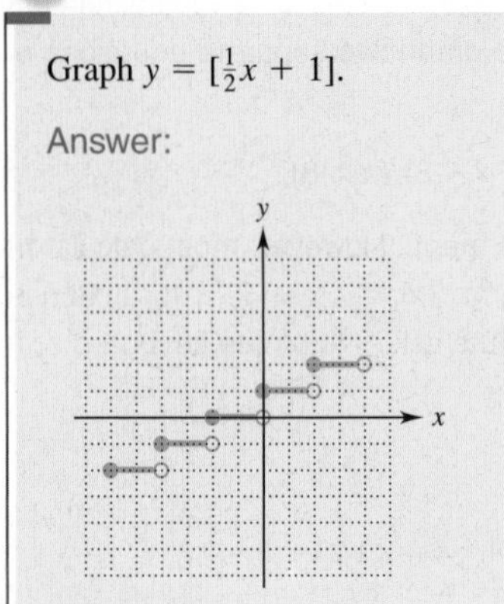

Step Functions

The **greatest integer function,** usually written $f(x) = [x]$, is defined by saying that $[x]$ denotes the largest integer that is less than or equal to x. For example, $[8] = 8$, $[7.45] = 7$, $[\pi] = 3$, $[-1] = -1$, $[-2.6] = -3$, and so on.

*Ohio Department of Public Safety.

$f(x) = [x]$

FIGURE 3.9

5

Assume that the post office charges 80¢ per ounce, or fraction of an ounce, to mail a letter to Europe. Graph the ordered pairs (ounces, cost).

Answer:

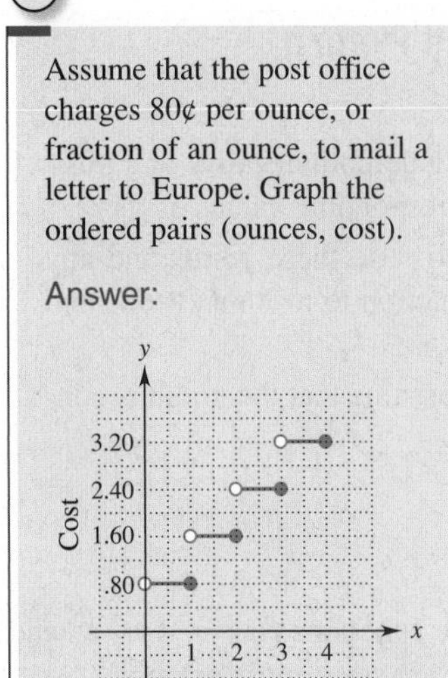

EXAMPLE 7 An overnight delivery service charges \$25 for a package weighing up to 2 pounds. For each additional pound or fraction of a pound there is an additional charge of \$3. Let $D(x)$ represent the cost to send a package weighing x pounds. Graph $D(x)$ for x in the interval $(0, 6]$.

For x in the interval $(0, 2]$, $y = 25$. For x in $(2, 3]$, $y = 25 + 3 = 28$. For x in $(3, 4]$, $y = 28 + 3 = 31$, and so on. The graph, which is that of a step function, is shown in Figure 3.10. ◀ 5

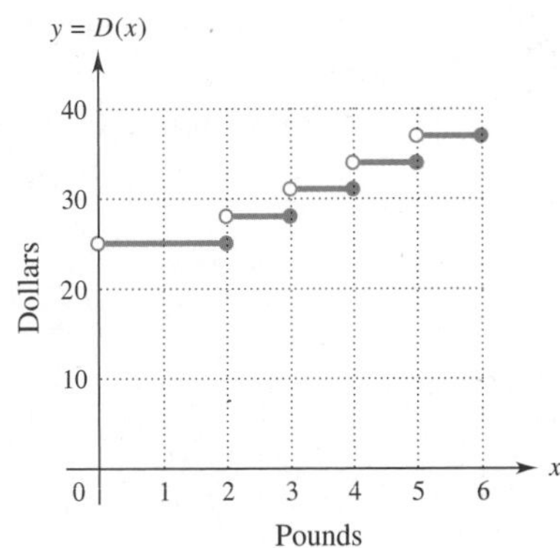

FIGURE 3.10

TECHNOLOGY TIP On most graphing calculators the greatest integer function is denoted INT or FLOOR (look on the MATH menu or its NUM submenu). Casio calculators use INTG for the greatest integer function and INT for a different function. When graphing these functions, put your calculator in "dot" graphing mode rather than the usual "connected" mode to avoid erroneous vertical line segments in the graph.

Other Functions

The graphs of many functions do not consist only of straight line segments. As a general rule when graphing functions by hand, you should follow the procedure introduced in Section 2.1 and summarized here.

Graphing a Function by Point Plotting

1. Determine the domain of the function.
2. Select a few numbers in the domain of f (include both negative and positive ones when possible) and compute the corresponding values of $f(x)$.
3. Plot the points $(x, f(x))$ computed in step two. Use these points and any other information you may have about the function to make an "educated guess" about the shape of the entire graph.
4. Unless you have information to the contrary, assume that the graph is continuous (unbroken) wherever it is defined.

6

Graph $f(x) = \sqrt{4 - x}$.

Answer:

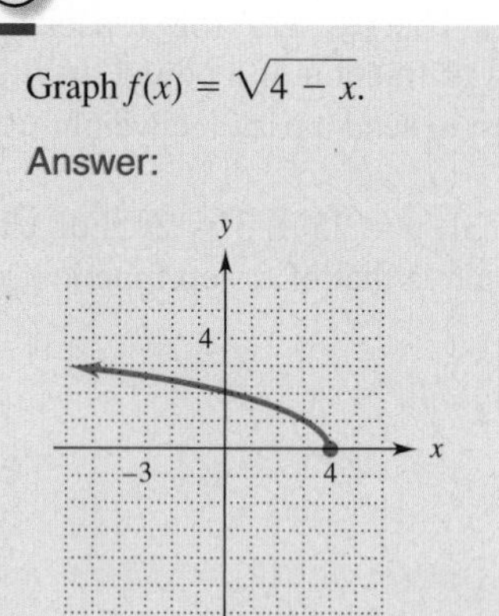

This method was used to find the graphs of the functions $f(x) = 4 - x^2$ and $g(x) = \sqrt{x}$ in Examples 3 and 4 of Section 2.1. Here are some more examples.

EXAMPLE 8 Graph $g(x) = \sqrt{x + 1}$.

Because the rule of the function is defined only when $x + 1 \geq 0$ (that is, when $x \geq -1$), the domain of g is the interval $[-1, \infty)$. Use a calculator to get a table of ordered pairs, such as the one in Figure 3.11. Plot the points and connect them in order (as x increases) to get the graph in Figure 3.11. ◀ 6

x	$g(x) = \sqrt{x + 1}$
−1	0
0	1
1	1.414
2	1.732
3	2
5	2.449
7	2.828
9	3.162

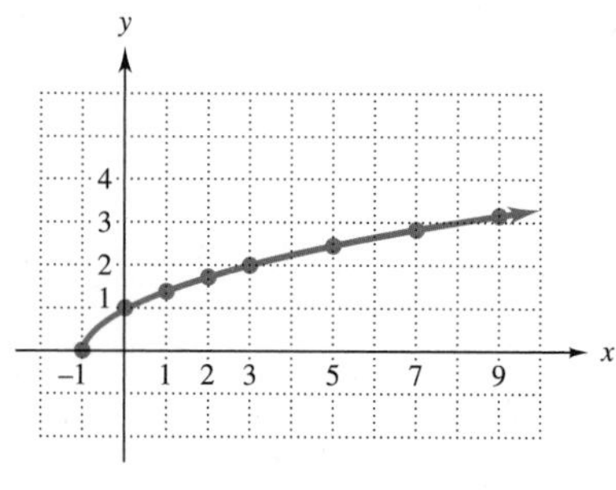

FIGURE 3.11

EXAMPLE 9 Graph the function whose rule is $f(x) = 2 - x^3/5$.

Make a table of values and plot the corresponding points. They suggest the graph in Figure 3.12. ◀

x	$f(x) = 2 - x^3/5$
−4	14.8
−3	7.4
−2	3.6
−1	2.2
0	2
1	1.8
2	.4
3	−3.4
4	−10.8

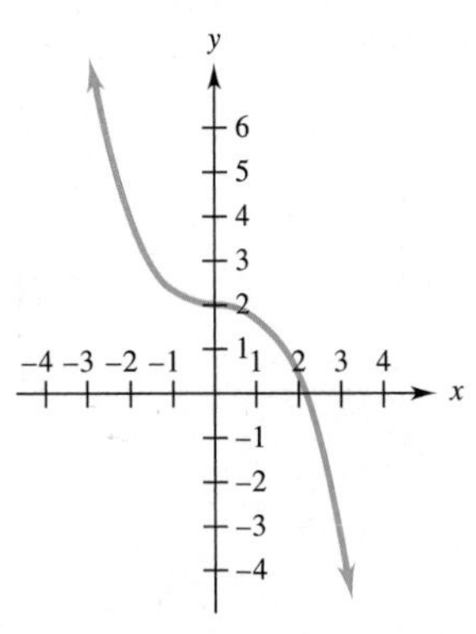

FIGURE 3.12

Many functions that arise in real-world applications are based on data and given in graphical form, so it is important to be able to read such graphs.

EXAMPLE 10 Figure 3.13 is the graph of the function h whose rule is $h(x)$ = Household debt payment as a percentage of after-tax income in year x.*

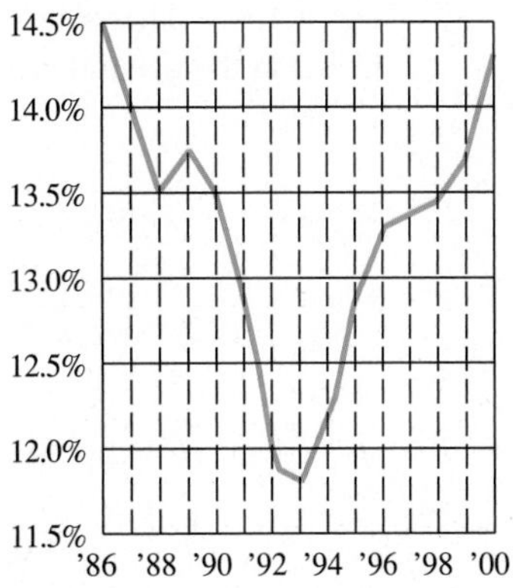

Source: © 2001 *The Plain Dealer.* All rights reserved. Reprinted with permission.

FIGURE 3.13

(a) Find the function values $h(1992)$, and $h(2000)$ and interptet them.

The point (1992, 12) is on the graph, which means that $h(1992) = 12$. Similarly, $h(2000) = 14.3$ because the point (2000, 14.3) is on the graph.† In early 1992, debt payment was 12% of after-tax income, and in early 2000 it was 14.3%.

(b) During what period was debt payment below 13%?

Look for points on the graph whose second coordinates are less than 13, that is, points below the horizontal line through 13. These points represent the period from late 1990 to mid-1995.

(c) During what period was debt payment above 13.5%?

*Federal Reserve Board.

†Since this small graph cannot be read with perfect accuracy, these results are approximate.

Look for points whose second coordinates are greater than 13.5 (above the horizontal line through 13.5). They occur during two periods: from 1986 to the beginning of 1990 and since the first quarter of 1998.

The Vertical Line Test

The following fact distinguishes function graphs from other graphs.

> **Vertical Line Test**
>
> No vertical line intersects the graph of a function more than once.

In other words, if a vertical line intersects a graph at more than one point, the graph is not the graph of a function. To see why this is true, consider the graph in Figure 3.14. The vertical line $x = 3$ intersects the graph at two points. If this were the graph of a function f, this would mean that $f(3) = 2$ (because (3, 2) is on the graph) and that $f(3) = -1$ (because (3, −1) is on the graph). This is impossible because a *function* can have only one value when $x = 3$ (each input determines exactly one output). Therefore, this cannot be the graph of a function. A similar argument works in the general case.

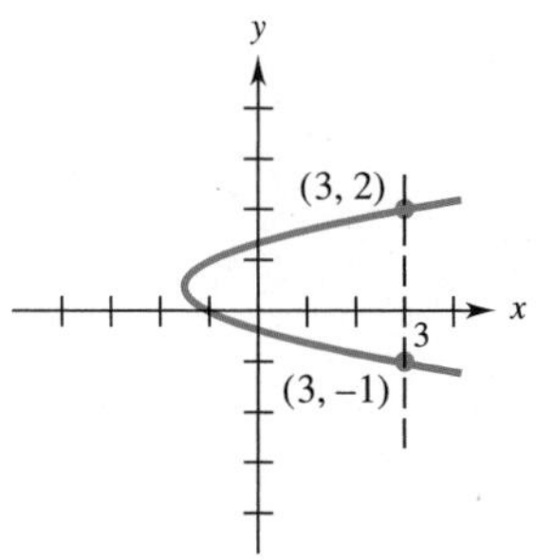

FIGURE 3.14

EXAMPLE 11 Which of the graphs in Figure 3.15 are the graphs of functions?

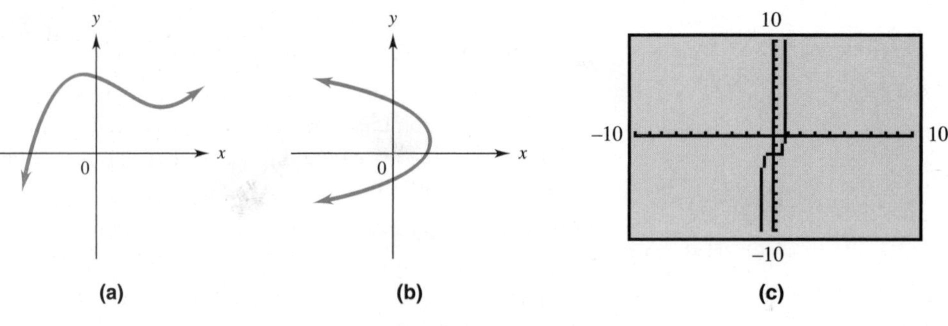

FIGURE 3.15

Find a viewing window that indicates the actual shape of the graph of the function $f(x) = 15x^{11} - 2$ of Example 10 near the point $(-1, -17)$.

Answer:

There are many correct answers, including, $-1.4 \le x \le -.6$ and $-30 \le y \le 0$.

Every vertical line intersects graph (a) in at most one point, so this graph is the graph of a function. Many vertical lines (including the y-axis) intersect graph (b) twice, so it is not the graph of a function.

Graph (c) appears to fail the Vertical Line Test near $x = 1$ and $x = -1$, indicating this is not the graph of a function. But this is misleading because of the low resolution of a calculator screen. The table in Figure 3.16 and the very narrow segment of the graph in Figure 3.17 show that the graph actually moves to the right as it moves upward. A similar fact is true near $x = -1$. So this graph *does* pass the Vertical Line Test and *is* the graph of a function. (Its rule is $f(x) = 15x^{11} - 2$.) The moral of this story is that you can't always trust a graphing calculator image. When in doubt, try other viewing windows or a table to see what's really going on. ◀ 7

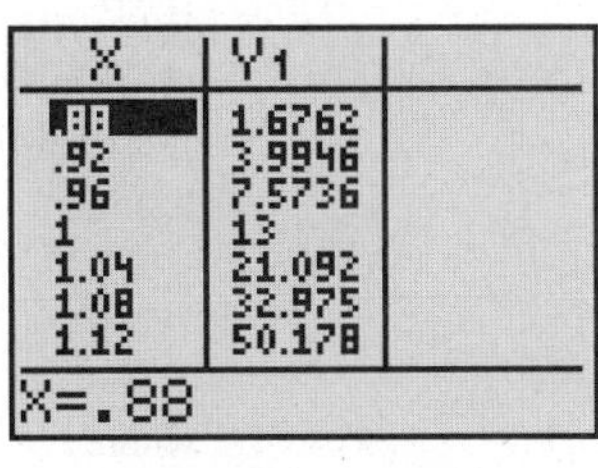

X	Y1
.88	1.6762
.92	3.9946
.96	7.5736
1	13
1.04	21.092
1.08	32.975
1.12	50.178

X=.88

FIGURE 3.16

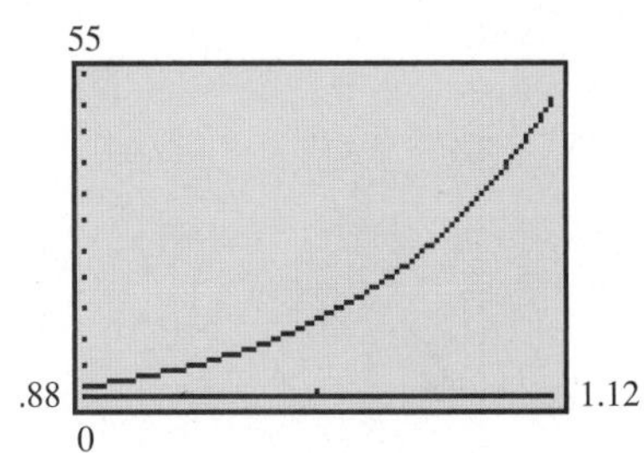

FIGURE 3.17

3.2 Exercises

Graph each function. (See Examples 1–4.)

1. $f(x) = -.5x + 2$
2. $g(x) = 3 - x$
3. $f(x) = \begin{cases} x + 2 & \text{if } x \le 1 \\ 3 & \text{if } x > 1 \end{cases}$
4. $g(x) = \begin{cases} 2x - 1 & \text{if } x < 0 \\ -1 & \text{if } x \ge 0 \end{cases}$
5. $y = \begin{cases} 3 - x & \text{if } x \le 0 \\ 2x + 3 & \text{if } x > 0 \end{cases}$
6. $y = \begin{cases} x + 5 & \text{if } x \le 1 \\ 2 - 3x & \text{if } x > 1 \end{cases}$
7. $f(x) = \begin{cases} |x| & \text{if } x \le 2 \\ -x & \text{if } x > 2 \end{cases}$
8. $g(x) = \begin{cases} -|x| & \text{if } x \le 1 \\ 2x & \text{if } x > 1 \end{cases}$
9. $f(x) = |x - 4|$
10. $g(x) = |4 - x|$
11. $f(x) = |3 - 4x|$
12. $g(x) = -|x|$
13. $y = -|x - 1|$
14. $f(x) = |x| - 2$
15. $y = |x| + 3$
16. $|x| + |y| = 1$ (*Hint:* This is not the graph of a function but is made up of four straight line segments. Find them by using the definition of absolute value in these four cases: $x \ge 0$ and $y \ge 0$; $x \ge 0$ and $y < 0$; $x < 0$ and $y \ge 0$; $x < 0$ and $y < 0$.)

Graph each of the following functions. (See Examples 6 and 7.)

17. $f(x) = [x - 3]$
18. $g(x) = [x + 2]$
19. $g(x) = [-x]$
20. $f(x) = -[x]$
21. $f(x) = [x] + [-x]$ (The graph contains horizontal segments but is *not* a horizontal line.)
22. Assume that postage rates are 37¢ for the first ounce, plus 23¢ for each additional ounce, and that each letter carries one 37¢ stamp and as many 23¢ stamps as necessary. Graph the *postage stamp function* whose rule is

$p(x) =$ the number of stamps on a letter weighing x ounces.

Graph each function. (See Examples 8 and 9.)

23. $f(x) = 3 - 2x^2$
24. $g(x) = 2 - x^2$
25. $h(x) = x^3/10 + 2$
26. $f(x) = x^3/20 - 3$

27. $g(x) = \sqrt{-x}$

28. $h(x) = \sqrt{x} - 1$

29. $f(x) = \sqrt[3]{x}$

30. $g(x) = \sqrt[3]{x - 4}$

Which of these are graphs of functions? (See Example 11.)

31.

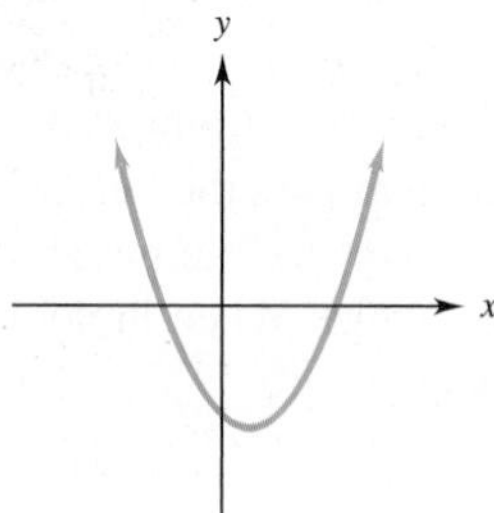

32.

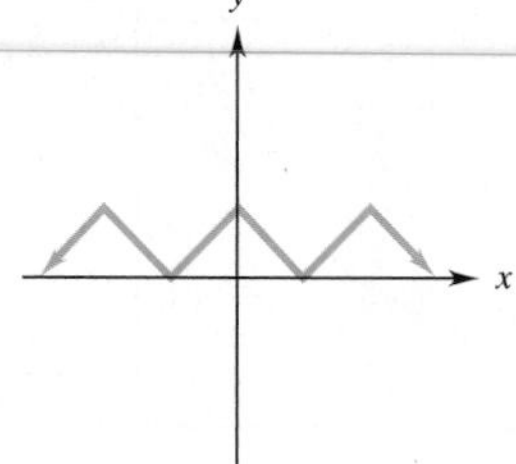

33.

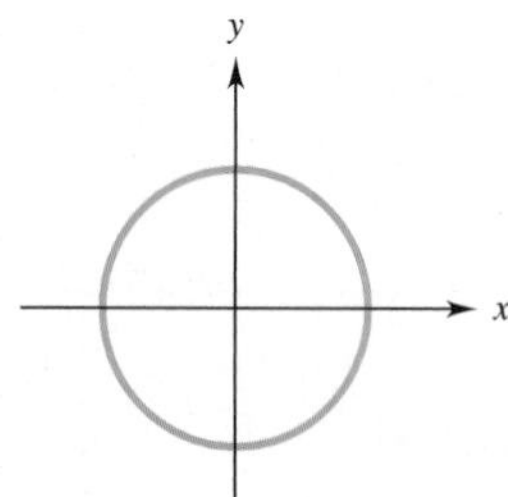

34.

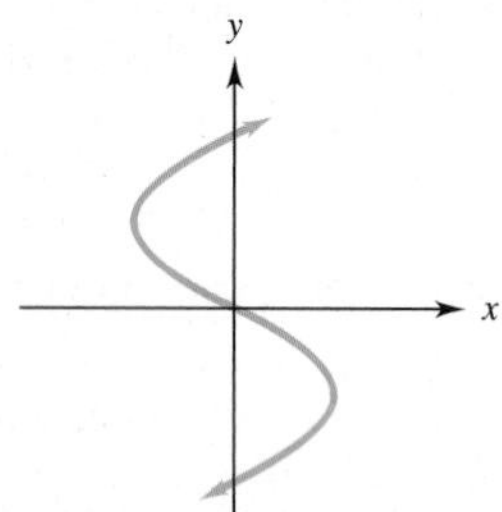

35.

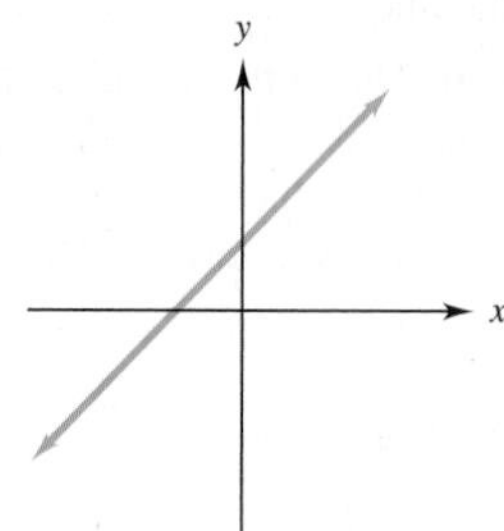

36.

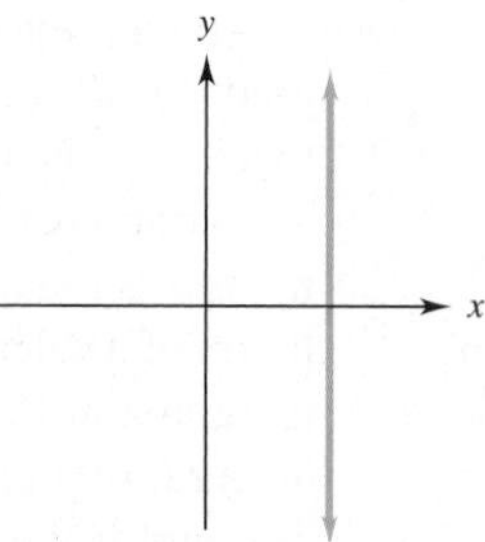

Work the following problems. (See Examples 3–5.)

37. Finance The maximum allowable yearly contribution to an individual retirement account (IRA) was \$2000 from 1981 to 2001 and has steadily increased since then. When the maximum is adjusted for inflation, however, the picture is a little different. The approximate maximum IRA contribution in 1981 dollars in year x is given by the function

$$f(x) = \begin{cases} -50x + 2050 & \text{if } 1 \le x < 22 \\ 84x - 348 & \text{if } 22 \le x < 29 \end{cases},$$

where $x = 0$ corresponds to 1980.

(a) Graph this function.

(b) What does the graph say about inflation from 1981 to 2001?

38. Finance The Connecticut state income tax in 2001 was given by the function

$$g(x) = \begin{cases} .03x & \text{if } 0 \le x \le 20{,}000 \\ .045x - 300 & \text{if } x > 20{,}000 \end{cases},$$

where x is the income in dollars. Graph this function for incomes from 0 to \$40,000.

39. Natural Science The snow depth in Michigan's Isle Royale National Park varies throughout the winter. In a typical winter, the snow depth in inches is approximated by the following function.

$$f(x) = \begin{cases} 6.5x & \text{if } 0 \le x \le 4 \\ -5.5x + 48 & \text{if } 4 < x \le 6 \\ -30x + 195 & \text{if } 6 < x \le 6.5 \end{cases}$$

Here, x represents the time in months with $x = 0$ representing the beginning of October, $x = 1$ representing the beginning of November, and so on.

(a) Graph $f(x)$.

(b) In what month is the snow deepest? What is the deepest snow depth?

(c) In what months does the snow begin and end?

40. Natural Science A factory begins emitting particulate matter into the atmosphere at 8 A.M. each workday, with the emissions continuing until 4 P.M. The level of pollutants, $P(t)$, measured by a monitoring station 1/2 mile

away is approximated as follows, where t represents the number of hours since 8 A.M.

$$P(t) = \begin{cases} 75t + 100 & \text{if } 0 \le t \le 4 \\ 400 & \text{if } 4 < t < 8 \\ -100t + 1200 & \text{if } 8 \le t \le 10 \\ -\dfrac{50}{7}t + \dfrac{1900}{7} & \text{if } 10 < t < 24 \end{cases}$$

Find the level of pollution at

(a) 9 A.M. **(b)** 11 A.M. **(c)** 5 P.M.
(d) 7 P.M. **(e)** Midnight.
(f) Graph $y = P(t)$.
(g) From the graph in part (f), at what time(s) is the pollution level highest? lowest?

41. Health The table shows the consumer price index (CPI) for medical care in selected years.*

Year	*Medical CPI*
1950	15.1
1975	47.5
1999	250.6

(a) Let $x = 0$ correspond to 1950. Find the rule of a piecewise linear function that models this data, that is, a piecewise linear function f with $f(0) = 15.1$, $f(25) = 47.5$, and $f(49) = 250.6$. Round all coefficients in your final answer to one decimal place.
(b) Graph the function f for $0 \le x \le 50$
(c) Use the function to estimate the medical CPI in 1995.
(d) Assume that this model remains accurate after 2000 and estimate the medical CPI in 2004.

42. Business The graph (from the Bureau of Public Debt) shows the federal debt from 1990 to 2000 (in billions of dollars).

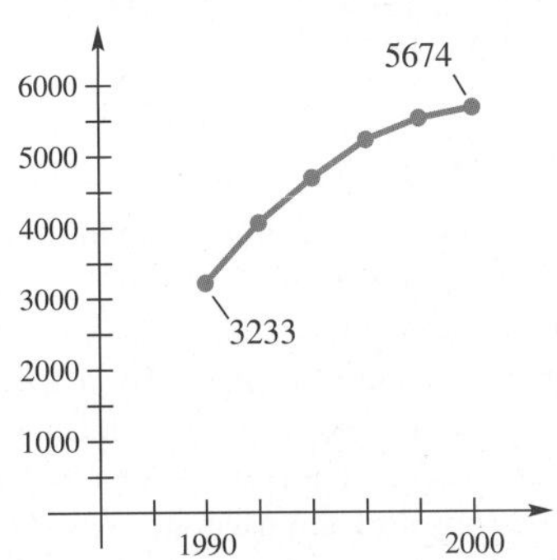

(a) Let $x = 0$ correspond to 1990. Find the rule of a linear function g that passes through the two points corresponding to 1990 and 2000. Draw the graph of g in the preceding figure.
(b) Is the function g a good model for the federal debt? Why?

Finance *The figure shows the annual percentage change in consumer price indexes (CPI). Use it for Exercises 43 and 44. (See Examples 5 and 10.)*

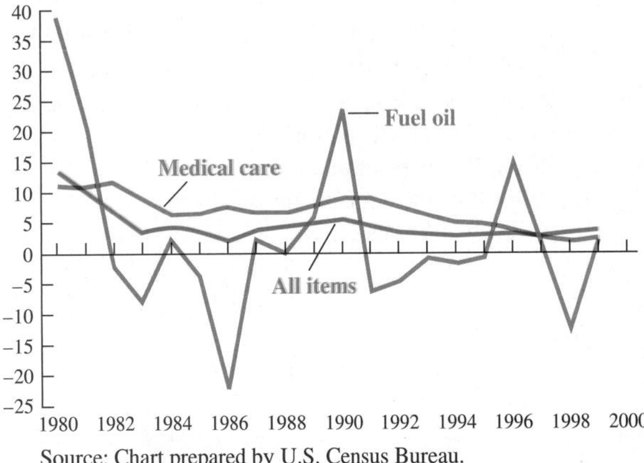

Source: Chart prepared by U.S. Census Bureau.

43. Answer each of these questions and explain how you obtained the answer from the graph.
(a) Was there any period between 1980 and 1998 when the CPI for all items was decreasing?
(b) During what years was the CPI for fuel oil increasing or constant?
(c) During what years was the CPI for fuel oil decreasing? How can you tell from the graph?

44. **(a)** When was the fuel oil CPI decreasing at the fastest rate?
(b) After 1982, when was the fuel oil CPI increasing at the fastest rate?
(c) Between 1985 and 1995, when were the medical care CPI and the fuel oil CPI increasing at the same rate? How can you tell from the graph?

45. Natural Science The table gives estimates of the percent of ozone change from 1985 for several years in the future, if chlorofluorocarbon production is reduced 80% globally.

Year	*Percent*
1985	0
2005	1.5
2025	3
2065	4
2085	5

*U.S. Bureau of Labor Statistics.

(a) Plot these ordered pairs on a grid.
(b) The points from part (a) should lie approximately on a straight line. Use the pairs (2005, 1.5) and (2085, 5) to write an equation of the line.
(c) Letting $f(x)$ represent the percent of ozone change and x represent the year, write your equation from part (b) as a rule that defines a function.
(d) Find $f(2065)$. Does it agree fairly closely with the number in the table that corresponds to 2065? Do you think the expression from part (c) describes this function adequately?

46. Business The graph below shows the number of cellular telephone accounts (in millions) since 1989.
(a) Is this the graph of a function?
(b) What does the domain represent?
(c) Estimate the range.

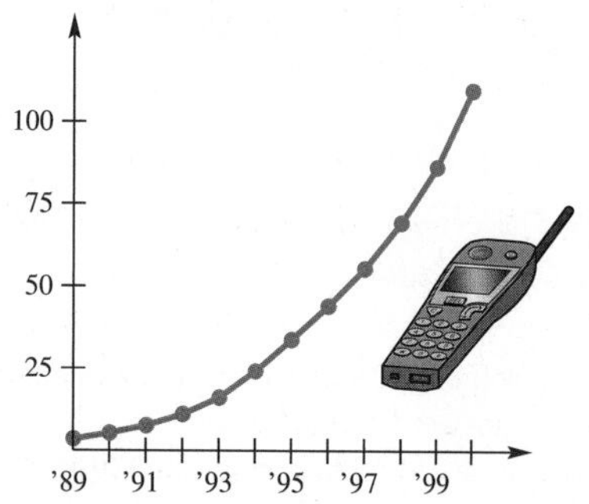

Business *Work these problems. (See Examples 6, 7, and 11.)*

47. Whenever postage rates change, some newspaper publishes a graph like this one, which shows the price of a first-class stamp from 1982 to 2003.

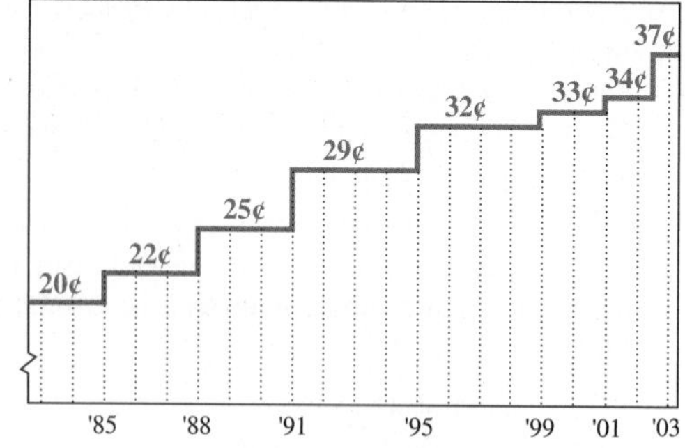

(a) Let f be the function whose rule is

$$f(x) = \text{cost of a first-class stamp in year } x.$$

Find $f(1990)$ and $f(2000)$.
(b) Explain why the graph in the figure is not the graph of the *function* f. What must be done to the figure to make it an accurate graph of the function f?

48. A chain-saw rental firm charges $7 per day or fraction of a day to rent a saw, plus a fixed fee of $4 for resharpening the blade. Let $S(x)$ represent the cost of renting a saw for x days. Find each of the following.

(a) $S\left(\frac{1}{2}\right)$ (b) $S(1)$ (c) $S\left(1\frac{1}{4}\right)$ (d) $S\left(3\frac{1}{2}\right)$
(e) What does it cost to rent a saw for $4\frac{9}{10}$ days?
(f) A portion of the graph of $y = S(x)$ is shown here. Explain how the graph could be continued.

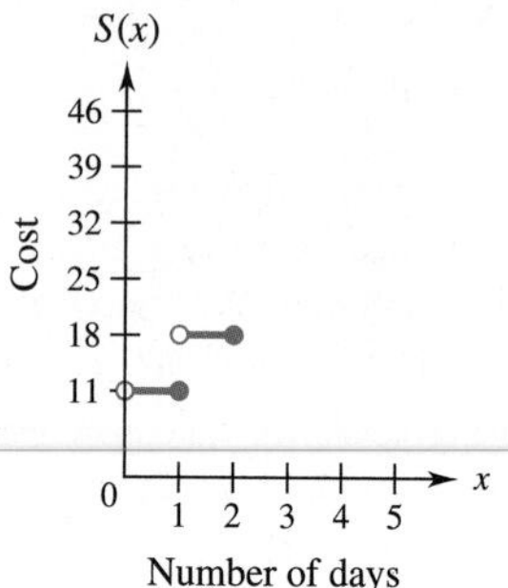

(g) What is the domain variable?
(h) What is the range variable?
(i) Write a sentence or two explaining what part (c) and its answer represent.
(j) We have left $x = 0$ out of the graph. Discuss why it should or shouldn't be included. If it were included, how would you define $S(0)$?

49. The charge to rent a Haul-It-Yourself Trailer is $25 plus $2 per hour or portion of an hour. Find the cost to rent a trailer for
(a) 2 hours; (b) 1.5 hours; (c) 4 hours;
(d) 3.7 hours.
(e) Graph the ordered pairs (hours, cost).

50. A delivery company charges $3 plus 50¢ per mile or part of a mile. Find the cost for a trip of
(a) 3 miles; (b) 3.4 miles; (c) 3.9 miles;
(d) 5 miles.
(e) Graph the ordered pairs (miles, cost).
(f) Is this a function?

Work these problems.

51. Natural Science A laboratory culture contains about one million bacteria at midnight. The culture grows very rapidly until noon, when a bactericide is introduced and the bacteria population plunges. By 4 P.M. the bacteria have adapted to the bactericide and the culture slowly increases in population until 9 P.M. when the culture is accidentally destroyed by the clean-up crew. Let $g(t)$ denote the bacteria population at time t (with $t = 0$ corresponding to midnight) and draw a plausible graph of the function g. (Many correct answers are possible.)

52. Transportation A plane flies from Austin, Texas, to Cleveland, Ohio, a distance of 1200 miles. Let f be the function whose rule is

$$f(t) = \text{distance (in miles) from Austin at time } t \text{ hours,}$$

with $t = 0$ corresponding to the 4 P.M. takeoff. In each of the following, draw a plausible graph of f under the given circumstances. (There are many correct answers for each part.)

(a) The flight is nonstop and takes between 3.5 and 4 hours.

(b) Bad weather forces the plane to land in Dallas (about 200 miles from Austin) at 5 P.M., remain overnight, and leave at 8 A.M. the next morning, flying nonstop to Cleveland.

(c) The plane flies nonstop, but due to heavy traffic it must fly in a holding pattern for an hour over Cincinnati (about 200 miles from Cleveland) and then go on to Cleveland.

3.3 APPLICATIONS OF LINEAR FUNCTIONS

Most of this section deals with the basic business relationships that were introduced in Section 1.2:

Revenue = (Price per item) × (Number of items)

Cost = Fixed Costs + Variable Costs

Profit = Revenue − Cost

The examples will use only linear functions, but the methods presented here also apply to more complicated functions.

Cost Analysis

Recall that the fixed costs are for such things as buildings, machinery, real estate taxes, and product design. Within broad limits, the fixed cost is constant for a particular product and does not change as more items are made. The variable costs are for labor, materials, shipping, and so on. The variable costs depend on the number of items made.

If $C(x)$ is the cost of making x items, then the fixed cost (the cost which occurs even when no items are produced) can be found by letting $x = 0$. For example, for the cost function $C(x) = 45x + 250{,}000$, the fixed cost is

$$C(0) = 45(0) + 250{,}000 = \$250{,}000.$$

In this case, the variable cost of making x items is $45x$, that is, \$45 per item manufactured.

EXAMPLE 1 An anticlot drug can be made for \$10 per unit. The total cost to produce 100 units is \$1500.

(a) Assume that the cost function is linear and find its rule.

Since the cost function $C(x)$ is linear, its rule is of the form $C(x) = mx + b$. We are given that m (the cost per item) is 10, so the rule is $C(x) = 10x + b$. To find b use the fact that it costs \$1500 to produce 100 units, which means that

$$\begin{aligned} C(100) &= 1500 \\ 10(100) + b &= 1500 \qquad C(x) = 10x + b \\ 1000 + b &= 1500 \\ b &= 500. \end{aligned}$$

So the rule of the cost function is $C(x) = 10x + 500$.

The total cost of producing 10 calculators is \$100. The variable costs per calculator are \$4. Find the rule of the linear cost function.

Answer:

$C(x) = 4x + 60$.

In side problem 1 above find the average cost per calculator when 100 are produced.

Answer:

\$4.60

(b) What are the fixed costs?

The fixed costs are $C(0) = 10(0) + 500 = \$500$. 

If $C(x)$ is the total cost to produce x items, then the **average cost** per item is given by

$$\overline{C}(x) = \frac{C(x)}{x}.$$

As more and more items are produced, the average cost per item typically decreases.

EXAMPLE 2 Find the average cost to produce 100 and 1000 units of the anticlot drug in Example 1.

The cost function is $C(x) = 10x + 500$, so the average cost of 100 units is

$$\overline{C}(100) = \frac{C(100)}{100} = \frac{10(100) + 500}{100} = \frac{1500}{100} = \$15.00 \text{ per unit.}$$

The average cost of 1000 units is

$$\overline{C}(1000) = \frac{C(1000)}{1000} = \frac{10(1000) + 500}{1000} = \frac{10500}{1000} = \$10.50 \text{ per unit.}$$ (2)

Rates of Change

The rate at which a quantity (such as revenue or profit) is changing can be quite important. For instance, if a company determines that the rate of change of its revenue is decreasing, then sales growth is slowing down, a trend that may require a response.

The rate of change of a linear function is easily determined. For example, suppose $f(x) = 3x + 5$ and consider the following table of values. The table shows that each time x changes by 1, the corresponding value of $f(x)$ changes by 3.

x	$f(x) = 3x + 5$
1	8
2	11
3	14
4	17
5	20

Thus, the rate of change of $f(x) = 3x + 5$ with respect to x is 3, which is the slope of the line $y = 3x + 5$. The same thing happens for any linear function:

The rate of change of a linear function $f(x) = mx + b$ is the slope m.

In particular, the rate of change of a linear function is constant. Rates of change of nonlinear functions, which are more complicated (and not constant), will be studied in Section 11.2.

In economics, the rate of change of the cost function is called the **marginal cost.** Marginal cost is important to management in making decisions in such areas as cost control, pricing, and production planning. When the cost function is linear, say, $C(x) = mx + b$, then the marginal cost is the number m (the slope of the graph of C). Marginal cost can also be thought of as the cost of producing one more item, as the next example demonstrates.

3

The cost in dollars to produce x kilograms of chocolate candy is given by $C(x)$, where

$$C(x) = 3.5x + 800.$$

Find each of the following.

(a) The fixed cost

(b) The total cost for 12 kilograms

(c) The marginal cost of the 40th kilogram

(d) The marginal cost per kilogram

Answers:

(a) \$800

(b) \$842

(c) \$3.50

(d) \$3.50

EXAMPLE 3 An electronics company manufactures handheld PCs. The cost function for one of their models is $C(x) = 160x + 750{,}000$.

(a) What are the fixed costs for this product?

The fixed costs are $C(0) = 160(0) + 750{,}000 = \$750{,}000$.

(b) What is the marginal cost?
The slope of $C(x) = 160x + 750{,}000$ is 160, so the marginal cost is \$160.

(c) After 50,000 units have been produced, what is the cost of producing one more?
The cost of producing 50,000 is

$$C(50{,}000) = 160(50{,}000) + 750{,}000 = \$8{,}750{,}000.$$

The cost of 50,001 units is

$$C(50{,}001) = 160(50{,}001) + 750{,}000 = \$8{,}750{,}160.$$

The cost of the additional unit is the difference

$$C(50{,}001) - C(50{,}000) = 8{,}750{,}160 - 8{,}750{,}000 = \$160.$$

Thus, the cost of one more item is the marginal cost. ◀ 3

Example 3 is typical of the general case, as summarized here.

In a **linear cost function** $C(x) = mx + b$, m represents the marginal cost and b the fixed cost. The marginal cost is the cost of producing one more item.

Conversely, if the fixed cost is b and the marginal cost is always the same constant m, then the cost function for producing x items is $C(x) = mx + b$.

Break-Even Analysis

A typical company must analyze its cost and the potential market for its product to determine when (or if) it will make a profit.

EXAMPLE 4 A company manufactures a particular model of DVD player that sells to retailers for \$168. The cost of making x of these DVD players is given by the function $C(x) = 118x + 800{,}000$.

(a) Find the function R that gives the revenue from selling x players.
Since revenue is the product of the price per item and the number of items, $R(x) = 168x$.

(b) What is the revenue from selling 40,00 players?
Evaluate the revenue function R at 40,000:

$$R(40{,}000) = 168(40{,}000) = \$6{,}720{,}000.$$

(c) Find the profit function P.
Since profit = revenue − cost,

$$P(x) = R(x) - C(x) = 168x - (118x + 800{,}000) = 50x - 800{,}000.$$

(d) What is the profit from selling 10,000 players?
Evaluate the profit function at 10,000:

$$P(10{,}000) = 50(10{,}000) - 800{,}000 = -300{,}000,$$

that is, a loss of \$300,000. ◀

A company can make a profit only if the revenue on a product exceeds the cost of manufacturing it. The number of units at which revenue equals cost (that is, profit is 0) is the **break-even point.**

For a certain magazine, the cost equation is $c = .70x + 1200$, where x is the number of magazines sold. The magazine sells for \$1 per copy. Find the break-even point.

Answer:

4000 magazines

EXAMPLE 5 Find the break-even point for the company in Example 4.

The company will break even when revenue equals cost, that is, when

$$\begin{aligned} R(x) &= C(x) \\ 168x &= 118x + 800{,}000 \\ 50x &= 800{,}000 \\ x &= 16{,}000 \end{aligned}$$

The company breaks even by selling 16,000 DVD players. The graphs of the revenue and cost functions and the break-even point (where $x = 16{,}000$) are shown in Figure 3.18. The company must sell more than 16,000 players ($x > 16{,}000$) in order to make a profit. ◀ (4)

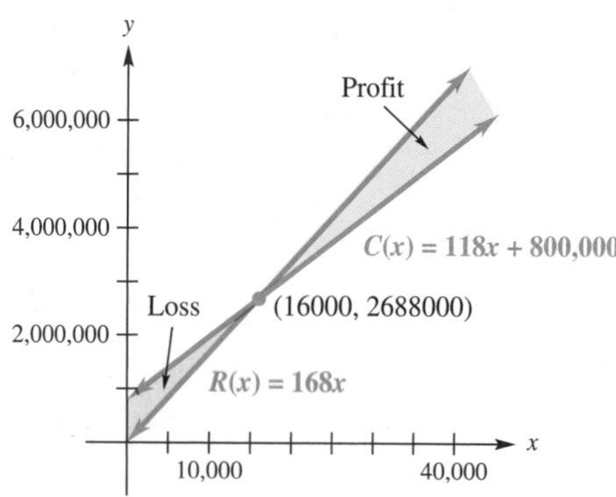

FIGURE 3.18

TECHNOLOGY TIP The break-even point in Example 5 can be found on a graphing calculator by graphing the cost and revenue functions on the same screen and using the calculator's intersection finder, as shown in Figure 3.19. Depending on the calculator, the intersection finder is in the CALC or G-SOLVE menu or in the MATH or FCN submenu of the GRAPH menu.

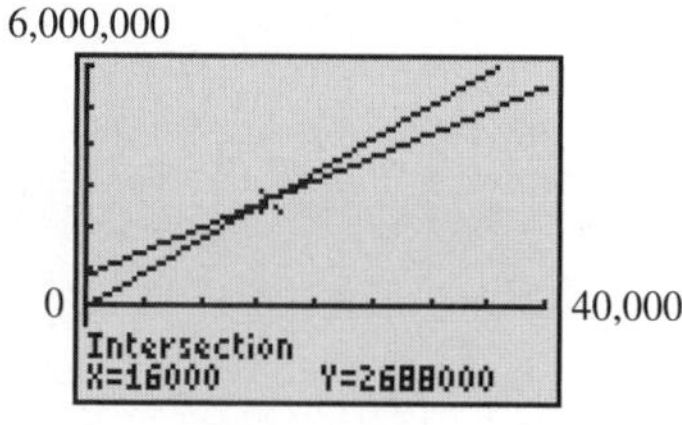

FIGURE 3.19

Supply and Demand

The supply of and demand for an item are usually related to its price. Producers will supply large numbers of the item at a high price, but consumer demand will be low. As the price of the item decreases, consumer demand increases, but producers are less willing to supply large numbers of the item. The curves that show the quantity that will be supplied at a given price and the quantity that will be demanded at a given price are called **supply and demand curves.** In supply and demand problems, we use p for price and q for quantity. We will discuss the economic concepts of supply and demand in more detail in later chapters.

EXAMPLE 6 Bill Cornett, an economist, has studied the supply and demand for aluminum siding and has determined that price per unit,* p, and the quantity demanded, q, are related by the linear equation

$$p = 60 - \frac{3}{4}q.$$

(a) Find the demand at a price of \$40 per unit.
Let $p = 40$.

$$p = 60 - \frac{3}{4}q$$

$$40 = 60 - \frac{3}{4}q \qquad \text{Let } p = 40.$$

$$-20 = -\frac{3}{4}q \qquad \text{Add } -60 \text{ on both sides.}$$

$$\frac{80}{3} = q \qquad \text{Multiply both sides by } -\frac{4}{3}.$$

At a price of \$40 per unit, 80/3 (or $26\frac{2}{3}$) units will be demanded.

(b) Find the price if the demand is 32 units.
Let $q = 32$.

$$p = 60 - \frac{3}{4}q$$

$$p = 60 - \frac{3}{4}(32) \qquad \text{Let } q = 32.$$

$$p = 60 - 24$$

$$p = 36$$

With a demand of 32 units, the price is \$36.

(c) Graph $p = 60 - \frac{3}{4}q$.

It is customary to use the horizontal axis for quantity q and the vertical axis for price p. In part (a) we saw that 80/3 units would be demanded at a price of \$40 per unit; this gives the ordered pair (80/3, 40). Part (b) shows that with a demand of 32

*An appropriate unit here might be, for example, one thousand square feet of siding.

Suppose price and demand are related by $p = 100 - 4q$.

(a) Find the price if the demand is 10 units.

(b) Find the demand if the price is $80.

(c) Write the corresponding ordered pairs.

Answers:

(a) $60

(b) 5 units

(c) (10, 60); (5, 80)

units, the price is $36, which gives the ordered pair (32, 36). Using the points (80/3, 40) and (32, 36) to get the demand graph shown in Figure 3.20. Only the portion of the graph in Quadrant I is shown, because supply and demand are meaningful only for positive values of p and q. (5)

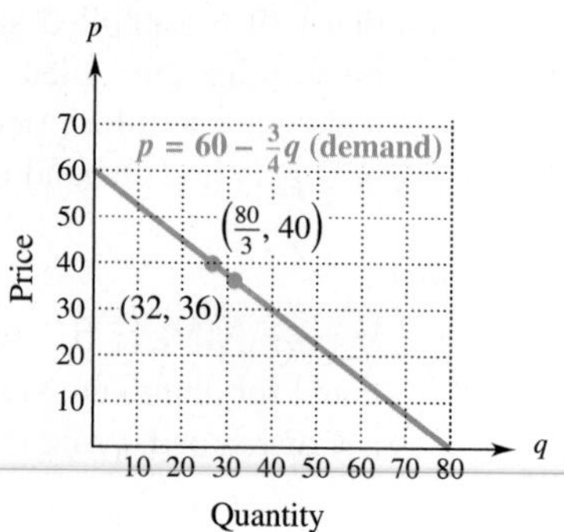

FIGURE 3.20

(d) From Figure 3.20, at a price of $30, what quantity is demanded?

Price is located on the vertical axis. Look for 30 on the p-axis and read across to where the line $p = 30$ crosses the demand graph. As the graph shows, this occurs where the demand is 40.

(e) At what price will 60 units be demanded?

Quantity is located on the horizontal axis. Find 60 on the q-axis and read up to where the vertical line $q = 60$ crosses the demand graph. This occurs where the price is about $15 per unit.

(f) What quantity is demanded at a price of $60 per unit?

The point (0, 60) on the demand graph shows that the demand is 0 at a price of $60 (that is, there is no demand at such a high price). ◀

EXAMPLE 7 Suppose the economist of Example 6 concludes that the supply q of siding is related to its price p by the equation

$$p = .85q.$$

(a) Find the supply if the price is $51 per unit.

$$51 = .85q \qquad \text{Let } p = 51.$$
$$60 = q$$

If the price is $51 per unit, then 60 units will be supplied to the marketplace.

(b) Find the price per unit if the supply is 20 units.

$$p = .85(20) = 17 \qquad \text{Let } q = 20.$$

If the supply is 20 units, then the price is $17 per unit.

(c) Graph the supply equation $p = .85q$.

As with demand, each point on the graph has quantity q as its first coordinate and the corresponding price p as its second coordinate. Part (a) shows that the ordered pair (60, 51) is on the graph of the supply equation, and part (b) shows that (20, 17) is on the graph. Using these points we obtain the supply graph in Figure 3.21.

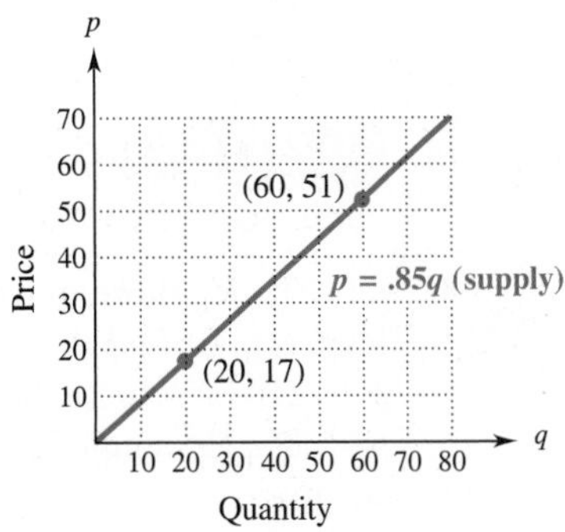

FIGURE 3.21

(d) Use the graph in Figure 3.21 to find the approximate price at which 35 units will be supplied. Then use algebra to find the exact price.

The point on the graph with first coordinate $q = 35$ is approximately (35, 30). Therefore, 35 units will be supplied when the price is approximately \$30. To determine the exact price algebraically, substitute $q = 35$ in the supply equation:

$$p = .85q = .85(35) = \$29.75$$

EXAMPLE 8 The supply and demand curves of Examples 6 and 7 are shown in Figure 3.22. Graphically determine whether there is a surplus or a shortage of supply at a price of \$40 per unit.

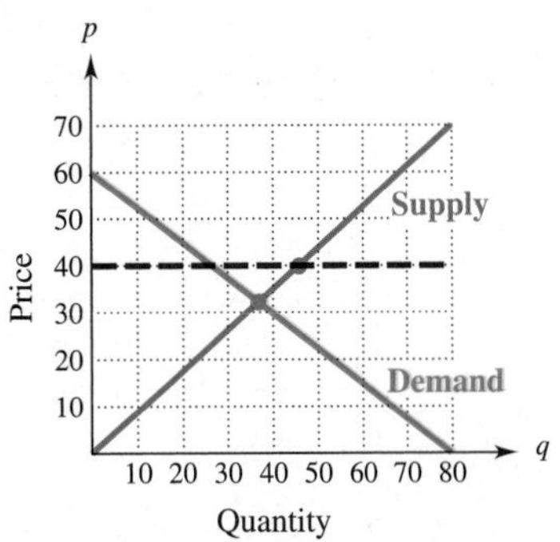

FIGURE 3.22

Find 40 on the vertical axis in Figure 3.22 and read across to the point where the horizontal line $p = 40$ crosses the supply graph (that is, the point corresponding to a price of \$40). This point lies above the demand graph, so supply is greater than demand at a price of \$40 and there is a surplus of supply.

Supply and demand are equal at the point where the supply curve intersects the demand curve. This is the **equilibrium point.** Its second coordinate is the **equilibrium price,** the price at which the same quantity will be supplied as is demanded. Its first coordinate is the quantity that will be demanded and supplied at the equilibrium price; this number is called the **equilibrium quantity.**

EXAMPLE 9 In the situation described in Examples 6–8, what is the equilibrium quantity? What is the equilibrium price?

The equilibrium point is where the supply and demand curves in Figure 3.22 intersect. To find the quantity q at which the price given by the demand equation $p = 60 - .75q$ (Example 6) is the same as that given by the supply equation $p = .85q$ (Example 7), set these two expressions for p equal and solve the resulting equation.

$$60 - .75q = .85q$$
$$60 = 1.6q$$
$$37.5 = q$$

Therefore, the equilibrium quantity is 37.5 units, the number of units for which supply will equal demand. Substituting $q = 37.5$ in either the demand or supply equation shows that

$$p = 60 - .75(37.5) = 31.875 \quad \text{or} \quad p = .85(37.5) = 31.875.$$

So the equilibrium price is \$31.875 (or \$31.88 rounded). (To avoid error, it's a good idea to substitute in both equations, as we did here, to be sure the same value of p results; if it doesn't, a mistake has been made.) In this case, the equilibrium point—the point whose coordinates are the equilibrium quantity and price—is (37.5, 31.875). ◀ 6

6

The demand for a certain commodity is related to the price by $p = 80 - (2/3)q$. The supply is related to the price by $p = (4/3)q$. Find

(a) the equilibrium quantity;

(b) the equilibrium price.

Answers:

(a) 40

(b) $160/3 \approx \$53.33$

TECHNOLOGY TIP The equilibrium point (37.5, 31.875) can be found on a graphing calculator by graphing the supply and demand curves on the same screen and using the calculator's intersection finder to locate their point of intersection.

3.3 Exercises

Business *Write a cost function for each of the following. Identify all variables used. (See Example 1.)*

1. A chain-saw rental firm charges \$12 plus \$1 per hour.
2. A trailer-hauling service charges \$45 plus \$2 per mile.
3. A parking garage charges 35¢ plus 30¢ per half hour.
4. For a 1-day rental, a car rental firm charges \$14 plus 6¢ per mile.

Business *Assume that each of the following can be expressed as a linear cost function. Find the appropriate cost function in each case. (See Example 1.)*

5. Fixed cost, \$100; 50 items cost \$1600 to produce.
6. Fixed cost, \$1000; 40 items cost \$2000 to produce.
7. Marginal cost, \$120; 100 items cost \$15,800 to produce.
8. Marginal cost, \$90; 150 items cost \$16,000 to produce.

Business *In each of the following, a cost function is given. Find the average cost per item when the required number of items is produced. (See Example 2.)*

9. $C(x) = 12x + 1800$; 50 items; 500 items; 1000 items

10. $C(x) = 50x + 12{,}000$; 100 items; 1000 items; 10,000 items

11. $C(x) = 6.5x + 9800$; 200 items; 2000 items; 5000 items

12. $C(x) = 8.75x + 16{,}500$; 1000 items; 10,000 items; 75,000 items

Business *Work these problems. (See Example 3.)*

13. The total cost (in dollars) to produce x college algebra books is $C(x) = 42.5x + 80{,}000$.
(a) What are the fixed costs?
(b) What is the marginal cost per book?
(c) What is the total cost of producing 1000 books? Of 32,000 books?
(d) What is the average cost per book when 1000 are produced? When 32,000 are produced?

14. The total cost (in dollars) to produce x compact disks is $C(x) = 6.80x + 450{,}000$.
(a) What are the fixed costs?
(b) What is the marginal cost per disk?
(c) What is the total cost of producing 50,000 disks? Of 600,000 disks?
(d) What is the average cost per disk when 50,000 are produced? When 500,000 are produced?

15. The manager of a restaurant found that the cost to produce 100 cups of coffee is \$11.02, while the cost to produce 400 cups is \$40.12. Assume the cost $C(x)$ is a linear function of x, the number of cups produced.
(a) Find a formula for $C(x)$.
(b) Find the total cost of producing 1000 cups.
(c) Find the total cost of producing 1001 cups.
(d) Find the marginal cost of the 1001st cup.
(e) What is the marginal cost of *any* cup?

16. In deciding whether to set up a new manufacturing plant, company analysts have decided that a linear function is a reasonable estimation for the total cost $C(x)$ in dollars to produce x items. They estimate the cost to produce 10,000 items as \$547,500 and the cost to produce 50,000 items as \$737,500.
(a) Find formula for $C(x)$.
(b) Find the total cost to produce 100,000 items.
(c) Find the marginal cost of the items to be produced in this plant.

Business *Assume that each of the following has a linear cost function. Find (a) the cost function; (b) the revenue function; (c) the profit function; (d) the profit on 100 items. (See Example 4.)*

	Fixed Cost	*Marginal Cost per Item*	*Item Sells For*
17.	\$500	\$10	\$35
18.	\$180	\$11	\$20
19.	\$250	\$18	\$28
20.	\$15,000	\$30	\$80
21.	\$18,000	\$12.50	\$25

22. Business In the profit-volume chart below, EF and GH represent the profit-volume graphs of a single-product company for 1989 and 1990, respectively.*

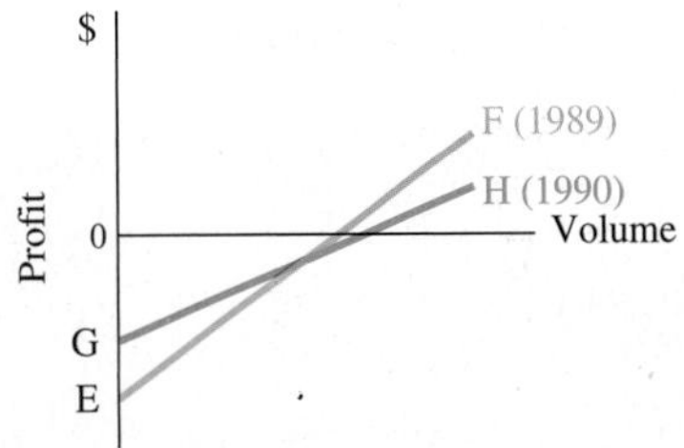

If 1989 and 1990 unit sales prices are identical, how did total fixed costs and unit variable costs of 1990 change compared to 1989? Choose one:

	1990 Total Fixed Costs	*1990 Unit Variable Costs*
(a)	Decreased	Increased
(b)	Decreased	Decreased
(c)	Increased	Increased
(d)	Increased	Decreased

Use algebra to find the intersection points of the graphs of the given equations. (See Examples 5 and 9.)

23. $2x - y = 7$ and $y = 8 - 3x$

24. $6x - y = 2$ and $y = 4x + 7$

25. $y = 3x - 7$ and $y = 7x + 4$

26. $y = 3x + 5$ and $y = 12 - 2x$

*Uniform CPA Examintion, May 1991. American Institute of Certified Public Accountants.

Business *Work the following problems. (See Example 5.)*

27. For x thousand policies, an insurance company claims that their monthly revenue in dollars is given by $R = 125x$ and their monthly cost in dollars is given by $C = 100x + 5000$.

(a) Find the break-even point.

(b) Graph the revenue and cost equations on the same axes.

(c) From the graph, estimate the revenue and cost when $x = 100$ (100 thousand policies).

28. The owners of a parking lot have determined that their weekly revenue and cost in dollars are given by $R = 80x$ and $C = 50x + 2400$, where x is the number of long-term parkers.

(a) Find the break-even point.

(b) Graph R and C on the same axes.

(c) From the graph, estimate the revenue and cost when there are 60 long-term parkers.

29. The revenue (in millions of dollars) from the sale of x units at a home supply outlet is given by $r = .21x$. The profit (in millions of dollars) from the sale of x units is given by $p = .084x - 1.5$.

(a) Find the cost equation.

(b) What is the cost of producing 7 units?

(c) What is the break-even point?

30. The profit (in millions of dollars) from the sale of x million units of Blue Glue is given by $p = .7x - 25.5$. The cost is given by $c = .9x + 25.5$.

(a) Find the revenue equation.

(b) What is the revenue from selling 10 million units?

(c) What is the break-even point?

Business *Suppose you are the manager of a firm. The accounting department has provided cost estimates and the sales department sales estimates on a new product. You must analyze the data they give you, determine what it will take to break even, and decide whether or not to go ahead with production of the new product. (See Example 5.)*

31. Cost estimate is given by $c = 80x + 7000$ and revenue estimate by $r = 95x$; no more than 400 units can be sold.

32. Cost is $c = 140x + 3000$ and revenue is $r = 125x$.

33. Cost is $c = 125x + 42{,}000$ and revenue is $r = 165.5x$; no more than 2000 units can be sold.

34. Cost is $c = 1750x + 95{,}000$ and revenue is $r = 1975x$; no more than 600 units can be sold.

35. Business The graphs show the hourly compensation (wages and benefits) in U.S. dollars of production workers in Austria and the United States since 1975.* Estimate the break-even point (the point at which workers in both countries had the same compensation).

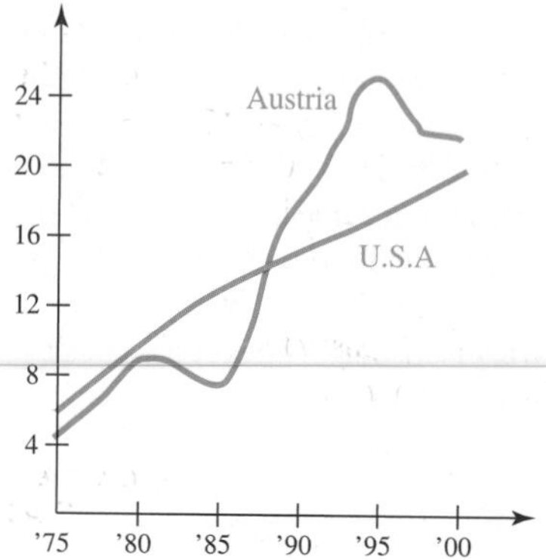

36. Business The graphs show U.S. exports and imports of goods and services (in billions of dollars) since 1965.† Estimate the break-even point (where the values of exports and imports were the same).

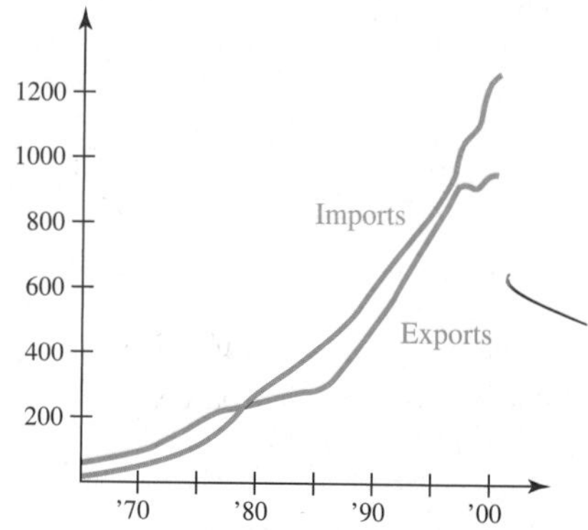

37. Business Direct U.S. investments in Ireland and Italy (in billions of dollars) in 1990 and 1999 are shown in the chart.‡

	1990	1999
Ireland	6.88	19.82
Italy	13.12	17.60

*U.S. Bureau of Labor Statistics.

†U.S. Bureau of Economic Analysis.

‡U.S. Department of Commerce.

(a). Let $x = 0$ correspond to 1990. Assume the change in investment in each case is linear and find linear functions that give the investment in year x for each country. Round all coefficients to two decimal places.
(b) Graph the functions from part (a) on the same coordinate axes.
(c) Find the intersection point of the graphs and interpret your answer.

38. Social Science The approximate median family income (in thousands of dollars) for white families in the United States is given by $f(x) = .15x + 37$, where x is the number of years since 1980. The approximate median family income for black families is given by $g(x) = .2x + 21$.
(a) Graph both functions on the same coordinate axes for $0 \leq x \leq 30$.
(b) Do the graphs intersect? What conclusions do you infer from this?

Business *Use the supply and demand curves graphed below to answer Exercises 39–42. (See Examples 6–9.)*

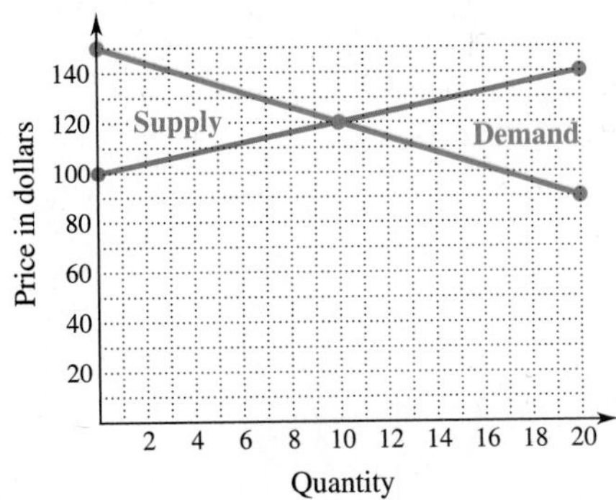

39. At what price are 20 items supplied?

40. At what price are 20 items demanded?

41. Find the equilibrium quantity.

42. Find the equilibrium price.

Business *Work the following exercises. (See Examples 6–9.)*

43. Suppose that the demand and price for a certain brand of shampoo are related by

$$p = 16 - \frac{5}{4}q,$$

where p is price, in dollars, and q is demand. Find the price for a demand of
(a) 0 units; **(b)** 4 units; **(c)** 8 units.

Find the demand for the shampoo at a price of
(d) \$6; **(e)** \$11; **(f)** \$16.
(g) Graph $p = 16 - (5/4)q$. Suppose the price and supply of the shampoo are related by

$$p = \frac{3}{4}q,$$

where q represents the supply and p the price. Find the supply when the price is
(h) \$0; **(i)** \$10; **(j)** \$20.
(k) Graph $p = (3/4)q$ on the same axes used for part (g).
(l) Find the equilibrium quantity.
(m) Find the equilibrium price.

44. Let the supply and demand for radial tires in dollars be given by

$$\text{supply: } p = \frac{3}{2}q \quad \text{and} \quad \text{demand: } p = 81 - \frac{3}{4}q.$$

(a) Graph these on the same axes.
(b) Find the equilibrium quantity.
(c) Find the equilibrium price.

45. Let the supply and demand for bananas in cents per pound be given by

$$\text{supply: } p = \frac{2}{5}q \quad \text{and} \quad \text{demand: } p = 100 - \frac{2}{5}q.$$

(a) Graph these on the same axes.
(b) Find the equilibrium quantity.
(c) Find the equilibrium price.
(d) On what interval does demand exceed supply?

46. Let the supply and demand for sugar be given by

$$\text{supply: } p = 1.4q - .6$$

and

$$\text{demand: } p = -2q + 3.2,$$

where p is in dollars.
(a) Graph these on the same axes.
(b) Find the equilibrium quantity.
(c) Find the equilibrium price.
(d) On what interval does supply exceed demand?

47. Explain why the graph of the (total) cost function is always above the x-axis and can never move downward as you go from left to right. Is the same thing true of the graph of the average cost function?

48. Explain why the graph of the profit function may rise or fall (as you go from left to right) and may be below the x-axis.

3.4 QUADRATIC FUNCTIONS

A **quadratic function** is a function whose rule is given by a quadratic polynomial, such as

$$f(x) = x^2, \qquad g(x) = 3x^2 + 30x + 67, \qquad \text{and} \quad h(x) = -x^2 + 4x.$$

Thus, a quadratic function is one whose rule can be written in the form

$$\boldsymbol{f(x) = ax^2 + bx + c}$$

for some constants a, b, c, with $a \neq 0$.

EXAMPLE 1 Graph each of these quadratic functions.

(a) $f(x) = x^2$

Choose several negative, zero, and positive values of x, find the values of $f(x)$, and plot the corresponding points. Connecting these points with a smooth curve, we obtain Figure 3.23 below.

1

Graph each quadratic function.

(a) $f(x) = x^2 - 4$

(b) $f(x) = -(x - 3)^2$

Answers:

(a)

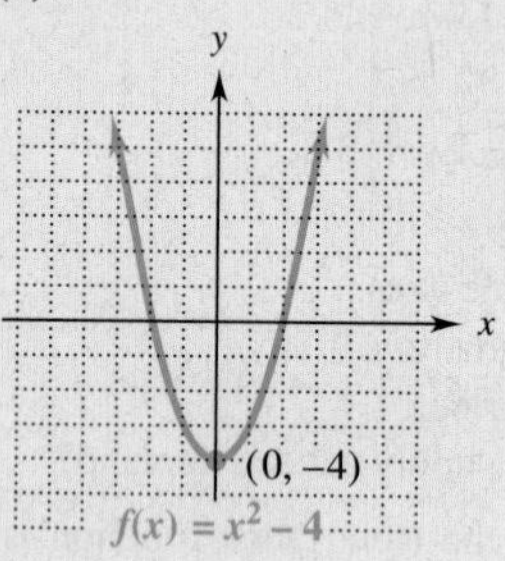

(b)

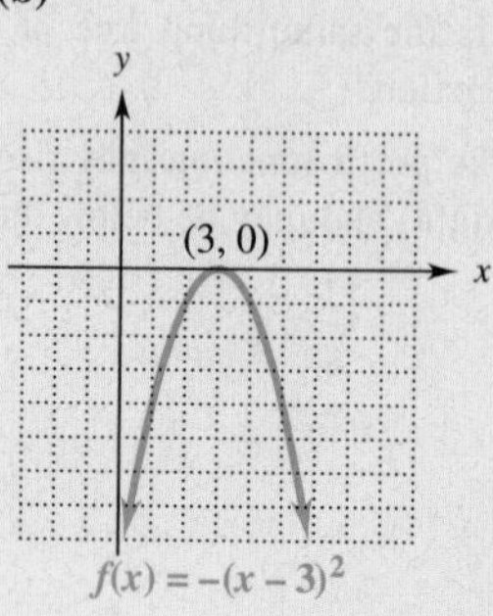

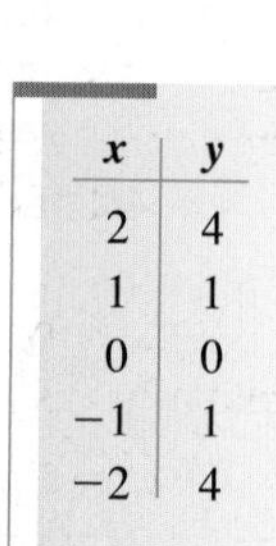

x	y
2	4
1	1
0	0
−1	1
−2	4

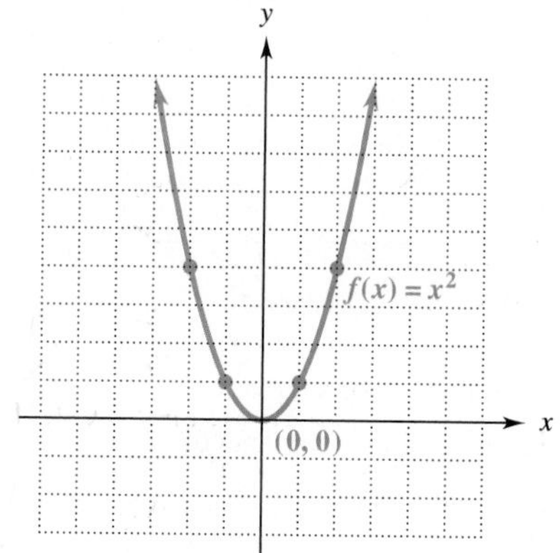

FIGURE 3.23

(b) $h(x) = -(x + 2)^2$

Since $h(x) = -(x + 2)^2 = -(x^2 + 4x + 4) = -x^2 - 4x - 4$, we see that h actually is a quadratic function. When $x = -2$, then

$$h(x) = h(-2) = -(-2 + 2)^2 = 0.$$

Therefore $(-2, 0)$ is on the graph. When $x \neq -2$, then $x + 2 \neq 0$, so that $(x + 2)^2$ is positive, and hence, $h(x) = -(x + 2)^2$ is negative. Thus, the graph lies below the x-axis whenever $x \neq -2$. Using these facts and plotting some points lead to Figure 3.24. ◀ 1

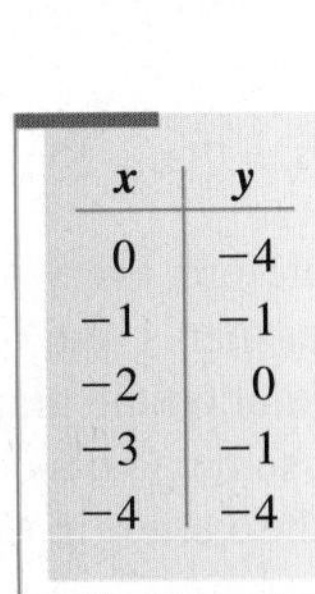

x	y
0	−4
−1	−1
−2	0
−3	−1
−4	−4

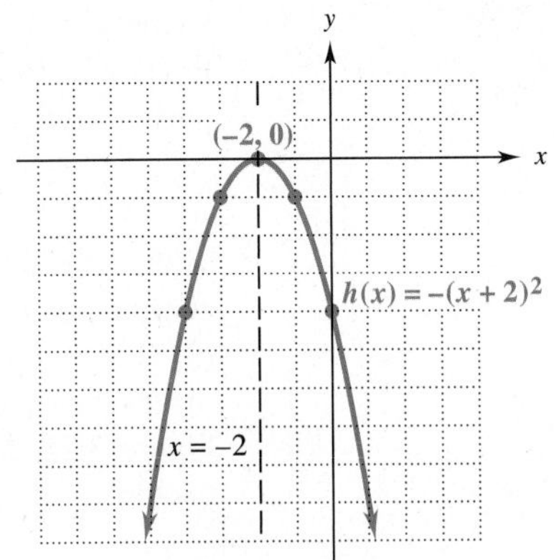

FIGURE 3.24

The curves in Figures 3.23 and 3.24 and in problem 1 in the margin are called **parabolas.** It can be shown that the graph of every quadratic function is a parabola. All parabolas have the same basic "cup" shape, though the cup may be broad or narrow. The parabola opens upward when the coefficient of the x^2-term is positive (as in Figure 3.23) and downward when this coefficient is negative (as in Figure 3.24).

Parabolas have many useful properties. Cross sections of radar dishes and spotlights form parabolas. Disks often visible on the sidelines of televised football games are microphones having reflectors with parabolic cross sections. These microphones are used by the television networks to pick up the shouted signals of the quarterbacks.

When a parabola opens upward (as in Figure 3.23), its lowest point is called the **vertex.** When a parabola opens downward (as in Figure 3.24), its highest point is called the **vertex.** The vertical line through the vertex of a parabola is called the **axis of the parabola.** For example, (0, 0) is the vertex of the parabola in Figure 3.23 and its axis is the y-axis. If you were to fold this graph along its axis, the two halves of the parabola would match exactly. This means that a parabola is *symmetric* about its axis.

The vertex of a parabola can be roughly approximated by using the trace feature or accurately approximated by using the minimum or maximum finder on a graphing calculator. However, there are algebraic techniques for finding the vertex precisely.

EXAMPLE 2 The function $g(x) = 2(x - 3)^2 + 1$ is quadratic because its rule can be written in the required form:

$$g(x) = 2(x - 3)^2 + 1 = 2(x^2 - 6x + 9) + 1 = 2x^2 - 12x + 19.$$

Its graph will be an upward-opening parabola. Note that

$$g(3) = 2(3 - 3)^2 + 1 = 1,$$

so that (3, 1) is on the graph. We claim that (3, 1) is the vertex—the lowest point on the graph. To see why this is true, note that when $x \neq 3$, the quantity $2(x - 3)^2$ is positive and hence

$$g(x) = 2(x - 3)^2 + 1 = (\text{a positive number}) + 1,$$

so that $g(x) > 1$. Therefore, every point $(x, g(x))$ on the graph with $x \neq 3$ has second coordinate $g(x)$ larger than 1 and hence lies *above* the point (3, 1). In other words, (3, 1) is the lowest point on the graph, the vertex of the parabola. Knowing this fact makes it easy to plot some points on both sides of the vertex and obtain the graph in Figure 3.25. Since the vertex is (3, 1), the vertical line $x = 3$ is the axis of the parabola.

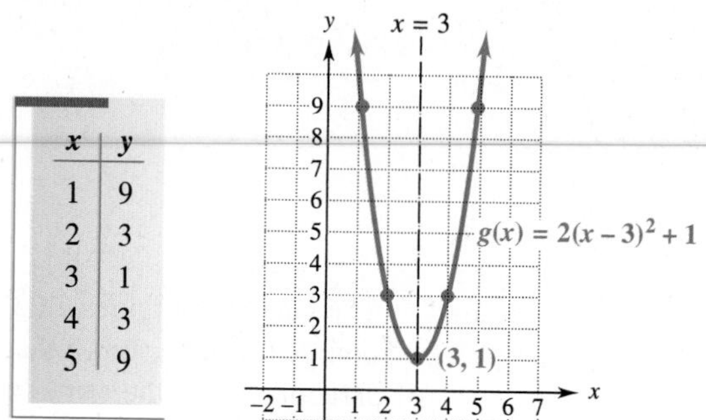

x	y
1	9
2	3
3	1
4	3
5	9

FIGURE 3.25

Example 2 illustrates some of the following facts, whose proofs are omitted.

If f is a quadratic function defined by $y = a(x - h)^2 + k$, then the graph of the function f is a parabola having its vertex at (h, k) and axis of symmetry $x = h$.

If $a > 0$, the parabola opens upward; if $a < 0$, it opens downward.

If $0 < |a| < 1$, the parabola is "broader" than $y = x^2$, while if $|a| > 1$, the parabola is "narrower" than $y = x^2$.

EXAMPLE 3 Determine algebraically whether the parabola opens upward or downward and find its vertex.

(a) $f(x) = -3(x - 4)^2 - 7$

The rule of the function is in the form $f(x) = a(x - h)^2 + k$ (with $a = -3$, $h = 4$, and $k = -7$). The parabola opens downward ($a < 0$) and its vertex is $(h, k) = (4, -7)$.

(b) $g(x) = 2(x + 3)^2 + 5$

Be careful here; the vertex is *not* (3, 5). To put the rule of $g(x)$ in the form $a(x - h)^2 + k$, we must rewrite it so that there is a minus sign inside the parentheses:

$$\begin{aligned} g(x) &= 2(x + 3)^2 + 5 \\ &= 2(x - (-3))^2 + 5. \end{aligned}$$

2

Determine the vertex of each parabola algebraically and graph it.

(a) $f(x) = (x + 4)^2 - 3$

(b) $f(x) = -2(x - 3)^2 + 1$

Answers:

(a)

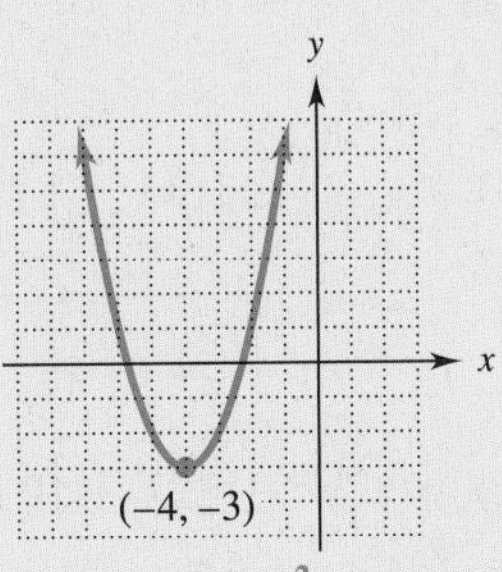

$f(x) = (x + 4)^2 - 3$

(b)

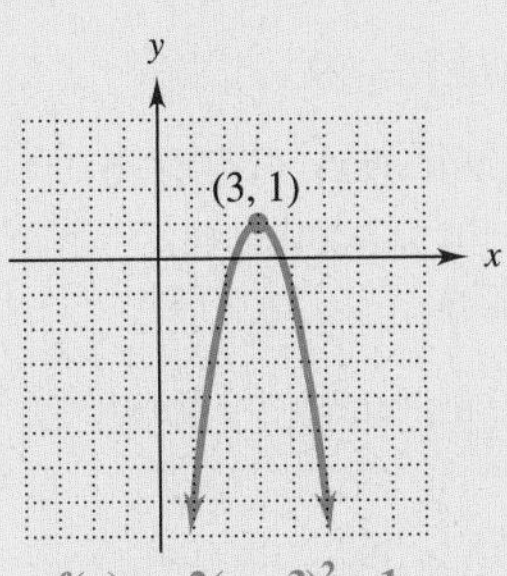

$f(x) = -2(x - 3)^2 + 1$

This is the required form with $a = 2$, $h = -3$, and $k = 5$. The parabola opens upward and its vertex is $(-3, 5)$. ◀ 2

EXAMPLE 4 Find the rule of a quadratic function whose graph has vertex (3, 4) and passes through the point (6, 22).

The graph of $f(x) = a(x - h)^2 + k$ has vertex (h, k). We want $h = 3$ and $k = 4$, so that $f(x) = a(x - 3)^2 + 4$. Since (6, 22) is on the graph, we must have $f(6) = 22$. Therefore

$$\begin{aligned} f(x) &= a(x - 3)^2 + 4 \\ f(6) &= a(6 - 3)^2 + 4 \\ 22 &= a(3)^2 + 4 \\ 9a &= 18 \\ a &= 2. \end{aligned}$$

Thus, the graph of $f(x) = 2(x - 3)^2 + 4$ is a parabola with vertex (3, 4) that passes through (6, 22). ◀

The vertex of each parabola in Examples 2 and 3 was easily determined because the rule of the function had the form

$$f(x) = a(x - h)^2 + k.$$

The rule of *any* quadratic function can be put in this form by using the technique of completing the square, which was discussed in Section 1.7.

EXAMPLE 5 By completing the square, determine the vertex of the graph of $f(x) = x^2 - 2x + 3$. Then graph the parabola.

In order to get the equation in the form $f(x) = a(x - h)^2 + k$, we first write $f(x) = x^2 - 2x + 3$ as

$$f(x) = (x^2 - 2x \quad\quad) + 3.$$

Now complete the square for the expression in parentheses. Take half the coefficient of x, namely, $(\frac{1}{2})(-2) = -1$, and square the result: $(-1)^2 = 1$. In order to complete the square we must add 1 inside the parentheses, but in order not to change the rule of the function we must also subtract 1:

$$f(x) = (x^2 - 2x + 1 - 1) + 3.$$

By the associative property,

$$f(x) = (x^2 - 2x + 1) + (-1 + 3).$$

Factor $x^2 - 2x + 1$ as $(x - 1)^2$, to get

$$f(x) = (x - 1)^2 + 2.$$

With the rule in this form, we can see that the graph is an upward-opening parabola with vertex (1, 2), as shown in Figure 3.26 on the next page. ◀ 3

3

Rewrite the rule of each function by completing the square and use this form to find the vertex of the graph.

(a) $f(x) = x^2 - 6x + 11$

(b) $g(x) = x^2 + 8x + 18$

Answers:

(a) $f(x) = (x - 3)^2 + 2$; (3, 2)

(b) $g(x) = (x + 4)^2 + 2$; (−4, 2)

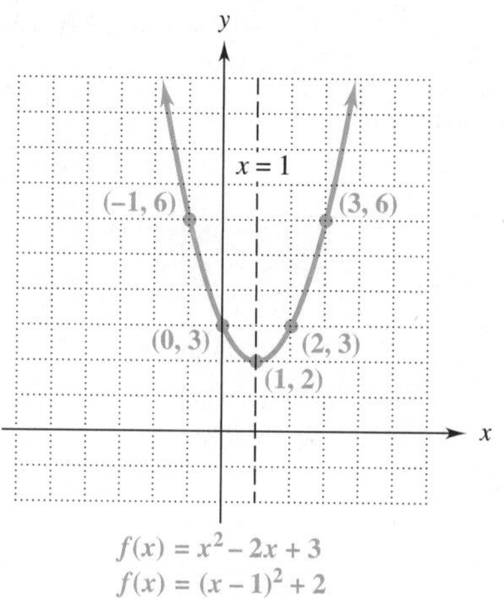

FIGURE 3.26

4

Complete the square, find the vertex, and graph each function.

(a) $f(x) = 3x^2 - 12x + 14$

(b) $f(x) = -x^2 + 6x - 12$

Answers:

(a) $f(x) = 3(x - 2)^2 + 2$

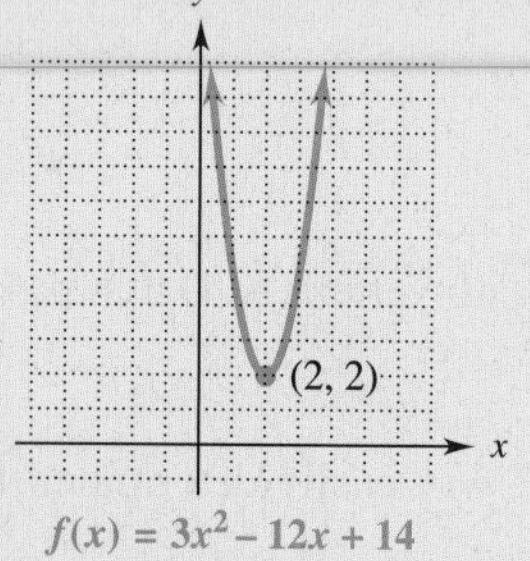

(b) $f(x) = -(x - 3)^2 - 3$

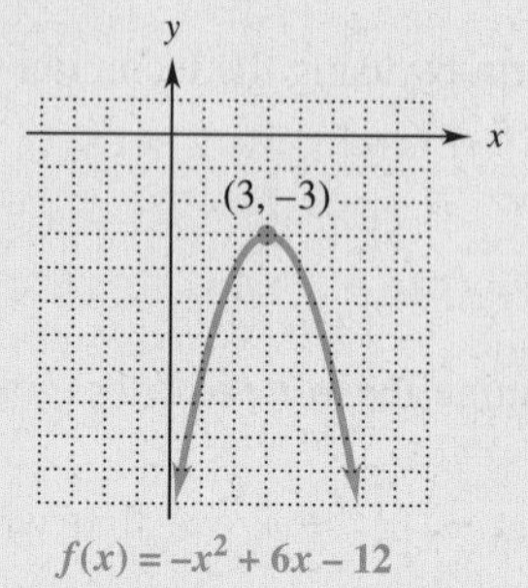

EXAMPLE 6 Find the vertex of the graph of $f(x) = -2x^2 + 12x - 19$ and graph the parabola.

Get $f(x)$ in the form $f(x) = a(x - h)^2 + k$ by first factoring -2 from the expression $-2x^2 + 12x$.

$$f(x) = -2x^2 + 12x - 19 = -2(x^2 - 6x) - 19$$

Working inside the parentheses, take half of -6 (coefficient of x): $(1/2)(-6) = -3$. Square this result: $(-3)^2 = 9$. Add and subtract 9 inside the parentheses.

$$y = -2(x^2 - 6x + 9 - 9) - 19$$

$$y = -2(x^2 - 6x + 9) + (-2)(-9) - 19 \qquad \text{Distributive property}$$

Simplify and factor to get

$$y = -2(x - 3)^2 - 1.$$

Since $a = -2$ is negative, the parabola opens downward. The vertex is at $(3, -1)$, and the parabola is narrower than $y = x^2$. Use these results and plot additional ordered pairs as needed to get the graph in Figure 3.27. ◀ 4

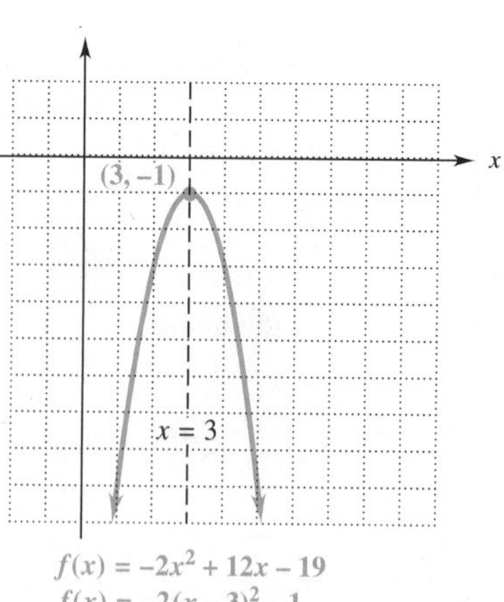

FIGURE 3.27

The technique of completing the square can be used to rewrite the general equation $f(x) = ax^2 + bx + c$ in the form $f(x) = a(x - h)^2 + k$, as is shown in Exercise 34. When this is done we obtain a formula for the coordinates of the vertex.

> The graph of $f(x) = ax^2 + bx + c$ is a parabola with vertex (h, k), where
>
> $$h = \frac{-b}{2a} \quad \text{and} \quad k = f(h).$$

EXAMPLE 7 Find the vertex, the axis, and the x- and y-intercepts of the graph of $f(x) = x^2 - x - 6$.

Since $a = 1$ and $b = -1$, the x-value of the vertex is

$$\frac{-b}{2a} = \frac{-(-1)}{2 \cdot 1} = \frac{1}{2}.$$

The y-value of the vertex is

$$f\left(\frac{1}{2}\right) = \left(\frac{1}{2}\right)^2 - \frac{1}{2} - 6 = -\frac{25}{4}.$$

The vertex is $(1/2, -25/4)$ and the axis of the parabola is $x = 1/2$, as shown in Figure 3.28. The intercepts are found by setting each variable equal to 0.

x-intercepts

Set $f(x) = y = 0$:

$$0 = x^2 - x - 6$$
$$0 = (x + 2)(x - 3)$$
$$x + 2 = 0 \quad \text{or} \quad x - 3 = 0$$
$$x = -2 \qquad\qquad x = 3$$

The x-intercepts are -2 and 3. ◀ (5)

y-intercept

Set $x = 0$:

$$f(x) = y = 0^2 - 0 - 6 = -6$$

The y-intercept is -6.

(5)

Find the vertex of the graph.

(a) $f(x) = 2x^2 - 7x + 12$

(b) $k(x) = -.4x^2 + 1.6x + 5$

Answers:

(a) (7/4, 47/8)

(b) (2, 6.6)

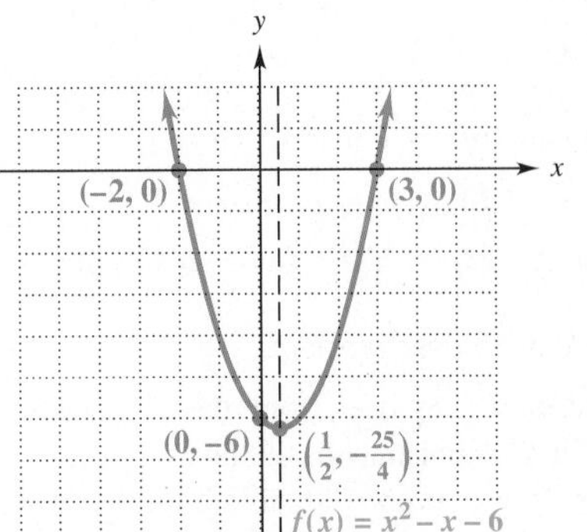

FIGURE 3.28

TECHNOLOGY TIP The maximum or minimum finder on a graphing calculator can approximate the vertex of a parabola with a high degree of accuracy. The max/min finder is in the CALC menu or in the MATH or FCN submenu of the GRAPH menu. Similarly, the calculator's graphical root finder can approximate the x-intercepts of a parabola.

3.4 Exercises

Without graphing, determine the vertex of each of the following parabolas and state whether it opens upward or downward. (See Examples 2, 3, 5–7.)

1. $f(x) = 3(x - 5)^2 + 2$ **2.** $g(x) = -6(x - 2)^2 - 5$

3. $f(x) = x^2 + 12x + 1$ **4.** $g(x) = x^2 - 10x + 3$

5. $f(x) = 2x^2 + 4x + 1$ **6.** $g(x) = -3x^2 + 6x + 5$

Without graphing, determine the x- and y-intercepts of each of the following parabolas. (See Example 7.)

7. $f(x) = 3(x - 2)^2 - 3$ **8.** $f(x) = x^2 - 4x - 1$

9. $g(x) = 2x^2 + 8x + 6$ **10.** $g(x) = x^2 - 10x + 20$

Find the rule of a quadratic function whose graph has the given vertex and passes through the given point. (See Example 4.)

11. Vertex (1, 2); point (5, 6)

12. Vertex (−3, 2); point (2, 1)

13. Vertex (−1, −2); point (1, 2)

14. Vertex (2, −4); point (5, 2)

Graph each parabola. Find the vertex and axis of symmetry of each. (See Examples 1–7.)

15. $f(x) = (x + 2)^2$ **16.** $f(x) = -(x + 5)^2$

17. $f(x) = (x - 1)^2 - 3$ **18.** $f(x) = (x - 2)^2 + 1$

19. $f(x) = x^2 - 4x + 6$ **20.** $f(x) = x^2 + 6x + 3$

21. $f(x) = 2x^2 - 4x + 5$ **22.** $f(x) = -3x^2 + 24x - 46$

Use a graphing calculator to work these problems.

23. **Transportation** According to data from the National Highway Traffic Safety Administration, the fatal accident rate per 1000 licensed drivers every 100 million miles can be approximated by the function $f(x) = .0031x^2 - .291x + 7.1$, where x is the age of the driver ($16 \le x \le 88$). At what age is the rate the lowest?

24. **Social Science** According to data from the U.S. Census Bureau, the population of Cleveland, Ohio (in thousands), in year x can be approximated by $g(x) = -.172x^2 + 16.82x + 487.62$, where $x = 0$ corresponds to 1900. In what year did Cleveland have its largest population?

25. **Social Science** Between 1992 and 1998, the percent of college freshmen who planned eventually to get some type of medical degree can be approximated by $f(x) = -.2369x^2 + 1.425x + 6.905$, where $x = 0$ corresponds to 1992. In what year did this percentage reach a maximum?

26. **Transportation** The money spent per year during the 1990s for research and development of "smart" highways is approximated by the function $g(x) = -19.321x^2 + 3608.7x - 168{,}310$, where $x = 0$ corresponds to 1900 and $g(x)$ is in millions of dollars. In what year was the maximum amount spent? What was that maximum amount?

In Exercises 27–32, graph the functions in parts (a)–(d) on the same set of axes; then answer part (e).

27. **(a)** $k(x) = x^2$
(b) $f(x) = 2x^2$
(c) $g(x) = 3x^2$
(d) $h(x) = 3.5x^2$
(e) Explain how the coefficient a in the function given by $f(x) = ax^2$ affects the shape of the graph, in comparison with the graph of $k(x) = x^2$, when $a > 1$.

28. **(a)** $k(x) = x^2$
(b) $f(x) = .8x^2$
(c) $g(x) = .5x^2$
(d) $h(x) = .3x^2$
(e) Explain how the coefficient a in the function given by $f(x) = ax^2$ affects the shape of the graph, in comparison with the graph of $k(x) = x^2$, when $0 < a < 1$.

29. (a) $k(x) = -x^2$
(b) $f(x) = -2x^2$
(c) $g(x) = -3x^2$
(d) $h(x) = -3.5x^2$
(e) Compare these graphs to the ones in Exercise 27 and explain how changing the sign of the coefficient a in the function given by $f(x) = ax^2$ affects the graph.

30. (a) $k(x) = x^2$
(b) $f(x) = x^2 + 2$
(c) $g(x) = x^2 + 3$
(d) $h(x) = x^2 + 5$
(e) Explain how the graph of $f(x) = x^2 + c$ (where c is a positive constant) can be obtained from the graph of $k(x) = x^2$.

31. (a) $k(x) = x^2$
(b) $f(x) = x^2 - 1$
(c) $g(x) = x^2 - 2$
(d) $h(x) = x^2 - 4$
(e) Explain how the graph of $f(x) = x^2 - c$ (where c is a positive constant) can be obtained from the graph of $k(x) = x^2$.

32. (a) $k(x) = x^2$
(b) $f(x) = (x + 2)^2$
(c) $g(x) = (x - 2)^2$
(d) $h(x) = (x - 4)^2$
(e) Explain how the graph of $f(x) = (x + c)^2$ and $f(x) = (x - c)^2$ (where c is a positive constant) can be obtained from the graph of $k(x) = x^2$.

33. Find the rule of a quadratic function whose graph is a parabola with vertex (0, 0) that includes the point (2, 12).

34. Verify that the right side of the equation

$$ax^2 + bx + c = a\left(x - \left(\frac{-b}{2a}\right)\right)^2 + \left(c - \frac{b^2}{4a}\right)$$

equals the left side. Since the right side of the equation is in the form $a(x - h)^2 + k$, we conclude that the vertex of the parabola $f(x) = ax^2 + bx + c$ has x-coordinate $h = -b/2a$.

3.5 APPLICATIONS OF QUADRATIC FUNCTIONS

The fact that the vertex of a parabola $y = ax^2 + bx + c$ is the highest or lowest point on the graph can be used in applications to find a maximum or a minimum value.

EXAMPLE 1 Anne Kelly owns and operates Aunt Emma's Blueberry Pies. She has hired a consultant to analyze her business operations. The consultant tells her that her profits, $P(x)$, from the sale of x cases of pies, are given by

$$P(x) = 120x - x^2.$$

How many cases of pies should she sell in order to maximize profit? What is the maximum profit?

The profit function can be rewritten as $P(x) = -x^2 + 120x$. Its graph is a downward-opening parabola (see Figure 3.29 on the next page). Its vertex, which may be found as in Section 3.4, is (60, 3600). For each point on the graph,

the x-coordinate is the number of cases;

the y-coordinate is the profit on that number of cases.

Only the portion of the graph in Quadrant I (where both coordinates are positive) is relevant here because she cannot sell a negative number of pies and she is not interested in a negative profit. *Maximum* profit occurs at the point with the largest y-coordinate, namely, the vertex, as shown in Figure 3.29. The maximum profit of \$3600 is obtained when 60 cases of pies are sold. ◀ 1

When a company sells x units of a product, its profit is $P(x) = -2x^2 + 40x + 280.$ Find

(a) the number of units that should be sold so that maximum profit is received;

(b) the maximum profit.

Answers:

(a) 10 units

(b) \$480

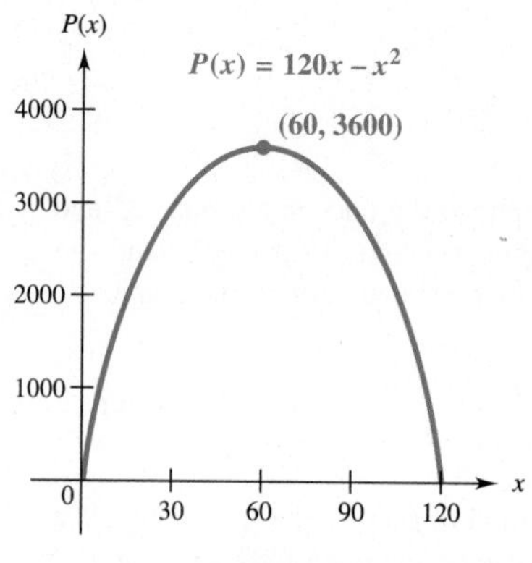

FIGURE 3.29

Supply and demand curves were introduced in Section 3.3. Here is a quadratic example.

EXAMPLE 2 Suppose that the price and demand for an item are related by

$$p = 150 - 6q^2, \qquad \text{Demand function}$$

where p is the price (in dollars) and q is the number of items demanded (in hundreds). The price and supply are related by

$$p = 10q^2 + 2q, \qquad \text{Supply function}$$

where q is the number of items supplied (in hundreds). Find the equilibrium quantity and the equilibrium price.

The graphs of both of these equations are parabolas (Figure 3.30). Only those portions of the graphs that lie in the first quadrant are included, because neither supply, demand, nor price can be negative.

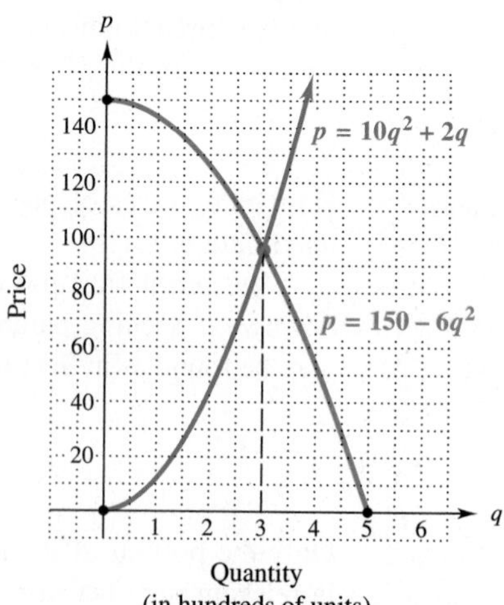

FIGURE 3.30

The point where the demand and supply curves intersect is the equilibrium point. Its first coordinate is the equilibrium quantity, and its second coordinate is the equilibrium price. These coordinates may be found in two ways.

Algebraic Method At the equilibrium point, the second coordinate of the demand curve must be the same as the second coordinate of the supply curve so that

$$150 - 6q^2 = 10q^2 + 2q.$$

Write this quadratic equation in standard form as follows.

$$0 = 16q^2 + 2q - 150 \qquad \text{Add } -150 \text{ and } 6q^2 \text{ to both sides.}$$

$$0 = 8q^2 + q - 75 \qquad \text{Multiply both sides by } \frac{1}{2}.$$

This equation can be solved by the quadratic formula given in Section 1.7. Here $a = 8$, $b = 1$, and $c = -75$.

$$q = \frac{-1 \pm \sqrt{1 - 4(8)(-75)}}{2(8)}$$

$$= \frac{-1 \pm \sqrt{1 + 2400}}{16} \qquad -4(8)(-75) = 2400$$

$$= \frac{-1 \pm 49}{16} \qquad \sqrt{1 + 2400} = \sqrt{2401} = 49$$

$$q = \frac{-1 + 49}{16} = \frac{48}{16} = 3 \quad \text{or} \quad q = \frac{-1 - 49}{16} = -\frac{50}{16} = -\frac{25}{8}$$

(2)

> The price and demand for an item are related by $p = 32 - x^2$, while price and supply are related by $p = x^2$. Find
>
> (a) the equilibrium quantity;
>
> (b) the equilibrium price.
>
> Answers:
>
> (a) 4
>
> (b) 16

It is not possible to make $-25/8$ units, so discard that answer and use only $q = 3$. Hence the equilibrium quantity is 300. Find the equilibrium price by substituting 3 for q in either the supply or the demand function (and check your answer by using the other one). Using the supply function gives

$$p = 10q^2 + 2q$$

$$p = 10 \cdot 3^2 + 2 \cdot 3 \qquad \text{Let } q = 3.$$

$$= 10 \cdot 9 + 6$$

$$p = 96.$$

Graphical Method Graph the two functions on a graphing calculator and use the intersection finder to determine that the equilibrium point is (3, 96), as in Figure 3.31. ◀ (2)

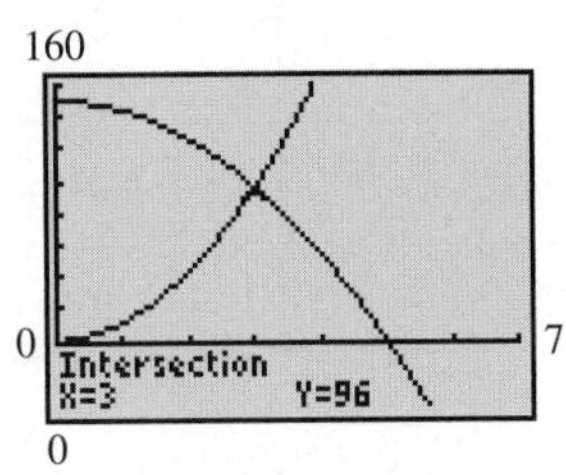

FIGURE 3.31

EXAMPLE 3 The rental manager of a small apartment complex with 16 units has found from experience that each \$40 increase in the monthly rent results in an empty apartment. All 16 apartments will be rented at a monthly rent of \$500. How many \$40 increases will produce maximum monthly income for the complex?

Let x represent the number of \$40 increases. Then the number of apartments rented will be $16 - x$. Also, the monthly rent per apartment will be $500 + 40x$. (There are x increases of \$40 for a total increase of $40x$.) The monthly income, $I(x)$, is given by the number of apartments rented times the rent per apartment, so

$$\begin{aligned} I(x) &= (16 - x)(500 + 40x) \\ &= 8000 + 640x - 500x - 40x^2 \\ &= 8000 + 140x - 40x^2. \end{aligned}$$

Since x represents the number of \$40 increases and each \$40 increase causes one empty apartment, x must be a whole number. Since there are only 16 apartments, $0 \leq x \leq 16$. Because there is a small number of possibilities, the value of x that produces maximum income may be found in several ways.

Brute Force Method Use a scientific calculator or the table feature of a graphing calculator (as in Figure 3.32) to evaluate $I(x)$ when $x = 1, 2, \ldots, 16$ and find the largest value.

X	Y_1
0	8000
1	8100
2	8120
3	8060
4	7920
5	7700
6	7400

X=0

X	Y_1
7	7020
8	6560
9	6020
10	5400
11	4700
12	3920
13	3060

X=7

X	Y_1
14	2120
15	1100
16	0
17	-1180
18	-2440
19	-3780
20	-5200

X=14

FIGURE 3.32

The tables show that a maximum income of \$8120 occurs when $x = 2$. So the manager should charge a rent of $500 + 2(40) = \$580$, leaving two apartments vacant.

Algebraic Method The graph of $I(x) = 8000 + 140x - 40x^2$ is a downward-opening parabola (why?) and the value of x that produces maximum income occurs at the vertex. The methods of Section 3.4 show that the vertex is (1.75, 8122.50). Since x must be a whole number, evaluate $I(x)$ at $x = 1$ and $x = 2$ to see which one gives the best result.

$$\text{If } x = 1, \text{ then } I(1) = -40(1)^2 + 140(1) + 8000 = 8100.$$
$$\text{If } x = 2, \text{ then } I(2) = -40(2)^2 + 140(2) + 8000 = 8120.$$

So maximum income occurs when $x = 2$. The manager should charge a rent of $500 + 2(40) = \$580$ and leave two apartments vacant.

Quadratic Models Real-world data can sometimes be used to construct a quadratic function that approximates the data. Such **quadratic models** can then be used (subject to limitations) to predict future behavior.

EXAMPLE 4 The number of women employed full-time in civilian jobs has increased dramatically in the past century. The table shows the number of employed women (in millions) in selected years.*

Year	1910	1920	1930	1940	1950	1960	1970	1980	1990	2000
Working Women	7.4	8.6	10.8	12.8	18.6	23.6	31.4	45.3	56.7	65.6

(a) Display this information graphically.

Let $x = 0$ correspond to 1900. Plot the points given by the table: (10, 7.4), (20, 8.6), and so on, as in Figure 3.33.

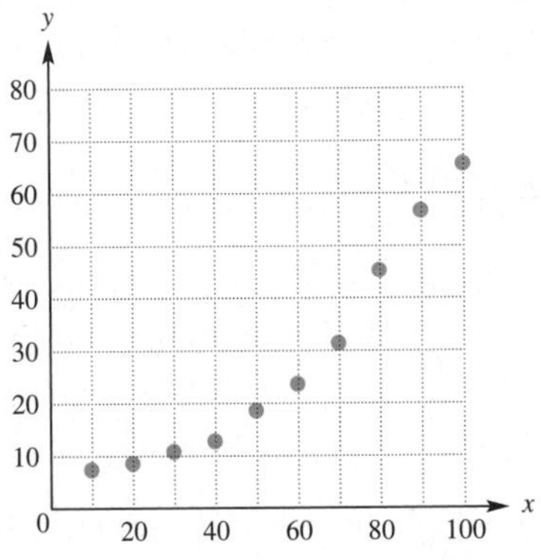

FIGURE 3.33

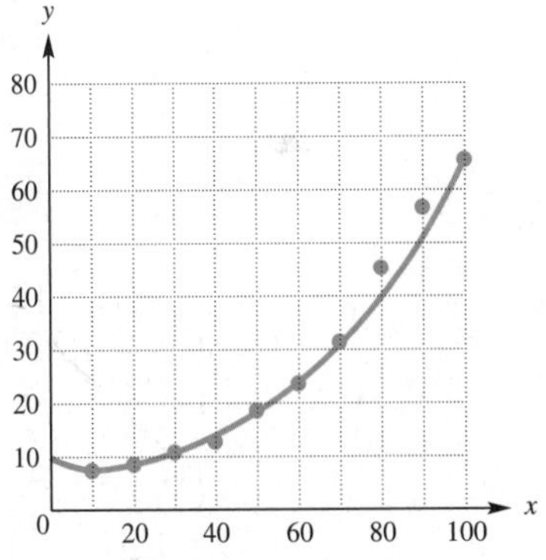

FIGURE 3.34

(b) The shape of the data points in Figure 3.33 resembles the right half of a parabola. Find a quadratic model $f(x) = a(x - h)^2 + k$ for this data.

Recall that when a quadratic function is written in this form, the vertex of its graph is (h, k). Based on Figure 3.33, let (10, 7.4) be the vertex so that

$$f(x) = a(x - 10)^2 + 7.4.$$

To find a, choose another data point, say, (100, 65.6). If this point is on the parabola, then $f(100) = 65.6$, so that

$$\begin{aligned} f(x) &= a(x - 10)^2 + 7.4 \\ 65.5 &= a(100 - 10)^2 + 7.4 && \text{Substitute 100 for } x \text{ and 65.6 for } f(x) \\ 65.6 &= 8100a + 7.4 \\ 8100a &= 58.2 \\ a &= \frac{58.2}{8100} \approx .007185 \end{aligned}$$

*U.S. Bureau of Labor Statistics.

Find another quadratic model in Example 4(b) by using (10, 7.4) as the vertex and (70, 31.4) as the other point.

Answer:

$f(x) = \frac{1}{150}(x - 10)^2 + 7.4$

Therefore $f(x) = .007185(x - 10)^2 + 7.4$ is a quadratic model for this data. The graph of $f(x)$ in Figure 3.34 on the preceding page appears to fit the data reasonably well.

(c) Use the quadratic model in part (b) to estimate the number of women in the work force in 2005.

The year 2005 corresponds to $x = 105$, so the number is approximately

$$f(105) = .007185(105 - 10)^2 + 7.4 \approx 72.2 \text{ million.}$$

Quadratic Regression

Linear regression was used in Section 2.3 to construct a linear function that modeled a set of data points. When the data points appear to lie on a parabola rather than on a straight line (as in Example 4), a similar least squares regression procedure is available on most graphing calculators and spreadsheet programs to construct a quadratic model for the data. Simply follow the same steps as in linear regression, with one exception: choose quadratic rather than linear regression (they're on the same menu).

EXAMPLE 5 Use a graphing calculator to do the following.

(a) Find a quadratic regression model for the data in Example 4.

Enter the first coordinates of the data points as list L_1 and the second coordinates as list L_2. Performing the quadratic regression, as in Figure 3.35, leads to the model

$$g(x) = .007996x^2 - .21667x + 9.21167.$$

The number R^2 in Figure 3.35 is very close to 1, which indicates that this model fits the data well. The graph of $g(x)$ and the data points in Figure 3.36 visually confirm this.

FIGURE 3.35

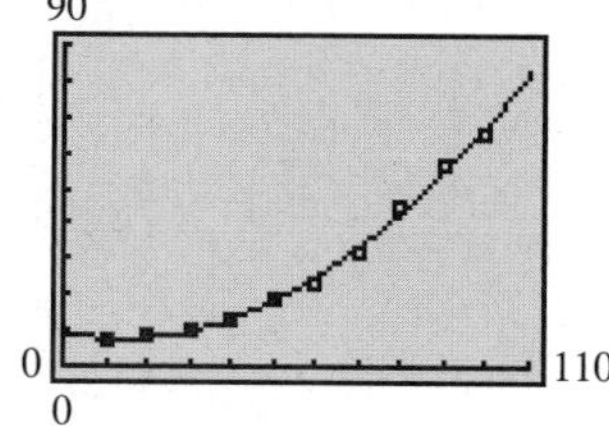

FIGURE 3.36

(b) Use the regression model to estimate the number of women in the workforce in 2005.

Evaluate $g(x)$ when $x = 105$:

$$g(105) = .007996(105^2) - .21667(105) + 9.21167 \approx 74.6 \text{ million.}$$

This estimate is about 2.4 million higher than the one in Example 4.

3.5 Exercises

Work these problems. (See Example 1.)

1. **Business** Shannise Cole makes and sells candy. She has found that the cost per box for making x boxes of candy is given by

$$C(x) = x^2 - 40x + 405.$$

   (a) How much does it cost per box to make 15 boxes? 18 boxes? 30 boxes?
   (b) Graph the cost function $C(x)$ and mark the points corresponding to 15, 18, and 30 boxes.
   (c) What point on the graph corresponds to the number of boxes that will make the cost per box as small as possible?
   (d) How many boxes should she make in order to keep the cost per box at a minimum? What is the minimum cost per box?

2. **Business** Greg Tobin sells bottled water. He has found that the average amount of time he spends with each customer is related to his weekly sales volume by the function

$$f(x) = x(60 - x),$$

   where x is the number of minutes per customer and $f(x)$ is the number of cases sold per week.
   (a) How many cases does he sell if he spends 10 minutes with each customer? 20 minutes? 45 minutes?
   (b) Choose an appropriate scale for the axes and sketch the graph of $f(x)$. Mark the points on the graph corresponding to 10, 20, and 45 minutes.
   (c) Explain what the vertex of the graph represents.
   (d) How long should Greg spend with each customer in order to sell as many cases per week as possible? If he does, how many cases will he sell?

3. **Natural Science** A researcher in physiology has decided that a good mathematical model for the number of impulses fired after a nerve has been stimulated is given by $y = -x^2 + 20x - 60$, where y is the number of responses per millisecond and x is the number of milliseconds since the nerve was stimulated.
   (a) When will the maximum firing rate be reached?
   (b) What is the maximum firing rate?

4. **Physical Science** If an object is thrown upward with an initial velocity of 32 feet per second, then its height, in feet, above the ground after t seconds is given by

$$h = 32t - 16t^2.$$

   Find the maximum height attained by the object. Find the number of seconds it takes for the object to hit the ground.

5. **Business** Colleen Davis owns a factory that manufactures souvenir key chains. Her weekly profit (in hundreds of dollars) is given by $P(x) = -2x^2 + 60x - 120$, where x is the number of cases of key chains sold.
   (a) What is the largest number of cases she can sell and still make a profit?
   (b) Explain how it is possible for her to lose money if she sells more cases than your answer in part (a).
   (c) How many cases should she make and sell in order to maximize her profits?

6. **Business** The manager of a bicycle shop has found that, at a price (in dollars) of $p(x) = 150 - \frac{x}{4}$ per bicycle, x bicycles will be sold.
   (a) Find an expression for the total revenue from the sale of x bicycles. (*Hint:* revenue = demand × price.)
   (b) Find the number of bicycle sales that leads to maximum revenue.
   (c) Find the maximum revenue.

Work the following problems. (See Example 2.)

7. **Business** Suppose the supply and demand for a certain textbook are given by

$$\text{supply}: p = \frac{1}{5}q^2; \qquad \text{demand}: p = -\frac{1}{5}q^2 + 40,$$

   where p is price and q is quantity. How many books are demanded at a price of
   (a) 10? (b) 20?
   (c) 30? (d) 40?

   How many books are supplied at a price of
   (e) 5? (f) 10?
   (g) 20? (h) 30?
   (i) Graph the supply and demand functions on the same axes.

8. **Business** Find the equilibrium quantity and the equilibrium price in Exercise 7.

9. **Business** Suppose the price p of widgets is related to the quantity q that is demanded by

$$p = 640 - 5q^2,$$

   where q is measured in hundreds of widgets. Find the price when the number of widgets demanded is
   (a) 0; (b) 5; (c) 10.

   Suppose the supply function for widgets is given by $p = 5q^2$, where q is the number of widgets (in hundreds) that are supplied at price p.
   (d) Graph the demand function $p = 640 - 5q^2$ and the supply function $p = 5q^2$ on the same axes.
   (e) Find the equilibrium quantity.
   (f) Find the equilibrium price.

10. Business The supply function for a commodity is given by $p = q^2 + 200$ and the demand function is given by $p = -10q + 3200$.
   (a) Graph the supply and demand functions on the same axes.
   (b) Find the equilibrium point.
   (c) What is the equilibrium quantity? The equilibrium price?

Business *Find equilibrium quantity and equilibrium price for the commodity whose supply and demand functions are given.*

11. Supply: $p = 45q$; Demand: $p = -q^2 + 10{,}000$
12. Supply: $p = q^2 + q + 10$; Demand: $p = -10q + 3060$
13. Supply: $p = q^2 + 20q$;
   Demand: $p = -2q^2 + 10q + 3000$
14. Supply: $p = .2q + 51$; Demand: $p = \dfrac{3000}{q + 5}$

Business *The revenue function R and the cost function C for a particular product are given. These functions are valid only for the specified range of values. Find the number of units that must be produced to break even.*

15. $R(x) = 200x - x^2$; $C(x) = 70x + 2200$; $0 \le x \le 100$
16. $R(x) = 300x - x^2$; $C(x) = 65x + 7000$; $0 \le x \le 150$
17. $R(x) = 400x - 2x^2$; $C(x) = -x^2 + 200x + 1900$; $0 \le x \le 100$
18. $R(x) = 500x - 2x^2$; $C(x) = -x^2 + 270x + 5125$; $0 \le x \le 125$

Business *Work each problem. (See Example 3.)*

19. A charter flight charges a fare of \$200 per person, plus \$4 per person for each unsold seat on the plane. If the plane holds 100 passengers and if x represents the number of unsold seats, find the following.
   (a) An expression for the total revenue received for the flight. (*Hint:* Multiply the number of people flying, $100 - x$, by the price per ticket.)
   (b) The graph for the expression of part (a)
   (c) The number of unsold seats that will produce the maximum revenue
   (d) The maximum revenue

20. The revenue of a charter bus company depends on the number of unsold seats. If 100 seats are sold, the price is \$50. Each unsold seat increases the price per seat by \$1. Let x represent the number of unsold seats.
   (a) Write an expression for the number of seats that are sold.
   (b) Write an expression for the price per seat.
   (c) Write an expression for the revenue.
   (d) Find the number of unsold seats that will produce the maximum revenue.
   (e) Find the maximum revenue.

21. Farmer Linton wants to find the best time to take her hogs to market. The current price is 88 cents per pound and her hogs weigh an average of 90 pounds. The hogs gain 5 pounds per week and the market price for hogs is falling each week by 2 cents per pound. How many weeks should Ms. Linton wait before taking her hogs to market in order to receive as much money as possible? At this time, how much money (per hog) will she get?

22. The manager of a peach orchard is trying to decide when to arrange for picking the peaches. If they are picked now, the average yield per tree will be 100 pounds, which can be sold for 40¢ per pound. Past experience shows that the yield per tree will increase about 5 pounds per week, while the price will decrease about 2¢ per pound per week.
   (a) Let x represent the number of weeks that the manager should wait. Find the income per pound.
   (b) Find the number of pounds per tree.
   (c) Find the total revenue from a tree.
   (d) When should the peaches be picked in order to produce maximum revenue?
   (e) What is the maximum revenue?

Unless directed otherwise by your instructor, do these problems as follows. If you do not have a graphing calculator or spreadsheet program, omit part (b). If you do have a graphing calculator or spreadsheet program, omit part (a). (See Examples 4 and 5.)

23. Health The table lists the total cumulative number of deaths in the United States to date known to have been caused by AIDS.*

Year	*Deaths*
1981	159
1985	12,652
1987	41,262
1989	90,218
1991	158,911
1993	245,662
1995	345,331
1997	403,206

   (a) Let $x = 0$ correspond to 1980. Use (1, 159) as the vertex and the data for 1997 to find a quadratic function $f(x) = a(x - h)^2 + k$ that models this data.
   (b) Let $x = 0$ correspond to 1980 and plot the data points. Use quadratic regression to find a function g that models this data.
   (c) Use the quadratic model from part (a) or (b) to estimate the cumulative number of deaths by 1992, 1996, and 2000. For comparison purposes, the actual figures for 1992 and 1996 are 200,391 and 382,261.

*U.S. Department of Health and Human Services, Centers for Disease Control and Prevention, *HIV/AIDS Surveillance Report,* vol. 11, no. 2.

24. Finance Federal government revenues (in billions of dollars) for selected years are shown in the table.*

Year	Revenue
1980	517.1
1985	734.1
1990	1031.3
1995	1351.5
2000	2025.0

(a) Let $x = 0$ correspond to 1980. Use (0, 517.1) as the vertex and the data for 2000 to find a quadratic function $f(x) = a(x - h)^2 + k$ that models this data.
(b) Let $x = 0$ correspond to 1980 and plot the data points. Use quadratic regression to find a function g that models this data.
(c) Use the quadratic model from part (a) or (b) to estimate government revenues in the years 1983, 1991, and 1996. How well do the estimates compare with the actual revenues of 600.6, 1054.3, and 1452.8?
(d) Estimate government revenue in 2001.

25. Transportation The table shows fuel consumption (in billions of gallons) by all nonmilitary motor vehicles in selected years.†

Year	Fuel Consumption
1975	109.0
1980	115.0
1985	121.3
1990	130.8
1995	143.8
1998	154.9

(a) Let $x = 0$ correspond to 1970. Use (5, 109) as the vertex and the data for 1998 to find a quadratic function $f(x) = a(x - h)^2 + k$ that models this data.
(b) Let $x = 0$ correspond to 1970 and plot the data points. Use quadratic regression to find a function g that models this data. Based on the value of r^2, how good a model is this?
(c) Use the quadratic model from part (a) or (b) to estimate a fuel consumption in 1993 and 1997. For comparison purposes, the actual figures for 1993 and 1997 are 137.3 and 150.3.
(d) Assuming these models remain accurate, estimate fuel consumption in 2000 and 2003.

26. Business The table shows the number of personal computers (in thousands) sold in selected years.*

Year	Unit Sales
1984	3,975
1985	3,200
1990	4,000
1995	8,400
1996	9,400
1998	12,800
1999	14,900
2000	16,800

(a) Let $x = 0$ correspond to 1980. Use (5, 3200) as the vertex and the data for 2000 to find a quadratic function $f(x) = a(x - h)^2 + k$ that models this data.
(b) Let $x = 0$ correspond to 1980 and plot the data points. Use quadratic regression to find a function g that models this data. Based on the value of r^2, how good a model is this?
(c) Use the quadratic model from part (a) or (b) to estimate the number of computers sold in 1992.
(d) Assuming these models remained accurate, how many computers were sold in 2002?

Work these problems.

27. Geometry A field bounded on one side by a river is to be fenced on three sides to form a rectangular enclosure. There are 320 ft of fencing available. What should the dimensions be to have an enclosure with the maximum possible area?

28. Geometry A rectangular garden bounded on one side by a river is to be fenced on the other three sides. Fencing material for the side parallel to the river costs $30 per foot and material for the other two sides costs $10 per foot. What are the dimensions of the garden of largest possible area, if $1200 is to be spent for fencing material?

Business *Recall that profit equals revenue minus cost. In Exercises 29 and 30 find the following.*

(a) *The break-even point (to the nearest tenth)*
(b) *The x-value that makes profit a maximum*
(c) *The maximum profit*
(d) *For what x-values will a loss occur?*
(e) *For what x-values will a profit occur?*

29. $R(x) = 400x - 2x^2$ and $C(x) = 200x + 2000$, with $0 \leq x \leq 100$

30. $R(x) = 900x - 3x^2$ and $C(x) = 450x + 5000$, with $0 \leq x \leq 150$

*Financial Management Service of the U.S. Department of the Treasury.

†U.S. Federal Highway Administration.

*Consumer Electronics Manufacturers Association.

3.6 POLYNOMIAL FUNCTIONS

A **polynomial function of degree *n*** is a function whose rule is given by a polynomial of degree n.* For example, linear functions, such as

$$f(x) = 3x - 2 \quad \text{and} \quad g(x) = 7x + 5,$$

are polynomial functions of degree 1 and quadratic functions, such as

$$f(x) = 3x^2 + 4x - 6 \quad \text{and} \quad g(x) = -2x^2 + 9$$

are polynomial functions of degree 2. Similarly, the function given by

$$f(x) = 2x^4 + 5x^3 - 6x^2 + x - 3$$

is a polynomial function of degree 4.

Graph $f(x) = x^3 - 5$.

Answer:

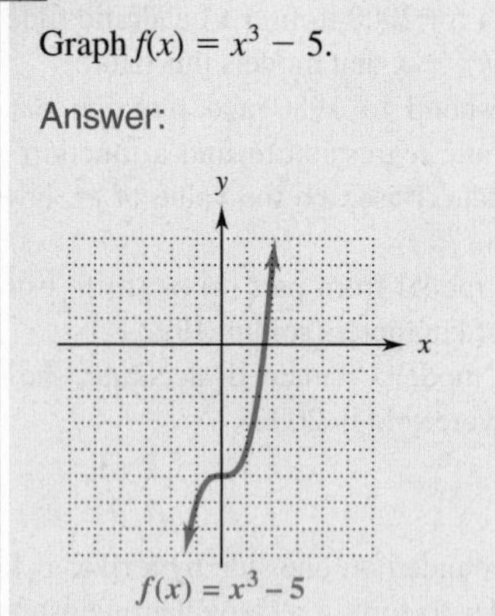

Basic Graphs

The simplest polynomial functions are those whose rules are of the form $f(x) = ax^n$ (where a is a constant).

EXAMPLE 1 Graph $f(x) = x^3$.

First, find several ordered pairs belonging to the graph. Be sure to choose some negative x-values, $x = 0$, and some positive x-values to get representative ordered pairs. Find as many ordered pairs as you need in order to see the shape of the graph. Then plot the ordered pairs and draw a smooth curve through them, getting the graph in Figure 3.37. ◀ 1

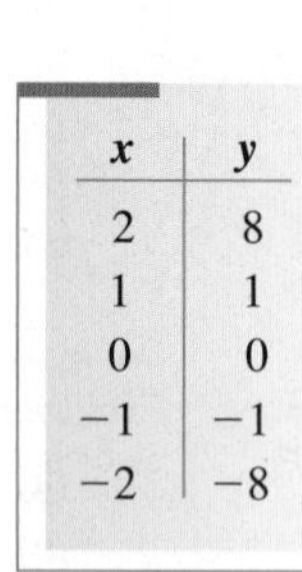

x	y
2	8
1	1
0	0
−1	−1
−2	−8

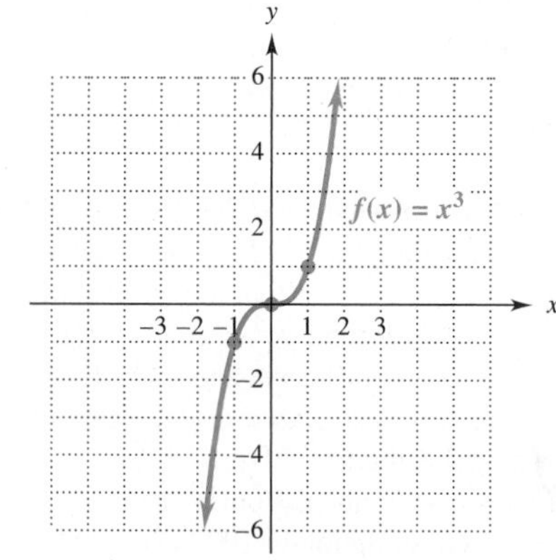

FIGURE 3.37

*The degree of a polynomial was defined on page 13.

2

Graph these functions.

(a) $f(x) = -.25x^5 - 2$

(b) $f(x) = -x^4 + 3$

Answers:

(a)

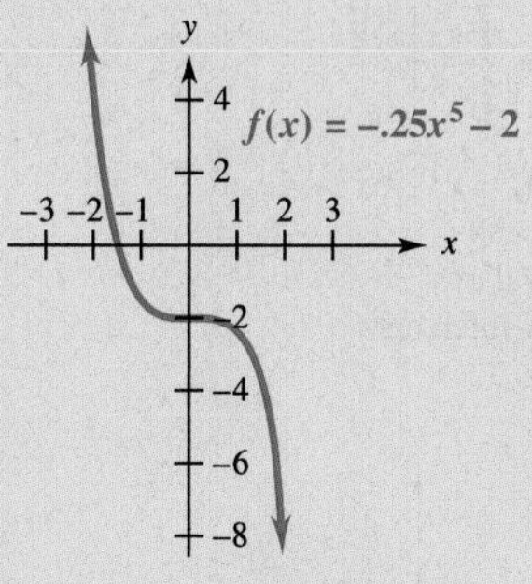

(b)

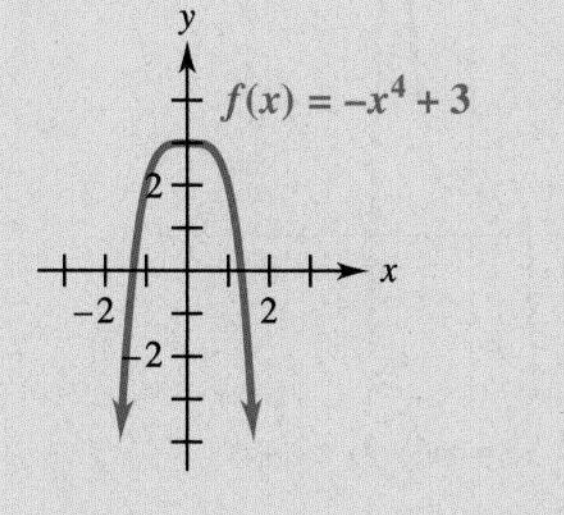

EXAMPLE 2 Graph $f(x) = (3/2)x^4$.

The following table gives some typical ordered pairs.

x	-2	-1	0	1	2
y	24	3/2	0	3/2	24

The graph is shown in Figure 3.38. 

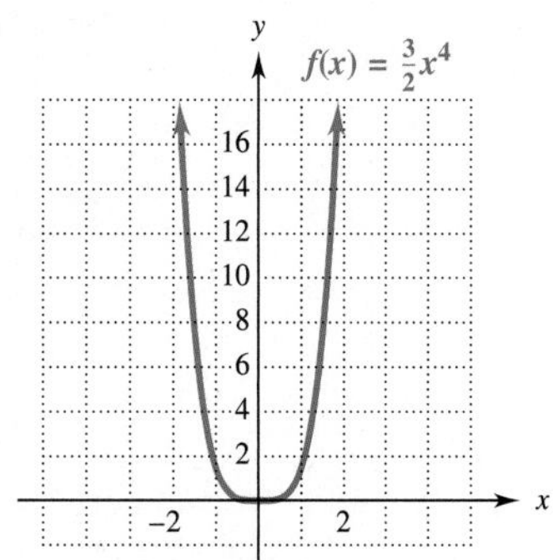

FIGURE 3.38

The graph of $f(x) = ax^n$ has the same general shape as one of the graphs in Figure 3.37 or 3.38.

Graph of $f(x) = ax^n$

If the exponent n is even, then the graph of $f(x) = ax^n$ is cup-shaped with the bottom of the cup at the origin and the y-axis through the middle; it opens upward when $a > 0$ and downward when $a < 0$.

If the exponent n is odd, then the graph of $f(x) = ax^n$ moves upward from left to right when $a > 0$ and downward when $a < 0$, with a single bend at the origin.

Properties of Polynomial Graphs

Unlike the graphs in the preceding figures, the graphs of more complicated polynomial functions may have several "peaks" and "valleys," as illustrated in Figure 3.39 on the next page. The location of the peaks and valleys can be accurately approximated by a maximum or minimum finder on a graphing calculator. Calculus is needed to determine their exact location.

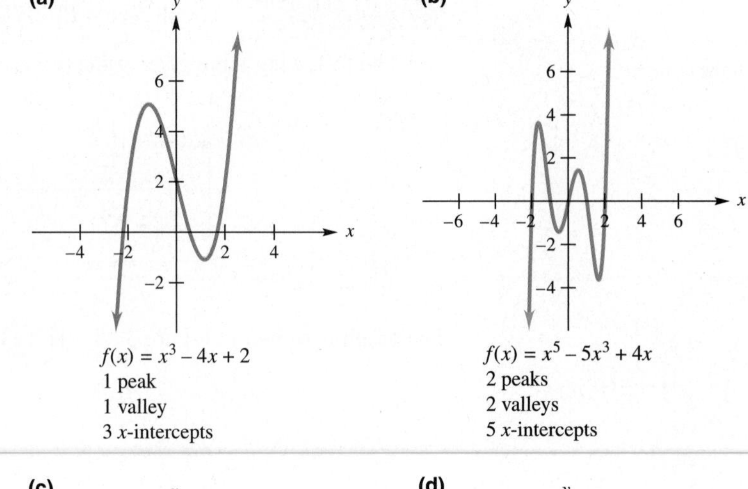

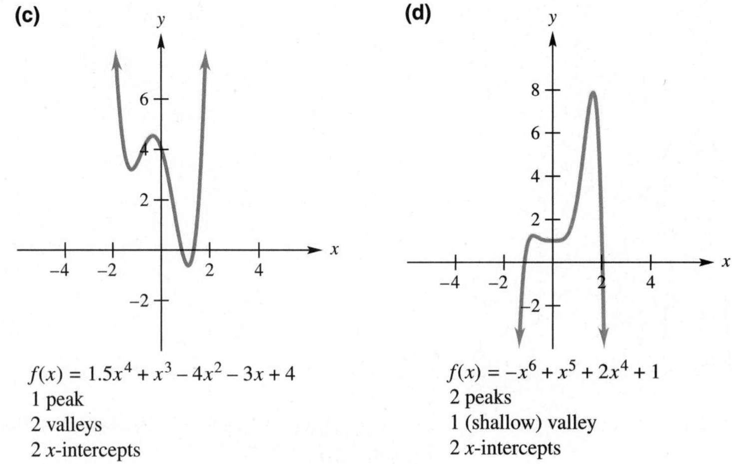

FIGURE 3.39

Figure 3.39 illustrates two important facts about polynomial functions.

1. The total number of peaks and valleys on the graph of a polynomial function of degree n is at most $n - 1$.
2. The number of x-intercepts on the graph of a polynomial function of degree n is at most n.

For instance, the polynomial in Figure 3.39(b) has degree $n = 5$ and the number of peaks and valleys on its graph is $n - 1 = 4$. The graph has 5 x-intercepts. Similarly, the polynomial in Figure 3.39(d) has degree $n = 6$, so that $n - 1 = 5$. The graph has 3 peaks and valleys and 2 x-intercepts.

The domain of every polynomial function is the set of all real numbers, which means that its graph extends forever to the left and right. The shape of the graph far from the center depends on the degree of the polynomial. For instance, compare the graph of $g(x) = 1.5x^4 + x^3 - 4x^2 - 3x + 4$ in Figure 3.39(c) with the graph of $f(x) = 1.5x^4$ in Figure 3.38. At the far left and far right both graphs shoot away from the x-axis in a similar fashion. When both functions are graphed in the same large viewing window on a graphing calculator (Figure 3.40), the ends of the graphs look almost the same. In other words, as you move out from the center, the graph of $g(x) = 1.5x^4 + x^3 - 4x^2 - 3x + 4$ looks very much like the graph of its highest degree term, $1.5x^4$. The same thing is true in the general case.

> The graph of a polynomial function is a smooth, unbroken curve that extends forever to the left and right. When $|x|$ is large, the graph resembles the graph of its highest degree term and moves sharply away from the x-axis.

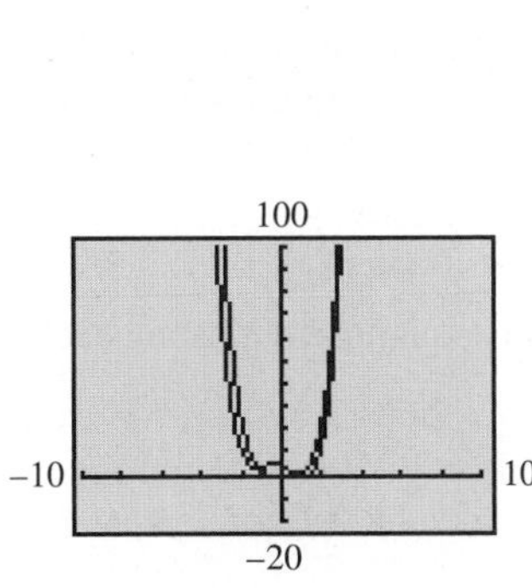

FIGURE 3.40

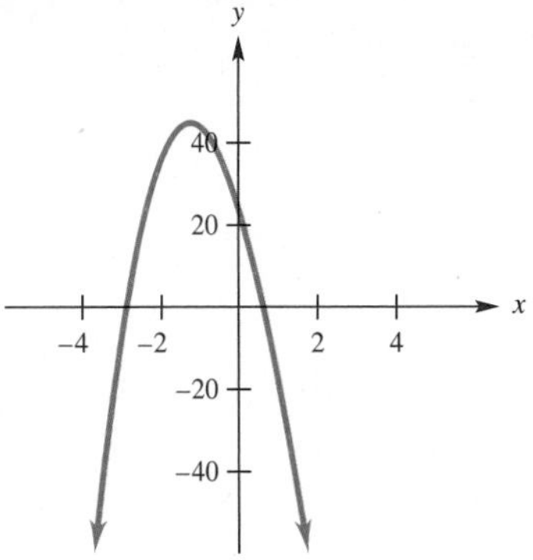

FIGURE 3.41

EXAMPLE 3 Let $g(x) = x^3 - 11x^2 - 32x + 24$.

(a) Is Figure 3.41 a complete graph of $g(x)$; that is, does it show all the important features of the graph?

The far ends of the graph of $g(x)$ should resemble the graph of its highest degree term x^3. The graph of $f(x) = x^3$ in Figure 3.37 moves upward at the far right, but Figure 3.41 does not. So Figure 3.41 is not a complete graph.

(b) Use a graphing calculator to find a complete graph of $g(x)$.

Since the graph of $g(x)$ must eventually start rising on the right side (as does the graph of x^3), a viewing window that shows a complete graph must extend beyond $x = 4$. By experimenting with various windows, we obtain Figure 3.42. This graph shows a total of 2 peaks and valleys and 3 x-intercepts (the maximum possible for a polynomial of degree 3). At the far ends, the graph of $g(x)$ in Figure 3.42 resembles the graph of $f(x) = x^3$. Therefore, Figure 3.42 is a complete graph of $g(x)$. ◀ (3)

(3) Find a viewing window on a graphing calculator that shows a complete graph of

$f(x) = -.7x^4 + 119x^2 + 400.$

(*Hint:* The graph has two x-intercepts and the maximum possible number of peaks and valleys.)

Answer:

Many correct answers, including $-15 \le x \le 15$ and $-2000 \le y \le 6000$.

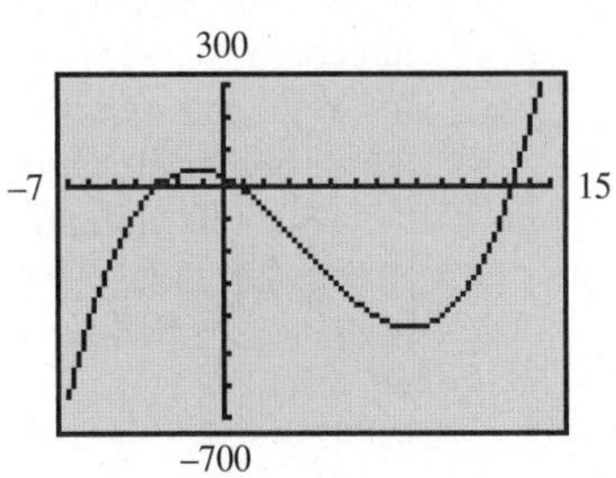

FIGURE 3.42

Graphing Techniques

Accurate graphs of first- and second-degree polynomial functions (lines and parabolas) are easily found algebraically, as we saw in Sections 2.2 and 3.4. All polynomial functions of degree 3, and some of higher degree, can be accurately graphed by hand by using calculus and algebra to locate the peaks and valleys. When a polynomial can be completely factored, the general shape of its graph can be determined algebraically by using the basic properties of polynomial graphs, as illustrated in Example 4 below. Accurate graphs of other polynomial functions generally require technology.

Multiply out the expression for $f(x)$ in Example 4 and determine its degree.

Answer:

$f(x) = 2x^3 + 5x^2 - x - 6$; degree 3.

EXAMPLE 4 Graph $f(x) = (2x + 3)(x - 1)(x + 2)$.

Note that $f(x)$ is a polynomial of degree 3 (if you don't see why, do problem 4 at the side). Begin by finding any x-intercepts. Set $f(x) = 0$ and solve for x:

$$f(x) = 0$$
$$(2x + 3)(x - 1)(x + 2) = 0$$

Solve this equation by placing each of the three factors equal to 0.

$$2x + 3 = 0 \quad \text{or} \quad x - 1 = 0 \quad \text{or} \quad x + 2 = 0$$
$$x = -\frac{3}{2} \qquad\qquad x = 1 \qquad\qquad x = -2$$

The three numbers, $-3/2$, 1, and -2, divide the x-axis into four intervals:

$$x < -2, \quad -2 < x < -\frac{3}{2}, \quad -\frac{3}{2} < x < 1, \quad \text{and} \quad 1 < x.$$

These intervals are shown in Figure 3.43.

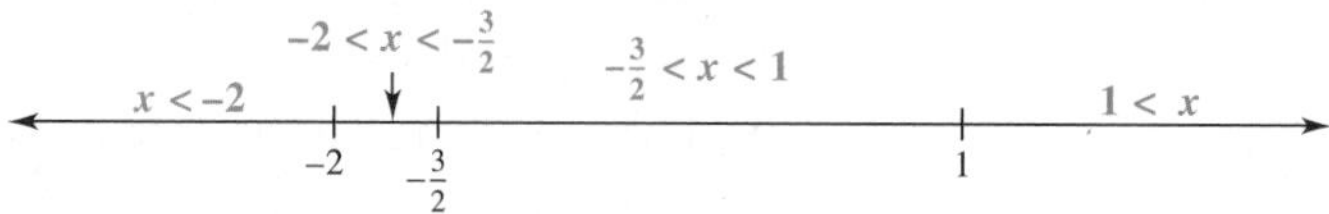

FIGURE 3.43

Since the graph is an unbroken curve, it can only change from above the x-axis to below it by passing through the x-axis. As we have seen, this occurs only at the x-intercepts $x = -2$, $-3/2$, and 1. Consequently, in the interval between two intercepts (or to the left of $x = -2$ or to the right of $x = 1$), the graph of $f(x)$ must lie entirely above or entirely below the x-axis.

We can determine where the graph lies over an interval, by evaluating $f(x) = (2x + 3)(x - 1)(x + 2)$ at a number in that interval. For example, $x = -3$ is in the interval where $x < -2$ and

$$f(-3) = (2(-3) + 3)(-3 - 1)(-3 + 2)$$
$$= -12.$$

Therefore, $(-3, -12)$ is on the graph. Since this point lies below the x-axis, all points in this interval (that is, all points with $x < -2$) must lie below the x-axis. By testing numbers in the other intervals, we obtain this chart.

Interval	$x < -2$	$-2 < x < -3/2$	$-3/2 < x < 1$	$x > 1$
Test Number	-3	$-7/4$	0	2
Value of f(x)	-12	11/32	-6	28
Sign of f(x)	Negative	Positive	Negative	Positive
Graph	Below x-axis	Above x-axis	Below x-axis	Above x-axis

5

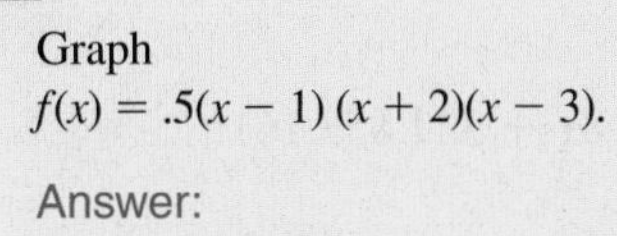

Graph
$f(x) = .5(x - 1)(x + 2)(x - 3)$.

Answer:

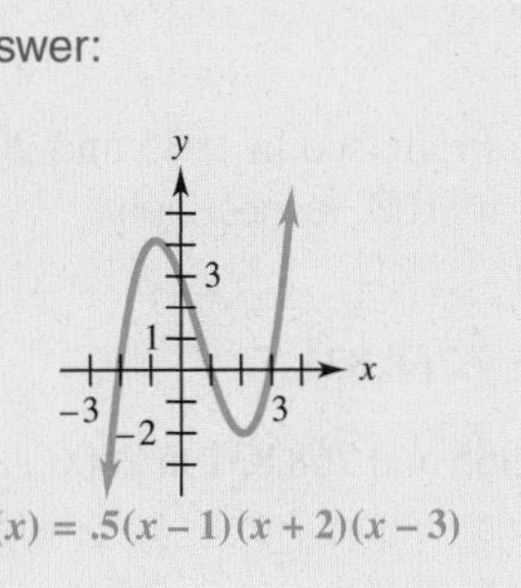

$f(x) = .5(x-1)(x+2)(x-3)$

Since the graph intersects the x-axis at the intercepts $x = -2$ and $x = -3/2$ and is above the x-axis between these intercepts, there must be at least one peak there. Similarly, there must be at least one valley between $x = -3/2$ and $x = 1$ because the graph is below the x-axis there. However, a polynomial function of degree 3 can have a total of at most $3 - 1 = 2$ peaks and valleys. So there must be exactly one peak and exactly one valley on this graph.

Furthermore, when $|x|$ is large, the graph must resemble the graph of $y = 2x^3$ (the highest degree term). The graph of $y = 2x^3$, like the graph of $y = x^3$ in Figure 3.37, moves upward to the right and downward to the left. Using these facts and plotting the x-intercepts shows that the graph must have the general shape shown in Figure 3.44. Plotting additional points leads to the reasonably accurate graph in Figure 3.45. We say "reasonably accurate" because we cannot be sure of the exact locations of the peaks and valleys on the graph without using calculus. ◀ 5

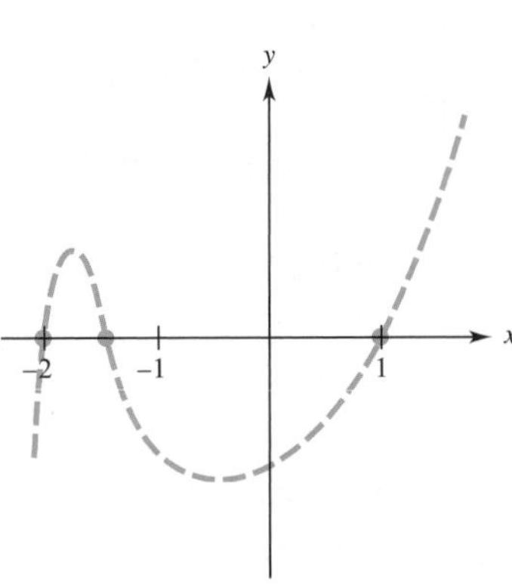

FIGURE 3.44

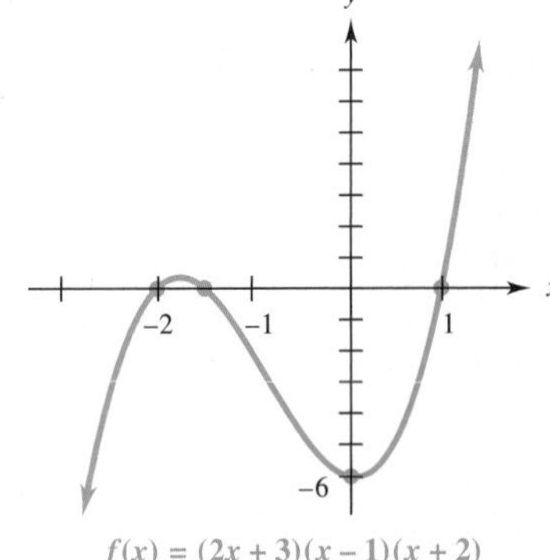

FIGURE 3.45

Polynomial Models

Regression procedures (similar to linear regression in Section 2.3 and quadratic regression in Section 3.5) can be used to find cubic and quartic (degree 4) polynomial models for appropriate data.

EXAMPLE 5 The table shows the population of San Francisco in selected years.

Year	1900	1950	1970	1980	1990	1998
Population	342,782	775,357	715,674	678,974	723,959	745,774

(a) Plot the data on graphing calculator, with $x = 0$ corresponding to 1900.

The points in Figure 3.46 suggest the general shape of a cubic polynomial graph.

(b) Use cubic regression to obtain a model for this data.

The procedure is the same as for linear regression (just choose "cubic" in place of "linear"). It produces the function

$$f(x) = 2.708x^3 - 493.867x^2 + 26{,}563.818x + 342{,}853.161^*$$

Its graph in Figure 3.47 appears to fit the data well.

(c) Use the model to estimate the population of San Francisco in 1985 and 2000.

Now 1985 and 2000 correspond to $x = 85$ and $x = 100$, respectively. Verify that

$$f(85) \approx 695{,}803 \quad \text{and} \quad f(100) \approx 768{,}831.$$

So the population was approximately 695,803 in 1985 and 768,831 in 2000.

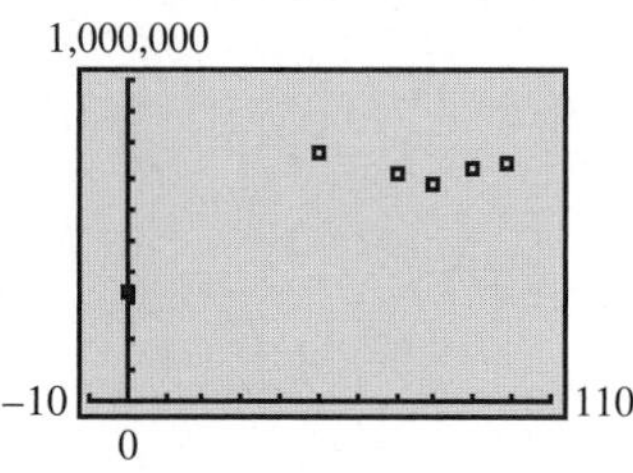

FIGURE 3.46

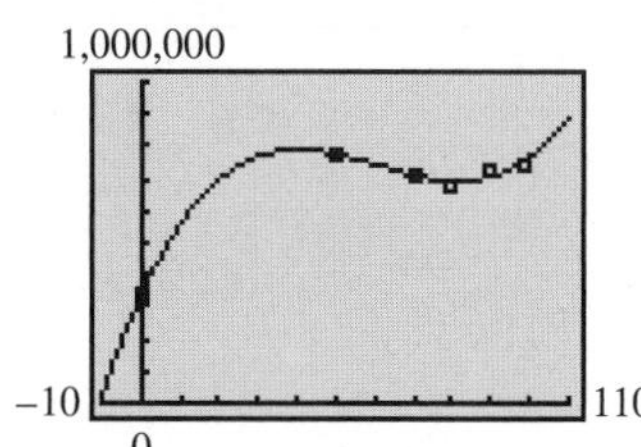

FIGURE 3.47

*Coefficients are rounded for easy reading, but the full coefficients are used for graphs and function evaluation.

3.6 Exercises

Graph each of the following polynomial functions. (See Examples 1 and 2.)

1. $f(x) = x^4$

2. $g(x) = -.5x^6$

3. $h(x) = -.2x^5$

4. $f(x) = x^7$

*In Exercises 5–8, state whether the graph could possibly be the graph of **(a)** some polynomial function; **(b)** a polynomial function of degree 3; **(c)** a polynomial function of degree 4; **(d)** a polynomial function of degree 5. (See Example 3.)*

5.

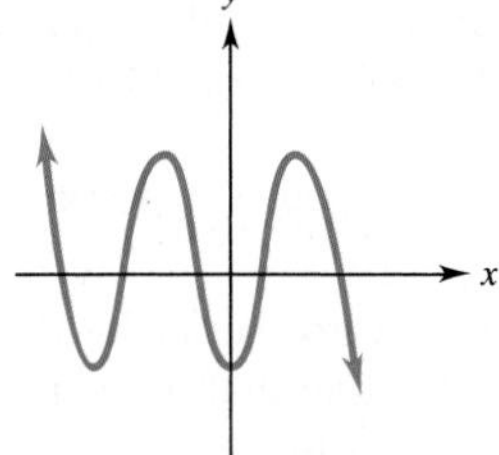

6.

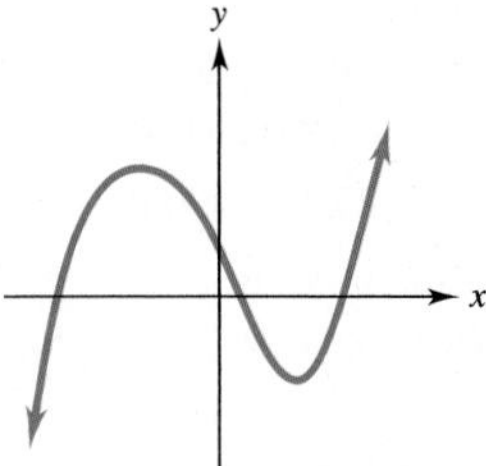

7.

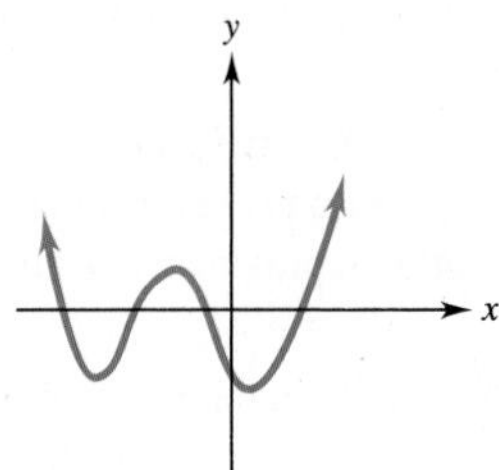

8.

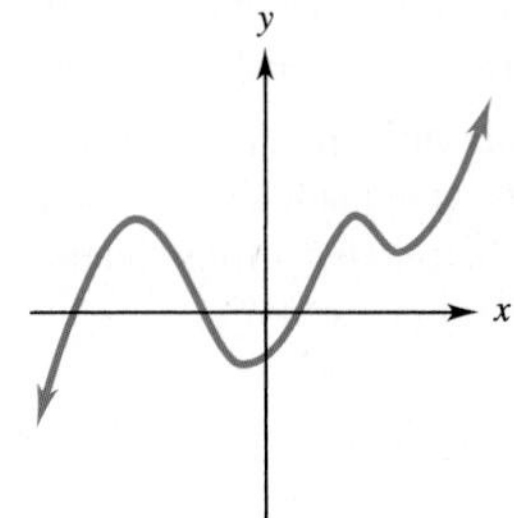

In Exercises 9–14, match the given polynomial function to its graph (A–F), without using a graphing calculator. (See Example 3 and the two boxes preceding it.)

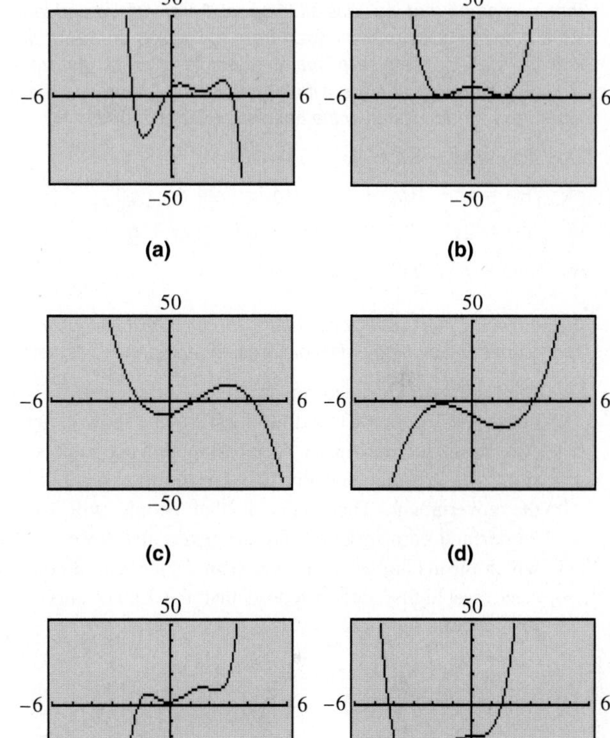

9. $f(x) = x^3 - 7x - 9$

10. $f(x) = -x^3 + 4x^2 + 3x - 8$

11. $f(x) = x^4 - 5x^2 + 7$

12. $f(x) = x^4 + 4x^3 - 20$

13. $f(x) = .7x^5 - 2.5x^4 - x^3 + 8x^2 + x + 2$

14. $f(x) = -x^5 + 4x^4 + x^3 - 16x^2 + 12x + 5$

Graph each of the following polynomial functions. (See Example 4.)

15. $f(x) = (x + 2)(x - 3)(x + 4)$

16. $f(x) = (x - 3)(x - 1)(x + 1)$

17. $f(x) = x^2(x - 2)(x + 3)$

18. $f(x) = x^2(x + 1)(x - 1)$
19. $f(x) = x^3 + x^2 - 6x$
20. $f(x) = x^3 - 2x^2 - 8x$
21. $f(x) = x^3 + 3x^2 - 4x$
22. $f(x) = x^4 - 5x^2$

Exercises 23–26 require a graphing calculator. Find a viewing window that shows a complete graph of the polynomial function (that is, a graph that includes all the peaks and valleys and indicates how the curve moves away from the x-axis at the far left and far right). There are many possible correct answers. Consider your answer correct if it shows all the features that appear in the window given in the answers. (See Example 3.)

23. $g(x) = x^3 - 3x^2 - 4x - 5$
24. $f(x) = x^4 - 10x^3 + 35x^2 - 50x + 24$
25. $f(x) = 2x^5 - 3.5x^4 - 10x^3 + 5x^2 + 12x + 6$
26. $g(x) = x^5 + 8x^4 + 20x^3 + 9x^2 - 27x - 7$

In Exercises 27–31, use a calculator to evaluate the functions. Do the graph by plotting points or by using a graphing calculator.

27. **Finance** An idealized version of the Laffer curve (originated by economist Arthur Laffer) is shown here. According to this theory, decreasing the tax rate, say, from x_2 to x_1, may actually increase the total revenue to the government. The theory is that people will work harder and earn more if they are taxed at a lower rate, which means higher total tax revenues than would be the case at a higher rate. Suppose that the Laffer curve is given by the function

$$f(x) = \frac{x(x - 100)(x - 160)}{240} \qquad (0 \le x \le 100),$$

where $f(x)$ is government revenue (in billions of dollars) from a tax rate of x percent. Find the revenue from the following tax rates.
(a) 20%
(b) 40%
(c) 50%
(d) 70%
(e) Graph $f(x)$

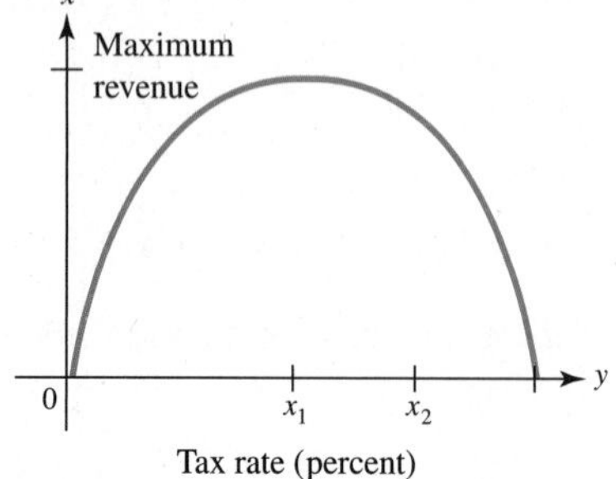

28. **Health** A technique for measuring cardiac output depends on the concentration of a dye after a known amount is injected into a vein near the heart. In a normal heart, the concentration of the dye at time x (in seconds) is given by the function defined by

$$g(x) = -.006x^4 + .140x^3 - .053x^2 + 1.79x.$$

(a) Find the following: $g(0)$; $g(1)$; $g(2)$; $g(3)$.
(b) Graph $g(x)$ for $x \ge 0$.

29. **Physical Science** The pressure of the oil in a reservoir tends to drop with time. By taking sample pressure readings for a particular oil reservoir, petroleum engineers have found that the change in pressure is given by

$$P(t) = t^3 - 18t^2 + 81t,$$

where t is time in years from the date of the first reading.
(a) Find the following: $P(0)$; $P(3)$; $P(7)$; $P(10)$.
(b) Graph $P(t)$.
(c) For what time period is the change in pressure (drop) increasing? decreasing?

30. **Natural Science** During the early part of the 20th century, the deer population of the Kaibab Plateau in Arizona experienced a rapid increase because hunters had reduced the number of natural predators. The increase in population depleted the food resources and eventually caused the population to decline. For the period from 1905 to 1930, the deer population was approximated by

$$D(x) = -.125x^5 + 3.125x^4 + 4000,$$

where x is time in years from 1905.
(a) Find the following: $D(0)$; $D(5)$; $D(10)$; $D(15)$; $D(20)$; $D(25)$.
(b) Graph $D(x)$.
(c) From the graph, over what period of time (from 1905 to 1930) was the population increasing? relatively stable? decreasing?

31. **Transportation** According to the U.S. Department of Transportation, the number of consumer complaints against top U.S. airlines in year x can be approximated by the polynomial function.

$$f(x) = .2196x^4 - 5.454x^3 + 50.258x^2 - 202.68x + 306.134 \qquad (5 \le x \le 9),$$

where $x = 5$ corresponds to 1995 and $f(x)$ is in thousands.
(a) How many complaints were made during 1996?
(b) How many complaints were made during 1999?

Assume this model remains valid in later years.
(c) How many complaints were made during 2000?
(d) In what year would the number of complaints reach 113,000? Does this seem plausible?

32. A partial graph of a cubic polynomial function $f(x)$ is shown on the calculator screen below. The coefficient of x^3 in the rule of $f(x)$ is negative. Sketch a graph that has the same shape as the entire graph of $f(x)$.

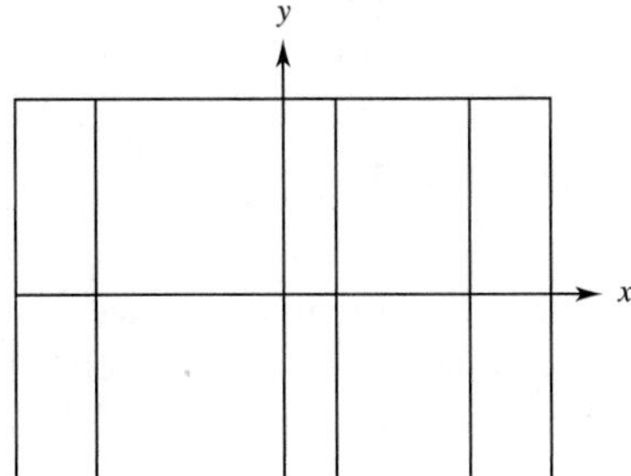

Use a graphing calculator to do the following problems. (See Example 5.)

33. Social Science The table shows the number of abandoned cars (in thousands) in New York City over a ten-year period.*

Year	*Abandoned Cars*
1986	82
1987	120
1988	140
1989	148
1990	135
1991	95
1992	75
1993	51
1994	38
1995	27
1996	20

(a) Plot the data points given by the table, with $x = 0$ corresponding to 1986.

(b) Use quartic regression to find a fourth-degree polynomial function $f(x)$ that models this data.

(c) Graph $f(x)$ on the same screen as the data points. Does the graph appear to fit the data well?

(d) Does this model predict an increase or a decrease in the number of abandoned cars after 1996?

New York Times, February 15, 1997.

34. Physical Science The table shows U.S. coal consumption (in millions of short tons) in selected years.*

Year	*Millions of Short Tons*
1950	494.1
1960	398.1
1970	523.2
1980	702.7
1985	818.0
1990	895.5
1995	962.0
1996	1005.6
1997	1027.1

(a) Let $x = 0$ correspond to 1900. Plot the data points in the viewing window with $50 \le x \le 100$ and $350 \le y \le 1100$.

(b) Use cubic regression to find a cubic polynomial function $g(x)$ that models this data.

(c) Graph $g(x)$ on the same screen as the data points. Does the graph appear to fit the data well?

(d) Assuming that this model remains valid, estimate the amount of coal consumed in 2003.

*U.S. Department of Energy, *Annual Energy Review* (1997).

3.7 RATIONAL FUNCTIONS

A **rational function** is a function whose rule is the quotient of two polynomials, such as

$$f(x) = \frac{2}{1+x}, \qquad g(x) = \frac{3x+2}{2x+4}, \qquad h(x) = \frac{x^2-2x-4}{x^3-2x^2+x}.$$

Thus a rational function is one whose rule can be written in the form

$$f(x) = \frac{P(x)}{Q(x)},$$

where $P(x)$ and $Q(x)$ are polynomials, with $Q(x) \neq 0$. The function is undefined for any values of x that make $Q(x) = 0$, so there are breaks in the graph at these numbers.

Linear Rational Functions

We begin with rational functions in which both numerator and denominator are first-degree or constant polynomials. Such functions are sometimes called **linear rational functions.**

EXAMPLE 1 Graph the rational function defined by $y = \dfrac{2}{1 + x}$.

This function is undefined for $x = -1$, since -1 leads to a 0 denominator. For this reason, the graph of this function will not intersect the vertical line $x = -1$. Since x can take on any value except -1, the values of x can approach -1 as closely as desired from either side of -1, as shown in the following table of values.

x approaches -1 ↓

x	-1.5	-1.2	-1.1	-1.01	$-.99$	$-.9$	$-.8$	$-.5$
$1 + x$	$-.5$	$-.2$	$-.1$	$-.01$	$.01$	$.1$	$.2$	$.5$
$\dfrac{2}{1 + x}$	-4	-10	-20	-200	200	20	10	4

↑ $|f(x)|$ gets larger and larger

1 Graph the following.

(a) $f(x) = \dfrac{3}{5 - x}$

(b) $f(x) = \dfrac{-4}{x + 4}$

Answers:

(a)

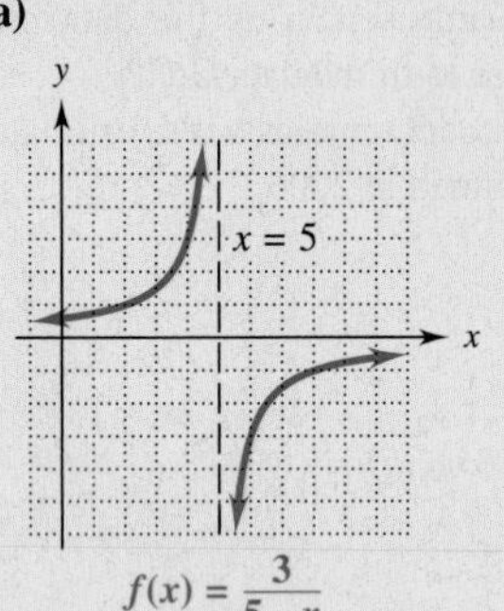

$f(x) = \dfrac{3}{5 - x}$

(b)

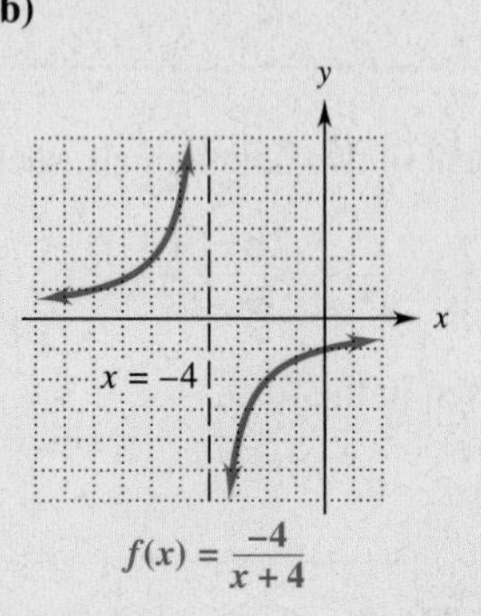

$f(x) = \dfrac{-4}{x + 4}$

The table above suggests that as x gets closer and closer to -1 from either side, the denominator $1 + x$ gets closer and closer to 0, and $|2/(1 + x)|$ gets larger and larger. The part of the graph near $x = -1$ in Figure 3.48 on the next page shows this behavior. The vertical line $x = -1$ that is approached by the curve is called a *vertical asymptote*. For convenience, the vertical asymptote is indicated by a dashed line in Figure 3.48, but this line is *not* part of the graph of the function.

As $|x|$ gets larger and larger so does the absolute value of the denominator $1 + x$. Hence $y = 2/(1 + x)$ gets closer and closer to 0, as shown in the table below.

x	-101	-11	-2	0	9	99
$1 + x$	-100	-10	-1	1	10	100
$\dfrac{2}{1 + x}$	$-.02$	$-.2$	-2	2	$.2$	$.02$

The horizontal line $y = 0$ is called a *horizontal asymptote* for this graph. Using the asymptotes and plotting the intercept and other points gives the graph of Figure 3.48. ◀ 1

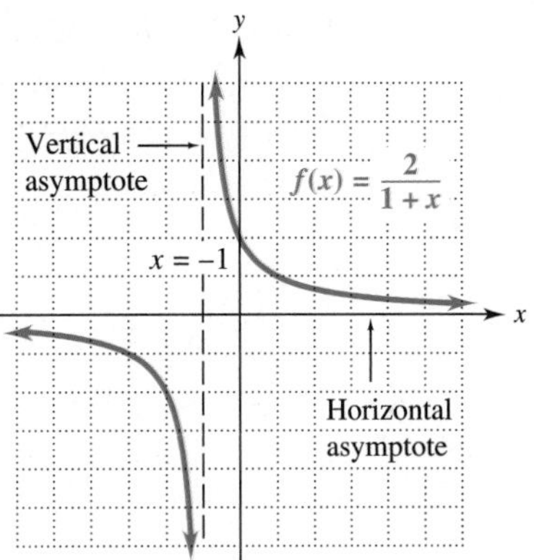

FIGURE 3.48

Example 1 suggests the following conclusion.

If a number c makes the denominator 0 but the numerator nonzero in the expression defining a rational function, then the line $x = c$ is a **vertical asymptote** for the graph of the function.

Also, wherever the values of y approach but do not equal some number k as $|x|$ gets larger and larger, the line $y = k$ is a **horizontal asymptote** for the graph.

EXAMPLE 2 Graph $f(x) = \dfrac{3x + 2}{2x + 4}$.

Find the vertical asymptote by setting the denominator equal to 0 and then solving for x.

$$2x + 4 = 0$$
$$x = -2$$

In order to see what the graph looks like when $|x|$ is very large, we rewrite the rule of the function. When $x \neq 0$, dividing both numerator and denominator by x does not change the value of the function:

$$f(x) = \frac{3x + 2}{2x + 4} = \frac{\dfrac{3x + 2}{x}}{\dfrac{2x + 4}{x}}$$
$$= \frac{\dfrac{3x}{x} + \dfrac{2}{x}}{\dfrac{2x}{x} + \dfrac{4}{x}} = \frac{3 + \dfrac{2}{x}}{2 + \dfrac{4}{x}}.$$

Now when $|x|$ is very large, the fractions $2/x$ and $4/x$ are very close to 0 (for instance, when $x = 200$, $4/x = 4/200 = .02$). Therefore the numerator of $f(x)$ is very

Graph the following.

(a) $f(x) = \dfrac{3x + 5}{x - 3}$

(b) $f(x) = \dfrac{2 - x}{x + 3}$

Answers:

(a)

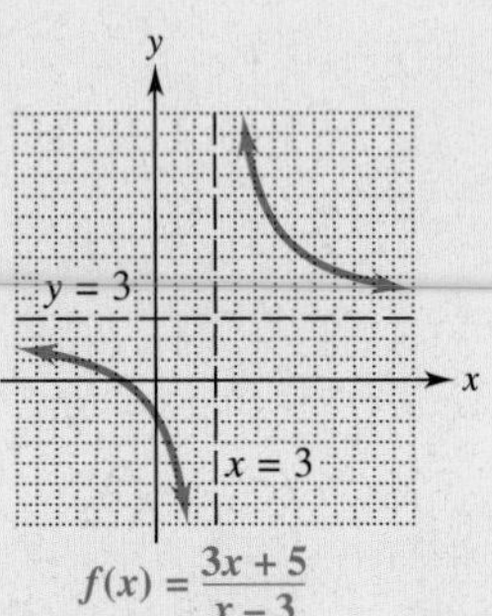

(b)

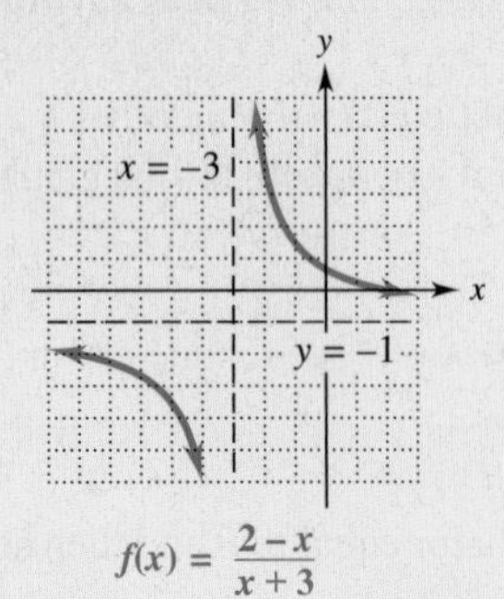

close to $3 + 0 = 3$ and the denominator is very close to $2 + 0 = 2$. Hence $f(x)$ is very close to 3/2 when $|x|$ is large, so the line $y = 3/2$ is the horizontal asymptote of the graph, as shown in Figure 3.49. ◀ (2)

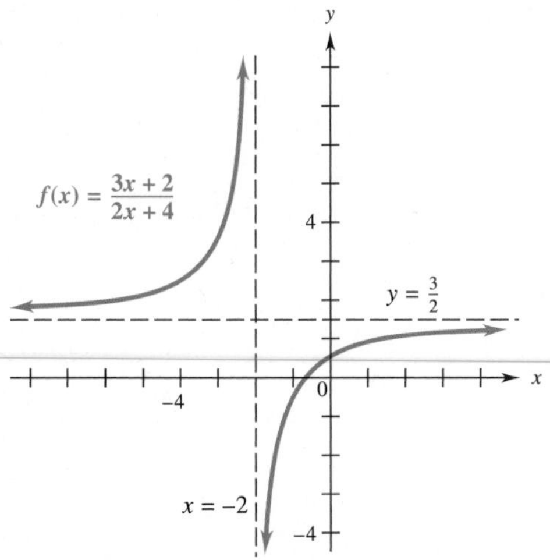

FIGURE 3.49

TECHNOLOGY TIP Depending on the viewing window, a graphing calculator may not accurately represent the graph of a rational function. For example, the graph of $f(x) = \dfrac{3x + 2}{2x + 4}$ in Figure 3.50, which should look like Figure 3.49, has an erroneous vertical line at the place where the graph has a vertical asymptote. This problem can usually be avoided by using a window that has the vertical asymptote at the center of the x-axis, as in Figure 3.51.

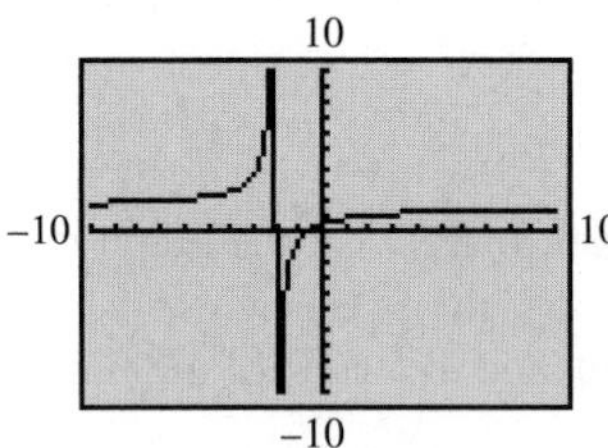

FIGURE 3.50

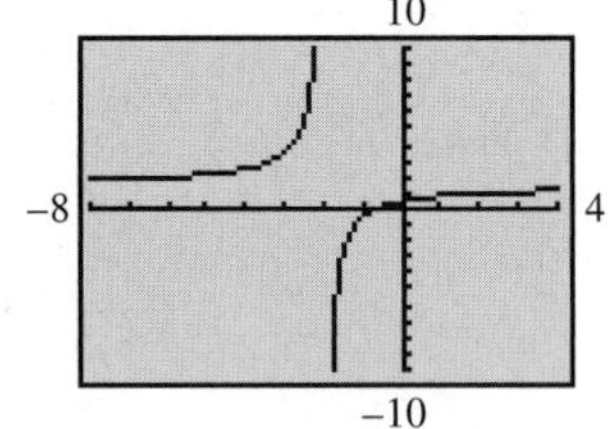

FIGURE 3.51

In Examples 1 and 2, each graph has a single vertical asymptote, determined by the root of the denominator, and a horizontal asymptote determined by the coefficients of x. In Example 2, for instance, $f(x) = \dfrac{3x + 2}{2x + 4}$ has horizontal asymptote $y = \dfrac{3}{2}$, and in Example 1, $f(x) = \dfrac{2}{1 + x} = \dfrac{0x + 2}{1x + 1}$ has horizontal asymptote $y = \dfrac{0}{1} = 0$ (the x-axis). Similar arguments work for any linear rational function.

The graph of $f(x) = \dfrac{ax + b}{cx + d}$ (where $c \neq 0$ and $ad \neq bc$) has a vertical asymptote at the root of the denominator and horizontal asymptote $y = \dfrac{a}{c}$.

Other Rational Functions

When the numerator or denominator of a rational function has degree greater than 1, its graph can be more complicated than those in Examples 1 and 2. The graph may have several vertical asymptotes, as well as peaks and valleys.

EXAMPLE 3 Graph $f(x) = \dfrac{2x^2}{x^2 - 4}$.

Find the vertical asymptotes by setting the denominator equal to 0 and solving for x.

$$x^2 - 4 = 0$$
$$(x + 2)(x - 2) = 0$$
$$x + 2 = 0 \quad \text{or} \quad x - 2 = 0$$
$$x = -2 \qquad\qquad x = 2$$

Since neither of these numbers makes the numerator 0, the lines $x = -2$ and $x = 2$ are vertical asymptotes of the graph. The horizontal asymptote can be determined by dividing both the numerator and denominator of $f(x)$ by x^2 (the highest power of x that appears in either one).

$$f(x) = \frac{2x^2}{x^2 - 4} = \frac{\frac{2x^2}{x^2}}{\frac{x^2 - 4}{x^2}} = \frac{\frac{2x^2}{x^2}}{\frac{x^2}{x^2} - \frac{4}{x^2}} = \frac{2}{1 - \frac{4}{x^2}}$$

When $|x|$ is very large, the fraction $4/x^2$ is very close to 0, so that the denominator is very close to 1 and $f(x)$ is very close to 2. Hence the line $y = 2$ is the horizontal asymptote of the graph. Using this information and plotting several points in each of the three regions determined by the vertical asymptotes, we obtain Figure 3.52 on the next page. ◀ (3)

(3)

List the vertical and horizontal asymptotes of the function.

(a) $f(x) = \dfrac{3x + 5}{x + 5}$

(b) $g(x) = \dfrac{2 - x^2}{x^2 - 4}$

Answers:

(a) Vertical $x = -5$; horizontal $y = 3$.

(b) Vertical $x = -2$ and $x = 2$; horizontal $y = -1$.

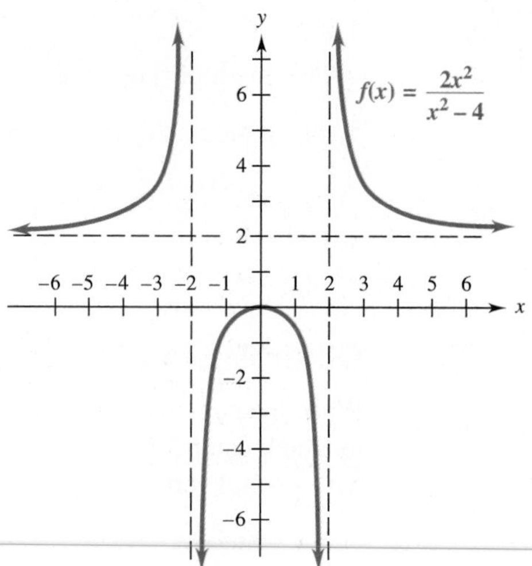

FIGURE 3.52

TECHNOLOGY TIP When a function whose graph has more than one vertical asymptote (as in Example 3) is graphed on a graphing calculator, erroneous vertical lines can sometimes be avoided by using a *decimal window* (with the *y*-range adjusted to show the graph). On TI, Sharp, and HP-39, use (Z)DECIMAL in the ZOOM or VIEWS menu. On Casio, use INIT in the V-WINDOW menu. Figure 3.53 shows the function of Example 3 graphed in a decimal window on a TI-83. (The *x*-range may be different on other calculators.)

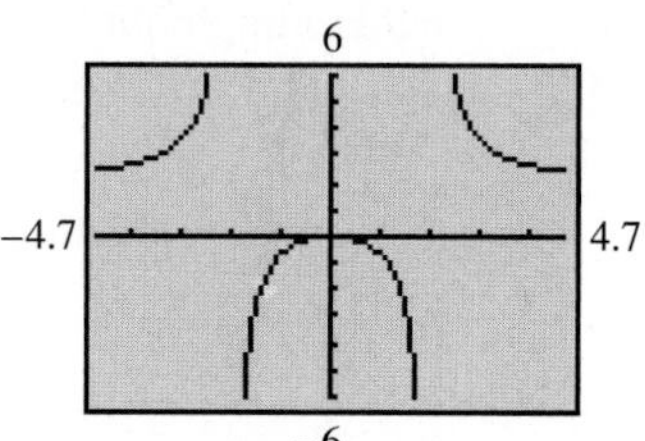

FIGURE 3.53

The arguments used to find the horizontal asymptotes in Examples 1–3 work in the general case and lead to this conclusion.

If the numerator of the rational function $f(x)$ has *smaller* degree than the denominator, then the x-axis (the line $y = 0$) is the horizontal asymptote of the graph. If the numerator and denominator have the *same* degree, say, $f(x) = \frac{ax^n + \cdots}{cx^n + \cdots}$, then the line $y = \frac{a}{c}$ is the horizontal asymptote.*

*When the numerator has larger degree than the denominator, the graph has no horizontal asymptote but may have non-horizontal lines or other curves as asymptotes; see Exercises 32 and 33 at the end of this section for examples.

Applications

Rational functions have a variety of applications, some of which are explored here.

EXAMPLE 4 In many situations involving environmental pollution, much of the pollutant can be removed from the air or water at a fairly reasonable cost, but the last, small part of the pollutant can be very expensive to remove.

Cost as a function of the percentage of pollutant removed from the environment can be calculated for various percentages of removal, with a curve fitted through the resulting data points. This curve then leads to a function that approximates the situation. Rational functions often are a good choice for these **cost-benefit functions.**

Using the function of Example 4, find the cost to remove the following percents of pollutants.

(a) 70%

(b) 85%

(c) 98%

Answers:

(a) $35,000

(b) About $73,000

(c) About $221,000

For example, suppose a cost-benefit function is given by

$$f(x) = \frac{18x}{106 - x},$$

where $f(x)$ or y is the cost (in thousands of dollars) of removing x percent of a certain pollutant. The domain of x is the set of all numbers from 0 to 100, inclusive; any amount of pollutant from 0% to 100% can be removed. To remove 100% of the pollutant here would cost

$$y = \frac{18(100)}{106 - 100} = 300,$$

or $300,000. Check that 95% of the pollutant can be removed for about $155,000, 90% for about $101,000, and 80% for about $55,000, as shown in Figure 3.54 (in which the displayed y-coordinates are rounded to the nearest integer). ◀ 4

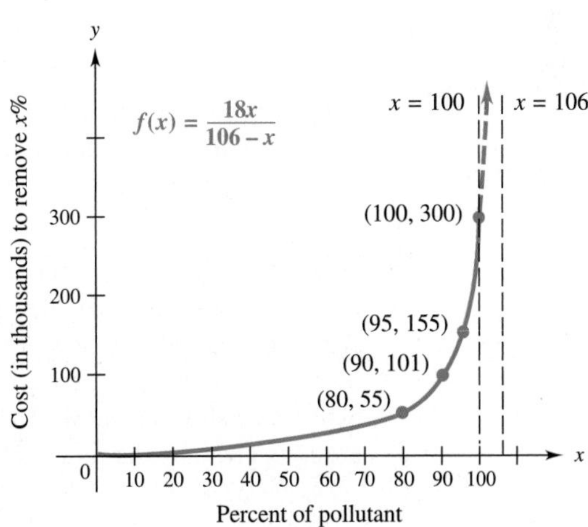

FIGURE 3.54

In management, **product-exchange functions** give the relationship between quantities of two items that can be produced by the same machine or factory. For example, an oil refinery can produce gasoline, heating oil, or a combination of the two; a winery can produce red wine, white wine, or a combination of the two. The next example discusses a product-exchange function.

EXAMPLE 5 The product-exchange function for the Golden Grape Winery for red wine x and white wine y, in tons, is

$$y = \frac{100{,}000 - 50x}{1000 + x}.$$

Graph the function and find the maximum quantity of each kind of wine that can be produced.

Only nonnegative values of x and y make sense in this situation, so we graph the function in the first quadrant (Figure 3.55). Note that the y-intercept of the graph (found by setting $x = 0$) is 100 and the x-intercept (found by setting $y = 0$ and solving for x) is 2000. Since we are interested only in the Quadrant I portion of the graph, we can find a few more points in Quadrant I and complete the graph as shown in Figure 3.55.

Rework Example 5 with the product-exchange function

$$y = \frac{60{,}000 - 10x}{60 + x}$$

to find the maximum amount of each wine that can be produced.

Answer:

6000 tons of red, 1000 tons of white

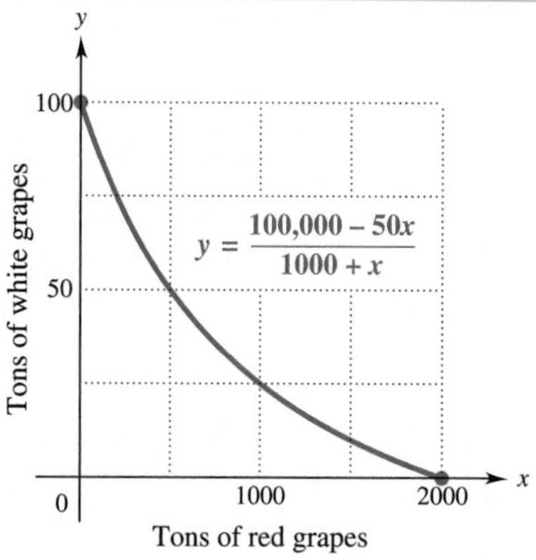

FIGURE 3.55

The maximum value of y occurs when $x = 0$, so the maximum amount of white wine that can be produced is 100 tons, given by the y-intercept. The x-intercept gives the maximum amount of red wine that can be produced, 2000 tons. ◀ 5

EXAMPLE 6 A retailer buys 2500 specialty lightbulbs from a distributor each year. In addition to the cost of each bulb, there is an order fee for each order, so she wants to order as few times as possible. On the other hand, storage costs are higher when there are fewer orders (and hence more bulbs per order to store). Past experience shows that the total annual cost (for the bulbs, ordering fees, and storage costs) is given by the rational function.

$$C(x) = \frac{.98x^2 + 1200x + 22{,}000}{x},$$

where x is the number of bulbs ordered each time. How many bulbs should be ordered each time in order to have the smallest possible cost?

Graph the cost function $C(x)$ in a window with $0 \leq x \leq 2500$ (because she can't order a negative number and needs only 2500 for the year).

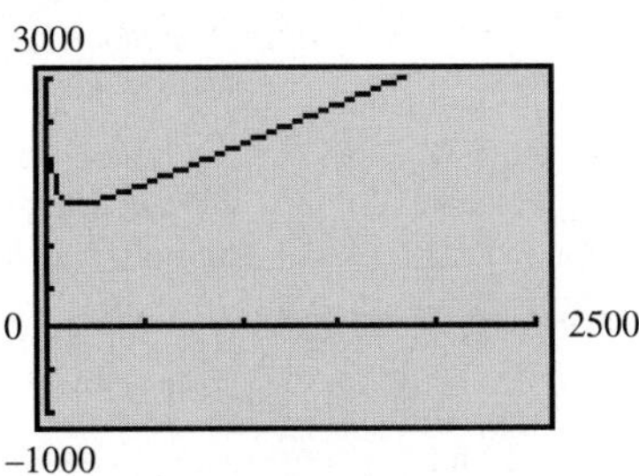

FIGURE 3.56

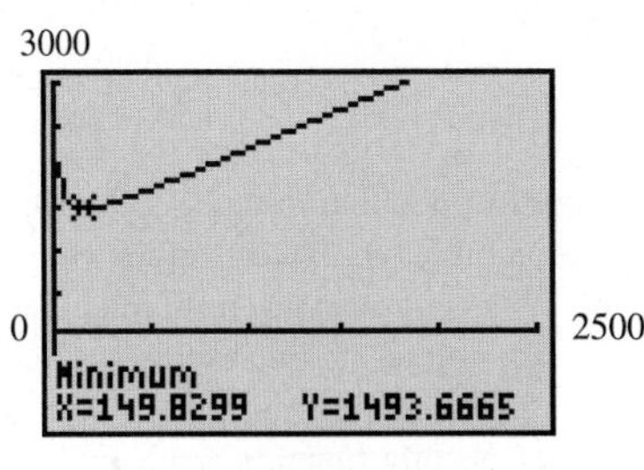

FIGURE 3.57

For each point on the graph in Figure 3.56,

the x-coordinate is the number of bulbs ordered each time;

the y-coordinate is the annual cost when x bulbs are ordered each time.

Use the minimum finder on a graphing calculator to find the point with smallest y-coordinate, which is approximately (149.83, 1493.67), as shown in Figure 3.57. Since the retailer can't order part of a lightbulb, she should order 150 bulbs each time for an approximate annual cost of $1494.

3.7 Exercises

Graph each function. Give the equations of the vertical and horizontal asymptotes (see Examples 1–3).

1. $f(x) = \dfrac{1}{x+5}$
2. $g(x) = \dfrac{-7}{x-6}$
3. $f(x) = \dfrac{-3}{2x+5}$
4. $h(x) = \dfrac{-4}{2-x}$
5. $f(x) = \dfrac{3x}{x-1}$
6. $g(x) = \dfrac{x-2}{x}$
7. $f(x) = \dfrac{x+1}{x-4}$
8. $f(x) = \dfrac{x-3}{x+5}$
9. $f(x) = \dfrac{2-x}{x-3}$
10. $g(x) = \dfrac{3x-2}{x+3}$
11. $f(x) = \dfrac{2x-1}{4x+2}$
12. $f(x) = \dfrac{3x-6}{6x-1}$
13. $h(x) = \dfrac{x+1}{x^2+3x-4}$
14. $g(x) = \dfrac{1}{x(x+1)^2}$
15. $f(x) = \dfrac{x^2+1}{x^2-1}$
16. $f(x) = \dfrac{x-1}{x^2-x-6}$

Find the equations of the vertical asymptotes of each of the following rational functions.

17. $f(x) = \dfrac{x-3}{x^2+x-2}$
18. $g(x) = \dfrac{x+2}{x^2-1}$
19. $g(x) = \dfrac{x^2+2x}{x^2-4x-5}$
20. $f(x) = \dfrac{x^2-2x-4}{x^3-2x^2+x}$

Work these problems. (See Example 4.)

21. **Natural Science** Suppose a cost-benefit model is given by
$$f(x) = \frac{4.3x}{100-x},$$
where $f(x)$ is the cost, in thousands of dollars, of removing x percent of a given pollutant. Find the cost of removing each of the following percents of pollutants.
(a) 50% **(b)** 70%
(c) 80% **(d)** 90%
(e) 95% **(f)** 98%
(g) 99%
(h) Is it possible, according to this model, to remove *all* the pollutant?
(i) Graph the function.

22. **Natural Science** Suppose a cost-benefit model is given by
$$f(x) = \frac{4.5x}{101-x},$$
where $f(x)$ is the cost, in thousands of dollars, of removing x percent of a certain pollutant. Find the cost of removing the following percents of pollutants.
(a) 0% **(b)** 50%
(c) 80% **(d)** 90%
(e) 95% **(f)** 99%
(g) 100% **(h)** Graph the function.

23. Natural Science The function

$$f(x) = \frac{\lambda x}{1 + (ax)^b}$$

is used in population models to give the size of the next generation ($f(x)$) in terms of the current generation (x).*

(a) What is a reasonable domain for this function, considering what x represents?
(b) Graph this function for $\lambda = a = b = 1$ and $x \geq 0$.
(c) Graph this function for $\lambda = a = 1$ and $b = 2$ and $x \geq 0$.
(d) What is the effect of making b larger?

24. Natural Science The function

$$f(x) = \frac{Kx}{A + x}$$

is used in biology to give the growth rate of a population in the presence of a quantity x of food. This is called Michaelis–Menten kinetics.†

(a) What is a reasonable domain for this function, considering what x represents?
(b) Graph this function for $K = 5, A = 2$ and $x \geq 0$.
(c) Show that $y = K$ is a horizontal asymptote.
(d) What do you think K represents?
(e) Show that A represents the quantity of food for which the growth rate is half of its maximum.

25. Social Science The average waiting time in a line (or queue) before getting served is given by

$$W = \frac{S(S - A)}{A},$$

where A is the average rate at which people arrive at the line and S is the average service time. At a certain fast food restaurant, the average service time is 3 minutes. Find W for each of the following average arrival times.

(a) 1 minute
(b) 2 minutes
(c) 2.5 minutes
(d) What is the vertical asymptote?
(e) Graph the equation on the interval (0, 3).
(f) What happens to W when $A > 3$? What does this mean?

*See J. Maynard Smith, *Models in Ecology* (Cambridge University Press, 1974).

†See Leah Edelstein-Keshet, *Mathematical Models in Biology* (Random House, 1988).

Business *Sketch the Quadrant I portion of the graph of each of the functions defined below, and then estimate the maximum quantities of each product that can be produced. (See Example 5.)*

26. The product-exchange function for gasoline, x, and heating oil, y, in hundreds of gallons per day, is

$$y = \frac{125{,}000 - 25x}{125 + 2x}.$$

27. A drug factory found the product-exchange function for a red tranquilizer, x, and a blue tranquilizer, y, is

$$y = \frac{900{,}000{,}000 - 30{,}000x}{x + 90{,}000}.$$

28. Physical Science The failure of several O-rings in field joints was the cause of the fatal crash of the *Challenger* space shuttle in 1986. NASA data from 24 successful launches prior to *Challenger* suggested that O-ring failure was related to launch temperature by a function similar to

$$N(t) = \frac{600 - 7t}{4t - 100} \qquad (50 \leq t \leq 85),$$

where t is the temperature (in °F) at launch and N is the approximate number of O-rings that fail. Assume that this function accurately models the number of O-ring failures that would occur at lower launch temperatures (an assumption NASA did not make).

(a) Does $N(t)$ have a vertical asymptote? At what value of t does it occur?
(b) Without graphing, what would you conjecture that the graph would look like just to the right of the vertical asymptote? What does this suggest about the number of O-ring failures that might be expected near that temperature? (The temperature at the *Challenger* launching was 31°.)
(c) Confirm your conjecture by graphing $N(t)$ between the vertical asymptote and $t = 85$.

29. Business A company has fixed costs of \$40,000 and marginal cost of \$2.60 per unit.

(a) Find the linear cost function.
(b) Find the average cost function. (Average cost was defined in Section 3.3.)
(c) Find the horizontal asymptote of the graph of the average cost function. Explain what the asymptote means in this situation (how low can the average cost possibly be?).

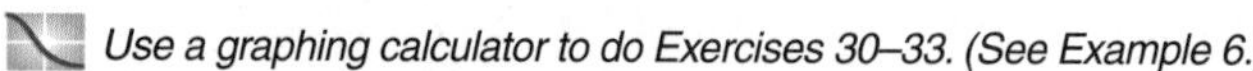

Use a graphing calculator to do Exercises 30–33. (See Example 6.)

30. Finance Another model of a Laffer curve (see Exercise 27 of Section 3.6) is given by

$$f(x) = \frac{300x - 3x^2}{10x + 200},$$

where $f(x)$ is government revenue (in billions of dollars) from a tax rate of x percent. Find the revenue from the following tax rates.

(a) 16% (b) 25%
(c) 40% (d) 55%
(e) Graph $f(x)$.
(f) What tax rate produces maximum revenue? What is the maximum revenue?

31. Business When no more than 110 units are produced, the cost of producing x units is given by

$$C(x) = .2x^3 - 25x^2 + 1531x + 25{,}000.$$

How many units should be produced in order to have the lowest possible average cost?

32. (a) Graph $f(x) = \dfrac{x^3 + 3x^2 + x + 1}{x^2 + 2x + 1}$

(b) Does the graph tote? Does the zontal straight li

(c) Graph $f(x)$ and t Does this line a of $f(x)$?

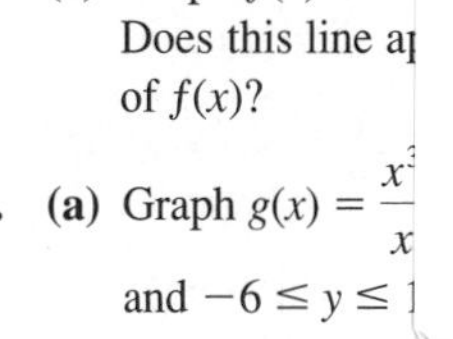

33. (a) Graph $g(x) = \dfrac{x^3}{x}$ and $-6 \le y \le$

(b) Graph $g(x)$ and the parabola $y = x^2 + x + 1$ on the same screen. How do the two graphs compare when $|x| \ge 2$?

CHAPTER 3 SUMMARY

Key Terms and Symbols

3.1 function
domain
range
functional notation

3.2 graph
linear function
piecewise linear function
absolute value function
greatest integer function
step function
vertical line test

3.3 fixed costs
variable cost
average cost
average rate of change
marginal cost
linear cost function
break-even point
supply and demand curves
equilibrium point
equilibrium price
equilibrium quantity

3.4 quadratic function
parabola
vertex
axis

3.5 quadratic model

3.6 polynomial function
graph of $f(x) = ax^n$
properties of polynomial graphs
polynomial models

3.7 rational function
linear rational function
vertical asymptote
horizontal asymptote

Key Concepts

A **function** consists of a set of inputs called the **domain,** a set of outputs called the **range,** and a rule by which each input determines exactly one output.

If a vertical line intersects a graph in more than one point, the graph is not that of a function.

A **linear cost function** has equation $C(x) = mx + b$ where m is the **marginal cost** (the cost of producing one more item) and b is the **fixed cost.**

If $p = f(q)$ gives the price per unit when x units can be supplied, and $p = g(q)$ gives the price per unit when q units are demanded, then the **equilibrium price** and **equilibrium quantity** occur at the q-value such that $f(q) = g(q)$.

The **quadratic function** defined by $f(x) = a(x - h)^2 + k$ has a graph that is a **parabola** with vertex (h, k) and axis of symmetry $x = h$. The parabola opens upward if $a > 0$, downward if $a < 0$.

If the equation is in the form $f(x) = ax^2 + bx + c$, the vertex is $\left(-\frac{b}{2a}, f\left(-\frac{b}{2a}\right)\right)$.

When $|x|$ is large, the graph of a **polynomial function** resembles the graph of its highest degree term ax^n. The graph of $f(x) = ax^n$ is described on page 169.

On the graph of a polynomial function of degree n:

The total number of peaks and valleys is at most $n - 1$.

The total number of x-intercepts is at most n.

If a number c makes the denominator of a **rational function** 0, but the numerator nonzero, then the line $x = c$ is a **vertical asymptote** for the graph.

Whenever the values of y approach, but never equal, some number k as $|x|$ gets larger and larger, the line $y = k$ is a **horizontal asymptote** for the graph.

If the numerator of a rational function has *smaller* degree than the denominator, then the x-axis is the horizontal asymptote of the graph.

If the numerator and denominator of a rational function have the *same* degree, say, $f(x) = \frac{ax^n + \cdots}{cx^n + \cdots}$, then the line $y = \frac{a}{c}$ is the horizontal asymptote of the graph.

Chapter 3 Review Exercises

Which of the following rules defines a function?

1.

x	3	2	1	0	1	2
y	8	5	2	0	−2	−5

2.

x	2	1	0	−1	−2
y	5	3	1	−1	−3

3. $y = \sqrt{x}$

4. $x = |y|$

5. $x = y^2 + 1$

6. $y = 5x - 2$

For each function, find

(a) $f(6)$ **(b)** $f(-2)$ **(c)** $f(p)$ **(d)** $f(r + 1)$.

7. $f(x) = 4x - 1$

8. $f(x) = 3 - 4x$

9. $f(x) = -x^2 + 2x - 4$

10. $f(x) = 8 - x - x^2$

11. Let $f(x) = 5x - 3$ and $g(x) = -x^2 + 4x$. Find each of the following.

(a) $f(-2)$ **(b)** $g(3)$ **(c)** $g(-k)$

(d) $g(3m)$ **(e)** $g(k - 5)$ **(f)** $f(3 - p)$

12. Let $f(x) = x^2 + x + 1$. Find each of the following.

(a) $f(3)$ **(b)** $f(1)$ **(c)** $f(4)$

(d) Based on your answers in parts (a)–(c), is it true that $f(a + b) = f(a) + f(b)$ for all real numbers a, b?

Graph each function.

13. $f(x) = |x| - 3$

14. $f(x) = -|x| - 2$

15. $f(x) = -|x + 1| + 3$

16. $f(x) = 2|x - 3| - 4$

17. $f(x) = [x - 3]$

18. $f(x) = \left[\frac{1}{2}x - 2\right]$

19. $f(x) = \begin{cases} -4x + 2 & \text{if } x \le 1 \\ 3x - 5 & \text{if } x > 1 \end{cases}$

20. $f(x) = \begin{cases} 3x + 1 & \text{if } x < 2 \\ -x + 4 & \text{if } x \ge 2 \end{cases}$

21. $f(x) = \begin{cases} |x| & \text{if } x < 3 \\ 6 - x & \text{if } x \ge 3 \end{cases}$

22. $f(x) = \sqrt{x^2}$

23. $g(x) = x^2/8 - 3$

24. $h(x) = \sqrt{x} + 2$

25. **Business** Let f be a function that gives the cost to rent a floor polisher for x hours. The cost is a flat \$3 for renting the polisher plus \$4 per day or fraction of a day for using the polisher.

(a) Graph f.

(b) Give the domain and range of f.

(c) David Fleming wants to rent a polisher, but he can spend no more than \$15. At most how many days can he use it?

26. Business A trailer hauling service charges \$45, plus \$2 per mile or part of a mile.
(a) Is \$90 enough for a 20-mile haul?
(b) Graph the ordered pairs (miles, cost).
(c) Give the domain and range.

27. Social Science The percent of children born to unmarried mothers since 1970 can be approximated by

$$f(x) = \begin{cases} .9x + 11 & \text{from 1970 to 1994} \\ 32.6 & \text{from 1994 to 2000} \end{cases}$$

Let $x = 0$ correspond to 1970 and graph the function. What does the graph suggest about births to unmarried mothers?

28. Social Science According to data from the U.S. Department of Agriculture, the annual per capita consumption of red meat in the United States has declined in a linear fashion from 131.7 pounds in 1970 to 107.7 in 2000.
(a) Let $x = 0$ correspond to 1900 and find the rule of a linear function that gives the approximate annual per capita consumption of red meat in year x.
(b) If this trend continues, in what year will per capita consumption be 100 pounds?

Business *In Exercises 29–32, find the following.*
(a) *the linear cost function*
(b) *the marginal cost*
(c) *the average cost per unit to produce 100 units*

29. Eight units cost \$300; fixed cost is \$60.
30. Fixed cost is \$2000; 36 units cost \$8480.
31. Twelve units cost \$445; 50 units cost \$1585.
32. Thirty units cost \$1500; 120 units cost \$5640.

33. Business The cost of producing x ink cartridges for a printer is given by $C(x) = 24x + 18{,}000$. Each cartridge can be sold for \$28.
(a) What are the fixed costs?
(b) Find the revenue function.
(c) Find the break-even point.
(d) If the company sells exactly the number of cartridges needed to break even, what is its revenue?

34. Business The prime time viewing share of network TV stations in year x is approximated by

$$f(x) = -1.2x + 68 \qquad (9 \le x \le 22),$$

where $x = 0$ corresponds to 1980. The share of basic cable stations is approximated by $g(x) = 2.3x - 3.6$.* In what year did network and cable stations have the same viewing share?

35. Business Suppose the demand and price for the HBO cable channel are related by $p = -.5q + 30.95$, where p is the monthly price in dollars and q is measured in millions of subscribers. If the price and supply are related by $p = .3q + 2.15$, what are the equilibrium quantity and price?

*Based on data from the National Cable Television Association.

36. Business Suppose the supply and price for prescription strength Tylenol are related by $p = .0015q + 1$, where p is the price (in dollars) of a 30-day prescription. If the demand is related to price by $p = -.0025q + 64.36$, what are the equilibrium quantity and price?

Without graphing, determine whether each of the following parabolas opens upward or downward and find its vertex.

37. $f(x) = 3(x - 2)^2 + 6$
38. $f(x) = 2(x + 3)^2 - 5$
39. $g(x) = -4(x + 1)^2 + 8$
40. $g(x) = -5(x - 4)^2 - 6$

Graph each of the following and label its vertex.

41. $f(x) = x^2 - 4$
42. $f(x) = 6 - x^2$
43. $f(x) = x^2 + 2x - 3$
44. $f(x) = -x^2 + 6x - 3$
45. $f(x) = -x^2 - 4x + 1$
46. $f(x) = 4x^2 - 8x + 3$
47. $f(x) = 2x^2 + 4x - 3$
48. $f(x) = -3x^2 - 12x - 8$

Determine whether each of the following functions has a minimum or a maximum value and find that value.

49. $f(x) = x^2 + 6x - 2$
50. $f(x) = x^2 + 4x + 5$
51. $g(x) = -4x^2 + 8x + 3$
52. $g(x) = -3x^2 - 6x + 3$

Solve each problem.

53. Business The commodity market is very unstable; money can be made or lost quickly when investing in soybeans, wheat, pork bellies, and the like. Suppose that an investor kept track of her total profit, P, at time t, measured in months, after she began investing and found that $P = -4t^2 + 32t - 20$. At what time is her profit largest? (*Hint:* $t > 0$ in this case.)

54. Physical Science The height h (in feet) of a rocket at time t seconds after liftoff is given by $h = -16t^2 + 800t$.
(a) How long does it take the rocket to reach 3200 feet?
(b) What is the maximum height of the rocket?

55. Business The manager of a large apartment complex has found that the profit is given by $P = -x^2 + 250x - 15{,}000$, where x is the number of units rented. For what value of x does the complex produce the largest profit?

56. Business A rectangular enclosure is to be built with three sides made out of redwood fencing at a cost of \$15 per running foot and the fourth side made out of cement blocks at a cost of \$30 per running foot. \$900 is available for the project. What are the dimensions of the enclosure with maximum possible area and what is this area?

57. Finance The table shows the expenditures (in billions of dollars) of the federal government on transportation in selected years.*

Year	1995	1997	1999	2000	2001
Expenditures	39	41	43	47	50

(a) Let $x = 0$ correspond to 1990. Find a quadratic function $f(x) = a(x - h)^2 + k$ that models this data, using (5, 39) as the vertex and the data point for 2001.
(b) Assuming the model remains valid, what are the transportation expenditures (to the nearest billion) in 2002?

58. Health The table shows the number of AIDS patients admitted to hospital in the state of New York in selected years.†

Year	1986	1988	1990	1992
Patients	11,875	23,440	40,940	57,500

Year	1994	1996	1998	1999
Patients	61,875	60,625	47,500	45,000

(a) Let $x = 0$ correspond to 1980 and plot the data points on a graphing calculator.
(b) Use quadratic regression to find a quadratic model for the data.
(c) Estimate the number of patients admitted in 1991.
(d) To the nearest year, in what year were the most patients admitted?

Graph each of the following polynomial functions.

59. $f(x) = x^4 - 2$

60. $g(x) = x^3 - x$

61. $f(x) = x(x - 2)(x + 3)$

62. $f(x) = (x - 1)(x + 2)(x - 3)$

63. $f(x) = 3x(3x + 2)(x - 1)$

64. $f(x) = x^3 - 3x^2 - 4x$

65. $f(x) = x^4 - 5x^2 - 6$

66. $f(x) = x^4 - 7x^2 - 8$

Use a graphing calculator to do Problems 67–70.

67. Business The demand equation for automobile oil filters is

$$p = -.000012q^3 - .00498q^2 + .1264q + 1508,$$

where p is in dollars and q is in thousands of items. The supply equation is

$$p = -.000001q^3 + .00097q^2 + 2q.$$

Find the equilibrium quantity and the equilibrium price.

68. Business The average cost (in dollars) per item of manufacturing x thousand cans of spray paint is given by

$$A(x) = -.000006x^4 + .0017x^3 + .03x^2 - 24x + 1110.$$

How many cans should be manufactured if the average cost is to be as low as possible? What is the average cost in that case?

69. Business Plastic racks for holding compact discs sell for \$23 each. The cost of manufacturing x racks is given by $C(x) = -.000006x^3 + .07x^2 + 2x + 1200$. At most 600 racks can be produced.
(a) Find the revenue and profit functions.
(b) What is the break-even point?
(c) What is the maximum number of racks that can be made without losing money?
(d) How many racks should be made in order to have as large a profit as possible? What is that profit?

70. Health The table shows the average remaining life expectancy (in years) of a male at selected ages.*

Current Age	Birth	20	40	60	80	90	100
Life Expectancy	73.6	54.7	36.2	19.4	7.5	4	2.4

(a) Let birth = 0 and plot the data points.
(b) Use quartic regression to find a fourth-degree polynomial $f(x)$ that models the data.
(c) What is the life expectancy of a man of age 25? Age 35? Age 50?
(d) What is the life expectancy of a man who is exactly your age?

List the vertical and horizontal asymptotes of each function and sketch its graph.

71. $f(x) = \dfrac{1}{x - 3}$

72. $f(x) = \dfrac{-2}{x + 4}$

73. $f(x) = \dfrac{-3}{2x - 4}$

74. $f(x) = \dfrac{5}{3x + 7}$

75. $g(x) = \dfrac{5x - 2}{4x^2 - 4x - 3}$

76. $g(x) = \dfrac{x^2}{x^2 - 1}$

77. Business The average cost per carton of producing x cartons of cocoa is given by

$$C(x) = \frac{650}{2x + 40}.$$

*U.S. Department of Treasury and the Office of Management and Budget.

†New York State Department of Health.

*National Center for Health Statistics.

Find the average cost per carton to make the given number of cartons.

(a) 10 cartons
(b) 50 cartons
(c) 70 cartons
(d) 100 cartons
(e) Graph $C(x)$.

78. Business The cost and revenue functions (in dollars) for a frozen yogurt shop are given by

$$C(x) = \frac{400x + 400}{x + 4} \quad \text{and} \quad R(x) = 100x,$$

where x is measured in hundreds of units.

(a) Graph $C(x)$ and $R(x)$ on the same axes.
(b) What is the break-even point for this shop?
(c) If the profit function is given by $P(x)$, does $P(1)$ represent a profit or a loss?
(d) Does $P(4)$ represent a profit or a loss?

79. Business The supply and demand functions for the yogurt shop in Exercise 78 are as follows:

$$\text{supply: } p = \frac{q^2}{4} + 25; \qquad \text{demand: } p = \frac{500}{q},$$

where p is the price in dollars for q hundred units of yogurt.

(a) Graph both functions on the same axes, and from the graph, estimate the equilibrium point.
(b) Give the q-intervals where supply exceeds demand.
(c) Give the q-intervals where demand exceeds supply.

80. Business A cost-benefit curve for pollution control is given by

$$y = \frac{9.2x}{106 - x},$$

where y is the cost in thousands of dollars of removing x percent of a specific industrial pollutant. Find y for each of the following values of x.

(a) $x = 50$
(b) $x = 98$
(c) What percent of the pollutant can be removed for $22,000?

Chapter 4

Exponential and Logarithmic Functions

Exponential and logarithmic functions play a key role in management, economics, the social and physical sciences, and engineering. They are used to study the growth of money and organizations; learning curves; the growth of human, animal, and bacterial populations; the spread of disease; and radioactive decay.

4.1 EXPONENTIAL FUNCTIONS

In polynomial functions, the variable is raised to various constant exponents; for instance, $f(x) = x^2 + 3x - 5$. In *exponential functions,* such as

$$f(x) = 10^x \quad \text{or} \quad g(x) = 2^{-x^2} \quad \text{or} \quad h(x) = 3^{.6x},$$

the variable is in the exponent and the base is a positive constant. We begin with the simplest type of exponential function.

If a is a positive constant other than 1 and k is any nonzero constant, then the function defined by

$$f(x) = a^{kx}$$

is called an **exponential function with base *a*.** Its domain is the set of all real numbers.

The function $f(x) = 1^x$ is the constant function $f(x) = 1$. Exponential functions with negative bases are not of interest because when a is negative, a^x may not be defined for some values of x; for instance, $(-4)^{1/2} = \sqrt{-4}$ is not a real number.

EXAMPLE 1 Graph these exponential functions and discuss the characteristics of their graphs.

(a) $f(x) = 2^x$

Either use a graphing calculator or graph by hand: make a table of values, plot the corresponding points, and join them by a smooth curve, as in Figure 4.1. The graph has the following properties.

The graph rises as x increases (that is, $f(x)$ is an increasing function).

The y-intercept is at 1.

The graph is always above the x-axis (because *every* power of 2 is positive), so there is no x-intercept.

There is no vertical asymptote, but the negative x-axis is a horizontal asymptote (the graph approaches but never touches it).

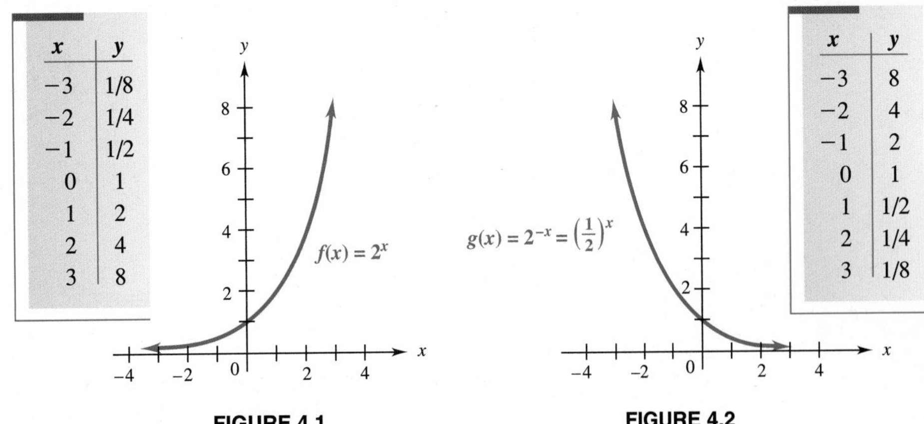

x	y
−3	1/8
−2	1/4
−1	1/2
0	1
1	2
2	4
3	8

x	y
−3	8
−2	4
−1	2
0	1
1	1/2
2	1/4
3	1/8

FIGURE 4.1

FIGURE 4.2

1 

Use a graphing calculator to graph $f(x) = 4^x$ and $g(x) = \left(\frac{1}{4}\right)^x$ on the same screen.

Answer:

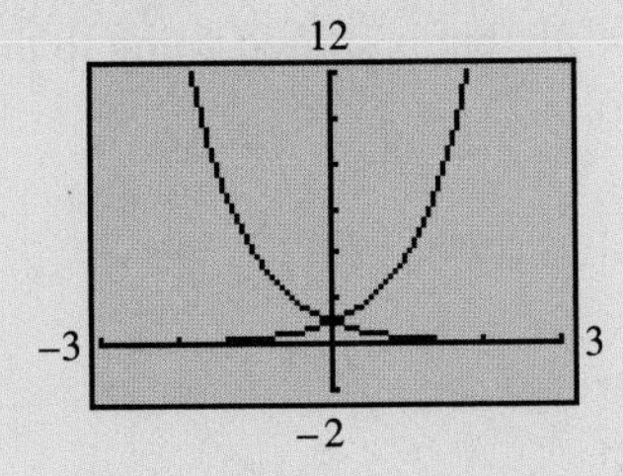

(b) $g(x) = 2^{-x}$

Again, use a graphing calculator or graph by hand, as in Figure 4.2. The graph has the following properties.

The graph falls as x increases (that is, $f(x)$ is a decreasing function).

The y-intercept is at 1.

The graph is always above the x-axis (because *every* power of 2 is positive), so there is no x-intercept.

There is no vertical asymptote, but the positive x-axis is a horizontal asymptote (the graph approaches but never touches it).

Notice that

$$2^{-x} = \frac{1}{2^x} = \left(\frac{1}{2}\right)^x,$$

by properties of exponents. The graph of $g(x) = (1/2)^x$ is the mirror image of the graph of $f(x) = 2^x$, with the y-axis as the mirror. ◀ 1

The graph of $f(x) = 2^x$ in Figure 4.1 illustrates **exponential growth,** which is far more explosive than polynomial growth. For example, if the graph in Figure 4.1 were extended on the same scale to include the point $(50, 2^{50})$, the graph would be more than 2 billion *miles* high! Similarly, the graph of $g(x) = 2^{-x} = (1/2)^x$ illustrates **exponential decay.**

When $a > 1$, the graph of the exponential function $h(x) = a^x$ has the same basic shape as the graph of $f(x) = 2^x$ in Figure 4.1. The larger the base a is, the more steeply the graph rises, as illustrated in Figure 4.3(a).

When $0 < a < 1$, the graph of $h(x) = a^x$ has the same basic shape as the graph of $g(x) = (1/2)^x$ in Figure 4.2. Graphs of some typical exponential functions with $0 < a < 1$ are shown in Figure 4.3(b).

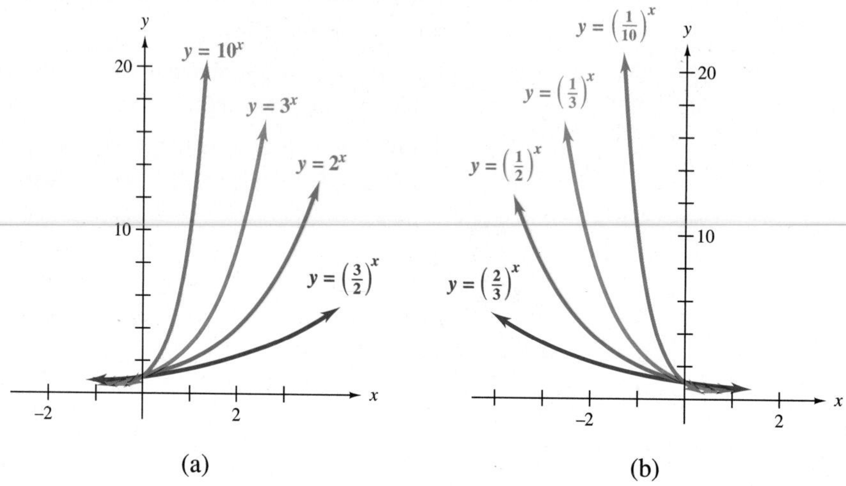

FIGURE 4.3

The graphs of exponential functions such as $f(x) = 3^{1-x}$ or $g(x) = 2^{.6x}$ or $h(x) = 3 \cdot 10^{2x+1}$ have the same general shape as the exponential graphs above. The only differences are that the graph may rise or fall at a different rate or the entire graph may be shifted vertically or horizontally.

EXAMPLE 2 Graph $f(x)$ and $g(x)$ as defined below and explain how the two graphs are related.

(a) $f(x) = 3^{1-x}$ and $g(x) = 3^{-x}$

Choose values of x that make the exponent positive, zero, and negative, and plot the corresponding points. The graphs are shown in Figure 4.4. The graph of $f(x) = 3^{1-x}$ has the same shape as the graph of $g(x) = 3^{-x}$ but is shifted 1 unit to the right, making the y-intercept (0, 3) rather than (0, 1).

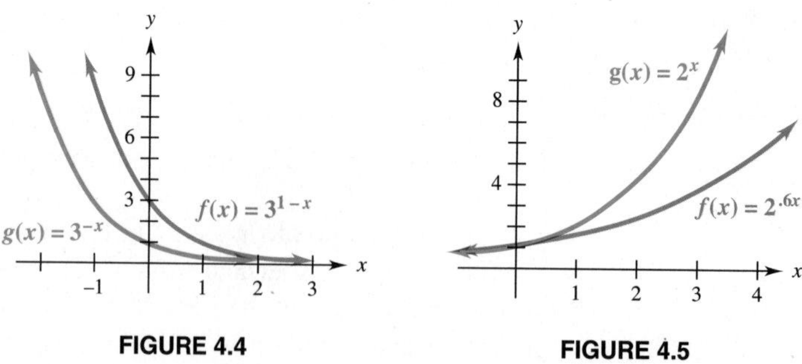

FIGURE 4.4 **FIGURE 4.5**

(b) $f(x) = 2^{.6x}$ and $g(x) = 2^x$

Comparing the graphs of $f(x) = 2^{.6x}$ and $g(x) = 2^x$ in Figure 4.5, we see that the graphs are both increasing but the graph of $f(x)$ rises at a slower rate. This happens because of the .6 in the exponent. If the coefficient of x were greater than 1, the graph would rise at a faster rate than the graph of $f(x) = 2^x$. ◀ (2)

(2)

Graph $f(x)= 2^{x+1}$.

Answer:

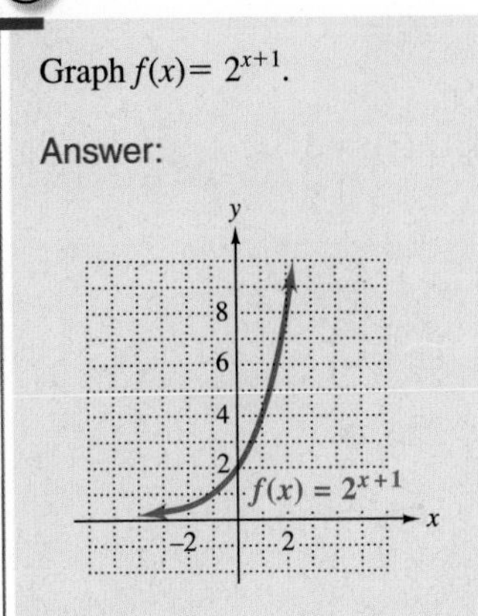

When the exponent involves a nonlinear expression in x, an exponential function graph may have a much different shape than the preceding ones.

EXAMPLE 3 Graph $f(x) = 2^{-x^2}$.

Either use a graphing calculator or plot points and connect them with a smooth curve, as in Figure 4.6. The graph is symmetric about the y-axis—that is, if the figure were folded on the y-axis, the two halves would match. Like the graphs above, this graph has the x-axis as a horizontal asymptote. The domain is still all real numbers, but here the range is $0 < y \leq 1$. Graphs such as this are important in probability, where the normal curve has an equation similar to $f(x)$ in this example. ◀ (3)

(3)

Graph $f(x)= \left(\frac{1}{2}\right)^{-x^2}$.

Answer:

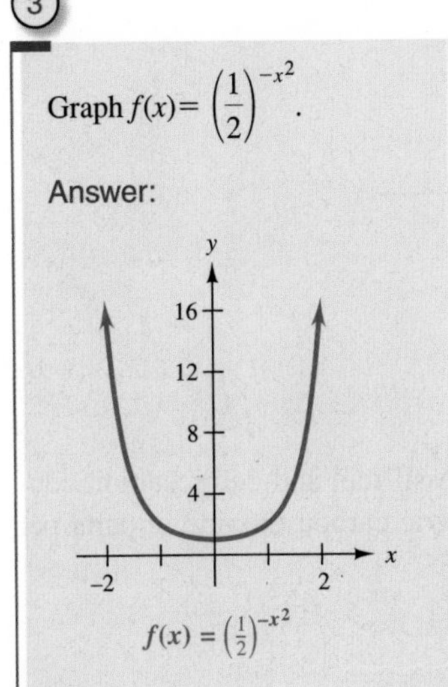

x	y
−2	1/16
−1.5	.21
−1	1/2
−.5	.84
0	1
.5	.84
1	1/2
1.5	.21
2	1/16

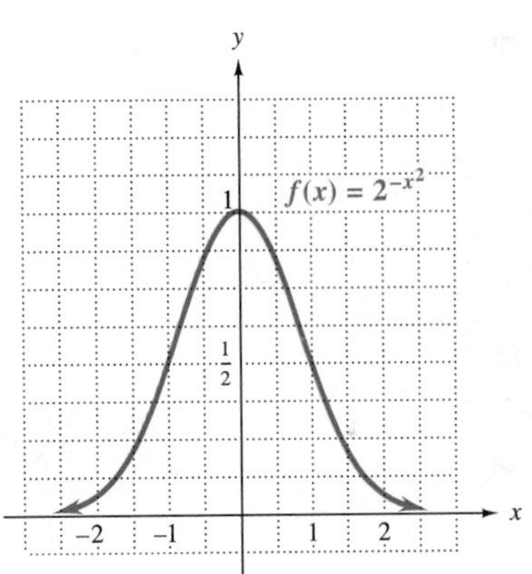

FIGURE 4.6

The Number *e*

There is an irrational number, denoted e, that arises naturally in a variety of mathematical situations. To 12 decimal places,

$$e \approx \mathbf{2.718281828459}.$$

Perhaps the single most useful exponential function is the function defined by $f(x) = e^x$.

TECHNOLOGY TIP To evaluate powers of e with a calculator, use the [e^x] key. On some calculators, you will need to use the two keys [INV] [LN] or [2nd] [LN]. For example, your calculator should show that $e^{.14} = 1.150273799$ to 8 decimal places. If these keys do not work, consult the instruction manual for your calculator. (4)

(4)

Evaluate the following powers of e.

(a) $e^{.06}$

(b) $e^{-.06}$

(c) $e^{2.30}$

(d) $e^{-2.30}$

Answers:

(a) 1.06184

(b) .94176

(c) 9.97418

(d) .10026

NOTE To display the decimal expansion of e on your calculator screen, calculate e^1.

In Figure 4.7 the functions defined by

$$g(x) = 2^x, \qquad f(x) = e^x, \qquad \text{and} \quad h(x) = 3^x$$

are graphed for comparison.

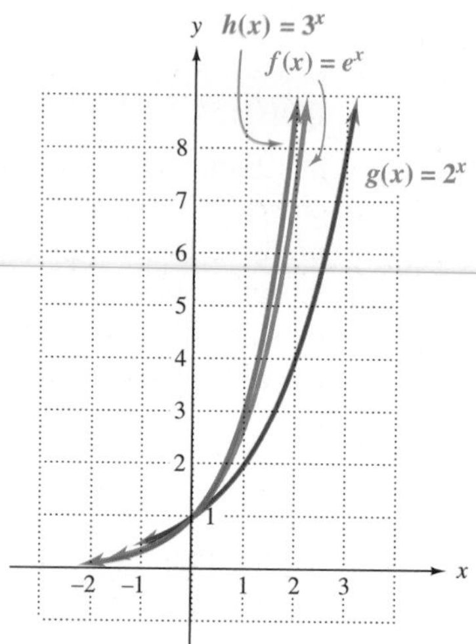

FIGURE 4.7

5

The number of organisms present at time t is given by $f(t) = 75e^{.458t}$.

(a) Is this a growth function or a decay function?

Find the number of organisms present at

(b) $t = 0$;

(c) $t = 2$;

(d) $t = 4$.

Answers:

(a) A growth function

(b) 75

(c) About 188

(d) About 469

EXAMPLE 4 If current trends of burning fossil fuel and deforestation continue, then the increase in the amounts of atmospheric carbon dioxide in parts per million (ppm) will be given by

$$f(t) = 375e^{.00609t},$$

where $t = 0$ corresponds to 2000.*

(a) What is the amount of carbon dioxide in 2000?

Since 2000 corresponds to $t = 0$, evaluate the function at $t = 0$:

$$f(0) = 375e^{.00609(0)} = 375e^0 = 375\,(1) = 375 \text{ ppm}.$$

(b) What will the amount of carbon dixode be in 2025?

Since 2025 corresponds to $t = 25$, the amount will be

$$f(25) = 375e^{.00609(25)} = 436.7 \text{ ppm}.$$

(c) Use a graphing calculator to find the year in which the amount of carbon dioxide will reach 500 ppm. In other words, solve the equation $375e^{.00609t} = 500$.

Graph $y = 375e^{.00609x}$ and $y = 500$ on the same screen. Find the intersection point, as in Figure 4.8, which shows that the 500 ppm level will be reached when t is about 47.2, that is, during 2047. ◀ 5

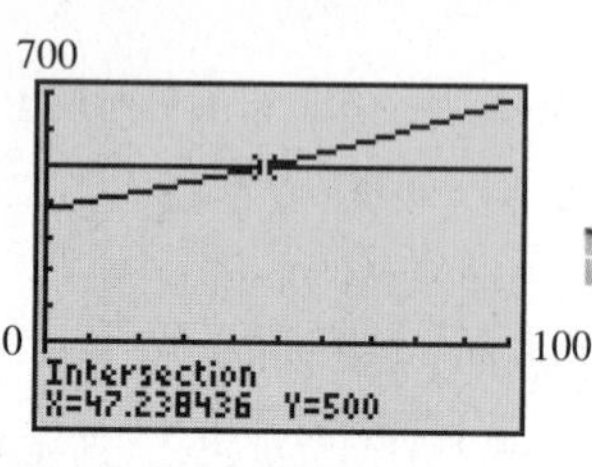

FIGURE 4.8

*Based on projections from the International Panel on Climate Change.

4.1 Exercises

Classify each function as linear, quadratic, or exponential.

1. $f(x) = 2x^2 + x - 3$

2. $g(x) = 5^{x-3}$

3. $h(x) = 3 \cdot 6^{2x^2+4}$

4. $f(x) = 3x + 1$

Without graphing,
(a) describe the shape of the graph of each function, and
(b) complete the ordered pairs (0,) and (1,) for each function.
(See Examples 1 and 2.)

5. $f(x) = .8^x$

6. $g(x) = 6^{-x}$

7. $f(x) = 5^{.4x}$

8. $g(x) = -e^x$

Graph each function. (See Examples 1 and 2.)

9. $f(x) = 3^x$

10. $g(x) = 3^{.5x}$

11. $f(x) = 2^{x/2}$

12. $g(x) = e^{x/4}$

13. $f(x) = (1/5)^x$

14. $g(x) = 2^{3x}$

15. Graph these functions on the same axes.
(a) $f(x) = 2^x$ **(b)** $g(x) = 2^{x+3}$ **(c)** $h(x) = 2^{x-4}$
(d) If c is a positive constant, explain how the graphs of $y = 2^{x+c}$ and $y = 2^{x-c}$ are related to the graph of $f(x) = 2^x$.

16. Graph these functions on the same axes.
(a) $f(x) = 3^x$ **(b)** $g(x) = 3^x + 2$
(c) $h(x) = 3^x - 4$
(d) If c is a positive constant, explain how the graphs of $y = 3^x + c$ and $y = 3^x - c$ are related to the graph of $f(x) = 3^x$.

The figure shows the graphs of $y = a^x$ for $a = 1.8, 2.3, 3.2, .4, .75$, and $.31$. They are identified by letter but not necessarily in the same order as the values of a just given. Use your knowledge of how the exponential function behaves for various powers of a to match each lettered graph with the correct value of a.

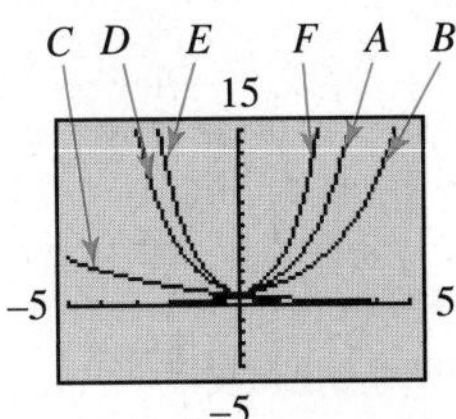

17. A **18.** B **19.** C **20.** D **21.** E **22.** F

In Exercises 23 and 24, the graph of an exponential function with base a is given. Follow the directions in parts (a)–(f) in each exercise.

23.

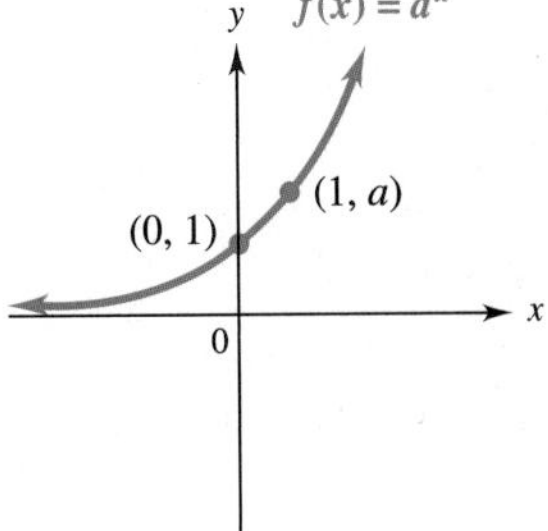

(a) Is $a > 1$ or is $0 < a < 1$?
(b) Give the domain and range of f.
(c) Sketch the graph of $g(x) = -a^x$.
(d) Give the domain and range of g.
(e) Sketch the graph of $h(x) = a^{-x}$.
(f) Give the domain and range of h.

24.

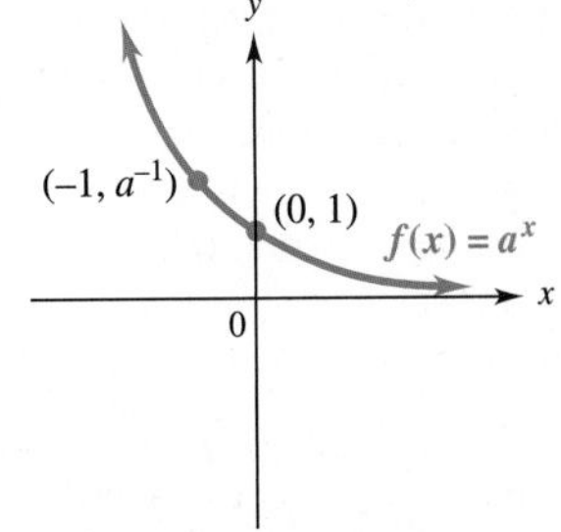

(a) Is $a > 1$ or is $0 < a < 1$?
(b) Give the domain and range of f.
(c) Sketch the graph of $g(x) = a^x + 2$.
(d) Give the domain and range of g.
(e) Sketch the graph of $h(x) = a^{x+2}$.
(f) Give the domain and range of h.

25. If $f(x) = a^x$ and $f(3) = 27$, find the following values of $f(x)$.
(a) $f(1)$ **(b)** $f(-1)$ **(c)** $f(2)$ **(d)** $f(0)$

26. Give a rule of the form $f(x) = a^x$ to define the exponential function whose graph contains the given point.
(a) $(3, 8)$ **(b)** $(-3, 64)$

Graph each function. (See Example 3.)

27. $f(x) = 2^{-x^2+2}$

28. $g(x) = 2^{x^2-2}$

29. $f(x) = x \cdot 2^x$

30. $f(x) = x^2 \cdot 2^x$

Work the following exercises.

31. Finance If $1 is deposited into an account paying 6% per year compounded annually, then after t years the account will contain

$$y = (1 + .06)^t = (1.06)^t$$

dollars.

(a) Use a calculator to complete the following table.

t	0	1	2	3	4	5	6	7	8	9	10
y	1					1.34					1.79

(b) Graph $y = (1.06)^t$.

32. Finance If money loses value at the rate of 3% per year, the value of $1 in t years is given by

$$y = (1 - .03)^t = (.97)^t.$$

(a) Use a calculator to complete the following table.

t	0	1	2	3	4	5	6	7	8	9	10
y	1					.86					.74

(b) Graph $y = (.97)^t$.

Work these problems. See Example 4.

33. Finance If money loses value, then it takes more dollars to buy the same item. Use the results of Exercise 32(a) to answer the following questions.
(a) Suppose a house costs $105,000 today. Estimate the cost of a similar house in 10 years. (*Hint:* Solve the equation $.74t = \$105{,}000$.)
(b) Estimate the cost of a $50 textbook in 8 years.

34. Natural Science Biologists studying salmon have found that the oxygen consumption of yearling salmon is given by $g(x) = 100e^{.7x}$, where x is the speed in feet per second. Find each of the following.
(a) The oxygen consumption when the fish are still
(b) The oxygen consumption at a speed of 2 feet per second

35. Social Science Cellular phone use increased exponentially in the 1990s. The number of U.S. cellular phone accounts (in millions) can be approximated by

$$f(x) = 25.49(1.2821)^x \qquad (0 \le x \le 7)$$

where $x = 0$ corresponds to 1994.* Use this function to estimate the number of cell phone accounts in
(a) 2000; (b) 2001.
(c) Will this model continue to be accurate over the long run?

*Based on data from the CTIA Semi-Annual Wireless Survey.

36. Business The monthly payment on a car loan at 12% interest per year on the unpaid balance is given by

$$f(n) = \frac{P}{\dfrac{1 - 1.01^{-n}}{.01}},$$

where P is the amount borrowed and n is the number of months over which the loan is paid back. Find the monthly payment for each of the following loans.
(a) $8000 for 48 months
(b) $8000 for 24 months
(c) $6500 for 36 months
(d) $6500 for 60 months

37. Natural Science The amount of plutonium remaining from one kilogram after x years is given by the function $W(x) = 2^{-x/24360}$. How much will be left after
(a) 1000 years?
(b) 10,000 years?
(c) 15,000 years?
(d) Estimate how long it will take for the one kilogram to decay to half its original weight. This may help to explain why nuclear waste disposal is a serious problem.

38. Health National health expenditures (in billions of dollars) in year x can be approximated by

$$f(x) = 976.8e^{.06062x},$$

where $x = 0$ corresponds to 1995.* Estimate the health care expenditures in the following years.
(a) 1999 (b) 2001 (c) 2005

39. Social Science There were fewer than a billion people on earth when Thomas Jefferson died in 1826, and there are now more than six billion. If world population continues to grow as expected, the population (in billions) in year t will be given by the function $P(t) = 4.834(1.01^{(t-1980)})$.† Estimate the world population in the following years.
(a) 2005 (b) 2010 (c) 2030
(d) What will the world population be when you reach 65 years old?

40. Social Science Using the function in Exercise 39 and a calculator, estimate the year in which the population will be double what it was in 2005. (*Hint:* If you have a graphing calculator, solve the appropriate equation, as in Example 4(c). If you have a scientific calculator, experiment with various values of t.)

Business *The scrap value of a machine is the value of the machine at the end of its useful life. By one method of calculating scrap value, where it is assumed a constant percentage of value is lost annually, the scrap value S is given by*

$$S = C(1 - r)^n,$$

*Based on projections of the U.S. Healthcare Financing Administration.

†Based on current data and projections by the U.S. Census Bureau.

where C is the original cost, n is the useful life of the machine in years, and r is the constant annual percentage of value lost. Find the scrap value for each of the following machines.

41. Original cost, \$54,000; life, 8 years; annual rate of value loss, 12%

42. Original cost, \$178,000; life, 11 years; annual rate of value loss, 14%

43. Use the graphs of $f(x) = 2^x$ and $g(x) = 2^{-x}$ (not a calculator) to explain why $2^x + 2^{-x}$ is approximately equal to 2^x when x is very large.

44. Business For a short period in 2001, Purina Mills' stock price could be approximated by the function

$$f(x) = 8.13(1.3084)^x,$$

where x is measured in months and $x = 0$ corresponds to March 1.

(a) What was the approximate stock price on April 1? On May 15? On June 21?

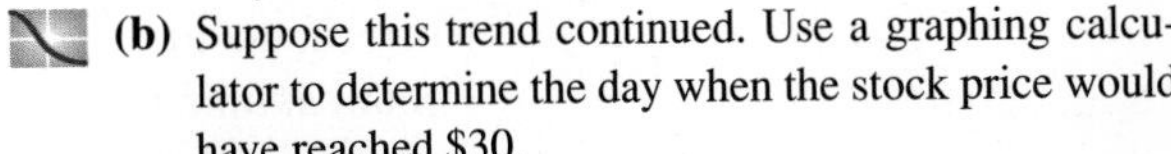

(b) Suppose this trend continued. Use a graphing calculator to determine the day when the stock price would have reached \$30.

(c) What price does the model predict for today's date? Does this price seem realistic? Check the stock price in today's paper (it's traded on the Nasdaq). What conclusions can you draw?

4.2 APPLICATIONS OF EXPONENTIAL FUNCTIONS

In many situations in biology, economics, and the social sciences, a quantity changes at a rate proportional to the quantity present. In such cases, the amount present at time t is a function of t, called the **exponential growth function.**

Exponential Growth Function

Under normal conditions, growth can be described by a function of the form

$$f(t) = y_0e^{kt} \quad \text{or} \quad f(t) = y_0b^t,$$

where $f(t)$ is the amount present at time t, y_0 is the amount present at time $t = 0$, and k and b are constants that depend on the rate of growth.

It is understood that this growth can involve either growing larger or growing smaller. Here is an example of growing smaller.

Suppose the number of bacteria in a culture at time t is

$$y = 500e^{.4t},$$

where t is measured in hours.

(a) How many bacteria are present initially?

(b) How many bacteria are present after 10 hours?

Answers:

(a) 500

(b) About 27,300

EXAMPLE 1 Radioactive lead-210 decays to polonium-210. The amount y of radioactive lead-210 at time t is given by $y = y_0e^{-.032t}$, where t is time in years. How much of an initial 500 grams of lead-210 will remain after 5 years?

Since the initial amount is 500 grams, $y_0 = 500$. Substitute 500 for y_0 and 5 for t in the definition of the function.

$$y = 500e^{-.032(5)} \approx 426$$

About 426 grams will remain. 

When a quantity is known to grow exponentially, it is sometimes possible to find a function that models this growth from a small amount of data.

EXAMPLE 2 When money is placed in a bank account that pays compound interest, the amount in the account grows exponentially, as we shall see in Chapter 5. Suppose such an account grows from \$1000 to \$1316 in seven years.

(a) Find a growth function of the form $f(t) = y_0b^t$ that gives the amount in the account at time t years.

The value of the account at time $t = 0$ and $t = 7$ are given, that is, $f(0) = 1000$ and $f(7) = 1316$. Solve the first of these equations for y_0:

$$f(0) = 1000$$
$$y_0b^0 = 1000 \qquad \text{Rule of } f$$
$$y_0 = 1000 \qquad b^0 = 1$$

So the rule of f has the form $f(t) = 1000b^t$. Now solve the equation $f(7) = 1316$ for b:

$$f(7) = 1316$$
$$1000b^7 = 1316 \qquad \text{Rule of } f$$
$$b^7 = 1.316 \qquad \text{Divide both sides by 1000.}$$
$$b = (1.316)^{1/7} \approx 1.04 \qquad \text{Take the seventh root of each side.}$$

So the rule of the function is $f(t) = 1000(1.04)^t$.

(b) How much is in the account after 12 years?

$$f(12) = 1000(1.04)^{12} = \$1601.03.$$ (2)

(2) Suppose an investment grows exponentially from \$500 to \$587.12 in three years.

(a) Find a function of the form $f(t) = y_0b^t$ that gives the value of the investment after t years.

(b) How much was the investment worth after 10 years?

Answers:

(a) $f(t) = 500(1.055)^t$

(b) \$854.07

EXAMPLE 3 Infant mortality rates in the United States are shown in the following table.*

Year	*Rate*	*Year*	*Rate*
1920	76.7	1970	20.0
1930	60.4	1980	12.6
1940	47.0	1985	10.6
1950	29.2	1990	9.2
1960	26.0	1995	7.6
		1998	7.2

(a) Let $t = 0$ correspond to 1920. Use the data for 1920 and 1998 to find a function of the form $f(t) = y_0b^t$ that models this data.

Since the rate is 76.7 when $t = 0$, we have $y_0 = 76.7$, so that $f(t) = 76.7b^t$. Since $t = 78$ corresponds to 1998, we have $f(78) = 7.2$, that is,

$$76.7b^{78} = 72 \qquad \text{Rule of } f$$
$$b^{78} = \frac{7.2}{76.7} \qquad \text{Divide both sides by 76.7.}$$
$$b = \left(\frac{7.2}{76.7}\right)^{\frac{1}{78}} \approx .97 \qquad \text{Take 78th roots on both sides.}$$

Therefore, the function is $f(t) = 76.7(.97)^t$.

*Infants under 1 year, deaths per 1000 live births. U.S. National Center for Health Statistics.

(b) Use exponential regression on a graphing calculator to find another model of this data.

The procedure for entering the data and finding the function is the same as for linear regression (just choose "exponential" instead of "linear"), as explained in Section 2.3. Depending on the calculator, one of these functions will be produced:

$$g(t) = 82.14(.9694)^t \quad \text{or} \quad h(t) = 82.14e^{-.03104t}.$$

Both functions give the same values (except for slight differences due to rounding the coefficients displayed above). They fit the data reasonably well, as shown in Figure 4.9.

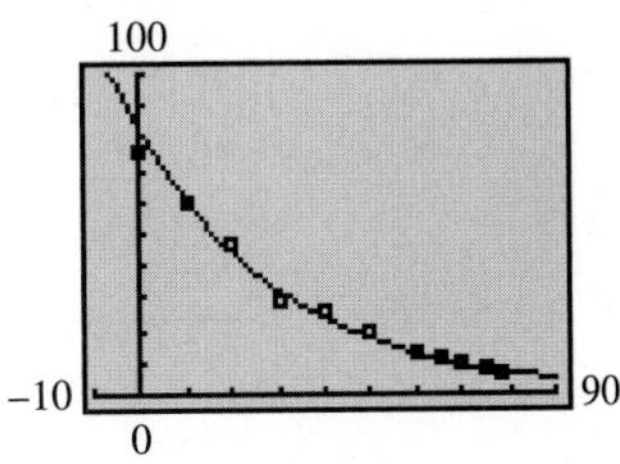

FIGURE 4.9

(c) Use the preceding models to estimate the mortality rate in 1986 and 2001.

Evaluating each of the models at $t = 66$ and $t = 81$ shows that the models in parts (a) and (b) give similar, but not identical, results.

t	$f(t)$	$g(t)$	$h(t)$
66	10.3	10.6	10.6
81	6.5	6.6	6.6

Other Exponential Models

When a quantity changes exponentially, but does not either grow very large or decrease practically to 0, as in Examples 1–3, different functions are needed.

EXAMPLE 4 Sales of a new product often grow rapidly at first and then begin to level off with time. Suppose the annual sales of an inexpensive can opener are given by

$$S(x) = 10{,}000(1 - e^{-.5x}),$$

where $x = 0$ corresponds to the time when the can opener went on the market.

(a) What were the sales in each of the first three years?

At the end of one year ($x = 1$), sales were

$$S(1) = 10{,}000(1 - e^{-.5(1)}) \approx 3935.$$

Sales in the next two years were

$$S(2) = 10{,}000(1 - e^{-.5(2)}) \approx 6321 \quad \text{and} \quad S(3) = 10{,}000(1 - e^{-.5(3)}) \approx 7769.$$

3

Suppose the value of the assets (in thousands of dollars) of a certain company after t years is given by

$$V(t) = 100 - 75e^{-.2t}.$$

(a) What is the initial value of the assets?

(b) What is the limiting value of the assets?

(c) Find the value after 10 years.

(d) Graph $V(t)$.

Answers:

(a) \$25,000

(b) \$100,000

(c) \$89,850

(d)

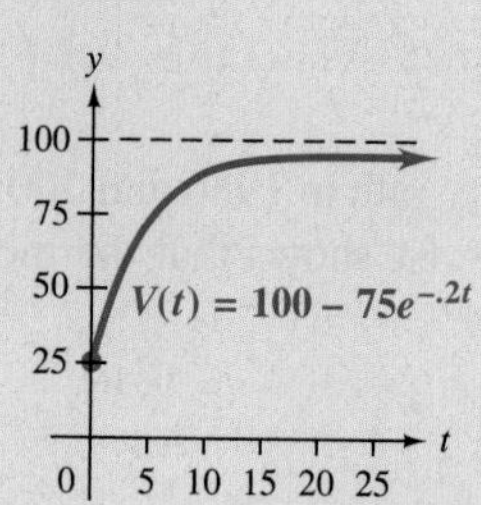

(b) What were the sales at the end of the tenth year?

$$S(10) = 10{,}000(1 - e^{-.5(10)}) \approx 9933.$$

(c) Graph the functions S. What does it suggest?

The graph can be obtained by plotting points and connecting them with a smooth curve, or by using a graphing calculator, as in Figure 4.10. The graph indicates that sales will level off after the tenth year to around 10,000 can openers per year. ◀ 3

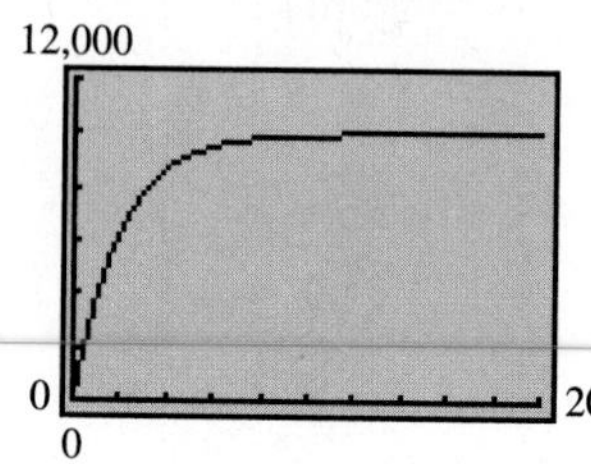

FIGURE 4.10

A variety of activities can be modeled by **logistic functions,** whose rules are the form

$$f(x) = \frac{c}{1 + ae^{kx}}.$$

The logistic function in the next example is sometimes called a **forgetting curve.**

EXAMPLE 5 Psychologists have measured people's ability to remember facts that they have memorized. In one such experiment, it was found that the number of facts $N(t)$ remembered after t days was given by

$$N(t) = \frac{10.003}{1 + .0003e^{.8t}}.$$

(a) How many facts were remembered at the beginning of the experiment?

When $t = 0$, then

$$N(0) = \frac{10.003}{1 + .0003e^{.8(0)}} = \frac{10.003}{1.0003} = 10.$$

So 10 facts were remembered at the beginning.

(b) How many facts were remembered after one week? After two weeks?

One and two weeks correspond to $t = 7$ and $t = 14$.

$$N(7) = \frac{10.003}{1 + .003e^{.8(7)}} = \frac{10.003}{1.0811} \approx 9.25$$

$$N(14) = \frac{10.003}{1 + .0003e^{.8(14)}} = \frac{10.003}{22.9391} \approx .44$$

In Example 5,

(a) find the number of facts remembered after 10 days;

(b) use the graph to estimate when just 1 fact will be remembered.

Answers:

(a) 5

(b) After about 12 days

So 9 facts were remembered after one week, but effectively none were remembered after two weeks. (.44 is less than "half a fact"). The graph of the function in Figure 4.11 gives a picture of this forgetting process. ◀ 4

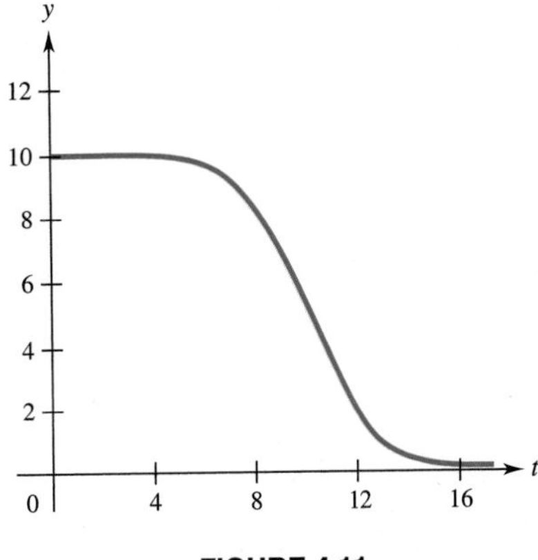

FIGURE 4.11

TECHNOLOGY TIP Many graphing calculators can find a logistic model for appropriate data.

4.2 Exercises

Work these problems. (See Example 1.)

1. **Physical Science** A sealed box contains radium. The number of grams present at time t is given by

$$Q(t) = 100e^{-.00043t},$$

where t is measured in years. Find the amount of radium in the box at the following times.

(a) $t = 0$
(b) $t = 800$
(c) $t = 1600$
(d) $t = 5000$

2. **Physical Science** The pressure of the atmosphere, $p(h)$, in pounds per square inch, is given by

$$p(h) = p_0e^{-.0000425h},$$

where h is the height above sea level and p_0 is a constant. The pressure at sea level is 15 pounds per square inch.
(a) Find p_0.
(b) Find the pressure at an altitude of 6000 feet.
(c) What would be the pressure encountered by a spaceship at an altitude of 150,000 feet?

3. **Business** The total revenues of the securities industry since 1980 are approximated by the function

$$f(t) = 21.53e^{.13t},$$

where $t = 0$ corresponds to 1980 and $f(t)$ is in billions of dollars. Find the revenues in the following years.
(a) 1987 **(b)** 1994 **(c)** 2000

4. **Natural Science** The graph on the next page shows how the risk of chromosomal abnormality in a child rises with the age of the mother.*
(a) Read from the graph the risk of chromosomal abnormality (per 1000) at ages 20, 35, 42, and 49.

New York Times, February 5, 1994, p. 24.

(b) Verify by substitution that the exponential equation $y = .590e^{.061t}$ "fits" the graph for ages 20 and 35.

(c) Does the equation in part (b) also fit the graph for ages 42 and 49? What does this mean?

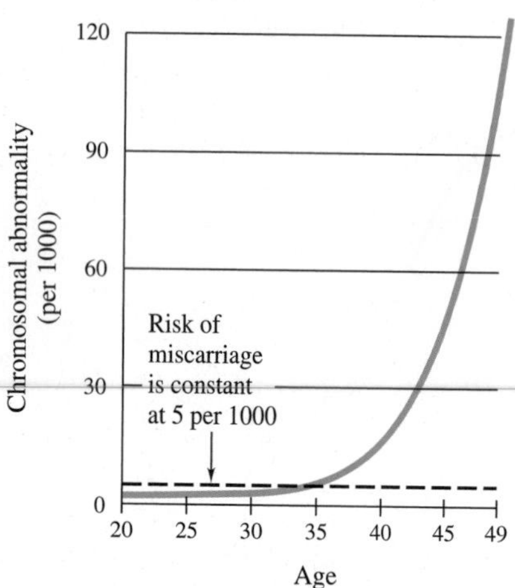

Source: American College of Obstetricians and Gynecologists.

In each of the following problems, find an exponential function of the form $f(t) = y_0b^t$ to model the data. (See Examples 2 and 3.)

5. **Social Sciences** The U.S. Census Bureau predicts that the Hispanic population in the United States will increase from 32.5 million in 2000 to 98.2 million in 2050.*
   (a) Find a model for this data, in which $t = 0$ corresponds to 2000.
   (b) What is the projected Hispanic population in 2010? In 2035?
   (c) By experimenting with different values of t (or by using a graphing calculator to solve an appropriate equation), estimate the year in which the Hispanic population will reach 54 million.

6. **Social Sciences** The U.S. Census Bureau predicts that the African-American population will increase from 35.3 million in 2000 to 59.2 million in 2050.*
   (a) Find a model for this data, in which $t = 0$ corresponds to 2000.
   (b) What is the projected African-American population in 2004? In 2030?
   (c) By experimenting with different values of t (or by using a graphing calculator to solve an appropriate equation), estimate the year in which the African-American population will reach 55 million.

7. **Finance** Suppose you owe $800 on your credit account and you stop making any payments or new purchases. In the unlikely event that the credit card company allows this situation to go on for a year, you will then owe $901.46.
   (a) Find a model for this data, in which t is measured in *months* and $t = 0$ is the present, when your balance is $800.
   (b) If your account does not get turned over to a collection agency (unlikely), how much will you owe after 18 months? After two years?

8. **Business** The number of Internet users in the United States was estimated to be 54 million in the fall of 1999 and 85 million in the fall of 2002.*
   (a) Find a model for this data, in which t is measured in years with $t = 0$ corresponding to fall 1999 and $f(t)$ is measured in millions.
   (b) Use the results of part (a) to determine $f(6)$. Interpret your result.

In the following exercises, find the exponential model as follows. If you do not have suitable technology, use the first and last data points to find a function. (See Examples 2 and 3.) If you have a graphing calculator or other suitable technology, use exponential regression to find a function. (See Example 3.)

9. **Business** The table shows the purchasing power of a dollar in various years, with the 1982–1984 average being the base.† For example, the entry .600 for 1999 means that a dollar in 1999 bought what 60 cents did in 1982–1984.

Year	1983	1985	1987	1989	1991
Purchasing Power of $1	1.003	.928	.880	.807	.734

Year	1993	1995	1997	1999
Purchasing Power of $1	.692	.656	.623	.600

   (a) Find an exponential model for this data, with $t = 0$ corresponding to 1983.
   (b) Find the purchasing power of a dollar in 2001, 2003, and 2005.
   (c) Use a graphing calculator (or trial and error) to determine the year in which the purchasing power of the 1982–84 dollar will drop to 40 cents.

10. **Physical Science** The table at the top of the next column shows the atmospheric pressure (in millibars) at various altitudes (in meters).

Statistical Abstracts of the United States, 2000.

*U.S. Department of Commerce.

†U.S. Bureau of Labor Statistics.

Altitude	Pressure
0	1,013
1000	899
2000	795
3000	701
4000	617
5000	541
6000	472
7000	411
8000	357
9000	308
10,000	265

(a) Find an exponential model for this data, in which t is measured in thousands (for instance, $t = 2$ is 2000 meters).

(b) Use the function in part (a) to estimate the atmospheric pressure at 1500 meters and 11,000 meters. Compare your results to the actual values of 846 millibars and 227 millibars, respectively.

11. Health The table shows age-adjusted death rates (per 100,000 population) for heart disease.*

Year	1970	1980	1990	1995	1997
Death Rate	253.6	202.2	152.0	138.3	130.5

(a) Find an exponential model for this data, with $t = 0$ corresponding to 1970.

(b) Estimate the death rate in 1985 and 2000.

(c) Assuming this model remains accurate, use a graphing calculator (or trial and error) to determine the year in which the rate will be 100.

12. Business The table shows outstanding consumer credit (in billions of dollars) in various years.†

Year	1980	1985	1990	1995
Credit	349.4	584.7	789.3	1096.0

Year	1996	1997	1998	1999
Credit	1182.4	1234.1	1300.5	1395.4

(a) Find an exponential model for this data, with $t = 0$ corresponding to 1980.

(b) If this model remains accurate, what will the outstanding consumer credit be in 2003 and 2006?

(c) In what year will consumer credit reach \$2500 billion?

Work the following problems. (See Example 4.)

13. Business Assembly line operations tend to have a high turnover of employees, forcing the companies involved to spend much time and effort in training new workers. It has been found that a worker new to the operation of a certain task on the assembly line will produce $P(t)$ items on day t, where

$$P(t) = 25 - 25e^{-.3t}.$$

(a) How many items will be produced on the first day?

(b) How many items will be produced on the eighth day?

(c) What is the maximum number of items, according to the function, the worker can produce?

14. Social Science The number of words per minute that an average typist can type is given by

$$W(t) = 60 - 30e^{-.5t},$$

where t is time in months after the beginning of a typing class. Find each of the following.

(a) $W(0)$

(b) $W(1)$

(c) $W(4)$

(d) $W(6)$

Natural Science *Newton's Law of Cooling says that the rate at which a body cools is proportional to the difference in temperature between the body and an environment into which it is introduced. Using calculus, the temperature F(t) of the body at time t after being introduced into an environment having constant temperature T_0 is*

$$F(t) = T_0 + Cb^t,$$

where C and b are constants. Use this result in Exercises 15 and 16.

15. Boiling water, at 100° Celsius, is placed in a freezer at 0° Celsius. The temperature of the water is 50° Celsius after 24 minutes. Find the temperature of the water after 96 minutes.

16. Paisley refuses to drink coffee cooler than 95°F. She makes coffee with a temperature of 170°F in a room with a temperature of 70°F. The coffee cools to 120°F in 10 minutes. What is the longest time she can let the coffee sit before she drinks the coffee?

*U.S. Department of Health and Human Services.

†*Federal Reserve Bulletin.*

17. **Social Science** A sociologist has shown that the fraction $y(t)$ of people in a group who have heard a rumor after t days is approximated by

$$y(t) = \frac{y_0 e^{kt}}{1 - y_0(1 - e^{kt})},$$

where y_0 is the fraction of people who have heard the rumor at time $t = 0$, and k is a constant. A graph of $y(t)$ for a particular value of k is shown in the figure.

(a) If $k = .1$ and $y_0 = .05$, find $y(10)$.

(b) If $k = .2$ and $y_0 = .10$, find $y(5)$.

(c) Assume the situation in part (b). How many *weeks* will it take for 65% of the people to have heard the rumor?

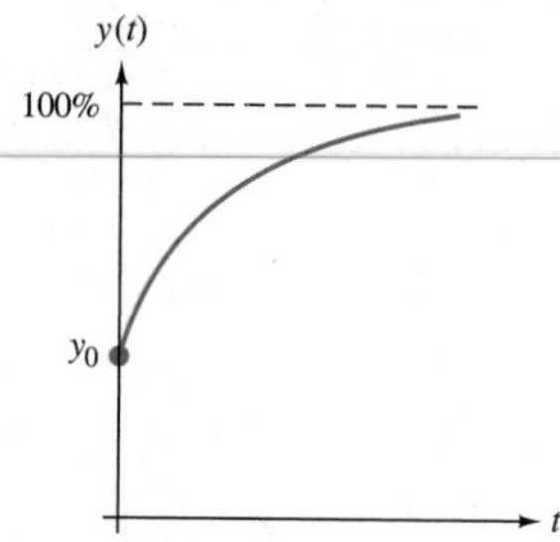

18. **Social Science** Data from the National Highway Traffic Safety Administration indicate that in year t, the approximate percentage of people in the United States who wear seat belts when driving is given by

$$g(t) = \frac{76.13}{1 + 10.1e^{-.3167t}} \qquad (3 \le t \le 18),$$

where $t = 0$ corresponds to 1980. What percentage used seat bealts in

(a) 1992; (b) 1994; (c) 1998?

Assuming this function is accurate after 1998, what percentage of people used seat belts in

(d) 2000; (e) 2002; (f) 2003?

(g) Graph the function. In California, seat belt usage was 87% in the late 1990s. Does the graph suggest that this will happen nationwide in the near future?

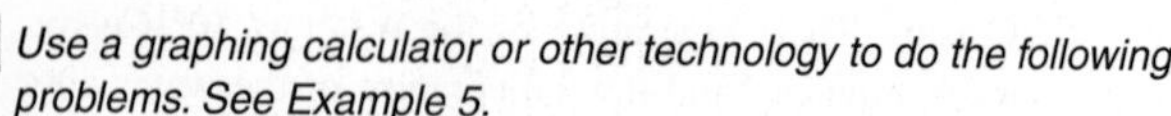
Use a graphing calculator or other technology to do the following problems. See Example 5.

19. **Business** The number of people who subscribe to basic cable TV can be approximated by

$$f(x) = \frac{76.7}{1 + 16(.8444)^x},$$

where $x = 0$ corresponds to 1970 and $f(x)$ is in millions.*

(a) Estimate the number of subscribers in 1990, and 2005.

(b) Graph $f(x)$ for the period 1970–2020.

(c) Use the graph to determine the year in which the number of subscribers reached 70 million.

(d) Based on the graph, will the number of subscribers reach 90 million in the foreseeable future?

20. **Natural Science** The population of fish in a certain lake at time t months is given by the function

$$p(t) = \frac{20{,}000}{1 + 24(2^{-.36t})}.$$

(a) Graph the population function from $t = 0$ to $t = 48$ (a four-year period).

(b) What was the population at the beginning?

(c) Use the graph to estimate the one-year period in which the population grew most rapidly.

(d) When do you think the population will reach 25,000? What factors in nature might explain your answer?

21. **Finance** Since 1980, the national debt can be approximated by the logistic model

$$f(x) = \frac{7070.4}{1 + 172.36e^{-.1666x}} \qquad (x \ge 20),$$

where $x = 0$ corresponds to 1960 and $f(x)$ is measured in billions of dollars.

(a) Estimate the national debt in 1995, 2000, and 2005.

(b) Graph the function for the period 1980–2030.

(c) Determine when the debt will reach 6.8 trillion dollars (that's 6800 billion).

(d) According to this model, will the national debt level off at any point in the future?

22. **Social Science** The probability P percent of having an automobile accident is related to the alcohol level of the driver's blood t by the function $P(t) = e^{21.459t}$.

(a) Graph $P(t)$ in a viewing window with $0 \le t \le .2$ and $0 \le P(t) \le 100$.

(b) At what blood alcohol level is the probability of an accident at least 50%? What is the legal blood alcohol level in your state?

*Based on data from *The Cable TV Financial Databook* and *The Pay TV Newsletter*.

4.3 LOGARITHMIC FUNCTIONS

Until the development of computers and calculators, logarithms were the only effective tool for large-scale numerical computations. They are no longer needed for this, but logarithmic functions still play a crucial role in calculus and in many applications.

Common Logarithms

Logarithms are simply *a new language for old ideas*—essentially a special case of exponents.

Definition of Common (Base 10) Logarithms

$$y = \log x \quad \text{means} \quad 10^y = x.$$

(1)

Find each common logarithm.

(a) log 100

(b) log 1000

(c) log .1

Answers:

(a) 2

(b) 3

(c) −1

"Log x," which is read "the logarithm of x," is the answer to the question

To what exponent must 10 be raised to produce x?

EXAMPLE 1 To find log 10,000 ask yourself, "To what exponent must 10 be raised to produce 10,000?" Since $10^4 = 10{,}000$, we see that $\log 10{,}000 = 4$. Similarly,

$$\log 1 = 0 \quad \text{because} \quad 10^0 = 1;$$

$$\log .01 = -2 \quad \text{because} \quad 10^{-2} = \frac{1}{10^2} = \frac{1}{100} = .01;$$

$$\log \sqrt{10} = 1/2 \quad \text{because} \quad 10^{1/2} = \sqrt{10}.$$ (1)

(2)

Find each common logarithm.

(a) log 27

(b) log 1089

(c) log .00426

Answers:

(a) 1.4314

(b) 3.0370

(c) −2.3706

EXAMPLE 2 Log(−25) is the exponent to which 10 must be raised to produce −25. But every power of 10 is positive! So there is no exponent that will produce −25. *Logarithms of negative numbers and 0 are not defined.*

```
log(359)
        2.555094449
10^(2.5551)
        359.004589
log(.026)
       -1.585026652
```

FIGURE 4.12

EXAMPLE 3 **(a)** We know that log 359 must be a number between 2 and 3 because $10^2 = 100$ and $10^3 = 1000$. By using the log key on a calculator, we find that log 359 (to four decimal places) is 2.5551. You can verify this by computing $10^{2.5551}$; the result (rounded) is 359. See Figure 4.12.

(b) When 10 is raised to a negative exponent, the result is a number less than 1. Consequently, the logarithms of numbers between 0 and 1 are negative. For instance, $\log .026 = -1.5850$, as shown in Figure 4.12. (2)

NOTE On scientific calculators, enter the number followed by the log key. On graphing calculators, press the log key followed by the number, as in Figure 4.12.

Natural Logarithms

Although common logarithms still have some uses (one of which is discussed in Section 4.4), the most widely used logarithms today are defined in terms of the number e (whose decimal expansion begins 2.71828 . . .) rather than 10. They have a special name and notation.

Definition of Natural (Base e) Logarithms

$$y = \ln x \quad \text{means} \quad e^y = x.$$

Thus the number **ln *x*** (which is sometimes read "el-en x") is the exponent to which e must be raised to produce the number x. For instance, $\ln 1 = 0$ because $e^0 = 1$. Although logarithms to base e may not seem as "natural" as common logarithms, there are several reasons for using them, some of which are discussed in Section 4.4.

3

Find the following.

(a) ln 6.1

(b) ln 20

(c) ln .8

(d) ln .1

Answers:

(a) 1.8083

(b) 2.9957

(c) −.2231

(d) −2.3026

EXAMPLE 4 **(a)** To find ln 85 use the LN key of your calculator. The result is 4.4427. Thus, 4.4427 is the exponent (to four decimal places) to which e must be raised to produce 85. You can verify this by computing $e^{4.4427}$; the answer (rounded) is 85.

(b) A calculator shows that $\ln .38 = -.9676$ (rounded), which means that $e^{-.9676} \approx .38$.

These result are shown in Figure 4.13. ◀ 3

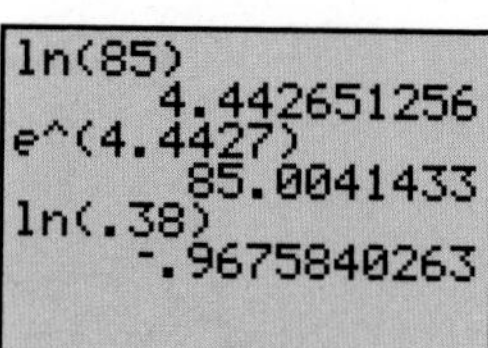

FIGURE 4.13

EXAMPLE 5 You don't need a calculator to find $\ln e^8$. Just ask yourself, "To what exponent must e be raised to produce e^8?" The answer, obviously, is 8. So $\ln e^8 = 8$. ◀

Other Logarithms

The procedure used to define common and natural logarithms can be carried out with any positive number $a \neq 1$ as the base in place of 10 or e.

Definition of Logarithms to the Base a

$$y = \log_a x \quad \text{means} \quad a^y = x.$$

Read $y = \log_a x$ as "y is the logarithm of x to the base a." For example, the exponential statement $2^4 = 16$ can be translated into the equivalent logarithmic statement $4 = \log_2 16$. Thus, $\log_a x$ is an *exponent;* it is the answer to the question

To what power must a be raised to produce x?

This key definition should be memorized. It is important to remember the location of the base and exponent in each part of the definition.

Exponent ↓

Logarithmic form: $y = \log_a x$

↑ Base

Exponent ↓

Exponential form: $a^y = x$

↑ Base

Common and natural logarithms are the special cases when $a = 10$ and when $a = e$, respectively. Both $\log u$ and $\log_{10} u$ mean the same thing. Similarly, $\ln u$ and $\log_e u$ mean the same thing.

4

Write the logarithmic form of

(a) $5^3 = 125$;

(b) $3^{-4} = 1/81$;

(c) $8^{2/3} = 4$.

Answers:

(a) $\log_5 125 = 3$

(b) $\log_3(1/81) = -4$

(c) $\log_8 4 = 2/3$

5

Write the exponential form of

(a) $\log_{16} 4 = 1/2$;

(b) $\log_3(1/9) = -2$;

(c) $\log_{16} 8 = 3/4$.

Answers:

(a) $16^{1/2} = 4$

(b) $3^{-2} = 1/9$

(c) $16^{3/4} = 8$

EXAMPLE 6 This example shows several statements written in both exponential and logarithmic forms.

Exponential Form	*Logarithmic Form*
(a) $3^2 = 9$	$\log_3 9 = 2$
(b) $(1/5)^{-2} = 25$	$\log_{1/5} 25 = -2$
(c) $10^5 = 100{,}000$	$\log_{10} 100{,}000$ (or $\log 100{,}000$) $= 5$
(d) $4^{-3} = 1/64$	$\log_4(1/64) = -3$
(e) $2^{-4} = 1/16$	$\log_2(1/16) = -4$
(f) $e^0 = 1$	$\log_e 1$ (or $\ln 1$) $= 0$

◀ 4 5

Properties of Logarithms

Some of the important properties of logarithms arise directly from their definition.

Let x and a be any positive real numbers, with $a \neq 1$, and r be any real number. Then

(a) $\log_a 1 = 0$ **(b)** $\log_a a = 1$

(c) $\log_a a^r = r$ **(d)** $a^{\log_a x} = x$

FIGURE 4.14

Property (a) was discussed in Example 1 (with $a = 10$). Property (c) was illustrated in Example 5 (with $a = e$ and $r = 8$). Property (b) is property (c) with $r = 1$. To understand property (d), recall that $\log_a x$ is the exponent to which a must be raised to produce x. Consequently, when you raise a to this exponent, the result is x, as illustrated for common and natural logarithms in Figure 4.14.

The following properties are part of the reason why logarithms are so useful. They will be used in the next section to solve exponential and logarithmic equations.

The Product, Quotient, and Power Properties

Let x, y, and a be any positive real numbers, with $a \neq 1$. Let r be any real number. Then

(a) $\log_a xy = \log_a x + \log_a y$ Product property

(b) $\log_a \frac{x}{y} = \log_a x - \log_a y$ Quotient property

(c) $\log_a x^r = r \log_a x$ Power property

To prove property (a), let

$$m = \log_a x \quad \text{and} \quad n = \log_a y.$$

Then, by the definition of logarithm,

$$a^m = x \quad \text{and} \quad a^n = y.$$

Multiply to get

$$a^m \cdot a^n = x \cdot y,$$

or, by a property of exponents,

$$a^{m+n} = xy.$$

Use the definition of logarithm to rewrite this last statement as

$$\log_a xy = m + n.$$

Replace m with $\log_a x$ and n with $\log_a y$ to get

$$\log_a xy = \log_a x + \log_a y.$$

Properties (b) and (c) can be proved similarly. You can also illustrate these properties on a calculator, as in Figure 4.15.

FIGURE 4.15

6

Write as a single logarithm, using the properties of logarithms.

(a) $\log_a 5x + \log_a 3x^4$

(b) $\log_a 3p - \log_a 5q$

(c) $4 \log_a k - 3 \log_a m$

Answers:

(a) $\log_a 15x^5$

(b) $\log_a(3p/5q)$

(c) $\log_a(k^4/m^3)$

EXAMPLE 7 Using the properties of logarithms, each of the following can be written as a single logarithm.*

(a) $\log_a x + \log_a(x - 1) = \log_a x(x - 1)$;

(b) $\log_a(x^2 + 4) - \log_a(x + 6) = \log_a \dfrac{x^2 + 4}{x + 6}$;

(c) $\log_a 9 + 5 \log_a x = \log_a 9 + \log_a x^5 = \log_a 9x^5$. ◀ 6

CAUTION There is no logarithm property that allows you to simplify the logarithm of a sum, such as $\log_a(x^2 + 4)$. In particular, $\log_a(x^2 + 4)$ is *not* equal to $\log_a x^2 + \log_a 4$. Property (a) of logarithms in Example 7 shows that $\log_a x^2 + \log_a 4 = \log_a 4x^2$.

7

Use the properties of logarithms to rewrite and evaluate each of the following, given $\log_3 7 \approx 1.77$ and $\log_3 5 \approx 1.46$.

(a) $\log_3 35$

(b) $\log_3 7/5$

(c) $\log_3 25$

(d) $\log_3 3$

(e) $\log_3 1$

Answers:

(a) 3.23

(b) .31

(c) 2.92

(d) 1

(e) 0

EXAMPLE 8 Assume that $\log_6 7 \approx 1.09$ and $\log_6 5 \approx .90$. Use the properties of logarithms to find each of the following.

(a) $\log_6 35 = \log_6(7 \cdot 5) = \log_6 7 + \log_6 5 \approx 1.09 + .90 = 1.99$;

(b) $\log_6 5/7 = \log_6 5 - \log_6 7 \approx .90 - 1.09 = -.19$;

(c) $\log_6 5^3 = 3 \log_6 5 \approx 3(.90) = 2.70$;

(d) $\log_6 6 = 1$;

(e) $\log_6 1 = 0$. ◀ 7

In Example 8 several logarithms to base 6 were given. However, they could have been found by using a calculator and the following formula.

Change of Base Theorem

For any positive numbers a and x (with $a \neq 1$),

$$\log_a x = \frac{\ln x}{\ln a}.$$

*Here and elsewhere, we assume that variable expressions represent positive numbers and that the base a is positive, with $a \neq 1$.

EXAMPLE 9 To find $\log_7 3$, use the theorem with $a = 7$ and $x = 3$:

$$\log_7 3 = \frac{\ln 3}{\ln 7} \approx \frac{1.0986}{1.9459} \approx .5646.$$

You can check this on your calculator by verifying that $7^{.5646} \approx 3$.

Logarithmic Functions

For a given *positive* value of x, the definition of logarithm leads to exactly one value of y, so that $y = \log_a x$ defines a logarithmic function of base a. (The base a must be positive, with $a \neq 1$.)

If $a > 0$ and $a \neq 1$, the **logarithmic function** with base a is defined as

$$f(x) = \log_a x.$$

The most important logarithmic function is the natural logarithmic function.

EXAMPLE 10 Graph $f(x) = \ln x$ and $g(x) = e^x$ on the same axes.

For each function, use a calculator to compute some ordered pairs. Then plot the corresponding points and connect them with a curve to obtain the graphs in Figure 4.16.

8

Graph $f(x) = \log x$ and $g(x) = 10^x$ on the same axes.

Answer:

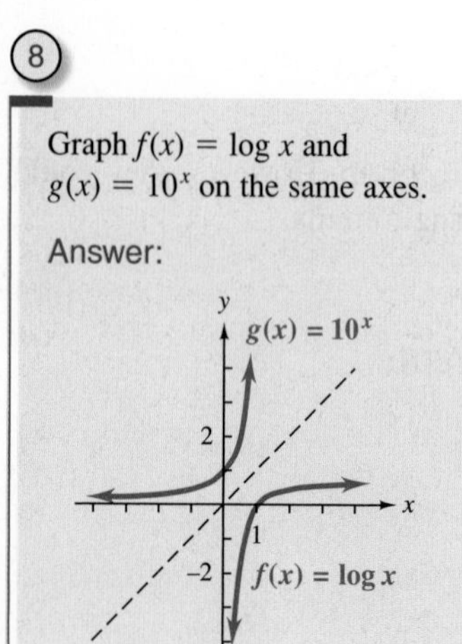

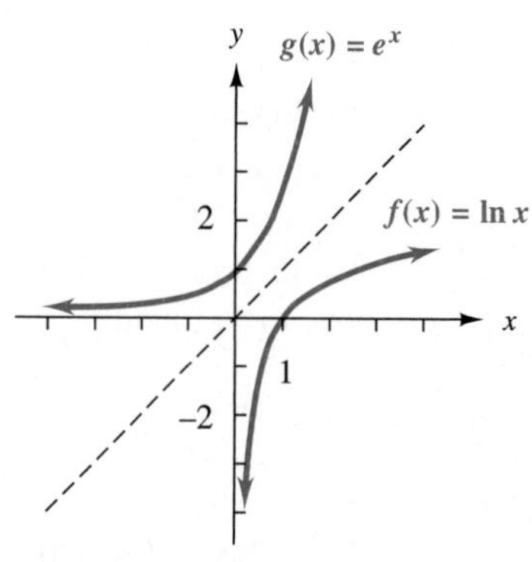

FIGURE 4.16

In Figure 4.16, the dashed straight line is the graph of $y = x$. Observe that the graph of $f(x) = \ln x$ is the mirror image of the graph of $g(x) = e^x$, with the line $y = x$ being the mirror. 8

When the base $a > 1$, the graph of $f(x) = \log_a x$ has the same basic shape as the graph of the natural logarithmic function in Figure 4.16, as summarized below.

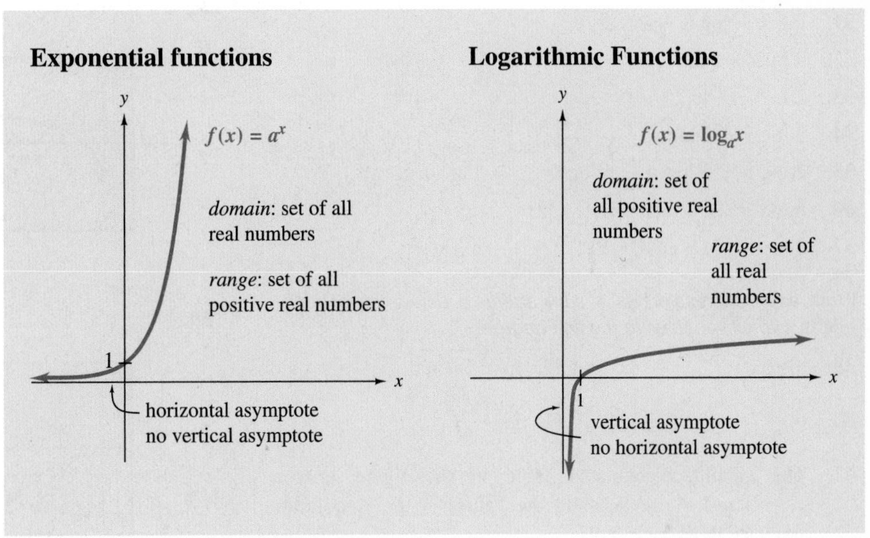

As the information in the box suggests, the graph of $f(x) = \log_a x$ is the mirror image of the graph of $g(x) = a^x$, with the line $y = x$ being the mirror. Functions whose graphs are related in this way are said to be **inverses** of each other. A more complete discussion of inverse functions is given in most standard college algebra books.

4.3 Exercises

Complete each statement in Exercises 1–4.

1. $y = \log_a x$ means $x =$ ________.

2. The statement $\log_5 125 = 3$ tells us that ________ is the power of ________ that equals ________.

3. What is wrong with the expression $y = \log_b$?

4. Logarithms of negative numbers are not defined because ________.

Translate each logarithmic statement into an equivalent exponential statement. (See Examples 1, 5, and 6.)

5. $\log 100{,}000 = 5$ **6.** $\log .001 = -3$

7. $\log_3 81 = 4$ **8.** $\log_2(1/4) = -2$

Translate each exponential statement into an equivalent logarithmic statement. (See Examples 5 and 6.)

9. $10^{1.8751} = 75$ **10.** $e^{3.2189} = 25$

11. $3^{-2} = 1/9$ **12.** $16^{1/2} = 4$

Without using a calculator, evaluate each of the following. (See Examples 1, 5, and 6.)

13. $\log 1000$ **14.** $\log .0001$

15. $\log_5 25$ **16.** $\log_9 81$

17. $\log_4 64$ **18.** $\log_6 216$

19. $\log_2 \frac{1}{4}$ **20.** $\log_3 \frac{1}{27}$

21. $\ln \sqrt{e}$ **22.** $\ln(1/e)$

23. $\ln e^{3.78}$ **24.** $\log 10^{56.9}$

Use a calculator to evaluate each logarithm to three decimal places. (See Examples 3 and 4.)

25. $\log 47$ **26.** $\log .004$

27. $\ln .351$ **28.** $\ln 2160$

29. Why does $\log_a 1$ always equal 0 for any valid base a?

Write each expression as the logarithm of a single number or expression. Assume all variables represent positive numbers. (See Example 7.)

30. $\log 15 - \log 3$

31. $\log 4 + \log 8 - \log 2$

32. $3 \ln 2 + 2 \ln 3$

33. $2 \ln 5 - \frac{1}{2} \ln 25$

34. $3 \log x - 2 \log y$

35. $2 \log u + 3 \log w - 6 \log v$

36. $\ln(3x + 2) + \ln(x + 4)$

37. $2 \ln(x + 1) - \ln(x + 2)$

Write each expression as a sum and/or a difference of logarithms with all variables to the first degree.

38. $\log 5x^2y^3$

39. $\ln \sqrt{6m^4n^2}$

40. $\ln \dfrac{3x}{5y}$

41. $\log \dfrac{\sqrt{xz}}{z^3}$

42. The calculator-generated table in the figure is for $y_1 = \log(4 - x)$. Why do the values in the y_1 column show ERROR for $x \geq 4$?

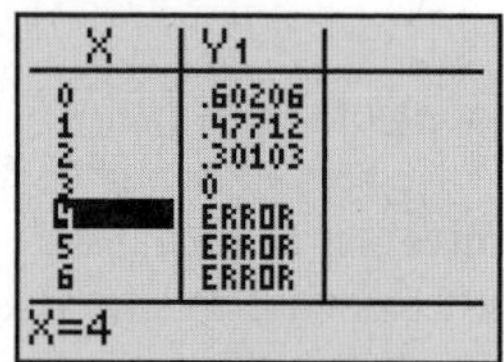

X	Y1
0	.60206
1	.47712
2	.30103
3	0
4	ERROR
5	ERROR
6	ERROR

X=4

Express each of the following in terms of u and v, where u = ln x and v = ln y. For example, ln x^3 = 3(ln x) = 3u.

43. $\ln(x^2y^5)$

44. $\ln(\sqrt{x} \cdot y^2)$

45. $\ln(x^3/y^2)$

46. $\ln(\sqrt{x}/y)$

Evaluate each expression. (See Example 9.)

47. $\log_6 543$

48. $\log_{20} 97$

49. $\log_{35} 6874$

50. $\log_5 50 - \log_{50} 5$

Find numerical values for b and c for which the given statement is FALSE.

51. $\log(b + c) = \log b + \log c$

52. $\dfrac{\ln b}{\ln c} = \ln\left(\dfrac{b}{c}\right)$

Graph each of the following. (See Example 10.)

53. $y = \ln(x + 2)$

54. $y = \ln x + 2$

55. $y = \log(x - 3)$

56. $y = \log x - 3$

57. Graph $f(x) = \log x$ and $g(x) = \log(x/4)$ for $-2 \leq x \leq 8$. How are these graphs related? How does the quotient rule support your answer?

In Exercises 58 and 59 the coordinates of a point on the graph of the indicated function are displayed at the bottom of the screen. Write the logarithmic and exponential equations associated with the display.

58. $f(x) = \log x$

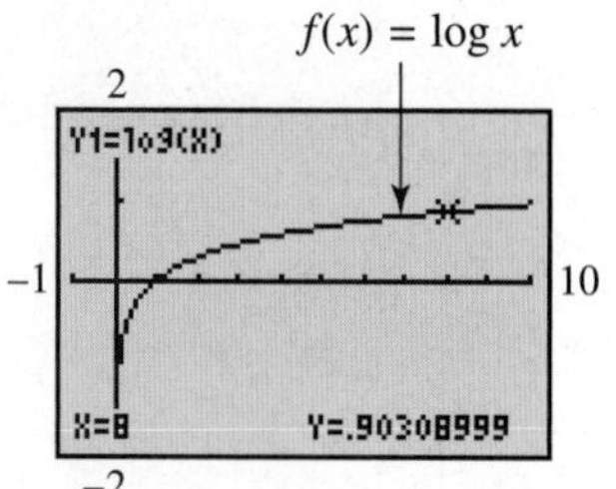

59. $g(x) = \ln x$

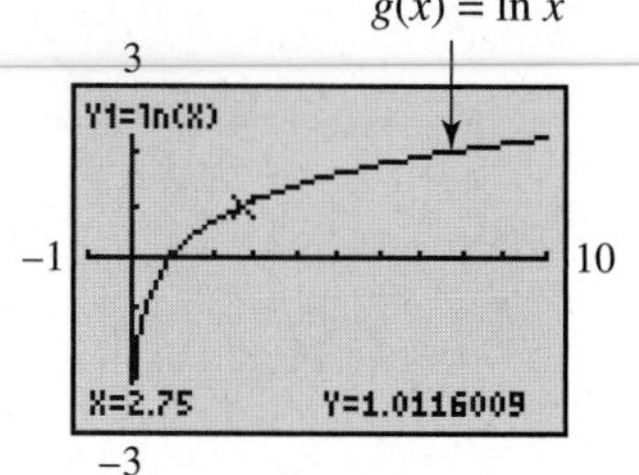

60. Match each equation with its graph. Each tick mark represents one unit.

(a) $y = \log x$ **(b)** $y = 10^x$

(c) $y = \ln x$ **(d)** $y = e^x$

A.

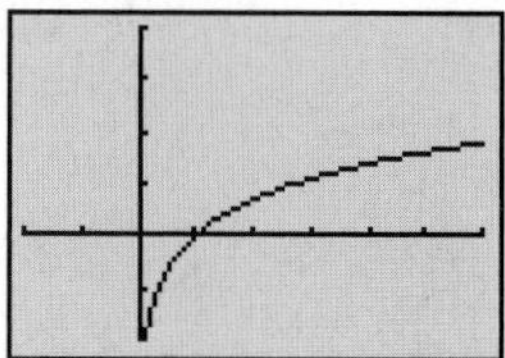

B.

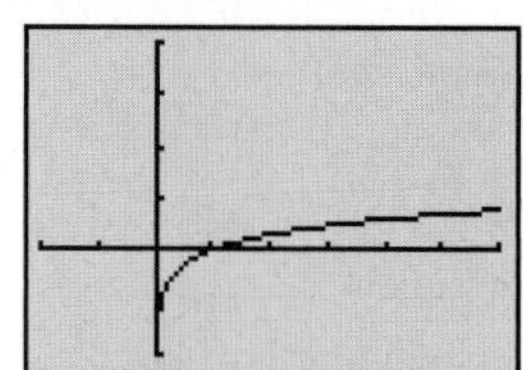

C.

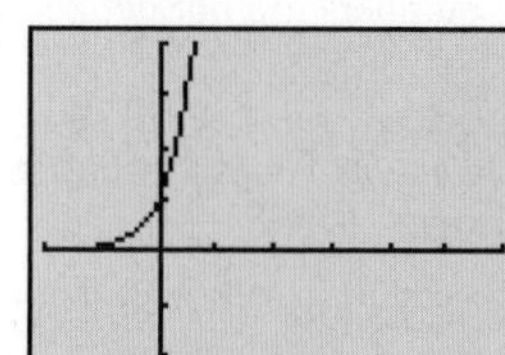

D.

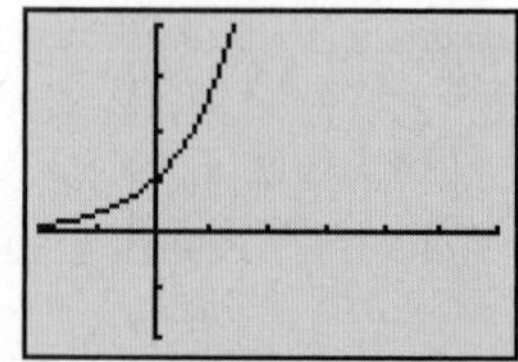

61. Finance The doubling function

$$D(r) = \frac{\ln 2}{\ln(1 + r)}$$

gives the number of years required to double your money when it is invested at interest rate r (expressed as a decimal), compounded annually. How long does it take to double your money at each of the following rates?

(a) 4% **(b)** 8%

(c) 18% **(d)** 36%

(e) Round each of your answers in parts (a)–(d) to the nearest year and compare them with these numbers: 72/4, 72/8, 72/18, 72/36. Use this evidence to state a "rule of thumb" for determining approximate doubling time without using the function D. This rule, which has long been used by bankers, is called the *rule of 72*.

62. Health Two people with the flu visited the campus of Big State U. The number of days T that it took for the flu virus to infect n people is given by

$$T = -1.43 \ln\left(\frac{10{,}000 - n}{4998n}\right).$$

How many days will it take for the virus to infect

(a) 500 people? **(b)** 5000 people?

63. Business The graph in the next column shows the percent increase in commercial rents in California from 1992 to 1999.

(a) Describe the growth in rents during the period shown in the graph.

(b) The graph can be approximated by the equation $f(x) = -650 + 143 \ln x$, where x is the number of years since 1900 and y is the corresponding percent change in rents. Find $f(92)$ and $f(99)$. Compare your results with the corresponding y-values from the graph.

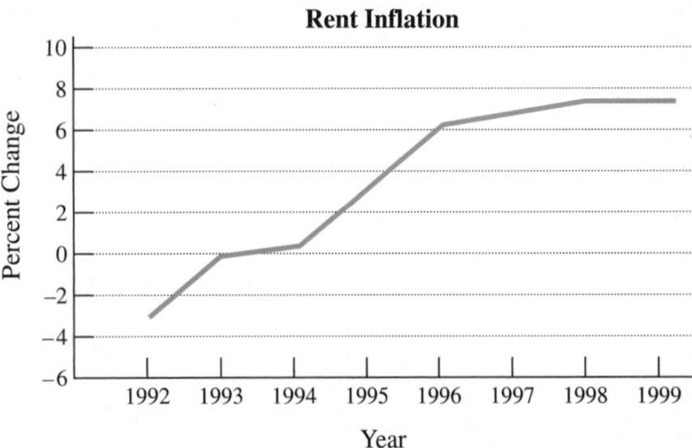

Source: CB Commercial/Torto Wheaton Research.

64. Physical Science The barometric pressure p (in inches of mercury) is related to the height h above sea level (in miles) by the equation

$$h = -5 \ln\left(\frac{p}{29.92}\right).$$

The following pressure readings were made by a weather balloon. At what heights were they made?

(a) 29.92 in. **(b)** 20.05 in. **(c)** 11.92 in.

(d) Use a graphing calculator to determine the pressure at a height of 3 miles.

4.4 LOGARITHMIC AND EXPONENTIAL EQUATIONS

Many applications involve solving logarithmic and exponential equations. So we begin with solution methods for such equations.

Logarithmic Equations

When an equation involves only logarithmic terms, use the logarithm properties to write each side as a single logarithm. Then use this fact.

Let u, v, and a be positive real numbers, with $a \neq 1$.

If $\log_a u = \log_a v$, then $u = v$.

EXAMPLE 1 Solve $\log x = \log (x + 3) - \log (x - 1)$

First, use the quotient property of logarithms to write the right side as a single logarithm.

$$\log x = \log (x + 3) - \log (x - 1)$$
$$\log x = \log\left(\frac{x + 3}{x - 1}\right)$$

The fact in the preceding box now shows that

$$x = \frac{x + 3}{x - 1}$$
$$x(x - 1) = x + 3$$
$$x^2 - x = x + 3$$
$$x^2 - 2x - 3 = 0$$
$$(x - 3)(x + 1) = 0$$
$$x = 3 \quad \text{or} \quad x = -1$$

Since $\log x$ is not defined when $x = -1$, the only possible solution is $x = 3$. Use a calculator to verify that 3 actually is a solution. ◀ 1

1

Solve each equation.

(a) $\log_2(p + 9) - \log_2 p = \log_2(p + 1)$

(b) $\log_3(m + 1) - \log_3(m - 1) = \log_3 m$

Answers:

(a) 3

(b) $1 + \sqrt{2} \approx 2.414$

When an equation involves constants and logarithmic terms, use algebra and the logarithm properties to write one side as a single logarithm and the other as a constant. Then use the following property of logarithms.

If a and u are positive real numbers, with $a \neq 1$, then

$$a^{\log_a u} = u$$

EXAMPLE 2 Solve each equation.

(a) $\log_5 (2x - 3) = 2$

Since the logarithm base is 5, raise 5 to the exponents given by the equation.

$$5^{\log_5 (2x-3)} = 5^2$$

Now use the fact in the preceding box (with $a = 5$ and $u = 2x - 3$) to conclude that

$$2x - 3 = 25$$
$$2x = 28$$
$$x = 14.$$

Verify that 14 is a solution of the original equation.

(b) $\log (x - 16) = 2 - \log (x - 1)$
First rearrange the terms to obtain a single logarithm on the left side.

$$\log (x - 16) + \log (x - 1) = 2$$
$$\log [(x - 16)(x - 1)] = 2 \qquad \text{Product property of logarithms}$$
$$\log (x^2 - 17x + 16) = 2$$

Since the logarithm base is 10, raise 10 to these powers.

$$10^{\log(x^2 - 17x + 16)} = 10^2$$

Apply the logarithm property in the preceding box (with $a = 10$ and $u = x^2 - 17x + 16$).

$$x^2 - 17x + 16 = 100$$
$$x^2 - 17x - 84 = 0$$
$$(x + 4)(x - 21) = 0$$
$$x = -4 \quad \text{or} \quad x = 21$$

When $x = -4$, then in the original equation, $\log (x - 16) = \log (-20)$, which is not defined. So -4 is not a solution. You should verify that 21 is a solution of the original equation.

(c) $\log_2 x - \log_2 (x - 1) = 1$
Proceed as above, with 2 as the logarithm base.

$$\log_2 \frac{x}{x - 1} = 1 \qquad \text{Quotient property of logarithms}$$
$$2^{\log_2 x/(x-1)} = 2^1 \qquad \text{Exponentiate to the base 2.}$$
$$\frac{x}{x - 1} = 2 \qquad \text{Use the fact in the preceding box.}$$
$$x = 2(x - 1) \qquad \text{Multiply both sides by } x - 1.$$
$$x = 2x - 2$$
$$-x = -2$$
$$x = 2$$

Verify that 2 is a solution of the original equation. 

2

Solve each equation.

(a) $\log_5 x + 2 \log_5 x = 3$

(b) $\log_6(a + 2) - \log_6 \dfrac{a - 7}{5} = 1$

Answers:

(a) 5

(b) 52

Exponential Equations

An equation in which all the variables are exponents is called an exponential equation. When such an equation can be written as two powers of the same base, it can be solved by using this fact.

Let a be a positive real number, with $a \neq 1$.

If $a^u = a^v$, then $u = v$.

EXAMPLE 3 Solve $9^x = 27$.

First, write both sides as powers of the same base. Since $9 = 3^2$ and $27 = 3^3$, we have

$$9^x = 27$$
$$(3^2)^x = 3^3$$
$$3^{2x} = 3^3$$

Apply the fact in the preceding box (with $a = 3$, $u = 2x$, and $v = 3$).

$$2x = 3$$
$$x = \frac{3}{2}$$

Verify that $x = 3/2$ is a solution of the original equation. 

3

Solve each equation.

(a) $8^{2x} = 4$

(b) $5^{3x} = 25^4$

(c) $36^{-2x} = 6$

Answers:

(a) 1/3

(b) 8/3

(c) −1/4

Exponential equations involving different bases can often be solved by using the power property of logarithms, which is repeated here for natural logarithms.

If u and r are real numbers, with u positive, then

$$\ln u^r = r \ln u.$$

Although natural logarithms are used in the following examples, logarithms to any base will produce the same solutions.

EXAMPLE 4 Solve $3^x = 5$.

Take logarithms on both sides.

$$\ln 3^x = \ln 5$$

Apply the power property of logarithms in the preceding box (with $r = x$) to the left side.

$$x \ln 3 = \ln 5$$
$$x = \frac{\ln 5}{\ln 3} \approx 1.465 \quad \text{Divide both sides by the constant ln 3.}$$

To check, evaluate $3^{1.465}$; the answer should be approximately 5, which verifies that the solution of the given equation is 1.465 to the nearest thousandth. 4

4

Solve each equation. Round solutions to the nearest thousandth.

(a) $2^x = 7$

(b) $5^m = 50$

(c) $3^y = 17$

Answers:

(a) 2.807

(b) 2.431

(c) 2.579

CAUTION Be careful; $\dfrac{\ln 5}{\ln 3}$ is *not* equal to $\ln\left(\dfrac{5}{3}\right)$ or $\ln 5 - \ln 3$.

EXAMPLE 5 Solve $3^{2x-1} = 4^{x+2}$.

Taking natural logarithms on both sides gives

$$\ln 3^{2x-1} = \ln 4^{x+2}.$$

Now use the power property of logarithms and the fact that ln 3 and ln 4 are constants to rewrite the equation.

$$(2x - 1)(\ln 3) = (x + 2)(\ln 4) \quad \text{Power property}$$
$$2x(\ln 3) - 1(\ln 3) = x(\ln 4) + 2(\ln 4) \quad \text{Distributive property}$$
$$2x(\ln 3) - x(\ln 4) = 2(\ln 4) + 1(\ln 3) \quad \text{Collect terms with } x \text{ on one side.}$$

Factor out x on the left side to get

$$[2(\ln 3) - \ln 4]x = 2(\ln 4) + \ln 3.$$

Divide both sides by the coefficient of x:

$$x = \frac{2(\ln 4) + \ln 3}{2(\ln 3) - \ln 4}.$$

Using a calculator to evaluate this last expression, we find that

$$x = \frac{2 \ln 4 + \ln 3}{2 \ln 3 - \ln 4} \approx 4.774.$$ (5)

Recall that $\ln e = 1$ (because 1 is the exponent to which e must be raised to produce e). This fact simplifies the solution of equations involving powers of e.

(5) Solve each equation. Round solutions to the nearest thousandth.

(a) $6^m = 3^{2m-1}$

(b) $5^{6a-3} = 2^{4a+1}$

Answers:

(a) 2.710

(b) .802

EXAMPLE 6 Solve $3e^{x^2} = 600$.

First divide each side by 3 to get

$$e^{x^2} = 200.$$

Now take natural logarithms on both sides; then use the power property of logarithms.

$$e^{x^2} = 200$$
$$\ln e^{x^2} = \ln 200$$
$$x^2 \ln e = \ln 200 \quad \text{Power property}$$
$$x^2 = \ln 200 \quad \ln e = 1$$
$$x = \pm\sqrt{\ln 200}$$
$$x \approx \pm 2.302$$

The solutions are ± 2.302, rounding to the nearest thousandth. (The symbol $\pm$ is used as a shortcut for writing the two solutions, 2.302 and -2.302.) (6)

(6) Solve each equation. Round solutions to the nearest thousandth.

(a) $e^{.1x} = 11$

(b) $e^{3+x} = .893$

(c) $e^{2x^2-3} = 9$

Answers:

(a) 23.979

(b) -3.113

(c) ± 1.612

TECHNOLOGY TIP Logarithmic and exponential equations can be solved on a graphing calculator. To solve $3^x = 5^{2x-1}$, for example, graph $y = 3^x$ and $y = 5^{2x-1}$ on the same screen. Then use the intersection finder to determine the x-coordinates of their intersection points. Alternatively, graph $y = 3^x - 5^{2x-1}$ and use the root finder to determine the x-intercepts of the graph.

Applications

Some of the most important applications of exponential and logarithmic functions arise in banking and finance. They will be thoroughly discussed in Chapter 5. The applications here are from other fields.

EXAMPLE 7 According to projections by the U.S. Census Bureau, the world population (in billions) in year x is approximated by the function $P(x) = 4.834(1.01^{x-1980})$. When will the population reach 7 billion?

You are asked to find the value of x for which $P(x) = 7$, that is, to solve the equation

$$4.834(1.01^{x-1980}) = 7$$

$$1.01^{x-1980} = \frac{7}{4.834} \quad \text{Divide both sides by 4.834.}$$

$$\ln 1.01^{x-1980} = \ln\left(\frac{7}{4.834}\right) \quad \text{Take logarithms on both sides.}$$

$$(x - 1980)\ln 1.01 = \ln\left(\frac{7}{4.834}\right) \quad \text{Power property of logarithms.}$$

$$x - 1980 = \frac{\ln(7/4.834)}{\ln 1.01} \quad \text{Divide both sides by ln 1.01.}$$

$$x = \frac{\ln(7/4.834)}{\ln 1.01} + 1980 \approx 2017.2$$

Hence, the world population will reach 7 billion around 2017. ◀ 7

> 7
>
> Use the function in Example 7 to determine when the earth's population will be 10 billion.
>
> Answer:
>
> In 2053

The **half-life** of a radioactive substance is the time it takes for a given quantity to decay to one-half its original mass. The half-life depends only on the substance, not on the size of the sample. It can be shown that the amount of a radioactive substance at time t is given by the function

$$f(t) = y_0\left(\frac{1}{2}\right)^{t/h} = y_0(.5^{t/h}),$$

where y_0 is the initial amount (at time $t = 0$) and h is the half-life of the substance.

Radioactive carbon-14 is found in every living plant and animal. After the plant or animal dies, its carbon-14 decays exponentially, with a half-life of 5730 years. This fact is the basis for a technique called *carbon dating* for determining the age of fossils.

EXAMPLE 8 A round wood table hanging in Winchester Castle (England) was alleged to have belonged to King Arthur, who lived in the fifth century. A recent chemical analysis showed that the wood had lost 9% of its carbon-14.* How old is the table?

*This is done by measuring the ratio of carbon-14 to nonradioactive carbon-12 in the table (which is approximately constant over long time periods) and comparing it to the ratio in living wood.

The decay function for carbon-14 is

$$f(t) = y_0(.5^{t/5730}),$$

where $t = 0$ corresponds to the time when the wood was cut to make the table (that's when the tree died). Since the wood has lost 9% of its carbon-14, the amount now in the table is 91% of the initial amount y_0, that is, $.91y_0$. We must find the value of t for which $f(t) = .91y_0$. So we must solve the equation

$$y_0(.5^{t/5730}) = .91y_0 \qquad \text{Definition of } f(t)$$

$$.5^{t/5730} = .91 \qquad \text{Divide both sides by } y_0.$$

$$\ln .5^{t/5730} = \ln .91 \qquad \text{Take logarithms on both sides.}$$

$$\left(\frac{t}{5730}\right)\ln .5 = \ln .91 \qquad \text{Power property of logarithms}$$

$$t \ln .5 = 5730 \ln .91 \qquad \text{Multiply both sides by 5730.}$$

$$t = \frac{5730 \ln .91}{\ln .5} \approx 779.63 \qquad \text{Divide both sides by ln.5.}$$

8

How old is a skeleton that has lost 65% of its carbon-14?

Answer:

About 8679 years

The table is about 780 years old and, therefore, could not have belonged to King Arthur. ◀ 8

Earthquakes are often in the news. What you may not have known is that the standard method of measuring their size, the **Richter scale,** is a logarithmic function (base 10).

EXAMPLE 9 The intensity $R(i)$ of an earthquake, measured on the Richter scale, is given by

$$R(i) = \log\left(\frac{i}{i_0}\right),$$

where i is the intensity of the ground motion of the earthquake and i_0 is the intensity of the ground motion of the so-called *zero earthquake* (the smallest detectable earthquake, against which others are measured). The 1989 San Francisco earthquake measured 7.1 on the Richter scale.

(a) How did the ground motion of this earthquake compare with that of the zero earthquake?

In this case $R(i) = 7.1$, that is, $\log(i/i_0) = 7.1$. Thus 7.1 is the exponent to which 10 must be raised to produce i/i_0. In other words,

$$10^{7.1} = \frac{i}{i_0}, \quad \text{or equivalently,} \quad i = 10^{7.1}i_0.$$

So this earthquake had $10^{7.1}$ (approximately 12.6 million) times more ground motion than the zero earthquake.

(b) What is the Richter scale intensity of an earthquake with 10 times as much ground motion as the 1989 San Francisco earthquake?

Using the result from (a), the ground motion of such a quake would be

$$i = 10(10^{7.1}i_0) = 10^1 \cdot 10^{7.1}i_0 = 10^{8.1}i_0$$

Find the Richter scale intensity of an earthquake whose ground motion is 100 times greater than the ground motion of the 1989 San Francisco earthquake discussed in Example 9.

Answer:

9.1

so that its Richter scale measure would be

$$R(i) = \log\left(\frac{i}{i_0}\right) = \log\left(\frac{10^{8.1}i_0}{i_0}\right) = \log 10^{8.1} = 8.1.$$

Therefore, a tenfold increase in ground motion increases the Richter scale intensity by just 1. ◀ 9

Earlier, we saw how rational functions describe cost-benefit functions. The final example illustrates another type of cost-benefit function.

EXAMPLE 10 One action that government could take to reduce carbon emissions into the atmosphere is to place a tax on fossil fuel. This tax would be based on the amount of carbon dioxide that is emitted into the air when the fuel is burned. The *cost-benefit* equation $\ln(1 - P) = -.0034 - .0053T$ describes the approximate relationship between a tax of T dollars per ton of carbon dioxide and the corresponding percent reduction P (in decimals) of emission of carbon dioxide.*

(a) Write P as a function of T.

We begin by writing the cost-benefit equation in exponential form.

$$\ln(1 - P) = -.0034 - .0053T$$
$$1 - P = e^{-.0034-.0053T}$$
$$P = P(T) = 1 - e^{-.0034-.0053T}$$

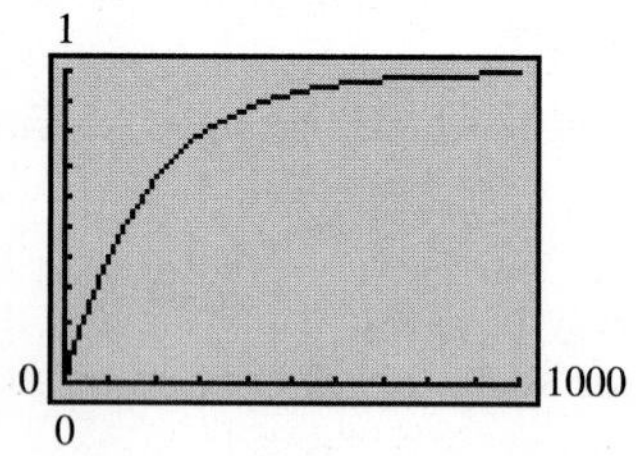

FIGURE 4.17

A calculator-generated graph of $P(T)$ is shown in Figure 4.17.

(b) Discuss the benefit of continuing to raise taxes on carbon dioxide emissions.

From the graph we see that initially there is a rapid reduction of carbon dioxide emissions. However, after a while there is little benefit in raising taxes further. ◀

*W. Nordhause, "To Slow or Not to Slow: The Economics of the Greenhouse Effect." Yale University, New Haven, Connecticut.

4.4 Exercises

Solve each logarithmic equation. (See Example 1.)

1. $\ln(3x - 1) - \ln(2 + x) = \ln 2$
2. $\ln(8k - 7) - \ln(3 + 4k) = \ln(9/11)$
3. $\ln x + 1 = \ln(x - 4)$
4. $\ln(4x - 2) = \ln 4 - \ln(x - 2)$
5. $2\ln(x - 3) = \ln(x + 5) + \ln 4$
6. $\ln(k + 5) + \ln(k + 2) = \ln 14k$

Solve each logarithmic equation. (See Example 2.)

7. $\log_3(5x + 1) = 2$
8. $\log_5(9x - 4) = 1$
9. $\log x - \log(x + 3) = -1$
10. $\log m - \log(m - 4) = -2$
11. $\log_3(y + 2) = \log_3(y - 7) + \log_3 4$
12. $\log_8(z - 6) = 2 - \log_8(z + 15)$
13. $\ln(x + 9) - \ln x = 1$
14. $\ln(2x + 1) - 1 = \ln(x - 2)$
15. $\log x + \log(x - 3) = 1$
16. $\log(x - 1) + \log(x + 2) = 1$

Solve each equation for c.

17. $\log(3 + b) = \log(4c - 1)$

18. $\ln(b + 7) = \ln(6c + 8)$

19. $2 - b = \log(6c + 5)$

20. $8b + 6 = \ln(2c) + \ln c$

21. Suppose you overhear the following statement: "I must reject any negative answer when I solve an equation involving logarithms." Is this correct? Write an explanation of why it is or is not correct.

22. What values of x could not possibly be solutions of the following equation?

$$\log_a(4x - 7) + \log_a(x^2 + 4) = 0$$

Solve these exponential equations without using logarithms. (See Example 3.)

23. $3^{x-1} = 9$

24. $16^{-x+1} = 8$

25. $25^{-2x} = 3125$

26. $81^{-2x} = 3^{x-1}$

27. $7^{-x} = 49^{x+3}$

28. $16^x = 64$

29. $\left(\frac{3}{4}\right)^x = \frac{16}{9}$

30. $2^{x^2-4x} = \frac{1}{16}$

Use logarithms to solve these exponential equations. (See Examples 4–6.)

31. $3^x = 5$

32. $5^x = 4$

33. $2^x = 3^{x-1}$

34. $4^{x+2} = 2^{x-1}$

35. $3^{1-2x} = 5^{x+5}$

36. $4^{3x-1} = 3^{x-2}$

37. $e^{2x} = 5$

38. $e^{-3x} = 2$

39. $2e^{5a+2} = 8$

40. $10e^{3z-7} = 5$

Solve each equation for c.

41. $10^{4c-3} = d$

42. $3 \cdot 10^{2c+1} = 4d$

43. $e^{2c-1} = b$

44. $3e^{5c-7} = b$

Solve these equations. (See Examples 1–6.)

45. $\log_5(r + 2) + \log_5(r - 2) = 1$

46. $\log_4(z + 3) + \log_4(z - 3) = 1$

47. $\log_3(a - 3) = 1 + \log_3(a + 1)$

48. $\log w + \log(3w - 13) = 1$

49. $\log_2 \sqrt{2y^2} - 1 = 1/2$

50. $\log_2(\log_2 x) = 1$

51. $\log z = \sqrt{\log z}$

52. $\log x^2 = (\log x)^2$

53. $5^{-2x} = \frac{1}{25}$

54. $5^{x^2+x} = 1$

55. $2^{|x|} = 16$

56. $5^{-|x|} = \frac{1}{25}$

57. $2^{x^2-1} = 12$

58. $3^{2-x^2} = 4$

59. $2(e^x + 1) = 10$

60. $5(e^{2x} - 2) = 15$

61. Explain why the equation $4^{x^2+1} = 2$ has no solutions.

62. Explain why the equation $\log(-x) = -4$ does have a solution and find that solution.

Work these problems. (See Example 7.)

63. **Business** The number of private cars per one million people in China is approximated by

$$g(x) = 751.635(1.2072^x),$$

where $x = 0$ corresponds to 1990.*

(a) What was the number of cars per million people in 1998 and 2003? (For comparison purposes, there were 768,333 cars per million people in the United States in 1998.)

(b) If this model remains accurate, when will there be 15,000 cars per million people?

64. **Business** According to data from the World Intellectual Property Organization, the number of patent applications filed worldwide (in millions) during the 1990s is approximated by

$$f(x) = 1.2174e^{.1885x},$$

where $x = 0$ corresponds to 1990.

(a) How many patent applications had been filled by 1995 and by 2000?

(b) If this model remains accurate, when will the number of patent applications reach 18 million?

65. **Health** A drug's effectiveness decreases over time. If each hour a drug is only 90% as effective as the previous hour, at some point the patient will not be receiving enough medication and must receive another dose. This situation can be modeled by an exponential function with $y = y_0(.90)^{t-1}$. In this equation, y_0 is the amount of the initial dose and y is the percent of medication still available t hours after the drug was administered. Suppose 200 mg of the drug is administered. How long will it take for this initial dose to reach the dangerously low level of 50 mg?

66. **Business** The U.S. gross domestic product (GDP) in billions of dollars is approximated by

$$f(x) = -6.168 + 5.069 \ln x \qquad (x \geq 10),$$

where $x = 10$ corresponds to 1990.† Assuming this model remains accurate, in what year will the GDP be 10 billion dollars?

Work these exercises. (See Example 8.)

67. **Natural Science** The amount of cobalt-60 (in grams) in a storage facility at time t is given by

$$C(t) = 25e^{-.14t},$$

where time is measured in years.

*Chinese Ministry of Communications and the *Economist* magazine.

†U.S. Bureau of Economic Analysis.

(a) How much cobalt-60 was present initially?
(b) What is the half-life of cobalt-60? (*Hint:* for what value of t is $C(t) = 12.5$?)

68. Natural Science A Native American mummy was found recently. It had 73.6% of the amount of radiocarbon present in living beings. Approximately how long ago did this person die?

69. Natural Science How old is a piece of ivory that has lost 36% of its radiocarbon?

70. Natural Science A sample from a refuse deposit near the Strait of Magellan had 60% of the carbon 14 of a contemporary living sample. How old was the sample?

Natural Science *Do the following problems. (See Example 9.)*

71. On July 14, 1991, Peshawar, Pakistan, was shaken by an earthquake that measured 6.6 on the Richter scale.
(a) Express this reading in terms of i_0.
(b) In February of the same year, a quake measuring 6.5 on the Richter scale killed about 900 people in the mountains of Pakistan and Afghanistan. Express the intensity of a 6.5 reading in terms of i_0.
(c) How many times more intense was the July earthquake than the February one?

72. Find the Richter scale intensity of earthquakes whose ground motion is
(a) $1000i_0$
(b) $100{,}000i_0$
(c) $10{,}000{,}000i_0$.
(d) Fill in the blank in this statement: Increasing the ground motion by a factor of 10^k increases the Richter intensity by ______ units.

73. The loudness of sound is measured in units called decibels. The decibel rating of a sound is given by

$$D(i) = 10 \cdot \log\left(\frac{i}{i_0}\right),$$

where i is the intensity of the sound and i_0 is the minimum intensity detectable by the human ear (the so-called *threshold sound*). Find the decibel rating of each of the following sounds whose intensities are given. Round answers to the nearest whole number.
(a) Whisper, $115i_0$
(b) Average sound level in the movie *Godzilla*, $10^{10}i_0$
(c) Jackhammer, $31{,}600{,}000{,}000i_0$
(d) Rock music, $895{,}000{,}000{,}000i_0$
(e) Jetliner at takeoff, $109{,}000{,}000{,}000{,}000i_0$

74. **(a)** How much more intense is a sound that measures 100 decibels than the threshold sound?
(b) How much more intense is a sound that measures 50 decibels than the threshold sound?
(c) How much more intense is a sound measuring 100 decibels than one measuring 50 decibels?

75. Natural Science Refer to Example 10.
(a) Determine the percent reduction in carbon dioxide when the tax is \$60.
(b) What tax will cause a 50% reduction in carbon dioxide emissions?

76. Social Science The number of years, $N(r)$, since two independently evolving languages split off from a common ancestral language is approximated by

$$N(r) = -5000 \ln r,$$

where r is the proportion of the words from the ancestral language that is common to both languages now. Find each of the following.
(a) $N(.9)$ **(b)** $N(.5)$ **(c)** $N(.3)$
(d) How many years have elapsed since the split if 70% of the words of the ancestral language are common to both languages today?
(e) If two languages split off from a common ancestral language about 1000 years ago, find r.

77. Natural Science In the central Sierra Nevada of California, the percent of moisture that falls as snow rather than rain is approximated reasonably well by

$$p = 86.3 \ln h - 680,$$

where p is the percent of moisture as snow at an altitude of h feet (with $3000 \le h < 8500$).
(a) Graph p.
(b) At what altitude is 50 percent of the moisture snow?

78. Physical Science The table gives some of the planets' average distances D from the sun and their period P of revolution around the sun in years. The distances have been normalized so that Earth is one unit from the sun. Thus, Jupiter's distance of 5.2 means that Jupiter's distance from the sun is 5.2 times farther than Earth's.*

Planet	*D*	*P*
Earth	1	1
Jupiter	5.2	11.9
Saturn	9.54	29.5
Uranus	19.2	84.0

(a) Plot the points (D, P) for these planets. Would a straight line or an exponential curve fit these points best?
(b) Plot the points $(\ln D, \ln P)$ for these planets. Do these points appear to lie on a line?

*C. Ronan, *The Natural History of the Universe*, (New York: Macmillan Publishing Co., 1991).

(c) Determine a linear equation that approximates the data points with $x = \ln D$ and $y = \ln P$. Use the first and last data points (rounded to 2 decimal places). Graph your line and the data on the same coordinate axes.

(d) Use the linear equation to predict the period of the planet Pluto if its distance is 39.5. Compare your answer to the true value of 248.5 years.

 *Use logarithmic regression for these exercises.*

79. Health The table shows the number of kidney transplants in selected years.*

Year	1985	1990	1995	1996	1997	1998
Transplants	7695	9880	11,901	12,152	12,307	13,139

(a) Find a logarithmic function that models this data with $x = 5$ corresponding to 1985.

(b) Estimate the number of transplants in 2002.

(c) If this model remains accurate, in what year will there be 14,500 transplants?

80. Social Science The table shows the number of visitors to U.S. national parks (in millions) in selected years.†

Year	1950	1960	1970	1980	1990	1994	1998
Visitors	14	28	46	47	47.7	63	64.5

(a) Find a logarithmic function that models this data with $x = 50$ corresponding to 1950.

(b) Assuming the model remains accurate, estimate the number of visitors in 2000, 2003, and 2005.

(c) In what year will the number of visitors reach 75 million?

*U.S. Department of Health and Human Services and the United Network for Organ Sharing.

†National Park Service.

CHAPTER 4 SUMMARY

Key Terms and Symbols

4.1 exponential function
exponential growth and decay
the number $e \approx 2.71828\ldots$

4.2 exponential growth function
logistic function

4.3 $\log x$ common (base 10 logarithm) of x
$\ln x$ natural (base e logarithm) of x
$\log_a x$ base a logarithm of x
product, quotient, and power properties of logarithms
change of base theorem
logarithmic function
inverses

4.4 logarithmic equations
exponential equations
half-life
Richter scale

Key Concepts

An important application of exponents is the **exponential growth function,** defined as $f(t) = y_0e^{kt}$ or $f(t) = y_0b^t$, where y_0 is the amount of a quantity present at time $t = 0$, $e \approx 2.71828$, and k and b are constants.

The **logarithm** of x to the base a is defined as follows. For $a > 0$ and $a \neq 1$, $y = \log_a x$ means $a^y = x$. Thus, $\log_a x$ is an *exponent,* the power to which a must be raised to produce x.

Properties of Logarithms

Let x, y, and a be positive real numbers, $a \neq 1$, and let r be any real number.

$\log_a 1 = 0$ $\qquad$ $\log_a a = 1$

$\log_a a^r = r$ $\qquad$ $a^{\log_a x} = x$

Product property: $\log_a xy = \log_a x + \log_a y$

Quotient property: $\log_a \dfrac{x}{y} = \log_a x - \log_a y$

Power property: $\log_a x^r = r \log_a x$

Solving Exponential and Logarithmic Equations

Let $a > 0$, $a \neq 1$.

If $\log_a u = \log_a v$, then $u = v$.

If $a^u = a^v$, $u = v$.

Chapter 4 Review Exercises

Match each equation with the letter of the graph that most closely resembles its graph. Assume that a > 1.

1. $y = a^{x+2}$ $\qquad$ **2.** $y = a^x + 2$

3. $y = -a^x + 2$ $\qquad$ **4.** $y = a^{-x} + 2$

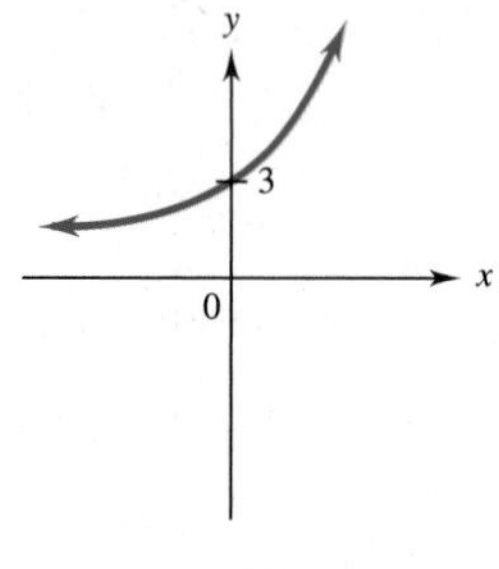

(a)

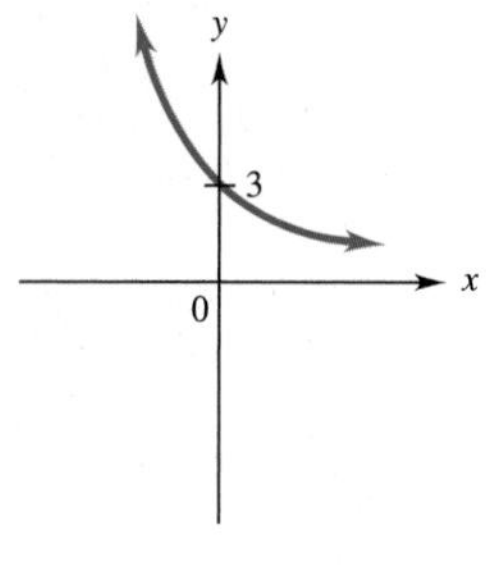

(b)

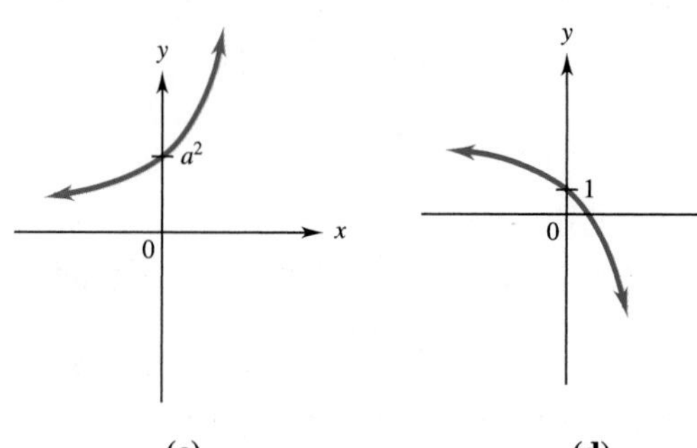

(c) $\qquad$ **(d)**

Consider the exponential function $y = f(x) = a^x$ graphed here. Answer each question based on the graph.

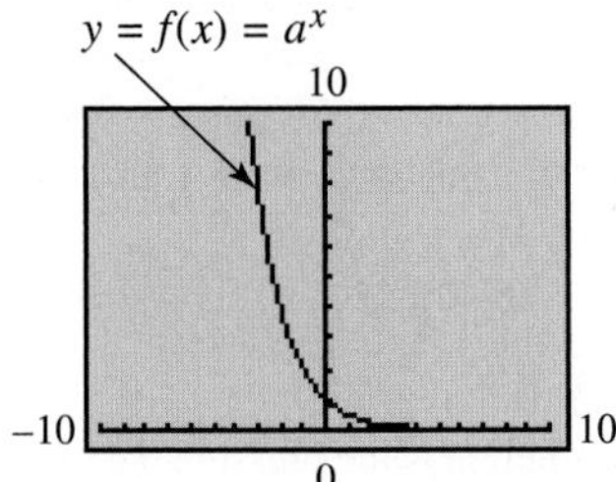

5. What is true about the value of a in comparison to 1?
6. What is the domain of f?
7. What is the range of f?
8. What is the value of $f(0)$?

Graph each function.

9. $f(x) = 4^x$
10. $g(x) = 4^{-x}$
11. $f(x) = \ln x + 5$
12. $g(x) = \log x - 3$

Work these problems.

13. **Business** India has become an important exporter of software to the United States. Since 1985 India's exports (in millions of dollars) can be approximated by the function

$$f(x) = 6.2(10^{-12})(1.4^x),$$

where x is the number of years since 1900.
   (a) According to this model, what was the value of software exports in 2000?
   (b) When will software exports double their value in 2000?

14. **Business** A person learning certain skills involving repetition tends to learn quickly at first. Then learning tapers off and approaches some upper limit. Suppose the number of symbols per minute a textbook typesetter can produce is given by $p(t) = 250 - 120(2.8)^{-.5t}$, where t is the number of months the typesetter has been in training. Find each of the following:
   (a) $p(2)$ (b) $p(4)$
   (c) $p(10)$. (d) Graph $y = p(t)$.

Translate each exponential statement into an equivalent logarithmic one.

15. $10^{1.6721} = 47$
16. $5^4 = 625$
17. $e^{3.6636} = 39$
18. $5^{1/2} = \sqrt{5}$

Translate each logarithmic statement into an equivalent exponential one.

19. $\log 1000 = 3$
20. $\log 16.6 = 1.2201$
21. $\ln 95.4 = 4.5581$
22. $\log_2 64 = 6$

Evaluate each expression without using a calculator.

23. $\ln e^3$
24. $\log \sqrt{10}$
25. $10^{\log 7.4}$
26. $\ln e^{4k}$
27. $\log_8 16$
28. $\log_{25} 5$

Write each expression as a single logarithm. Assume all variables represent positive quantities.

29. $\log 4k + \log 5k^3$
30. $4 \log x - 2 \log x^3$
31. $2 \log b - 3 \log c$
32. $4 \ln x - 2(\ln x^3 + 4 \ln x)$

Solve each equation. Round to the nearest thousandth.

33. $\ln(m + 3) - \ln m = \ln 2$
34. $2 \ln(y + 1) = \ln(y^2 - 1) + \ln 5$
35. $\log(m + 2) = 1$
36. $\log x^2 = 2$
37. $\log_2(3k - 2) = 4$
38. $\log_5\left(\frac{5z}{z - 2}\right) = 2$
39. $\log x + \log(x + 3) = 1$
40. $\log_2 r + \log_2(r - 2) = 3$
41. $2^{3x} = \frac{1}{8}$
42. $\left(\frac{9}{16}\right)^x = \frac{3}{4}$
43. $9^{2y-1} = 27^y$
44. $\frac{1}{2} = \left(\frac{b}{4}\right)^{1/4}$
45. $8^p = 19$
46. $3^z = 11$
47. $5 \cdot 2^{-m} = 35$
48. $2 \cdot 15^{-k} = 18$
49. $e^{-5-2x} = 5$
50. $e^{3x-1} = 12$
51. $6^{2-m} = 2^{3m+1}$
52. $5^{3r-1} = 6^{2r+5}$
53. $(1 + .003)^k = 1.089$
54. $(1 + .094)^z = 2.387$

Work these problems.

55. **Business** Suppose the gross national product (GNP) of a small country (in millions of dollars) is approximated by the function, $G(t) = 15 + 2 \log t$, where t is time in years, for $1 \le t \le 6$. Find the GNP at the following times.
   (a) 1 year (b) 2 years (c) 5 years

56. **Natural Science** A population is increasing according to the growth law $y = 2e^{.02t}$, where y is in millions and t is in years. Match each of the questions (a), (b), (c), and (d) with one of the solutions (A), (B), (C), or (D).

(a) How long will it take for the population to triple?	(A) Evaluate $2e^{.02(1/3)}$.
(b) When will the population reach 3 million?	(B) Solve $2e^{.02t} = 3 \cdot 2$ for t.
(c) How large will the population be in 3 years?	(C) Evaluate $2e^{.02(3)}$.
(d) How large will the population be in 4 months?	(D) Solve $2e^{.02t} = 3$ for t.

57. **Natural Science** The amount of polonium (in grams) present after t days is given by

$$A(t) = 10e^{-.00495t}.$$

(a) How much polonium was present initially?
(b) What is the half-life of polonium?
(c) How long will it take for the polonium to decay to 3 grams?

58. Natural Science One earthquake measures 4.6 on the Richter scale. A second earthquake has ground motion 1000 times greater than the first. What does the second one measure on the Richter scale?

Natural Science *Here is another form of Newton's Law of Cooling (see Section 4.2, Exercises 15 and 16): $F(t) = T_0 + Ce^{-kt}$, where C and k are constants.*

59. A piece of metal is heated to 300° Celsius and then placed in a cooling liquid at 50° Celsius. After 4 minutes the metal has cooled to 175° Celsius. Find its temperature after 12 minutes.

60. A frozen pizza has a temperature of 3.4° Celsius when it is taken from the freezer and left out in a room at 18° Celsius. After half an hour its temperature is 7.2° Celsius. How long will it take for the pizza to thaw to 10° Celsius?

In Exercises 61 and 62, do part (a) and skip part (b) if you do not have a graphing calculator. If you have a graphing calculator, then do part (b) and skip part (a).

61. Business The number of music cassette tapes sold each year (in millions) has been steadily dropping, as shown in the table.

Year	1994	1995	1996	1997	1998	1999
Cassettes Sold	345.4	272.6	225.3	172.6	158.5	123.6

(a) Let $x = 0$ correspond to 1994. Use the data points from 1994 and 1999 to find a function of the form $f(x) = a(b^x)$ that models this data.
(b) Use exponential regression to find a function $g(x)$ that models this data, with $x = 0$ corresponding to 1994.
(c) Estimate the number of cassetes sold in 2002.
(d) According to this model, when will cassette sales drop below 50 million?

62. Physical Science The atmospheric pressure (in millibars) at a given altitude (in thousands of meters) is listed in the table.

Altitude	0	2000	4000	6000	8000	10,000
Pressure	1013	795	617	472	357	265

(a) Use the data points for altitudes 0 and 10,000 to find a function of the form $f(x) = a(b^x)$ that models this data, where altitude x is measured in thousands of meters (for example, $x = 4$ means 4000 meters).
(b) Use exponential regression to find a function $g(x)$ that models this data, with x measured in thousands of meters.
(c) Estimate the pressure at 1500 m and 11,000 m and compare the results to the actual values of 846 and 227 millibars.
(d) At what height is the pressure 500 millibars?

Use a graphing calculator to do the following exercises.

63. Business The table shows the number of Southwest Airlines passengers (in millions) in selected years.

Year	1988	1990	1992	1994	1996	2000
Passengers	14.9	19.8	27.8	42.7	49.6	63.7

(a) Let $x = 0$ correspond to 1980. If your calculator has the capability, find a logistic model for this data.
(b) If this model remains accurate, when will the number of passengers reach 70 million?

64. Physical Science The power of personal computers has increased dramatically as a result of the ability to place an increasing number of transistors on a single processor chip. The table lists the number of transistors on some popular computer chips by Intel.*

Year	*Chip*	*Transistors*
1971	4004	2,300
1986	386DX	275,000
1989	486DX	1,200,000
1993	Pentium	3,100,000
1995	P6	5,500,000
2000	Pentium 4	42,000,000

(a) Let x be the year, where $x = 0$ corresponds to 1971, and y be the number of transistors. Plot the data.
(b) Would a linear, exponential, or logarithmic function fit this data best?
(c) Use regression to find a model for this data.
(d) Graph the function from part (c) with the data points. Does the function fit the data?
(e) If this model remains accurate, when will the number of transistors be 70 million?

*Data provided by Intel.

Case 3

Characteristics of the Monkeyface Prickleback*

The monkeyface prickleback (*Cebidichthys violaceus*), known to anglers as the monkeyface "eel," is found in rocky intertidal and subtidal habitats ranging from San Quintin Bay, Baja California, to Brookings, Oregon. Pricklebacks are prime targets of the few sports anglers who "poke pole" in the rocky intertidal zone at low tide. Little is known about the life history of this species. The results of a study of the length, weight, and age of this species are discussed in this case.

Data on standard length (*SL*) and total length (*TL*) were collected. Early in the study only *TL* was measured, so a conversion to *SL* was necessary. The equation relating the two lengths, calculated from 177 observations for which both lengths had been measured, is

$$SL = TL(.931) + 1.416.$$

Ages (determined by standard aging techniques) were used to estimate parameters of the von Bertanfany growth model

$$L_t = L_x(1 - e^{-kt}) \qquad (1)$$

where

L_t = length at age t,
L_x = asymptotic age of the species,
k = growth completion rate, and
t_0 = theoretical age at zero length.

The constants a and b in the model

$$W = aL^b \qquad (2)$$

where

W = weight in g,
L = standard length in cm,

were determined using 139 fish ranging from 27 cm and 145 g to 60 cm and 195 g.

Growth curves giving length as a function of age are shown in the accompanying figure. For the data marked opercle, the lengths were computed from the ages using equation (1).

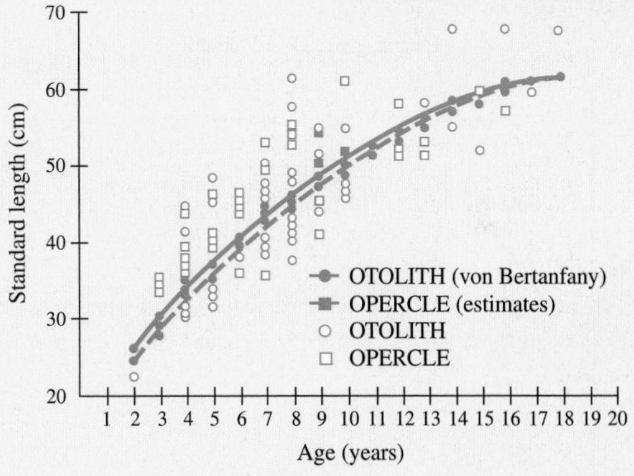

Estimated length from equation (1) at a given age was larger for males than females after age eight. See the table below.

Structure/ Sex	*Age (yr)*	*Length (cm)*	L_x	k	t_0	n
Otolith						
Est.	2–18	23–67	72	.10	−1.89	91
S.D.			8	.03	1.08	
Opercle						
Est.	2–18	23–67	71	.10	−2.63	91
S.D.			8	.04	1.31	
Opercle–Females						
Est.	0–18	15–62	62	.14	−1.95	115
S.D.			2	.02	.28	
Opercle–Males						
Est.	0–18	13–67	70	.12	−1.91	74
S.D.			5	.02	.29	

*From William H. Marshall and Tina Wyllie Echeverria, "Characteristics of the Monkeyface Prickleback," *California Fish & Game*, Vol. 78, No. 2, Spring 1992. Copyright 1992. American Association for the Advancement of Science. Reprinted with permission.

Weight/length relationships found with equation (2) are shown in the accompanying figure, along with data from other studies.

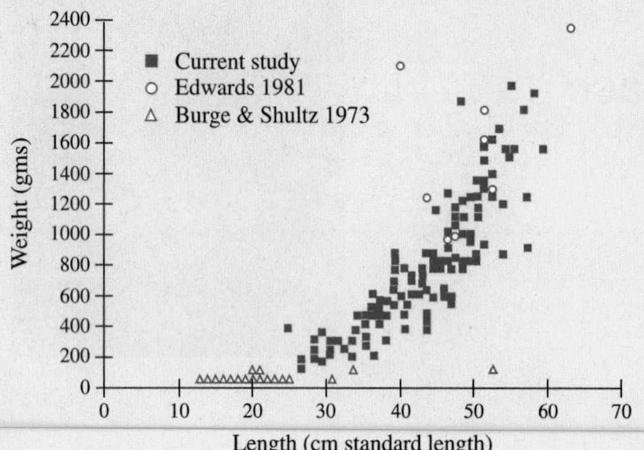

Exercises

1. Use equation (1) to estimate the lengths at ages 4, 11, and 17. Let $L_x = 71.5$ and $k = .1$. Compare your answers with the results in the first figure. What do you find?
2. Use equation (2) with $a = .01289$ and $b = 2.9$ to estimate the weights for lengths of 25 cm, 40 cm, and 60 cm. Compare with the results in the second figure. Are your answers reasonable compared to the curve?

Chapter 5

Mathematics of Finance

It is important for both businesspersons and consumers to understand the mathematics of finance in order to make sound financial decisions. Eventually, almost everyone takes out a loan to buy a car, a household appliance, or a home. We must all plan for eventual retirement, which involves investing in various types of investments, including Individual Retirement Accounts (IRAs), stocks, bonds, and annuities. The interest formulas for borrowing or investing money are introduced in this chapter.

5.1 SIMPLE INTEREST AND DISCOUNT

Interest is the fee paid to use someone else's money. Interest on loans of a year or less is frequently calculated as **simple interest,** which is paid only on the amount borrowed or invested and not on past interest. The amount borrowed or deposited is called the **principal.** The **rate** of interest is given as a percent per year, expressed as a decimal. For example, $6\% = .06$ and $11\frac{1}{2}\% = .115$. The time the money is earning interest is calculated in years. Simple interest is the product of the principal, rate, and time.

Simple Interest

The simple interest, I, on P dollars at a rate of interest r per year for t years is

$$I = Prt.$$

It is customary in financial problems to round interest to the nearest cent.

Find the simple interest for each loan.

(a) \$2000 at 8.5% for 10 months

(b) \$3500 at 10.5% for $1\frac{1}{2}$ years

Answers:

(a) \$141.67

(b) \$551.25

EXAMPLE 1 To furnish her new apartment, Eleanor Chin borrowed \$6000 at 9.2% interest for 11 months. How much interest will she pay?

Use the formula, $I = Prt$, with $P = 6000$, $r = .092$, and $t = 11/12$ years.

$$I = 6000(.092)(11/12) = 506,$$

or \$506.00 

A deposit of P dollars today at a rate of interest r for t years produces interest of $I = Prt$. The interest, added to the original principal P, gives

$$P + Prt = P(1 + rt).$$

This amount is called the **future value** of P dollars at an interest rate r for time t in years. When loans are involved, the future value is often called the **maturity value** of the loan. This idea is summarized as follows.

Future or Maturity Value for Simple Interest

The **future value** or **maturity value,** A, of P dollars for t years at a rate of interest r per year is

$$\mathbf{A = P(1 + rt).}$$

EXAMPLE 2 Find each maturity value and the amount of interest paid.

(a) Rick borrows \$20,000 from his parents at 5.25% to add a room on his house. He plans to repay the loan in 9 months with a bonus he expects to receive at that time.

The loan is for 9 months, or 9/12 of a year, so $t = .75$, $P = 20{,}000$, and $r = .0525$. Use the formula.

$$\begin{aligned} A &= P(1 + rt) \\ &= 20{,}000[1 + .0525(.75)] \\ &\approx 20{,}787.5, \qquad \text{Use a calculator.} \end{aligned}$$

Find each future value.

(a) \$1000 at 4.6% for 6 months

(b) \$8970 at 11% for 9 months

(c) \$95,106 at 9.8% for 76 days

Answers:

(a) \$1023

(b) \$9710.03

(c) \$97,073.64

or \$20,787.50. The maturity value is the sum of the principal and interest, so the interest on the loan is

$$\$20{,}787.50 - \$20{,}000 = \$787.50.$$

(b) A loan of \$11,280 for 85 days at 11% interest

It is common to assume 360 days in a year when working with simple interest. We shall usually make such an assumption in this book. The maturity value in this example is

$$A = 11{,}280\left[1 + .11\left(\frac{85}{360}\right)\right] \approx 11{,}280[1.0259722] \approx 11{,}572.97,$$

or \$11,572.97. The interest is

$$\$11{,}572.97 - \$11{,}280 = \$292.97.$$ 

TECHNOLOGY TIP The table feature of a graphing calculator makes it easy to compare the effect on the future value of a loan of a small difference in the interest rate. The screen in Figure 5.1 shows the portion of a year in the X column and the future value of a \$2500 loan with interest rates of 12.1% in the Y_1 column and 11.6% in the Y_2 column. From the table we see that when X = .75 the future value is \$9.50 less at 11.6% than at 12.1%.

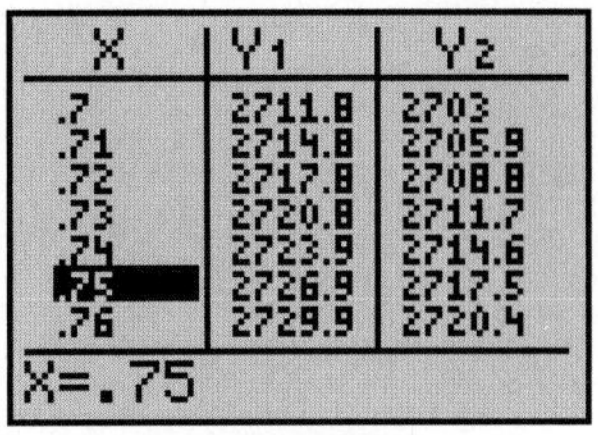

X	Y_1	Y_2
.7	2711.8	2703
.71	2714.8	2705.9
.72	2717.8	2708.8
.73	2720.8	2711.7
.74	2723.9	2714.6
.75	2726.9	2717.5
.76	2729.9	2720.4

X=.75

FIGURE 5.1

Present Value

A sum of money that can be deposited today to yield some larger amount in the future is called the **present value** of that future amount. Present value refers to the principal to be invested or loaned, so we use the same variable P as we did for principal. In interest problems, P always represents the amount at the beginning of the time period, and A always represents the amount at the end of the time period. To find a formula for P, we begin with the future value formula

$$A = P(1 + rt).$$

Dividing each side by $1 + rt$ gives the following formula for present value.

Present Value for Simple Interest

The **present value** P of a future amount of A dollars at a simple interest rate r for t years is

$$P = \frac{A}{1 + rt}.$$

3

Find the present value of the following future amounts. Assume 6% interest.

(a) \$7500 in 1 year

(b) \$89,000 in 5 months

(c) \$164,200 in 125 days

Answers:

(a) \$7075.47

(b) \$86.829.27

(c) \$160,848.98

EXAMPLE 3 Find the present value of \$32,000 in 4 months at 9% interest.

$$P = \frac{32{,}000}{1 + (.09)\left(\frac{4}{12}\right)} = \frac{32{,}000}{1.03} = 31{,}067.96$$

A deposit of \$31,067.96 today, at 9% interest, would produce \$32,000 in 4 months. These two sums, \$31,067.96 today and \$32,000.00 in 4 months, are equivalent (at 9%) because the first amount becomes the second amount in 4 months. ◀ 3

Jerrell Davis is owed \$19,500 by Christine O'Brien. The money will be paid in 11 months, with no interest. If the current interest rate is 10%, how much should Davis be willing to accept today in settlement of the debt?

Answer:

\$17,862.60

EXAMPLE 4 Because of a court settlement, Carter Fenton owes \$5000 to Jay Beckenstein. The money must be paid in 10 months, with no interest. Suppose Fenton wishes to pay the money today and Beckenstein can invest it at an annual rate of 5%. What amount should he be willing to accept?

The amount that Beckenstein should be willing to accept is given by the present value:

$$P = \frac{5000}{1 + (.05)\left(\frac{10}{12}\right)} = 4800.00.$$

Beckenstein should be willing to accept \$4800 today in settlement of the obligation. (4)

EXAMPLE 5 Suppose you borrow \$40,000 today and are required to pay \$41,400 in 4 months to pay off the loan and interest. What is the simple interest rate?

We can use the future value formula, with $P = 40{,}000$, $A = 41{,}400$, and $t = 4/12 = 1/3$, and solve for r.

$$A = P(1 + rt)$$

$$41{,}400 = 40{,}000\left(1 + r \cdot \frac{1}{3}\right)$$

$$41{,}400 = 40{,}000 + \frac{40{,}000r}{3}$$

$$1400 = \frac{40{,}000r}{3}$$

$$40{,}000r = 3 \cdot 1400 = 4200$$

$$r = \frac{4200}{40{,}000} = .105$$

Therefore, the interest rate is 10.5%. (5)

You lend a friend \$500. She agrees to pay you \$520 in 6 months. What is the interest rate?

Answer:

8%

Simple Discount Notes

The loans discussed up to this point are called **simple interest notes,** where interest on the face value of the loan is added to the loan and paid at maturity. Another common type of note, called a **simple discount note,** has the interest deducted in advance from the amount of a loan before giving the *balance* to the borrower. The *full* value of the note must be paid at maturity. The money that is deducted is called the **bank discount** or just the **discount,** and the money actually received by the borrower is called the **proceeds.**

For example, consider a loan of \$3000 at 6% interest for 9 months. We can compare the two types of loan arrangements as follows.

	Simple Interest Note	*Bank Discount Note*
Interest on the note	3000(.06)(9/12) = \$135	3000(.06)(9/12) = \$135
Borrower receives	\$3000	\$2865
Borrower pays back	\$3135	\$3000

Kelly Bell signs an agreement at her bank to pay the bank \$25,000 in 5 months. The bank charges a 13% discount rate. Find the amount of the discount and the amount Bell actually receives.

Answer:

\$1354.17; \$23,645.83

EXAMPLE 6 Theresa Vasquez needs a loan from her bank and agrees to pay \$8500 to her banker in 9 months. The banker subtracts a discount of 12% and gives the balance to Vasquez. Find the amount of the discount and the proceeds.

As shown above, the discount is found in the same way that simple interest is found, except that it is based on the amount to be repaid.

$$\text{Discount} = 8500(.12)\left(\frac{9}{12}\right) = 765.00$$

The proceeds are found by subtracting the discount from the original amount.

$$\text{Proceeds} = \$8500 - \$765.00 = \$7735.00$$ 

In Example 6, the borrower was charged a discount of 12%. However, 12% is *not* the interest rate paid, since 12% applies to the \$8500, while the borrower actually received only \$7735. In the next example, we find the rate of interest actually paid by the borrower.

7

Refer to problem 6 and find the actual rate of interest paid by Bell.

Answer:

13.7% (to the nearest tenth)

EXAMPLE 7 Find the actual rate of interest paid by Vasquez in Example 6.

Use the formula for simple interest, $I = Prt$, with r the unknown. Since the borrower received only \$7735 and must repay \$8500, $I = 8500 - 7735 = 765$. Here, $P = 7735$ and $t = 9/12 = .75$. Substitute these values into $I = Prt$.

$$I = Prt$$
$$765 = 7735(r)(.75)$$
$$\frac{765}{7735(.75)} = r$$
$$.132 \approx r$$

The actual interest rate paid by the borrower is about 13.2%.  

Let D represent the amount of discount on a loan. Then $D = Art$, where A is the maturity value of the loan (the amount borrowed plus interest) and r is the stated rate of interest. The amount actually received, the proceeds, can be written as $P = A - D$, or $P = A - Art = A(1 - rt)$.

The formulas for discount are summarized below.

Discount

If D is the discount on a loan having a maturity value A at a rate of interest r for t years, and if P represents the proceeds, then

$$\boldsymbol{P = A - D} \quad \text{or} \quad \boldsymbol{P = A(1 - rt)}.$$

EXAMPLE 8 John Cross owes \$4250 to Jane Fleming. The loan is payable in 1 year at 10% interest. Fleming needs cash to buy a new car, so 3 months before the loan is payable she goes to the bank to have the loan discounted. That is, she sells the loan (note) to the bank. The bank charges an 11% discount fee. Find the amount of cash she will receive from the bank.

First find the maturity value of the loan, the amount (with interest) Cross must pay Fleming. By the formula for maturity value,

$$\begin{aligned} A &= P(1 + rt) \\ &= 4250[1 + (.10)(1)] \\ &= 4250(1.10) = 4675, \end{aligned}$$

or $4675.00.

The bank applies its discount rate to this total:

$$\textbf{Amount of discount} = 4675(.11)(3/12) \approx \textbf{128.56}.$$

(Remember that the loan was discounted 3 months before it was due.) Fleming actually receives

$$\$4675 - \$128.56 = \textbf{\$4546.44}$$

in cash from the bank. Three months later, the bank will get $4675 from Cross.

8

8 A firm accepts a $21,000 note due in 7 months with interest of 10.5%. Suppose the firm discounts the note at a bank 75 days before it is due. Find the amount the firm would receive if the bank charges a 12.4% discount rate. (Use 360 days in a year.)

Answer:

$21,710.52

5.1 Exercises

1. What factors determine the amount of interest earned on a fixed principal?

Find the simple interest. (See Example 1.)

2. $25,000 at 7% for 9 months
3. $3850 at 9% for 8 months
4. $1974 at 6.3% for 7 months
5. $3724 at 8.4% for 11 months

Find the simple interest. Assume a 360-day year and a 30-day month. (See Example 2.)

6. $5147.18 at 10.1% for 58 days
7. $2930.42 at 11.9% for 123 days
8. $7980 at 10%; loan made on May 7 and due September 19
9. $5408 at 12%; loan made on August 16 and due December 30

Find the simple interest. Assume 365 days in a year and use the exact number of days in a month. (Assume 28 days in February.)

10. $7800 at 11%; made on July 7 and due October 25
11. $11,000 at 10%; made on February 19 and due May 31
12. $2579 at 9.6%; made on October 4 and due March 15
13. $37,098 at 11.2%; made on September 12 and due July 30
14. In your own words, describe the *maturity value* of a loan.
15. What is meant by the *present value* of money?

Find the present value of each future amount. Assume 360 days in a year. (See Examples 3 and 4.)

16. $15,000 for 8 months; money earns 6%
17. $48,000 for 9 months; money earns 5%
18. $15,402 for 125 days; money earns 6.3%
19. $29,764 for 310 days; money earns 7.2%

Find the proceeds for each loan. Assume 360 days in a year. (See Example 6.)

20. $7150; discount rate 12%; length of loan 11 months
21. $9450; discount rate 10%; length of loan 7 months
22. $35,800; discount rate 9.1%; length of loan 183 days
23. $50,900; discount rate 8.2%; length of loan 238 days
24. Why is the discount rate charged on a simple discount note different from the actual interest rate paid on the proceeds?

Find the interest rate to the nearest tenth on the proceeds for the following simple discount notes. (See Example 7.)

25. $6200; discount rate 10%; length of loan 8 months
26. $5000; discount rate 8.1%; length of loan 6 months
27. $58,000; discount rate 10.8%; length of loan 9 months
28. $43,000; discount rate 9%; length of loan 4 months

Finance *Work the following applied problems.*

29. Anne Kelly borrowed \$25,900 from her father to start a flower shop. She repaid him after 11 months, with interest of 8.4%. Find the total amount she repaid.

30. An accountant for a corporation forgot to pay the firm's income tax of \$725,896.15 on time. The government charged a penalty of 12.7% interest for the 34 days the money was late. Find the total amount (tax and penalty) that was paid. (Use a 365-day year.)

31. Mike Branson invested his summer earnings of \$3000 in a savings account for college. The account pays 2.5% interest. How much will this amount to in 9 months?

32. To pay for textbooks, a student borrows \$450 dollars from a credit union at 6.5% simple interest. He will repay the loan in 38 days, when he expects to be paid for tutoring. How much interest will he pay? (Assume a 360-day year.)

33. An account invested in a money market fund grew from \$67,081.20 to \$67,359.39 in a month. What was the interest rate to the nearest tenth?

34. A \$100,000 certificate of deposit held for 60 days is worth \$101,133.33. To the nearest tenth of a percent, what interest rate was earned?

35. A stock that sold for \$22 at the beginning of the year was selling for \$24 at the end of the year. If the stock paid a dividend of \$.50 per share, what is the simple interest rate on an investment in this stock? (*Hint:* Consider the interest to be the increase in value plus the dividend.)

36. A bond with a face value of \$10,000 in 10 years can be purchased now for \$5988.02. What is the simple interest rate?

37. Tuition of \$1769 will be due when the spring term begins in 4 months. What amount should a student deposit today, at 6.25%, to have enough to pay the tuition?

38. A firm of accountants has ordered 7 new computers at a cost of \$5104 each. The machines will not be delivered for 7 months. What amount could the firm deposit in an account paying 6.42% to have enough to pay for the machines?

39. John Sun Yee needs \$6000 to pay for remodeling work on his house. A contractor agrees to do the work in 10 months. How much should Yee deposit at 4.6% to accumulate the \$6000 at that time?

40. Lorie Reilly decides to go back to college. For transportation, she borrows money from her parents to buy a small car for \$7200. She plans to repay the loan in 7 months. What amount can she deposit today at 5.25% to have enough to pay off the loan?

41. John Matthews signs a \$4200 note at the bank. The bank charges a 12.2% discount rate. Find the net proceeds if the note is for 10 months. Find the actual interest rate (to the nearest hundredth) charged by the bank.

42. A building contractor gives a \$13,500 note to a plumber. (The plumber loans \$13,500 to the contractor.) The note is due in 9 months, with interest of 9%. Three months after the note is signed, the plumber discounts it at the bank. The bank charges a 10.1% discount rate. How much will the plumber receive? Will it be enough to pay a bill for \$13,582?

43. Shalia Johnson owes \$7000 to the Eastside Music Shop. She has agreed to pay the amount in 7 months at an interest rate of 10%. Two months before the loan is due, the store needs \$7350 to pay a wholesaler's bill. The bank will discount the note at a rate of 10.5%. How much will the store receive? Is it enough to pay the bill?

44. Fay, Inc., received a \$30,000, six-month, 12% interest-bearing note from a customer.* The note was discounted the same day by Carr National Bank at 15%. The amount of cash received by Fay from the bank was

(a) \$30,000 **(b)** \$29,550

(c) \$29,415 **(d)** \$27,750.

45. Many credit card companies target entering college freshmen with attractive interest rates. What they do not emphasize, however, is that the low rate will be in effect for only a short time, changing automatically to a much higher rate, often in just a few months. Freshman Michael Branson accepts a credit card offer at 8% annual interest on any unpaid charges. The rate will increase to 15% annually after 6 months. To furnish his dorm room, he charges a small refrigerator, some bedding, and personal supplies amounting to \$457.80. When he receives his first month's bill, he finds that he can pay just a minimum amount of \$87.50.

(a) If Michael pays the minimum amount and makes no additional charges, how much will he owe the following month?

(b) Suppose Michael continues to pay the same minimum amount for 4 more months. What will he owe at the end of the next month?

(c) The minimum amount normally is reduced as the total charge is reduced. Suppose each month, the minimum is reduced by \$15 and Michael continues to pay only the minimum. Can be pay off the charges in 6 months? If not, what will he owe at the end of 6 months?

*Uniform CPA Examination, May 1989, American Institute of Certified Public Accountants.

5.2 COMPOUND INTEREST

Simple interest is normally used for loans or investments of a year or less. For longer periods *compound interest* is used. With **compound interest,** interest is charged (or paid) on interest as well as on principal. For example, if \$1000 is deposited at 5% compounded annually, then the interest for the first year is \$1000(.05) = \$50, just as with simple interest, so that the account balance is \$1050 at the end of the year. During the second year, interest is paid on the entire \$1050 (not just on the original \$1000 as with simple interest), so the amount in the account at the end of the second year is \$1050 + \$1050(.05) = \$1102.50. This is more than simple interest would produce. (1)

Use the formula

$$A = P(1 + rt)$$

to find the amount in the account after 2 years at 5% simple interest.

Answer:

\$1100

To find a formula for compound interest, suppose that P dollars are deposited at interest rate r per year. The amount A on deposit after 1 year is found by the simple interest formula.

$$A = P[1 + r(1)] = P(1 + r)$$

If the deposit earns compound interest, the interest for the second year is paid on the total amount on deposit at the end of the first year, $P(1 + r)$. Using the formula $A = P(1 + rt)$ again, with $P(1 + r)$ (in place of P) and $t = 1$ gives the total amount on deposit at the end of the second year.

$$A = [P(1 + r)](1 + r \cdot 1) = P(1 + r)^2$$

In the same way, the total amount on deposit at the end of the third year is

$$P(1 + r)^3.$$

Continuing in this way, the total amount on deposit after t years is

$$A = P(1 + r)^t,$$

which is called the **compound amount.**

NOTE Compare this formula for compound interest with the formula for simple interest from the previous section.

Compound interest	$A = P(1 + r)^t$
Simple interest	$A = P(1 + rt)$

The important distinction between the two formulas is that in the compound interest formula, the number of years t is an *exponent,* so that money grows much more rapidly when interest is compounded.

Spreadsheets are ideal for performing financial calculations. Figure 5.2 shows a Microsoft Excel spreadsheet with the formulas for compound and simple interest used to create columns B and C, respectively, when \$1000 is invested at an annual rate of 10%. Notice how rapidly the compound amound increases compared to the maturity value with simple interest. For more details on the use of spreadsheets in the mathematics of finance, see *The Spreadsheet Manual* that is available with this book.

	A	B	C
1	period	compound	simple
2	1	1100	1100
3	2	1210	1200
4	3	1331	1300
5	4	1464.1	1400
6	5	1610.51	1500
7	6	1771.561	1600
8	7	1948.7171	1700
9	8	2143.58881	1800
10	9	2357.947691	1900
11	10	2593.74246	2000
12	11	2853.116706	2100
13	12	3138.428377	2200
14	13	3452.271214	2300
15	14	3797.498336	2400
16	15	4177.248169	2500
17	16	4594.972986	2600
18	17	5054.470285	2700
19	18	5559.917313	2800
20	19	6115.909045	2900
21	20	6727.499949	3000

FIGURE 5.2

Interest can be compounded more than once a year. Common **compounding periods** include

semiannually (two periods per year),
quarterly (four periods per year),
monthly (twelve periods), and
daily (usually 365 periods per year).

To find the *interest rate per period, i,* we *divide* the annual interest rate r by the number of compounding periods, m, per year. The total number of compounding periods, n, is found by *multiplying* the number of years t by the number of compounding periods per year, m. Then the general formula for compound interest can be derived in much the same way as the formula given earlier.

Compound Interest

If P dollars are deposited for n compounding periods at a rate of interest i per period, the compound amount (or future value) A is

$$A = P(1 + i)^n.$$

In particular, if the annual interest rate r is compounded m times per year and the number of years is t, then $i = r/m$ and $n = mt$.

Suppose \$17,000 is deposited at 4% compounded annually for 11 years.

(a) Find the compound amount.

(b) Find the amount of interest earned.

Answers:

(a) \$26,170.72

(b) \$9170.72

EXAMPLE 1 Suppose \$1000 is deposited for 6 years in an account paying 8.31% per year compounded annually.

(a) Find the compound amount.

In the formula above, $P = 1000$, $i = .0831$, and $n = 6$. The compound amount is

$$A = P(1 + i)^n$$
$$A = 1000(1.0831)^6$$
$$A = \$1614.40.$$

(b) Find the amount of interest earned.

Subtract the initial deposit from the compound amount.

Amount of interest = \$1614.40 − \$1000 = \$614.40 

Find the compound amount.

(a) \$10,000 at 8% compounded quarterly for 7 years

(b) \$36,000 at 6% compounded monthly for 2 years

Answers:

(a) \$17,410.24

(b) \$40,577.75

EXAMPLE 2 Find the amount of interest earned by a deposit of \$2450 for 6.5 years at 5.25% compounded quarterly.

Interest compounded quarterly is compounded 4 times a year. In 6.5 years, there are 6.5(4) = 26 periods. Thus, $n = \mathbf{26}$. Interest of 5.25% per year is 5.25/4 per quarter, so $i = \mathbf{.0525/4}$. Now use the formula for compound amount.

$$A = P(1 + i)^n$$
$$A = 2450(1 + \mathbf{.0525/4})^{\mathbf{26}}$$
$$= 3438.784190$$

Rounded to the nearest cent, the compound amount is \$3438.78, so the interest is \$3438.78 − \$2450 = \$988.78. ◀ 3

CAUTION As shown in Example 2, compound interest problems involve two rates—the nominal or stated annual rate r and the rate i per compounding period. Be sure you understand the distinction between them. When interest is compounded annually, these rates are the same. In all other cases, $i \neq r$.

The more often interest is compounded within a given time period, the more interest will be earned. Surprisingly, however, there is a limit on the amount of interest, no matter how often it is compounded. To see this, suppose that \$1 is invested at 100% interest per year, compounded n times per year. Then the interest rate (in decimal form) is 1.00 and the interest rate per period is $1/n$. According to the formula (with $P = 1$), the compound amount at the end of 1 year will be

$$A = \left(1 + \frac{1}{n}\right)^n.$$

A computer gives the following results for various values of n.

Interest Is Compounded	n	$\left(1+\frac{1}{n}\right)^n$
Annually	1	$\left(1+\frac{1}{1}\right)^1 = 2$
Semiannually	2	$\left(1+\frac{1}{2}\right)^2 = 2.25$
Quarterly	4	$\left(1+\frac{1}{4}\right)^4 \approx 2.4414$
Monthly	12	$\left(1+\frac{1}{12}\right)^{12} \approx 2.6130$
Daily	365	$\left(1+\frac{1}{365}\right)^{365} \approx 2.71457$
Hourly	8760	$\left(1+\frac{1}{8760}\right)^{8760} \approx 2.718127$
Every minute	525,600	$\left(1+\frac{1}{525{,}600}\right)^{525{,}600} \approx 2.7182792$
Every second	31,536,000	$\left(1+\frac{1}{31{,}536{,}000}\right)^{31{,}536{,}000} \approx 2.7182818$

Because interest is rounded to the nearest penny, the compound amount never exceeds \$2.72, no matter how big n is. (If you try computing the values in the table with your calculator, you may notice that your answers do not agree exactly. This is because of round-off error.)

The table above suggests that as n takes larger and larger values, then the corresponding values of $\left(1+\frac{1}{n}\right)^n$ get closer and closer to a specific real number, whose decimal expansion begins 2.71828. . . . This is indeed the case, as is shown in calculus, and the number 2.71828 . . . is denoted e.

The preceding example is typical of what happens when interest is compounded n times per year, with larger and larger values of n. It can be shown that no matter what interest rate or principal is used, there is always an upper limit on the compound amount, which is called the compound amount from **continuous compounding.**

Continuous Compounding

The compound amount A for a deposit of P dollars at interest rate r per year compounded continuously for t years is given by

$$A = Pe^{rt}.*$$

Most calculators have an e^x key for computing powers of e. See Chapter 4 for details on using a calculator to evaluate e^x.

*Other applications of the exponential function $f(x) = e^x$ are discussed in Chapter 4.

Find the compound amount, assuming continuous compounding.

(a) \$12,000 at 10% for 5 years

(b) \$22,867 at 7.2% for 9 years

Answers:

(a) \$19,784.66

(b) \$43,715.15

EXAMPLE 3 Suppose \$5000 is invested at an annual rate of 4% compounded continuously for 5 years. Find the compound amount.

In the formula for continuous compounding, let $P = 5000$, $r = .04$, and $t = 5$. Then a calculator with an e^x key shows that

$$A = 5000e^{(.04)5} = 5000e^{.2} \approx \$6107.01.$$

You can readily verify that daily compounding would have produced a compound amount about 6¢ less. (4)

Effective Rate

If \$1 is deposited at 4% compounded quarterly, a calculator can be used to find that at the end of one year, the compound amount is \$1.0406, an increase of 4.06% over the original \$1. The actual increase of 4.06% in the money is somewhat higher than the stated increase of 4%. To differentiate between these two numbers, 4% is called the **nominal** or **stated rate** of interest, while 4.06% is called the **effective rate.** To avoid confusion between stated rates and effective rates, we shall continue to use r for the stated rate and we will use r_E for the effective rate.*

Find the effective rate corresponding to a nominal rate of

(a) 12% compounded monthly;

(b) 8% compounded quarterly.

Answers:

(a) 12.68%

(b) 8.24%

EXAMPLE 4 Find the effective rate corresponding to a stated rate of 6% compounded semiannually.

A calculator shows that \$100 at 6% compounded semiannually will grow to

$$A = 100\left(1 + \frac{.06}{2}\right)^2 = 100(1.03)^2 = \$106.09.$$

Thus, the actual amount of compound interest is \$106.09 − \$100 = \$6.09. Now if you earn \$6.09 interest on \$100 in 1 year with annual compounding, your rate is 6.09/100 = .0609 = 6.09%. Thus, the effective rate is $r_E = 6.09\%$. (5)

If you examine Example 4, you see that the effective rate is $\frac{106.09 - 100}{100}$, that is,

$$\frac{100\left(1 + \frac{.06}{2}\right)^2 - 100}{100} = \frac{100\left[\left(1 + \frac{.06}{2}\right)^2 - 1\right]}{100} = \left(1 + \frac{.06}{2}\right)^2 - 1$$

The same argument works with any values of P, r, and m in place of 100, .06 and 2 and leads to this formula.

Effective Rate

The effective rate corresponding to a stated rate of interest r compounded m times per year is

$$r_E = \left(1 + \frac{r}{m}\right)^m - 1.$$

*When applied to consumer finance, the effective rate is called the annual percentage rate, APR, or annual percentage yield, APY.

Using the effective rate formula, a graphing calculator generated the table in Figure 5.3, in which the interest rate r is 10%, Y_1 denotes the effective rate, and X is the number of compoundings. The table shows how the effective rate increases with more frequent compounding.

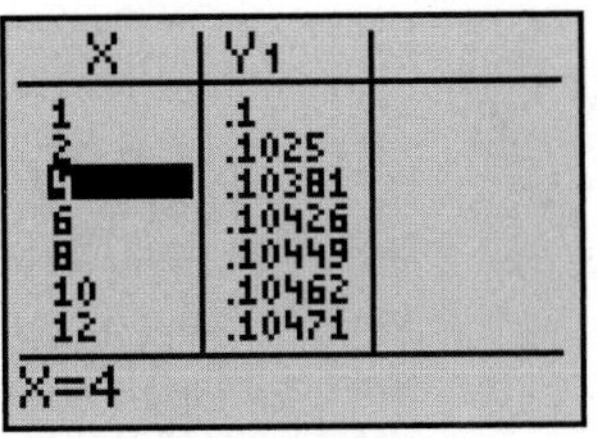

FIGURE 5.3

6

Find the effective rate corresponding to a nominal rate of

(a) 8% compounded monthly;

(b) 10% compounded quarterly.

Answers:

(a) 8.30%

(b) 10.38%

EXAMPLE 5 A bank pays interest of 4.9% compounded monthly. Find the effective rate.

Use the formula given above with $r = .049$ and $m = 12$. The effective rate is

$$r_E = \left(1 + \frac{.049}{12}\right)^{12} - 1$$

$$= 1.050115575 - 1 \approx .0501,$$

or 5.01%. 

EXAMPLE 6 Bank A is now lending money at 10% interest compounded annually. The rate at Bank B is 9.6% compounded monthly, and the rate at Bank C is 9.7% compounded quarterly. If you need to borrow money, at which bank will you pay the least interest?

Compare the effective rates.

$$\text{Bank A:} \quad \left(1 + \frac{.10}{1}\right)^{1} - 1 = .10 = 10\%$$

$$\text{Bank B:} \quad \left(1 + \frac{.096}{12}\right)^{12} - 1 \approx .10034 = 10.034\%$$

$$\text{Bank C:} \quad \left(1 + \frac{.097}{4}\right)^{4} - 1 \approx .10057 = 10.057\%$$

The lowest effective interest rate is at Bank A, which has the highest nominal rate.

Present Value for Compound Interest

The formula for compound interest, $A = P(1 + i)^n$, has four variables, A, P, i, and n. Given the values of any three of these variables, the value of the fourth can be found. In particular, if A (the future amount), i, and n are known, then P can be found. Here P is the amount that should be deposited today to produce A dollars in n periods.

EXAMPLE 7 Keisha Jones must pay a lump sum of \$6000 in 5 years. What amount deposited today at 6.2% compounded annually will amount to \$6000 in 5 years?

Here $A = 6000$, $i = .062$, $n = 5$, and P is unknown. Substituting these values into the formula for the compound amount gives

$$6000 = P(1.062)^5$$

$$P = \frac{6000}{(1.062)^5} = 4441.49,$$

or \$4441.49. If Jones leaves \$4441.49 for 5 years in an account paying 6.2% compounded annually, she will have \$6000 when she needs it. To check your work, use the compound interest formula with $P = \$4441.49$, $i = .062$, and $n = 5$. You should get $A = \$6000.00$. ◀ 7

7

Find P in Example 7 if the interest rate is

(a) 6%;

(b) 10%.

Answers:

(a) \$4483.55

(b) \$3725.53

As Example 7 shows, \$6000 in 5 years is (approximately) the same as \$4441.49 today (if money can be deposited at 6.2% annual interest). An amount that can be deposited today to yield a given amount in the future is called the *present value* of the future amount. By solving $A = P(1 + i)^n$ for P, we get the following general formula for present value.

Present Value for Compound Interest

The **present value** of A dollars compounded at an interest rate i per period for n periods is

$$\mathbf{P = \frac{A}{(1 + i)^n} \quad \text{or} \quad P = A(1 + i)^{-n}.}$$

NOTE Since this is just the compound interest formula solved for P, it is not necessary to remember a new formula for present value. You can use the compound interest formula, if you understand what each of the variables represents.

EXAMPLE 8 Find the present value of \$16,000 in 9 years if money can be deposited at 6% compounded semiannually.

In 9 years there are $2 \cdot 9 = 18$ semiannual periods. A rate of 6% per year is 3% in each semiannual period. Apply the formula with $A = 16{,}000$, $i = .03$, and $n = 18$.

$$P = \frac{A}{(1 + i)^n} = \frac{16{,}000}{(1 + .03)^{18}} \approx \frac{16{,}000}{1.702433} \approx 9398.31$$

A deposit of \$9398.31 today, at 6% compounded semiannually, will produce a total of \$16,000 in 9 years. ◀ 8

Find the present value in Example 8 if money is deposited at 10% compounded semiannually.

Answer:

\$6648.33

The formula for compound amount can also be solved for n.

EXAMPLE 9 Suppose the general level of inflation in the economy averages 8% per year. Find the number of years it would take for the overall level of prices to double.

To find the number of years it will take for \$1 worth of goods or services to cost \$2, we must solve for n in the equation

$$2 = 1(1 + .08)^n,$$

where $A = 2$, $P = 1$, and $i = .08$. This equation simplifies to

$$2 = (1.08)^n.$$

To solve for n, we will use base 10 logarithms as in Chapter 4.

$$\log 2 = \log (1.08)^n \quad \text{Take the logarithm of each side.}$$
$$\log 2 = n \log 1.08 \quad \text{Property (c) of logarithms}$$
$$n = \frac{\log 2}{\log 1.08} \quad \text{Divide both sides by log 1.08.}$$
$$n \approx 9.01$$

To check, use a calculator to get $1.08^{9.01} = 2.00$ to the nearest hundredth. 

Using a calculator, find the number of years it will take for \$500 to increase to \$750 in an account paying 6% interest compounded semiannually.

Answer:

About 7 years

When interest is compounded continuously, the present value can be found by solving the continuous compounding formula $A = Pe^{rt}$ for P.

EXAMPLE 10 Assuming a continuous inflation rate of 3.5%, how much did a \$100 item today cost 4 years ago?

Here the future value is $A = 100$, $r = .035$, and $t = 4$. We want to find the present value P.

$$A = Pe^{rt}$$
$$100 = Pe^{(.035)4}$$
$$100 = Pe^{.14}$$
$$P = \frac{100}{e^{.14}} \approx \$86.94$$

The item cost \$86.94 four years ago. ◀ (10)

(10)

Find the value of a \$100 item in 3 years at a continuous inflation rate of 4%.

Answer:

\$112.75

Summary

At this point, it seems helpful to summarize the notation and the most important formulas for simple and compound interest. We use the following variables.

P = principal or present value
A = future or maturity value
r = annual (stated or nominal) interest rate
t = number of years
m = number of compounding periods per year
i = interest rate per period ($i = r/m$)

n = total number of compounding periods ($n = tm$)

r_E = effective rate

Simple Interest	*Compound Interest*	*Continuous Compounding*
$A = P(1 + rt)$	$A = P(1 + i)^n$	$A = Pe^{rt}$
$P = \frac{A}{1 + rt}$	$P = \frac{A}{(1 + i)^n} = A(1 + i)^{-n}$	$P = \frac{A}{e^{rt}}$
	$r_E = \left(1 + \frac{r}{m}\right)^m - 1$	

5.2 Exercises

1. In the compound interest formula in the box on page 239, what is the difference between r and i? between t and n?
2. Explain the difference between simple interest and compound interest.
3. What factors determine the amount of interest earned on a fixed principal?
4. In your own words, describe the *maturity value* of a loan.
5. What is meant by the *present value* of money?
6. If interest is compounded more than once per year, which rate is higher, the stated rate or the effective rate?

Find the compound amount for each of the following deposits. (See Examples 1 and 2.)

7. \$1000 at 6% compounded annually for 8 years
8. \$1000 at 7% compounded annually for 10 years
9. \$470 at 10% compounded semiannually for 12 years
10. \$15,000 at 6% compounded semiannually for 11 years
11. \$6500 at 12% compounded quarterly for 6 years
12. \$9100 at 8% compounded quarterly for 4 years

Find the amount of interest earned by each of the following deposits. (See Examples 1 and 2.)

13. \$26,000 at 7% compounded annually for 5 years
14. \$32,000 at 5% compounded annually for 10 years
15. \$8000 at 4% compounded semiannually for 6.4 years
16. \$2500 at 4.5% compounded semiannually for 8 years
17. \$5124.98 at 6.3% compounded quarterly for 5.2 years
18. \$27,630.35 at 7.1% compounded quarterly for 3.7 years

Find the compound amount if \$25,000 is invested at 6% compounded continuously for the following number of years. (See Example 3.)

19. 1 **20.** 5 **21.** 10

22. 15 **23.** 20

24. How do the nominal or stated interest rate and the effective interest rate differ?

Find the effective rate corresponding to the following nominal rates. (See Examples 4–6.)

25. 4% compounded semiannually
26. 8% compounded quarterly
27. 8% compounded semiannually
28. 10% compounded semiannually
29. 12% compounded semiannually
30. 12% compounded quarterly

Find the present value of the following future amounts. (See Examples 7 and 8.)

31. \$12,000 at 5% compounded annually for 6 years
32. \$8500 at 6% compounded annually for 9 years
33. \$4253.91 at 6.8% compounded semiannually for 4 years
34. \$27,692.53 at 4.6% compounded semiannually for 5 years
35. \$17,230 at 4% compounded quarterly for 10 years
36. \$5240 at 8% compounded quarterly for 8 years
37. If money can be invested at 8% compounded quarterly, which is larger: \$1000 now or \$1210 in 5 years? Use present value to decide.

38. If money can be invested at 6% compounded annually, which is larger: \$10,000 now or \$15,000 in 10 years? Use present value to decide.

Finance *Work the following applied problems.*

39. A small business borrows \$50,000 for expansion at 12% compounded monthly. The loan is due in 4 years. How much interest will the business pay?

40. A developer needs \$80,000 to buy land. He is able to borrow the money at 10% per year compounded quarterly. How much will the interest amount to if he pays off the loan in 5 years?

41. A company has agreed to pay \$2.9 million in 5 years to settle a lawsuit. How much must they invest now in an account paying 8% compounded monthly to have that amount when it is due?

42. Bill Poole wants to have \$20,000 available in 5 years for a down payment on a house. He has inherited \$15,000. How much of the inheritance should he invest now to accumulate the \$20,000, if he can get an interest rate of 8% compounded quarterly?

43. An advertisement for WingspanBank.com boasted the "highest CD yield in the universe," with an APY (or effective yield) of 7.35%.* The actual rate was not stated. Given that interest was compounded monthly, find the actual rate.

44. According to a financial Web site, on October 16, 2000, DeepGreen Bank of Cleveland, Ohio, paid 6.96% interest, compounded daily, on a one-year CD, while Flagstar Bank of Bloomfield Hills, Michigan, paid 7.01% compounded quarterly.† What are the effective rates for the two CDs, and which bank pays a higher effective rate?

45. A New York bank offered the following special on CD (certificate of deposit) rates. The rates are annual percentage yield, or effective rates, which are higher than the corresponding nominal rates. Assume quarterly compounding. Solve for r to approximate the corresponding nominal rates to the nearest hundredth.

Term	6 mo	1 yr	18 mo	2 yr	3 yr
APY (%)	5.00	5.30	5.45	5.68	5.75

46. The pie graph shows the percent of baby boomers ages 46 to 49 who said they had investments with a total value as shown in each category.‡ Note that 30% say they have saved less than \$10,000 and 28% don't know or gave no answer. Assume the money is invested at an average rate of 8% compounded quarterly for 20 years, when this age group will be ready for retirement. Find the range of amounts each group (except the "don't know or no answer" group) in the graph will have saved for retirement if no more is added.

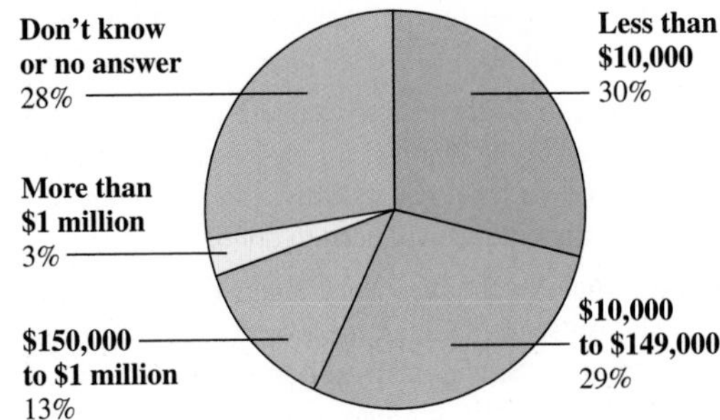

Note: Figures add to more than 100% because of rounding.

Sources: Census Bureau (age distribution); Merrill Lynch Baby Boom Retirement Index (investments); William M. Mercer Inc. (life expectancy).

47. Scott Silva borrowed \$5200 from his friend Joe Vetere to buy computer equipment. He repaid the loan 10 months later with simple interest at 7%. Vetere then invested the proceeds in a 5-year certificate of deposit paying 6.3% compounded quarterly. How much will he have at the end of 5 years? (*Hint:* You need to use both simple and compound interest.)

48. Lora Reilly has inherited \$10,000 from her uncle's estate. She will invest the money for 2 years. She is considering two investments: a money market fund that pays a guaranteed 5.25% interest compounded monthly or a 2-year Treasury note at 6% annual interest. Which investment pays the most interest over the 2-year period?

49. You decide to invest a \$16,000 bonus in a money market fund that guarantees a 5.5% annual interest rate compounded monthly for 7 years. A one-time fee of \$30 is charged to set up the account. In addition there is an annual administrative charge of 1.25% of the balance in the account at the end of each year.
   (a) How much is in the account at the end of the first year?
   (b) How much is in the account at the end of the seventh year?

50. Two partners agree to invest equal amounts in their business. One will contribute \$10,000 immediately. The other plans to contribute an equivalent amount in 3 years, when she expects to acquire a large sum of money. How much should she contribute at that time to match her partner's investment now, assuming an interest rate of 6% compounded semi-annually?

51. As the prize in a contest, you are offered \$1000 now or \$1210 in 5 years. If money can be invested at 6% compounded annually, which is larger?

**New York Times*, October 15, 2000, p. 39.

†www.bankrate.com.

‡*New York Times*, December 31, 1995, sec. 3, p.5.

Use the approach in Example 9 to find the time it would take for the general level of prices in the economy to double at the average annual inflation rates in Exercises 52 and 53.

52. 4% **53.** 5%

54. The consumption of electricity has increased historically at 6% per year. If it continues to increase at this rate indefinitely, find the number of years before the electric utilities will need to double their generating capacity.

55. Suppose a conservation campaign coupled with higher rates causes the demand for electricity to increase at only 2% per year, as it has recently. Find the number of years before the utilities will need to double generating capacity.

56. An article in the *New York Times* discussed how long it will take for Bill Gates, the world's second richest person (behind the Sultan of Brunei), to become the world's first trillionaire.* His birthday is October 28, 1955, and on July 16, 1997, he was worth $42 billion. (*Note:* A trillion dollars is 1000 billion dollars.)

(a) Assume that Bill Gates' fortune grows at an annual rate of 58%, the historical growth rate through 1997 of Microsoft stock, which made up most of his wealth in 1997. Find the age at which he becomes a trillionaire. (*Hint:* Use the formula for interest compounded annually, $A = P(1 + i)^n$, with $P = 42$. Graph the future value as a function of n on a graphing calculator and find where the graph crosses the line $y = 1000$.)

(b) Repeat part (a) using 10.9% growth, the average return on all stocks since 1926.

(c) What rate of growth would be necessary for Bill Gates to become a trillionaire by the time he is eligible for Social Security on January 1, 2022, after he has turned 66?

(d) An article on September 19, 1999, gave Bill Gates's wealth as roughly $90 billion.† What was the rate of growth of his wealth between the 1997 and 1999 articles?

57. In 1995, O. G. McClain of Houston, Texas, mailed a $100 check to a descendant of Texas independence hero Sam Houston to repay a $100 debt of McClain's great-great-grandfather, who died in 1835, to Sam Houston.‡ A bank estimated the interest on the loan to be $420 million for the 160 years it was due. Find the interest rate the bank was using, assuming interest is compounded annually.

58. In the New Testament, Jesus commends a widow who contributed 2 mites to the temple treasury (Mark 12:42–44). A mite was worth roughly 1/8 of a cent. Suppose the temple had invested those 2 mites at 4% interest compounded quarterly. How much would the money be worth 2000 years later?

59. Suppose $10,000 is invested at an annual rate of 5% for 10 yr. Find the future value if interest is compounded as follows.

(a) annually **(b)** quarterly
(c) monthly **(d)** daily (365 days)

60. In Exercise 59, notice that as the money is compounded more often, the compound amount becomes larger and larger. Is it possible to compound often enough so that the compound amount is $17,000 after 10 years? Explain.

The following exercises are from professional examinations.

61. On January 1, 1980, Jack deposited $1000 into Bank X to earn interest at the rate of j per annum compounded semiannually. On January 1, 1985, he transferred his account to Bank Y to earn interest at the rate of k per annum compounded quarterly. On January 1, 1988, the balance at Bank Y was $1990.76. If Jack could have earned interest at the rate of k per annum compounded quarterly from January 1, 1980, through January 1, 1988, his balance would have been $2203.76. Which of the following represents the ratio k/j?*

(a) 1.25 **(b)** 1.30 **(c)** 1.35 **(d)** 1.40
(e) 1.45

62. On January 1, 1987, Tone Company exchanged equipment for a $200,000 noninterest bearing note due on January 1, 1990. The prevailing rate of interest for a note of this type at January 1, 1987, was 10%. The present value of $1 at 10% for three periods is 0.75. What amount of interest revenue should be included in Tone's 1988 income statement?†

(a) $7500 **(b)** $15,000 **(c)** $16,500
(d) $20,000

**New York Times*, July 20, 1997, sec. 4, p. 2.

†*New York Times*, September 19, 1999, WK rev., p. 2.

‡*New York Times*, March 30, 1995.

*Problem from "Course 140 Examination, Mathematics of Compound Interest" of the Education and Examination Committee of The Society of Actuaries. Reprinted by permission of The Society of Actuaries.

†Uniform CPA Examination, May 1989, American Institute of Certified Public Accountants.

5.3 FUTURE VALUE OF AN ANNUITY AND SINKING FUNDS

So far in this chapter, only lump sum deposits and payments have been discussed. Many financial situations, however, involve a sequence of equal payments at regular intervals, such as weekly deposits in a savings account or monthly payments on a mortgage or car loan. To develop formulas to deal with periodic payments like these, we first discuss *sequences*.

Geometric Sequences

If a and r are fixed nonzero numbers, then the infinite list of numbers, a, ar, ar^2, ar^3, ar^4, . . . is called a **geometric sequence.** For instance, if $a = 5$ and $r = 2$, we have the sequence

$$5, 5 \cdot 2, 5 \cdot 2^2, 5 \cdot 2^3, 5 \cdot 2^4, \ldots,$$

or

$$5, 10, 20, 40, 80, \ldots.$$

In the sequence $a, ar, ar^2, ar^3, ar^4, \ldots$, the number a is the *first term* of the sequence, ar the *second term,* ar^2 the *third term*, ar^3 the *fourth term,* and so on. Thus, for any $n \geq 1$,

$$ar^{n-1} \text{ is the } n\text{th term of the sequence.}$$

Each term in the sequence is r times the preceding term. The number r is called the **common ratio** of the sequence.

EXAMPLE 1 Find the seventh term of the geometric sequence with $a = 6$ and $r = 4$. Using ar^{n-1}, with $n = 7$, the seventh term is

$$ar^6 = 6(4)^6 = 6(4096) = 24{,}576.$$

TECHNOLOGY TIP Some graphing calculators have the ability to produce a list of the terms of a sequence, given the expression for the nth term. With other calculators you can use the TABLE feature by entering the expression for the nth term as a function. The first four terms of the sequence discussed in Example 1 are shown in Figure 5.4.

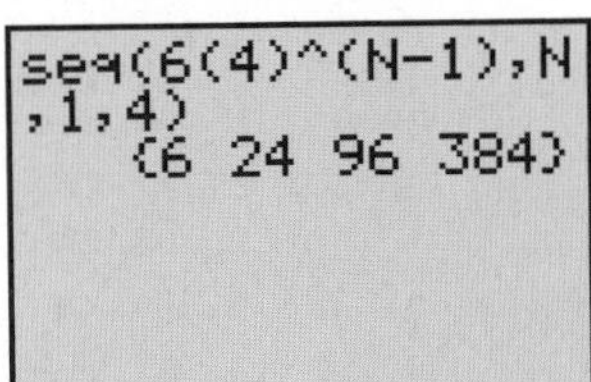

FIGURE 5.4

EXAMPLE 2 \$100 is deposited in an account that pays interest of 10% compounded annually. How much is in the account at the beginning of each of the first five years?

Write the first four terms of the geometric sequence with $a = 5$ and $r = -2$. Then find the seventh term.

Answer:

5, −10, 20, −40; 320

According to the compound interest formula (with $P = 100$ and $i = .1$), the value of the account at the beginning of each of the first five years is given by the first five terms of the geometric sequence with $a = 100$ and $r = 1.1$. The first five terms are

$$100, \quad 100(1.1), \quad 100(1.1)^2, \quad 100(1.1)^3, \quad 100(1.1)^4,$$

so the value of the account at the beginning of each of the five years is

\$100, \$110, \$121, \$133.10, \$146.41. 

The sum S_n of the first n terms of a geometric sequence can be written as

$$S_n = a + ar + ar^2 + ar^3 + \cdots + ar^{n-1}. \qquad (1)$$

If $r = 1$, this sum is

$$S_n = \underbrace{a + a + a + \cdots + a}_{n \text{ terms}} = na.$$

To derive a formula for S_n when $r \neq 1$, multiply both sides of equation (1) by r, obtaining

$$rS_n = ar + ar^2 + ar^3 + ar^4 + \cdots + ar^n. \qquad (2)$$

Now subtract corresponding sides of equation (1) from equation (2), and solve the equation for S_n.

$$\begin{aligned} rS_n &= \quad ar + ar^2 + ar^3 + \cdots + ar^{n-1} + ar^n \\ -S_n &= -(a + ar + ar^2 + ar^3 + \cdots + ar^{n-1}) \\ \hline rS_n - S_n &= ar^n - a \\ S_n(r-1) &= a(r^n - 1) \\ S_n &= \frac{a(r^n - 1)}{r - 1} \end{aligned}$$

Sum of Terms

If a geometric sequence has first term a and common ratio r, then the **sum of the first n terms** is given by

$$S_n = \frac{a(r^n - 1)}{r - 1}, \qquad r \neq 1.$$

(2)

(a) Find S_4 and S_7 for the geometric sequence 5, 15, 45, 135, ...

(b) Find S_5 for the geometric sequence having $a = -3$ and $r = -5$.

Answers:

(a) $S_4 = 200, S_7 = 5465$

(b) $S_5 = -1563$

EXAMPLE 3 Find the sum of the first six terms of the geometric sequence 3, 12, 48, ...

Here $a = 3$ and $r = 4$. Find S_6 by the result above.

$$S_6 = \frac{3(4^6 - 1)}{4 - 1} \qquad \text{Let } n = 6, a = 3, r = 4.$$

$$= \frac{3(4096 - 1)}{3}$$

$$= 4095$$ ◀ (2)

TECHNOLOGY TIP Graphing calculators with sequence capability can also find the sum of the first n terms of a sequence, given the expression for the nth term. In Example 3, for instance, the sequence is given by $ar^{n-1} = 3(4^{n-1})$. The sum of the first six terms is shown in Figure 5.5.

FIGURE 5.5

Ordinary Annuities

A sequence of equal payments made at equal periods of time is called an **annuity.** If the payments are made at the end of the time period, and if the frequency of payments is the same as the frequency of compounding, the annuity is called an **ordinary annuity.** The time between payments is the **payment period,** and the time from the beginning of the first payment period to the end of the last period is called the **term of the annuity.** The **future value of the annuity,** the final sum on deposit, is defined as the sum of the compound amounts of all the payments, compounded to the end of the term.

Two common uses of annuities are to accumulate funds for some goal or to withdraw funds from an account. For example, an annuity may be used to save money for a large purchase, such as an automobile, an expensive trip, or a down payment on a home. An annuity also may be used to provide monthly payments for retirement. We explore these options in this and the next section.

For example, suppose \$1500 is deposited at the end of the year for the next 6 years in an account paying 8% per year compounded annually. Figure 5.6 on the next page shows this annuity schematically.

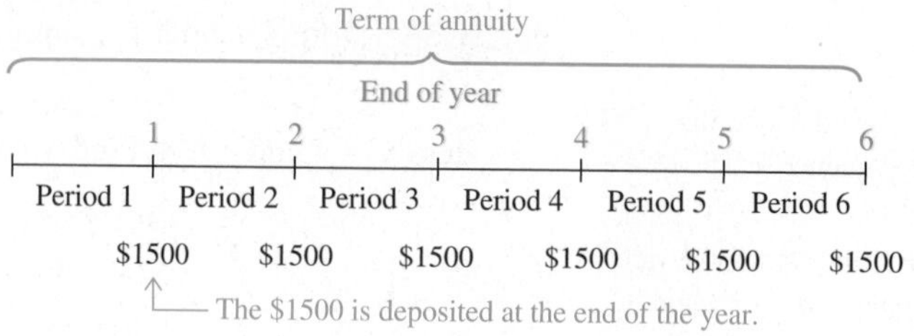

Term of annuity

End of year

1 2 3 4 5 6

Period 1 Period 2 Period 3 Period 4 Period 5 Period 6

\$1500 \$1500 \$1500 \$1500 \$1500 \$1500

The \$1500 is deposited at the end of the year.

FIGURE 5.6

To find the future value of this annuity, look separately at each of the \$1500 payments. The first of these payments will produce a compound amount of

$$1500(1 + .08)^5 = 1500(1.08)^5.$$

Use 5 as the exponent instead of 6 since the money is deposited at the *end* of the first year and earns interest for only 5 years. The second payment of \$1500 will produce a compound amount of $1500(1.08)^4$.

The computation of the compound amounts for the remaining payments is shown in Figure 5.7. (Note that the last payment earns no interest.)

Complete these steps for an annuity of \$2000 at the end of each year for 3 years. Assume interest of 6% compounded annually.

(a) The first deposit of \$2000 produces a total of ______.

(b) The second deposit becomes ______.

(c) No interest is earned on the third deposit, so the total in the account is ______.

Answers:

(a) \$2247.20

(b) \$2120.00

(c) \$6367.20

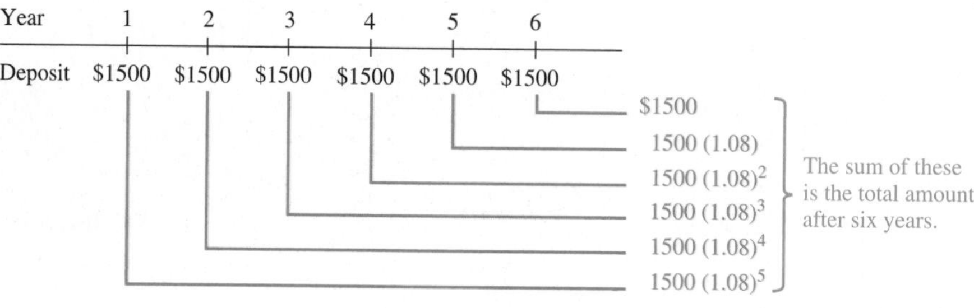

FIGURE 5.7

Figure 5.7 shows that the total amount after six years is the sum

$$1500 + 1500 \cdot 1.08^1 + 1500 \cdot 1.08^2 + 1500 \cdot 1.08^3 + 1500 \cdot 1.08^4 + 1500 \cdot 1.08^5,$$

which is the sum of the first six terms of a geometric sequence with $a = 1500$ and $r = 1.08$. Applying the sum formula with $n = 6$, we see that the sum is

$$\frac{a(r^n - 1)}{r - 1} = \frac{1500[(1.08)^6 - 1]}{1.08 - 1} = \$11{,}003.89.$$ (3)

To generalize this result, suppose that payments of R dollars each are deposited into an account at the *end of each period* for *n periods,* at a rate of interest *i per period.* The first payment of R dollars will produce a compound amount of $R(1 + i)^{n-1}$ dollars, the second payment will produce $R(1 + i)^{n-2}$ dollars, and so on; the final payment earns no interest and contributes just R dollars to the total. If S represents the future value of the annuity, then (as shown in Figure 5.8),

$$S = R + R(1 + i)^1 + R(1 + i)^2 + \cdots + R(1 + i)^{n-1}.$$

Solving this dependent system by back substitution, we have

$$y = 16 - \frac{8}{3}z$$

$$x = 44 - 2y - 6z = 44 - 2\left(16 - \frac{8}{3}z\right) - 6z = 12 - \frac{2}{3}z,$$

so that all solutions of the system are given by $\left(12 - \frac{2}{3}z, 16 - \frac{8}{3}z, z\right)$. The only solutions that apply in this situation, however, are those given by $z = 0, 3,$ or 6 because all other values of z lead to fractions or negative numbers (you can't deliver a negative number of boxes or part of a box). Hence, there are three ways to have a fully loaded van.

Solution	*Van Load*
(12, 16, 0)	12 EZ, 16 compact, 0 commercial
(10, 8, 3)	10 EZ, 8 compact, 3 commercial
(8, 0, 6)	8 EZ, 0 compact, 6 commercial

6.1 Exercises

Determine whether the given ordered list of numbers is a solution of the system of equations.

1. $(-1, 3)$
$$\begin{aligned} 2x + y &= 1 \\ -3x + 2y &= 9 \end{aligned}$$

2. $(2, 1.5, -.5)$
$$\begin{aligned} 3x + 4y - 2z &= -.5 \\ .5x + 8z &= -3 \\ x - 3y + 5z &= -5 \end{aligned}$$

Solve each of the following systems of two equations in two variables. (See Examples 1–3.)

3. $\begin{aligned} x - 2y &= 5 \\ 2x + y &= 3 \end{aligned}$

4. $\begin{aligned} 3x - y &= 1 \\ -x + 2y &= 4 \end{aligned}$

5. $\begin{aligned} 2x - 2y &= 12 \\ -2x + 3y &= 10 \end{aligned}$

6. $\begin{aligned} 3x + 2y &= -4 \\ 4x - 2y &= -10 \end{aligned}$

7. $\begin{aligned} x + 3y &= -1 \\ 2x - y &= 5 \end{aligned}$

8. $\begin{aligned} 4x - 3y &= -1 \\ x + 2y &= 19 \end{aligned}$

9. $\begin{aligned} 2x + 3y &= 15 \\ 8x + 12y &= 40 \end{aligned}$

10. $\begin{aligned} 2x + 5y &= 8 \\ 6x + 15y &= 18 \end{aligned}$

11. $\begin{aligned} 2x - 8y &= 2 \\ 3x - 12y &= 3 \end{aligned}$

12. $\begin{aligned} 3x - 2y &= 4 \\ 6x - 4y &= 8 \end{aligned}$

13. $\begin{aligned} 3x + 2y &= 5 \\ 6x + 4y &= 8 \end{aligned}$

14. $\begin{aligned} 9x - 5y &= 1 \\ -18x + 10y &= 1 \end{aligned}$

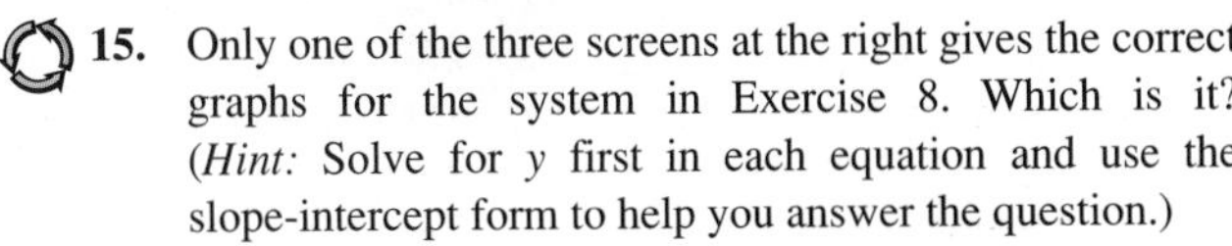

15. Only one of the three screens at the right gives the correct graphs for the system in Exercise 8. Which is it? (*Hint:* Solve for y first in each equation and use the slope-intercept form to help you answer the question.)

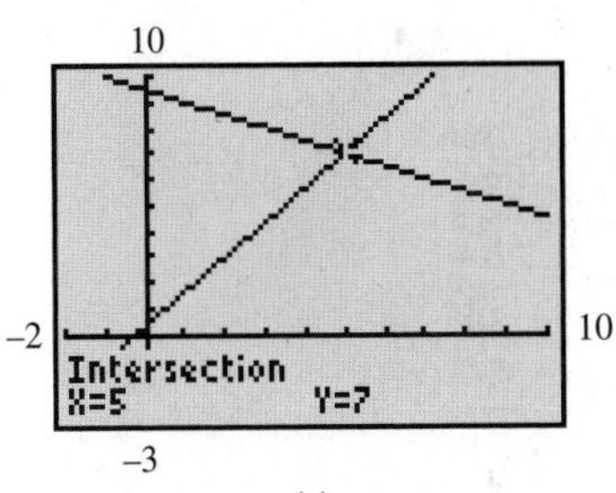

(a)

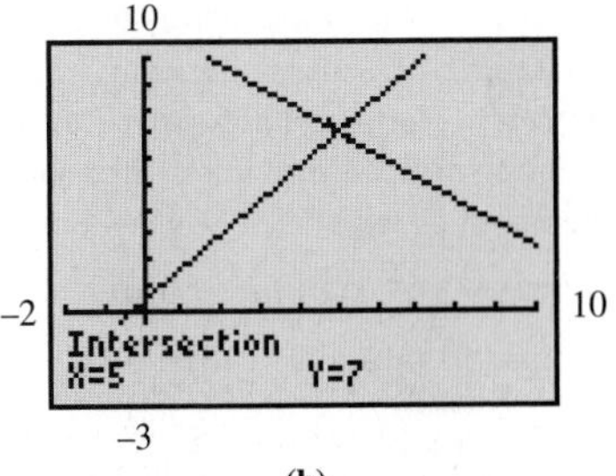

(b)

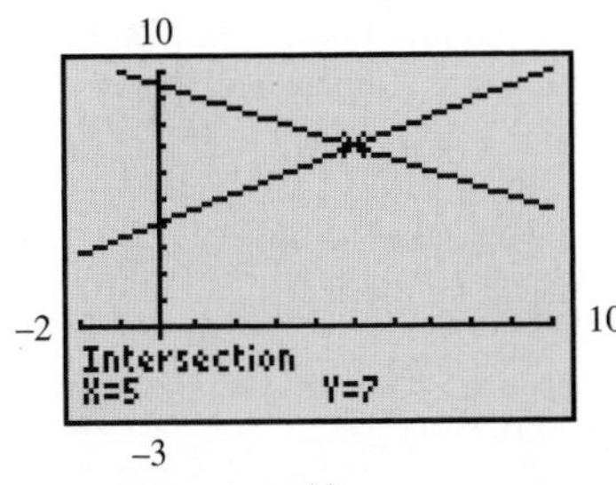

(c)

In Exercises 16 and 17, multiply both sides of each equation by a common denominator to eliminate the fractions. Then solve the system.

16. $\frac{x}{2} + \frac{y}{3} = 8$
$\frac{2x}{3} + \frac{3y}{2} = 17$

17. $\frac{x}{5} + 3y = 31$
$2x - \frac{y}{5} = 8$

18. Business When Neil Simon opens a new play, he has to decide whether to open the show on Broadway or off Broadway. For example, he decided to open his play *London Suite* off Broadway. From information provided by Emanuel Azenberg, his producer, the following equations were developed:

$$43{,}500x - y = 1{,}295{,}000$$
$$27{,}000x - y = 440{,}000,$$

where x represents the number of weeks that the show has run and y represents the profit or loss from the show (first equation is Broadway and second equation is off Broadway).*

(a) Solve this system of equations to determine when the profit/loss from the show will be equal for each venue. What is the profit?

(b) Discuss which venue is favorable for the show.

19. Social Sciences One of the factors that contribute to the success or failure of a particular army during war is its ability to get new troops ready for service. It is possible to analyze the rate of change in the number of troops of two hypothetical armies with the following simplified model,

$$\text{Rate of increase (Red Army)} = 200{,}000 - .5r - .3b$$
$$\text{Rate of increase (Blue Army)} = 350{,}000 - .5r - .7b,$$

where r is the number of soldiers in the Red Army at a given time and b is the number of soldiers in the Blue Army at a given time. The factors .5 and .7 represent each army's efficiency at bringing new soldiers to the fight.†

(a) Solve this system of equations to determine the number of soldiers in each army when the rate of increase for each is zero.

(b) Describe what might be going on in a war when the rate of increase is zero.

Write the augmented matrix of each of the following systems. Do not solve the systems.

20. $2x + y + z = 3$
$3x - 4y + 2z = -5$
$x + y + z = 2$

21. $3x + 4y - 2z - 3w = 0$
$x - 3y + 7z + 4w = 9$
$2x + 5z - 6w = 0$

Write the system of equations associated with the following augmented matrices. Do not solve the systems.

22. $\left[\begin{array}{ccc|c} 2 & 3 & 8 & 20 \\ 1 & 4 & 6 & 12 \\ 0 & 3 & 5 & 10 \end{array}\right]$

23. $\left[\begin{array}{ccc|c} 3 & 2 & 6 & 18 \\ 2 & -2 & 5 & 7 \\ 1 & 0 & 5 & 20 \end{array}\right]$

Use the indicated row operation to transform each matrix.

24. Interchange R_2 and R_3.
$\left[\begin{array}{ccc|c} 1 & 2 & 3 & -1 \\ 6 & 5 & 4 & 6 \\ 2 & 0 & 7 & -4 \end{array}\right]$

25. Replace R_3 by $-3R_1 + R_3$.
$\left[\begin{array}{cccc|c} 1 & 5 & 2 & 0 & -1 \\ 8 & 5 & 4 & 6 & 6 \\ 3 & 0 & 7 & 1 & -4 \end{array}\right]$

26. Replace R_2 by $2R_1 + R_2$.
$\left[\begin{array}{cccc|c} -4 & -3 & 1 & -1 & 2 \\ 8 & 2 & 5 & 0 & 6 \\ 0 & -2 & 9 & 4 & 5 \end{array}\right]$

27. Replace R_3 by $\frac{1}{4}R_3$.
$\left[\begin{array}{ccc|c} 2 & 5 & 1 & -1 \\ -4 & 0 & 4 & 6 \\ 6 & 0 & 8 & -4 \end{array}\right]$

In Exercises 28–33, perform row operations on the augmented matrix as far as necessary to determine whether the system is independent, dependent, or inconsistent. (See Examples 7–9.)

28. $x + 2y = 0$
$y - z = 2$
$x + y + z = -2$

29. $x + 2y + z = 0$
$y + 2z = 0$
$x + y - z = 0$

30. $x + 2y + 4z = 6$
$y + z = 1$
$x + 3y + 5z = 10$

31. $x + y + 2z + 3w = 1$
$2x + y + 3z + 4w = 1$
$3x + y + 4z + 5w = 2$

32. $a - 3b - 2c = -3$
$3a + 2b - c = 12$
$-a - b + 4c = 3$

33. $2x + 2y + 2z = 6$
$3x - 3y - 4z = -1$
$x + y + 3z = 11$

Write the augmented matrix of the system and use the matrix method to solve the system. If the system has an infinite number of solutions, express them in terms of the parameter z. (See Examples 4–9.)

34. $x + y + z = 2$
$2x + y - z = 5$
$x - y + z = -2$

35. $2x + y + z = 9$
$-x - y + z = 1$
$3x - y + z = 9$

36. $x + 3y + 4z = 14$
$2x - 3y + 2z = 10$
$3x - y + z = 9$
$4x + 2y + 5z = 23$

37. $4x - y + 3z = -2$
$3x + 5y - z = 15$
$-2x + y + 4z = 14$
$x + 6y + 3z = 29$

---

*Albert Goetz, "Basic Economics: Calculating against Theatrical Disaster," *Mathematics Teacher* 89, no. 1 (January 1996): 30–32. Reprinted with permission, ©1996 by the National Council of Teachers of Mathematics. All rights reserved.

†Ian Bellamy, "Modelling War," *Journal of Peace Research* 36, no. 6 (1999): 729–739. ©1999 Sage Publications, Ltd.

38. $$\begin{aligned} x + 2y + 3z &= 8 \\ 3x - y + 2z &= 5 \\ -2x - 4y - 6z &= 5 \end{aligned}$$

39. $$\begin{aligned} 3x - 2y - 8z &= 1 \\ 9x - 6y - 24z &= -2 \\ x - y + z &= 1 \end{aligned}$$

40. $$\begin{aligned} 2x - 4y + z &= -4 \\ x + 2y - z &= 0 \\ -x + y + z &= 6 \\ 2x - y + z &= 2 \end{aligned}$$

41. $$\begin{aligned} 4x - 3y + z &= 9 \\ 3x + 2y - 2z &= 4 \\ x - y + 3z &= 5 \\ 2x + 3y - 5z &= -1 \end{aligned}$$

42. $$\begin{aligned} 5x + 3y + 4z &= 19 \\ 3x - y + z &= -4 \end{aligned}$$

43. $$\begin{aligned} 3x + y - z &= 0 \\ 2x - y + 3z &= -7 \end{aligned}$$

44. $$\begin{aligned} 11x + 10y + 9z &= 5 \\ x + 2y + 3z &= 1 \\ 3x + 2y + z &= 1 \end{aligned}$$

45. $$\begin{aligned} x + y &= 3 \\ 5x - y &= 3 \\ 9x - 4y &= 1 \end{aligned}$$

46. Find constants a, b, c such that the points $(2, 3)$, $(-1, 0)$, and $(-2, 2)$ lie on the graph of the equation $y = ax^2 + bx + c$. (*Hint:* Since $(2, 3)$ is on the graph, we must have $3 = a(2^2) + b(2) + c$, that is, $4a + 2b + c = 3$. Similarly, the other two points lead to two more equations. Solve the resulting system for a, b, c.)

47. **(a)** Find the equation of the straight line through $(1, 2)$ and $(3, 4)$.
(b) Find the equation of the line through $(-1, 1)$ with slope 3.
(c) Find a point that lies on both the lines in (a) and (b).

48. Graph the equations in the following system. Then explain why the graphs show that the system is inconsistent.

$$\begin{aligned} 2x + 3y &= 8 \\ x - y &= 4 \\ 5x + y &= 7 \end{aligned}$$

49. Explain why a system with more variables than equations cannot have a unique solution (that is, be an independent system). (*Hint:* When you apply the elimination method to such a system, what must necessarily happen?)

Work the following problems by writing and solving a system of equations. (See Example 10.)

50. **Physical Science** Linear systems occur in the design of roof trusses for new homes and buildings. The simplest type of roof truss is a triangle. The truss shown in the figure at the top of the next column is used to frame roofs of small buildings. If a 100-pound force is applied at the peak of the truss, then the forces or weights W_1 and W_2 exerted parallel to each rafter of the truss are determined by the following system of linear equations.

$$\frac{\sqrt{3}}{2}(W_1 + W_2) = 100$$

$$W_1 - W_2 = 0$$

Solve the system to find W_1 and W_2.*

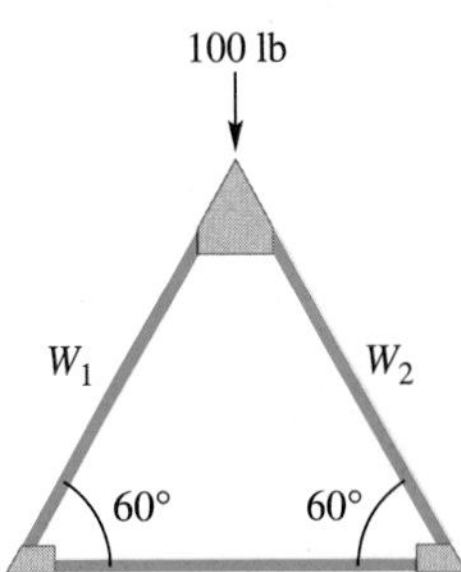

51. **Physical Science** (Refer to Exercise 50.) Use the following system of equations to determine the forces or weights W_1 and W_2 exerted on each rafter for the truss shown in the figure.

$$W_1 + \sqrt{2}W_2 = 300$$

$$\sqrt{3}W_1 - \sqrt{2}W_2 = 0$$

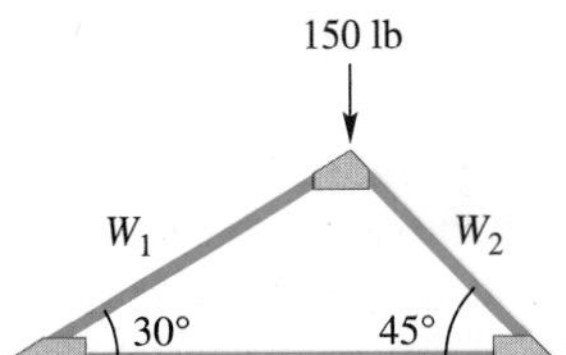

52. **Finance** Shirley Cicero has \$16,000 invested in Boeing and GE stock. The Boeing stock currently sells for \$30 a share and the GE stock for \$70 a share. If GE stock triples in value and Boeing stock goes up 50%, her stock will be worth \$34,500. How many shares of each stock does she own?

53. **Business** An apparel shop sells skirts for \$45 and blouses for \$35. Its entire stock is worth \$51,750. But sales are slow and only half the skirts and two-thirds of the blouses are sold, for a total of \$30,600. How many skirts and blouses are left in the store?

54. **Business** A company produces two models of bicycles, model 201 and model 301. Model 201 requires 2 hours of assembly time, and model 301 requires 3 hours of assembly time. The parts for model 201 cost \$25 per bike, and the parts for model 301 cost \$30 per bike. If the company has a total of 34 hours of assembly time and \$365 available per day for these two models, how many of each can be made in a day?

55. **Social Science** The relationship between a professional basketball player's height H (in inches) and weight W (in pounds) was modeled using two different samples of players. The resulting equations that modeled each sample were $W = 7.46H - 374$ and $W = 7.93H - 405$.

*R. Hibbeler, *Structural Analysis* (Englewood Cliffs, N.J.: Prentice-Hall, 1995).

(a) Use both equations to predict the weight of a 6′11″ professional basketball player.
(b) According to each model, what change in weight is associated with a 1-inch increase in height?
(c) Determine the weight and height where the two models agree.

56. Finance An investor wants to invest \$30,000 in corporate bonds that are rated AAA, A, and B. The lower rated ones pay higher interest but are higher risk as well. The average yield is 5% on AAA bonds, 6% on A bonds and 10% on B bonds. Being conservative, the investor wants to have twice as much in AAA bonds as in B bonds. How much should she invest in each type of bond to have an interest income of \$2000?

57. Business A candy store sells cashews for \$4.40 per pound and peanuts for \$1.20 per pound. If you want to buy exactly 3 pounds of nuts for \$6, how many pounds of each kind should you buy? (*Hint:* If you buy x pounds of cashews and y pounds of peanuts, then $x + y = 3$. Find a second equation by considering cost; then solve the resulting system.)

58. Business A winemaker has two large casks of wine. One wine is 8% alcohol and the other is 18% alcohol. How many gallons of each wine should be mixed to produce 30 gallons of wine that is 12% alcohol? (*Hint:* If x is the number of gallons of the 8% wine and y the number of gallons of the 18% wine, then $x + y =$? Use the fact that the amount of alcohol in the mixture is "8% of x + 18% of y" to find a second equation.)

59. Health Computer Aided Tomography (CAT) scanners take X-rays of a part of the body from different directions and put the information together to create a picture of a cross section of the body.* The amount by which the energy of the X-ray decreases, measured in linear-attenuation units, tells whether the X-ray has passed through healthy tissue, tumorous tissue, or bone, based on the following table.

Type of Tissue	*Linear-Attenuation Values*
Healthy tissue	.1625–.2977
Tumorous tissue	.2679–.3930
Bone	.3857–.5108

The part of the body to be scanned is divided into cells. If an X-ray passes through more than one cell, the total linear-attenuation value is the sum of the values for the cells. For example, in the figure at the top of the next column, let a, b, and c be the values for cells A, B, and C. The attenuation value for beam 1 is $a + b$ and for beam 2 is $a + c$.

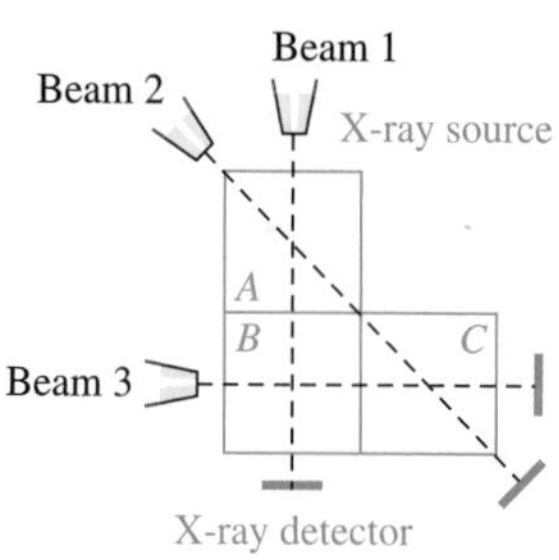

(a) Find the attenuation value for beam 3.
(b) Suppose that the attenuation values are .8, .55, and .65 for beams 1, 2, and 3, respectively. Set up and solve the system of three equations for a, b, and c. What can you conclude about cells A, B, and C?

60. Health (Refer to Exercise 59.) Four X-ray beams are aimed at four cells, as shown in the following figure.

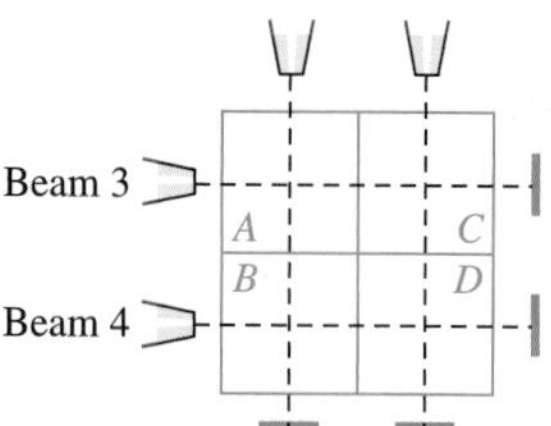

(a) Suppose the attenuation values for beams 1, 2, 3, and 4 are .60, .75, .65, and .70, respectively. Do we have enough information to determine the values of a, b, c, and d? Explain.
(b) Suppose we have the data from part (a), as well as the following values for d. Find the values for a, b, and c and make conclusions about cells A, B, C, and D in each case.
(i) .33 (ii) .43
(c) Two X-ray beams are added, as shown in the figure. In addition to the data in part (a), we now have attenuation values for beams 5 and 6 of .85 and .50. Find the values for a, b, c, and d and make conclusions about cells A, B, C, and D.

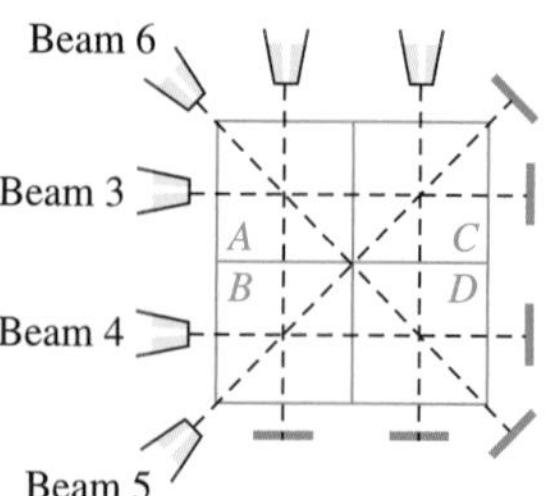

*Exercises 59 and 60 are based on the article "Medical Applications of Linear Equations" by David Jabon, Gail Nord, Bryce W. Wilson, and Penny Coffman, *Mathematics Teacher* 89, no. 5 (May 1996) p. 98: 398.

61. Transportation At rush hours, substantial traffic congestion is encountered at the traffic intersections shown in the figure. (The streets are one-way, as shown by the arrows.)

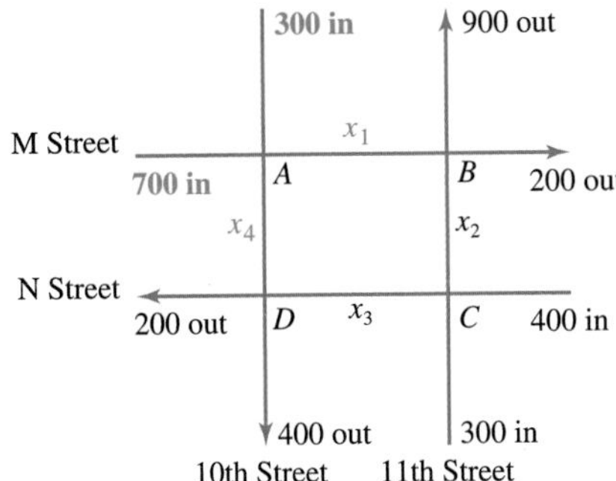

The city wishes to improve the signals at these corners to speed the flow of traffic. The traffic engineers first gather data. As the figure shows, 700 cars per hour come down M Street to intersection A, and 300 cars per hour come down 10th Street to intersection A. A total of x_1 of these cars leave A on M Street, and x_4 cars leave A on 10th Street. The number of cars entering A must equal the number leaving, so that

$$x_1 + x_4 = 700 + 300$$

or

$$x_1 + x_4 = 1000.$$

For intersection B, x_1 cars enter on M street and x_2 on 11th Street. The figure shows that 900 cars leave B on 11th and 200 on M. We have

$$x_1 + x_2 = 900 + 200$$
$$x_1 + x_2 = 1100.$$

(a) Write two equations representing the traffic entering and leaving intersections C and D.
(b) Solve the system of four equations using x_4 as the parameter.
(c) Based on your solution to part (b) what are the largest and smallest possible values for the number of cars leaving intersection A on 10th Street?
(d) Answer the question in part (c) for the other three variables.
(e) Verify that you could have discarded any one of the four original equations without changing the solution. What does this tell you about the original problem?

62. Transportation The diagram shows the traffic flow at four intersections during rush hour, as in Exercise 61.

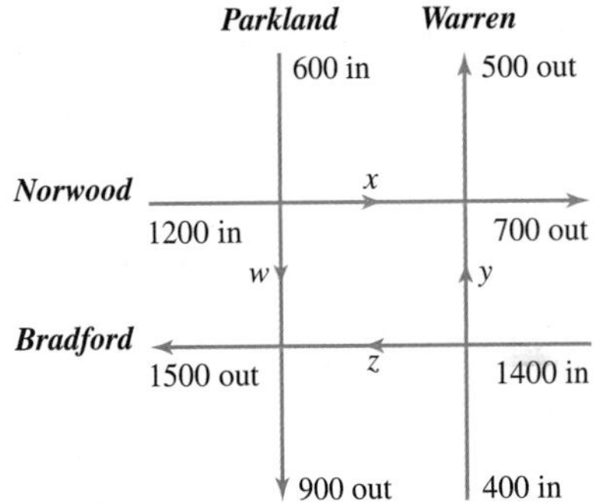

(a) What are the possible values x, y, z, w in order to avoid any congestion? (Use w as the parameter.)
(b) What are the possible values of w?

6.2 THE GAUSS-JORDAN METHOD

In Example 5 of the previous section, we used matrix methods to rewrite the system

$$x - 2y = 6 - 4z$$
$$x + 13z = 6 - y$$
$$-2x + 6y - z = -10$$

as an augmented matrix. We carried out the steps of the matrix method until the final matrix was

$$\left[\begin{array}{ccc|c} 1 & -2 & 4 & 6 \\ 0 & 1 & 3 & 0 \\ 0 & 0 & 1 & 2 \end{array}\right].$$

1

Use the Gauss-Jordan method to solve the system

$$x + 2y = 11$$
$$-4x + y = -8,$$

as follows. Give the shorthand notation and the new matrix in (b)–(d).

(a) Set up the augmented matrix.

(b) Get 0 in row two, column one.

(c) Get 1 in row two, column two.

(d) Finally, get 0 in row one, column two.

(e) The solution for the system is _____.

Answers:

(a) $\left[\begin{array}{cc|c} 1 & 2 & 11 \\ -4 & 1 & -8 \end{array}\right]$

(b) $4R_1 + R_2$; $\left[\begin{array}{cc|c} 1 & 2 & 11 \\ 0 & 9 & 36 \end{array}\right]$

(c) $\frac{1}{9}R_2$; $\left[\begin{array}{cc|c} 1 & 2 & 11 \\ 0 & 1 & 4 \end{array}\right]$

(d) $-2R_2 + R_1$; $\left[\begin{array}{cc|c} 1 & 0 & 3 \\ 0 & 1 & 4 \end{array}\right]$

(e) (3, 4)

We then used back substitution to solve it. In the **Gauss-Jordan method,** additional elimination of variables replaces back substitution, as follows.

$$\left[\begin{array}{ccc|c} 1 & -2 & 4 & 6 \\ 0 & 1 & 0 & -6 \\ 0 & 0 & 1 & 2 \end{array}\right] \quad -3R_3 + R_2$$

$$\left[\begin{array}{ccc|c} 1 & -2 & 0 & -2 \\ 0 & 1 & 0 & -6 \\ 0 & 0 & 1 & 2 \end{array}\right] \quad -4R_3 + R_1$$

$$\left[\begin{array}{ccc|c} 1 & 0 & 0 & -14 \\ 0 & 1 & 0 & -6 \\ 0 & 0 & 1 & 2 \end{array}\right] \quad 2R_2 + R_1$$

The solution of the system is now obvious; it is (−14, −6, 2). Note that this solution is the last column of the final augmented matrix. 1

A matrix, such as the last one above, is said to be in **reduced row echelon form** if it is in echelon form *and* every column containing a leading 1 has zeros in all its entries. In the Gauss-Jordan method, row operations may be performed in any order, but it is best to transform the matrix systematically. Either follow the procedure in the preceding example (which first puts the system in a form where back substitution could be used and then eliminates additional variable terms) or work column by column from left to right, as in the next example.

EXAMPLE 1 Use the Gauss-Jordan method to solve the system

$$\begin{aligned} x \quad\quad + 5z &= -6 + y \\ 3x + 3y \quad\quad &= 10 + z \\ x + 3y + 2z &= 5. \end{aligned}$$

The system must first be rewritten in proper form as follows.

$$\begin{aligned} x - y + 5z &= -6 \\ 3x + 3y - z &= 10 \\ x + 3y + 2z &= 5 \end{aligned}$$

Begin the solution by writing the augmented matrix of the linear system.

$$\left[\begin{array}{ccc|c} 1 & -1 & 5 & -6 \\ 3 & 3 & -1 & 10 \\ 1 & 3 & 2 & 5 \end{array}\right]$$

The first element in column one is already 1. Get 0 for the second element in column one by multiplying each element in the first row by −3 and adding the results to the corresponding elements in row two.

$$\left[\begin{array}{ccc|c} 1 & -1 & 5 & -6 \\ 0 & 6 & -16 & 28 \\ 1 & 3 & 2 & 5 \end{array}\right] \quad -3R_1 + R_2$$

Continue the solution of the system in Example 1 as follows. Give the shorthand notation and the matrix for each step.

(a) Get 1 in row two, column two.

(b) Get 0 in row one, column two.

(c) Now get 0 in row three, column two.

Answers:

(a) $\frac{1}{6}R_2$;

$$\left[\begin{array}{ccc|c} 1 & -1 & 5 & -6 \\ 0 & 1 & -\frac{8}{3} & \frac{14}{3} \\ 0 & 4 & -3 & 11 \end{array}\right]$$

(b) $R_2 + R_1$;

$$\left[\begin{array}{ccc|c} 1 & 0 & \frac{7}{3} & -\frac{4}{3} \\ 0 & 1 & -\frac{8}{3} & \frac{14}{3} \\ 0 & 4 & -3 & 11 \end{array}\right]$$

(c) $-4R_2 + R_3$;

$$\left[\begin{array}{ccc|c} 1 & 0 & \frac{7}{3} & -\frac{4}{3} \\ 0 & 1 & -\frac{8}{3} & \frac{14}{3} \\ 0 & 0 & \frac{23}{3} & -\frac{23}{3} \end{array}\right]$$

(Solution continued in the text.)

Now, change the first element in row three to 0 by multiplying each element of the first row by -1 and adding the results to the corresponding elements of the third row.

$$\left[\begin{array}{ccc|c} 1 & -1 & 5 & -6 \\ 0 & 6 & -16 & 28 \\ 0 & 4 & -3 & 11 \end{array}\right] \quad -1R_1 + R_3$$

This transforms the first column. Transform the second column in a similar manner, as directed in side problem 2. (2)

Complete the solution by transforming the third column of the matrix in part (c) of side problem 2.

$$\left[\begin{array}{ccc|c} 1 & 0 & \frac{7}{3} & -\frac{4}{3} \\ 0 & 1 & -\frac{8}{3} & \frac{14}{3} \\ 0 & 0 & 1 & -1 \end{array}\right] \quad \frac{3}{23}R_3$$

$$\left[\begin{array}{ccc|c} 1 & 0 & 0 & 1 \\ 0 & 1 & -\frac{8}{3} & \frac{14}{3} \\ 0 & 0 & 1 & -1 \end{array}\right] \quad -\frac{7}{3}R_3 + R_1$$

$$\left[\begin{array}{ccc|c} 1 & 0 & 0 & 1 \\ 0 & 1 & 0 & 2 \\ 0 & 0 & 1 & -1 \end{array}\right] \quad \frac{8}{3}R_3 + R_2$$

The linear system associated with this last augmented matrix is

$$\begin{aligned} x \quad\quad &= 1 \\ y \quad &= 2 \\ z &= -1, \end{aligned}$$

and the solution is $(1, 2, -1)$. ◀ (3)

Use the Gauss-Jordan method to solve

$$\begin{aligned} x + y - z &= 6 \\ 2x - y + z &= 3 \\ -x + y + z &= -4. \end{aligned}$$

Answer:

$(3, 1, -2)$

Most graphing calculators can transform a matrix into reduced row echelon form in a single keystroke, as illustrated in the next example.

EXAMPLE 2 Use a graphing calculator to solve the system

$$\begin{aligned} x - y + 5z &= -6 \\ 3x + 3y - z &= 10 \\ x - 5y + 8z &= -17 \\ x + 3y + 2z &= 5. \end{aligned}$$

Enter the augmented matrix in the calculator (Figure 6.6 on the next page). Then use the RREF key (in the MATH or OPS submenu of the MATRIX menu) or a REFF program (in the Graphing Calculator Appendix that is available at www.aw.com/MWA8) to put this matrix in reduced row echelon form (Figure 6.7 on the next page).

4 

Use a graphing calculator to solve the system.

$$\begin{aligned} x + 3y &= 4 \\ 4x + 8y &= 4 \\ 6x + 12y &= 6 \end{aligned}$$

Answer:

$(-5, 3)$

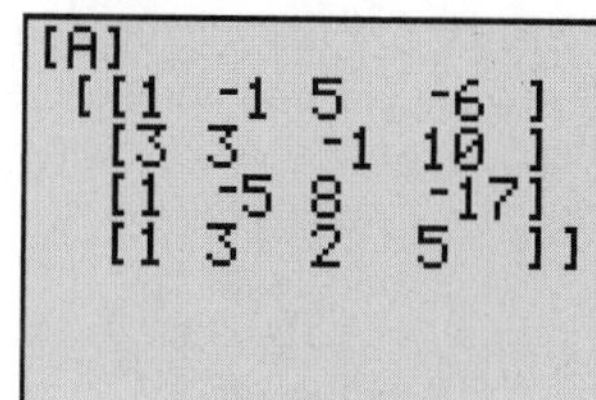

FIGURE 6.6

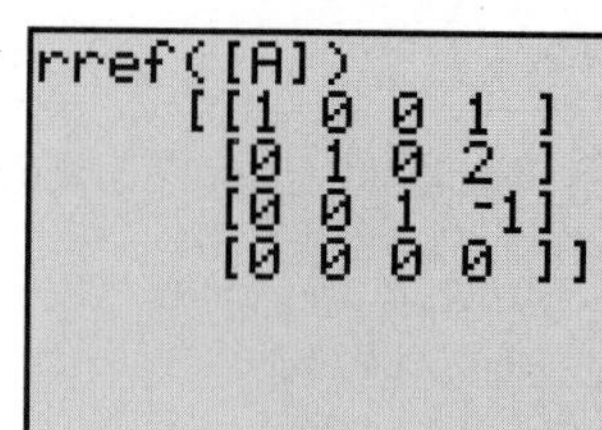

FIGURE 6.7

The solutions can now be read from the last column of the matrix in Figure 6.7: $x = 1, y = 2, z = -1$. ◀ 4

EXAMPLE 3 Use the Gauss-Jordan method to solve the system

$$\begin{aligned} 2x + 4y &= 4 \\ 3x + 6y &= 8 \\ 2x + y &= 7. \end{aligned}$$

There are several possible ways to proceed.

Manual Method Write the augmented matrix and perform row operations to obtain a first column whose entries (from top to bottom) are 1, 0, 0.

$$\left[\begin{array}{cc|c} 2 & 4 & 4 \\ 3 & 6 & 8 \\ 2 & 1 & 7 \end{array}\right]$$

$$\left[\begin{array}{cc|c} 1 & 2 & 2 \\ 3 & 6 & 8 \\ 2 & 1 & 7 \end{array}\right] \quad \frac{1}{2}R_1$$

$$\left[\begin{array}{cc|c} 1 & 2 & 2 \\ 0 & 0 & 2 \\ 2 & 1 & 7 \end{array}\right] \quad -3R_1 + R_2$$

5

Solve each system.

(a) $\begin{aligned} x - y &= 4 \\ -2x + 2y &= 1 \end{aligned}$

(b) $\begin{aligned} 3x - 4y &= 0 \\ 2x + y &= 0 \end{aligned}$

Answers:

(a) No solution

(b) $(0, 0)$

Stop! The second row of the matrix denotes the equation $0x + 0y = 2$. Since the left side of this equation is always 0 and the right side is 2, it has no solution. Therefore, the original system has no solution.

Calculator Method Enter the augmented matrix into a graphing calculator and use the RREF key to put it in reduced row echelon form, as in Figure 6.8. The last row of that matrix corresponds to $0x + 0y = 1$, which has no solution. Hence the original system has no solution. ◀ 5

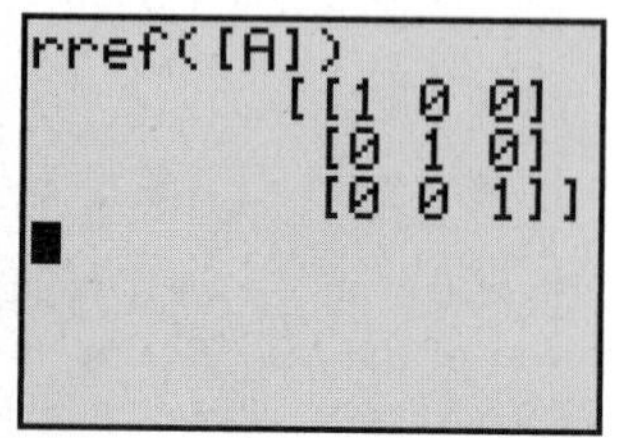

FIGURE 6.8

NOTE Whenever the Gauss-Jordan method produces a row whose elements are all 0 except the last one, as in Example 3, the system is inconsistent and has no solutions. On the other hand, if a row with *every* element 0 is produced, the system may have solutions. In that case, continue carrying out the Gauss-Jordan method.

EXAMPLE 4 Use the Gauss-Jordan method to solve the system

$$\begin{aligned} x + 2y - z &= 0 \\ 3x - y + z &= 6. \end{aligned}$$

When working manually, start with the augmented matrix and use row operations to obtain a first column whose entries (from top to bottom) are 1, 0.

$$\left[\begin{array}{rrr|r} 1 & 2 & -1 & 0 \\ 3 & -1 & 1 & 6 \end{array}\right]$$

$$\left[\begin{array}{rrr|r} 1 & 2 & -1 & 0 \\ 0 & -7 & 4 & 6 \end{array}\right] \quad -3R_1 + R_2$$

Now use row operations to obtain a second column whose entries (from top to bottom) are 0, 1.

$$\left[\begin{array}{rrr|r} 1 & 2 & -1 & 0 \\ 0 & 1 & -\frac{4}{7} & -\frac{6}{7} \end{array}\right] \quad -\frac{1}{7}R_2$$

$$\left[\begin{array}{rrr|r} 1 & 0 & \frac{1}{7} & \frac{12}{7} \\ 0 & 1 & -\frac{4}{7} & -\frac{6}{7} \end{array}\right] \quad -2R_2 + R_1$$

The RREF key on a graphing calculator produces the same matrix (but see the Technology Tip below). In any case, this last matrix is the augmented matrix of the system

$$\begin{aligned} x + \frac{1}{7}z &= \frac{12}{7} \\ y - \frac{4}{7}z &= -\frac{6}{7}. \end{aligned}$$

6

Use the Gauss-Jordan method to solve the following.

(a) $3x + 9y = -6$
$-x - 3y = 2$

(b) $2x + 9y = 12$
$4x + 18y = 5$

Answers:

(a) y arbitrary,
$x = -3y - 2$
or $(-3y - 2, y)$

(b) No solution

Solving the first equation for x and the second for y gives the solution

$$\begin{aligned} &z \text{ arbitrary} \\ &y = -\frac{6}{7} + \frac{4}{7}z \\ &x = \frac{12}{7} - \frac{1}{7}z, \end{aligned}$$

or $(12/7 - z/7, -6/7 + 4z/7, z)$. ◀ 6

TECHNOLOGY TIP When the RREF key is used, a calculator may produce a matrix with some long decimal entries. Such entries can often be displayed as fractions by using the FRAC key on TI calculators (or changing the number format mode to "fraction" on HP-39). Figure 6.9 illustrates this for the matrix in Example 4.

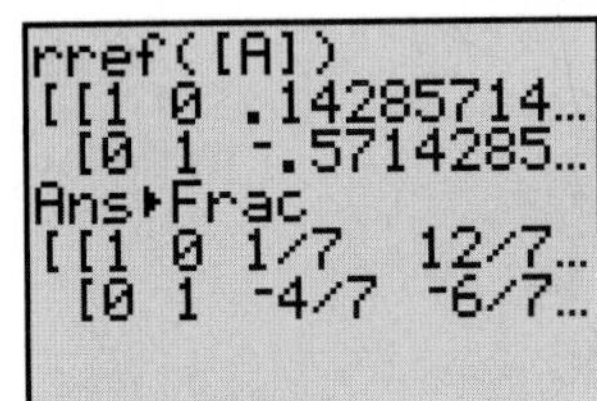

FIGURE 6.9

The techniques used in Example 1–4 can be summarized as follows:

The Gauss-Jordan Method for Solving a System of Linear Equations

1. Arrange the equations with the variable terms in the same order on the left of the equal sign and the constants on the right.
2. Write the augmented matrix of the system.
3. Use row operations or a graphing calculator to transform the augmented matrix into reduced row echelon form.
4. Stop the process in Step 3 if you obtain a row whose elements are all zero except the last one. In that case, the system is inconsistent and has no solutions. Otherwise, finish Step 3 and read the solutions of the system from the final matrix.

Applications

The key to solving applied problems is to determine the system of equations that must be solved.

EXAMPLE 5 An animal feed is to be made from corn, soybeans, and cottonseed. Determine how many units of each ingredient are needed to make a feed that supplies 1800 units of fiber, 2800 units of fat, and 2200 units of protein, given that one unit of each ingredient provides the numbers of units shown in the table below. The table states, for example, that a unit of corn provides 10 units of fiber, 30 units of fat, and 20 units of protein.

	Corn	*Soybeans*	*Cottonseed*	*Totals*
Units of Fiber	10	20	30	1800
Units of Fat	30	20	40	2800
Units of Protein	20	40	25	2200

Let x represent the required number of units of corn, y the number of units of soybeans, and z the number of units of cottonseed. Since the total amount of fiber is to be 1800,

$$10x + 20y + 30z = 1800.$$

The feed must supply 2800 units of fat, so

$$30x + 20y + 40z = 2800.$$

Finally, since 2200 units of protein are required,

$$20x + 40y + 25z = 2200.$$

List a sequence of row operations that will transform the augmented matrix of system (1) in Example 5 into the augmented matrix of system (2).

Answer:

Many sequences are possible, including this one:

replace R_1 by $\frac{1}{10}R_1$;

replace R_2 by $\frac{1}{10}R_2$;

replace R_3 by $\frac{1}{5}R_3$;

replace R_2 by $-3R_1 + R_2$;

replace R_3 by $-4R_1 + R_3$;

replace R_2 by $-\frac{1}{4}R_2$;

replace R_3 by $-\frac{1}{7}R_3$.

Thus we must solve this system of equations.

$$\begin{aligned} 10x + 20y + 30z &= 1800 \\ 30x + 20y + 40z &= 2800 \\ 20x + 40y + 25z &= 2200 \end{aligned} \qquad (1)$$

Now solve this system, either manually or by using technology.

Manual Method Write the augmented matrix and use row operations to transform it into row echelon form, as in problem 7 at the side. The resulting matrix represents the following system. (7)

$$\begin{aligned} x + 2y + 3z &= 180 \\ y + \frac{5}{4}z &= 65 \\ z &= 40 \end{aligned} \qquad (2)$$

Back substitution now shows that

$$z = 40, \quad y = 65 - \frac{5}{4}(40) = 15, \quad \text{and} \quad x = 180 - 2(15) - 3(40) = 30.$$

Thus, the feed should contain 30 units of corn, 15 units of soybeans, and 40 units of cottonseed.

Calculator Method Enter the augmented matrix of the system into the calculator (top of Figure 6.10). Use the RREF key or a REFF program to transform it into reduced row echelon form (bottom of Figure 6.10), which shows that $x = 30$, $y = 15$, and $z = 40$. ◀

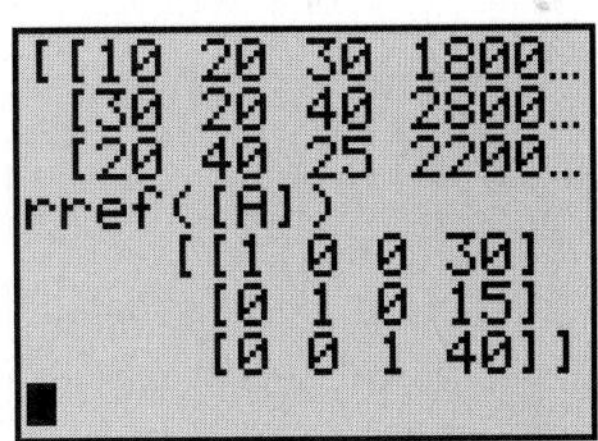

FIGURE 6.10

EXAMPLE 6 The U-Drive Rent-a-Truck Company plans to spend 3 million dollars on 200 new vehicles. Each van will cost \$10,000; each small truck, \$15,000; and each large truck, \$25,000. Past experience shows that they need twice as many vans as small trucks. How many of each kind of vehicle can they buy?

Let x be the number of vans, y the number of small trucks, and z the number of large trucks. Then

$$\begin{pmatrix}\text{Number of}\\ \text{vans}\end{pmatrix} + \begin{pmatrix}\text{Number of}\\ \text{small trucks}\end{pmatrix} + \begin{pmatrix}\text{Number of}\\ \text{large trucks}\end{pmatrix} = \text{Total number of vehicles}$$

$$x + y + z = 200. \qquad (1)$$

Similarly,

$$\begin{pmatrix}\text{Cost of } x\\ \text{vans}\end{pmatrix} + \begin{pmatrix}\text{Cost of } y\\ \text{small trucks}\end{pmatrix} + \begin{pmatrix}\text{Cost of } z\\ \text{large trucks}\end{pmatrix} = \text{Total cost}$$

$$10{,}000x + 15{,}000y + 25{,}000z = 3{,}000{,}000$$

Dividing both sides by 5000 produces the equivalent equation

$$2x + 3y + 5z = 600 \qquad (2)$$

Finally, the number of vans is twice the number of small trucks, that is, $x = 2y$, or equivalently,

$$x - 2y = 0. \qquad (3)$$

We must solve the system given by equations (1)–(3).

$$\begin{aligned} x + y + z &= 200 \\ 2x + 3y + 5z &= 600 \\ x - 2y &= 0 \end{aligned}$$

Form the augmented matrix and transform it into reduced row echelon form (either manually, as shown below, or by using technology).

$$\left[\begin{array}{ccc|c} 1 & 1 & 1 & 200 \\ 2 & 3 & 5 & 600 \\ 1 & -2 & 0 & 0 \end{array}\right]$$

$$\left[\begin{array}{ccc|c} 1 & 1 & 1 & 200 \\ 0 & 1 & 3 & 200 \\ 0 & -3 & -1 & -200 \end{array}\right] \quad \begin{array}{l} \\ -2R_1 + R_2 \\ -R_1 + R_3 \end{array}$$

$$\left[\begin{array}{ccc|c} 1 & 0 & -2 & 0 \\ 0 & 1 & 3 & 200 \\ 0 & 0 & 8 & 400 \end{array}\right] \quad \begin{array}{l} -R_2 + R_1 \\ \\ 3R_2 + R_3 \end{array}$$

$$\left[\begin{array}{ccc|c} 1 & 0 & -2 & 0 \\ 0 & 1 & 3 & 200 \\ 0 & 0 & 1 & 50 \end{array}\right] \quad \begin{array}{l} \\ \\ \frac{1}{8}R_3 \end{array}$$

$$\left[\begin{array}{ccc|c} 1 & 0 & 0 & 100 \\ 0 & 1 & 0 & 50 \\ 0 & 0 & 1 & 50 \end{array}\right] \quad \begin{array}{l} 2R_3 + R_1 \\ -3R_3 + R_2 \\ \\ \end{array}$$

The final matrix corresponds to the system

$$\begin{aligned} x &= 100 \\ y &= 50 \\ z &= 50. \end{aligned}$$

Therefore, U-Drive should buy 100 vans, 50 small trucks, and 50 large trucks. ◀ 8

8

In Example 6, suppose that the U-Drive Company can spend only 2 million dollars on 150 new vehicles and that they need three times as many vans as small trucks. Write a system of equations to express these conditions.

Answer:

$$\begin{aligned} x + y + z &= 150 \\ 2x + 3y + 5z &= 400 \\ x - 3y &= 0 \end{aligned}$$

6.2 Exercises

In Exercises 1–4, the reduced row echelon form of the augmented matrix of a system of equations is given. Find the solutions of the system.

1. $\begin{bmatrix} 1 & 0 & 0 & 0 & 3/2 \\ 0 & 1 & 0 & 0 & 5 \\ 0 & 0 & 1 & 0 & -2 \\ 0 & 0 & 0 & 1 & 0 \end{bmatrix}$

2. $\begin{bmatrix} 1 & 0 & 0 & 0 & 0 & 5 \\ 0 & 1 & 0 & 0 & 0 & 4 \\ 0 & 0 & 1 & 0 & 0 & 3 \\ 0 & 0 & 0 & 0 & 1 & 2 \\ 0 & 0 & 0 & 0 & 0 & 1 \end{bmatrix}$

3. $\begin{bmatrix} 1 & 0 & 0 & 1 & 2 \\ 0 & 1 & 0 & 2 & -3 \\ 0 & 0 & 1 & 0 & 4 \\ 0 & 0 & 0 & 0 & 0 \end{bmatrix}$

4. $\begin{bmatrix} 1 & 0 & 0 & 0 & 7 \\ 0 & 1 & 0 & 0 & 1 \\ 0 & 0 & 1 & 0 & -5 \\ 0 & 0 & 0 & 1 & 4 \\ 0 & 0 & 0 & 0 & 0 \\ 0 & 0 & 0 & 0 & 0 \end{bmatrix}$

Use the Gauss-Jordan method to solve each of the following systems of equations. (See Examples 1–4.)

5. $\begin{aligned} x + 2y + z &= 5 \\ 2x + y - 3z &= -2 \\ 3x + y + 4z &= -5 \end{aligned}$

6. $\begin{aligned} 3x - 2y + z &= 6 \\ 3x + y - z &= -4 \\ -x + 2y - 2z &= -8 \end{aligned}$

7. $\begin{aligned} x + 3y - 6z &= 7 \\ 2x - y + 2z &= 0 \\ x + y + 2z &= -1 \end{aligned}$

8. $\begin{aligned} x &= 1 - y \\ 2x &= z \\ 2z &= -2 - y \end{aligned}$

9. $\begin{aligned} 3x + 5y - z &= 0 \\ 4x - y + 2z &= 1 \\ -6x - 10y + 2z &= 0 \end{aligned}$

10. $\begin{aligned} x + y &= -1 \\ y + z &= 4 \\ x + z &= 1 \end{aligned}$

11. $\begin{aligned} x + y - z &= 6 \\ 2x - y + z &= -9 \\ x - 2y + 3z &= 1 \end{aligned}$

12. $\begin{aligned} y &= x - 1 \\ y &= 6 + z \\ z &= -1 - x \end{aligned}$

13. $\begin{aligned} x - 2y + z &= 5 \\ 2x + y - z &= 2 \\ -2x + 4y - 2z &= 2 \end{aligned}$

14. $\begin{aligned} 2x + 3y + z &= 9 \\ 4x + y - 3z &= -7 \\ 6x + 2y - 4z &= -8 \end{aligned}$

15. $\begin{aligned} -8x - 9y &= 11 \\ 24x + 34y &= 2 \\ 16x + 11y &= -57 \end{aligned}$

16. $\begin{aligned} 2x + y &= 7 \\ x - y &= 3 \\ x + 3y &= 4 \end{aligned}$

17. $\begin{aligned} x + y - z &= -20 \\ 2x - y + z &= 11 \end{aligned}$

18. $\begin{aligned} 4x + 3y + z &= 1 \\ -2x - y + 2z &= 0 \end{aligned}$

19. $\begin{aligned} 2x + y + 3z - 2w &= -6 \\ 4x + 3y + z - w &= -2 \\ x + y + z + w &= -5 \\ -2x - 2y + 2z + 2w &= -10 \end{aligned}$

20. $\begin{aligned} x + y + z + w &= -1 \\ -x + 4y + z - w &= 0 \\ x - 2y + z - 2w &= 11 \\ -x - 2y + z + 2w &= -3 \end{aligned}$

21. $\begin{aligned} x + 2y - z &= 3 \\ 3x + y + w &= 4 \\ 2x - y + z + w &= 2 \end{aligned}$

22. $\begin{aligned} x - 2y - z - 3w &= -3 \\ -x + y + z &= 2 \\ 4y + 3z - 6w &= -2 \end{aligned}$

Use systems of equations to answer these questions. (See Examples 5 and 6.)

23. Business McFrugal Snack Shops plan to hire two public relations firms to survey 500 customers by phone, 750 by mail, and 250 by in-person interviews. The Garcia firm has personnel to do 10 phone surveys, 30 mail surveys, and 5 interviews per hour. The Wong firm can handle 20 phone surveys, 10 mail surveys, and 10 interviews per hour. For how many hours should each firm be hired to produce the exact number of surveys needed?

24. Business A knitting shop ordered yarn from three suppliers, I, II, and III. One month the shop ordered a total of 100 units of yarn from these suppliers. The delivery costs were \$80, \$50, and \$65 per unit for the orders from suppliers I, II, and III, respectively, with total delivery costs of \$5990. The shop ordered the same amount from suppliers I and III. How many units were ordered from each supplier?

25. Business Pretzels cost \$3 per pound, dried fruit \$4 per pound, and nuts \$8 per pound. How many pounds of each should be used to produce 140 pounds of trail mix costing \$6 per pound in which there are twice as many pretzels (by weight) as dried fruit?

26. Business An auto manufacturer sends cars from two plants, I and II, to dealerships A and B located in a midwestern city. Plant I has a total of 28 cars to send, and plant II has 8. Dealer A needs 20 cars, and dealer B needs 16. Transportation costs based on the distance of each dealership from each plant are \$220 from I to A, \$300 from I to B, \$400 from II to A, and \$180 from II to B. The manufacturer wants to limit transportation costs to \$10,640. How many cars should be sent from each plant to each of the two dealerships?

27. Physical Sciences The stopping distance of a car traveling 25 mph is 61.7 feet, and for a car traveling 35 mph it is 106 feet.* The stopping distance in feet can be described by the equation $y = ax^2 + bx$, where x is the speed in mph.

(a) Find the values of a and b.

(b) Use your answers from part (a) to find the stopping distance for a car traveling 55 mph.

**National Traffic Safety Institute Student Workbook,* 1993, p. 7.

28. Natural Science An animal breeder can buy four types of tiger food. Each case of Brand A contains 25 units of fiber, 30 units of protein, and 30 units of fat. Each case of Brand B contains 50 units of fiber, 30 units of protein, and 20 units of fat. Each case of Brand C contains 75 units of fiber, 30 units of protein, and 20 units of fat. Each case of Brand D contains 100 units of fiber, 60 units of protein, and 30 units of fat. How many cases of each brand should the breeder mix together to obtain a food that provides 1200 units of fiber, 600 units of protein, and 400 units of fat?

29. Finance An investor plans to invest $70,000 in a mutual fund, corporate bonds, and a fast food franchise. She plans to put twice as much in bonds as in the mutual fund. Based on past performance, she expects the mutual fund to pay a 2% dividend, the bonds 10%, and the franchise 6%. She would like a dividend income of $4800. How much should she put in each of three investments?

30. Business According to data from a Texas agricultural report, the amount of nitrogen (in lbs/acre), phosphate (in lbs/acre), and labor (in hr/acre) needed to grow honeydews, yellow onions, and lettuce is given by the following table.*

	Honeydews	*Yellow Onions*	*Lettuce*
Nitrogen	120	150	180
Phosphate	180	80	80
Labor	4.97	4.45	4.65

(a) If the farmer has 220 acres, 29,100 pounds of nitrogen, 32,600 pounds of phosphate, and 480 hours of labor, is it possible to use all resources completely? If so, how many acres should he allot for each crop?

(b) Suppose everything is the same as in part (a), except that 1061 hours of labor are available. Is it possible to use all resources completely? If so, how many acres should he allot for each crop?

A graphing calculator or other technology is recommended for the following exercises.

31. Health The table shows the calories, sodium, and protein in one cup of various kinds of soup.

	Progresso Roasted Chicken Rotini	*Healthy Choice Hearty Chicken*	*Campbell's Chunky Chicken Noodle*
Calories	100	130	130
Sodium (mg)	970	480	880
Protein (g)	6	8	8

How many cups of each kind of soup should be mixed together to produce ten servings of soup, each of which provides 203 calories, 1190 milligrams of sodium, and 12.4 grams of protein? What is the serving size (in cups)? (*Hint:* In ten servings, there must be 2030 calories, 11,900 milligrams of sodium, and 124 grams of protein.)

32. Health The table shows the calories, sodium, and fat in one ounce of various snack foods (all produced by Planters).

	Cashews	*Dry Roasted Honey Peanuts*	*Cajun Crunch Trail Mix*
Calories	170	150	130
Sodium (mg)	120	95	10
Fat (g)	6	7	3

How many ounces of each kind of snack should be combined to produce ten servings, each of which provides 288 calories, 115 milligrams of sodium, and 9.6 grams of fat? What is the serving size?

33. Finance An investment firm recommends that a client invest in AAA, A, and B rated bonds. The average yield on AAA bonds is 6%, on A bonds 7%, and on B bonds 10%. The client wants to invest twice as much in AAA bonds as in B bonds. How much should be invested in each type of bond under the following conditions?

(a) The total investment is $25,000, and the investor wants an annual return of $1810 on the three investments.

(b) The values in part (a) are changed to $30,000 and $2150, respectively.

(c) The values in part (a) are changed to $40,000 and $2900, respectively.

34. Business An electronics company produces transistors, resistors, and computer chips. Each transistor requires 3 units of copper, 1 unit of zinc, and 2 units of glass. Each resistor requires 3, 2, and 1 units of the three materials, and each computer chip requires 2, 1, and 2 units of these materials, respectively. How many of each product can be made with the following amounts of materials?

(a) 810 units of copper, 410 units of zinc, and 490 units of glass

(b) 765 units of copper, 385 units of zinc, and 470 units of glass

(c) 1010 units of copper, 500 units of zinc, and 610 units of glass

35. Business At a pottery factory, fuel consumption for heating the kilns varies with the size of the order being fired. In the past, the company recorded the following figures.

*Miguel Paredes, Mohammad Fatehi, and Richard Hinthorn, "The Transformation of an Inconsistent Linear System into a Consistent System," *AMATYC Review*, 13, no. 2 (spring 1992).

x = *Number of Platters*	y = *Fuel Cost per Platter*
6	\$2.80
8	2.48
10	2.24

(a) Find an equation of the form $y = ax^2 + bx + c$ whose graph contains the three points corresponding to the data in the table.

(b) How many platters should be fired at one time in order to minimize the fuel cost per platter? What is the minimum fuel cost per platter?

36. Business The business analyst for Melcher Manufacturing wants to find an equation that can be used to project sales of a relatively new product. For the years 2000, 2001, and 2002, sales were \$15,000, \$32,000, and \$123,000, respectively.

(a) Graph the sales for the years 2000, 2001, and 2002, letting the year 2000 equal 0 on the x-axis. Let the values on the vertical axis be in thousands. (For example, the point (2001, 32,000) will be graphed as (1, 32).)

(b) Find the equation of the straight line $ax + by = c$ through the points for 2000 and 2002.

(c) Find the equation of the parabola $y = ax^2 + bx + c$ through the three given points.

(d) Find the projected sales for 2005 first by using the equation from part (b) and then by using the equation from part (c). If you were to estimate sales of the product in 2005, which result would you choose? Why?

37. Natural Science Determining the amount of carbon dioxide in the atmosphere is important because carbon dioxide is known to be a greenhouse gas. Carbon dioxide concentrations (in parts per million) have been measured at Mauna Loa, Hawaii, over the past 45 years. The concentrations have increased quadratically.* The table lists readings for three years.

Year	CO_2
1958	315
1978	335
1998	367

(a) If the relationship between the carbon dioxide concentration C and the year t is expressed as $C = at^2 + bt + c$, where $t = 0$ corresponds to 1958, use a linear system of equations to determine the constants a, b, and c.

(b) Predict the year when the amount of carbon dioxide in the atmosphere will double from its 1958 level.

38. Physical Science For certain aircraft there exists a quadratic relationship between an airplane's maximum speed S (in knots) and its ceiling C, its highest altitude possible (in thousands of feet).* The table lists three airplanes that conform to this relationship.

Airplane	*Maximum Speed*	*Ceiling*
Hawkeye	320	33
Corsair	600	40
Tomcat	1283	50

(a) If the relationship between C and S is written as $C = aS^2 + bS + c$, use a linear system of equations to determine the constants a, b, and c.

(b) A new aircraft of this type has a ceiling of 45,000 feet. Predict its top speed.

39. Physical Science Estimates of the total amount of lead emissions in the United States are made by the Environmental Protection Agency's Office of Air Quality Planning and Standards each year. The table lists lead estimates for four years.†

Year	*Lead Emissions (in short tons)*
1988	7053
1992	3808
1994	4043
1997	3915

(a) If the relationship between the lead emissions L and the year t is expressed as $L = at^2 + bt + c$, where $t = 0$ corresponds to 1988, use data from 1988, 1992, and 1997 and a linear system of equations to determine the constants a, b, and c.

(b) Use the equation from part (a) to predict the emissions in 1994 and compare the result with the actual data.

(c) If the relationship between the lead emissions L and the year t is expressed as $L = at^3 + bt^2 + ct + d$, where $t = 0$ corresponds to 1988, use all four data points and a linear system of equations to determine the constants a, b, c, and d.

(d) Discuss the appropriateness of the functions used in parts (a) and (b) to model this data.

*C. D. Keeling and T. P. Whorf, Scripps Institution of Oceanography.

*D. Sanders, *Statistics: A First Course*, Fifth Edition (McGraw Hill, 1995).

†U.S. Environmental Protection Agency.

6.3 BASIC MATRIX OPERATIONS

Until now we have used matrices only as a convenient shorthand to solve systems of equations. However, matrices are also important in the fields of management, natural science, engineering, and social science as a way to organize data, as Example 1 demonstrates.

EXAMPLE 1 The EZ Life Company manufactures sofas and armchairs in three models, A, B, and C. The company has regional warehouses in New York, Chicago, and San Francisco. In its August shipment, the company sends 10 model A sofas, 12 model B sofas, 5 model C sofas, 15 model A chairs, 20 model B chairs, and 8 model C chairs to each warehouse.

This data might be organized by first listing it as follows.

Sofas	10 model A	12 model B	5 model C
Chairs	15 model A	20 model B	8 model C

Alternatively, we might tabulate the data in a chart.

		MODEL		
		A	*B*	*C*
FURNITURE	Sofa	10	12	5
	Chair	15	20	8

With the understanding that the numbers in each row refer to the furniture type (sofa, chair) and the numbers in each column refer to the model (A, B, C), the same information can be given by a matrix, as follows.

$$M = \begin{bmatrix} 10 & 12 & 5 \\ 15 & 20 & 8 \end{bmatrix}$$ ◀ 1

Rewrite this information in a matrix with three rows and two columns.

Answer:

$$\begin{bmatrix} 10 & 15 \\ 12 & 20 \\ 5 & 8 \end{bmatrix}$$

A matrix with m rows and n columns has dimensions, or size, $m \times n$. The number of rows is always given first.

EXAMPLE 2 **(a)** The matrix $\begin{bmatrix} 6 & 5 \\ 3 & 4 \\ 5 & -1 \end{bmatrix}$ is a 3×2 matrix.

(b) $\begin{bmatrix} 5 & 8 & 9 \\ 0 & 5 & -3 \\ -4 & 0 & 5 \end{bmatrix}$ is a 3×3 matrix. **(c)** $[1 \quad 6 \quad 5 \quad -2 \quad 5]$ is a 1×5 matrix.

(d) A graphing calculator displays a 4×1 matrix like this.

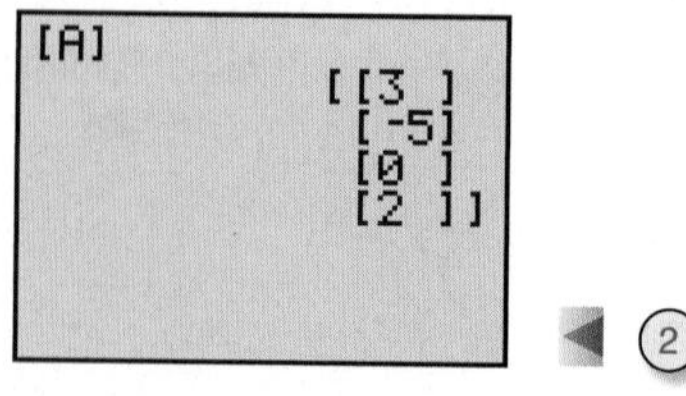

◀ 2

Give the size of each of the following matrices.

(a) $\begin{bmatrix} 2 & 1 & -5 & 6 \\ 3 & 0 & 7 & -4 \end{bmatrix}$

(b) $\begin{bmatrix} 1 & 2 & 3 \\ 4 & 5 & 6 \\ 9 & 8 & 7 \end{bmatrix}$

Answers:

(a) 2×4

(b) 3×3

3

Use the numbers 2, 5, −8, 4 to write

(a) a row matrix;

(b) a column matrix;

(c) a square matrix.

Answers:

(a) $[2 \quad 5 \quad -8 \quad 4]$

(b) $\begin{bmatrix} 2 \\ 5 \\ -8 \\ 4 \end{bmatrix}$

(c) $\begin{bmatrix} 2 & 5 \\ -8 & 4 \end{bmatrix}$ or $\begin{bmatrix} 2 & -8 \\ 5 & 4 \end{bmatrix}$

(Other answers are possible.)

A matrix with only one row, as in Example 2(c), is called a **row matrix** or **row vector.** A matrix with only one column, as in Example 2(d), is called a **column matrix** or **column vector.** A matrix with the same number of rows as columns is called a **square matrix.** The matrix in Example 2(b) above is a square matrix, as are

$$A = \begin{bmatrix} -5 & 6 \\ 8 & 3 \end{bmatrix} \quad \text{and} \quad B = \begin{bmatrix} 0 & 0 & 0 & 0 \\ -2 & 4 & 1 & 3 \\ 0 & 0 & 0 & 0 \\ -5 & -4 & 1 & 8 \end{bmatrix}.$$ 3

When a matrix is denoted by a single letter, such as the matrix A above, then the element in row i and column j is denoted a_{ij}. For example, $a_{21} = 8$ (the element in row 2, column 1). Similarly, in matrix B above, $b_{42} = -4$ (the element in row 4, column 2).

Addition

The matrix given in Example 1,

$$M = \begin{bmatrix} 10 & 12 & 5 \\ 15 & 20 & 8 \end{bmatrix},$$

shows the August shipment from the EZ Life plant to each of its warehouses. If matrix N below gives the September shipment to the New York warehouse, what is the total shipment for each item of furniture to the New York warehouse for these two months?

$$N = \begin{bmatrix} 45 & 35 & 20 \\ 65 & 40 & 35 \end{bmatrix}$$

If 10 model A sofas were shipped in August and 45 in September, then altogether 55 model A sofas were shipped in the 2 months. Adding the other corresponding entries gives a new matrix, Q, that represents the total shipment to the New York warehouse for the 2 months.

$$Q = \begin{bmatrix} 55 & 47 & 25 \\ 80 & 60 & 43 \end{bmatrix}$$

It is convenient to refer to Q as the "sum" of M and N.

The way these two matrices were added illustrates the following definition of addition of matrices.

The **sum** of two $m \times n$ matrices X and Y is the $m \times n$ matrix $X + Y$ in which each element is the sum of the corresponding elements of X and Y.

It is important to remember that only matrices that are the same size can be added.

(4)

Find each sum when possible.

(a) $\begin{bmatrix} 2 & 5 & 7 \\ 3 & -1 & 4 \end{bmatrix} + \begin{bmatrix} -1 & 2 & 0 \\ 10 & -4 & 5 \end{bmatrix}$

(b) $\begin{bmatrix} 1 \\ 2 \\ 3 \end{bmatrix} + \begin{bmatrix} 2 & -1 \\ 4 & 5 \\ 6 & 0 \end{bmatrix}$

(c) $[5 \;\; 4 \;\; -1] + [-5 \;\; 2 \;\; 3]$

Answers:

(a) $\begin{bmatrix} 1 & 7 & 7 \\ 13 & -5 & 9 \end{bmatrix}$

(b) Not possible

(c) $[0 \;\; 6 \;\; 2]$

EXAMPLE 3 Find each sum if possible.

(a) $\begin{bmatrix} 5 & -6 \\ 8 & 9 \end{bmatrix} + \begin{bmatrix} -4 & 6 \\ 8 & -3 \end{bmatrix} = \begin{bmatrix} 5 + (-4) & -6 + 6 \\ 8 + 8 & 9 + (-3) \end{bmatrix} = \begin{bmatrix} 1 & 0 \\ 16 & 6 \end{bmatrix}$

(b) The matrices

$$A = \begin{bmatrix} 5 & 8 \\ 6 & 2 \end{bmatrix} \quad \text{and} \quad B = \begin{bmatrix} 3 & 9 & 1 \\ 4 & 2 & 5 \end{bmatrix}$$

are different sizes, so it is not possible to find the sum $A + B$. ◀ (4)

TECHNOLOGY TIP Spreadsheets and graphing calculators can find matrix sums, as illustrated in Figure 6.11.

```
[A]
      [[-5 4 6]
       [2  0 4]]
[B]
      [[-8 3 2 ]
       [6  2 -4]]
```

```
[A]+[B]
      [[-13 7 8]
       [8   2 0]]
```

FIGURE 6.11

EXAMPLE 4 The September shipments of the three models of sofas and chairs from the EZ Life Company to the New York, San Francisco, and Chicago warehouses are given in matrices N, S, and C below.

$$N = \begin{bmatrix} 45 & 35 & 20 \\ 65 & 40 & 35 \end{bmatrix}, \quad S = \begin{bmatrix} 30 & 32 & 28 \\ 43 & 47 & 30 \end{bmatrix}, \quad C = \begin{bmatrix} 22 & 25 & 38 \\ 31 & 34 & 35 \end{bmatrix}$$

What was the total amount shipped to the three warehouses in September?

The total of the September shipments is represented by the sum of the three matrices N, S, and C.

$$N + S + C = \begin{bmatrix} 45 & 35 & 20 \\ 65 & 40 & 35 \end{bmatrix} + \begin{bmatrix} 30 & 32 & 28 \\ 43 & 47 & 30 \end{bmatrix} + \begin{bmatrix} 22 & 25 & 38 \\ 31 & 34 & 35 \end{bmatrix}$$

$$= \begin{bmatrix} 97 & 92 & 86 \\ 139 & 121 & 100 \end{bmatrix}$$

For example, from this sum the total number of model C sofas shipped to the three warehouses in September was 86. ◀ (5)

(5)

From the result of Example 4, find the total number of the following shipped to the three warehouses.

(a) Model A chairs

(b) Model B sofas

(c) Model C chairs

Answers:

(a) 139

(b) 92

(c) 100

As mentioned in Section 1.1, the additive inverse of the real number a is $-a$; a similar definition is given for the additive inverse of a matrix.

The **additive inverse** (or *negative*) of a matrix X is the matrix $-X$ in which each element is the additive inverse of the corresponding element of X.

If

$$A = \begin{bmatrix} 1 & 2 & 3 \\ 0 & -1 & 5 \end{bmatrix} \quad \text{and} \quad B = \begin{bmatrix} -2 & 3 & 0 \\ 1 & -7 & 2 \end{bmatrix},$$

then by the definition of the additive inverse of a matrix,

$$-A = \begin{bmatrix} -1 & -2 & -3 \\ 0 & 1 & -5 \end{bmatrix} \quad \text{and} \quad -B = \begin{bmatrix} 2 & -3 & 0 \\ -1 & 7 & -2 \end{bmatrix}.$$

TECHNOLOGY TIP A graphing calculator gives the additive inverse of a matrix by preceding the matrix with a negative sign. See Figure 6.12.

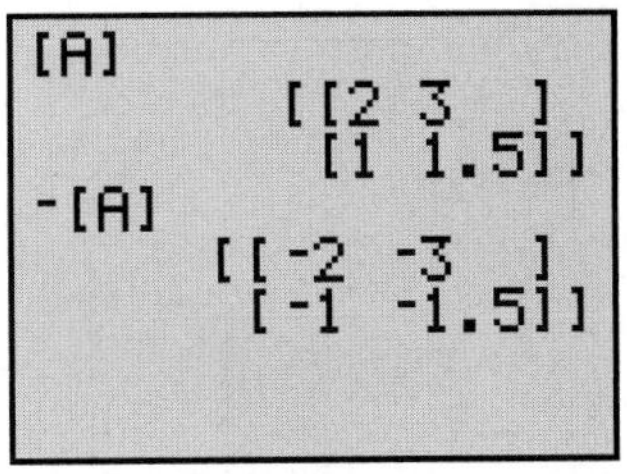

```
[A]
        [[2 3  ]
         [1 1.5]]
-[A]
      [[-2 -3  ]
       [-1 -1.5]]
```

FIGURE 6.12

By the definition of matrix addition, for each matrix X, the sum $X + (-X)$ is a **zero matrix,** O, whose elements are all zeros. There is an $m \times n$ zero matrix for each pair of values of m and n.

$$\begin{bmatrix} 0 & 0 \\ 0 & 0 \end{bmatrix} \qquad \begin{bmatrix} 0 & 0 & 0 & 0 \\ 0 & 0 & 0 & 0 \end{bmatrix}$$

2×2 zero matrix $\qquad$ 2×4 zero matrix

Zero matrices have the following *identity property.*

If O is the $m \times n$ zero matrix, and A is any $m \times n$ matrix, then

$$\boldsymbol{A + O = O + A = A.}$$

Compare this with the identity property for real numbers: for any real number a, $a + 0 = 0 + a = a$.

Subtraction

The **subtraction** of matrices can be defined in a manner comparable to subtraction for real numbers.

For two $m \times n$ matrices X and Y, the **difference** of X and Y is the $m \times n$ matrix $X - Y$ in which each element is the difference of the corresponding elements of X and Y, or, equivalently,

$$X - Y = X + (-Y).$$

According to this definition, matrix subtraction can be performed by subtracting corresponding elements. For example,

$$\begin{bmatrix} 1 & 2 & 3 \\ 0 & -1 & 5 \end{bmatrix} - \begin{bmatrix} -2 & 3 & 0 \\ 1 & -7 & 2 \end{bmatrix} = \begin{bmatrix} 1-(-2) & 2-3 & 3-0 \\ 0-1 & -1-(-7) & 5-2 \end{bmatrix}$$
$$= \begin{bmatrix} 3 & -1 & 3 \\ -1 & 6 & 3 \end{bmatrix}.$$

6

Find each of the following differences when possible.

(a) $\begin{bmatrix} 2 & 5 \\ -1 & 0 \end{bmatrix} - \begin{bmatrix} 6 & 4 \\ 3 & -2 \end{bmatrix}$

(b) $\begin{bmatrix} 1 & 5 & 6 \\ 2 & 4 & 8 \end{bmatrix} - \begin{bmatrix} 2 & 1 \\ 10 & 3 \end{bmatrix}$

(c) $[5 \quad -4 \quad 1] - [6 \quad 0 \quad -3]$

Answers:

(a) $\begin{bmatrix} -4 & 1 \\ -4 & 2 \end{bmatrix}$

(b) Not possible

(c) $[-1 \quad -4 \quad 4]$

EXAMPLE 5 (a) $[8 \quad 6 \quad -4] - [3 \quad 5 \quad -8] = [5 \quad 1 \quad 4]$

(b) The matrices

$$\begin{bmatrix} -2 & 5 \\ 0 & 1 \end{bmatrix} \quad \text{and} \quad \begin{bmatrix} 3 \\ 5 \end{bmatrix}$$

are different sizes and cannot be subtracted.

(c) Spreadsheets and graphing calculators can do matrix subtraction, as shown in Figure 6.13. ◀ 6

```
[B]
   [[-4 5 0  2]
    [6  3 -7 1]]
[C]
   [[4 12 8 -5]
    [2 8  5 0 ]]
```

```
[B]-[C]
  [[-8 -7 -8  7]
   [4  -5 -12 1]]
```

FIGURE 6.13

EXAMPLE 6 During September the Chicago warehouse of the EZ Life Company shipped out the following numbers of each model.

$$K = \begin{bmatrix} 5 & 10 & 8 \\ 11 & 14 & 15 \end{bmatrix}$$

What was the Chicago warehouse inventory on October 1, taking into account only the number of items received and sent out during the month?

The number of each kind of item received during September is given by matrix C from Example 4; the number of each model sent out during September is given by matrix K above. The October 1 inventory will be represented by the matrix $C - K$ as shown below.

$$C - K = \begin{bmatrix} 22 & 25 & 38 \\ 31 & 34 & 35 \end{bmatrix} - \begin{bmatrix} 5 & 10 & 8 \\ 11 & 14 & 15 \end{bmatrix} = \begin{bmatrix} 17 & 15 & 30 \\ 20 & 20 & 20 \end{bmatrix}$$

EXAMPLE 7 A drug company is testing 200 patients to see if Painoff (a new headache medicine) is effective. Half the patients receive Painoff and half receive a placebo. The data on the first 50 patients is summarized in this matrix.

	Pain Relief Obtained Yes	No
Patient took Painoff	22	3
Patient took placebo	8	17

For example, row 2 shows that of the people who took the placebo, 8 got relief but 17 did not. The test was repeated on three more groups of 50 patients each, with the results summarized by these matrices.

$$\begin{bmatrix} 21 & 4 \\ 6 & 19 \end{bmatrix} \quad \begin{bmatrix} 19 & 6 \\ 10 & 15 \end{bmatrix} \quad \begin{bmatrix} 23 & 2 \\ 3 & 22 \end{bmatrix}$$

The total results of the test can be obtained by adding these four matrices.

$$\begin{bmatrix} 22 & 3 \\ 8 & 17 \end{bmatrix} + \begin{bmatrix} 21 & 4 \\ 6 & 19 \end{bmatrix} + \begin{bmatrix} 19 & 6 \\ 10 & 15 \end{bmatrix} + \begin{bmatrix} 23 & 2 \\ 3 & 22 \end{bmatrix} = \begin{bmatrix} 85 & 15 \\ 27 & 73 \end{bmatrix}$$

Because 85 of 100 patients got relief with Painoff and only 27 of 100 with the placebo, it appears that Painoff is effective. 7

7

Later it was discovered that the data in the last group of 50 patients in Example 7 was invalid. Use a matrix to represent the total test results after those data were eliminated.

Answer:

$$\begin{bmatrix} 62 & 13 \\ 24 & 51 \end{bmatrix}$$

Suppose one of the EZ Life Company warehouses receives the following order, written in matrix form, where the entries have the same meaning as given earlier.

$$\begin{bmatrix} 5 & 4 & 1 \\ 3 & 2 & 3 \end{bmatrix}$$

Later, the store that sent the order asks the warehouse to send six more of the same order. The six new orders can be written as one matrix by multiplying each element in the matrix by 6, giving the product

$$6\begin{bmatrix} 5 & 4 & 1 \\ 3 & 2 & 3 \end{bmatrix} = \begin{bmatrix} 30 & 24 & 6 \\ 18 & 12 & 18 \end{bmatrix}.$$

In work with matrices, a real number, like the 6 in the multiplication above, is called a **scalar.**

The **product** of a scalar k and a matrix X is the matrix kX, each of whose elements is k times the corresponding element of X.

8

Find each product.

(a) $-3\begin{bmatrix} 4 & -2 \\ 1 & 5 \end{bmatrix}$

(b) $4\begin{bmatrix} 2 & 4 & 7 \\ 8 & 2 & 1 \\ 5 & 7 & 3 \end{bmatrix}$

Answers:

(a) $\begin{bmatrix} -12 & 6 \\ -3 & -15 \end{bmatrix}$

(b) $\begin{bmatrix} 8 & 16 & 28 \\ 32 & 8 & 4 \\ 20 & 28 & 12 \end{bmatrix}$

Here are two examples, one done by hand and the other on a graphing calculator (Figure 6.14).

$$(-3)\begin{bmatrix} 2 & -5 \\ 1 & 7 \\ 4 & -6 \end{bmatrix} = \begin{bmatrix} -6 & 15 \\ -3 & -21 \\ -12 & 18 \end{bmatrix}$$

```
[A]
              [[3  -4]
               [5  2 ]]
2*[A]
             [[6   -8]
              [10  4 ]]
```

FIGURE 6.14

8

6.3 Exercises

Find the size of each of the following. Identify any square, column, or row matrices. (See Example 2.) Give the additive inverse of each matrix.

1. $\begin{bmatrix} 7 & -8 & 4 \\ 0 & 13 & 9 \end{bmatrix}$

2. $\begin{bmatrix} -7 & 23 \\ 5 & -6 \end{bmatrix}$

3. $\begin{bmatrix} -3 & 0 & 11 \\ 1 & \frac{1}{4} & -7 \\ 5 & -3 & 9 \end{bmatrix}$

4. $[6 \quad -4 \quad \frac{2}{3} \quad 12 \quad 2]$

5. $\begin{bmatrix} 7 \\ 11 \end{bmatrix}$

6. $[-5]$

7. If A is a 5×3 matrix and $A + B = A$, what do you know about B?

8. If C is a 3×3 matrix and D is a 3×4 matrix, then $C + D$ is ______.

Perform the indicated operations where possible. (See Examples 3–6.)

9. $\begin{bmatrix} 1 & 2 & 5 & -1 \\ 3 & 0 & 2 & -4 \end{bmatrix} + \begin{bmatrix} 8 & 10 & -5 & 3 \\ -2 & -1 & 0 & 0 \end{bmatrix}$

10. $\begin{bmatrix} 1 & 5 \\ 2 & -3 \\ 3 & 7 \end{bmatrix} + \begin{bmatrix} 2 & 3 \\ 8 & 5 \\ -1 & 9 \end{bmatrix}$

11. $\begin{bmatrix} 1 & 5 & 7 \\ 2 & 2 & 3 \end{bmatrix} + \begin{bmatrix} 4 & 8 & -7 \\ 1 & -1 & 5 \end{bmatrix}$

12. $\begin{bmatrix} 2 & 4 \\ -8 & 1 \end{bmatrix} + \begin{bmatrix} 9 & -3 \\ 8 & 5 \end{bmatrix}$

13. $\begin{bmatrix} 4 & -2 & 5 \\ 3 & 7 & 0 \end{bmatrix} - \begin{bmatrix} 1 & 5 & -2 \\ -1 & 3 & 8 \end{bmatrix}$

14. $\begin{bmatrix} 9 & 1 \\ 0 & -3 \\ 4 & 10 \end{bmatrix} - \begin{bmatrix} 1 & 9 & -4 \\ -1 & 1 & 0 \end{bmatrix}$

Let $A = \begin{bmatrix} -2 & 4 \\ 0 & 3 \end{bmatrix}$ *and* $B = \begin{bmatrix} -6 & 2 \\ 4 & 0 \end{bmatrix}$. *Find each of the following.*

15. $2A$ **16.** $-3B$ **17.** $-4B$

18. $5A$ **19.** $-4A + 5B$ **20.** $3A - 10B$

Let $A = \begin{bmatrix} 1 & -2 \\ 4 & 3 \end{bmatrix}$ *and* $B = \begin{bmatrix} 2 & -1 \\ 0 & 5 \end{bmatrix}$. *Find a matrix X satisfying the given equation.*

21. $2X = 2A + 3B$ **22.** $3X = A - 3B$

Using matrices

$$O = \begin{bmatrix} 0 & 0 \\ 0 & 0 \end{bmatrix}, P = \begin{bmatrix} m & n \\ p & q \end{bmatrix}, T = \begin{bmatrix} r & s \\ t & u \end{bmatrix}, \text{ and}$$

$$X = \begin{bmatrix} x & y \\ z & w \end{bmatrix},$$

verify that the statements in Exercises 23–28 are true.

23. $X + T$ is a 2×2 matrix.

24. $X + T = T + X$ (Commutative property of addition of matrices)

25. $X + (T + P) = (X + T) + P$ (Associative property of addition of matrices)

26. $X + (-X) = O$ (Inverse property of addition of matrices)

27. $P + O = P$ (Identity property of addition of matrices)

28. Which of the above properties are valid for matrices that are not square?

Work the following exercises. (See Examples 1 and 7.)

29. Business When ticket holders fail to attend, major league sports teams lose the money these fans would have spent on refreshments, souvenirs, etc. The percentage of fans who don't show up is 16% in basketball and hockey, 20% in football, and 18% in baseball. The lost revenue per fan is \$18.20 in basketball, \$18.25 in hockey, \$19 in football, and \$15.40 in baseball. The total annual lost revenue is \$22.7 million in basketball, \$35.8 million in hockey, \$51.9 million in football, and \$96.3 million in baseball.* Express this information in matrix form; specify what the rows and columns represent.

30. Finance 67% of undergraduate students had credit cards in 1998 and 78% in 2000. The average credit card debt was \$1879 in 1998 and \$2748 in 2000. The percentage of students with balances over \$7000 was 10% in 1998 and 9% in 2000. The average number of credit cards per student was 3.5 in 1998 and 3 in 2000.* Write this information first as a 4×2 matrix and then as a 2×4 matrix.

31. Health The shortage of organs for transplants is a continuing problem in the United States. In 1990 there were 1788 people waiting for a heart transplant, 308 for a lung transplant, 1237 for a liver transplant, and 17,883 for a kidney transplant. Corresponding figures for 1998 were 4121, 3171, 12,070, and 38,270. In mid-2000, the figures were 4143, 3614, 15,359, and 45,273.† Express this information as a matrix, labeling rows and columns.

Work these exercises. (See Examples 1, 4, 6, and 7.)

32. Management There are three convenience stores in Gambier. This week, Store I sold 88 loaves of bread, 48 quarts of milk, 16 jars of peanut butter, and 112 pounds of cold cuts. Store II sold 105 loaves of bread, 72 quarts of milk, 21 jars of peanut butter, and 147 pounds of cold cuts. Store III sold 60 loaves of bread, 40 quarts of milk, no peanut butter, and 50 pounds of cold cuts.

(a) Use a 3×4 matrix to express the sales information for the three stores.

(b) During the following week, sales on these products at Store I increased by 25%; sales at Store II increased by 1/3; and sales at Store III increased by 10%. Write the sales matrix for that week.

(c) Write a matrix that represents total sales over the two-week period.

33. Social Science Some of the fastest growing college sports for both men and women are soccer, rowing, and equestrian sports. In 1981–82, there were 80 women's college soccer teams, 43 rowing teams, and 7 equestrian teams. The numbers of men's college teams in these sports were 744, 48, and 2. In 1998–99, there were 926 women's soccer teams, 122 rowing teams, and 41 equestrian teams. The numbers of men's teams in these sports were 879, 70, and 33.‡

(a) Express the information for women's teams as a 3×2 matrix A.

(b) Express the information for men's teams as a 3×2 matrix B.

(c) Find $A + B$ and explain what it means.

(d) Suppose the number of women's teams in each sport has increased by 10% since 1998–99 (rounded to the nearest whole number). Write a matrix C that expresses the new numbers.

*American Demographics.

*Nellie Mae.

†United Network of Organ Sharing and National Organ Procurement and Transplantation Network.

‡General Accounting Office.

34. Social Sciences The following table gives the educational attainment of the U.S. population 25 years and older in various years.*

	MALE		FEMALE	
	Four Years of High School or More	*Four Years of College or More*	*Four Years of High School or More*	*Four Years of College or More*
1940	22.7%	5.5%	26.3%	3.8%
1950	32.6	7.3	36.0	5.2
1960	39.5	9.7	42.5	5.8
1970	51.9	13.5	52.8	8.1
1980	67.3	20.1	65.8	12.8
1990	77.7	24.4	77.5	18.4
1999	83.4	27.5	83.4	23.1

(a) Write a matrix for the educational attainment of males.

(b) Write a matrix for the educational attainment of females.

(c) Use the matrices from parts (a) and (b) to write a matrix showing how much more (or less) education males have attained than females.

35. Transportation The tables in the next column give the death rates (per million person trips) for male and female drivers for various ages and number of passengers.†

MALE DRIVERS

Age	*Number of Passengers*			
	0	1	2	≥ 3
16	2.61	4.39	6.29	9.08
17	1.63	2.77	4.61	6.92
30–59	.92	.75	.62	.54

FEMALE DRIVERS

Age	*Number of Passengers*			
	0	1	2	≥ 3
16	1.38	1.72	1.94	3.31
17	1.26	1.48	2.82	2.28
30–59	.41	.33	.27	.40

(a) Write a matrix A for the death rate of male drivers.

(b) Write a matrix B for the death rate of female drivers.

(c) Use the matrices from parts (a) and (b) to write a matrix showing the difference between the death rate of males and females.

36. Transportation Use matrix operations on the matrices found in Exercise 35(a)(b) to find one matrix that gives the combined death rates for males and females (per million person trips) of drivers of various ages, with varying numbers of passengers. (*Hint:* Consider $\frac{1}{2}(A + B)$.)

*U.S. Census Bureau.

†Li-Hui Chen, Susan Baker, Elisa Braver, and Guohua Li, "Carrying Passengers as a Risk Factor for Crashes Fatal to 16- and 17-Year-Old Drivers," JAMA 283, no. 12 (March 22/29, 2000): 1578–1582.

6.4 MATRIX PRODUCTS AND INVERSES

In the previous section we showed how to multiply a matrix by a scalar. Finding the product of two matrices is more involved but is important in solving practical problems. To understand the reasoning behind the definition of matrix multiplication, look again at the EZ Life Company. Suppose sofas and chairs of the same model are often sold as sets with matrix W showing the number of each model set in each warehouse.

$$\begin{array}{l} \\ \text{New York} \\ \text{Chicago} \\ \text{San Francisco} \end{array} \begin{array}{c} \begin{array}{ccc} \text{A} & \text{B} & \text{C} \end{array} \\ \begin{bmatrix} 10 & 7 & 3 \\ 5 & 9 & 6 \\ 4 & 8 & 2 \end{bmatrix} \end{array} = W$$

If the selling price of a model A set is \$800, of a model B set \$1000, and of a model C set \$1200, find the total value of the sets in the New York warehouse as follows.

1

In this example of the EZ Life Company, find the total value of the New York sets if model A sets sell for $1200, model B for $1600, and model C for $1300.

Answer:

$27,100

Type	*Number of Sets*		*Price of Set*		*Total*
A	10	×	$ 800	=	$ 8,000
B	7	×	1000	=	7,000
C	3	×	1200	=	3,600
			Total for New York		$18,600

The total value of the three kinds of sets in New York is $18,600. (1)

The work done in the table above is summarized as follows:

$$10(\$800) + 7(\$1000) + 3(\$1200) = \$18{,}600.$$

In the same way, the Chicago sets have a total value of

$$5(\$800) + 9(\$1000) + 6(\$1200) = \$20{,}200,$$

and in San Francisco, the total value of the sets is

$$4(\$800) + 8(\$1000) + 2(\$1200) = \$13{,}600.$$

The selling prices can be written as a column matrix, P, and the total value in each location as a column matrix, V.

$$P = \begin{bmatrix} 800 \\ 1000 \\ 1200 \end{bmatrix} \quad \text{and} \quad V = \begin{bmatrix} 18{,}600 \\ 20{,}200 \\ 13{,}600 \end{bmatrix}$$

Consider how the first row of the matrix W and the single column P lead to the first entry of V.

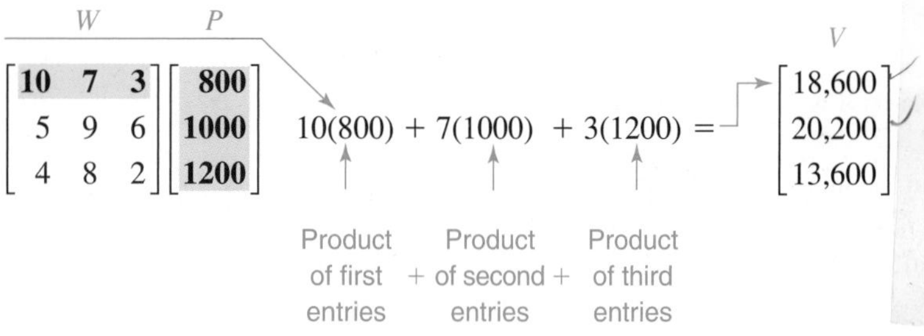

Similarly, adding the products of corresponding entries in the second row of W and the column P produces the second entry in V. The third entry in V is obtained in the same way by using the third row of W and column P. This suggests that it is reasonable to *define* the product WP to be V.

$$WP = \begin{bmatrix} 10 & 7 & 3 \\ 5 & 9 & 6 \\ 4 & 8 & 2 \end{bmatrix} \begin{bmatrix} 800 \\ 1000 \\ 1200 \end{bmatrix} = \begin{bmatrix} 18{,}600 \\ 20{,}200 \\ 13{,}600 \end{bmatrix} = V$$

Note the sizes of the matrices here: the product of a 3×3 matrix and a 3×1 matrix is a 3×1 matrix.

Multiplying Matrices

We first define the **product of a row of a matrix and a column of a matrix** (with the same number of entries in each) to be the *number* obtained by multiplying the corresponding entries (first by first, second by second, etc.) and adding the results. For instance,

$$[3 \quad -2 \quad 1] \cdot \begin{bmatrix} 4 \\ 5 \\ 0 \end{bmatrix} = 3 \cdot 4 + (-2) \cdot 5 + 1 \cdot 0 = 12 - 10 + 0 = 2.$$

Now **matrix multiplication** is defined as follows.

> Let A be an $m \times n$ matrix and let B be an $n \times k$ matrix. The **product matrix** AB is the $m \times k$ matrix whose entry in the ith row and jth column is
>
> the product of the ith row of A and the jth column of B.

CAUTION Be careful when multiplying matrices. Remember that the number of *columns* of A must equal the number of *rows* of B in order to get the product matrix AB. The final product will have as many rows as A and as many columns as B.

EXAMPLE 1 Suppose matrix A is 2×2 and matrix B is 2×4. Can the product AB be calculated? What is the size of the product?

The following diagram helps decide the answers to these questions.

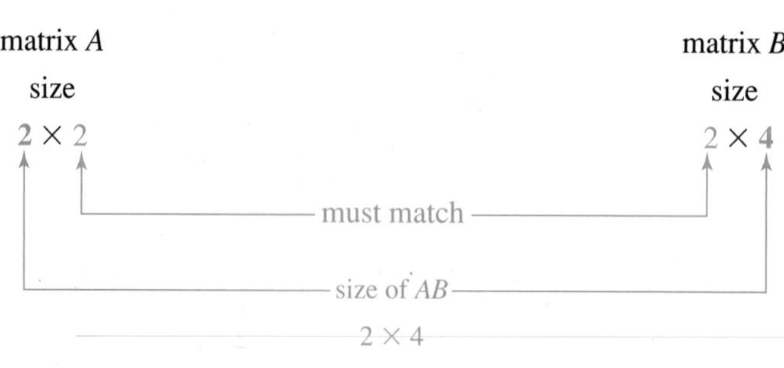

The product AB can be calculated because A has two columns and B has two rows. The product will be a 2×4 matrix. ◀ ②

> ②
>
> Matrix A is 4×6 and matrix B is 2×4.
>
> **(a)** Can AB be found? If so, give its size.
>
> **(b)** Can BA be found? If so, give its size.
>
> Answers:
>
> **(a)** No
>
> **(b)** Yes; 2×6

EXAMPLE 2 Find the product CD when

$$C = \begin{bmatrix} -3 & 4 & 2 \\ 5 & 0 & 4 \end{bmatrix} \quad \text{and} \quad D = \begin{bmatrix} -6 & 4 \\ 2 & 3 \\ 3 & -2 \end{bmatrix}.$$

Here matrix C is 2×3 and matrix D is 3×2, so matrix CD can be found and will be 2×2.

Step 1: row 1, column 1

$$\begin{bmatrix} \mathbf{-3} & \mathbf{4} & \mathbf{2} \\ 5 & 0 & 4 \end{bmatrix}\begin{bmatrix} \mathbf{-6} & 4 \\ \mathbf{2} & 3 \\ \mathbf{3} & -2 \end{bmatrix} \quad \mathbf{(-3)} \cdot \mathbf{(-6)} + \mathbf{4} \cdot \mathbf{2} + \mathbf{2} \cdot \mathbf{3} = 32$$

Hence, 32 is the entry in row 1 column 1 of CD, as shown in Step 5 below.

Step 2: row 1, column 2

$$\begin{bmatrix} \mathbf{-3} & \mathbf{4} & \mathbf{2} \\ 5 & 0 & 4 \end{bmatrix}\begin{bmatrix} -6 & \mathbf{4} \\ 2 & \mathbf{3} \\ 3 & \mathbf{-2} \end{bmatrix} \quad \mathbf{(-3)} \cdot \mathbf{4} + \mathbf{4} \cdot \mathbf{3} + \mathbf{2} \cdot \mathbf{(-2)} = -4$$

So -4 is the entry in row 1, column 2 of CD, as shown in Step 5. Continue in this manner to find the remaining entries of CD.

Step 3: row 2, column 1

$$\begin{bmatrix} -3 & 4 & 2 \\ \mathbf{5} & \mathbf{0} & \mathbf{4} \end{bmatrix}\begin{bmatrix} \mathbf{-6} & 4 \\ \mathbf{2} & 3 \\ \mathbf{3} & -2 \end{bmatrix} \quad \mathbf{5} \cdot \mathbf{(-6)} + \mathbf{0} \cdot \mathbf{2} + \mathbf{4} \cdot \mathbf{3} = -18$$

Step 4: row 2, column 2

$$\begin{bmatrix} -3 & 4 & 2 \\ \mathbf{5} & \mathbf{0} & \mathbf{4} \end{bmatrix}\begin{bmatrix} -6 & \mathbf{4} \\ 2 & \mathbf{3} \\ 3 & \mathbf{-2} \end{bmatrix} \quad \mathbf{5} \cdot \mathbf{4} + \mathbf{0} \cdot \mathbf{3} + \mathbf{4} \cdot \mathbf{(-2)} = 12$$

Step 5: The product is

$$CD = \begin{bmatrix} -3 & 4 & 2 \\ 5 & 0 & 4 \end{bmatrix}\begin{bmatrix} -6 & 4 \\ 2 & 3 \\ 3 & -2 \end{bmatrix} = \begin{bmatrix} 32 & -4 \\ -18 & 12 \end{bmatrix}.$$ ◀ (3)

(3)

Find the product CD given

$$C = \begin{bmatrix} 1 & 3 & 5 \\ 2 & -4 & -1 \end{bmatrix}$$

and

$$D = \begin{bmatrix} 2 & -1 \\ 4 & 3 \\ 1 & -2 \end{bmatrix}.$$

Answer:

$$CD = \begin{bmatrix} 19 & -2 \\ -13 & -12 \end{bmatrix}$$

▶ **EXAMPLE 3** Find BA given

$$A = \begin{bmatrix} 1 & -3 \\ 7 & 2 \end{bmatrix} \quad \text{and} \quad B = \begin{bmatrix} 1 & 0 & -1 \\ 3 & 1 & 4 \end{bmatrix}.$$

Since B is a 2×3 matrix and A is a 2×2 matrix, the product BA cannot be found. ◀ (4)

Give the size of each of the following products that can be found.

(a) $\begin{bmatrix} 2 & 4 \\ 6 & 8 \end{bmatrix}\begin{bmatrix} 1 & 2 & 3 \\ 0 & -1 & 2 \end{bmatrix}$

(b) $\begin{bmatrix} 1 & 2 \\ 5 & 10 \\ 12 & 7 \end{bmatrix}\begin{bmatrix} 2 & 4 \\ 3 & 6 \\ 9 & 1 \end{bmatrix}$

(c) $\begin{bmatrix} 5 \\ 2 \\ 4 \end{bmatrix}[1 \quad 0 \quad 6]$

Answers:

(a) 2×3

(b) Not possible

(c) 3×3

TECHNOLOGY TIP Graphing calculators can find matrix products. However, if you use a graphing calculator to try to find the product in Example 3, the calculator will display an error message.

Matrix multiplication has some similarities with the multiplication of numbers.

For any matrices $A, B, C,$ such that all the indicated sums and products exist, matrix multiplication is associative and distributive.

$$A(BC) = (AB)C \qquad A(B + C) = AB + AC \qquad (B + C)A = BA + CA$$

However, there are important differences between matrix multiplication and multiplication of numbers. (See Exercises 19–22 at the end of this section.) In particular, matrix multiplication is *not* commutative.

If A and B are matrices such that the products AB and BA exist,

AB may not equal BA.

Figure 6.15 shows an example of this situation.

```
[A]
          [[1  -4]
           [3  2 ]]
[B]
          [[-2  4]
           [5   3]]
```

```
[A][B]
          [[-22  -8]
           [4    18]]
[B][A]
          [[10  16 ]
           [14  -14]]
```

FIGURE 6.15

EXAMPLE 4 A contractor builds three kinds of houses, models A, B, and C, with a choice of two styles, Spanish or contemporary. Matrix P shows the number of each kind of house planned for a new 100-home subdivision.

$$\begin{array}{c} \\ \text{Model A} \\ \text{Model B} \\ \text{Model C} \end{array}\begin{array}{c} \begin{array}{cc} \text{Spanish} & \text{Contemporary} \end{array} \\ \begin{bmatrix} 0 & 30 \\ 10 & 20 \\ 20 & 20 \end{bmatrix} \end{array} = P$$

The amounts for each of the exterior materials used depend primarily on the style of the house. These amounts are shown in matrix Q. (Concrete is in cubic yards, lumber in units of 1000 board feet, brick in 1000s, and shingles in units of 100 square feet.)

$$\begin{array}{c} \\ \text{Spanish} \\ \text{Contemporary} \end{array}\begin{array}{c} \begin{array}{cccc} \text{Concrete} & \text{Lumber} & \text{Brick} & \text{Shingles} \end{array} \\ \begin{bmatrix} 10 & 2 & 0 & 2 \\ 50 & 1 & 20 & 2 \end{bmatrix} \end{array} = Q$$

Matrix R gives the cost for each kind of material.

$$\begin{array}{l} \text{Concrete} \\ \text{Lumber} \\ \text{Brick} \\ \text{Shingles} \end{array} \overset{\text{Cost per Unit}}{\begin{bmatrix} 20 \\ 180 \\ 60 \\ 25 \end{bmatrix}} = R$$

(a) What is the total cost for each model house?

First find PQ. The product PQ shows the amount of each material needed for each model house.

$$PQ = \begin{bmatrix} 0 & 30 \\ 10 & 20 \\ 20 & 20 \end{bmatrix} \begin{bmatrix} 10 & 2 & 0 & 2 \\ 50 & 1 & 20 & 2 \end{bmatrix}$$

$$PQ = \overset{\begin{array}{cccc} \text{Concrete} & \text{Lumber} & \text{Brick} & \text{Shingles} \end{array}}{\begin{bmatrix} 1500 & 30 & 600 & 60 \\ 1100 & 40 & 400 & 60 \\ 1200 & 60 & 400 & 80 \end{bmatrix}} \begin{array}{l} \text{Model A} \\ \text{Model B} \\ \text{Model C} \end{array}$$

Now multiply PQ and R, the cost matrix, to get the total cost for each model house.

$$\begin{bmatrix} 1500 & 30 & 600 & 60 \\ 1100 & 40 & 400 & 60 \\ 1200 & 60 & 400 & 80 \end{bmatrix} \begin{bmatrix} 20 \\ 180 \\ 60 \\ 25 \end{bmatrix} = \overset{\text{Cost}}{\begin{bmatrix} 72{,}900 \\ 54{,}700 \\ 60{,}800 \end{bmatrix}} \begin{array}{l} \text{Model A} \\ \text{Model B} \\ \text{Model C} \end{array}$$

(b) How much of each of the four kinds of material must be ordered?

The totals of the columns of matrix PQ will give a matrix whose elements represent the total amounts of each material needed for the subdivision. Call this matrix T and write it as a row matrix.

$$T = [3800 \quad 130 \quad 1400 \quad 200]$$

(c) What is the total cost for material?

Find the total cost of all the materials by taking the product of matrix T, the matrix showing the total amounts of each material, and matrix R, the cost matrix. (To multiply these and get a 1×1 matrix, representing total cost, we must multiply a 1×4 matrix by a 4×1 matrix. This is why T was written as a row matrix in (b) above.)

$$TR = [3800 \quad 130 \quad 1400 \quad 200] \begin{bmatrix} 20 \\ 180 \\ 60 \\ 25 \end{bmatrix} = [188{,}400]$$

(d) Suppose the contractor builds the same number of homes in five subdivisions. Calculate the total amount of each material for each model for all five subdivisions. Multiply PQ by the scalar 5, as follows.

$$5\begin{bmatrix} 1500 & 30 & 600 & 60 \\ 1100 & 40 & 400 & 60 \\ 1200 & 60 & 400 & 80 \end{bmatrix} = \begin{bmatrix} 7500 & 150 & 3000 & 300 \\ 5500 & 200 & 2000 & 300 \\ 6000 & 300 & 2000 & 400 \end{bmatrix}$$

Let matrix A be

$$\begin{array}{cc} & \textit{Vitamin} \\ & \begin{array}{ccc} \text{C} & \text{E} & \text{K} \end{array} \\ \textit{Brand} \begin{array}{c} \text{X} \\ \text{Y} \end{array} & \begin{bmatrix} 2 & 7 & 5 \\ 4 & 6 & 9 \end{bmatrix} \end{array}$$

and matrix B be

$$\begin{array}{cc} & \textit{Cost} \\ & \begin{array}{cc} \text{X} & \text{Y} \end{array} \\ \textit{Vitamin} \begin{array}{c} \text{C} \\ \text{E} \\ \text{K} \end{array} & \begin{bmatrix} 12 & 14 \\ 18 & 15 \\ 9 & 10 \end{bmatrix} \end{array}.$$

(a) What quantities do matrices A and B represent?

(b) What quantities does the product AB represent?

(c) What quantities does the product BA represent?

Answers:

(a) A = brand/vitamin, B = vitamin/cost

(b) AB = brand/cost

(c) Not meaningful, although the product BA can be found

We can introduce a notation to help keep track of the quantities a matrix represents. For example, we can say that matrix P, from Example 4, represents models/styles, matrix Q represents styles/materials, and matrix R represents materials/cost. In each case, the meaning of the rows is written first and the columns second. When we found the product PQ in Example 4, the rows of the matrix represented models and the columns represented materials. Therefore, we can say the matrix product PQ represents models/materials. The common quantity, styles, in both P and Q was eliminated in the product PQ. Do you see that the product $(PQ)R$ represents models/cost?

In practical problems this notation helps decide in what order to multiply two matrices so that the results are meaningful. In Example 4(c) we could have found either product RT or product TR. However, since T represents subdivisions/materials and R represents materials/cost, the product TR gives subdivisions/cost. (5)

Identity and Inverse Matrices

Recall from Section 1.1 that the real number 1 is the identity element for multiplication of real numbers: for any real number a, $a \cdot 1 = 1 \cdot a = a$. In this section, an **identity matrix I** is defined that has properties similar to those of the number 1.

If I is to be the identity matrix, the products AI and IA must both equal A. The 2×2 identity matrix that satisfies these conditions is

$$I = \begin{bmatrix} 1 & 0 \\ 0 & 1 \end{bmatrix}. \quad (6)$$

To check that I is really the 2×2 identity matrix, let

$$A = \begin{bmatrix} a & b \\ c & d \end{bmatrix}.$$

Then AI and IA should both equal A.

$$AI = \begin{bmatrix} a & b \\ c & d \end{bmatrix}\begin{bmatrix} 1 & 0 \\ 0 & 1 \end{bmatrix} = \begin{bmatrix} a(1) + b(0) & a(0) + b(1) \\ c(1) + d(0) & c(0) + d(1) \end{bmatrix} = \begin{bmatrix} a & b \\ c & d \end{bmatrix} = A$$

$$IA = \begin{bmatrix} 1 & 0 \\ 0 & 1 \end{bmatrix}\begin{bmatrix} a & b \\ c & d \end{bmatrix} = \begin{bmatrix} 1(a) + 0(c) & 1(b) + 0(d) \\ 0(a) + 1(c) & 0(b) + 1(d) \end{bmatrix} = \begin{bmatrix} a & b \\ c & d \end{bmatrix} = A$$

This verifies that I has been defined correctly. (It can also be shown that I is the only 2×2 identity matrix.)

The identity matrices for 3×3 matrices and 4×4 matrices, respectively, are

$$I = \begin{bmatrix} 1 & 0 & 0 \\ 0 & 1 & 0 \\ 0 & 0 & 1 \end{bmatrix} \quad \text{and} \quad I = \begin{bmatrix} 1 & 0 & 0 & 0 \\ 0 & 1 & 0 & 0 \\ 0 & 0 & 1 & 0 \\ 0 & 0 & 0 & 1 \end{bmatrix}.$$

By generalizing, an identity matrix can be found for any n by n matrix: this identity matrix will have 1's on the main diagonal from upper left to lower right, with all other entries equal to 0.

Let $A = \begin{bmatrix} 3 & -2 \\ 4 & -1 \end{bmatrix}$ and $I = \begin{bmatrix} 1 & 0 \\ 0 & 1 \end{bmatrix}$.

Find IA and AI.

Answer:

$IA = \begin{bmatrix} 3 & -2 \\ 4 & -1 \end{bmatrix} = A$ and

$AI = \begin{bmatrix} 3 & -2 \\ 4 & -1 \end{bmatrix} = A$

TECHNOLOGY TIP An $n \times n$ identity matrix can be displayed on most graphing calculators by using IDENTITY *n* or IDENT *n* or IDENMAT(*n*). Look in the MATH or OPS submenu of the TI MATRIX menu, or the OPTN MAT menu of Casio, or the MATRIX submenu of the HP-39 MATH menu.

Recall that for every nonzero real number a, the equation $ax = 1$ has a solution, namely, $x = 1/a = a^{-1}$. Similarly, for a square matrix A, we consider the matrix equation $AX = I$. This equation does not always have a solution, but when it does, we use special terminology. If there is a matrix A^{-1} satisfying

$$AA^{-1} = I,$$

(that is, A^{-1} is a solution of $AX = I$), then A^{-1} is called the **inverse matrix** of A. In this case it can be proved that $A^{-1}A = I$ and that A^{-1} is unique (that is, a square matrix has no more than one inverse). When a matrix has an inverse, it can be found by using the row operations given in Section 6.2, as we shall see below.

CAUTION Only square matrices have inverses, but not every square matrix has one. A matrix that does not have an inverse is called a **singular matrix.** Note that the symbol A^{-1} (read A-inverse) does *not* mean $1/A$ or I/A; the symbol A^{-1} is just the notation for the inverse of matrix A. There is no such thing as matrix division.

Given $A = \begin{bmatrix} 2 & 3 \\ 1 & 8 \end{bmatrix}$ and $B = \begin{bmatrix} -1 & 3 \\ 1 & -2 \end{bmatrix}$, determine if they are inverses.

Answer:

No, because $AB \neq I$.

EXAMPLE 5 Given matrices A and B below, determine whether B is the inverse of A.

$$A = \begin{bmatrix} 1 & 2 \\ 4 & 6 \end{bmatrix} \qquad B = \begin{bmatrix} -3 & 1 \\ 2 & -\frac{1}{2} \end{bmatrix}$$

B is the inverse of A if $AB = I$ and $BA = I$, so we find those products.

$$AB = \begin{bmatrix} 1 & 2 \\ 4 & 6 \end{bmatrix}\begin{bmatrix} -3 & 1 \\ 2 & -\frac{1}{2} \end{bmatrix} = \begin{bmatrix} 1 & 0 \\ 0 & 1 \end{bmatrix} = I$$

$$BA = \begin{bmatrix} -3 & 1 \\ 2 & -\frac{1}{2} \end{bmatrix}\begin{bmatrix} 1 & 2 \\ 4 & 6 \end{bmatrix} = \begin{bmatrix} 1 & 0 \\ 0 & 1 \end{bmatrix} = I$$

Therefore, B is inverse of A, that is, $A^{-1} = B$. (It is also true that A is the inverse of B, that is, $B^{-1} = A$.) ◀ 7

To see how to find the multiplicative inverse of a matrix, let us look for the inverse of

$$A = \begin{bmatrix} 2 & 4 \\ 1 & -1 \end{bmatrix}.$$

Let the unknown inverse matrix be

$$A^{-1} = \begin{bmatrix} x & y \\ z & w \end{bmatrix}.$$

By the definition of matrix inverse, $AA^{-1} = I$, or

$$AA^{-1} = \begin{bmatrix} 2 & 4 \\ 1 & -1 \end{bmatrix} \begin{bmatrix} x & y \\ z & w \end{bmatrix} = \begin{bmatrix} 1 & 0 \\ 0 & 1 \end{bmatrix}.$$

Use matrix multiplication to get

$$\begin{bmatrix} 2x + 4z & 2y + 4w \\ x - z & y - w \end{bmatrix} = \begin{bmatrix} 1 & 0 \\ 0 & 1 \end{bmatrix}.$$

Setting corresponding elements equal gives the system of equations

$$2x + 4z = 1 \qquad (1)$$
$$2y + 4w = 0 \qquad (2)$$
$$x - z = 0 \qquad (3)$$
$$y - w = 1. \qquad (4)$$

Since equations (1) and (3) involve only x and z, while equations (2) and (4) involve only y and w, these four equations lead to two systems of equations,

$$\begin{aligned} 2x + 4z &= 1 \\ x - z &= 0 \end{aligned} \quad \text{and} \quad \begin{aligned} 2y + 4w &= 0 \\ y - w &= 1. \end{aligned}$$

Writing the two systems as augmented matrices gives

$$\left[\begin{array}{cc|c} 2 & 4 & 1 \\ 1 & -1 & 0 \end{array}\right] \quad \text{and} \quad \left[\begin{array}{cc|c} 2 & 4 & 0 \\ 1 & -1 & 1 \end{array}\right].$$

Note that the row operations needed to transform both matrices are the same because the first two columns of both matrices are identical. Consequently, we can save time by combining these matrices into the single matrix

$$\left[\begin{array}{cc|cc} 2 & 4 & 1 & 0 \\ 1 & -1 & 0 & 1 \end{array}\right] \qquad (5)$$

Columns 1–3 represent the first system and columns 1, 2, and 4 represent the second system. Now use row operations as follows.

$$\left[\begin{array}{cc|cc} 1 & -1 & 0 & 1 \\ 2 & 4 & 1 & 0 \end{array}\right] \quad \text{Interchange } R_1, R_2$$

$$\left[\begin{array}{cc|cc} 1 & -1 & 0 & 1 \\ 0 & 6 & 1 & -2 \end{array}\right] \quad -2R_1 + R_2$$

$$\left[\begin{array}{cc|cc} 1 & -1 & 0 & 1 \\ 0 & 1 & \frac{1}{6} & -\frac{1}{3} \end{array}\right] \quad \frac{1}{6}R_2$$

$$\left[\begin{array}{cc|cc} 1 & 0 & \frac{1}{6} & \frac{2}{3} \\ 0 & 1 & \frac{1}{6} & -\frac{1}{3} \end{array}\right] \quad R_2 + R_1 \qquad (6)$$

The left half of the augmented matrix (6) is the identity matrix, so the Gauss-Jordan process is finished and the solutions can be read from the right half of the augmented matrix. The numbers in the first column to the right of the vertical bar give the values of x and z. The second column to the right of the bar gives the values of y and w. That is,

(a) Find A^{-1} if

$$A = \begin{bmatrix} 2 & 2 \\ 4 & 1 \end{bmatrix}.$$

(b) Check your answer by finding AA^{-1}.

Answers:

(a) $\begin{bmatrix} -\frac{1}{6} & \frac{1}{3} \\ \frac{2}{3} & -\frac{1}{3} \end{bmatrix}$

(b) $\begin{bmatrix} 1 & 0 \\ 0 & 1 \end{bmatrix}$

$$\left[\begin{array}{cc|cc} 1 & 0 & x & y \\ 0 & 1 & z & w \end{array}\right] = \left[\begin{array}{cc|cc} 1 & 0 & \frac{1}{6} & \frac{2}{3} \\ 0 & 1 & \frac{1}{6} & -\frac{1}{3} \end{array}\right]$$

so that

$$A^{-1} = \begin{bmatrix} x & y \\ z & w \end{bmatrix} = \begin{bmatrix} \frac{1}{6} & \frac{2}{3} \\ \frac{1}{6} & -\frac{1}{3} \end{bmatrix}.$$

Check by multiplying A and A^{-1}. The result should be I.

$$AA^{-1} = \begin{bmatrix} 2 & 4 \\ 1 & -1 \end{bmatrix}\begin{bmatrix} \frac{1}{6} & \frac{2}{3} \\ \frac{1}{6} & -\frac{1}{3} \end{bmatrix} = \begin{bmatrix} \frac{1}{3} + \frac{2}{3} & \frac{4}{3} - \frac{4}{3} \\ \frac{1}{6} - \frac{1}{6} & \frac{2}{3} + \frac{1}{3} \end{bmatrix} = \begin{bmatrix} 1 & 0 \\ 0 & 1 \end{bmatrix} = I$$ (8)

Thus, the original augmented matrix (5) has A as its left half and the identity matrix as its right half, while the final augmented matrix (6), at the end of the Gauss-Jordan process, has the identity matrix as its left half and the inverse matrix A^{-1} as its right half.

$$[A \mid I] \rightarrow [I \mid A^{-1}]$$

This procedure for finding the inverse of a matrix can be generalized as follows.

To obtain an **inverse matrix** A^{-1} for any $n \times n$ matrix A for which A^{-1} exists, follow these steps.

1. Form the augmented matrix $[A \mid I]$ where I is the $n \times n$ identity matrix.
2. Perform row operations on $[A \mid I]$ to get a matrix of the form $[I \mid B]$.
3. Matrix B is A^{-1}.

EXAMPLE 6 Find A^{-1} if $A = \begin{bmatrix} 1 & 0 & 1 \\ 2 & -2 & -1 \\ 3 & 0 & 0 \end{bmatrix}$.

First write the augmented matrix $[A \mid I]$.

$$[A \mid I] = \left[\begin{array}{ccc|ccc} 1 & 0 & 1 & 1 & 0 & 0 \\ 2 & -2 & -1 & 0 & 1 & 0 \\ 3 & 0 & 0 & 0 & 0 & 1 \end{array}\right]$$

The augmented matrix already has 1 in the upper left-hand corner as needed, so begin by selecting the row operation that will result in a 0 for the first element in row two. Multiply row one by -2 and add the result to row two. This gives

$$\left[\begin{array}{ccc|ccc} 1 & 0 & 1 & 1 & 0 & 0 \\ 0 & -2 & -3 & -2 & 1 & 0 \\ 3 & 0 & 0 & 0 & 0 & 1 \end{array}\right]. \quad -2R_1 + R_2$$

Get 0 for the first element in row three by multiplying row one by -3 and adding to row three as directed in Side Problem 9. (9)

9

(a) Complete this step.

(b) Write this row transformation as ______.

Answers:

(a)

$$\left[\begin{array}{ccc|ccc} 1 & 0 & 1 & 1 & 0 & 0 \\ 0 & -2 & -3 & -2 & 1 & 0 \\ 0 & 0 & -3 & -3 & 0 & 1 \end{array}\right]$$

(b) $-3R_1 + R_3$

10

(a) Complete these steps.

(b) Write these row transformations as ______.

Answers:

(a)

$$\left[\begin{array}{ccc|ccc} 1 & 0 & 0 & 0 & 0 & \frac{1}{3} \\ 0 & 1 & 0 & -\frac{1}{2} & -\frac{1}{2} & \frac{1}{2} \\ 0 & 0 & 1 & 1 & 0 & -\frac{1}{3} \end{array}\right]$$

(b) $-1R_3 + R_1$; $-\frac{3}{2}R_3 + R_2$

Get 1 for the second element in row two by multiplying row two of the matrix found in problem 9 at the side by $-1/2$, obtaining the new matrix

$$\left[\begin{array}{ccc|ccc} 1 & 0 & 1 & 1 & 0 & 0 \\ 0 & 1 & \frac{3}{2} & 1 & -\frac{1}{2} & 0 \\ 0 & 0 & -3 & -3 & 0 & 1 \end{array}\right]. \qquad -\frac{1}{2}R_2$$

Get 1 for the third element in row three by multiplying row three by $-1/3$, with the result

$$\left[\begin{array}{ccc|ccc} 1 & 0 & 1 & 1 & 0 & 0 \\ 0 & 1 & \frac{3}{2} & 1 & -\frac{1}{2} & 0 \\ 0 & 0 & 1 & 1 & 0 & -\frac{1}{3} \end{array}\right]. \qquad -\frac{1}{3}R_3$$

Now do side problem 10 to get 0's for the third elements in rows one and two. (10)

The answer for side problem 10(a) gives the desired inverse:

$$A^{-1} = \begin{bmatrix} 0 & 0 & \frac{1}{3} \\ -\frac{1}{2} & -\frac{1}{2} & \frac{1}{2} \\ 1 & 0 & -\frac{1}{3} \end{bmatrix}.$$

Verify that AA^{-1} is I. ◀

A graphing calculator provides a fast way to find matrix inverses.

EXAMPLE 7 Use a graphing calculator to find the inverse of the following matrices (if they have inverses).

$$A = \begin{bmatrix} 2 & 1 & -1 \\ 1 & 3 & 2 \\ 1 & 1 & 1 \end{bmatrix} \quad \text{and} \quad B = \begin{bmatrix} 2 & 4 \\ 3 & 6 \end{bmatrix}$$

Enter matrix A in the calculator (Figure 6.16(a)). Then use the x^{-1} key to find the inverse matrix, as in Figure 6.16(b). (Using $\wedge$ and -1 for the inverse results in an error message on most calculators.)

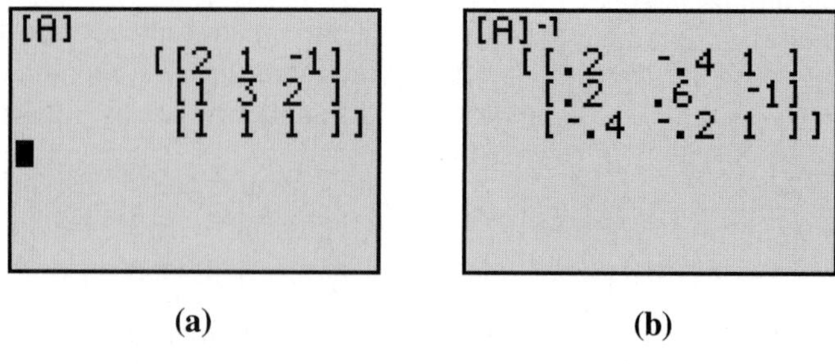

FIGURE 6.16

Use a graphing calculator to find the inverses of these matrices (if they exist).

(a) $A = \begin{bmatrix} 1 & 2 & 3 \\ 4 & -1 & 0 \\ 5 & 1 & 3 \end{bmatrix}$

(b) $B = \begin{bmatrix} 2 & 3 \\ -1 & 4 \end{bmatrix}$

Answers:

(a) No inverse

(b) Use the FRAC key (if you have one) to simplify the answer, which is

$$B^{-1} = \begin{bmatrix} 4/11 & -3/11 \\ 1/11 & 2/11 \end{bmatrix}.$$

Now enter matrix B in the calculator and use the x^{-1} key. The result is an error message (Figure 6.17), which indicates that the matrix is singular; it does not have an inverse. (If you were working by hand, you would have found that the appropriate system of equations has no solution.) ◀ (11)

FIGURE 6.17

TECHNOLOGY TIP Because of round-off error, a graphing calculator may sometimes display an "inverse" for a matrix that doesn't actually have one. So always verify your results by multiplying A and A^{-1}. If the product is not the identity matrix, then A does not have an inverse.

6.4 Exercises

In Exercises 1–6, the sizes of two matrices A and B are given. Find the sizes of the product AB and the product BA whenever these products exist. (See Example 1.)

1. A is 2×2 and B is 2×2.
2. A is 3×3 and B is 3×3.
3. A is 3×5 and B is 5×2.
4. A is 4×3 and B is 3×6.
5. A is 4×2 and B is 3×4.
6. A is 7×3 and B is 2×7.
7. To find the product matrix AB, the number of ______ of A must be the same as the number of ______ of B.
8. The product matrix AB has the same number of ______ as A and the same number of ______ as B.

Find each of the following matrix products. (See Examples 2–4.)

9. $\begin{bmatrix} 1 & 2 \\ 3 & 4 \end{bmatrix}\begin{bmatrix} -1 \\ 7 \end{bmatrix}$

10. $\begin{bmatrix} -1 & 5 \\ 7 & 0 \end{bmatrix}\begin{bmatrix} 6 \\ 2 \end{bmatrix}$

11. $\begin{bmatrix} 2 & 2 & -1 \\ 3 & 0 & 1 \end{bmatrix}\begin{bmatrix} 0 & 2 \\ -1 & 4 \\ 0 & 2 \end{bmatrix}$

12. $\begin{bmatrix} -9 & 2 & 1 \\ 3 & 0 & 0 \end{bmatrix}\begin{bmatrix} 2 \\ -1 \\ 4 \end{bmatrix}$

13. $\begin{bmatrix} -4 & 1 \\ 2 & -3 \end{bmatrix}\begin{bmatrix} 1 & 0 \\ 0 & 1 \end{bmatrix}$

14. $\begin{bmatrix} 1 & 0 \\ 0 & 1 \end{bmatrix}\begin{bmatrix} 3 & -2 \\ 1 & -5 \end{bmatrix}$

15. $\begin{bmatrix} 1 & 0 & 0 \\ 0 & 1 & 0 \\ 0 & 0 & 1 \end{bmatrix}\begin{bmatrix} 3 & -5 & 7 \\ -2 & 1 & 6 \\ 0 & -3 & 4 \end{bmatrix}$

16. $\begin{bmatrix} -8 & 9 \\ 3 & -4 \\ -1 & 6 \end{bmatrix}\begin{bmatrix} 1 & 0 & 0 \\ 0 & 1 & 0 \end{bmatrix}$

17. $\begin{bmatrix} 1 & 2 & 3 \\ 4 & 5 & 6 \\ 7 & 8 & 9 \end{bmatrix}\begin{bmatrix} -1 & 5 \\ 7 & 0 \\ 1 & 2 \end{bmatrix}$

18. $\begin{bmatrix} -2 & 0 & 3 \\ 5 & -3 & -1 \end{bmatrix}\begin{bmatrix} 2 & 0 & -1 & 3 \\ 0 & 1 & 0 & -1 \\ 4 & 2 & 5 & -4 \end{bmatrix}$

In Exercises 19–21, use the matrices

$$A = \begin{bmatrix} -3 & -9 \\ 2 & 6 \end{bmatrix} \quad \text{and} \quad B = \begin{bmatrix} 4 & 6 \\ 2 & 3 \end{bmatrix}.$$

19. Show that $AB \neq BA$. Hence, matrix multiplication is not commutative.

20. Show that $(A + B)^2 \neq A^2 + 2AB + B^2$.

21. Show that $(A + B)(A - B) \neq A^2 - B^2$.

22. Show that $D^2 = D$, where

$$D = \begin{bmatrix} 1 & 0 & 0 \\ \frac{1}{2} & 0 & \frac{1}{2} \\ 0 & 0 & 1 \end{bmatrix}.$$

Given matrices

$$P = \begin{bmatrix} m & n \\ p & q \end{bmatrix}, \quad X = \begin{bmatrix} x & y \\ z & w \end{bmatrix}, \quad T = \begin{bmatrix} r & s \\ t & u \end{bmatrix},$$

verify that the statements in Exercises 23–26 are true.

23. $(PX)T = P(XT)$ (Associative property)

24. $P(X + T) = PX + PT$ (Distributive property)

25. $k(X + T) = kX + kT$ for any real number k

26. $(k + h)P = kP + hP$ for any real numbers k and h

Determine whether the given matrices are inverses of each other by computing their product. (See Example 5.)

27. $\begin{bmatrix} 5 & 2 \\ 3 & -1 \end{bmatrix}$ and $\begin{bmatrix} -1 & 2 \\ 3 & -4 \end{bmatrix}$

28. $\begin{bmatrix} 0 & 1 \\ 1 & 0 \end{bmatrix}$ and $\begin{bmatrix} 3 & 5 \\ 7 & 9 \end{bmatrix}$

29. $\begin{bmatrix} 3 & -1 \\ -4 & 2 \end{bmatrix}$ and $\begin{bmatrix} 1 & \frac{1}{2} \\ 2 & \frac{3}{2} \end{bmatrix}$

30. $\begin{bmatrix} 1 & 1 \\ .1 & .2 \end{bmatrix}$ and $\begin{bmatrix} 2 & -10 \\ -1 & 10 \end{bmatrix}$

31. $\begin{bmatrix} 1 & 1 & 1 \\ 2 & 3 & 0 \\ 1 & 2 & 1 \end{bmatrix}$ and $\begin{bmatrix} 1.5 & .5 & -1.5 \\ -1 & 0 & 1 \\ .5 & -2 & 2 \end{bmatrix}$

32. $\begin{bmatrix} 2 & 5 & 4 \\ 1 & 4 & 3 \\ 1 & 3 & 2 \end{bmatrix}$ and $\begin{bmatrix} 1 & 2 & 1 \\ -5 & 8 & 2 \\ 7 & -11 & -3 \end{bmatrix}$

Find the inverse, if it exists, for each of the following matrices. (See Example 6.)

33. $\begin{bmatrix} 2 & 3 \\ 1 & 2 \end{bmatrix}$

34. $\begin{bmatrix} -1 & 2 \\ 1 & -1 \end{bmatrix}$

35. $\begin{bmatrix} 2 & 4 \\ 3 & 6 \end{bmatrix}$

36. $\begin{bmatrix} -3 & -5 \\ 6 & 10 \end{bmatrix}$

37. $\begin{bmatrix} 2 & 6 \\ 1 & 4 \end{bmatrix}$

38. $\begin{bmatrix} 1 & 2 \\ 3 & 4 \end{bmatrix}$

39. $\begin{bmatrix} 1 & -1 & 1 \\ 0 & 2 & -1 \\ 2 & 3 & 0 \end{bmatrix}$

40. $\begin{bmatrix} 1 & 2 & 3 \\ 1 & 1 & 2 \\ 0 & 1 & 2 \end{bmatrix}$

41. $\begin{bmatrix} 1 & 4 & 3 \\ 1 & -3 & -2 \\ 2 & 5 & 4 \end{bmatrix}$

42. $\begin{bmatrix} 1 & 2 & 0 \\ 3 & -1 & 2 \\ -2 & 3 & -2 \end{bmatrix}$

43. $\begin{bmatrix} 1 & -3 & 4 \\ 2 & -5 & 7 \\ 0 & -1 & 1 \end{bmatrix}$

44. $\begin{bmatrix} 5 & 0 & 2 \\ 2 & 2 & 1 \\ -3 & 1 & -1 \end{bmatrix}$

Use a graphing calculator to find the inverse of each matrix.

45. $\begin{bmatrix} 2 & 4 & 6 \\ -1 & -4 & -3 \\ 0 & 1 & -1 \end{bmatrix}$

46. $\begin{bmatrix} 2 & 2 & -4 \\ 2 & 6 & 0 \\ -3 & -3 & 5 \end{bmatrix}$

47. $\begin{bmatrix} 1 & -2 & 3 & 0 \\ 0 & 1 & -1 & 1 \\ -2 & 2 & -2 & 4 \\ 0 & 2 & -3 & 1 \end{bmatrix}$

48. $\begin{bmatrix} 1 & 1 & 0 & 2 \\ 2 & -1 & 1 & -1 \\ 3 & 3 & 2 & -2 \\ 1 & 2 & 1 & 0 \end{bmatrix}$

A graphing calculator or other technology is recommended for part (c) of Exercises 49–51.

49. Social Sciences The average birth and death rates per million for several regions of the world and their populations (in millions) are given below.*

Region	*Births*	*Deaths*
Asia	.024	.008
Latin America	.025	.007
North America	.015	.009
Europe	.011	.011

Year	*Asia*	*Latin America*	*North America*	*Europe*
1970	1996	286	226	460
1980	2440	365	252	484
1990	2906	455	277	499
2000	3683	519	310	729
2025 (projected)	4723	697	364	702

(a) Write the information in the first table as a 4×2 matrix R.

(b) Write the information in the second table as a 5×4 matrix P.

(c) Find the product PR.

(d) Explain what PR represents.

(e) From matrix PR, what was the total number of births in 2000? What total number of deaths is projected for 2025?

*U.S. Bureau of Census and the United Nations Population Fund.

50. Business The first table shows the number of telecommunications devices sold (in millions) over a four-year period. The second gives the average price of each such device (in dollars) over the same period.*

Year	1998	1999	2000	2001
Cordless Phone	31.6	32.6	34.0	35.0
Answering Machine	18.3	18.9	19.5	20.1
Home Fax Machine	2.5	2.6	2.7	2.8

Year	1998	1999	2000	2001
Cordless Phone	71.20	68.71	66.47	64.51
Answering Machine	67.76	64.29	62.05	60.20
Home Fax Machine	262	240.38	229.63	219.64

(a) Write the information in the first table as a 3×4 matrix A.

(b) Write the information in the second table as a 4×3 matrix B.

(c) Find the product AB.

(d) Explain what the following entries in AB represent: row 1, column 1; row 2, column 2; row 3, column 3. What do the other entries represent?

(e) From matrix AB, what were the total sales (in dollars) of answering machines over this four-year period?

51. Health The first table shows the number of live births (in thousands). The second shows the infant mortality rates (deaths per 1000 live births) in those years.†

Year	1995	1996	1997	1998
Black	394	595	601	610
White	2753	3093	3085	3119

Year	1995	1996	1997	1998
Black	15.1	14.2	13.7	14.1
White	6.3	6.0	6.0	6.0

(a) Write the information in the first table as a 4×2 matrix C.

(b) Write the information in the second table as a 2×4 matrix D.

(c) Find the product DC.

(d) Which entry in DC gives the total number of black infant deaths in 1995–98?

(e) Which entry in DC gives the total number of white infant deaths in 1995–98?

52. Business Burger Barn's three locations sell hamburgers, fries, and soft drinks. Barn I sells 900 burgers, 600 orders of fries, and 750 soft drinks each day. Barn II sells 1500 burgers a day and Barn III sells 1150. Soft drink sales number 900 a day at Barn II and 825 a day at Barn III. Barn II sells 950 and Barn III sells 800 orders of fries per day.

(a) Write a 3×3 matrix S that displays daily sales figures for all locations.

(b) Burgers cost $1.50 each, fries $.90 an order, and soft drinks $.60 each. Write a 1×3 matrix P that displays the prices.

(c) What matrix product displays the daily revenue at each of the three locations?

(d) What is the total daily revenue from all locations?

53. Business The four departments of Stagg Enterprises need to order the following amounts of the same products.

	Paper	*Tape*	*Printer Ribbon*	*Memo Pads*	*Pens*
Department 1	10	4	3	5	6
Department 2	7	2	2	3	8
Department 3	4	5	1	0	10
Department 4	0	3	4	5	5

The unit price (in dollars) of each product is given below for two suppliers.

Product	*Supplier A*	*Supplier B*
Paper	2	3
Tape	1	1
Printer Ribbon	4	3
Memo Pads	3	3
Pens	1	2

(a) Use matrix multiplication to get a matrix showing the comparative costs for each department for the products from the two suppliers.

(b) Find the total cost to buy products from each supplier. From which supplier should the company make the purchase?

54. Business The Perulli Candy Company makes three types of chocolate candy: Cheery Cherry, Mucho Mocha, and Almond Delight. The company produces its products in San Diego, Mexico City, and Managua using two main ingredients: chocolate and sugar.

*Multimedia Telecommunications Association.

†U.S. National Center for Health Statistics.

(a) Each kilogram of Cheery Cherry requires .5 kg of sugar and .2 kg of chocolate; each kilogram of Mucho Mocha requires .4 kg of sugar and .3 kg of chocolate; and each kilogram of Almond Delight requires .3 kg of sugar and .3 kg of chocolate. Put this information into a 2×3 matrix, labeling the rows and columns.

(b) The cost of 1 kg of sugar is \$3 in San Diego, \$2 in Mexico City, and \$1 in Managua. The cost of 1 kg of chocolate is \$3 in San Diego, \$3 in Mexico City, and \$4 in Managua. Put this information into a matrix in such a way that when you multiply it with your matrix from part (a), you get a matrix representing the ingredient cost of producing each type of candy in each city.

(c) Multiply the matrices in parts (a) and (b), labeling the product matrix.

(d) From part (c) what is the combined sugar-and-chocolate cost to produce 1 kg of Mucho Mocha in Managua?

(e) Perulli Candy needs to quickly produce a special shipment of 100 kg of Cheery Cherry, 200 kg of Mucho Mocha, and 500 kg of Almond Delight, and it decides to select one factory to fill the entire order. Use matrix multiplication to determine in which city the total sugar-and-chocolate cost to produce the order is the smallest.

6.5 APPLICATIONS OF MATRICES

This section gives a variety of applications of matrices.

(1)

Write the matrix of coefficients, the matrix of variables, and the matrix of constants for the system

$$2x + 6y = -14$$
$$-x - 2y = 3.$$

Answers:

$$A = \begin{bmatrix} 2 & 6 \\ -1 & -2 \end{bmatrix},$$

$$X = \begin{bmatrix} x \\ y \end{bmatrix},$$

$$B = \begin{bmatrix} -14 \\ 3 \end{bmatrix}.$$

Solving Systems with Matrices

Consider this system of linear equations.

$$2x - 3y = 4$$
$$x + 5y = 2$$

Let

$$A = \begin{bmatrix} 2 & -3 \\ 1 & 5 \end{bmatrix}, \quad X = \begin{bmatrix} x \\ y \end{bmatrix}, \quad B = \begin{bmatrix} 4 \\ 2 \end{bmatrix}.$$

Since

$$AX = \begin{bmatrix} 2 & -3 \\ 1 & 5 \end{bmatrix}\begin{bmatrix} x \\ y \end{bmatrix} = \begin{bmatrix} 2x - 3y \\ x + 5y \end{bmatrix} \quad \text{and} \quad B = \begin{bmatrix} 4 \\ 2 \end{bmatrix},$$

the original system is equivalent to the single matrix equation $AX = B$. Similarly, any system of linear equations can be written as a matrix equation $AX = B$. The matrix A is called the **coefficient matrix.** (1)

A matrix equation $AX = B$ can be solved if A^{-1} exists. Assuming A^{-1} exists and using the facts that $A^{-1}A = I$ and $IX = X$ along with the associative property of multiplication of matrices gives

$$\begin{aligned} AX &= B \\ A^{-1}(AX) &= A^{-1}B && \text{Multiply both sides by } A^{-1}. \\ (A^{-1}A)X &= A^{-1}B && \text{Associative property} \\ IX &= A^{-1}B && \text{Inverse property} \\ X &= A^{-1}B. && \text{Identity property} \end{aligned}$$

When multiplying by matrices on both sides of a matrix equation, be careful to multiply in the same order on both sides of the equation, since multiplication of matrices is not commutative (unlike multiplication of real numbers). This discussion is summarized below.

A system of equations $AX = B$, where A is the matrix of coefficients, X is the matrix of variables, and B is the matrix of constants, is solved by first finding A^{-1}. Then, if A^{-1} exists, $X = A^{-1}B$.

This method is very efficient when technology is available. The elimination method (the matrix version) or the Gauss-Jordan method is the best choice when solving a system by hand.

EXAMPLE 1 Use the inverse of the coefficient matrix to solve the system

$$\begin{aligned} -x - 2y + 2z &= 9 \\ 2x + y - z &= -3 \\ 3x - 2y + z &= -6. \end{aligned}$$

We use the fact that $X = A^{-1}B$, with

$$A = \begin{bmatrix} -1 & -2 & 2 \\ 2 & 1 & -1 \\ 3 & -2 & 1 \end{bmatrix} \quad \text{and} \quad B = \begin{bmatrix} 9 \\ -3 \\ -6 \end{bmatrix}.$$

Using either the method explained in Section 6.4 or a graphing calculator,

$$A^{-1} = \begin{bmatrix} \frac{1}{3} & \frac{2}{3} & 0 \\ \frac{5}{3} & \frac{7}{3} & -1 \\ \frac{7}{3} & \frac{8}{3} & -1 \end{bmatrix}$$

and

$$X = A^{-1}B = \begin{bmatrix} \frac{1}{3} & \frac{2}{3} & 0 \\ \frac{5}{3} & \frac{7}{3} & -1 \\ \frac{7}{3} & \frac{8}{3} & -1 \end{bmatrix} \begin{bmatrix} 9 \\ -3 \\ -6 \end{bmatrix} = \begin{bmatrix} 1 \\ 14 \\ 19 \end{bmatrix}.$$

Thus, the solution is (1, 14, 19). ◀ (2)

(2)

Use the inverse matrix to solve the system in Example 1, if the constants for the three equations are 12, 0, and 8, respectively.

Answer:

(4, 12, 20)

EXAMPLE 2 Use the inverse of the coefficient matrix to solve the system

$$\begin{aligned} x + 1.5y &= 8 \\ 2x + \ \ 3y &= 10. \end{aligned}$$

The coefficient matrix is $A = \begin{bmatrix} 1 & 1.5 \\ 2 & 3 \end{bmatrix}$. A graphing calculator will indicate that A^{-1} does not exist. If we try to carry out the row operations, we see why.

$$\left[\begin{array}{cc|cc} 1 & 1.5 & 1 & 0 \\ 2 & 3 & 0 & 1 \end{array}\right]$$

$$\left[\begin{array}{cc|cc} 1 & 1.5 & 1 & 0 \\ 0 & 0 & -2 & 1 \end{array}\right]$$

The next step cannot be performed because of the zero in the second row, second column. Verify that the original system has no solution. ◀ (3)

(3)

Solve the system in Example 2 if the constants are, respectively, 3 and 6.

Answer:

$(3 - 1.5y, y)$ for all real numbers y

Input-Output Analysis

An interesting application of matrix theory to economics was developed by Nobel Prize winner Wassily Leontief. His application of matrices to the interdependencies in an economy is called **input-output** analysis. In practice, input-output analysis is very complicated with many variables. We shall discuss only simple examples with a few variables.

Input-output models are concerned with the production and flow of goods (and perhaps services). In an economy with n basic commodities (or sectors), the production of each commodity uses some (perhaps all) of the commodities in the economy as inputs. The amounts of each commodity used in the production of 1 unit of each commodity can be written as an $n \times n$ matrix A, called the **technological** or **input-output matrix** of the economy.

EXAMPLE 3 Suppose a simplified economy involves just three commodity categories: agriculture, manufacturing, and transportation, all in appropriate units. Production of 1 unit of agriculture requires 1/2 unit of manufacturing and 1/4 unit of transportation. Production of 1 unit of manufacturing requires 1/4 unit of agriculture and 1/4 unit of transportation; while production of 1 unit of transportation requires 1/3 unit of agriculture and 1/4 unit of manufacturing. Write the input-output matrix of this economy.

The matrix is shown below.

		Output Agriculture	Manufacturing	Transportation
Input	Agriculture	0	$\frac{1}{4}$	$\frac{1}{3}$
	Manufacturing	$\frac{1}{2}$	0	$\frac{1}{4}$
	Transportation	$\frac{1}{4}$	$\frac{1}{4}$	0

$$\begin{bmatrix} 0 & \frac{1}{4} & \frac{1}{3} \\ \frac{1}{2} & 0 & \frac{1}{4} \\ \frac{1}{4} & \frac{1}{4} & 0 \end{bmatrix} = A$$

The first column of the input-output matrix represents the amount of each of the three commodities consumed in the production of 1 unit of agriculture. The second column gives the corresponding amounts required to produce 1 unit of manufacturing, and the last column gives the amounts needed to produce 1 unit of transportation. (Although it is unrealistic, perhaps, that production of 1 unit of a commodity requires none of that commodity, the simpler matrix involved is useful for our purposes.) ◀ (4)

(4)

Write a 2×2 technological matrix in which 1 unit of electricity requires 1/2 unit of water and 1/3 unit of electricity, while 1 unit of water requires no water but 1/4 unit of electricity.

Answer:

$$\begin{array}{c} \\ \text{Elec.} \\ \text{Water} \end{array} \begin{array}{c} \begin{array}{cc} \text{Elec.} & \text{Water} \end{array} \\ \begin{bmatrix} \frac{1}{3} & \frac{1}{4} \\ \frac{1}{2} & 0 \end{bmatrix} \end{array}$$

Another matrix used with the input-output matrix is a matrix giving the amount of each commodity produced, called the **production matrix,** or the **vector of gross output.** In an economy producing n commodities, the production matrix can be represented by a column matrix x with entries $x_1, x_2, x_3, \ldots, x_n$.

EXAMPLE 4 In Example 3, suppose the production matrix is

$$X = \begin{bmatrix} 60 \\ 52 \\ 48 \end{bmatrix}.$$

Then 60 units of agriculture, 52 units of manufacturing, and 48 units of transportation are produced. As 1/4 unit of agriculture is used for each unit of manufacturing produced, $1/4 \times 52 = 13$ units of agriculture must be used up in the "production" of manufacturing. Similarly, $1/3 \times 48 = 16$ units of agriculture will be used up in the "production" of transportation. Thus $13 + 16 = 29$ units of agriculture are used for production in the economy. Look again at the matrices A and X. Since X gives the number of units of each commodity produced and A gives the amount (in units) of each commodity used to produce 1 unit of the various commodities, the matrix product AX gives the amount of each commodity used up in production.

$$AX = \begin{bmatrix} 0 & \frac{1}{4} & \frac{1}{3} \\ \frac{1}{2} & 0 & \frac{1}{4} \\ \frac{1}{4} & \frac{1}{4} & 0 \end{bmatrix} \begin{bmatrix} 60 \\ 52 \\ 48 \end{bmatrix} = \begin{bmatrix} 29 \\ 42 \\ 28 \end{bmatrix}$$

This product shows that 29 units of agriculture, 42 units of manufacturing, and 28 units of transportation are used to produce 60 units of agriculture, 52 units of manufacturing, and 48 units of transportation. ◀

(5)

(a) Write a 2×1 matrix X to represent gross production of 9000 units of electricity and 12,000 units of water.

(b) Find AX using A from the last side problem.

(c) Find D using $D = X - AX$.

Answers:

(a) $\begin{bmatrix} 9000 \\ 12{,}000 \end{bmatrix}$

(b) $\begin{bmatrix} 6000 \\ 4500 \end{bmatrix}$

(c) $\begin{bmatrix} 3000 \\ 7500 \end{bmatrix}$

We have seen that the matrix product AX represents the amount of each commodity used in the production process. The remainder (if any) must be enough to satisfy the demand for the various commodities from outside the production system. In an n-commodity economy, this demand can be represented by a **demand matrix** D with entries $d_1, d_2, \ldots, d_n$. The difference between the production matrix, X, and the amount, AX, used in the production process must equal the demand, D, or

$$D = X - AX.$$

In Example 4,

$$D = \begin{bmatrix} 60 \\ 52 \\ 48 \end{bmatrix} - \begin{bmatrix} 29 \\ 42 \\ 28 \end{bmatrix} = \begin{bmatrix} 31 \\ 10 \\ 20 \end{bmatrix}.$$

This result shows that production of 60 units of agriculture, 52 units of manufacturing, and 48 units of transportation would satisfy a demand of 31, 10, and 20 units of each, respectively. (5)

In practice, A and D usually are known and X must be found. That is, we need to decide what amounts of production are necessary to satisfy the required demands. Matrix algebra can be used to solve the equation $D = X - AX$ for X.

$$\begin{aligned} D &= X - AX \\ D &= IX - AX && \text{Identity property} \\ D &= (I - A)X && \text{Distributive property} \end{aligned}$$

If the matrix $I - A$ has an inverse, then

$$\boldsymbol{X = (I - A)^{-1}D.}$$

EXAMPLE 5 Suppose, in the 3-commodity economy of Examples 3 and 4, there is a demand for 516 units of agriculture, 258 units of manufacturing, and 129 units of transportation. What should production of each commodity be?

The demand matrix is

$$D = \begin{bmatrix} 516 \\ 258 \\ 129 \end{bmatrix}.$$

Find the production matrix by first calculating $I - A$.

$$I - A = \begin{bmatrix} 1 & 0 & 0 \\ 0 & 1 & 0 \\ 0 & 0 & 1 \end{bmatrix} - \begin{bmatrix} 0 & \frac{1}{4} & \frac{1}{3} \\ \frac{1}{2} & 0 & \frac{1}{4} \\ \frac{1}{4} & \frac{1}{4} & 0 \end{bmatrix} = \begin{bmatrix} 1 & -\frac{1}{4} & -\frac{1}{3} \\ -\frac{1}{2} & 1 & -\frac{1}{4} \\ -\frac{1}{4} & -\frac{1}{4} & 1 \end{bmatrix}$$

Using a calculator with matrix capability or row operations, find the inverse of $I - A$.

$$(I - A)^{-1} = \begin{bmatrix} 1.3953 & .4961 & .5891 \\ .8372 & 1.3643 & .6202 \\ .5581 & .4651 & 1.3023 \end{bmatrix}$$

(The entries are rounded to four decimal places.)* Since $X = (I - A)^{-1}D$,

$$X = \begin{bmatrix} 1.3953 & .4961 & .5891 \\ .8372 & 1.3643 & .6202 \\ .5581 & .4651 & 1.3023 \end{bmatrix} \begin{bmatrix} 516 \\ 258 \\ 129 \end{bmatrix} = \begin{bmatrix} 924 \\ 864 \\ 576 \end{bmatrix}$$

(rounded to the nearest whole numbers).

From the last result, we see that production of 924 units of agriculture, 864 units of manufacturing, and 576 units of transportation is required to satisfy demands of 516, 258, and 129 units, respectively. ◀

EXAMPLE 6 An economy depends on two basic products, wheat and oil. To produce 1 metric ton of wheat requires .25 metric ton of wheat and .33 metric

---

*Although we show the matrix $(I - A)^{-1}$ with entries rounded to four decimal places, we did not round off in calculating $(I - A)^{-1}D$. If the rounded figures are used, the numbers in the product may vary slightly in the last digit.

ton of oil. Production of 1 metric ton of oil consumes .08 metric ton of wheat and .11 metric ton of oil. Find the production that will satisfy a demand of 500 metric tons of wheat and 1000 metric tons of oil.

The input-output matrix, A, and the matrix $I - A$ are

$$A = \begin{bmatrix} .25 & .08 \\ .33 & .11 \end{bmatrix} \quad \text{and} \quad I - A = \begin{bmatrix} .75 & -.08 \\ -.33 & .89 \end{bmatrix}.$$

Next, calculate $(I - A)^{-1}$.

$$(I - A)^{-1} = \begin{bmatrix} 1.3882 & .1248 \\ .5147 & 1.1699 \end{bmatrix} \quad \text{(rounded)}$$

To find the production matrix X, use the equation $X = (I - A)^{-1}D$, with

$$D = \begin{bmatrix} 500 \\ 1000 \end{bmatrix}.$$

The production matrix is

$$X = \begin{bmatrix} 1.3882 & .1248 \\ .5147 & 1.1699 \end{bmatrix} \begin{bmatrix} 500 \\ 1000 \end{bmatrix} = \begin{bmatrix} 819 \\ 1427 \end{bmatrix}.$$

The production numbers were rounded to the nearest whole number. Production of 819 metric tons of wheat and 1427 metric tons of oil is required to satisfy the indicated demand. ◀ (6)

(6)

A simple economy depends on just two products, beer and pretzels.

(a) Suppose 1/2 unit of beer and 1/2 unit of pretzels are needed to make 1 unit of beer, and 3/4 unit of beer is needed to make 1 unit of pretzels. Write the technological matrix A for the economy.

(b) Find $I - A$.

(c) Find $(I - A)^{-1}$.

(d) Find the gross production X that will be needed to get a net production of

$$D = \begin{bmatrix} 100 \\ 1000 \end{bmatrix}.$$

Answers:

(a) $\begin{bmatrix} \frac{1}{2} & \frac{3}{4} \\ \frac{1}{2} & 0 \end{bmatrix}$

(b) $\begin{bmatrix} \frac{1}{2} & -\frac{3}{4} \\ -\frac{1}{2} & 1 \end{bmatrix}$

(c) $\begin{bmatrix} 8 & 6 \\ 4 & 4 \end{bmatrix}$

(d) $\begin{bmatrix} 6800 \\ 4400 \end{bmatrix}$

TECHNOLOGY TIP If you are using a graphing calculator to determine X, you can calculate $(I - A)^{-1}D$ in one step without finding the intermediate matrices $I - A$ and $(I - A)^{-1}$.

Code Theory

Governments need sophisticated methods of coding and decoding messages. One example of such an advanced code uses matrix theory. Such a code takes the letters in the words and divides them into groups. (Each space between words is treated as a letter; punctuation is disregarded.) Then, numbers are assigned to the letters of the alphabet. For our purposes, let the letter *a* correspond to 1, *b* to 2, and so on. Let the number 27 correspond to a space between words.

For example, the message

mathematics is for the birds

can be divided into groups of three letters each.

mat hem ati cs– is– for –th e–b ird s––

(We used – to represent a space between words.) We now write a column matrix for each group of three symbols using the corresponding numbers, as determined above, instead of letters. For example, the letters *mat* can be encoded as

$$\begin{bmatrix} 13 \\ 1 \\ 20 \end{bmatrix}.$$

Write the message "*when*" using 2×1 matrices.

Answer:

$$\begin{bmatrix} 23 \\ 8 \end{bmatrix}, \begin{bmatrix} 5 \\ 14 \end{bmatrix}$$

Use the matrix given below to find the 2×1 matrices to be transmitted for the message you encoded in side problem 7.

$$\begin{bmatrix} 2 & 1 \\ 5 & 0 \end{bmatrix}$$

Answer:

$$\begin{bmatrix} 54 \\ 115 \end{bmatrix}, \begin{bmatrix} 24 \\ 25 \end{bmatrix}$$

The coded message then consists of the 3×1 column matrices:

$$\begin{bmatrix} 13 \\ 1 \\ 20 \end{bmatrix}, \begin{bmatrix} 8 \\ 5 \\ 13 \end{bmatrix}, \begin{bmatrix} 1 \\ 20 \\ 9 \end{bmatrix}, \begin{bmatrix} 3 \\ 19 \\ 27 \end{bmatrix}, \begin{bmatrix} 9 \\ 19 \\ 27 \end{bmatrix}, \begin{bmatrix} 6 \\ 15 \\ 18 \end{bmatrix}, \begin{bmatrix} 27 \\ 20 \\ 8 \end{bmatrix}, \begin{bmatrix} 5 \\ 27 \\ 2 \end{bmatrix}, \begin{bmatrix} 9 \\ 18 \\ 4 \end{bmatrix}, \begin{bmatrix} 19 \\ 27 \\ 27 \end{bmatrix}.$$ (7)

We can further complicate the code by choosing a matrix that has an inverse (in this case a 3×3 matrix, call it M) and finding the products of this matrix and each of the above column matrices. The size of each group, the assignment of numbers to letters, and the choice of matrix M must all be predetermined.

Suppose we choose

$$M = \begin{bmatrix} 1 & 3 & 3 \\ 1 & 4 & 3 \\ 1 & 3 & 4 \end{bmatrix}.$$

If we find the products of M and the column matrices above, we have a new set of column matrices,

$$\begin{bmatrix} 1 & 3 & 3 \\ 1 & 4 & 3 \\ 1 & 3 & 4 \end{bmatrix}\begin{bmatrix} 13 \\ 1 \\ 20 \end{bmatrix} = \begin{bmatrix} 76 \\ 77 \\ 96 \end{bmatrix}, \quad \text{and so on.}$$

The entries of these matrices can then be transmitted to an agent as the message 76, 77, 96, and so on. (8)

When the agent receives the message, it is divided into groups of three numbers with each group formed into a column matrix. After multiplying each column matrix by the matrix M^{-1}, the message can be read. For example,

$$M^{-1} \cdot \begin{bmatrix} 76 \\ 77 \\ 96 \end{bmatrix} = \begin{bmatrix} 13 \\ 1 \\ 20 \end{bmatrix}.$$

Although this type of code is relatively simple, it is actually difficult to break. Many complications are possible. For example, a long message might be placed in groups of 20, thus requiring a 20×20 matrix for coding and decoding. Finding the inverse of such a matrix would require an impractical amount of time if calculated by hand. For this reason some of the largest computers are used by government agencies involved in coding.

Routing

The diagram in Figure 6.18 shows the roads connecting four cities. Another way of representing this information is shown in matrix A, where the entries represent the number of roads connecting two cities without passing through another city.* For example, from the diagram we see that there are two roads connecting city 1 to city

*From *Matrices with Applications,* section 3.2, example 5, by Hugh G. Campbell. Copyright © 1968, pp. 50–51. Adapted by permission of Prentice-Hall, Englewood Cliffs, New Jersey.

4 without passing through either city 2 or 3. This information is entered in row one, column four and again in row four, column one, of matrix A.

$$A = \begin{bmatrix} 0 & 1 & 2 & 2 \\ 1 & 0 & 1 & 0 \\ 2 & 1 & 0 & 1 \\ 2 & 0 & 1 & 0 \end{bmatrix}$$

Note that there are zero roads connecting each city to itself. Also, there is one road connecting cities 3 and 2.

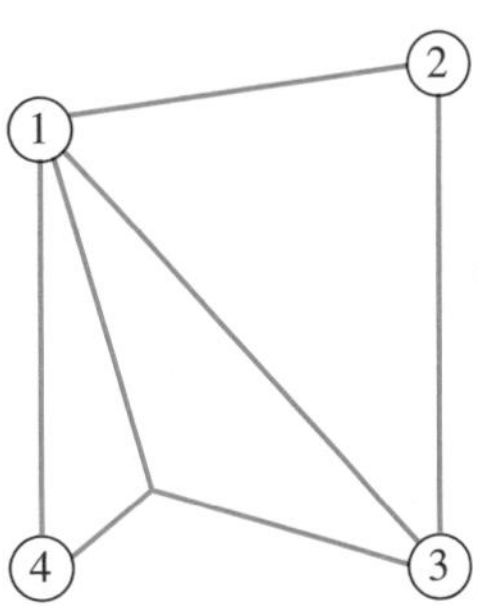

FIGURE 6.18

9

Use a graphing calculator to find the following

(a) A^3

(b) $A + A^2$

Answers:

(a)

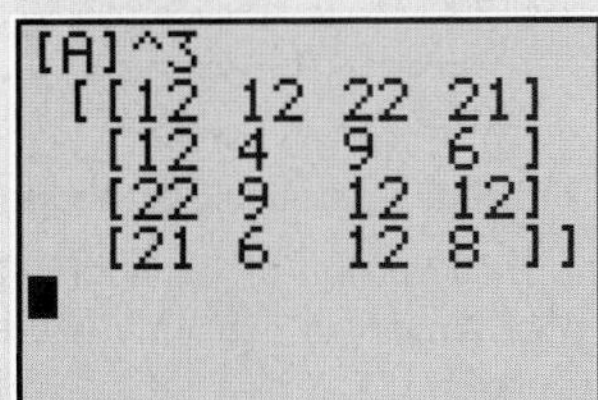

(b)

[A]+[A]²
[[9 3 5 4]
[3 2 3 3]
[5 3 6 5]
[4 3 5 5]]

How many ways are there to go from city 1 to city 2, for example, by going through exactly one other city? Because we must go through one other city, we must go through either city 3 or city 4. On the diagram in Figure 6.18, we see that we can go from city 1 to city 2 through city 3 in two ways. We can go from city 1 to city 3 in two ways and then from city 3 to city 2 in one way, so there are $2 \cdot 1 = 2$ ways to get from city 1 to city 2 through city 3. It is not possible to go from city 1 to city 2 through city 4, because there is no direct route between cities 4 and 2.

The matrix A^2 gives the number of ways to travel between any two cities by passing through exactly one other city. Multiply matrix A by itself, to get A^2. Let the first row, second column, entry of A^2 be b_{12}. (We use a_{ij} to denote the entry in the ith row and jth column of matrix A.) The entry b_{12} is found as follows.

$$\begin{aligned} b_{12} &= a_{11}a_{12} + a_{12}a_{22} + a_{13}a_{32} + a_{14}a_{42} \\ &= 0 \cdot 1 + 1 \cdot 0 + 2 \cdot 1 + 2 \cdot 0 \\ &= 2 \end{aligned}$$

The first product $0 \cdot 1$ in the calculations above represents the number of ways to go from city 1 to city 1 (that is, 0) and then from city 1 to city 2 (that is, 1). The 0 result indicates that such a trip does not involve a third city. The only nonzero product $(2 \cdot 1)$ represents the two routes from city 1 to city 3 and the one route from city 3 to city 2, which result in the $2 \cdot 1$ or 2 routes from city 1 to city 2 by going through city 3.

Similarly, A^3 gives the number of ways to travel between any two cities by passing through exactly two cities. Also, $A + A^2$ represents the total number of ways to travel between two cities with at most one intermediate city. 9

The diagram can be given many other interpretations. For example, the lines could represent lines of mutual influence between people or nations, or they could represent communication lines such as telephone lines.

6.5 Exercises

Solve the matrix equation AX = B for X. (See Example 1.)

1. $A = \begin{bmatrix} 1 & -1 \\ 5 & -6 \end{bmatrix}, B = \begin{bmatrix} 2 \\ 4 \end{bmatrix}$

2. $A = \begin{bmatrix} 3 & -2 \\ -1 & 1 \end{bmatrix}, B = \begin{bmatrix} -3 \\ 5 \end{bmatrix}$

3. $A = \begin{bmatrix} 3 & 1 \\ 4 & 2 \end{bmatrix}, B = \begin{bmatrix} 3 & 4 \\ 5 & 6 \end{bmatrix}$

4. $A = \begin{bmatrix} 7 & -3 \\ -2 & 1 \end{bmatrix}, B = \begin{bmatrix} 0 & 8 \\ 4 & 1 \end{bmatrix}$

5. $A = \begin{bmatrix} 1 & -2 & -3 \\ -1 & 4 & 6 \\ 1 & -1 & -2 \end{bmatrix}, B = \begin{bmatrix} 2 \\ 7 \\ 4 \end{bmatrix}$

6. $A = \begin{bmatrix} 3 & -1 & 0 \\ 0 & 1 & 2 \\ 6 & 0 & 5 \end{bmatrix}, B = \begin{bmatrix} -6 \\ 12 \\ 15 \end{bmatrix}$

Use the inverse of the coefficient matrix to solve each system of equations. (The inverses for Exercises 9–14 were found in Exercises 41 and 44–48 of Section 6.4.) (See Example 1.)

7. $\begin{aligned} x + 2y + 3z &= 5 \\ 2x + 3y + 2z &= 2 \\ -x - 2y - 4z &= -1 \end{aligned}$

8. $\begin{aligned} x + y - 3z &= 4 \\ 2x + 4y - 4z &= 8 \\ -x + y + 4z &= -3 \end{aligned}$

9. $\begin{aligned} x + 4y + 3z &= -12 \\ x - 3y - 2z &= 0 \\ 2x + 5y + 4z &= 7 \end{aligned}$

10. $\begin{aligned} 5x \qquad + 2z &= 3 \\ 2x + 2y + z &= 4 \\ -3x + y - z &= 5 \end{aligned}$

11. $\begin{aligned} 2x + 4y + 6z &= 4 \\ -x - 4y - 3z &= 8 \\ y - z &= -4 \end{aligned}$

12. $\begin{aligned} 2x + 2y - 4z &= 12 \\ 2x + 6y \qquad &= 16 \\ -3x - 3y + 5z &= -20 \end{aligned}$

13. $\begin{aligned} x - 2y + 3z \qquad &= 4 \\ y - z + w &= -8 \\ -2x + 2y - 2z + 4w &= 12 \\ 2y - 3z + w &= -4 \end{aligned}$

14. $\begin{aligned} x + y \qquad + 2w &= 3 \\ 2x - y + z - w &= 3 \\ 3x + 3y + 2z - 2w &= 5 \\ x + 2y + z \qquad &= 3 \end{aligned}$

Use matrix algebra to solve the following matrix equations for X. Then use the given matrices to find X and check your work.

15. $N = X - MX, N = \begin{bmatrix} 8 \\ -12 \end{bmatrix}, M = \begin{bmatrix} 0 & 1 \\ -2 & 1 \end{bmatrix}$

16. $A = BX + X, A = \begin{bmatrix} 4 & 6 \\ -2 & 2 \end{bmatrix}, B = \begin{bmatrix} -2 & -2 \\ 3 & 3 \end{bmatrix}$

Find the production matrix for the given input-output and demand matrices. (See Examples 3–6.)

17. $A = \begin{bmatrix} \frac{1}{2} & \frac{2}{5} \\ \frac{1}{4} & \frac{1}{5} \end{bmatrix}, D = \begin{bmatrix} 2 \\ 4 \end{bmatrix}$

18. $A = \begin{bmatrix} .1 & .03 \\ .07 & .6 \end{bmatrix}, D = \begin{bmatrix} 5 \\ 10 \end{bmatrix}$

19. **Business** In his work *Input-Output Economics,* Leontief provides an example of a simplified economy with just three sectors: agriculture, manufacturing, and households (that is, the sector of the economy that produces labor.)* It has the following input-output matrix.

	Agriculture	Manufacturing	Households
Agriculture	.25	.40	.133
Manufacturing	.14	.12	.100
Households	.80	3.60	.133

He also gives the demand matrix

$$D = \begin{bmatrix} 35 \\ 38 \\ 40 \end{bmatrix}$$

Find the amount of each commodity that should be produced.

20. **Business** A much simplified version of Leontief's 42-sector analysis of the 1947 American economy has the following input-output matrix:†

	Agriculture	Manufacturing	Households
Agriculture	.245	.102	.051
Manufacturing	.099	.291	.279
Households	.433	.372	.011

The demand matrix (in billions of dollars) is

$$D = \begin{bmatrix} 2.88 \\ 31.45 \\ 30.91 \end{bmatrix}$$

Find the amount of each commodity that should be produced.

---

*Wassily Leontief, *Input-Output Economics,* Second Edition (Oxford University Press, 1966), pp. 20–27.

†Ibid, pp. 6–9.

21. Business An analysis of the 1958 Israeli economy is simplified here by grouping the economy into three sectors, with the following input-output matrix.*

	Agriculture	Manufacturing	Energy
Agriculture	.293	0	0
Manufacturing	.014	.207	.017
Energy	.044	.010	.216

The demand (in thousands of Israeli pounds) as measured by exports is

$$D = \begin{bmatrix} 138{,}213 \\ 17{,}597 \\ 1786 \end{bmatrix}$$

Find the amount of each commodity that should be produced.

22. Business The 1981 Chinese economy can be simplified to three sectors: agriculture, industry and construction, and transportation and commerce.† The input-output matrix is given below.

	Agriculture	Industry/Constr.	Trans./Commerce
Agriculture	.158	.156	.009
Industry/Constr.	.136	.432	.071
Trans./Commerce	.013	.041	.011

The demand (in 100,000 RMB, the unit of money in China) is

$$D = \begin{bmatrix} 106{,}674 \\ 144{,}739 \\ 26{,}725 \end{bmatrix}$$

(a) Find the amount of each commodity that should be produced.

(b) Interpret the economic value of an increase in demand of 1 RMB in agriculture exports.

Write a system of equations and use the inverse of the coefficient matrix to solve the system.

23. Business Felsted Furniture makes dining room furniture. A buffet requires 30 hours for construction and 10 hours for finishing. A chair requires 10 hours for construction and 10 hours for finishing. A table requires 10 hours for construction and 30 hours for finishing. The construction department has 350 hours of labor and the finishing department 150 hours of labor available each week. How many pieces of each type of furniture should be produced each week if the factory is to run at full capacity?

*Ibid, pp. 174–175.

†*Input-Output Tables of China, 1981*, China Statistical Information and Consultancy Service Centre, 1987, pp. 17–19.

24. Natural Science **(a)** A hospital dietician is planning a special diet for a certain patient. The total amount per meal of food groups A, B, and C must equal 400 grams. The diet should include one-third as much of group A as of group B, and the sum of the amounts of group A and group C should equal twice the amount of group B. How many grams of each food group should be included?

(b) Suppose we drop the requirement that the diet include one-third as much of group A as of group B. Describe the set of all possible solutions.

(c) Suppose that, in addition to the conditions given in part (a), foods A and B cost 2 cents per gram and food C costs 3 cents per gram and that a meal must cost \$8. Is a solution possible?

25. Natural Science Three species of bacteria are fed three foods, I, II, and III. A bacterium of the first species consumes 1.3 units each of foods I and II and 2.3 units of food III each day. A bacterium of the second species consumes 1.1 units of food I, 2.4 units of food II, and 3.7 units of food III each day. A bacterium of the third species consumes 8.1 units of I, 2.9 units of II, and 5.1 units of III each day. If 16,000 units of I, 28,000 units of II, and 44,000 units of III are supplied each day, how many of each species can be maintained in this environment?

26. Business A company produces three combinations of mixed vegetables which sell in 1-kilogram packages. Italian style combines .3 kilogram of zucchini, .3 of broccoli, and .4 of carrots. French style combines .6 kilogram of broccoli and .4 of carrots. Oriental style combines .2 kilogram of zucchini, .5 of broccoli, and .3 of carrots. The company has a stock of 16,200 kilograms of zucchini, 41,400 kilograms of broccoli, and 29,400 kilograms of carrots. How many packages of each style should they prepare to use up their supplies?

27. Business A national chain of casual clothing stores recently sent shipments of jeans, jackets, sweaters, and shirts to its stores in various cities. The number of items shipped to each city and their total wholesale cost are shown in the table below. Find the wholesale price of one pair of jeans, one jacket, one sweater, and one shirt.

City	*Jeans*	*Jackets*	*Sweaters*	*Shirts*	*Total Cost*
Cleveland	3000	3000	2200	4200	\$507,650
St. Louis	2700	2500	2100	4300	459,075
Seattle	5000	2000	1400	7500	541,225
Phoenix	7000	1800	600	8000	571,500

28. **Health** A 100-bed nursing home provides two levels of long-term care: regular and maximum. Patients at each level have a choice of a private room or a less expensive semiprivate room. The tables below show the number of patients in each category at various times last year. The total daily costs for all patients were $18,824 in January, $18,738 in April, $18,606 in July, and $18,824 in October. Find the daily cost of each of the following: a private room (regular care), a private room (maximum care), a semiprivate room (regular care), and a semiprivate room (maximum care).

	REGULAR CARE PATIENTS	
Month	*Semiprivate*	*Private*
January	22	8
April	26	8
July	24	14
October	20	10

	MAXIMUM CARE PATIENTS	
Month	*Semiprivate*	*Private*
January	60	10
April	54	12
July	56	6
October	62	8

Exercises 29 and 30 refer to Example 6.

29. **Business** If the demand is changed to 690 metric tons of wheat and 920 metric tons of oil, how many units of each commodity should be produced?

30. **Business** Change the technological matrix so that production of 1 metric ton of wheat requires 1/5 metric ton of oil (and no wheat) and the production of 1 metric ton of oil requires 1/3 metric ton of wheat (and no oil). To satisfy the same demand matrix, how many units of each commodity should be produced?

Work these problems. (See Examples 3–6.)

31. **Business** A simplified economy has only two industries, the electric company and the gas company. Each dollar's worth of the electric company's output requires $.40 of its own output and $.50 of the gas company's output. Each dollar's worth of the gas company's output requires $.25 of its own output and $.60 of the electric company's output. What should the production of electricity and gas be (in dollars) if there is a $12 million demand for gas and a $15 million demand for electricity?

32. **Business** A two-segment economy consists of manufacturing and agriculture. To produce one unit of manufacturing output requires .40 unit of its own output and .20 unit of agricultural output. To produce one unit of agricultural output requires .30 unit of its own output and .40 unit of manufacturing output. If there is a demand of 240 units of manufacturing and 90 units of agriculture, what should be the output of each segment?

33. **Business** A primitive economy depends on two basic goods, yams and pork. Production of 1 bushel of yams requires 1/4 bushel of yams and 1/2 of a pig. To produce 1 pig requires 1/6 bushel of yams. Find the amount of each commodity that should be produced to get
(a) 1 bushel of yams and 1 pig;
(b) 100 bushels of yams and 70 pigs.

34. **Business** A simplified economy is based on agriculture, manufacturing, and transportation. Each unit of agricultural output requires .4 unit of its own output, .3 unit of manufacturing, and .2 unit of transportation output. One unit of manufacturing output requires .4 unit of its own output, .2 unit of agricultural, and .3 unit of transportation output. One unit of transportation output requires .4 unit of its own output, .1 unit of agricultural, and .2 unit of manufacturing output. There is demand for 35 units of agricultural, 90 units of manufacturing, and 20 units of transportation output. How many units should each segment of the economy produce?

35. **Business** The use of input-output analysis can also be used to model how changes in one city can affect cities that are connected with it in some way.* For example, if a large manufacturing company shuts down in one city, it is very likely that the economic welfare of all of the cities around it will suffer. Consider three Pennsylvania communities: Sharon, Farrell, and Hermitage. Due to their proximity to each other, residents of these three communities regularly spend time and money in the other communities. Suppose that we have gathered information in the form of an input-output matrix

$$\begin{bmatrix} .2 & .1 & .1 \\ .1 & .1 & 0 \\ .5 & .6 & .7 \end{bmatrix}.$$

This matrix can be thought of as the likelihood that a person from a particular community will spend money in each of the communities.
(a) Treat this matrix like an input-output matrix and calculate $(I - A)^{-1}$, where A is the input-output matrix above.
(b) Interpret the entries of this inverse matrix.

*The idea for this problem came from an example created by Thayer Watkins, Department of Economics, San Jose State University.

Work these coding problems.

36. Social Science Use the method discussed in the text to encode the message

Anne is home.

Break the message into groups of two letters and use the matrix

$$M = \begin{bmatrix} 1 & 3 \\ 2 & 7 \end{bmatrix}$$

37. Social Science Use the matrix of Exercise 36 to encode the message

Head for the hills!

38. Social Science Decode the following message, which was encoded by using the matrix M of Exercise 36.

$$\begin{bmatrix} 90 \\ 207 \end{bmatrix}, \begin{bmatrix} 39 \\ 87 \end{bmatrix}, \begin{bmatrix} 26 \\ 57 \end{bmatrix}, \begin{bmatrix} 66 \\ 145 \end{bmatrix}, \begin{bmatrix} 61 \\ 142 \end{bmatrix}, \begin{bmatrix} 89 \\ 205 \end{bmatrix},$$

Work these routing problems.

39. Social Science Use matrix A in the discussion on routing in the text to find A^2. Then answer the following questions. How many ways are there to travel from

(a) City 1 to city 3 by passing through exactly one city?
(b) City 2 to city 4 by passing through exactly one city?
(c) City 1 to city 3 by passing through at most one city?
(d) City 2 to city 4 by passing through at most one city?

40. Social Sciences The matrix A^3 (See Exercise 39) was found in side problem 9 of the text. Use it to answer the following questions.

(a) How many ways are there to travel between cities 1 and 4 by passing through exactly two cities?
(b) How many ways are there to travel between cities 1 and 4 by passing through at most two cities?

41. Business A small telephone system connects three cities. There are four lines between cities 3 and 2, three lines connecting city 3 with city 1, and two lines between cities 1 and 2.

(a) Write a matrix B to represent this information.
(b) Find B^2.
(c) How many lines that connect cities 1 and 2 go through exactly one other city (city 3)?
(d) How many lines that connect cities 1 and 2 go through at most one other city?

42. Transportation The figure shows four southern cities served by Supersouth Airlines.

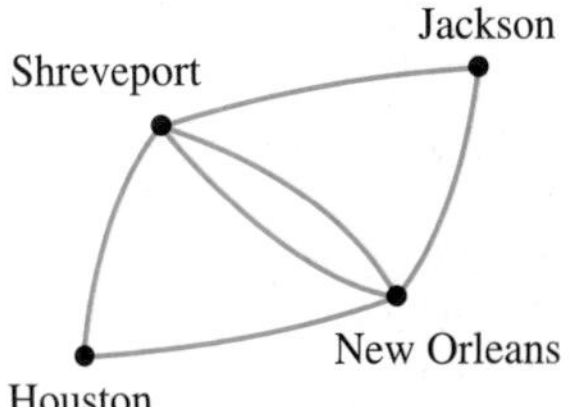

(a) Write a matrix to represent the number of nonstop routes between cities.
(b) Find the number of one-stop flights between Houston and Jackson.
(c) Find the number of flights between Houston and Shreveport that require at most one stop.
(d) Find the number of one-stop flights between New Orleans and Houston.

43. Natural Science The figure shows a food web. The arrows indicate the food sources of each population. For example, cats feed on rats and on mice.

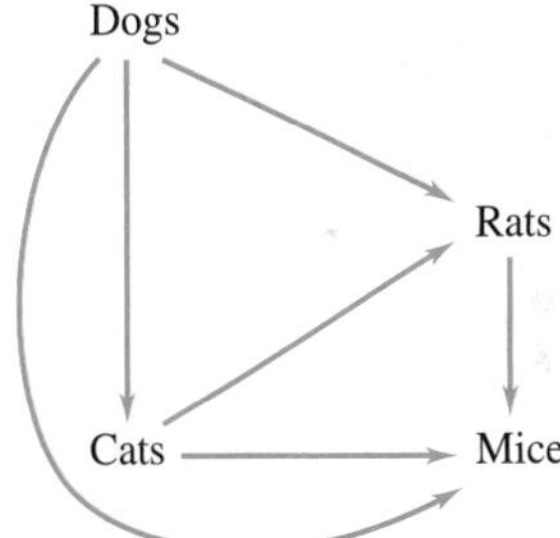

(a) Write a matrix C in which each row and corresponding column represent a population in the food chain. Enter a 1 when the population in a given row feeds on the population in the given column.
(b) Calculate and interpret C^2.

CHAPTER 6 SUMMARY

Key Terms and Symbols

6.1 linear equation
system of linear equations
solution of a system
independent system
dependent system
inconsistent system
elimination method
equivalent systems
elementary operations
row
column
matrix (matrices)
element (entry)
augmented matrix
row operations
row echelon form
parameter

6.2 Gauss-Jordan method
reduced row echelon form

6.3 row matrix (row vector)
column matrix (column vector)
square matrix
additive inverse of a matrix
zero matrix
scalar
product of a scalar and a matrix

6.4 product matrix
identity matrix
inverse matrix
singular matrix

6.5 coefficient matrix
input-output model
technological matrix (input-output matrix)
production matrix (vector of gross output)
demand matrix
code theory
routing theory

Key Concepts

Solving Systems of Equations

The following **elementary operations** are used to transform a system of equations into a simpler equivalent system.

1. Interchange any two equations.
2. Multiply both sides of an equation by a nonzero constant.
3. Replace an equation by the sum of itself and a constant multiple of another equation in the system.

The **elimination method** is a systematic way of using elementary operations to transform a system into an equivalent system that can be solved by **back substitution.** See Section 6.1 for details.

The matrix version of the elimination method uses the following **matrix row operations** that correspond to using elementary row operations with back substitution on a system of equations.

1. Interchange any two rows.
2. Multiply each element of a row by a nonzero constant.
3. Replace a row by the sum of itself and a constant multiple of another row in the matrix.

The **Gauss-Jordan method** is an extension of the elimination method for solving a system of linear equations. It uses row operations on the augmented matrix of the system. See Section 6.2 for details.

Operations on Matrices

The **sum** of two $m \times n$ matrices X and Y is the $m \times n$ matrix $X + Y$ in which each element is the sum of the corresponding elements of X and Y. The **difference** of two $m \times n$ matrices X and Y is the $m \times n$ matrix $X - Y$ in which each element is the difference of the corresponding elements of X and Y.

The **product** of a scalar k and a matrix X is the matrix kX, with each element k times the corresponding element of X.

The **product matrix** AB of an $m \times n$ matrix A and an $n \times k$ matrix B is the $m \times k$ matrix whose entry in the ith row and jth column is the product of the ith row of A and the jth column of B.

The **inverse matrix** A^{-1} for any $n \times n$ matrix A for which A^{-1} exists is found as follows. Form the augmented matrix $[A \mid I]$; perform row operations on $[A \mid I]$ to get the matrix $[I \mid A^{-1}]$.

Chapter 6 Review Exercises

Use the elimination or matrix method to solve each of the following systems. Identify any dependent or inconsistent systems.

1. $\begin{aligned} -5x - 3y &= 4 \\ 2x + y &= -3 \end{aligned}$

2. $\begin{aligned} 3x - y &= 6 \\ 2x + 3y &= 7 \end{aligned}$

3. $\begin{aligned} 3x - 5y &= 10 \\ 4x - 3y &= 6 \end{aligned}$

4. $\begin{aligned} \frac{1}{4}x - \frac{1}{3}y &= -\frac{1}{4} \\ \frac{1}{10}x + \frac{2}{5}y &= \frac{2}{5} \end{aligned}$

5. $\begin{aligned} x - 2y &= 1 \\ 4x + 4y &= 2 \\ 10x + 8y &= 4 \end{aligned}$

6. $\begin{aligned} x + y - 4z &= 0 \\ 2x + y - 3z &= 2 \end{aligned}$

7. $\begin{aligned} 3x + y - z &= 13 \\ x + 2z &= 9 \\ -3x - y + 2z &= 9 \end{aligned}$

8. $\begin{aligned} 4x - y - 2z &= 4 \\ x - y - \frac{1}{2}z &= 1 \\ 2x - y - z &= 8 \end{aligned}$

9. Business An office supply manufacturer makes two kinds of paper clips, standard and extra large. To make 1000 standard paper clips requires 1/4 hour on a cutting machine and 1/2 hour on a machine that shapes the clips. One thousand extra large paper clips require 1/3 hour on each machine. The manager of paper clip production has 4 hours per day available on the cutting machine and 6 hours per day on the shaping machine. How many of each kind of clip can he make?

10. Business Gretchen Schmidt plans to buy shares of two stocks. One costs \$32 per share and pays dividends of \$1.20 per share. The other costs \$23 per share and pays dividends of \$1.40 per share. She has \$10,100 to spend and wants to earn dividends of \$540. How many shares of each stock should she buy?

11. Business Joyce Pluth has money in two investment funds. Last year the first fund paid a dividend of 8% and the second a dividend of 2% and Joyce received a total of \$780. This year the first fund paid a 10% dividend and the second only 1% and Joyce received \$810. How much does she have invested in each fund?

12. Finance You are given \$144 in one, five, and ten dollar bills. There are 35 bills. There are two more ten dollar bills than five dollar bills. How many bills of each type are there?

13. Social Science A social service agency provides counseling, meals, and shelter to clients referred by sources I, II, and III. Clients from source I require an average of \$100 for food, \$250 for shelter, and no counseling. Source II clients require an average of \$100 for counseling, \$200 for food, and nothing for shelter. Source III clients require an average of \$100 for counseling, \$150 for food, and \$200 for shelter. The agency has funding of \$25,000 for counseling, \$50,000 for food, and \$32,500 for shelter. How many clients from each source can be served?

14. Business The Waputi Indians make woven blankets, rugs, and skirts. Each blanket requires 24 hours for spinning the yarn, 4 hours for dying the yarn, and 15 hours for weaving. Rugs require 30, 5, and 18 hours and skirts 12, 3, and 9 hours, respectively. If there are 306, 59, and 201 hours available for spinning, dying, and weaving, respectively, how many of each item can be made? (*Hint:* Simplify the equations you write, if possible, before solving the system.)

Use the Gauss-Jordan method to solve the following systems.

15. $\begin{aligned} x - z &= -3 \\ y + z &= 6 \\ 2x - 3z &= -9 \end{aligned}$

16. $\begin{aligned} 2x - y + 4z &= -1 \\ -3x + 5y - z &= 5 \\ 2x + 3y + 2z &= 3 \end{aligned}$

17. $\begin{aligned} 5x - 8y + z &= 1 \\ 3x - 2y + 4z &= 3 \\ 10x - 16y + 2z &= 3 \end{aligned}$

18. $\begin{aligned} x - 2y + 3z &= 4 \\ 2x + y - 4z &= 3 \\ -3x + 4y - z &= -2 \end{aligned}$

19. $\begin{aligned} 3x + 2y - 6z &= 9 \\ x + y + 2z &= 4 \\ 2x + 2y + 5z &= 0 \end{aligned}$

20. Business Each week at a furniture factory, there are 2000 work hours available in the construction department, 1400 work hours in the painting department, and 1300 work hours in the packing department. Producing a chair requires 2 hours of construction, 1 hour of painting, and 2 hours for packing. Producing a table requires 4 hours of construction, 3 hours of painting, and 3 hours for packing. Producing a chest requires 8 hours of construction, 6 hours of painting, and 4 hours for packing. If all available time is used in every department, how many of each item are produced each week?

For each of the following, find the dimensions of the matrix and identify any square, row, or column matrices.

21. $\begin{bmatrix} 2 & 3 \\ 5 & 9 \end{bmatrix}$

22. $\begin{bmatrix} 2 & -1 \\ 4 & 6 \\ 5 & 7 \end{bmatrix}$

23. $[12 \quad 4 \quad -8 \quad -1]$

24. $\begin{bmatrix} -7 & 5 & 6 \\ 3 & 2 & -1 \\ -1 & 12 & 8 \end{bmatrix}$

25. $\begin{bmatrix} 6 & 8 & 10 \\ 5 & 3 & -2 \end{bmatrix}$

26. $\begin{bmatrix} -9 \\ 15 \\ 4 \end{bmatrix}$

27. **Natural Science** The activities of a grazing animal can be classified roughly into three categories: grazing, moving, and resting. Suppose horses spend 8 hours grazing, 8 moving, and 8 resting; cattle spend 10 grazing, 5 moving, and 9 resting; sheep spend 7 grazing, 10 moving, and 7 resting; and goats spend 8 grazing, 9 moving, and 7 resting. Write this information as a 4×3 matrix.

28. **Business** The New York Stock Exchange reports in the daily newspapers give the dividend, price-to-earnings ratio, sales (in hundreds of shares), last price, and change in price for each company. Write the following stock reports as a 4×5 matrix. American Telephone & Telegraph: 5, 7, 2532, $52\frac{3}{8}$, $-\frac{1}{4}$. General Electric: 3, 9, 1464, 56, $+\frac{1}{8}$. Gulf Oil: 2.50, 5, 4974, 41, $-1\frac{1}{2}$. Sears: 1.36, 10, 1754, 18, $+\frac{1}{2}$.

Given the matrices

$$A = \begin{bmatrix} 4 & 10 \\ -2 & -3 \\ 6 & 9 \end{bmatrix}, \quad B = \begin{bmatrix} 2 & 3 & -2 \\ 2 & 4 & 0 \\ 0 & 1 & 2 \end{bmatrix}, \quad C = \begin{bmatrix} 5 & 0 \\ -1 & 3 \\ 4 & 7 \end{bmatrix},$$

$$D = \begin{bmatrix} 6 \\ 1 \\ 0 \end{bmatrix}, \quad E = [1 \quad 3 \quad -4], \quad F = \begin{bmatrix} -1 & 4 \\ 3 & 7 \end{bmatrix},$$

$$G = \begin{bmatrix} 2 & 5 \\ 1 & 6 \end{bmatrix},$$

find each of the following (if possible).

29. $-B$

30. $-D$

31. $3A - 2C$

32. $F + 3G$

33. $2B - 5C$

34. $G - 2F$

35. **Business** Refer to Exercise 28. Write a 4×2 matrix using the sales and price changes for the four companies. The next day's sales and price changes for the same four companies were 2310, 1258, 5061, 1812 and $-1/4$, $-1/4$, $+1/2$, $+1/2$, respectively. Write a 4×2 matrix using these new sales and price change figures. Use matrix addition to find the total sales and price changes for the two days.

36. **Business** An oil refinery in Tulsa sent 110,000 gallons of oil to a Chicago distributor, 73,000 to a Dallas distributor, and 95,000 to an Atlanta distributor. Another refinery in New Orleans sent the following amounts to the same three distributors: 85,000, 108,000, 69,000. The next month the two refineries sent the same distributors new shipments of oil as follows: from Tulsa, 58,000 to Chicago, 33,000 to Dallas, and 80,000 to Atlanta; from New Orleans, 40,000, 52,000, and 30,000, respectively.
   (a) Write the monthly shipments from the two distributors to the three refineries as 3×2 matrices.
   (b) Use matrix addition to find the total amounts sent to the refineries from each distributor.

Use the matrices given above Exercise 29 to find each of the following (if possible).

37. AG

38. EB

39. GF

40. CA

41. AGF

42. **Health** In a recent study, the numbers of head and neck injuries among hockey players wearing full face shields and half face shields were compared. The following table provides the rates per 1000 athlete-exposures for specific injuries that caused a player wearing either shield to miss one or more events.*

	Half Shield	*Full Shield*
Head and Face Injuries excluding Concussions	3.54	1.41
Concussions	1.53	1.57
Neck Injuries	.34	.29
Other	7.53	6.21

If an equal number of players in a large league wear each type of shield and the total number of athlete-exposures for the league in a season is 8000, use matrix operations to estimate the total number of injuries of each type.

43. **Business** An office supply manufacturer makes two kinds of paper clips, standard and extra large. To make a unit of standard paper clips requires 1/4 hour on a cutting machine and 1/2 hour on a machine that shapes the clips. A unit of extra large paper clips requires 1/3 hour on each machine.
   (a) Write this information as a 2×2 matrix (size/machine).
   (b) If 48 units of standard and 66 units of extra large clips are to be produced, use matrix multiplication to find out how many hours each machine will operate. (*Hint:* Write the units as a 1×2 matrix.)

*Brian Benson, Nicholas Nohtadi, Sarah Rose, and Willem Meeuwisse, "Head and Neck Injuries among Ice Hockey Players Wearing Full Face Shields vs. Half Face Shields," *JAMA*, 282, no. 24, (December 22/29, 1999): 2328–2332.

44. Business Theresa DePalo buys shares of three stocks. Their cost per share and dividend earnings per share are \$32, \$23, and \$54, and \$1.20, \$1.49, and \$2.10, respectively. She buys 50 shares of the first stock, 20 shares of the second, and 15 shares of the third.
(a) Write the cost per share and earnings per share of the stocks as a 3×2 matrix.
(b) Write the number of shares of each stock as a 1×3 matrix.
(c) Use matrix multiplication to find the total cost and total dividend earnings of these stocks.

45. If $A = \begin{bmatrix} 3 & 0 \\ 2 & 1 \end{bmatrix}$, find a matrix B such that both AB and BA are defined and $AB \neq BA$.

46. Is it possible to do Exercise 45 if $A = \begin{bmatrix} 4 & 0 \\ 0 & 4 \end{bmatrix}$? Explain.

Find the inverse of each of the following matrices that has an inverse.

47. $\begin{bmatrix} -4 & 2 \\ 0 & 3 \end{bmatrix}$

48. $\begin{bmatrix} 2 & 1 \\ 5 & 3 \end{bmatrix}$

49. $\begin{bmatrix} 6 & 4 \\ 3 & 2 \end{bmatrix}$

50. $\begin{bmatrix} 2 & 0 \\ -1 & 5 \end{bmatrix}$

51. $\begin{bmatrix} 2 & 0 & 4 \\ 1 & -1 & 0 \\ 0 & 1 & -2 \end{bmatrix}$

52. $\begin{bmatrix} 2 & -1 & 0 \\ 1 & 0 & 1 \\ 1 & -2 & 0 \end{bmatrix}$

53. $\begin{bmatrix} 2 & 3 & 5 \\ -2 & -3 & -5 \\ 1 & 4 & 2 \end{bmatrix}$

54. $\begin{bmatrix} 1 & 3 & 6 \\ 4 & 0 & 9 \\ 5 & 15 & 30 \end{bmatrix}$

55. $\begin{bmatrix} 1 & 3 & -2 & -1 \\ 0 & 1 & 1 & 2 \\ -1 & -1 & 1 & -1 \\ 1 & -1 & -3 & -2 \end{bmatrix}$

56. $\begin{bmatrix} 3 & 2 & 0 & -1 \\ 2 & 0 & 1 & 2 \\ 1 & 2 & -1 & 0 \\ 2 & -1 & 1 & 1 \end{bmatrix}$

Refer again to the matrices given above Exercise 29 to find each of the following (if possible).

57. F^{-1}

58. G^{-1}

59. $(G - F)^{-1}$

60. $(F + G)^{-1}$

61. B^{-1}

62. Explain why the matrix $\begin{bmatrix} a & 0 \\ c & 0 \end{bmatrix}$, where a and c are nonzero constants, cannot possibly have an inverse.

Solve each of the following matrix equations $AX = B$ for X.

63. $A = \begin{bmatrix} 2 & 4 \\ -1 & -3 \end{bmatrix}$, $B = \begin{bmatrix} 8 \\ 3 \end{bmatrix}$

64. $A = \begin{bmatrix} 1 & 3 \\ -2 & 4 \end{bmatrix}$, $B = \begin{bmatrix} 15 \\ 10 \end{bmatrix}$

65. $A = \begin{bmatrix} 1 & 0 & 2 \\ -1 & 1 & 0 \\ 3 & 0 & 4 \end{bmatrix}$, $B = \begin{bmatrix} 8 \\ 4 \\ -6 \end{bmatrix}$

66. $A = \begin{bmatrix} 2 & 4 & 0 \\ 1 & -2 & 0 \\ 0 & 0 & 3 \end{bmatrix}$, $B = \begin{bmatrix} 72 \\ -24 \\ 48 \end{bmatrix}$

Use the method of matrix inverses to solve each of the following systems.

67. $\begin{aligned} x + y &= 4 \\ 2x + 3y &= 10 \end{aligned}$

68. $\begin{aligned} 5x - 3y &= -2 \\ 2x + 7y &= -9 \end{aligned}$

69. $\begin{aligned} 2x + y &= 5 \\ 3x - 2y &= 4 \end{aligned}$

70. $\begin{aligned} x - 2y &= 7 \\ 3x + y &= 7 \end{aligned}$

71. $\begin{aligned} x + y + z &= 1 \\ 2x - y &= -2 \\ 3y + z &= 2 \end{aligned}$

72. $\begin{aligned} x &= -3 \\ y + z &= 6 \\ 2x - 3z &= -9 \end{aligned}$

73. $\begin{aligned} 3x - 2y + 4z &= 4 \\ 4x + y - 5z &= 2 \\ -6x + 4y - 8z &= -2 \end{aligned}$

74. $\begin{aligned} x + 2y &= -1 \\ 3y - z &= -5 \\ x + 2y - z &= -3 \end{aligned}$

Solve each of the following problems by any method.

75. Business A wine maker has two large casks of wine. One is 9% alcohol and the other is 14% alcohol. How many liters of each wine should be mixed to produce 40 liters of wine that is 12% alcohol?

76. Business A gold merchant has some 12 carat gold (12/24 pure gold), and some 22 carat gold (22/24 pure). How many grams of each could be mixed to get 25 grams of 15 carat gold?

77. Natural Science A chemist has some 40% acid solution and some 60% solution. How many liters of each should be used to get 40 liters of a 45% solution?

78. Business How many pounds of tea worth \$4.60 a pound should be mixed with tea worth \$6.50 a pound to get 10 pounds of a mixture worth \$5.74 a pound?

79. Business A machine in a pottery factory takes 3 minutes to form a bowl and 2 minutes to form a plate. The material for a bowl costs \$.25 and the material for a plate costs \$.20. If the machine runs for 8 hours and exactly \$44 is spent for material, how many bowls and plates can be produced?

80. Transportation A boat travels at a constant speed a distance of 57 kilometers downstream in 3 hours, then turns around and travels 55 kilometers upstream in 5 hours. What is the speed of the boat and of the current?

81. Business Ms. Tham invests \$50,000 three ways—at 8%, $8\frac{1}{2}$%, and 11%. In total, she receives \$4436.25 per year in interest. The interest from the 11% investment is

\$80 more than the interest on the 8% investment. Find the amount she has invested at each rate.

82. Business Tickets to a band concert cost \$2 for children, \$3 for teenagers, and \$5 for adults. 570 people attended the concert and total ticket receipts were \$1950. Three-fourths as many teenagers as children attended. How many children, teenagers, and adults were at the concert?

83. Given the input-output matrix $A = \begin{bmatrix} 0 & \frac{1}{4} \\ \frac{1}{2} & 0 \end{bmatrix}$ and the demand matrix $D = \begin{bmatrix} 2100 \\ 1400 \end{bmatrix}$, find each of the following.

(a) $I - A$ **(b)** $(I - A)^{-1}$

(c) the production matrix X

84. Business The 1987 economy of the state of Washington has been simplified to four sectors: natural resource, manufacturing, trade and services, and personal consumption. The input-output matrix is given below.*

	natural resources	manu-facturing	trade and services	personal consumption
natural resources	.1045	.0428	.0029	.0031
manufacturing	.0826	.1087	.0584	.0321
trade and services	.0867	.1019	.2032	.3555
personal consumption	.6253	.3448	.6106	.0798

Suppose the demand (in millions of dollars) is

$$D = \begin{bmatrix} 450 \\ 300 \\ 125 \\ 100 \end{bmatrix}.$$

Find the amount of each commodity that should be produced.

85. Business The 1963 economy of the state of Nebraska has been condensed to six sectors; livestock, crops, food products, mining and manufacturing, households, and other. The input-output matrix is in the next column.[†]

$$\begin{bmatrix} .178 & .018 & .411 & 0 & .005 & 0 \\ .143 & .018 & .088 & 0 & .001 & 0 \\ .089 & 0 & .035 & 0 & .060 & .003 \\ .001 & .010 & .012 & .063 & .007 & .014 \\ .141 & .252 & .088 & .089 & .402 & .124 \\ .188 & .156 & .103 & .255 & .008 & .474 \end{bmatrix}$$

(a) Find the matrix $(I - A)^{-1}$ and interpret the value in row two, column one, of this matrix.

(b) Suppose the demand (in millions of dollars) is

$$D = \begin{bmatrix} 1980 \\ 650 \\ 1750 \\ 1000 \\ 2500 \\ 3750 \end{bmatrix}.$$

Find the dollar amount of each commodity that should be produced.

86. Business An economy depends on two commodities, goats and cheese. It takes 2/3 of a unit of goats to produce 1 unit of cheese and 1/2 unit of cheese to produce 1 unit of goats.

(a) Write the input-output matrix for this economy.

(a) Find the production required to satisfy a demand of 400 units of cheese and 800 units of goats.

87. Business In a simple economic model, a country has two industries: agriculture and manufacturing. To produce \$1 of agricultural output requires \$.10 of agricultural output and \$.40 of manufacturing output. To produce \$1 of manufacturing output requires \$.70 of agricultural output and \$.20 of manufacturing output. If agricultural demand is \$60,000 and manufacturing demand is \$20,000, what must each industry produce? (Round answers to the nearest whole number.)

88. Business The matrix below represents the number of direct flights between four cities.

	A	B	C	D
A	0	1	0	1
B	1	0	0	1
C	0	0	0	1
D	1	1	1	0

(a) Find the number of one-stop flights between cities A and C.

(b) Find the total number of flights between cities B and C that are either direct or one-stop.

(c) Find the matrix that gives the number of two-stop flights between these cities.

89. Social Science **(a)** Use the matrix $M = \begin{bmatrix} 2 & 6 \\ 1 & 4 \end{bmatrix}$ to encode the message "leave now."

(b) What matrix should be used to decode this message?

*Robert Chase, Philip Bourque, and Richard Conway, Jr., "The 1987 Washington State Input-Output Study," Report to the Graduate School of Business Administration, University of Washington, September 1993.

†F. Charles Lamphear and Theodore Roesler, "1970 Nebraska Input-Output Tables," *Nebraska Economic and Business Report No. 10,* Bureau of Business Research, University of Nebraska-Lincoln.

Case 5

Matrix Operations and Airline Route Maps

Airline route maps are usually published on airline web sites, as well as in in-flight magazines. The purpose of these maps is to show what cities are connected to each other by non-stop flights provided by the airline. We can think of these maps as another type of **graph,** and we can use matrix operations to answer questions of interest about the graph. In order to study these graphs, a bit of terminology will be helpful. A **graph** is a set of points (called **vertices,** or **nodes**) and a set of lines called **edges** connecting some pairs of vertices. Two vertices connected by an edge are said to be **adjacent.** Consider for example the northern route map for Cape Air from May 2001, which is Figure 1 below.* Here the vertices are the cities to which Cape Air flies, and two vertices are connected if there is a non-stop flight between them.

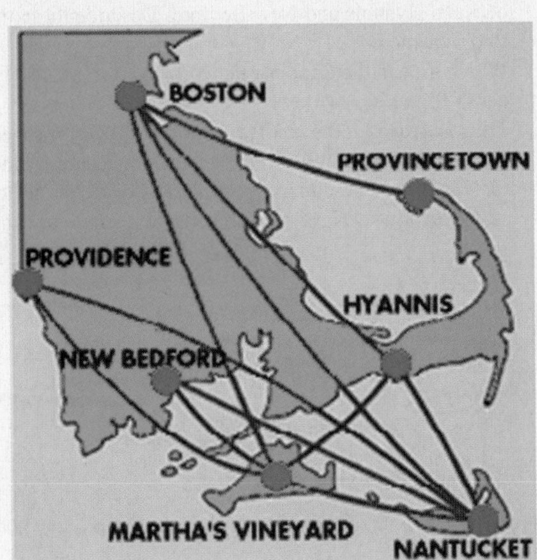

FIGURE 1 Cape Air Northern Route Map—May 2001

Some natural questions arise about graphs. It might be important to know if two vertices are connected by a sequence of two edges, even if they are not connected by a single edge. In the route map, Provincetown and Hyannis are connected by a two-edge sequence, meaning that a passenger would have to stop in Boston while flying between those cities on Cape Air. It might be important to know if it is possible to get from a vertex to another vertex in a given number of flights. In the example, a passenger on Cape Air can get from any city in their network to any other city given enough flights. But how many flights are enough? This is another issue of interest: what is the minimum number of steps to get from one vertex to another? What is the minimum number of steps to get from **any** vertex on the graph to any other? While these questions are relatively easy to answer for a small graph, as the number of vertices and edges grows, it becomes harder to keep track of all the different ways the vertices are connected. Matrix notation and computation can help to answer these questions.

The **adjacency matrix** for a graph with n vertices is an $n \times n$ matrix whose (i, j) entry is 1 if the ith vertex and jth are connected and 0 if they are not. If the vertices in the Cape Air graph respectively correspond to Boston (B), Hyannis (H), Martha's Vineyard (M), Nantucket (N), New Bedford (NB), Providence (P), and Provincetown (PT), then the adjacency matrix for Cape Air is as follows.

$$
\begin{array}{c} \\ A = \end{array}
\begin{array}{c}
\begin{array}{ccccccc} \text{B} & \text{H} & \text{M} & \text{N} & \text{NB} & \text{P} & \text{PT} \end{array} \\
\begin{bmatrix}
0 & 1 & 1 & 1 & 0 & 0 & 1 \\
1 & 0 & 1 & 1 & 0 & 0 & 0 \\
1 & 1 & 0 & 1 & 1 & 1 & 0 \\
1 & 1 & 1 & 0 & 1 & 1 & 0 \\
0 & 0 & 1 & 1 & 0 & 0 & 0 \\
0 & 0 & 1 & 1 & 0 & 0 & 0 \\
1 & 0 & 0 & 0 & 0 & 0 & 0
\end{bmatrix}
\end{array}
\begin{array}{l} \\ \text{B} \\ \text{H} \\ \text{M} \\ \text{N} \\ \text{NB} \\ \text{P} \\ \text{PT} \end{array}
$$

Adjacency matrices can be used to address the questions about graphs raised earlier. Which vertices are connected by a two-edge sequence? How many different two-edge sequences connect each pair of vertices? Consider the matrix A^2, which is A multiplied by itself. For example, let the (6, 2) entry in the matrix A^2 be named b_{62}. This entry in A^2 is the product of the 6th row of A and the 2nd column of A, or

$$
\begin{aligned}
b_{62} &= a_{61}a_{12} + a_{62}a_{22} + a_{63}a_{32} + a_{64}a_{42} + a_{65}a_{52} + a_{66}a_{62} + a_{67}a_{72} \\
&= 0 \cdot 1 + 0 \cdot 0 + 1 \cdot 1 + 1 \cdot 1 + 0 \cdot 0 + 0 \cdot 0 + 0 \cdot 0 \\
&= 2,
\end{aligned}
$$

*Compliments of Cape Air. See www.flycapeair.com/rout_cc_frameset.htm

which happens to be the number of two-flight sequences which connect city 6 (Providence) and city 2 (Hyannis). A careful look at Figure 1 confirms this fact. This calculation works because in order for a two-flight sequence to occur between Providence and Hyannis, Providence and Hyannis must both connect to an intermediate city. Since Providence connects to Martha's Vineyard (city 3) and Martha's Vineyard connects to Hyannis, $a_{63}a_{32} = 1 \cdot 1 = 1$. Thus there is 1 two-flight sequence from Providence to Hyannis that passes through Martha's Vineyard. Since Providence does not connect to Boston (city 1) but Boston does connect with Hyannis, $a_{61}a_{12} = 0 \cdot 1 = 0$. Thus there is no two-flight sequence from Providence to Hyannis that passes through Boston. To find the total number of two-flight sequences between Providence and Hyannis, simply sum over all intermediate points. Notice that this sum, which is

$$a_{61}a_{12} + a_{62}a_{22} + a_{63}a_{32} + a_{64}a_{42} + a_{65}a_{52} + a_{66}a_{62} + a_{67}a_{72},$$

is just the b_{62}, the (6, 2) entry in the matrix A^2. And so it is seen that the number of two-step sequences between vertex i and vertex j in a graph with adjacency matrix A is the (i, j) entry in A^2. A more general result is the following:

The number of k-step sequences between vertex i and vertex j in a graph with adjacency matrix A is the (i, j) entry in A^k.

If A is the adjacency matrix for Figure 1,

$$A^2 = \begin{array}{c} \begin{matrix} \text{B} & \text{H} & \text{M} & \text{N} & \text{NB} & \text{P} & \text{PT} \end{matrix} \\ \left[\begin{matrix} 4 & 2 & 2 & 2 & 2 & 2 & 0 \\ 2 & 3 & 2 & 2 & 2 & 2 & 1 \\ 2 & 2 & 5 & 4 & 1 & 1 & 1 \\ 2 & 2 & 4 & 5 & 1 & 1 & 1 \\ 2 & 2 & 1 & 1 & 2 & 2 & 0 \\ 2 & 2 & 1 & 1 & 2 & 2 & 0 \\ 0 & 1 & 1 & 1 & 0 & 0 & 1 \end{matrix}\right] \begin{matrix} \text{B} \\ \text{H} \\ \text{M} \\ \text{N} \\ \text{NB} \\ \text{P} \\ \text{PT} \end{matrix} \end{array}$$

and

$$A^3 = \begin{array}{c} \begin{matrix} \text{B} & \text{H} & \text{M} & \text{N} & \text{NB} & \text{P} & \text{PT} \end{matrix} \\ \left[\begin{matrix} 6 & 8 & 12 & 12 & 4 & 4 & 4 \\ 8 & 6 & 11 & 11 & 4 & 4 & 2 \\ 12 & 11 & 10 & 11 & 9 & 9 & 2 \\ 12 & 11 & 11 & 10 & 9 & 9 & 2 \\ 4 & 4 & 9 & 9 & 2 & 2 & 2 \\ 4 & 4 & 9 & 9 & 2 & 2 & 2 \\ 4 & 2 & 2 & 2 & 2 & 2 & 0 \end{matrix}\right] \begin{matrix} \text{B} \\ \text{H} \\ \text{M} \\ \text{N} \\ \text{NB} \\ \text{P} \\ \text{PT} \end{matrix} \end{array}$$

Since the (6,3) entry in A^2 is 1, there is 1 two-step sequence from Providence to Martha's Vineyard. Likewise, there are 4 three-step sequences between Hyannis and New Bedford since the (2,5) entry in A^3 is 4.

In observing the figures, note that some two-step or three step sequences may not be meaningful. On the Cape Air route map, Nantucket is reachable in two steps from Boston (via Hyannis or Martha's Vineyard), but in reality this does not matter, since there is a non-stop flight between the two cities. A better question to ask of a graph might be, "What is the least number of edges that must be traversed to go from vertex i to vertex j?"

To answer this question, consider the matrix $S_k = A + A^2 + \cdots + A^k$. The (i, j) entry in this matrix tallies the number of ways to get from vertex i to vertex j in k steps or less. If such a trip is impossible, this entry will be zero. Thus to find the shortest number of steps between the vertices, continue to compute S_k as k increases; the first k for which the (i, j) entry in S_k is nonzero is the shortest number of steps between i and j. Note that although the shortest number of steps may be computed, the method does not determine what those steps are.

If you are interested in other airlines' route maps, visit the web site dir.yahoo.com/Business_and_Economy/Shopping_and_Services/Travel_and_Transportation/Airlines/. This includes links to many obscure and smaller airlines, as well as the more well known carriers.

Exercises

1. Which Cape Air cities may be reached by a two-flight sequence from New Bedford? Which may be reached by a three-flight sequence?
2. It was shown above that there are 4 three-step sequences between Hyannis and New Bedford. Describe these three-step sequences.
3. Which trips in the Cape Air network take the greatest number of flights?
4. The route map for the northern routes of Big Sky Airlines for May 2001 is given in Figure 2.* Produce an adjacency matrix for this map. The current route map is available at hosts3.in-tch.com/www.bigskyair.com/body_northern_routes. html.

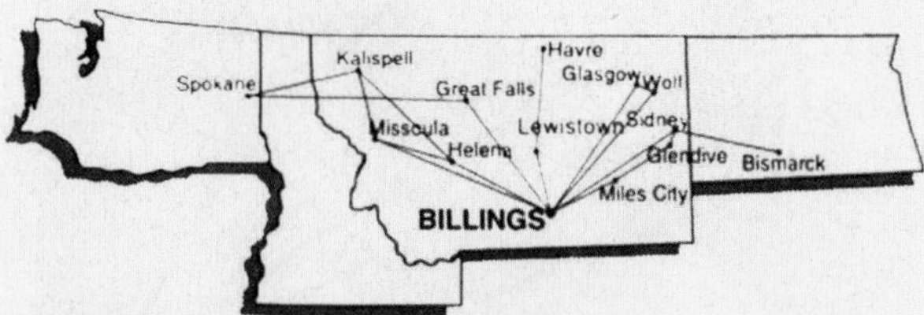

FIGURE 2 Big Sky Airlines Northern Route Map—May 2001

5. Which Big Sky cities may be reached by a three-flight sequence from Helena? For which cities does it take *at least* three flights to get from them to Helena?
6. Which trips in the Big Sky network take the largest number of flights? How many flights do these trips take?

*Airline route map courtesy of Big Sky Airlines, Billings, MT, www.bigskyair.com.

Chapter 7

Linear Programming

Many realistic problems involve inequalities. For example, a factory may have no more than 200 workers on a shift and must manufacture at least 3000 units at a cost of no more than \$35 each. How many workers should it have per shift in order to produce the required units at minimal cost? *Linear programming* is a method for finding the optimal (best possible) solution for such problems, if there is one.

In this chapter we shall study two methods of solving linear programming problems: the graphical method and the simplex method. The graphical method requires a knowledge of **linear inequalities,** those involving only first-degree polynomials in x and y. So we begin with a study of such inequalities.

7.1 GRAPHING LINEAR INEQUALITIES IN TWO VARIABLES

Examples of linear inequalities in two variables include

$$x + 2y < 4, \qquad 3x + 2y > 6, \qquad \text{and} \quad 2x - 5y \geq 10.$$

A solution of a linear inequality is an ordered pair that satisfies the inequality. For example (4, 4) is a solution of

$$3x - 2y \leq 6.$$

(Check by substituting 4 for x and 4 for y.) A linear inequality has an infinite number of solutions, one for every choice of a value for x. The best way to show these solutions is with a graph that consists of all the points in the plane whose coordinates satisfy the inequality.

EXAMPLE 1 Graph the inequality $3x - 2y \le 6$.

First, solve the inequality for y.

$$3x - 2y \le 6$$
$$-2y \le -3x + 6$$
$$y \ge \frac{3}{2}x - 3 \qquad \text{Multiply by } -1/2; \text{ reverse the inequality.}$$
$$y \ge 1.5x - 3$$

This inequality has the same solutions as the original one. To solve it, note that the points on the line $y = 1.5x - 3$ certainly satisfy $y \ge 1.5x - 3$. Plot some points and graph this line, as in Figure 7.1.

1

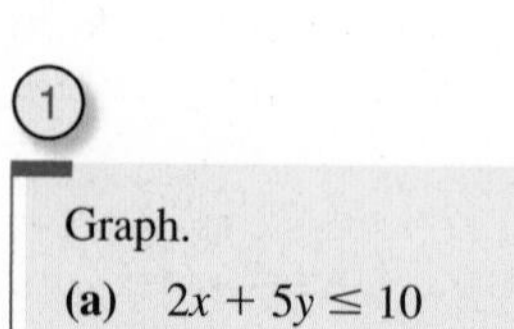
Graph.

(a) $2x + 5y \le 10$

(b) $x - y \ge 4$

Answers:

(a)

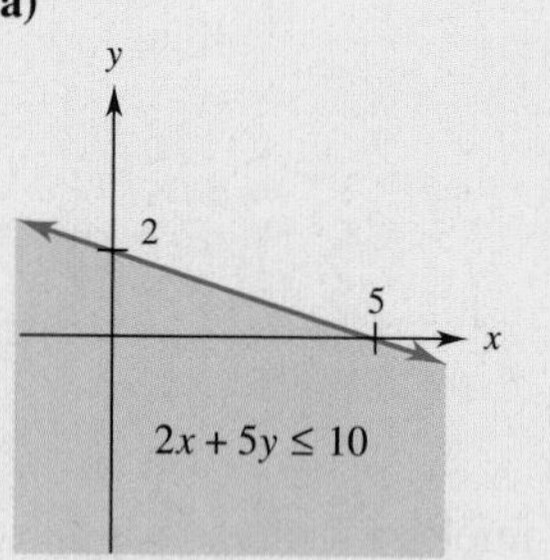

(b)

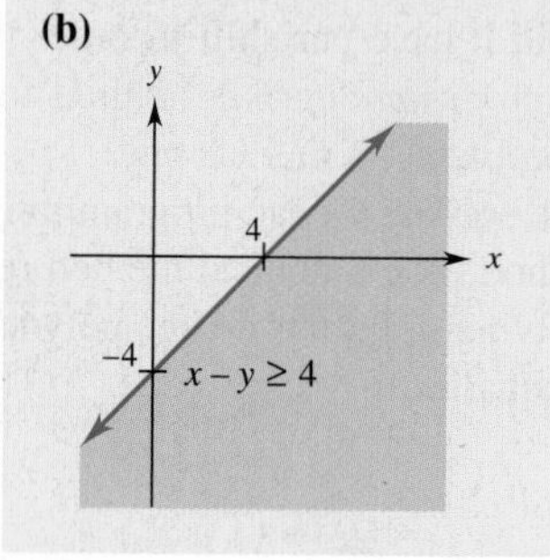

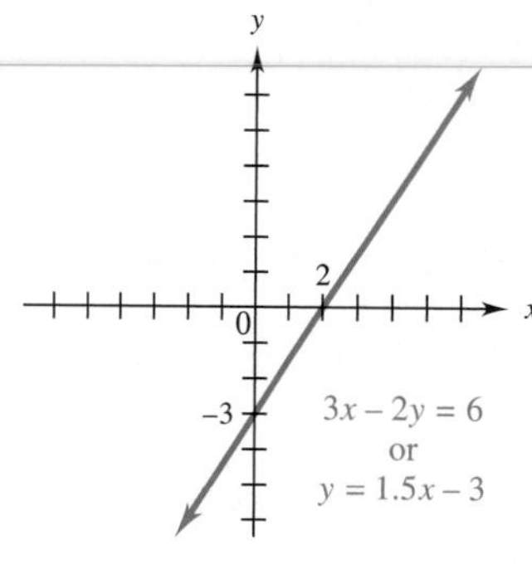

FIGURE 7.1

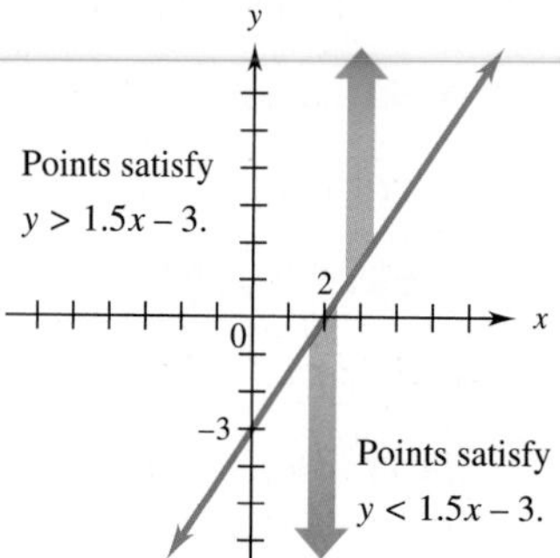

FIGURE 7.2

The points on the line satisfy "y *equals* $1.5x - 3$." The points satisfying "y is *greater than* $1.5x - 3$" are the points *above* the line (because they have larger second coordinates than the points on the line). Similarly, the points satisfying

$$y < 1.5x - 3$$

lie below the line (because they have smaller second coordinates), as shown in Figure 7.2. The line $y = 1.5x - 3$ is the **boundary line.**

In summary, the solutions of $y \ge 1.5x - 3$ are all points *on or above* the line $y = 1.5x - 3$. The line and the shaded region of Figure 7.3 make up the graph of the inequality $y \ge 1.5x - 3$. ◀ 1

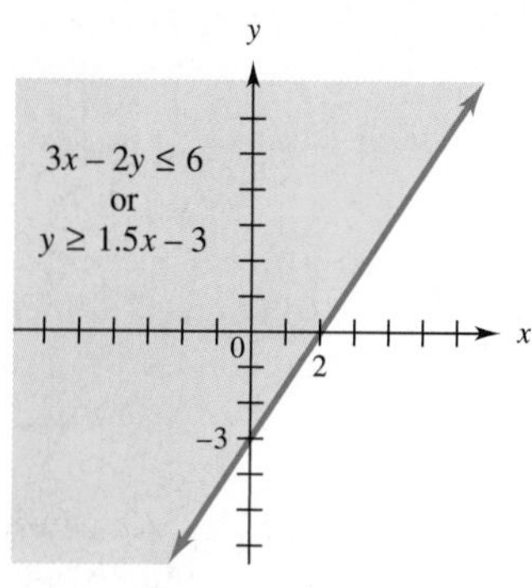

FIGURE 7.3

2

Graph.

(a) $2x + 3y > 12$

(b) $3x - 2y < 6$

Answers:

(a)

(b)

EXAMPLE 2 Graph $x + 4y < 4$.

First obtain an equivalent inequality by solving for y.

$$4y < -x + 4$$

$$y < -\frac{1}{4}x + 1$$

$$y < -.25x + 1$$

The boundary line is $y = -.25x + 1$, but *it* is *not* part of the solution since points on the line do not satisfy $y < -.25x + 1$. To indicate this, the line is drawn dashed in Figure 7.4. The points *below* the boundary line are the solutions of $y < -.25x + 1$ because they have smaller second coordinates than the points on the line $y = -.25x + 1$. The shaded region in Figure 7.4 (excluding the dashed line) is the graph of the inequality $y < -.25x + 1$. ◀ 2

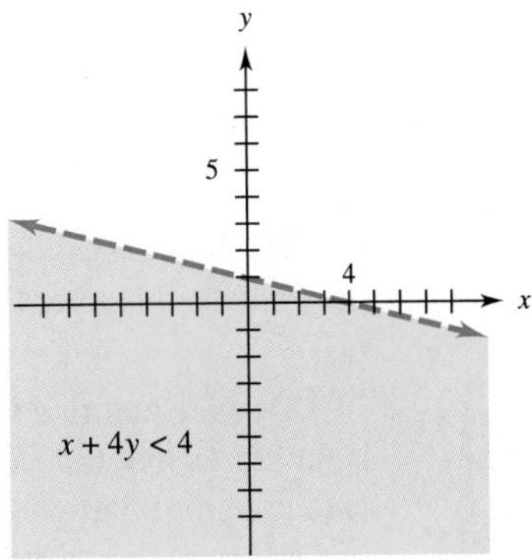

FIGURE 7.4

Examples 1 and 2 show that the solutions of a linear inequality form a **half-plane** consisting of all points on one side of the boundary line (and possibly the line itself). When an inequality is solved for y, the inequality symbol immediately tells whether the points above (>), on (=), or below (<) the boundary line satisfy the inequality, as summarized here.

Inequality	*Solution Consists of All Points*
$y \geq mx + b$	*on or above* the line $y = mx + b$
$y > mx + b$	*above* the line $y = mx + b$
$y \leq mx + b$	*on or below* the line $y = mx + b$
$y < mx + b$	*below* the line $y = mx + b$

When graphing by hand, draw the boundary line $y = mx + b$ solid when it is included in the solution (≥ or ≤ inequalities) and dashed when it is not part of the solution (> or < inequalities).

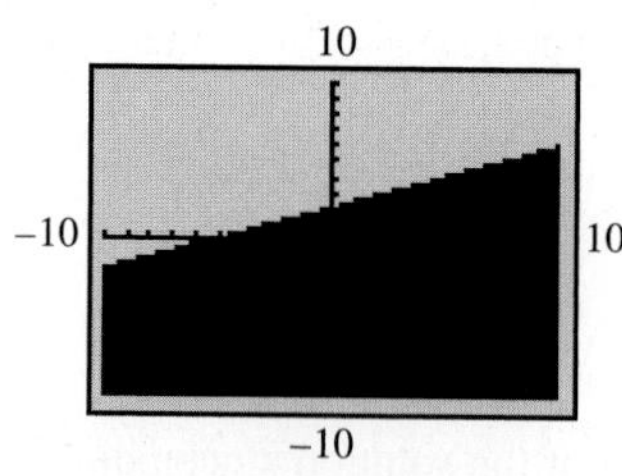

FIGURE 7.5

(3)

Use a graphing calculator to graph $2x < y$.

Answer:

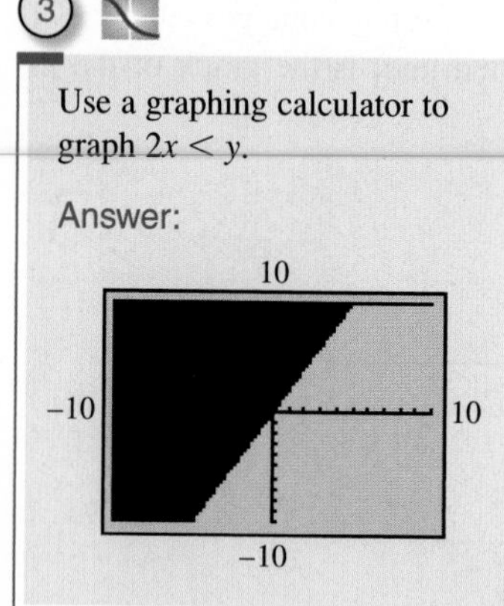

(4)

Graph each of the following.

(a) $x \geq 3$

(b) $y - 3 \leq 0$

Answers:

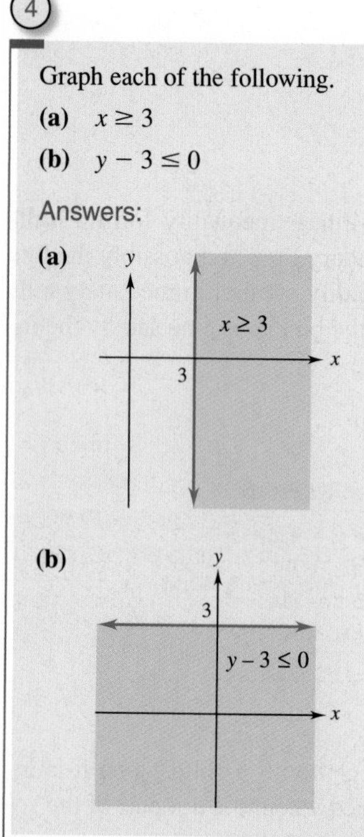

EXAMPLE 3 Graph $5y - 2x \leq 10$.

Solve the inequality for y.

$$5y \leq 2x + 10$$

$$y \leq \frac{2}{5}x + 2$$

$$y \leq .4x + 2$$

The graph consists of all points on or below the boundary line $y \leq .4x + 2$, as shown in Figure 7.5. (Check your instruction manual for directions on shading calculator generated graphs.) (3)

CAUTION You cannot tell from a calculator-produced graph whether or not the boundary line is included. It is included in Figure 7.5 but not in the answer to problem 3 at the side.

EXAMPLE 4 Graph each of the following.

(a) $y \geq 2$

The boundary line is the horizontal line $y = 2$. The graph consists of all points on or above this line (Figure 7.6).

(b) $x \leq -1$

This inequality does not fit the pattern discussed above, but it can be solved by a similar technique. Here the boundary line is the vertical line $x = -1$ and it is included in the solution. The points satisfying $x < -1$ are all points to the left of this line (because they have x-coordinates smaller than -1). So the graph consists of the points that are *on or to the left of* the vertical line $x = -1$, as shown in Figure 7.7. (4)

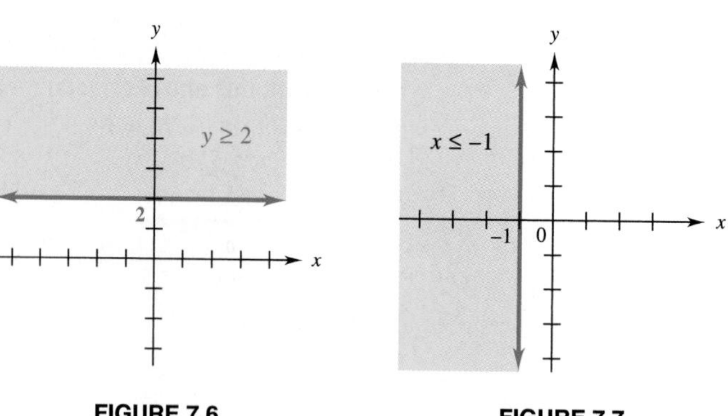

FIGURE 7.6

FIGURE 7.7

An alternative technique for solving inequalities, that does not require solving for y is illustrated in the next example. Feel free to use it, if you find it easter than the technique presented in Examples 1–4.

EXAMPLE 5 Graph $4y - 2x > 6$.

The boundary line is $4y - 2x = 6$, which can be graphed by finding its intercepts.

x-intercept: let $y = 0$	y-intercept: let $x = 0$
$4(0) - 2x = 6$	$4y - 2(0) = 6$
$x = -3$	$y = \frac{6}{4} = 1.5$

The graph is the half-plane above or below this line. To determine which, choose a test point—any point not on the boundary line, say, (0, 0). Letting $x = 0$ and $y = 0$ in the inequality produces

$$4(0) - 2(0) > 6, \quad \text{a } \textit{false} \text{ statement.}$$

Therefore, (0, 0) is not in the solution. So the solution is the half-plane that does *not* include (0, 0), as shown in Figure 7.8.

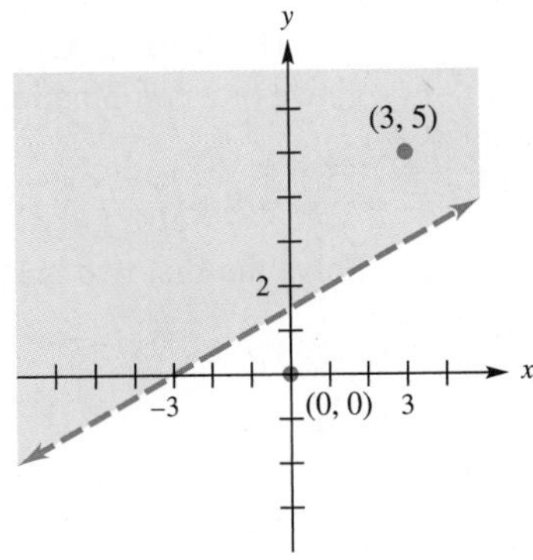

FIGURE 7.8

If a different test point is used, say (3, 5), then substituting $x = 3$ and $y = 5$ in the inequality produces

$$4(5) - 2(3) > 6, \quad \text{a } \textit{true} \text{ statement.}$$

Therefore, the solution of the inequality is the half-plane containing (3, 5), as shown in Figure 7.8. ◀

Systems of Inequalities

Realistic problems often involve many inequalities. For example, a manufacturing problem might produce inequalities resulting from production requirements, as well as inequalities about cost requirements. A set of at least two inequalities is called a **system of inequalities.** The **graph** of a system of inequalities is made up of all those points that satisfy all the inequalities of the system at the same time.

EXAMPLE 6 Graph the system

$$y < -3x + 12$$
$$x < 2y.$$

First, graph the solution of $y < -3x + 12$. The boundary, the line with equation $y = -3x + 12$, is dashed in Figure 7.9 on the next page. The points below the

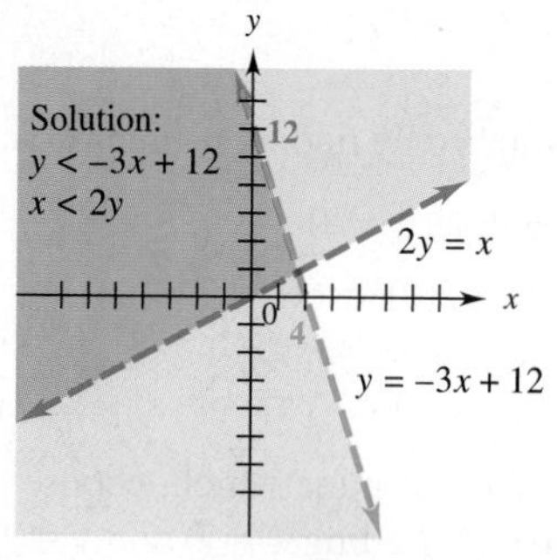

FIGURE 7.9

boundary satisfy $y < -3x + 12$. Now graph the solution of $x < 2y$ on the same axes. Again, the boundary line $x = 2y$ is dashed. Use a test point to see that the region above the boundary satisfies $x < 2y$. The purple region in Figure 7.8 shows all the points that satisfy both inequalities of the system. ◀ (5)

The purple region of Figure 7.9 is sometimes called the **region of feasible solutions,** or just the **feasible region,** since it is made up of all the points that satisfy (are feasible for) each inequality of the system.

(5)

Graph the system
$x + y \le 6, 2x + y \ge 4.$

Answer:

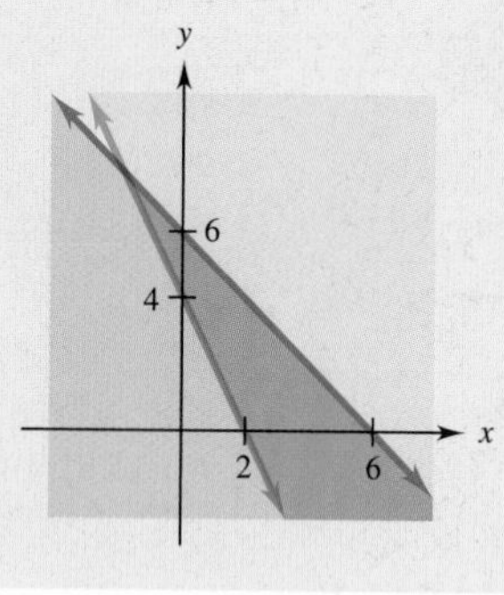

EXAMPLE 7 Graph the feasible region for the system

$$2x - 5y \le 10$$
$$x + 2y \le 8$$
$$x \ge 0, y \ge 0.$$

The feasible region may be found by using either of the methods illustrated above or a combination of them.

"Solve for y" Method

Solve the first two inequalities for y:

$$2x - 5y \le 10$$
$$-5y \le -2x + 10$$
$$y \ge .4x - 2$$

and

$$x + 2y \le 8$$
$$2y \le -x + 8$$
$$y \le -.5x + 4.$$

The inequalities $x \ge 0$ and $y \ge 0$ restrict the graph to the first quadrant. Therefore, the feasible region consists of all points in the first quadrant that are on or above the line $y = .4x - 2$ and on or below the line $y = -.5x + 4$. A calculator-generated graph, with the feasible region shaded by hand, is shown in Figure 7.10.

Test Point Method

Graph $2x - 5y = 10$ by finding its intercepts, which give the points (5, 0) and (0, −2). Choose a test point and determine that the solutions of $2x - 5y \le 10$ are the points on or above the boundary line.

Similarly, determine the intercepts of $x + 2y = 8$, which show that the line goes through (8, 0) and (0, 4). Choose a test point and conclude that the solutions of $x + 2y \le 8$ are the points that lie on or below the boundary line. The inequalities $x \ge 0$ and $y \ge 0$ restrict the graph to the first quadrant. Hence, the feasible region consists of all points in the first quadrant that are on or above $2x - 5y = 10$ and on or below $x + 2y = 8$, as shown in Figure 7.11. ◀ (6)

(6)

Graph the feasible region of the system
$$x + 4y \le 8$$
$$x - y \ge 3$$
$$x \ge 0, y \ge 0.$$

Answer:

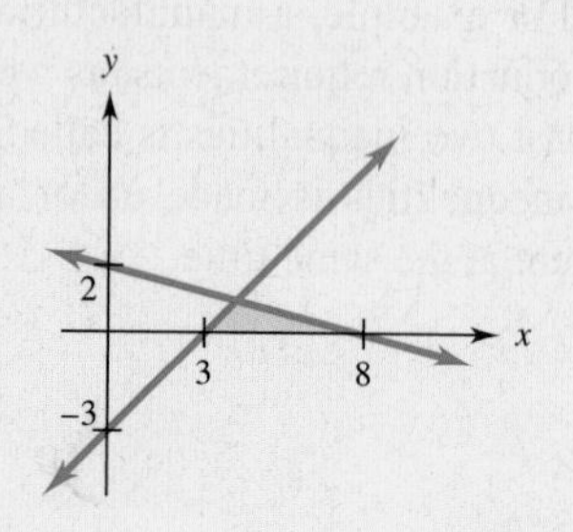

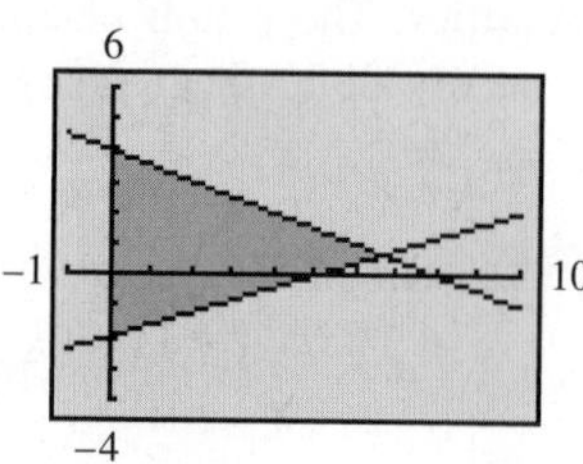

FIGURE 7.10

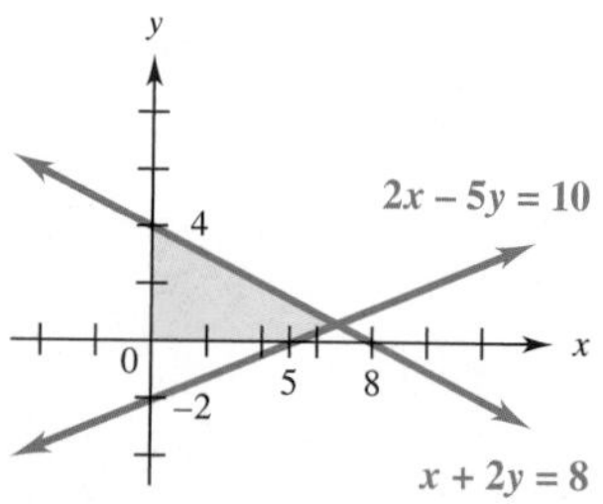

FIGURE 7.11

Applications

As we shall see in the rest of this chapter, many realistic problems lead to systems of linear inequalities. The next example is typical of such problems.

EXAMPLE 8 Midtown Manufacturing Company makes plastic plates and cups, both of which require time on two machines. A unit of plates requires 1 hour on machine A and 2 on machine B, while a unit of cups requires 3 hours on machine A and 1 on machine B. Each machine is operated for at most 15 hours per day. Write a system of inequalities expressing these conditions and graph the feasible region.

Let x represent the number of units of plates to be made and y represent the number of units of cups. Then make a chart that summarizes the given information.

	Number of Units	TIME ON MACHINE A	TIME ON MACHINE B
Plates	x	1	2
Cups	y	3	1
Maximum Time Available		15	15

On machine A, x units of plates require a total of $1 \cdot x = x$ hours while y units of cups require $3 \cdot y = 3y$ hours. Since machine A is available no more than 15 hours a day,

$$x + 3y \le 15. \qquad \text{Machine A}$$

The requirement that machine B be used no more than 15 hours a day gives

$$2x + y \le 15. \qquad \text{Machine B}$$

It is not possible to produce a negative number of cups or plates, so that

$$x \ge 0 \quad \text{and} \quad y \ge 0.$$

The feasible region for this system of inequalities is shown in Figure 7.12.

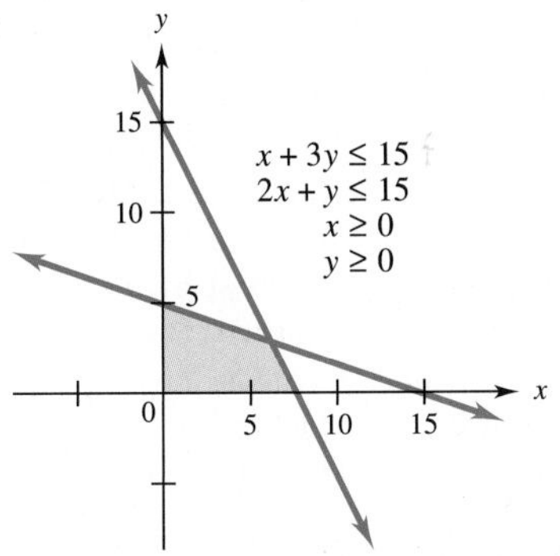

FIGURE 7.12

7.1 Exercises

Graph each of the following linear inequalities. (See Examples 1–5.)

1. $y < 5 - 2x$
2. $y > x + 3$
3. $3x - 2y \ge 18$
4. $2x + 5y \le 10$
5. $2x - y \le 4$
6. $4x - 3y \ge 24$
7. $y \le -4$
8. $x \ge -2$
9. $x + 4y \le 2$
10. $3x + 2y \ge 6$
11. $4x + 3y > -3$
12. $5x + 3y > 15$
13. $2x - 4y < 3$
14. $4x - 3y < 12$
15. $x \le 5y$
16. $2x \ge y$
17. $-3x < y$
18. $-x > 6y$
19. $y < x$
20. $y > -2x$

21. In your own words, explain how to determine whether the boundary of an inequality is solid or dashed.

22. When graphing $y \leq 3x - 6$, would you shade above or below the line $y = 3x - 6$? Explain your answer.

Graph the feasible region for the following systems of inequalities. (See Examples 6 and 7.)

23. $x - y \geq 1$, $x \leq 3$

24. $2x + y \leq 5$, $x + 2y \leq 5$

25. $4x + y \geq 9$, $2x + 3y \leq 7$

26. $2x + y > 8$, $4x - y < 3$

27. $x + y > 5$, $x - 2y < 2$

28. $3x - 4y < 6$, $2x + 5y > 15$

29. $2x - y < 1$, $3x + y < 6$

30. $x + 3y \leq 6$, $2x + 4y \geq 7$

31. $-x - y < 5$, $2x - y < 4$

32. $6x - 4y > 8$, $3x + 2y > 4$

33. $3x + y \geq 6$, $x + 2y \geq 7$, $x \geq 0$, $y \geq 0$

34. $2x + 3y \geq 12$, $x + y \geq 4$, $x \geq 0$, $y \geq 0$

35. $-2 < x < 3$, $-1 \leq y \leq 5$, $2x + y < 6$

36. $-2 < x < 2$, $y > 1$, $x - y > 0$

37. $2y - x \geq -5$, $y \leq 3 + x$, $x \geq 0$, $y \geq 0$

38. $2x + 3y \leq 12$, $2x + 3y > -6$, $3x + y < 4$, $x \geq 0$, $y \geq 0$

39. $3x + 4y > 12$, $2x - 3y < 6$, $0 \leq y \leq 2$, $x \geq 0$

40. $0 \leq x \leq 9$, $x - 2y \geq 4$, $3x + 5y \leq 30$, $y \geq 0$

In Exercises 41 and 42, find a system of inequalities whose feasible region is the interior of the given polygon.

41. Rectangle with vertices (2, 3), (2, −1), (7, 3), (7, −1)

42. Triangle with vertices (2, 4), (−4, 0), (2, −1)

43. **Business** Cindi Herring and Kent Merrill produce handmade shawls and afghans. They spin the yarn, dye it, and then weave it. A shawl requires 1 hour of spinning, 1 hour of dyeing, and 1 hour of weaving. An afghan needs 2 hours of spinning, 1 of dyeing, and 4 of weaving. Together, they spend at most 8 hours spinning, 6 hours dyeing, and 14 hours weaving.

(a) Complete the following chart.

	Number	Hours Spinning	Hours Dyeing	Hours Weaving
Shawls	x			
Afghans	y			
Maximum Number of Hours Available		8	6	14

(b) Use the chart to write a system of inequalities that describes the situation.

(c) Graph the feasible region of this system of inequalities.

44. **Business** An electric shaver manufacturer makes two models, the regular and the flex. Because of demand, the number of regular shavers made is never more than half the number of flex shavers. The factory's production cannot exceed 1200 shavers per week.

(a) Write a system of inequalities that describes the possibilities for making x regular and y flex shavers per week.

(b) Graph the feasible region of this system of inequalities.

In each of the following, write a system of inequalities that describes all the conditions and graph the feasible region of the system. (See Example 8.)

45. **Business** Southwestern Oil supplies two distributors located in the Northwest. One distributor needs at least 3000 barrels of oil, and the other needs at least 5000 barrels. Southwestern can send out at most 10,000 barrels. Let x = the number of barrels of oil sent to distributor 1 and y = the number sent to distributor 2.

46. **Business** The California Almond Growers have 2400 boxes of almonds to be shipped from their plant in Sacramento to Des Moines and San Antonio. The Des Moines market needs at least 1000 boxes, while the San Antonio market must have at least 800 boxes. Let x = the number of boxes to be shipped to Des Moines and y = the number of boxes to be shipped to San Antonio.

47. **Business** A cement manufacturer produces at least 3.2 million barrels of cement annually. He is told by the Environmental Protection Agency that his operation emits 2.5 pounds of dust for each barrel produced. The EPA has ruled that annual emissions must be reduced to 1.8 million pounds. To do this, the manufacturer plans to replace the present dust collectors with two types of electronic precipitators. One type would reduce emissions to .5 pound per barrel and would cost 16¢ per barrel. The other would reduce the dust to .3 pound per barrel and would cost 20¢ per barrel. The manufacturer does not want to spend more than .8 million dollars on the precipitators. He needs to know how many barrels he should produce with each type. Let x = the number of barrels in millions produced with the first type and y = the number of barrels in millions produced with the second type.

48. **Health** A dietician is planning a snack package of fruit and nuts. Each ounce of fruit will supply 1 unit of protein, 2 units of carbohydrates, and 1 unit of fat. Each ounce of nuts will supply 1 unit of protein, 1 unit of carbohydrates, and 1 unit of fat. Every package must provide at least 7 units of protein, at least 10 units of carbohydrates, and no more than 9 units of fat. Let x be the ounces of fruit and y the ounces of nuts to be used in each package.

7.2 LINEAR PROGRAMMING: THE GRAPHICAL METHOD

Many problems in business, science, and economics involve finding the optimal value of a function (for instance, the maximum value of the profit function or the minimum value of the cost function) subject to various **constraints** (such as transportation costs, environmental protection laws, availability of parts, interest rates, etc.). **Linear programming** deals with such situations in which the function to be optimized, called the **objective function,** is linear and the constraints are given by linear inequalities. Linear programming problems that involve only two variables can be solved by the graphical method, explained in Example 1.

EXAMPLE 1 Find the maximum and minimum values of the objective function $z = 2x + 5y$, subject to the following constraints.

$$\begin{aligned} 3x + 2y &\le 6 \\ -2x + 4y &\le 8 \\ x + y &\ge 1 \\ x \ge 0, y &\ge 0 \end{aligned}$$

First, graph the feasible region of the system of inequalities (Figure 7.13). The points in this region or on its boundaries are the only ones that satisfy all the constraints. However, each such point may produce a different value of the objective function. For instance, the points (.5, 1) and (1, 0) in the feasible region lead to the values

$$z = 2(.5) + 5(1) = 6 \quad \text{and} \quad z = 2(1) + 5(0) = 2.$$

We must find the points that produce the maximum and minimum values of z.

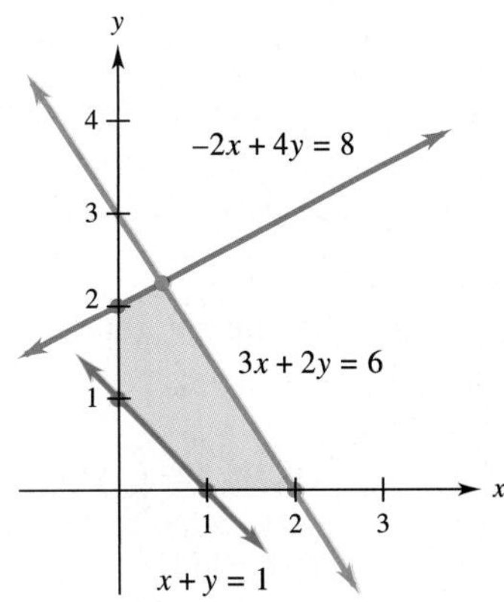

FIGURE 7.13

To find the maximum value, consider various possible values for z. For instance, when $z = 0$, then the objective function is $0 = 2x + 5y$, whose graph is a straight line. Similarly, when z is 5, 10, and 15, the objective function becomes (in turn)

$$5 = 2x + 5y, \qquad 10 = 2x + 5y, \qquad 15 = 2x + 5y.$$

These four lines are graphed in Figure 7.14 on the next page. (All the lines are parallel because they have the same slope.) The figure shows that z cannot take on the value 15 because the graph for $z = 15$ is entirely outside the feasible region. The maximum possible value of z will be obtained from a line parallel to the others and between the lines representing the objective function when $z = 10$ and $z = 15$. The value of z will be as large as possible and all constraints will be satisfied if this line just touches the feasible region. This occurs with the green line through point A.

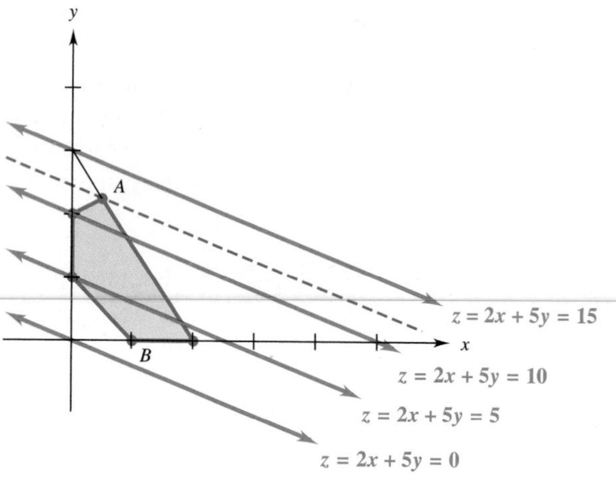

FIGURE 7.14

Suppose the objective function in Example 1 is changed to $z = 5x + 2y$.

(a) Sketch the graphs of the objective function when $z = 0$, $z = 5$, and $z = 10$ on the region of feasible solutions given in Figure 7.13.

(b) From the graph, decide what values of x and y will maximize the objective function.

Answers:

(a)

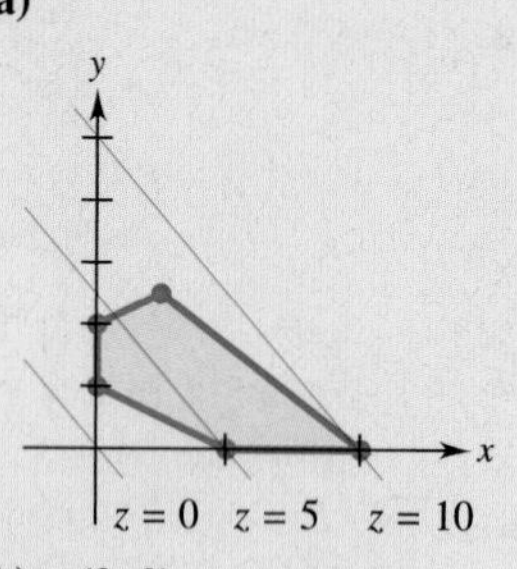

(b) (2, 0)

The point A is the intersection of the graphs of $3x + 2y = 6$ and $-2x + 4y = 8$. Its coordinates can be found either algebraically or graphically (using a graphing calculator).

Algebraic Method

Solve the system

$$3x + 2y = 6$$
$$-2x + 4y = 8,$$

as in Section 6.1, to get $x = 1/2$ and $y = 9/4$. Hence A has coordinates $(1/2, 9/4) = (.5, 2.25)$.

Graphical Method

Solve the two equations for y.

$$y = -1.5x + 3$$
$$y = .5x + 2$$

Graph both equations on the same screen and use the intersection finder to find that the coordinates of the intersection point A are $(.5, 2.25)$.

The value of z at the point A is

$$z = 2x + 5y = 2(.5) + 5(2.25) = 12.25.$$

Thus, the maximum possible value of z is 12.25. Similarly, the minimum value of z occurs at point B, which has coordinates (1, 0). The minimum value of z is $2(1) + 5(0) = 2$. ◀ 1

Points such as A and B in Example 1 are called corner points. A **corner point** is a point in the feasible region where the boundary lines of two constraints cross. The feasible region in Figure 7.13 is **bounded,** because the region is enclosed by boundary lines on all sides. Linear programming problems with bounded regions always have solutions. However, if Example 1 did not include the constraint $3x + 2y \leq 6$, the feasible region would be **unbounded,** and there would be no way to *maximize* the value of the objective function.

Some general conclusions can be drawn from the method of solution used in Example 1. Figure 7.15 shows various feasible regions and the lines that result from various values of z. (Figure 7.15 shows the situation in which the lines are in order from left to right as z increases.) In part (a) of the figure, the objective function takes on its minimum value at corner point Q and its maximum value at P. The minimum is again at Q in part (b), but the maximum occurs at P_1 or P_2, or any point on the line segment connecting them. Finally, in part (c), the minimum value occurs at Q, but the objective function has no maximum value because the feasible region is unbounded.

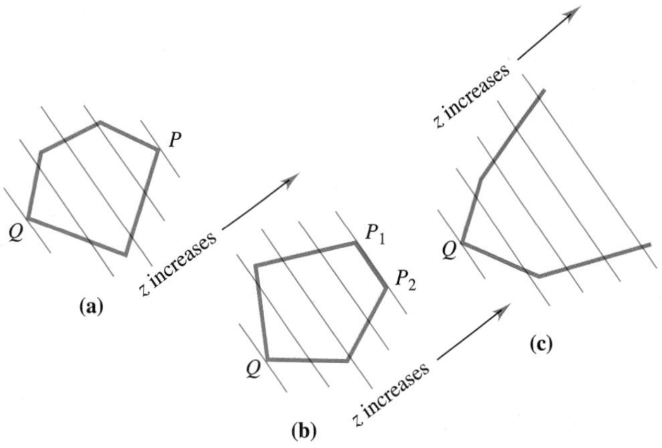

FIGURE 7.15

The preceding discussion suggests the truth of the **corner point theorem.**

Corner Point Theorem

If the feasible region is bounded, then the objective function has both a maximum and a minimum value and each occurs at one or more corner points.

If the feasible region is unbounded, the objective function may not have a maximum or minimum. But if a maximum or minimum value exists, it will occur at one or more corner points.

This theorem simplifies the job of finding an optimum value: First, graph the feasible region and find all corner points. Then test each point in the objective function. Finally, identify the corner point producing the optimum solution.

With the theorem, the problem in Example 1 could have been solved by identifying the five corner points of Figure 7.13: (0, 1), (0, 2), (.5, 2.25), (2, 0), and (1, 0). Then, substituting each of these points into the objective function $z = 2x + 5y$ would identify the corner points that produce the maximum and the minimum values of z.

Corner Point	*Value of* $z = 2x + 5y$
(0, 1)	$2(0) + 5(1) = 5$
(0, 2)	$2(0) + 5(2) = 10$
(.5, 2.25)	$2(.5) + 5(2.25) = 12.25$ (maximum)
(2, 0)	$2(2) + 5(0) = 4$
(1, 0)	$2(1) + 5(0) = 2$ (minimum)

2

(a) Identify the corner points in the graph.

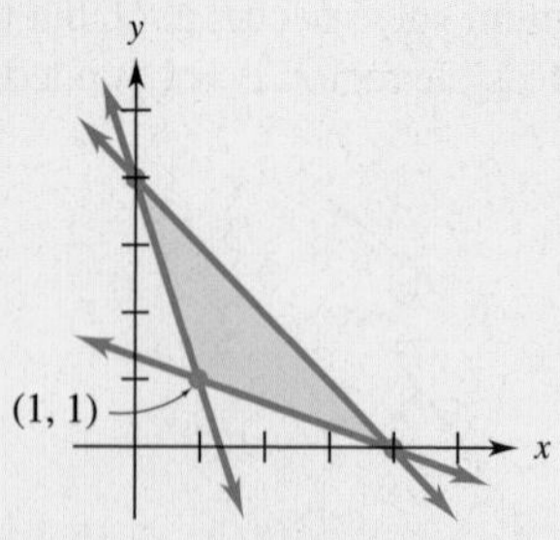

(b) Which corner point would minimize $z = 2x + 3y$?

Answers:

(a) (0, 4), (1, 1), (4, 0)

(b) (1, 1)

From these results, the corner point (.5, 2.25) yields the maximum value of 12.25 and the corner point (1, 0) gives the minimum value of 2. These are the same values found earlier. (2)

A summary of the steps in solving a linear programming problem by the graphical method is given here.

Solving a Linear Programming Problem Graphically

1. Write the objective function and all necessary constraints.
2. Graph the feasible region.
3. Determine the coordinates of each of the corner points.
4. Find the value of the objective function at each corner point.
5. If the feasible region is bounded, the solution is given by the corner point producing the optimum value of the objective function.
6. If the feasible region is an unbounded region in the first quadrant and both coefficients of the objective function are positive,* then the minimum value of the objective function occurs at a corner point and there is no maximum value.

EXAMPLE 2 Sketch the feasible region for the following set of constraints:

$$3y - 2x \geq 0$$
$$y + 8x \leq 52$$
$$y - 2x \leq 2$$
$$x \geq 3.$$

Then find the maximum and minimum values of the objective function $z = 5x + 2y$.

The graph in Figure 7.16(a) shows that the feasible region is bounded. The corner points are found by solving systems of two equations either algebraically, using the methods of Chapter 6, or graphically, using the intersection finder on a graphing calculator. Figure 7.16(b) shows the calculator graphs of the first three inequalities. With the graphical method, corner points on the line $x = 3$ are found by observation.

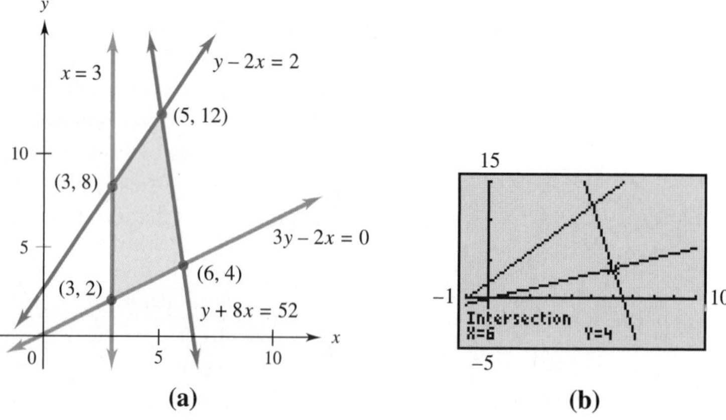

FIGURE 7.16

*This is the only case of an unbounded region that occurs in the applications considered here.

(3)

Use the region of feasible solutions in the sketch to find the following.

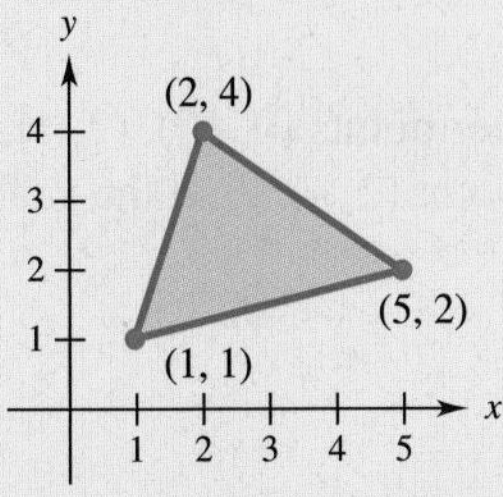

(a) The values of x and y that maximize $z = 2x - y$.

(b) The maximum value of $z = 2x - y$.

(c) The values of x and y that minimize $z = 4x + 3y$.

(d) The minimum value of $z = 4x + 3y$.

Answers:

(a) (5, 2)

(b) 8

(c) (1, 1)

(d) 7

The sketch shows a feasible region. Let $z = x + 3y$. Use the sketch to find the values of x and y that

(a) minimize z;

(b) maximize z.

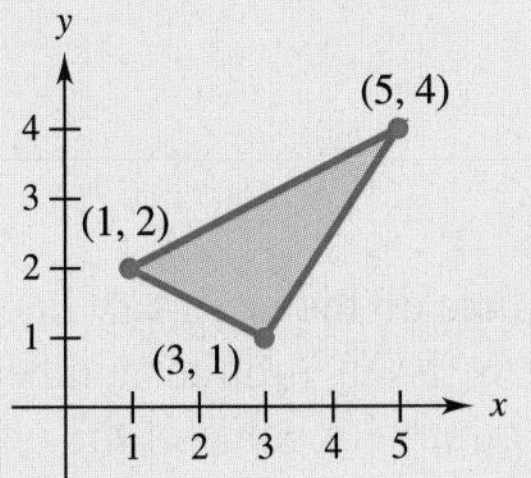

Answers:

(a) (3, 1)

(b) (5, 4)

Use the corner points from the graph to find the maximum and minimum values of the objective function.

Corner Point	*Value of* $z = 5x + 2y$
(3, 2)	$5(3) + 2(2) = 19$ (minimum)
(6, 4)	$5(6) + 2(4) = 38$
(5, 12)	$5(5) + 2(12) = 49$ (maximum)
(3, 8)	$5(3) + 2(8) = 31$

The minimum value of $z = 5x + 2y$ is 19 at the corner point (3, 2). The maximum value is 49 at (5, 12). (3)

EXAMPLE 3 Solve the following linear programming problem.

$$\begin{aligned} \text{Minimize} \quad & z = x + 2y \\ \text{subject to:} \quad & x + y \le 10 \\ & 3x + 2y \ge 6 \\ & x \ge 0, y \ge 0. \end{aligned}$$

The feasible region is shown in Figure 7.17. From the figure, the corner points are (0, 3), (0, 10), (10, 0), and (2, 0). These corner points give the following values of z.

Corner Point	*Value of* $z = x + 2y$
(0, 3)	$0 + 2(3) = 6$
(0, 10)	$0 + 2(10) = 20$
(10, 0)	$10 + 2(0) = 10$
(2, 0)	$2 + 2(0) = 2$ (minimum)

The minimum value of z is 2; it occurs at (2, 0). (4)

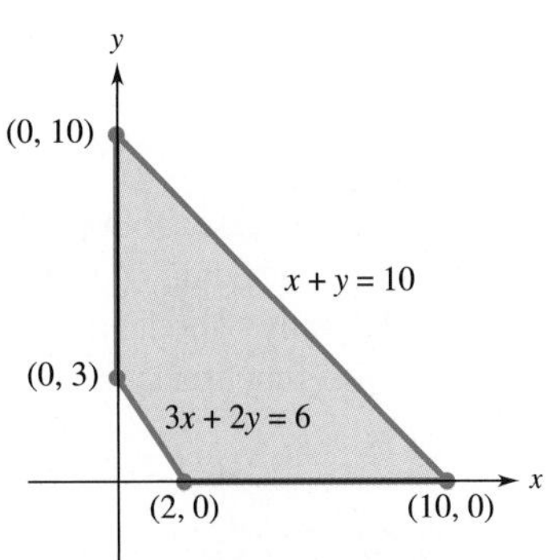

FIGURE 7.17

EXAMPLE 4 Solve the following linear programming problem.

$$\begin{aligned} \text{Minimize} \quad & z = 2x + 4y \\ \text{subject to:} \quad & x + 2y \geq 10 \\ & 3x + y \geq 10 \\ & x \geq 0, y \geq 0. \end{aligned}$$

Figure 7.18 shows the hand-drawn graph with corner points (0, 10), (2, 4), and (10, 0), as well as the calculator graph with the corner point (2, 4). Find the value of z for each point.

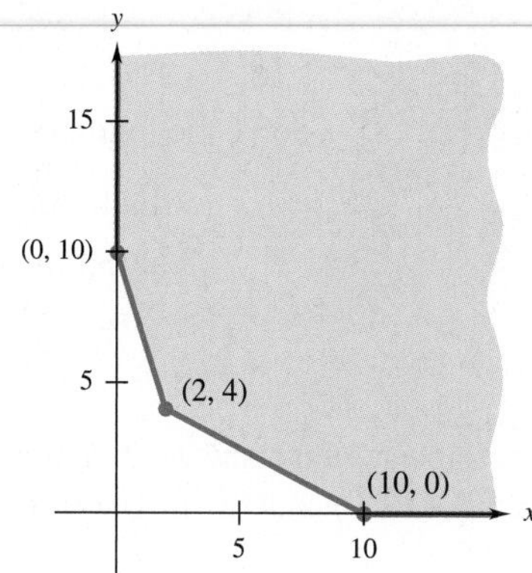

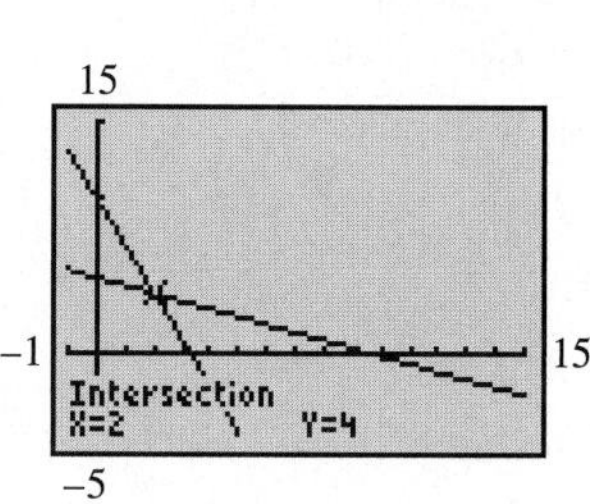

FIGURE 7.18

Corner Point	*Value of $z = 2x + 4y$*
(0, 10)	$2(0) + 4(10) = 40$
(2, 4)	$2(2) + 4(4) = 20$ (minimum)
(10, 0)	$2(10) + 4(0) = 20$ (minimum)

In this case, both (2, 4) and (10, 0), as well as all the points on the boundary line between them, give the same optimum value of z. There is an infinite number of equally "good" values of x and y which give the same minimum value of the objective function $z = 2x + 4y$. The minimum value is 20. ◀ (5)

(5)

The sketch below shows a region of feasible solutions. From the sketch decide what ordered pair would minimize $z = 2x + 4y$.

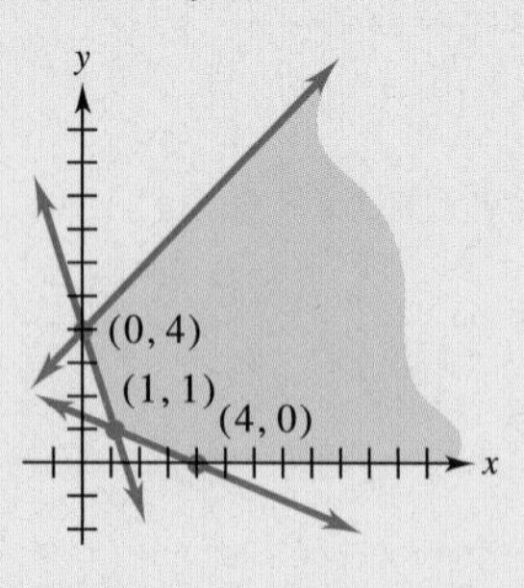

Answer:

(1, 1)

7.2 Exercises

Exercises 1–6 show regions of feasible solutions. Use these regions to find maximum and minimum values of each given objective function. (See Examples 1 and 2.)

1. $z = 3x + 5y$

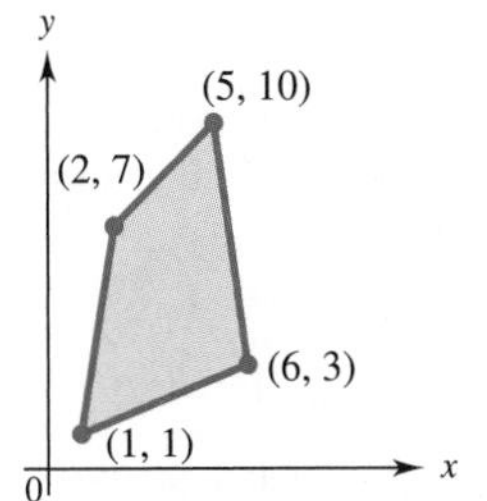

2. $z = 6x + y$

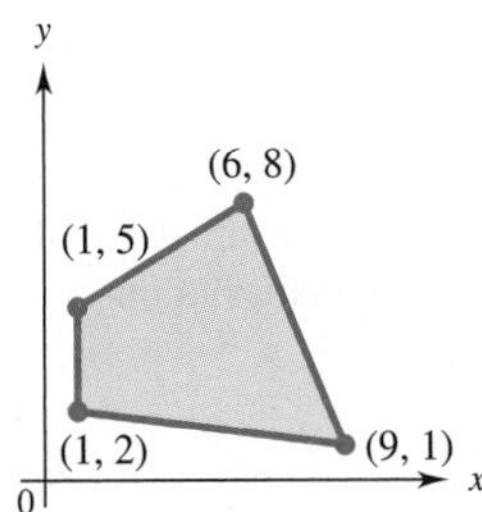

3. $z = .40x + .75y$

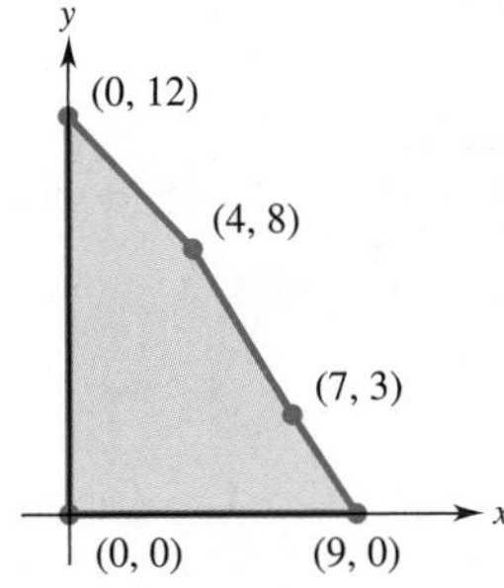

4. $z = .35x + 1.25y$

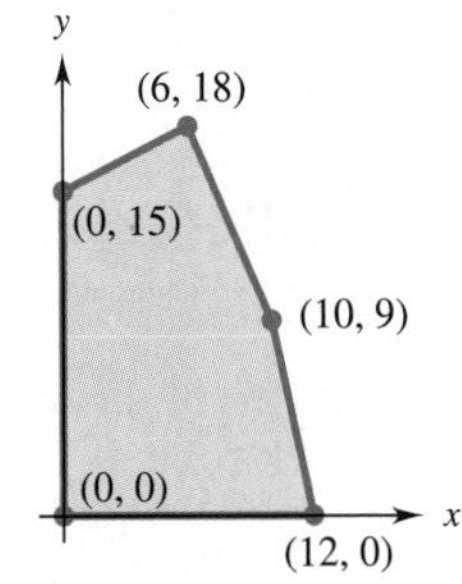

5. **(a)** $z = 4x + 2y$
(b) $z = 2x + 3y$
(c) $z = 2x + 4y$
(d) $z = x + 4y$

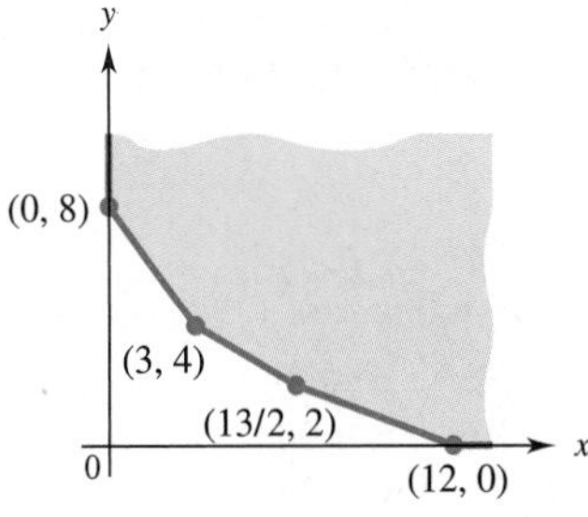

6. **(a)** $z = 4x + y$
(b) $z = 5x + 6y$
(c) $z = x + 2y$
(d) $z = x + 6y$

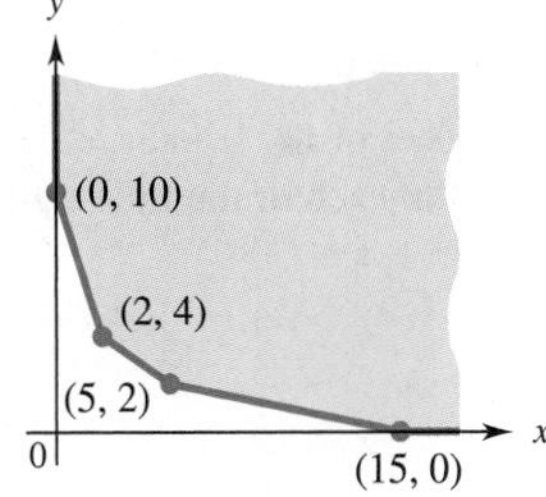

Use graphical methods to solve Exercises 7–12. (See Examples 2–4.)

7. Maximize $z = 5x + 2y$
subject to:
$2x + 3y \le 6$
$4x + y \le 6$
$x \ge 0, y \ge 0.$

8. Minimize $z = x + 3y$
subject to:
$2x + y \le 10$
$5x + 2y \ge 20$
$-x + 2y \ge 0$
$x \ge 0, y \ge 0.$

9. Minimize $z = 2x + y$
subject to:
$3x - y \ge 12$
$x + y \le 15$
$x \ge 2, y \ge 3.$

10. Maximize $z = x + 3y$
subject to:
$2x + 3y \le 100$
$5x + 4y \le 200$
$x \ge 10, y \ge 20.$

11. Maximize $z = 4x + 2y$
subject to:
$x - y \le 10$
$5x + 3y \le 75$
$x \ge 0, y \ge 0.$

12. Maximize $z = 4x + 5y$
subject to: $10x - 5y \leq 100$
$20x + 10y \geq 150$
$x \geq 0, y \geq 0.$

Find the minimum and maximum values of $z = 3x + 4y$ (if possible) for each of the following sets of constraints. (See Examples 2–4.)

13. $3x + 2y \geq 6$
$x + 2y \geq 4$
$x \geq 0, y \geq 0$

14. $2x + y \leq 20$
$10x + y \geq 36$
$2x + 5y \geq 36$

15. $x + y \leq 6$
$-x + y \leq 2$
$2x - y \leq 8$

16. $-x + 2y \leq 6$
$3x + y \geq 3$
$x \geq 0, y \geq 0$

17. Find values of $x \geq 0$ and $y \geq 0$ that maximize $z = 10x + 12y$ subject to each of the following sets of constraints.

(a) $x + y \leq 20$
$x + 3y \leq 24$

(b) $3x + y \leq 15$
$x + 2y \leq 18$

(c) $x + 2y \geq 10$
$2x + y \geq 12$
$x - y \leq 8$

18. Find values of $x \geq 0$ and $y \geq 0$ that minimize $z = 3x + 2y$ subject to each of the following sets of constraints.

(a) $10x + 7y \leq 42$
$4x + 10y \geq 35$

(b) $6x + 5y \geq 25$
$2x + 6y \geq 15$

(c) $2x + 5y \geq 22$
$4x + 3y \leq 28$
$2x + 2y \leq 17$

19. Explain why it is impossible to maximize the function $z = 3x + 4y$ subject to the constraints:
$x + y \geq 8$
$2x + y \leq 10$
$x + 2y \leq 8$
$x \geq 0, \ y \geq 0$

20. You are given the following linear programming problem:*

Maximize $z = c_1x_1 + c_2x_2$
subject to: $2x_1 + x_2 \leq 11$
$-x_1 + 2x_2 \leq 2$
$x_1 \geq 0, x_2 \geq 0.$

If $c_2 > 0$, determine the range of c_1/c_2 for which $(x_1, x_2) = (4, 3)$ is an optimal solution.

(a) $[-2, 1/2]$

(b) $[-1/2, 2]$

(c) $[-11, -1]$

(d) $[1, 11]$

(e) $[-11, 11]$

*Problem from "Course 130 Examination Operations Research" of the *Education and Examination Committee of The Society of Actuaries*. Reprinted by permission of The Society of Actuaries.

7.3 APPLICATIONS OF LINEAR PROGRAMMING

In this section we show several applications of linear programming with two variables.

EXAMPLE 1 A 4-H Club member raises geese and pigs. She wants to raise no more than 16 animals including no more than 10 geese. She spends \$15 to raise a goose and \$45 to raise a pig, and she has \$540 available for this project. Find the 0maximum profit she can make if each goose produces a profit of \$7 and each pig a profit of \$20.

The total profit is determined by the number of geese and pigs. So let x be the number of geese to be produced and let y be the number of pigs. Then summarize the information of the problem in a table.

	Number	*Cost to Raise*	*Profit Each*
Geese	x	\$ 15	\$ 7
Pigs	y	\$ 45	\$20
Maximum Available	16	\$540	

Use this table to write the necessary constraints. Since the total number of animals cannot exceed 16, the first constraint is

$$x + y \le 16.$$

"No more than 10 geese" leads to

$$x \le 10.$$

The cost to raise x geese at \$15 per goose is $15x$ dollars, while the cost for y pigs at \$45 each is $45y$ dollars. Only \$540 is available, so

$$15x + 45y \le 540.$$

Dividing both sides by 15 gives the equivalent inequality

$$x + 3y \le 36.$$

The number of geese and pigs cannot be negative, so

$$x \ge 0, \quad y \ge 0.$$

The 4-H Club member wants to know the number of geese and the number of pigs that should be raised for maximum profit. Each goose produces a profit of \$7 and each pig, \$20. If z represents total profit, then

$$z = 7x + 20y$$

is the objective function, which is to be maximized.

We must solve the following linear programming problem.

$$\begin{aligned} \text{Maximize} \quad & z = 7x + 20y && \text{Objective function} \\ \text{subject to:} \quad & \left.\begin{aligned} x + y &\le 16 \\ x &\le 10 \\ x + 3y &\le 36 \\ x \ge 0, y &\ge 0. \end{aligned}\right\} && \text{Constraints} \end{aligned}$$

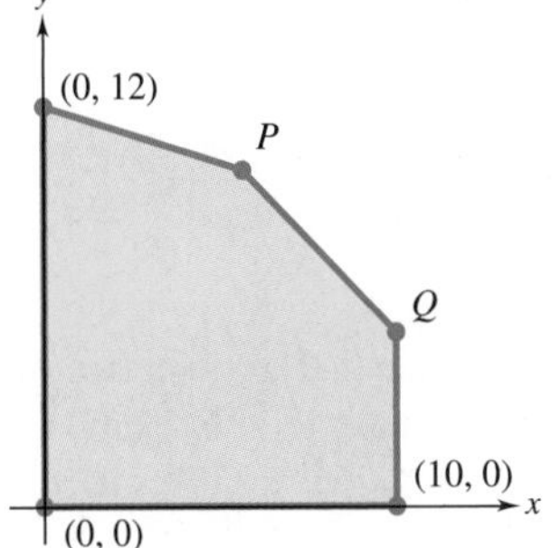

FIGURE 7.19

Using the methods of the previous section, graph the feasible region for the system of inequalities given by the constraints, as in Figure 7.19.

Find the corner points P and Q in Figure 7.19.

Answer:

$P = (6, 10)$
$Q = (10, 6)$

The corner points (0, 12), (0, 0), and (10, 0) can be read directly from the graph. Find the coordinates of the other corner points by solving a system of equations or with a graphing calculator. (1)

Test each corner point in the objective function to find the maximum profit.

Corner Point	$z = 7x + 20y$
(0, 12)	$7(0) + 20(12) = 240$
(6, 10)	$7(6) + 20(10) = 242$ (maximum)
(10, 6)	$7(10) + 20(6) = 190$
(10, 0)	$7(10) + 20(0) = 70$
(0, 0)	$7(0) + 20(0) = 0$

The maximum value for z of 242 occurs at (6, 10). Thus 6 geese and 10 pigs will produce a maximum profit of $242.

EXAMPLE 2 An office manager needs to purchase new filing cabinets. He knows that Ace cabinets cost $40 each, require 6 square feet of floor space, and hold 8 cubic feet of files. On the other hand, each Excello cabinet costs $80, requires 8 square feet of floor space, and holds 12 cubic feet. His budget permits him to spend no more than $560 on files, while the office has room for no more than 72 square feet of cabinets. The manager desires the greatest storage capacity within the limitations imposed by funds and space. How many of each type of cabinet should he buy?

Let x represent the number of Ace cabinets to be bought and let y represent the number of Excello cabinets. The information given in the problem can be summarized as follows.

	Number	*Cost of Each*	*Space Required*	*Storage Capacity*
Ace	x	$ 40	6 sq ft	8 cu ft
Excello	y	$ 80	8 sq ft	12 cu ft
Maximum Available		$560	72 sq ft	

The constraints imposed by cost and space are

$$40x + 80y \leq 560 \quad \text{Cost}$$
$$6x + 8y \leq 72. \quad \text{Floor space}$$

The number of cabinets cannot be negative, so $x \geq 0$ and $y \geq 0$. The objective function to be maximized gives the amount of storage capacity provided by some combination of Ace and Excello cabinets. From the information in the chart, the objective function is

$$\text{Storage space} = z = 8x + 12y.$$

2

Find the corner point labeled P on the region of feasible solutions given below.

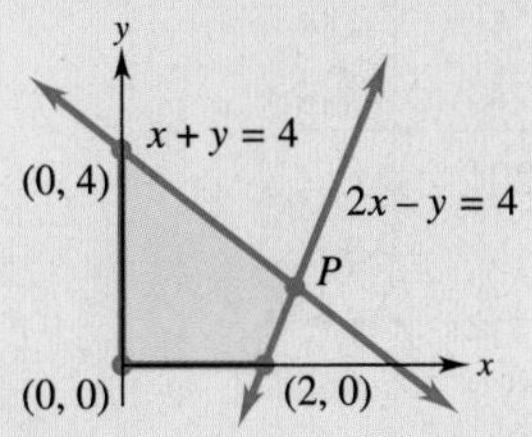

Answer:

(8/3, 4/3)

3

A popular cereal combines oats and corn. At least 27 tons of the cereal is to be made. For the best flavor, the amount of corn should be no more than twice the amount of oats. Oats cost $300 per ton and corn costs $200 per ton. How much of each grain should be used to minimize the cost?

(a) Make a chart to organize the information given in the problem.

(b) Write an equation for the objective function.

(c) Write four inequalities for the constraints.

Answers:

(a)

	Number of Tons	*Cost/ Ton*
Oats	x	$300
Corn	y	$200
	27	

(b) $z = 300x + 200y$

(c) $x + y \geq 27$
$y \leq 2x$
$x \geq 0$
$y \geq 0$

In summary, the given problem has produced the following linear programming problem.

$$\begin{aligned} \text{Maximize} \quad & z = 8x + 12y \\ \text{subject to:} \quad & 40x + 80y \leq 560 \\ & 6x + 8y \leq 72 \\ & x \geq 0, y \geq 0. \end{aligned}$$

A graph of the feasible region is shown in Figure 7.20. Three of the corner points can be identified from the graph as (0, 0), (0, 7), and (12, 0). The fourth corner point, labeled Q in the figure, can be found algebraically or with a graphing calculator to be (8, 3). (2)

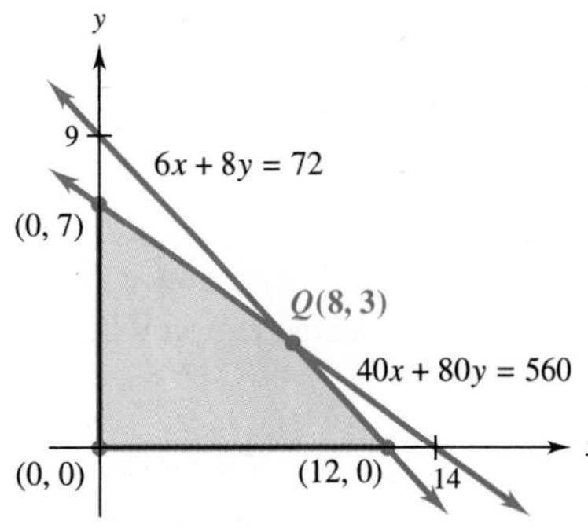

FIGURE 7.20

Use the corner point theorem to find the maximum value of z.

Corner Point	*Value of* $z = 8x + 12y$
(0, 0)	0
(0, 7)	84
(12, 0)	96
(8, 3)	100 (maximum)

The objective function, which represents storage space, is maximized when $x = 8$ and $y = 3$. The manager should buy 8 Ace cabinets and 3 Excello cabinets. (3)

EXAMPLE 3 Certain laboratory animals must have at least 30 grams of protein and at least 20 grams of fat per feeding period. These nutrients come from food A, which costs 18¢ per unit and supplies 2 grams of protein and 4 of fat, and food B, with 6 grams of protein and 2 of fat, costing 12¢ per unit. Food B is bought under a long-term contract requiring that at least 2 units of B be used per serving. How much of each food must be bought to produce minimum cost per serving?

Let x represent the amount of food A needed and y the amount of food B. Use the given information to produce the following table.

Food	*Number of Units*	*Grams of Protein*	*Grams of Fat*	*Cost*
A	x	2	4	18¢
B	y	6	2	12¢
Minimum Required		30	20	

The linear programming problem can be stated as follows.

$$\begin{aligned} \text{Minimize} \quad & z = .18x + .12y \\ \text{subject to:} \quad & 2x + 6y \geq 30 \\ & 4x + 2y \geq 20 \\ & y \geq 2 \\ & x \geq 0, y \geq 0. \end{aligned}$$

(The constraint $y \geq 0$ is redundant because of the constraint $y \geq 2$.) A graph of the feasible region with the corner points identified is shown in Figure 7.21.

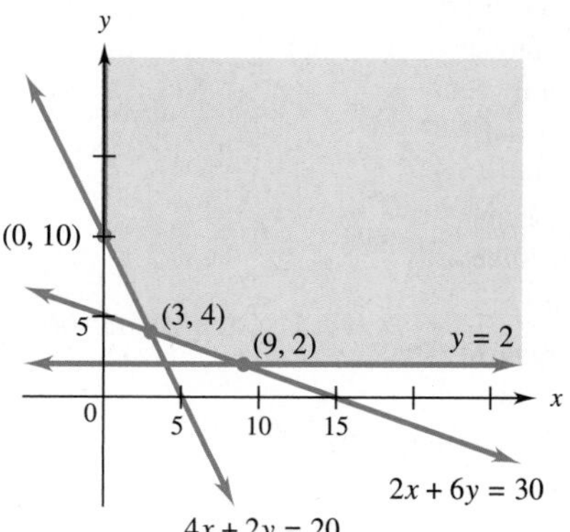

FIGURE 7.21

Use the corner point theorem to find the minimum value of z as shown in the chart below.

Corner Points	$z = .18x + .12y$
(0, 10)	$.18(0) + .12(10) = 1.20$
(3, 4)	$.18(3) + .12(4) = 1.02$ **(minimum)**
(9, 2)	$.18(9) + .12(2) = 1.86$

The minimum value of 1.02 occurs at (3, 4). Thus, 3 units of A and 4 units of B will produce a minimum cost of $1.02 per serving. ◀ (4)

The feasible region in Figure 7.21 is an unbounded feasible region—the region extends indefinitely to the upper right. With this region it would not be possible to *maximize* the objective function, because the total cost of the food could always be increased by encouraging the animals to eat more.

(4)

Use the information in side problem 3 to do the following.

(a) Graph the feasible region and find the corner points.

(b) Determine the minimum value of the objective function and the point where it occurs.

(c) Is there a maximum cost?

Answers:

(a)

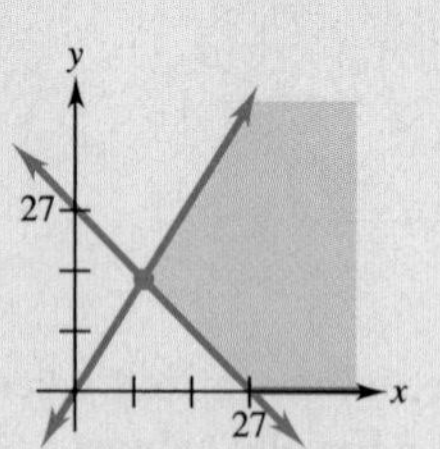

Corner points: (27, 0), (9, 18)

(b) $6300 at (9, 18)

(c) No

7.3 Exercises

Write the constraints in Exercises 1–5 as linear inequalities and identify all variables used. In some instances, not all of the information is needed to write the constraints. (See Examples 1–3.)

1. A canoe requires 6 hours of fabrication and a rowboat 4 hours. The fabrication department has at most 90 hours of labor available each week.
2. Doug Gilbert needs at least 2400 mg of Vitamin C per day. Each Supervite pill provides 250 mg and each Vitahealth pill provides 350 mg.
3. A candidate can afford to spend no more than $8500 on radio and TV advertising. Each radio spot costs $150 and each TV ad costs $750.
4. A hospital dietician has two meal choices, one for patients on solid food that costs $2.25 and one for patients on liquids that costs $3.75. There are a maximum of 400 patients in the hospital.

Solve these linear programming problems, which are somewhat simpler than the examples in the text.

5. **Business** A chain saw requires 4 hours of assembly and a wood chipper 6 hours. A maximum of 48 hours of assembly time are available. The profit is $150 on a chain saw and $220 on a chipper. How many of each should be assembled for maximum profit?
6. **Health** Don Garcia needs at least 24 units of vitamin A per day. Each green pill provides 4 units and each red pill provides 1 unit. Green pills cost 3.5¢ each and red pills 1¢ each. How many of each kind of pill should he take to minimize his cost?
7. **Business** Deluxe coffee is to be mixed with regular coffee to make at least 50 pounds of a blended coffee. The mixture must contain at least 10 pounds of deluxe coffee. Deluxe coffee costs $6 per pound and regular coffee $5 per pound. How many pounds of each kind of coffee should be used to minimize costs?
8. **Business** Pauline Wong spends 4 hours selling a used car and 6 hours selling a new car. She works no more than 40 hours per week. In order to receive a bonus, she must sell at least one used car and four new cars each week. In that case, she receives a bonus of $180 for each used car and $290 for each new car. How many used and how many new cars should she try to sell to maximize her bonus?

Solve the following linear programming problems. (See Examples 1–3.)

9. **Business** A company is considering two insurance plans with the types of coverage and premiums shown in the table below.

	Policy A	*Policy B*
Fire/Theft	$10,000	$15,000
Liability	$180,000	$120,000
Premium	$50	$40

(For example, this means that $50 buys one unit of plan A, consisting of $10,000 fire and theft insurance and $180,000 of liability insurance.)

(a) The company wants at least $300,000 fire/theft insurance and at least $3,000,000 liability insurance from these plans. How many units should be purchased from each plan to minimize the cost of the premiums? What is the minimum premium?

(b) Suppose the premium for policy A is reduced to $25. Now how many units should be purchased from each plan to minimize the cost of the premiums? What is the minimum premium?

10. **Business** A manufacturer of refrigerators must ship at least 100 refrigerators to its two West Coast warehouses. Each warehouse holds a maximum of 100 refrigerators. Warehouse A holds 25 refrigerators already, and warehouse B has 20 on hand. It costs $12 to ship a refrigerator to warehouse A and $10 to ship one to warehouse B. Union rules require that at least 300 workers be hired. Shipping a refrigerator to Warehouse A requires 4 workers, while shipping a refrigerator to Warehouse B requires 2 workers. How many refrigerators should be shipped to each warehouse to minimize costs? What is the minimum cost?
11. **Health** Sing, who is dieting, requires two food supplements, I and II. He can get these supplements from two different products, A and B, as shown in the table below.

		GRAMS OF SUPPLEMENT PER SERVING	
		I	II
PRODUCT	A	3	2
	B	2	4

Sing's physician has recommended that he include at least 15 g of each supplement in his daily diet. If product A costs 25¢ per serving and product B costs 40¢ per serving, how can he satisfy his dietetic requirements most economically? Find the minimum cost.

12. **Business** The manufacturing process requires that oil refineries must manufacture at least 2 gallons of gasoline for every gallon of fuel oil. To meet the winter demand for fuel oil, at least 3 million gallons a day must be produced. The demand for gasoline is no more than 6.4 million gallons per day. It takes .25 hour to ship each million gallons of gasoline and 1 hour to ship each million gallons of fuel oil out of the warehouse. No more than 4.65 hours are available for shipping. If the refinery sells gasoline for $1.25 per gallon and fuel oil for $1 per gallon, how much of each should be produced to maximize revenue? Find the maximum revenue.

13. **Business** A machine shop manufactures two types of bolts. The bolts require time on each of three groups of machines, but the time required on each group differs, as shown in the table below.

		MACHINE GROUP		
		I	*II*	*III*
BOLTS	Type 1	.1 min	.1 min	.1 min
	Type 2	.1 min	.4 min	.02 min

Production schedules are made up one day at a time. In a day there are 240, 720, and 160 minutes available, respectively, on these machines. Type 1 bolts sell for 10¢ and type 2 bolts for 12¢. How many of each type of bolt should be manufactured per day to maximize revenue? What is the maximum revenue?

14. **Health** Kim Walrath has a nutritional deficiency and is told to take at least 2400 mg of iron, 2100 mg of Vitamin B-1, and 1500 mg of Vitamin B-2. One Maxivite pill contains 40 mg of iron, 10 mg of B-1, and 5 mg of B-2 and costs 6¢. One Healthovite pill provides 10 mg of iron, 15 mg of B-1, and 15 mg of B-2 and costs 8¢. What combination of Maxivite and Healthovite pills will meet the requirement at lowest cost? What is the minimum cost?

15. **Social Sciences** An anthropology article describes a hypothetical situation that could be described by a linear programming model.* Suppose a population gathers plants and animals for survival. They need at least 360 units of energy, 300 units of protein, and 8 hides during some time period. One unit of plants provides 30 units of energy, 10 units of protein, and no hides. One animal provides 20 units of energy, 25 units of protein, and 1 hide. Only 25 units of plants and 25 animals are available. It costs the population 30 hours of labor to gather one unit of a plant and 15 hours for an animal. Find how many units of plants and how many animals should be gathered to meet the requirements with a minimum number of hours of labor.

16. **Business** The Miers Company produces small engines for several manufacturers. The company receives orders from two assembly plants for their Topflight engine. Plant I needs at least 50 engines, and plant II needs at least 27 engines. The company can send at most 85 engines to these two assembly plants. It costs $20 per engine to ship to plant I and $35 per engine to ship to plant II. Plant I gives Miers $15 in rebates toward its products for each engine they buy, while plant II gives similar $10 rebates. Miers estimates that they need at least $1110 in rebates to cover products they plan to buy from the two plants. How many engines should be shipped to each plant to minimize shipping costs? What is the minimum cost?

17. **Business** A greeting card manufacturer has 370 boxes of a particular card in warehouse I and 290 boxes of the same card in warehouse II. A greeting card shop in San Jose orders 350 boxes of the card, and another shop in Memphis orders 300 boxes. The shipping costs per box to these shops from the two warehouses are shown in the following table.

		DESTINATION	
		San Jose	*Memphis*
WAREHOUSE	I	$.25	$.22
	II	$.23	$.21

How many boxes should be shipped to each city from each warehouse to minimize shipping costs? What is the minimum cost? (*Hint:* Use x, $350 - x$, y, and $300 - y$ as the variables.)

18. **Business** Hotnews Magazine publishes a U.S. and a Canadian edition each week. There are 30,000 subscribers in the United States and 20,000 in Canada. Other copies are sold at newsstands. Postage and shipping costs average $80 per thousand copies for the U.S. and $60 per thousand copies for Canada. Surveys show that no more than 120,000 copies of each issue can be sold (including subscriptions) and that the number of copies of the Canadian edition should not exceed twice the number of copies of the U.S. edition. The publisher can spend at most $8400 a month on postage and shipping. If the profit is $200 for each thousand copies of the U.S. edition and $150 for each thousand copies of the Canadian edition,

*Van A. Reidhead, "Linear Programming Models in Archaeology," *Annual Review of Anthropology* 8 (1979): 543–578.

how many copies of each version should be printed to earn as large a profit as possible? What is that profit?

19. Finance A pension fund manager decides to invest at most \$40 million in U.S. Treasury Bonds paying 8% annual interest and in mutual funds paying 12% annual interest. He plans to invest at least \$20 million in bonds and at least \$15 million in mutual funds. Bonds have an initial fee of \$300 per million dollars, while the fee for mutual funds is \$100 per million. The fund manager is allowed to spend no more than \$8400 on fees. How much should be invested in each to maximize annual interest? What is the maximum annual interest?

20. Natural Science A certain predator requires at least 10 units of protein and 8 units of fat per day. One prey of Species I provides 5 units of protein and 2 units of fat; one prey of Species II provides 3 units of protein and 4 units of fat. Capturing and digesting each Species II prey requires 3 units of energy, and capturing and digesting each Species I prey requires 2 units of energy. How many of each prey would meet the predator's daily food requirements with the least expenditure of energy? Are the answers reasonable? How could they be interpreted?

21. Social Science Students at Upscale U are required to take at least 3 humanities and 4 science courses. The maximum allowable number of science courses is 12. Each humanities course carries 4 credits and each science course 5 credits. The total number of credits in science and humanities cannot exceed 80. Quality points for each course are assigned in the usual way: the number of credit hours times 4 for an A grade; times 3 for a B grade; times 2 for a C grade. Susan Katz expects to get B's in all her science courses. She expects to get C's in half her humanities courses, B's in one-fourth of them, and A's in the rest. Under these assumptions, how many courses of each kind should she take in order to earn the maximum possible number of quality points?

22. Social Science In Exercise 21, find Susan's grade point average (the total number of quality points divided by the total number of credit hours) at each corner point of the feasible region. Does the distribution of courses that produces the highest number of quality points also yield the highest grade point average? Is this a contradiction?

*The importance of linear programming is shown by the inclusion of linear programming problems on most qualification examinations for Certified Public Accountants. Exercises 23–25 are reprinted from one such examination.**

The Random Company manufactures two products, Zeta and Beta. Each product must pass through two processing operations. All materials are introduced at the start of Process No. 1. There are no work-in-process inventories. Random may produce either one product exclusively or various combinations of both products subject to the following constraints.

	Process No. 1	*Process No. 2*	*Contribution Margin per Unit*
Hours required to produce 1 unit of:			
Zeta	1 hour	1 hour	\$4.00
Beta	2 hours	3 hours	\$5.25
Total capacity in hours per day	1000 hours	1275 hours	

A shortage of technical labor has limited Beta production to 400 units per day. There are no constraints on the production of Zeta other than the hour constraints in the above schedule. Assume that all the relationships between capacity and production are linear.

23. Given the objective to maximize total contribution margin, what is the production constraint for Process No. 1?
(a) Zeta + Beta $\le$ 1000
(b) Zeta + 2 Beta $\le$ 1000
(c) Zeta + Beta $\ge$ 1000
(d) Zeta + 2 Beta $\ge$ 1000

24. Given the objective to maximize total contribution margin, what is the labor constraint for production of Beta?
(a) Beta $\le$ 400 (b) Beta $\ge$ 400
(c) Beta $\le$ 425 (d) Beta $\ge$ 425

25. What is the objective function of the data presented?
(a) Zeta + 2 Beta = \$9.25
(b) \$4.00 Zeta + 3(\$5.25) Beta = total contribution margin
(c) \$4.00 Zeta + \$5.25 Beta = total contribution margin
(d) 2(\$4.00) Zeta + 3(\$5.25) Beta = total contribution margin

*Material from *Uniform CPA Examinations and Unofficial Answers*, copyright © 1973, 1974, 1975 by the American Institute of Certified Public Accountants, Inc., is reprinted with permission.

7.4 THE SIMPLEX METHOD: MAXIMIZATION

For linear programming problems with more than two variables or with two variables and many constraints, the graphical method is usually inefficient or impossible, so the **simplex method** is used. The simplex method, which is introduced here, was developed for the U.S. Air Force by George B. Danzig in 1947. It was used successfully during the Berlin airlift in 1948–49 to maximize the amount of cargo delivered under very severe constraints and is widely used today in a variety of industries.

Because the simplex method is used for problems with many variables, it usually is not convenient to use letters such as x, y, z, or w as variable names. Instead, the symbols x_1 (read "x-sub-one"), x_2, x_3, and so on, are used. In the simplex method, all constraints must be expressed in the linear form

$$a_1x_1 + a_2x_2 + a_3x_3 + \cdots \leq b,$$

where $x_1, x_2, x_3, \ldots$ are variables, $a_1, a_2, a_3, \ldots$ are coefficients, and b is a constant.

We first discuss the simplex method for linear programming problems in *standard maximum form.*

Standard Maximum Form

A linear programming problem is in **standard maximum form** if

1. the objective function is to be maximized;
2. all variables are nonnegative ($x_i \geq 0$, $i = 1, 2, 3, \ldots$);
3. all constraints involve $\leq$;
4. the constants on the right side in the constraints are all nonnegative ($b \geq 0$).

Problems that do not meet all of these conditions are considered in Sections 7.6 and 7.7.

The "mechanics" of the simplex method are demonstrated in Examples 1–5. Although the procedures to be followed will be made clear, as will the fact that they result in an optimal solution, the reasons why these procedures are used may not be immediately apparent. Examples 6 and 7 will supply these reasons and explain the connection between the simplex method and the graphical method used in Section 7.3.

Setting Up the Problem

The first step is to convert each constraint, a linear inequality, into a linear equation. This is done by adding a nonnegative variable, called a **slack variable,** to each constraint. For example, convert the inequality $x_1 + x_2 \leq 10$ into an equation by adding the slack variable x_3, to get

$$x_1 + x_2 + x_3 = 10, \quad \text{where } x_3 \geq 0.$$

The inequality $x_1 + x_2 \le 10$ says that the sum $x_1 + x_2$ is less than or perhaps equal to 10. The variable x_3 "takes up any slack" and represents the amount by which $x_1 + x_2$ fails to equal 10. For example, if $x_1 + x_2$ equals 8, then x_3 is 2. If $x_1 + x_2 = 10$, the value of x_3 is 0.

CAUTION A different slack variable must be used for each constraint.

EXAMPLE 1 Restate the following linear programming problem by introducing slack variables.

$$\begin{aligned} \text{Maximize} \quad & z = 2x_1 + 3x_2 + x_3 \\ \text{subject to:} \quad & x_1 + x_2 + 4x_3 \le 100 \\ & x_1 + 2x_2 + x_3 \le 150 \\ & 3x_1 + 2x_2 + x_3 \le 320 \\ \text{with} \quad & x_1 \ge 0, x_2 \ge 0, x_3 \ge 0. \end{aligned}$$

Rewrite the three constraints as equations by introducing nonnegative slack variables x_4, x_5, and x_6, one for each constraint. Then the problem can be restated as

$$\begin{aligned} \text{Maximize} \quad & z = 2x_1 + 3x_2 + x_3 \\ \text{subject to:} \quad & x_1 + x_2 + 4x_3 + x_4 = 100 \\ & x_1 + 2x_2 + x_3 + x_5 = 150 \\ & 3x_1 + 2x_2 + x_3 + x_6 = 320 \\ \text{with} \quad & x_1 \ge 0, x_2 \ge 0, x_3 \ge 0, x_4 \ge 0, x_5 \ge 0, x_6 \ge 0. \end{aligned}$$

(1)

1

Rewrite the following set of constraints as equations by adding nonnegative slack variables.

$$\begin{aligned} x_1 + x_2 + x_3 &\le 12 \\ 2x_1 + 4x_2 &\le 15 \\ x_2 + 3x_3 &\le 10 \end{aligned}$$

Answer:

$$\begin{aligned} x_1 + x_2 + x_3 + x_4 &= 12 \\ 2x_1 + 4x_2 + x_5 &= 15 \\ x_2 + 3x_3 + x_6 &= 10 \end{aligned}$$

Adding slack variables to the constraints converts a linear programming problem into a system of linear equations. These equations should have all variables on the left of the equals sign and all constants on the right. All the equations of Example 1 satisfy this condition except for the objective function, $z = 2x_1 + 3x_2 + x_3$, which may be written with all variables on the left as

$$-2x_1 - 3x_2 - x_3 + z = 0.$$

Now the equations of Example 1 (with the constraints listed first) can be written as the following augmented matrix.

$$\begin{array}{ccccccc|c} x_1 & x_2 & x_3 & x_4 & x_5 & x_6 & z & \\ 1 & 1 & 4 & 1 & 0 & 0 & 0 & 100 \\ 1 & 2 & 1 & 0 & 1 & 0 & 0 & 150 \\ 3 & 2 & 1 & 0 & 0 & 1 & 0 & 320 \\ \hline -2 & -3 & -1 & 0 & 0 & 0 & 1 & 0 \end{array}$$

Indicators

This matrix is the initial **simplex tableau.** Except for the last entries, the 1 and 0 on the right end, the numbers in the bottom row of a simplex tableau are called **indicators.** (2)

2

Set up the initial simplex tableau for the following linear programming problem:

$$\begin{aligned} \text{Maximize} \quad & z = 2x_1 + 3x_2 \\ \text{subject to:} \quad & x_1 + 2x_2 \le 85 \\ & 2x_1 + x_2 \le 92 \\ & x_1 + 4x_2 \le 104 \\ \text{with} \quad & x_1 \ge 0, x_2 \ge 0. \end{aligned}$$

Locate and label the indicators.

Answer:

$$\begin{array}{cccccc|c} x_1 & x_2 & x_3 & x_4 & x_5 & z & \\ 1 & 2 & 1 & 0 & 0 & 0 & 85 \\ 2 & 1 & 0 & 1 & 0 & 0 & 92 \\ 1 & 4 & 0 & 0 & 1 & 0 & 104 \\ \hline -2 & -3 & 0 & 0 & 0 & 1 & 0 \end{array}$$

Indicators

This simplex tableau represents a system of 4 linear equations in 7 variables. Since there are more variables than equations, the system is dependent and has infinitely many solutions. Our goal is to find a solution in which all the variables are nonnegative and z is as large as possible. This will be done by using row operations to replace the given system by an equivalent one in which certain variables are eliminated from some of the equations. The process will be repeated until the optimum solution can be read from the matrix, as explained below.

Selecting the Pivot

Recall how row operations are used to eliminate variables in the Gauss-Jordan method. A particular nonzero entry in the matrix is chosen and changed to a 1; then all other entries in that column are changed to zeros. A similar process is used in the simplex method. The chosen entry is called the **pivot.** If we were only interested in solving the system, we could choose the various pivots in many different ways, as in Chapter 6. Here, however, it's not enough just to find a solution. We must find one that is nonnegative (so that it satisfies all the constraints) *and* makes z as a large as possible. Consequently, the pivot must be chosen carefully, as explained in the next example. The reasons why this procedure is used and why it works are discussed in Example 7.

EXAMPLE 2 Determine the pivot in the simplex tableau for the problem in Example 1.

Look at the indicators (the last row of the tableau) and choose the most negative one.

$$\begin{array}{ccccccc} x_1 & x_2 & x_3 & x_4 & x_5 & x_6 & z \\ \end{array}$$
$$\left[\begin{array}{ccccccc|c} 1 & 1 & 4 & 1 & 0 & 0 & 0 & 100 \\ 1 & 2 & 1 & 0 & 1 & 0 & 0 & 150 \\ 3 & 2 & 1 & 0 & 0 & 1 & 0 & 320 \\ \hline -2 & \boxed{-3} & -1 & 0 & 0 & 0 & 1 & 0 \end{array}\right]$$

Most negative indicator (points to -3)

The most negative indicator identifies the variable that is to be eliminated from all but one of the equations (rows), in this case x_2. The column containing the most negative indicator is called the **pivot column.** Now for each *positive* entry in the pivot column, divide the number in the far right column of the same row by the positive number in the pivot column.

$$\begin{array}{ccccccc} x_1 & x_2 & x_3 & x_4 & x_5 & x_6 & z \\ \end{array}$$
$$\left[\begin{array}{ccccccc|c} 1 & \boxed{1} & 4 & 1 & 0 & 0 & 0 & \mathbf{100} \\ 1 & \boxed{2} & 1 & 0 & 1 & 0 & 0 & \mathbf{150} \\ 3 & \boxed{2} & 1 & 0 & 0 & 1 & 0 & \mathbf{320} \\ \hline -2 & -3 & -1 & 0 & 0 & 0 & 1 & 0 \end{array}\right]
\begin{array}{l} \text{Quotients} \\ 100/1 = 100 \\ 150/2 = 75 \leftarrow \text{Smallest} \\ 320/2 = 160 \\ \\ \end{array}$$

The row with the smallest quotient (in this case, the second row) is called the **pivot row.** The entry in the pivot row and pivot column is the pivot.

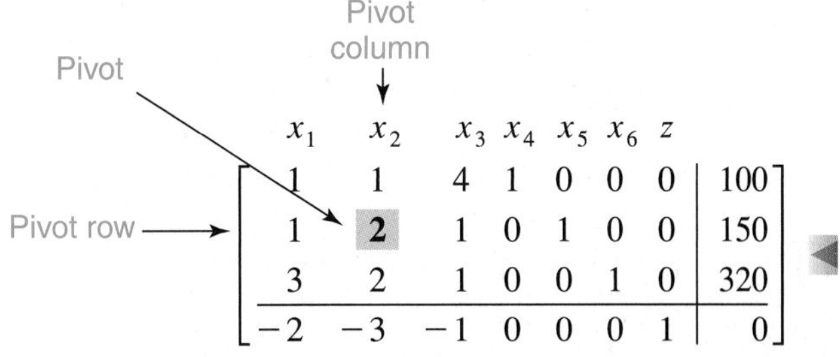

x_1	x_2	x_3	x_4	x_5	x_6	z	
1	1	4	1	0	0	0	100
1	**2**	1	0	1	0	0	150
3	2	1	0	0	1	0	320
−2	−3	−1	0	0	0	1	0

3

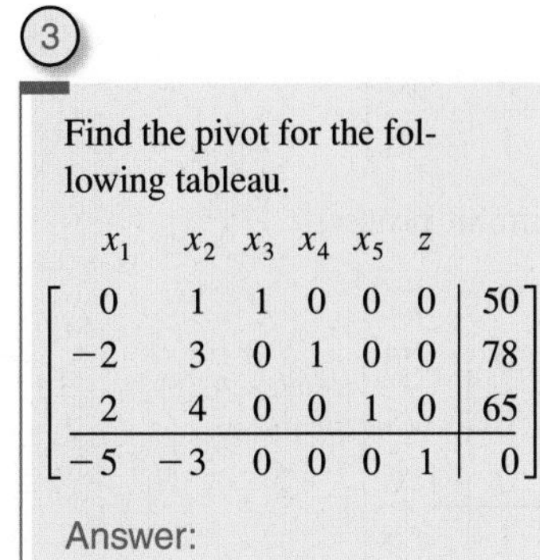

Find the pivot for the following tableau.

x_1	x_2	x_3	x_4	x_5	z	
0	1	1	0	0	0	50
−2	3	0	1	0	0	78
2	4	0	0	1	0	65
−5	−3	0	0	0	1	0

Answer:

2 (in first column)

CAUTION In some simplex tableaus the pivot column may contain zeros or negative entries. Only the positive entries in the pivot column should be used to form the quotients and determine the pivot row. If there are no positive entries in the pivot column (so that a pivot row cannot be chosen), then no maximum solution exists. (3)

Pivoting

Once the pivot has been selected, row operations are used to replace the initial simplex tableau by another simplex tableau in which the pivot column variable is eliminated from all but one of the equations. Since this new tableau is obtained by row operations, it represents an equivalent system of equations (that is, a system with the same solutions as the original system). This process, which is called **pivoting,** is explained in the next example.

EXAMPLE 3 Use the indicated pivot, 2, to perform the pivoting on the simplex tableau of Example 2:

x_1	x_2	x_3	x_4	x_5	x_6	z	
1	1	4	1	0	0	0	100
1	**2**	1	0	1	0	0	150
3	2	1	0	0	1	0	320
−2	−3	−1	0	0	0	1	0

Start by multiplying each entry of row two by 1/2 in order to change the pivot to 1.

x_1	x_2	x_3	x_4	x_5	x_6	z		
1	1	4	1	0	0	0	100	
$\frac{1}{2}$	**1**	$\frac{1}{2}$	0	$\frac{1}{2}$	0	0	75	$\frac{1}{2}R_2$
3	2	1	0	0	1	0	320	
−2	−3	−1	0	0	0	1	0	

Now use row operations to make the entry in row one, column two, a 0.

$$\begin{array}{ccccccc|c} x_1 & x_2 & x_3 & x_4 & x_5 & x_6 & z & \\ \frac{1}{2} & 0 & \frac{7}{2} & 1 & -\frac{1}{2} & 0 & 0 & 25 \\ \frac{1}{2} & 1 & \frac{1}{2} & 0 & \frac{1}{2} & 0 & 0 & 75 \\ 3 & 2 & 1 & 0 & 0 & 1 & 0 & 320 \\ \hline -2 & -3 & -1 & 0 & 0 & 0 & 1 & 0 \end{array} \quad -R_2 + R_1$$

Change the 2 in row three, column two, to a 0 by a similar process.

$$\begin{array}{ccccccc|c} x_1 & x_2 & x_3 & x_4 & x_5 & x_6 & z & \\ \frac{1}{2} & 0 & \frac{7}{2} & 1 & -\frac{1}{2} & 0 & 0 & 25 \\ \frac{1}{2} & 1 & \frac{1}{2} & 0 & \frac{1}{2} & 0 & 0 & 75 \\ 2 & 0 & 0 & 0 & -1 & 1 & 0 & 170 \\ \hline -2 & -3 & -1 & 0 & 0 & 0 & 1 & 0 \end{array} \quad -2R_2 + R_3$$

Finally, add 3 times row 2 to the last row in order to change the indicator -3 to 0.

$$\begin{array}{ccccccc|c} x_1 & x_2 & x_3 & x_4 & x_5 & x_6 & z & \\ \frac{1}{2} & 0 & \frac{7}{2} & 1 & -\frac{1}{2} & 0 & 0 & 25 \\ \frac{1}{2} & 1 & \frac{1}{2} & 0 & \frac{1}{2} & 0 & 0 & 75 \\ 2 & 0 & 0 & 0 & -1 & 1 & 0 & 170 \\ \hline -\frac{1}{2} & 0 & \frac{1}{2} & 0 & \frac{3}{2} & 0 & 1 & 225 \end{array} \quad -3R_2 + R_4$$

The pivoting is now complete because the pivot column variable x_2 has been eliminated from all equations except the one represented by the pivot row. The initial simplex tableau has been replaced by a new simplex tableau, which represents an equivalent system of equations.

4 For the simplex tableau below,

(a) find the pivot.

(b) Perform the pivoting and write the new tableau.

$$\begin{array}{cccccc|c} x_1 & x_2 & x_3 & x_4 & x_5 & z & \\ 1 & 2 & 6 & 1 & 0 & 0 & 16 \\ 1 & 3 & 0 & 0 & 1 & 0 & 25 \\ \hline -1 & -4 & -3 & 0 & 0 & 1 & 0 \end{array}$$

Answers:

(a) 2

(b)

$$\begin{array}{cccccc|c} x_1 & x_2 & x_3 & x_4 & x_5 & z & \\ \frac{1}{2} & 1 & 3 & \frac{1}{2} & 0 & 0 & 8 \\ -\frac{1}{2} & 0 & -9 & -\frac{3}{2} & 1 & 0 & 1 \\ \hline 1 & 0 & 9 & 2 & 0 & 1 & 32 \end{array}$$

CAUTION During pivoting, do not interchange rows of the matrix. Make the pivot entry 1 by multiplying the pivot row by an appropriate constant, as in Example 3. (4)

When at least one of the indicators in the last row of a simplex tableau is negative (as is the case with the tableau obtained in Example 3), the simplex method requires that a new pivot be selected and the pivoting be performed again. This procedure is repeated until a simplex tableau with no negative indicators in the last row is obtained or a tableau is reached in which no pivot row can be chosen.

EXAMPLE 4 In the simplex tableau obtained in Example 3, select a new pivot and perform the pivoting.

First, locate the pivot column by finding the most negative indicator in the last row. Then locate the pivot row by computing the necessary quotients and finding the smallest one, as shown here.

$$
\begin{array}{c}
\text{Pivot row} \rightarrow \\ \\ \\ \\
\end{array}
\begin{array}{c}
\begin{array}{ccccccc|c}
x_1 & x_2 & x_3 & x_4 & x_5 & x_6 & z & \\
\end{array} \\
\left[\begin{array}{ccccccc|c}
\frac{1}{2} & 0 & \frac{7}{2} & 1 & -\frac{1}{2} & 0 & 0 & 25 \\
\frac{1}{2} & 1 & \frac{1}{2} & 0 & \frac{1}{2} & 0 & 0 & 75 \\
2 & 0 & 0 & 0 & -1 & 1 & 0 & 170 \\ \hline
-\frac{1}{2} & 0 & \frac{1}{2} & 0 & \frac{3}{2} & 0 & 1 & 225
\end{array}\right]
\end{array}
\begin{array}{l}
\text{Quotients} \\
\frac{25}{1/2} = 50 \quad \text{Smallest} \\
\frac{75}{1/2} = 150 \\
170/2 = 85 \\
\end{array}
$$

Pivot column (arrow under first column)

So the pivot is the number 1/2 in row one, column one. Begin the pivoting by multiplying every entry in row one by 2. Then continue as indicated below, to obtain the following simplex tableau.

$$
\begin{array}{c}
\begin{array}{ccccccc|c}
x_1 & x_2 & x_3 & x_4 & x_5 & x_6 & z & \\
\end{array} \\
\left[\begin{array}{ccccccc|c}
1 & 0 & 7 & 2 & -1 & 0 & 0 & 50 \\
0 & 1 & -3 & -1 & 1 & 0 & 0 & 50 \\
0 & 0 & -14 & -4 & 1 & 1 & 0 & 70 \\ \hline
0 & 0 & 4 & 1 & 1 & 0 & 1 & 250
\end{array}\right]
\end{array}
\begin{array}{l}
2R_1 \\
-\frac{1}{2}R_1 + R_2 \\
-2R_1 + R_3 \\
\frac{1}{2}R_1 + R_4
\end{array}
$$

Since there are no negative indicators in the last row, no further pivoting is necessary and we call this the **final simplex tableau.** ◀

Reading the Solution

The next example shows how to read an optimal solution of the original linear programming problem from the final simplex tableau.

▶ **EXAMPLE 5** Solve the linear programming problem introduced in Example 1.

Look at the final simplex tableau for this problem, which was obtained in Example 4.

$$
\begin{array}{c}
\begin{array}{ccccccc|c}
x_1 & x_2 & x_3 & x_4 & x_5 & x_6 & z & \\
\end{array} \\
\left[\begin{array}{ccccccc|c}
1 & 0 & 7 & 2 & -1 & 0 & 0 & 50 \\
0 & 1 & -3 & -1 & 1 & 0 & 0 & 50 \\
0 & 0 & -14 & -4 & 1 & 1 & 0 & 70 \\ \hline
0 & 0 & 4 & 1 & 1 & 0 & 1 & 250
\end{array}\right]
\end{array}
$$

The last row of this matrix represents the equation

$$4x_3 + x_4 + x_5 + z = 250, \quad \text{or equivalently,} \quad z = 250 - 4x_3 - x_4 - x_5.$$

If x_3, x_4, and x_5 are all 0, then the value of z is 250. If any one of x_3, x_4, or x_5 is positive, then z will have a smaller value than 250 (why?). Consequently, since we want a solution for this system in which all the variables are nonnegative and z is as large as possible, we must have

$$x_3 = 0, \qquad x_4 = 0, \qquad x_5 = 0.$$

When these values are substituted in the first equation (represented by the first row of the final simplex tableau), the result is

$$x_1 + 7 \cdot 0 + 2 \cdot 0 - 1 \cdot 0 = 50, \quad \text{that is,} \quad x_1 = 50.$$

Similarly, substituting 0 for x_3, x_4, and x_5 in the last three equations represented by the final simplex tableau shows that

$$x_2 = 50, \qquad x_6 = 70, \qquad z = 250.$$

Therefore, the maximum value of $z = 2x_1 + 3x_2 + x_3$ occurs when

$$x_1 = 50, \qquad x_2 = 50, \qquad x_3 = 0,$$

in which case $z = 2 \cdot 50 + 3 \cdot 50 + 0 = 250$. (The values of the slack variables are irrelevant in stating the solution of the original problem.) ◀

A linear programming problem with slack variables x_4 and x_5 has the final simplex tableau shown below. What is the optimal solution?

x_1	x_2	x_3	x_4	x_5	z	
0	3	1	5	2	0	9
1	−2	0	4	1	0	6
0	5	0	1	0	1	21

Answer:

$z = 21$ when $x_1 = 6$, $x_2 = 0$, $x_3 = 9$.

In any simplex tableau, some columns look like columns of an identity matrix (one entry is 1, the rest are 0). The variables corresponding to these columns are called **basic variables** and the variables corresponding to the other columns **nonbasic variables.** In the tableau of Example 5, for instance, the basic variables are x_1, x_2, x_6, and z (shown in blue below), and the nonbasic variables are x_3, x_4, and x_5.

x_1	x_2	x_3	x_4	x_5	x_6	z	
1	0	7	2	−1	0	0	50
0	1	−3	−1	1	0	0	50
0	0	−14	−4	1	1	0	70
0	0	4	1	1	0	1	250

The optimal solution in Example 5 was obtained from the final simplex tableau by setting the nonbasic variables equal to 0 and solving for the basic variables. Furthermore, the values of the basic variables are easy to read off the matrix: find the 1 in the column representing a basic variable; the last entry in that row is the value of that basic variable in the optimal solution. In particular, *the entry in the lower right-hand corner of the final simplex tableau is the maximum value of z.* (5)

CAUTION If there are two identical columns in a tableau, each of which is a column in an identity matrix, only one of the variables corresponding to these columns can be a basic variable. The other is treated as a nonbasic variable. You may choose either one to be the basic variable, unless one of them is z, in which case z must be the basic variable.

The steps involved in solving a standard maximum linear programming problem by the simplex method have been illustrated in Examples 1–5 and are summarized here.

Simplex Method

1. Determine the objective function.
2. Write all necessary constraints.
3. Convert each constraint into an equation by adding slack variables.

continued

4. Set up the initial simplex tableau.
5. Locate the most negative indicator. If there are two such indicators, choose one. This indicator determines the pivot column.
6. Use the positive entries in the pivot column to form the quotients necessary for determining the pivot. If there are no positive entries in the pivot column, no maximum solution exists. If two quotients are equally the smallest, let either determine the pivot.*
7. Multiply every entry in the pivot row by the reciprocal of the pivot to change the pivot to 1. Then use row operations to change all other entries in the pivot column to 0 by adding suitable multiples of the pivot row to the other rows.
8. If the indicators are all positive or 0, this is the final tableau. If not, go back to Step 5 above and repeat the process until a tableau with no negative indicators is obtained.†
9. In the final tableau, the *basic* variables correspond to the columns that have one entry 1 and the rest 0. The *nonbasic* variables correspond to the other columns. Set each nonbasic variable equal to 0 and solve the system for the basic variables. The maximum value of the objective function is the number in the lower right-hand corner of the final tableau.

6

A linear programming problem has the initial tableau given below. Use the simplex method to solve the problem.

$$\begin{array}{c} \begin{matrix} x_1 & x_2 & x_3 & x_4 & z & \end{matrix} \\ \left[\begin{array}{ccccc|c} 1 & 1 & 1 & 0 & 0 & 40 \\ 2 & 1 & 0 & 1 & 0 & 24 \\ \hline -300 & -200 & 0 & 0 & 1 & 0 \end{array}\right] \end{array}$$

Answer:

$x_1 = 0, x_2 = 24, x_3 = 16,$
$x_4 = 0, z = 4800$

The solution found by the simplex method may not be unique, especially when choices are possible in steps 5, 6, or 9. There may be other solutions that produce the same maximum value of the objective function. (See Exercises 37 and 38 at the end of this section) 6

The Simplex Method with Technology

In the preceding discussion and in most of the examples that follow, the simplex method is carried out by hand. This allows the reader to become familiar with the procedure and to understand how it works.

Unfortunately, it's easy to make an error in manual computations, and even a small error can make a big difference in the final result. Furthermore, most real-life applications involve a large number of variables and constraints, which makes the manual approach impractical. Consequently,

we recommend that you use some kind of technology for doing the simplex method.

Unless marked otherwise, the exercises can be done manually, but using technology will make your life much easier and your answers more accurate. Readily available technology includes the following.

*It may be that the first choice of a pivot does not produce a solution. In that case try the other choice.

†Some special circumstances are noted at the end of Section 7.7.

GRAPHING CALCULATORS Any graphing calculator that has matrix capabilities can do row operations. Doing the simplex method this way is an improvement on manual computation but still rather cumbersome. It is much more effective to use the simplex method programs in the Graphing Calculator Appendix. You can install these programs manually in your calculator or use a computer to download them from www.aw.com/MWA8 if you have the appropriate hardware for transferring the programs to the calculator.

SPREADSHEETS Most spreadsheets have a built-in simplex method program. Figure 7.22 shows the Solver of Microsoft Excel. Spreadsheets also provide a "sensitivity analysis," which allows you to see how much the constraints could be varied without changing the maximal solution.

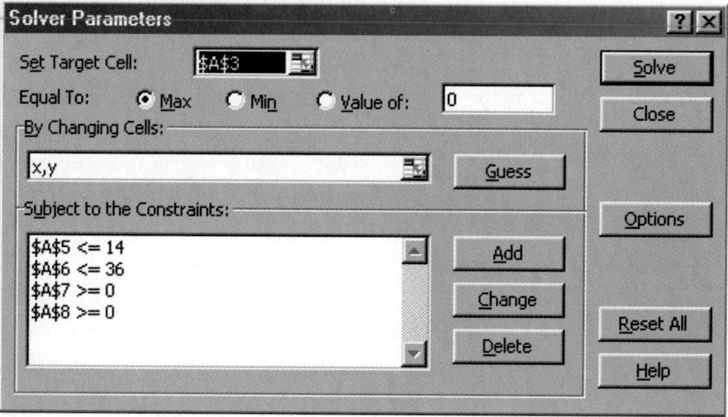

FIGURE 7.22

Geometric Interpretation of the Simplex Method

Although it may not be immediately apparent, the simplex method is based on the same geometrical considerations as the graphical method. This can be seen by looking at a problem that can be readily solved by both methods.

EXAMPLE 6 In Example 2 of Section 7.3 the following problem was solved graphically (using x, y instead of x_1, x_2):

$$\begin{aligned} \text{Maximize} \quad & z = 8x_1 + 12x_2 \\ \text{subject to:} \quad & 40x_1 + 80x_2 \le 560 \\ & 6x_1 + 8x_2 \le 72 \\ & x_1 \ge 0, x_2 \ge 0. \end{aligned}$$

Graphing the feasible region (Figure 7.23) and evaluating z at each corner point shows that the maximum value of z occurs at (8, 3).

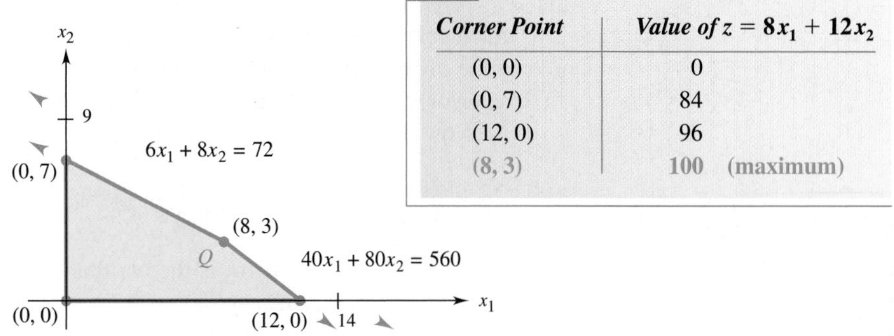

Corner Point	*Value of* $z = 8x_1 + 12x_2$
(0, 0)	0
(0, 7)	84
(12, 0)	96
(8, 3)	100 (maximum)

FIGURE 7.23

To solve the same problem by the simplex method, add a slack variable to each constraint.

$$\begin{aligned} 40x_1 + 80x_2 + x_3 \qquad &= 560 \\ 6x_1 + \ 8x_2 \qquad + x_4 &= 72 \end{aligned}$$

Then write the initial simplex tableau.

$$\begin{array}{c} \begin{array}{ccccc} x_1 & x_2 & x_3 & x_4 & z \end{array} \\ \left[\begin{array}{rrrrr|r} 40 & 80 & 1 & 0 & 0 & 560 \\ 6 & 8 & 0 & 1 & 0 & 72 \\ \hline -8 & -12 & 0 & 0 & 1 & 0 \end{array}\right] \end{array}$$

In this tableau the basic variables are x_3, x_4, and z (why?). By setting the nonbasic variables (namely, x_1 and x_2) equal to 0 and solving for the basic variables, we obtain the following solution (which will be called a **basic feasible solution**):

$$x_1 = 0, \qquad x_2 = 0, \qquad x_3 = 560, \qquad x_4 = 72, \qquad z = 0.$$

Since $x_1 = 0$ and $x_2 = 0$, this solution corresponds to the corner point at the origin in the graphical solution (Figure 7.23). The value $z = 0$ at the origin is obviously not maximal and pivoting in the simplex method is designed to improve it.

The most negative indicator in the initial tableau is -12 and the necessary quotients are

$$\frac{560}{80} = 7 \quad \text{and} \quad \frac{72}{8} = 9.$$

The smaller quotient is 7, giving 80 as the pivot. Performing the pivoting leads to this tableau.

$$\begin{array}{c} \begin{array}{ccccc} x_1 & x_2 & x_3 & x_4 & z \end{array} \\ \left[\begin{array}{rrrrr|r} \frac{1}{2} & 1 & \frac{1}{80} & 0 & 0 & 7 \\ 2 & 0 & -\frac{1}{10} & 1 & 0 & 16 \\ \hline -2 & 0 & \frac{3}{20} & 0 & 1 & 84 \end{array}\right] \end{array} \begin{array}{l} \frac{1}{80}R_1 \\ -8R_1 + R_2 \\ 12R_1 + R_3 \end{array}$$

The basic variables here are x_2, x_4, and z and the basic feasible solution (found by setting the nonbasic variables equal to 0 and solving for the basic variables) is

$$x_1 = 0, \qquad x_2 = 7, \qquad x_3 = 0, \qquad x_4 = 16, \qquad z = 84,$$

which corresponds to the corner point (0, 7) in Figure 7.23. Note that the new value of the pivot variable x_2 is precisely the smallest quotient, 7, that was used to select the pivot row. Although this value of z is better, further improvement is possible.

Now the most negative indicator is -2 and the quotients are

$$\frac{7}{1/2} = 14 \quad \text{and} \quad \frac{16}{2} = 8.$$

Since 8 is smaller, the pivot is the number 2 in row two, column one. Pivoting produces the final tableau.

$$\begin{array}{c} \begin{matrix} x_1 & x_2 & x_3 & x_4 & z & \end{matrix} \\ \left[\begin{array}{ccccc|c} 0 & 1 & \frac{3}{80} & -\frac{1}{4} & 0 & 3 \\ 1 & 0 & -\frac{1}{20} & \frac{1}{2} & 0 & 8 \\ \hline 0 & 0 & \frac{1}{20} & 1 & 1 & 100 \end{array}\right] \end{array} \begin{array}{l} -\frac{1}{2}R_2 + R_1 \\ \frac{1}{2}R_2 \\ 2R_2 + R_3 \end{array}$$

Here the basic feasible solution is

$$x_1 = 8, \qquad x_2 = 3, \qquad x_3 = 0, \qquad x_4 = 0, \qquad z = 100,$$

which corresponds to the corner point (8, 3) in Figure 7.23. Once again, the new value of the pivot variable x_1 is the smallest quotient, 8, that was used to select the pivot. From the graphical method we know that this solution provides the maximum value of the objective function. This fact can also be seen algebraically by using an algebraic argument similar to the one used in Example 5. Thus, there is no need to move to another corner point and the simplex method ends. ◀

As illustrated in Example 6, the basic feasible solution obtained from a simplex tableau corresponds to a corner point of the feasible region. Pivoting, which replaces one tableau with another, is a systematic way of moving from one corner point to another, each time improving the value of the objective function. The simplex method ends when a corner point that produces the maximum value of the objective function is reached (or when it becomes clear that the problem has no maximum solution).

When there are three or more variables in a linear programming problem, it may be difficult or impossible to draw a picture, but it can be proved that the optimal value of the objective function occurs at a basic feasible solution (corresponding to a corner point in the two variable case). The simplex method provides a means of moving from one basic feasible solution to another until one that produces the optimal value of the objective function is reached.

Explanation of Pivoting

The rules for selecting the pivot in the simplex method can be understood by examining how the first pivot was chosen in Example 6.

EXAMPLE 7 The initial simplex tableau of Example 6 provides a basic feasible solution with $x_1 = 0, x_2 = 0$.

$$\begin{array}{ccccc|c} x_1 & x_2 & x_3 & x_4 & z & \\ \hline 40 & 80 & 1 & 0 & 0 & 560 \\ 6 & 8 & 0 & 1 & 0 & 72 \\ \hline -8 & -12 & 0 & 0 & 1 & 0 \end{array}$$

This solution certainly does not give a maximum value for the objective function $z = 8x_1 + 12x_2$. Since x_2 has the largest coefficient, z will be increased most if x_2 is increased. In other words, the most negative indicator in the tableau (which corresponds to the largest coefficient in the objective function) identifies the variable that will provide the greatest change in the value of z.

To determine how much x_2 can be increased without leaving the feasible region, look at the first two equations

$$\begin{aligned} 40x_1 + 80x_2 + x_3 \qquad &= 560 \\ 6x_1 + \ \ 8x_2 \qquad + x_4 &= 72 \end{aligned}$$

and solve for the basic variables x_3 and x_4.

$$\begin{aligned} x_3 &= 560 - 40x_1 - 80x_2 \\ x_4 &= 72 - 6x_1 - 8x_2 \end{aligned}$$

Now x_2 is to be increased while x_1 is to keep the value 0. Hence

$$\begin{aligned} x_3 &= 560 - 80x_2 \\ x_4 &= 72 - 8x_2. \end{aligned}$$

Since $x_3 \geq 0$ and $x_4 \geq 0$, we must have:

$$\begin{array}{rclcrcl} 0 \leq 560 - 80x_2 & & & \text{and} & 0 \leq 72 - 8x_2 & & \\ 80x_2 \leq 560 & & & & 8x_2 \leq 72 & & \\ x_2 \leq \dfrac{560}{80} = 7 & & & & x_2 \leq \dfrac{72}{8} = 9. & & \end{array}$$

The right sides of these last inequalities are the quotients used to select the pivot row. Since x_2 must satisfy both inequalities, x_2 can be at most 7. In other words, the smallest quotient formed from positive entries in the pivot column identifies the value of x_2 that produces the largest change in z while remaining in the feasible region. By pivoting with the pivot determined in this way, we obtain the second tableau and a basic feasible solution in which $x_2 = 7$, as was shown in Example 6. ◀

An analysis similar to that in Example 7 applies to each occurrence of pivoting in the simplex method. The idea is to improve the value of the objective function by adjusting one variable at a time. The most negative indicator identifies the variable that will account for the largest increase in z. The smallest quotient determines the largest value of that variable that will produce a feasible solution. Pivoting leads to a solution in which the selected variable has this largest value.

The simplex method is easily implemented on a computer and some graphing calculators. Technology is essential for the simplex method in any situation where there are a large number of variables and constraints (and hence, an enormous number of corner points to check).

7.4 Exercises

In Exercises 1–4, (a) *determine the number of slack variables needed;* (b) *name them;* (c) *use the slack variables to convert each constraint into a linear equation. (See Example 1.)*

1. Maximize $z = 32x_1 + 9x_2$
subject to: $4x_1 + 2x_2 \le 20$
$5x_1 + x_2 \le 50$
$2x_1 + 3x_2 \le 25$
$x_1 \ge 0, x_2 \ge 0.$

2. Maximize $z = 3.7x_1 + 4.3x_2$
subject to: $2.4x_1 + 1.5x_2 \le 10$
$1.7x_1 + 1.9x_2 \le 15$
$x_1 \ge 0, x_2 \ge 0.$

3. Maximize $z = 8x_1 + 3x_2 + x_3$
subject to: $3x_1 - x_2 + 4x_3 \le 95$
$7x_1 + 6x_2 + 8x_3 \le 118$
$4x_1 + 5x_2 + 10x_3 \le 220$
$x_1 \ge 0, x_2 \ge 0, x_3 \ge 0.$

4. Maximize $z = 12x_1 + 15x_2 + 10x_3$
subject to: $2x_1 + 2x_2 + x_3 \le 8$
$x_1 + 4x_2 + 3x_3 \le 12$
$x_1 \ge 0, x_2 \ge 0, x_3 \ge 0.$

Introduce slack variables as necessary and then write the initial simplex tableau for each of these linear programming problems.

5. Maximize $z = 5x_1 + x_2$
subject to: $2x_1 + 3x_2 \le 6$
$4x_1 + x_2 \le 6$
$5x_1 + 2x_2 \le 15$
$x_1 \ge 0, x_2 \ge 0.$

6. Maximize $z = 5x_1 + 3x_2 + 4x_3$
subject to: $4x_1 + 3x_2 + 2x_3 \le 60$
$3x_1 + 4x_2 + x_3 \le 24$
$x_1 \ge 0, x_2 \ge 0, x_3 \ge 0.$

7. Maximize $z = x_1 + 5x_2 + 10x_3$
subject to: $x_1 + 2x_2 + 3x_3 \le 10$
$2x_1 + x_2 + x_3 \le 8$
$3x_1 + 2x_3 \le 6$
$x_1 \ge 0, x_2 \ge 0, x_3 \ge 0.$

8. Maximize $z = 5x_1 - x_2 + 3x_3$
subject to: $3x_1 + 2x_2 + x_3 \le 36$
$x_1 + 4x_2 + x_3 \le 24$
$x_1 - x_2 - x_3 \le 32$
$x_1 \ge 0, x_2 \ge 0, x_3 \ge 0.$

Find the pivot in each of the following simplex tableaus. (See Example 2.)

9.

$$\begin{array}{cccccc|c} x_1 & x_2 & x_3 & x_4 & x_5 & z & \\ \hline 2 & 2 & 0 & 3 & 1 & 0 & 15 \\ 3 & 4 & 1 & 6 & 0 & 0 & 20 \\ \hline -2 & -1 & 0 & 1 & 0 & 1 & 10 \end{array}$$

10.

$$\begin{array}{cccccc|c} x_1 & x_2 & x_3 & x_4 & x_5 & z & \\ \hline 0 & 2 & 1 & 1 & 3 & 0 & 5 \\ 1 & -5 & 0 & 1 & 2 & 0 & 8 \\ \hline 0 & -2 & 0 & -1 & 1 & 1 & 10 \end{array}$$

11.

$$\begin{array}{ccccccc|c} x_1 & x_2 & x_3 & x_4 & x_5 & x_6 & z & \\ \hline 6 & 2 & 1 & 3 & 0 & 0 & 0 & 8 \\ 0 & 2 & 0 & 1 & 0 & 1 & 0 & 7 \\ 2 & 1 & 0 & 3 & 1 & 0 & 0 & 6 \\ \hline -3 & -2 & 0 & 2 & 0 & 0 & 1 & 12 \end{array}$$

12.

$$\begin{array}{ccccccc|c} x_1 & x_2 & x_3 & x_4 & x_5 & x_6 & z & \\ \hline 0 & 2 & 0 & 1 & 2 & 2 & 0 & 3 \\ 0 & 3 & 1 & 0 & 1 & 2 & 0 & 2 \\ 1 & 4 & 0 & 0 & 3 & 5 & 0 & 5 \\ \hline 0 & -4 & 0 & 0 & 4 & -3 & 1 & 20 \end{array}$$

In Exercises 13–16, use the indicated entry as the pivot and perform the pivoting. (See Examples 3 and 4.)

13.

$$\begin{array}{cccccc|c} x_1 & x_2 & x_3 & x_4 & x_5 & z & \\ \hline 1 & 2 & 4 & 1 & 0 & 0 & 56 \\ 2 & \boxed{2} & 1 & 0 & 1 & 0 & 40 \\ \hline -1 & -3 & -2 & 0 & 0 & 1 & 0 \end{array}$$

14.

$$\begin{array}{ccccccc|c} x_1 & x_2 & x_3 & x_4 & x_5 & x_6 & z & \\ \hline 2 & 2 & \boxed{1} & 1 & 0 & 0 & 0 & 12 \\ 1 & 2 & 3 & 0 & 1 & 0 & 0 & 45 \\ 3 & 1 & 1 & 0 & 0 & 1 & 0 & 20 \\ \hline -2 & -1 & -3 & 0 & 0 & 0 & 1 & 0 \end{array}$$

15.

$$\begin{array}{ccccccc|c} x_1 & x_2 & x_3 & x_4 & x_5 & x_6 & z & \\ \hline 1 & 1 & 1 & 1 & 0 & 0 & 0 & 60 \\ 3 & 1 & \boxed{2} & 0 & 1 & 0 & 0 & 100 \\ 1 & 2 & 3 & 0 & 0 & 1 & 0 & 200 \\ \hline -1 & -1 & -2 & 0 & 0 & 0 & 1 & 0 \end{array}$$

16.

$$\begin{array}{ccccccc|c} x_1 & x_2 & x_3 & x_4 & x_5 & x_6 & z & \\ \hline 4 & 2 & 3 & 1 & 0 & 0 & 0 & 22 \\ 2 & 2 & \boxed{5} & 0 & 1 & 0 & 0 & 28 \\ 1 & 3 & 2 & 0 & 0 & 1 & 0 & 45 \\ \hline -3 & -2 & -4 & 0 & 0 & 0 & 1 & 0 \end{array}$$

For each simplex tableau in Exercises 17–20, (a) *list the basic and the nonbasic variables;* (b) *find the basic feasible solution determined by setting the nonbasic variables equal to 0;* (c) *decide whether this is a maximum solution. (See Examples 5 and 6.)*

17.

$$\begin{array}{cccccc|c} x_1 & x_2 & x_3 & x_4 & x_5 & z & \\ \hline 3 & 2 & 0 & -3 & 1 & 0 & 29 \\ 4 & 0 & 1 & -2 & 0 & 0 & 16 \\ \hline -5 & 0 & 0 & -1 & 0 & 1 & 11 \end{array}$$

18.

$$\begin{array}{ccccccc} x_1 & x_2 & x_3 & x_4 & x_5 & x_6 & z & \\ \end{array}$$
$$\left[\begin{array}{ccccccc|c} -3 & 0 & \frac{1}{2} & 1 & -2 & 0 & 0 & 22 \\ 2 & 0 & -3 & 0 & 1 & 1 & 0 & 10 \\ 4 & 1 & 4 & 0 & \frac{3}{4} & 0 & 0 & 17 \\ \hline -1 & 0 & 0 & 0 & 1 & 0 & 1 & 120 \end{array}\right]$$

19.

$$\begin{array}{ccccccc} x_1 & x_2 & x_3 & x_4 & x_5 & x_6 & z \end{array}$$
$$\left[\begin{array}{ccccccc|c} 1 & 0 & 2 & \frac{1}{2} & 0 & \frac{1}{3} & 0 & 6 \\ 0 & 1 & -1 & 5 & 0 & -1 & 0 & 13 \\ 0 & 0 & 1 & \frac{3}{2} & 1 & -\frac{1}{3} & 0 & 21 \\ \hline 0 & 0 & 2 & \frac{1}{2} & 0 & 3 & 1 & 18 \end{array}\right]$$

20.

$$\begin{array}{cccccccc} x_1 & x_2 & x_3 & x_4 & x_5 & x_6 & x_7 & z \end{array}$$
$$\left[\begin{array}{cccccccc|c} -1 & 0 & 0 & 1 & 0 & 3 & -2 & 0 & 47 \\ 2 & 0 & 1 & 0 & 0 & 2 & -\frac{1}{2} & 0 & 37 \\ 3 & 0 & 0 & 0 & 1 & -1 & 6 & 0 & 43 \\ \hline 4 & 1 & 0 & 0 & 0 & 6 & 0 & 1 & 86 \end{array}\right]$$

Use the simplex method to solve Exercises 21–36.

21. Maximize $z = x_1 + 3x_2$
subject to:
$x_1 + x_2 \leq 10$
$5x_1 + 2x_2 \leq 20$
$x_1 + 2x_2 \leq 36$
$x_1 \geq 0, x_2 \geq 0.$

22. Maximize $z = 5x_1 + x_2$
subject to:
$2x_1 + 3x_2 \leq 8$
$4x_1 + 8x_2 \leq 12$
$5x_1 + 2x_2 \leq 30$
$x_1 \geq 0, x_2 \geq 0.$

23. Maximize $z = 2x_1 + x_2$
subject to:
$x_1 + 3x_2 \leq 12$
$2x_1 + x_2 \leq 10$
$x_1 + x_2 \leq 4$
$x_1 \geq 0, x_2 \geq 0.$

24. Maximize $z = 4x_1 + 2x_2$
subject to:
$-x_1 - x_2 \leq 12$
$3x_1 - x_2 \leq 15$
$x_1 \geq 0, x_2 \geq 0.$

25. Maximize $z = 5x_1 + 4x_2 + x_3$
subject to:
$-2x_1 + x_2 + 2x_3 \leq 3$
$x_1 - x_2 + x_3 \leq 1$
$x_1 \geq 0, x_2 \geq 0, x_3 \geq 0.$

26. Maximize $z = 3x_1 + 2x_2 + x_3$
subject to:
$2x_1 + 2x_2 + x_3 \leq 10$
$x_1 + 2x_2 + 3x_3 \leq 15$
$x_1 \geq 0, x_2 \geq 0, x_3 \geq 0.$

27. Maximize $z = 2x_1 + x_2 + x_3$
subject to:
$x_1 - 3x_2 + x_3 \leq 3$
$x_1 - 2x_2 + 2x_3 \leq 12$
$x_1 \geq 0, x_2 \geq 0, x_3 \geq 0.$

28. Maximize $z = 4x_1 + 5x_2 + x_3$
subject to:
$x_1 + 2x_2 + 4x_3 \leq 10$
$2x_1 + 2x_2 + x_3 \leq 10$
$x_1 \geq 0, x_2 \geq 0, x_3 \geq 0.$

29. Maximize $z = 2x_1 + 2x_2 - 4x_3$
subject to:
$3x_1 + 3x_2 - 6x_3 \leq 51$
$5x_1 + 5x_2 + 10x_3 \leq 99$
$x_1 \geq 0, x_2 \geq 0, x_3 \geq 0.$

30. Maximize $z = 4x_1 + x_2 + 3x_3$
subject to:
$x_1 + 3x_3 \leq 6$
$6x_1 + 3x_2 + 12x_3 \leq 40$
$x_1 \geq 0, x_2 \geq 0, x_3 \geq 0.$

31. Maximize $z = 300x_1 + 200x_2 + 100x_3$
subject to:
$x_1 + x_2 + x_3 \leq 100$
$2x_1 + 3x_2 + 4x_3 \leq 320$
$2x_1 + x_2 + x_3 \leq 160$
$x_1 \geq 0, x_2 \geq 0, x_3 \geq 0.$

32. Maximize $z = x_1 + 5x_2 - 10x_3$
subject to:
$8x_1 + 4x_2 + 12x_3 \leq 18$
$x_1 + 6x_2 + 2x_3 \leq 45$
$5x_1 + 7x_2 + 3x_3 \leq 60$
$x_1 \geq 0, x_2 \geq 0, x_3 \geq 0.$

33. Maximize $z = 4x_1 - 3x_2 + 2x_3$
subject to:
$2x_1 - x_2 + 8x_3 \leq 40$
$4x_1 - 5x_2 + 6x_3 \leq 60$
$2x_1 - 2x_2 + 6x_3 \leq 24$
$x_1 \geq 0, x_2 \geq 0, x_3 \geq 0.$

34. Maximize $z = 3x_1 + 2x_2 - 4x_3$
subject to:
$x_1 - x_2 + x_3 \leq 10$
$2x_1 - x_2 + 2x_3 \leq 30$
$-3x_1 + x_2 + 3x_3 \leq 40$
$x_1 \geq 0, x_2 \geq 0, x_3 \geq 0.$

35. Maximize $z = x_1 + 2x_2 + x_3 + 5x_4$
subject to:
$x_1 + 2x_2 + x_3 + x_4 \leq 50$
$3x_1 + x_2 + 2x_3 + x_4 \leq 100$
$x_1 \geq 0, x_2 \geq 0, x_3 \geq 0, x_4 \geq 0.$

36. Maximize $z = x_1 + x_2 + 4x_3 + 5x_4$
subject to:
$x_1 + 2x_2 + 3x_3 + x_4 \leq 115$
$2x_1 + x_2 + 8x_3 + 5x_4 \leq 200$
$x_1 + x_3 \leq 50$
$x_1 \geq 0, x_2 \geq 0, x_3 \geq 0, x_4 \geq 0.$

37. The initial simplex tableau of a linear programming problem is given below.

$$\begin{array}{cccccc} x_1 & x_2 & x_3 & x_4 & x_5 & z \end{array}$$
$$\left[\begin{array}{cccccc|c} 1 & 1 & 1 & 1 & 0 & 0 & 12 \\ 2 & 1 & 2 & 0 & 1 & 0 & 30 \\ \hline -2 & -2 & -1 & 0 & 0 & 1 & 0 \end{array}\right]$$

(a) Use the simplex method to solve the problem, with column one as the first pivot column.

(b) Now use the simplex method to solve the problem, with column two as the first pivot column.

(c) Does this problem have a unique maximum solution? Why?

38. The final simplex tableau of a linear programming problem is given here.

$$\begin{array}{ccccc} x_1 & x_2 & x_3 & x_4 & z \end{array}$$
$$\left[\begin{array}{ccccc|c} 1 & 1 & 2 & 0 & 0 & 24 \\ 2 & 0 & 2 & 1 & 0 & 8 \\ \hline 4 & 0 & 0 & 0 & 1 & 40 \end{array}\right]$$

(a) What is the solution given by this tableau?

(b) Even though all the indicators are nonnegative, perform one more round of pivoting on this tableau, using column three as the pivot column and choosing the pivot row by forming quotients in the usual way.

(c) Show that there is more than one solution to the linear programming problem by comparing your answer in part (a) to the basic feasible solution given by the tableau found in part (b). Does it give the same value of z as the solution in part (a)?

7.5 MAXIMIZATION APPLICATIONS

Applications of the simplex method are considered in this section. First, however, we make a slight change in notation. You have noticed that the column representing the variable z in a simplex tableau never changes during pivoting. (Since all the entries except the last one in this column are 0, performing row operations has no effect on these entries—they remain 0.) Consequently, this column is unnecessary and can be omitted without causing any difficulty.

Hereafter in this text, the column corresponding to the variable z (representing the objective function) will be omitted from all simplex tableaus.

EXAMPLE 1 A farmer has 110 acres of available land he wishes to plant with a mixture of potatoes, corn, and cabbage. It costs him \$400 to produce an acre of potatoes, \$160 to produce an acre of corn, and \$280 to produce an acre of cabbage. He has a maximum of \$20,000 to spend. He makes a profit of \$120 per acre of potatoes, \$40 per acre of corn, and \$60 per acre of cabbage.

(a) How many acres of each crop should he plant to maximize his profit? Begin by summarizing the given information as follows.

Crop	*Number of Acres*	*Cost per Acre*	*Profit per Acre*
Potatoes	x_1	\$400	\$120
Corn	x_2	\$160	\$ 40
Cabbage	x_3	\$280	\$ 60
Maximum Available	110	\$20,000	

If the number of acres allotted to each of the three crops is represented by x_1, x_2, and x_3, respectively, then the constraints can be expressed as

$$\begin{aligned} x_1 + \quad x_2 + \quad x_3 &\le 110 && \text{Number of acres} \\ 400x_1 + 160x_2 + 280x_3 &\le 20{,}000 && \text{Production costs} \end{aligned}$$

where x_1, x_2, and x_3 are all nonnegative. The first of these constraints says that $x_1 + x_2 + x_3$ is less than or perhaps equal to 110. Use x_4 as the slack variable, giving the equation

$$x_1 + x_2 + x_3 + x_4 = 110.$$

Here x_4 represents the amount of the farmer's 110 acres that will not be used. (x_4 may be 0 or any value up to 110.)

In the same way, the constraint $400x_1 + 160x_2 + 280x_3 \le 20{,}000$ can be converted into an equation by adding a slack variable, x_5:

$$400x_1 + 160x_2 + 280x_3 + x_5 = 20{,}000.$$

The slack variable x_5 represents any unused portion of the farmer's \$20,000 capital. (Again, x_5 may have any value from 0 to 20,000.)

The farmer's profit on potatoes is the product of the profit per acre (\$120) and the number x_1 of acres, that is, $120x_1$. His profit on corn and cabbage is computed similarly. Hence his total profit is given by

$$\begin{aligned} z &= \text{profit on potatoes} + \text{profit on corn} + \text{profit on cabbage} \\ z &= 120x_1 + 40x_2 + 60x_3. \end{aligned}$$

The linear programming problem can now be stated as follows:

$$\begin{aligned} &\text{Maximize} && z = 120x_1 + 40x_2 + 60x_3 \\ &\text{subject to:} && x_1 + x_2 + x_3 + x_4 = 110 \\ & && 400x_1 + 160x_2 + 280x_3 + x_5 = 20{,}000 \\ &\text{with} && x_1 \ge 0, x_2 \ge 0, x_3 \ge 0, x_4 \ge 0, x_5 \ge 0. \end{aligned}$$

The initial simplex tableau (without the z column) is

$$\begin{array}{c} \begin{matrix} x_1 & x_2 & x_3 & x_4 & x_5 \end{matrix} \\ \left[\begin{array}{rrrrr|r} 1 & 1 & 1 & 1 & 0 & 110 \\ 400 & 160 & 280 & 0 & 1 & 20{,}000 \\ \hline -120 & -40 & -60 & 0 & 0 & 0 \end{array}\right]. \end{array}$$

The most negative indicator is -120; column one is the pivot column. The quotients needed to determine the pivot row are $110/1 = 110$ and $20{,}000/400 = 50$. So the pivot is 400 in row two, column one. Multiplying row two by 1/400 and completing the pivoting leads to the final simplex tableau.

$$\begin{array}{c} \begin{matrix} x_1 & x_2 & x_3 & x_4 & x_5 \end{matrix} \\ \left[\begin{array}{rrrrr|r} 0 & .6 & .3 & 1 & -.0025 & 60 \\ 1 & .4 & .7 & 0 & .0025 & 50 \\ \hline 0 & 8 & 24 & 0 & .3 & \mathbf{6000} \end{array}\right] \begin{array}{l} -1R_2 + R_1 \\ \frac{1}{400}R_2 \\ 120R_2 + R_3 \end{array} \end{array}$$

Setting the nonbasic variables x_2, x_3, and x_5 equal to 0, solving for the basic variables x_1 and x_4, and remembering that the value of z is in the lower right-hand corner leads to this maximum solution:

$$x_1 = 50, \quad x_2 = 0, \quad x_3 = 0, \quad x_4 = 60, \quad x_5 = 0, \quad z = 6000.$$

Therefore, the farmer will make a maximum profit of \$6000 by planting 50 acres of potatoes, and no corn or cabbage.

(b) If the farmer maximizes his profit, how much land will remain unplanted? What is the explanation for this?

Since 50 of 110 acres are planted, 60 acres will remain unplanted. Alternatively, note that the unplanted acres of land are represented by x_4, the slack variable in the "number of acres" constraint. In the maximal solution found in part (a), $x_4 = 60$, which means that 60 acres are left unplanted.

The amount of unused cash is represented by x_5, the slack variable in the "production costs" constraint. Since $x_5 = 0$, all the available money has been used. By using the maximal solution in part (a), the farmer has used his \$20,000 most effectively. If he had more cash, he would plant more crops and make a larger profit. ◀

EXAMPLE 2 Ana Pott, who is a candidate for the state legislature, has \$96,000 to buy TV advertising time. Ads cost \$400 per minute on a local cable channel, \$4000 per minute on a regional independent channel, and \$12,000 per minute on a national network channel. Because of existing contracts the TV stations can provide at most 30 minutes of advertising time, with a maximum of 6 minutes on the national network channel. At any given time during the evening, approximately 100,000 people watch the cable channel, 200,000 the independent channel, and 600,000 the network channel. To get maximum exposure, how much time should Ana buy from each station?

(a) Set up the initial simplex tableau for this problem.

Let x_1 be the number of minutes of ads on the cable channel, x_2 the number of minutes on the independent channel, and x_3 the number of minutes on the network channel. Exposure is measured in viewer-minutes. For instance, 100,000 people watching x_1 minutes of ads on the cable channel produces $100{,}000x_1$ viewer-minutes. The amount of exposure is given by the total number of viewer-minutes for all three channels, namely,

$$100{,}000x_1 + 200{,}000x_2 + 600{,}000x_3.$$

Since 30 minutes are available,

$$x_1 + x_2 + x_3 \leq 30.$$

The fact that only 6 minutes can be used on the network channel means that

$$x_3 \leq 6.$$

Expenditures are limited to \$96,000, so

$$\begin{aligned} \text{Cable cost} + \text{independent cost} + \text{network cost} &\leq 96{,}000 \\ 400x_1 \quad + \quad 4000x_2 \quad + \quad 12{,}000x_3 &\leq 96{,}000. \end{aligned}$$

Therefore, Ana must solve the following linear programming problem:

$$\begin{aligned} \text{Maximize} \quad z = 100{,}000x_1 + 200{,}000x_2 + 600{,}000x_3 \\ \text{subject to:} \qquad x_1 + x_2 + x_3 &\leq 30 \\ x_3 &\leq 6 \\ 400x_1 + 4000x_2 + 12{,}000x_3 &\leq 96{,}000 \\ \text{with} \quad x_1 \geq 0, x_2 \geq 0, x_3 \geq 0. \end{aligned}$$

Introducing slack variables x_4, x_5, and x_6 (one for each constraint), rewriting the constraints as equations, and expressing the objective function as

$$-100{,}000x_1 - 200{,}000x_2 - 600{,}000x_3 + z = 0$$

leads to the initial simplex tableau:

$$\begin{array}{c} \\ \\ \\ \\ \end{array}\left[\begin{array}{cccccc|c} x_1 & x_2 & x_3 & x_4 & x_5 & x_6 & \\ 1 & 1 & 1 & 1 & 0 & 0 & 30 \\ 0 & 0 & 1 & 0 & 1 & 0 & 6 \\ 400 & 4000 & 12{,}000 & 0 & 0 & 1 & 96{,}000 \\ \hline -100{,}000 & -200{,}000 & -600{,}000 & 0 & 0 & 0 & 0 \end{array}\right].$$

(b) Use the simplex method to find the final simplex tableau.

Work by hand, or use a graphing calculator simplex program or a spreadsheet, to obtain this final tableau.

$$\left[\begin{array}{cccccc|c} x_1 & x_2 & x_3 & x_4 & x_5 & x_6 & \\ 1 & 0 & 0 & \frac{10}{9} & \frac{20}{9} & -\frac{25}{90{,}000} & 20 \\ 0 & 0 & 1 & 0 & 1 & 0 & 6 \\ 0 & 1 & 0 & -\frac{1}{9} & -\frac{29}{9} & \frac{25}{90{,}000} & 4 \\ \hline 0 & 0 & 0 & \frac{800{,}000}{9} & \frac{1{,}600{,}000}{9} & \frac{250}{9} & 6{,}400.000 \end{array}\right].$$

Therefore, the optimal solution is

$$x_1 = 20, \quad x_2 = 4, \quad x_3 = 6, \quad x_4 = 0, \quad x_5 = 0, \quad x_6 = 0.$$

Ana should buy 20 minutes of time on the cable channel, 4 minutes on the independent channel, and 6 minutes on the network channel.

(c) What do the values of the slack variables in the optimal solution tell you?

All three slack variables are 0. This means that all the available minutes have been used ($x_4 = 0$ in the first constraint), the maximum possible six minutes on the national network have been used ($x_5 = 0$ in the second constraint), and all of the \$96,000 has been spent ($x_6 = 0$ in the third constraint). ◀ 1

1

In Example 2, what is the number of viewer-minutes in the optimal solution?

Answer:

$z = 6{,}400{,}000$

EXAMPLE 3 A chemical plant makes three products: glaze, solvent, and clay, each of which brings in different revenue per truckload. Production is limited, first by the number of air pollution units the plant is allowed to produce each day and second by the time available in the evaporation tank. The plant manager wants to maximize the daily revenue. Using information not given here, he sets up an initial simplex tableau and uses the simplex method to produce the following final simplex tableau.

$$\left[\begin{array}{ccccc|c} -10 & -25 & 0 & 1 & -1 & 60 \\ 3 & 4 & 1 & 0 & .1 & 24 \\ \hline 7 & 13 & 0 & 0 & .4 & 96 \end{array}\right]$$

The three variables represent the number of truckloads of glaze, solvent, and clay, respectively. The first slack variable comes from the air pollution constraint and the second slack variable from the time constraint on the evaporation tank. The revenue function is given in hundreds of dollars.

(a) What is the optimal solution?

$$x_1 = 0, \quad x_2 = 0, \quad x_3 = 24, \quad x_4 = 60, \quad x_5 = 0, \quad \text{and} \quad z = 96.$$

(b) Interpret this solution. What do the variables represent and what does it mean?

The variable x_1 is the number of truckloads of glaze, x_2 the number of truckloads of solvent, x_3 the number of truckloads of clay to be produced and z is the revenue produced (in hundreds of dollars). The plant should produce 24 truckloads of clay and no glaze or solvent for a maximum revenue of \$9600. The first slack variable x_4 represents the number of air pollution units below the maximum number allowed. Since $x_4 = 60$, air pollution will be 60 units less than the allowable maximum. The second slack variable x_5 represents the unused time in the evaporation tank. Since $x_5 = 0$, the evaporation tank is fully used.

7.5 Exercises

Set up the initial simplex tableau for each of the following problems.

1. Business A cat breeder has the following amounts of cat food: 90 units of tuna, 80 units of liver, and 50 units of chicken. To raise a Siamese cat, the breeder must use 2 units of tuna, 1 of liver, and 1 of chicken per day, while raising a Persian cat requires 1, 2, and 1 units, respectively, per day. If a Siamese cat sells for \$12 while a Persian cat sells for \$10, how many of each should be raised in order to obtain maximum gross income? What is the maximum gross income?

2. Business Banal, Inc., produces art for motel rooms. Its painters can turn out mountain scenes, seascapes, and pictures of clowns. Each painting is worked on by three different artists, T, D, and H. Artist T works only 25 hours per week, while D and H work 45 and 40 hours per week, respectively. Artist T spends 1 hour on a mountain scene, 2 hours on a seascape, and 1 hour on a clown. Corresponding times for D and H are 3, 2, and 2 hours, and 2, 1, and 4 hours, respectively. Banal makes \$20 on a mountain scene, \$18 on a seascape, and \$22 on a clown. The head painting packer can't stand clowns, so that no more than 4 clown paintings may be done in a week. Find the number of each type of painting that should be made weekly in order to maximize profit. Find the maximum possible profit.

3. Health A biologist has 500 kg of nutrient A, 600 kg of nutrient B, and 300 kg of nutrient C. These nutrients will be used to make 4 types of food, whose contents (in percent of nutrient per kilogram of food) and whose "growth values" are as shown in the table.

	P	Q	R	S
A	0	0	37.5	62.5
B	0	75	50	37.5
C	100	25	12.5	0
Growth Value	90	70	60	50

How many kilograms of each food should be produced in order to maximize total growth value? Find the maximum growth value.

4. Natural Science A lake is stocked each spring with three species of fish, A, B, and C. The average weights of the fish are 1.62, 2.12, and 3.01 kilograms for species A, B, and C, respectively. Three foods, I, II, and III, are available in the lake. Each fish of species A requires 1.32 units of food I, 2.9 units of food II, and 1.75 units of food III on the average each day. Species B fish each require 2.1 units of food I, .95 units of food II, and .6 units of food III daily. Species C fish require .86, 1.52, and 2.01 units of I, II, and III per day, respectively. If 490 units of food I, 897 units of food II, and 653 units of food III are available daily, how should the lake be stocked to maximize the weight of the fish supported by the lake?

In each of the following exercises: (a) *Use the simplex method to solve the problem;* (b) *explain what the values of the slack variables in the optimal solution mean in the context of the problem. (See Examples 1–3).*

5. Business A manufacturer of bicycles builds one-, three-, and ten-speed models. The bicycles need both aluminum and steel. The company has available 91,800 units of steel and 42,000 units of aluminum. The one-, three-, and ten-speed models need, respectively, 20, 30, and 40 units of steel and 12, 21, and 16 units of aluminum. How many of each type of bicycle should be made in order to maximize profit if the company makes \$8 per one-speed bike, \$12 per three-speed, and \$24 per ten-speed? What is the maximum possible profit?

6. Social Science Jayanta is working to raise money for the homeless by sending information letters and making follow-up calls to local labor organizations and church groups. She discovered that each church group requires 2 hours of letter writing and 1 hour of follow-up, while for each labor union she needs 2 hours of letter writing and 3 hours of follow-up. Jayanta can raise \$100 from each church group and \$200 from each union local, and she has a maximum of 16 hours of letter-writing time and a maximum of 12 hours of follow-up time available per month. Determine the most profitable mixture of groups she should contact and the most money she can raise in a month.

7. Social Science A political party is planning a half-hour television show. The show will have 3 minutes of direct requests for money from viewers. Three of the party's politicians will be on the show—a senator, a congresswoman, and a governor. The senator, a party "elder statesman," demands that he be on screen at least twice as long as the governor. The total time taken by the senator and the governor must be at least twice the time taken by the congresswoman. Based on a pre-show survey, it is believed that 40, 60, and 50 (in thousands) viewers will watch the program for each minute the senator, congresswoman, and governor, respectively, are on the air. Find the time that should be allotted to each politician in order to get the maximum number of viewers. Find the maximum number of viewers.

8. Business The Cut-Right Company sells sets of kitchen knives. The Basic Set consists of 2 utility knives and 1 chef's knife. The Regular Set consists of 2 utility knives, 1 chef's knife, and 1 slicer. The Deluxe Set consists of 3 utility knives, 1 chef's knife, and 1 slicer. Their profit is \$30 on a Basic Set, \$40 on a Regular Set, and \$60 on a Deluxe Set. The factory has on hand 800 utility knives, 400 chef's knives, and 200 slicers. Assuming that all sets will be sold, how many of each type should be made up in order to maximize profit? What is the maximum profit?

Use the simplex method to solve the following problems. (See Examples 1–3.)

9. Business The Fancy Fashions Store has \$8000 available each month for advertising. Newspaper ads cost \$400 each and no more than 20 can be run per month. Radio ads cost \$200 each and no more than 30 can run per month. TV ads cost \$1200 each, with a maximum of 6 available each month. Approximately 2000 women will see each newspaper ad, 1200 will hear each radio commercial, and 10,000 will see each TV ad. How much of each type of advertising should be used if the store wants to maximize its ad exposure?

10. Business Caroline's Quality Candy Confectionery is famous for fudge, chocolate cremes, and pralines. Its candy-making equipment is set up to make 100-pound batches at a time. Currently there is a chocolate shortage and the company can get only 120 pounds of chocolate in the next shipment. On a week's run, the confectionery's cooking and processing equipment is available for a total of 42 machine hours. During the same period the employees have a total of 56 work hours available for packaging. A batch of fudge requires 20 pounds of chocolate while a batch of cremes uses 25 pounds of chocolate. The cooking and processing take 120 minutes for fudge, 150 minutes for chocolate cremes, and 200 minutes for pralines. The packaging times measured in minutes per 1-pound box are 1, 2, 3, respectively, for fudge, cremes, and pralines. Determine how many batches of each type of candy the confectionery should make, assuming that the profit per pound box is 50¢ on fudge, 40¢ on chocolate cremes, and 45¢ on pralines. Also, find the maximum profit for the week.

11. Finance A political party is planning its fund-raising activities for a coming election. They plan to raise money through large fund-raising parties, letters requesting funds, and dinner parties where people can meet the candidate personally. Each large fund-raising party costs \$3000, each mailing costs \$1000, and each dinner party costs \$12,000. They can spend up to \$102,000 for these activities. From experience they know that each large party will raise \$200,000, each letter campaign will raise \$100,000. and each dinner party will raise \$600,000. They are able to carry out as many as 25 of these activities.
   (a) How many of each should they plan in order to raise the maximum amount of money? What is the maximum amount?
   (b) Dinner parties are more expensive than letter campaigns, yet the optimum solution found in part (a) includes dinner parties but no letter campaigns. Explain how this is possible.

12. Business A baker has 60 units of flour, 132 units of sugar, and 102 units of raisins. A loaf of raisin bread requires 1 unit of flour, 1 unit of sugar, and 2 units of

raisins, while a raisin cake needs 2, 4, and 1 units, respectively. If raisin bread sells for \$3 a loaf and a raisin cake for \$4, how many of each should be baked so that the gross income is maximized? What is the maximum gross income?

Business *The next two problems come from past CPA examinations.* Select the appropriate answer for each question.*

13. The Ball Company manufactures three types of lamps, labeled A, B, and C. Each lamp is processed in two departments, I and II. Total available man-hours per day for departments I and II are 400 and 600, respectively. No additional labor is available. Time requirements and profit per unit for each lamp type are as follows.

	A	B	C
Man-Hours in I	2	3	1
Man-Hours in II	4	2	3
Profit per Unit	\$5	\$4	\$3

The company has assigned you as the accounting member of its profit planning committee to determine the numbers of types of A, B, and C lamps that it should produce in order to maximize its total profit from the sale of lamps. The following questions relate to a linear programming model that your group has developed.

(a) The coefficients of the objective function would be
(1) 4, 2, 3;
(2) 2, 3, 1;
(3) 5, 4, 3;
(4) 400,600.

(b) The constraints in the model would be
(1) 2, 3, 1;
(2) 5, 4, 3;
(3) 4, 2, 3;
(4) 400,600.

(c) The constraint imposed by the available man-hours in department I could be expressed as
(1) $4X_1 + 2X_2 + 3X_3 \leq 400$;
(2) $4X_1 + 2X_2 + 3X_3 \geq 400$;
(3) $2X_1 + 3X_2 + 1X_3 \leq 400$;
(4) $2X_1 + 3X_2 + 1X_3 \geq 400$.

14. The Golden Hawk Manufacturing Company wants to maximize the profits on products A, B, and C. The contribution margin for each product follows.

Product	*Contribution Margin*
A	\$2
B	5
C	4

The production requirements and departmental capacities, by departments, are as follows.

	PRODUCTION REQUIREMENTS BY PRODUCT (HOURS)			DEPARTMENTAL CAPACITY (TOTAL HOURS)
Department	A	B	C	
Assembling	2	3	2	30,000
Painting	1	2	2	38,000
Finishing	2	3	1	28,000

(a) What is the profit-maximization formula for the Golden Hawk Company?
(1) \$2A + \$5B + \$4C = X (where X = profit)
(2) 5A + 8B + 5C ≤ 96,000
(3) \$2A + \$5B + \$4C ≤ X
(4) \$2A + \$5B + \$4C = 96,000

(b) What is the constraint for the Painting Department of the Golden Hawk Company?
(1) 1A + 2B + 2C ≥ 38,000
(2) \$2A + \$5B + \$4C ≥ 38,000
(3) 1A + 2B + 2C ≤ 38,000
(4) 2A + 3B + 2C ≤ 30,000

15. Solve the problem in Exercise 1.

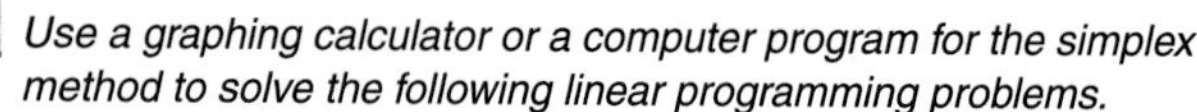

Use a graphing calculator or a computer program for the simplex method to solve the following linear programming problems.

16. Exercise 2. Your final answer should consist of whole numbers (Banal can't sell half a painting).

17. Exercise 3

18. Exercise 4

---

*Material from *Uniform CPA Examination Questions and Unofficial Answers,* copyright © 1973, 1974, 1975 by the American Institute of Certified Public Accountants, Inc., is reprinted with permission.

7.6 THE SIMPLEX METHOD: DUALITY AND MINIMIZATION

In this section the simplex method is extended to linear programming problems satisfying the following conditions.

1. The objective function is to be *minimized.*
2. All the coefficients of the objective function are nonnegative.
3. All constraints involve $\geq$.
4. All variables are nonnegative.

The method of solving minimization problems presented here is based on an interesting connection between maximizing and minimizing problems: any solution of a maximizing problem produces the solution of an associated minimizing problem, or vice-versa. Each of the associated problems is called the **dual** of the other. Thus, duals enable us to solve minimization problems of the type described above by the simplex method introduced in Section 7.4. (An alternative approach for solving minimization problems is given in the next section.)

When dealing with minimization problems, we use y_1, y_2, y_3, etc., as variables and denote the objective function by w. An example will explain the idea of a dual.

EXAMPLE 1 Minimize $w = 8y_1 + 16y_2$

subject to:
$$y_1 + 5y_2 \geq 9$$
$$2y_1 + 2y_2 \geq 10$$
$$y_1 \geq 0,\ y_2 \geq 0.$$

Before considering slack variables, write the augmented matrix of the system of inequalities and include the coefficients of the objective function (not their negatives) as the last row in the matrix.

Constants

$$\left[\begin{array}{cc|c} 1 & 5 & 9 \\ 2 & 2 & 10 \\ \hline 8 & 16 & 0 \end{array}\right] \leftarrow \text{Objective function}$$

Look now at the following new matrix, obtained from the one above by interchanging rows and columns.

Constants

$$\left[\begin{array}{cc|c} 1 & 2 & 8 \\ 5 & 2 & 16 \\ \hline 9 & 10 & 0 \end{array}\right] \leftarrow \text{Objective function}$$

The *rows* of the first matrix (for the minimizing problem) are the *columns* of the second matrix.

The entries in this second matrix could be used to write the following maximizing problem in standard form (again ignoring the fact that the numbers in the last row are not negative).

Use the corner points in Figure 7.24(a) to find the minimum value of $w = 8y_1 + 16y_2$ and where it occurs.

Answer:

48 when $y_1 = 4, y_2 = 1$

Use Figure 7.24(b) to find the maximum value of $z = 9x_1 + 10x_2$ and where it occurs.

Answer:

48 when $x_1 = 2, x_2 = 3$

$$\begin{aligned} \text{Maximize} \quad & z = 9x_1 + 10x_2 \\ \text{subject to:} \quad & x_1 + 2x_2 \le 8 \\ & 5x_1 + 2x_2 \le 16 \\ & x_1 \ge 0, x_2 \ge 0. \end{aligned}$$

Figure 7.24(a) shows the region of feasible solutions for the minimization problem given above, while Figure 7.24(b) shows the region of feasible solutions for the maximization problem produced by exchanging rows and columns. ◀ 1 2

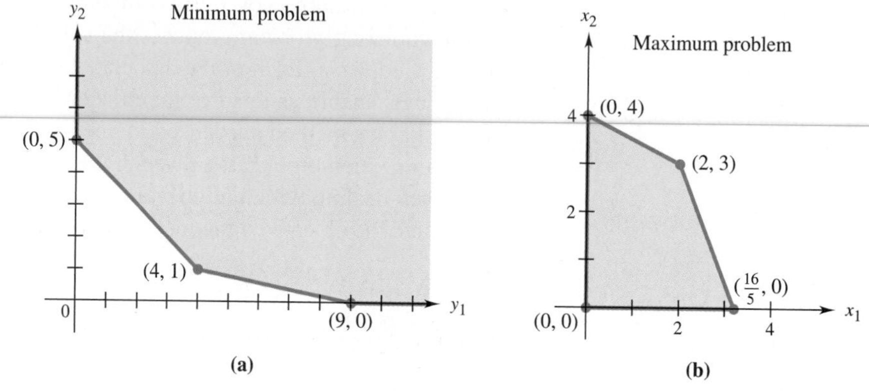

FIGURE 7.24

The two feasible regions in Figure 7.24 are different and the corner points are different, but the values of the objective functions found in Problems 1 and 2 at the side are equal—both are 48. An even closer connection between the two problems is shown by using the simplex method to solve the maximization problem given above.

Maximization problem

$$\begin{array}{c} \begin{matrix} x_1 & x_2 & x_3 & x_4 & \end{matrix} \\ \left[\begin{array}{cccc|c} 1 & \boxed{2} & 1 & 0 & 8 \\ 5 & 2 & 0 & 1 & 16 \\ \hline -9 & -10 & 0 & 0 & 0 \end{array}\right] \end{array}$$

$$\begin{array}{c} \begin{matrix} x_1 & x_2 & x_3 & x_4 & \end{matrix} \\ \left[\begin{array}{cccc|c} \frac{1}{2} & 1 & \frac{1}{2} & 0 & 4 \\ \boxed{4} & 0 & -1 & 1 & 8 \\ \hline -4 & 0 & 5 & 0 & 40 \end{array}\right] \end{array} \quad \begin{array}{l} \frac{1}{2}R_1 \\ -2R_1 + R_2 \\ 10R_1 + R_3 \end{array}$$

$$\begin{array}{c} \begin{matrix} x_1 & x_2 & x_3 & x_4 & \end{matrix} \\ \left[\begin{array}{cccc|c} 0 & 1 & \frac{5}{8} & -\frac{1}{8} & 3 \\ 1 & 0 & -\frac{1}{4} & \frac{1}{4} & 2 \\ \hline 0 & 0 & 4 & 1 & 48 \end{array}\right] \end{array} \quad \begin{array}{l} -\frac{1}{2}R_2 + R_1 \\ \frac{1}{4}R_2 \\ 4R_2 + R_3 \end{array}$$

The maximum is 48 when $x_1 = 2, x_2 = 3$.

Notice that the solution to the *minimization problem* (namely, $y_1 = 4, y_2 = 1$) is found in the bottom row and slack variable columns of the final simplex tableau for the maximization problem. This result suggests that a minimization problem can be solved by forming the dual maximization problem, solving it by the simplex method, and then reading the solution for the minimization problem from the bottom row of the final simplex tableau.

Before using this method to solve a minimization problem, let us find the duals of some typical linear programming problems. The process of exchanging the rows and columns of a matrix, which is used to find the dual, is called **transposing** the matrix, and each of the two matrices is the **transpose** of the other.

Give the transpose of each matrix.

(a) $\begin{bmatrix} 2 & 4 \\ 6 & 3 \\ 1 & 5 \end{bmatrix}$

(b) $\begin{bmatrix} 4 & 7 & 10 \\ 3 & 2 & 6 \\ 5 & 8 & 12 \end{bmatrix}$

Answers:

(a) $\begin{bmatrix} 2 & 6 & 1 \\ 4 & 3 & 5 \end{bmatrix}$

(b) $\begin{bmatrix} 4 & 3 & 5 \\ 7 & 2 & 8 \\ 10 & 6 & 12 \end{bmatrix}$

EXAMPLE 2 Find the transpose of each matrix.

(a) $A = \begin{bmatrix} 2 & -1 & 5 \\ 6 & 8 & 0 \\ -3 & 7 & -1 \end{bmatrix}$

Write the rows of matrix A as the columns of the transpose.

$$\text{Transpose of } A = \begin{bmatrix} 2 & 6 & -3 \\ -1 & 8 & 7 \\ 5 & 0 & -1 \end{bmatrix}$$

(b) The transpose of $\begin{bmatrix} 1 & 2 & 4 & 0 \\ 2 & 1 & 7 & 6 \end{bmatrix}$ is $\begin{bmatrix} 1 & 2 \\ 2 & 1 \\ 4 & 7 \\ 0 & 6 \end{bmatrix}$. ◀ 3

TECHNOLOGY TIP Most graphing calculators can find the transpose of a matrix. Look for this feature in the MATRIX MATH menu (TI) or the OPTN MAT menu (Casio) or the MATH MATRIX menu (HP-39). The transpose of matrix A from Example 2(a) is shown in Figure 7.25.

```
[[2  -1 5 ]
 [6   8 0 ]
 [-3 7  -1]]
AnsT
[[2  6 -3]
 [-1 8 7 ]
 [5  0 -1]]
```

FIGURE 7.25

EXAMPLE 3 Write the duals of the following minimization linear programming problems.

(a) Minimize $w = 10y_1 + 8y_2$

subject to: $y_1 + 2y_2 \geq 2$

$y_1 + y_2 \geq 5$

$y_1 \geq 0, y_2 \geq 0.$

Begin by writing the augmented matrix for the given problem.

$$\left[\begin{array}{cc|c} 1 & 2 & 2 \\ 1 & 1 & 5 \\ \hline 10 & 8 & 0 \end{array}\right]$$

Form the transpose of this matrix to get

$$\left[\begin{array}{cc|c} 1 & 1 & 10 \\ 2 & 1 & 8 \\ \hline 2 & 5 & 0 \end{array}\right].$$

Write the dual of the following linear programming problem.

Minimize

$w = 2y_1 + 5y_2 + 6y_3$

subject to:

$$\begin{aligned} 2y_1 + 3y_2 + \ y_3 &\geq 15 \\ y_1 + \ y_2 + 2y_3 &\geq 12 \\ 5y_1 + 3y_2 \qquad &\geq 10 \\ y_1 \geq 0, y_2 \geq 0, y_3 &\geq 0. \end{aligned}$$

Answer:

Maximize

$z = 15x_1 + 12x_2 + 10x_3$

subject to:

$$\begin{aligned} 2x_1 + \ x_2 + 5x_3 &\leq 2 \\ 3x_1 + \ x_2 + 3x_3 &\leq 5 \\ x_1 + 2x_2 \qquad &\leq 6 \\ x_1 \geq 0, x_2 \geq 0, x_3 &\geq 0. \end{aligned}$$

The dual problem is stated from this second matrix as follows (using x instead of y).

$$\begin{aligned} \text{Maximize} \quad z &= 2x_1 + 5x_2 \\ \text{subject to:} \quad x_1 + x_2 &\leq 10 \\ 2x_1 + x_2 &\leq 8 \\ x_1 \geq 0, x_2 &\geq 0. \end{aligned}$$

(b) Minimize $w = 7y_1 + 5y_2 + 8y_3$

subject to:

$$\begin{aligned} 3y_1 + 2y_2 + y_3 &\geq 10 \\ y_1 + \ y_2 + y_3 &\geq 8 \\ 4y_1 + 5y_2 \qquad &\geq 25 \\ y_1 \geq 0, y_2 \geq 0, y_3 &\geq 0. \end{aligned}$$

The dual problem is stated as follows.

$$\begin{aligned} \text{Maximize} \quad z &= 10x_1 + 8x_2 + 25x_3 \\ \text{subject to:} \quad 3x_1 + x_2 + 4x_3 &\leq 7 \\ 2x_1 + x_2 + 5x_3 &\leq 5 \\ x_1 + x_2 \qquad &\leq 8 \\ x_1 \geq 0, x_2 \geq 0, x_3 &\geq 0. \end{aligned}$$ ◀ 4

In Example 3, all the constraints of the minimization problems were $\geq$ inequalities, while all those in the dual maximization problems were $\leq$ inequalities. This is generally the case; inequalities are reversed when the dual problem is stated.

The following table shows the close connection between a problem and its dual.

Given Problem	*Dual Problem*
m variables	n variables
n constraints	m constraints (m slack variables)
Coefficients from objective function	Constraint constants
Constraint constants	Coefficients from objective function

The next theorem, whose proof requires advanced methods, guarantees that a minimization problem can be solved by forming a dual maximization problem.

Theorem of Duality

The objective function w of a minimizing linear programming problem takes on a minimum value if and only if the objective function z of the corresponding dual maximizing problem takes on a maximum value. The maximum value of z equals the minimum value of w.

This method is illustrated in the following example.

EXAMPLE 4 Minimize $w = 3y_1 + 2y_2$

subject to:

$$\begin{aligned} y_1 + 3y_2 &\geq 6 \\ 2y_1 + y_2 &\geq 3 \\ y_1 \geq 0,\ y_2 &\geq 0. \end{aligned}$$

Use the given information to write the matrix.

$$\left[\begin{array}{cc|c} 1 & 3 & 6 \\ 2 & 1 & 3 \\ \hline 3 & 2 & 0 \end{array}\right]$$

Transpose to get the following matrix for the dual problem.

$$\left[\begin{array}{cc|c} 1 & 2 & 3 \\ 3 & 1 & 2 \\ \hline 6 & 3 & 0 \end{array}\right]$$

Write the dual problem from this matrix, as follows.

Maximize $z = 6x_1 + 3x_2$

subject to:

$$\begin{aligned} x_1 + 2x_2 &\leq 3 \\ 3x_1 + x_2 &\leq 2 \\ x_1 \geq 0, x_2 &\geq 0. \end{aligned}$$

Solve this standard maximization problem using the simplex method. Start by introducing slack variables to give the system

$$\begin{aligned} x_1 + 2x_2 + x_3 \qquad\qquad &= 3 \\ 3x_1 + x_2 \qquad + x_4 \qquad &= 2 \\ -6x_1 - 3x_2 - 0x_3 - 0x_4 + z &= 0 \end{aligned}$$

with $x_1 \geq 0, x_2 \geq 0, x_3 \geq 0, x_4 \geq 0.$

The initial tableau for this system is given below with the pivot as indicated.

$$\begin{array}{c} \begin{matrix} x_1 & x_2 & x_3 & x_4 & & \end{matrix} \\ \left[\begin{array}{cccc|c} 1 & 2 & 1 & 0 & 3 \\ \boxed{3} & 1 & 0 & 1 & 2 \\ \hline -6 & -3 & 0 & 0 & 0 \end{array}\right] \end{array} \quad \begin{array}{l} \text{Quotients} \\ 3/1 = 3 \\ 2/3 \\ \end{array}$$

The simplex method gives the following final tableau.

$$
\begin{array}{cccc}
x_1 & x_2 & x_3 & x_4 & \\
\end{array}
$$

$$
\left[\begin{array}{cccc|c}
0 & 1 & \frac{3}{5} & -\frac{1}{5} & \frac{7}{5} \\
1 & 0 & -\frac{1}{5} & \frac{2}{5} & \frac{1}{5} \\
\hline
0 & 0 & \frac{3}{5} & \frac{9}{5} & \frac{27}{5}
\end{array}\right]
$$

$$
\begin{array}{ccc}
y_1 & y_2 & w
\end{array}
$$

Minimize $w = 10y_1 + 8y_2$
subject to:

$$y_1 + 2y_2 \geq 2$$
$$y_1 + \ \ y_2 \geq 5$$
$$y_1 \geq 0, y_2 \geq 0.$$

Answer:

$y_1 = 0, y_2 = 5$, for a minimum of 40

This tableau shows that the solution of the dual maximization problem is $x_1 = 1/5$, $x_2 = 7/5$ and the maximum value of z is 27/5. But that is *not* the answer to the original *minimization* problem. To find the minimization answer, we must read the final tableau differently (as suggested in the discussion after Example 1).

As indicated in blue below the final tableau, the last entries in the columns corresponding to the slack variables (x_3 and x_4) give the values of the original variables y_1 and y_2 that produce the minimal value of w. This minimal value of w appears in the lower right-hand corner (and is the same as the maximal value of z in the dual problem). So the solution of the given minimization problems is as follows.

> The minimum value of $w = 3y_1 + 2y_2$, subject to the given constraints, is 27/5 and occurs when $y_1 = 3/5$ and $y_2 = 9/5$. ◀ 5

A minimizing problem that meets the conditions listed at the beginning of the section can be solved by the method of duals, as illustrated in Examples 1–4 and summarized here.

Solving Minimization Problems with Duals

1. Find the dual standard maximum problem.*
2. Use the simplex method to solve the dual maximum problem.
3. Read the optimal solution of the original minimization problem from the final simplex tableau:

   y_1 is the last entry in the column corresponding to the first slack variable;

   y_2 is the last entry in the column corresponding to the second slack variable; and so on.

These values of y_1, y_2, y_3, etc., produce the minimum value of w, which is the entry in the lower right-hand corner of the tableau.

EXAMPLE 5 A minimization problem in three variables was solved by using duals. The final simplex tableau for the dual maximization problem is shown here.

$$
\begin{array}{c}
\begin{array}{cccccc}
x_1 & x_2 & x_3 & x_4 & x_5 & x_6
\end{array}\\
\left[\begin{array}{cccccc|c}
3 & 1 & 1 & 0 & 9 & 0 & 1 \\
13 & -1 & 0 & 1 & -2 & 0 & 10 \\
9 & 10 & 0 & 0 & 7 & 1 & 7 \\
\hline
5 & 1 & 0 & 4 & 1 & 7 & 28
\end{array}\right]
\end{array}
$$

*The coefficients of the objective function in the minimization problem are the constants on the right side of the constraints in the dual maximization problem. So when all these coefficients are nonnegative (condition 2), the dual problem is in standard maximum form.

(a) What is the optimal solution of the minimization problem?

Since there are three variables in the minimization problem, there must be three slack variables in the dual maximization problem, namely, x_4, x_5, and x_6. Looking at the bottom of the columns corresponding to these slack variables, we see that the solution of the minimization problem is

$$y_1 = 4, \qquad y_2 = 1, \qquad y_3 = 7, \quad \text{with a minimal value of } w = 28.$$

(b) What is the optimal solution of the dual maximization problem?

Since x_4, x_5, and x_6 are slack variables by part (a), the variables in the dual problem are x_1, x_2, and x_3. Read the solution from the final tableau, as in Sections 7.4 and 7.5:

$$x_1 = 0, \qquad x_2 = 0 \qquad x_3 = 1, \quad \text{with a maximal value of } z = 28.$$

Further Uses of the Dual

The dual is useful not only in solving minimization problems but also in seeing how small changes in one variable will affect the value of the objective function. For example, suppose an animal breeder needs at least 6 units per day of nutrient A and at least 3 units of nutrient B and that the breeder can choose between two different feeds, feed 1 and feed 2. Find the minimum cost for the breeder if each bag of feed 1 costs \$3 and provides 1 unit of nutrient A and 2 units of B, while each bag of feed 2 costs \$2 and provides 3 units of nutrient A and 1 of B.

If y_1 represents the number of bags of feed 1 and y_2 represents the number of bags of feed 2, the given information leads to

$$\begin{aligned} \text{Minimize} \quad & w = 3y_1 + 2y_2 \\ \text{subject to:} \quad & y_1 + 3y_2 \geq 6 \\ & 2y_1 + y_2 \geq 3 \\ & y_1 \geq 0, y_2 \geq 0. \end{aligned}$$

This minimization linear programming problem is the one we solved in Example 4 of this section. In that example, we formed the dual and reached the following final tableau.

$$\begin{array}{cccc|c} x_1 & x_2 & x_3 & x_4 & \\ \hline 0 & 1 & \frac{3}{5} & -\frac{1}{5} & \frac{7}{5} \\ 1 & 0 & -\frac{1}{5} & \frac{2}{5} & \frac{1}{5} \\ \hline 0 & 0 & \frac{3}{5} & \frac{9}{5} & \frac{27}{5} \end{array}$$

This final tableau shows that the breeder will obtain minimum feed costs by using 3/5 bag of feed 1 and 9/5 bags of feed 2 per day, for a daily cost of 27/5 = 5.40 dollars.

Now look at the data from the feed problem shown in the table below.

	UNITS OF NUTRIENT (PER BAG)		*Cost*
	A	B	*per Bag*
Feed 1	1	2	\$3
Feed 2	3	1	\$2
Minimum Nutrient Needed	6	3	

If x_1 and x_2 are the cost per unit of nutrients A and B, the constraints of the dual problem can be stated as follows.

$$\text{Cost of feed 1:} \quad x_1 + 2x_2 \le 3$$
$$\text{Cost of feed 2:} \quad 3x_1 + x_2 \le 2$$

The solution of the dual problem, which maximizes nutrients, also can be read from the final tableau:

$$x_1 = \frac{1}{5} = .20 \quad \text{and} \quad x_2 = \frac{7}{5} = 1.40,$$

which means that a unit of nutrient A costs 1/5 of a dollar = \$.20, while a unit of nutrient B costs 7/5 dollars = \$1.40. The minimum daily cost, \$5.40, is found by the following procedure.

(\$.20 per unit of A) × (6 units of A)	= \$1.20
+ (\$1.40 per unit of B) × (3 units of B)	= \$4.20
Minimum daily cost	= \$5.40

The numbers .20 and 1.40 are called the **shadow costs** of the nutrients. These two numbers from the dual, \$.20 and \$1.40, also allow the breeder to estimate feed costs for "small" changes in nutrient requirements. For example, an increase of 1 unit in the requirement for each nutrient would produce total cost as follows.

\$5.40	6 units of A, 3 of B
.20	1 extra unit of A
1.40	1 extra unit of B
\$7.00	Total cost per day (6)

(6) The final tableau of the dual of the problem about filing cabinets in Example 2, Section 7.3, and Example 6, Section 7.4 is given below.

$$\begin{array}{cccc} x_1 & x_2 & x_3 & x_4 \end{array}$$
$$\left[\begin{array}{cccc|c} 0 & 1 & \frac{1}{2} & -\frac{1}{4} & 1 \\ 1 & 0 & -\frac{1}{20} & -\frac{1}{80} & \frac{1}{20} \\ 0 & 0 & 8 & 3 & 100 \end{array}\right]$$

(a) What are the imputed amounts of storage for each unit of cost and floor space?

(b) What are the shadow values of the cost and the floor space?

Answers:

(a) Cost: 28 sq ft
floor space: 72 sq ft

(b) $\frac{1}{20}$, 1

7.6 Exercises

Find the transpose of each matrix. (See Example 2.)

1. $\begin{bmatrix} 3 & -4 & 5 \\ 1 & 10 & 7 \\ 0 & 3 & 6 \end{bmatrix}$

2. $\begin{bmatrix} 3 & -5 & 9 & 4 \\ 1 & 6 & -7 & 0 \\ 4 & 18 & 11 & 9 \end{bmatrix}$

3. $\begin{bmatrix} 3 & 0 & 14 & -5 & 3 \\ 4 & 17 & 8 & -6 & 1 \end{bmatrix}$

4. $\begin{bmatrix} 15 & -6 & -2 \\ 13 & -1 & 11 \\ 10 & 12 & -3 \\ 24 & 1 & 0 \end{bmatrix}$

State the dual problem for each of the following, but do not solve it. (See Example 3.)

5. Minimize $w = 3y_1 + 5y_2$
subject to: $3y_1 + y_2 \ge 4$
$-y_1 + 2y_2 \ge 6$
$y_1 \ge 0, y_2 \ge 0.$

6. Minimize $w = 4y_1 + 7y_2$
subject to: $y_1 + y_2 \ge 17$
$3y_1 + 6y_2 \ge 21$
$2y_1 + 4y_2 \ge 19$
$y_1 \ge 0, y_2 \ge 0.$

7. Minimize $w = 2y_1 + 8y_2$
subject to: $y_1 + 7y_2 \ge 18$
$4y_1 + y_2 \ge 15$
$5y_1 + 3y_2 \ge 20$
$y_1 \ge 0, y_2 \ge 0.$

8. Minimize $w = 5y_1 + y_2 + 3y_3$
subject to: $7y_1 + 6y_2 + 8y_3 \ge 18$
$4y_1 + 5y_2 + 10y_3 \ge 20$
$y_1 \ge 0, y_2 \ge 0, y_3 \ge 0.$

9. Minimize $w = y_1 + 2y_2 + 6y_3$
subject to: $3y_1 + 4y_2 + 6y_3 \geq 8$
$y_1 + 5y_2 + 2y_3 \geq 12$
$y_1 \geq 0, y_2 \geq 0, y_3 \geq 0.$

10. Minimize $w = 4y_1 + 3y_2 + y_3$
subject to: $y_1 + 2y_2 + 3y_3 \geq 115$
$2y_1 + y_2 + 8y_3 \geq 200$
$y_1 - y_3 \geq 50$
$y_1 \geq 0, y_2 \geq 0, y_3 \geq 0.$

11. Minimize $w = 8y_1 + 9y_2 + 3y_3$
subject to: $y_1 + y_2 + y_3 \geq 5$
$y_1 + y_2 \geq 4$
$2y_1 + y_2 + 3y_3 \geq 15$
$y_1 \geq 0, y_2 \geq 0, y_3 \geq 0.$

12. Minimize $w = y_1 + 2y_2 + y_3 + 5y_4$
subject to: $y_1 + y_2 + y_3 + y_4 \geq 50$
$3y_1 + y_2 + 2y_3 + y_4 \geq 100$
$y_1 \geq 0, y_2 \geq 0, y_3 \geq 0, y_4 \geq 0.$

Use duality to solve the following problems. (See Example 4.)

13. Minimize $w = 2y_1 + y_2 + 3y_3$
subject to: $y_1 + y_2 + y_3 \geq 100$
$2y_1 + y_2 \geq 50$
$y_1 \geq 0, y_2 \geq 0, y_3 \geq 0.$

14. Minimize $w = 2y_1 + 4y_2$
subject to: $4y_1 + 2y_2 \geq 10$
$4y_1 + y_2 \geq 8$
$2y_1 + y_2 \geq 12$
$y_1 \geq 0, y_2 \geq 0.$

15. Minimize $w = 3y_1 + y_2 + 4y_3$
subject to: $2y_1 + y_2 + y_3 \geq 6$
$y_1 + 2y_2 + y_3 \geq 8$
$2y_1 + y_2 + 2y_3 \geq 12$
$y_1 \geq 0, y_2 \geq 0, y_3 \geq 0.$

16. Minimize $w = y_1 + y_2 + 3y_3$
subject to: $2y_1 + 6y_2 + y_3 \geq 8$
$y_1 + 2y_2 + 4y_3 \geq 12$
$y_1 \geq 0, y_2 \geq 0, y_3 \geq 0.$

17. Minimize $w = 6y_1 + 4y_2 + 2y_3$
subject to: $2y_1 + 2y_2 + y_3 \geq 2$
$y_1 + 3y_2 + 2y_3 \geq 3$
$y_1 + y_2 + 2y_3 \geq 4$
$y_1 \geq 0, y_2 \geq 0, y_3 \geq 0.$

18. Minimize $w = 12y_1 + 10y_2 + 7y_3$
subject to: $2y_1 + y_2 + y_3 \geq 7$
$y_1 + 2y_2 + y_3 \geq 4$
$y_1 \geq 0, y_2 \geq 0, y_3 \geq 0.$

19. Minimize $w = 20y_1 + 12y_2 + 40y_3$
subject to: $y_1 + y_2 + 5y_3 \geq 20$
$2y_1 + y_2 + y_3 \geq 30$
$y_1 \geq 0, y_2 \geq 0, y_3 \geq 0.$

20. Minimize $w = 4y_1 + 5y_2$
subject to: $10y_1 + 5y_2 \geq 100$
$20y_1 + 10y_2 \geq 150$
$y_1 \geq 0, y_2 \geq 0.$

21. Minimize $w = 4y_1 + 2y_2 + y_3$
subject to: $y_1 + y_2 + y_3 \geq 4$
$3y_1 + y_2 + 3y_3 \geq 6$
$y_1 + y_2 + 3y_3 \geq 5$
$y_1 \geq 0, y_2 \geq 0, y_3 \geq 0.$

22. Minimize $w = 3y_1 + 2y_2$
subject to: $2y_1 + 3y_2 \geq 60$
$y_1 + 4y_2 \geq 40$
$y_1 \geq 0, y_2 \geq 0.$

23. Health Glenn Russell, who is dieting, requires two food supplements, I and II. He can get these supplements from two different products, A and B, as shown in the following table.

		SUPPLEMENT (GRAMS PER SERVING)	
		I	II
PRODUCT	A	4	2
	B	2	5

Glenn's physician has recommended that he include at least 20 grams of supplement I and 18 grams of supplement II in his diet. If product A costs 24¢ per serving and product B costs 40¢ per serving, how can he satisfy these requirements most economically?

24. Business An animal food must provide at least 54 units of vitamins and 60 calories per serving. One gram of soybean meal provides at least 2.5 units of vitamins and 5 calories. One gram of meat by-products provides at least 4.5 units of vitamins and 3 calories. One gram of grain provides at least 5 units of vitamins and 10 calories. If a gram of soybean meal costs 8¢, a gram of meat by-products 9¢, and a gram of grain 10¢, what mixture of these three ingredients will provide the required vitamins and calories at minimum cost?

25. Business A brewery produces regular beer and a lower-carbohydrate "light" beer. Steady customers of the brewery buy 12 units of regular beer and 10 units of light beer monthly. While setting up the brewery to produce the beers, the management decides to produce extra beer, beyond the need to satisfy the customers. The cost per unit of regular beer is $36,000 and the cost per unit of light beer is $48,000. Every unit of regular beer brings in $100,000 in revenue, while every unit of light beer brings in $300,000 in revenue. The brewery wants at least \$7,000,000 in revenue. At least 20 additional units of beer can be sold. How much of each type of beer should be made so as to minimize total production costs?

26. Business Joan McKee has a part-time job conducting public opinion interviews. She has found that a political interview takes 45 minutes and a market interview takes 55 minutes. She needs to minimize the time she spends doing interviews to allow more time for her full-time job. Unfortunately, to keep her part-time job, she must complete at least 8 interviews each week. Also, she must earn at least \$60 per week at this job: she earns \$8 for each political interview and \$10 for each market interview. Finally, to stay in good standing with her supervisor, she must earn at least 40 bonus points per week; she receives 6 bonus points for each political interview and 5 points for each market interview. How many of each interview should she do each week to minimize the time spent?

27. You are given the following linear programming problem (P):*

$$\begin{aligned} \text{Minimize} \quad & z = x_1 + 2x_2 \\ \text{subject to:} \quad & -2x_1 + x_2 \geq 1 \\ & x_1 - 2x_2 \geq 1 \\ & x_1 \geq 0, x_2 \geq 0. \end{aligned}$$

The dual of (P) is (D). Which of the statements below is true?

(a) (P) has no feasible solution and the objective function of (D) is unbounded.

(b) (D) has no feasible solution and the objective function of (P) is unbounded.

(c) The objective functions of both (P) and (D) are unbounded.

(d) Both (P) and (D) have optimal solutions.

(e) Neither (P) nor (D) has feasible solutions.

28. Business Refer to the end of this section, to the text on minimizing the daily cost of feeds.

(a) Find a combination of feeds that will cost \$7.00 and give 7 units of A and 4 units of B.

(b) Use the dual variables to predict the daily cost of feed if the requirements change to 5 units of A and 4 units of B. Find a combination of feeds to meet these requirements at the predicted price.

29. Business A small toy manufacturing firm has 200 squares of felt, 600 ounces of stuffing, and 90 feet of trim available to make two types of toys, a small bear and a monkey. The bear requires 1 square of felt and 4 ounces of stuffing. The monkey requires 2 squares of felt, 3 ounces of stuffing, and 1 foot of trim. The firm makes \$1 profit on each bear and \$1.50 profit on each monkey. The linear program to maximize profit is

$$\begin{aligned} \text{Maximize} \quad & z = x_1 + 1.5x_2 \\ \text{subject to:} \quad & x_1 + 2x_2 \leq 200 \\ & 4x_1 + 3x_2 \leq 600 \\ & x_2 \leq 90 \\ & x_1 \geq 0, x_2 \geq 0. \end{aligned}$$

The final simplex tableau is

$$\left[\begin{array}{cccccc|c} 1 & 0 & -.6 & .4 & 0 & & 120 \\ 0 & 0 & -.8 & .2 & 1 & & 50 \\ 0 & 1 & .8 & -.2 & 0 & & 40 \\ \hline 0 & 0 & .6 & .1 & 0 & & 180 \end{array}\right].$$

(a) What is the corresponding dual problem?

(b) What is the optimal solution to the dual problem?

(c) Use the shadow values to estimate the profit the firm will make if their supply of felt increases to 210 squares.

(d) How much profit will the firm make if their supply of stuffing is cut to 590 ounces and their supply of trim is cut to 80 feet?

30. Refer to Example 1 in Section 7.5.

(a) Give the dual problem.

(b) Use the shadow values to estimate the farmer's profit if land is cut to 90 acres but capital increases to \$21,000.

(c) Suppose the farmer has 110 acres but only \$19,000. Find the optimum profit and the planting strategy that will produce this profit.

*Problem 2 from "November 1989 Course 130 Examination Operations Research" of the *Education and Examination Committee of The Society of Actuaries*. Reprinted by permission of The Society of Actuaries.

7.7 THE SIMPLEX METHOD: NONSTANDARD PROBLEMS

So far the simplex method has been used to solve problems in which the variables are nonnegative and all the other constraints are of one type (either all $\leq$ or all $\geq$). Now we extend the simplex method to linear programming problems with nonnegative variables and mixed constraints ($\leq$, $=$, and $\geq$).

The solution method to be used here requires that all inequality constraints be written so that the constant on the right side is nonnegative. For instance, the inequality

$$4x_1 + 5x_2 - 12x_3 \leq -30$$

can be replaced by the equivalent one obtained by multiplying both sides by -1 and reversing the direction of the inequality sign:

$$-4x_1 - 5x_2 + 12x_3 \geq 30.$$

Maximization with ≤ and ≥ Constraints

As is always the case when the simplex method is involved, each inequality constraint must be written as an equation. Constraints involving ≤ are converted to equations by adding a nonnegative slack variable, as in Section 7.4. Similarly, constraints involving ≥ are converted to equations by *subtracting* a nonnegative **surplus variable.** For example, the inequality $2x_1 - x_2 + 5x_3 \geq 12$ is written as

$$2x_1 - x_2 + 5x_3 - x_4 = 12,$$

where $x_4 \geq 0$. The surplus variable x_4 represents the amount by which $2x_1 - x_2 + 5x_3$ exceeds 12.

EXAMPLE 1 Restate the following problem in terms of equations and write its initial simplex tableau.

$$\begin{aligned} \text{Maximize} \quad & z = 4x_1 + 10x_2 + 6x_3 \\ \text{subject to:} \quad & x_1 + 4x_2 + 4x_3 \geq 8 \\ & x_1 + 3x_2 + 2x_3 \leq 6 \\ & 3x_1 + 4x_2 + 8x_3 \leq 22 \\ & x_1 \geq 0, x_2 \geq 0, x_3 \geq 0. \end{aligned}$$

In order to write the constraints as equations, subtract a surplus variable from the ≥ constraint and add a slack variable to each ≤ constraint. So the problem becomes

$$\begin{aligned} \text{Maximize} \quad & z = 4x_1 + 10x_2 + 6x_3 \\ \text{subject to:} \quad & x_1 + 4x_2 + 4x_3 - x_4 = 8 \\ & x_1 + 3x_2 + 2x_3 + x_5 = 6 \\ & 3x_1 + 4x_2 + 8x_3 + x_6 = 22 \\ & x_1 \geq 0, x_2 \geq 0, x_3 \geq 0, x_4 \geq 0, x_5 \geq 0, x_6 \geq 0. \end{aligned}$$

Write the objective function as $z - 4x_1 - 10x_2 - 6x_3 = 0$ and use the coefficients of the four equations to write the initial simplex tableau (omitting the z column):

$$\begin{array}{cccccc|c} x_1 & x_2 & x_3 & x_4 & x_5 & x_6 & \\ 1 & 4 & 4 & -1 & 0 & 0 & 8 \\ 1 & 3 & 2 & 0 & 1 & 0 & 6 \\ 3 & 4 & 8 & 0 & 0 & 1 & 22 \\ \hline -4 & -10 & -6 & 0 & 0 & 0 & 0 \end{array}$$ ◀ 1

1

(a) Restate this problem in terms of equations:

$$\begin{aligned} \text{Maximize} \quad & z = 3x_1 - 2x_2 \\ \text{subject to:} \quad & 2x_1 + 3x_2 \leq 8 \\ & 6x_1 - 2x_2 \geq 3 \\ & x_1 + 4x_2 \geq 1 \\ & x_1 \geq 0, x_2 \geq 0. \end{aligned}$$

(b) Write the initial simplex tableau.

Answers:

(a) Maximize $z = 3x_1 - 2x_2$ subject to:

$$\begin{aligned} & 2x_1 + 3x_2 + x_3 = 8 \\ & 6x_1 - 2x_2 - x_4 = 3 \\ & x_1 + 4x_2 - x_5 = 1 \\ & x_1 \geq 0, x_2 \geq 0, x_3 \geq 0, \\ & x_4 \geq 0, x_5 \geq 0. \end{aligned}$$

(b)

$$\begin{array}{ccccc|c} x_1 & x_2 & x_3 & x_4 & x_5 & \\ 2 & 3 & 1 & 0 & 0 & 8 \\ 6 & -2 & 0 & -1 & 0 & 3 \\ 1 & 4 & 0 & 0 & -1 & 1 \\ \hline -3 & 2 & 0 & 0 & 0 & 0 \end{array}$$

State the basic solution given by each tableau. Is it feasible?

(a)

$$\begin{array}{ccccc} x_1 & x_2 & x_3 & x_4 & x_5 & \\ \end{array}$$
$$\left[\begin{array}{ccccc|c} 3 & -5 & 1 & 0 & 0 & 12 \\ 4 & 7 & 0 & 1 & 0 & 6 \\ 1 & 3 & 0 & 0 & -1 & 5 \\ \hline -7 & 4 & 0 & 0 & 0 & 0 \end{array}\right]$$

(b)

$$\begin{array}{ccccc} x_1 & x_2 & x_3 & x_4 & x_5 & \\ \end{array}$$
$$\left[\begin{array}{ccccc|c} 9 & 8 & -1 & 1 & 0 & 12 \\ -5 & 3 & 0 & 0 & 1 & 7 \\ \hline 4 & 2 & 3 & 0 & 0 & 0 \end{array}\right]$$

Answers:

(a) $x_1 = 0, x_2 = 0, x_3 = 12, x_4 = 6, x_5 = -5$; no.

(b) $x_1 = 0, x_2 = 0, x_3 = 0, x_4 = 12, x_5 = 7$; yes.

The tableau in Example 1 resembles those that have appeared previously, and similar terminology is used. The variables whose columns have one entry ± 1 and the rest 0 will be called **basic variables;** the other variables are nonbasic. A solution obtained by setting the nonbasic variables equal to 0 and solving for the basic variables (by looking at the constants in the right-hand column) will be called a **basic solution.** A basic solution that is feasible is called a **basic feasible solution.** In the tableau of Example 1, for instance, the basic variables are x_4, x_5, and x_6, and the basic solution is:

$$x_1 = 0, \quad x_2 = 0, \quad x_3 = 0, \quad x_4 = -8, \quad x_5 = 6, \quad x_6 = 22.$$

However, because one variable is negative, this solution is not feasible. (2)

The solution method for problems such as the one in Example 1 consists of two stages. **Stage I** consists of finding a basic *feasible* solution that can be used as the starting point for the simplex method. (This stage is unnecessary in a standard maximization problem because the solution given by the initial tableau is always feasible.) There are many systematic ways of finding a feasible solution, all of which depend on the fact that row operations (such as pivoting) produce a tableau that represents a system with the same solutions as the original one. One such technique is explained in the next example. Since the immediate goal is to find a feasible solution, not necessarily an optimal one, the procedures for choosing pivots differ from those in the ordinary simplex method.

EXAMPLE 2 Find a basic feasible solution for the problem in Example 1, whose initial tableau is

$$\begin{array}{cccccc} x_1 & x_2 & x_3 & x_4 & x_5 & x_6 \end{array}$$
$$\left[\begin{array}{cccccc|c} 1 & 4 & 4 & -1 & 0 & 0 & 8 \\ 1 & 3 & 2 & 0 & 1 & 0 & 6 \\ 3 & 4 & 8 & 0 & 0 & 1 & 22 \\ \hline -4 & -10 & -6 & 0 & 0 & 0 & 0 \end{array}\right].$$

In the basic solution given by this tableau, x_4 has a negative value. The only nonzero entry in its column is the -1 in row one. Choose any *positive* entry in row one except the entry on the far right. The column that the chosen entry is in will be the pivot column. We choose the first positive entry in row one, the 1 in column one. The pivot row is determined in the usual way by considering quotients (constant at the right end of the row divided by the positive entry in the pivot column) in each row except the objective row:

$$8/1 = 8, \qquad 6/1 = 6, \qquad 22/3 = 7\frac{1}{3}.$$

The smallest quotient is 6, so the pivot is the 1 in row two, column one. Pivoting in the usual way leads to this tableau

$$\begin{array}{cccccc} x_1 & x_2 & x_3 & x_4 & x_5 & x_6 \end{array}$$
$$\left[\begin{array}{cccccc|c} 0 & 1 & 2 & -1 & -1 & 0 & 2 \\ 1 & 3 & 2 & 0 & 1 & 0 & 6 \\ 0 & -5 & 2 & 0 & -3 & 1 & 4 \\ \hline 0 & 2 & 2 & 0 & 4 & 0 & 24 \end{array}\right] \begin{array}{l} -R_2 + R_1 \\ \\ -3R_2 + R_3 \\ 4R_2 + R_4 \end{array}$$

and the basic solution

$$x_1 = 6, \quad x_2 = 0, \quad x_3 = 0, \quad x_4 = -2, \quad x_5 = 0, \quad x_6 = 4.$$

Since the basic variable x_4 is negative, this solution is not feasible. So we repeat the pivoting process described above. The x_4 column has a -1 in row one, so we choose a positive entry in that row, namely, the 1 in row one, column two. This choice makes column two the pivot column. The pivot row is determined by the quotients $2/1 = 2$ and $6/3 = 2$ (negative entries in the pivot column and the entry in the objective row are not used). Since there is a tie, we can choose either row one or row two. We choose row one and use the 1 in row one, column 2, as the pivot. Pivoting produces this tableau

$$\begin{array}{c} \\ \\ \\ \\ \end{array}\begin{array}{l} x_1 \; x_2 \; x_3 \; x_4 \; x_5 \; x_6 \end{array}$$

$$\left[\begin{array}{cccccc|c} 0 & 1 & 2 & -1 & -1 & 0 & 2 \\ 1 & 0 & -4 & 3 & 4 & 0 & 0 \\ 0 & 0 & 12 & -5 & -8 & 1 & 14 \\ \hline 0 & 0 & -2 & 2 & 6 & 0 & 20 \end{array}\right] \quad \begin{array}{l} \\ -3R_1 + R_2 \\ 5R_1 + R_3 \\ -2R_1 + R_4 \end{array}$$

and the basic *feasible* solution

$$x_1 = 0, \quad x_2 = 2, \quad x_3 = 0, \quad x_4 = 0, \quad x_5 = 0, \quad x_6 = 14.$$ ◀

Once a basic feasible solution has been found, Stage I is ended. The procedures used in Stage I are summarized below.*

Finding a Basic Feasible Solution

1. **If** any basic variable has a negative value, locate the -1 in that variable's column and note the row it is in.
2. In the row determined in Step 1, choose a positive entry (other than the one at the far right) and note the column it is in. This is the pivot column.
3. Use the positive entries in the pivot column (except in the objective row) to form quotients and select the pivot.
4. Pivot as usual, which results in the pivot column's having one entry 1 and the rest 0's.
5. Repeat Steps 1–4 until every basic variable is nonnegative, so that the basic solution given by the tableau is feasible. If it ever becomes impossible to continue, then the problem has no feasible solution.

One way to make the required choices systematically is to choose the first possibility in each case (going from the top for rows or from the left for columns). However, any choice meeting the required conditions may be used. For maximum efficiency, it is usually best to choose the pivot column in Step 2 so that the pivot is in the same row chosen in Step 1, if this is possible. (3)

(3)

The initial tableau of a maximization problem is given below. Use column one as the pivot column for carrying out Stage I and state the basic feasible solution that results.

$$\begin{array}{ccccc} x_1 & x_2 & x_3 & x_4 & \end{array}$$
$$\left[\begin{array}{cccc|c} 1 & 3 & 1 & 0 & 70 \\ 2 & 4 & 0 & -1 & 50 \\ \hline -8 & -10 & 0 & 0 & 0 \end{array}\right]$$

Answer:

$$\begin{array}{ccccc} x_1 & x_2 & x_3 & x_4 & \end{array}$$
$$\left[\begin{array}{cccc|c} 0 & 1 & 1 & \frac{1}{2} & 45 \\ 1 & 2 & 0 & -\frac{1}{2} & 25 \\ \hline 0 & 6 & 0 & -4 & 200 \end{array}\right]$$

$x_1 = 25, x_2 = 0, x_3 = 45, x_4 = 0$.

*Except in rare cases that do not occur in this book, this method eventually produces a basic feasible solution or shows that one does not exist. The *two-phase method* using artificial variables, which is discussed in more advanced texts, works in all cases and often is more efficient.

In **Stage II,** the simplex method is applied as usual to the tableau that produced the basic feasible solution in Stage I. Just as in Section 7.4, each round of pivoting replaces the basic feasible solution of one tableau with the basic feasible solution of a new tableau in such a way that the value of the objective function is increased, until an optimal value is obtained (or it becomes clear that no optimal solution exists).

EXAMPLE 3 Solve the linear programming problem in Example 1.

A basic feasible solution for this problem was found in Example 2 by using the tableau shown below. However, this solution is not maximal because there is a negative indicator in the objective row. So we use the simplex method: the most negative indicator determines the pivot column and the usual quotients determine that the number 2 in row one, column three, is the pivot.

$$\begin{array}{cccccc|c|l} x_1 & x_2 & x_3 & x_4 & x_5 & x_6 & & \text{Quotients} \\ 0 & 1 & \boxed{2} & -1 & -1 & 0 & 2 & 2/2 \leftarrow \text{Smallest} \\ 1 & 0 & -4 & 3 & 4 & 0 & 0 & \\ 0 & 0 & 12 & -5 & -8 & 1 & 14 & 14/12 \\ \hline 0 & 0 & -2 & 2 & 6 & 0 & 20 & \end{array}$$

(Arrow under -2: Most negative indicator)

Pivoting leads to the final tableau.

$$\begin{array}{cccccc|c|l} x_1 & x_2 & x_3 & x_4 & x_5 & x_6 & & \\ 0 & \frac{1}{2} & 1 & -\frac{1}{2} & -\frac{1}{2} & 0 & 1 & \frac{1}{2}R_1 \\ 1 & 0 & -4 & 3 & 4 & 0 & 0 & \\ 0 & 0 & 12 & -5 & -8 & 1 & 14 & \\ \hline 0 & 0 & -2 & 2 & 6 & 0 & 20 & \end{array}$$

$$\begin{array}{cccccc|c|l} x_1 & x_2 & x_3 & x_4 & x_5 & x_6 & & \\ 0 & \frac{1}{2} & 1 & -\frac{1}{2} & -\frac{1}{2} & 0 & 1 & \\ 1 & 2 & 0 & 1 & 2 & 0 & 4 & 4R_1 + R_2 \\ 0 & -6 & 0 & 1 & -2 & 1 & 2 & -12R_1 + R_3 \\ \hline 0 & 1 & 0 & 1 & 5 & 0 & 22 & 2R_1 + R_4 \end{array}$$

Complete Stage II and find an optimal solution for side problem 3 on page 399. What is the optimal value of the objective function z?

Answer:

The optimal value $z = 560$ occurs when $x_1 = 70$, $x_2 = 0$, $x_3 = 0$, $x_4 = 90$.

Therefore, the maximum value of z occurs when $x_1 = 4$, $x_2 = 0$, and $x_3 = 1$, in which case $z = 22$. ◀ (4)

Minimization Problems

The two-stage method for maximization problems illustrated in Examples 1–3 also provides a means of solving minimization problems. To see why, consider this simple fact: when a number t gets smaller, then $-t$ gets larger, and vice versa. For instance, if t goes from 6 down to -8, then $-t$ goes from -6 up to 8. Thus, if w is the objective function of a linear programming problem, the feasible solution that produces the minimum value of w also produces the maximum value of $-w$, and vice versa. Therefore, to solve a minimization problem with objective function w, we need only solve the maximization problem with the same constraints and objective function $z = -w$.

EXAMPLE 4 Minimize $w = 2y_1 + y_2 - y_3$

subject to:
$$\begin{aligned} -y_1 - y_2 + y_3 &\le -4 \\ y_1 + 3y_2 + 3y_3 &\ge 6 \\ y_1 \ge 0, y_2 \ge 0, y_3 &\ge 0. \end{aligned}$$

Make the constant in the first constraint positive by multiplying both sides by -1. Then solve this maximization problem:

Maximize $z = -w = -2y_1 - y_2 + y_3$

subject to:
$$\begin{aligned} y_1 + y_2 - y_3 &\ge 4 \\ y_1 + 3y_2 + 3y_3 &\ge 6 \\ y_1 \ge 0, y_2 \ge 0, y_3 &\ge 0. \end{aligned}$$

Convert the constraints to equations by subtracting surplus variables and set up the first tableau.

$$\begin{array}{c} \begin{matrix} y_1 & y_2 & y_3 & y_4 & y_5 & \end{matrix} \\ \left[\begin{array}{ccccc|c} 1 & 1 & -1 & -1 & 0 & 4 \\ 1 & 3 & 3 & 0 & -1 & 6 \\ \hline 2 & 1 & -1 & 0 & 0 & 0 \end{array}\right] \end{array}$$

The basic solution given by this tableau, $y_1 = 0, y_2 = 0, y_3 = 0, y_4 = -4, y_5 = -6$, is not feasible, so the procedures of Stage I must be used to find a basic feasible solution. In the column of the negative basic variable y_4, there is a -1 in row one; we choose the first positive entry in that row, so that column one will be the pivot column. The quotients $4/1 = 4$ and $6/1 = 6$ show that the pivot is the 1 in row one, column one. Pivoting produces this tableau:

$$\begin{array}{c} \begin{matrix} y_1 & y_2 & y_3 & y_4 & y_5 & \end{matrix} \\ \left[\begin{array}{ccccc|c} 1 & 1 & -1 & -1 & 0 & 4 \\ 0 & 2 & 4 & 1 & -1 & 2 \\ \hline 0 & -1 & 1 & 2 & 0 & -8 \end{array}\right]. \end{array} \quad \begin{array}{l} \\ -R_1 + R_2 \\ -2R_1 + R_3 \end{array}$$

The basic solution $y_1 = 4, y_2 = 0, y_3 = 0, y_4 = 0, y_5 = -2$ is not feasible because y_5 is negative, so we repeat the process. We choose the first positive entry in row two (the row containing the -1 in the y_5 column), which is in column two, so that column two is the pivot column. The relevant quotients are $4/1 = 4$ and $2/2 = 1$, so the pivot is the 2 in row two, column two. Pivoting produces a new tableau.

$$\begin{array}{c} \begin{matrix} y_1 & y_2 & y_3 & y_4 & y_5 & \end{matrix} \\ \left[\begin{array}{ccccc|c} 1 & 1 & -1 & -1 & 0 & 4 \\ 0 & 1 & 2 & \frac{1}{2} & -\frac{1}{2} & 1 \\ \hline 0 & -1 & 1 & 2 & 0 & -8 \end{array}\right] \end{array} \quad \begin{array}{l} \\ \frac{1}{2}R_2 \\ \\ \end{array}$$

$$\begin{array}{c} \begin{matrix} y_1 & y_2 & y_3 & y_4 & y_5 & \end{matrix} \\ \left[\begin{array}{ccccc|c} 1 & 0 & -3 & -\frac{3}{2} & \frac{1}{2} & 3 \\ 0 & 1 & 2 & \frac{1}{2} & -\frac{1}{2} & 1 \\ \hline 0 & 0 & 3 & \frac{5}{2} & -\frac{1}{2} & -7 \end{array}\right] \end{array} \quad \begin{array}{l} -R_2 + R_1 \\ \\ R_2 + R_3 \end{array}$$

The basic solution $y_1 = 3, y_2 = 1, y_3 = 0, y_4 = 0, y_5 = 0$ is feasible, so Stage I is complete. However, this solution is not optimal because the objective row contains the negative indicator $-1/2$ in column five. According to the simplex method, column five is the next pivot column. The only positive ratio $3/\frac{1}{2} = 6$ is in row one, so the pivot is 1/2 in row one, column five. Pivoting produces the final tableau.

$$\begin{array}{ccccc|c} y_1 & y_2 & y_3 & y_4 & y_5 & \\ 2 & 0 & -6 & -3 & 1 & 6 \\ 0 & 1 & 2 & \frac{1}{2} & -\frac{1}{2} & 1 \\ \hline 0 & 0 & 3 & \frac{5}{2} & -\frac{1}{2} & -7 \end{array} \quad 2R_1$$

$$\begin{array}{ccccc|c} y_1 & y_2 & y_3 & y_4 & y_5 & \\ 2 & 0 & -6 & -3 & 1 & 6 \\ 1 & 1 & -1 & -1 & 0 & 4 \\ \hline 1 & 0 & 0 & 1 & 0 & -4 \end{array} \quad \begin{array}{l} \\ \frac{1}{2}R_1 + R_2 \\ \frac{1}{2}R_1 + R_3 \end{array}$$

Since there are no negative indicators, the solution given by this tableau ($y_1 = 0$, $y_2 = 4$, $y_3 = 0$, $y_4 = 0$, $y_5 = 6$) is optimal. The maximum value of $z = -w$ is -4. Therefore, the minimum value of the original objective function w is $-(-4) = 4$, which occurs when $y_1 = 0$, $y_2 = 4$, $y_3 = 0$. ◀ 5

5

Minimize $w = 2y_1 + 3y_2$
subject to:

$$y_1 + y_2 \geq 10$$
$$2y_1 + y_2 \geq 16$$
$$y_1 \geq 0, y_2 \geq 0.$$

Answer:

$y_1 = 10, y_2 = 0; w = 20$

Equation Constraints

Recall that for any real numbers a and b,

$$a = b \quad \text{exactly when } a \geq b \text{ and simultaneously } a \leq b.$$

Thus, an equation such as $y_1 + 3y_2 + 3y_3 = 6$ is equivalent to this pair of inequalities:

$$y_1 + 3y_2 + 3y_3 \geq 6$$
$$y_1 + 3y_2 + 3y_3 \leq 6.$$

In a linear programming problem, each equation constraint should be replaced in this way by a pair of inequality constraints. Then the problem can be solved by the two-stage method.

EXAMPLE 5 Minimize $w = 2y_1 + y_2 - y_3$

subject to:
$$-y_1 - y_2 + y_3 \leq -4$$
$$y_1 + 3y_2 + 3y_3 = 6$$
$$y_1 \geq 0, y_2 \geq 0, y_3 \geq 0.$$

Multiply the first inequality by -1 and replace the equation by an equivalent pair of inequalities, as explained above, to obtain this problem:

Maximize $z = -w = -2y_1 - y_2 + y_3$

subject to:
$$y_1 + y_2 - y_3 \geq 4$$
$$y_1 + 3y_2 + 3y_3 \geq 6$$
$$y_1 + 3y_2 + 3y_3 \leq 6$$
$$y_1 \geq 0, y_2 \geq 0, y_3 \geq 0.$$

Convert the constraints to equations by subtracting surplus variables y_4 and y_5 from the first two and adding a slack variable y_6 to the third. Then the first tableau is

$$\begin{array}{cccccc|c} y_1 & y_2 & y_3 & y_4 & y_5 & y_6 & \\ \hline 1 & 1 & -1 & -1 & 0 & 0 & 4 \\ 1 & 3 & 3 & 0 & -1 & 0 & 6 \\ 1 & 3 & 3 & 0 & 0 & 1 & 6 \\ \hline 2 & 1 & -1 & 0 & 0 & 0 & 0 \end{array}$$

The basic solution given by this tableau is $y_1 = 0$, $y_2 = 0$, $y_3 = 0$, $y_4 = -4$, $y_5 = -6$, $y_6 = 6$, which is not feasible. So we carry out the procedures of Stage I.

$$\begin{array}{cccccc|c} y_1 & y_2 & y_3 & y_4 & y_5 & y_6 & \\ \hline 1 & 1 & -1 & -1 & 0 & 0 & 4 \\ 0 & 2 & 4 & 1 & -1 & 0 & 2 \\ 0 & 2 & 4 & 1 & 0 & 1 & 2 \\ \hline 0 & -1 & 1 & 2 & 0 & 0 & -8 \end{array}$$

$$\begin{array}{cccccc|c} y_1 & y_2 & y_3 & y_4 & y_5 & y_6 & \\ \hline 1 & 0 & -3 & -1.5 & .5 & 0 & 3 \\ 0 & 1 & 2 & .5 & -.5 & 0 & 1 \\ 0 & 0 & 0 & 0 & 1 & 1 & 0 \\ \hline 0 & 0 & 3 & 2.5 & -.5 & 0 & -7 \end{array}$$

This tableau gives the basic feasible solution $y_1 = 3$, $y_2 = 1$, $y_3 = 0$, $y_4 = 0$, $y_5 = 0$, $y_6 = 0$, so Stage I is complete. Now apply the simplex method. One round of pivoting produces the final tableau.

$$\begin{array}{cccccc|c} y_1 & y_2 & y_3 & y_4 & y_5 & y_6 & \\ \hline 1 & 0 & -3 & -1.5 & 0 & -.5 & 3 \\ 0 & 1 & 2 & .5 & 0 & .5 & 1 \\ 0 & 0 & 0 & 0 & 1 & 1 & 0 \\ \hline 0 & 0 & 3 & 2.5 & 0 & .5 & -7 \end{array}$$

Therefore, the minimum value of $w = -z$ is $w = -(-7) = 7$, which occurs when $y_1 = 3$, $y_2 = 1$, and $y_3 = 0$. ◀

You may have noticed that Example 5 is just Example 4, with the last inequality constraint replaced by an equation constraint. Note, however, that the optimal solutions are different in the two examples. The minimal value of w found in Example 4 is smaller than the one found in Example 5, but does not satisfy the equation constraint in Example 5.

The two-stage method used in Examples 1–5 is summarized here.

Solving Nonstandard Problems

1. **Replace** each equation constraint by an equivalent pair of inequality constraints.
2. **If** necessary, write each constraint with a positive constant.

continued

3. Convert a minimization problem to maximization problem by letting $z = -w$.
4. Add slack variables and subtract surplus variables as needed to convert the constraints into equations.
5. Write the initial simplex tableau.
6. Find a basic feasible solution for the problem, if one exists (Stage I).
7. When a basic feasible solution is found, use the simplex method to solve the problem (Stage II).

NOTE It may happen that the tableau that gives the basic feasible solution in Stage I has no negative indicators in its last row. In this case, the solution found is already optimal and Stage II is not necessary.

Applications

Many real-world applications of linear programming involve mixed constraints. Since they typically involve a large number of variables and constraints, technology is normally required to solve such problems.

EXAMPLE 6 A college textbook publisher has received orders from two colleges, C_1 and C_2. C_1 needs 500 books, and C_2 needs 1000. The publisher can supply the books from either of two warehouses. Warehouse W_1 has 900 books available and warehouse W_2 has 700. The costs to ship a book from each warehouse to each college are given below.

		TO	
		C_1	C_2
FROM	W_1	\$1.20	\$1.80
	W_2	\$2.10	\$1.50

How many books should be sent from each warehouse to each college to minimize the shipping costs?

To begin, let

$$y_1 = \text{the number of books shipped from } W_1 \text{ to } C_1;$$
$$y_2 = \text{the number of books shipped from } W_2 \text{ to } C_1;$$
$$y_3 = \text{the number of books shipped from } W_1 \text{ to } C_2;$$
$$y_4 = \text{the number of books shipped from } W_2 \text{ to } C_2.$$

C_1 needs 500 books, so $y_1 + y_2 = 500$, which is equivalent to this pair of inequalities:

$$y_1 + y_2 \geq 500$$
$$y_1 + y_2 \leq 500.$$

Similarly, $y_3 + y_4 = 1000$, which is equivalent to

$$y_3 + y_4 \geq 1000$$
$$y_3 + y_4 \leq 1000.$$

Since W_1 has 900 books availble and W_2 has 700 vailable,

$$y_1 + y_3 \leq 900 \quad \text{and} \quad y_2 + y_4 \leq 700.$$

The company wants to minimize shipping costs, so the objective function is

$$w = 1.20y_1 + 2.10y_2 + 1.80y_3 + 1.50y_4.$$

Now write the problem as a system of linear equations, adding slack or surplus variables as needed, and let $z = -w$.

$$\begin{array}{llllllllllll}
y_1 + & y_2 & & & - y_5 & & & & & & & = 500 \\
y_1 + & y_2 + & & & & y_6 & & & & & & = 500 \\
& & y_3 + & y_4 & & & - y_7 & & & & & = 1000 \\
& & y_3 + & y_4 + & & & & y_8 & & & & = 1000 \\
y_1 + & & y_3 + & & & & & & y_9 & & & = 900 \\
& y_2 + & & y_4 + & & & & & & y_{10} & & = 700 \\
\multicolumn{4}{l}{1.20y_1 + 2.10y_2 + 1.80y_3 + 1.50y_4} & & & & & & & + z & = 0
\end{array}$$

Set up the initial simplex tableau.

$$\begin{array}{c}
\begin{matrix} y_1 & y_2 & y_3 & y_4 & y_5 & y_6 & y_7 & y_8 & y_9 & y_{10} & \end{matrix} \\
\left[\begin{array}{cccccccccc|c}
1 & 1 & 0 & 0 & -1 & 0 & 0 & 0 & 0 & 0 & 500 \\
1 & 1 & 0 & 0 & 0 & 1 & 0 & 0 & 0 & 0 & 500 \\
0 & 0 & 1 & 1 & 0 & 0 & -1 & 0 & 0 & 0 & 1000 \\
0 & 0 & 1 & 1 & 0 & 0 & 0 & 1 & 0 & 0 & 1000 \\
1 & 0 & 1 & 0 & 0 & 0 & 0 & 0 & 1 & 0 & 900 \\
0 & 1 & 0 & 1 & 0 & 0 & 0 & 0 & 0 & 1 & 700 \\
\hline
1.20 & 2.10 & 1.80 & 1.50 & 0 & 0 & 0 & 0 & 0 & 0 & 0
\end{array}\right]
\end{array}$$

The basic solution here is not feasible because $y_5 = -500$ and $y_7 = -1000$. Stages I and II could be done by hand here, but because of the large size of the matrix, it's more efficient to use technology such, as the program in the Graphing Calculator Appendix. Stage I takes four rounds of pivoting and produces the feasible solution in Figure 7.26. Because of the small size of a calculator screen, you must scroll to the right to see the entire matrix.

$y_1 \quad y_2 \; y_3 \; y_4 \; y_5 \quad y_6 \; y_7$

```
[[1    1 0 0 -1   0 0…
 [0    0 0 0 1    1 0…
 [-1   0 0 1 0    0 -…
 [0    0 0 0 0    0 1…
 [1    0 1 0 0    0 0…
 [0    0 0 0 1    0 1…
 [-1.2 0 0 0 2.1  0 1…
```

$y_7 \quad y_8 \; y_9 \quad y_{10}$

```
… 0   0 0  0 500  ]
… 0   0 0  0 0    ]
… -1  0 -1 0 100  ]
… 1   1 0  0 0    ]
… 0   0 1  0 900  ]
… 1   0 1  1 100  ]
… 1.5 0 -.3 0 -2820]]
```

FIGURE 7.26

Now Stage II begins. Two rounds of pivoting produce the final tableau (Figure 7.27).

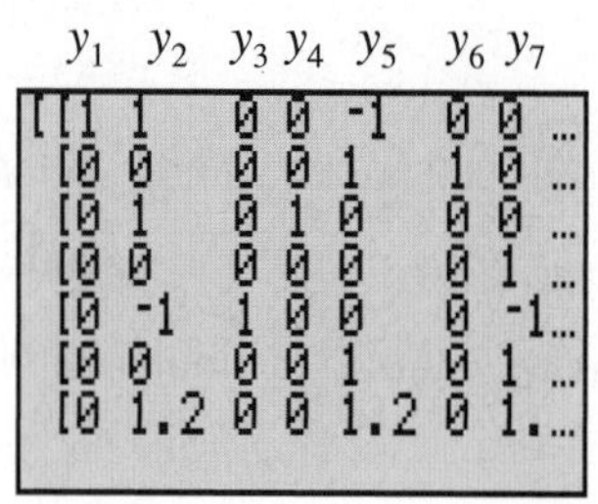

FIGURE 7.27

The optimal solution is $y_1 = 500$, $y_2 = 0$, $y_3 = 300$, $y_4 = 700$, which results in a minimum shipping cost of \$2190 (remember that the optimal value for the original minimization problem is the negative of the optimal value of the associated maximization problem).

7.7 Exercises

In Exercises 1–4, (a) restate the problem in terms of equations by introducing slack and surplus variables; (b) write the initial simplex tableau. (See Example 1.)

1. Maximize $z = 5x_1 + 2x_2 - x_3$
subject to:
$$2x_1 + 3x_2 + 5x_3 \geq 8$$
$$4x_1 - x_2 + 3x_3 \leq 7$$
$$x_1 \geq 0, x_2 \geq 0, x_3 \geq 0.$$

2. Maximize $z = x_1 + 4x_2 + 6x_3$
subject to:
$$5x_1 + 8x_2 - 5x_3 \leq 10$$
$$6x_1 + 2x_2 + 3x_3 \geq 7$$
$$x_1 \geq 0, x_2 \geq 0, x_3 \geq 0.$$

3. Maximize $z = 2x_1 - 3x_2 + 4x_3$
subject to:
$$x_1 + x_2 + x_3 \leq 100$$
$$x_1 + x_2 + x_3 \geq 75$$
$$x_1 + x_2 \geq 27$$
$$x_1 \geq 0, x_2 \geq 0, x_3 \geq 0.$$

4. Maximize $z = -x_1 + 5x_2 + x_3$
subject to:
$$2x_1 + x_3 \leq 40$$
$$x_1 + x_2 \geq 18$$
$$x_1 + x_3 = 20$$
$$x_1 \geq 0, x_2 \geq 0, x_3 \geq 0.$$

Convert Exercises 5–8 into maximization problems with positive constants on the right side of each constraint and write the initial simplex tableau. (See Examples 4 and 5.)

5. Minimize $w = 2y_1 + 5y_2 - 3y_3$
subject to:
$$y_1 + 2y_2 + 3y_3 \geq 115$$
$$2y_1 + y_2 + y_3 \leq 200$$
$$y_1 + y_3 \geq 50$$
$$y_1 \geq 0, y_2 \geq 0, y_3 \geq 0.$$

6. Minimize $w = 7y_1 + 6y_2 + y_3$
subject to:
$$y_1 + y_2 + y_3 \geq 5$$
$$-y_1 + y_2 \leq -4$$
$$2y_1 + y_2 + 3y_3 \geq 15$$
$$y_1 \geq 0, y_2 \geq 0, y_3 \geq 0.$$

7. Minimize $w = y_1 - 4y_2 + 2y_3$
subject to:
$$-7y_1 + 6y_2 - 8y_3 \leq -18$$
$$4y_1 + 5y_2 + 10y_3 \geq 20$$
$$y_1 \geq 0, y_2 \geq 0, y_3 \geq 0.$$

8. Minimize $w = y_1 + 2y_2 + y_3 + 5y_4$
subject to:
$$-y_1 + y_2 + y_3 + y_4 \leq -50$$
$$3y_1 + y_2 + 2y_3 + y_4 = 100$$
$$y_1 \geq 0, y_2 \geq 0, y_3 \geq 0, y_4 \geq 0.$$

Use the two-stage method to solve Exercises 9–20. (See Examples 1–5.)

9. Maximize $z = 12x_1 + 10x_2$
subject to:
$$x_1 + 2x_2 \geq 24$$
$$x_1 + x_2 \leq 40$$
$$x_1 \geq 0, x_2 \geq 0.$$

10. Find $x_1 \geq 0$, $x_2 \geq 0$, and $x_3 \geq 0$ such that
$$x_1 + x_2 + x_3 \leq 150$$
$$x_1 + x_2 + x_3 \geq 100$$
and $z = 2x_1 + 5x_2 + 3x_3$ is maximized.

11. Find $x_1 \geq 0$, $x_2 \geq 0$, and $x_3 \geq 0$ such that
$$x_1 + x_2 + 2x_3 \leq 38$$
$$2x_1 + x_2 + x_3 \geq 24$$
and $z = 3x_1 + 2x_2 + 2x_3$ is maximized.

12. Maximize $z = 6x_1 + 8x_2$
subject to: $3x_1 + 12x_2 \geq 48$
$2x_1 + 4x_2 \leq 60$
$x_1 \geq 0, x_2 \geq 0.$

13. Find $x_1 \geq 0$ and $x_2 \geq 0$ such that
$$x_1 + 2x_2 \leq 18$$
$$x_1 + 3x_2 \geq 12$$
$$2x_1 + 2x_2 \leq 30$$
and $z = 5x_1 + 10x_2$ is maximized.

14. Find $y_1 \geq 0, y_2 \geq 0$ such that
$$10y_1 + 5y_2 \geq 100$$
$$20y_1 + 10y_2 \geq 160$$
and $w = 4y_1 + 5y_2$ is minimized.

15. Minimize $w = 3y_1 + 2y_2$
subject to: $2y_1 + 3y_2 \geq 60$
$y_1 + 4y_2 \geq 40$
$y_1 \geq 0, y_2 \geq 0.$

16. Minimize $w = 3y_1 + 4y_2$
subject to: $y_1 + 2y_2 \geq 10$
$y_1 + y_2 \geq 8$
$2y_1 + y_2 \leq 22$
$y_1 \geq 0, y_2 \geq 0.$

17. Maximize $z = 3x_1 + 2x_2$
subject to: $x_1 + x_2 = 50$
$4x_1 + 2x_2 \geq 120$
$5x_1 + 2x_2 \leq 200$
with $x_1 \geq 0, x_2 \geq 0.$

18. Maximize $z = 10x_1 + 9x_2$
subject to: $x_1 + x_2 = 30$
$x_1 + x_2 \geq 25$
$2x_1 + x_2 \leq 40$
with $x_1 \geq 0, x_2 \geq 0.$

19. Minimize $w = 32y_1 + 40y_2$
subject to: $20y_1 + 10y_2 = 200$
$25y_1 + 40y_2 \leq 500$
$18y_1 + 24y_2 \geq 300$
with $y_1 \geq 0, y_2 \geq 0.$

20. Minimize $w = 15y_1 + 12y_2$
subject to: $y_1 + 2y_2 \leq 12$
$3y_1 + y_2 \geq 18$
$y_1 + y_2 = 10$
with $y_1 \geq 0, \quad y_2 \geq 0.$

In Exercises 21–24, set up the initial simplex tableau, but do not solve the problem. (See Example 6.)

21. Business A company is developing a new additive for gasoline. The additive is a mixture of three liquid ingredients, I, II, and III. For proper performance, the total amount of additive must be at least 10 ounces per barrel of gasoline. However, for safety reasons, the amount of additive should not exceed 15 ounces per barrel of gasoline. At least 1/4 ounce of ingredient I must be used for every ounce of ingredient II and at least 1 ounce of ingredient III must be used for every ounce of ingredient I. If the costs of I, II, and III are \$.30, \$.09, and \$.27 per ounce, respectively, find the mixture of the three ingredients that produces the minimum cost of the additive. What is the minimum cost?

22. Business A popular soft drink called Sugarlo, which is advertised as having a sugar content of no more than 10%, is blended from five ingredients, each of which has some sugar content. Water may also be added to dilute the mixture. The sugar content of the ingredients and their costs per gallon are given below.

	INGREDIENT					
	1	2	3	4	5	*Water*
Sugar content (%)	.28	.19	.43	.57	.22	0
Cost (\$/gal.)	.48	.32	.53	.28	.43	.04

At least .01 of the content of Sugarlo must come from ingredients 3 or 4, .01 must come from ingredients 2 or 5, and .01 from ingredients 1 or 4. How much of each ingredient should be used in preparing at least 15,000 gallons of Sugarlo to minimize the cost? What is the minimum cost?

23. Business The manufacturer of a popular personal computer has orders from two dealers. Dealer D_1 wants 32 computers, and dealer D_2 wants 20 computers. The manufacturer can fill the orders from either of two warehouses, W_1 or W_2. W_1 has 25 of the computers on hand, and W_2 has 30. The costs (in dollars) to ship one computer to each dealer from each warehouse are given below

		TO	
		D_1	D_2
FROM	W_1	\$14	\$22
	W_2	\$12	\$10

How should the orders be filled to minimize shipping costs? What is the minimum cost?

24. Natural Science Mark, who is ill, takes vitamin pills. Each day he must have at least 16 units of vitamin A, 5 units of vitamin B_1, and 20 units of vitamin C. He can choose between pill #1 which costs 10¢ and contains 8 units of A, 1 of B_1, and 2 of C, and pill #2 which costs 20¢ and contains 2 units of A, 1 of B_1, and 7 of C. How many of each pill should he buy in order to minimize his cost?

Use the two-stage method to solve Exercises 23–30. (See Examples 5 and 6.)

25. Transportation Southwestern Oil supplies two distributors in the Northwest from two outlets, S_1 and S_2.

Distributor D_1 needs at least 3000 barrels of oil, and distributor D_2 needs at least 5000 barrels. The two outlets can each furnish up to 5000 barrels of oil. The costs per barrel to ship the oil are given in the table.

		DISTRIBUTORS	
		D_1	D_2
OUTLETS	S_1	\$30	\$20
	S_2	\$25	\$22

There is also a shipping tax per barrel as given in the table below. Southwestern Oil is determined to spend no more than \$40,000 on shipping tax.

	D_1	D_2
S_1	\$2	\$6
S_2	\$5	\$4

How should the oil be supplied to minimize shipping costs?

26. Transportation Change Exercise 25 so that the two outlets each furnish exactly 5000 barrels of oil, with everything else the same. Solve the problem as in Example 5.

27. Business Topgrade Turf lawn seed mixture contains three types of seeds: bluegrass, rye, and Bermuda. The costs per pound of the three types of seed are 12¢, 15¢, and 5¢, respectively. In each batch there must be at least 20% bluegrass seed, and the amount of Bermuda must be no more than 2/3 the amount of rye. To fill current orders, the company must make at least 5000 pounds of the mixture. How much of each kind of seed should be used to minimize cost?

28. Business Change Exercise 27 so that the company must make exactly 5000 pounds of the mixture. Solve the problem as in Example 5.

29. Finance A bank has set aside a maximum of \$25 million for commercial and home loans. Every million dollars in commercial loans requires 2 lengthy application forms, while every million dollars in home loans requires 3 lengthy application forms. The bank cannot process more than 72 application forms at this time. The bank's policy is to loan at least four times as much for home loans as for commercial loans. Because of prior commitments, at least \$10 million will be used for these two types of loans. The bank earns 12% on home loans and 10% on commercial loans. What amount of money should be allotted for each type of loan to maximize the interest income?

30. Finance Virginia Keleske has decided to invest a \$100,000 inheritance in government securities that earn 7% per year, municipal bonds that earn 6% per year, and mutual funds that earn an average of 10% per year. She will spend at least \$40,000 on government securities, and she wants at least half the inheritance to go to bonds and mutual funds. Government securities have an initial fee of 2%, municipal bonds have an initial fee of 1%, and mutual funds have an initial fee of 3%. Virginia has \$2400 available to pay initial fees. How much should be invested in each way to maximize the interest yet meet the constraints? What is the maximum interest she can earn?

31. Business A brewery produces regular beer and a lower-carbohydrate "light" beer. Steady customers of the brewery buy 12 units of regular beer and 10 units of light beer. While setting up the brewery to produce the beers, the management decides to produce extra beer, beyond that needed to satisfy the steady customers. The cost per unit of regular beer is \$36,000 and the cost per unit of light beer is \$48,000. The number of units of light beer should not exceed twice the number of units of regular beer. At least 20 additional units of beer can be sold. How much of each type beer should be made so as to minimize total production costs?

32. Business The chemistry department at a local college decides to stock at least 800 small test tubes and 500 large test tubes. It wants to buy at least 1500 test tubes to take advantage of a special price. Since the small tubes are broken twice as often as the larger, the department will order at least twice as many small tubes as large. If the small test tubes cost 15¢ each and the large ones, made of a cheaper glass, cost 12¢ each, how many of each size should they order to minimize cost?

Business *Use technology to solve the following exercises, whose initial tableaus were set up in Exercises 21–23.*

33. Exercise 21

34. Exercise 22

35. Exercise 23

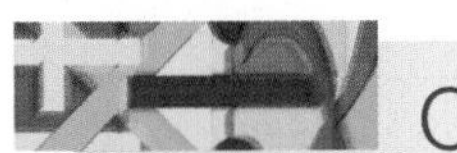

CHAPTER 7 SUMMARY

Key Terms and Symbols

7.1 linear inequality
boundary
half-plane
system of inequalities
region of feasible solutions (feasible region)

7.2 linear programming
objective function
constraints
corner point
bounded feasible region
unbounded feasible region
corner point theorem

7.4 standard maximum form
slack variable
simplex tableau
indicator
pivot and pivoting
final simplex tableau
basic variables
nonbasic variables
basic feasible solution

7.6 dual
transpose of a matrix
theorem of duality
shadow costs

7.7 surplus variable
basic variables
basic solution
basic feasible solution
Stage I
Stage II

Key Concepts

Graphing a Linear Inequality

Graph the boundary line as a solid line if the inequality includes "or equal," a dashed line otherwise. Shade the half-plane that includes a test point that makes the inequality true. The graph of a system of inequalities, called the **region of feasible solutions,** includes all points that satisfy all the inequalities of the system at the same time.

Solving Linear Programming Problems

Graphically: Determine the objective function and all necessary constraints. Graph the region of feasible solutions. The maximum or minimum value will occur at one or more of the corner points of this region.

Simplex Method: Determine the objective function and all necessary constraints. Convert each constraint into an equation by adding slack variables. Set up the initial simplex tableau. Locate the most negative indicator. Form the quotients to determine the pivot. Use row operations to change the pivot to 1 and all other numbers in that column to 0. If the indicators are all positive or 0, this is the final tableau. If not, choose a new pivot and repeat the process until no indicators are negative. Read the solution from the final tableau. The optimum value of the objective function is the number in the lower right corner of the final tableau. For problems with **mixed constraints,** replace each equation constraint by a pair of inequality constraints. Then add slack variables and subtract surplus variables as needed to convert each constraint into an equation. In Stage I, use row operations to transform the matrix until the solution is feasible. In Stage II, use the simplex method as described above. For **minimum** problems, let the objective function be w and set $-w = z$. Then proceed as with mixed constraints.

Solving Minimum Problems with Duals

Find the dual maximum problem. Solve the dual using the simplex method. The minimum value of the objective function w is the maximum value of the dual objective function z. The optimal solution is found in the entries in the bottom row of the columns corresponding to the slack variables.

Chapter 7 Review Exercises

Graph each of the following linear inequalities.

1. $y \le 3x + 2$
2. $2x - y \ge 6$
3. $3x + 4y \ge 12$
4. $y \le 4$

Graph the solution of each of the following systems of inequalities.

5. $x + y \le 6$
$2x - y \ge 3$
6. $4x + y \ge 8$
$2x - 3y \le 6$
7. $2 \le x \le 5$
$1 \le y \le 7$
$x - y \le 3$
8. $x + 2y \le 4$
$2x - 3y \le 6$
$x \ge 0$
$y \ge 0$

Set up a system of inequalities for each of the following problems; then graph the region of feasible solutions.

9. **Business** A bakery makes both cakes and cookies. Each batch of cakes requires 2 hours in the oven and 3 hours in the decorating room. Each batch of cookies needs $1\frac{1}{2}$ hours in the oven and 2/3 of an hour in the decorating room. The oven is available no more than 15 hours a day, while the decorating room can be used no more than 13 hours a day.

10. **Business** A company makes two kinds of pizza, special and basic. The special has toppings of cheese, tomatoes, and vegetables. Basic has just cheese and tomatoes. The company sells at least 6 units a day of the special pizza and 4 units a day of the basic. The cost of the vegetables (including tomatoes) is \$2 per unit for special and \$1 per unit for basic. No more than \$32 per day can be spent on vegetables (including tomatoes). The cheese used for the special is \$5 per unit, and the cheese for the basic is \$4 per unit. The company can spend no more than \$100 per day on cheese.

Use the given regions to find the maximum and minimum values of the objective function $z = 2x + 4y$.

11.

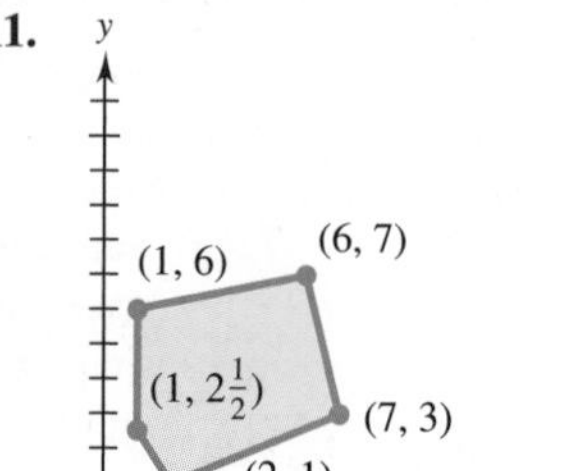

12.

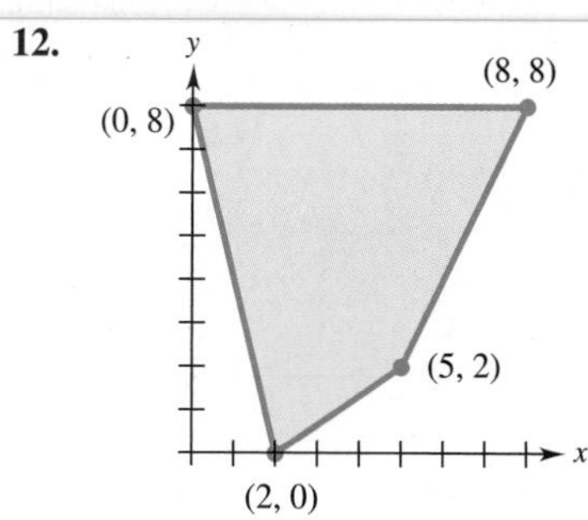

Use the graphical method to solve Exercises 13–16.

13. Maximize $z = 3x + 2y$
subject to: $2x + 7y \le 14$
$2x + 3y \le 10$
$x \ge 0, y \ge 0.$

14. Find $x \ge 0$ and $y \ge 0$ such that
$8x + 9y \ge 72$
$6x + 8y \ge 72$
and $w = 4x + 12y$ is minimized.

15. Find $x \ge 0$ and $y \ge 0$ such that
$x + y \le 50$
$2x + y \ge 20$
$x + 2y \ge 30$
and $w = 8x + 3y$ is minimized.

16. Maximize $z = 2x - 5y$
subject to: $3x + 2y \le 12$
$5x + y \ge 5$
$x \ge 0, y \ge 0.$

17. **Business** How many batches of cakes and cookies should the bakery of Exercise 9 make in order to maximize profits if cookies produce a profit of \$20 per batch and cakes produce a profit of \$30 per batch?

18. **Business** How many units of each kind of pizza should the company of Exercise 10 make in order to maximize revenue if special sells for \$20 per unit and basic for \$15 per unit?

For Exercises 19–22, (a) *select appropriate variables,* (b) *write the objective function,* (c) *write the constraints as inequalities.*

19. **Business** Roberta Hernandez sells three items, A, B, and C, in her gift shop. Each unit of A costs her \$2 to buy, \$1 to sell, and \$2 to deliver. For each unit of B, the costs are \$3, \$2, and \$2, respectively, and for each unit of C the costs are \$6, \$2, and \$4, respectively. The profit on A is \$4, on B it is \$3, and on C, \$3. How many of each should she order to maximize her profit if she can spend \$1200 to buy, \$800 on selling costs, and \$500 on delivery costs?

20. **Business** An investor is considering three types of investment: a high-risk venture into oil leases with a potential return of 15%, a medium-risk investment in bonds with a 9% return, and a relatively safe stock investment with a 5% return. He has \$50,000 to invest. Because of the risk, he will limit his investment in oil leases and bonds to 30% and his investment in oil leases and stock to 50%. How much should he invest in each to maximize his return, assuming investment returns are as expected?

21. **Business** The Aged Wood Winery makes two white wines, Fruity and Crystal, from two kinds of grapes and sugar. The wines require the following amounts of each ingredient per gallon and produce a profit per gallon as shown below.

	Grape A (bushels)	*Grape B (bushels)*	*Sugar (pounds)*	*Profit (dollars)*
Fruity	2	2	2	12
Crystal	1	3	1	15

The winery has available 110 bushels of grape A, 125 bushels of grape B, and 90 pounds of sugar. How much of each wine should be made to maximize profit?

22. **Business** A company makes three sizes of plastic bags: 5 gallon, 10 gallon, and 20 gallon. The production time in hours for cutting, sealing, and packaging a unit of each size is shown below.

Size	*Cutting*	*Sealing*	*Packaging*
5 gallon	1	1	2
10 gallon	1.1	1.2	3
20 gallon	1.5	1.3	4

There are at most 8 hours available each day for each of the three operations. If the profit per unit is \$1 for 5-gallon bags, \$.90 for 10-gallon bags, and \$.95 for 20-gallon bags, how many of each size should be made per day to maximize the profit?

23. When is it necessary to use the simplex method rather than the graphical method?

24. What types of problems can be solved using slack variables and surplus variables?

25. What kind of problem can be solved using the method of duals?

26. In solving a linear programming problem, you are given the following initial tableau.

$$\left[\begin{array}{cccccc|c} 4 & 2 & 3 & 1 & 0 & 0 & 9 \\ 5 & 4 & 1 & 0 & 1 & 0 & 10 \\ \hline -6 & -7 & -5 & 0 & 0 & 1 & 0 \end{array}\right]$$

(a) What is the problem being solved?

(b) If the 1 in row 1, column 4, were a -1 rather than a 1, how would it change your answer to part (a)?

(c) After several steps of the simplex algorithm, the following tableau results.

$$\left[\begin{array}{cccccc|c} 3 & 0 & 5 & 2 & -1 & 0 & 8 \\ 11 & 10 & 0 & -1 & 3 & 0 & 21 \\ \hline 47 & 0 & 0 & 13 & 11 & 10 & 227 \end{array}\right]$$

What is the solution? (List only the values of the original variables and the objective function. Do not include slack or surplus variables.)

(d) What is the dual of the problem you found in part (a)?

(e) What is the solution of the dual you found in part (d)? (Do not perform any steps of the simplex algorithm; just examine the tableau given in part (c).)

For each of the following problems, (a) *add slack variables and* (b) *set up the initial simplex tableau.*

27. Maximize $z = 2x_1 + 7x_2$
subject to: $3x_1 + 5x_2 \le 47$
$x_1 + x_2 \le 25$
$5x_1 + 2x_2 \le 35$
$2x_1 + x_2 \le 30$
$x_1 \ge 0, x_2 \ge 0.$

28. Maximize $z = 15x_1 + 10x_2$
subject to: $2x_1 + 5x_2 \le 50$
$x_1 + 3x_2 \le 25$
$4x_1 + x_2 \le 18$
$x_1 + x_2 \le 12$
$x_1 \ge 0, x_2 \ge 0.$

29. Maximize $z = 4x_1 + 6x_2 + 3x_3$
subject to: $x_1 + x_2 + x_3 \le 100$
$2x_1 + 3x_2 \le 500$
$x_1 + 2x_3 \le 350$
$x_1 \ge 0, x_2 \ge 0, x_3 \ge 0.$

30. Maximize $z = x_1 + 4x_2 + 2x_3$
subject to: $x_1 + x_2 + x_3 \le 90$
$2x_1 + 5x_2 + x_3 \le 120$
$x_1 + 3x_2 \le 80$
$x_1 \ge 0, x_2 \ge 0, x_3 \ge 0.$

For each of the following, use the simplex method to solve the maximization linear programming problems with initial tableaus as given.

31.
$$\begin{array}{ccccc|c} x_1 & x_2 & x_3 & x_4 & x_5 & \\ 1 & 2 & 3 & 1 & 0 & 28 \\ 2 & 4 & 8 & 0 & 1 & 32 \\ \hline -5 & -2 & -3 & 0 & 0 & 0 \end{array}$$

32.
$$\begin{array}{cccc|c} x_1 & x_2 & x_3 & x_4 & \\ 2 & 1 & 1 & 0 & 10 \\ 9 & 3 & 0 & 1 & 15 \\ \hline -2 & -3 & 0 & 0 & 0 \end{array}$$

33.
$$\begin{array}{cccccc|c} x_1 & x_2 & x_3 & x_4 & x_5 & x_6 & \\ 1 & 2 & 2 & 1 & 0 & 0 & 50 \\ 4 & 24 & 0 & 0 & 1 & 0 & 20 \\ 1 & 0 & 2 & 0 & 0 & 1 & 15 \\ \hline -5 & -3 & -2 & 0 & 0 & 0 & 0 \end{array}$$

34.
$$\begin{array}{ccccc|c} x_1 & x_2 & x_3 & x_4 & x_5 & \\ 1 & -2 & 1 & 0 & 0 & 38 \\ 1 & -1 & 0 & 1 & 0 & 12 \\ 2 & 1 & 0 & 0 & 1 & 30 \\ \hline -1 & -2 & 0 & 0 & 0 & 0 \end{array}$$

Convert the following problems into maximization problems without using duals.

35. Minimize $w = 18y_1 + 10y_2$
subject to: $y_1 + y_2 \geq 17$
$5y_1 + 8y_2 \geq 42$
$y_1 \geq 0, y_2 \geq 0.$

36. Minimize $w = 12y_1 + 20y_2 - 8y_3$
subject to: $y_1 + y_2 + 2y_3 \geq 48$
$y_1 + y_2 \geq 12$
$y_3 \geq 10$
$3y_1 + y_3 \geq 30$
$y_1 \geq 0, y_2 \geq 0, y_3 \geq 0.$

37. Minimize $w = 6y_1 - 3y_2 + 4y_3$
subject to: $2y_1 + y_2 + y_3 \geq 112$
$y_1 + y_2 + y_3 \geq 80$
$y_1 + y_2 \geq 45$
$y_1 \geq 0, y_2 \geq 0, y_3 \geq 0.$

Use the simplex method to solve the following mixed constraint problems.

38. Maximize $z = 2x_1 + 4x_2$
subject to: $3x_1 + 2x_2 \leq 12$
$5x_1 + x_2 \geq 5$
$x_1 \geq 0, x_2 \geq 0.$

39. Minimize $w = 4y_1 - 8y_2$
subject to: $y_1 + y_2 \leq 50$
$2y_1 - 4y_2 \geq 20$
$y_1 - y_2 \leq 22$
$y_1 \geq 0, y_2 \geq 0.$

The following tableaus are the final tableaus of minimizing problems solved by letting $w = -z$. Give the solution and the minimum value of the objective function for each problem.

40.
$$\left[\begin{array}{cccccc|c} 0 & 1 & 0 & 2 & 5 & 0 & 17 \\ 0 & 0 & 1 & 3 & 1 & 1 & 25 \\ 1 & 0 & 0 & 4 & 2 & \frac{1}{2} & 8 \\ \hline 0 & 0 & 0 & 2 & 5 & 0 & -427 \end{array}\right]$$

41.
$$\left[\begin{array}{ccccccc|c} 0 & 0 & 2 & 1 & 0 & 6 & 6 & 92 \\ 1 & 0 & 3 & 0 & 0 & 0 & 2 & 47 \\ 0 & 1 & 0 & 0 & 0 & 1 & 0 & 68 \\ 0 & 0 & 4 & 0 & 1 & 0 & 3 & 35 \\ \hline 0 & 0 & 5 & 0 & 0 & 2 & 9 & -1957 \end{array}\right]$$

The tableaus in Exercises 42–44 are the final tableaus of minimizing problems solved by the method of duals. State the solution and the minimum value of the objective function for each problem.

42.
$$\left[\begin{array}{cccccc|c} 1 & 0 & 0 & 3 & 1 & 2 & 12 \\ 0 & 0 & 1 & 4 & 5 & 3 & 5 \\ 0 & 1 & 0 & -2 & 7 & -6 & 8 \\ \hline 0 & 0 & 0 & 5 & 7 & 3 & 172 \end{array}\right]$$

43.
$$\left[\begin{array}{cccccc|c} 0 & 0 & 1 & 6 & 3 & 1 & 2 \\ 1 & 0 & 0 & 4 & -2 & 2 & 8 \\ 0 & 1 & 0 & 10 & 7 & 0 & 12 \\ \hline 0 & 0 & 0 & 9 & 5 & 8 & 62 \end{array}\right]$$

44.
$$\left[\begin{array}{cccc|c} 1 & 0 & 7 & -1 & 100 \\ 0 & 1 & 1 & 3 & 27 \\ \hline 0 & 0 & 7 & 2 & 640 \end{array}\right]$$

Business *Solve the following maximization problems, which were begun in Exercises 19–22.*

45. Exercise 19

46. Exercise 20

47. Exercise 21

48. Exercise 22

Business *Solve the following minimization problems.*

49. Cauchy Canners produces canned corn, beans, and carrots. Demand for vegetables requires it to produce at least 1000 cases per month. Based on past sales, it should produce at least twice as many cases of corn as of beans and at least 340 cases of carrots. It costs \$10 to produce a case of corn, \$15 to produce a case of beans, and \$25 to produce a case of carrots. How many cases of each vegetable should be produced to minimize costs? What is the minimum cost?

50. A contractor builds boathouses in two basic models, the atlantic and pacific. Each atlantic model requires 1000 feet of framing lumber, 3000 cubic feet of concrete, and \$2000 for advertising. Each pacific model requires 2000 feet of framing lumber, 3000 cubic feet of concrete, and

$3000 for advertising. Contracts call for using at least 8000 feet of framing lumber, 18,000 cubic feet of concrete, and $15,000 worth of advertising. If the total spent on each atlantic model is $3000 and the total spent on each pacific model is $4000, how many of each model should be built to minimize costs?

Business *Solve these mixed constraint problems.*

51. Brand X Cannery produces canned whole tomatoes and tomato sauce. This season, they have available 3,000,000 kilograms of tomatoes for these two products. To meet the demands of regular customers, they must produce at least 80,000 kilograms of sauce and 800,000 kilograms of whole tomatoes. The cost per kilogram is $4 to produce canned whole tomatoes and $3.25 to produce tomato sauce. Labor agreements require that at least 110,000 person hours be used. Each one-kilogram can of sauce requires 3 minutes for one worker, and each one-kilogram can of whole tomatoes requires 6 minutes for one worker. How many kilograms of tomatoes should Brand X use for each product to minimize cost? (For simplicity, assume production of y_1 kilograms of canned whole tomatoes and y_2 kilograms of tomato sauce requires $y_1 + y_2$ kilograms of tomatoes.)

52. A steel company produces two types of alloys. A run of type I requires 3000 pounds of molybdenum and 2000 tons of iron ore pellets as well as $2000 in advertising. A run of type II requires 3000 pounds of molybdenum and 1000 tons of iron ore pellets as well as $3000 in advertising. Total costs are $15,000 on a run of type I and $6000 on a run of type II. The company has on hand 18,000 pounds of molybdenum and 7000 tons of iron ore pellets and wants to use all of it. They plan to spend at least $14,000 on advertising. How much of each type should be produced to minimize cost? What is the minimum cost?

Case 6

Cooking with Linear Programming

Constructing a nutritious recipe can be a difficult task. The recipe must produce food that tastes good, and it must also balance the nutrients that each ingredient brings to the dish. This balancing of nutrients is very important in several diet plans that are currently popular. Many of these plans restrict the intake of certain nutrients (usually fat) while allowing for large amounts of other nutrients (protein and carbohydrates are popular choices). The number of calories in the dish is also often minimized. Linear programming can be used to help create recipes that balance nutrients.

In order to develop solutions to this type of problem we will need to know nutritional data for the ingredients in our recipes. This data is available from the United States Department of Agriculture in the USDA Nutrient Database for Standard Reference. This database, which is available at www.nal.udsa.gov/fnic/foodcomp/data/sr13.html, contains the nutrient levels for hundreds of basic foodstuffs. The nutrient levels are given per 100 grams of foodstuff. Unfortunately, grams are not often used in recipes; instead, kitchen measures like cups, tablespoons, and fractions of vegetables are used. Table 1 shows the conversion factors from grams to more familiar kitchen units and also gives serving sizes for various foodstuffs.

Table 1 **Serving Sizes of Various Foodstuffs**

Foodstuff	*Serving Size*
Beef	6 oz = 170 g
Egg	1 egg = 61 g
Feta Cheese	1/4 cup = 38 g
Lettuce	1/2 cup = 28 g
Milk	1 cup = 244 g
Oil	1 Tbsp = 13.5 g
Onion	1 onion = 110 g
Salad Dressing	1 cup = 250 g
Soy Sauce	1 Tbsp = 18 g
Spinach	1 cup = 180 g
Tomato	1 tomato = 123 g

Consider creating a recipe for a spinach omelet from eggs, milk, vegetable oil, and spinach. The nutrients of interest will be protein, fat, and carbohydrates. Calories will also be monitored. The amounts of the nutrients and calories for these ingredients are given in Table 2:

Table 2 Nutritional Values per 100 g of Foodstuff

Nutrient (units)	*Eggs*	*Milk*	*Oil*	*Spinach*
Calories (kcal)	152	61.44	884	23
Protein (g)	10.33	3.29	0	2.9
Fat (g)	11.44	3.34	100	.26
Carbohydrates (g)	1.04	4.66	0	3.75

Let x_1 be the number of 100 gram units of eggs to use in the recipe, x_2 be the number of 100 gram units of milk, x_3 be the number of 100 gram units of oil, and x_4 be the number of 100 gram units of spinach. We will want to minimize the amount of calories in the dish, while providing at least 4 grams of protein, 15 grams of carbohydrates, and 20 grams of fat. The cooking technique specifies that at least 1/8 of a cup of milk (30.5 grams) must be used in the recipe. We should thus minimize the objective function (using 100 gram units of foodstuff):

$$z = 152x_1 + 61.44x_2 + 884x_3 + 23x_4$$

subject to

$$\begin{aligned} 10.33x_1 + 3.29x_2 + 0x_3 + 2.90x_4 &\geq 4 \\ 11.44x_1 + 3.34x_2 + 100x_3 + .26x_4 &\geq 20 \\ 1.04x_1 + 4.66x_2 + 0x_3 + 3.75x_4 &\geq 15 \\ 0x_1 + 1x_2 + 0x_3 + 0x_4 &\geq .305. \end{aligned}$$

Of course, all variables are subject to nonnegativity constraints:

$$x_1 \geq 0, \qquad x_2 \geq 0, \qquad x_3 \geq 0, \qquad x_4 \geq 0.$$

Using a graphing calculator or a computer with linear programming software, we get the following solution:

$$x_1 = 1.2600, \qquad x_2 = .3050, \qquad x_3 = .0448, \qquad x_4 = .338.$$

This recipe produces an omelet with 257.63 calories. In grams, the amounts of each ingredient are 126 grams of eggs, 30.5 grams of milk, 4.48 grams of oil, and 33.8 grams of spinach. Converting to kitchen units using Table 1, we find the recipe to be approximately 2 eggs, 1/8 cup milk, 1 teaspoon oil, and 1/4 cup spinach.

Exercises

1. Consider preparing a high-carbohydrate Greek salad using feta cheese, lettuce, salad dressing, and tomato. The amount of carbohydrates in the salad should be maximized. In addition the salad should have less than 260 calories, over 210 milligrams of calcium, and over 6 grams of protein. The salad should also weigh in total less than 400 grams and be dressed with at least 2 tablespoons (1/8 cup) of salad dressing. The amounts of the nutrients and calories for these ingredients are given in Table 3:

Table 3 Nutritional Values per 100 g of Foodstuff

Nutrient (units)	*Feta Cheese*	*Lettuce*	*Salad Dressing*	*Tomato*
Calories (kcal)	263	14	448.8	21
Calcium (mg)	492.5	36	0	5
Protein (g)	10.33	1.62	0	.85
Carbohydrates (g)	4.09	2.37	2.5	4.64

Use linear programming to find the number of 100 gram units of each ingredient to be used in such a Greek salad and convert to kitchen units using Table 1.

2. Consider preparing a stir fry using beef, oil, onion, and soy sauce. A low-calorie stir fry is desired, which also contains less than 10 grams of carbohydrates, more than 50 grams of protein, and more than 3.5 grams of vitamin C. In order for the wok to function correctly, at least one teaspoon (or 4.5 grams) of oil must be used in the recipe. The amounts of the nutrients and calories for these ingredients are given in Table 4:

Table 4 Nutritional Values per 100 g of Foodstuff

Nutrient (units)	*Beef*	*Oil*	*Onion*	*Soy Sauce*
Calories (kcal)	215	884	38	60
Protein (g)	26	0	1.16	10.51
Carbohydrates (g)	0	1	8.63	5.57
Vitamin C (g)	0	0	6.4	0

Use linear programming to find the number of 100 gram units of each ingredient to be used in the stir fry and convert to kitchen units using Table 1.

Chapter 8

Sets and Probability

Federal officials cannot predict exactly how traffic deaths are affected by the trends toward fewer drunken drivers and increased use of seat belts. Economists cannot tell exactly how stricter federal regulations on bank loans affect the U.S. economy. The number of traffic deaths and the growth of the economy are subject to many factors that cannot be predicted precisely.

Probability theory enables us to deal with uncertainty. The basic concepts of probability are discussed in this chapter, and applications of probability are discussed in the next chapter. Sets and set operations are the basic tools for the study of probability, so we begin with them.

8.1 SETS

Think of a set as a well-defined collection of objects. A set of coins might include one of each type of coin now put out by the U.S. government. Another set might be made up of all the students in your English class. On the other hand, a collection of "young adults" does not constitute a set unless the designation young adult is clearly defined. For example, this set might be defined as those aged 18–29.

In mathematics sets are often made up of numbers. The set consisting of the numbers 3, 4, and 5 is written

$$\{3, 4, 5\},$$

where **set braces,** { }, are used to enclose the numbers belonging to the set. The numbers, 3, 4, and 5 are called the **elements** or **members** of this set. To show that 4 is an element of the set {3, 4, 5}, we use the symbol $\in$ and write

$$4 \in \{3, 4, 5\},$$

read "4 is an element of the set containing 3, 4, and 5."

Write *true* or *false*.

(a) $9 \in \{8, 4, -3, -9, 6\}$

(b) $4 \notin \{3, 9, 7\}$

(c) If $M = \{0, 1, 2, 3, 4\}$, then $0 \in M$.

Answers:

(a) False

(b) True

(c) True

Also, $5 \in \{3, 4, 5\}$. Place a slash through the symbol $\in$ to show that 8 is *not* an element of this set.

$$8 \notin \{3, 4, 5\}$$

This is read "8 is not an element of the set $\{3, 4, 5\}$."

Sets are often named with capital letters, so that if

$$B = \{5, 6, 7\},$$

then, for example, $6 \in B$ and $10 \notin B$. (1)

Sometimes a set has no elements. Some examples are the set of female presidents of the United States in the period 1788–2004, the set of counting numbers less than 1, and the set of men more than 10 feet tall. A set with no elements is called the **empty set.** The symbol $\emptyset$ is used to represent the empty set.

CAUTION Be careful to distinguish between the symbols 0, $\emptyset$ and $\{0\}$. The symbol 0 represents a *number;* $\emptyset$ represents a *set* with no elements; and $\{0\}$ represents a *set* with one element, the number 0. Do not confuse the empty set symbol $\emptyset$ with the zero Ø on a computer screen or printout.

Two sets are **equal** if they contain exactly the same elements. The sets $\{5, 6, 7\}$, $\{7, 6, 5\}$, and $\{6, 5, 7\}$ all contain exactly the same elements and are equal. In symbols,

$$\{5, 6, 7\} = \{7, 6, 5\} = \{6, 5, 7\}.$$

This means that the ordering of the elements in a set is unimportant. Sets that do not contain exactly the same elements are *not equal*. For example, the sets $\{5, 6, 7\}$ and $\{5, 6, 7, 8\}$ do not contain exactly the same elements and are not equal. We show this by writing

$$\{5, 6, 7\} \neq \{5, 6, 7, 8\}.$$

List the elements in the following sets.

(a) $\{x \mid x$ is a counting number more than 5 and less than $8\}$

(b) $\{x \mid x$ is an integer, $-3 < x \leq 1\}$

Answers:

(a) $\{6, 7\}$

(b) $\{-2, -1, 0, 1\}$

Sometimes we describe a set by a common property of its elements rather than by a list of elements. This common property can be expressed with **set-builder notation;** for example,

$$\{x \mid x \text{ has property } P\}$$

(read "the set of all elements x such that x has property P") represents the set of all elements x having some property P.

EXAMPLE 1 List the elements belonging to each of the following sets.

(a) $\{x \mid x$ is a natural number less than $5\}$

The natural numbers less than 5 make up the set $\{1, 2, 3, 4\}$.

(b) $\{x \mid x$ is a state that borders Florida$\}$ = {Alabama, Georgia}

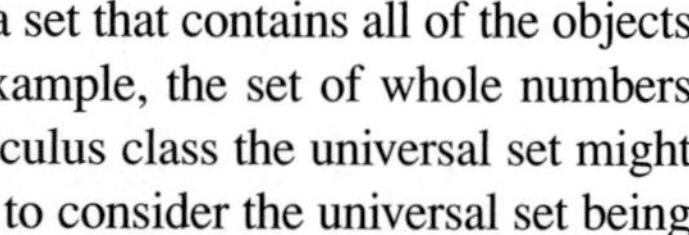

The **universal set** in a particular discussion is a set that contains all of the objects being discussed. In grade-school arithmetic, for example, the set of whole numbers might be the universal set, whereas in a college calculus class the universal set might be the set of all real numbers. When it is necessary to consider the universal set being used, it will be clearly specified or easily understood from the context of the problem.

Sometimes every element of one set also belongs to another set. For example, if

$$A = \{3, 4, 5, 6\}$$

and

$$B = \{2, 3, 4, 5, 6, 7, 8\},$$

then every element of A is also an element of B. This is an example of the following definition.

A set A is a **subset** of a set B (written $A \subseteq B$) provided that every element of A is also an element of B.

3

Write *true* or *false*.

(a) $\{3, 4, 5\} \subseteq \{2, 3, 4, 6\}$

(b) $\{x \mid x \text{ is an automobile}\} \subseteq \{x \mid x \text{ is a motor vehicle}\}$

(c) $\{3, 6, 9, 10\} \subseteq \{3, 9, 11, 13\}$

Answers:

(a) False

(b) True

(c) False

EXAMPLE 2 Decide whether or not $M \subseteq N$.

(a) M is the set of all small businesses with less than 20 employees. N is the set of all businesses.

Each business with less than 20 employees is also a business, so $M \subseteq N$.

(b) M is the set of all fourth-grade students in a school at the end of the school year, and N is the set of all nine-year-old students in the school at the end of the school year.

By the end of the school year, some fourth-grade students are ten years old, so there are elements in M that are not in N. Thus, M is not a subset of N, written $M \not\subseteq N$. ◀ 3

Every set A is a subset of itself because the statement "every element of A is also an element of A" is always true. It is also true that the empty set is a subset of every set.*

For any set A,

$$\emptyset \subseteq A \quad \text{and} \quad A \subseteq A.$$

EXAMPLE 3 List all possible subsets for each of the following sets.

(a) $\{7, 8\}$

There are 4 subsets of $\{7, 8\}$:

$$\emptyset, \quad \{7\}, \quad \{8\}, \quad \{7, 8\}.$$

*This fact is not intuitively obvious to most people. If you wish, you can think of it as a convention that we agree to adopt in order to simplify the statements of several results later.

List all subsets of {w, x, y, z}.

Answer:

∅, {w}, {x}, {y}, {z}, {w, x}, {w, y}, {w, z}, {x, y}, {x, z}, {y, z}, {w, x, y}, {w, x, z}, {w, y, z}, {x, y, z}, {w, x, y, z}

(b) {a, b, c}

There are 8 subsets of {a, b, c}:

∅, {a}, {b}, {c}, {a, b}, {a, c}, {b, c}, {a, b, c}.

A good way to find the subsets of {7, 8} and the subsets of {a, b, c} in Example 3 is to use a **tree diagram,** a systematic way of listing all the subsets of a given set. Figures 8.1(a) and (b) show tree diagrams for finding the subsets of {7, 8} and {a, b, c}. (4)

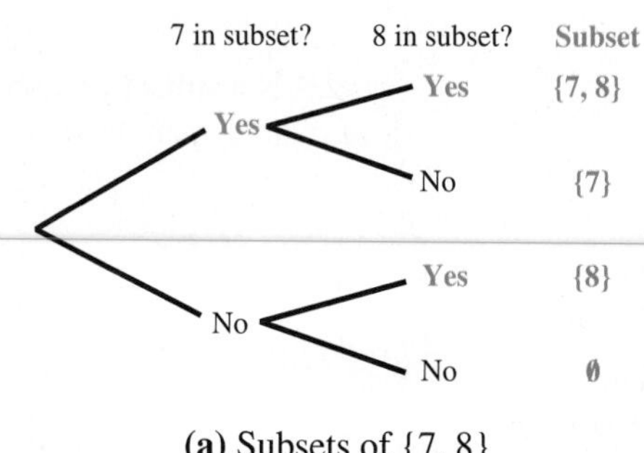

(a) Subsets of {7, 8}

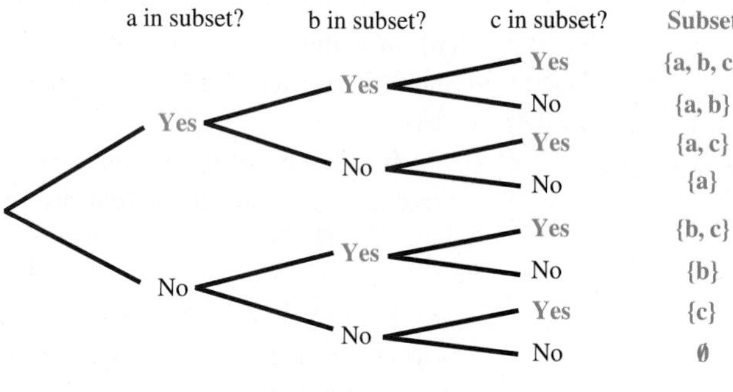

(b) Subsets of {a, b, c}

FIGURE 8.1

By using the fact that there are two possibilities for each element (either it is in the subset or it is not) we have found that a set with 2 elements has 4 ($= 2^2$) subsets and a set with 3 elements has 8 ($= 2^3$) subsets. Similar arguments work for any finite set and lead to this conclusion.

A set of n distinct elements has 2^n subsets.

EXAMPLE 4 Find the number of subsets for each of the following sets.

(a) {3, 4, 5, 6, 7}

Since this set has 5 elements, it has 2^5 or 32 subsets.

Find the number of subsets for each of the following sets.

(a) $\{x \mid x$ is a season of the year$\}$

(b) $\{-6, -5, -4, -3, -2, -1, 0\}$

(c) $\{6\}$

Answers:

(a) 16

(b) 128

(c) 2

6

Refer to sets A, B, C, and U in the diagram.

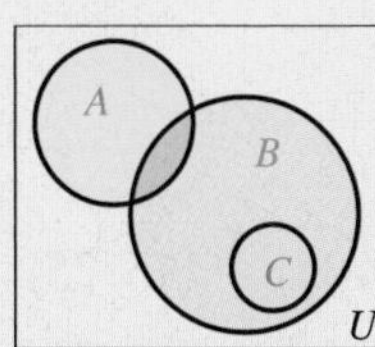

(a) Is $A \subseteq B$?

(b) Is $C \subseteq B$?

(c) Is $C \subseteq U$?

(d) Is $\emptyset \subseteq A$?

Answers:

(a) No

(b) Yes

(c) Yes

(d) Yes

Let $U = \{a, b, c, d, e, f, g\}$, with $K = \{c, d, f, g\}$ and $R = \{a, c, d, e, g\}$. Find

(a) K';

(b) R'.

Answers:

(a) $\{a, b, e\}$

(b) $\{b, f\}$

(b) $\{x \mid x$ is a day of the week$\}$
This set has 7 elements and therefore has $2^7 = 128$ subsets.

(c) $\emptyset$
Since the empty set has 0 elements, it has $2^0 = 1$ subset, $\emptyset$ itself. ◀ 5

Figure 8.2 shows a set A, which is a subset of a set B, because A is entirely in B. (The areas of the regions are not meant to be proportional to the size of the corresponding sets.) The rectangle represents the universal set, U. Such diagrams, called **Venn diagrams,** are used to illustrate relationships among sets. 6

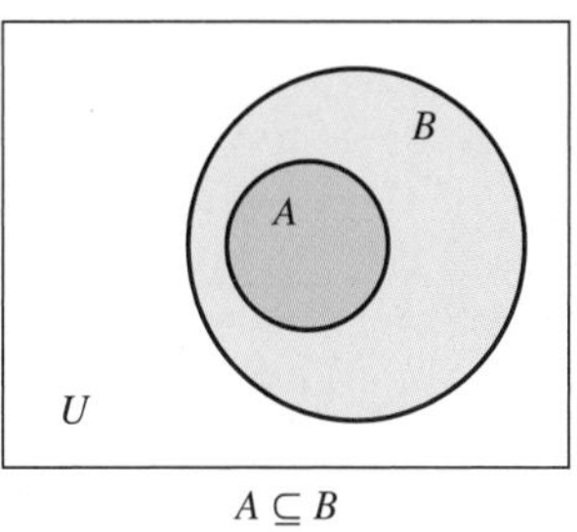

$A \subseteq B$

FIGURE 8.2

We can operate on sets to get other sets, just as operations on numbers (such as finding the negative, addition, and multiplication) produce other numbers. For example, given a set A and a universal set U, the set of all elements of U that do *not* belong to A is called the **complement** of set A. For example, if set A is the set of all the female students in your class and U is the set of all students in the class, then the complement of A would be the set of all male students in the class. The complement of set A is written A' (read "A-prime"). The Venn diagram of Figure 8.3 shows a set B. Its complement, B', is shown in color.

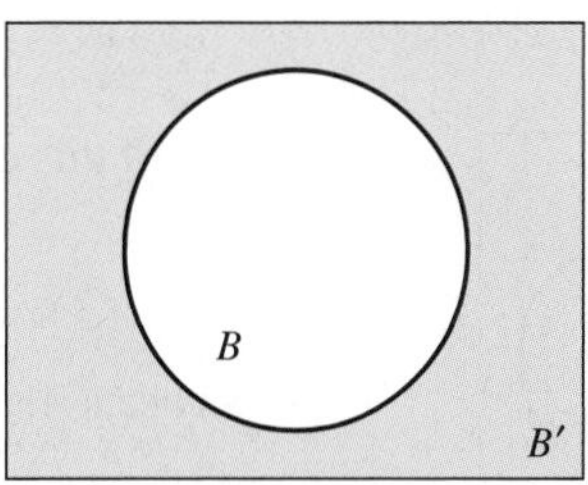

FIGURE 8.3

EXAMPLE 5 Let $U = \{1, 2, 3, 4, 5, 6, 7\}$, $A = \{1, 3, 5, 7\}$, and $B = \{3, 4, 6\}$. Find the following sets.

(a) A'
Set A' contains the elements of U that are not in A.

$$A' = \{2, 4, 6\}$$

(b) $B' = \{1, 2, 5, 7\}$

(c) $\emptyset' = U$ and $U' = \emptyset$ ◀ 7

Given two sets A and B, the set of all elements belonging to *both* set A and set B is called the **intersection** of the two sets, written $A \cap B$. For example, the elements that belong to both $A = \{1, 2, 4, 5, 7\}$ and $B = \{2, 4, 5, 7, 9, 11\}$ are 2, 4, 5, and 7, so

$$\begin{aligned} \text{A and B} &= A \cap B \\ &= \{1, 2, 4, 5, 7\} \cap \{2, 4, 5, 7, 9, 11\} \\ &= \{2, 4, 5, 7\}. \end{aligned}$$

The Venn diagram of Figure 8.4 shows two sets A and B with their intersection, $A \cap B$, shown in color.

8

Find the following.

(a) $\{1, 2, 3, 4\} \cap \{3, 5, 7, 9\}$

(b) Suppose set K is the set of all blue-eyed blondes in a class and J is the set of all blue-eyed brunettes in the class. Let $P = \{x \mid x \text{ is a brown-eyed redhead}\}$. If the class has only blondes or brunettes, find $K \cap P$.

Answers:

(a) $\{3\}$

(b) $\emptyset$

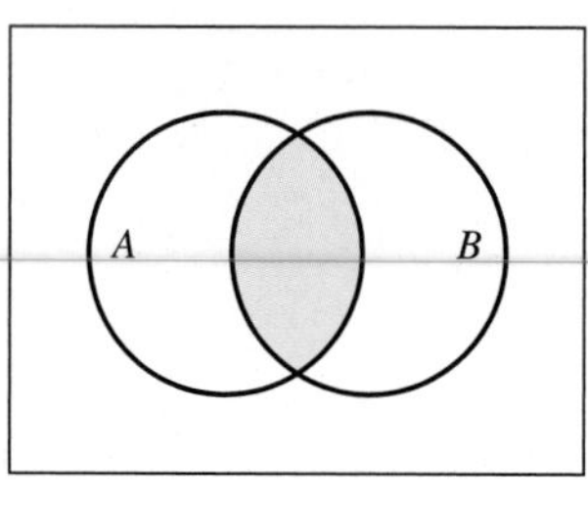

$A \cap B$

FIGURE 8.4

EXAMPLE 6 **(a)** $\{9, 15, 25, 36\} \cap \{15, 20, 25, 30, 35\} = \{15, 25\}$

The elements 15 and 25 are the only ones belonging to both sets.

(b) $\{x \mid x \text{ is a teen-ager}\} \cap \{x \mid x \text{ is a senior citizen}\}$ is an empty set. ◀ 8

Two sets that have no elements in common are called **disjoint sets.** For example, there are no elements common to both {50, 51, 54} and {52, 53, 55, 56}, so these two sets are disjoint, and

$$\{50, 51, 54\} \cap \{52, 53, 55, 56\} = \emptyset.$$

The result of this example can be generalized:

For any sets A and B,

if A and B are disjoint sets, then $A \cap B = \emptyset$.

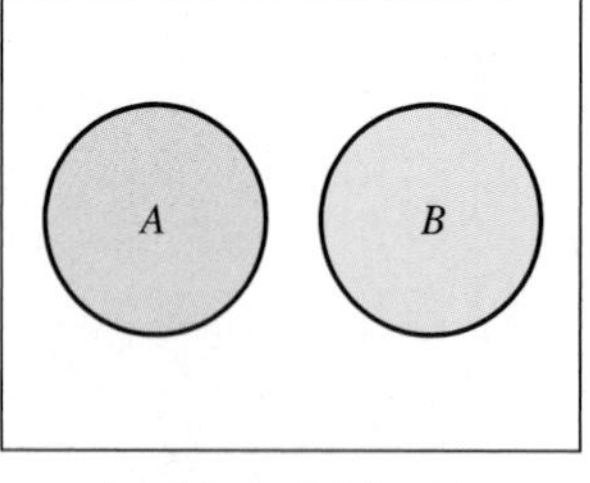

A and *B* are disjoint sets.

FIGURE 8.5

Figure 8.5 is a Venn diagram of disjoint sets.

The set of all elements belonging to set A or to set B, or to both sets, is called the **union** of the two sets, written $A \cup B$. For example, for sets $A = \{1, 3, 5\}$ and $B = \{3, 5, 7, 9\}$,

$$\begin{aligned} \text{A or B} &= A \cup B \\ &= \{1, 3, 5\} \cup \{3, 5, 7, 9\} \\ &= \{1, 3, 5, 7, 9\}. \end{aligned}$$

The Venn diagram of Figure 8.6 shows two sets A and B, with their union $A \cup B$ shown in color.

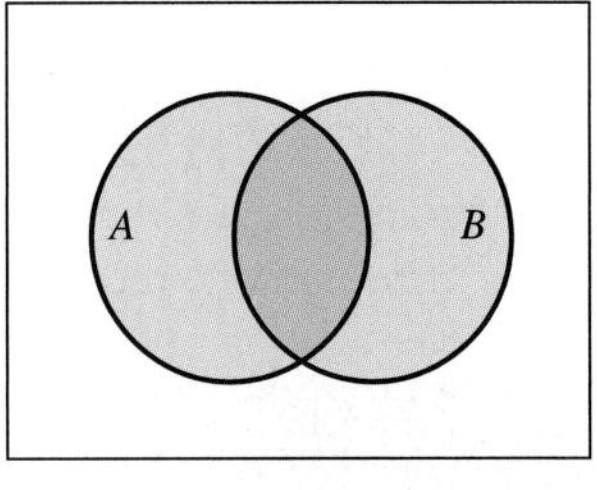

$A \cup B$

FIGURE 8.6

9

Find the following.

(a) $\{a, b, c\} \cup \{a, c, e\}$

(b) Describe $K \cup J$ in words for the sets given in side problem 8(b).

Answers:

(a) $\{a, b, c, e\}$

(b) All members of the class with blue eyes

EXAMPLE 7 **(a)** Find the union of $\{1, 2, 5, 9, 14\}$ and $\{1, 3, 4, 8\}$.

Begin by listing the elements of the first set, $\{1, 2, 5, 9, 14\}$. Then include any elements from the second set *that are not already listed.* Doing this gives

$$\{1, 2, 5, 9, 14\} \cup \{1, 3, 4, 8\} = \{1, 2, 3, 4, 5, 8, 9, 14\}.$$

(b) {terriers, spaniels, chows, dalmatians} ∪ {spaniels, collies, bulldogs} = {terriers, spaniels, chows, dalmatians, collies, bulldogs} 9

Finding the complement of a set, the intersection of two sets, or the union of two sets are examples of *set operations.* The set operations are summarized below.

Operations on Sets

Let A and B be any sets with U the universal set. Then

the **complement** of A, written A', is

$$A' = \{x \mid x \notin A \text{ and } x \in U\};$$

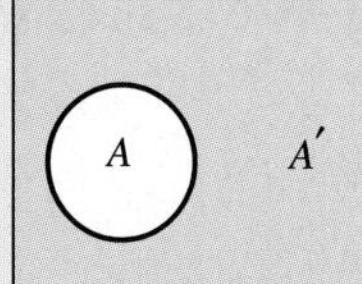

the **intersection** of A and B is

$$A \cap B = \{x \mid x \in A \textbf{ and } x \in B\};$$

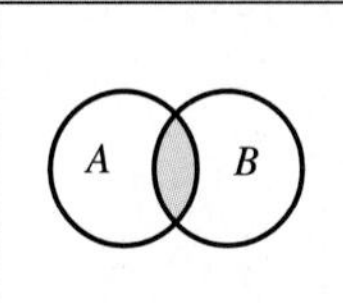

the **union** of A and B is

$$A \cup B = \{x \mid x \in A \textbf{ or } x \in B \textbf{ or both}\}.$$

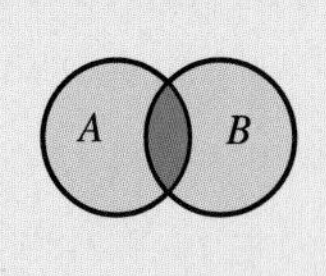

CAUTION As shown in the definitions given above, an element is in the *intersection* of sets A and B if it is in *both A and B*. On the other hand, an element is in the *union* of sets A and B if it is in *A or B or both*.

EXAMPLE 8 The table gives the annual high and low prices, the last price, and the weekly change for six stocks, as listed on the New York Stock Exchange at the end of a recent week.*

Stock	*High*	*Low*	*Last*	*Change*
BiotechT	206.00	92.51	138.20	+3.10
Cisco	70.00	13.19	20.49	+1.64
Disney	43.00	26.00	31.85	+.13
GenElec	60.50	36.42	48.14	−.86
Lucent	67.19	5.50	8.40	+.41

Let set A include all stocks with a high price greater than \$100; B all stocks with a last price between \$20 and \$40; C all stocks with a positive price change. Find the following.

10 **(a)** B'

Set B' contains all the listed stocks that are not in set B, so

$$B' = \{\text{BiotechT, GenElec, Lucent}\}.$$

(b) $A \cap C$

The intersection, A and C, contains those stocks with a high price greater than \$100 and a positive price change.

$$A \cap C = \{\text{BiotechT}\}$$

(c) $A \cup B$

The union of A and B includes all stocks with a high price greater than \$100 or a last price between \$20 and \$40.

$$A \cup B = \{\text{BiotechT, Cisco, Disney}\}$$  10

Find all stocks with

(a) a last price between \$20 and \$40 and with a negative price change.

(b) a last price between \$20 and \$40 or with a negative price change.

Answers:

(a) ∅

(b) {Cisco, Disney, GenElec}

**Sacramento Bee*, June 9, 2001.

8.1 Exercises

Write true *or* false *for each statement.*

1. $3 \in \{2, 5, 7, 9, 10\}$

2. $6 \in \{-2, 6, 9, 5\}$

3. $9 \notin \{2, 1, 5, 8\}$

4. $3 \notin \{7, 6, 5, 4\}$

5. $\{2, 5, 8, 9\} = \{2, 5, 9, 8\}$

6. $\{3, 7, 12, 14\} = \{3, 7, 12, 14, 0\}$

7. {all whole numbers greater than 7 and less than 10} = {8, 9}

8. {all counting numbers not greater than 3} = {0, 1, 2}

9. $\{x \mid x \text{ is an odd integer}, 6 \le x \le 18\} = \{7, 9, 11, 15, 17\}$

10. $\{x \mid x \text{ is a vowel}\} = \{\text{a, e, i, o, u}\}$

11. The elements of a set may be sets themselves, as in $\{1, \{1, 3\}, \{2\}, 4\}$. Explain why the set $\{\emptyset\}$ is not the same set as $\{0\}$.

12. What is set-builder notation? Give an example.

Let $A = \{-3, 0, 3\}$, $B = \{-2, -1, 0, 1, 2\}$, $C = \{-3, -1\}$, $D = \{0\}$, $E = \{-2\}$, and $U = \{-3, -2, -1, 0, 1, 2, 3\}$. Insert $\subseteq$ or $\not\subseteq$ to make the following statements true. (See Example 2.)

13. A ____ U

14. E ____ A

15. A ____ E

16. B ____ C

17. $\emptyset$ ____ A

18. $\{0, 2\}$ ____ D

19. D ____ B

20. A ____ C

Find the number of subsets for each set. (See Example 4.)

21. $\{A, B, C\}$

22. {red, yellow, blue, black, white}

23. $\{x \mid x \text{ is an integer strictly between 0 and 7}\}$

24. $\{x \mid x \text{ is a whole number less than 4}\}$

Find the complement of each set. (See Example 5.)

25. The set in Exercise 23 if U is the set of all integers.

26. The set in Exercise 24 if U is the set of all whole numbers.

27. Describe the intersection and union of sets. How do they differ?

Insert $\cap$ or $\cup$ to make each statement true. (See Examples 6 and 7.)

28. $\{5, 7, 9, 19\}$ ____ $\{7, 9, 11, 15\} = \{7, 9\}$

29. $\{8, 11, 15\}$ ____ $\{8, 11, 19, 20\} = \{8, 11\}$

30. $\{2, 1, 7\}$ ____ $\{1, 5, 9\} = \{1\}$

31. $\{6, 12, 14, 16\}$ ____ $\{6, 14, 19\} = \{6, 14\}$

32. $\{3, 5, 9, 10\}$ ____ $\emptyset = \emptyset$

33. $\{3, 5, 9, 10\}$ ____ $\emptyset = \{3, 5, 9, 10\}$

34. $\{1, 2, 4\}$ ____ $\{1, 2, 4\} = \{1, 2, 4\}$

35. $\{1, 2, 4\}$ ____ $\{1, 2\} = \{1, 2, 4\}$

36. Is it possible for two nonempty sets to have the same intersection and union? If so, give an example.

Let $U = \{2, 3, 4, 5, 7, 9\}$; $X = \{2, 3, 4, 5\}$; $Y = \{3, 5, 7, 9\}$; and $Z = \{2, 4, 5, 7, 9\}$.

List the members of each of the following sets, using set braces. (See Examples 5–7.)

37. $X \cap Y$

38. $X \cup Y$

39. X'

40. Y'

41. $X' \cap Y'$

42. $X' \cap Z$

43. $X \cup (Y \cap Z)$

44. $Y \cap (X \cup Z)$

Let U = {all students in this school}; M = {all students taking this course}; N = {all students taking accounting}; P = {all students taking zoology}.

Describe each of the following sets in words.

45. M'

46. $M \cup N$

47. $N \cap P$

48. $N' \cap P'$

49. Refer to the sets listed in the directions for Exercises 13–20. Which pairs of sets are disjoint?

50. Refer to the sets listed in the directions for Exercises 37–44. Which pairs are disjoint?

Refer to Example 8 in the text. Describe each of the sets in Exercises 51–54 in words; then list the elements of each set.

51. C'

52. $A \cap B$

53. $(A \cup B)'$

54. $(B \cap C)'$

Business *An electronics store classifies credit applicants by sex, marital status, and employment status. Let the universal set be the set of all applicants, M be the set of male applicants, S be the set of all single applicants, and E be the set of employed applicants. Describe the following sets in words.*

55. $M \cap E$

56. $M' \cap S$

57. $M' \cup S'$

Business *The total U.S. advertising volume (in millions of dollars) spent by medium in 1998 and 1999 is shown in the table.**

Medium	*1998*	*1999*
Newspapers	44,292	46,648
Magazines	10,518	11,433
Television	47,720	50,440
Radio	15,073	17,215
Yellow Pages	11,990	12,652
Direct Mail	39,620	41,403
Internet	1,050	1,904

List the elements of each set.

58. The set of all media that collected more than \$40,000 million in both 1998 and 1999.

59. The set of all media that collected less than \$12,000 million in 1998 or 1999 (or both years).

60. The dollar amounts spent for magazine ads in each year shown.

**Advertising Age*, May 22, 2000. www.adage.com.

Business *The top five pay-cable services in 1999 are listed here.* Use this information for Exercises 61–66.*

Network	*Subscribers (in millions)*	*Content*
The Disney Channel	42.9	Movies, cartoons
HBO/Cinemax	35.7	Movies, variety, sports, documentaries
Encore	23.1	On-demand movies
Showtime/The Movie Channel	22.3	Movies, variety, comedy, sports
Spice	11.0	Adult movies

List the elements of the following sets.

61. F, the set of networks with more than 25 million subscribers.

62. G, the set of networks that show sports.

63. H, the set of networks that show comedy.

64. $F \cap G$ **65.** $H \cup F$ **66.** F'

*National Cable Television Association, *Cable TV Developments* (Spring 2000).

Health *The table below shows some symptoms of an overactive thyroid and an underactive thyroid.*

Underactive Thyroid	*Overactive Thyroid*
Sleepiness, s	Insomnia, i
Dry hands, d	Moist hands, m
Intolerance of cold, c	Intolerance of heat, h
Goiter, g	Goiter, g

Let U be the smallest possible set that includes all the symptoms listed, N be the set of symptoms for an underactive thyroid, and O be the set of symptoms for an overactive thyroid. Use the lower case letters in the table to list the elements of each set.

67. O' **68.** N'

69. $N \cap O$ **70.** $N \cup O$

8.2 APPLICATIONS OF VENN DIAGRAMS

We used Venn diagrams in the last section to illustrate set union and intersection. The rectangular region in a Venn diagram represents the universal set, U. Including only a single set, A, inside the universal set, as in Figure 8.7, divides U into two nonoverlapping regions. Region 1 represents A', those elements outside set A, while region 2 represents those elements belonging to set A. (The numbering of these regions is arbitrary.)

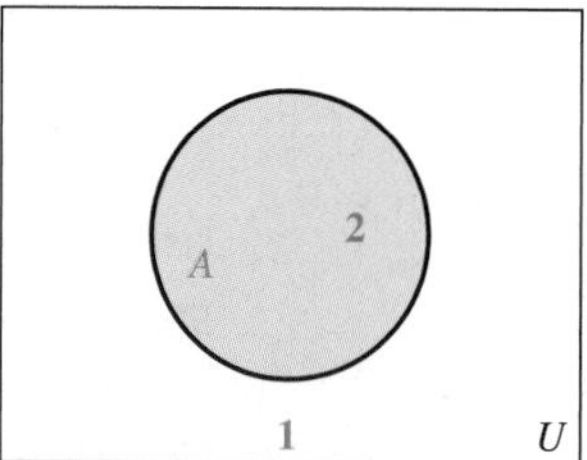

One set leads to 2 regions. (numbering is arbitrary).

FIGURE 8.7

Two sets lead to 4 regions. (numbering is arbitrary).

FIGURE 8.8

The Venn diagram of Figure 8.8 shows two sets inside U. These two sets divide the universal set into four nonoverlapping regions. As labeled in Figure 8.8, region 1 includes those elements outside both set A and set B. Region 2 includes those elements belonging to A and not to B. Region 3 includes those elements belonging

Draw Venn diagrams for the following.

(a) $A \cup B'$

(b) $A' \cap B'$

Answers:

(a)

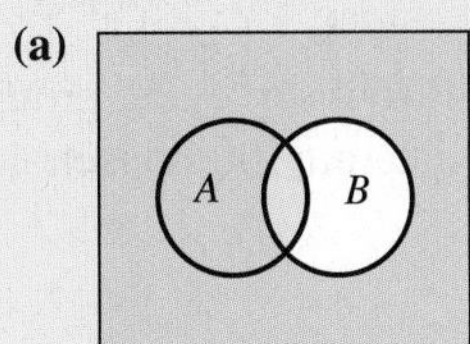

$A' \cup B'$

(b)

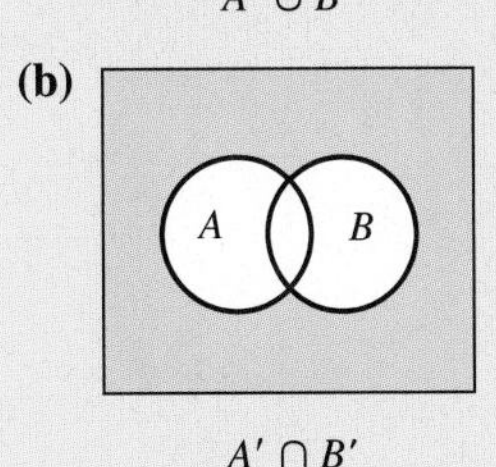

$A' \cap B'$

to both A and B. Which elements belong to region 4? (Again, the numbering is arbitrary.)

EXAMPLE 1 Draw Venn diagrams similar to Figure 8.8 and shade the regions representing the following sets.

(a) $A' \cap B$

Set A' contains all the elements outside set A. As labeled in Figure 8.8, A' is represented by regions 1 and 4. Set B is represented by the elements in regions 3 and 4. The intersection of sets A' and B, the set $A' \cap B$, is given by the region common to regions 1 and 4 and regions 3 and 4. The result, region 4, is shaded in Figure 8.9.

(b) $A' \cup B'$

Again, set A' is represented by regions 1 and 4 and set B' by regions 1 and 2. To find $A' \cup B'$, identify the region that represents the set of all elements in A', B', or both. The result, which is shaded in Figure 8.10, includes regions 1, 2, and 4. ◀ 1

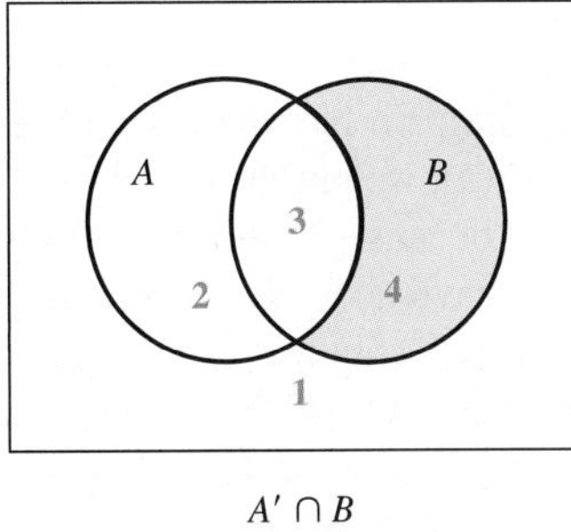

$A' \cap B$

FIGURE 8.9

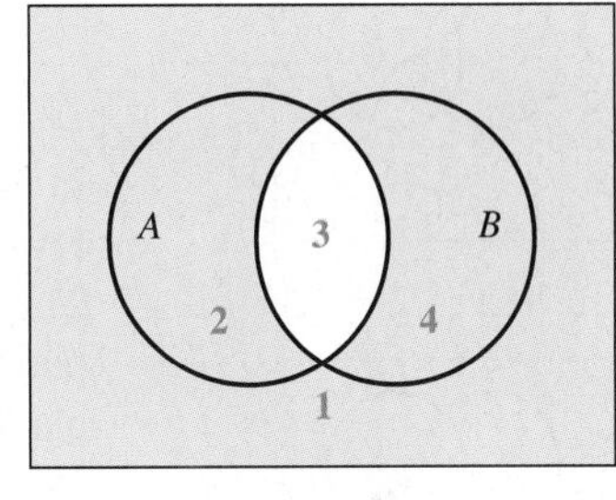

$A' \cup B'$

FIGURE 8.10

Venn diagrams also can be drawn with three sets inside U. These three sets divide the universal set into eight nonoverlapping regions which can be numbered (arbitrarily) as in Figure 8.11.

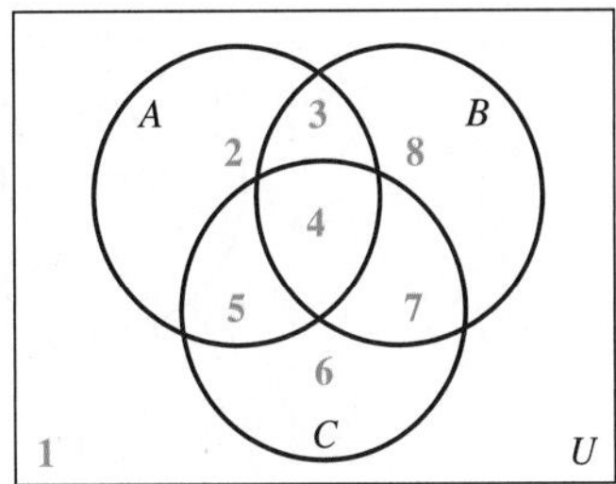

Three sets leads to 8 regions.

FIGURE 8.11

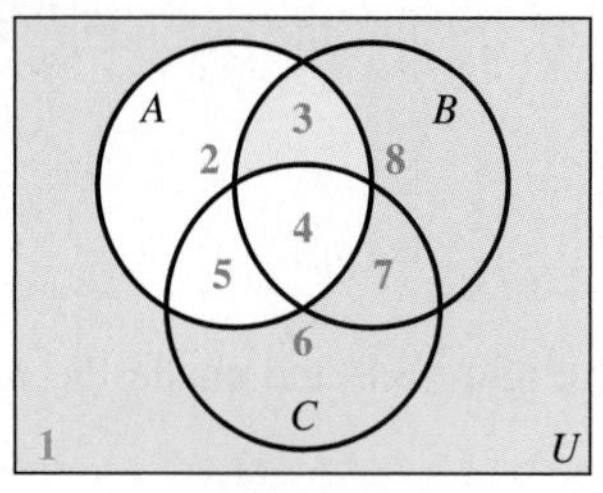

$A' \cup (B \cap C')$

FIGURE 8.12

EXAMPLE 2 Shade $A' \cup (B \cap C')$ in a Venn diagram.

First find $B \cap C'$. See Figure 8.12. Set B is represented by regions 3, 4, 7, and 8, and C' by regions 1, 2, 3, and 8. The overlap of these regions, regions 3 and 8, represents the set $B \cap C'$. Set A' is represented by regions 1, 6, 7, and 8. The union of regions 3 and 8 and regions 1, 6, 7, and 8 includes regions 1, 3, 6, 7, and 8, which are shaded in Figure 8.12. ◀ (2)

(2)

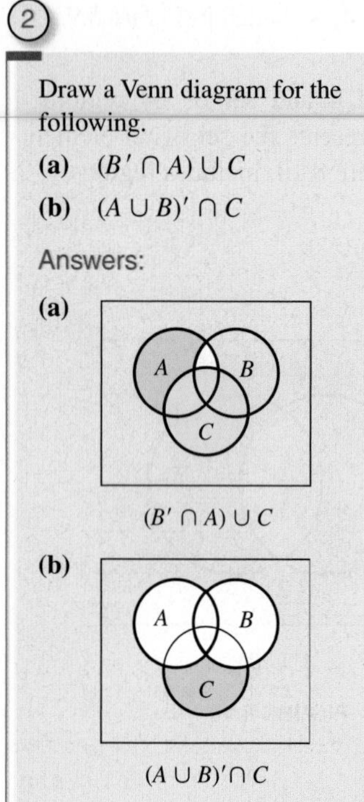

Draw a Venn diagram for the following.

(a) $(B' \cap A) \cup C$

(b) $(A \cup B)' \cap C$

Answers:

(a)

$(B' \cap A) \cup C$

(b)

$(A \cup B)' \cap C$

Venn diagrams can be used to solve problems that result from surveying groups of people. As an example, suppose a researcher collecting data on 100 households finds that

21 have a DVD player;

56 have a videocassette recorder (VCR); and

12 have both.

The researcher wants to answer the following questions.

(a) How many do not have a VCR?

(b) How many have neither a DVD player nor a VCR?

(c) How many have a DVD player but not a VCR?

A Venn diagram like the one in Figure 8.13 will help sort out the information. In Figure 8.13(a), we put the number 12 in the region common to both a VCR and a DVD player, because 12 households have both. Of the 21 with a DVD player, $21 - 12 = 9$ have no VCR, so in Figure 8.13(b) we put 9 in the region for a DVD player but no VCR. Similarly, $56 - 12 = 44$ households have a VCR but not a DVD player, so we put 44 in that region. Finally, the diagram shows that $100 - 44 - 12 - 9 = 35$ households have neither a VCR nor a DVD player. Now we can answer the questions:

(a) $35 + 9 = 44$ do not have a VCR;

(b) 35 have neither;

(c) 9 have a DVD player but not a VCR. 

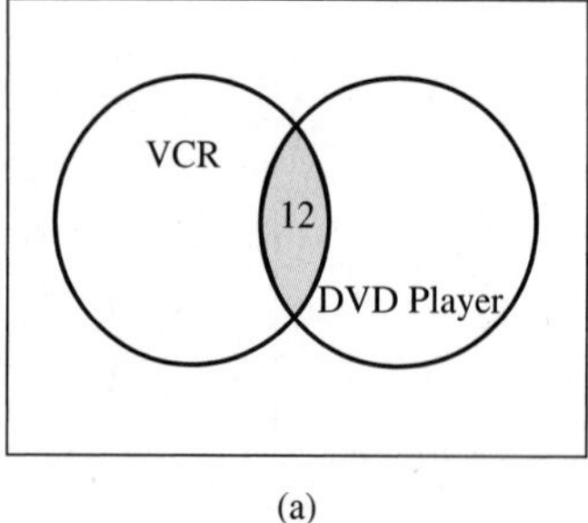

(a)

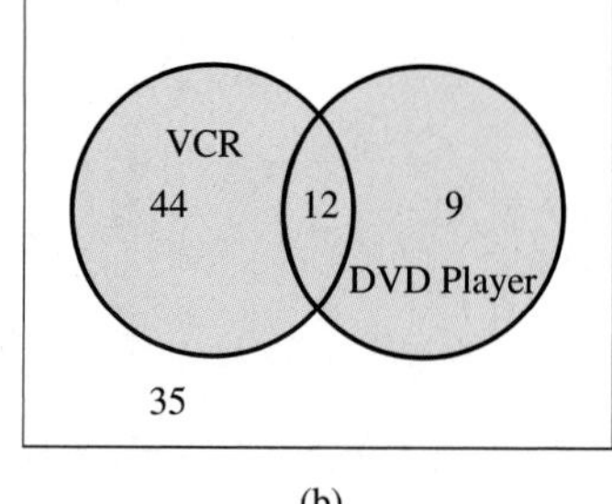

(b)

FIGURE 8.13

3

(a) Place numbers in the regions on a Venn diagram if the data on the 100 households showed

29 DVD players;

63 VCRs;

20 with both.

(b) How many have a VCR but not a DVD player?

Answers:

(a)

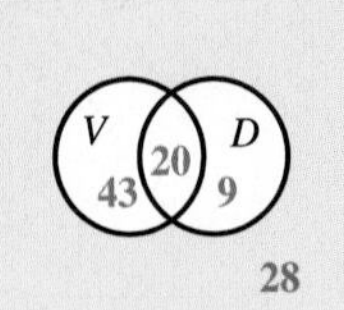

(b) 43

EXAMPLE 3 A group of 60 freshman business students at a large university was surveyed, with the following results.

19 of the students read *Business Week;*

18 read the *Wall Street Journal;*

50 read *Fortune;*

13 read *Business Week* and the *Journal;*

11 read the *Journal* and *Fortune;*

13 read *Business Week* and *Fortune;*

9 read all three.

Use this data to answer the following questions.

(a) How many students read none of the publications?

(b) How many read only *Fortune?*

(c) How many read *Business Week* and the *Journal* but not *Fortune?*

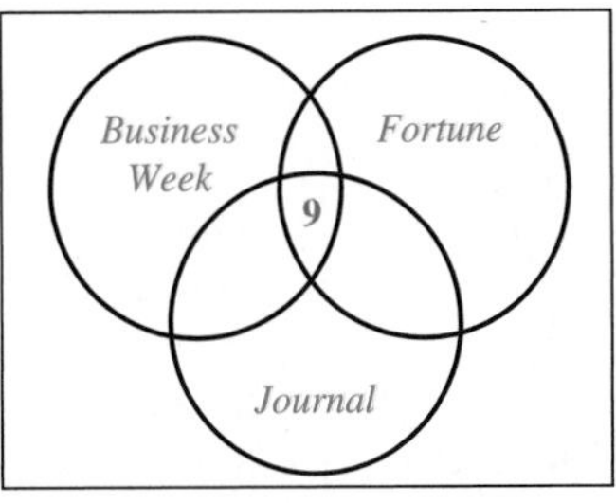

FIGURE 8.14(a)

Once again, use a Venn diagram to represent the data. Since 9 students read all three publications, begin by placing 9 in the area in Figure 8.14(a) that belongs to all three regions.

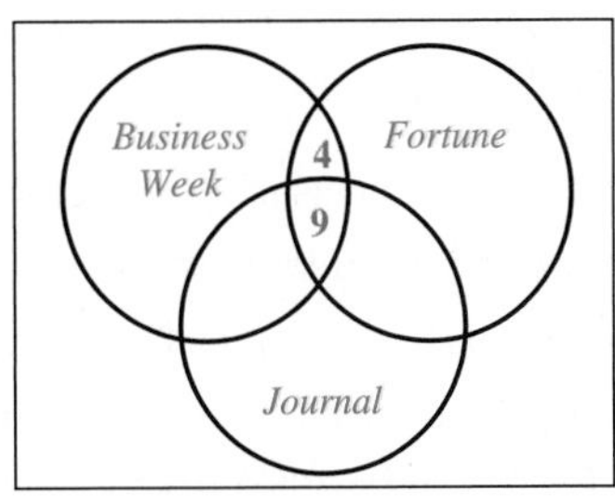

FIGURE 8.14(b)

Of the 13 students who read *Business Week* and *Fortune,* 9 also read the *Journal.* Therefore only $13 - 9 = 4$ students read just *Business Week* and *Fortune.* So place a 4 in the region common only to *Business Week* and *Fortune* readers, as in Figure 8.14(b).

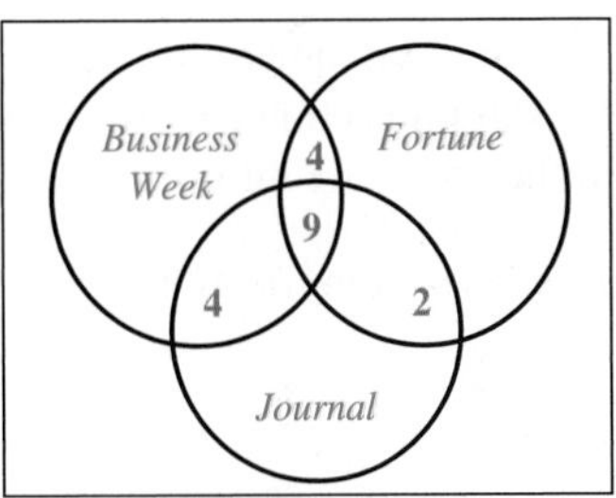

FIGURE 8.14(c)

In the same way, place a 4 in the region of Figure 8.14(c) common only to *Business Week* and the *Journal,* and 2 in the region common only to *Fortune* and the *Journal.*

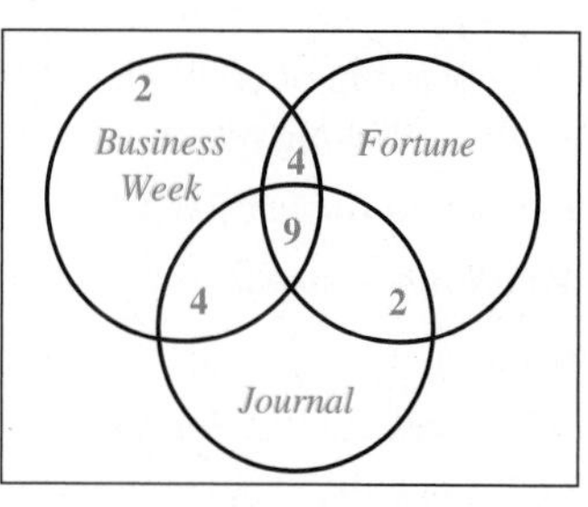

FIGURE 8.14(d)

The data shows that 19 students read *Business Week*. However, $4 + 9 + 4 = 17$ readers have already been placed in the *Business Week* region. The balance of this region in Figure 8.14(d) will contain only $19 - 17 = 2$ students. These 2 students read *Business Week* only—not *Fortune* and not the *Journal*.

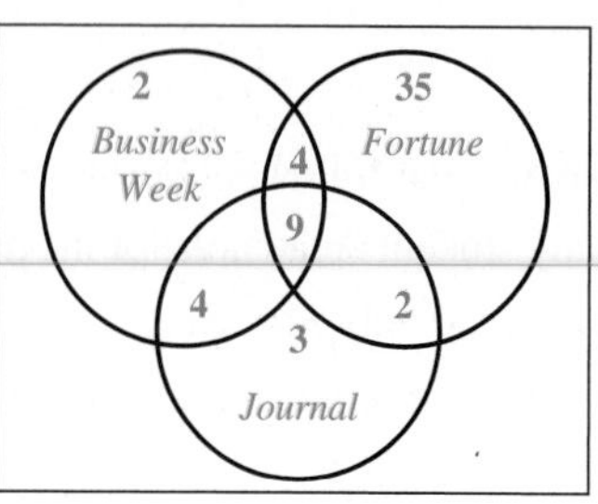

FIGURE 8.14(e)

In the same way, 3 students read only the *Journal* and 35 read only *Fortune*, as shown in Figure 8.14(e).

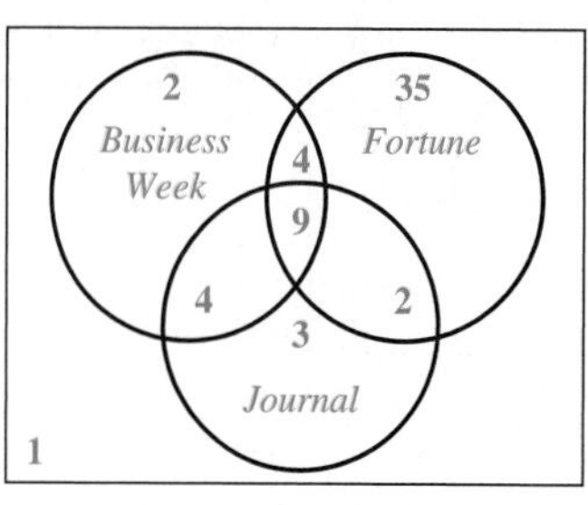

FIGURE 8.14(f)

A total of $2 + 4 + 3 + 4 + 9 + 2 + 35 = 59$ students are placed in the various regions of Figure 8.14(e). Since 60 students were surveyed, $60 - 59 = 1$ student reads none of the three publications and 1 is placed outside the other regions in Figure 8.14(f).

In the example about the three publications, how many students read exactly

(a) 1 of the publications?

(b) 2 of the publications?

Answers:

(a) 40

(b) 10

Figure 8.14(f) can now be used to answer the questions asked above.

(a) Only 1 student reads none of the publications.

(b) There are 35 students who read only *Fortune*.

(c) The overlap of the regions representing *Business Week* and the *Journal* shows that 4 students read *Business Week* and the *Journal* but not *Fortune*. ◀ (4)

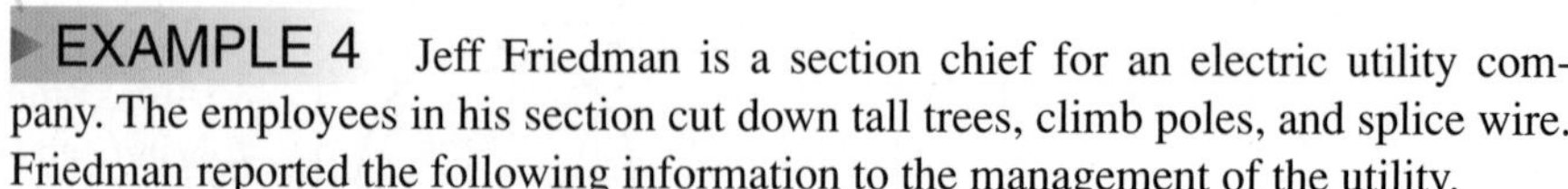

EXAMPLE 4 Jeff Friedman is a section chief for an electric utility company. The employees in his section cut down tall trees, climb poles, and splice wire. Friedman reported the following information to the management of the utility.

Of the 100 employees in my section,

45 can cut tall trees;

50 can climb poles;

57 can splice wire;

28 can cut trees and climb poles;

5

In Example 4, suppose 46 employees can cut tall trees. Then how many

(a) can only cut tall trees?

(b) can cut trees or climb poles?

(c) can cut trees or climb poles or splice wire?

Answers:

(a) 4

(b) 68

(c) 91

20 can climb poles and splice wire;

25 can cut trees and splice wire;

11 can do all three;

9 can't do any of the three (management trainees).

The data supplied by Friedman leads to the numbers shown in Figure 8.15. Add the numbers from all the regions to get the total number of Friedman's employees.

$$9 + 3 + 14 + 23 + 11 + 9 + 17 + 13 = 99$$

Friedman claimed to have 100 employees, but his data indicates only 99. The management decided that Friedman didn't qualify as a section chief and reassigned him as a nightshift meter reader in Guam. (Moral: He should have taken this course.) ◀ 5

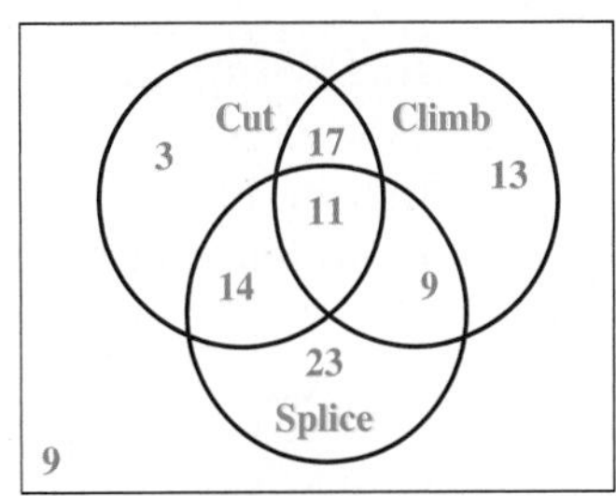

FIGURE 8.15

NOTE In all the examples above, we started in the innermost region with the intersection of the categories. This is usually the best way to begin solving problems of this type.

We use the symbol $n(A)$ to denote the *number* of elements in A. For instance, if $A = \{w, x, y, z\}$, then $n(A) = 4$. The following useful fact is proved below.

Union Rule for Counting

$$n(A \cup B) = n(A) + n(B) - n(A \cap B)$$

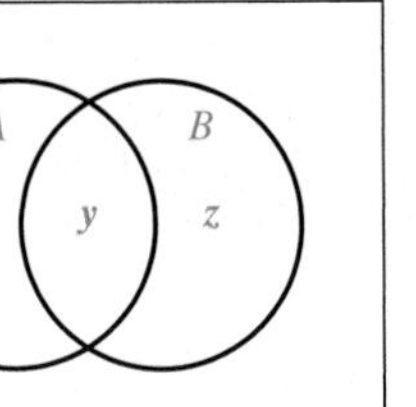

FIGURE 8.16

For example, if $A = \{r, s, t, u, v\}$ and $B = \{r, t, w\}$, then $A \cap B = \{r, t\}$, so that $n(A) = 5$, $n(B) = 3$, and $n(A \cap B) = 2$. By the formula in the box, $n(A \cup B) = 5 + 3 - 2 = 6$, which is certainly true since $A \cup B = \{r, s, t, u, v, w\}$.

Here is a proof of the statement in the box: let x be the number of elements in A that are not in B, y be the number of elements in $A \cap B$, and z be the number of elements in B that are not in A, as indicated in Figure 8.16. That diagram shows that $n(A \cup B) = x + y + z$. It also shows that $n(A) = x + y$ and $n(B) = y + z$, so that

$$\begin{aligned} n(A) + n(B) - n(A \cap B) &= (x + y) + (z + y) - y \\ &= x + y + z \\ &= n(A \cup B). \end{aligned}$$

EXAMPLE 5 A group of 10 students meets to plan a school function. All are majoring in accounting or economics or both. Five of the students are economics majors and 7 are majors in accounting. How many major in both subjects?

Let A represent the set of accounting majors and B represent the set of economics majors. Use the union rule, with $n(A) = 5$, $n(B) = 7$, and $n(A \cup B) = 10$. We must find $n(A \cap B)$.

$$n(A \cup B) = n(A) + n(B) - n(A \cap B)$$
$$10 = 5 + 7 - n(A \cap B),$$

so

$$n(A \cap B) = 5 + 7 - 10 = 2. \quad ⑥$$

⑥

If $n(A) = 10$, $n(B) = 7$, and $n(A \cap B) = 3$, find $n(A \cup B)$.

Answer:

14

EXAMPLE 6 Suppose a random sample of 200 voters was selected in the 1994 elections for the U.S. House of Representatives. The approximate numbers from the East, Midwest, South, or West who voted Democrat or Republican is given by the following table.*

	East (E)	*Midwest (M)*	*South (S)*	*West (W)*	*Total*
Democrat (*D*)	24	22	27	26	99
Republican (*R*)	22	28	33	18	101
Total	46	50	60	44	200

Using the letters given in the table, find the number of people in each of the following sets.

(a) $D \cap S$

The set $D \cap S$ consists of all those who voted Democrat *and* were from the South. From the table, we see that there were 27 such people.

(b) $D \cup S$

The set $D \cup S$ consists of all those who voted Democrat *or* were from the South. We include all 99 who voted Democrat, plus the 33 who were from the South and did not vote Democrat, for a total of 132. Alternatively, we could use the formula $n(D \cup S) = n(D) + n(S) - n(D \cap S) = 99 + 60 - 27 = 132$.

(c) $(E \cup W) \cap R'$

Begin with the set $E \cup W$, which is everyone from the East or the West. This consists of the four categories with 24, 22, 26, and 18 people. Of this set, take those who did *not* vote Republican, for a total of $24 + 26 = 50$ people. This is the number of people from the East or the West who did not vote Republican. ⑦

⑦

Find the number of people in each set.

(a) $R \cup M$

(b) $(R \cap W) \cup E'$

Answers:

(a) 123

(b) 154

*New York Times, November 13, 1994, p. 24.

8.2 Exercises

Sketch a Venn diagram like the one below and use shading to show each of the following sets. (See Example 1.)

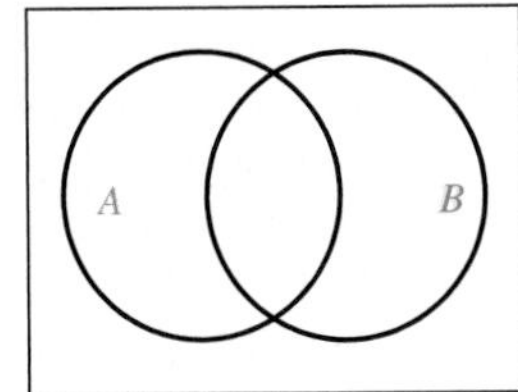

1. $B \cap A'$ 2. $A \cup B'$ 3. $A' \cup B$ 4. $A' \cap B'$
5. $B' \cup (A' \cap B')$ 6. $(A \cap B) \cup B'$
7. U' 8. $\emptyset'$
9. Three sets divide the universal set into at most _____ regions.
10. What does the notation $n(A)$ represent?

Sketch a Venn diagram like the one shown and use shading to show each of the following sets. (See Example 2.)

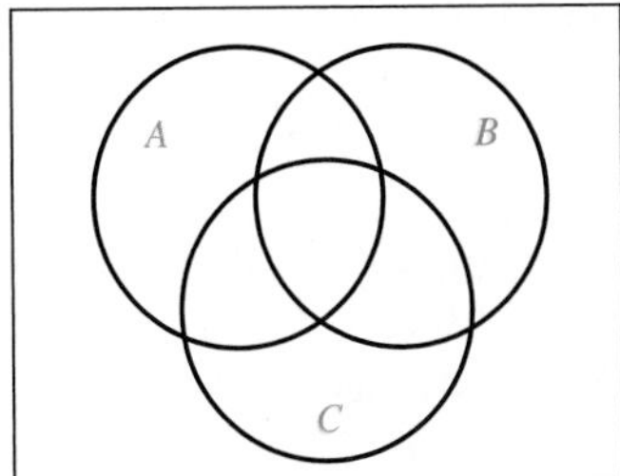

11. $(A \cap C') \cup B$ 12. $A \cap (B \cup C')$
13. $A' \cap (B \cap C)$ 14. $(A' \cap B') \cap C$
15. $(A \cap B') \cup C$ 16. $(A \cap B') \cap C$

Use Venn diagrams to answer the following questions. (See Examples 2 and 4.)

17. Social Science In 2000, the percent of white children under 18 years of age who lived with both parents was 75.3; the percent of white children under 18 years of age who lived with their father only was 4.3; and the percent of white children under 18 years of age who lived with neither parent was 3.1.* What percent of white children under age 18 lived with their mother only?

18. Social Science In 2000, there were 3842 (in thousands) children under the age of 18 living with their grandparents. Of these children, 531 had both parents also living with them, 1732 had only their mother living with them, and 220 had only their father living with them.* How many children lived with their grandparents only?

19. Social Science A survey of people attending a Lunar New Year celebration in Chinatown yielded the following results:

    120 were women;
    150 spoke Cantonese;
    170 lit firecrackers;
    108 of the men spoke Cantonese;
    100 of the men did not light firecrackers;
    18 of the non-Cantonese-speaking women lit firecrackers;
    78 non-Cantonese-speaking men did not light firecrackers;
    30 of the women who spoke Cantonese lit firecrackers.

    (a) How many attended?
    (b) How many of those who attended did not speak Cantonese?
    (c) How many women did not light firecrackers?
    (d) How many of those who lit firecrackers were Cantonese-speaking men?

20. Business Jeff Friedman, of Example 4 in the text, was again reassigned, this time to the home economics department of the electric utility. He interviewed 140 people in a suburban shopping center to find out some of their cooking habits. He obtained the following results. Should he be reassigned yet one more time?

    58 use microwave ovens;
    63 use electric ranges;
    58 use gas ranges;
    19 use microwave ovens and electric ranges;
    17 use microwave ovens and gas ranges;
    4 use both gas and electric ranges;
    1 uses all three;
    2 cook only with solar energy.

21. Natural Science After a genetics experiment, the number of pea plants having certain characteristics was tallied, with the results as follows.

    22 were tall;
    25 had green peas;
    39 had smooth peas;
    9 were tall and had green peas;
    17 were tall and had smooth peas;
    20 had green peas and smooth peas;

*U.S. Bureau of the Census, *America's Families and Living Arrangements 2000.*

*U.S. Bureau of the Census.

6 had all three characteristics;
4 had none of the characteristics.

(a) Find the total number of plants counted.
(b) How many plants were tall and had peas that were neither smooth nor green?
(c) How many plants were not tall but had peas that were smooth and green?

22. Health Human blood can contain either no antigens, the A antigen, the B antigen, or both the A and B antigens. A third antigen, called the Rh antigen, is important in human reproduction, and again may or may not be present in an individual. Blood is called type A-positive if the individual has the A and Rh antigens, but not the B antigen. A person having only the A and B antigens is said to have type AB-negative blood. A person having only the Rh antigen has type O-positive blood. Other blood types are defined in a similar manner. Identify the blood type of the individuals in regions (a)–(g) of the Venn diagram.

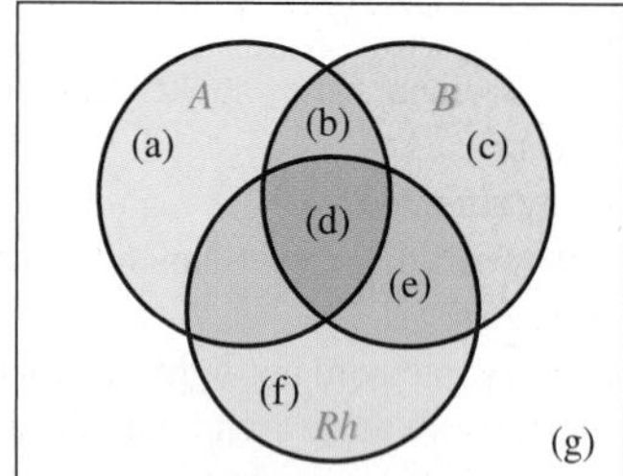

23. Natural Science Use the diagram from Exercise 22. In a certain hospital, the following data was recorded.

25 patients had the A antigen;
17 had the A and B antigens;
27 had the B antigen;
22 had the B and Rh antigens;
30 had the Rh antigen;
12 had none of the antigens;
16 had the A and Rh antigens;
15 had all three antigens.

How many patients
(a) were represented?
(b) had exactly one antigen?
(c) had exactly two antigens?
(d) had O-positive blood?
(e) had AB-positive blood?
(f) had B-negative blood?
(g) had O-negative blood?
(h) had A-positive blood?

24. Social Science At a pow-wow in Arizona, Native Americans from all over the Southwest came to participate in the ceremonies. A coordinator of the pow-wow took a survey and found that

15 families brought food, costumes, and crafts;
25 families brought food and crafts;
42 families brought food;
20 families brought costumes and food;
6 families brought costumes and crafts but not food;
4 families brought crafts but neither food nor costumes;
10 families brought none of the three items;
18 families brought costumes but not crafts.

(a) How many families were surveyed?
(b) How many families brought costumes?
(c) How many families brought crafts, but not costumes?
(d) How many families did not bring crafts?
(e) How many families brought food or costumes?

25. Social Science The following table lists the number of head and neck injuries for 319 ice hockey players wearing either a full shield or half shield in the Canadian Inter-University Athletics Union during the 1997–98 season.* Using the letters given in the table, find the number of injuries in each set.

	Half Shield (*H*)	*Full Shield* (*F*)
Head and Face Injuries (A)	95	34
Concussions (B)	41	38
Neck Injuries (C)	9	7
Other Injuries (D)	202	150

(a) $A \cap F$ (b) $C \cap (H \cup F)$
(c) $D \cup F$ (d) $B' \cap C'$

26. Social Science The number of female military personnel on active duty in September 1998 is given in the table.†

	Army (*A*)	*Air Force* (*B*)	*Navy* (*C*)	*Marines* (*D*)	*Total*
Officers (O)	10,367	11,971	7777	854	30,969
Enlisted (E)	60,787	53,542	42,261	8928	165,518
Cadets and Midshipmen (M)	624	653	656	0	1,933
Total	71,778	66,166	50,694	9782	198,420

Use this information and the letters given to find the number of female military personnel in each of the following sets.

(a) $A \cup B$ (b) $E \cup (C \cup D)$ (c) $O' \cap M'$

*Brian Benson, Nicholas Nohtaki, M. Sarah Rose, and Willem Meeuwisse, "Head and Neck Injuries Among Ice Hockey Players Wearing Full Face Shields vs. Half Face Shields," *JAMA* 282, no. 24 December 22/29, pp. 2328–2332.

†Borga Brunner, ed., *Time Almanac 2000*, p. 395.

27. Social Science The U.S. population by age and race in 2000 (in millions) is given in the table.*

	Non-Hispanic White (A)	Hispanic (B)	African-American (C)	Asian-American (D)	Totals
Under 45 (E)	120.1	25.7	24.4	7.6	177.8
45–64 (F)	47.5	4.9	6.3	2.2	60.9
65 and Over (G)	29.1	1.9	2.8	0.8	34.6
Totals	196.7	32.5	33.5	10.6	273.3

Using the letters given in the table, find the number of people in each set. (See Example 6.)

(a) $A \cap E$ **(b)** $F \cup B$ **(c)** $F \cup (C \cap G)$
(d) $E' \cap (B \cup G)$ **(e)** $G' \cup D$ **(f)** $F' \cap (A' \cap C')$

28. Health The table shows the numbers of breast cancers diagnosed among women in the California Teachers Study from 1996 through 1998.

Race (A)	No Breast Cancer (B)	Breast Cancer (C)	Total
White (D)	99,649	1236	100,885
Black (E)	3,083	36	3,119
Hispanic (F)	5,024	35	5,059
Asian/Pacific Islander (G)	4,167	45	4,212
Other (H)	2,500	21	2,521
Total	114,423	1373	115,796

Using the letters given in the table, find the number of women in each set. (See Example 6.)

(a) $G \cap C$ **(b)** $B \cup F$
(c) $(D \cup E) \cap B$ **(d)** $(D' \cap C)'$

*U.S. Bureau of the Census, *Current Population Reports* (Projections).

29. Restate the union rule in words.

Use Venn diagrams to answer the following questions. (See Example 5.)

30. If $n(A) = 5$, $n(B) = 8$, and $n(A \cap B) = 4$, what is $n(A \cup B)$?

31. If $n(A) = 12$, $n(B) = 27$, and $n(A \cup B) = 30$, what is $n(A \cap B)$?

32. Suppose $n(B) = 7$, $n(A \cap B) = 3$, and $n(A \cup B) = 20$. What is $n(A)$?

33. Suppose $n(A \cap B) = 5$, $n(A \cup B) = 35$, and $n(A) = 13$. What is $n(B)$?

Draw a Venn diagram and use the given information to fill in the number of elements for each region.

34. $n(U) = 38$, $n(A) = 16$, $n(A \cap B) = 12$, $n(B') = 20$

35. $n(A) = 26$, $n(B) = 10$, $n(A \cup B) = 30$, $n(A') = 17$

36. $n(A \cup B) = 17$, $n(A \cap B) = 3$, $n(A) = 8$, $n(A' \cup B') = 21$

37. $n(A') = 28$, $n(B) = 25$, $n(A' \cup B') = 45$, $n(A \cap B) = 12$

38. $n(A) = 28$, $n(B) = 34$, $n(C) = 25$, $n(A \cap B) = 14$, $n(B \cap C) = 15$, $n(A \cap C) = 11$, $n(A \cap B \cap C) = 9$, $n(U) = 59$

39. $n(A) = 54$, $n(A \cap B) = 22$, $n(A \cup B) = 85$, $n(A \cap B \cap C) = 4$, $n(A \cap C) = 15$, $n(B \cap C) = 16$, $n(C) = 44$, $n(B') = 63$

*In Exercises 40–43, show that the statements are true by drawing Venn diagrams and shading the regions representing the sets on each side of the equals signs.**

40. $(A \cup B)' = A' \cap B'$

41. $(A \cap B)' = A' \cup B'$

42. $A \cap (B \cup C) = (A \cap B) \cup (A \cap C)$

43. $A \cup (B \cap C) = (A \cup B) \cap (A \cup C)$

*The statements in Exercises 40 and 41 are known as De Morgan's laws. They are named for the English mathematician Augustus De Morgan (1806–71).

8.3 INTRODUCTION TO PROBABILITY

If you go to a supermarket and buy 5 pounds of peaches at $1.26 per pound, you can easily find the *exact* price of your purchase: $6.30. On the other hand, the produce manager of the market is faced with the problem of ordering peaches. The manager may have a good estimate of the number of pounds of peaches that will be sold during the day, but it is impossible to predict the *exact* amount. The number of pounds that customers will purchase during a day is *random:* the quantity cannot be predicted exactly. A great many problems that come up in applications of mathematics involve random phenomena—those for which exact prediction is impossible. The best that we can do is determine the *probability* of the possible outcomes.

Sample Spaces

In probability, an **experiment** is an activity or occurrence with an observable result. Each repetition of an experiment is called a **trial.** The possible results of each trial are called **outcomes.** The set of all possible outcomes for an experiment is the **sample space** for that experiment. A sample space for the experiment of tossing a coin is made up of the outcomes heads (h) and tails (t). If S represents this sample space, then

$$S = \{h, t\}.$$

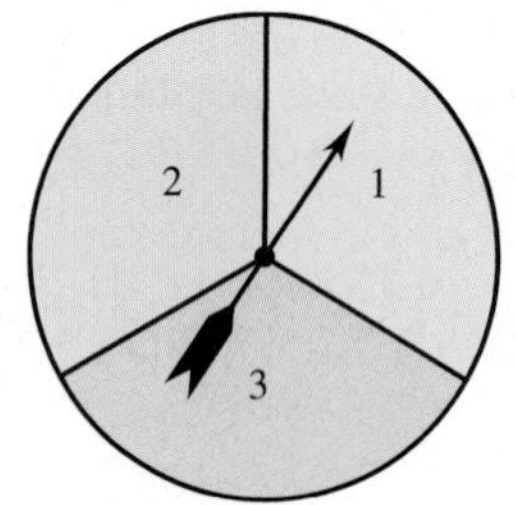

FIGURE 8.17

EXAMPLE 1 Give the sample space for each experiment.

(a) A spinner like the one in Figure 8.17 is spun.
The three outcomes are 1, 2, or 3, so the sample space is

$$\{1, 2, 3\}$$

(b) For the purposes of a public opinion poll, respondents are classified as young, middle-aged, or senior and as male or female.

A sample space for this poll could be written as a set of orderend pairs:

{(young, male), (young, female), (middle-aged, male),
(middle-aged, female), (senior, male), (senior, female)}

(c) An experiment consists of studying the numbers of boys and girls in families with exactly 3 children. Let b represent *boy* and g represent *girl.*

A three-child family can have 3 boys, written *bbb,* 3 girls, *ggg,* or various combinations, such as *bgg.* A sample space with four outcomes (not equally likely) is

$$S_1 = \{3 \text{ boys}, 2 \text{ boys and } 1 \text{ girl}, 1 \text{ boy and } 2 \text{ girls}, 3 \text{ girls}\}.$$

Notice that a family with 3 boys or 3 girls can occur in just one way, but a family of two boys and 1 girl or 1 boy and 2 girls can occur in more than one way. If the order of the births is considered, so that *bgg* is different from *gbg* or *ggb,* for example, another sample space is

$$S_2 = \{bbb, bbg, bgb, gbb, bgg, gbg, ggb, ggg\}.$$

The second sample space, S_2, has equally likely outcomes if we assume that having a boy or a girl is equally likely. This assumption, while not quite true, is approximately true, so we will use it throughout this book. The outcomes in S_1 are not equally likely, since there is more than one way to get a family with 2 boys and 1 girl (*bbg, bgb,* or *gbb*) or a family with 2 girls and 1 boy (*ggb, gbg,* or *bgg*), but only one way to get 3 boys (*bbb*) or 3 girls (*ggg*).

1

(a) Write an equally likely sample space for the experiment of tossing 2 fair coins.

(b) Write a sample space for the experiment of tossing 2 fair coins, if we are interested only in the number of heads. Are the outcomes in this sample space equally likely?

Answers:

(a) {hh, ht, th, tt}

(b) {0, 1, 2}; no

CAUTION An experiment may have more than one sample space, as shown in Example 1(c). The most convenient sample spaces have equally likely outcomes, but it is not always possible to choose such a sample space. (1)

Events

An **event** is a subset of a sample space. If the sample space for tossing a coin is $S = \{h, t\}$, then one event is $E = \{h\}$, which represents the outcome "heads."

An ordinary die is a cube whose six different faces show the following numbers of dots: 1, 2, 3, 4, 5, and 6. If the die is fair (not "loaded" to favor certain faces over others), then any one of the faces is equally likely to come up when the die is rolled. The sample space for the experiment of rolling a single fair die is $S = \{1, 2, 3, 4, 5, 6\}$. Some possible events are listed below.

The die shows an even number: $E_1 = \{2, 4, 6\}$.

The die shows a 1: $E_2 = \{1\}$.

The die shows a number less than 5: $E_3 = \{1, 2, 3, 4\}$.

The die shows a multiple of 3: $E_4 = \{3, 6\}$.

Suppose a die is tossed. Write the following events.

(a) The number showing is less than 3.

(b) The number showing is 5.

(c) The number showing is 8.

Answers:

(a) {1, 2}

(b) {5}

(c) ∅

EXAMPLE 2 For the sample space S_2 in Example 1(c), write the following events.

(a) Event H: the family has exactly 2 girls.
Families with three children can have exactly 2 girls with either *bgg*, *gbg*, or *ggb*, so that event H is

$$H = \{bgg, gbg, ggb\}.$$

(b) Event K: the three children are the same sex.
Two outcomes satisfy this condition, all boys or all girls.

$$K = \{bbb, ggg\}$$

(c) Event J: the family has three girls.
Only *ggg* satisfies this condition, so

$$J = \{ggg\}.$$ (2)

In Example 2(c), event J had only one possible outcome, *ggg*. Such an event, with only one possible outcome, is a **simple event.** If an event E equals the sample space S, then E is a **certain event.** If event $E = \emptyset$, then E is an **impossible event.**

EXAMPLE 3 Suppose a fair die is rolled. As shown above, the sample space is $\{1, 2, 3, 4, 5, 6\}$.

(a) The event "the die shows a 4," $\{4\}$, has only one possible outcome. It is a simple event.

(b) The event "the number showing is less than 10" equals the sample space $S = \{1, 2, 3, 4, 5, 6\}$. This event is a certain event; if a die is rolled the number showing (either 1, 2, 3, 4, 5, or 6) must be less than 10.

(3)

Which of the events listed in side problem 2 is

(a) simple?

(b) certain?

(c) impossible?

Answers:

(a) Part (b)

(b) None

(c) Part (c)

(c) The event "the die shows a 7" is the empty set, $\emptyset$; this is an impossible event. (3)

Since events are sets, we can use set operations to find unions, intersections, and complements of events. A summary of the set operations for events is given below.

Set Operations for Events

Let E and F be events for a sample space S.

$E \cap F$ occurs when both E **and** F occur;

$E \cup F$ occurs when E **or** F **or both** occur;

E' occurs when E does **not** occur.

EXAMPLE 4 A study of workers earning the minimum wage grouped such workers into various categories, which can be interpreted as events when a worker is selected at random.* Consider the following events:

E: worker is under 20;

F: worker is white;

G: worker is female.

Give the following events for the experiment of rolling a fair die if $E = \{1, 3\}$ and $F = \{2, 3, 4, 5\}$.

(a) $E \cap F$

(b) $E \cup F$

(c) E'

Answers:

(a) $\{3\}$

(b) $\{1, 2, 3, 4, 5\}$

(c) $\{2, 4, 5, 6\}$

Describe each of the following events in words.

(a) E'

E' is the event that the worker is 20 or over.

(b) $F \cap G'$

$F \cap G'$ is the event that the worker is white and not a female, that is, the worker is a white male.

(c) $E \cup G$

$E \cup G$ is the event that the worker is under 20 or is female. Note that this event includes all workers under 20, both male and female, and all female workers of any age. (4)

Two events that cannot both occur at the same time, such as getting both a head and a tail on the same toss of a coin, are called **mutually exclusive events.**

Mutually Exclusive Events

Events E and F are mutually exclusive events if $E \cap F = \emptyset$.

For any event E, E and E' are mutually exclusive. By definition, mutually exclusive events are disjoint sets.

*_New York Times_, April 19, 1996, p. A26.

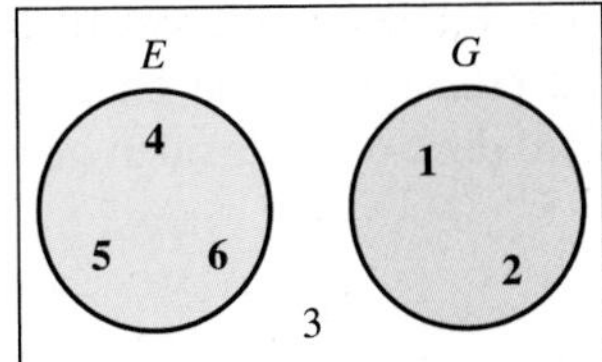

FIGURE 8.18

EXAMPLE 5 Let $S = \{1, 2, 3, 4, 5, 6\}$, the sample space for tossing a die. Let $E = \{4, 5, 6\}$ and let $G = \{1, 2\}$. Then E and G are mutually exclusive events because they have no outcomes in common; $E \cap G = \emptyset$. See Figure 8.18. ◀ 5

In Example 5, let $F = \{2, 4, 6\}$, $K = \{1, 3, 5\}$, and G remain the same. Are the following events mutually exclusive?

(a) F and K

(b) F and G

Answers:

(a) Yes

(b) No

Probability

For sample spaces with *equally likely* outcomes, the probability of an event is defined as follows.

Basic Probability Principle

Let S be a sample space of equally likely outcomes and let event E be a subset of S. Then the **probability that event E occurs** is

$$P(E) = \frac{n(E)}{n(S)}.$$

By this definition, the **probability of an event** is a number that indicates the relative likelihood of the event.

CAUTION The basic probability principle only applies when the outcomes are equally likely.

EXAMPLE 6 Suppose a single fair die is rolled. Use the sample space $S = \{1, 2, 3, 4, 5, 6\}$ and give the probability of each of the following events.

(a) E: the die shows an even number.
Here, $E = \{2, 4, 6\}$, a set with three elements. Because S contains six elements,

$$P(E) = \frac{3}{6} = \frac{1}{2}.$$

(b) F: the die shows a number less than 10.
Event F is a certain event, with

$$F = \{1, 2, 3, 4, 5, 6\},$$

so that

$$P(F) = \frac{6}{6} = 1.$$

(c) G: the die shows an 8.
This event is impossible, so

$$P(G) = \frac{0}{6} = 0.$$ ◀ 6

A fair die is rolled. Find the probability of rolling

(a) an odd number;

(b) 2, 4, 5, or 6;

(c) a number greater than 5;

(d) the number 7.

Answers:

(a) 1/2

(b) 2/3

(c) 1/6

(d) 0

A standard deck of 52 cards has four suits: hearts (♥), clubs (♣), diamonds (♦), and spades (♠), with 13 cards in each suit. The hearts and diamonds are red, and the spades and clubs are black. Each suit has an ace (A), a king (K), a queen (Q), a jack (J), and cards numbered from 2 to 10. The jack, queen, and king are called face cards and for many purposes can be thought of as having values 11, 12, and 13, respectively. The ace can be thought of as the low card (value 1) or the high card (value 14). See Figure 8.19. We will refer to this standard deck of cards often in our discussion of probability.

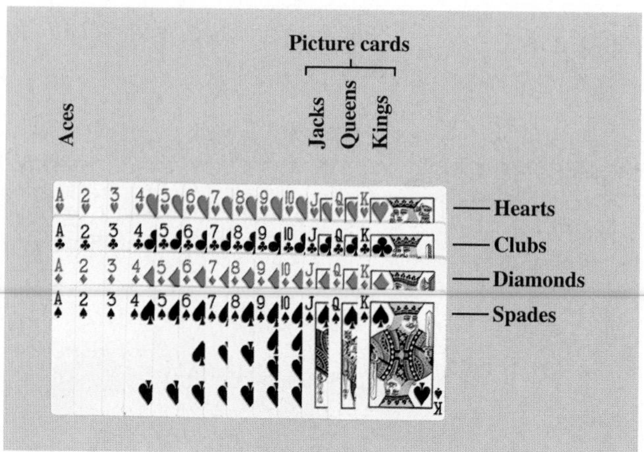

FIGURE 8.19

EXAMPLE 7 If a single card is drawn at random from a standard, well-shuffled, 52-card deck, find the probability of each of the following events.

(a) Drawing an ace
There are 4 aces in the deck. The event "drawing an ace" is

$$\{\text{heart ace, diamond ace, club ace, spade ace}\}.$$

Therefore,

$$P(\text{ace}) = \frac{4}{52} = \frac{1}{13}.$$

(b) Drawing a face card
Since there are 12 face cards,

$$P(\text{face card}) = \frac{12}{52} = \frac{3}{13}.$$

(c) Drawing a spade
The deck contains 13 spades, so

$$P(\text{spade}) = \frac{13}{52} = \frac{1}{4}.$$

(d) Drawing a spade or a heart
Besides the 13 spades, the deck contains 13 hearts, so

$$P(\text{spade or heart}) = \frac{26}{52} = \frac{1}{2}.$$ ◀ 7

7

A single playing card is drawn at random from an ordinary 52-card deck. Find the probability of drawing

(a) a queen;

(b) a diamond;

(c) a red card.

Answers:

(a) 1/13

(b) 1/4

(c) 1/2

In the preceding examples, the probability of each event was a number between 0 and 1, inclusive. The same thing is true in general. Any event E is the subset of the sample space S, so $0 \le n(E) \le n(S)$. Since $P(E) = n(E)/n(S)$, it follows that $0 \le P(E) \le 1$.

For any event E,

$$\mathbf{0 \le P(E) \le 1.}$$

EXAMPLE 8 The table gives the number of years of service of senators in the 107th Congress of the United States of America, which convened on January 3, 2001.*

Years of Service	*Number of Senators*
0–9	54
10–19	24
20–29	16
30–39	4
40 or more	2

8

Find the probability that a randomly selected senator of the 107th Congress had served 30 or more years when Congress convened.

Answer:

.06

Find the probability that a randomly selected senator of the 107th Congress served 20–29 years when Congress convened.

This probability is found by dividing the number of senators who served 20–29 years by the total number of senators. Thus,

$$P(20\text{–}29 \text{ years}) = \frac{16}{100} = .16.$$ 8

World Almanac and Book of Facts 2002, p. 83.

8.3 Exercises

1. What is meant by a "fair" coin or die?
2. What is the sample space for an experiment?

Write sample spaces for the experiments in Exercises 3–8. (See Example 1.)

3. A month of the year is chosen for a wedding.
4. A day in April is selected for a bicycle race.
5. A student is asked how many points she earned on a recent 80-point test.
6. A person is asked the number of hours (to the nearest hour) he watched television yesterday.
7. The management of an oil company must decide whether to go ahead with a new oil shale plant or to cancel it.
8. A coin is tossed and a die is rolled.

9. Define an event.

10. What is a simple event?

For the experiments in Exercises 11–16, write out an equally likely sample space, and then write the indicated events in set notation. (See Examples 2 and 3.)

11. A marble is drawn at random from a bowl containing 3 yellow, 4 white, and 8 blue marbles.
    (a) A yellow marble is drawn.
    (b) A blue marble is drawn.
    (c) A white marble is drawn.
    (d) A black marble is drawn.

12. Slips of paper marked with the numbers 1, 2, 3, 4, and 5 are placed in a box. After they are mixed, two slips are drawn.
    (a) Both slips are marked with even numbers.
    (b) One slip is marked with an odd number and the other is marked with an even number.
    (c) Both slips are marked with the same number.

13. An unprepared student takes a three-question true/false quiz in which he guesses the answers to all three questions.
    (a) The student gets three answers wrong.
    (b) The student gets exactly two answers correct.
    (c) The student gets only the first answer correct.

14. A die is tossed twice, with the tosses recorded as ordered pairs.
    (a) The first die shows a 3.
    (b) The sum of the numbers showing is 8.
    (c) The sum of the numbers showing is 13.

15. A jar contains four balls, labeled 1, 2, 3, and 4. A second jar contains five balls, labeled 1, 2, 3, 4, and 5. An experiment consists of taking one ball from the first jar, and then taking a ball from the second jar
    (a) The number on the first ball is even.
    (b) The number on the second ball is even.
    (c) The sum of the numbers on the two balls is 5.
    (d) The sum of the numbers on the two balls is 1.

16. From 5 employees, Strutz, Martin, Hampton, Williams, and Ewing, 2 are selected to attend a conference.
    (a) Hampton is selected.
    (b) Strutz and Martin are not both selected.
    (c) Both Williams and Ewing are selected.

A single fair die is rolled. Find the probabilities of the following events. (See Example 6.)

17. Getting a 5

18. Getting a number less than 4

19. Getting a number greater than 4

20. Getting a 2 or a 5

21. Getting a multiple of 3

22. Getting any number except 3

A card is drawn from a well-shuffled deck of 52 cards. Find the probability of drawing each of the following. (See Example 7.)

23. A 9

24. A black card

25. A black 9

26. A heart

27. The 9 of hearts

28. A 2 or a queen

29. A black 7 or a red 8

30. A red face card

A jar contains 5 red, 4 black, 7 purple, and 9 green marbles. If a marble is drawn at random, what is the probability that the marble is

31. red?

32. black?

33. green?

34. purple?

35. not black?

36. not purple?

37. red or black?

38. black or purple?

Solve each problem. (See Example 8.)

39. **Finance** As of July 2000, the Janus Mercury fund invested in equities throughout the world, as shown below.*

Region	*% of Equities*
Europe	18.69
Pacific Rim	6.44
United States	73.31
Latin America	1.56

Find the probability that a randomly selected equity would be from each of the following regions.
    (a) The Pacific Rim
    (b) Europe
    (c) The United States

40. **Health** For a medical experiment, people are classified as to whether they smoke, have a family history of heart disease, or are overweight. Define events E, F, and G as follows.

E: person smokes

F: person has a family history of heart disease

G: person is overweight

Describe each of the following events in words.
(a) G' **(b)** $F \cap G$ **(c)** $E \cup G'$

41. **Health** Refer to Exercise 40. Describe each of the following events in words.
(a) $E \cup F$ **(b)** $E' \cap F$ **(c)** $F' \cup G'$

*www.janus.com

42. Social Sciences The population of the United States by race in 2000, and the projected population by race for the year 2025 are given in the table at the right (in thousands).* Find the probability of a randomly selected person being each of the following.

(a) Hispanic in 2000

(b) Hispanic in 2025

(c) African-American in 2000

(d) African-American in 2025

Race	*2000*	*2025*
White	196,700	209,900
Hispanic	32,500	56,900
African-American	33,500	44,700
Asian-American	10,600	24,000
Other	2,100	2,800

*"Projections of the Total Resident Population by 5-Year Age Groups, Race, and Hispanic Origin, 1999 to 2000," from U.S. Bureau of the Census, *Current Population Reports*.

8.4 BASIC CONCEPTS OF PROBABILITY

We determine the probability of more complex events in this section.

To find the probability of the union of two sets E and F in a sample space, we use the union rule for counting given in Section 8.2.

$$n(E \cup F) = n(E) + n(F) - n(E \cap F).$$

Dividing both sides by $n(S)$ shows that

$$\frac{n(E \cup F)}{n(S)} = \frac{n(E)}{n(S)} + \frac{n(F)}{n(S)} - \frac{n(E \cap F)}{n(S)}$$

$$P(E \cup F) = P(E) + P(F) - P(E \cap F).$$

This discussion is summarized below.

Union Rule for Probability

For any events E and F from a sample space S,

$$\mathbf{P(E \cup F) = P(E) + P(F) - P(E \cap F).}$$

(Although the union rule applies to any events E and F from any sample space, the derivation we have given is valid only for sample spaces with equally likely simple events.)

EXAMPLE 1 If a single card is drawn from an ordinary deck of cards, find the probability that it will be red or a face card.

Let R represent the event "red card" and F the event "face card." There are 26 red cards in the deck, so $P(R) = 26/52$. There are 12 face cards in the deck, so

A single card is drawn from an ordinary deck. Find the probability that it is black or a 9.

Answer:

7/13

$P(F) = 12/52$. Since there are 6 red face cards in the deck, $P(R \cap F) = 6/52$. By the union rule, the probability that the card will be red or a face card is

$$P(R \cup F) = P(R) + P(F) - P(R \cap F)$$
$$= \frac{26}{52} + \frac{12}{52} - \frac{6}{52} = \frac{32}{52} = \frac{8}{13}.$$ ◀ 1

CAUTION Recall from Section 8.1, the word "or" always indicates a *union*.

EXAMPLE 2 Suppose two fair dice (plural of *die*) are rolled. Find each of the following probabilities.

(a) The first die shows a 2 or the sum of the results is 6 or 7.

The sample space for the throw of two dice is shown in Figure 8.20, where 1-1 represents the event "the first die shows a 1 and the second die shows a 1," 1-2 represents "the first die shows a 1 and the second die shows a 2," and so on. Let A represent the event "the first die shows a 2" and B represent the event "the sum of the results is 6 or 7." These events are indicated in Figure 8.20. From the diagram, event A has 6 elements, B has 11 elements, and the sample space has 36 elements. Thus,

$$P(A) = \frac{6}{36}, \quad P(B) = \frac{11}{36}, \quad \text{and} \quad P(A \cap B) = \frac{2}{36}.$$

By the union rule,

$$P(A \cup B) = P(A) + P(B) - P(A \cap B),$$
$$P(A \cup B) = \frac{6}{36} + \frac{11}{36} - \frac{2}{36} = \frac{15}{36} = \frac{5}{12}.$$

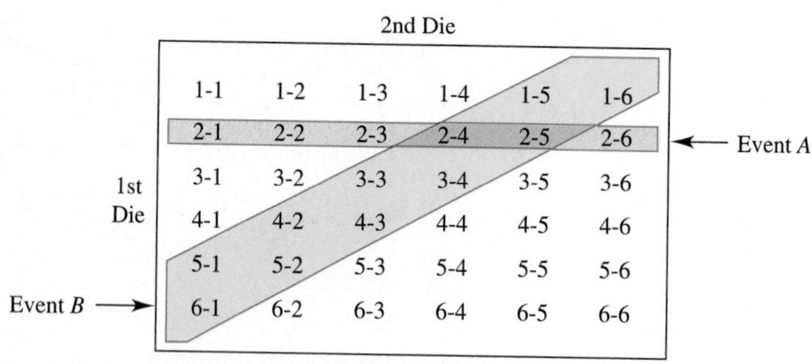

FIGURE 8.20

2

In the experiment of Example 2, find the following probabilities.

(a) The sum is 5 or the second die shows a 3.

(b) Both dice show the same number or the sum is at least 11.

Answers:

(a) 1/4

(b) 2/9

(b) The sum is 11 or the second die shows a 5.

$P(\text{sum is } 11) = 2/36$, $P(\text{second die shows } 5) = 6/36$, and $P(\text{sum is 11 and second die shows } 5) = 1/36$, so

$$P(\text{sum is 11 or second die shows } 5) = \frac{2}{36} + \frac{6}{36} - \frac{1}{36} = \frac{7}{36}.$$ ◀ 2

CAUTION You may wonder why we did not use $S = \{2,3,4,5, \ldots, 12\}$ as the sample space in Example 2. Remember, we prefer to use a sample space with equally likely outcomes. The outcomes in set S above are not equally likely—a sum of 2 can occur in just one way, a sum of 3 in two ways, a sum of 4 in three ways, and so on, as shown in Figure 8.20.

If events E and F are mutually exclusive, then $E \cap F = \emptyset$ by definition; hence; $P(E \cap F) = 0$. Applying the union rule yields this useful fact.

Union Rule for Mutually Exclusive Events

For mutually exclusive events E and F,

$$\mathbf{P(E \cup F) = P(E) + P(F).}$$

EXAMPLE 3 Assume that the probability of a couple having a boy is the same as the probability of their having a girl. If the couple has 3 children, find the probability that at least 2 of them are girls.

The event of having at least 2 girls is the union of the mutually exclusive events E = "the family has exactly 2 girls" and F = "the family has exactly 3 girls." Using the equally likely sample space

$$\{ggg, ggb, gbg, bgg, gbb, bgb, bbg, bbb\},$$

we see that $P(2 \text{ girls}) = 3/8$ and $P(3 \text{ girls}) = 1/8$. Therefore,

$$P(\text{at least 2 girls}) = P(2 \text{ girls}) + P(3 \text{ girls})$$
$$= \frac{3}{8} + \frac{1}{8} = \frac{1}{2}.$$ ◀ 3

3

In Example 3, find the probability of no more than 2 girls.

Answer:

7/8

By definition of E', for any event E from a sample space S,

$$E \cup E' = S \quad \text{and} \quad E \cap E' = \emptyset.$$

Because $E \cap E' = \emptyset$, events E and E' are mutually exclusive, so that

$$P(E \cup E') = P(E) + P(E').$$

However, $E \cup E' = S$, the sample space, and $P(S) = 1$. Thus

$$P(E \cup E') = P(E) + P(E') = 1.$$

Rearranging these terms gives the following useful rule.

Complement Rule

For any event E,

$$\mathbf{P(E') = 1 - P(E)} \quad \text{and} \quad \mathbf{P(E) = 1 - P(E').}$$

(a) Let $P(K) = 2/3$. Find $P(K')$.

(b) If $P(X') = 3/4$, find $P(X)$.

Answers:

(a) 1/3

(b) 1/4

EXAMPLE 4 If a fair die is rolled, what is the probability that any number but 5 will come up?

If E is the event that 5 comes up, then E' is the event that any number but 5 comes up. $P(E) = 1/6$, so we have $P(E') = 1 - 1/6 = 5/6$. ◀ 4

In Example 5, find the probability that the sum of the numbers rolled is at least 5.

Answer:

5/6

EXAMPLE 5 If two fair dice are rolled, find the probability that the sum of the numbers showing is greater than 3.

To calculate this probability directly, we must find the probabilities that the sum is 4, 5, 6, 7, 8, 9, 10, 11, or 12 and then add them. It is much simpler to first find the probability of the complement, the event that the sum is less than or equal to 3.

$$P(\text{sum} \leq 3) = P(\text{sum is } 2) + P(\text{sum is } 3)$$
$$= \frac{1}{36} + \frac{2}{36} = \frac{3}{36} = \frac{1}{12}$$

Now use the fact that $P(E) = 1 - P(E')$ to get

$$P(\text{sum} > 3) = 1 - P(\text{sum} \leq 3) = 1 - \frac{1}{12} = \frac{11}{12}.$$ ◀ 5

Odds

Sometimes probability statements are given in terms of **odds,** a comparison of $P(E)$ with $P(E')$. For example, suppose $P(E) = \frac{4}{5}$. Then $P(E') = 1 - \frac{4}{5} = \frac{1}{5}$. These probabilities predict that E will occur 4 out of 5 times and E' will occur 1 out of 5 times. Then we say the **odds in favor** of E are 4 to 1 or 4:1.

Odds

The **odds in favor** of an event E are defined as the ratio of $P(E)$ to $P(E')$, or

$$\frac{P(E)}{P(E')}, \qquad P(E') \neq 0.$$

Suppose $P(E) = 9/10$. Find the odds

(a) in favor of E;

(b) against E.

Answers:

(a) 9 to 1

(b) 1 to 9

EXAMPLE 6 Suppose the weather forecaster says that the probability of rain tomorrow is 1/3. Find the odds in favor of rain tomorrow.

Let E be the event "rain tomorrow." Then E' is the event "no rain tomorrow." Since $P(E) = 1/3$, $P(E') = 2/3$. By the definition of odds, the odds in favor of rain are

$$\frac{1/3}{2/3} = \frac{1}{2}, \quad \text{written 1 to 2} \quad \text{or} \quad 1:2.$$

On the other hand, the odds that it will *not* rain, or the odds *against* rain, are

$$\frac{2/3}{1/3} = \frac{2}{1}, \quad \text{written 2 to 1.}$$ ◀ 6

If the odds in favor of an event are, say, 3 to 5, then the probability of the event is 3/8, while the probability of the complement of the event is 5/8. (Odds of 3 to 5 indicate 3 outcomes in favor of the event out of a total of 8 outcomes.) This example suggests the following generalization.

If the odds favoring event E are m to n, then

$$P(E) = \frac{m}{m+n} \quad \text{and} \quad P(E') = \frac{n}{m+n}.$$

EXAMPLE 7 The odds that a particular bid will be the low bid are 4 to 5.

(a) Find the probability that the bid will be the low bid.

Odds of 4 to 5 show 4 favorable chances out of $4 + 5 = 9$ chances altogether.

$$P(\text{bid will be low bid}) = \frac{4}{4+5} = \frac{4}{9}$$

If the odds in favor of event E are 1 to 5, find

(a) $P(E)$;

(b) $P(E')$.

Answers:

(a) 1/6

(b) 5/6

(b) Find the odds against that bid being the low bid.

There is a 5/9 chance that the bid will not be the low bid, so the odds against a low bid are

$$\frac{P(\text{bid will not be low})}{P(\text{bid will be low})} = \frac{5/9}{4/9} = \frac{5}{4}$$

or 5:4. ◀ (7)

Empirical Probability

In many real-life problems, it is not possible to establish exact probabilities for events. Instead, useful approximations are often found by drawing on past experience as a guide to the future. The next example shows one approach to such **empirical probabilities.**

EXAMPLE 8 The table lists U.S. advertising volume in thousands of dollars by medium in 1998.*

Medium	*Expenditures*
Magazines	13,780,249
Newspapers	16,130,928
Network Television	16,271,972
Sport Television	15,486,766
Syndicated Television	2,691,648
Cable Television	6,671,978
Network Radio	824,007
Sunday Magazines	1,029,447

We can find the empirical probability that a dollar of advertising is spent on each medium by first finding the total spent and then dividing the amount spent on each medium by the total. Verify that the amounts in the table sum to 72,886,995.

**The World Almanac and Book of Facts 2000*, p. 192.

The probability that a dollar is spent on newspapers, for example, is P(newspapers) = 16,130,928/72,886,995 $\approx$ .221. Similarly, we could divide each amount by 72,886,995, with the results (rounded to three decimal places) shown below.

Medium	*Probabilities*
Magazines	.189
Newspapers	.221
Network Television	.223
Sport Television	.212
Syndicated Television	.037
Cable Television	.092
Network Radio	.011
Sunday Magazines	.014

8

A traffic engineer gathered the following data on the number of accidents at four busy intersections.

Intersection	*Number of Accidents*
1	25
2	32
3	28
4	35

If an accident occurs at one of these four intersections, what is the probability (to 2 decimal places) that it occurs at intersections 1 or 3?

Answer:

.44

The numbers in this table sum to .999. In theory, they should total 1.000, but this does not always occur with rounded numbers.

The categories in the table are mutually exclusive simple events. Thus, to find the probability that an advertising dollar is spent on newspapers or sport television, we use the union rule to calculate

$$P(\text{newspapers or sport television}) = .221 + .212 = .433.$$

We could get this same result by summing the amount spent on newspapers and sport television and dividing the total by 72,886,995.

Thus, nearly half of all advertising dollars are spent on these two media, a figure that should be of interest to both advertisers and the owners of the various media. ◀ 8

A table of probabilities, as in Example 8, sets up a probability distribution; that is, for each possible outcome of an experiment, a number, called the probability of that outcome, is assigned. This assignment may be done in any reasonable way (on a empirical basis, as in Example 8, or by theoretical reasoning, as in Section 8.3), provided that it satisfies the following conditions.

Properties of Probability

Let S be a sample space consisting of n distinct outcomes $s_1, s_2, \ldots, s_n$. An acceptable probability assignment consists of assigning to each outcome s_i a number p_i (the probability of s_i) according to these rules.

1. The probability of each outcome is a number between 0 and 1.

$$0 \le p_1 \le 1, \quad 0 \le p_2 \le 1, \ldots, \quad 0 \le p_n \le 1$$

2. The sum of the probabilities of all possible outcomes is 1.

$$p_1 + p_2 + p_3 + \cdots + p_n = 1$$

Probability distributions are discussed further in the next chapter.

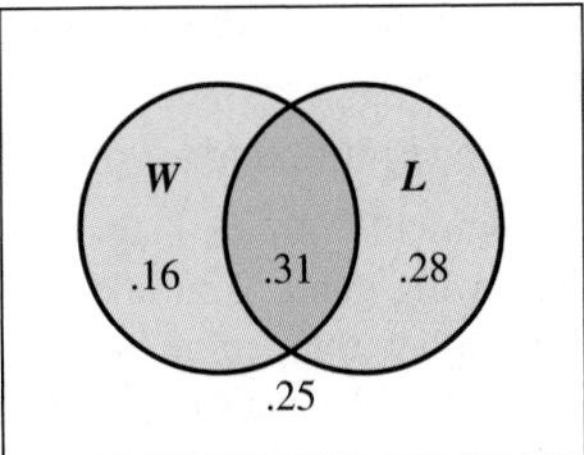

FIGURE 8.21

EXAMPLE 9 Susan is a college student who receives heavy sweaters from her aunt at the first sign of cold weather. She estimates that the probability that a sweater is the wrong size is .47, the probability that it is a loud color is .59, and the probability that it is both the wrong size and a loud color is .31.

(a) Find the probability that the sweater is the correct size and not a loud color.

Let W be the event "wrong size" and L be the event "loud color." Place the given information on a Venn diagram, starting with .31 in the intersection of the regions W and L (see Figure 8.21). As stated earlier, event W has probability .47. Since .31 has already been placed inside the intersection of W and L,

$$.47 - .31 = .16$$

goes inside region W but outside the intersection of W and L. In the same way,

$$.59 - .31 = .28$$

goes inside the region for L and outside the overlap.

Using regions W and L, the event we want is $W' \cap L'$. From the Venn diagram in Figure 8.21, the labeled regions have a total probability of

$$.16 + .31 + .28 = .75$$

Since the entire region of the Venn diagram must have probability 1, the region outside W and L, or $W' \cap L'$, has probability

$$1 - .75 = .25.$$

The probability is .25 that the sweater is the correct size and not a loud color.

(b) Find the probability that the sweater is the correct size *or* not a loud color.

The corresponding region, $W' \cup L'$, has probability

$$.25 + .16 + .28 = .69.$$ ◀ 9

Find the probability that the sweater in Example 9 is

(a) not a loud color;

(b) the correct size *and* a loud color.

Answers:

(a) .41

(b) .28

8.4 Exercises

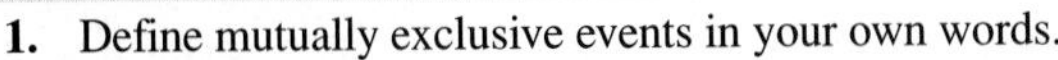

1. Define mutually exclusive events in your own words.

Decide whether the events in Exercises 2–7 are mutually exclusive.

2. Owning a car and owning a truck

3. Wearing glasses and wearing sandals

4. Being married and being over 30 years old

5. Being a teenager and being over 30 years old

6. Rolling a die once and getting a 4 and an odd number

7. Being a male and being a postal worker

Two dice are rolled. Find the probabilities of rolling the given sums. (See Examples 2, 4, and 5.)

8. **(a)** 2 **(b)** 4 **(c)** 5 **(d)** 6

9. **(a)** 8 **(b)** 9 **(c)** 10 **(d)** 13

10. **(a)** 9 or more **(b)** Less than 7
(c) Between 5 and 8 (exclusive)

11. **(a)** Not more than 5 **(b)** Not less than 8
(c) Between 3 and 7 (exclusive)

Two dice are rolled. Find the probabilities of the following events.

12. The first die is 3 or the sum is 8.

13. The second die is 5 or the sum is 10.

14. Three unusual dice, A, B, and C, are constructed such that die A has the numbers 3, 3, 4, 4, 8, 8; die B has the numbers, 1, 1, 5, 5, 9, 9; and die C has the numbers 2, 2, 6, 6, 7, 7.

(a) If dice A and B are rolled, find the probability that B beats A, that is, the number that appears on die B is greater than the number that appears on die A.

(b) Which die is better? Explain.

One card is drawn from an ordinary deck of 52 cards. Find the probabilities of drawing the following cards. (See Example 1.)

15. **(a)** A 3 or a king
(b) An ace or a spade
(c) A red card or a 7

16. **(a)** A 9 or 10 **(b)** A red card or a 3
(c) A 9 or a black 10 **(d)** A heart or a black card
(e) A face card or a diamond

17. **(a)** Less than a 4 (count aces as ones)
(b) A diamond or a 7 **(c)** A black card or an ace
(d) A heart or a jack **(e)** A red card or a face card

Ms. Elliott invites 10 relatives to a party: her mother, 2 aunts, 3 uncles, 2 brothers, 1 male cousin, and 1 female cousin. If the chances of any 1 guest arriving first are equally likely, find the probabilities that the first guest to arrive is as follows.

18. **(a)** A brother or an uncle
(b) A brother or a cousin
(c) A brother or her mother

19. **(a)** An uncle or a cousin
(b) A male or a cousin
(c) A female or a cousin

The numbers 1, 2, 3, 4, and 5 are written on slips of paper, and two slips are drawn at random one at a time without replacement. Find each of the probabilities in Exercises 20 and 21.

20. **(a)** The sum of the numbers is 9.
(b) The sum of the numbers is 5 or less.
(c) The first number is 2 or the sum is 6.

21. **(a)** Both numbers are even.
(b) One of the numbers is even or greater than 3.
(c) The sum is 5 or the second number is 2.

Use Venn diagrams to work Exercises 22 and 23.

22. Suppose $P(E) = .26$, $P(F) = .41$, and $P(E \cap F) = .14$. Find each of the following.
(a) $P(E \cup F)$ **(b)** $P(E' \cap F)$
(c) $P(E \cap F')$ **(d)** $P(E' \cup F')$

23. Let $P(Z) = .42$, $P(Y) = .35$, and $P(Z \cup Y) = .61$. Find each of the following probabilities.
(a) $P(Z' \cap Y')$ **(b)** $P(Z' \cup Y')$
(c) $P(Z' \cup Y)$ **(d)** $P(Z \cap Y')$

24. Define what is meant by odds.

A single fair die is rolled. Find the odds in favor of getting the results in Exercises 25–28. (See Examples 6 and 7.)

25. 5 **26.** 3, 4, or 5
27. 1, 2, 3, or 4 **28.** Some number less than 2

29. A marble is drawn from a box containing 3 yellow, 4 white, and 8 blue marbles. Find the odds in favor of drawing the following.
(a) A yellow marble **(b)** A blue marble
(c) A white marble

30. Find the odds of *not* drawing a white marble in Exercise 29.

31. Two dice are rolled. Find the odds of rolling a 7 or 11.

32. In the "Ask Marilyn" column of *Parade* magazine, a reader wrote about the following game: You and I each roll a die. If your die is higher than mine, you win. Otherwise, I win. The reader thought that the probability that each player wins is 1/2. Is this correct? If not, what is the probability that each player wins?*

33. On page 134 of Roger Staubach's autobiography, *First Down, Lifetime to Go*. Staubach makes the following statement regarding his experience in Vietnam:

> Odds against a direct hit are very low but when your life is in danger, you don't worry too much about the odds.

Is this wording consistent with our definition of odds, for and against? How could it have been said so as to be technically correct?

34. The table gives the odds that a particular event will occur.† Convert each odds to the probability that the event will occur.

Event	*Odds for the Event*
You will eat out today.	1 to 2
The next bottled water you buy will be nothing more than tap water.	1 to 4
The earth will be struck by a huge meteor during your lifetime.	1 to 9000
You will go to Disney World this year.	1 to 9
You'll regain weight you lost by dieting.	9 to 10

Which of Exercises 35–42 are examples of empirical probability?

35. The probability of heads on 5 consecutive tosses of a coin

36. The probability that a freshman entering college will graduate with a degree

37. The probability that a person is allergic to penicillin

38. The probability of drawing an ace from a standard deck of 52 cards

39. The probability that a person will get lung cancer from smoking cigarettes

---

**Parade* Magazine, November 6, 1994, p. 10. Reprinted by permission of the William Morris Agency, Inc. on behalf of the author. Copyright © 1994 by Marilyn vos Savant.

†The Forum for Investor Advice: Krantz, Les, *What the Odds Are* (Harper Perennial, 1992); and Larry Laudan, *Danger Ahead: The Risks You Really Face on Life's Highway* (John Wiley & Sons, 1997).

40. A weather forecast that predicts a 70% chance of rain tomorrow

41. A gambler's claim that on a roll of a fair die, $P(\text{even}) = 1/2$

42. A surgeon's prediction that a patient has a 90% chance of full recovery

43. What is a probability distribution?

An experiment is conducted for which the sample space is $S = \{s_1, s_2, s_3, s_4, s_5\}$. Which of the probability assignments in Exercises 44–49 is possible for this experiment? If an assignment is not possible, tell why.

44.

Outcomes	s_1	s_2	s_3	s_4	s_5
Probabilities	.09	.32	.21	.25	.13

45.

Outcomes	s_1	s_2	s_3	s_4	s_5
Probabilities	.92	.03	0	.02	.03

46.

Outcomes	s_1	s_2	s_3	s_4	s_5
Probabilities	1/3	1/4	1/6	1/8	1/10

47.

Outcomes	s_1	s_2	s_3	s_4	s_5
Probabilities	1/5	1/3	1/4	1/5	1/10

48.

Outcomes	s_1	s_2	s_3	s_4	s_5
Probabilities	.64	−.08	.30	.12	.02

49.

Outcomes	s_1	s_2	s_3	s_4	s_5
Probabilities	.05	.35	.5	.2	−.3

One way to solve a probability problem is to repeat the experiment many times, keeping track of the results. Then the probability can be approximated using the basic definition of the probability of an event E: $P(E) = n(E)/n(S)$, where E occurs $n(E)$ times out of $n(S)$ trials of an expeirment. This is called the Monte Carlo method of finding probabilities. If physically repeating the experiment is too tedious, it may be simulated using a random number generator, available on most computers and scientific or graphing calculators. To simulate a coin toss or the roll of a die on a graphing calculator, change the setting to fixed decimal mode with 0 digits displayed. To simulate multiple tosses of a coin, press RAND (or RANDOM or RND#) in the PROB submenu of the MATH (or OPTN) menu and then press ENTER repeatedly. Interpret 0 as a head and 1 as a tail. To simulate multiple rolls of a die, press RAND × 6 + .5 and then press ENTER repeatedly.

50. Suppose two dice are rolled. Use the Monte Carlo method with at least 50 repetitions to approximate the following probabilities. Compare with the results of Exercise 11.
(a) P(the sum is not more than 5)
(b) P(the sum is not less than 8)

51. Suppose two dice are rolled. Use the Monte Carlo method with at least 50 repetitions to approximate the following probabilities. Compare with the results of Exercise 10.
(a) P(the sum is 9 or more)
(b) P(the sum is less than 7)

52. Suppose three dice are rolled. Use the Monte Carlo method with at least 100 repetitions to approximate the following probabilities.
(a) P(the sum is 5 or less)
(b) P(neither a 1 nor a 6 is rolled)

53. Suppose a coin is tossed 5 times. Use the Monte Carlo method with at least 50 repetitions to approximate the following probabilities.
(a) P(exactly 4 heads) **(b)** P(2 heads and 3 tails)

54. Business Suppose that 8% of a certain batch of calculators have a defective case and that 11% have defective batteries. Also, 3% have both a defective case and defective batteries. A calculator is selected from the batch at random. Find the probability that the calculator has a good case and good batteries.

55. Business The table shows the probabilities of a person accumulating specific amounts of credit card charges over a 12-month period. Find the probabilities that a person's total charges during the period are the following.
(a) $500 or more **(b)** Less than $1000
(c) $500 to $2999 **(d)** $3000 or more

Charges	*Probability*
Under $100	.31
$100–$499	.18
$500–$999	.18
$1000–$1999	.13
$2000–$2999	.08
$3000–$4999	.05
$5000–$9999	.06
$10,000 or more	.01

Business *The table below shows the probability that a customer of a department store will make a purchase in the indicated range.*

Amount Spent	*Probability*
Below $10	.07
$10–$24.99	.18
$25–$49.99	.21
$50–$74.99	.16
$75–$99.99	.11
$100–$199.99	.09
$200–$349.99	.07
$350–$499.99	.08
$500 or more	.03

Find the probabilities that a customer makes a purchase in the following ranges.

56. **(a)** Less than \$25
(b) More than \$24.99
(c) \$50 to \$199.99

57. **(a)** Less than \$350
(b) \$75 or more
(c) \$200 or more

58. Business The probability that a company will make a profit this year is .74. Find the odds against the company making a profit.

59. Natural Science A study on body types gave the following results: 45% were short; 25% were short and overweight; and 24% were tall and not overweight. Find the probabilities that a person is the following.
(a) Overweight
(b) Short but not overweight
(c) Tall and overweight

60. Natural Science Color blindness is an inherited characteristic that is more common in males than in females. If M represents male and C represents red-green color blindness, we use the relative frequencies of the incidences of males and red-green color blindness as probabilities to get

$$P(C) = .039, \quad P(M \cap C) = .035, \quad P(M \cup C) = .495.*$$

Find the following probabilities.
(a) $P(C')$ **(b)** $P(M)$ **(c)** $P(M')$
(d) $P(M' \cap C')$ **(e)** $P(C \cap M')$ **(f)** $P(C \cup M')$

61. Natural Science Gregor Mendel, an Austrian monk, was the first to use probability in the study of genetics. In an effort to understand the mechanism of character transmittal from one generation to the next in plants, he counted the number of occurrences of various characteristics. Mendel found that the flower color in certain pea plants obeyed this scheme:

Pure red crossed with pure white produces red.

From its parents, the red offspring received genes for both red (R) and white (W), but in this case red is *dominant* and white *recessive,* so the offspring exhibits the color red. However, the offspring still caries both genes, and when two such offspring are crossed, several things can happen in the third generation. The table in the next column, which is called a *Punnet square,* shows the equally likely outcomes.

		SECOND PARENT	
		R	***W***
FIRST PARENT	***R***	*RR*	*RW*
	W	*WR*	*WW*

Use the fact that red is dominant over white to find each of the following. Assume that there are an equal number of red and white genes in the population.
(a) P(a flower is red) **(b)** P(a flower is white)

62. Natural Science Mendel found no dominance in snapdragons, with one red gene and one white gene producing pink-flowered offspring. These second-generation pinks, however, still carry one red and one white gene, and when they are crossed, the next generation still yields the Punnet square shown above. Find each of the following probabilities.
(a) P(red) **(b)** P(pink) **(c)** P(white)

(Mendel verified these probability ratios experimentally and did the same for many characteristics other than flower color. His work, published in 1866, was not recognized until 1890.)

63. Natural Science In most animals and plants, it is very unusual for the number of main parts of the organism (such as arms, legs, toes, or flower petals) to vary from generation to generation. Some species, however, have *meristic variability,* in which the number of certain body parts varies from generation to generation. One researcher studied the front feet of certain guinea pigs and produced the following probabilities.*

$$P(\text{only four toes, all perfect}) = .77$$
$$P(\text{one imperfect toe and four good ones}) = .13$$
$$P(\text{exactly five good toes}) = .10$$

Find the probability of each of the following events.
(a) No more than four good toes
(b) Five toes, whether perfect or not

64. Business The following data was gathered for 130 adult U.S. workers: 55 were women; 3 women earned more than \$40,000; and 62 men earned less than or equal \$40,000. Find the probabilities that an individual is
(a) a woman earning less than or equal \$40,000;
(b) a man earning more than \$40,000;
(c) a man or is earning more than \$40,000;
(d) a woman or is earning less than or equal \$40,000.

*The probabilities of a person being male or female are from *The World Almanac and Book of Facts*, 2002. The probabilities of a male and female being color-blind are from *Parsons' Diseases of the Eye* (18th ed.) by Stephen J. H. Miller (Churchill Livingston, 1990), p. 269. This reference gives a range of 3 to 4% for the probability of gross color blindness in men; we used the midpoint of this range.

*J. R. Wright, "An Analysis of Variability in Guinea Pigs," *Genetics*, Vol. 19, pp. 506–536.

65. Business A survey of 100 people about their music expenditures gave the following information: 38 bought rock music; 20 were teenagers who bought rock music; and 26 were teenagers. Find the probabilities that a person is
    (a) a teenager who buys nonrock music;
    (b) someone who buys rock music or is a teenager;
    (c) not a teenager;
    (d) not a teenager but a buyer of rock music.

66. Social Science In a refugee camp in southern Mexico, it was found that 90% of the refugees came to escape political oppression, 80% came to escape abject poverty, and 70% came to escape both. What is the probability that a refugee in the camp was not poor nor seeking political asylum?

67. Social Science There were 198,420 female military personnel on active duty in September 1998 in various ranks and military branches, as listed in the table below.*

	Army (*A*)	*Air Force* (*B*)	*Navy* (*C*)	*Marines* (*D*)
Officers (*O*)	10,367	11,971	7,777	854
Enlisted (*E*)	60,787	53,542	42,261	8928
Cadets and Midshipmen (*M*)	624	653	656	0

(a) Convert the numbers in the table to probabilities.
(b) Find the probability that a randomly selected woman is enlisted in the Army.
(c) Find the probability that a randomly selected woman is an officer in the Navy or Marine Corps.
(d) $P(A \cup B)$
(e) $P(E \cup (C \cup D))$

*Borga Brunner, ed., *Time Almanac 2000*, p. 395.

68. Social Science Research has been carried out to measure the amount of intolerance that U.S. citizens have for left-wing Communists and right-wing Fascists, as indicated in the table below. Note that the numbers are given as percents and each row sums to 100 (except for rounding).*
    (a) Find the probability that a randomly chosen U.S. citizen would have at least some intolerance of right-wing Fascists.
    (b) Find the probability that a randomly chosen U.S. citizen would have at least some intolerance of left-wing Communists.
    (c) Compare your answers to parts (a) and (b) and provide possible reasons for these numbers.

	None at All	*Don't Know*	*Not Very Much*	*Somewhat*	*Extremely*
Left-Wing Communists	13.0	2.7	33.0	34.2	17.1
Right-Wing Fascists	10.1	3.3	20.7	43.1	22.9

*J. L. Gibson, "Putting Up with Fellow Russians: An Analysis of Political Tolerance in the Fledging Russian Democracy," *Political Research Quarterly*, vol. 51, no. 6, March 1998, pp. 37–38.

8.5 CONDITIONAL PROBABILITY AND INDEPENDENT EVENTS

The training manager for a large brokerage firm has noticed that some of the firm's brokers use the firm's research advice, while other brokers tend to go with their own beliefs about which stocks will go up. To see whether the research department performs better than the beliefs of the brokers, the manager conducted a survey of 100 brokers, with the results shown in the following table.

1

Use the data in the table to find

(a) $P(B)$,

(b) $P(A')$,

(c) $P(B')$.

Answers:

(a) .45

(b) .4

(c) .55

	Picked a Stock That Went Up	*Didn't Pick A Stock That Went Up*	*Totals*
Used Research	30	15	45
Didn't Use Research	30	25	55
Totals	60	40	100

Letting A be the event "picked a stock that went up" and letting B be the event "used research," $P(A)$, $P(A')$, $P(B)$, and $P(B')$ can be found. For example, the chart shows that a total of 60 brokers picked stocks that went up, so $P(A) = 60/100 = .6$. (1)

Suppose we want to find the probability that a broker using research will pick a stock that goes up. From the table above, of the 45 brokers who use research, there are 30 who picked stocks that went up, so

$$P(\text{broker who uses research picks stocks that go up}) = \frac{30}{45} \approx .667.$$

This is a different number than the probability that a broker picks a stock that goes up, .6, because *we have additional information* (the broker uses research) *that has reduced the sample space.* In other words, we found the probability that a broker picks a stock that goes up, A, given the additional information that the broker uses research, B. This is called the *conditional probability* of event A, given that event B has occurred, written $P(A \mid B)$. ($P(A \mid B)$ may also be read as "the probability of A given B.")

In the example above,

$$P(A \mid B) = \frac{30}{45}.$$

If we divide the numerator and denominator by 100 (the size of the sample space), this can be written as

$$P(A \mid B) = \frac{\frac{30}{100}}{\frac{45}{100}} = \frac{P(A \cap B)}{P(B)},$$

where $P(A \cap B)$ represents, as usual, the probability that both A and B will occur.

To generalize this result, assume that E and F are two events for a particular experiment. Assume that the sample space S for this experiment has n possible equally likely outcomes. Suppose event F has m elements and $E \cap F$ has k elements ($k \leq m$). Using the fundamental principle of probability,

$$P(F) = \frac{m}{n} \quad \text{and} \quad P(E \cap F) = \frac{k}{n}.$$

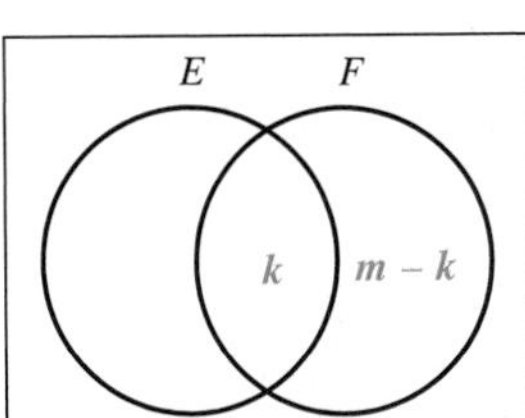

Event F has a total of m elements.

FIGURE 8.22

We now want to find $P(E \mid F)$, the probability that E occurs given that F has occurred. Since we assume F has occurred, reduce the sample space to F: look only at the m elements inside F. (See Figure 8.22.) Of these m elements, there are k elements where E also occurs, because $E \cap F$ has k elements. This makes

$$P(E \mid F) = \frac{k}{m}.$$

Divide numerator and denominator by n to get

$$P(E \mid F) = \frac{k/n}{m/n} = \frac{P(E \cap F)}{P(F)}.$$

The last result motivates the definition of conditional probability.

The **conditional probability** of an event E given event F, written $P(E \mid F)$, is

$$\mathbf{P(E \mid F) = \frac{P(E \cap F)}{P(F)}, \qquad P(F) \neq 0.}$$

This definition tells us that, for equally likely outcomes, conditional probability is found by *reducing the sample space to event F*, and then finding the number of outcomes in F that are also in event E. Thus,

$$\mathbf{P(E \mid F) = \frac{n(E \cap F)}{n(F)}.}$$

Although the definition of conditional probability was motivated by an example with equally likely outcomes, it is valid in all cases. For an intuitive explanation, think of the formula as giving the probability that both E and F occur compared with the entire probability of F.

EXAMPLE 1 Use the information given in the table on the preceding page to find the following probabilities.

(a) $P(B \mid A)$

This represents the probability that the broker used research, given that the broker picked a stock that went up. Reduce the sample space to A. Then find $n(A \cap B)$ and $n(A)$.

$$P(B \mid A) = \frac{n(A \cap B)}{n(A)} = \frac{30}{60} = \frac{1}{2}$$

If a broker picked a stock that went up, then the probability is 1/2 that the broker used research.

(b) $P(A' \mid B)$

In words, this is the probability that a broker picks a stock that does not go up, even though he used research.

$$P(A' \mid B) = \frac{n(A' \cap B)}{n(B)} = \frac{15}{45} = \frac{1}{3}$$

The table shows the results of a survey of a buffalo herd.

	Males	*Females*	*Totals*
Adults	500	1300	1800
Calves	520	500	1020
Totals	1020	1800	2820

Let M represent "male" and A represent "adult." Find each of the following.

(a) $P(M \mid A)$

(b) $P(M' \mid A)$

(c) $P(A \mid M')$

(d) $P(A' \mid M)$

(e) State the probability in part (d) in words.

Answers:

(a) 5/18

(b) 13/18

(c) 13/18

(d) 26/51

(e) The probability that a buffalo is a calf given that it is a male

Find $P(F \mid E)$ if $P(E) = .3$, $P(F) = .4$, and $P(E \cup F) = .6$.

Answer:

1/3

(c) $P(B' \mid A')$

Here, we want the probability that a broker who picked a stock that did not go up did not use research.

$$P(B' \mid A') = \frac{n(B' \cap A')}{n(A')} = \frac{25}{40} = \frac{5}{8}$$ ②

Venn diagrams can be used to illustrate problems in conditional probability. A Venn diagram for Example 1, in which the probabilities are used to indicate the number in the set defined by each region, is shown in Figure 8.23. In the diagram, $P(B \mid A)$ is found by *reducing the sample space to just set A*. Then $P(B \mid A)$ is the ratio of the number in that part of set B which is also in A to the number in set A, or $.3/.6 = .5$.

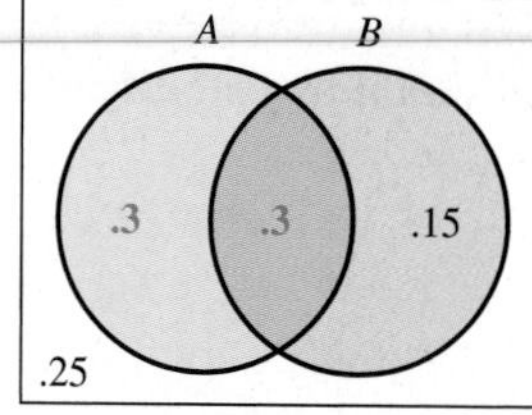

FIGURE 8.23

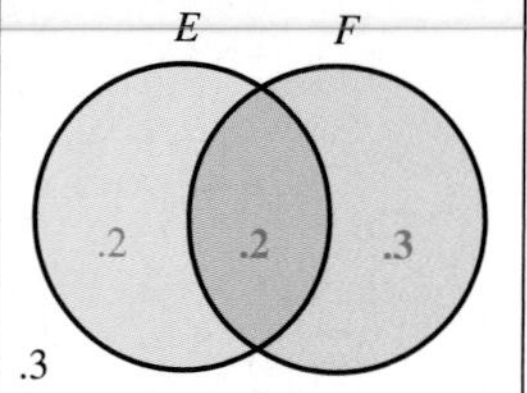

FIGURE 8.24

EXAMPLE 2 Given $P(E) = .4$, $P(F) = .5$, and $P(E \cup F) = .7$, find $P(E \mid F)$.

Find $P(E \cap F)$ first. Then use a Venn diagram to find $P(E \mid F)$. By the union rule,

$$P(E \cup F) = P(E) + P(F) - P(E \cap F)$$
$$.7 = .4 + .5 - P(E \cap F)$$
$$P(E \cap F) = .2.$$

Now use the probabilities to indicate the number in each region of the Venn diagram in Figure 8.24. $P(E \mid F)$ is the ratio of the probability of that part of E which is in F to the probability of F or

$$P(E \mid F) = \frac{P(E \cap F)}{P(F)} = \frac{.2}{.5} = \frac{2}{5} = .4.$$ ③

EXAMPLE 3 Two fair coins were tossed, and it is known that at least one was a head. Find the probability that both were heads.

The sample space has four equally likely outcomes, $S = \{hh, ht, th, tt\}$. Define two events:

$$E_1 = \text{at least 1 head} = \{hh, ht, th\}$$

and

$$E_2 = \text{2 heads} = \{hh\}.$$

Because there are four equally likely outcomes, $P(E_1) = 3/4$. Also, $P(E_1 \cap E_2) = 1/4$. We want the probability that both were heads, given that at least one was a head: that

4

In Example 3, find the probability that exactly one coin showed a head, given that at least one was a head.

Answer:

2/3

is, we want to find $P(E_2 \mid E_1)$. Because of the condition that at least one coin was a head, the reduced sample space is

$$\{hh, ht, th\}.$$

Since only one outcome in this reduced sample space is 2 heads,

$$P(E_2 \mid E_1) = \frac{1}{3}.$$

Alternatively, use the definition given above.

$$P(E_2 \mid E_1) = \frac{P(E_2 \cap E_1)}{P(E_1)} = \frac{1/4}{3/4} = \frac{1}{3}$$ ◀ 4

It is important not to confuse $P(A \mid B)$ with $P(B \mid A)$. For example, in a criminal trial, a prosecutor may point out to the jury that the probability of the defendant's DNA profile matching that of a sample taken at the scene of the crime, given that the defendant is innocent, is very small. What the jury must decide, however, is the probability that the defendant is innocent, given that the defendant's DNA profile matches the sample. Confusing the two is an error sometimes called "the prosecutor's fallacy," and the 1990 conviction of a rape suspect in England was overturned by a panel of judges, who ordered a retrial, because the fallacy made the original trial unfair.*

In the next section, we will see how to compute $P(A \mid B)$ when we know $P(B \mid A)$.

Product Rule

If $P(E) \neq 0$ and $P(F) \neq 0$, then the definition of conditional probability shows that

$$P(E \mid F) = \frac{P(E \cap F)}{P(F)} \quad \text{and} \quad P(F \mid E) = \frac{P(F \cap E)}{P(E)}.$$

Using the fact that $P(E \cap F) = P(F \cap E)$, and solving each of these equations for $P(E \cap F)$, we obtain the following rule.

Product Rule of Probability

If E and F are events, then $P(E \cap F)$ may be found by either of these formulas.

$$\mathbf{P(E \cap F) = P(F) \cdot P(E \mid F)} \quad \text{or} \quad \mathbf{P(E \cap F) = P(E) \cdot P(F \mid E).}$$

The **product rule** gives a method for finding the probability that events E and F both occur. Here is a simple way to remember the ordering of E and F in the probability rule.

$$P(E \cap F) = P(F) \cdot P(E \mid F) \quad \text{or} \quad P(E \cap F) = P(E) \cdot P(F \mid E)$$

*David Pringle, "Who's the DNA Fingerprinting Pointing At?" *New Scientist*, January 29, 1994, pp. 51–52.

In a litter of puppies, 3 were female and 4 were male. Half the males were black. Find the probability that a puppy chosen at random from the litter would be a black male.

Answer:

2/7

EXAMPLE 4 In a class with 2/5 women and 3/5 men, 25% of the women are business majors. Find the probability that a student chosen at random from the class is a female business major.

Let B and W represent the events "business major" and "woman," respectively. We want to find $P(B \cap W)$. By the product rule,

$$P(B \cap W) = P(W) \cdot P(B \mid W).$$

From the given information, $P(W) = 2/5 = .4$ and the probability that a woman is a business major is $P(B \mid W) = .25$. Then

$$P(B \cap W) = .4(.25) = .10.$$ ◀ 5

In Section 8.1 we used a tree diagram to find the number of subsets of a given set. By including the probabilities for each branch of a tree diagram, we convert it to a **probability tree.** The next examples show how a probability tree is used with the product rule to find the probability of a sequence of events.

EXAMPLE 5 A company needs to hire a new director of advertising. It has decided to try to hire either person A or person B, who are assistant advertising directors for its major competitor. To decide between A and B, the company does research on the campaigns managed by A or B (none are managed by both) and finds that A is in charge of twice as many advertising campaigns as B. Also, A's campaigns have satisfactory results three out of four times, while B's campaigns have satisfactory results only two out of five times. Suppose one of the competitor's advertising campaigns (managed by A or B) is selected randomly.

We can represent this situation schematically as follows. Let A denote the event "Person A does the job" and B the event "Person B does the job." Let S be the event "satisfactory results" and U the event "unsatisfactory results." Then the given information can be summarized in the probability tree in Figure 8.25. Since A does twice as many jobs as B, $P(A) = 2/3$ and $P(B) = 1/3$, as noted on the first-stage branches of the tree. When A does a job, the probability of satisfactory results is 3/4 and of unsatisfactory results 1/4, as noted on the second-stage branches. Similarly, the probabilities when B does the job are noted on the remaining second-stage branches. The composite branches labeled 1–4 represent the four mutually exclusive possibilities for the running and outcome of the campaign.

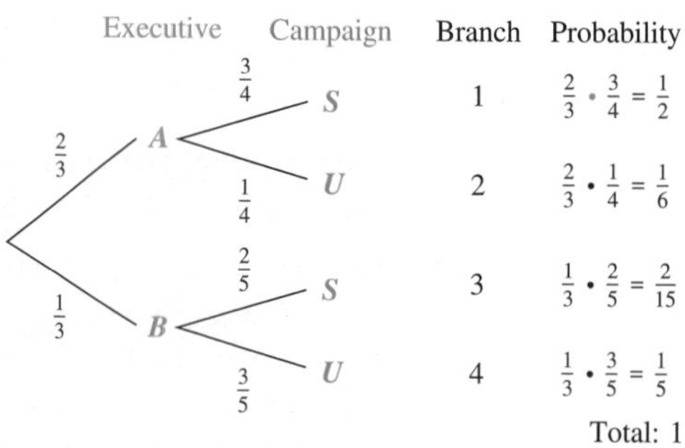

FIGURE 8.25

(a) Find the probability that A is in charge of a campaign that produces satisfactory results.

We are asked to find $P(A \cap S)$. We know that when A does the job, the probability of success is 3/4, that is, $P(S \mid A) = 3/4$. Hence, by the product rule,

$$P(A \cap S) = P(A) \cdot P(S \mid A) = \frac{2}{3} \cdot \frac{3}{4} = \frac{1}{2}.$$

The event $A \cap S$ is represented by branch 1 of the tree, and as we have just seen, its probability is the product of the probabilities that make up that branch.

(b) Find the probability that B runs a campaign that produces satisfactory results.

We must find $P(B \cap S)$. The event is represented by branch 3 of the tree and, as before, its probability is the product of the probabilities of the pieces of that branch:

$$P(B \cap S) = P(B) \cdot P(S \mid B) = \frac{1}{3} \cdot \frac{2}{5} = \frac{2}{15}.$$

(c) What is the probability that the selected campaign is satisfactory?

The event S is the union of the mutually exclusive events $A \cap S$ and $B \cap S$, which are represented by branches 1 and 3 of the tree diagram. By the union rule,

$$P(S) = P(A \cap S) + P(B \cap S) = \frac{1}{2} + \frac{2}{15} = \frac{19}{30}.$$

Thus, the probability of an event that appears on several branches is the sum of the probabilities of each of these branches.

(d) What is the probability that the selected campaign is unsatisfactory?

$P(U)$ can be read from branches 2 and 4 of the tree.

$$P(U) = \frac{1}{6} + \frac{1}{5} = \frac{11}{30}$$

Alternatively, because U is the complement of S,

$$P(U) = 1 - P(S) = 1 - \frac{19}{30} = \frac{11}{30}.$$

(e) Find the probability that either A runs the campaign or the results are satisfactory (or possibly both).

Event A combines branches 1 and 2, while event S combines branches 1 and 3, so use branches 1, 2, and 3.

$$P(A \cup S) = \frac{1}{2} + \frac{1}{6} + \frac{2}{15} = \frac{4}{5}$$ ◀ 6

6

Find each of the following probabilities for Example 5.

(a) $P(U \mid A)$

(b) $P(U \mid B)$

Answers:

(a) 1/4

(b) 3/5

▶ **EXAMPLE 6** From a box containing 1 red, 3 white, and 2 green marbles, two marbles are drawn one at a time without replacing the first before the second is drawn. Find the probability that one white and one green marble are drawn.

A probability tree showing the various possible outcomes is given in Figure 8.26 on the next page. In this diagram, W represents the event "drawing a white marble" and G represents "drawing a green marble." On the first draw, $P(W \text{ on the 1st}) = 3/6 = 1/2$ because three of the six marbles in the box are white. On the second draw, $P(G \text{ on the 2nd} \mid W \text{ on the 1st}) = 2/5$. One white marble has been removed, leaving 5, of which 2 are green.

Find the probability of drawing a green marble and then a white marble.

Answer:

1/5

In Example 6, find the probability of drawing 1 white and 1 red marble.

Answer:

1/5

We want to find the probability of drawing exactly one white marble and exactly one green marble. Two events satisfy this condition: drawing a white marble first and then a green one (branch 2 of the tree) or drawing a green marble first and then a white one (branch 4). For branch 2,

$$P(W \text{ on 1st}) \cdot P(G \text{ on 2nd} \mid W \text{ on 1st}) = \frac{1}{2} \cdot \frac{2}{5} = \frac{1}{5}$$ (7)

For branch 4, where the green marble is drawn first,

$$P(G \text{ first}) \cdot P(W \text{ second} \mid G \text{ first}) = \frac{1}{3} \cdot \frac{3}{5} = \frac{1}{5}.$$

Since these two events are mutually exclusive, the final probability is the sum of the two probabilities.

$$P(\text{one } W, \text{one } G) = P(W \text{ on 1st}) \cdot P(G \text{ on 2nd} \mid W \text{ on 1st}) + P(G \text{ on 1st}) \cdot P(W \text{ on 2nd} \mid G \text{ on 1st}) = \frac{2}{5}$$ (8)

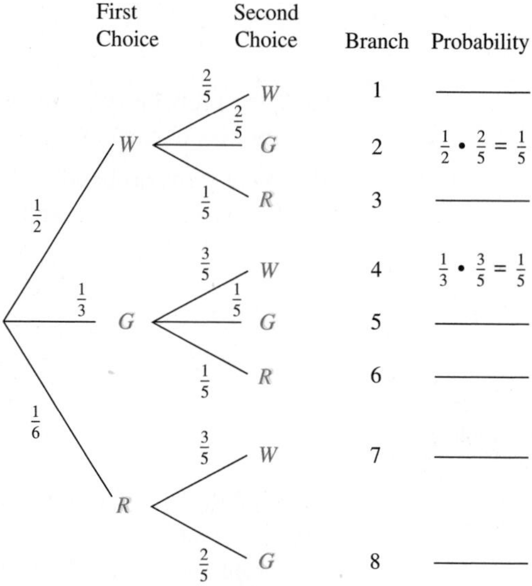

FIGURE 8.26

The product rule is often used when dealing with *stochastic processes,* which are mathematical models that evolve over time in a probabilistic manner. For example, drawing different colored marbles from a box (without replacing them) is such a process, in which the probabilities change with each successive draw. (Particular stochastic processes are studied further in Section 9.5.)

EXAMPLE 7 Two cards are drawn without replacement from an ordinary deck (52 cards). Find the probability that the first card is a heart and the second card is red.

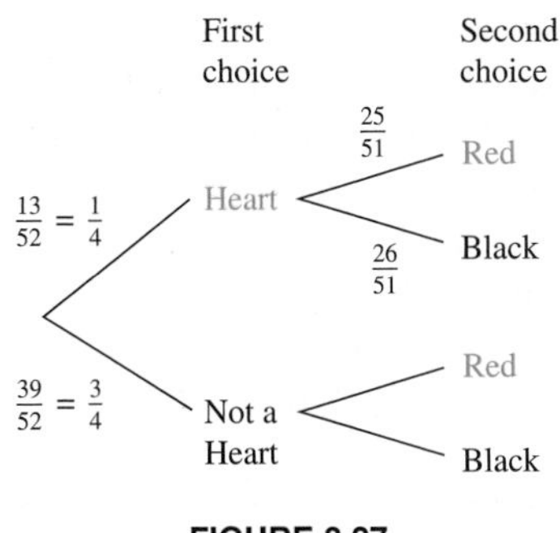

FIGURE 8.27

Start with the probability tree of Figure 8.27. (You may wish to refer to the deck of cards shown in Figure 8.19.) On the first draw, since there are 13 hearts in the 52 cards, the probability of drawing a heart first is $13/52 = 1/4$. On the second draw, since a (red) heart has been drawn already, there are 25 red cards in the remaining 51 cards. Thus the probability of drawing a red card on the second draw, given that the first is a heart, is 25/51. By the product rule of probability,

$$
\begin{aligned}
&P(\text{heart on 1st and red on 2nd}) \\
&\quad = P(\text{heart on 1st}) \cdot P(\text{red on 2nd} \mid \text{heart on 1st}) \\
&\quad = \frac{1}{4} \cdot \frac{25}{51} = \frac{25}{204} \approx .1225.
\end{aligned}
$$

◀ 9

Find the probability of drawing a heart on the first draw and a black card on the second, if two cards are drawn without replacement.

Answer:

$13/102 \approx .1275$

EXAMPLE 8 Three cards are drawn, without replacement, from an ordinary deck. Find the probability that exactly 2 of the cards are red.

Here we need a probability tree with three stages, as shown in Figure 8.28. The three branches indicated with arrows produce exactly 2 red cards from the draws. Multiply the probabilities along each of these branches and then add.

$$
\begin{aligned}
P(\text{exactly 2 red cards}) &= \frac{26}{52} \cdot \frac{25}{51} \cdot \frac{26}{50} + \frac{26}{52} \cdot \frac{26}{51} \cdot \frac{25}{50} + \frac{26}{52} \cdot \frac{26}{51} \cdot \frac{25}{50} \\
&= \frac{50{,}700}{132{,}600} = \frac{13}{34} \approx .382
\end{aligned}
$$

◀ 10

Use the tree in Example 8 to find the probability that exactly one of the cards is red.

Answer:

$13/34 \approx .382$

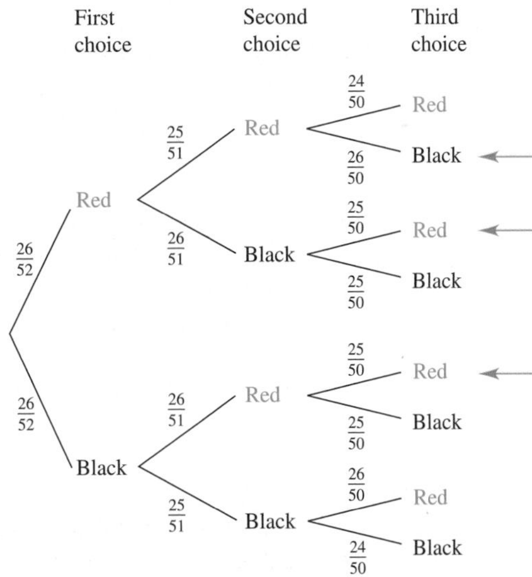

FIGURE 8.28

Independent Events

Suppose, in Example 7 that we draw the two cards *with* replacement rathen than without replacement (that is, we put the first card back before drawing the second card). If the first card is a heart, then the probability of drawing a red card on the second draw is 26/52, rather than 25/51 because there are still 52 cards in the deck, 26 of them red. In this case, P(red second | heart first) is the same as P(red second). The value of the second card is not affected by the value of the first card. We say that the event that the second card is red is *independent* of the event that the first card is a heart since the knowledge of the first card does not influence what happens to the second card. On the other hand, when we draw without replacement, the events that the first card is a heart and the second is red are *dependent* events. The fact that the first card is a heart means there is one less red card in the deck, influencing the probability that the second card is red.

As another example, consider tossing a fair coin twice. If the first toss shows heads, the probability that the next toss is heads is still 1/2. Coin tosses are independent events, since the outcome of one toss does not influence the outcome of the next toss. Similarly, rolls of a fair die are independent events. On the other hand, the events "the milk is old" and "the milk is sour" are dependent events; if the milk is old, there is an increased chance that it is sour. Also, in the example at the beginning of this section, the events A (broker picked a stock that went up) and B (broker used research) are dependent events, because information about the use of research affects the probability of picking stocks that go up. That is, $P(A \mid B)$ is different from $P(A)$.

If events E and F are independent, then the knowledge that E has occurred gives no (probability) information about the occurrence or nonoccurrence of event F. That is, $P(F)$ is exactly the same as $P(F \mid E)$, or

$$P(F \mid E) = P(F).$$

This, in fact, this is the formal definition of independent events.

E and F are **independent events** if

$$\boldsymbol{P(F \mid E) = P(F)} \quad \text{or} \quad \boldsymbol{P(E \mid F) = P(E).}$$

If the events are not independent, they are **dependent events.**

When E and F are independent events, then $P(F \mid E) = P(F)$ and the product rule becomes

$$P(E \cap F) = P(E) \cdot P(F \mid E) = P(E) \cdot P(F).$$

Conversely, if this equation holds, it follows that $P(F) = P(F \mid E)$. Consequently, we have this useful fact.

Product Rule for Independent Events

E and F are independent events if and only if

$$\boldsymbol{P(E \cap F) = P(E) \cdot P(F).}$$

EXAMPLE 9 A calculator requires a key-stroke assembly and a logic circuit. Assume that 99% of the key-stroke assemblies are satisfactory and 97% of the logic circuits are satisfactory. Find the probability that a finished calculator will be satisfactory.

Find the probability of getting 4 successive heads on 4 tosses of a fair coin.

Answer:

1/16

If the failure of a key-stroke assembly and the failure of a logic circuit are independent events, then

$$P(\text{satisfactory calculator})$$
$$= P(\text{satisfactory key-stroke assembly}) \cdot P(\text{satisfactory logic circuit})$$
$$= (.99)(.97) \approx .96.$$ (11)

CAUTION It is common for students to confuse the ideas of *mutually exclusive* events and *independent* events. Events E and F are mutually exclusive if $E \cap F = \emptyset$. For example, if a family has exactly one child, the only possible outcomes are $B = \{\text{boy}\}$ and $G = \{\text{girl}\}$. These two events are mutually exclusive. However, the events are *not* independent, since $P(G \mid B) = 0$ (if a family with only one child has a boy, the probability it has a girl is then 0). Since $P(G \mid B) \neq P(G)$, the events are not independent. Of all the families with exactly *two* children, the events $G_1 = \{\text{first child is a girl}\}$ and $G_2 = \{\text{second child is a girl}\}$ are independent, because $P(G_2 \mid G_1)$ equals $P(G_2)$. However, G_1 and G_2 are not mutually exclusive, since $G_1 \cap G_2 = \{\text{both children are girls}\} \neq \emptyset$.

To show that two events E and F are independent, we can show that $P(F \mid E) = P(F)$ or that $P(E \mid F) = P(E)$ or that $P(E \cap F) = P(E) \cdot P(F)$. Another way is to observe that knowledge of one outcome does not influence the probability of the other outcome, as we did for coin tosses.

NOTE In some cases, it may not be apparent from the physical description of the problem whether two events are independent or not. For example, it is not obvious whether the event that a baseball player gets a hit tomorrow is independent of the event that he got a hit today. In such cases, it is necessary to use the definition and calculate whether $P(F \mid E) = P(F)$ or, equivalently, whether $P(E \mid F) = P(E) \cdot P(F)$.

In the U.S. population, the probability of being Hispanic is .11, the probability of living in California is .12, and the probability of being a Hispanic living in California is .04. Are the events being Hispanic and living in California independent?

Answer:

No

EXAMPLE 10 On a typical January day in Manhattan the probability of snow is .10, the probability of a traffic jam is .80, and the probability of snow or a traffic jam (or both) is .82. Are the event "it snows" and the event "a traffic jam occurs" independent?

Let S represent the event "it snows" and T represent the event "a traffic jam occurs." We must determine whether

$$P(T \mid S) = P(T) \quad \text{or} \quad P(S \mid T) = P(S).$$

We know $P(S) = .10$, $P(T) = .8$, and $P(S \cup T) = .82$. We can use the union rule (or a Venn diagram) to find $P(S \cap T) = .08$, $P(T \mid S) = .8$, and $P(S \mid T) = .1$. Since

$$P(T \mid S) = P(T) = .8 \quad \text{and} \quad P(S \mid T) = P(S) = .1,$$

the events "it snows" and "a traffic jam occurs" are independent. (12)

Although we showed $P(T \mid S) = P(T)$ and $P(S \mid T) = P(S)$ in Example 10, only one of these results is needed to establish independence. It is also important to note that independence of events does not necessarily follow intuition; it is established from the mathematical definition of independence.

8.5 Exercises

If a single fair die is rolled, find the probability of rolling the following. (See Examples 1 and 2.)

1. 3, given that the number rolled was odd
2. 5, given that the number rolled was even
3. An odd number, given that the number rolled was 3

If two fair dice are rolled (recall the 36-outcome sample space), find the probability of rolling the following.

4. A sum of 8, given the sum was greater than 7
5. A sum of 6, given the roll was a "double" (two identical numbers)
6. A double, given that the sum was 9

If two cards are drawn without replacement from an ordinary deck (see Example 7), find the following probabilities.

7. The second is a heart, given that the first is a heart
8. The second is black, given that the first is a spade
9. The second is a face card, given that the first is a jack
10. The second is an ace, given that the first is not an ace
11. A jack and a 10 are drawn.
12. An ace and a 4 are drawn.
13. Two black cards are drawn.
14. Two hearts are drawn.
15. In your own words, explain how to find the conditional probability $P(E \mid F)$.
16. Your friend asks you to explain how the product rule for independent events differs from the product rule for dependent events. How would you respond?
17. Another friend asks you to explain how to tell whether two events are dependent or independent. How would you reply? (Use your own words.)
18. A student reasons that the probability in Example 3 of both coins being heads is just the probability that the other coin is a head, that is, 1/2. Explain why this reasoning is wrong.
19. In a two-child family, if we assume that the probabilities of a male child and a female child are each .5, are the events *each child is the same sex* and *at most one male* independent? Are they independent for a three-child family?
20. Let A and B be independent events with $P(A) = \frac{1}{4}$ and $P(B) = \frac{1}{5}$. Find $P(A \cap B)$ and $P(A \cup B)$.

Business *Use a probability tree or Venn diagram in Exercises 21–24 (See Examples 2 and 5–9. A shop that produces custom kitchen cabinets has two employees, Sitlington and Capek. 95% of Capek's work is satisfactory and 10% of Sitlington's work is unsatisfactory. 60% of the shop's cabinets are made by Capek (the rest by Sitlington). Find the following probabilities.*

21. An unsatisfactory cabinet was made by Capek. (*Hint:* Consider which event came first.)
22. A finished cabinet is unsatisfactory.
23. Social Science Marriages between cousins are very common in some countries, where about 50% of marriages are consanguineous (between first cousins or people even more closely related). A recent study in Pakistan has shown that 16% of children from unrelated marriages died by age 10, while 21% of children from consanguineous marriages died by age 10. Find the following probabilities.
   (a) a child survives
   (b) a child from a consanguineous marriage survives
24. Business Among users of automated teller machines (ATMs), 92% use ATMs to withdraw cash, and 32% use them to check their account balance.* Suppose that 96% use ATMs to either withdraw cash or check their account balance (or both). Find the probability that
   (a) an ATM user withdraws funds if that person uses the ATM to check an account balance.
   (b) an ATM user checks an account balance given that the ATM is used to withdraw funds.

25. Health The table at the top of the next column, based on data from the Centers for Disease Control, gives the number of new cases of the AIDS virus for men and women in the United States in 1998 by method of transmission.†

**Chicago Tribune*, Sec. 4, p. 1, December 18, 1995.

†*Health United States*, 1999, CDC, National Center for HIV, STD, and TB Prevention, Division of HIV/AIDS Prevention and *The World Almanac and Book of Facts 2000*, p. 903.

Method of Transmission	*Male*	*Female*	*Total*
Homosexual Contact	8388	0	8388
Heterosexual Contact	1146	1806	2952
Intravenous Drug Use	3652	1541	5193
Other	5237	2054	7291
Total	18,423	5401	23,824

(a) Find the probability that a male resident who was newly diagnosed with AIDS in 1998 contracted it via homosexual contact.

(b) Find the probability that a female resident who was newly diagnosed with AIDS in 1998 contracted it via intravenous drug use.

(c) Find the probability that a newly diagnosed person with AIDS is a female.

(d) Find the probability that a newly diagnosed person who contracted AIDS via heterosexual contact was a female.

Health *The table lists the number of head and neck injuries for* 319 *ice hockey players' exposures wearing either a full shield or half shield in the Canadian Inter-University Athletics Union during the* 1997–1998 *season.**

	Half Shield (H)	*Full Shield (F)*	*Total*
Head and Face Injuries (*A*)	95	34	129
Concussions (*B*)	41	38	79
Neck Injuries (*C*)	9	7	16
Other Injuries (*D*)	202	150	352
Total	347	229	576

For a randomly selected injury, find the following probabilities.

26. $P(A)$

27. $P(C \mid F)$

28. $P(A \mid H)$

29. $P(B' \mid H')$

30. Natural Science In a letter to the journal *Nature*, Robert A. J. Matthews gives the following table of outcomes of forecast and weather over 1000 1-hour walks, based on the United Kingdom's Meteorological Office's 83% accuracy in 24-hour forecasts.†

	Rain	*No Rain*	*Sum*
Forecast of Rain	66	156	222
Forecast of No Rain	14	764	778
Sum	80	920	1000

(a) Verify that the probability that the forecast called for rain, given that there was rain, is indeed 83%. Also verify that the probability that the forecast called for no rain, given that there was no rain, is also 83%.

(b) Calculate the probability that there was rain, given that the forecast called for rain.

(c) Calculate the probability that there was no rain, given that the forecast called for no rain.

(d) Observe that your answer to part (c) is higher than 83% and that your answer to part (b) is much lower. Discuss which figure best describes the accuracy of the weather forecast in recommending whether or not you should carry an umbrella.

Business *The table shows* 1999 *employment figures (in thousands) in two occupational categories.**

	Managerial/ Professional	*Technical/Sales/ Adminis. Support*	*Totals*
Women	14,100	24,800	38,900
Men	20,400	20,000	40,400
Totals	34,500	44,800	79,300

Let A represent managerial/professional workers, B represent Technical/Sales/Administrative Support workers, C represent women, and D represent men. Express each of the following probabilities in words and find its value. (See Example 1.)

31. $P(A \mid D)$

32. $P(C \mid A)$

33. $P(B \mid C)$

34. $P(D \cap A)$

Natural Science *The following table shows frequencies for red-green color blindness, where M represents male and C represents color-blind.*

	M	*M'*	*Totals*
C	.042	.007	.049
C'	.485	.466	.951
Totals	.527	.473	1.000

Use this table to find the following probabilities.

35. $P(M)$

36. $P(C)$

37. $P(M \cap C)$

38. $P(M \cup C)$

39. $P(M \mid C)$

40. $P(M' \mid C)$

*Brian Benson, Nicholas Nohtaki, M. Sarah Rose, and Willem Meeuwisse, "Head and Neck Injuries among Ice Hockey Players Wearing Full Face Shields vs. Half Face Shields," *JAMA* 282, no. 24 (December 22/29, 1999): 2328–2332.

†Robert A. J. Matthews, Nature 382, August 29, 1996: p 3.

*Bureau of Labor Statistics, U.S. Department of Labor.

41. Are the events C and M described above dependent? Recall that two events E and F are dependent if $P(E \mid F) \neq P(E)$. (See Example 10.)

42. **Natural Science** A scientist wishes to determine if there is any dependence between color blindness (C) and deafness (D). Given the probabilities listed in the table below, what should his findings be? (See Example 10.)

	D	*D'*	*Totals*
C	.0004	.0796	.0800
C'	.0046	.9154	.9200
Totals	.0050	.9950	1.0000

Social Science *The Motor Vehicle Department has found that the probability of a person passing the test for a driver's license on the first try is .75. The probability that an individual who fails on the first test will pass on the second try is .80, and the probability that an individual who fails the first and second tests will pass the third time is .70. Find the probability that an individual*

43. fails both the first and second tests;

44. will fail three times in a row;

45. will require at least two tries to pass the test.

46. **Social Science** A smooth-talking young man has a 1/3 probability of talking a policeman out of giving him a speeding ticket. The probability that he is stopped for speeding during a given weekend is 1/2. Find the probabilities of the events in parts (a) and (b).
    (a) He will receive no speeding tickets on a given weekend.
    (b) He will receive no speeding tickets on 3 consecutive weekends.
    (c) We have assumed that what happens on the second or third weekend is the same as what happened on the first weekend. Is this realistic? Will driving habits remain the same after getting a ticket?

Natural Science *Four different medications, C, D, E, and F, may be used to control high blood pressure. A physician usually prescribes C first because it is least likely to cause side effects. If blood pressure remains high, the patient is switched to D. If this fails to work, the patient is switched to E, and if necessary to F. The probability that C will work is .7. If C fails, the probability that D will work is .8. If D fails, the probability that E will work is .62. If E fails, the probability that F will work is .45. Find the probability that*

47. A patient's blood pressure will not be reduced by any of the medications.

48. A patient will have to take at least two medications and will have his or her blood pressure reduced.

49. If medications C and D fail, what is the probability that a patient's blood pressure will be reduced by medication E or F?

Business *The number of new passenger cars in thousands from major worldwide producers in selected years is shown in the table.*

	U.S.	*Canada*	*Germany*	*Japan*
1993	10,898	1468	184	1597
1994	12,263	1591	188	1593
1995	11,985	1678	207	1387
1996	11,799	1688	234	1191

Find the following probabilities for a vehicle selected at random.

50. It was made in the United States.

51. It was made in 1993.

52. It was made in Japan in 1995.

53. It was made in 1996, given that it was made in Germany.

The probability that the first record by a singing group will be a hit is .32. If their first record is a hit, so are all their subsequent records. If their first record is not a hit, the probability of their second record and all subsequent ones being hits is .16. If the first two records are not hits, the probability that the third is a hit is .08. The probability of a hit continues to decrease by half with each successive nonhit record. Find the probability that

54. the group will have at least one hit in their first four records.

55. the group will have exactly one hit in their first three records.

56. the group will have a hit in their first six records if the first three are not hits.

Work the following problems on independent events. (See Examples 9 and 10.)

57. **Business** Corporations such as banks, where a computer is essential to day-to-day operations, often have a second, backup computer in case of failure by the main computer. Suppose that there is a .003 chance that the main computer will fail in a given time period and a .005 chance that the backup computer will fail while the main computer is being repaired. Assume these failures represent independent events and find the fraction of the time that the corporation can assume it will have computer service. How realistic is our assumption of independence?

58. **Business** According to a booklet put out by Eastwest Airlines, 98% of all scheduled Eastwest flights actually take place. (The other flights are canceled due to weather, equipment problems, and so on.) Assume that the event that a given flight takes place is independent of the event that another flight takes place.
    (a) Elisabeta Guervara plans to visit her company's branch offices; her journey requires 3 separate flights on Eastwest Airlines. What is the probability that all of these flights will take place?
    (b) Based on the reasons we gave for a flight to be canceled, how realistic is the assumption of independence that we made?

59. Social Science The probability that a key component of a space rocket will fail is .03.

(a) How many such components must be used as backups to ensure that the probability of at least one of the components' working is .999999?

(b) Is it reasonable to assume independence here?

60. Natural Science A medical experiment showed that the probability that a new medicine is effective is .75, the probability that a patient will have a certain side effect is .4, and the probability that both events occur is .3. Decide whether these events are dependent or independent.

61. Social Science A teacher has found that the probability that a student studies for a test is .6, the probability that a student gets a good grade on a test is .7, and the probability that both occur is .52. Are these events independent?

62. Business Refer to Exercises 31–34. Are the events being a woman and being employed in the managerial/professional category independent?

8.6 BAYES' FORMULA

Suppose the probability that a person gets lung cancer, given that the person smokes a pack or more of cigarettes daily, is known. For a research project, it might be necessary to know the probability that a person smokes a pack or more of cigarettes daily, given that the person has lung cancer. More generally, if $P(E \mid F)$ is known for two events E and F, can $P(F \mid E)$ be found? The answer is yes, we can find $P(F \mid E)$ using the formula to be developed in this section. To develop this formula, we can use a probability tree to find $P(F \mid E)$. Since $P(E \mid F)$ is known, the first outcome is either F or F'. Then for each of these outcomes, either E or E' occurs, as shown in Figure 8.29.

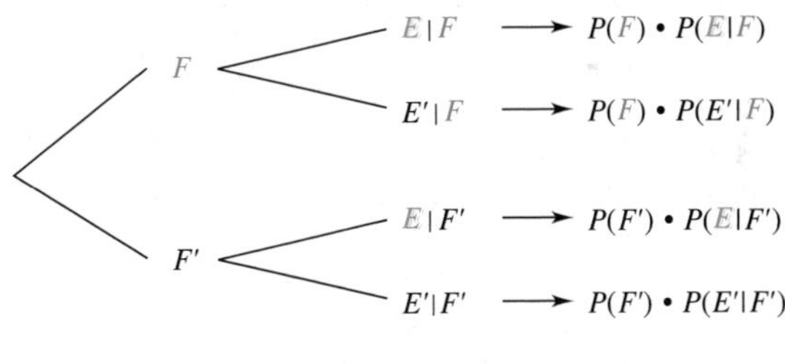

FIGURE 8.29

The four cases have the probabilities shown on the right. By the definition of conditional probability and the product rule,

$$P(E) = P(F \cap E) + P(F' \cap E),$$

$$P(F \cap E) = P(F) \cdot P(E \mid F), \quad \text{and} \quad P(F' \cap E) = P(F') \cdot P(E \mid F').$$

By substitution,

$$P(E) = P(F) \cdot P(E \mid F) + P(F') \cdot P(E \mid F'),$$

and

$$P(F \mid E) = \frac{P(F \cap E)}{P(E)} = \frac{P(F) \cdot P(E \mid F)}{P(F) \cdot P(E \mid F) + P(F') \cdot P(E \mid F')}.$$

We have proved a special case of Bayes' formula, which is generalized later in this section. (1)

(1) Use the special case of Bayes' formula to find $P(F \mid E)$ if $P(F) = .2$, $P(E \mid F) = .1$, and $P(E \mid F') = .3$. (*Hint:* $P(F') = 1 - P(F)$.)

Answer:

$1/13 \approx .077$

Bayes' Formula (Special Case)

$$P(F \mid E) = \frac{P(F) \cdot P(E \mid F)}{P(F) \cdot P(E \mid F) + P(F') \cdot P(E \mid F')}.$$

EXAMPLE 1 For a fixed length of time, the probability of worker error on a certain production line is .1, the probability that an accident will occur when there is a worker error is .3, and the probability that an accident will occur when there is no worker error is .2. Find the probability of a worker error if there is an accident.

Let E represent the event of an accident, and let F represent the event of worker error. From the information above,

$$P(F) = .1, \quad P(E \mid F) = .3, \quad \text{and} \quad P(E \mid F') = .2.$$

These probabilities are shown on the probability tree in Figure 8.30.

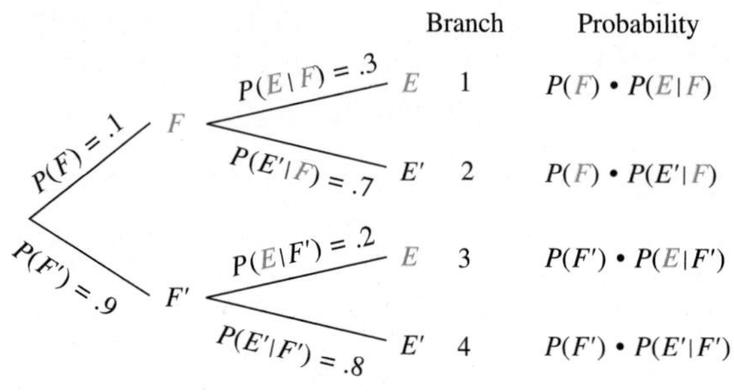

FIGURE 8.30

Find $P(F \mid E)$ by dividing the probability that both E and F occur, given by branch 1, by the probability that E occurs, given by the sum of branches 1 and 3.

$$P(F \mid E) = \frac{P(F) \cdot P(E \mid F)}{P(F) \cdot P(E \mid F) + P(F') \cdot P(E \mid F')}$$

$$= \frac{(.1)(.3)}{(.1)(.3) + (.9)(.2)} \approx .143 \quad ②$$

②

In Example 1, find $P(F' \mid E)$.

Answer:

$6/7 \approx .857$

The special case of Bayes' formula can be generalized to more than two possibilities with the probability tree of Figure 8.31. This diagram shows the paths that can produce an event E. We assume that events $F_1, F_2, \ldots, F_n$ are pairwise mutually exclusive events (that is, events which, taken two at a time, are disjoint), whose union is the sample space, and E is an event that has occurred. See Figure 8.32.

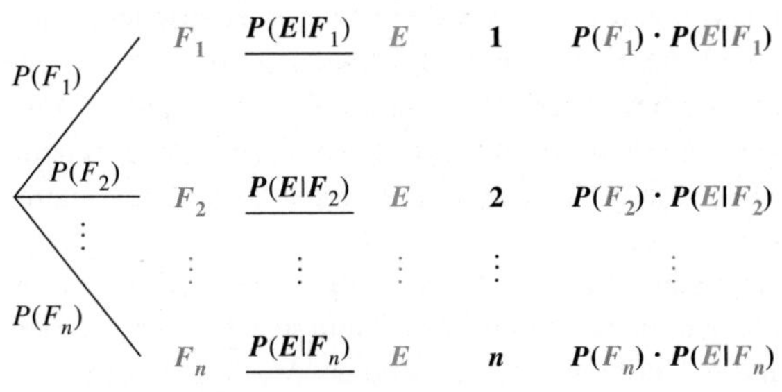

FIGURE 8.31

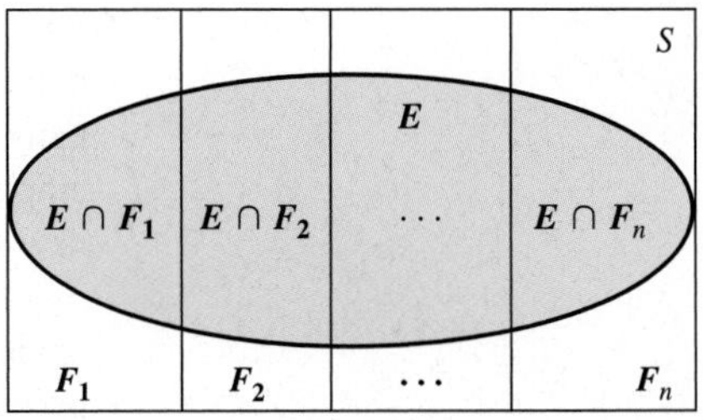

S
E
$E \cap F_1$ $E \cap F_2$ ... $E \cap F_n$
F_1 F_2 ... F_n

FIGURE 8.32

The probability $P(F_i \mid E)$, where $1 \leq i \leq n$, can be found by dividing the probability for the branch containing $P(E \mid F_i)$ by the sum of the probabilities of all the branches producing event E.

Bayes' Formula

$$P(F_i \mid E) = \frac{P(F_i) \cdot P(E \mid F_i)}{P(F_1) \cdot P(E \mid F_1) + \cdots + P(F_n) \cdot P(E \mid F_n)}.$$

This result is known as **Bayes' formula,** after the Reverend Thomas Bayes (1702–61), whose paper on probability was published about two hundred years ago.

The statement of Bayes' formula can be daunting. Actually, it is easier to remember the formula by thinking of the probability tree that produced it. Go through the following steps.

Using Bayes' Formula

Step 1 Start a probability tree with branches representing events F_1, $F_2, \ldots, F_n$. Label each branch with its corresponding probability.

Step 2 From the end of each of these branches, draw a branch for event E. Label this branch with the probability of getting to it, or $P(E \mid F_i)$.

Step 3 There are now n different paths that result in event E. Next to each path, put its probability—the product of the probabilities that the first branch occurs, $P(F_i)$, and that the second branch occurs, $P(E \mid F_i)$: that is, $P(F_i) \cdot P(E \mid F_i)$.

Step 4 $P(F_i \mid E)$ is found by dividing the probability of the branch for F_i by the sum of the probabilities of all the branches producing event E.

EXAMPLE 2 Based on past experience, a company knows that an experienced machine operator (one or more years of experience) will produce a defective item 1% of the time. Operators with some experience (up to one year) have a 2.5% defect rate, while new operators have a 6% defect rate. At any one time, the company has 60% experienced employees, 30% with some experience, and 10% new employees. Find the probability that a particular defective item was produced by a new operator.

Let E represent the event "item is defective," with F_1 representing "item was made by an experienced operator," F_2 "item was made by an operator with some experience," and F_3 "item was made by a new operator." Then

$$P(F_1) = .60 \qquad P(E \mid F_1) = .01$$
$$P(F_2) = .30 \qquad P(E \mid F_2) = .025$$
$$P(F_3) = .10 \qquad P(E \mid F_3) = .06.$$

We need to find $P(F_3 \mid E)$, the probability that an item was produced by a new operator, given that it is defective. First, draw a probability tree using the given information, as in Figure 8.33. The steps leading to event E are shown.

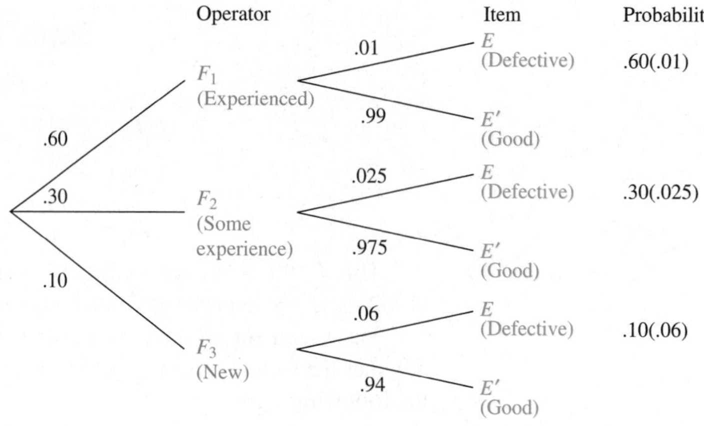

FIGURE 8.33

3 In Example 2, find
(a) $P(F_1 \mid E)$;
(b) $P(F_2 \mid E)$.

Answers:

(a) $4/13 \approx .3077$
(b) $5/13 \approx .3846$

Find $P(F_3 \mid E)$ using the bottom branch of the tree shown in Figure 8.33: divide the probability for this branch by the sum of the probabilities of all the branches leading to E.

$$P(F_3 \mid E) = \frac{.10(.06)}{.60(.01) + .30(.025) + .10(.06)} = \frac{.006}{.0195} \approx .3077 \quad ③$$

After working problem 3 at the side, check that $P(F_1 \mid E) + P(F_2 \mid E) + P(F_3 \mid E) = 1$. (That is, the defective item was made by *someone*.)

EXAMPLE 3 A manufacturer buys items from six different suppliers. The fraction of the total number of items obtained from each supplier, along with the probability that an item purchased from that supplier is defective, is shown in the following table.

Supplier	*Fraction of Total Supplied*	*Probability of Defective*
1	.05	.04
2	.12	.02
3	.16	.07
4	.23	.01
5	.35	.03
6	.09	.05

Find the probability that a defective item came from supplier 5.

Let F_1 be the event that an item came from supplier 1, with F_2, F_3, F_4, F_5, and F_6 defined in a similar manner. Let E be the event that an item is defective. We want to find $P(F_5 \mid E)$. Use the probabilities in the table above to prepare a probability tree. By Bayes' formula,

$$P(F_5 \mid E) = \frac{(.35)(.03)}{(.05)(.04) + (.12)(.02) + (.16)(.07) + (.23)(.01) + (.35)(.03) + (.09)(.05)}$$

$$= \frac{.0105}{.0329} \approx .319.$$

There is about a 32% chance that a defective item came from supplier 5. Even though supplier 5 has only 3% defectives, his probability of being "guilty" is relatively high, about 32%, because of the large fraction supplied by 5. ◀ (4)

(4)

In Example 3, find the probability that the defective item came from

(a) supplier 3;

(b) supplier 6.

Answers:

(a) .340

(b) .137

8.6 Exercises

For two events, M and N, $P(M) = .4$, $P(N \mid M) = .3$, and $P(N \mid M') = .4$. Find each of the following. (See Example 1.)

1. $P(M \mid N)$

2. $P(M' \mid N)$

For mutually exclusive events R_1, R_2, R_3, $P(R_1) = .05$, $P(R_2) = .6$, and $P(R_3) = .35$. In addition, $P(Q \mid R_1) = .40$, $P(Q \mid R_2) = .30$, and $P(Q \mid R_3) = .60$. Find each of the following. (See Examples 2 and 3.)

3. $P(R_1 \mid Q)$

4. $P(R_2 \mid Q)$

5. $P(R_3 \mid Q)$

6. $P(R_1' \mid Q)$

Suppose three jars have the following contents: 2 black balls and 1 white ball in the first; 1 black ball and 2 white balls in the second; 1 black ball and 1 white ball in the third. If the probability of selecting one of the three jars is 1/2, 1/3, and 1/6, respectively, find the probability that if a white ball is drawn, it came from the

7. second jar;

8. third jar.

Social Science *In 1990, 1% of the U.S. population was Native American. The probability that a Native American lives in the West or South is .76. The probability is .56 that a person who is not Native American lives in the West or South. Find the following probabilities.*

9. A person living in the West or South is Native American.

10. A person who does not live in the West or South is not Native American.

Social Science *A federal study showed that in 1990, 49% of all those involved in a fatal car crash wore seat belts. Of those in a fatal crash who wore seat belts, 44% were injured and 27% were killed. For those not wearing seat belts, the comparable figures were 41% and 50%, respectively.**

*National Highway Traffic Safety Administration, Office of Driver and Pedestrian Research: "Occupant Protection Trends in 19 Cities" (November 1989) and "Use of Automatic Safety Belt Systems in 19 Cities" (February 1991).

11. Find the probability that a randomly selected person who was killed in a car crash was wearing a seat belt.

12. Find the probability that a randomly selected person who was unharmed in a fatal crash was not wearing a seat belt.

Business *A building contractor buys 70% of his cement from supplier A, and 30% from supplier B. A total of 90% of the bags from A arrive undamaged, while 95% of the bags from B arrive undamaged. Give the probabilities that a damaged bag is from each of the following sources.*

13. Supplier A **14.** Supplier B

Business *The probability that a customer of a local department store will be a "slow pay" is .02. The probability that a "slow pay" will make a large down payment when buying a refrigerator is .14. The probability that a person who is not a "slow pay" will make a large down payment when buying a refrigerator is .50. Suppose a customer makes a large down payment on a refrigerator. Find the probability that the customer is*

15. a "slow pay"; **16.** not a "slow pay."

17. Social Science A 1995 survey showed that of Baby Boomers (Americans born between 1946 and 1960), 56% have a family income of more than $40,000.* The probability that a Baby Boomer's family own their own home if their income is over $40,000 is .51, while the probability that they own their home if the family income is less than or equal to $40,000 is .27. Find the probability that a Baby Boomer has family income over $40,000, if it is known that the family owns its home.

18. Social Science During the murder trial of O. J. Simpson, Alan Dershowitz, an adviser to the defense team, stated on television that only about .1% of men who batter their wives actually murder them. Statistician I. J. Good observed that even if, given that a husband is a batterer, the probability he is guilty of murdering his wife is .001, what we really want to know is the probability that the husband is guilty, given that the wife was murdered.† Good estimates the probability of a battered wife being murdered, given that her husband is not guilty, as .001. The probability that she is murdered if her husband is guilty is 1, of course. Using these numbers and Dershowitz's .001 probability of the husband being guilty, find the probability that the husband is guilty, given that the wife was murdered.

19. Management The following information pertains to three shipping terminals operated by Krag Corp.:‡

Terminal	*Percentage of Cargo Handled*	*Percentage of Error*
Land	50	2
Air	40	4
Sea	10	14

Krag's internal auditor randomly selects one set of shipping documents, ascertaining that the set selected contains an error. Which of the following gives the probability that the error occurred in the Land Terminal?
(a) .02 **(b)** .10 **(c)** .25 **(d)** .50

Health *In a test for toxemia, a disease that affects pregnant women, the woman lies on her left side and then rolls over on her back. The test is considered positive if there is a 20 mm rise in her blood pressure within one minute. The results have produced the following probabilities, where T represents having toxemia at some time during the pregnancy and N represents a negative test.*

$$P(T' \mid N) = .90 \quad \text{and} \quad P(T \mid N') = .75$$

Assume that $P(N') = .11$, *and find each of the following.*

20. $P(N \mid T)$

21. $P(N' \mid T)$

22. Health In 2000, 7.3% of Americans were diabetic, 19.8% of Americans were obese, and 80% of diabetics were obese.* Find the probability that an American is diabetic given that person is obese.

23. Health In 1987, it was found that one woman in 10,000 was infected with the AIDS virus. The most commonly used test always detected an infected woman, but one healthy woman out of 20,000 was incorrectly identified as being infected with the AIDS virus.†
(a) Find the probability that a woman who tested positive in 1987 was not infected.
(b) It has been argued that everyone should be tested for AIDS. Based on the results of part (a), how useful would the results of such testing be?

24. Health The probability that a person with certain symptoms has hepatitis is .8. The blood test used to confirm this diagnosis gives positive results for 90% of people with the disease and 5% of those without the disease. What is the probability that an individual who has the symptoms and who reacts positively to the test actually has hepatitis?

**New York Times*, sec. 3, p. 1, December 31, 1995.

†I. J. Good, "When Batterer Turns Murderer," *Nature* 375, no. 15 (June 15, 1995), p. 541.

‡Uniform CPA Examination, November 1989.

**The Sacramento Bee*, September 12, 2001, p. A20.

†*New York Times*, September 5, 1987.

25. Health The sensitivity of a medical test is defined as the probability that a test will be positive given that a person has a disease, written $P(T^+ \mid D^+)$. The specificity of a test is defined as the probability that a test will be negative given that the person does not have the disease, written $P(T^- \mid D^-)$. For example, the sensitivity and specificity for breast cancer during a clinical breast examination by a trained expert are approximately .54 and .94, respectively.*

(a) If 2% of U.S. women have breast cancer,† find the probability that a woman who tests positive during a clinical breast examination actually has breast cancer.

(b) Given that a women tests negative during a clinical breast examination, find the probability that she does not have breast cancer.

(c) How many false positives could be expected for every 1000 clinical breast examinations? (A false positive refers to a person who does not have the disease, but tests positive.)

26. Health A new test for the virus that causes AIDS, developed by Octopus Diagnostics Research of Hantsport. Nova Scotia, shows the presence or absence of HIV in a drop of blood in two minutes, compared with five days for a conventional test.‡ Preliminary results indicate a false positive rate (an indication that the HIV virus is present when it is not) of less than 2% and a false negative rate (a failure to detect the presence of the HIV virus) of up to 5%. Assume for this exercise that these rates are exactly 2% and 5%. In 1996, there were 780,000 people in North America with the HIV virus, out of a population of 295 million.§ Suppose a resident of North America is chosen at random and given this test. If the result is positive, what is the probability that the person actually has the HIV virus?

27. Social Science A recent study by the Harvard School of Public Health reported that 86% of male students who live in a fraternity house are binge drinkers. The figure for fraternity members who are not residents of a fraternity house is 71%, while the figure for men who do not belong to a fraternity is 45%.¶ Suppose that 10% of U.S. students live in a fraternity house, 15% belong to a fraternity, but do not live in a fraternity house, and 75% do not belong to a fraternity.

(a) What is the probability that a randomly selected male student is a binge drinker?

(b) If a randomly selected male student is a binge drinker, what is the probability that he lives in a fraternity house?

28. Finance A bank finds that the relationship between mortgage defaults and the size of the down payment is given by this table.

Down Payment (%)	5%	10%	20%	25%
Number of Mortgages with this Down Payment	1260	700	560	280
Probability of Default	.05	.03	.02	.01

If a default occurs, what is the probability that it is on a mortgage with a 5% down payment? (See Examples 2 and 3.)

Business *The table shows the percent of U.S. households in each region and the percent per region using central air conditioning.*

	Northeast	*Midwest*	*South*	*West*
Region	19.4	23.7	35.2	21.5
Uses Central AC	22.1	51.3	69.4	26.9

Find the probability that a household using central air conditioning is

29. in the South

30. in the Midwest

Social Science *The following tables give the proportions of men and women in the U.S. population, and the proportions of men and women who have never married, in a recent year.* Use them for Exercises 31–33.*

	MEN	
Age	*Proportion in Population*	*Proportion Never Married*
18–24	.133	.874
25–34	.205	.397
35–44	.232	.187
45–64	.287	.075
65 or Over	.142	.038

*Mary B. Barton, Russell Harris, and Suzanne Fletcher, "Does This Patient Have Breast Cancer? The Screening Clinical Breast Examinations: Should It Be Done? How?" *JAMA,* 282 no. 13 (October 6, 1999): 1270–1280.

†*The World Almanac and Book of Facts 2000,* p. 902.

‡*Maclean's,* February 17, 1997, p. 70.

§*The World Almanac Book of Facts 1997.*

¶*New York Times,* December 6, 1995, p. B16.

*"Marital Status and Living Arrangements of Adults 18 Years and Over," March, 1998, Table B. U.S. Bureau of the Census.

WOMEN		
Age	*Proportion in Population*	*Proportion Never Married*
18–24	.123	.775
25–34	.194	.298
35–44	.219	.121
45–64	.284	.062
65 or Over	.181	.047

31. Find the probability that a randomly selected man who has never married is between 35 and 44 years old (inclusive).

32. Find the probability that a randomly selected woman who has been married is between 18 and 24 (inclusive).

33. Find the probability that a randomly selected woman who has never been married is between 45 and 64 (inclusive).

Social Science *The following table gives the proportions of people over 18 who are in various age categories, along with the probabilities that a person in a given category will vote in a general election.*

Age	*Percent of Voting Age Population*	*Probability of a Person of This Age Voting*
18–21	11.0%	.48
22–24	7.6%	.53
25–44	37.6%	.68
45–64	28.3%	.64
65 or Over	15.5%	.74

Suppose a voter is picked at random. Find the probabilities that the voter is in the following age categories.

34. 18–21

35. 65 or over

36. Find the probability that a person who did not vote was in the age category 45–64.

CHAPTER 8 SUMMARY

Key Terms and Symbols

{ } set braces
∈ is an element of
∉ is not an element of
∅ empty set
⊆ is a subset of
⊄ is not a subset of
A' complement of set A
∩ set intersection
∪ set union
$P(E)$ probability of event E
$P(F \mid E)$ probabilty of F, given that E has occurred

8.1 set
element (member)
empty set
set-builder notation
universal set
subset
set operations
tree diagram
Venn diagram
complement
intersection
disjoint sets
union

8.2 union rule for counting

8.3 experiment
trial
outcome
sample space
event
simple event
certain event
impossible event
mutually exclusive events
basic probability principle

8.4 union rule for probability
complement rule
odds
empirical probability
probability distribution

8.5 conditional probability
product rule of probability
probability tree
independent events
dependent events

8.6 Bayes' formula

Key Concepts

Sets

Set A is a **subset** of set B if every element of A is also an element of B.

A set of n elements has 2^n subsets.

Let A and B be any sets with universal set U.

The **complement** of A is $A' = \{x \mid x \notin A \text{ and } x \in U\}$.

The **intersection** of A and B is $A \cap B = \{x \mid x \in A \text{ and } x \in B\}$.

The **union** of A and B is $A \cup B = \{x \mid x \in A \text{ or } x \in B \text{ or both}\}$.

$n(A \cup B) = n(A) + n(B) - n(A \cap B)$, where $n(X)$ is the number of elements in set X

Probability Summary

Basic Probability Principle

Let S be a sample space of equally likely outcomes and let event E be a subset of S. Then the probability that event E occurs is

$$P(E) = \frac{n(E)}{n(S)}.$$

Union Rule

For any events E and F from a sample space S.

$$P(E \cup F) = P(E) + P(F) - P(E \cap F).$$

For mutually exclusive events E and F,

$$P(E) \cup F) = P(E) + P(F).$$

Complement Rule

$P(E) = 1 - P(E)$ and $P(E') = 1 - P(E)$

Odds

The odds in favor of event E are $\frac{P(E)}{P(E')}$, $P(E') \neq 0$.

Properties of Probability

1. For any event E in sample space S, $0 \leq P(E) \leq 1$.
2. The sum of the probabilities of all possible distinct outcomes is 1.

Conditional Probability

The conditional probability of event E, given that event F has occurred, is

$$P(E \mid F) = \frac{P(E \cap F)}{P(F)}, \quad \text{where} \quad P(F) \neq 0.$$

For equally likely outcomes, conditional probability is found by reducing the sample space to event F; then

$$P(E \mid F) = \frac{n(E \cap F)}{n(F)}.$$

Product Rule of Probability

If E and F are events, then $P(E \cap F)$ may be found by either of these formulas.

$$P(E \cap F) = P(F) \cdot P(E \mid F) \quad \text{or} \quad P(E \cap F) = P(E) \cdot P(F \mid E)$$

If E and F are independent events, then $P(E \cap F) = P(E) \cdot P(F)$.

Bayes' Formula

$$P(F_i \mid E) = \frac{P(F_i) \cdot P(E \mid F_i)}{P(F_1) \cdot P(E \mid F_1) + P(F_2) \cdot P(E \mid F_2) + \cdots + P(F_n) \cdot P(E \mid F_n)}$$

Chapter 8 Review Exercises

Write true *or* false *for each of the following.*

1. $9 \in \{8, 4, -3, -9, 6\}$
2. $4 \in \{3, 9, 7\}$
3. $2 \notin \{0, 1, 2, 3, 4\}$
4. $0 \notin \{0, 1, 2, 3, 4\}$
5. $\{3, 4, 5\} \subseteq \{2, 3, 4, 5, 6\}$
6. $\{1, 2, 5, 8\} \subseteq \{1, 2, 5, 10, 11\}$
7. $\emptyset \subseteq \{1\}$
8. $0 \subseteq \emptyset$

List the elements in the following sets.

9. $\{x \mid x \text{ is a national holiday}\}$
10. $\{x \mid x \text{ is an integer}, -3 \leq x < 1\}$
11. $\{\text{all counting numbers less than } 5\}$
12. $\{x \mid x \text{ is a leap year between 1989 and 2003}\}$

Let $U = \{\text{Vitamins } A, B_1, B_2, B_3, B_6, B_{12}, C, D, E\}$, $M = \{A, C, D, E\}$, and $N = \{A, B_1, B_2, C, E\}$. Find the following.

13. M'
14. N'
15. $M \cap N$
16. $M \cup N$
17. $M \cup N'$
18. $M' \cap N$

Let U = {all students in a class}, A = {all male students}, B = {all A students}, C = {all students with red hair}, and D = {all students younger than 21}. Describe the following sets in words.

19. $A \cap C$
20. $B \cap D$
21. $A \cup D$
22. $A' \cap D$
23. $B' \cap C'$

Draw a Venn diagram and shade the given set.

24. $B \cup A'$
25. $A' \cap B$
26. $A' \cap (B' \cap C)$
27. $(A \cup B)' \cap C$

Social Science *A survey of a group of military personnel revealed the following information.*

20 officers
27 minorities
19 women
5 women officers
8 minority women
10 minority officers
3 women minority officers
6 Caucasian male enlisted personnel

28. How many were interviewed?
29. How many were enlisted minority women?
30. How many were male minority officers?

Write sample spaces for the following.

31. A die is rolled and the number of points showing is noted.
32. A card is drawn from a deck containing only 4 aces.
33. A color is selected from the set {red, blue, green}, and then a number is chosen from the set {10, 20, 30}.

A jar contains 5 discs labeled 2, 4, 6, 8, 10, and another jar contains 2 blue and 3 yellow balls. One disc is drawn and then a ball is drawn. Give the following.

34. The sample space
35. Event F, the ball is blue.
36. Event E, the disc shows a number greater than 5.
37. Are the outcomes in this sample space equally likely?

Business *A company sells computers and copiers. Let E be the event "a customer buys a computer," and let F be the event "a customer buys a copier." In Exercises 38 and 39, write each of the following using ∩, ∪, or ′ as necessary.*

38. A customer buys neither.
39. A customer buys at least one.
40. A student gives the answer to a probability problem as 6/5. Explain why this answer must be incorrect.
41. Describe what is meant by disjoint sets and give an example.
42. Describe what is meant by mutually exclusive events and give an example.
43. How are disjoint sets and mutually exclusive events related?

A single card is drawn from an ordinary deck. Find the probability of drawing each of the following.

44. A black king
45. A face card
46. A red card or a face card
47. A black card, given it is a 2
48. A jack, given it is a face card
49. A face card, given it is a jack

Find the odds in favor of drawing the following.

50. A spade

51. A red queen

52. A black face card or a 7

Business *A sample shipment of five electric motors is chosen at random. The probability of exactly* 0, 1, 2, 3, 4, *or* 5 *motors being defective is given in the following table.*

Number defective	0	1	2	3	4	5
Probability	.31	.25	.18	.12	.08	.06

Find the probability that

53. no more than 3 are defective.

54. at least 3 are defective.

Health *The partial table shows the four possible (equally likely) combinations when both parents are carriers of the sickle cell anemia trait. Each carrier parent has normal cells (N) and trait cells (T).*

		2nd Parent	
		N_2	T_2
1st Parent	N_1		N_1T_2
	T_1		

55. Complete the table.

56. If the disease occurs only when two trait cells combine, find the probability that a child born to these parents will have sickle cell anemia.

57. The child will carry the trait but not have the disease if a normal cell combines with a trait cell. Find this probability.

58. Find the probability that the child is neither a carrier nor has the disease.

Find the probabilities for the following sums when two fair dice are rolled.

59. 8

60. At least 10

61. No more than 5

62. Odd and greater than 8

63. 12, given the sum is greater than 10

64. 7, given that at least one die shows a 4

Suppose P(E) = .51, P(F) = .37, and P(E ∩ F) = .22. Find each of the following probabilities.

65. $P(E \cup F)$

66. $P(E \cap F')$

67. $P(E' \cup F)$

68. $P(E' \cap F')$

69. For the events E and F, $P(E) = .2$, $P(E \mid F) = .3$, and $P(F) = .4$. Find each of the following.

(a) $P(E' \mid F)$ **(b)** $P(E \mid F')$

70. Define independent events and give an example.

71. Are independent events always mutually exclusive? Are they ever mutually exclusive? Give examples.

Business *Of the appliance repair shops listed in the phone book,* 80% *are competent and* 20% *are not. A competent shop can repair an appliance correctly* 95% *of the time; an incompetent shop can repair an appliance correctly* 60% *of the time. Suppose an appliance was repaired correctly. Find the probability that it was repaired by*

72. a competent shop;

73. an incompetent shop.

Suppose an appliance was repaired incorrectly. Find the probability that it was repaired by

74. a competent shop;

75. an incompetent shop.

76. Business A manufacturer buys items from four different suppliers. The fraction of the total number of items that is obtained from each supplier, along with the probability that an item purchased from that supplier is defective, is shown in the following table.

Supplier	*Fraction of Total Supplied*	*Probability of Defective*
1	.17	.04
2	.39	.02
3	.35	.07
4	.09	.03

(a) Find the probability that a defective item came from supplier 4.

(b) Find the probability that a defective item came from supplier 2.

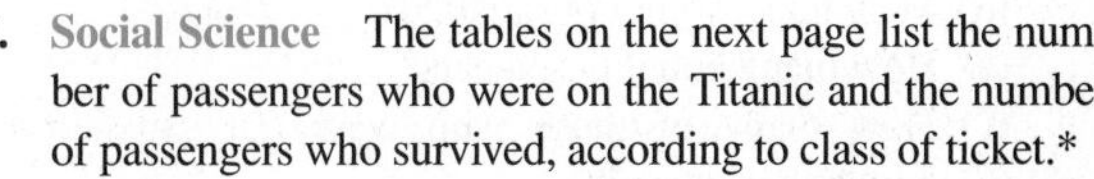

77. Social Science The tables on the next page list the number of passengers who were on the Titanic and the number of passengers who survived, according to class of ticket.*

*Sandra L. Takis, "Titanic: A Statistical Exploration," *Mathematics Teacher* 92, no. 8, (November 1999): p. 660–664. Reprinted with permission. ©1999 by the National Council of Teachers of Mathematics. All rights reserved.

	CHILDREN		WOMEN	
	On	*Survived*	*On*	*Survived*
First Class	6	6	144	140
Second Class	24	24	165	76
Third Class	79	27	93	80
Total	109	57	402	296

	MEN		TOTALS	
	On	*Survived*	*On*	*Survived*
First Class	175	57	325	203
Second Class	168	14	357	114
Third Class	462	75	634	182
Total	805	146	1316	499

Use this information to determine the following (round answers to two decimal places).

(a) What is the probability that a randomly selected passenger was in second class?

(b) What is the overall probability of surviving?

(c) What is the probability of a first-class passenger surviving?

(d) What is the probability of a child who was also in third class surviving?

(e) Given that the survivor is from first class, what is the probability that she was a woman?

(f) Given that a male has survived, what is the probability that he was in third class?

(g) Are the events third-class survival and male survival independent events? What does this imply?

78. Business The table below shows the results of a survey of buyers of a certain model of car.

Car Type	*Satisfied*	*Not Satisfied*	*Totals*
New	300	100	
Used	450		600
Totals		250	

(a) Complete the table.

(b) How many buyers were surveyed?

(c) How many bought a new car and were satisfied?

(d) How many were not satisfied?

(e) How many bought used cars?

(f) How many of those who were not satisfied had bought a used car?

(g) Rewrite the event stated in part (f) using the expression "given that."

(h) Find the probability of the outcome in parts (f) and (g).

(i) Find the probability that a used-car buyer is not satisfied.

(j) You should have different answers in parts (h) and (i). Explain why.

Additional Probability Review Exercises

Use these exercises for practice deciding which rule, principle, or formula to apply.

1. Suppose $P(E) = .4$, $P(F) = .22$, and $P(E \cup F) = .52$. Find

(a) $P(E \cup F')$ **(b)** $P(E \cap F')$ **(c)** $P(E' \cup F)$

2. A jar contains 2 white, 3 orange, 5 yellow, and 8 black marbles. If a marble is drawn at random, find the probability that it is the following.

(a) White **(c)** Not black

(b) Orange **(d)** Orange or yellow

3. Finance GT Global Theme Funds has worldwide assets distributed as follows.*

Geographic Location	*Percent*
Africa and Middle East	4.2
Latin America	6.0
Asia-Pacific	18.7
Europe	23.1
United States and Canada	48.0

If an asset is chosen randomly, what is the probability that it is from the following regions?

(a) Asia-Pacific

(b) North or South America

(c) Not from Europe

4. Finance GT Global Health Care Fund has invested in the following sectors.*

Sector	*Percent*
Biotechnology	30.7
Pharmaceuticals	25.4
Medical Technology and Supplies	22.3
Health Care Services	8.2
Short-Term and Other	13.4

*Exercises 3 and 4 from GT Global Theme Funds Annual Report, October 31, 1996.

Find the probability that an investment selected at random from this fund is as follows.

(a) In biotechnology or health care services

(b) In pharmaceuticals or medical technology and supplies

(c) Not in biotechnology

5. A single fair die is rolled. Find the following probabilities, if the die shows

(a) a 2, given that the number was odd;

(b) a 4, given that the number was even;

(c) an even number given that the number was 6.

6. Finance According to an article in *Business Week*, in 1989 funding for university research in the United States totaled $15 billion. Support came from various sources, as shown in the table below.*

Source	*Amount (in billions of dollars)*
Federal government	9.0
State and local government	1.2
Institutional	2.7
Industry	1.0
Other	1.1

Find the probability that funds for a particular project came from each of the following sources.

(a) Federal government
(b) Industry
(c) The institution

7. Social Science The population of the United States by race in 1995, and the projected population by race for the year 2025 are given below.†

U.S. Population by Race (in thousands)

Race	*1995*	*2025*
White	193,900	209,900
Hispanic	26,800	50,900
African-American	31,600	44,700
Asian-American	9,200	24,000
Other	1,900	2,800

Find the probability of a randomly selected person being each of the following.

(a) Hispanic in 1995
(b) Hispanic in 2025
(c) African-American in 1995
(d) African-American in 2025

8. Given $P(A) = .2$, $P(B \mid A) = .5$, $P(B \mid A') = .4$, find $P(A \mid B)$.

Business *Companies A, B, and C produce 15%, 40%, and 45%, respectively, of the major appliances sold in a certain area. In that area, 1% of the Company A appliances, $1\frac{1}{2}$% of the Company B appliances, and 2% of the Company C appliances need service within the first year. Suppose an appliance that needs service within the first year is chosen at random; find the probability that it was manufactured by Company*

9. A; **10.** B.

11. Social Science In one area, 4% of the population drives luxury cars. However, 17% of the CPAs drive luxury cars. Are the events "person drives a luxury car" and "person is a CPA" independent?

12. Suppose $P(E) = .05$, $P(F) = .1$, and $P(E \cap F) = .02$. Find

(a) $P(E' \cap F)$ **(b)** $P(E' \cup F')$ **(c)** $P(E \cap F')$.

13. One orange and four red slips of paper are placed in a box. Two red and three orange slips are placed in a second box. A box is chosen at random, and a slip of paper is selected from it. The probability of choosing the first box is 3/8. If the selected slip of paper is orange, what is the probability that it came from the first box?

14. Find the probability that the slip of paper in Exercise 13 came from the second box, given that it is red.

15. Business A manufacturing firm finds that 70% of its new hires turn out to be good workers and 30% poor workers. All current workers are given a reasoning test. Of the good workers, 80% pass it; 40% of the poor workers pass it. Assume that these figures will hold true in the future. If the company makes the test part of its hiring procedure and only hires people who meet the previous requirements and pass the test, what percent of the new hires will turn out to be good workers?

Social Science *A survey has shown that 52% of the women in a certain community work outside the home. Of these women, 64% are married, while 86% of the women who do not work outside the home are married. Find the probability that a woman in that community is*

16. married;

17. a single woman working outside the home.

Social Science *A study showed that, in 1991, 31.6% of men and 35.0% of women were obese.* Given that 48.7% of*

*"University Research: The Squeeze Is On," *Business Week*, May 20, 1991. Reprinted by special permission. Copyright © 1991 by McGraw-Hill, Inc.

†Projections of Hispanic and Non-Hispanic Populations by Age and Sex: 1995 to 2025, from U.S. Bureau of the Census, *Current Population Reports*.

*U.S. Bureau of the Census: Statistical Abstract of the United States, 1992.

Americans are men and 51.3% are women, find the probability that a randomly selected adult fits the following description.

18. An obese man

19. Obese

Business *On a given weekend in the fall, a tire company can buy television advertising time for a college football game, a baseball game, or a professional football game. If the company sponsors the college game, there is a 70% chance of a high rating, a 50% chance if they sponsor a baseball game, and a 60% chance if they sponsor a professional football game. The probability of the company sponsoring these various games is .5, .2, and .3, respectively. Suppose the company does get a high rating; find the probability that it sponsored*

20. a college game;

21. a professional football game.

22. Social Science A nationwide survey showed that the odds that an individual uses a seat belt when in a car's front seat are 17:8.* What is the probability that such an individual does not use a seat belt?

23. Social Science The survey mentioned in Exercise 22 also showed that the odds that a driver does not drink or does not drive after drinking are 4:1.* What is the probability that a driver was drinking?

24. If a marble is drawn from a bag containing 2 yellow, 5 red, and 3 blue marbles, what are the probabilities of the following?
(a) The marble is red.
(b) The marble is either yellow or blue.
(c) The marble is yellow or red.
(d) The marble is green.

25. The law firm of Alam, Bartolini, Chinn, Dickinson, and Ellsberg has two senior partners, Alam and Bartolini. Two of the attorneys are to be selected to attend a conference. Assuming that all are equally likely to be selected, find the following probabilities.
(a) Chinn is selected.
(b) Ellsberg is not selected.
(c) Alam and Dickinson are selected.
(d) At least one senior partner is selected.

*Exercises 22 and 23 based on an article in the *Sacramento Bee*, June 4, 1992.

Case 7

Medical Diagnosis

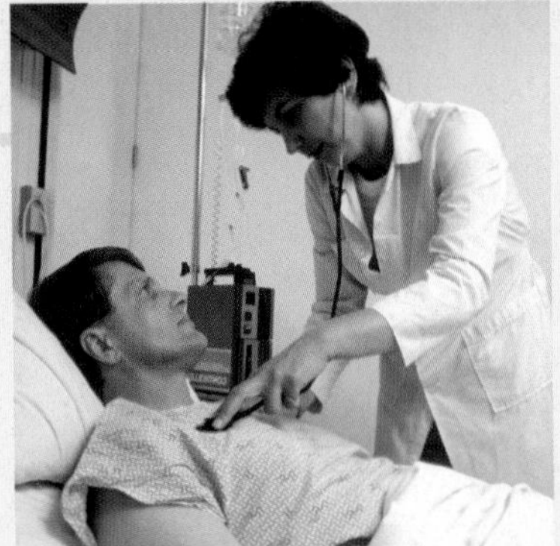

When a patient is examined, information, typically incomplete, is obtained about his or her state of health. Probability theory provides a mathematical model appropriate for this situation, as well as a procedure for quantitatively interpreting such partial information to arrive at a reasonable diagnosis.*

To do this, list the states of health that can be distinguished in such a way that the patient can be in one and only one state at the time of the examination. Each state of health H is associated with a number $P(H)$ between 0 and 1 such that the sum of all these numbers is 1. This number $P(H)$ represents the probability, before exam, that a patient is in the state of health H, and $P(H)$ may be chosen subjectively from medical experience, using any information available prior to the examination. The probability may be most conveniently established from clinical records; that is, a mean probability is established for patients in general, although the number would vary from patient to patient.

For example, limiting the discussion to the condition of a patient's heart, suppose there are exactly 3 states of health, with probabilities as follows.

	State of Health H	*P(H)*
H_1	patient has a normal heart	.8
H_2	patient has minor heart irregularities	.15
H_3	patient has a severe heart condition	.05

Having selected $P(H)$, the information from the examination is processed. First, the results of the examination must be classified. The examination itself consists of observing the state of a number of characteristics of the patient. Let us assume that the examination for a heart condition consists of a stethoscope examination and a cardiogram. The outcome of such an examination, C, might be one of the following:

C_1 = stethoscope shows normal heart and cardiogram shows normal heart;

C_2 = stethoscope shows normal heart and cardiogram shows minor irregularities;

and so on.

*From "Probabilistic Medical Diagnosis," Roger Wright, *Some Mathematical Models in Biology*, Robert M. Thrall, ed. (The University of Michigan, 1967), by permission of Robert M. Thrall.

It remains to assess for each state of health H the conditional probability $P(C \mid H)$ of each examination outcome C using only the knowledge that a patient is in a given state of health. (This may be based on the medical knowledge and clinical experience of the doctor.) The conditional probabilities $P(C \mid H)$ will not vary from patient to patient, so that they may be built into a diagnostic system, although they should be reviewed periodically.

Suppose the result of the examination is C_1. Let us assume the following probabilities.

$$P(C_1 \mid H_1) = .9$$
$$P(C_1 \mid H_2) = .4$$
$$P(C_1 \mid H_3) = .1$$

Now, for a given patient, the appropriate probability associated with each state of health H, after examination, is $P(H \mid C)$ where C is the outcome of the examination. This can be calculated by using Bayes' formula. For example, to find $P(H_1 \mid C_1)$—that is, the probability that the patient has a normal heart given that the examination showed a normal stethoscope examination and a normal cardiogram—we use Bayes' formula as follows.

$$P(H_1 \mid C_1)$$
$$= \frac{P(C_1 \mid H_1)P(H_1)}{P(C_1 \mid H_1)P(H_1) + P(C_1 \mid H_2)P(H_2) + P(C_1 \mid H_3)P(H_3)}$$
$$= \frac{(.9)(.8)}{(.9)(.8) + (.4)(.15) + (.1)(.05)} \approx .92$$

Hence, the probability is about .92 that the patient has a normal heart on the basis of the examination results. This means that in 8 out of 100 patients, some abnormality will be present and not be detected by the stethoscope or the cardiogram.

Exercises

1. Find $P(H_2 \mid C_1)$.
2. Assuming the following probabilities, find $P(H_1 \mid C_2)$:

   $P(C_2 \mid H_1) = .2, \quad P(C_2 \mid H_2) = .8, \quad P(C_2 \mid H_3) = .3.$
3. Assuming the probabilities of Exercise 2, find $P(H_3 \mid C_2)$.

Chapter 9

Counting, Probability Distributions, and Further Topics in Probability

A survey by *Money* magazine found that supermarket scanners overcharged customers at 30% of stores. If you shop at 3 supermarkets that use scanners, what is the probability that you will be overcharged? In this chapter, we introduce additional techniques for finding the probabilities of events such as this.

To use the basic definition of probability, $P(E) = n(E)/n(S)$ (where S is a sample space with equally likely outcomes), up to now we have simply listed the outcomes in S and in E. However, when S has many outcomes, listing them all becomes very tedious. In the first section of this chapter, we introduce methods for counting the number of outcomes in a set without actually listing them, and then we use this approach in later sections to find probabilities.

9.1 PERMUTATIONS AND COMBINATIONS

We begin with a simple example. If there are three roads from town A to town B and 2 roads from town B to town C, in how many ways can someone travel from A to C by way of B? For each of the 3 roads from A there are 2 different routes leading from B to C, making $3 \cdot 2 = 6$ different trips, as shown in Figure 9.1 on the next page. This example illustrates a general principle of counting called the *multiplication principle*.

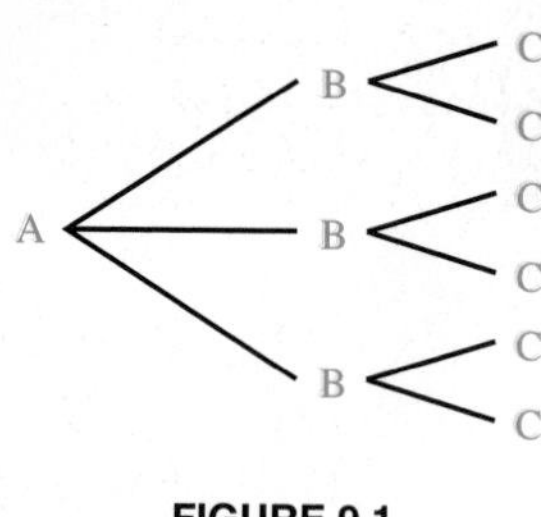

FIGURE 9.1

Multiplication Principle

Suppose n choices must be made, with

$$m_1 \text{ ways to make choice } 1,$$

and for each of these,

$$m_2 \text{ ways to make choice } 2,$$

and so on, with

$$m_n \text{ ways to make choice } n.$$

Then there are

$$m_1 \cdot m_2 \cdot \cdots \cdot m_n$$

different ways to make the entire sequence of choices.

EXAMPLE 1 A combination lock can be set to open to any 3-letter sequence. How many such sequences are possible?

Since there are 26 letters in the alphabet, there are 26 choices for each of the 3 letters, and, by the multiplication principle, $26 \cdot 26 \cdot 26 = 17{,}576$ different sequences.

(a) In how many ways can 6 business tycoons line up their golf carts at the country club?

(b) How many ways can 4 pupils be seated in a row with 4 seats?

Answers:

(a) $6 \cdot 5 \cdot 4 \cdot 3 \cdot 2 \cdot 1 = 720$

(b) $4 \cdot 3 \cdot 2 \cdot 1 = 24$

EXAMPLE 2 A teacher has 5 different books to be arranged side by side. How many different arrangements are possible?

Five choices will be made, 1 for each space that will hold a book. Any of the 5 possible books could be chosen for the first space. There are 4 possible choices for the second space, since 1 book has already been placed in the first space, 3 possible choices for the third space, and so on. By the multiplication principle, the number of different possible arrangements (sequence of choices) is $5 \cdot 4 \cdot 3 \cdot 2 \cdot 1 = 120$.

The use of the multiplication principle often leads to products such as $5 \cdot 4 \cdot 3 \cdot 2 \cdot 1$, the product of all the natural numbers from 5 down to 1. If n is a natural number, the symbol $n!$ (read "*n factorial*") denotes the product of all the natural numbers from n down to 1. If $n = 1$, this formula is understood to give $1! = 1$.

n-Factorial

For any natural number n,

$$n! = n(n-1)(n-2)\ldots(3)(2)(1).$$

Also, 0! is defined to be the number 1.

(2)

Evaluate:

(a) 4!

(b) 6!

(c) 1!

(d) 6!/4!

Answers:

(a) 24

(b) 720

(c) 1

(d) 30

With this symbol, the product $5 \cdot 4 \cdot 3 \cdot 2 \cdot 1$ can be written as 5!. Also, $3! = 3 \cdot 2 \cdot 1 = 6$. The definition of $n!$ could be used to show that $n! = n \cdot (n - 1)!$ for all natural numbers $n \geq 2$. It is helpful if this result also holds for $n = 1$. This can only happen if 0! equals 1, as defined above. (2)

Almost all calculators have an $n!$ key. A calculator with a 10-digit display and scientific notation capability will usually give the exact value of $n!$ for $n \leq 13$ and approximate values of $n!$ for $14 \leq n \leq 69$. The value of 70! is approximately $1.198 \cdot 10^{100}$, which is too large for most calculators.

TECHNOLOGY TIP The factorial key on a graphing calculator is usually located in the PRB or PROB submenu of the MATH or OPTN menu.

EXAMPLE 3 Morse code uses a sequence of dots and dashes to represent letters and words. How many sequences are possible with at most 3 symbols?

"At most 3" means "1 or 2 or 3." Each symbol may be either a dot or a dash. Thus, the following numbers of sequences are possible in each case.

Number of Symbols	*Number of Sequences*
1	2
2	$2 \cdot 2 = 4$
3	$2 \cdot 2 \cdot 2 = 8$

(3)

How many sequences are possible with at most 4 symbols?

Answer:

30

Altogether, $2 + 4 + 8 = 14$ different sequences of at most 3 symbols are possible. Because there are 26 letters in the alphabet, some letters must be represented by sequences of 4 symbols in Morse code. (3)

In how many ways can 3 of 7 items be arranged?

Answer:

$7 \cdot 6 \cdot 5 = 210$

EXAMPLE 4 Suppose the teacher in Example 2 wishes to place only 3 of the 5 books on his desk. How many arrangements of 3 books are possible?

The teacher again has 5 ways to fill the first space, 4 ways to fill the second space, and 3 ways to fill the third. Because he wants to use only 3 books, there are only 3 spaces to be filled giving $5 \cdot 4 \cdot 3 = 60$ arrangements. (4)

Permutations

The answer 60 in Example 4 is called the number of *permutations* of 5 things taken 3 at a time. A **permutation** of r elements (where $r \geq 1$) from a set of n elements is any arrangement, *without repetition,* of the r elements. The number of permutations of n things taken r at a time (with $r \leq n$) is written ${}_nP_r$.* Based on the work in Example 4,

$${}_5P_3 = 5 \cdot 4 \cdot 3 = 60,$$

which can be expressed as

$$5 \cdot 4 \cdot 3 = 5 \cdot 4 \cdot 3 \cdot \frac{2 \cdot 1}{2 \cdot 1} = \frac{5 \cdot 4 \cdot 3 \cdot 2 \cdot 1}{2 \cdot 1} = \frac{5!}{2!} = \frac{5!}{(5-3)!}$$

*Another notation that is sometimes used is $P(n, r)$.

This example illustrates the following general rule for permutations.

Permutations

If ${}_nP_r$ (where $r \le n$) is the number of permutations of n elements taken r at a time, then

$$ {}_nP_r = \frac{n!}{(n-r)!}. $$

5

Find the number of permutations of

(a) 5 things taken 2 at a time;

(b) 9 things taken 3 at a time.

Find each of the following.

(c) ${}_3P_1$

(d) ${}_7P_3$

(e) ${}_{12}P_2$

Answers:

(a) 20

(b) 504

(c) 3

(d) 210

(e) 132

TECHNOLOGY TIP The permutation function on a graphing calculator is in the same menu as the factorial key. As with $n!$, for large values of n and r, the calculator display for ${}_nP_r$ may be an approximation.

To find ${}_nP_r$, we can use either the rule above or direct application of the multiplication principle, as the following example shows.

EXAMPLE 5 Early in 1996, eight candidates sought the Republican nomination for president. In how many ways could voters rank their first, second, and third choices?

This is the same as finding the number of permutations of 8 elements taken 3 at a time. Since there are 3 choices to be made, the multiplication principle gives ${}_8P_3 = 8 \cdot 7 \cdot 6 = 336$. Alternatively, by the formula for ${}_nP_r$,

$$ {}_8P_3 = \frac{8!}{(8-3)!} = \frac{8!}{5!} = \frac{8 \cdot 7 \cdot 6 \cdot 5 \cdot 4 \cdot 3 \cdot 2 \cdot 1}{5 \cdot 4 \cdot 3 \cdot 2 \cdot 1} = 8 \cdot 7 \cdot 6 = 336. $$

Figure 9.2 shows this result on a TI-83 graphing calculator. ◀ 5

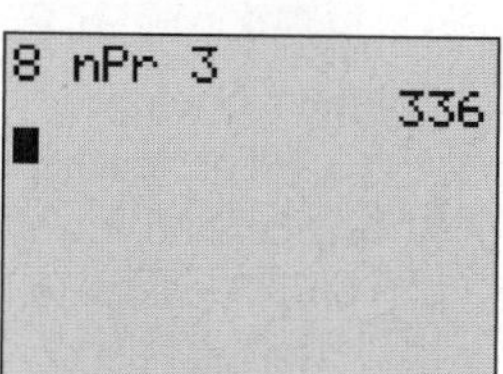

FIGURE 9.2

EXAMPLE 6 Find each of the following.

(a) The number of permutations of the letters A, B, and C

By the formula for ${}_nP_r$ with both n and r equal to 3,

$$ {}_3P_3 = \frac{3!}{(3-3)!} = \frac{3!}{0!} = \frac{3!}{1} = 3! = 6. $$

The 6 permutations (or arrangements) are

ABC, ACB, BAC, BCA, CAB, CBA.

6

Find the number of permutations of the letters C, O, D, and E

(a) using all the letters;

(b) using 2 of the 4 letters;

(c) using 3 of the 4 letters.

Answers:

(a) 24

(b) 12

(c) 24

(b) The number of permutations using just 2 of the letters A, B, and C
Find ${}_3P_2$:

$$ {}_3P_2 = \frac{3!}{(3-2)!} = \frac{3!}{1!} = 3! = 6. $$

This result is exactly the same answer as in part (a). This is because, in the case of ${}_3P_3$, after the first 2 choices are made, the third is already determined, as shown in the table below.

First Two Letters	AB	AC	BA	BC	CA	CB
Third Letter	C	B	C	A	B	A

◀ 6

EXAMPLE 7 A televised talk show will include 4 women and 3 men as panelists.

(a) In how many ways can the panelists be seated in a row of 7 chairs?
Find ${}_7P_7$, the total number of ways to seat 7 panelists in 7 chairs.

$$ {}_7P_7 = \frac{7!}{(7-7)!} = \frac{7!}{0!} = \frac{7!}{1} = 7 \cdot 6 \cdot 5 \cdot 4 \cdot 3 \cdot 2 \cdot 1 = 5040 $$

There are 5040 ways to seat the 7 panelists.

A collection of 3 paintings by one artist and 2 by another is to be displayed. In how many ways can the paintings be shown

(a) in a row?

(b) if the works of the artists are to be alternated?

(c) if one painting by each artist is displayed?

Answers:

(a) 120

(b) 12

(c) 6

(b) In how many ways can the panelists be seated if the men and women are to be alternated?

Use the multiplication principle. In order to alternate men and women, a woman must be seated in the first chair (since there are 4 women and only 3 men), any of the men next, and so on. Thus, there are 4 ways to fill the first seat, 3 ways to fill the second seat, 3 ways to fill the third seat (with any of the 3 remaining women), and so on.

$$ 4 \cdot 3 \cdot 3 \cdot 2 \cdot 2 \cdot 1 \cdot 1 = 144 $$

There are 144 ways to seat the panelists.

(c) In how many ways can the panelists be seated if the men must sit together and the women must sit together?

Use the multiplication principle. We first must decide how to arrange the two groups (men and women). There are 2! ways of doing this. Next, there are 4! ways of arranging the women and 3! ways of arranging the men, for a total of

$$ 2!\,4!\,3! = 2 \cdot 24 \cdot 6 = 288 $$

ways. ◀ 7

Combinations

In Example 4, we found there are 60 ways that a teacher can arrange 3 of 5 different books on a desk. That is, there are 60 permutations of 5 things taken 3 at a time. Suppose now that the teacher does not wish to arrange the books on his desk but rather wishes to choose, at random, any 3 of the 5 books to give to a book sale to raise money for his school. In how many ways can he do this?

At first glance, we might say 60 again, but this is incorrect. The number 60 counts all possible *arrangements* of 3 books chosen from 5. However, the following arrangements would all lead to the same set of 3 books being given to the book sale.

mystery-biography-textbook	biography-textbook-mystery
mystery-textbook-biography	textbook-biography-mystery
biography-mystery-textbook	textbook-mystery-biography

The list shows 6 different *arrangements* of 3 books but only one *subset* of 3 books. A subset of items selected *without regard to order* is called a **combination.** The number of combinations of 5 things taken 3 at a time is written $\binom{5}{3}$ or ${}_5C_3$. Since they are subsets, combinations are *not ordered.*

To evaluate $\binom{5}{3}$, start with the $5 \cdot 4 \cdot 3$ *permutations* of 5 things taken 3 at a time. Combinations are unordered; therefore, we find the number of combinations by dividing the number of permutations by the number of ways each group of 3 can be ordered—that is, by 3!.

$$\binom{5}{3} = \frac{5 \cdot 4 \cdot 3}{3!} = \frac{5 \cdot 4 \cdot 3}{3 \cdot 2 \cdot 1} = 10$$

There are 10 ways that the teacher can choose 3 books at random for the book sale.

Generalizing this discussion gives the formula for the number of combinations of n elements taken r at a time, written $\binom{n}{r}$ or ${}_nC_r$. In general, a set of r elements can be ordered in $r!$ ways, so we divide ${}_nP_r$ by $r!$ to get $\binom{n}{r} = {}_nC_r$.

8

Evaluate $\frac{{}_nP_r}{r!}$ for the following values.

(a) $n = 6, r = 2$

(b) $n = 8, r = 4$

(c) $n = 7, r = 0$

Answers:

(a) 15

(b) 70

(c) 1

$$\binom{n}{r} = {}_nC_r = \frac{{}_nP_r}{r!}$$

$$= {}_nP_r \frac{1}{r!}$$

$$= \frac{n!}{(n-r)!} \cdot \frac{1}{r!} \qquad \text{Definition of } {}_nP_r$$

$$= \frac{n!}{(n-r)!r!}$$

This last form is the most useful for setting up the calculation. ◀ 8

Combinations

The number of combinations of n elements taken r at a time, where $r \le n$, is

$$\binom{n}{r} = {}_nC_r = \frac{n!}{(n-r)!\,r!}.$$

Replacing r by $n - r$ in the combinations formula gives

$$\binom{n}{n-r} = \frac{n!}{(n - [n - r])!\,(n - r)!}$$
$$= \frac{n!}{r!\,(n - r)!} = \frac{n!}{(n - r)!\,r!}.$$

Therefore,

$$\binom{n}{r} = \binom{n}{n-r}, \quad \text{or equivalently,} \quad {}_nC_r = {}_nC_{n-r}.$$

For example, by this result,

$$\binom{5}{3} = \binom{5}{2} \quad \text{and} \quad \binom{10}{4} = \binom{10}{6}.$$

9

Use $\dfrac{n!}{(n-r)!r!}$ to evaluate $\binom{n}{r}$.

(a) $\binom{6}{2}$

(b) $\binom{8}{4}$

(c) $\binom{7}{0}$

Compare your answers with the answers for side problem 8.

Answers:

(a) 15

(b) 70

(c) 1

EXAMPLE 8 From a group of 8 professors, a committee is selected to develop a new curriculum. How many committees can be formed with the following numbers of members?

(a) 3

A committee is an unordered group, so find $\binom{8}{3}$. By the formula for combinations,

$$\binom{8}{3} = \frac{8!}{5!\,3!} = \frac{8 \cdot 7 \cdot 6 \cdot 5 \cdot 4 \cdot 3 \cdot 2 \cdot 1}{5 \cdot 4 \cdot 3 \cdot 2 \cdot 1 \cdot 3 \cdot 2 \cdot 1} = \frac{8 \cdot 7 \cdot 6}{3 \cdot 2 \cdot 1} = 56.$$

(b) 5

We want $\binom{8}{5}$. By the result above,

$$\binom{8}{5} = \binom{8}{3} = 56.$$

This makes sense; since each group of 3 leaves a corresponding group of 5, the number of ways to choose 5 people from 8 is the same as the number of ways to choose 3 people from 8. ◀ 9

TECHNOLOGY TIP The combination function on a graphing calculator, designated ${}_nC_r$, is in the same menu as ${}_nP_r$. Figure 9.3 shows the result of computing ${}_8C_3$.

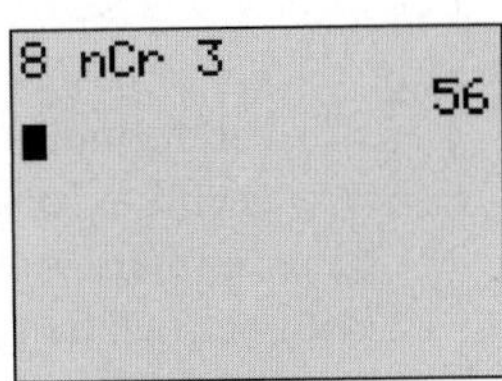

FIGURE 9.3

EXAMPLE 9 Three managers are to be selected from a group of 30 to work on a special project.

(a) In how many different ways can the managers be selected?

Here we wish to know the number of 3-element combinations that can be formed from a set of 30 elements. (We want combinations, not permutations, since order within the group of 3 does not matter.)

$$\binom{30}{3} = \frac{30!}{27!\,3!} = 4060$$

There are 4060 ways to select the project group.

(b) In how many ways can the group of 3 be selected if a certain manager must work on the project?

Since 1 manager has already been selected for the project, the problem is reduced to selecting 2 more from the remaining 29 managers.

$$\binom{29}{2} = \frac{29!}{27!\,2!} = 406$$

In this case, the project group can be selected in 406 ways.

(c) In how many ways can a nonempty group of at most 3 managers be selected from these 30 managers?

The group is to be nonempty; therefore, "at most 3" means "1 or 2 or 3." Find the number of ways for each case.

Five orchids from a collection of 20 are to be selected for a flower show.

(a) In how many ways can this be done?

(b) In how many different ways can the group of 5 be selected if 2 particular orchids must be included?

(c) In how many ways can at least 1 and at most 5 orchids be selected? (*Hint:* Use a calculator.)

Answers:

(a) $\binom{20}{5} = 15{,}504$

(b) $\binom{18}{3} = 816$

(c) 21,699

Case	*Number of Ways*
1	$\binom{30}{1} = \frac{30!}{29!\,1!} = \frac{30 \cdot 29!}{29!\,1!} = 30$
2	$\binom{30}{2} = \frac{30!}{28!\,2!} = \frac{30 \cdot 29 \cdot 28!}{28! \cdot 2 \cdot 1} = 435$
3	$\binom{30}{3} = \frac{30!}{27!\,3!} = \frac{30 \cdot 29 \cdot 28 \cdot 27!}{27! \cdot 3 \cdot 2 \cdot 1} = 4060$

The total number of ways to select at most 3 managers will be the sum

$$30 + 435 + 4060 = 4525.$$ (10)

The formulas for permutations and combinations given in this section will be very useful in solving probability problems in later sections. Any difficulty in using these formulas usually comes from being unable to differentiate between them. Both permutations and combinations give the number of ways to choose r objects from a set of n objects. The differences between permutations and combinations are outlined in the following summary.

Permutations	**Combinations**
Different orderings or arrangements of the r objects are different permutations.	Each choice or subset of r objects gives 1 combination. Order within the r objects does not matter.
$${}_nP_r = \frac{n!}{(n-r)!}$$	$$\binom{n}{r} = {}_nC_r = \frac{n!}{(n-r)!\,r!}$$
Clue words: arrangement, schedule, order	Clue words: group, committee, set, sample
Order matters!	Order does not matter!

In the next examples, concentrate on recognizing which of the formulas should be applied.

EXAMPLE 10 For each of the following problems, tell whether permutations or combinations should be used to solve the problem.

(a) How many 4-digit code numbers are possible if no digits are repeated?

Since changing the order of the 4 digits results in a different code, we use permutations.

(b) A sample of 3 lightbulbs is randomly selected from a batch of 15 bulbs. How many different samples are possible?

The order in which the 3 lightbulbs are selected is not important. The sample is unchanged if the bulbs are rearranged, so combinations should be used.

(c) In a basketball conference with 8 teams, how many games must be played so that each team plays every other team exactly once?

Selection of 2 teams for a game is an *unordered* subset of 2 from the set of 8 teams. Use combinations again.

(d) In how many ways can 4 patients be assigned to 6 hospital rooms so that each patient has a private room?

The room assignments are an *ordered* selection of 4 rooms from the 6 rooms. Exchanging the rooms of any 2 patients within a selection of 4 rooms gives a different assignment, so permutations should be used. ◀ 11

11

Solve the problems in Example 10.

Answers:

(a) 5040

(b) 455

(c) 28

(d) 360

EXAMPLE 11 A manager must select 4 employees for promotion. 12 employees are eligible.

(a) In how many ways can the 4 be chosen?

Because there is no reason to consider the order in which the 4 are selected, we use combinations.

$$\binom{12}{4} = \frac{12!}{4!\,8!} = 495$$

A mailman has special delivery mail for 7 customers.

(a) In how many ways can he arrange his schedule to deliver to all 7?

(b) In how many ways can he schedule deliveries if he can deliver to only 4 of the 7?

Answers:

(a) 5040

(b) 840

(b) In how many ways can 4 employees be chosen (from 12) to be placed in 4 different jobs?

In this case, once a group of 4 is selected, they can be assigned in many different ways (or arrangements) to the 4 jobs. Therefore, this problem requires permutations.

$$_{12}P_4 = \frac{12!}{8!} = 11{,}880$$ (12)

The following problems involve a standard deck of 52 playing cards, shown in Figure 8.19 in Chapter 8.

In how many ways can 4 aces and any other card be dealt?

Answer:

48

EXAMPLE 12 In how many ways can a full house of aces and eights (3 aces and 2 eights) be dealt in a 5-card poker hand?

The order of the 3 aces and the 2 eights does not matter, so use combinations and the multiplication principle. There are $\binom{4}{3}$ ways to get 3 aces from the 4 aces in the deck and $\binom{4}{2}$ ways to get 2 eights. By the multiplication principle, the number of ways to get 3 aces *and* 2 eights is

$$\binom{4}{3} \cdot \binom{4}{2} = 4 \cdot 6 = 24.$$ (13)

EXAMPLE 13 Five cards are dealt from a standard 52-card deck.

(a) How many such hands have all face cards?

The face cards are the king, queen, and jack of each suit. There are 4 suits, so there are 12 face cards. The order of the 5 cards is not important, so use combinations to get

$$\binom{12}{5} = \frac{12!}{5!\,7!} = 792.$$

(b) How many such hands have exactly 2 hearts?

There are 13 hearts in the deck, so the 2 hearts will be selected from those 13 cards. The other 3 cards must come from the remaining 39 cards cards that are not hearts. Use combinations and the multiplication principle to get

$$\binom{13}{2}\binom{39}{3} = 78 \cdot 9139 = 712{,}842.$$

Notice that the two top numbers in the combinations add up to 52, the total number of cards, and the two bottom numbers add up to 5, the number of cards in a hand.

39. Use a computer or a graphing calculator and the Monte Carlo method with $n = 50$ to estimate the probabilities of the following hands at poker. (See the directions for Exercises 50–53 on page 449.) Assume aces are either high or low. Since each hand has 5 cards, you will need $50 \cdot 5 = 250$ random numbers to "look at" 50 hands. Compare these experimental results with the theoretical results.
(a) A pair of aces
(b) Any two cards of the same value
(c) Three of a kind

40. Use a computer or a graphing calculator and the Monte Carlo method with $n = 20$ to estimate the probabilities of the following 13-card bridge hands. Since each hand has 13 cards, you will need $20 \cdot 13 = 260$ random numbers to "look at" 20 hands.
(a) No aces
(b) 2 kings and 2 aces
(c) No cards of any one suit—that is, only 3 suits represented

9.3 BINOMIAL PROBABILITY

Many probability problems are concerned with repetitions of the same experiments. Some examples include finding the probability of getting 7 heads in 8 tosses of a coin, of hitting a target 6 times out of 6, and of finding 1 defective item in a sample of 15 items. Probability problems of this kind are called **Bernoulli trials** problems, or **Bernoulli processes,** and are named after the Swiss mathematician Jakob Bernoulli (1654–1705). In each case, some outcome is designated a success, and any other outcome is considered a failure. Thus, if the probability of a success in a single trial is p, the probability of failure will be $1 - p$. A Bernoulli trials problem, or **binomial experiment,** must satisfy the following conditions.

Binomial Experiment

1. The same experiment is repeated several times.
2. There are only two possible outcomes, success or failure.
3. The repeated trials are independent so that the probability of each outcome remains the same for each trial.

Consider a typical problem of this type: find the probability of getting 5 ones on 5 rolls of a die. Here, the experiment, rolling a die, is repeated 5 times. If getting a one is designated as a success, then getting any other outcome is a failure. The 5 trials are independent; the probability of success (getting a one) is 1/6 on each trial, while the probability of a failure (any other result) is 5/6. Thus, the required probability is

$$P(5 \text{ ones in } 5 \text{ rolls}) = P(1) \cdot P(1) \cdot P(1) \cdot P(1) \cdot P(1) = \left(\frac{1}{6}\right)^5$$

$$\approx .00013.$$

Now suppose the problem is changed to that of finding the probability of getting a one exactly 4 times in 5 rolls of the die. Again, a success on any roll of the die is defined as getting a one, and the trials are independent, with the same probability of success each time. The desired outcome for this experiment can occur in more than one way, as shown below, where s represents a success (getting a one), and f represents a failure (getting any other result), and each row represents a different outcome.

$$\begin{matrix} s & s & s & s & f \\ s & s & s & f & s \\ s & s & f & s & s \\ s & f & s & s & s \\ f & s & s & s & s \end{matrix}$$

Using combinations, the total number of ways in which 4 successes (and 1 failure) can occur is $\binom{5}{4} = 5$. The probability of each of these 5 outcomes is

$$P(1) \cdot P(1) \cdot P(1) \cdot P(1) \cdot P(\text{not } 1) = \left(\frac{1}{6}\right)^4 \left(\frac{5}{6}\right).$$

Find the probability of rolling a one

(a) exactly twice in 5 rolls of a die;

(b) exactly once in 5 rolls of a die.

(c) Do the probabilities found in the text, plus the two probabilities found here, cover all possible outcomes for this experiment? If not, what other outcomes are possible?

Answers:

(a) .161

(b) .402

(c) No; 0 ones

Since the 5 outcomes represent mutually exclusive alternative events, add the 5 probabilities, or multiply by $\binom{5}{4} = 5$.

$$P(4 \text{ ones in } 5 \text{ rolls}) = \binom{5}{4}\left(\frac{1}{6}\right)^4\left(\frac{5}{6}\right)^{5-4} = 5\left(\frac{1}{6}\right)^4\left(\frac{5}{6}\right)^1 = \frac{5^2}{6^5} \approx .0032$$

The probability of rolling a one exactly 3 times in 5 rolls of a die can be computed in the same way. The probability of any one way of achieving 3 successes and 2 failures will be

$$\left(\frac{1}{6}\right)^3\left(\frac{5}{6}\right)^2.$$

Again the desired outcome can occur in more than one way. The number of ways in which 3 successes (and 2 failures) can occur, using combinations, is $\binom{5}{3} = 10$.

$$P(3 \text{ ones in } 5 \text{ rolls}) = \binom{5}{3}\left(\frac{1}{6}\right)^3\left(\frac{5}{6}\right)^{5-3} = 10\left(\frac{1}{6}\right)^3\left(\frac{5}{6}\right)^2 = \frac{250}{6^5} \approx .032$$ (1)

A similar argument works in the general case.

Binomial Probability

If p is the probability of success in a single trial of a binomial experiment, the probability of x successes and $n - x$ failures in n independent repeated trials of the experiment is

$$\binom{n}{x} p^x (1-p)^{n-x}.$$

This formula plays an important role in the study of *binomial distributions,* which are discussed in Section 10.4.

TECHNOLOGY TIP On the TI-83 calculator, use "binompdf(n,p,x)" in the DISTR menu to compute the probability of exactly x successes in n trials (where p is the probability of success in a single trial). Use "binomcdf(n,p,x)" to compute the probability of at most x successes in n trials. Figure 9.6 shows the probability of exactly three ones in five rolls of a die and the probability of at most three ones in five rolls of a die.

FIGURE 9.6

EXAMPLE 1 The advertising agency that handles the Diet Supercola account believes that 40% of all consumers prefer this product over its competitors. Suppose a random sample of 6 people is chosen. Assume that all responses are independent of each other. Find the probability of the following.

(a) Exactly 4 of the 6 people prefer Diet Supercola.

We can think of the 6 responses as 6 independent trials. A success occurs if a person prefers Diet Supercola. Then this is a binomial experiment with

$$p = P(\text{success}) = P(\text{prefer Diet Supercola}) = .4.$$

The sample is made up of 6 people, so $n = 6$. To find the probability that exactly 4 people prefer this drink, we let $x = 4$ and use the result in the box.

$$\begin{aligned} P(\text{exactly } 4) &= \binom{6}{4}(.4)^4(1 - .4)^{6-4} \\ &= 15(.4)^4(.6)^2 \\ &= 15(.0256)(.36) \quad \text{Use a calculator.} \\ &= .13824 \end{aligned}$$

(b) None of the 6 people prefers Diet Supercola.

Let $x = 0$.

$$\begin{aligned} P(\text{exactly } 0) &= \binom{6}{0}(.4)^0(1 - .4)^6 \\ &= 1(1)(.6)^6 \\ &\approx .0467 \end{aligned}$$

2

2

Eighty percent of all students at a certain school ski. If a sample of 5 students at this school is selected, and if their responses are independent, find the probability that exactly

(a) 1 of the 5 students skis;

(b) 4 of the 5 students ski.

Answers:

(a) $\binom{5}{1}(.8)^1(.2)^4 = .0064$

(b) $\binom{5}{4}(.8)^4(.2)^1 = .4096$

Find the probability of getting 2 fours in 8 tosses of a die.

Answer:

$\binom{8}{2}\left(\frac{1}{6}\right)^2\left(\frac{5}{6}\right)^6 \approx .2605$

EXAMPLE 2 Find the probability of getting exactly 7 heads in 8 tosses of a fair coin.

The probability of success, getting a head in a single toss, is 1/2, so the probability of failure, getting a tail, is 1/2.

$$P(\text{7 heads in 8 tosses}) = \binom{8}{7}\left(\frac{1}{2}\right)^7\left(\frac{1}{2}\right)^1 = 8\left(\frac{1}{2}\right)^8 = .03125$$ 3

EXAMPLE 3 The chance that an American falls asleep with the TV on at least three nights a week is 1/4.*

(a) Suppose a researcher selects 15 Americans at random and is interested in the probability that 3 are "TV sleepers."

The experiment, selecting a person, is repeated 15 times. If selecting a TV sleeper is labeled a success, then getting a non–TV sleeper is labeled a failure. Researchers consider such trials to be independent. Thus, the probability that 3 persons in the sample of 15 are sleepers is

$$\binom{15}{3}\left(\frac{1}{4}\right)^3\left(\frac{3}{4}\right)^{12} \approx .225.$$

Five percent of the clay pots fired in a certain way are defective. Assume independence and find the probability of getting each result in a sample of 12 pots.

(a) 2 defective pots

(b) At most 2 defective pots

Answer:

(a) $\binom{12}{2}(.05)^2(.95)^{10} \approx .0988$

(b) .980

(b) Find the probability that at most 3 of the 15 people in the sample are TV sleepers.

"At most 3" means 0 or 1 or 2 or 3. We must find the probability for each case and then use the union rule for mutually exclusive events.

$$P(\text{0 TV sleepers}) = \binom{15}{0}\left(\frac{1}{4}\right)^0\left(\frac{3}{4}\right)^{15} \approx .0134$$

$$P(\text{1 TV sleeper}) = \binom{15}{1}\left(\frac{1}{4}\right)^1\left(\frac{3}{4}\right)^{14} \approx .0668$$

$$P(\text{2 TV sleepers}) = \binom{15}{2}\left(\frac{1}{4}\right)^2\left(\frac{3}{4}\right)^{13} \approx .156$$

By the union rule,

$$P(\text{at most 3 TV sleepers}) \approx .0134 + .0668 + .156 + .225 \approx .461.$$ 

In the next example we return to the problem posed at the beginning of this chapter.

EXAMPLE 4 A survey by *Money* magazine found that supermarket scanners are overcharging customers at 30% of stores.† Assume all such scanners consistently overcharge. If you shop at 3 supermarkets that use the scanners, what is the probability that you will be overcharged in at least one store?

**Harper's Magazine,* March 1996, p. 13.

†"Don't Get Cheated by Supermarket Scanners," pp. 132–138, by Vanessa O'Connell reprinted from the April 1993 issue of *Money* by special permission (reprinted in the *Chicago Tribune* as "Beware, When Scanners Err, You Pay Price"). Copyright © 1993, Time Inc.

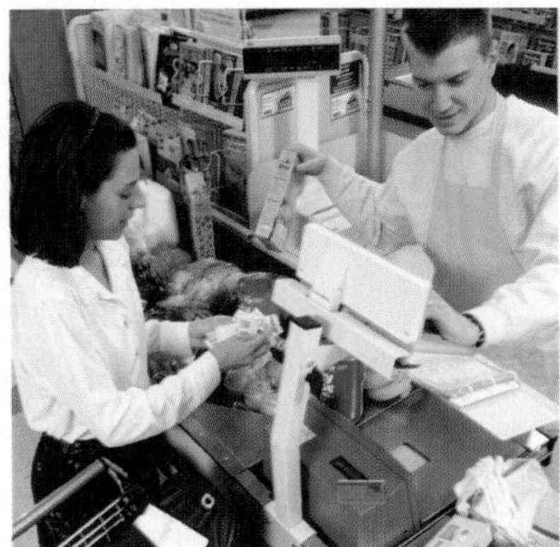

5

In Example 4, find the probability that

(a) you are overcharged in more than 1 store.

(b) you are overcharged in less than 1 store.

Answers:

(a) .216

(b) .343

We can treat this as a binomial experiment, letting $n = 3$ and $p = .3$. At least 1 of 3 means 1 or 2 or 3. It will be simpler here to find the probability of being overcharged in none of the 3 stores, that is, P(0 overcharges), and then find the probability of at least 1 overcharge, which is the number $1 - P$(0 overcharges).

$$\begin{aligned} P(0 \text{ overcharges}) &= \binom{3}{0}(.3)^0(.7)^3 \\ &= 1(1)(.343) \\ &= .343 \\ P(\text{at least } 1) &= 1 - P(0 \text{ overcharges}) \\ &= 1 - .343 \\ &= .657 \end{aligned}$$

5

EXAMPLE 5 If each member of a 9-person jury acts independently of the other members and makes the correct determination of guilt or innocence with probability .65, find the probability that the majority of jurors will reach a correct verdict.*

Since the jurors in this particular situation act independently, we can treat this as a binomial experiment. Thus, the probability that the majority of the jurors will reach the correct verdict is given by

$$\begin{aligned} P(\text{at least } 5) &= \binom{9}{5}(.65)^5(.35)^4 + \binom{9}{6}(.65)^6(.35)^3 + \binom{9}{7}(.65)^7(.35)^2 \\ &\quad + \binom{9}{8}(.65)^8(.35)^1 + \binom{9}{9}(.65)^9 \\ &\approx .2194 + .2716 + .2162 + .1004 + .0207 \\ &= .8283. \end{aligned}$$

*Bernard Grofman, "A Preliminary Model of Jury Decision Making as a Function of Jury Size, Effective Jury Decision Rule, and Mean Juror Judgmental Competence," *Frontiers in Economics* (1979), pp. 98–110.

TECHNOLOGY TIP Some spreadsheets provide binomial probabilities. In Microsoft Excel, for example, the command "5BINOMDIST (5, 9, .65, 0)" gives .21939, which is the probability for $x = 5$ in Example 5. Alternatively, the command "5BINOMDIST (4, 9, .65, 1)" gives .17172 as the probability that 4 or fewer jurors will make the correct decision. Subtract .17172 from 1 to get .82828 as the probability that the majority of the jurors will make the correct decision. This value agrees with the value found in Example 5.

9.3 Exercises

In Exercises 1–39, see Examples 1–5.

Social Science *In 1995 12.7% of the U.S. population was 65 or older, an increase from 4.1% in 1900.* Find the probabilities that the following number of persons selected at random from 20 U.S. residents in 1995 were 65 or older.*

1. Exactly 5

2. Exactly 1

3. None

4. All

5. At least 1

6. At most 3

A die is rolled 12 times. Find the probability of rolling

7. Exactly 12 ones;

8. Exactly 6 ones;

9. Exactly 1 one;

10. Exactly 2 ones;

11. No more than 3 ones;

12. No more than 1 one.

A coin is tossed 5 times. Find the probability of getting

13. All heads;

14. Exactly 3 heads;

15. No more than 3 heads;

16. At least 3 heads.

17. How do you identify a probability problem that involves a binomial experiment?

18. Why do combinations occur in the binomial probability formula?

Natural Science *Alzheimer's disease affects an estimated 4 million Americans. The probability that an American aged 65 or older has Alzheimer's is .0127.* A group of 10 people aged 65 or older is selected randomly. Find the following probabilities.*

19. 2 people have the disease.

20. No one has the disease.

21. At most 1 person has the disease.

22. 8 people do not have the disease.

Business *The survey discussed in Example 4 also found that customers overpay for 1 out of every 10 items, on average. Suppose a customer purchases 15 items. Find the following probabilities.*

23. A customer overpays on 3 items.

24. A customer does not overpay for any item.

25. A customer overpays on at least one item.

26. A customer overpays on at least 2 items.

27. A customer overpays on at most 2 items.

Natural Science *The probability that a birth will result in twins is .027.* Assuming independence (perhaps not a valid assumption), what are the probabilities that out of 100 births in a hospital, there will be the following numbers of sets of twins?*

28. Exactly 2 sets of twins

29. At most 2 sets of twins

30. Natural Science In an experiment on the effects of a radiation dose on cells, a beam of radioactive particles is aimed at a group of 10 cells. Find the probability that 8 of the cells will be hit by the beam if the probability that any single cell will be hit is .6. (Assume independence.)

31. Natural Science The probability of a mutation of a given gene under a dose of 1 roentgen of radiation is approximately 2.5×10^{-7}. What is the probability that in 10,000 genes, at least 1 mutation occurs?

32. Social Science Refer to Example 3. The probability that an American falls asleep with the TV on at least three nights a week is 1/4. Suppose a researcher selects 5 Americans at random and is interested in the probability that one or more is a "TV sleeper." Find the following probabilities.

(a) All 5 are TV sleepers.
(b) Three of the 5 are TV sleepers.

33. Natural Science Children in a certain school are monitored for a year and it is found that 30% of those who do not brush their teeth regularly don't have a cavity during the year. A child who brushes regularly has a 70% chance of having no cavities during the year.

**The World Almanac and Book of Facts*, 2001, p. 721.

**The World Almanac and Book of Facts*, 2001, p. 873.

(a) If 10 children brush regularly, what is the probability that at least 9 will have no cavities during the year?

(b) 150 children who brush regularly are divided into 15 groups of 10 each. What is the probability that in at least one group of 10, at least 9 will have no cavities during the year?

34. Social Science A recent study found that 33% of women would prefer to work part time rather than full time if money were not a concern.* Find the probability that if 10 women are selected at random, at least 3 of them would prefer to work part time.

35. Social Science The fatal accident involvement rate for drivers younger than 20 is .0006.† Find the probability that in 10 unrelated fatal accidents, the following number of accidents involved a driver younger than 20.

(a) 1 **(b)** at most 1

Social Science *Refer to Example 5. Find the probability of each result.*

36. All jurors reach a correct verdict.

37. No jurors reach a correct verdict.

38. At least 8 jurors reach a correct verdict.

39. 7 jurors reach a correct verdict.

40. Social Science According to a poll conducted by *U.S. News and World Report,* 84% of college students believe they need to cheat to get ahead in the world today.‡

(a) Do the results of this poll indicate that 84% of all college students cheat? Explain.

(b) If this result is accurate and 100 college students are asked if they believe that cheating is necessary to get ahead in the world, what is the probability that 90 or more of the students will answer affirmatively to the question?

41. Social Science In the "Numbers" section of a recent *Time* magazine, it was reported that 15.2% of low-birth-weight babies graduate from high school by age 19. On the other hand, it was reported that 57.5% of the normal-birth-weight siblings graduated from high school.§

(a) If 40 low-birth-weight babies were tracked through high school, what is the probability that fewer than 15 will graduate from high school by age 19?

(b) What are some of the factors that may contribute to the wide difference in high school success between these siblings? Do you believe that low birth weight is the primary cause of the difference? What other information do you need to better answer these questions?

42. Natural Science In placebo-controlled trials of Pravachol®, a drug that is prescribed to lower cholesterol, 7.3% of the patients who were taking the drug experienced nausea/vomiting, whereas 7.1% of the patients who were taking the placebo experienced nausea/vomiting.*

(a) If 100 patients who are taking Pravachol® are selected, what is the probability that 10 or more will experience nausea/vomiting?

(b) If a second group of 100 patients receives a placebo, what is the probability that 10 or more will experience nausea/vomiting?

(c) Since 7.3% is larger than 7.1%, do you believe that the Pravachol® causes more people to experience nausea/vomiting than a placebo? Explain.

43. Natural Science The use of DNA has become an integral part of many court cases. When DNA is extracted from cells and body fluids, genetic information is represented by bands of information, which look similar to a bar code at a grocery store. It is generally accepted that in unrelated people, the probability of a particular band matching is 1 in 4.†

(a) If 5 bands are compared in unrelated people, what is the probability that all 5 of the bands match? (Express your answer in terms of "1 chance in ?".)

(b) If 20 bands are compared in unrelated people, what is the probability that all 20 of the bands match? (Express your answer in terms of "1 chance in ?".)

(c) If 20 bands are compared in unrelated people, what is the probability that 16 or more bands match? (Express your answer in terms of "1 chance in ?".)

(d) If you were deciding paternity and there were 16 matches out of 20 bands compared, would you believe that the person being tested was the father? Explain.

44. Social Science In England, a woman was found guilty of smothering her two infant children. Much of the Crown's case against her was based on the testimony from a pediatrician who indicated that the chances of two crib deaths occurring in both siblings was only about 1 in 73 million. This number was calculated by assuming that the probability of a single crib death is 1 in 8500 and the probability of two crib deaths is 1 in 8500^2 (i.e., binomial).‡ Why is the use of binomial probability not correct in this situation?

---

*Cathleen Ferraro, "Feelings of the Working Woman," *Sacramento Bee,* May 11, 1995, pp. Al, A22.

†National Safety Council; National Highway Traffic Safety Administration.

‡Carolyn Kleiner and Mary Lord, "The Cheating Game," *U.S. News and World Report,* November 22, 1999, pp. 55–66.

§"Numbers." *Time,* July 17, 2000, p. 21.

*Advertisement in *Time,* July 17, 2000, for Pravachol®, developed and marketed by Bristol-Myers Squibb Company.

†"Genetic Fingerprinting Worksheet." Centre for Innovation in Mathematics Teaching, http://www.ex.ac.uk/cimt/resource/fgrprnts.htm.

‡Stephen J. Watkins, "Conviction by Mathematical Error?" *British Medical Journal* 320. no. 7226 (January 1, 2000): 2–3.

The area of the bar above $x = 0$ in Figure 9.7 is the product of 1 and 14/55, or $1 \cdot 14/55 = 14/55$. Since each bar has a width of 1, its area is equal to the probability that corresponds to that value of x. The probability that a particular value will occur is thus given by the area of the appropriate bar of the graph. For example, the probability that one or more monitors is broken is the sum of the areas for $x = 1$, $x = 2$, and $x = 3$. This area, shown in red in Figure 9.8, corresponds to 41/55 of the total area, since

$$\begin{aligned} P(x \geq 1) &= P(x = 1) + P(x = 2) + P(x = 3) \\ &= 28/55 + 12/55 + 1/55 \\ &= 41/55. \end{aligned}$$

FIGURE 9.8

1

(a) Give the probability distribution for the number of heads showing when 3 coins are tossed.

(b) Draw a histogram for the distribution in part (a). Find the probability that no more than 1 coin comes up heads.

Answers:

(a)

x	$P(x)$
0	1/8
1	3/8
2	3/8
3	1/8

(b) 1/2

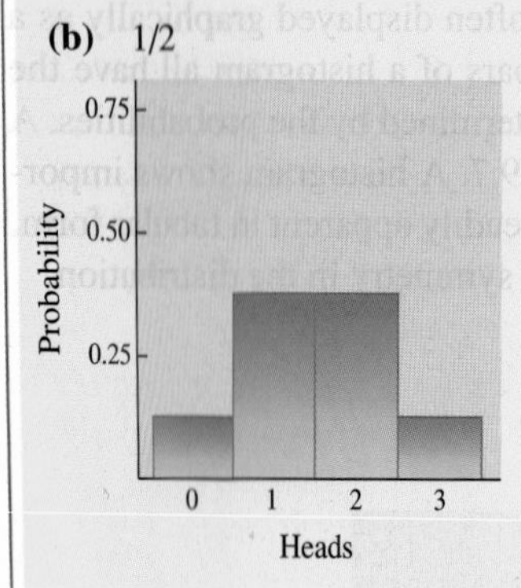

EXAMPLE 1

(a) Give the probability distribution for the number of heads showing when two coins are tossed.

Let x represent the random variable "number of heads." Then x can take on the values 0, 1, or 2. Now find the probability of each outcome. The results are shown in Table 2.

Table 2

x	0	1	2
$P(x)$	1/4	1/2	1/4

(b) Draw a histogram for the distribution in Table 2. Find the probability that at least one coin comes up heads.

The histogram is shown in Figure 9.9. The portion in red represents

$$\begin{aligned} P(x \geq 1) &= P(x = 1) + P(x = 2) \\ &= \frac{3}{4}. \end{aligned}$$

1

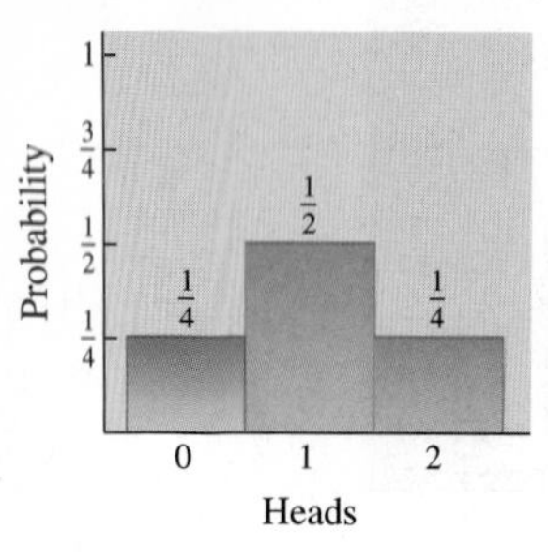

FIGURE 9.9

TECHNOLOGY TIP Virtually all graphing calculators can produce histograms. The procedures differ on various calculators, but it usually requires you to enter the outcomes in one list and the corresponding frequencies in a second list. For specific details, check your instruction manual under "statistics graphs" or "statistical plotting." To get the histogram in Figure 9.9 with a TI-83 calculator, we entered the outcomes 0, 1, and 2 in the first list and entered the probabilities .25, .5 and .25 in a second list. Two versions of the histogram are shown in Figure 9.10. They differ slightly because different viewing windows were used. With some calculators, the probabilities must be entered as integers, so make the entries in the second list 1, 2, and 1 (corresponding to 1/4, 2/4, and 1/4) and use a window with $0 \le y \le 4$.

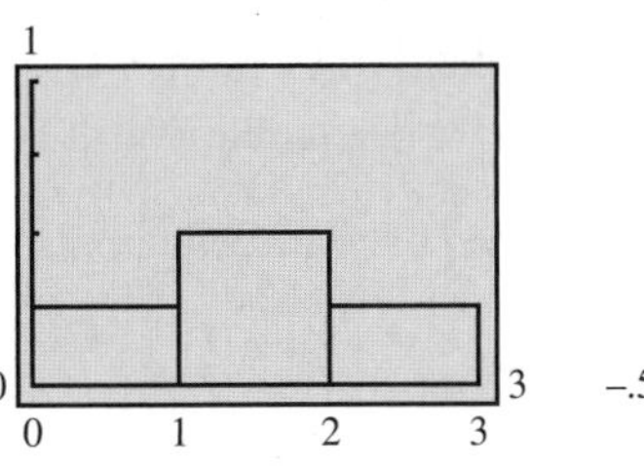

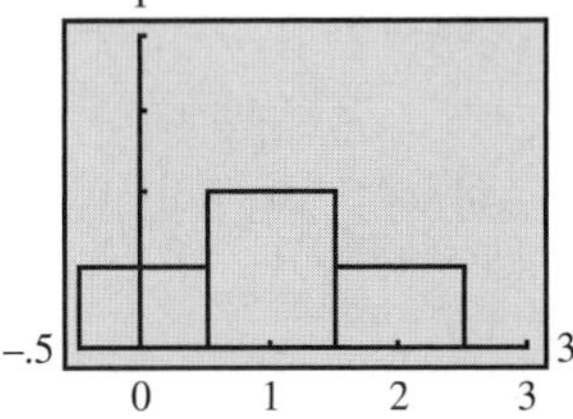

FIGURE 9.10

Expected Value

In working with probability distributions, it is useful to have a concept of the typical or average value that the random variable takes on. In Example 1, for instance, it seems reasonable that, on the average, one head shows when two coins are tossed. This does not tell what will happen the next time we toss two coins; we may get two heads, or we may get none. If we tossed two coins many times, however, we would expect that, in the long run, we would average about one head for each toss of two coins.

A way to solve such problems in general is to imagine flipping two coins 4 times. Based on the probability distribution in Example 1, we would expect that 1 of the 4 times we would get 0 heads, 2 of the 4 times we would get 1 head, and 1 of the 4 times we would get 2 heads. The total number of heads we would get, then, is

$$0 \cdot 1 + 1 \cdot 2 + 2 \cdot 1 = 4.$$

The expected number of heads per toss is found by dividing the total number of heads by the total number of tosses, or

$$\frac{0 \cdot 1 + 1 \cdot 2 + 2 \cdot 1}{4} = 0 \cdot \frac{1}{4} + 1 \cdot \frac{1}{2} + 2 \cdot \frac{1}{4} = 1.$$

Notice that the expected number of heads turns out to be the sum of the three values of the random variable x multiplied by their corresponding probabilities. We can use this idea to define the *expected value* of a random variable as follows.

Expected Value

Suppose the random variable x can take on the n values $x_1, x_2, x_3, \ldots, x_n$. Also, suppose the probabilities that these values occur are, respectively, $p_1, p_2, p_3, \ldots, p_n$. Then the **expected value** of the random variable is

$$E(x) = x_1p_1 + x_2p_2 + x_3p_3 + \cdots + x_np_n.$$

EXAMPLE 2 In the example with the computer monitors, find the expected number of broken monitors that the shipping manager finds.

2

Find the expected value of the number of heads showing when four coins are tossed.

Answer:

2

Multiply each outcome in Table 1 by its probability, and sum these products.

$$E(x) = 0 \cdot \frac{14}{55} + 1 \cdot \frac{28}{55} + 2 \cdot \frac{12}{55} + 3 \cdot \frac{1}{55} = 1$$

On the average, the shipping manager will find 1 broken monitor in the sample of 4. On reflection, this seems natural; 3 of the 12 monitors, or 1/4 of the total, are broken. We should expect, then, that 1/4 of the sample of 4 monitors are broken. ◀ 2

Physically, the expected value of a probability distribution represents a balance point. If we think of the histogram in Figure 9.7 as a series of weights with magnitudes represented by the heights of the bars, then the system would balance if supported at the point corresponding to the expected value.

EXAMPLE 3 Suppose a local symphony decides to raise money by raffling a microwave oven worth \$400, a dinner for two worth \$80, and 2 books worth \$20 each. A total of 2000 tickets are sold at \$1 each. Find the expected value of winning for a person who buys 1 ticket in the raffle.

Here the random variable represents the possible amounts of net winnings, where net winnings = amount won − cost of ticket. The net winnings of the person winning the oven are \$400 (amount won) − \$1 (cost of ticket) = \$399. The net winnings for each losing ticket are \$0 − \$1 = −\$1.

The net winnings of the various prizes, as well as their respective probabilities, are shown in Table 3. The probability of winning \$19 is 2/2000 because there are 2 prizes worth \$20. We have not reduced the fractions in order to keep all the denominators equal. Because there are 4 winning tickets, there are 1996 losing tickets, so the probability of winning −\$1 is 1996/2000.

Table 3

x	\$399	\$79	\$19	−\$1
$P(x)$	1/2000	1/2000	2/2000	1996/2000

Suppose you buy 1 of 10,000 tickets at $1 each in a lottery where the prize is $5,000. What are your expected net winnings? What does this answer mean?

Answer:

−$ .50. On the average you lose $ .50 per ticket purchased.

The expected winnings for a person buying one ticket are

$$399\left(\frac{1}{2000}\right) + 79\left(\frac{1}{2000}\right) + 19\left(\frac{2}{2000}\right) + (-1)\left(\frac{1996}{2000}\right) = -\frac{1480}{2000}$$
$$= -.74.$$

On the average, a person buying 1 ticket in the raffle will lose $.74, or 74¢.

It is not possible to lose 74¢ in this raffle: either you lose $1, or you win a prize worth $400, $80, or $20, minus the $1 you pay to play. But if you bought tickets in many such raffles over a long period of time, you would lose 74¢ per ticket, on the average. It is important to note that the expected value of a random variable may be a number that can never occur in any one trial of the experiment. ◀ (3)

NOTE An alternative way to compute expected value in this and other examples is to calculate the expected amount won and then subtract the cost of the ticket afterward. The amount won is either $400 (with probability 1/2000), $80 (with probability 1/2000), $20 (with probability 2/2000), or $0 (with probability 1996/2000). The expected winnings for a person buying one ticket are then

$$400\left(\frac{1}{2000}\right) + 80\left(\frac{1}{2000}\right) + 20\left(\frac{2}{2000}\right) + 0\left(\frac{1996}{2000}\right) - 1 = -\frac{1480}{2000}$$
$$= -.74.$$

EXAMPLE 4 Each day Donna and Mary toss a coin to see who buys coffee (80 cents a cup). One tosses and the other calls the outcome. If the person who calls the outcome is correct, the other buys the coffee; otherwise the caller pays. Find Donna's expected winnings.

Assume that an honest coin is used, that Mary tosses the coin, and that Donna calls the outcome. The possible results and corresponding probabilities are shown below.

	Possible Results			
Result of Toss	Heads	Heads	Tails	Tails
Call	Heads	Tails	Heads	Tails
Caller Wins?	Yes	No	No	Yes
Probability	1/4	1/4	1/4	1/4

(4)

Find Mary's expected winnings.

Answer:

0

Donna wins an 80¢ cup of coffee whenever the results and calls match, and she loses an 80¢ cup when there is no match. Her expected winnings are

$$(.80)\left(\frac{1}{4}\right) + (-.80)\left(\frac{1}{4}\right) + (-.80)\left(\frac{1}{4}\right) + (.80)\left(\frac{1}{4}\right) = 0.$$

On the average, over the long run, Donna neither wins nor loses. ◀ (4)

A game with an expected value of 0 (such as the one in Example 4) is called a **fair game.** Casinos do not offer fair games. If they did, they would win (on the average) \$0 and have a hard time paying the help! Casino games have expected winnings for the house that vary from 1.5 cents per dollar to 60 cents per dollar. Exercises 30–35 at the end of the section ask you to find the expected winnings for certain games of chance.

The idea of expected value can be very useful in decision making, as shown by the next example.

5

A person can take one of two jobs. With job A, there is a 50% chance of making \$60,000 per year after 5 years and a 50% chance of making \$30,000. With job B, there is a 30% chance of making \$90,000 per year after 5 years and a 70% chance of making \$20,000. Based strictly on expected value, which job should be taken?

Answer:

Job A has an expected salary of \$45,000; job B, \$41,000. Take job A.

EXAMPLE 5 At age 50, you receive a letter from Mutual of Mauritania Insurance Company. According to the letter, you must tell the company immediately which of the following two options you will choose: take \$20,000 at age 60 (if you are alive, \$0 otherwise) or \$30,000 at age 70 (again, if you are alive, \$0 otherwise). Based *only* on the idea of expected value, which should you choose?

Life insurance companies have constructed elaborate tables showing the probability of a person living a given number of years into the future. From a recent such table, the probability of living from age 50 to age 60 is .88, while the probability of living from age 50 to 70 is .64. The expected values of the two options are given below.

First Option: $(20{,}000)(.88) + (0)(.12) = 17{,}600$

Second Option: $(30{,}000)(.64) + (0)(.36) = 19{,}200$

Based strictly on expected values, choose the second option. 

EXAMPLE 6 According to the National Center for Education Statistics, 79% of the U.S. holders of bachelor's degrees in education in 1992–1993 were women.* Suppose 5 holders of bachelor's degrees in education from 1992–93 are picked at random.

(a) Find the probability distribution for the number that are female.

We first note that each of the 5 people in the sample is either female (with probability .79) or male (with probability .21). As in the previous section, we may assume that the probability for each member of the sample is independent of that of any other. Such a situation is described by binomial probability with $n = 5$ and $p = .79$, for which we use the binomial probability formula

$$\binom{n}{x} \cdot p^x \cdot (1-p)^{n-x},$$

where x is the number of females in the sample. For example,

$$P(x = 0) = \binom{5}{0}(.79)^0(.21)^5 \approx .0004.$$

**New York Times*, January 7, 1996. Education Life, p. 24.

Similarly, we could calculate the probability that x is any value from 0 to 5, resulting in the probability distribution in Table 4 (with all probabilities rounded to four places).

Table 4

x	0	1	2	3	4	5
$P(x)$	.0004	.0077	.0578	.2174	.4090	.3077

(b) Find the expected number of females in the sample of 5 people.
Using the formula for expected value, we have

$$E(x) = 0(.0004) + 1(.0077) + 2(.0578) + 3(.2174) + 4(.4090) + 5(.3077) = 3.95.$$

On the average, 3.95 of the people in the sample of 5 will be female.

There is another way to get the answer in part (b) of the previous example. Because 79% of the U.S. holders of bachelor's degrees in education from 1992–93 are female, it is reasonable to expect 79% of our sample to be female. Thus, 79% of 5 is $5(.79) = 3.95$. Notice that what we have done is to multiply n by p. It can be shown that this method always gives the expected value for binomial probability.

Expected Value for Binomial Probability

For binomial probability, $E(x) = np$. In other words, the expected number of successes is the number of trials times the probability of success in each trial.

EXAMPLE 7 Suppose a family has 3 children.

(a) Find the probability distribution for the number of girls.
Assuming girls and boys are equally likely, the probability distribution is binomial with $n = 3$ and $p = 1/2$. Letting x be the number of girls in the formula for binomial probability, we find, for example,

$$P(x = 0) = \binom{3}{0}\left(\frac{1}{2}\right)^0\left(\frac{1}{2}\right)^3 = \frac{1}{8}.$$

The other values are found similarly, and the results are shown in Table 5.

Table 5

x	0	1	2	3
$P(x)$	1/8	3/8	3/8	1/8

We can verify this by noticing that in the sample space S of all 3-child families, there are eight equally likely outcomes: $S = \{ggg, ggb, gbg, gbb, bgg, bgb, bbg, bbb\}$. One of the outcomes has 0 girls, three have 1 girl, three have 2 girls, and one has 3 girls.

(b) Find the expected number of girls in a 3-child family using the distribution from part (a).

Using the formula for expected value, we have

$$\text{Expected number of girls} = 0\left(\frac{1}{8}\right) + 1\left(\frac{3}{8}\right) + 2\left(\frac{3}{8}\right) + 3\left(\frac{1}{8}\right)$$
$$= \frac{12}{8} = 1.5.$$

On average, a 3-child family will have 1.5 girls. This result agrees with our intuition that, on the average, half the children born will be girls.

(c) Find the expected number of girls in a 3-child family using the formula for expected value for binomial probability.

Using the formula $E(x) = np$ with $n = 3$ and $p = 1/2$, we have

$$\text{Expected number of girls} = 3\left(\frac{1}{2}\right) = 1.5.$$

This agrees with our answer from part (b), as it must. ◀ 6

We will return to the concepts in this section in Chapter 10.

6

The probability that a can of soda from a certain plant is defective is .005. A sample of 4 cans is selected at random. Write a distribution for the number of defective cans in the sample, and give its expected value.

Answers:

x	$P(x)$
0	.98
1	.0197
2	.00015
3	.0000005
4	6.25×10^{-10}

$E(x) = .02$

9.4 Exercises

For each of the experiments described below, let x determine a random variable and use your knowledge of probability to prepare a probability distribution. (See Example 1.)

1. Four coins are tossed, and the number of heads is noted.
2. Two dice are rolled, and the total number of points is recorded.
3. Three cards are drawn from a deck. The number of aces is counted.
4. Two balls are drawn from a bag in which there are 4 white balls and 2 black balls. The number of black balls is counted.

Draw a histogram for each of the following and shade the region that gives the indicated probability. (See Example 1.)

5. Exercise 1; $P(x \le 2)$
6. Exercise 2; $P(x \ge 11)$
7. Exercise 3; P(at least one ace)
8. Exercise 4; P(at least one black ball)

Find the expected value for each random variable. (See Example 2.)

9.

x	2	3	4	5
$P(x)$	.1	.4	.3	.2

10.

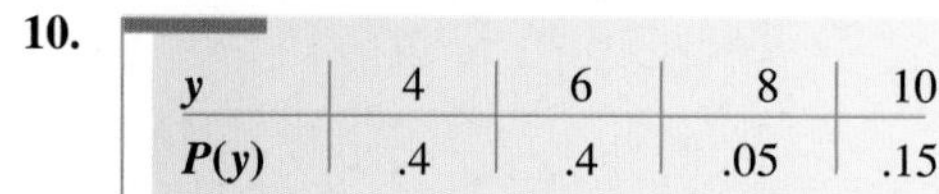

y	4	6	8	10
$P(y)$	.4	.4	.05	.15

11.

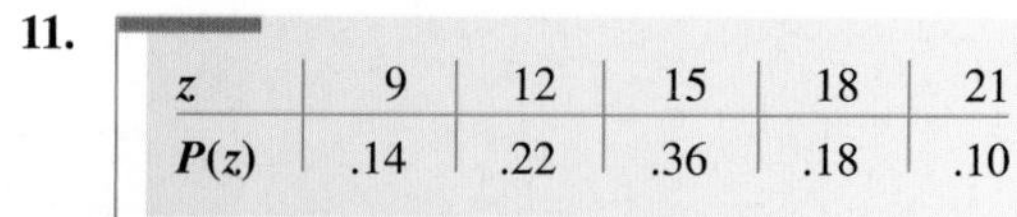

z	9	12	15	18	21
$P(z)$	.14	.22	.36	.18	.10

12.

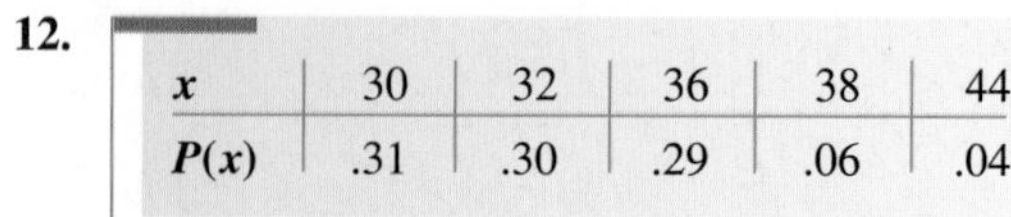

x	30	32	36	38	44
$P(x)$	.31	.30	.29	.06	.04

Find the expected values for the random variables x having the probability functions graphed below.

13.

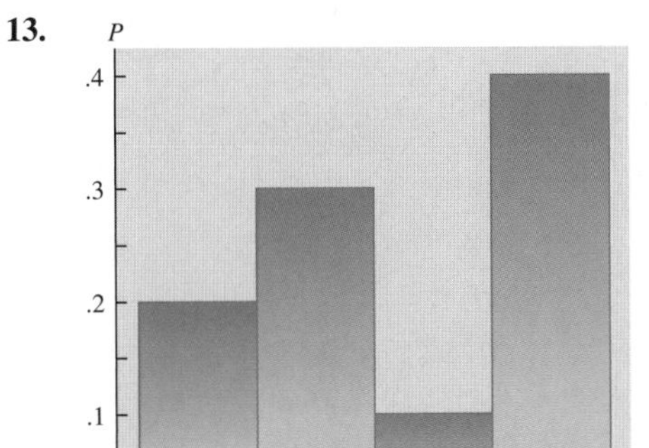

14.

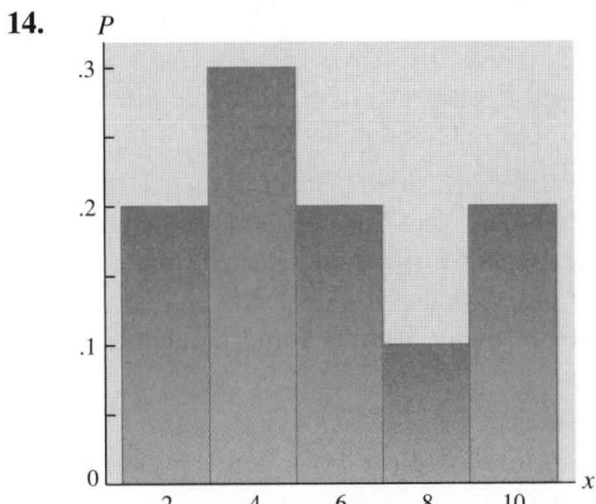

15.

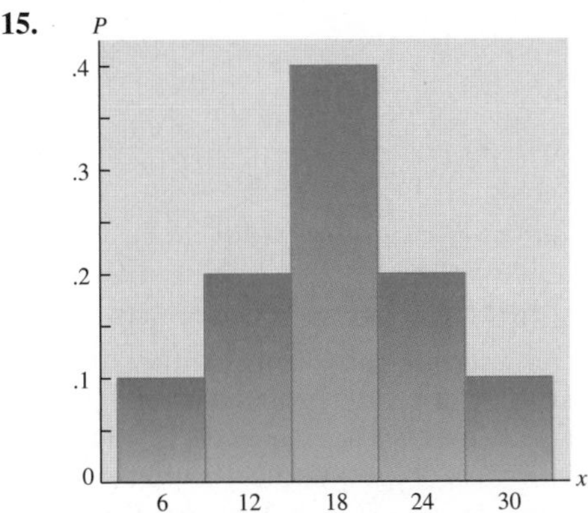

16.

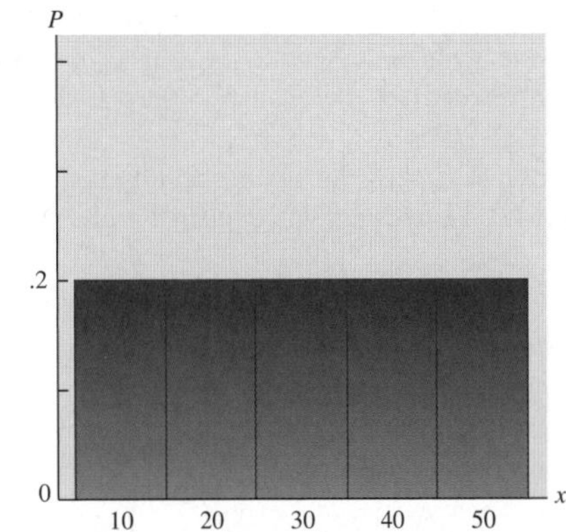

17. For the game in Example 4, suppose one day Mary brings a 2-headed coin and uses it to toss for the coffee. Since Mary tosses, Donna calls. Is this still a fair game?

18. In Exercise 17,

(a) what is Donna's expected gain if she calls heads?

(b) what is Donna's expected gain if she calls tails?

Solve each exercise. Many of these exercises require the use of combinations. (See Examples 3–5.)

19. Suppose 3 marbles are drawn from a bag containing 3 yellow and 4 white marbles.

(a) Draw a histogram for the number of yellow marbles in the sample.

(b) What is the expected number of yellow marbles in the sample?

20. Suppose 5 apples in a barrel of 25 apples are known to be rotten.

(a) Draw a histogram for the number of rotten apples in a sample of 2 apples.

(b) What is the expected number of rotten apples in a sample of 2 apples?

21. A delegation of 3 is selected from a city council made up of 5 liberals and 4 conservatives.

(a) What is the expected number of liberals in the delegation?

(b) What is the expected number of conservatives in the delegation?

22. From a group of 2 women and 5 men, a delegation of 2 is selected. Find the expected number of women in the delegation.

23. In a club with 20 senior and 10 junior members, what is the expected number of junior members on a 3-member committee?

24. If 2 cards are drawn at one time from a deck of 52 cards, what is the expected number of diamonds?

25. Suppose someone offers to pay you $5 if you draw 2 diamonds in the game in Exercise 24. He says that you should pay 50 cents for the chance to play. Is this a fair game?

26. Your friend missed class the day probability distributions were discussed. How would you explain a probability distribution to him?

27. Explain what expected value means in your own words.

28. A raffle offers a first prize of $100 and 2 second prizes of $40 each. One ticket costs $1, and 500 tickets are sold. Find the expected winnings for a person who buys 1 ticket. Is this a fair game?

29. A raffle offers a first prize of $1000, 2 second prizes of $300 each, and 20 third prizes of $10 each. If 10,000 tickets are sold at 50¢ each, find the expected winnings for a person buying 1 ticket. Is this a fair game?

Find the expected winnings for the games of chance described in Exercises 30–35.

30. A lottery requires you to choose 4 cards from an ordinary deck: 1 heart, 1 club, 1 diamond, and 1 spade in that order from the 13 cards in each suit. If all four choices are selected by the lottery, you win $5000. It costs $1 to play.

31. If exactly 3 of the 4 choices in Exercise 30 are selected, the player wins $200. (Ignore the possibility that all 4 choices are selected. It still costs $1 to play.)

32. In one form of roulette, you bet $1 on "even." If 1 of the 18 even numbers comes up, you get your dollar back, plus another one. If 1 of the 20 noneven (18 odd, 0, and 00) numbers comes up, you lose your dollar.

33. Repeat Exercise 32 if there are only 19 noneven numbers (no 00).

34. *Numbers* is a game in which you bet $1 on any three-digit number from 000 to 999. If your number comes up, you get $500.

35. In one form of the game Keno, the house has a pot containing 80 balls, each marked with a different number from 1 to 80. You buy a ticket for $1 and mark one of the 80 numbers on it. The house then selects 20 numbers at random. If your number is among the 20, you get $3.20 (for a net winning of $2.20).

36. Business A magazine distributor offers a first prize of $100,000, two second prizes of $40,000 each, and two third prizes of $10,000 each. A total of 2,000,000 entries are received in the contest. Find the expected winnings if you submit one entry to the contest. If it would cost you 50¢ in paper, and stamps to enter, would it be worth it?

37. Business A contest at a fast-food restaurant offered the following cash prizes and probabilities of winning on one visit.

Prize	*Probability*
$100,000	1/176,402,500
$25,000	1/39,200,556
$5000	1/17,640,250
$1000	1/1,568,022
$100	1/282,244
$5	1/7056
$1	1/588

Suppose you spend $1 to buy a bus pass that lets you go to 25 different restaurants in the chain and pick up entry forms. Find your expected value.

38. Business According to officials of Mars, the makers of M & M Plain Chocolate Candies, 20% of the candies in each bag are red.* 4 candies are selected from a bag and the number of red candies is recorded.
(a) Give a probability distribution for the number of red candies.
(b) Find the expected value.

39. Business A local used-car dealer gets complaints about his cars as shown in the following table.

Number of Complaints per Day	0	1	2	3	4	5	6
Probability	.01	.05	.15	.26	.33	.14	.06

Find the expected number of complaints per day.

40. Finance An insurance company has written 100 policies of $10,000, 500 of $5000, and 1000 of $1000 for people of age 20. If experience shows that the probability that a person will die at age 20 is .001, how much can the company expect to pay out during the year the policies were written?

41. Business Levi Strauss and Company uses expected value to help its salespeople rate their accounts.† For each account, a salesperson estimates potential additional vol-

*NCTM News Bulletin, February 1995, p. 5.

†This example was supplied by James McDonald, Levi Strauss and Company, San Francisco.

Account Number	Existing Volume	Potential Additional Volume	Probability of Additional Volume	Expected Value of Potential	Existing Volume + Expected Value of Potential	Class
1	$15,000	$10,000	.25	$2,500	$17,500	C
2	40,000	0	—	—	40,000	C
3	20,000	10,000	.20			
4	50,000	10,000	.10			
5	5,000	50,000	.50			
6	0	100,000	.60			
7	30,000	20,000	.80			

ume and the probability of getting it. The product of these figures gives the expected value of the potential, which is added to the existing volume. The totals are then classified as A, B, or C as follows: $40,000 or below, class C; above $40,000 up to and including $55,000, class B; above $55,000, class A. Complete the chart above for one salesperson.

42. Natural Science In a certain animal species, the probability that a healthy adult female will have no offspring in a given year is .31, while the probabilities of 1, 2, 3, or 4 offspring are, respectively, .21, .19, .17, and .12. Find the expected number of offspring.

43. Natural Science Otitis media, or middle ear infection, is initially treated with an antibiotic. Researchers have compared two antibiotics, amoxicillin and cefaclor, for their cost effectiveness. Amoxicillin is inexpensive, safe, and effective. Cefaclor is also safe. However, it is considerably more expensive and it is generally more effective. Use the tree diagram below (where the costs are estimated as the total cost of medication, office visit, ear check, and hours of lost work) to answer the following.*
(a) Find the expected cost of using each antibiotic to treat a middle ear infection.
(b) To minimize the total expected cost, which antibiotic should be chosen?

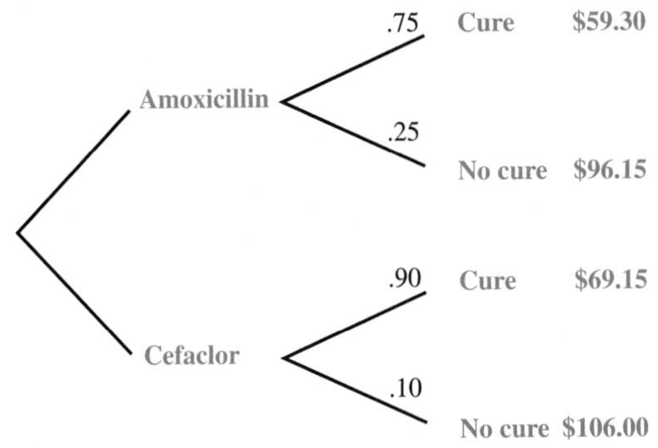

44. Physical Science One of the few methods that can be used in an attempt to cut the severity of a hurricane is to *seed* the storm. In this process, silver iodide crystals are dropped into the storm. Unfortunately, silver iodide crystals sometimes cause the storm to *increase* its speed. Wind speeds may also increase or decrease even with no seeding. Use the tree diagram below to answer the following.*
(a) Find the expected amount of damage under each option, "seed" and "do not seed."
(b) To minimize total expected damage, what option should be chosen?

Change in wind speed
Property damage (millions of dollars)
Seed
0.038 +32% 335.8
0.143 +16% 191.1
0.392 0 100.0
0.255 −16% 46.7
0.172 −34% 16.3
Do not seed
0.054 +32% 335.8
0.206 +16% 191.1
0.480 0 100.0
0.206 −16% 46.7
0.054 −34% 16.3

45. At the end of play in a major golf tournament, two players, an "old pro" and a "new kid," are tied. Suppose the first prize is $80,000 and second prize is $20,000. Find the expected winnings for the old pro if
(a) both players are of equal ability;
(b) the new kid will freeze up, giving the old pro a 3/4 chance of winning.

*Jeffrey Weiss and Shoshana Melman, "Cost Effectiveness in the Choice of Antibiotics for the Initial Treatment of Otitis Media in Children: A Decision Analysis Approach." *Journal of Pediatric Infectious Disease* vol. 7, no. 1 (1998): 23–26.

*Data from "The Decision to Seed Hurricanes," by R. A. Howard from *Science*, Vol. 176, pp. 1191–1202, Copyright 1972 by the American Association for the Advancement of Science.

46. Kimberly Workman has four cats: Riley, Abby, Beastie, and Sylvester. Each cat has a 30% probability of climbing into the chair in which Kimberly is sitting, independent of how many cats are already in the chair with Kimberly.

(a) Find the probability distribution for the number of cats in the chair with Kimberly.

(b) Find the expected number of cats in the chair with Kimberly using the probability distribution in part (a).

(c) Find the expected number of cats in the chair with Kimberly using the formula for expected value of the binomial distribution.

47. Social Science Mr. Statistics (a feature in *Fortune* magazine) investigated the claim of the United States Postal Service that 83% of first class mail in New York City arrives by the next day.* (The figure is 87% nationwide.) He mailed a letter to himself on 10 consecutive days; only 4 were delivered by the next day.

(a) Find the probability distribution for the number of letters delivered by the next day if the overall probability of next-day delivery is 83%.

(b) Using your answer to part (a), find the probability that 4 or fewer out of 10 letters would be delivered by the next day.

(c) Based on your answer to part (b), do you think it is likely that the 83% figure is accurate? Explain.

(d) Find the number of letters out of 10 that you would expect to be delivered by the next day if the 83% figure is accurate.

48. After a team scores a touchdown, it can either attempt to kick an extra point or attempt a two-point conversion. During the 1999–2000 NFL season, two-point conversions were successful 37% of the time and the extra-point kicks were successful 94% of the time.*

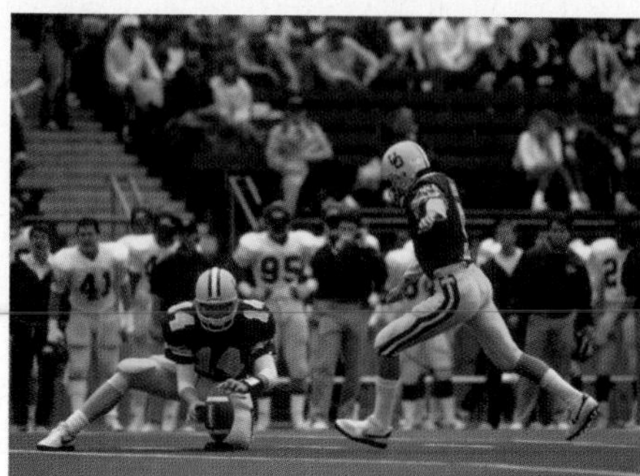

(a) Calculate the expected value of each strategy.

(b) Which strategy, over the long run, will maximize the number of points scored?

(c) Using this information, should a team always only use one strategy? Explain.

*Daniel Seligman, "Ask Mr. Statistics," *Fortune*, July 24, 1995, pp. 170–171.

*David Leonhardt, "In Football, 6 + 2 Often Equals 6," *New York Times*, Sunday, January 16, 2000. p. 4–2.

9.5 MARKOV CHAINS

In Section 8.5, we touched on **stochastic processes,** mathematical models that evolve over time in a probabilistic manner. In this section we study a special kind of stochastic process called a **Markov chain,** where the outcome of an experiment depends only on the outcome of the previous experiment. In other words, the next **state** of the system depends only on the present state, not on preceding states. Such experiments are common enough in applications to make their study worthwhile. Markov chains are named after the Russian mathematician A. A. Markov, 1856–1922, who started the theory of stochastic processes. To see how Markov chains work, we look at an example.

EXAMPLE 1 A small town has only two dry cleaners, Johnson and NorthClean. Johnson's manager hopes to increase the firm's market share by an extensive advertising campaign. After the campaign, a market research firm finds that there is a probability of .8 that a Johnson customer will bring his next batch of dirty items to Johnson and a .35 chance that a NorthClean customer will switch to Johnson for his next batch. Assume that the probability that a customer comes to a given cleaners depends only on where the last load of clothes was taken. If there is an .8 chance that a Johnson customer will return to Johnson, then there must be a $1 - .8 = .2$ chance that the customer will switch to NorthClean. In the same way, there is a $1 - .35 = .65$ chance that a NorthClean customer will return to NorthClean. If an individual bringing a load to Johnson is said to be in state 1 and an individual bringing a load to NorthClean is said to be in state 2, then these probabilities of change from one cleaner to the other are as shown in the following table.

	State	SECOND LOAD 1	2
FIRST LOAD	1	.8	.2
	2	.35	.65

1

Given the transition matrix

$$\text{State} \begin{matrix} 1 \\ 2 \end{matrix} \overset{\begin{matrix} \text{State} \\ 1 \quad 2 \end{matrix}}{\begin{bmatrix} .3 & .7 \\ .1 & .9 \end{bmatrix}}.$$

(a) What is the probability of changing from state 1 to state 2?

(b) What does the number .1 represent?

(c) Draw a transition diagram for this information.

Answers:

(a) .7

(b) The probability of changing from state 2 to state 1

(c)

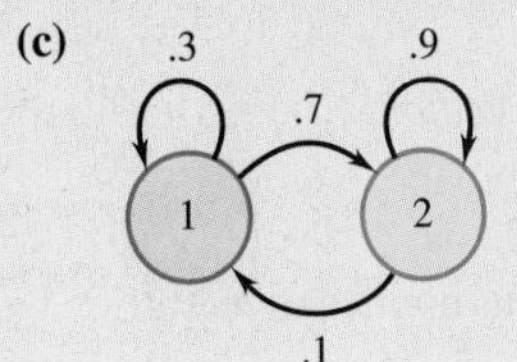

The information from the table can be written in other forms. Figure 9.11 is a **transition diagram** that shows the two states and the probabilities of going from one state to another.

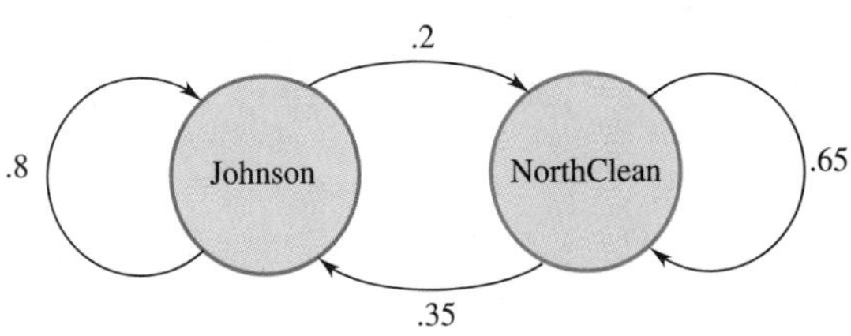

FIGURE 9.11

In a **transition matrix,** the states are indicated at the side and top, as follows.

$$\textit{First Load} \begin{matrix} \text{Johnson} \\ \text{NorthClean} \end{matrix} \overset{\begin{matrix} \textit{Second Load} \\ \text{Johnson} \quad \text{NorthClean} \end{matrix}}{\begin{bmatrix} .8 & .2 \\ .35 & .65 \end{bmatrix}}$$ 1

A **transition matrix** has the following features.

1. It is square, since all possible states must be used both as rows and as columns.
2. All entries are between 0 and 1, inclusive, because all entries represent probabilities.
3. The sum of the entries in any row must be 1, because the numbers in the row give the probability of changing from the state at the left to one of the states indicated across the top.

EXAMPLE 2 Suppose that when the new promotional campaign began, Johnson had 40% of the market and NorthClean had 60%. Use the probability tree in Figure 9.12 to find how these proportions would change after another week of advertising.

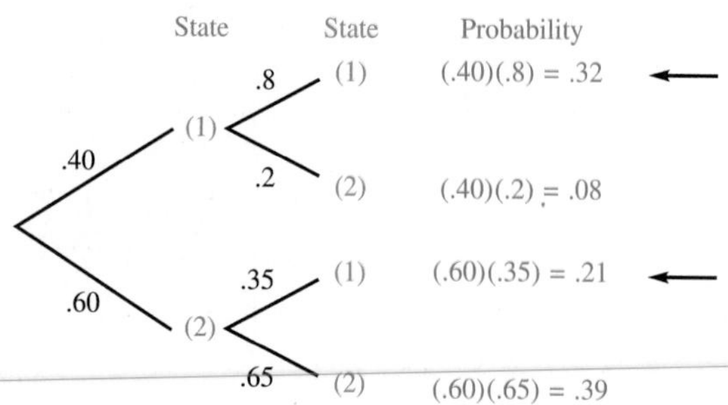

State 1: taking cleaning to Johnson
State 2: taking cleaning to NorthClean

FIGURE 9.12

Add the numbers indicated with arrows to find the proportion of people taking their cleaning to Johnson after 1 week.

$$.32 + .21 = .53$$

Similarly, the proportion taking their cleaning to NorthClean is

$$.08 + .39 = .47.$$

The initial distribution of 40% and 60% becomes, after 1 week, 53% and 47%.

These distributions can be written as the *probability vectors*

$$[.40 \quad .60] \quad \text{and} \quad [.53 \quad .47].$$

A **probability vector** is a matrix of only one row, having nonnegative entries with the sum of the entries equal to 1.

The results from the probability tree above are exactly the same as the result of multiplying the initial probability vector by the transition matrix. (Multiplication of matrices was discussed in Section 6.4.)

$$[.4 \quad .6]\begin{bmatrix} .8 & .2 \\ .35 & .65 \end{bmatrix} = [.53 \quad .47]$$

If v denotes the original probability vector [.4 .6] and P denotes the transition matrix, then the market share vector after one week is vP = [.53 .47]. To find the market share after 2 weeks multiply the vector vP = [.53 .47] by P, which amounts to finding vP^2. (2)

Problem 2 at the side shows that after two weeks, the market share vector is vP^2 = [.59 .41]. To get the market share after 3 weeks, multiply this vector by P, that is, find vP^3. Do not use the rounded answer from Problem 2. (3)

2

Find the product

$$[.53 \quad .47]\begin{bmatrix} .8 & .2 \\ .35 & .65 \end{bmatrix}.$$

Answer:

[.59 .41] (rounded)

Find each cleaner's market share after 3 weeks.

Answer:

[.61 .39] (rounded)

Continuing this process gives each cleaner's share of the market after additional weeks.

Weeks after Start	*Johnson*	*NorthClean*	
0	.4	.6	v
1	.53	.47	vP^1
2	.59	.41	vP^2
3	.61	.39	vP^3
4	.63	.37	vP^4
12	.64	.36	vP^{12}
13	.64	.36	vP^{13}

The results seem to approach the probability vector [.64 .36].

What happens if the initial probability vector is different from [.4 .6]? Suppose [.75 .25] is used; the same powers of the transition matrix as above give the following results.

Week after Start	*Johnson*	*NorthClean*	
0	.75	.25	v
1	.69	.31	vP^1
2	.66	.34	vP^2
3	.65	.35	vP^3
4	.64	.36	vP^4
5	.64	.36	vP^5
6	.64	.36	vP^6

The results again seem to be approaching the numbers in the probability vector [.64 .36], the same numbers approached with the initial probability vector [.4 .6]. In either case, the long-range trend is for a market share of about 64% for Johnson and 36% for NorthClean. The example above suggests that this long-range trend does not depend on the initial distribution of market shares. This means that if the initial market share for Johnson was less than 64%, the advertising campaign has paid off in terms of a greater long-range market share. If the initial share was more than 64%, the campaign did not pay off.

Regular Transition Matrices

One of the many applications of Markov chains is in finding these long-range predictions. It is not possible to make long-range predictions with all transition matrices, but for a large set of transition matrices, long-range predictions *are* possible. Such predictions are always possible with **regular transition matrices.** A transition matrix is **regular** if some power of the matrix contains all positive entries. A Markov chain is a **regular Markov chain** if its transition matrix is regular.

EXAMPLE 3 Decide whether the following transition matrices are regular.

(a) $A = \begin{bmatrix} .75 & .25 & 0 \\ 0 & .5 & .5 \\ .6 & .4 & 0 \end{bmatrix}$

Square A.

$$A^2 = \begin{bmatrix} .5625 & .3125 & .125 \\ .3 & .45 & .25 \\ .45 & .35 & .2 \end{bmatrix}$$

Since all entries in A^2 are positive, matrix A is regular.

Decide if the following transition matrices are regular.

(a) $\begin{bmatrix} 0 & 1 \\ 1 & 0 \end{bmatrix}$

(b) $\begin{bmatrix} .3 & .7 \\ 1 & 0 \end{bmatrix}$

Answers:

(a) No

(b) Yes

(b) $B = \begin{bmatrix} .5 & 0 & .5 \\ 0 & 1 & 0 \\ 0 & 0 & 1 \end{bmatrix}$

Find various powers of B.

$$B^2 = \begin{bmatrix} .25 & 0 & .75 \\ 0 & 1 & 0 \\ 0 & 0 & 1 \end{bmatrix} \quad B^3 = \begin{bmatrix} .125 & 0 & .875 \\ 0 & 1 & 0 \\ 0 & 0 & 1 \end{bmatrix} \quad B^4 = \begin{bmatrix} .0625 & 0 & .9375 \\ 0 & 1 & 0 \\ 0 & 0 & 1 \end{bmatrix}$$

Notice that all of the powers of B shown here have zeros in the same locations. Thus, further powers of B will still give the same zero entries, so that no power of matrix B contains all positive entries. For this reason, B is not regular. ◀ 4

NOTE If a transition matrix P has some zero entries, and P^2 does as well, you may wonder how far you must compute P^k to be certain that the matrix is not regular. The answer is that if zeros occur in the identical places in both P^k and P^{k+1} for any k, they will appear in those places for all higher powers of P, so P is not regular.

Suppose that v is any probability vector. It can be shown that for a regular Markov chain with a transition matrix P, there exists a single vector V that does not depend on v, such that $v \cdot P^n$ gets closer and closer to V as n gets larger and larger.

Equilibrium Vector of a Markov Chain

If a Markov chain with transition matrix P is regular, then there is a unique vector V such that, for any probability vector v and for large values of n,

$$v \cdot P^n \approx V.$$

Vector V is called the equilibrium vector or the fixed vector of the Markov chain.

In the example with Johnson Cleaners, the equilibrium vector V is approximately $[.64 \quad .36]$. Vector V can be determined by finding P^n for larger and larger values of n and then looking for a vector that the product $v \cdot P^n$ approaches. Such an approach can be very tedious, however, and is prone to error. To find a better way, start with the fact that for a large value of n,

$$v \cdot P^n \approx V,$$

as mentioned above. From this result, $v \cdot P^n \cdot P \approx V \cdot P$, so that

$$v \cdot P^n \cdot P = v \cdot P^{n+1} \approx VP.$$

Since $v \cdot P^n \approx V$ for large values of n, it is also true that $v \cdot P^{n+1} \approx V$ for large values of n (the product $v \cdot P^n$ approaches V, so that $v \cdot P^{n+1}$ must also approach V). Thus, $v \cdot P^{n+1} \approx V$ and $v \cdot P^{n+1} \approx VP$, which suggests that

$$VP = V.$$

If a Markov chain with transition matrix P is regular, then the equilibrium vector V satisties

$$VP = V.$$

The equilibrium vector V can be found by solving a system of linear equations, as shown in the next examples.

EXAMPLE 4 Find the long-range trend for the Markov chain in Examples 1 and 2, with transition matrix

$$P = \begin{bmatrix} .8 & .2 \\ .35 & .65 \end{bmatrix}.$$

This matrix is regular since all entries are positive. Let P represent this transition matrix and let V be the probability vector $[v_1 \quad v_2]$. We want to find V such that

$$VP = V,$$

or

$$[v_1 \quad v_2]\begin{bmatrix} .8 & .2 \\ .35 & .65 \end{bmatrix} = [v_1 \quad v_2].$$

Multiply on the left to get

$$[.8v_1 + .35v_2 \quad .2v_1 + .65v_2] = [v_1 \quad v_2].$$

Set corresponding entries from the two matrices equal to get

$$.8v_1 + .35v_2 = v_1 \qquad .2v_1 + .65v_2 = v_2.$$

Simplify each of these equations,

$$-.2v_1 + .35v_2 = 0 \qquad .2v_1 - .35v_2 = 0$$

These last two equations are really the same. (The equations in the system obtained from $VP = V$ are always dependent.) To find the values of v_1 and v_2, recall that $V = [v_1 \quad v_2]$ is a probability vector, so that

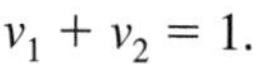

$$v_1 + v_2 = 1.$$

Find v_1 and v_2 by solving the system

$$\begin{aligned} -.2v_1 + .35v_2 &= 0 \\ v_1 + \quad v_2 &= 1. \end{aligned}$$

From the second equation, $v_1 = 1 - v_2$. Substitute for v_1 in the first equation:

$$-.2(1 - v_2) + .35v_2 = 0.$$ (5)

Since $v_2 = .364$ (from problem 5 at the side) and $v_1 = 1 - v_2$, then $v_1 = 1 - .364 = .636$ and the equilibrium vector is $[.636 \quad .364] \approx [.64 \quad .36]$. ◀

(5)

Solve the equation for v_2. Round to the nearest thousandth.

Answer:

.364

EXAMPLE 5 The probability that a complex assembly line works correctly depends on whether or not the line worked correctly the last time it was used. The various probabilities are as given in the following transition matrix.

	Works Properly Now	Does Not
Worked Properly Before	.9	.1
Did Not	.7	.3

Find the long-range probability that the assembly line will work properly.

Begin by finding the equilibrium vector $[v_1 \quad v_2]$, where

$$[v_1 \quad v_2]\begin{bmatrix} .9 & .1 \\ .7 & .3 \end{bmatrix} = [v_1 \quad v_2].$$

Multiplying on the left and setting corresponding entries equal gives the equations

$$.9v_1 + .7v_2 = v_1 \text{and} \quad .1v_1 + .3v_2 = v_2$$

or

$$-.1v_1 + .7v_2 = 0 \quad \text{and} \quad .1v_1 - .7v_2 = 0.$$

Substitute $v_1 = 1 - v_2$ in the first of these equations to get

$$\begin{aligned} -.1(1 - v_2) + .7v_2 &= 0 \\ -.1 + .1v_2 + .7v_2 &= 0 \\ -.1 + .8v_2 &= 0 \\ .8v_2 &= .1 \\ v_2 &= \frac{1}{8}, \end{aligned}$$

and $v_1 = 1 - 1/8 = 7/8$. The equilibrium vector is $[7/8 \quad 1/8]$. In the long run, the company can expect the assembly line to run properly 7/8 of the time. ◀ (6)

(6)

In Example 5, suppose the company modifies the line so that the transition matrix becomes

$$\begin{bmatrix} .95 & .05 \\ .8 & .2 \end{bmatrix}.$$

Find the long-range probability that the assembly line will work properly.

Answer:

16/17

EXAMPLE 6 Find the equilibrium vector for the transition matrix

$$K = \begin{bmatrix} .2 & .6 & .2 \\ .1 & .1 & .8 \\ .3 & .3 & .4 \end{bmatrix}.$$

Matrix K has all positive entries and thus is regular. For this reason, an equilibrium vector V must exist such that $VK = V$. Let $V = [v_1 \quad v_2 \quad v_3]$.

Then

$$[v_1 \quad v_2 \quad v_3] \begin{bmatrix} .2 & .6 & .2 \\ .1 & .1 & .8 \\ .3 & .3 & .4 \end{bmatrix} = [v_1 \quad v_2 \quad v_3].$$

Use matrix multiplication on the left.

$$[.2v_1 + .1v_2 + .3v_3 \quad .6v_1 + .1v_2 + .3v_3 \quad .2v_1 + .8v_2 + .4v_3] = [v_1 \quad v_2 \quad v_3]$$

Set corresponding entries equal.

$$\begin{aligned} .2v_1 + .1v_2 + .3v_3 &= v_1 \\ .6v_1 + .1v_2 + .3v_3 &= v_2 \\ .2v_1 + .8v_2 + .4v_3 &= v_3 \end{aligned}$$

Simplify these equations.

$$\begin{aligned} -.8v_1 + .1v_2 + .3v_3 &= 0 \\ .6v_1 - .9v_2 + .3v_3 &= 0 \\ .2v_1 + .8v_2 - .6v_3 &= 0 \end{aligned}$$

Since V is a probability vector,

$$v_1 + v_2 + v_3 = 1.$$

This gives a system of four equations in three unknowns:

$$\begin{aligned} v_1 + \quad v_2 + \quad v_3 &= 1 \\ -.8v_1 + .1v_2 + .3v_3 &= 0 \\ .6v_1 - .9v_2 + .3v_3 &= 0 \\ .2v_1 + .8v_2 - .6v_3 &= 0. \end{aligned}$$

This system can be solved with the Gauss-Jordan method in Section 6.2. Start with the augmented matrix

$$\left[\begin{array}{ccc|c} 1 & 1 & 1 & 1 \\ -.8 & .1 & .3 & 0 \\ .6 & -.9 & .3 & 0 \\ .2 & .8 & -.6 & 0 \end{array}\right].$$

The solution of this system is $v_1 = 5/23$, $v_2 = 7/23$, $v_3 = 11/23$, and

$$V = \begin{bmatrix} \frac{5}{23} & \frac{7}{23} & \frac{11}{23} \end{bmatrix} \approx [.22 \quad .30 \quad .48].$$ ◀ 7

7

Find the equilibrium vector for the transition matrix

$$P = \begin{bmatrix} .3 & .7 \\ .5 & .5 \end{bmatrix}.$$

Answer:

[5/12 7/12]

In Example 4, we found that [.64 .36] was the equilibrium vector for the regular transition matrix.

$$P = \begin{bmatrix} .8 & .2 \\ .35 & .65 \end{bmatrix}.$$

Observe what happens when you take powers of the matrix P (displayed enries have been rounded for easy reading, but the full decimals were used in the calculations).

$$P^2 = \begin{bmatrix} .71 & .29 \\ .51 & .49 \end{bmatrix} \qquad P^3 = \begin{bmatrix} .67 & .33 \\ .58 & .42 \end{bmatrix} \qquad P^4 = \begin{bmatrix} .65 & .35 \\ .61 & .39 \end{bmatrix}$$

$$P^5 = \begin{bmatrix} .64 & .36 \\ .62 & .38 \end{bmatrix} \qquad P^6 = \begin{bmatrix} .64 & .36 \\ .63 & .37 \end{bmatrix} \qquad P^{10} = \begin{bmatrix} .64 & .36 \\ .64 & .36 \end{bmatrix}$$

As these results suggest, higher and higher powers of the transition matrix P approach a matrix having all rows identical: these identical rows have as entries the entries of the equilibrium vector V.

If you have the technology to compute matrix powers easily (such as a graphing calculator), you can approximate the equilibrium vector by taking higher and higher powers of the transition matrix until all its rows are identical. Figure 9.13 shows part of this process for the transition matrix

$$B = \begin{bmatrix} .9 & .1 \\ .7 & .3 \end{bmatrix}$$

of Example 5.

```
[B]^4
 [[.8752 .1248]
  [.8736 .1264]]
[B]^8
[[.87500032 .12…
 [.87499776 .12…
```

```
[B]^12
[[.8750000005 :…
 [.8749999964 :…
[B]^16
   [[.875 .125]
    [.875 .125]]
```

FIGURE 9.13

Figure 9.13 indicates that the equilibrium vector is [.875 .125] = [7/8 1/8], which is what was found algebraically in Example 5.

The results of this section can be summarized as follows.

Properties of Regular Markov Chains

Suppose a regular Markov chain has a transition matrix P.

1. As n gets larger and larger, the product $v \cdot P^n$ approaches a unique vector V for any initial probability vector v. Vector V is called the *equilibrium vector or fixed vector*.
2. Vector V has the property that $VP = V$.
3. To find V, solve a system of equations obtained from the matrix equation $VP = V$ and from the fact that the sum of the entries of V is 1.
4. The powers P^n come closer and closer to a matrix whose rows are made up of the entries of the equilibrium vector V.

9.5 Exercises

Decide which of the following could be a probability vector.

1. $\begin{bmatrix} \frac{1}{3} & \frac{2}{3} \end{bmatrix}$

2. $\begin{bmatrix} \frac{3}{4} & \frac{1}{4} \end{bmatrix}$

3. $[0 \quad 1]$

4. $[.4 \quad .2 \quad 0]$

5. $[.3 \quad -.1 \quad .6]$

6. $\begin{bmatrix} \frac{1}{4} & \frac{1}{8} & \frac{5}{8} \end{bmatrix}$

Decide which of the following could be a transition matrix. Sketch a transition diagram for any transition matrices.

7. $\begin{bmatrix} .7 & .1 \\ .5 & .5 \end{bmatrix}$

8. $\begin{bmatrix} \frac{1}{4} & \frac{3}{4} \\ 0 & 1 \end{bmatrix}$

9. $\begin{bmatrix} \frac{2}{3} & \frac{1}{3} \\ \frac{1}{5} & \frac{4}{5} \end{bmatrix}$

10. $\begin{bmatrix} 0 & 1 & 0 \\ \frac{1}{3} & \frac{1}{3} & \frac{1}{3} \\ 1 & 0 & 0 \end{bmatrix}$

11. $\begin{bmatrix} \frac{1}{2} & \frac{1}{4} & 1 \\ \frac{2}{3} & 0 & \frac{1}{3} \\ \frac{1}{3} & 1 & 0 \end{bmatrix}$

12. $\begin{bmatrix} .2 & .3 & .5 \\ 0 & 0 & 1 \\ .1 & .9 & 0 \end{bmatrix}$

In Exercises 13–15, write any transition diagrams as transition matrices.

13.

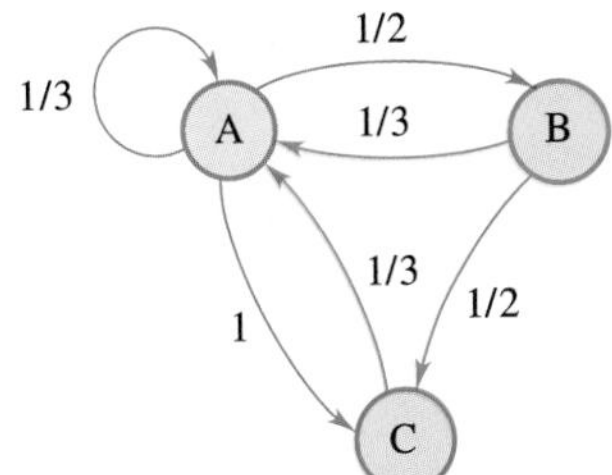

14.

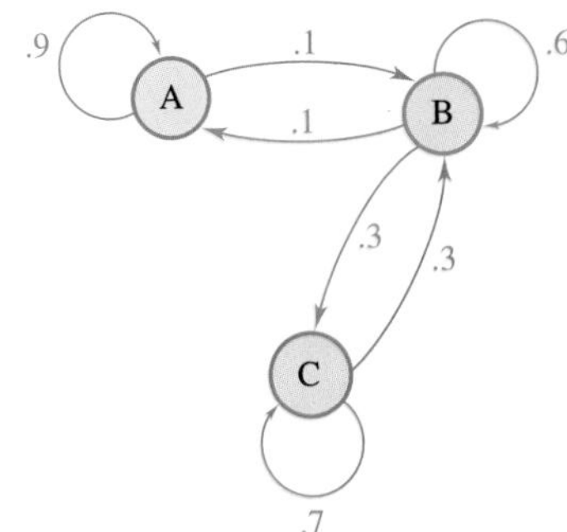

15.

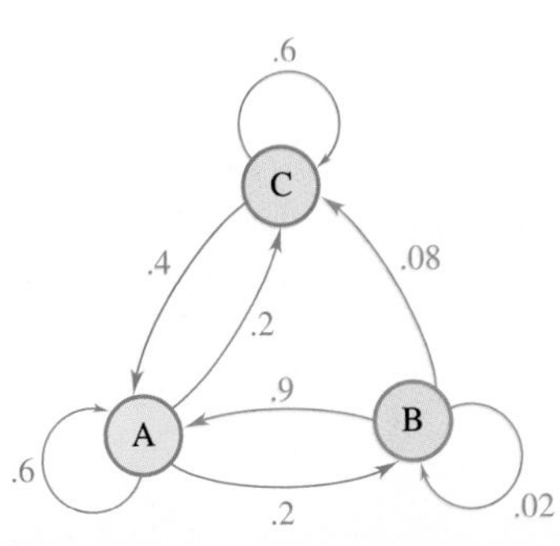

Decide whether the following transition matrices are regular. (See Example 3.)

16. $\begin{bmatrix} 1 & 0 \\ .6 & .4 \end{bmatrix}$

17. $\begin{bmatrix} .2 & .8 \\ .9 & .1 \end{bmatrix}$

18. $\begin{bmatrix} .3 & .5 & .2 \\ 1 & 0 & 0 \\ .5 & .1 & .4 \end{bmatrix}$

19. $\begin{bmatrix} 0 & 1 & 0 \\ .4 & .2 & .4 \\ 1 & 0 & 0 \end{bmatrix}$

20. $\begin{bmatrix} .25 & .40 & .30 & .05 \\ .18 & .23 & .59 & 0 \\ 0 & .15 & .36 & .49 \\ .28 & .32 & .24 & .16 \end{bmatrix}$

21. $\begin{bmatrix} .12 & .68 & 0 & .20 \\ 0 & .33 & .40 & .27 \\ 0 & 0 & 1 & 0 \\ .52 & 0 & .43 & .05 \end{bmatrix}$

Find the equilibrium vector for each of the following transition matrices. (See Examples 4 and 5.)

22. $\begin{bmatrix} .3 & .7 \\ .4 & .6 \end{bmatrix}$

23. $\begin{bmatrix} .8 & .2 \\ .1 & .9 \end{bmatrix}$

24. $\begin{bmatrix} \frac{1}{4} & \frac{3}{4} \\ \frac{1}{2} & \frac{1}{2} \end{bmatrix}$

25. $\begin{bmatrix} \frac{2}{3} & \frac{1}{3} \\ \frac{1}{8} & \frac{7}{8} \end{bmatrix}$

26. $\begin{bmatrix} .25 & .35 & .4 \\ .1 & .3 & .6 \\ .55 & .4 & .05 \end{bmatrix}$

27. $\begin{bmatrix} .16 & .28 & .56 \\ .43 & .12 & .45 \\ .86 & .05 & .09 \end{bmatrix}$

28. $\begin{bmatrix} .1 & .1 & .8 \\ .4 & .4 & .2 \\ .1 & .2 & .7 \end{bmatrix}$

29. $\begin{bmatrix} .5 & .2 & .3 \\ .1 & .4 & .5 \\ .2 & .2 & .6 \end{bmatrix}$

For each of the following transition matrices, use a graphing calculator or computer to find the first five powers of the matrix. Then find the probability that state 2 changes to state 4 after 5 repetitions of the experiment.

30. $\begin{bmatrix} .1 & .2 & .2 & .3 & .2 \\ .2 & .1 & .1 & .2 & .4 \\ .2 & .1 & .4 & .2 & .1 \\ .3 & .1 & .1 & .2 & .3 \\ .1 & .3 & .1 & .1 & .4 \end{bmatrix}$

31. $\begin{bmatrix} .3 & .2 & .3 & .1 & .1 \\ .4 & .2 & .1 & .2 & .1 \\ .1 & .3 & .2 & .2 & .2 \\ .2 & .1 & .3 & .2 & .2 \\ .1 & .1 & .4 & .2 & .2 \end{bmatrix}$

32. **Social Science** The chart on the next page shows the percent of the poor, middle class, and affluent that change into another class. The first graph shows the figures for 1967–1979, and the second those for 1980–1991.* For each time period, assume the number of people who move directly from poor to affluent or from affluent to poor is essentially 0.

**New York Times*, June 4, 1995, p. E4.

(a) Find the long-range percent of poor, middle class, and affluent people if the 1967–1979 trends were to continue.

(b) Find the long-range percent of poor, middle class, and affluent people if the 1980–1991 trends were to continue.

The poor are more likely to stay poor and the affluent are more likely to stay affluent.

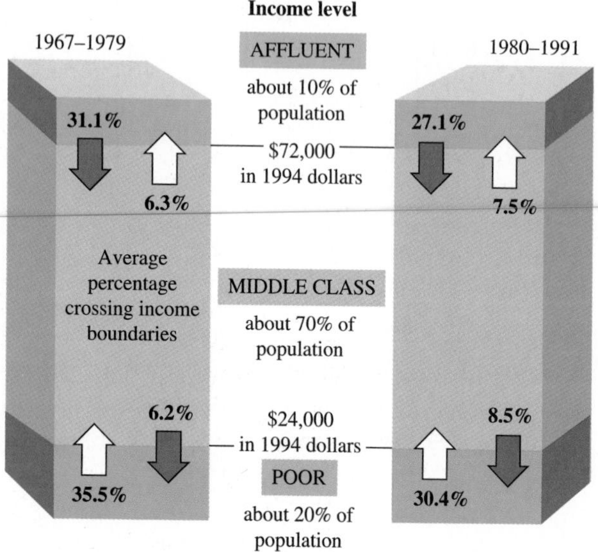

All figures are for household after-tax income, including wages, salaries, and some Government assistance programs like food stamps.

Sources: Greg Duncan, Northwestern University; Timothy Smeeding, Syracuse University

33. Finance In 1992, many homeowners refinanced their mortgages to take advantage of lower interest rates. Most of the mortgages could be classified into three groups: 30-year fixed-rate, 15-year fixed-rate, and adjustable rate. (For this exercise, the small number of loans that are not of those three types will be classified with the adjustable rate loans.) Sometimes when a homeowner refinanced, the new loan was the same type as the old loan, and sometimes it was different. The breakdown of the percent in each category is shown in the following table.*

	NEW LOAN		
OLD LOAN	*30-Year Fixed*	*15-Year Fixed*	*Adjustable*
30-year fixed	.444	.479	.077
15-year fixed	.150	.802	.048
Adjustable	.463	.367	.170

New York Times, November 18, 1992, p. D2.

If these conversion rates were to persist, find the long-range trend for the percent of loans of each type.

34. Business The probability that a complex assembly line works correctly depends on whether the line worked correctly the last time it was used. There is a .95 chance that the line will work correctly if it worked correctly the time before, and a .7 chance that it will work correctly if it did *not* work correctly the time before. Set up a transition matrix with this information and find the long-run probability that the line will work correctly. (See Example 5.)

35. Business Suppose improvements are made in the assembly line of Exercise 34, so that the transition matrix becomes

$$\begin{array}{r} \\ \text{Works} \\ \text{Doesn't Work} \end{array} \begin{array}{c} \begin{array}{cc} \text{Works} & \text{Doesn't Work} \end{array} \\ \begin{bmatrix} .95 & .05 \\ .85 & .15 \end{bmatrix} \end{array}.$$

Find the new long-run probability that the line will work properly.

36. Natural Science In Exercises 61 and 62 of Section 8.4 we discussed the effect on flower color of cross-pollinating pea plants. As shown there, since the gene for red is dominant and the gene for white is recessive, 75% of these pea plants have red flowers and 25% have white flowers, because plants with 1 red and 1 white gene appear red. If a red-flowered plant is crossed with a red-flowered plant known to have 1 red and 1 white gene, then 75% of the offspring will be red and 25% will be white. Crossing a red-flowered plant that has 1 red and 1 white gene with a white-flowered plant produces 50% red-flowered offspring and 50% white-flowered offspring.

(a) Write a transition matrix using this information.

(b) Write a probability vector for the initial distribution of colors.

(c) Find the distribution of colors after 4 generations.

(d) Find the long-range distribution of colors.

37. Natural Science Snapdragons with 1 red gene and 1 white gene produce pink-flowered offspring. If a red snapdragon is crossed with a pink snapdragon, the probabilities that the offspring will be red, pink, or white are 1/2, 1/2, and 0, respectively. If 2 pink snapdragons are crossed, the probabilities of red, pink, or white offspring are 1/4, 1/2, and 1/4, respectively. For a cross between a white and a pink snapdragon, the corresponding probabilities are 0, 1/2, and 1/2. Set up a transition matrix and find the long-range prediction for the fraction of red, pink, and white snapdragons.

38. Many phenomena can be viewed as examples of a random walk. Consider the following simple example. A security guard can stand in front of any one of three doors 20 feet apart in front of a building, and every minute he decides whether to move to another door chosen at random. If he is at the middle door, he is equally likely to stay where he is, move to the door to the left, or move to the door to the right. If he is at the door on either end, he is equally likely to stay where he is or move to the middle door.

(a) Verify that the transition matrix is given by

$$\begin{array}{c} \\ 1 \\ 2 \\ 3 \end{array}\begin{array}{c} \begin{array}{ccc} 1 & 2 & 3 \end{array} \\ \begin{bmatrix} \frac{1}{2} & \frac{1}{2} & 0 \\ \frac{1}{3} & \frac{1}{3} & \frac{1}{3} \\ 0 & \frac{1}{2} & \frac{1}{2} \end{bmatrix} \end{array}.$$

(b) Find the long-range trend for the fraction of time the guard spends in front of each door.

39. Social Science The probability that a homeowner will become a renter in five years is .03. The probability that a renter will become a homeowner in five years is .1. Suppose the proportions in the population are 64% homeowners (O), 35.5% renters (R), and .5% homeless (H), with the following transition matrix. Assume that these figures continue to apply.

	O	R	H
O	.97	.03	0
R	.1	.899	.001
H	0	.4	.6

Find the long-range probabilities for the three categories.

40. Business An insurance company classifies its drivers into three groups: G_0 (no accidents), G_1 (one accident), and G_2 (more than one accident). The probability that a driver in G_0 will stay in G_0 after 1 year is .85, that he will become a G_1 is .10, and that he will become a G_2 is .05. A driver in G_1 cannot move to G_0 (this insurance company has a long memory). There is an .80 probability that a G_1 driver will stay in G_1 and a .20 probability that he will become a G_2. A driver in G_2 must stay in G_2.

(a) Write a transition matrix using this information.

Suppose that the company accepts 50,000 new policyholders, all of whom are in group G_0. Find the number in each group

(b) after 1 year; (c) after 2 years;
(d) after 3 years; (e) after 4 years.
(f) Find the equilibrium vector here. Interpret your result.

41. Business The difficulty with the mathematical model of Exercise 40 is that no "grace period" is provided; there should be a certain probability of moving from G_1 or G_2 back to G_0 (say, after 4 years with no accidents). A new system with this feature might produce the following transition matrix.

$$\begin{bmatrix} .85 & .10 & .05 \\ .15 & .75 & .10 \\ .10 & .30 & .60 \end{bmatrix}$$

Suppose that when this new policy is adopted, the company has 50,000 policyholders in group G_0. Find the number of these in each group

(a) after 1 year; (b) after 2 years;
(c) after 3 years.
(d) Find the equilibrium vector here. Interpret your result.

42. Business Research done by the Gulf Oil Corporation produced the following transition matrix for the probability that a person with one form of home heating would switch to another.*

		Will Switch to		
		Oil	Gas	Electric
	Oil	.825	.175	0
Now Has	Gas	.060	.919	.021
	Electric	.049	0	.951

The current share of the market held by these three types of heat is given by [.26 .60 .14]. Find the share of the market held by each type of heat after

(a) 1 year; (b) 2 years; (c) 3 years.
(d) What is the long-range prediction?

43. The results of cricket matches between England and Australia have been found to be modeled by a Markov chain.† The probability that England wins, loses, or draws is based on the result of the previous game, with the following transition matrix.

	Wins	Loses	Draws
Wins	.443	.364	.193
Loses	.277	.436	.287
Draws	.266	.304	.430

(a) Compute the transition matrix for the game after the next one, based on the result of the last game.
(b) Use your answer from part (a) to find the probability that, if England won the last game, England will win the game after the next one.
(c) Use your answer from part (a) to find the probability that, if Australia won the last game, England will win the game after the next one.

44. Social Science At one liberal arts college, students are classified as humanities majors, science majors, or undecided. There is a 20% chance that a humanities major will change to a science major from one year to the next and a 45% chance that a humanities major will change to undecided. A science major will change to humanities with probability .15 and to undecided with probability .35. An undecided will switch to humanities or science with probabilities of .5 and .3, respectively. Find the long-range prediction for the fraction of students in each of these three majors.

*From "Forecasting Market Shares of Alternative Home-Heating Units by Markov Process Using Transition Probabilities Estimates from Aggregate Time Series Data" by Ali Ezzati, from *Management Science*, Vol. 21, No. 4, December 1974. Copyright © 1974 The Institute of Management Sciences. Reprinted by permission.

†From Derek Colwell, Brian Jones, and Jack Gillett, "A Markov Chain in Cricket," *Mathematical Gazette*, June 1991.

45. Business In the queuing chain, we assume that people are queuing up to be served by, say, a bank teller. For simplicity, let us assume that once two people are in line, no one else can enter the line. Let us further assume that one person is served every minute, as long as someone is in line. Assume further that in any minute, there is a probability of $\frac{1}{2}$ that no one enters the line, a probability of $\frac{1}{3}$ that exactly one person enters the line, and a probability of $\frac{1}{6}$ that exactly two people enter the line, assuming there is room. If there is not enough room for two people, then the probability that one person enters the line is $\frac{1}{2}$. Let the state be given by the number of people in line.

(a) Verify that the transition matrix is

$$\begin{array}{c} \\ 0 \\ 1 \\ 2 \end{array}\begin{array}{c} \begin{array}{ccc} 0 & 1 & 2 \end{array} \\ \begin{bmatrix} \frac{1}{2} & \frac{1}{3} & \frac{1}{6} \\ \frac{1}{2} & \frac{1}{3} & \frac{1}{6} \\ 0 & \frac{1}{2} & \frac{1}{2} \end{bmatrix} \end{array}.$$

(b) Find the transition matrix for a two-minute period.

(c) Use your result from part (b) to find the probability that a queue with no one in line has two people in line two minutes later.

Use a graphing calculator or computer for Exercises 46 and 47.

46. Business A company with a new training program classified each employee in one of four states: s_1, never in the program; s_2, currently in the program; s_3, discharged; s_4, completed the program. The transition matrix for this company is given below.

$$\begin{array}{c} \\ s_1 \\ s_2 \\ s_3 \\ s_4 \end{array}\begin{array}{c} \begin{array}{cccc} s_1 & s_2 & s_3 & s_4 \end{array} \\ \begin{bmatrix} .4 & .2 & .05 & .35 \\ 0 & .45 & .05 & .5 \\ 0 & 0 & 1 & 0 \\ 0 & 0 & 0 & 1 \end{bmatrix} \end{array}$$

(a) What percent of employees who had never been in the program (state s_1) completed the program (state s_4) after the program had been offered five times?

(b) If the initial percent of employees in each state was $[.5 \quad .5 \quad 0 \quad 0]$, find the corresponding percents after the program had been offered four times.

47. Business Find the long-range prediction for the percent of employees in each state for the company training program from Exercise 46.

9.6 DECISION MAKING

John F. Kennedy once remarked he had assumed that as president it would be difficult to choose between distinct, opposite alternatives when a decision needed to be made. Actually, however, he found that such decisions were easy to make; the hard decisions came when he was faced with choices that were not as clear-cut. Most decisions fall in this last category—decisions that must be made under conditions of uncertainty. In Section 9.4 we saw how to use expected values to help make a decision. These ideas are extended in this section, where we consider decision making in the face of uncertainty. We begin with an example.

EXAMPLE 1 Freezing temperatures are endangering the orange crop in central California. A farmer can protect his crop by burning smudge pots—the heat from the pots keeps the oranges from freezing. However, burning the pots is expensive; the cost is $4000. The farmer knows that if he burns smudge pots he will be able to sell his crop for a net profit (after smudge pot costs are deducted) of $10,000, provided that the freeze does develop and wipes out other orange crops in California. If he does nothing he will either lose $2000 already invested in the crop if it does freeze or make a profit of $9600 if it does not freeze. (If it does not freeze, there will be a large supply of oranges, and thus his profit will be lower than if there was a small supply.) What should the farmer do?

He should begin by carefully defining the problem. First, he must decide on the **states of nature,** the possible alternatives over which he has no control. Here there are two: freezing temperatures, or no freezing temperatures. Next, the farmer should list the things he can control—his actions or **strategies.** He has two possible strategies; to use smudge pots or not. The consequences of each action under each state

Explain how each of the following payoffs in the matrix were obtained.

(a) −$2000

(b) $10,000

Answers:

(a) If it freezes and smudge pots are not used, the farmer's profit is −$2000 for labor costs.

(b) If it freezes and smudge pots are used, the farmer makes a profit of $10,000.

What should the farmer do if the probability of a freeze is .6? What is his expected profit?

Answer:

Use smudge pots; $8240

of nature, called **payoffs,** are summarized in a **payoff matrix,** as shown below. The payoffs in this case are the profits for each possible combination of events.

Strategies of Farmer		*States of Nature* Freeze	No Freeze
	Use Smudge Pots	$10,000	$5600
	Do Not Use Pots	−$2000	$9600

To get the $5600 entry in the payoff matrix, use the profit if there is no freeze, $9600, and subtract the $4000 cost of using the pots. (1)

Once the farmer makes the payoff matrix, what then? The farmer might be an optimist (some might call him a gambler); in this case he might assume that the best will happen and go for the biggest number of the matrix ($10,000). For this profit, he must adopt the strategy "use smudge pots."

On the other hand, if the farmer is a pessimist, he would want to minimize the worst thing that could happen. If he uses smudge pots, the worst thing that could happen to him would be profit of $5600, which will result if there is no freeze. If he does not use smudge pots, he might face a loss of $2000. To minimize the worst, he once again should adopt the strategy "use smudge pots."

Suppose the farmer decides that he is neither an optimist nor a pessimist but would like further information before choosing a strategy. For example, he might call the weather forecaster and ask for the probability of a freeze. Suppose the forecaster says that this probability is only .1. What should the farmer do? He should recall the discussion of expected value and work out the expected profit for each of his two possible strategies. If the probability of a freeze is .1, then the probability that there is no freeze is .9. This information leads to the following expected values.

If smudge pots are used: $10{,}000(.1) + 5600(.9) = 6040$

If no smudge pots are used: $-2000(.1) + 9600(.9) = 8440$

Here the maximum expected profit, $8440, is obtained if smudge pots are not used. (2)

As the example shows, the farmer's beliefs about the probabilities of a freeze affect his choice of strategies.

EXAMPLE 2 A small manufacturer of Christmas cards must decide in February about the type of cards she should emphasize in her fall line. She has three possible strategies: emphasize modern cards, emphasize old-fashioned cards, or a mixture of the two. Her success is dependent on the state of the economy in December. If the economy is strong, she will do well with her modern cards, while in a weak economy people long for the old days and buy old-fashioned cards. In an in-between economy, her mixture of lines would do the best. She first prepares a payoff matrix for all three possibilities. The numbers in the matrix represent her profits in thousands of dollars.

Strategies		*States of Nature* Weak Economy	In-between	Strong Economy
	Modern	40	85	120
	Old-fashioned	106	46	83
	Mixture	72	90	68

(a) What would an optimist do?

If the manufacturer is an optimist, she should aim for the biggest number on the matrix, 120 (representing $120,000 in profit). Her strategy in this case would be to produce modern cards.

(b) How would a pessimist react?

A pessimistic manufacturer wants to find the best of the worst of all bad things that can happen. If she produces modern cards, the worst that can happen is a profit of $40,000. For old-fashioned cards, the worst is a profit of $46,000. From a mixture, the worst is a profit of $68,000. Her strategy here is to use a mixture.

(c) Suppose the manufacturer reads in a business magazine that leading experts believe there is a 50% chance of a weak economy at Christmas, a 20% chance of an in-between economy, and a 30% chance of a strong economy. How might she use this information?

The manufacturer can now find her expected profit for each possible strategy.

Modern:	$40(.50) + 85(.20) + 120(.30) = 73$
Old-fashioned:	$106(.50) + 46(.20) + 83(.30) = 87.1$
Mixture:	$72(.50) + 90(.20) + 68(.30) = 74.4$

Here the best strategy is old-fashioned cards; the expected profit is 87.1, or $87,100. ◀ 3

3

Suppose the manufacturer reads another article which gives the following predictions: 35% chance of a weak economy, 25% chance of an in-between economy, and a 40% chance of a strong economy. What is the best strategy now? What is the expected profit?

Answer:

Modern; $83,250

9.6 Exercises

1. Business A developer has $100,000 to invest in land. He has a choice of two parcels (at the same price), one on the highway and one on the coast. With both parcels, his ultimate profit depends on whether he faces light opposition from environmental groups or heavy opposition. He estimates that the payoff matrix is as follows (the numbers represent his profit).

	Opposition Light	Heavy
Highway	$70,000	$30,000
Coast	$150,000	−$40,000

What should the developer do if he is

(a) an optimist? **(b)** a pessimist?

(c) Suppose the probability of heavy opposition is .8. What is his best strategy? What is the expected profit?

(d) What is the best strategy if the probability of heavy opposition is only .4?

2. Business Hillsdale College has sold out all tickets for a jazz concert to be held in the stadium. If it rains, the show will have to be moved to the gym, which has a much smaller seating capacity. The dean must decide in advance whether to set up the seats and the stage in the gym or in the stadium, or both, just in case. The payoff matrix below shows the net profit in each case.

		States of Nature Rain	No Rain
Strategies	Set up in Stadium	−$1550	$1500
	Set up in Gym	$1000	$1000
	Set up Both	$750	$1400

What strategy should the dean choose if she is

(a) an optimist? **(b)** a pessimist?

(c) If the weather forecaster predicts rain with a probability of .6, what strategy should she choose to maximize expected profit? What is the maximum expected profit?

3. Business An analyst must decide what fraction of the items produced by a certain machine are defective. He has already decided that there are three possibilities for the fraction of defective items: .01, .10, and .20. He may recommend two courses of action: repair the machine or make no repairs. The payoff matrix below represents the *costs* to the company in each case, in hundreds of dollars.

		Defectives		
		.01	.10	.20
Strategies	Repair	130	130	130
	No Repair	25	200	500

What strategy should the analyst recommend if he is

(a) an optimist? **(b)** a pessimist?

(c) Suppose the analyst is able to estimate probabilities for the three states of nature as follows.

Fraction of Defectives	*Probability*
.01	.70
.10	.20
.20	.10

Which strategy should he recommend? Find the expected cost to the company if this strategy is chosen.

4. Business The research department of the Allied Manufacturing Company has developed a new process that it believes will result in an improved product. Management must decide whether or not to go ahead and market the new product. The new product may be better than the old or it may not be better. If the new product is better and the company decides to market it, sales should increase by $50,000. If it is not better and they replace the old product with the new product on the market, they will lose $25,000 to competitors. If they decide not to market the new product they will lose $40,000 if it is better, and research costs of $10,000 if it is not.

(a) Prepare a payoff matrix.

(b) If management believes the probability that the new product is better to be .4, find the expected profits under each strategy and determine the best action.

5. Business A businessman is planning to ship a used machine to his plant in Nigeria. He would like to use it there for the next 4 years. He must decide whether or not to overhaul the machine before sending it. The cost of overhaul is $2600. If the machine fails when in operation in Nigeria, it will cost him $6000 in lost production and repairs. He estimates the probability that it will fail at .3 if he does not overhaul it and .1 if he does overhaul it. Neglect the possibility that the machine might fail more than once in the 4 years.

(a) Prepare a payoff matrix.

(b) What should the businessman do to minimize his expected costs?

6. Business A contractor prepares to bid on a job. If all goes well, his bid should be $30,000, which will cover his costs plus his usual profit margin of $4500. However, if a threatened labor strike actually occurs, his bid should be $40,000 to give him the same profit. If there is a strike and he bids $30,000, he will lose $5500. If his bid is too high, he may lose the job entirely, while if it is too low, he may lose money.

(a) Prepare a payoff matrix.

(b) If the contractor believes that the probability of a strike is .6, how much should he bid?

7. Natural Science A community is considering an anti-smoking campaign.* The city council will choose one of three possible strategies: a campaign for everyone over age 10 in the community, a campaign for youths only, or no campaign at all. The two states of nature are a true cause-effect relationship between smoking and cancer and no cause-effect relationship. The costs to the community (including loss of life and productivity) in each case are as shown at the top of the next column.

	States of Nature	
Strategies	Cause-Effect Relationship	No Cause-Effect Relationship
Campaign for All	$100,000	$800,000
Campaign for Youth	$2,820,000	$20,000
No Campaign	$3,100,100	$0

What action should the city council choose if it is

(a) optimistic? **(b)** pessimistic?

(c) If the Director of Public Health estimates that the probability of a true cause-effect relationship is .8, which strategy should the city council choose?

8. Business An investor has $20,000 to invest in stocks. She has two possible strategies: buy conservative blue-chip stocks or buy highly speculative stocks. There are two states of nature: the market goes up or the market goes down. The following payoff matrix shows the net amounts she will have under the various circumstances.

	Market Up	Market Down
Buy Blue-Chip	$25,000	$18,000
Buy Speculative	$30,000	$11,000

What should the investor do if she is

(a) an optimist? **(b)** a pessimist?

(c) Suppose there is a .7 probability of the market going up. What is the best strategy? What is the expected profit?

(d) What is the best strategy if the probability of a market rise is .2?

Sometimes the numbers (or payoffs) in a payoff matrix do not represent money (profits or costs, for example), but utility. A utility is a number that measures the satisfaction (or lack of it) that results from a certain action. The numbers must be assigned by each individual, depending on how he or she feels about a

*This problem is based on an article by B. G. Greenberg in the September 1969 issue of the *Journal of the American Statistical Association*.

situation. For example, one person might assign a utility of +20 for a week's vacation in San Francisco, with −6 being assigned if the vacation were moved to Sacramento. Work the following problems in the same way as those above.

9. Social Science A politician must plan her reelection strategy. She can emphasize jobs or she can emphasize the environment. The voters can be concerned about jobs or about the environment. A payoff matrix showing the utility of each possible outcome is shown below.

		Voters	
		Jobs	Environment
Candidate	Jobs	+25	−10
	Environment	−15	+30

The political analysts feel that there is a .35 chance that the voters will emphasize jobs. What strategy should the candidate adopt? What is its expected utility?

10. In an accounting class, the instructor permits the students to bring a calculator or a reference book (but not both) to an examination. The examination itself can emphasize either numbers or definitions. In trying to decide which aid to take to an examination, a student first decides on the utilities shown in the following payoff matrix.

		Exam Emphasizes	
		Numbers	Definitions
Student Chooses	Calculator	+50	0
	Book	+10	+40

(a) What strategy should the student choose if the probability that the examination will emphasize numbers is .6? What is the expected utility in this case?

(b) Suppose the probability that the examination emphasizes numbers is .4. What strategy should be chosen by the student?

CHAPTER 9 SUMMARY

Key Terms and Symbols

9.1 $n!$ n factorial
multiplication principle
permutations
combinations

9.3 Bernoulli trials (processes)
binomial experiment
binomial probability

9.4 random variable
probability distribution
histogram
expected value
fair game

9.5 stochastic processes
Markov chain
state
transition diagram
transition matrix
probability vector
regular transition matrix
regular Markov chain
equilibrium vector (fixed vector)

9.6 states of nature
strategies
payoffs
payoff matrix

Key Concepts

Multiplication Principle: If there are m_1 ways to make a first choice, m_2 ways to make a second choice, and so on, where none of the choices depend on any of the others, then there are $m_1 m_2 \cdots m_n$ different ways to make the entire sequence of choices.

The number of **permutations** of n elements taken r at a time is ${}_nP_r = \dfrac{n!}{(n-r)!}$.

The number of **combinations** of n elements taken r at a time is

$${}_nC_r = \binom{n}{r} = \frac{n!}{(n-r)!r!}.$$

Binomial experiments have the following characteristics: The same experiment is repeated several times. There are only *two* outcomes, labeled success and failure. The trials are independent so that the probability of success is the same for each trial. If the probability of success in a single trial is p, then the probability of x successes in n trials is

$$\binom{n}{x}p^x(1-p)^{n-x}.$$

Expected Value of a Probability Distribution: For a random variable x with values $x_1, x_2, \ldots, x_n$ and probabilities $p_1, p_2, \ldots, p_n$, respectively, the expected value is

$$E(x) = x_1p_1 + x_2p_2 + \cdots + x_np_n.$$

Markov Chains: A **transition matrix** must be square, with all entries between 0 and 1 inclusive, and the sum of the entries in any row must be 1. A Markov chain is *regular* if some power of its transition matrix P contains all positive entries. The long-range probabilities for a regular Markov chain are given by the **equilibrium** or **fixed vector** V, where for any initial probability vector v, the products vP^n approach V as n gets larger and larger, and $VP = V$. To find V, solve the system of equations formed by $VP = V$ and the fact that the sum of the entries of V is 1.

Decision Making: A **payoff matrix** which includes all available strategies and states of nature is used in decision making to define the problem and the possible solutions. The expected value of each strategy can help to determine the best course of action.

Chapter 9 Review Exercises

1. In how many ways can 5 shuttle vans line up at the airport?
2. How many variations in first-, second-, and third-place finishes are possible in a 100-yard dash with 7 runners?
3. In how many ways can a sample of 3 pears be taken from a basket containing a dozen pears?
4. If 2 of the pears in Exercise 3 are spoiled, in how many ways can the sample of 3 include the following?
   (a) 1 spoiled pear **(b)** No spoiled pears
   (c) At most 1 spoiled pear
5. In how many ways can 3 pictures, selected from a group of 6 pictures, be arranged in a row on a wall?
6. In how many ways can the 6 pictures in Exercise 5 be arranged in a row, if a certain one must be first?
7. In how many ways can the 6 pictures in Exercise 5 be arranged if 3 are landscapes and 3 are puppies, and if
   (a) like types must be kept together?
   (b) landscapes and puppies are alternated?
8. A representative is to be selected from each of 3 departments in a large company. There are 7 people in the first department, 5 in the second department, and 8 in the third department.
   (a) How many different groups of 3 representatives are possible?
   (b) How many groups are possible, if any number (at least 1) up to 3 representatives can form a group?
9. Explain under what circumstances a permutation should be used in a probability problem and under what circumstances a combination should be used.
10. Discuss under what circumstances the binomial probability formula should be used in a probability problem.

Suppose 2 cards are drawn without replacement from an ordinary deck of 52 cards. Find the probabilities of the following results.

11. Both cards are black.
12. Both cards are hearts.
13. Exactly 1 is a face card.
14. At most 1 is an ace.

A collection of golf balls contains 4 yellow, 2 blue, and 6 white balls. A golfer selects 3 balls at random. Find the probability that the selection includes the following.

15. All white balls
16. All yellow balls
17. At least 1 blue ball
18. 1 ball of each color
19. In this exercise we study the connection between sets (from Chapter 8) and combinations.
   (a) Given a set with n elements, what is the number of subsets of size 0? of size 1? of size 2? of size n?
   (b) Using your answer from part (a), give an expression for the total number of subsets of a set with n elements.

(c) Using your answer from part (b) and a result from Chapter 8, explain why the following equation must be true:

$$\binom{n}{0} + \binom{n}{1} + \binom{n}{2} + \cdots \binom{n}{n} = 2^n.$$

(d) Verify the equation in part (c) for $n = 4$ and $n = 5$.

Business *A bottle capping machine has an error rate of .01. A random sample of 20 bottles is selected. Find the following probabilities.*

20. Exactly 4 bottles are improperly capped.

21. No more than 3 bottles are improperly capped.

22. At least 1 bottle is improperly capped.

Business *A certain machine that is used to manufacture screws produces a defect rate of .01. A random sample of 20 screws is selected. Find the probabilities that the sample contains the following.*

23. Exactly 4 defective screws

24. Exactly 3 defective screws

25. No more than 4 defective screws

26. Set up the probability that the sample has 12 or more defective screws. (Do not evaluate.)

27. **Business** A developer can buy a piece of property that will produce a profit of \$16,000 with probability .7 or a loss of \$9000 with probability .3. What is the expected profit?

In Exercises 28–32, (a) give a probability distribution, (b) sketch its histogram, and (c) find the expected value.

28. A pair of dice is rolled and the sum of the results for each roll is recorded.

29. A coin is tossed 4 times and the number of tails is recorded.

30. **Social Science** In 1992, the Big 10 collegiate sports conference moved to have women compose at least 40% of its athletes within 5 years.* Suppose they exactly achieved the 40% figure and that 5 athletes are picked at random from Big 10 universities. The number of women is recorded.

31. **Social Science** In the mathematics honors society at a college, 2 of the 8 members are African American. Three members are selected at random to be interviewed by the student newspaper, and the number of African Americans is noted.

32. **Social Science** In a small class of 10 students, 3 did not do their homework. The professor selects half of the class to present solutions to homework problems on the board, and records how many of those selected did not do their homework.

Solve the following problems.

33. Suppose someone offers to pay you \$100 if you draw 3 cards from a standard deck of 52 cards and all the cards are clubs. What should you pay for the chance to win if it is a fair game?

34. You pay \$6 to play in a game where you will roll a die, with payoffs as follows: \$8 for a 6, \$7 for a 5, and \$4 for any other results. What are your expected winnings? Is the game fair?

35. A lottery has a first prize of \$5000, two second prizes of \$1000 each, and two \$100 third prizes. Ten thousand tickets are sold, at \$1 each. Find the expected winnings of a person buying 1 ticket.

36. Find the expected number of girls in a family of 5 children.

37. Three cards are drawn from a standard deck of 52 cards.
(a) What is the expected number of aces?
(b) What is the expected number of clubs?

Decide whether each of the following is a regular transition matrix.

38. $\begin{bmatrix} 0 & 1 \\ .8 & .2 \end{bmatrix}$

39. $\begin{bmatrix} -.1 & .4 \\ .3 & .7 \end{bmatrix}$

40. $\begin{bmatrix} .3 & 0 & .7 \\ .5 & .1 & .4 \\ 1 & 0 & 0 \end{bmatrix}$

41. $\begin{bmatrix} .2 & .3 & .5 \\ .4 & .4 & .2 \\ .5 & .1 & .4 \end{bmatrix}$

42. **Business** A credit card company classified its customers in three groups: nonusers in a given month, light users, and heavy users. The transition matrix for these states is

	Nonuser	Light	Heavy
Nonuser	.8	.15	.05
Light	.25	.55	.2
Heavy	.04	.21	.75

Suppose the initial distribution for the three states is [.4 .4 .2]. Find the distribution after
(a) 1 month; (b) 2 months.
(c) What is the long-range prediction for the distribution of users?

43. **Business** Savmor Investments starts a heavy advertising campaign. At the start of the campaign, Savmor sells 35% of all mutual funds sold in the area, while 65% are sold by the Highrate Company. The campaigns produces the following transition matrix.

		After campaign	
		Savmor	Highrate
Before Campaign	Savmor	.8	.2
	Highrate	.4	.6

(a) Find the market share for each company after the campaign.
(b) Find the share of the market for each company after three such campaigns.
(c) Predict the long-range market share for Savmor.

NCTM News Bulletin, February 1995, p. 5.

44. Business In labor-management relations, both labor and management can adopt either a friendly or a hostile attitude. The results are shown in the following payoff matrix. The numbers give the wage gains made by an average worker.

		Management Friendly	Hostile
Labor	Friendly	\$600	\$800
	Hostile	\$400	\$950

(a) Suppose the chief negotiator for labor is an optimist. What strategy should he choose?

(b) What strategy should he choose if he is a pessimist?

(c) The chief negotiator for labor feels that there is a 70% chance that the company will be hostile. What strategy should he adopt? What is the expected payoff?

(d) Just before negotiations begin, a new management is installed in the company. There is only a 40% chance that the new management will be hostile. What strategy should be adopted by labor?

45. Social Science A candidate for city council can come out in favor of a new factory, be opposed to it, or waffle on the issue. The change in votes for the candidate depends on what her opponent does, with payoffs as shown.

		Opponent Favors	Waffles	Opposes
Candidate	Favors	0	−1000	−4000
	Waffles	1000	0	−500
	Opposes	5000	2000	0

(a) What should the candidate do if she is an optimist?

(b) What should she do if she is a pessimist?

(c) Suppose the candidate's campaign manager feels there is a 40% chance that the opponent will favor the plant and a 35% chance that he will waffle. What strategy should the candidate adopt? What is the expected change in the number of votes?

(d) The opponent conducts a new poll that shows strong opposition to the new factory. This changes the probability he will favor the factory to 0 and the probability he will waffle to .7. What strategy should our candidate adopt? What is the expected change in the number of votes now?

*Exercises 46 and 47 are taken from actuarial examinations given by the Society of Actuaries.**

46. Business A company is considering the introduction of a new product that is believed to have probability .5 of being successful and probability .5 of being unsuccessful. Successful products pass quality control 80% of the time. Unsuccessful products pass quality control 25% of the time. If the product is successful, the net profit to the company will be \$40 million; if unsuccessful, the net loss will be \$15 million. Determine the expected net profit if the product passes quality control.

(a) \$23 million **(b)** \$24 million
(c) \$25 million **(d)** \$26 million
(e) \$27 million

47. Business A merchant buys boxes of fruit from a grower and sells them. Each box of fruit is either Good or Bad. A Good box contains 80% excellent fruit and will earn \$200 profit on the retail market. A Bad box contains 30% excellent fruit and will produce a loss of \$1000. The a priori probability of receiving a Good box of fruit is .9. Before the merchant decides to put the box on the market, he can sample one piece of fruit to test whether it is excellent. Based on that sample, he has the option of rejecting the box without paying for it. Determine the expected value of the right to sample. (*Hint:* If the merchant samples the fruit, what are the probabilities of accepting a Good box, accepting a Bad box, and not accepting the box? What are these probabilities if he does not sample the fruit?)

(a) 0 **(b)** \$16 **(c)** \$34
(d) \$72 **(e)** \$80

48. Business The March 1982 issue of *Mathematics Teacher* included "Overbooking Airline Flights," an article by Joe Dan Austin. In this article, Austin developed a model for the expected income for an airline flight. With appropriate assumptions, the probability that exactly x of n people with reservations show up at the airport to buy a ticket is given by the binomial probability formula. Assume the following: 6 reservations have been accepted for 3 seats, $p = .6$ is the probability that a person with a reservation will show up, a ticket costs \$100, and the airline must pay \$100 to anyone with a reservation who does not get a ticket. Complete the following table.

Number Who Show Up (x)	0	1	2	3	4	5	6
Airline's Income							
$P(x)$							

(a) Use the table to find $E(I)$, the expected airline income from the 3 seats.

(b) Find $E(I)$ for $n = 3$, $n = 4$, and $n = 5$. Compare these answers with $E(I)$ for $n = 6$. For these values of n, how many reservations should the airline book for the 3 seats in order to maximize the expected revenue?

*Problem from "Course 130 Examination, Operations Research," of the *Education and Examination Committee of the Society of Actuaries*. Reprinted by permission of the Society of Actuaries.

Case 8

Optimal Inventory for a Service Truck

For many different items it is difficult or impossible to take the item to a central repair facility when service is required. Washing machines, large television sets, office copiers, and computers are only a few examples of such items. Service for items of this type is commonly performed by sending a repair person to the item, with the person driving to the location in a truck containing various parts that might be required in repairing the item. Ideally, the truck should contain all the parts that might be required. However, most parts would be needed only infrequently, so that inventory costs for the parts would be high.

An optimum policy for deciding on which parts to stock on a truck would require that the probability of not being able to repair an item without a trip back to the warehouse for needed parts be as low as possible, consistent with minimum inventory costs. An analysis similar to the one below was developed at the Xerox Corporation.*

To set up a mathematical model for deciding on the optimum truck-stocking policy, let us assume that a broken machine might require one of 5 different parts (we could assume any number of different parts—we use 5 to simplify the notation). Suppose also that the probability that a particular machine requires part 1 is p_1; that it requires part 2 is p_2; and so on. Assume also that failures of different part types are independent, and that at most one part of each type is used on a given job.

Suppose that, on the average, a repair person makes N service calls per time period. If the repair person is unable to make a repair because at least one of the parts is unavailable, there is a penalty cost, L, corresponding to wasted time for the repair person, an extra trip to the parts depot, customer unhappiness, and so on. For each of the parts carried on the truck, an average inventory cost is incurred. Let H_i be the average inventory cost for part i, where $1 \le i \le 5$.

Let M_1 represent a policy of carrying only part 1 on the repair truck, M_{24} represent a policy of carrying only parts 2 and 4, with M_{12345} and M_0 representing policies of carrying all parts and no parts, respectively.

For policy M_{35}, carrying parts 3 and 5 only, the expected cost per time period per repair person, written $C(M_{35})$, is

$$C(M_{35}) = (H_3 + H_5) + NL[1 - (1 - p_1)(1 - p_2)(1 - p_4)].$$

(The expression in brackets represents the probability of needing at least one of the parts not carried, 1, 2, or 4 here.) As further examples,

$$C(M_{125}) = (H_1 + H_2 + H_5) + NL[1 - (1 - p_3)(1 - p_4)],$$

while

$$C(M_{12345}) = (H_1 + H_2 + H_3 + H_4 + H_5) + NL[1 - 1]$$
$$= H_1 + H_2 + H_3 + H_4 + H_5,$$

and

$$C(M_0) = NL[1 - (1 - p_1)(1 - p_2)(1 - p_3)(1 - p_4)(1 - p_5)].$$

To find the best policy, evaluate $C(M_0)$, $C(M_1)$, . . . , $C(M_{12345})$, and choose the smallest result. (A general solution method is in the *Management Science* paper.)

EXAMPLE 1 Suppose that for a particular item, only 3 possible parts might need to be replaced. By studying past records of failures of the item, and finding necessary inventory costs, suppose that the following values have been found.

p_1	p_2	p_3
.09	.24	.17

H_1	H_2	H_3
\$15	\$40	\$9

*Reprinted by permission of Stephen Smith, John Chambers, and Eli Shlifer. "Optimal Inventories Based on Job Completion Rate for Repairs Requiring Multiple Items." *Management Science*, Vol. 26, No. 8, August 1980, copyright © 1980 by The Institute of Management Sciences.

Suppose $N = 3$ and L is \$54. Then, as an example,

$$\begin{aligned} C(M_1) &= H_1 + NL[1 - (1 - p_2)(1 - p_3)] \\ &= 15 + 3(54)[1 - (1 - .24)(1 - .17)] \\ &= 15 + 3(54)[1 - (.76)(.83)] \\ &\approx 15 + 59.81 \\ &= 74.81. \end{aligned}$$

Thus, if policy M_1 is followed (carrying only part 1 on the truck), the expected cost per repair person per time period is \$74.81. Also,

$$\begin{aligned} C(M_{23}) &= H_2 + H_3 + NL[1 - (1 - p_1)] \\ &= 40 + 9 + 3(54)(.09) \\ &= 63.58, \end{aligned}$$

so that M_{23} is a better policy than M_1. By finding the expected values for all other possible policies (see the exercises), the optimum policy may be chosen. ◀

Exercises

1. Refer to the example and find each of the following.
   (a) $C(M_0)$ **(b)** $C(M_2)$ **(c)** $C(M_3)$
   (d) $C(M_{12})$ **(e)** $C(M_{13})$ **(f)** $C(M_{123})$
2. Which policy leads to the lowest expected cost?
3. In Example 1, $p_1 + p_2 + p_3 = .09 + .24 + .17 = .50$. Why is it not necessary that the probabilities add up to 1?
4. Suppose an item to be repaired might need one of n different parts. How many different policies would then need to be evaluated?

Chapter 10

Introduction to Statistics

Statistics is a branch of mathematics that deals with the collection and summarization of data. Methods of statistical analysis make it possible to draw conclusions about a population based on data from a sample of the population. Statistical models have become increasingly useful in manufacturing, government, agriculture, medicine, and the social sciences and in all types of research. An Indianapolis racecar team is using statistics to improve performance by gathering data on each run around the track. They sample data 300 times a second and use computers to process the data. In this chapter we give a brief introduction to some of the key topics from statistical theory.

10.1 FREQUENCY DISTRIBUTIONS AND MEASURES OF CENTRAL TENDENCY

Often, a researcher wishes to learn something about a characteristic of a population, but because the population is very large or mobile, it is not possible to examine all of its elements. Instead, a limited sample drawn from the population is studied to determine the characteristics of the population. For these inferences to be correct, the sample chosen must be a **random sample.** Random samples are representative of the population because they are chosen so that every element of the population is equally likely to be selected. For example, a hand dealt from a well-shuffled deck of cards is a random sample.

A random sample can be difficult to obtain in real life. For example, suppose you want to take a random sample of voters in your congressional district to see which candidate they prefer in the next election. If you do a telephone survey, you have a random sample of people who are at home to answer the telephone, underrepresenting those who work a lot of hours and are rarely home to answer the phone, or those who have an unlisted number, or those who cannot afford a telephone, or those who refuse to answer telephone surveys. Such people may have a different opinion than those you interview.

A famous example of an inaccurate poll was made by the *Literary Digest* in 1936. Their survey indicated that Alfred Landon would win the presidential election; in fact, Franklin Roosevelt won with 62% of the popular vote. The *Digest's* major error was mailing their surveys to a sample of those listed in telephone directories. During the depression, many poor people did not have telephones, and the poor voted overwhelmingly for Roosevelt. Modern pollsters use sophisticated techniques to ensure that their sample is as random as possible.

Once a sample has been chosen and all data of interest are collected, the data must be organized so that conclusions may be more easily drawn. One method of organization is to group the data into intervals; equal intervals are usually chosen.

EXAMPLE 1 A survey asked a random sample of 30 business executives for their recommendations as to the number of college units in management that a business major should have. The results are shown below. Group the data into intervals and find the frequency of each interval.

3	25	22	16	0	9	14	8	34	21
15	12	9	3	8	15	20	12	28	19
17	16	23	19	12	14	29	13	24	18

The highest number in the list is 34 and the lowest is 0; one convenient way to group the data is in intervals of size 5, starting with 0–4 and ending with 30–34. This gives an interval for each number in the list and results in seven equal intervals of a convenient size. Too many intervals of smaller size would not simplify the data enough, while too few intervals of larger size would conceal information that the data might provide. A rule of thumb is to use from six to fifteen intervals.

First tally the number of college units falling into each interval. Then total the tallies in each interval, as in the table below. This table is an example of a **grouped frequency distribution.**

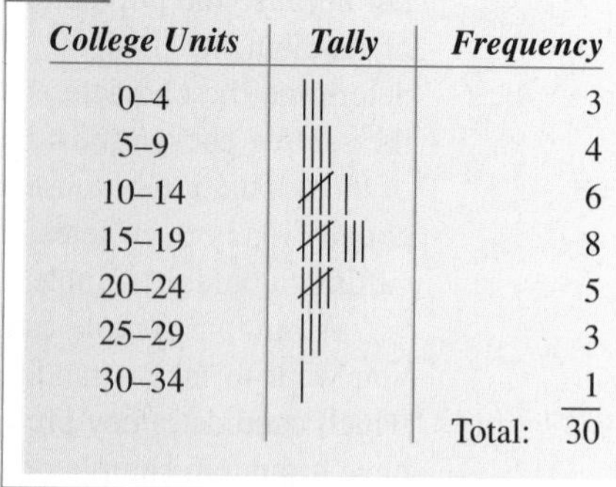

College Units	*Tally*	*Frequency*
0–4	|||	3
5–9	||||	4
10–14	~~||||~~ |	6
15–19	~~||||~~ |||	8
20–24	~~||||~~	5
25–29	|||	3
30–34	|	1
		Total: 30

1 An accounting firm selected 24 complex tax returns prepared by a certain tax preparer. The number of errors per return were as follows.

8	12	0	6	10	8	0	14
8	12	14	16	4	14	7	11
9	12	7	15	11	21	22	19

Prepare a grouped frequency distribution for this data. Use intervals 0–4, 5–9, and so on.

Answer:

Interval	*Frequency*
0–4	3
5–9	7
10–14	9
15–19	3
20–24	2
	Total: 24

The frequency distribution in Example 1 shows information about the data that might not have been noticed before. For example, the interval with the largest number of units is 15–19, and 19 executives (more than half) had between 9 and 25 units. Also, the frequency in each interval increases rather evenly (up to 8) and then decreases at about the same pace. However, some information has been lost; for example, we no longer know how many executives had 12 units.

Picturing Data

The information in a grouped frequency distribution can be displayed in a histogram similar to the ones for probability distributions in the previous chapter. The intervals determine the widths of the bars; if equal intervals are used, all the bars have the same width. The heights of the bars are determined by the frequencies.

A **frequency polygon** is another form of graph that illustrates a grouped frequency distribution. The polygon is formed by joining consecutive midpoints of the tops of the histogram bars with straight line segments. Sometimes the midpoints of the first and last bars are joined to endpoints on the horizontal axis where the next midpoint would appear. (See Figure 10.1)

(2)

Make a histogram and a frequency polygon for the distribution found in side problem 1.

Answer:

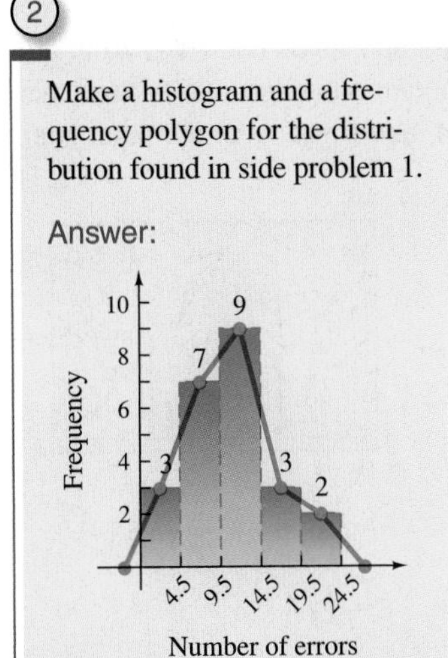

EXAMPLE 2 A grouped frequency distribution of suggested college units was found in Example 1. Draw a histogram and a frequency polygon for this distribution.

First, draw a histogram, shown in blue in Figure 10.1. To get a frequency polygon, connect consecutive midpoints of the tops of the bars. The frequency polygon is shown in red. ◀ (2)

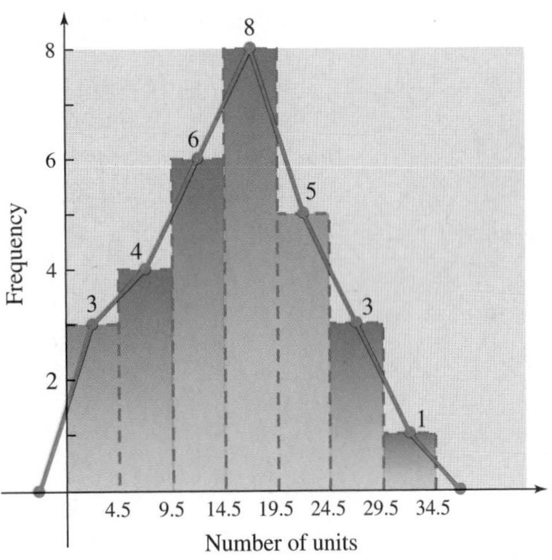

FIGURE 10.1

TECHNOLOGY TIP As noted in Section 9.4, most graphing calculators can display histograms. Many will also display frequency polygons (which are usually labeled LINE or xyLINE in calculator menus). When dealing with grouped frequency distributions, however, certain adjustments must be made on a calculator.

1. *A calculator list of outcomes must consist of single numbers, not intervals.* The table in Example 1, for example, cannot be entered as shown. To convert the first column of the table for calculator use, choose one number in each interval, say 2 in the interval 0–4, 7 in the interval 5–9, 12 in the interval 10–14, etc. Then use 2, 7, 12, . . . as the list of outcomes to be entered into the calculator. The frequency list (the last column of the table) remains the same.
2. *The histogram bar width affects the shape of the graph.* If you use a bar width of 4 in Example 1, the calculator may produce a histogram with gaps in it. To avoid this, use the interval $0 \le x < 5$ in place of $0 \le x \le 4$, and similarly for the other intervals and make 5 the bar width.

Following this procedure, we obtain the calculator-generated histogram and frequency polygon in Figure 10.2 for the data from Example 1. Note that the width of each histogram bar is 5. Some calculators cannot display both the histogram and the frequency polygon on the same screen as is done here.

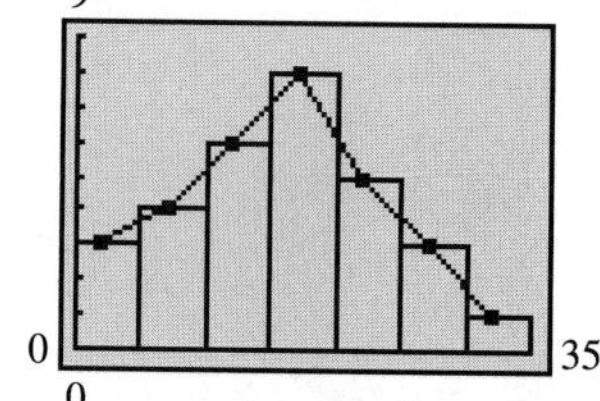

FIGURE 10.2

Stem-and-leaf plots allow us to organize the data in a distribution without the disadvantage of losing the original information. In a **stem-and-leaf plot,** we separate the digits in each of the data into two parts consisting of the first one or two digits (the stem) and the other digits (the leaf).

EXAMPLE 3 Construct a stem-and-leaf plot for the data in Example 1.

Since the data are two-digit numbers, we use the first digit for the stems: 0, 1, 2, and 3. The second digits provide the leaves. For example, as shown below, the first stem 0 has leaves 0, 3, 3 corresponding to the entries 3 and 0 in the first row of the data list and 3 in the second row. Here, each stem corresponds to an interval in the frequency table. The stems and leaves are separated by a vertical line.

Stem	*Leaves*
0	033
0	8899
1	222344
1	55667899
2	01234
2	589
3	4

If we turn the page on its side, the distribution looks like a histogram but still retains each of the original values. We used each stem digit twice, because, as with a histogram, too few intervals conceal useful information about the shape of the distribution. ◀ (3)

(3)

Make a stem-and-leaf plot for the data in Example 1 using one stem for 0, 1, 2, and 3.

Answer:

Stem	Leaves
0	0338899
1	22234455667899
2	01234589
3	4

EXAMPLE 4 List the original data for the stem-and-leaf plot shown below.

Stem	Leaves
15	258
16	23469
17	145668
18	0137

The first stem and its leaves correspond to the data 152, 155, and 158. Similarly, the rest of the data are 162, 163, 164, 166, 169, 171, 174, 175, 176, 176, 178, 180, 181, 183, and 187. ◀ (4)

(4)

List the original data for the following stem-and-leaf plot.

Stem	Leaves
4	36
5	2589
6	244678
7	1248
8	5

Answer:

43, 46, 52, 55, 58, 59, 62, 64, 64, 66, 67, 68, 71, 72, 74, 78, 85

NOTE The remainder of this section deals with topics that are generally referred to as "measures of central tendency." Computing these various measures is greatly simplified by the statistical capabilities of most scientific and graphing calculators. Calculators vary considerably in how data is entered, so read your instruction manual to learn how to enter lists of data and the corresponding frequencies. On scientific calculators with statistical capabilities, there are keys for finding most of the measures of central tendency discussed below. On graphing calculators, most or all of these measures can be obtained with a single keystroke (look for *one-variable statistics,* which is often labeled 1-VAR, in the STAT menu or its CALC submenu).

Mean

Three measures of central tendency, or "averages," are used with frequency distributions: the mean, the median, and the mode. The most important of these is the mean, which is similar to the expected value of a probability distribution. The **arithmetic mean** (the mean) of a set of numbers is the sum of the numbers, divided by the total number of numbers. We write the sum of n numbers $x_1, x_2, x_3, \ldots, x_n$ in a compact way using **summation notation,** also called **sigma notation.** With the Greek letter Σ (sigma), the sum

$$x_1 + x_2 + x_3 + \cdots + x_n$$

is written

$$x_1 + x_2 + x_3 + \cdots + x_n = \sum_{i=1}^{n} x_i.$$

In statistics, $\Sigma_{i=1}^{n} x_i$ is often abbreviated as just Σx. The symbol $\bar{x}$ (read "x-bar") is used to represent the mean of a sample.

Mean

The mean of the n numbers $x_1, x_2, x_3, \ldots, x_n$ is

$$\bar{x} = \frac{x_1 + x_2 + \cdots + x_n}{n} = \frac{\Sigma x}{n}.$$

EXAMPLE 5 The numbers of bankruptcy petitions (in thousands) filed in the United States in the fiscal years (1994–1999) are given in the table below.* Find the mean number of bankruptcy petitions filed annually during this period.

Year	*Petitions Filed (in thousands)*
1994	845
1995	858
1996	1042
1997	1317
1998	1429
1999	1392

5

Find the mean of the following list of sales at a boutique.

\$25.12	\$42.58
\$76.19	\$32
\$81.11	\$26.41
\$19.76	\$59.32
\$71.18	\$21.03

Answer:

\$45.47

Let $x_1 = 845$, $x_2 = 858$, and so on. Here, $n = 6$, since there are 6 numbers in the list.

$$\bar{x} = \frac{845 + 858 + 1042 + 1317 + 1429 + 1392}{6} \approx 1147$$

The mean number of bankruptcy petitions filed during the given years is 1,147,000.

TECHNOLOGY TIP The mean of the six numbers in Example 5 is easily found by using the $\bar{x}$ key on a scientific calculator or the one-variable statistics key on a graphing calculator. A graphing calculator will also display additional information, which will be discussed in the next section.

*Administrative Offices of the U.S. Courts, *Statistical Tables for the Federal Judiciary.*

The mean of data that have been arranged in a frequency distribution is found in a similar way. For example, suppose the following data is collected.

Value	*Frequency*
84	2
87	4
88	7
93	4
99	3
	Total: 20

The value 84 appears twice, 87 four times, and so on. To find the mean, first add 84 two times, 87 four times, and so on; or get the same result faster by multiplying 84 by 2, 87 by 4, and so on, and then by adding the results. Dividing the sum by 20, the total of the frequencies, gives the mean.

$$\begin{aligned}\bar{x} &= \frac{(84 \cdot 2) + (87 \cdot 4) + (88 \cdot 7) + (93 \cdot 4) + (99 \cdot 3)}{20} \\ &= \frac{168 + 348 + 616 + 372 + 297}{20} \\ &= \frac{1801}{20} \\ \bar{x} &= 90.05\end{aligned}$$

Verify that your calculator gives the same result.

EXAMPLE 6 Find the mean for the data shown in the following frequency distribution.

Value	*Frequency*	*Value × Frequency*
30	6	30 · 6 = 180
32	9	32 . 9 = 288
33	7	33 . 7 = 231
37	12	37 . 12 = 444
42	6	42 . 6 = 252
	Total: 40	Total: 1395

The value 30 appears six times, 32 nine times, and so on. To find the mean, first multiply 30 times 6, 32 times 9, and so on, to get the new column "Value × Frequency,"

Find $\bar{x}$ for the following frequency distribution.

Value	Frequency
7	2
9	3
11	6
13	4
15	1
17	4

Answer:

$\bar{x} = 12.1$

that has been added to the frequency distribution. Adding the products from this column gives a total of 1395. The total from the frequency column is 40. The mean is

$$\bar{x} = \frac{1395}{40} = 34.875.$$ ◀ 6

The mean of grouped data is found in a similar way. For grouped data, intervals are used, rather than single values. To calculate the mean, it is assumed that all these values are located at the midpoint of the interval. The letter x is used to represent the midpoints and f represents the frequencies, as shown in the next example.

EXAMPLE 7 Listed below is the grouped frequency distribution for the 30 business executives described in Example 1. Find the mean from the grouped frequency distribution.

Interval	Midpoint, x	Frequency, f	Product, xf
0–4	2	3	6
5–9	7	4	28
10–14	12	6	72
15–19	17	8	136
20–24	22	5	110
25–29	27	3	81
30–34	32	1	32
		Total: 30	Total: 465

A column for the midpoint of each interval has been added. The numbers in this column are found by adding the endpoints of each interval and dividing by 2. For the interval 0–4, the midpoint is $(0 + 4)/2 = 2$. The numbers in the product column on the right are found by multiplying each frequency by its corresponding midpoint. Finally, we divide the total of the product column by the total of the frequency column to get

$$\bar{x} = \frac{465}{30} = 15.5$$

Find the mean of the following grouped frequency distribution.

Interval	Frequency
0–4	6
5–9	4
10–14	7
15–19	3

Answer:

8.75

Notice that this mean is slightly different from the earlier mean of 15.93. The reason for this difference is that we have acted as if each piece of data is at the midpoint, which is not true here and is not true in most cases. Information is always lost when the data are grouped. It is more accurate to use the original data, rather than the grouped frequency, when calculating the mean, but the original data might not be available. Furthermore, the mean based upon the grouped data is typically not too different from the mean based upon the original data, and there may be situations in which the extra accuracy is not worth the extra effort. ◀ 7

NOTE **1.** The midpoint of an interval in a grouped frequency distribution may be a value that the data does not assume. For example, if we grouped the data for the 30 business executives into the intervals 0–5, 6–11, 12–17, 18–23, 24–29, and 30–35, the midpoints would be 2.5, 8.5, 14.5, 20.5, 26.5, and 32.5, even though all the data are whole numbers.

2. If we used different intervals in Example 7, the mean would come out to be a slightly different number. Verify that with the intervals 0–5, 6–11, 12–17, 18–23, 24–29, and 30–35, the mean in Example 7 is 16.1.

The formula for the mean of a grouped frequency distribution is given below.

Mean of a Grouped Distribution

The mean of a distribution where x represents the midpoints, f the frequencies, and $n = \Sigma f$, is

$$\bar{x} = \frac{\Sigma(xf)}{n}.$$

The mean of a random sample is a random variable, and for this reason it is sometimes called the **sample mean.** The sample mean is a random variable because it assigns a number to the experiment of taking a random sample. If a different random sample were taken, the mean would probably have a different value, with some values more probable than others. If another set of 30 business executives were selected in Example 1, the mean number of college units in management recommended for a business major might be 13.22 or 17.69. It is unlikely that the mean would be as small as 1.21 or as large as 32.75, although these values are remotely possible.

We saw in Section 9.4 how to calculate the expected value of a random variable when we know its probability distribution. The expected value is sometimes called the **population mean,** denoted by the Greek letter μ. In other words,

$$E(x) = \mu.$$

Furthermore, it can be shown that the expected value of $\bar{x}$ is also equal to μ; that is,

$$E(\bar{x}) = \mu.$$

For instance, consider again the 30 business executives in Example 1. We found that $\bar{x} = 15.93$, but the value of μ, the average for all possible business executives, is unknown. If a good estimate of μ were needed, the best guess (based on this data) is 15.93. Although $E(x) = \mu$ and $E(\bar{x}) = \mu$, $E(x)$ and $E(\bar{x})$ are not the same. $E(x)$ is the population mean, while $E(\bar{x})$ is the expected value of all possible sample means.

Median

Asked by a reporter to give the average height of the players on his team, the Little League coach lined up his 15 players by increasing height. He picked out the player in the middle and pronounced this player to be of average height. This kind of average, called the **median,** is defined as the middle entry in a set of data arranged in either increasing or decreasing order. If there is an even number of entries, the median is defined to be the mean of the two middle entries. The following table shows how to find the median for two sets of data: {8, 7, 4, 3, 1} and {2, 3, 4, 7, 9, 12}.

Odd Number of Entries	*Even Number of Entries*
8	2
7	3
Median = 4	4 } Median = $\frac{4+7}{2} = 5.5$
3	7 }
1	9
	12

NOTE As shown in the table above, when there are an even number of entries, the median is not always equal to one of the data entries.

EXAMPLE 8 Find the median for the following lists of numbers.

(a) 11, 12, 17, 20, 23, 28, 29

The median is the middle number, in this case 20. (Note that the numbers are already arranged in numerical order.) In this list, three numbers are smaller than 20 and three are larger.

(b) 15, 13, 7, 11, 19, 30, 39, 5, 10

First arrange the numbers in numerical order, from smallest to largest or vice versa.

5, 7, 10, 11, 13, 15, 19, 30, 39

The middle number can now be determined; the median is 13.

(c) 47, 59, 32, 81, 74, 153

Write the numbers in numerical order.

32, 47, 59, 74, 81, 153

There are six numbers here; the median is the mean of the two middle numbers, or

$$\text{Median} = \frac{59 + 74}{2} = \frac{133}{2} = 66.5.$$

CAUTION Remember, the data must be arranged in numerical order before locating the median. (8)

(8)

Find the median for each of the following lists of numbers.

(a) 12, 15, 17, 19, 35, 42, 58

(b) 28, 68, 7, 15, 47, 59, 13, 74, 32, 25

Answers:

(a) 19

(b) 30

TECHNOLOGY TIP Many graphing calculators (including TI-82/83/86, HP-39, Casio 9850) display the median when doing one-variable statistics. You may have to scroll down to a second screen to find it.

Both the mean and the median are examples of a **statistic,** which is simply a number that gives information about a sample. In some situations, the median gives a truer representative or typical element of the data than the mean. For example, suppose in an office there are 10 salespersons, 4 secretaries, the sales manager, and Ms. Daly, who owns the business. Their annual salaries are as follows: support staff, \$30,000 each; salespersons, \$50,000 each; manager, \$70,000; and owner, \$400,000. The mean salary is

$$\bar{x} = \frac{(30{,}000)4 + (50{,}000)10 + 70{,}000 + 400{,}000}{16} = \$68{,}125.$$

However, since 14 people earn less than \$68,125 and only 2 earn more, this does not seem very representative. The median salary is found by ranking the salaries by size: \$30,000, \$30,000, \$30,000, \$30,000, \$50,000, \$50,000, . . . , \$400,000. There are 16 salaries (an even number) in the list, so the mean of the 8th and 9th entries will give the value of the median. The 8th and 9th entries are both \$50,000, so the median is \$50,000. In this example, the median is more representative of the distribution than the mean.

Mode

Sue's scores on ten class quizzes include one 7, two 8's, six 9's and one 10. She claims that her average grade on quizzes is 9, because most of her scores are 9's. This kind of "average," found by selecting the most frequent entry, is called the **mode.**

9

Find the mode for each of the following lists of numbers.

(a) 29, 35, 29, 18, 29, 56, 48

(b) 13, 17, 19, 20, 20, 13, 25, 27, 13, 20

(c) 512, 546, 318, 729, 854, 253

Answers:

(a) 29

(b) 13 and 20

(c) No mode

EXAMPLE 9 Find the mode for each list of numbers.

(a) 57, 38, 55, 55, 80, 87, 98

The number 55 occurs more often than any other, so it is the mode. It is not necessary to place the numbers in numerical order when looking for the mode.

(b) 182, 185, 183, 185, 187, 187, 189

Both 185 and 187 occur twice. This list has *two* modes, so it is bimodal.

(c) 10,708, 11,519, 10,972, 17,546, 13,905, 12,182

No number occurs more than once. This list has no mode. 9

The mode has the advantages of being easily found and not being influenced by data that are extreme values. It is often used in samples where the data to be "averaged" are not numerical. A major disadvantage of the mode is that there may be more than one, in case of ties, or there may be no mode at all when all entries occur with the same frequency.

The mean is the most commonly used measure of central tendency. Its advantages are that it is easy to compute, it takes all the data into consideration, and it is reliable—that is, repeated samples are likely to give very similar means. A disadvantage of the mean is that it is influenced by extreme values, as illustrated in the salary example above.

The median can be easy to compute and is influenced very little by extremes. Like the mode, the median can be found in situations where the data are not numerical. A disadvantage of the median is the need to rank the data in order; this can be tedious when the number of items is large and difficult or not possible when the data is not numerical.

EXAMPLE 10 Seeds that are dried, placed in an airtight container, and stored in a cool, dry place remain ready to be planted for a long time. The following table gives the amount of time that each type of seed can be stored and still remain viable for planting.*

Vegetable	*Years*
Beans	3
Cabbage	4
Carrots	1
Cauliflower	4
Corn	2
Cucumbers	5
Melons	4
Peppers	2
Pumpkin	4
Tomatoes	3

Find the mean, median, and mode of the information in the table.

The mean amount of time that the seeds can be stored is

$$\bar{x} = \frac{3 + 4 + 1 + 4 + 2 + 5 + 4 + 2 + 4 + 3}{10} = 3.2 \text{ years}$$

After the numbers are arranged in order from smallest to largest, the middle number, or median, is found; it is 3.5.

The number 4 occurs more often than any other, so it is the mode.

TECHNOLOGY TIP Most scientific calculators have some statistical capability and can calculate the mean of a set of data; graphing calculators can often calculate the median as well. For example, Figure 10.3 shows the mean and the median for the data above calculated on a TI-83, where the data is stored in the list L_1. This calculator does not include a command for finding the mode.

```
mean(L1)
                 3.2
median(L1)
                 3.5
```

FIGURE 10.3

Spreadsheets can also be used to find the mean, median, and mode. Using Microsoft Excel, place the data in cells A1 through A10. To find the mean of this data, type "=mean (A1 . . A10)" in cell A11, or any other unused cell, and then press Enter. The result of 3.2 will appear in cell A11. To find the median of this data, type "=median (A1 . . A10)" in cell A12, or any other unused cell, and press Enter. The result of 3.5 will appear in cell A12. To find the mode of this data, type "=mode (A1 . . A10)" in cell A13, or any other unused cell, and press Enter. The result of 4 will appear in cell A13.

**The Handy Science Answer Book*, Second Edition, The Carnegie Library of Pittsburgh, p. 247.

10.1 Exercises

For Exercises 1–4, (a) *group the data as indicated;* (b) *prepare a frequency distribution with columns for intervals and frequencies;* (c) *construct a histogram;* (d) *construct a frequency polygon. (See Examples 1 and 2.)*

1. Use six intervals, starting with 0–24.

74	133	4	127	20	30
103	27	139	118	138	121
149	132	64	141	130	76
42	50	95	56	65	104
4	140	12	88	119	64

2. Use seven intervals, starting with 30–39.

79	71	78	87	69	50	63	51	60	46
65	65	56	88	94	56	74	63	87	62
84	76	82	67	59	66	57	81	93	93
54	88	55	69	78	63	63	48	89	81
98	42	91	66	60	70	64	70	61	75
82	65	68	39	77	81	67	62	73	49
51	76	94	54	83	71	94	45	73	95
72	66	71	77	48	51	54	57	69	87

3. Use 70–74 as the first interval.

79	84	88	96	102	104	110	108	106	106	
104	99	97	92	94	90	82	74	72	83	
84	92	100	99	101	107	111	102	97	94	92

4. Use 140–149 as the first interval.

174	190	172	182	179	186	171	152	174	185
180	170	160	173	163	177	165	157	149	167
169	182	178	158	182	169	181	173	183	176
170	162	159	147	150	192	179	165	148	188

Construct a stem-and-leaf plot for the data in the indicated exercise. (Use a 2-digit stem. See Example 3.)

5. Exercise 3 (Use Stems 07, 08, . . . ,11)

6. Exercise 4

7. How does a frequency polygon differ from a histogram?

8. Discuss the advantages and disadvantages of the mean as a measure of central tendency.

Find the mean for each list. Round to the nearest tenth. (See Example 5.)

9. 21,900, 22,850, 24,930, 29,710, 28,340, 40,000

10. 38,500, 39,720, 42,183, 21,982, 43,250

11. 9.4, 11.3, 10.5, 7.4, 9.1, 8.4, 9.7, 5.2, 1.1, 4.7

12. 30.1, 42.8, 91.6, 51.2, 88.3, 21.9, 43.7, 51.2

Find the mean for each distribution. Round to the nearest tenth. (See Examples 5, 6, and 7.)

13.

Value	Frequency
3	4
5	2
9	1
12	3

14.

Value	Frequency
9	3
12	5
15	1
18	1

15.

Value	Frequency
12	4
13	2
15	5
19	3
22	1
23	5

16.

Value	Frequency
25	1
26	2
29	5
30	4
32	3
33	5

Find the median for each of the following lists of numbers. (See Example 8.)

17. 12, 18, 32, 51, 58, 92, 106

18. 596, 604, 612, 683, 719

19. 100, 114, 125, 135, 150, 172

20. 1072, 1068, 1093, 1042, 1056, 1005, 1009

21. 28.4, 9.1, 3.4, 27.6, 59.8, 32.1, 47.6, 29.8

22. .6, .4, .9, 1.2, .3, 4.1, 2.2, .4, .7, .1

Find the mode or modes for each of the following lists of numbers. (See Example 9.)

23. 4, 9, 8, 6, 9, 2, 1, 3

24. 21, 32, 46, 32, 49, 32, 49

25. 74, 68, 68, 68, 75, 75, 74, 74, 70

26. 158, 162, 165, 162, 165, 157, 163

27. 6.8, 6.3, 6.3, 6.9, 6.7, 6.4, 6.1, 6.0

28. 12.75, 18.32, 19.41, 12.75, 18.30, 19.45, 18.33

29. When is the median the most appropriate measure of central tendency?

30. Under what circumstances would the mode be an appropriate measure of central tendency?

For grouped data, the modal class *is the interval containing the most data values. Give the mean and modal class for each of the following collections of grouped data. (See Example 7.)*

31. The distribution in Exercise 3.

32. The distribution in Exercise 4.

33. To predict the outcome of the next congressional election, you take a survey of your friends. Is this a random sample of the voters in your congressional district? Explain why or why not.

Work each problem. (See Example 10.)

34. Social Science The number of nations participating in the winter Olympic games, from 1972 to 1998, is given below.*

Year	Nations Participating
1972	35
1976	37
1980	37
1984	49
1988	57
1992	64
1994	67
1998	72

Find the following statistics for the data.
(a) Mean **(b)** Median **(c)** Mode

35. Business The total pay (in thousands of dollars) for the 10 highest paid CEOs in 1997 is given in the following table.†

Person, Company	Total Pay
Sanford Weill, Travelers Group	230,725
Roberto Goizueta, Coca-Cola	111,832
Richard Scrushy, Health South	106,790
Ray Irani, Occidental Petroleum	101,505
Eugene Isenberg, Nabors Industries	84,547
Joseph Costelo, Cadence Design Systems	66,842
Andrew Grove, Intel	52,214
Charles McCall, HBO Co.	51,409
Philip Purcell, Morgan Stanley Dean Witter	50,807
Robert Shapiro, Monsanto	49,326

(a) Find the mean compensation for this group of people.
(b) Find the median compensation for this group of people.
(c) What might account for the difference between these values?
(d) Write the salary for Shapiro.

36. Business The top ten box-office receipts (in millions of dollars) for the weekend of 6/15/2001 through 6/17/2001 at the movies are given here.*

Movie (Weeks in release)	Weekend Receipts (millions)
Lara Croft: Tomb Raider (1)	47.7
Atlantis: The Lost Empire (2)	20.3
Shrek (5)	13.2
Swordfish (2)	12.7
Pearl Harbor (4)	9.9
Evolution (2)	6.6
The Animal (3)	5.8
Moulin Rouge (5)	5
What's the Worst Thing That Could Happen? (3)	3
The Mummy Returns (7)	2.7

(a) Find the mean weekend receipts in dollars for this group.
(b) Find the median weekend receipts in dollars for this group.

**The New York Times Almanac, 2001.*

†John W. Wright, *The American Almanac of Jobs and Salaries,* 2000–2001 ed., p. 321.

*Exhibitor Relations Co.

37. Natural Science The number of recognized blood types varies by species, as indicated by the following table.* Find the mean, median, and mode of this data.

Animal	*Number of Blood Types*
Pig	16
Cow	12
Chicken	11
Horse	9
Human	8
Sheep	7
Dog	7
Rhesus Monkey	6
Mink	5
Rabbit	5
Mouse	4
Rat	4
Cat	2

38. Social Science The number of weddings in each month of 1996 is given in the following table.†

Month	*Number (in thousands)*
January	110
February	155
March	118
April	172
May	241
June	242
July	235
August	239
September	225
October	231
November	171
December	184

(a) Calculate the mean and median for this data.
(b) Which month is closest to the mean?

Physical Science *The table below gives the average monthly temperatures in degrees Fahrenheit for a certain area.*

Month	*Maximum*	*Minimum*
January	39	16
February	39	18
March	44	21
April	50	26
May	60	32
June	69	37
July	79	43
August	78	42
September	70	37
October	51	31
November	47	24
December	40	20

Find the mean and median for each of the following.

39. The maximum temperature

40. The minimum temperature

Business *U.S. wheat prices and production figures for a recent decade are given below.**

Year	*Price ($ per bushel)*	*Production (millions of bushels)*
1989	3.72	2037
1990	2.61	2730
1991	3.00	1980
1992	3.24	2467
1993	3.26	2396
1994	3.45	2321
1995	4.55	2183
1996	4.30	2285
1997	3.38	2482
1998	2.65	2550

Find the mean and median for each of the following.

41. Price per bushel of wheat

42. Wheat production

**The Handy Science Answer Book,* Carnegie Library of Pittsburgh, Pennsylvania, p. 264.

†*The Amazing Almanac,* Blackbirch Press, Inc., p. 245.

**The World Almanac and Book of Facts 2000,* pp. 141–142.

43. Social Science The total household income for full-time African-American workers making under $100,000 in 1997 is given in the table below.* (See Example 7.)

Income Range	*Midpoint Salary*	*Frequency (in thousands)*
Under $5,000	$2500	923
$5,000–$9,999	$7500	1746
$10,000–$14,999	$12,500	1310
$15,000–$24,999	$20,000	2233
$25,000–$34,999	$30,000	1771
$35,000–$49,999	$42,500	1859
$50,000–$74,999	$62,500	1634
$75,000–$99,999	$87,500	574

Use this table to estimate the mean household income for full-time African-American workers in 1997.

44. Social Science The total household income for full-time white American workers making under $100,000 in 1997 is given in the following table.* (See Example 7.)

Income Range	*Midpoint Salary*	*Frequency (in thousands)*
Under $5,000	$2500	2411
$5,000–$9,999	$7500	5769
$10,000–$14,999	$12,500	6716
$15,000–$24,999	$20,000	12,571
$25,000–$34,999	$30,000	11,366
$35,000–$49,999	$42,500	14,207
$50,000–$74,999	$62,500	16,188
$75,000–$99,999	$87,500	8180

(a) Use this table to estimate the mean household income for full-time white American workers in 1997.

(b) Compare this estimate with the estimate found in Exercise 43. Does this provide some evidence that full-time white American workers have higher household earnings than full-time African-American workers?

45. Social Science The histogram below shows estimates of the percent of the U.S. population in each age group in the year 2000.† What percent of the population is estimated to be in each of the following age groups?

(a) 10–19
(b) 60–69
(c) What age range has the largest percent of the population?

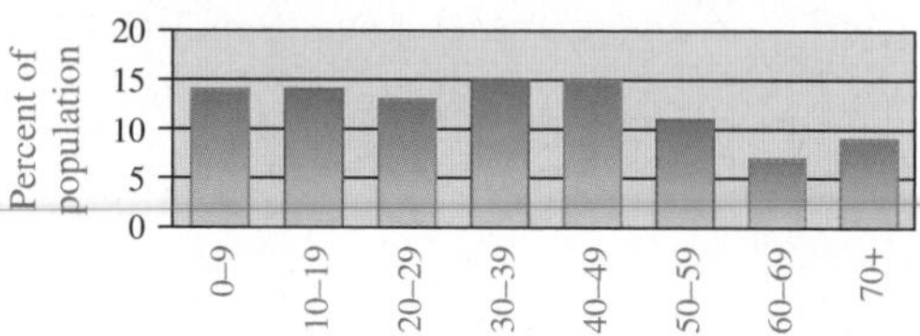

46. Social Science The histogram below shows estimates of the percent of the U.S. population in each age group in the year 2025.† What percent of the population is estimated to be in each of the following age groups then?

(a) 20–29
(b) 70+
(c) What age group will have the smallest percent of the population?
(d) Compare the histogram in Exercise 45 with the histogram below. What seems to be true of the U.S. population?

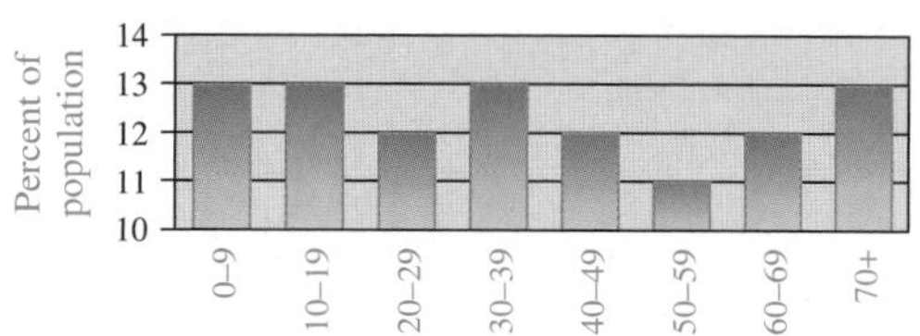

**Time Almanac 2000*, Time Inc., p. 829.

†U.S. Census Bureau, January 13, 2000.

10.2 MEASURES OF VARIATION

The mean, median, and mode are measures of central tendency of a list of numbers but tell nothing about the *spread* of the numbers in the list. For example, look at the following three lists.

I	3	5	6	3	3
II	4	4	4	4	4
III	10	1	0	0	9

Each of these three lists has a mean of 4, and yet they are quite different; the amount of dispersion or variation within the lists is different. Therefore, in addition to a measure of central tendency, another kind of measure is needed that describes how much the numbers vary.

The largest number in sample I is 6, while the smallest is 3, a difference of 3. In sample II this difference is 0; in sample III it is 10. The difference between the largest and smallest number in a sample is called the **range,** one instance of a measure of variation. The range of sample I is 3, of sample II is 0, and of sample III is 10. The range has the advantage of being very easy to compute and gives a rough estimate of the variation among the data in the sample. However, it depends only on the two extremes and tells nothing about how the other data are distributed between the extremes.

EXAMPLE 1 Find the range for each list of numbers.

(1)

Find the range for the numbers 159, 283, 490, 390, 375, 297.

Answer:

331

(a) 12, 27, 6, 19, 38, 9, 42, 15

The highest number here is 42; the lowest is 6. The range is the difference of these numbers, or

$$42 - 6 = 36.$$

(b) 74, 112, 59, 88, 200, 73, 92, 175

$$\text{Range} = 200 - 59 = 141$$ (1)

TECHNOLOGY TIP Many graphing calculators list the largest and smallest numbers in a list when displaying one-variable statistics, usually on the second screen of the display.

To find a useful measure of variation, we begin by finding the **deviations from the mean,** the differences found by subtracting the mean from each number in a distribution.

EXAMPLE 2 Find the deviations from the mean for the numbers

$$32, \quad 41, \quad 47, \quad 53, \quad 57.$$

Adding these numbers and dividing by 5 gives a mean of 46. To find the deviations from the mean, subtract 46 from each number in the list. For example, the first deviation from the mean is $32 - 46 = -14$; the last is $57 - 46 = 11$.

Number	*Deviation from Mean*
32	−14
41	−5
47	1
53	7
57	11

Find the deviations from the mean for each set of numbers.

(a) 19, 25, 36, 41, 52, 61

(b) 6, 9, 5, 11, 3, 2

Answers:

(a) Mean is 39; deviations are −20, −14, −3, 2, 13, 22.

(b) Mean is 6; deviations from the mean are 0, 3, −1, 5, −3, −4.

To check your work, find the sum of these deviations. It should always equal 0. (The answer is always 0 because the positive and negative deviations cancel each other.) ◀ 2

To find a measure of variation, we might be tempted to use the mean of the deviations. However, as mentioned above, this number is always 0, no matter how widely the data are dispersed. To avoid the problem of the positive and negative deviations averaging 0, we take absolute values and find $\Sigma|x - \bar{x}|$, then divide it by n to get the *mean deviation*. Statisticians consider a better measure of variation to be squaring each deviation to get nonnegative numbers, then taking the square root of the mean of the squared variations to preserve the units of the original data (such as inches, pounds, and so on). The squared deviations for the data in Example 2 are shown in the table.

Number	*Deviation from Mean*	*Square of Deviation*
32	−14	196
41	−5	25
47	1	1
53	7	49
57	11	121

In this case the mean of the squared deviations is

$$\frac{196 + 25 + 1 + 49 + 121}{5} = \frac{392}{5} = 78.4.$$

This number is called the **population variance,** because the sum was divided by $n = 5$, the number of items in the original list.

Because the deviations from the mean must add up to 0, if we know any 4 of the 5 deviations, the 5th will be determined. That is, only $n - 1$ of the deviations are free to vary, so we really have only $n - 1$ independent pieces of information, or *degrees of freedom*. Using $n - 1$ as the divisor in the formula for the mean gives

$$\frac{196 + 25 + 1 + 49 + 121}{5 - 1} = \frac{392}{4} = 98.$$

This number 98 is called the **sample variance** of the distribution and is denoted s^2 because it is found by averaging a list of squares. In this case, the population and sample variances differ by quite a bit. But when n is relatively large, as is the case in real-life applications, the difference between them is rather small.

Sample Variance

The variance of a sample of n numbers $x_1, x_2, x_3, \ldots, x_n$, with mean $\bar{x}$, is

$$s^2 = \frac{\Sigma(x - \bar{x})^2}{n - 1}.$$

When computing the sample variance by hand, it is often convenient to use the following shortcut formula, which can be derived algebraically from the definition in the box above.

$$s^2 = \frac{\Sigma x^2 - n\bar{x}^2}{n - 1}$$

To find the sample variance, we square the deviations from the mean, so the variance is in squared units. To return to the same units as the data, we use the *square root* of the variance, called the **sample standard deviation,** denoted s.

Sample Standard Deviation

The standard deviation of a sample of n numbers $x_1, x_2, x_3, \ldots, x_n$, with mean $\bar{x}$, is

$$s = \sqrt{\frac{\Sigma(x - \bar{x})^2}{n - 1}}.$$

NOTE The **population standard deviation** is

$$\sigma = \sqrt{\frac{\Sigma(x - \bar{x})^2}{n}},$$

where n is the population size.

TECHNOLOGY TIP When a graphing calculator computes one-variable statistics for a list of data, it usually displays the following information (not necessarily in this order and sometimes on two screens), and possibly other information as well.

Information	*Notation*
Number of data entries	n or $N\Sigma$
Mean	$\bar{x}$ or mean Σ
Sum of all data entries	Σx or TOT Σ
Sum of the squares of all data entries	Σx^2
Sample standard deviation	Sx or sx or $x\sigma_{n-1}$ or SSDEV
Population standard deviation	σx or $x\sigma_n$ or PSDEV
Largest/smallest data entries	maxX/minX or MAXΣ/MINΣ
Median	Med or MEDIAN

NOTE In the rest of this section, we shall deal exclusively with the sample variance and the sample standard deviation. So whenever standard deviation is mentioned, it means "sample standard deviation," not population standard deviation.

As its name indicates, the standard deviation is the most commonly used measure of variation. The standard deviation is a measure of the variation from the mean. The size of the standard deviation indicates how spread out the data are from the mean.

EXAMPLE 3 Find the standard deviation of the numbers

$$7,\quad 9,\quad 18,\quad 22,\quad 27,\quad 29,\quad 32,\quad 40$$

by hand, using the shortcut variance formula at the top of page 563.

Arrange the work in columns, as shown in the table.

Number	*Square of the Number*
7	49
9	81
18	324
22	484
27	729
29	841
32	1024
40	1600
184	5132

Now find the mean.

$$\bar{x} = \frac{\Sigma x}{8} = \frac{184}{8} = 23$$

The total of the second column gives $\Sigma x^2 = 5132$. The variance is

$$s^2 = \frac{\Sigma x^2 - n\bar{x}^2}{n - 1}$$
$$= \frac{5132 - 8(23)^2}{8 - 1}$$
$$= 128.6 \quad \text{(rounded)},$$

and the standard deviation is

$$s \approx \sqrt{128.6} \approx 11.3.$$ (3)

(3)

Find the standard deviation of each set of numbers. The deviations from the mean were found earlier in problem 2 at the side.

(a) 19, 25, 36, 41, 52, 61

(b) 6, 9, 5, 11, 3, 2

Answers:

(a) 15.9

(b) 3.5

TECHNOLOGY TIP The screens in Figure 10.5 show two ways to find variance and standard deviation on a TI-83 calculator, using the LIST menu and the STAT menu. The data are first entered in a list, here L_5. See your instruction book for details.

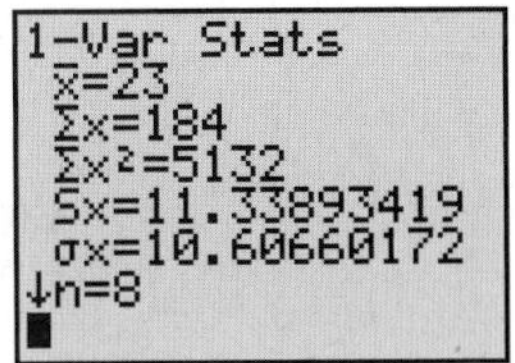

FIGURE 10.4

With a spreadsheet, the data are entered in cells A1 through A8. Then, in cell A9, type "=VAR (A1..A8)" and press Enter. The standard deviation can be calculated by either taking the square root of cell A9 or by typing "=STDEV (A1..A8)" in cell A10 and pressing Enter.

CAUTION We must be careful to divide by $n - 1$, not n, when calculating the standard deviation of a sample. Many calculators are equipped with statistical keys that compute the variance and standard deviation. Some of these calculators use $n - 1$ and others use n for these computations; some may have keys for both. Check your calculator's instruction book before using a statistical calculator for the exercises.

One way to interpret the standard deviation uses the fact that, for many populations, most of the data is within three standard deviations of the mean. (See Section 10.3.) This implies that, in Example 3, most of the population data from which this sample is taken are between

$$\bar{x} - 3s = 23 - 3(11.3) = -10.9$$

and

$$\bar{x} + 3s = 23 + 3(11.3) = 56.9.$$

This has important implications for quality control. If the sample in Example 3 represents measurements of a product that the manufacturer wants to be between 5 and 45, the standard deviation is too large, even though all the numbers are within these bounds.

For data in a grouped frequency distribution, a slightly different formula for the standard deviation is used.

Standard Deviation for a Grouped Distribution

The standard deviation for a sample distribution with mean $\bar{x}$, where x is an interval midpoint with frequency f, and $n = \Sigma f$, is

$$s = \sqrt{\frac{\Sigma fx^2 - n\bar{x}^2}{n-1}}.$$

The formula indicates that the product fx^2 is to be found for each interval. Then these products are summed, n times the square of the mean is subtracted, and the difference is divided by one less than the total frequency; that is, by $n - 1$. The square root of this result is s, the standard deviation. The standard deviation found by this formula may (probably will) differ somewhat from the standard deviation found from the original data.

CAUTION In calculating the standard deviation for either a grouped or ungrouped distribution, using a rounded value for the mean or variance may produce an inaccurate value.

EXAMPLE 4 Find s for the grouped data of Example 7, Section 10.1.

Begin by including columns for x^2 (where x is the midpoint of the interval) and fx^2.

Interval	x	x^2	f	fx^2
0–4	2	4	3	12
5–9	7	49	4	196
10–14	12	144	6	864
15–19	17	289	8	2312
20–24	22	484	5	2420
25–29	27	729	3	2187
30–34	32	1024	1	1024
		Total:	30	Total: 9015

4

Find the standard deviation for the grouped data that follows. (*Hint:* $\bar{x} = 28.5$)

Value	*Frequency*
20–24	3
25–29	2
30–34	4
35–39	1

Answer:

5.3

Recall from Example 7 of Section 10.1 that $\bar{x} = 15.5$. Use the formula above with $n = 30$ to find s.

$$s = \sqrt{\frac{\Sigma fx^2 - n\bar{x}^2}{n - 1}}$$

$$= \sqrt{\frac{9015 - 30(15.5)^2}{30 - 1}}$$

$$\approx 7.89$$ 4

NOTE A calculator is almost a necessity for finding a standard deviation. With a non-graphing calculator, a good procedure to follow is to first calculate $\bar{x}$. Then for each x, square that number, and multiply the result by the appropriate frequency. If your calculator has a key that accumulates a sum, use it to accumulate the total in the last column of the table. With a graphing calculator, simply enter the midpoints and the frequencies, and then ask for the 1-variable statistics.

10.2 Exercises

1. How are the variance and the standard deviation related?
2. Why can't we use the sum of the deviations from the mean as a measure of dispersion of a distribution?

Find the range and standard deviation for each of the following sets of numbers. (See Examples 1 and 3.)

3. 6, 8, 9, 10, 12
4. 12, 15, 19, 23, 26
5. 7, 6, 12, 14, 18, 15
6. 4, 3, 8, 9, 7, 10, 1
7. 42, 38, 29, 74, 82, 71, 35
8. 122, 132, 141, 158, 162, 169, 180
9. 241, 248, 251, 257, 252, 287
10. 51, 58, 62, 64, 67, 71, 74, 78, 82, 93

Find the standard deviation for the grouped data in Exercises 11 and 12. (See Example 4.)

11. (From Exercise 1, Section 10.1)

College Units	*Frequency*
0–24	4
25–49	3
50–74	6
75–99	3
100–124	5
125–149	9

12. (From Exercise 2, Section 10.1)

Scores	*Frequency*
30–39	1
40–49	6
50–59	13
60–69	22
70–79	17
80–89	13
90–99	8

13. Natural Science Twenty-five laboratory rats, used in an experiment to test the food value of a new product, made the following weight gains in grams.

5.25	5.03	4.90	4.97	5.03
5.12	5.08	5.15	5.20	4.95
4.90	5.00	5.13	5.18	5.18
5.22	5.04	5.09	5.10	5.11
5.23	5.22	5.19	4.99	4.93

Find the mean gain and the standard deviation of the gains.

14. Business An assembly-line machine turns out washers with the following thicknesses (in millimeters).

1.20	1.01	1.25	2.20	2.58	2.19	1.29	1.15
2.05	1.46	1.90	2.03	2.13	1.86	1.65	2.27
1.64	2.19	2.25	2.08	1.96	1.83	1.17	2.24

Find the mean and standard deviation of these thicknesses.

An application of standard deviation is given by Chebyshev's Theorem. *(P. L. Chebyshev was a Russian mathematician who lived from 1821 to 1894.) This theorem applies to any distribution of numerical data. It states:*

> *For any distribution of numerical data, at least* $1 - 1/k^2$ *of the numbers lie within k standard deviations of the mean.*

Example *For any distribution, at least*

$$1 - \frac{1}{3^2} = 1 - \frac{1}{9} = \frac{8}{9}$$

of the numbers lie within 3 standard deviations of the mean. Find the fraction of all the numbers of a data set lying within the following numbers of standard deviations from the mean.

15. 2
16. 4
17. 1.5

In a certain distribution of numbers, the mean is 50 with a standard deviation of 6. Use Chebyshev's theorem to tell what percent of the numbers are

18. between 32 and 68;
19. between 26 and 74;
20. less than 38 or more than 62;
21. less than 32 or more than 68;
22. less than 26 or more than 74.
23. Business The Britelite Company conducted tests on the life of its lightbulbs and those of a competitor (Brand X) with the following results for samples of 10 bulbs of each brand.

	Hours of Use (in 100s)									
Britelite	20	22	22	25	26	27	27	28	30	35
Brand X	15	18	19	23	25	25	28	30	34	38

(a) Compute the mean and standard deviation for each sample. Compare the means and standard deviations of the two brands and then answer the questions below.
(b) Which bulbs have a more uniform life in hours?
(c) Which bulbs have the highest average life in hours?

24. Business The weekly wages of the six employees of Harold's Hardware Store are \$300, \$320, \$380, \$420, \$500, and \$2000.
(a) Find the mean and standard deviation of this distribution.
(b) How many of the employees earn within one standard deviation of the mean? How many earn within two standard deviations of the mean?

25. Social Science The number of unemployed workers in the United States in recent years (in millions) is given below.*

Year	Number Unemployed
1989	6.53
1990	7.05
1991	8.63
1992	9.61
1993	8.94
1994	8.00
1995	7.40
1996	7.24
1997	6.74
1998	6.21

Use sample statistics to find the following.
(a) Find the mean number unemployed (in millions) in this period. Which year has unemployment closest to the mean?
(b) Find the standard deviation for the data.
(c) In how many of these years is unemployment within 1 standard deviation of the mean?
(d) In how many of these years is unemployment within 3 standard deviations of the mean?

26. Life Science A medical laboratory tested 21 samples of human blood for acidity on the pH scale, with the following results.

7.1	7.5	7.3	7.4	7.6	7.2	7.3
7.4	7.5	7.3	7.2	7.4	7.3	7.5
7.5	7.4	7.4	7.1	7.3	7.4	7.4

(a) Find the mean and standard deviation.
(b) What percent of the data is within 2 standard deviations of the mean?

The World Almanac and Book of Facts 2000, p. 145.

27. Life Science The number of recognized blood types between species is given in the following table.* In Exercise 37 of the previous section, the mean was found to be 7.38.

Animal	Number of Blood Types
Pig	16
Cow	12
Chicken	11
Horse	9
Human	8
Sheep	7
Dog	7
Rhesus Monkey	6
Mink	5
Rabbit	5
Mouse	4
Rat	4
Cat	2

(a) Find the variance and the standard deviation of these data.
(b) How many of these animals have blood types that are within 1 standard deviation of the mean?

28. Social Science In 1999, fourteen state governors earned at least \$115,000 annually (not counting expense allowances) as listed below. (Salaries are given in thousands of dollars and are rounded to the nearest \$1000.)†

State	Salary
California	165
Georgia	116
Illinois	140
Maryland	120
Michigan	127
Minnesota	120
Nevada	117
New York	179
Ohio	119
Texas	115
Vermont	116
Virginia	125
Washington	132
Wisconsin	116

(a) Find the mean salary of these governors. Which state has the governor with the salary closest to the mean?

**The Handy Science Answer Book*, Carnegie Library of Pittsburgh, Pennsylvania, p. 264.

†*The World Almanac and Book of Facts 2000*, p. 98.

(b) Find the standard deviation for the data.
(c) What percent of the governors have salaries within 1 standard deviation of the mean?
(d) What percent of the governors have salaries within 3 standard deviations of the mean?

29. Health The amounts of time that it takes for various slow-growing tumors to double in size are listed in the table below.*

Type of Cancer	*Doubling Time (days)*
Breast cancer	84
Rectal cancer	91
Synovioma	128
Skin cancer	131
Lip cancer	143
Testicular cancer	153
Esophageal cancer	164

(a) Find the mean and standard deviation of these data.
(b) How many of these cancers have doubling times that are within 2 standard deviations of the mean?
(c) If a person had a nonspecified tumor that was doubling every 200 days, discuss if this particular tumor is growing at a rate that would be expected.

30. Business The Quaker Oats Company conducted a survey to determine if a proposed premium, to be included in their cereal, was appealing enough to generate new sales.† Four cities were used as test markets, where the cereal was distributed with the premium, and four cities as control markets, where the cereal was distributed without the premium. The eight cities were chosen on the basis of their similarity in terms of population, per capita income, and total cereal purchase volume. The results were as follows.

		Percent Change in Average Market Shares per Month
Test Cities	1	+18
	2	+15
	3	+7
	4	+10
Control Cities	1	+1
	2	−8
	3	−5
	4	0

(a) Find the mean of the change in market share for the four test cities.
(b) Find the mean of the change in market share for the four control cities.
(c) Find the standard deviation of the change in market share for the test cities.
(d) Find the standard deviation of the change in market share for the control cities.
(e) Find the difference between the mean of part (a) and the mean of part (b). This represents the estimate of the percent change in sales due to the premium.
(f) The two standard deviations from part (c) and part (d) were used to calculate an "error" of ± 7.95 for the estimate in part (e). With this amount of error what is the smallest and largest estimate of the increase in sales? (*Hint:* Use the answer to part (e).)

On the basis of the results of Exercise 30, the company decided to mass produce the premium and distribute it nationally.

31. Business The following table gives 10 samples of three measurements, made during a production run.

SAMPLE NUMBER									
1	*2*	*3*	*4*	*5*	*6*	*7*	*8*	*9*	*10*
2	3	−2	−3	−1	3	0	−1	2	0
−2	−1	0	1	2	2	1	2	3	0
1	4	1	2	4	2	2	3	2	2

(a) Find the mean $\bar{x}$ for each sample of three measurements.
(b) Find the standard deviation s for each sample of three measurements.
(c) Find the mean $\bar{\bar{x}}$ of the sample means.
(d) Find the mean $\bar{s}$ of the sample standard deviations.
(e) The upper and lower control limits of the sample means here are $\bar{\bar{x}} \pm 1.954\bar{s}$. Find these limits. If any of the measurements are outside these limits, the process is out of control. Decide if this production process is out of control.

32. Discuss what the standard deviation tells us about a distribution.

Social Science *The reading scores of a second-grade class given individualized instruction are shown below. The table also shows the reading scores of a second-grade class given traditional instruction in the same school.*

Scores	*Individualized Instruction*	*Traditional Instruction*
50–59	2	5
60–69	4	8
70–79	7	8
80–89	9	7
90–99	8	6

*Vincent Collins, R. Kenneth Lodffer, and Harold Tivey, "Observations on Growth Rates of Human Tumors," *American Journal of Roentgen*, 76, no. 5 (November 1956): 988–1000.

†This example was supplied by Jeffrey S. Berman, Senior Analyst, Marketing Information, Quaker Oats Company.

33. Find the mean and standard deviation for the individualized instruction scores.

34. Find the mean and standard deviation for the traditional instruction scores.

35. Discuss a possible interpretation of the differences in the means and the standard deviations in Exercises 33 and 34.

36. Business Refer to Section 10.1. Exercise 35, which gives total pay for CEOs.
(a) Calculate the standard deviation of the data.
(b) What percent of the CEOs listed have compensation more than 2 standard deviations from the mean?
(c) What does your answer to part (b) suggest?

10.3 NORMAL DISTRIBUTIONS

Suppose a bank is interested in improving its services to customers. The manager decides to begin by finding the amount of time tellers spend on each transaction, rounded to the nearest minute. The times for 75 different transactions are recorded, with the results shown in the following table. The frequencies listed in the second column are divided by 75 to find the empirical probabilities.

Time	*Frequency*	*Probability*
1	3	$3/75 = .04$
2	5	$5/75 \approx .07$
3	9	$9/75 = .12$
4	12	$12/75 = .16$
5	15	$15/75 = .20$
6	11	$11/75 \approx .15$
7	10	$10/75 \approx .13$
8	6	$6/75 = .08$
9	3	$3/75 = .04$
10	1	$1/75 \approx .01$

Figure 10.5(a) shows a histogram and frequency polygon for the data. The heights of the bars are the empirical probabilities, rather than the frequencies. The transaction times are given to the nearest minute. Theoretically at least, they could have been timed to the nearest tenth of a minute, or hundredth of a minute, or even more precisely. In each case, a histogram and frequency polygon could be drawn. If the times are measured with smaller and smaller units, there are more bars in the histogram and the frequency polygon begins to look more and more like the curve in Figure 10.5(b) instead of a polygon. Actually, it is possible for the transaction times to take on any real number value greater than 0. A distribution in which the outcomes can take any real number value within some interval is a **continuous distribution.** The graph of a continuous distribution is a curve.

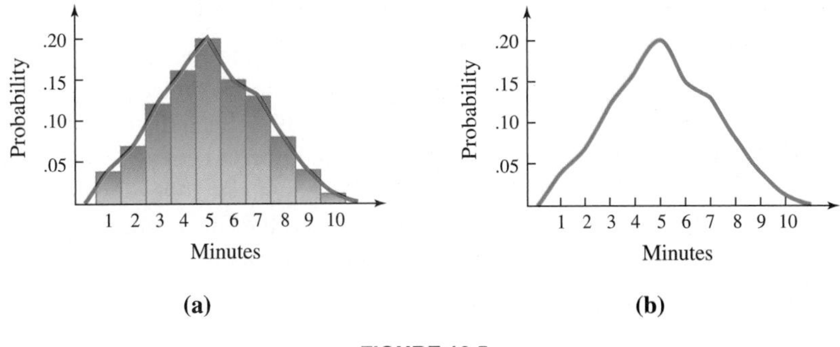

FIGURE 10.5

The distribution of heights (in inches) of college women is another example of a continuous distribution, since these heights include infinitely many possible measurements, such as 53, 58.5, 66.3, 72.666, . . . , and so on. Figure 10.6 shows the continuous distribution of heights of college women. Here the most frequent heights occur near the center of the interval shown.

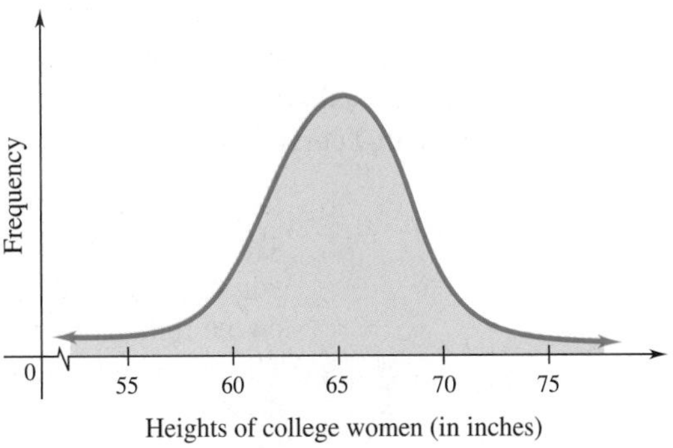

FIGURE 10.6

Another continuous curve, which approximates the distribution of yearly incomes in the United States, is given in Figure 10.7. The graph shows that the most frequent incomes are grouped near the low end of the interval. This kind of distribution, where the peak is not at the center, is called **skewed.**

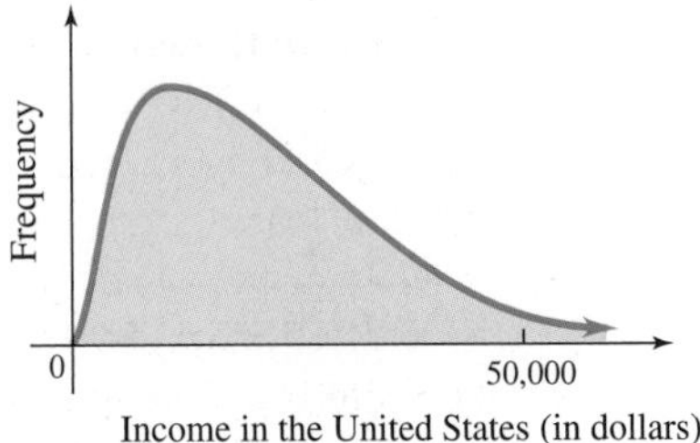

FIGURE 10.7

Normal Distributions

Many natural and social phenomena produce continuous probability distributions whose graphs are approximated very well by bell-shaped curves, such as those shown in Figure 10.8. Such distributions are called **normal distributions** and their graphs are called **normal curves.** Examples of distributions that are approximately normal are the heights of college women and the errors made in filling 1-pound cereal boxes. We use the Greek letters μ (mu) to denote the mean and σ (sigma) to denote the standard deviation of a normal distribution.

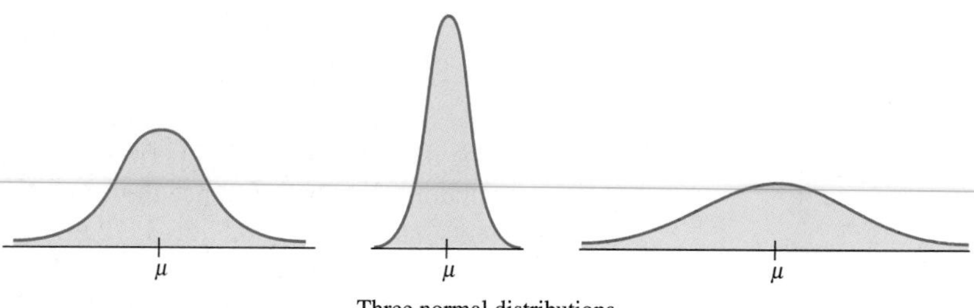

Three normal distributions

FIGURE 10.8

There are many normal distributions. Some of the corresponding normal curves are tall and thin and others short and wide, as shown in Figure 10.8. But every normal curve has the following properties.

1. Its peak occurs directly above the mean μ.
2. The curve is symmetric about the vertical line through the mean (that is, if you fold the graph along this line, the left half of the graph will fit exactly on the right half).
3. The curve never touches the x-axis—it extends indefinitely in both directions.
4. The area under the curve (and above the horizontal axis) is 1. (As can be shown with calculus, this is a consequence of the fact that the sum of the probabilities in any distribution is 1.)

A normal distribution is completely determined by its mean μ and standard deviation σ.* A small standard deviation leads to a tall, narrow curve like the one in the center of Figure 10.8, because most of the data are close to the mean. A large standard deviation means the data are very spread out, producing a flat wide curve like the one on the right in Figure 10.8.

Since the area under a normal curve is 1, parts of this area can be used to determine certain probabilities. For instance, Figure 10.9(a) is the probability distribution of the annual rainfall in a certain region. The probability that the annual rainfall will

*As shown in more advanced courses, its graph is the graph of the function

$$f(x) = \frac{1}{\sigma\sqrt{2\pi}} e^{-(x-\mu)^2/(2\sigma^2)},$$

where $e \approx 2.71828$ is the real number introduced in Section 4.1

be between 25 and 35 inches is the area under the curve from 25 to 35. The general case, shown in Figure 10.9(b), can be stated as follows.

The area of the shaded region under the normal curve from a to b is the probability that an observed data value will be between a and b.

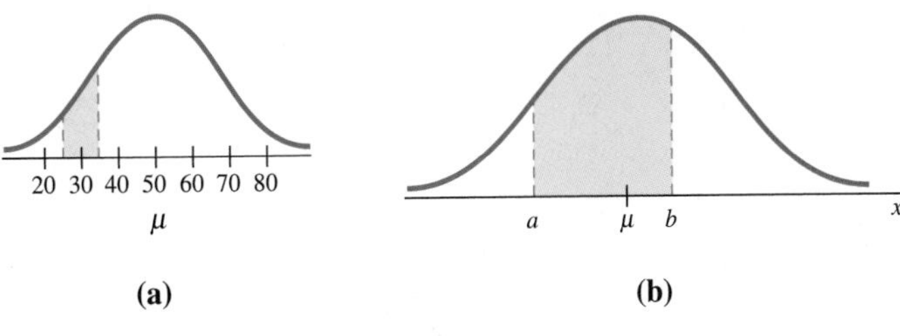

(a) (b)

FIGURE 10.9

To use normal curves effectively we must be able to calculate areas under portions of these curves. These calculations have already been done for the normal curve with mean $\mu = 0$ and standard deviation $\sigma = 1$ (which is called the **standard normal curve**) and are available in Table 2 at the back of the book. The following examples demonstrate how to use Table 2 to find such areas. Later we shall see how the standard normal curve may be used to find areas under any normal curve.

The horizontal axis of the standard normal curve is usually labeled z. Since the standard deviation of the standard normal curve is 1, the numbers along the horizontal axis (the z-values) measure the number of standard deviations above or below the mean $z = 0$.

TECHNOLOGY TIP Some graphing calculators (such as the TI-83 and Casio 9850) have the ability to graph a normal distribution given its mean and standard deviation and to find areas under the curve between two x-values. For an area under the curve, some calculators will give the corresponding z-value. For details, see your instruction book. (Look for "distribution" or "probability distribution.") A calculator-generated graph of the standard normal curve is shown in Figure 10.10.

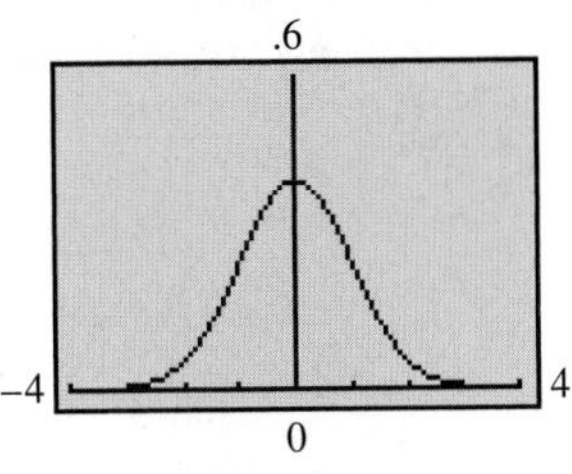

FIGURE 10.10

Find the percent of area between the mean and

(a) $z = 1.51$;

(b) $z = -2.04$.

(c) Find the percent of area in the shaded region.

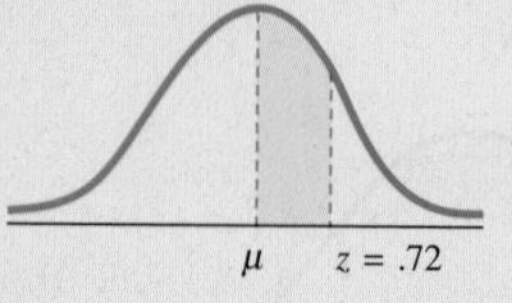

Answers:

(a) 43.45%

(b) 47.93%

(c) 26.42%

If your calculator can graph probability distributions and find area, use it to find the areas required in Example 1.

Answers:

(a) 34.13%

(b) 49.25%

EXAMPLE 1 Find the following areas under the standard normal curve.

(a) The area between $z = 0$ and $z = 1$, the shaded region in Figure 10.11

Find the entry 1 in the z-column of Table 2. The entry next to it in the A-column is .3413, which means that the area between $z = 0$ and $z = 1$ is .3413. Since the total area under the curve is 1, the shaded area in Figure 10.11 is 34.13% of the total area under the normal curve.

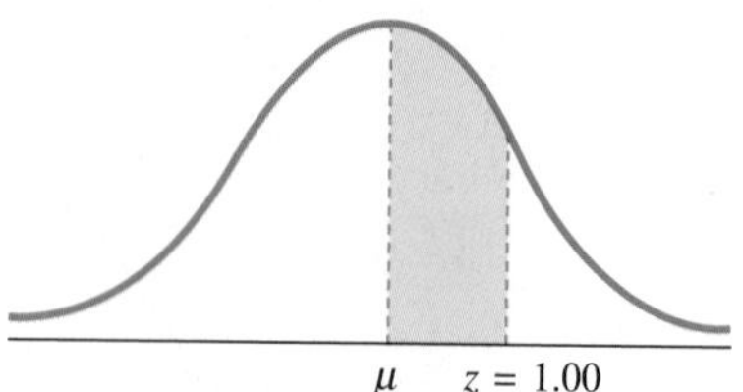

FIGURE 10.11

(b) The area between $z = -2.43$ and $z = 0$

Table 2 lists only positive values of z. But the normal curve is symmetric around the mean $z = 0$, so the area between $z = 0$ and $z = -2.43$ is the same as the area between $z = 0$ and $z = 2.43$. Find 2.43 in the z-column of Table 2. The entry next to it in the A-column shows that the area is .4925. Hence, the shaded area in Figure 10.12 is 49.25% of the total area under the curve. ◀ (1) (2)

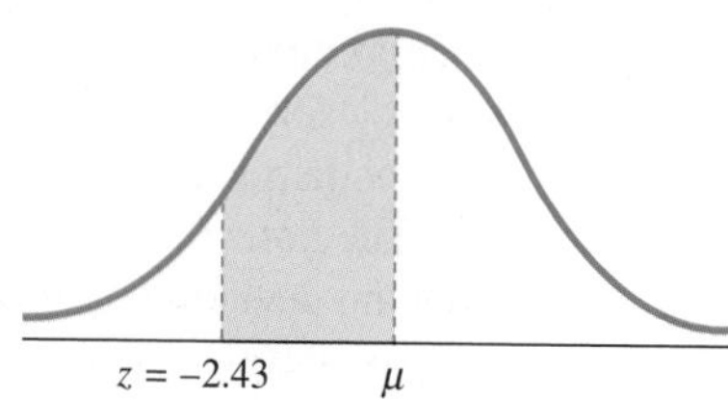

FIGURE 10.12

TECHNOLOGY TIP Because of convenience and accuracy, graphing calculators and computers have made normal curve tables less important. Figure 10.13 shows how part (b) of Example 1 can be done on a TI-83 calculator using a command from the DISTR menu. The second result in the calculator screen gives the area between $-\infty$ and $z = -2.43$; the entry -1E99 represents $-1 \cdot 10^{99}$, which is used to approximate $-\infty$.

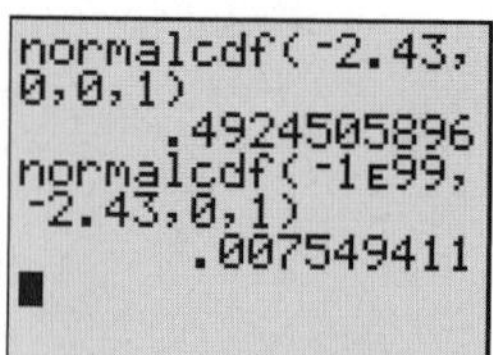

FIGURE 10.13

Many statistical software packages are widely used today. All of these packages are set up in a way that is similar to a spreadsheet, and they all can be used to generate normal curve values. In addition, most spreadsheets can also perform a wide range of statistical calculations.

EXAMPLE 2 Use technology or Table 2 to find the percent of the total area for the following areas under the standard normal curve.

(a) The area between 1.41 standard deviations *below* the mean and 2.25 standard deviations *above* the mean (that is, between $z = -1.41$ and $z = 2.25$)

First, draw a sketch showing the desired area, as in Figure 10.14. From Table 2, the area between the mean and 1.41 standard deviations below the mean is .4207. Also, the area from the mean to 2.25 standard deviations above the mean is .4878. As the figure shows, the total desired area can be found by *adding* these numbers.

$$\begin{array}{r} .4207 \\ + .4878 \\ \hline .9085 \end{array}$$

The shaded area in Figure 10.14 represents 90.85% of the total area under the normal curve.

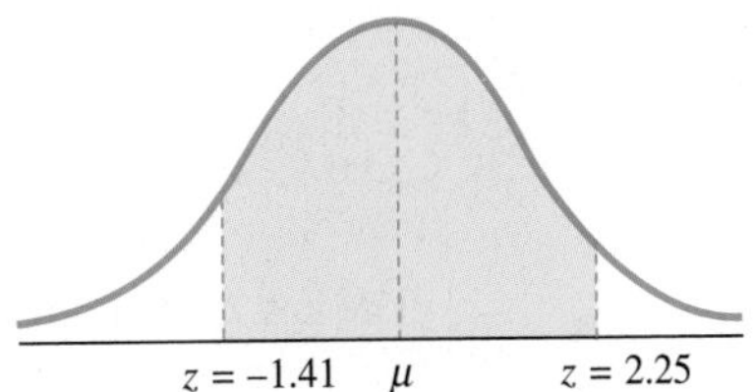

FIGURE 10.14

(b) The area between .58 standard deviation above the mean and 1.94 standard deviations above the mean

Figure 10.15 shows the desired area. The area between the mean and .58 standard deviation above the mean is .2190. The area between the mean and 1.94 standard deviations above the mean is .4738. As the figure shows, the desired area is found by *subtracting* one area from the other.

$$\begin{array}{r} .4738 \\ - .2190 \\ \hline .2548 \end{array}$$

The shaded area of Figure 10.15 represents 25.48% of the total area under the normal curve.

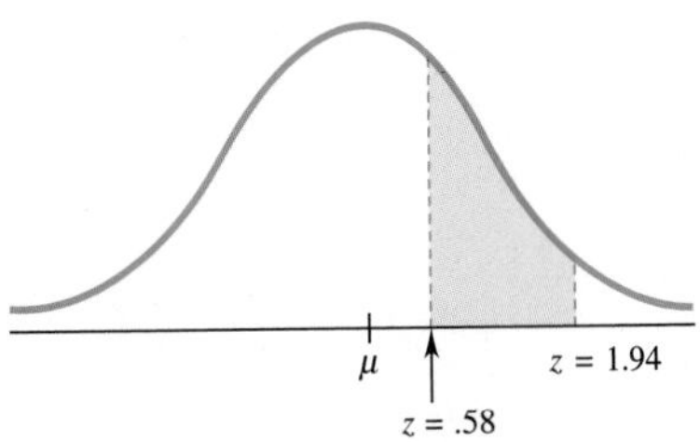

FIGURE 10.15

(c) The area to the right of 2.09 standard deviations above the mean

The total area under a normal curve is 1. Thus, the total area to the right of the mean is 1/2, or .5000. From Table 2, the area from the mean to 2.09 standard

Find the following standard normal curve areas as percents of the total area.

(a) Between .31 standard deviations below the mean and 1.01 standard deviations above the mean

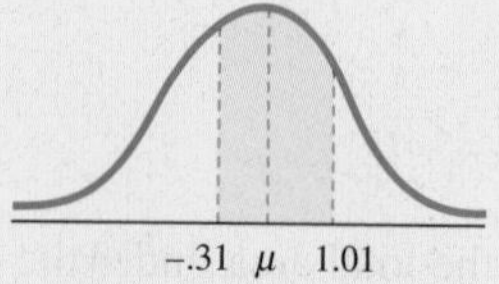

(b) Between .38 and 1.98 standard deviations below the mean

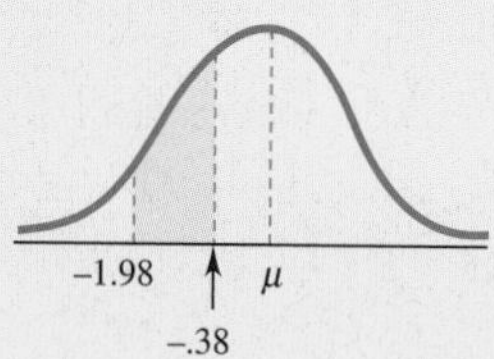

(c) To the right of 1.49 standard deviations above the mean

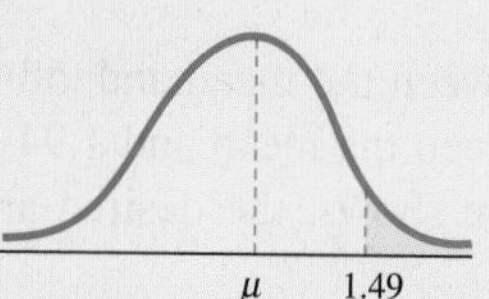

(d) What percent of the area is within 1 standard deviation of the mean? within 2 standard deviations of the mean? within 3 standard deviations of the mean? What can you conclude from the last answer?

Answers:

(a) 46.55%

(b) 32.82%

(c) 6.81%

(d) 68.3%, 95.47%, 99.7% Almost all the data lies within 3 standard deviations of the mean.

deviations above the mean is .4817. The area to the right of 2.09 standard deviations is found by subtracting .4817 from .5000.

$$\begin{array}{r} .5000 \\ -\ .4817 \\ \hline .0183 \end{array}$$

A total of 1.83% of the total area is to the right of 2.09 standard deviations above the mean. Figure 10.16 (which is not to scale) shows the desired area. ◀ 3

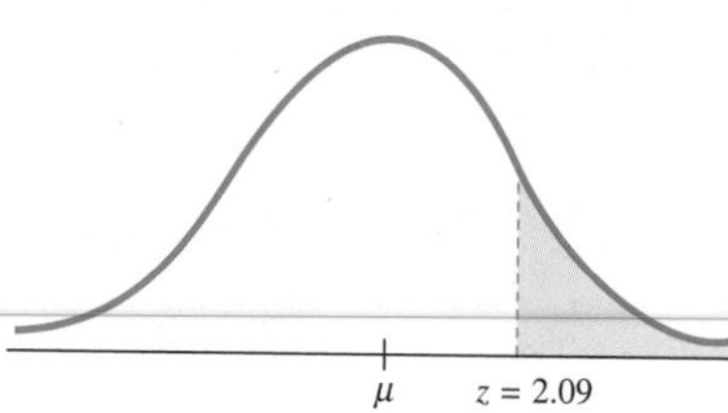

FIGURE 10.16

The key to finding areas under *any* normal curve is to express each number x on the horizontal axis in terms of standard deviations above or below the mean. The **z-score** for x is the number of standard deviations that x lies from the mean (positive if x is above the mean, negative if x is below the mean).

EXAMPLE 3 If a normal distribution has mean 50 and standard deviation 4, find the following z-scores.

(a) The z-score for $x = 46$

Since 46 is 4 units below 50 and the standard deviation is 4, 46 is 1 standard deviation below the mean. So, its z-score is -1.

(b) The z-score for $x = 60$

The z-score is 2.5 because 60 is 10 units above the mean (since $60 - 50 = 10$) and 10 units is 2.5 standard deviations (since $10/4 = 2.5$). ◀ 4

In Example 3(b) we found the z-score by taking the difference between 60 and the mean and dividing this difference by the standard deviation. The same procedure works in the general case.

If a normal distribution has mean μ and standard deviation σ, then the z-score for the number x is

$$z = \frac{x - \mu}{\sigma}.$$

The importance of z-scores is the following fact, whose proof is omitted.

4

Find each z-score using the information in Example 3.

(a) $x = 36$

(b) $x = 55$

Answers:

(a) -3.5

(b) 1.25

Area under a Normal Curve

The area under a normal curve between $x = a$ and $x = b$ is the same as the area under the standard normal curve between the z-score for a and the z-score for b.

Therefore, by converting to z-scores and using a graphing calculator or Table 2 for the standard normal curve, we can find areas under any normal curve. Since these areas are probabilities (as explained earlier), we can now handle a variety of applications.

Graphing calculators, computer programs, and CAS programs (such as DERIVE) can be used to find areas under the normal curve and hence, probabilities. The equation of the standard normal curve, with $\mu = 0$ and $\sigma = 1$, is

$$f(x) = (1/\sqrt{2\pi})e^{-x^2/2}.$$

A good approximation of the area under this curve (and above $y = 0$) can be found by using the x-interval $[-4, 4]$. See Chapter 13 for more information on finding such areas.

EXAMPLE 4 Dixie Office Supplies finds that its sales force drives an average of 1200 miles per month per person, with a standard deviation of 150 miles. Assume that the number of miles driven by a salesperson is closely approximated by a normal distribution.

(a) Find the probability that a salesperson drives between 1200 and 1600 miles per month.

Here $\mu = 1200$ and $\sigma = 150$, and we must find the area under the normal distribution curve between $x = 1200$ and $x = 1600$. We begin by finding the z-score for $x = 1200$.

$$z = \frac{x - \mu}{\sigma} = \frac{1200 - 1200}{150} = \frac{0}{150} = 0$$

The z-score for $x = 1600$ is

$$z = \frac{x - \mu}{\sigma} = \frac{1600 - 1200}{150} = \frac{400}{150} = 2.67.*$$

So the area under the curve from $x = 1200$ to $x = 1600$ is the same as the area under the standard normal curve from $z = 0$ to $z = 2.67$, as indicated in Figure 10.17. A graphing calculator or Table 2 shows that this area is .4962. Therefore, the probability that a salesperson drives between 1200 and 1600 miles per month is .4962.

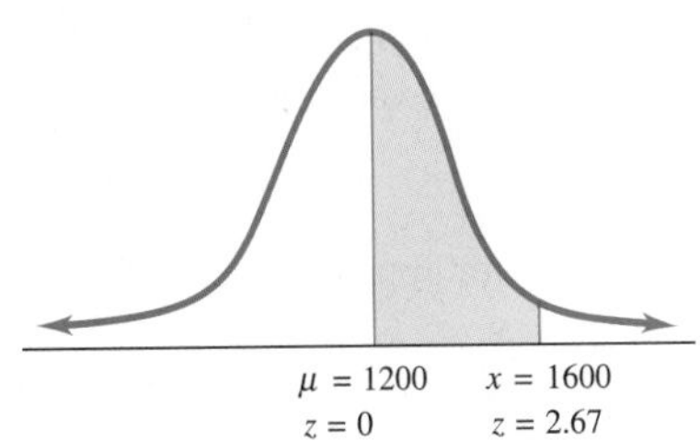

FIGURE 10.17

*All z-scores here are rounded to two decimal places.

The heights of female sophomore college students at one school have $\mu = 172$ centimeters, with $\sigma = 10$ centimeters. Find the probability that the height of such a student is

(a) between 172 cm and 185 cm;

(b) between 160 cm and 180 cm;

(c) less than 165 cm.

Answers:

(a) .4032

(b) .6730

(c) .2420

(b) Find the probability that a salesperson drives between 1000 and 1500 miles per month.

As shown in Figure 10.18, z-scores for both $x = 1000$ and $x = 1500$ are needed.

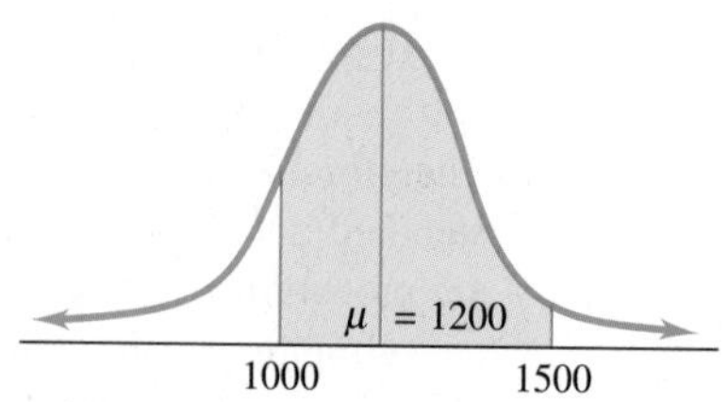

FIGURE 10.18

For $x = 1000$,

$$z = \frac{1000 - 1200}{150} = \frac{-200}{150} = -1.33.$$

For $x = 1500$,

$$z = \frac{1500 - 1200}{150} = \frac{300}{150} = 2.00.$$

From the table $z = 1.33$ leads to an area of .4082, while $z = 2.00$ corresponds to .4773. A total of $.4082 + .4773 = .8855$, or 88.55%, of all drivers travel between 1000 and 1500 miles per month. From this, the probability that a driver travels between 1000 and 1500 miles per month is .8855. ◀ 5

EXAMPLE 5 A tire store finds that the tread life of its tires is normally distributed, with a mean of 26,640 miles and a standard deviation of 4000 miles. The store sold 9000 tires this month. How many of them can be expected to last more than 35,000 miles?

Here $\mu = 26{,}640$ and $\sigma = 4000$. The probability that a tire will last more than 35,000 miles is the area under the normal curve to the right of $x = 35{,}000$. The z-score for $x = 35{,}000$ is

$$z = \frac{x - \mu}{\sigma} = \frac{35{,}000 - 26{,}640}{4000} = \frac{8360}{4000} = 2.09.$$

Example 2(c) and Figure 10.16 show that the area to the right of $z = 2.09$ is .0183, which is 1.83% of the total area under the curve. Therefore, 1.83% of the tires can be expected to last more than 35,000 miles. Thus,

$$1.83\% \text{ of } 9000 = .0183 \cdot 9000 = 164.7,$$

or approximately 165 tires can be expected to last more than 35,000 miles. ◀

NOTE Notice in Example 5 that $P(z \geq 2.09) = P(z > 2.09.)$. The area under the curve is the same, whether we include the endpoint or not. Notice also that $P(z = 2.09) = 0$, because no area is included.

CAUTION When calculating normal probability, it is wise to draw a normal curve with the mean and the z-scores every time. This will avoid confusion as to whether you should add or subtract probabilities.

As mentioned earlier, z-scores are standard deviations, so $z = 1$ corresponds to 1 standard deviation above the mean, and so on. As found in side problem 3(d) of this section, 68.3% of the area under a normal curve lies within 1 standard deviation of the mean. Also, 95.47% lies within 2 standard deviations of the mean, and 99.7% lies within 3 standard deviations of the mean. These results, summarized in Figure 10.19, can be used to get a quick estimate of results when working with normal curves.

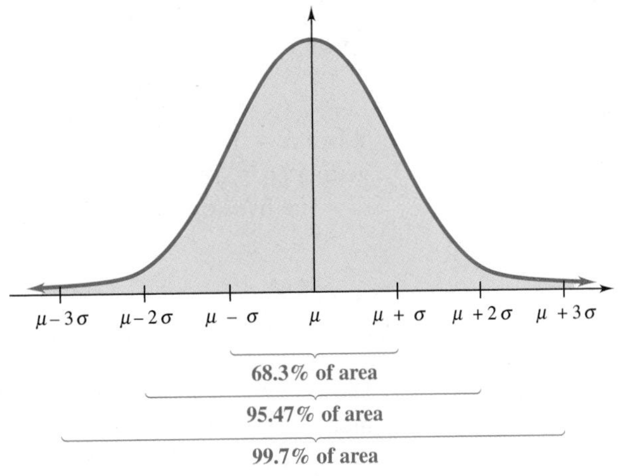

FIGURE 10.19

Boxplots

The normal curve is useful because you can easily read various characteristics of the data from the picture. Boxplots are another graphical means of presenting key characteristics of a data set. The idea is to arrange the data in increasing order and choose three numbers Q_1, Q_2, and Q_3 that divide it into four equal parts, as indicated schematically here.

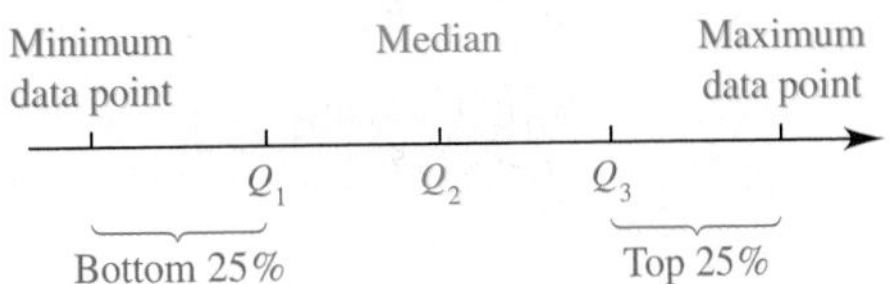

The number Q_1 is called the **first quartile,** the median Q_2 is called the **second quartile,** and Q_3 is called the **third quartile.** These numbers are used to construct a boxplot, as illustrated in the following examples.

EXAMPLE 6 Construct a boxplot for the Britelite data in Exercise 23 of Section 10.2:

20, 22, 22, 25, 26, 27, 27, 28, 30, 35.

The data is already ranked. The minimum score is 20 and the maximum score is 35. Since there are an even number of scores, the median score Q_2 is 26.5 (halfway between the two center entries). To find Q_1, which separates the bottom 25% of the data from the rest, we first calculate

25% of n (rounded up to the nearest integer),

where n is the number of data. Here $n = 10$ and $.25(10) = 2.5$, which rounds to 3. Now count to the *third* score, to get $Q_1 = 22$.

Similarly, since Q_3 separates the bottom 75% of the data from the rest, we calculate

75% of n (rounded up to the nearest integer).

When $n = 10$, then $.75(10) = 7.5$, which rounds to 8. Count to the *eighth* score, getting $Q_3 = 28$.

The five key numbers for constructing the boxplot are

$$\begin{array}{ccccc} \text{Minimum}, & Q_1, & Q_2, & Q_3, & \text{Maximum} \\ \searrow & \downarrow & \downarrow & \downarrow & \swarrow \\ 20, & 22, & 26.5, & 28, & 35 \end{array}$$

Draw a horizontal line parallel to the number axis, from the minimum to the maximum score. Around this line, construct a box whose ends are at Q_1 and Q_3 and mark the location of Q_2 by a vertical line in the box, as shown in Figure 10.20.

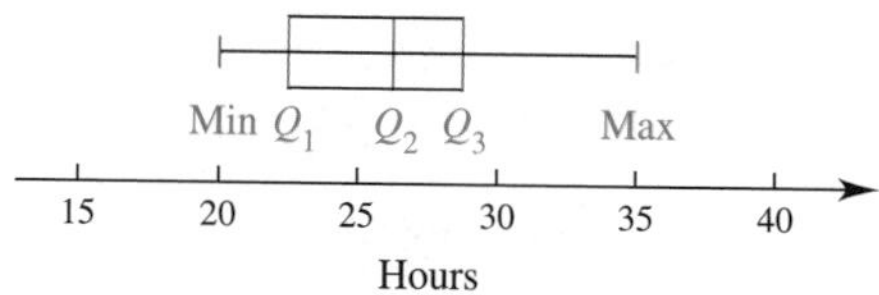

FIGURE 10.20

Because they don't show the detail of a histogram or stem-and-leaf plots, boxplots are not as useful for analyzing a single data set. However, they can be useful for comparing two sets of data, as long as the same scale is used.

EXAMPLE 7 In Exercise 23 of Section 10.2, the following data was given for Brand X:

15, 18, 19, 23, 25, 25, 28, 30, 34, 38.

Construct a boxplot for this data and compare it to the one for Britelite in Example 6.

The minimum is 15, the maximum is 38, and the median is $Q_2 = 25$. Again $n = 10$ and

25% of $n = .25(10) = 2.5$, which rounds up to 3

so that $Q_1 = 19$ (the *third* data score). Similarly,

$$75\% \text{ of } n = .75(10) = 7.5, \text{ which rounds up to } 8,$$

so that $Q_3 = 30$ (the *eighth* data score). The minimum, maximum, and the three quartiles (namely, 15, 19, 25, 30, and 38) are used to get the boxplot in Figure 10.21.

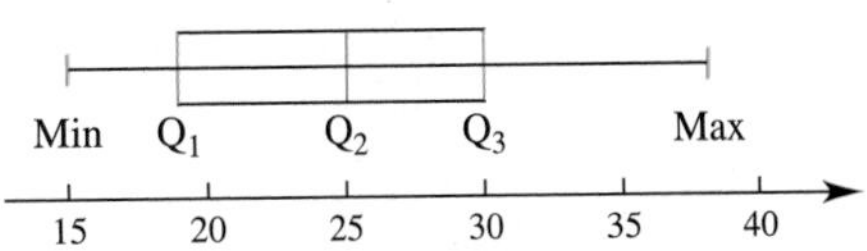

FIGURE 10.21

6

Construct a box plot for the number of blood types data given in Section 10.2, Exercise 27.

Answer:

0 4 8 12 16

Comparing the graphs in Figures 10.20 and 10.21 shows that the range of hours in use is less for Brand X bulbs than for Britelite bulbs, the width of the boxplot indicates where the middle 50% of the distribution lie, and the median indicates how that 50% is distributed. The boxplot for Britelite bulbs is narrower than that for Brand X, showing that they are more consistent in number of hours of use, and the higher placement of the median indicates that they tend to last longer than brand X. ◀ 6

TECHNOLOGY TIP Many graphing calculators can graph boxplots for single-variable data. The procedure is similar to plotting data points using the STAT PLOT menu.

10.3 Exercises

1. The peak in a normal curve occurs directly above ______.
2. The total area under a normal curve (above the horizontal axis) is ______.
3. How are z-scores found for normal distributions where $\mu \neq 0$ or $\sigma \neq 1$?
4. How is the standard normal curve used to find probabilities for normal distributions?

Find the percent of the area under a normal curve between the mean and the following number of standard deviations from the mean. (See Example 2.)

5. 1.75
6. .23
7. −.43
8. −2.1

Find the percent of the total area under the standard normal curve between the following z-scores. (See Examples 1 and 2.)

9. $z = 1.41$ and $z = 2.83$
10. $z = .64$ and $z = 2.11$
11. $z = -2.48$ and $z = -.05$
12. $z = -1.74$ and $z = -1.02$
13. $z = -3.11$ and $z = 1.44$
14. $z = -2.94$ and $z = -.43$

Find a z-score satisfying each of the following conditions. (Hint: Use Table 2 backward or a graphing calculator.)

15. 5% of the total area is to the right of z.
16. 1% of the total area is to the left of z.
17. 15% of the total area is to the left of z.
18. 25% of the total area is to the right of z.
19. For any normal distribution, what is the value of $P(x \leq \mu)$? $P(x \geq \mu)$?
20. Compare the probability that a number will lie within 2 standard deviations of the mean of a probability distribution using Chebyshev's theorem and using the normal

distribution. (See Exercises 15–22, Section 10.2) Explain what you observe.

21. Repeat Exercise 20 using 3 standard deviations.

Assume the following distributions are all normal. (See Example 4.)

22. **Business** According to the label, a regular can of Campbell's soup holds an average of 305 grams with a standard deviation of 4.2 grams. What is the probability that a can will be sold that holds more than 306 grams?

23. **Business** A jar of Adams Old Fashioned Peanut Butter contains 453 grams with a standard deviation of 10.1 grams. Find the probability that one of these jars contains less than 450 grams.

24. **Business** A General Electric soft white 3-way bulb has an average life of 1200 hours with a standard deviation of 50 hours. Find the probability that the life of one of these bulbs will be between 1150 and 1300 hours.

25. **Business** A 100-watt lightbulb has an average brightness of 1640 lumens with a standard deviation of 62 lumens. What is the probability that a 100-watt bulb will have a brightness between 1600 and 1700 lumens?

26. **Social Science** The scores on a standardized test in a suburban high school have a mean of 80 with a standard deviation of 12. What is the probability that a student will have a score less than 60?

27. **Health** The mean clotting time of blood is 7.45 seconds with a standard deviation of 3.6 seconds. What is the probability that an individual's blood clotting time will be less than 7 seconds or greater than 8 seconds?

28. **Social Science** The average time to complete the test in Exercise 26 is 40 minutes with a standard deviation of 5.2 minutes. Find the probability that a student will require more than 50 minutes to finish the test.

29. **Social Science** A 120-minute videotape has a standard deviation of .7 minute. Find the probability that a 118-minute movie will not be entirely copied.

30. **Natural Science** The distribution of low temperatures for a city has an average of 44° with a standard deviation of 6.7°. What is the probability that the low temperature will be above 55°?

Business *A certain type of lightbulb has an average life of 500 hours, with a standard deviation of 100 hours. The length of life of the bulb can be closely approximated by a normal curve. An amusement park buys and installs 10,000 such bulbs. Find the total number that can be expected to last for each of the following periods of time. (See Example 5.)*

31. At least 500 hours

32. Less than 500 hours

33. Between 650 and 780 hours

34. Between 290 and 540 hours

35. Less than 740 hours

36. More than 300 hours

37. Find the shortest and longest lengths of life for the middle 80% of the bulbs.

Social Science *New studies by Federal Highway Administration traffic engineers suggest that speed limits on many thoroughfares are set arbitrarily and often are artificially low. According to traffic engineers, the ideal limit should be the "85th percentile speed." This means the speed at or below which 85 percent of the traffic moves. Assuming speeds are normally distributed, find the 85th percentile speed for roads with the following conditions.*

38. The mean speed is 50 mph with a standard deviation of 10 mph.

39. The mean speed is 30 mph with a standard deviation of 5 mph.

Social Science *One professor uses the following system for assigning letter grades in a course.*

Grade	*Total Points*
A	Greater than $\mu + \frac{3}{2}\sigma$
B	$\mu + \frac{1}{2}\sigma$ to $\mu + \frac{3}{2}\sigma$
C	$\mu - \frac{1}{2}\sigma$ to $\mu + \frac{1}{2}\sigma$
D	$\mu - \frac{3}{2}\sigma$ to $\mu - \frac{1}{2}\sigma$
F	Below $\mu - \frac{3}{2}\sigma$

What percent of the students receive the following grades?

40. A

41. B

42. C

43. Do you think the system in Exercises 40–42 would be more likely to be fair in a large freshman class in psychology or in a graduate seminar of five students? Why?

Health *In nutrition, the recommended daily allowance of vitamins is a number set by the government as a guide to an individual's daily vitamin intake. Actually, vitamin needs vary* *drastically from person to person, but the needs are very closely approximated by a normal curve. To calculate the recommended daily allowance, the government first finds the average need for vitamins among people in the population and then the standard deviation. The recommended daily allowance is defined as the mean plus 2.5 times the standard deviation.*

44. What percent of the population will receive adequate amounts of vitamins under this plan?

Find the recommended daily allowance for the following vitamins.

45. Mean = 500 units, standard deviation = 50 units

46. Mean = 1800 units, standard deviation = 140 units

47. Mean = 159 units, standard deviation = 12 units

48. Mean = 1200 units, standard deviation = 92 units

Social Science *The mean performance score of a large group of fifth-grade students on a math achievement test is 88. The scores are known to be normally distributed. What percent of the students had scores as follows?*

49. More than 1 standard deviation above the mean

50. More than 2 standard deviations above the mean

Social Science *Studies have shown that women are charged an average of $500 more than men for cars.* Assume a normal distribution of overcharges with a mean of $500 and a standard deviation of $60. Find the probability of a woman's paying the following additional amounts for a car.*

51. Less than $500

52. At least $600

53. Between $400 and $600

*"From repair shops to cleaners, women pay more," by Bob Dart as appeared in *The Chicago Tribune,* May 27, 1993. Reprinted by permission of the author.

Social Science *Women earn an average of $.73 for every $1 earned by a man, a difference of $.27.* Assume differences in pay are normally distributed with a mean of $.27 and a standard deviation of $.08. Find the probability that a woman earns the following amounts less than a man in the same job.*

54. Between $.20 and $.30 **55.** More than $.19

56. At most $.27 **57.** At least $.25

58. Business The table gives the percent of U.S. agricultural exports to selected nations.† Use this information in the following exercises. (See Example 6.)

	1990	*1998*
Japan	20.6	18.5
South Korea	6.7	5.4
Taiwan	4.2	4
China	2.1	1.8
Hong Kong	1.8	2.5
European Union	17.4	13.7
Mexico	6.5	11.7
Canada	10.7	14.6
Africa	4.9	4.5

(a) Construct a boxplot for the 1990 data.
(b) Construct a boxplot for the 1998 data.
(c) Use the graphs in parts (a) and (b) to compare the 1990 and 1998 distributions.

*Bureau of the Census, U.S. Department of Commerce.

†U.S. Department of Agriculture, *Agricultural Statistics,* annual.

10.4 NORMAL APPROXIMATION TO THE BINOMIAL DISTRIBUTION

As we saw in Section 9.3, many practical experiments have only two possible outcomes, sometimes referred to as success or failure. Such experiments are called Bernoulli trials or Bernoulli processes. Examples of Bernoulli trials include flipping a coin (with heads being a success, for instance, and tails a failure) or testing a computer chip coming off the assembly line to see whether or not it is defective. A binomial experiment consists of repeated independent Bernoulli trials, such as flipping a coin 10 times or taking a random sample of 20 computer chips from the assembly line. In Section 9.4 we found the probability distribution for several binomial experiments, such as sampling five people with bachelor's degrees in education and counting how many are women. The probability distribution for a binomial experiment is known as a **binomial distribution.**

As another example, it is reported that 29% of drivers on Interstate 10 in Texas exceed the 70 mph speed limit.* Suppose a state trooper wants to verify this statistic and records the speed of 10 randomly selected drivers. The trooper finds that 4 out of 10, or 40%, exceed the speed limit. How likely is this if the 29% figure is accurate? We can answer this question with the binomial probability formula

$$\binom{n}{x} \cdot p^x \cdot (1 - p)^{n-x},$$

where n is the size of the sample (10 in this case), x is the number of speeders (4 in this example), and p is the probability that a driver is a speeder (.29). This gives

$$P(x = 4) = \binom{10}{4} \cdot .29^4 \cdot (1 - .29)^6$$

$$= 210(.007073)(.1281) \approx .1903.$$

The probability is almost 20%, so this result is not unusual.

Suppose that the state trooper takes a larger random sample of 100 drivers. What is the probability that 40 or more drivers speed if the 29% figure is accurate? Calculating $P(x = 40) + P(x = 41) + \cdots + P(x = 100)$ is a formidable task. One solution is provided by graphing calculators or computers. On the TI-83, for example, we can first calculate the probability that 39 or fewer drivers exceed the speed limit using the DISTR menu command binomcdf (100, .29, 39). Subtracting the answer from 1 gives a probability of .0119. But this high-tech method fails as n becomes larger; the command binomcdf (1000, .29, 300) gives an error message. On the other hand, there is a low-tech method that works regardless of the size of n. It has further interest because it connects two different distributions: the normal and the binomial. The normal distribution is continuous, since the random variable can take on any real number. The binomial distribution is *discrete,* because the random variable can take on only integer values between 0 and n. Nevertheless, the normal distribution can be used to give a good approximation to binomial probability.

In order to use the normal approximation, we first need to know the mean and standard deviation of the binomial distribution. Recall from Section 9.4 that for the binomial distribution, $E(x) = np$. In Section 10.1, we referred to $E(x)$ as μ, and that notation will be used here. It is shown in more advanced courses in statistics that the standard deviation of the binomial distribution is given by $\sigma = \sqrt{np(1 - p)}$.

Mean and Standard Deviation for the Binomial Distribution

For the binomial distribution, the mean and standard deviation are given by

$$\boldsymbol{\mu = np} \quad \text{and} \quad \boldsymbol{\sigma = \sqrt{np(1 - p)}},$$

where n is the number of trials and p is the probability of success on a single trial. (1)

1

Find μ and σ for a binomial distribution having $n = 120$ and $p = 1/6$.

Answer:

$\mu = 20$; $\sigma = 4.08$

Time, May 13, 1996, p. 34.

EXAMPLE 1 Suppose a fair coin is flipped 15 times.

(a) Find the mean and standard deviation for the number of heads.
Using $n = 15$ and $p = 1/2$, the mean is

$$\mu = np = 15\left(\frac{1}{2}\right) = 7.5.$$

The standard deviation is

$$\sigma = \sqrt{np(1-p)} = \sqrt{15\left(\frac{1}{2}\right)\left(1-\frac{1}{2}\right)}$$
$$= \sqrt{15\left(\frac{1}{2}\right)\left(\frac{1}{2}\right)} = \sqrt{3.75} \approx 1.94.$$

We expect, on average, to get 7.5 heads out of 15 tosses. Most of the time, the number of heads will be within three standard deviations of the mean, or between $7.5 - 3(1.94) = 1.68$ and $7.5 + 3(1.94) = 13.32$.

(b) Find the probability distribution for the number of heads and draw a histogram of the probabilities.

The probability distribution is found by putting $n = 15$ and $p = 1/2$ into the formula for binomial probability. For example, the probability of 9 heads is given by

$$P(x = 9) = \binom{15}{9}\left(\frac{1}{2}\right)^9\left(1-\frac{1}{2}\right)^6 \approx .15274.$$

Probabilities for the other values of x between 0 and 15, as well as a histogram of the probabilities, are shown in Figure 10.22.

x	$P(x)$
0	.00003
1	.00046
2	.00320
3	.01389
4	.04166
5	.09164
6	.15274
7	.19638
8	.19638
9	.15274
10	.09164
11	.04166
12	.01389
13	.00320
14	.00046
15	.00003

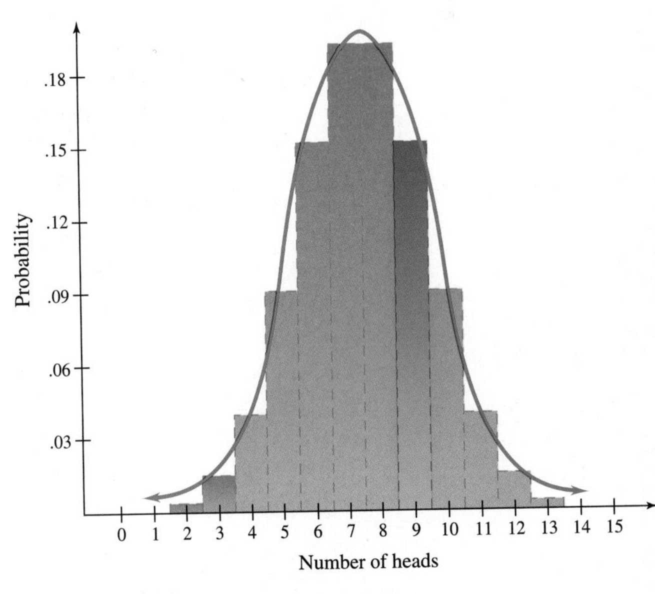

FIGURE 10.22

In Figure 10.22, we have superimposed the normal curve with $\mu = 7.5$ and $\sigma = 1.94$ over the histogram of the distribution. Notice how well the normal distribution fits the binomial distribution. This approximation was first discovered in 1718 by Abraham De Moivre (1667–1754) for the case $p = 1/2$. The result was generalized by the French mathematician Pierre-Simon Laplace (1749–1827) in a book published in 1812. As n becomes larger and larger, a histogram for the binomial distribution looks more and more like a normal curve. Figures 10.23 (a) and (b) show histograms of the binomial distribution with $p = .3$, using $n = 8$ and $n = 50$, respectively.

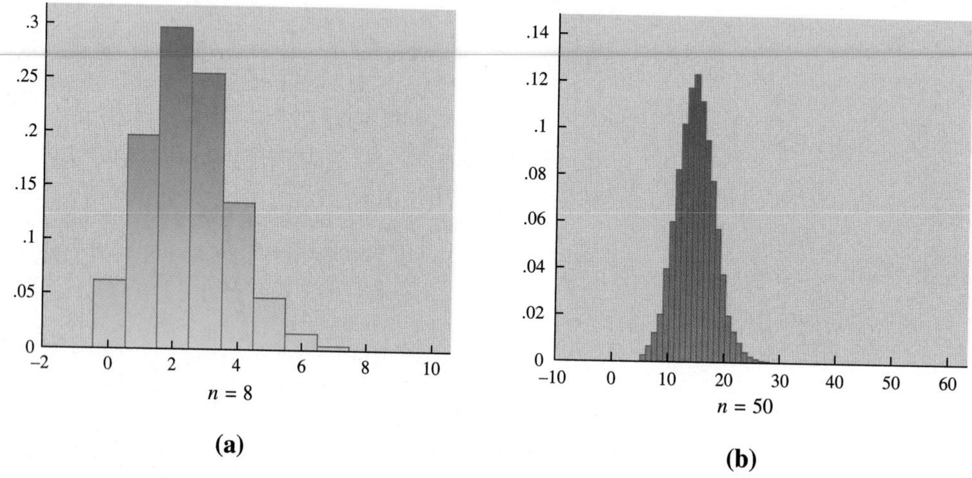

FIGURE 10.23

The probability of getting exactly 9 heads in the 15 tosses, or .15274, is the same as the area of the blue bar in Figure 10.22. As the graph suggests, the blue area is approximately equal to the area under the normal curve from $x = 8.5$ to $x = 9.5$. The normal curve is higher than the top of the bar in the left half but lower in the right half.

To find the area under the normal curve from $x = 8.5$ to $x = 9.5$, first find z-scores, as in the previous section. Use the mean and the standard deviation for the distribution, which we have already calculated, to get z-scores for $x = 8.5$ and $x = 9.5$.

For $x = 8.5$,

$$z = \frac{8.5 - 7.5}{1.94} = \frac{1.00}{1.94}$$

$$z \approx .52.$$

For $x = 9.5$,

$$z = \frac{9.5 - 7.5}{1.94} = \frac{2.00}{1.94}$$

$$z \approx 1.03.$$

2

Use the normal distribution to find the probability of getting exactly the following number of heads in 15 tosses of a coin.

(a) 7

(b) 10

Answers:

(a) .1985

(b) .0909

From Table 2, $z = .52$ gives an area of .1985, and $z = 1.03$ gives .3485. The difference between these two numbers is the desired result.

$$.3485 - .1985 = .1500$$

This answer (.1500) is not far from the more accurate answer of .15274 found above. ◀ 2

CAUTION The normal curve approximation to a binomial distribution is quite accurate *provided that n* is large and p is not close to 0 or 1. As a rule of thumb, the normal curve approximation can be used as long as both np and $n(1 - p)$ are at least 5.

EXAMPLE 2 Consider the random sample discussed earlier of 100 drivers on Interstate 10 in Texas, where 29% of the drivers exceed the 70 mph speed limit.

(a) Use the normal distribution to approximate the probability that at least 40 drivers exceed the speed limit.

First find the mean and the standard deviation using $n = 100$ and $p = .29$.

$$\mu = 100(.29) = 29 \qquad \sigma = \sqrt{100(.29)(1 - .29)} = \sqrt{100(.29)(.71)} = \sqrt{20.59} \approx 4.54$$

As the graph in Figure 10.24 shows, we need to find the area to the right of $x = 39.5$ (since we want 40 or more speeders). The z-score corresponding to $x = 39.5$ is

$$z = \frac{39.5 - 29}{4.54} \approx 2.31.$$

From the table, $z = 2.31$ leads to an area of .4896, so

$$P(z > 2.31) = .5 - .4896 = .0104.$$

This value is close to the value of .0119 found earlier with the help of a graphing calculator. Either method tells us there is roughly a 1% chance of finding 40 or more speeders out of a random sample of 100. If the trooper found this many in his sample, he might suspect that either his sample is not truly random or the 29% figure is too low.

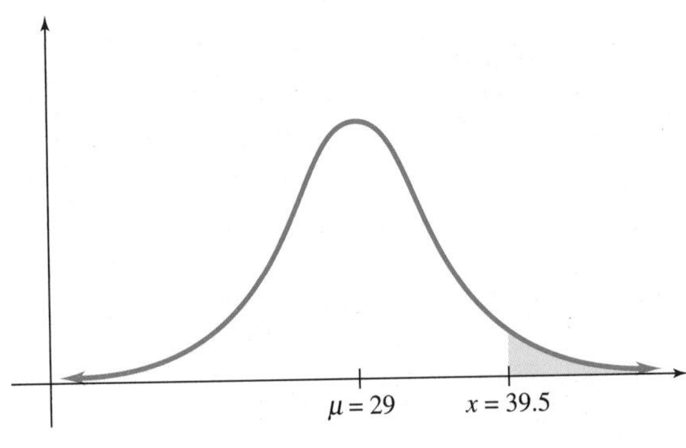

FIGURE 10.24

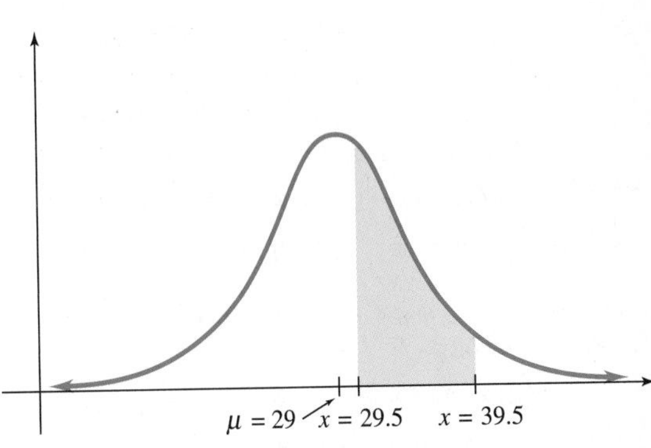

FIGURE 10.25

About 9% of the transistors produced by a certain factory are defective. Find the approximate probability that in a sample of 200, the following numbers of transistors will be defective. (*Hint:* $\sigma = 4.05$)

(a) Exactly 11

(b) 16 or fewer

(c) More than 14

Answers:

(a) .0226

(b) .3557

(c) .8051

(b) Find the probability of finding between 30 and 35 speeders in a random sample of 100.

As Figure 10.25 shows, we need to find the area between $x = 29.5$ and $x = 35.5$.

$$\text{If } x = 29.5, \text{ then } z = \frac{29.5 - 29}{4.54} \approx .11.$$

$$\text{If } x = 35.5, \text{ then } z = \frac{35.5 - 29}{4.54} \approx 1.43.$$

Use the table to find that $z = .11$ gives an area of .0458 and $z = 1.43$ yields .4236. The final answer is the difference of these numbers.

$$P(.11 \le z \le 1.43) = .4236 - .0458 = .3798.$$

The probability of finding between 30 and 35 speeders is about .3798. 

10.4 Exercises

1. What must be known to find the mean and standard deviation of a binomial distribution?
2. What is the rule of thumb for using the normal distribution to approximate a binomial distribution?

Suppose 16 coins are tossed. Find the probability of getting each of the following results (a) *using the binomial probability formula, and* (b) *using the normal curve approximation. (See Examples 1 and 2.)*

3. Exactly 8 heads
4. Exactly 7 heads
5. More than 12 tails
6. Fewer than 5 tails

For the remaining exercises in this section, use the normal curve approximation to the binomial distribution.

Suppose 1000 coins are tossed. Find the probability of getting each of the following results.

7. Exactly 500 heads
8. Exactly 510 heads
9. 480 heads or more
10. Fewer than 470 tails

A die is tossed 120 times. Find the probability of getting each of the following results.

11. Exactly twenty 5's
12. Exactly twenty-four 6's
13. More than eighteen 3's
14. Fewer than twenty-two 6's
15. A reader asked Mr. Statistics (a feature in *Fortune* magazine) about the game of 26 once played in the bars of Chicago.* The player chooses a number between 1 and 6 and then rolls a cup full of 10 dice 13 times. Out of the 130 numbers rolled, if the number chosen appears at least 26 times, the player wins. Calculate the probability of winning.
16. Social Science According to the National Center for Education Statistics, 79% of the U.S. holders of bachelor's degrees in education in 1992–93 were women.† Suppose five holders of bachelor's degrees in education are picked at random.
    (a) Find the probability that four of the five are women.
    (b) Find the probability that at least two of the five are women.
17. Natural Science For certain bird species, with appropriate assumptions, the number of nests escaping predation has a binomial distribution.‡ Suppose the probability of success (that is, a nest escaping predation) is .3. Find the probability that at least half of 26 nests escape predation.
18. Natural Science Under certain appropriate assumptions, the probability of a competing young animal eating

*Daniel Seligman and Patty De Llosa, "Ask Mr. Statistics," *Fortune,* May 1, 1995, p. 141.

†*The New York Times,* January 7, 1996, Education Life, p. 24.

‡From H. M. Wilbur, *American Naturalist,* vol. 111.

x units of food is binomially distributed, with n equal to the maximum number of food units the animal can acquire and p equal to the probability per time unit that an animal eats a unit of food.* Suppose $n = 120$ and $p = .6$.

(a) Find the probability that an animal consumes 80 units of food.

(b) Suppose the animal must consume at least 70 units of food to survive. What is the probability that this happens?

19. Business A recent study of minimum wage earners found that 25.6% of them are teenagers.† Suppose a random sample of 600 minimum wage earners is selected. What is the probability that more than 160 of them are teenagers?

20. Social Science In Exercise 19, what is the probability that less than 140 of them are teenagers?

21. Health A study found that 23.6% of people on kidney dialysis in the United States die, more than in any other industrial country.‡ A hospital surveyed 40 of its patients who had been on dialysis that year and found that 15 died so far. What is the probability that 15 or more people out of 40 would die if this hospital's patients are typical of those in the United States?

22. Natural Science A flu vaccine has a probability of 80% of preventing a person who is inoculated from getting the flu. A county health office inoculates 134 people. Find the probabilities of the following.

(a) Exactly 10 of the people inoculated get the flu.

(b) No more than 10 of the people inoculated get the flu.

(c) None of the people inoculated get the flu.

23. Natural Science The probability that a male will be color blind is .042. Find the probabilities that in a group of 53 men, the following will be true.

(a) Exactly 5 are color blind.

(b) No more than 5 are color blind.

(c) At least 1 is color blind.

24. Business The probability that a certain machine turns out a defective item is .05. Find the probabilities that in a run of 75 items, the following results are obtained.

(a) Exactly 5 defectives

(b) No defectives

(c) At least 1 defective

25. Business A company is taking a survey to find out whether people like its product. Their last survey indicated that 70% of the population likes the product. Based on that, of a sample of 58 people, find the probabilities of the following.

(a) All 58 like the product.

(b) From 28 to 30 (inclusive) like the product.

26. Natural Science The blood types B− and AB− are the rarest of the eight human blood types, representing 1.5% and .6% of the population, respectively.*

(a) If the blood types of a random sample of 1000 blood donors are recorded, what is the probability that 10 or more of the samples are AB−?

(b) If the blood types of a random sample of 1000 blood donors are recorded, what is the probability that 20 to 40 inclusive of the samples are B−?

(c) If a particular city had a blood drive in which 500 people gave blood and 3% of the donations were B−, would we have reason to believe that this town has a higher than normal number of donors who are B−? (*Hint:* Calculate the probability of 15 or more donors being B− for a random sample of 500 and then consider the probability obtained.)

27. Social Science In Washington State, there were 27.4 motorcycle rider fatalities per 1000 motorcycle crashes in 1989.† If this rate is a true reflection of the nationwide proportion of motorcycle crashes that result in death, find the probability that for a state that has 3000 motorcycle accidents in a given year, there will be between 80 and 100 fatalities (inclusive).

28. Social Science In one state, 55% of the voters expect to vote for Bill Poole. Suppose 1400 people are asked the name of the person for whom they expect to vote. Find the probability that at least 750 people will say that they expect to vote for Poole.

29. Social Science A poll of 2000 teenagers found that 1 in 8 reported carrying a weapon for protection.‡ In a typical high school with 1200 students, what is the probability that more than 120 students, but fewer than 180, carry a weapon?

30. Social Science Only 1 out of 12 American parents requires that children do their homework before watching TV.§ If your neighborhood is typical, what is the probability that that out of 51 parents, 5 or fewer require their children to do homework before watching TV?

31. In the 1989 U.S. Open, four golfers each made a hole in one on the same par-3 hole on the same day. *Sports Illustrated* writer R. Reilly stated the probability of a hole

*From G. deJong, *American Naturalist*, vol. 110.

†*Chicago Tribune*, December 28, 1995, pp. 1, 14.

‡*New York Times*, December 4, 1995, p. A1.

**The Handy Science Answer Book*, The Carnegie Library, Pittsburgh, Pennsylvania, 2000, p. 332.

†Jefferson Rowland, Frederick Rivara, Philip Salzberg, Robert Soderberg, Ronald Maier, and Thomas Keopsell, "Motorcycle Helmet Use and Injury Outcome and Hospitalization Costs from Crashes in Washington State," *American Journal of Public Health* 86, no. 1 (January 1996): 41–45.

‡*New York Times*, January 12, 1996, p. A6.

§"Harper's Index," *Harper's*, September 1996: 15.

in one for a given golf pro on a given par-3 hole to be 1/3709.*

(a) For a specific par-3 hole, use the binomial distribution to find the probability that 4 or more of the 156 golf pros in the tournament field shoot a hole in one.†

(b) For a specific par-3 hole, use the normal approximation to the binomial distribution to find the probability that 4 or more of the 156 golf pros in the tournament field shoot a hole in one. Why must we be very cautious when using this approximation for this application?

(c) If the probability of a hole in one remains constant and is 1/3709 for any par-3 hole, find the probability that in 20,000 attempts by golf pros, there will be 4 or more hole in ones. Discuss whether this assumption is reasonable.

*R. Reilly, "King of the Hill," *Sports Illustrated,* June 1989, pp. 20–25.

†Bonnie Litwiller and David Duncan, "The Probability of a Hole in One," *School Science and Mathematics* 91, no. 1, (January 1991): 30.

CHAPTER 10 SUMMARY

Key Terms and Symbols

10.1 Σ summation (sigma) notation
$\bar{x}$ sample mean
μ population mean
random sample
grouped frequency distribution
frequency polygon
(arithmetic) mean
median
statistic
mode

10.2 s^2 sample variance
s sample standard deviation
σ population standard deviation
range
deviations from the mean
variance
standard deviation

10.3 μ mean of a continuous distribution
σ standard deviation of a normal distribution
continuous distribution
skewed distribution
normal distributions
normal curves
standard normal curve
z-score
boxplot
quartile

10.4 binomial distribution

Key Concepts

To organize the data from a sample, we use a **grouped frequency distribution,** a set of intervals with their corresponding frequencies. The same information can be displayed with a histogram, a type of bar graph with a bar for each interval. Each bar has width 1 and height equal to the probability of the corresponding interval. A **stem-and-leaf plot** presents the individual data in a similar form, so it can be viewed as bars. Another way to display this information is with a **frequency polygon,** which is formed by connecting the midpoints of consecutive bars of the histogram with straight line segments.

The **mean** $\bar{x}$ of a frequency distribution is the expected value.

For n numbers $x_1, x_2, \ldots, x_n$,

$$\bar{x} = \frac{\Sigma x}{n}.$$

For a grouped distribution,

$$\bar{x} = \frac{\Sigma(xf)}{n}.$$

The **median** is the middle entry in a set of data arranged in either increasing or decreasing order.

The **mode** is the most frequent entry in a set of numbers.

The **range** of a distribution is the difference between the largest and smallest numbers in the distribution.

The **sample standard deviation** s is the square root of the sample **variance.**

For n numbers,

$$s = \sqrt{\frac{\Sigma x^2 - n\bar{x}^2}{n - 1}}.$$

For a grouped distribution,

$$s = \sqrt{\frac{\Sigma fx^2 - n\bar{x}^2}{n - 1}}.$$

A **normal distribution** is a continuous distribution with the following properties: The highest frequency is at the mean; the graph is symmetric about a vertical line through the mean; the total area under the curve, above the x-axis, is 1. If a normal distribution has mean μ and standard deviation σ, then the z-score for the number x is $z = \dfrac{x - \mu}{\sigma}$.

A **boxplot** organizes a list of data using the minimum and maximum values, the median, and the first and third quartiles to give a visual overview of the distribution.

Area under a Normal Curve The area under a normal curve between $x = a$ and $x = b$ gives the probability that an observed data value will be between a and b.

The **binomial distribution** is a distribution with the following properties: For n independent repeated trials, where the probability of success in a single trial is p, the probability of x successes is $\binom{n}{x}p^x(1 - p)^{n-x}$. The mean is $\mu = np$ and the standard deviation is

$$\sigma = \sqrt{np(1 - p)}.$$

Chapter 10 Review Exercises

1. Discuss some reasons for organizing data into a grouped frequency distribution.

2. What is the rule of thumb for an appropriate interval in a grouped frequency distribution?

In Exercises 3 and 4, (a) *write a frequency distribution;* (b) *draw a histogram;* (c) *draw a frequency polygon.* (d) *In Exercise 4, also construct a stem-and-leaf plot, using two separate intervals for 9 and 10.*

3. The following numbers give the sales in dollars for the lunch hour at a local hamburger store for the last twenty Fridays. (Use intervals 450–474, 475–499, and so on.)

480	451	501	478	512	473	509	515	458	566
516	535	492	558	488	547	461	475	492	471

4. The number of units carried in one semester by the students in a business mathematics class was as follows. (Use intervals of 9–10, 11–12, 13–14, 15–16.)

10	9	16	12	13	15	13	16	15	11	13
12	12	15	12	14	10	12	14	15	15	13

Find the mean for each of the following.

5. 41, 60, 67, 68, 72, 74, 78, 83, 90, 97

6. 105, 108, 110, 115, 106, 110, 104, 113, 117

7.

Interval	*Frequency*
10–19	6
20–29	12
30–39	14
40–49	10
50–59	8

8.

Interval	*Frequency*
40–44	2
45–49	5
50–54	7
55–59	10
60–64	4
65–69	1

9. What do the mean, median, and mode of a distribution have in common? How do they differ? Describe each in a sentence or two.

Find the median and the mode (or modes) for each of the following.

10. 32, 35, 36, 44, 46, 46, 59

11. 38, 36, 42, 44, 38, 36, 48, 35

The modal class is the interval containing the most data values. Find the modal class for the distributions of

12. Exercise 7 above.

13. Exercise 8 above.

14. What is meant by the range of a distribution?

15. How are the variance and the standard deviation of a distribution related? What is measured by the standard deviation?

Find the range and standard deviation for each of the following distributions.

16. 14, 17, 18, 19, 32

17. 26, 43, 51, 29, 37, 56, 29, 82, 74, 93

Find the standard deviation for the following.

18. Exercise 7 above.

19. Exercise 8 above.

20. Describe the characteristics of a normal distribution.

21. What is meant by a skewed distribution?

Find the following areas under the standard normal curve.

22. Between $z = 0$ and $z = 1.27$

23. To the left of $z = .41$

24. Between $z = -1.88$ and $z = 2.10$

25. Between $z = 1.53$ and $z = 2.82$

26. Find a z-score such that 8% of the area under the curve is to the right of z.

27. Why is the normal distribution not a good approximation of a binomial distribution that has a value of p close to 0 or 1?

Business *The table gives the probability distribution of U.S. buyers of recorded music in 1999 by age.**

Age	*Probability*
10–14	.085
15–19	.126
20–24	.126
25–29	.105
30–34	.101
35–39	.104
40–44	.093
45 & up	.247

28. **(a)** Draw a histogram for this distribution.
(b) From the shape of the histogram, is this a normal distribution?

29. As shown in the distribution above, 24.7% of buyers of recorded music are age 45 or over. Assume a binomial distribution with $n = 6$ and $p = .247$.
(a) Find the probability that 4 of 6 randomly selected recorded music buyers are age 45 or over.
(b) Find the mean and standard deviation of the binomial distribution. Interpret your results.

30. Finance The annual returns of two stocks for three years are given below.

	2000	**2001**	**2002**
Stock I	11%	−1%	14%
Stock II	9%	5%	10%

(a) Find the mean and standard deviation for each stock over the three-year period.
(b) If you are looking for security with an 8% return, which of these two stocks would you choose using your answers from part (a) to decide?

31. Natural Science The weight gains of two groups of 10 rats fed on two different experimental diets were as follows.

	Weight Gains									
Diet A	1	0	3	7	1	1	5	4	1	4
Diet B	2	1	1	2	3	2	1	0	1	0

Compute the mean and standard deviation for each group and compare them to answer the questions below.
(a) Which diet produced the greatest mean gain?
(b) Which diet produced the most consistent gain?

32. Natural Science Refer to Exercise 31.
(a) Construct boxplots for each diet data.
(b) Use the boxplots to compare weight gains for the two diets.

33. Health Between 1990 and 1999 HMO enrollment in the United States increased as shown below.*

Location	*Gains*
South	16.8%
Midwest	20.7%
Northeast	22.1%
West	18.2%

(a) Compute the mean and standard deviation of the gains.
(b) How many standard deviations from the mean is the greatest gain? the smallest gain?

*Recording Industry Association of America.

*U.S. Department of Health and Human Services, *Health, United States, 2000.*

34. Social Science On standard IQ tests, the mean is 100, with a standard deviation of 15. The results are very close to fitting a normal curve. Suppose an IQ test is given to a very large group of people. Find the percent of people whose IQ score is
(a) more than 130;
(b) less than 85;
(c) between 85 and 115.

35. Business A machine that fills quart orange juice cartons is set to fill them with 32.1 oz. If the actual contents of the cartons vary normally, with a standard deviation of .1 oz, what percent of the cartons contains less than a quart (32 oz)?

36. Business About 60% of the bags of blended coffee produced by a certain company contain cheap coffee. Assume a binomial distribution and use the normal distribution to find the probability that, in a sample of 500 bags of coffee, the following is true.
(a) 310 or fewer contain cheap coffee.
(b) Exactly 300 contain cheap coffee.

37. Natural Science An area infested with fruit flies is to be sprayed with a chemical that is known to be 98% effective for each application. A sample of 100 flies is checked. Assume a binomial distribution and use the normal distribution to approximate the following probabilities.
(a) 95% of the flies are killed in one application.
(b) At least 95% of the flies are killed in one application.
(c) At least 90% of the flies are killed in one application.
(d) All the flies are killed in one application. Compare your answer with the probability found using the binomial distribution.

38. Natural Science Twenty percent of third-world women would like to avoid pregnancy but are not using birth control. For 50 such women, use the normal curve approximation to find the probability of each of the following. Assume a binomial distribution and use the normal distribution approximation.
(a) Exactly 10 are not using birth control.
(b) At least 20 are using birth control.

39. Health A survey of injuries during practice or performance given to 313 Broadway performers produced the demographics in the table.*

	Mean	*Standard Deviation*
Dancer's Age (female)	28.0	5.5
Dancer's Age (male)	32.2	8.4
Height in m (female)	1.64	.08
Duration as Professional in Years (female)	9.5	6.0
Total No. of Injuries as Performer (female)	3.0	2.2

*Randolph Evans, Richard Evans, Scott Carvajal, and Susan Perry, "A Survey of Injuries among Broadway Performers," *American Journal of Public Health* 86, no.1 (January 1996): 77–80.

Assume that all of these demographics follow a normal distribution.
(a) Find the probability that a female dancer is 35 years old or older.
(b) Find the probability that a male dancer is 35 years old or older.
(c) Compare your answers to parts (a) and (b).
(d) Find the probability that a female performer is 1.4 meters tall or taller.
(e) Find the probability that a female performer has a career duration that is more than 1.5 standard deviations from the mean.
(f) Would a female who has more than 6 injuries during her career be considered a rare event? Explain.

40. Much of our work in Chapters 9 and 10 is interrelated. Note the similarities in the following parallel treatments of a frequency distribution and a probability distribution on the next page.

Frequency Distribution

Complete the table below for the following data. (Recall that x is the midpoint of the interval.)

14, 7, 1, 11, 2, 3, 11, 6, 10, 13, 11, 11, 16, 12, 9, 11, 9, 10, 7, 12, 9, 6, 4, 5, 9, 16, 12, 12, 11, 10, 14, 9, 13, 10, 15, 11, 11, 1, 12, 12, 6, 7, 8, 2, 9, 12, 10, 15, 9, 3

Interval	*Tally*	x	f	$x \cdot f$
1–3				
4–6				
7–9				
10–12				
13–15				
16–18				

Probability Distribution
A binomial distribution has $n = 10$ and $p = .5$. Complete the table to the right.

x	$P(x)$	$x \cdot P(x)$
0	.001	
1	.010	
2	.044	
3	.117	
4		
5		
6		
7		
8		
9		
10		

For each distribution, do the following for (a)–(c).

(a) Find the mean (or expected value).
(b) Find the standard deviation.
(c) Use Chebyshev's theorem to find the interval that contains 75% of the distribution. (See Section 10.2 exercises.)
(d) Use the normal approximation of the binomial probability distribution to find the interval that contains 95.44% of that distribution.
(e) Why can't we use the normal distribution to answer probability questions about the frequency distribution?

Case 9

Statistics in the Law—The *Castaneda* Decision

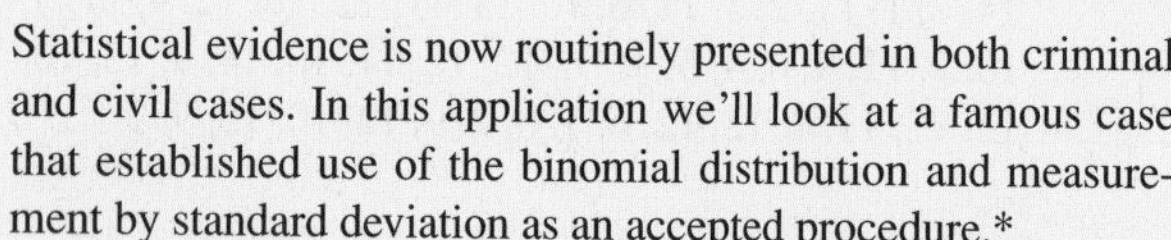

Statistical evidence is now routinely presented in both criminal and civil cases. In this application we'll look at a famous case that established use of the binomial distribution and measurement by standard deviation as an accepted procedure.*

Defendants who are convicted in criminal cases sometimes appeal their conviction on the grounds that the jury that indicted or convicted them was drawn from a pool of jurors that does not represent the population of the district in which they live. These appeals almost always cite the Supreme Court's decision in *Castaneda v. Partida* [430 U.S. 482], a case that dealt with the selection of grand juries in the state of Texas. The decision summarizes the facts this way:

> After respondent, a Mexican-American, had been convicted of a crime in a Texas District Court and had exhausted his state remedies on his claim of discrimination in the selection of the grand jury that had indicted him, he filed a habeas corpus petition in the Federal District Court, alleging a denial of due process and equal protection under the Fourteenth Amendment, because of gross underrepresentation of Mexican-Americans on the county grand juries.

The case went to the Appeals Court, which noted that "the county population was 79% Mexican-American, but, over an 11-year period, only 39% of those summoned for grand jury service were Mexican-American," and concluded that together with other testimony about the selection process, "the proof offered by respondent was sufficient to demonstrate a prima facie case of intentional discrimination in grand jury selection. . . ."

The state appealed to the Supreme Court, and the Supreme Court needed to decide whether the underrepresentation of Mexican-Americans on grand juries was indeed too ex-

*The Castaneda case and many other interesting applications of statistics in law are discussed in Michael O. Finkelstein and Bruce Levin, *Statistics for Lawyers*, New York, Springer-Verlag, 1990. U.S. Supreme Court decisions are online at http://www.findlaw.com/casecode/supreme.html, and most states now have important state court decisions online.

treme to be an effect of chance. To do so, they invoked the binomial distribution. Here is the argument:

> Given that 79.1% of the population is Mexican-American, the expected number of Mexican-Americans among the 870 persons summoned to serve as grand jurors over the 11-year period is approximately 688. The observed number is 339. Of course, in any given drawing some fluctuation from the expected number is predicted. The important point, however, is that the statistical model shows that the results of a random drawing are likely to fall in the vicinity of the expected value. . . .
>
> The measure of the predicted fluctuations from the expected value is the standard deviation, defined for the binomial distribution as the square root of the product of the total number in the sample (here 870) times the probability of selecting a Mexican-American (.791) times the probability of selecting a non-Mexican-American (.209). . . . Thus, in this case the standard deviation is approximately 12. As a general rule for such large samples, if the difference between the expected value and the observed number is greater than two or three standard deviations, then the hypothesis that the jury drawing was random would be suspect to a social scientist. The 11-year data here reflect a difference between the expected and observed number of Mexican-Americans of approximately 29 standard deviations. A detailed calculation reveals that the likelihood that such a substantial departure from the expected value would occur by chance is less than 1 in 10^{140}.

The Court decided that the statistical evidence supported the conclusion that jurors were not randomly selected, and that it was up to the state to show that its selection process did not discriminate against Mexican-Americans. The Court concluded:

> The proof offered by respondent was sufficient to demonstrate a prima facie case of discrimination in grand jury selection. Since the State failed to rebut the presumption of purposeful discrimination by competent testimony, despite two opportunities to do so, we affirm the Court of Appeals' holding of a denial of equal protection of the law in the grand jury selection process in respondent's case.

Exercises

1. Check the Court's calculation of 29 standard deviations as the difference between the expected number of Mexican-Americans and the number actually chosen.
2. Where do you think the Court's figure of 1 in 10^{140} came from?
3. The *Castaneda* decison also presents data from a $2\frac{1}{2}$-year period during which the state district judge supervised the selection process. During this period, 220 persons were called to serve as grand jurors, and only 100 of these were Mexican-American.
   (a) Considering the 220 jurors as a random selection from a large population, what is the expected number of Mexican-Americans, using the 79.1% population figure?
   (b) If we model the drawing of jurors as a sequence of 220 independent Bernoulli trials, what is the standard deviation of the number of Mexican-Americans?
   (c) About how many standard deviations is the actual number of Mexican-Americans drawn (100) from the expected number that you calculated in (a)?
   (d) What does the normal distribution table at the back of the book tell you about this result?
4. The following information is from an appeal brought by Hy-Vee stores before the Iowa Supreme Court, appealing a ruling by the Iowa Civil Rights Commission in favor of a female employee of one of their grocery stores.

   > In 1985, there were 112 managerial positions in the ten Hy-Vee stores located in Cedar Rapids. Only 6 of these managers were women. During that same year there were 294 employees; 206 were men and 88 were women.

   (a) How far from the expected number of women in management was the actual number, assuming that gender had nothing to do with promotion? Measure the difference in standard deviations.
   (b) Does this look like evidence of purposeful discrimination?

Chapter 11

Differential Calculus

The algebraic problems considered in earlier chapters dealt with *static* situations.

What is the revenue when x items are sold?
How much interest is earned in 2 years?
What is the equilibrium price?

Calculus, on the other hand, deals with *dynamic* situations.

At what rate is the economy growing?
How fast is a rocket going at any instant after liftoff?
How quickly can production be increased without adversely affecting profits?

The techniques of calculus will allow us to answer many questions like these that deal with rates of change.

The key idea underlying the development of calculus is the concept of limit. So we begin by studying limits.

11.1 LIMITS

We have often dealt with a problem like this: Find the value of the function $f(x)$ when $x = a$. The underlying idea of "limit," however, is to examine what the function does *near* $x = a$, rather than what it does *at* $x = a$.

EXAMPLE 1 The function

$$f(x) = \frac{2x^2 - 3x - 2}{x - 2}$$

is not defined when $x = 2$ (why?). What happens to the values of $f(x)$ when x is *very close* to 2?

Evaluate f at several numbers that are very close to $x = 2$, as in the following table.

x	1.99	1.999	2	2.0001	2.001
$f(x)$	4.98	4.998	—	5.0002	5.002

1 Use a calculator to estimate $\lim_{x\to 1} \frac{x^3 + x^2 - 2x}{x - 1}$ by completing the following table.

x	$f(x)$
.9	
.99	
.999	
1.0001	
1.001	
1.01	
1.1	

Answer:

2.61; 2.9601; 2.996; 3.0004; 3.004; 3.0401; 3.41; the limit appears to be 3.

The table suggests that

> as x gets closer and closer to 2, from either direction, the corresponding value of $f(x)$ gets closer and closer to 5.

In fact, by experimenting further, you can convince yourself that the values of $f(x)$ can be made *as close as you want* to 5 by taking values of x close enough to 2. This situation is usually described by saying "the *limit* of $f(x)$ as x approaches 2 is the number 5," which is written symbolically as

$$\lim_{x\to 2} f(x) = 5, \quad \text{or equivalently,} \quad \lim_{x\to 2} \frac{2x^2 - 3x - 2}{x - 2} = 5.$$ ◀ 1

The following informal definition of "limit" is similar to the situation in Example 1, but now f is any function, and a and L are fixed real numbers (in Example 1, $a = 2$ and $L = 5$).

Limit of a Function

Let f be a function and let a and L be real numbers. Assume that $f(x)$ is defined for all x near $x = a$. Suppose that

> as x takes values very close (but not equal) to a (on both sides of a), the corresponding values of $f(x)$ are very close (and possibly equal) to L;

and that

> the values of $f(x)$ can be made as close as you want to L for all values of x that are close enough to a.

Then the number L is the **limit** of the function $f(x)$ as x approaches a, which is written

$$\lim_{x\to a} f(x) = L.$$

This definition is *informal* because the expressions "near," "very close," and "as close as you want" have not been precisely defined. In particular, the tables used in Example 1 and the following examples provide strong intuitive evidence of what the limits must be, but not a proof.

EXAMPLE 2 If $f(x) = x^2 + x + 1$, what is $\lim_{x \to 3} f(x)$?

Make a table showing the values of the function at numbers very close to 3.

	x approaches 3 from the left →			3 ←	x approaches 3 from the right		
x	2.9	2.99	2.9999	3	3.0001	3.01	3.1
$f(x)$	12.31	12.9301	12.9993 . . .		13.0007 . . .	13.0701	13.71

The table suggests that as x approaches 3 from either direction, $f(x)$ gets closer and closer to 13 and, hence, that

$$\lim_{x \to 3} f(x) = 13, \quad \text{or equivalently,} \quad \lim_{x \to 3} (x^2 + x + 1) = 13.$$

Note that the function $f(x)$ is defined when $x = 3$ and $f(3) = 3^2 + 3 + 1 = 13$. So in this case, the limit of $f(x)$ as x approaches 3 is $f(3)$, the value of the function at 3.

EXAMPLE 3 Use a graphing calculator to find

$$\lim_{x \to 3} \frac{x - 3}{e^{x-3} - 1}.$$

There are two ways to estimate the limit.

Graphical Method Graph $f(x) = \dfrac{x - 3}{e^{x-3} - 1}$ in a very narrow window near $x = 3$. Use the trace feature to move along the graph and observe the y-coordinates as x gets very close to 3 from either side. Figure 11.1 suggests that $\lim_{x \to 3} f(x) = 1$.

Numerical Method Use the table feature to make a table of values for $f(x)$ when x is very close to 3. Figure 11.2 shows that when x is very close to 3, then $f(x)$ is very close to 1. (The table displays "error" at $x = 3$ because the function is not defined when $x = 3$.) Thus, it appears that

$$\lim_{x \to 3} \frac{x - 3}{e^{x-3} - 1} = 1.$$

The function has the limit 1 as x approaches 3, even though $f(3)$ is not defined.

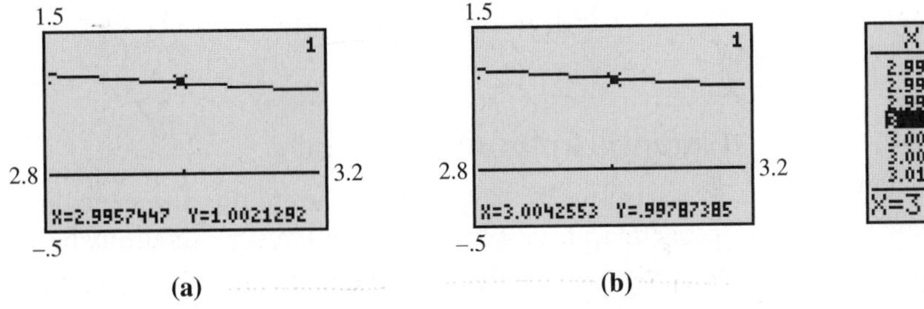

FIGURE 11.1

FIGURE 11.2

EXAMPLE 4 Find $\lim_{x \to 4} f(x)$, where f is the function whose rule is

$$f(x) = \begin{cases} 0 & \text{if } x \text{ is an integer} \\ 1 & \text{if } x \text{ is not an integer} \end{cases}$$

and whose graph is shown in Figure 11.3.

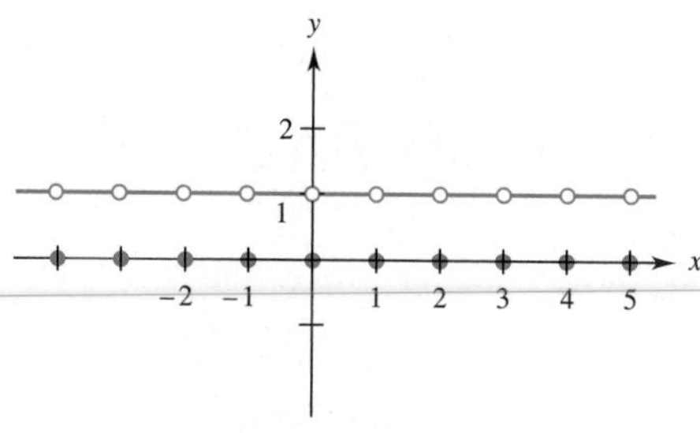

FIGURE 11.3

The definition of the limit as x approaches 4 involves only values of x that are close to, but not equal to 4—corresponding to the part of the graph on either side of 4, but not at 4 itself. Now $f(x) = 1$ for all these numbers (because the numbers very near 4, such as 3.99995 or 4.00002, are not integers). Thus for all x very close to 4, the corresponding value of $f(x)$ is 1, so $\lim_{x \to 4} f(x) = 1$. However, since 4 is an integer, $f(4) = 0$. Therefore, $\lim_{x \to 4} f(x) \neq f(4)$.

Examples 1–4 illustrate the following facts.

Limits and Function Values

If the limit of a function $f(x)$ as x approaches a exists, then there are three possibilities:

1. $f(a)$ is not defined, but $\lim f(x)$ is defined. (Examples 1 and 3)
2. $f(a)$ is defined and $\lim f(x) = f(a)$. (Example 2)
3. $f(a)$ is defined, but $\lim f(x) \neq f(a)$. (Example 4)

Finding Limits Algebraically

As we have seen, tables are very useful for estimating limits. However, it is often more efficient and accurate to find limits algebraically by using the following properties of limits.

Properties of Limits

Let a, k, A, and B be real numbers and let f and g be functions such that

$$\lim_{x \to a} f(x) = A \text{ and } \lim_{x \to a} g(x) = B.$$

1. $\lim\limits_{x \to a} k = k$ (for any constant k)

   (The limit of a constant is the constant.)

2. $\lim\limits_{x \to a} x = a$ (for any real number a)

3. $\lim\limits_{x \to a} [f(x) \pm g(x)] = A \pm B = \lim\limits_{x \to a} f(x) \pm \lim\limits_{x \to a} g(x)$

   (The limit of a sum or difference is the sum or difference of the limits.)

4. $\lim\limits_{x \to a} [f(x) \cdot g(x)] = A \cdot B = \lim\limits_{x \to a} f(x) \cdot \lim\limits_{x \to a} g(x)$

   (The limit of a product is the product of the limits.)

5. $\lim\limits_{x \to a} \dfrac{f(x)}{g(x)} = \dfrac{A}{B} = \dfrac{\lim\limits_{x \to a} f(x)}{\lim\limits_{x \to a} g(x)}$ $(B \neq 0)$

   (The limit of a quotient is the quotient of the limits, provided the limit of the denominator is nonzero.)

6. For any real number r for which A^r exists,

$$\lim_{x \to a} [f(x)]^r = A^r = [\lim_{x \to a} f(x)]^r.$$

Although we won't prove these properties (a rigorous definition of limit is needed for that), you should find most of them very plausible. For instance, if the values of $f(x)$ get very close to A and the values of $g(x)$ get very close to B when x approaches a, it is reasonable to expect that the corresponding values of $f(x) + g(x)$ will get very close to $A + B$ (Property 3) and that the corresponding values of $f(x)g(x)$ will get very close to AB (Property 4).

EXAMPLE 5 Find $\lim\limits_{x \to 2} (3x^2 + 5x - 1)$.

$\lim\limits_{x \to 2} (3x^2 + 5x - 1)$

$= \lim\limits_{x \to 2} 3x^2 + \lim\limits_{x \to 2} 5x + \lim\limits_{x \to 2} (-1)$ — Property 3

$= \lim\limits_{x \to 2} 3 \cdot \lim\limits_{x \to 2} x^2 + \lim\limits_{x \to 2} 5 \cdot \lim\limits_{x \to 2} x + \lim\limits_{x \to 2} (-1)$ — Property 4

$= \lim\limits_{x \to 2} 3 \cdot [\lim\limits_{x \to 2} x]^2 + \lim\limits_{x \to 2} 5 \cdot \lim\limits_{x \to 2} x + \lim\limits_{x \to 2} (-1)$ — Property 6

$= 3 \cdot 2^2 + 5 \cdot 2 + (-1) = 21$ — Properties 1 and 2

Example 5 shows that $\lim_{x\to 2} f(x) = 21$, where $f(x) = 3x^2 + 5x - 1$. Note that $f(2) = 3 \cdot 2^2 + 5 \cdot 2 - 1 = 21$. In other words, the limit as x approaches 2 is the value of the function at 2, that is

$$\lim_{x\to 2} f(x) = f(2).$$

The same analysis used in Example 5 works with any polynomial function and leads to the following conclusion.

2

If $f(x) = 2x^4 - 4x^3 + 3x$, find

(a) $\lim_{x\to 2} f(x)$;

(b) $\lim_{x\to -1} f(x)$.

Answers:

(a) 6

(b) 3

Polynomial Limits

If $f(x)$ is a polynomial function and a is a real number, then

$$\lim_{x\to a} f(x) = f(a).$$

This property will be used frequently. (2)

EXAMPLE 6 Find each limit.

(a) $\lim_{x\to 2} [(x^2 + 1) + (x^3 - x + 3)]$

$$\begin{aligned}\lim_{x\to 2} &[(x^2 + 1) + (x^3 - x + 3)] \\ &= \lim_{x\to 2} (x^2 + 1) + \lim_{x\to 2} (x^3 - x + 3) && \text{Property 3} \\ &= (2^2 + 1) + (2^3 - 2 + 3) = 5 + 9 = 14 && \text{Polynomial limit}\end{aligned}$$

(b) $\lim_{x\to -1} (x^3 + 4x)(2x^2 - 3x)$

$$\begin{aligned}\lim_{x\to -1} &(x^3 + 4x)(2x^2 - 3x) \\ &= \lim_{x\to -1} (x^3 + 4x) \cdot \lim_{x\to -1} (2x^2 - 3x) && \text{Property 4} \\ &= [(-1)^3 + 4(-1)] \cdot [2(-1)^2 - 3(-1)] && \text{Polynomial limit} \\ &= (-1 - 4)(2 + 3) = -25\end{aligned}$$

(c) $\lim_{x\to -1} 5(3x^2 + 2)$

$$\begin{aligned}\lim_{x\to -1} 5(3x^2 + 2) &= \lim_{x\to -1} 5 \cdot \lim_{x\to -1} (3x^2 + 2) && \text{Property 4} \\ &= 5[3(-1)^2 + 2] && \text{Property 1 and polynomial limit} \\ &= 25\end{aligned}$$

(d) $\lim_{x\to 4} \dfrac{x}{x + 2}$

$$\begin{aligned}\lim_{x\to 4} \frac{x}{x + 2} &= \frac{\lim_{x\to 4} x}{\lim_{x\to 4} (x + 2)} && \text{Property 5} \\ &= \frac{4}{4 + 2} = \frac{2}{3} && \text{Polynomial limit}\end{aligned}$$

3

Use the limit properties to find the following.

(a) $\lim_{x\to 4} (3x - 9)$

(b) $\lim_{x\to -1} (2x^2 - 4x + 1)$

(c) $\lim_{x\to 2} \frac{x - 1}{3x + 2}$

(d) $\lim_{x\to 2} \sqrt{3x + 3}$

Answers:

(a) 3

(b) 7

(c) 1/8

(d) 3

(e) $\lim_{x\to 9} \sqrt{4x - 11}$

Begin by writing the square root in exponential form.

$$\begin{aligned}\lim_{x\to 9} \sqrt{4x - 11} &= \lim_{x\to 9} [4x - 11]^{1/2} \\ &= [\lim_{x\to 9} (4x - 11)]^{1/2} && \text{Property 6} \\ &= [4 \cdot 9 - 11]^{1/2} && \text{Polynomial limit} \\ &= [25]^{1/2} = \sqrt{25} = 5.\end{aligned}$$

3

The definition of the limit as x approaches a involves only the values of the function when x is *near a,* but not the value of the function *at a.* So two functions that agree for all values of x, except possibly at $x = a$, will necessarily have the same limit when x approaches a. Thus, we have this fact:

Limit Theorem

If f and g are functions that have limits as x approaches a, and $f(x) = g(x)$ for all x near a, then

$$\lim_{x\to a} f(x) = \lim_{x\to a} g(x).$$

EXAMPLE 7 Find $\lim_{x\to 2} \frac{x^2 + x - 6}{x - 2}$.

Property 5 cannot be used here, because

$$\lim_{x\to 2} (x - 2) = 0.$$

We can, however, simplify the function by rewriting the fraction as

$$\frac{x^2 + x - 6}{x - 2} = \frac{(x + 3)(x - 2)}{x - 2}.$$

When $x \neq 2$, the quantity $x - 2$ is nonzero and may be cancelled, so that

$$\frac{x^2 + x - 6}{x - 2} = x + 3 \quad \text{for all } x \neq 2.$$

Now the Limit Theorem can be used.

$$\lim_{x\to 2} \frac{x^2 + x - 6}{x - 2} = \lim_{x\to 2} (x + 3) = 2 + 3 = 5$$

4

4

Find $\lim_{x\to 1} \frac{2x^2 + x - 3}{x - 1}$.

Answer:

5

EXAMPLE 8 Find $\lim_{x\to 4} \frac{\sqrt{x} - 2}{x - 4}$.

As $x \to 4$, the numerator approaches 0 and the denominator also approaches 0, giving the meaningless expression 0/0. To change the form of the expression,

Find the following.

(a) $\lim\limits_{x\to1} \dfrac{\sqrt{x}-1}{x-1}$

(b) $\lim\limits_{x\to9} \dfrac{\sqrt{x}-3}{x-9}$

Answers:

(a) 1/2

(b) 1/6

algebra can be used to rationalize the numerator by multiplying both the numerator and the denominator by $\sqrt{x}+2$. This gives

$$\frac{\sqrt{x}-2}{x-4} = \frac{\sqrt{x}-2}{x-4}\cdot\frac{\sqrt{x}+2}{\sqrt{x}+2} = \frac{\sqrt{x}\cdot\sqrt{x}+2\sqrt{x}-2\sqrt{x}-4}{(x-4)(\sqrt{x}+2)}$$

$$= \frac{x-4}{(x-4)(\sqrt{x}+2)} = \frac{1}{\sqrt{x}+2} \quad \text{for all } x \neq 4.$$

Now use the Limit Theorem and the properties of limits.

$$\lim_{x\to4}\frac{\sqrt{x}-2}{x-4} = \lim_{x\to4}\frac{1}{\sqrt{x}+2} = \frac{1}{\sqrt{4}+2} = \frac{1}{2+2} = \frac{1}{4}$$ ◀ 5

Existence of Limits

It is possible that $\lim\limits_{x\to a} f(x)$ may not exist, that is, there may be no number L satisfying the definition of $\lim\limits_{x\to a} f(x) = L$. This can happen in many ways, two of which are illustrated here.

6

Let $f(x) = \dfrac{x^2+9}{x-3}$. Find the following.

(a) $\lim\limits_{x\to3} f(x)$

(b) $\lim\limits_{x\to0} f(x)$

Answers:

(a) Does not exist

(b) -3

EXAMPLE 9 Let $g(x) = \dfrac{x^2+4}{x-2}$ and find $\lim\limits_{x\to2} g(x)$.

Property 5 cannot be used since $\lim\limits_{x\to2}(x-2) = 0$. Furthermore, there is no way to simplify $g(x)$ algebraically (because x^2+4 doesn't factor). Hence, the Limit Theorem cannot be used. So try to estimate the limit. Make a table of values, such as the one below, and graph the function. Remember that negative numbers far from 0 (such as -1000 or -5000) are very small numbers (even though their absolute values may be large).

	x approaches 2 from the left →				2	← x approaches 2 from the right		
x	1.8	1.9	1.99	1.999	2	2.001	2.01	2.05
$g(x)$	−36.2	−76.1	−796	−7996		8004	804	164
	$g(x)$ gets smaller and smaller					$g(x)$ gets larger and larger		

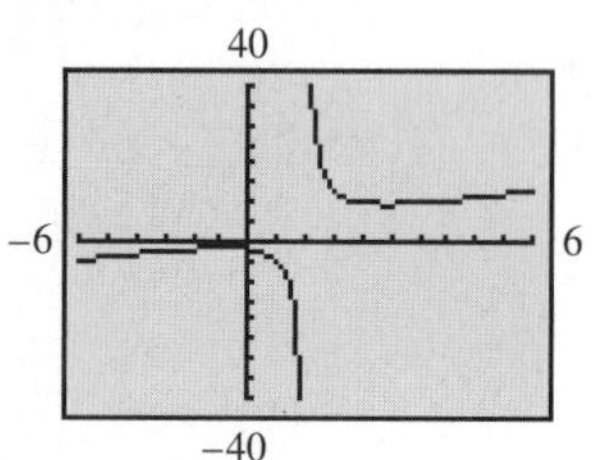

FIGURE 11.4

The table above and the graph of $g(x)$ in Figure 11.4 show that as x approaches 2 from the left, $g(x)$ gets smaller and smaller, but as x approaches 2 from the right, $g(x)$ gets larger and larger. Since $g(x)$ does not get closer and closer to a single real number as x approaches 2 from either side,

$$\lim_{x\to2}\frac{x^2+4}{x-2} \text{ does not exist.}$$ ◀ 6

EXAMPLE 10 What is $\lim_{x \to 0} \frac{|x|}{x}$?

The function $f(x) = \frac{|x|}{x}$ is not defined when $x = 0$. Recall the definition of absolute value:

$$|x| = \begin{cases} x & \text{if } x \geq 0 \\ -x & \text{if } x < 0 \end{cases}.$$

Consequently, when $x > 0$,

$$f(x) = \frac{|x|}{x} = \frac{x}{x} = 1$$

and when $x < 0$,

$$f(x) = \frac{|x|}{x} = \frac{-x}{x} = -1.$$

The graph of f is shown in Figure 11.5. As x approaches 0 from the right, x is always positive and the corresponding value of $f(x)$ is 1. But as x approaches 0 from the left, x is always negative and the corresponding value of $f(x)$ is -1. Thus, as x approaches 0 from *both* sides, the corresponding values of $f(x)$ do not get closer and closer to a *single* real number. Therefore, the limit does not exist.*

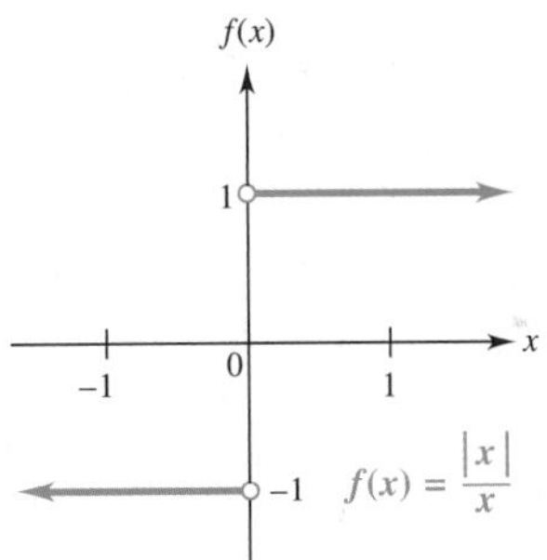

FIGURE 11.5

Examples 9 and 10 illustrate the following facts.

Existence of Limits

The limit of a function f as x approaches a may fail to exist if

1. $f(x)$ becomes infinitely large in absolute value as x approaches a from either side (Example 9); or
2. $f(x)$ gets closer and closer to a number L as x approaches a from the left, but $f(x)$ gets closer and closer to a different number M as x approaches a from the right (Example 10).

---

*In a situation like this, one sometimes says that -1 is the *limit of $f(x)$ from the left* and that 1 is the *limit of $f(x)$ from the right*. The concept of limit as we have defined it is sometimes called a *two-sided limit*.

The function f whose graph is shown in Figure 11.6 illustrates various facts about limits that were discussed in this section.

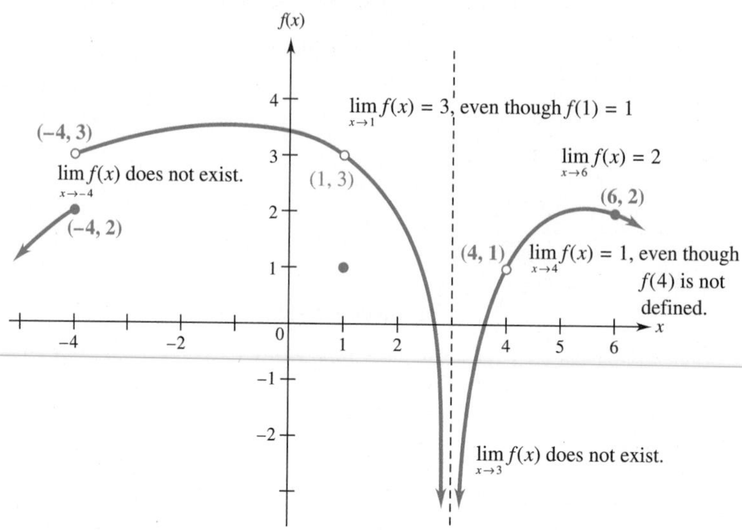

FIGURE 11.6

11.1 Exercises

In each of the following, use the graph to determine the value of the indicated limits. (See Examples 3, 4, 9, and 10 and Figure 11.6.)

1. **(a)** $\lim_{x\to 3} f(x)$ **(b)** $\lim_{x\to -1.5} f(x)$

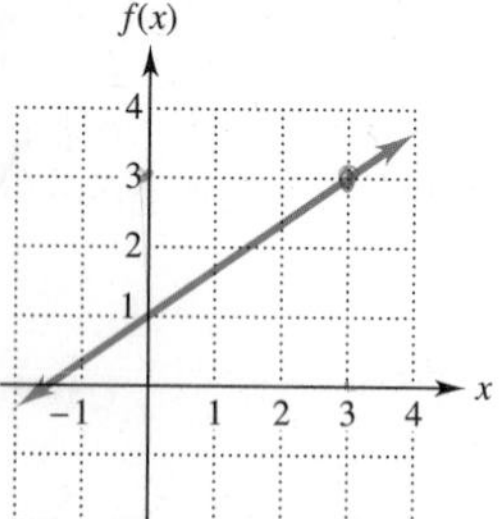

2. **(a)** $\lim_{x\to 2} F(x)$ **(b)** $\lim_{x\to -1} F(x)$

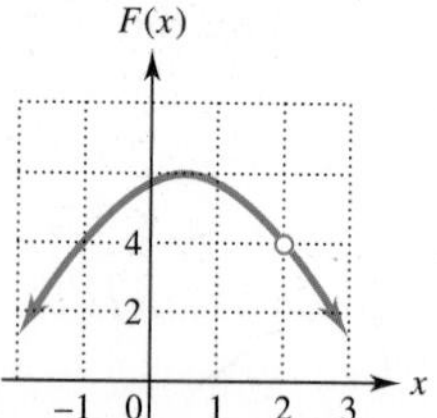

3. **(a)** $\lim_{x\to -2} f(x)$ **(b)** $\lim_{x\to 1} f(x)$

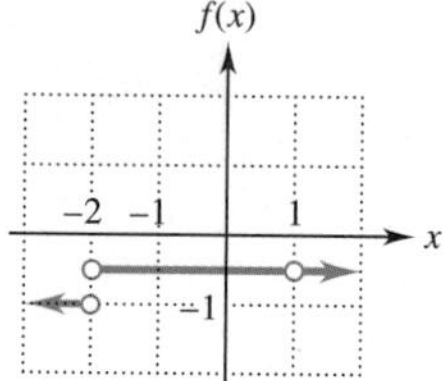

4. **(a)** $\lim_{x\to -1} g(x)$ **(b)** $\lim_{x\to 3} g(x)$

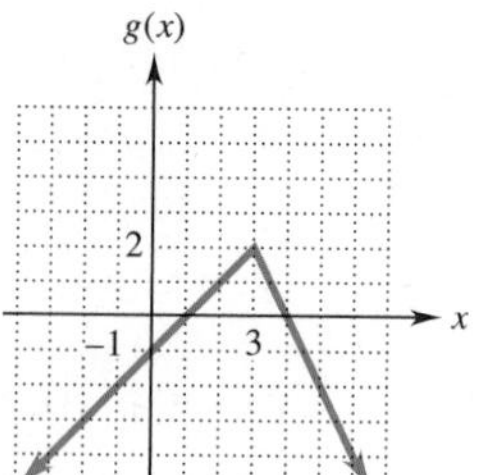

5. **(a)** $\lim_{x \to 0} f(x)$ **(b)** $\lim_{x \to -1} f(x)$

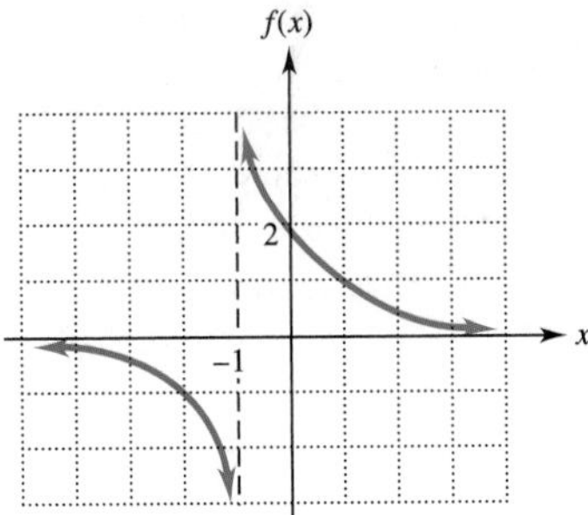

6. **(a)** $\lim_{x \to 1} h(x)$ **(b)** $\lim_{x \to 2} h(x)$

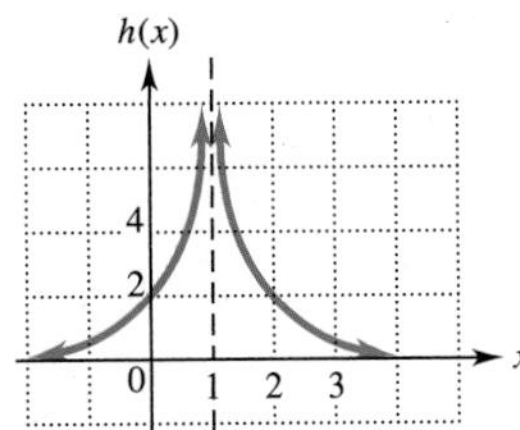

7. **(a)** $\lim_{x \to 1} g(x)$ **(b)** $\lim_{x \to -1} g(x)$

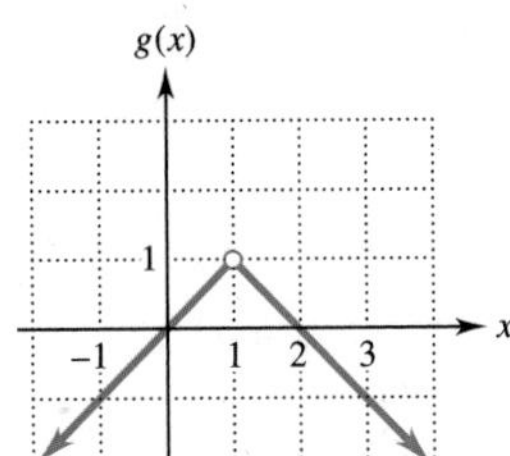

8. **(a)** $\lim_{x \to 3} f(x)$ **(b)** $\lim_{x \to 0} f(x)$

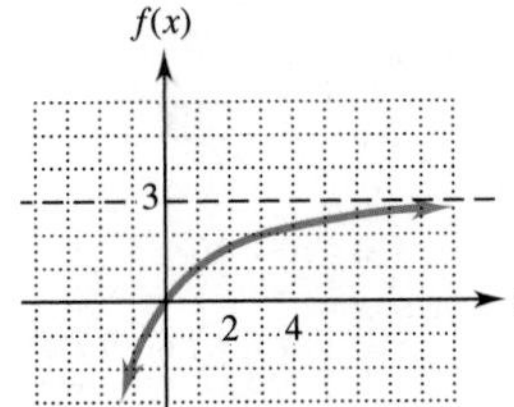

9. Explain why $\lim_{x \to 2} F(x)$ in Exercise 2(a) exists, but $\lim_{x \to -2} f(x)$ in Exercise 3(a) does not.

10. In Exercise 7(a), why does $\lim_{x \to 1} g(x)$ exist even though $g(1)$ is not defined?

Use a calculator to estimate the limit. (See Examples 1–3.)

11. $\lim_{x \to 1} \dfrac{\ln x}{x - 1}$

12. $\lim_{x \to 3} \dfrac{\ln x - \ln 3}{x - 3}$

13. $\lim_{x \to 0} \dfrac{e^{2x} - 1}{x}$

14. $\lim_{x \to 0} \dfrac{x}{\ln |x|}$

15. $\lim_{x \to 0} (x \ln |x|)$

16. $\lim_{x \to 0} \dfrac{x}{e^x - 1}$

17. $\lim_{x \to 3} \dfrac{x^3 - 3x^2 - x + 3}{x - 3}$

18. $\lim_{x \to 4} \dfrac{.1x^4 - .8x^3 + 1.6x^2 + 2x - 8}{x - 4}$

19. $\lim_{x \to -2} \dfrac{x^4 + 2x^3 - x^2 + 3x + 1}{x + 2}$

20. $\lim_{x \to 0} \dfrac{e^{2x} + e^x - 2}{e^x - 1}$

Suppose $\lim_{x \to 4} f(x) = 16$ and $\lim_{x \to 4} g(x) = 8$. Use the limit properties to find the following limits.

21. $\lim_{x \to 4} [f(x) - g(x)]$

22. $\lim_{x \to 4} [g(x) \cdot f(x)]$

23. $\lim_{x \to 4} \dfrac{f(x)}{g(x)}$

24. $\lim_{x \to 4} [3 \cdot f(x)]$

25. $\lim_{x \to 4} \sqrt{f(x)}$

26. $\lim_{x \to 4} [g(x)]^3$

27. $\lim_{x \to 4} \dfrac{f(x) + g(x)}{2g(x)}$

28. $\lim_{x \to 4} \dfrac{5g(x) + 2}{1 - f(x)}$

29. **(a)** Graph the function f whose rule is

$$f(x) = \begin{cases} 3 - x & \text{if } x < -2 \\ x + 2 & \text{if } -2 \le x < 2. \\ 1 & \text{if } x \ge 2 \end{cases}$$

Use the graph in part (a) to find these limits.

(b) $\lim_{x \to -2} f(x)$ **(c)** $\lim_{x \to 1} f(x)$ **(d)** $\lim_{x \to 2} f(x)$

30. **(a)** Graph the function g whose rule is

$$g(x) = \begin{cases} x^2 & \text{if } x < -1 \\ x + 2 & \text{if } -1 \le x < 1. \\ 3 - x & \text{if } x \ge 1 \end{cases}$$

Use the graph in part (a) to find these limits.

(b) $\lim_{x \to -1} g(x)$ **(c)** $\lim_{x \to 0} g(x)$ **(d)** $\lim_{x \to 1} g(x)$

Use algebra and the properties of limits as needed to find the following limits. If the limit does not exist, say so. (See Examples 5–10.)

31. $\lim_{x \to 2} (2x^3 + 5x^2 + 2x + 1)$

32. $\lim_{x \to -1} (4x^3 - x^2 + 3x - 1)$

33. $\lim_{x \to 3} \dfrac{5x - 6}{2x + 1}$

34. $\lim_{x \to -2} \dfrac{2x + 1}{3x - 4}$

35. $\lim_{x \to 3} \dfrac{x^2 - 9}{x - 3}$

36. $\lim_{x \to -2} \dfrac{x^2 - 4}{x + 2}$

37. $\lim_{x \to -2} \dfrac{x^2 - x - 6}{x + 2}$

38. $\lim_{x \to 5} \dfrac{x^2 - 3x - 10}{x - 5}$

39. $\lim_{x \to 2} \dfrac{x^2 - 5x + 6}{x^2 - 6x + 8}$

40. $\lim_{x \to -2} \dfrac{x^2 + 3x + 2}{x^2 - x - 6}$

41. $\lim_{x\to 4} \frac{(x+4)^2(x-5)}{(x-4)(x+4)^2}$

42. $\lim_{x\to -3} \frac{(x+3)(x-3)(x+4)}{(x+8)(x+3)(x-4)}$

43. $\lim_{x\to 3} \sqrt{x^2-4}$

44. $\lim_{x\to 3} \sqrt{x^2-5}$

45. $\lim_{x\to 4} \frac{-6}{(x-4)^2}$

46. $\lim_{x\to -2} \frac{3x}{(x+2)^3}$

47. $\lim_{x\to 0} \frac{[1/(x+3)] - 1/3}{x}$

48. $\lim_{x\to 0} \frac{[-1/(x+2)] + 1/2}{x}$

49. $\lim_{x\to 25} \frac{\sqrt{x}-5}{x-25}$

50. $\lim_{x\to 36} \frac{\sqrt{x}-6}{x-36}$

51. $\lim_{x\to 5} \frac{\sqrt{x}-\sqrt{5}}{x-5}$

52. **(a)** Approximate $\lim_{x\to 0}(1+x)^{1/x}$ to five decimal places. (Evaluate the function at numbers closer and closer to 0 until successive approximations agree in the first five places.)
(b) Find the decimal expansion of the number e to as many places as your calculator can manage.
(c) What do parts (a) and (b) suggest about the exact value of $\lim_{x\to 0}(1+x)^{1/x}$?

53. **Business** A company training program has determined that a new employee can do an average of $P(s)$ pieces of work per day after s days of on-the-job training, where

$$P(s) = \frac{90s}{s+6}.$$

Find the following.
(a) $P(1)$ **(b)** $P(11)$ **(c)** $\lim_{s\to 11} P(s)$

54. **Natural Science** The concentration of a drug in a patient's bloodstream h hours after it was injected is given by

$$A(h) = \frac{.2h}{h^2+2}.$$

Find the following.
(a) $A(.5)$ **(b)** $A(1)$ **(c)** $\lim_{h\to 1} A(h)$

55. **Business** The cost of manufacturing a particular videotape is

$$c(x) = 20{,}000 + 5x,$$

where x is the number of tapes produced. The average cost per tape, denoted by $\overline{c}(x)$, is found by dividing $c(x)$ by x. Find the following.
(a) $\overline{c}(1000)$ **(b)** $\overline{c}(100{,}000)$
(c) $\lim_{x\to 10{,}000} \overline{c}(x)$

56. **Business** When the price of an essential commodity (such as gasoline) rises rapidly, consumption drops slowly at first. If the price continues to rise, however, a "tipping" point may be reached, at which consumption takes a sudden, substantial drop. Suppose the accompanying graph shows the consumption of gasoline, $G(t)$, in millions of gallons, in a certain area. We assume that the price is rising rapidly. Here t is time in months after the price began rising. Use the graph to find the following.
(a) $\lim_{t\to 12} G(t)$ **(b)** $\lim_{t\to 16} G(t)$ **(c)** $G(16)$
(d) The tipping point (in months)

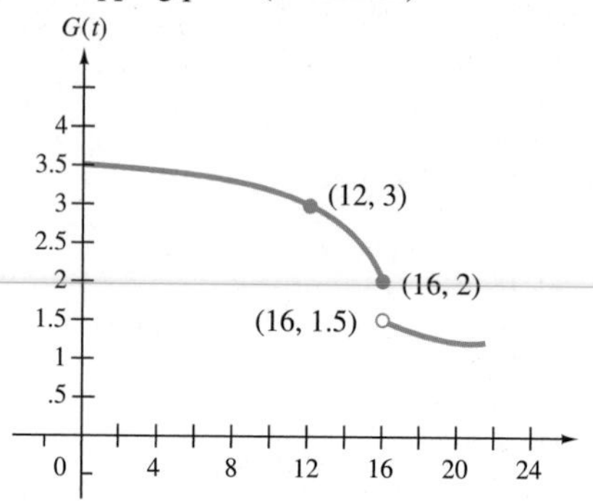

57. **Social Sciences** The figure shows the graphs of the functions C and I whose rules are

$C(x)$ = population of China (in billions) in year x,
$I(x)$ = population of India (in billions) in year x.*

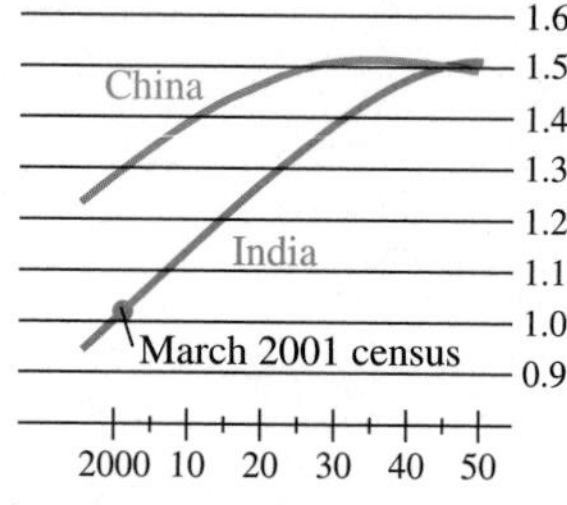

Sources: Indian Census; UN

Find each of these limits.
(a) $\lim_{x\to 2030} C(x)$ **(b)** $\lim_{x\to 2015} I(x)$
(c) $\lim_{x\to 2045} C(x) - I(x)$ **(d)** $\lim_{x\to 2045} C(x) + I(x)$

58. **Business** The graph on the next page shows the profit from the daily production of x thousand kilograms of an industrial chemical. Use the graph to find the following limits.
(a) $\lim_{x\to 6} P(x)$ **(b)** $\lim_{x\to 10} P(x)$ **(c)** $\lim_{x\to 15} P(x)$

*Graph appeared in the March 31, 2001, edition of the *Economist* magazine.

(d) Use the graph to estimate the number of units of the chemical that must be produced before the second shift is beneficial.

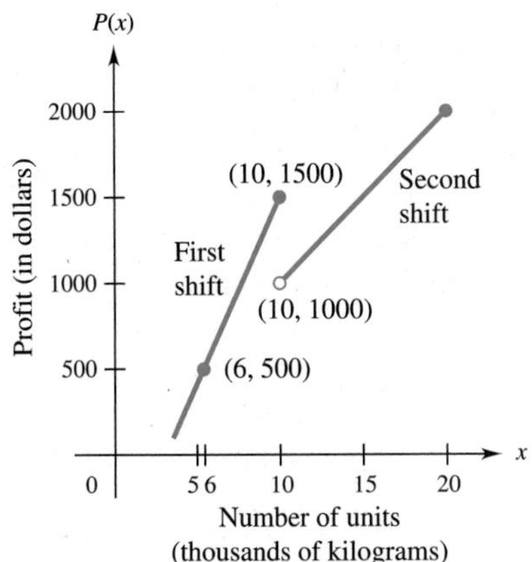

59. Business A long-distance phone service advertises that "all calls up to 20 minutes are 99 cents, and only 7 cents per minute [or fraction thereof] after that." So a 20-minute call costs .99, a 20.1-minute call costs \$1.06, a 21-minute call costs \$1.06, and so on. Let $P(x)$ be the price of a phone call lasting x minutes.
(a) Write the rule of the function P for $0 \le x \le 23$.
(b) Graph $P(x)$ for $0 \le x \le 23$.
Find the following limits.
(c) $\lim_{x \to 10} P(x)$ **(d)** $\lim_{x \to 20} P(x)$ **(e)** $\lim_{x \to 22.5} P(x)$

11.2 RATES OF CHANGE

One of the main applications of calculus is determining how one variable changes in relation to another. A person in business wants to know how profit changes with respect to changes in advertising, while a person in medicine wants to know how a patient's reaction to a drug changes with respect to changes in the dose.

We begin the discussion with a familiar situation. A driver makes the 168-mile trip from Cleveland to Columbus, Ohio, in 3 hours. The following table shows how far the driver has traveled from Cleveland at various times.

Time (in hours)	0	.5	1	1.5	2	2.5	3
Distance (in miles)	0	22	52	86	118	148	168

If f is the function whose rule is

$$f(x) = \text{distance from Cleveland at time } x,$$

then the table shows, for example, that $f(2) = 118$ and $f(3) = 168$. So the distance traveled from time $x = 2$ to $x = 3$ is $168 - 118$, that is, $f(3) - f(2)$. In a similar fashion, we obtain the other entries in the following chart.

Time Interval	*Distance Traveled*
$x = 2$ to $x = 3$	$f(3) - f(2) = 168 - 118 = 50$
$x = 1$ to $x = 3$	$f(3) - f(1) = 168 - 52 = 116$
$x = 0$ to $x = 2.5$	$f(2.5) - f(0) = 148 - 0 = 148$
$x = .5$ to $x = 1$	$f(1) - f(.5) = 52 - 22 = 30$
$x = a$ to $x = b$	$f(b) - f(a)$

The last line of the chart shows how to find the distance traveled in any time interval ($0 \le a < b \le 3$).

Since distance = average speed × time,

$$\text{Average speed} = \frac{\text{distance traveled}}{\text{time interval}}.$$

In the chart above, you can compute the length of each time interval by taking the difference between the two times. Thus, for example, from $x = 1$ to $x = 3$ is a time interval of length $3 - 1 = 2$ hours and, hence, the average speed over this interval is $116/2 = 58$ mph. Similarly, we have the following information.

Time Interval	*Average Speed* = $\dfrac{\textit{Distance Traveled}}{\textit{Time Interval}}$
$x = 2$ to $x = 3$	$\dfrac{f(3) - f(2)}{3 - 2} = \dfrac{168 - 118}{3 - 2} = \dfrac{50}{1} = 50$ mph
$x = 1$ to $x = 3$	$\dfrac{f(3) - f(1)}{3 - 1} = \dfrac{168 - 52}{3 - 1} = \dfrac{116}{2} = 58$ mph
$x = 0$ to $x = 2.5$	$\dfrac{f(2.5) - f(0)}{2.5 - 0} = \dfrac{148 - 0}{2.5 - 0} = \dfrac{148}{2.5} = 59.2$ mph
$x = .5$ to $x = 1$	$\dfrac{f(1) - f(.5)}{1 - .5} = \dfrac{52 - 22}{1 - .5} = \dfrac{30}{.5} = 60$ mph
$x = a$ to $x = b$	$\dfrac{f(b) - f(a)}{b - a}$ mph

The last line of the chart shows how to compute the average speed over any time interval ($0 \le a < b \le 3$). (1)

1

Find the average speed

(a) from $t = 1.5$ to $t = 2$;

(b) from $t = s$ to $t = r$.

Answers:

(a) 64 mph

(b) $\dfrac{f(r) - f(s)}{r - s}$ mph

Now speed (miles per hour) is simply the *rate of change* of distance with respect to time and what was done for the distance function f in the preceding discussion can be done with any function.

Quantity	*Meaning for the Distance Function*	*Meaning for an Arbitrary Function f*
$b - a$	Time interval = change in time from $x = a$ to $x = b$	Change in x from $x = a$ to $x = b$
$f(b) - f(a)$	Distance traveled = corresponding change in distance as time changes from a to b	Corresponding change in $f(x)$ as x changes from a to b
$\dfrac{f(b) - f(a)}{b - a}$	Average speed = average rate of change of distance with respect to time as time changes from a to b	**Average rate of change** of $f(x)$ with respect to x as x changes from a to b (where $a < b$)

Find the average rate of change of $f(x)$ in Example 1 when x changes from

(a) 0 to 4;

(b) 2 to 7.

Answers:

(a) 8

(b) 13

EXAMPLE 1 If $f(x) = x^2 + 4x + 5$, find the average rate of change of $f(x)$ with respect to x as x changes from -2 to 3.

This is the situation described in the last line of the chart above, with $a = -2$ and $b = 3$. The average rate of change is

$$\frac{f(3) - f(-2)}{3 - (-2)} = \frac{26 - 1}{5} = \frac{25}{5} = 5.$$

2

EXAMPLE 2 Suppose $P(t)$ denotes the production cost (in millions of dollars) for a high-budget movie in the United States in year t. Figure 11.7 shows the graph of P.*

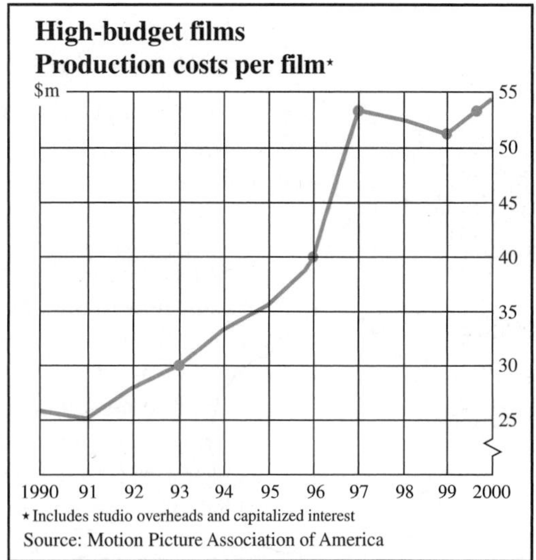

FIGURE 11.7

Approximate the average rate of change of price with respect to time over the following intervals.

(a) From the beginning of 1993 to the beginning of 1996.

The graph shows that costs were 30 million at the start of 1993 and 40 million at the start of 1996, that is, $P(1993) = 30$ and $P(1996) = 40$. Therefore the average rate of change over this period is

$$\frac{P(1996) - P(1993)}{1996 - 1993} = \frac{40 - 30}{3} = \frac{10}{3} = 3\frac{1}{3}.$$

Thus, production costs were increasing during this period at the rate of $3\frac{1}{3}$ million dollars per year.

*From the *Economist,* March 31, 2001.

Use Figure 11.7 to find the average rate of change of cost with respect to time

(a) From the beginning of 1993 to the middle of 1996.

(b) From the beginning of 1990 to the beginning of 1991.

Answers:

(a) Increasing at the rate of approximately 4.286 million dollars per year

(b) Graph is difficult to read accurately. Costs are decreasing at a rate between .5 and 1 million dollars per year.

(b) From the beginning of 1997 to the beginning of 1999.

The graph shows that $P(1997) \approx 53.5$ and $P(1999) \approx 51.5$, so that the average rate of change is

$$\frac{P(1999) - P(1997)}{1999 - 1997} = \frac{51.5 - 53.5}{2} = \frac{-2}{2} = -1.$$

The negative number means that costs were *decreasing* at the rate of 1 million dollars per year during this period.

(c) From the beginning of 1997 to the middle of 1999.

Mid-1999 corresponds to 1999.5 and the graph shows that $P(1999.5) \approx 53.5$. Hence, the average rate of change is

$$\frac{P(1999.5) - P(1997)}{1999.5 - 1997} = \frac{53.5 - 53.5}{2.5} = \frac{0}{2.5} = 0.$$

The average rate of change is 0, which means that, on the average during this period, costs did not change. ◀ 3

Instantaneous Rate of Change

Suppose a car is stopped at a traffic light. When the light turns green, the car begins to move along a straight road. Assume that the distance traveled by the car is given by the function

$$s(t) = 2t^2 \qquad (0 \le t \le 30),$$

where time t is measured in seconds and the distance $s(t)$ at time t is measured in feet. We know how to find the *average* speed of the car over any time interval, so we now turn to a different problem: determining the *exact* speed of the car at a particular instant, say $t = 10$.*

The intuitive idea is that the exact speed at $t = 10$ is very close to the average speed over a very short time interval near $t = 10$. If we take shorter and shorter time intervals near $t = 10$, the average speeds over these intervals should get closer and closer to the exact speed at $t = 10$. In other words,

> the exact speed at $t = 10$ is the limit of the average speeds over shorter and shorter time intervals near $t = 10$.

The following chart illustrates this idea.

Interval	*Average Speed*
$t = 10$ to $t = 10.1$	$\frac{s(10.1) - s(10)}{10.1 - 10} = \frac{204.02 - 200}{.1} = 40.2$
$t = 10$ to $t = 10.01$	$\frac{s(10.01) - s(10)}{10.01 - 10} = \frac{200.4002 - 200}{.01} = 40.02$
$t = 10$ to $t = 10.001$	$\frac{s(10.001) - s(10)}{10.001 - 10} = \frac{200.040002 - 200}{.001} = 40.002$

The chart suggests that the exact speed at $t = 10$ is 40 ft/sec. We can confirm this intuition by computing the average speed from $t = 10$ to $t = 10 + h$, where h is any

*As distance is measured in feet and time in seconds here, speed is measured in feet per second. It may help to know that 15 mph is equivalent to 22 ft/sec and 60 mph to 88 ft/sec.

very small nonzero number. (The chart does this for $h = .1$, $h = .01$, and $h = .001$.) The average speed from $t = 10$ to $t = 10 + h$ is

$$\begin{aligned}\frac{s(10+h)-s(10)}{(10+h)-10} &= \frac{s(10+h)-s(10)}{h}\\ &= \frac{2(10+h)^2 - 2\cdot 10^2}{h}\\ &= \frac{2(100+20h+h^2)-200}{h}\\ &= \frac{200+40h+2h^2-200}{h}\\ &= \frac{40h+2h^2}{h} = \frac{h(40+2h)}{h} \qquad (h \neq 0)\\ &= 40+2h.\end{aligned}$$

Saying that the time interval from 10 to $10 + h$ gets shorter and shorter is equivalent to saying h gets closer and closer to 0. Hence, the exact speed at $t = 10$ is the limit as h approaches 0 of the average speeds over the intervals from $t = 10$ to $t = 10 + h$; that is,

$$\begin{aligned}\lim_{h\to 0}\frac{s(10+h)-s(10)}{h} &= \lim_{h\to 0}(40+2h)\\ &= 40 \text{ ft/sec}.\end{aligned}$$

In the preceding example, the car moved in one direction along a straight line. Now suppose that an object is moving back and forth along a number line, with its position at time t given by the function $s(t)$. The **velocity** of the object is the rate of motion in the direction in which it is moving. **Speed** is the absolute value of velocity. For example, a velocity of -30 ft/sec indicates a speed of 30 ft/sec in the negative direction, while a velocity of 40 ft/sec indicates a speed of 40 ft/sec in the positive direction.

Hereafter, we deal with **average velocity** and **instantaneous velocity** instead of average speed and exact speed. When "velocity" is used alone, it means instantaneous velocity.

Let a be a fixed number. By replacing 10 by a in the discussion above, we see that the average velocity of the object from time $t = a$ to time $t = a + h$ is the quotient

$$\frac{s(a+h)-s(a)}{(a+h)-a} = \frac{s(a+h)-s(a)}{h}.$$

The instantaneous velocity at time a is the limit of this quotient as h approaches 0.

Velocity

If an object moves along a straight line, with position $s(t)$ at time t, then the **velocity** of the object at $t = a$ is

$$\lim_{h\to 0}\frac{s(a+h)-s(a)}{h},$$

provided this limit exists.

EXAMPLE 3 The distance in feet of an object from a starting point is given by $s(t) = 2t^2 - 5t + 40$, where t is time in seconds.

(a) Find the average velocity of the object from 2 seconds to 4 seconds.
The average velocity is

$$\frac{s(4) - s(2)}{4 - 2} = \frac{52 - 38}{2} = \frac{14}{2} = 7$$

feet per second.

(b) Find the instantaneous velocity at 4 seconds.
For $t = 4$, the instantaneous velocity is

$$\lim_{h \to 0} \frac{s(4 + h) - s(4)}{h}$$

feet per second. We have

$$\begin{aligned} s(4 + h) &= 2(4 + h)^2 - 5(4 + h) + 40 \\ &= 2(16 + 8h + h^2) - 20 - 5h + 40 \\ &= 32 + 16h + 2h^2 - 20 - 5h + 40 \\ &= 2h^2 + 11h + 52 \end{aligned}$$

and

$$s(4) = 2(4)^2 - 5(4) + 40 = 52.$$

Thus,

$$s(4 + h) - s(4) = (2h^2 + 11h + 52) - 52 = 2h^2 + 11h$$

and the instantaneous velocity at $t = 4$ is

$$\begin{aligned} \lim_{h \to 0} \frac{2h^2 + 11h}{h} &= \lim_{h \to 0} \frac{h(2h + 11)}{h} \\ &= \lim_{h \to 0} (2h + 11) = 11 \text{ ft/sec.} \end{aligned}$$

4

In Example 3, if $s(t) = s^2 + 3$, find

(a) the average velocity from 1 second to 5 seconds;

(b) the instantaneous velocity at 5 seconds.

Answers:

(a) 6 ft per second

(b) 10 ft per second

EXAMPLE 4 The velocity of blood cells is of interest to physicians; a slower velocity than normal might indicate a constriction, for example. Suppose the position of a red blood cell in a capillary is given by

$$s(t) = 1.2t + 5,$$

where $s(t)$ gives the position of a cell in millimeters from some reference point and t is time in seconds. Find the velocity of this cell at time $t = a$.

Evaluate the limit given above. To find $s(a + h)$, substitute $a + h$ for the variable t in $s(t) = 1.2t + 5$.

$$s(a + h) = 1.2(a + h) + 5$$

Now use the definition of velocity.

$$v(t) = \lim_{h\to0} \frac{s(a+h) - s(a)}{h}$$

$$= \lim_{h\to0} \frac{1.2(a+h) + 5 - (1.2a + 5)}{h}$$

$$= \lim_{h\to0} \frac{1.2a + 1.2h + 5 - 1.2a - 5}{h} = \lim_{h\to0} \frac{1.2h}{h} = 1.2$$

The velocity of the blood cell at $t = a$ is 1.2 millimeters per second, regardless of the value of a. In other words, the blood velocity is a constant 1.2 millimeters per second at any time. ◀ 5

5

Repeat Example 4 with $s(t) = .3t - 2$.

Answer:

The velocity is .3 millimeter per second.

The ideas underlying the concept of the velocity of a moving object can be extended to any function $f(x)$. In place of average velocity at time t, we have the average rate of change of $f(x)$ with respect to x as x changes from one value to another. Taking limits leads to this definition.

The **instantaneous rate of change** for a function f when $x = a$ is

$$\lim_{h\to0} \frac{f(a+h) - f(a)}{h},$$

provided this limit exists.

EXAMPLE 5 A company determines that the cost (in hundreds of dollars) of manufacturing x cases of computer mice is

$$C(x) = -.2x^2 + 8x + 40 \qquad (0 \le x \le 20).$$

(a) Find the average rate of change of cost per case for manufacturing between 5 and 10 cases.

Use the formula for average rate of change. The cost to manufacture 5 cases is

$$C(5) = -.2(5^2) + 8(5) + 40 = 75,$$

or \$7500. The cost to manufacture 10 cases is

$$C(10) = -.2(10^2) + 8(10) + 40 = 100,$$

or \$10,000. The average rate of change of cost is

$$\frac{C(10) - C(5)}{10 - 5} = \frac{100 - 75}{5} = 5.$$

Thus, on the average, cost is increasing at the rate of \$500 per case when production is increased from 5 to 10 cases.

(b) Find the instantaneous rate of change with respect to the number of cases produced when 5 cases are produced.

The instantaneous rate of change when $x = 5$ is given by

$$\lim_{h\to 0} \frac{C(5+h) - C(5)}{h}$$

$$= \lim_{h\to 0} \frac{[-.2(5+h)^2 + 8(5+h) + 40] - [-.2(5^2) + 8(5) + 40]}{h}$$

$$= \lim_{h\to 0} \frac{[-5 - 2h - .2h^2 + 40 + 8h + 40] - [75]}{h}$$

$$= \lim_{h\to 0} \frac{6h - .2h^2}{h} \qquad \text{Combine terms.}$$

$$= \lim_{h\to 0} (6 - .2h) \qquad \text{Divide by } h.$$

$$= 6. \qquad \text{Calculate the limit.}$$

When 5 cases are manufactured, the cost is increasing at the rate of \$600 per case. ◀

The rate of change of the cost function is called the **marginal cost.*** Similarly, **marginal revenue** and **marginal profit** are the rates of change of the revenue and profit functions, respectively. Part (b) of Example 5 shows that the marginal cost when 5 cases are manufactured is \$600.

▶ **EXAMPLE 6** Total revenue (in billions of dollars) in year x from the sales of organic fruits and vegetables in the United States is approximated by

$$f(x) = 1.9(1.14^x) \qquad (0 \le x \le 11),$$

where $x = 0$ corresponds to 1994.†

(a) Find the marginal revenue in 1998.

Since $x = 4$ corresponds to 1998, we must find the instantaneous rate of change of $f(x)$ at $x = 4$. The algebraic techniques used in preceding examples won't work with an exponential function, but the rate of change can be approximated by a graphing calculator or spreadsheet program. The average rate of change from 4 to $4 + h$ is

$$\frac{f(4+h) - f(4)}{h} = \frac{1.9(1.14^{4+h}) - 1.9(1.14^4)}{h}.$$

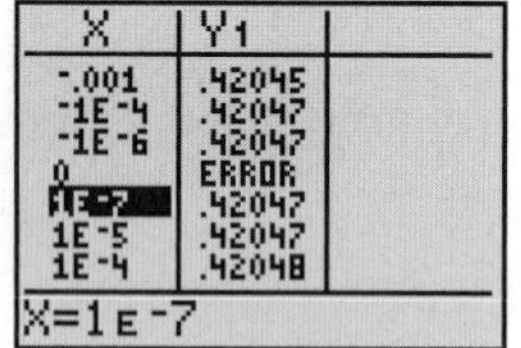

FIGURE 11.8

To approximate the instantaneous rate of change, we evaluate this quantity for very small values of h, as shown in Figure 11.8 (in which X is used in place of h and Y_1 is the average rate of change shown above). The table suggests that at $x = 4$,

$$\text{Marginal revenue} = \lim_{h\to 0} \frac{1.9(1.14^{4+h}) - 1.9(1.14^4)}{h} \approx .42047.$$

Thus, revenue was increasing at a rate of about .42047 billion dollars per year in 1998.

---

*Marginal cost for linear cost functions was discussed in Section 3.3.

†Based on data and projections from www.datamonitor.com, published in the St. Louis *Post-Dispatch,* June 18, 2001.

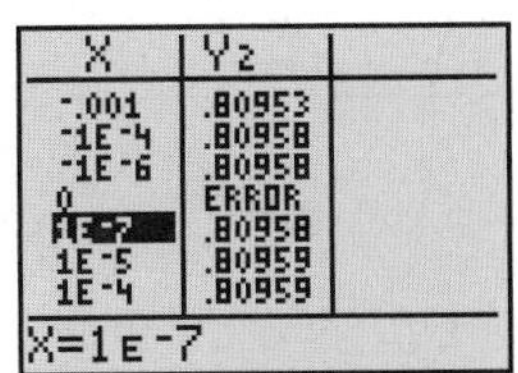

FIGURE 11.9

(b) Find the marginal revenue in 2003.

The marginal revenue, which is the instantaneous rate of change of $f(x)$ when $x = 9$, is

$$\lim_{h\to 0} \frac{f(9+h) - f(9)}{h} = \lim_{h\to 0} \frac{1.9(1.14^{9+h}) - 1.9(1.14^{9})}{h} \approx .80958,$$

as shown in Figure 11.9. In 2003, therefore, revenue is increasing at a rate of about .80958 billion dollars per year, almost twice the rate in 1998. ◀

11.2 Exercises

Find the average rate of change for the following functions (See Example 1.)

1. $f(x) = x^2 + 2x$ between $x = 0$ and $x = 5$
2. $f(x) = -4x^2 - 6$ between $x = 2$ and $x = 5$
3. $f(x) = 2x^3 - 4x^2 + 6x$ between $x = -1$ and $x = 2$
4. $f(x) = -3x^3 + 2x^2 - 4x + 1$ between $x = 0$ and $x = 1$
5. $f(x) = \sqrt{x}$ between $x = 1$ and $x = 4$
6. $f(x) = \sqrt{3x - 2}$ between $x = 1$ and $x = 3$
7. $f(x) = \dfrac{1}{x - 1}$ between $x = -2$ and $x = 0$
8. $f(x) = .4525\, e^{1.556\sqrt{x}}$ between $x = 4$ and $x = 4.5$
9. **Business** The graph shows the total sales in thousands of dollars from the distribution of x thousand catalogs. Find and interpret the average rate of change of sales with respect to the number of catalogs distributed for the following changes in x.
   (a) 10 to 20 **(b)** 20 to 30 **(c)** 30 to 40
   (d) What is happening to the average rate of change of sales as the number of catalogs distributed increases?
   (e) Explain why part (d) might happen.

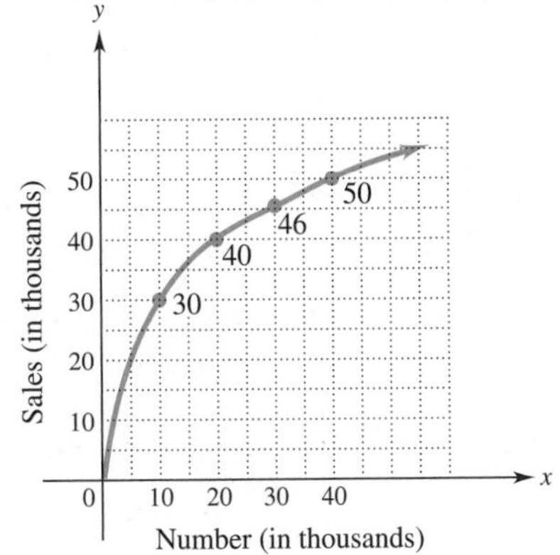

10. **Business** The graph shows annual sales (in appropriate units) of a computer game. Find the average annual rate of change in sales for the following changes in years.
    (a) 1 to 4 **(b)** 4 to 7 **(c)** 7 to 12
    (d) What do your answers for parts (a) to (c) tell you about the sales of this product?
    (e) Give an example of another product that might have such a sales curve.

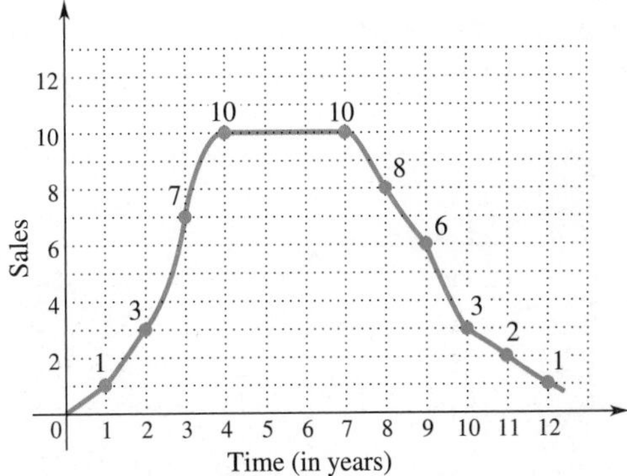

11. **Finance** The graph shows the money remaining in the Medicare trust fund at the end of the calendar year, adjusted for inflation in 2000 dollars.*

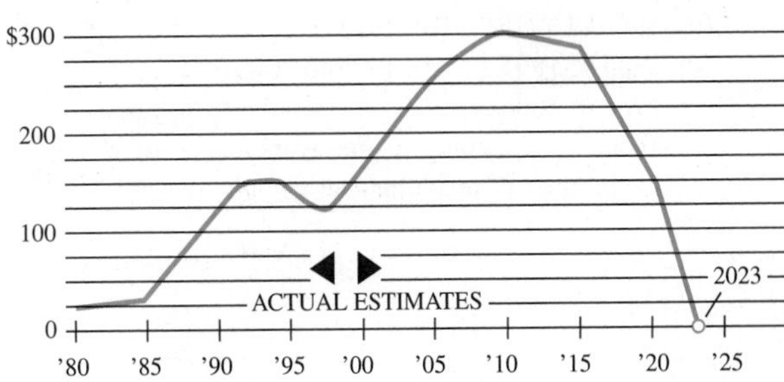

*Social Security Administration; Department of Health and Human Services.

Find the approximate average change in the trust fund for each time period.

(a) from 1994 (the peak) to 1998 (the low point)
(b) from 1998 to the estimated value for 2010
(c) from 1990 to 1998

12. Business The light vehicle market share (in percent) in the United States for domestic cars is shown in the figure.

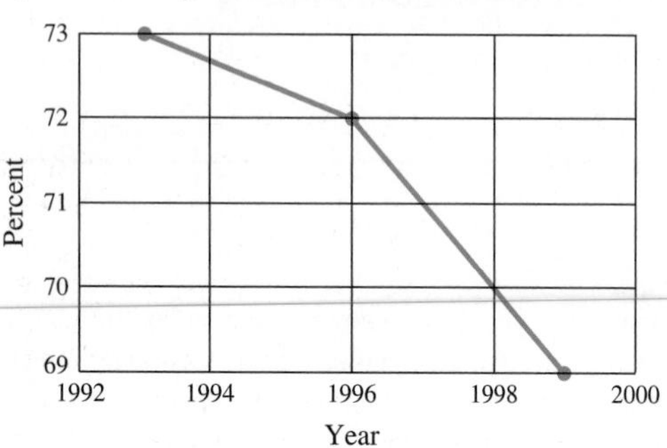

Source: J.D. Powers & Associates

Use the points marked on the graph to find and interpret the rate of change of market share
(a) from 1993 to 1996; (b) from 1996 to 1999.

13. Business The graph at the top of the next column shows the number of McDonald's restaurants in the United States as a function of time.*

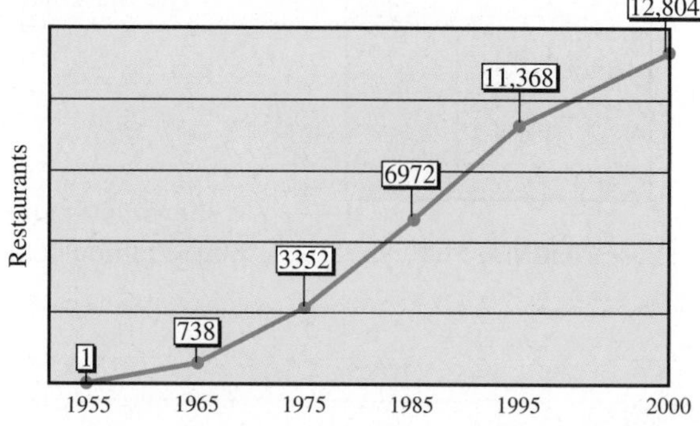

Find and interpret the average rate of change in the number of restaurants with respect to time over the following intervals.

(a) 1965 to 1975 (b) 1975 to 1985
(c) 1985 to 1995 (d) 1995 to 2000
(e) 1965 to 2000
(f) What is happening to the average rate of change in the number of restaurants as time increases?

14. Business The graph below shows the projected annual revenue (in billions of dollars) from sales of all organic food products (fruits, vegetables, meats, juices, ready meals, etc.).*

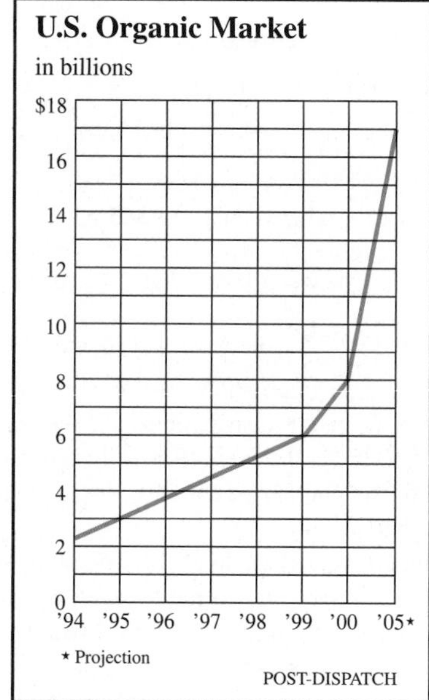

Find the average of change of revenue in the following periods.

(a) From 1995 to 2000
(b) From 2000 to 2005
(c) From 1995 to 2005
(d) How can the answer to part (c) be obtained from the answers to parts (a) and (b)?

15. Transportation Because of growing congestion, more people are using public transportation. The graph on the next page shows the number of trips annually on U.S. mass transportation systems.† Find and interpret the following average rates of change.

*Adapted from the *Chicago Tribune,* January 20, 1996, sec. 5, p. 1.

*St. Louis *Post-Dispatch,* June 18, 2001.

†*U.S. News and World Report,* April 30, 2001.

(a) From 1992 to 1995 (b) From 1995 to 2000
(c) Find the approximate average rate of change from 1992 to the last quarter of 1997. Does this answer accurately reflect what occurred during those years?

Annual Trips on U.S. Mass Transportation Systems

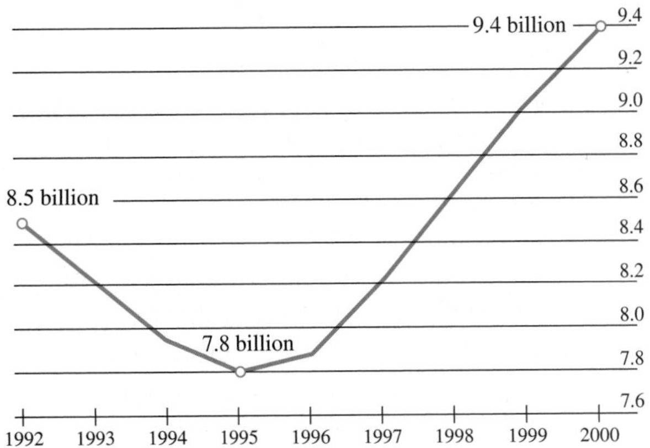

16. Social Science The future size of the world population depends on how soon it reaches replacement-level fertility, the point at which each woman bears on average about 2.1 children. The graphs show projections for reaching that point in different years.* Estimate the average rate of change in population for each projection from 1990 to 2050. Which projection shows the smallest rate of change in world population? (Tick marks on the horizontal axis are 5 years apart.)

Ultimate World Population Size Under Different Assumptions

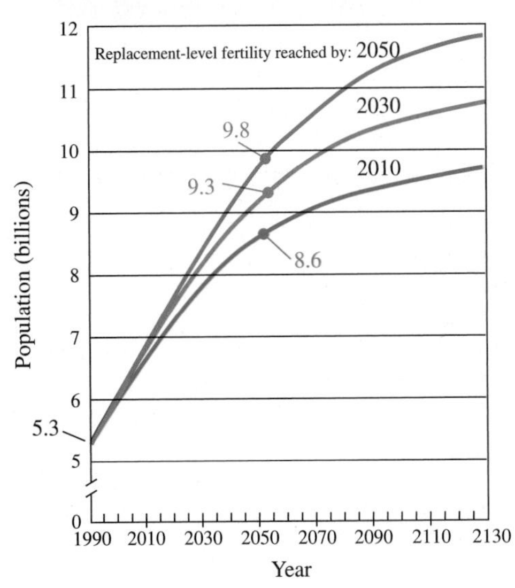

*Carl Haub, Population Reference Bureau.

17. Explain the difference between the average rate of change of $y = f(x)$ as x changes from a to b, and the instantaneous rate of change of y at $x = a$.

18. If the instantaneous rate of change of $f(x)$ with respect to x is positive when $x = 1$, is f increasing or decreasing there?

Exercises 19–21 deal with a car moving along a straight road, as discussed on pages 612–613. At time t seconds the distance of the car (in feet) from the starting point is $s(t) = 2t^2$. Find the instantaneous velocity (speed) of the car at

19. $t = 5$; 20. $t = 20$.

21. What was the average speed of the car during the first 30 seconds?

An object moves along a straight line; its distance (in feet) from a fixed point at time t seconds is $s(t) = t^2 + 5t + 2$. Find the instantaneous velocity of the object at the following times. (See Example 3.)

22. $t = 6$ 23. $t = 1$ 24. $t = 10$

25. Physical Science A car is moving along a straight test track. The position in feet of the car, $s(t)$, at various times t is measured, with the following results.

t (seconds)	0	2	4	6	8	10
s(t) (feet)	0	10	14	20	30	36

Find and interpret the average velocities for the following changes in t.
(a) 0 to 2 seconds (b) 2 to 4 seconds
(c) 4 to 6 seconds (d) 6 to 8 seconds
(e) Estimate the instantaneous velocity at 4 seconds
(i) by using the formula for estimating instantaneous rate (with $h = 2$), and
(ii) by averaging the answers for the average velocity in the two seconds before and the two seconds afterwards (that is, the answers to parts (b) and (c)).
(f) Estimate the instantaneous velocity at 6 seconds using the two methods in part (e).

In each of the following exercises, find: (a) $f(a + h)$; (b) $\frac{f(a+h) - f(a)}{h}$; (c) *the instantaneous rate of change of f when a* = 5. *(See Examples 3–5.)*

26. $f(x) = x^2 + x$ 27. $f(x) = x^2 - x - 1$
28. $f(x) = x^2 + 2x + 2$ 29. $f(x) = x^3$
30. $f(x) = x^3 - x$

Solve Exercises 31 and 32 by algebraic methods. (See Examples 3–5.)

31. Business The revenue (in thousands of dollars) from producing x units of an item is

$$R(x) = 10x - .002x^2.$$

(a) Find the average rate of change of revenue when production is increased from 1000 to 1001 units.
(b) Find the marginal revenue when 1000 units are produced.
(c) Find the additional revenue if production is increased from 1000 to 1001 units.
(d) Compare your answers for parts (a) and (c). What do you find?

32. Business Suppose customers in a hardware store are willing to buy $N(p)$ boxes of nails at p dollars per box, as given by

$$N(p) = 80 - 5p^2, \qquad 1 \le p \le 4.$$

(a) Find the average rate of change of demand for a change in price from \$2 to \$3.
(b) Find the instantaneous rate of change of demand when the price is \$2.
(c) Find the instantaneous rate of change of demand when the price is \$3.
(d) As the price is increased from \$2 to \$3, how is demand changing? Is the change to be expected?

33. Health Epidemiologists in College Station, Texas, estimate that t days after the flu begins to spread in town, the percent of the population infected by the flu is approximated by

$$p(t) = t^2 + t$$

for $0 \le t \le 5$.
(a) Find the average rate of change of p with respect to t over the the interval from 1 to 4 days.
(b) Find the instantaneous rate of change of p with respect to t at $t = 3$.

Use technology to work Exercises 34–36. (See Example 6.)

34. Finance Outstanding consumer credit (in billions of dollars) in year x can be approximated by

$$f(x) = 378.63(1.0726^x),$$

where $x = 0$ corresponds to 1980.* Estimate the rate of which consumer credit is increasing in
(a) 1995 **(b)** 2002.

35. Business The number of people (in millions) who subscribe to basic cable TV can be approximated by

$$f(x) = \frac{76.7}{1 + 16(.8444^x)},$$

where $x = 0$ corresponds to 1970.† Estimate the rate at which the number of subscribers was changing in
(a) 1990 **(b)** 2004.
(c) What factors might explain why the answers to parts (a) and (b) differ significantly?

36. Social Science The number of visitors to U.S. national parks (in millions) is approximately given by

$$g(x) = -270.8 + 73.21 \ln x \qquad (50 \le x \le 110),$$

where $x = 50$ corresponds to 1950.‡ Estimate the rate at which the number of visitors was changing in
(a) 1980 **(b)** 2000.

Federal Reserve Bulletin.

†Based on data from *The Cable TV Financial Databook* and *The Pay TV Newsletter.*

‡National Park Service.

11.3 TANGENT LINES AND DERIVATIVES

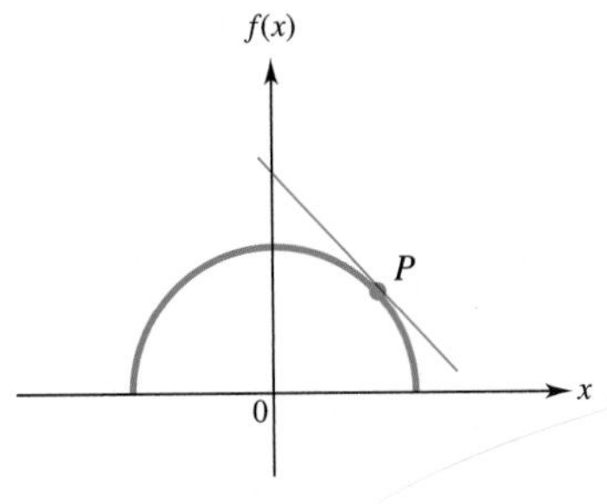

FIGURE 11.10

We now develop a geometric interpretation of the rates of change considered in the previous section. In geometry, a tangent line to a circle at a point P is defined to be the line through P that is perpendicular to the radius OP, as in Figure 11.10 (which shows only the top half of the circle). If you think of this circle as a road on which you are driving at night, then the tangent line indicates the direction of the light beam from your headlights as you pass through the point P. This suggests a way of extending the idea of a tangent line to any curve: the tangent line to the curve at a point P indicates the "direction" of the curve as it passes through P. Using this intuitive idea of direction, we see, for example, that the lines through P_1 and P_3 in Figure 11.11 appear to be tangent lines, whereas the lines through P_2 and P_4 do not.

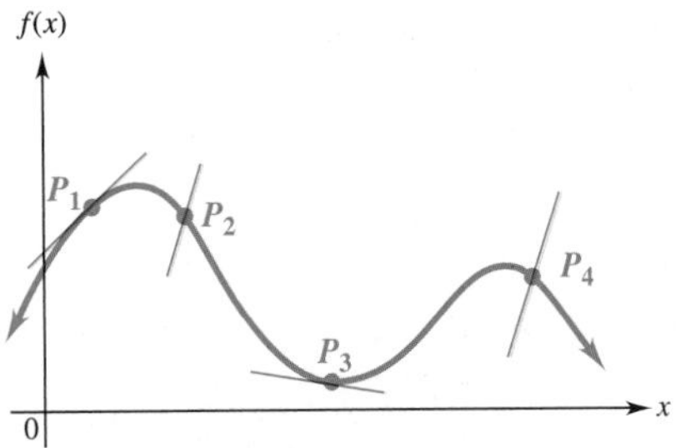

FIGURE 11.11

We can use these ideas to develop a precise definition of the tangent line to the graph of a function f at the point R. As shown in Figure 11.12, choose a second point S on the graph and draw the line through R and S: this line is called a **secant line**. You can think of this secant line as a rough approximation of the tangent line.

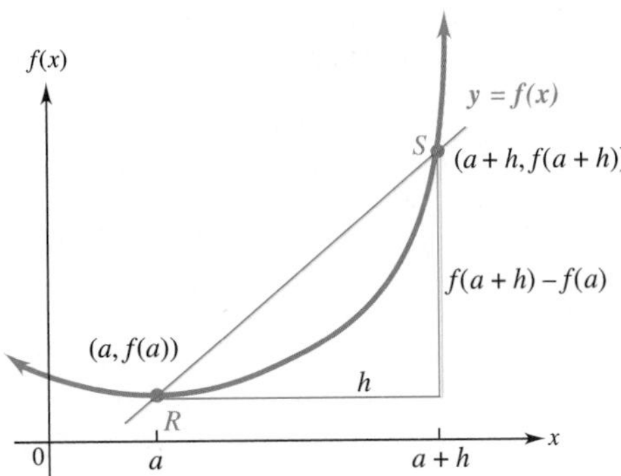

FIGURE 11.12

Now suppose that the point S slides down the curve closer to R. Figure 11.13 shows successive positions S_2, S_3, S_4 of the point S. The closer S gets to R, the better the secant line RS approximates our intuitive idea of the tangent line at R.

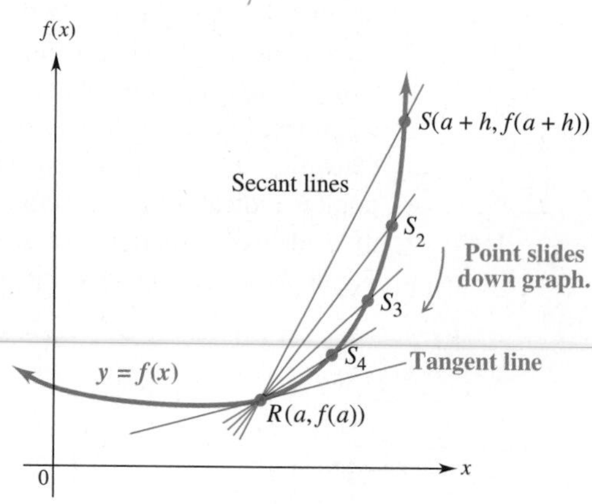

FIGURE 11.13

In particular, the closer S gets to R, the closer the slope of the secant line gets to the slope of the tangent line. Informally we say that

The slope of tangent line at R = The limit of the slope of secant line RS as S gets closer and closer to R.

In order to make this more precise, suppose the first coordinate of R is a. Then the first coordinate of S can be written as $a + h$ for some number h (in Figure 11.12, h is the distance on the x-axis between the two first coordinates). Thus R has coordinates $(a, f(a))$ and S has coordinates $(a + h, f(a + h))$, as shown in Figures 11.12 and 11.13. Consequently, the slope of the secant line RS is

$$\frac{f(a+h) - f(a)}{(a+h) - a} = \frac{f(a+h) - f(a)}{h}.$$

Now as S moves closer to R, their first coordinates move closer to each other, that is, h gets smaller and smaller. Hence

Slope of tangent line at R = The limit of slope of secant line RS as S gets closer and closer to R;

= The limit of $\dfrac{f(a+h) - f(a)}{h}$ as h gets closer and closer to 0;

$$= \lim_{h \to 0} \frac{f(a+h) - f(a)}{h}.$$

This intuitive development suggests the following formal definition.

Tangent Line

The **tangent line** to the graph of $y = f(x)$ at the point $(a, f(a))$ is the line through this point having slope

$$\lim_{h \to 0} \frac{f(a+h) - f(a)}{h},$$

provided this limit exists. If this limit does not exist, then there is no tangent line at the point.

The slope of the tangent line at a point is also called the **slope of the curve** at that point. Since the slope of a line indicates its direction (see the table on page 81), the slope of the tangent line at a point indicates the direction of the curve at that point.

EXAMPLE 1 Find the slope of the tangent line to the graph of $y = x^2 + 2$ when $x = -1$. Find the equation of the tangent line.

Use the definition above, with $f(x) = x^2 + 2$ and $a = -1$. The slope of the tangent line is

$$\begin{aligned}
\text{Slope of tangent} &= \lim_{h \to 0} \frac{f(a+h) - f(a)}{h} \\
&= \lim_{h \to 0} \frac{[(-1+h)^2 + 2] - [(-1)^2 + 2]}{h} \\
&= \lim_{h \to 0} \frac{[1 - 2h + h^2 + 2] - [1 + 2]}{h} \\
&= \lim_{h \to 0} \frac{-2h + h^2}{h} = \lim_{h \to 0} (-2 + h) = -2.
\end{aligned}$$

The slope of the tangent line at $(-1, f(-1)) = (-1, 3)$ is -2. The equation of the tangent line can be found with the point-slope form of the equation of a line from Chapter 2.

$$\begin{aligned}
y - y_1 &= m(x - x_1) \\
y - 3 &= -2[x - (-1)] \\
y - 3 &= -2(x + 1) \\
y - 3 &= -2x - 2 \\
y &= -2x + 1
\end{aligned}$$

Figure 11.14 on the next page shows a graph of $f(x) = x^2 + 2$, along with a graph of the tangent line at $x = -1$. ◀ 1

Let $f(x) = x^2 + 2$.
Find the equation of the tangent line to the graph at the point where $x = 1$.

Answer:

$y = 2x + 1$

Use a graphing calculator to confirm your answer to side problem 1 on the previous page by graphing $f(x) = x^2 + 2$ and the tangent line at the point where $x = 1$ on the same screen.

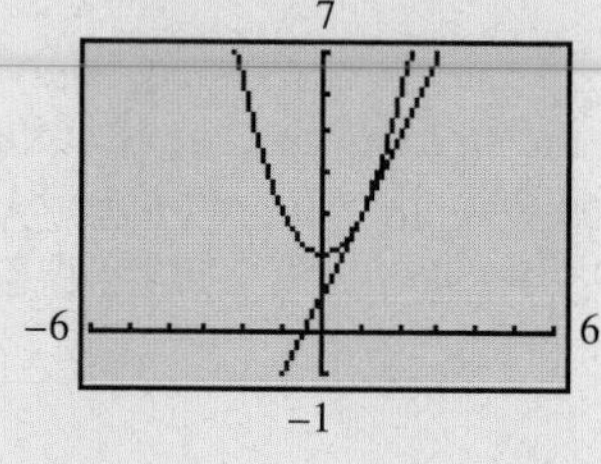

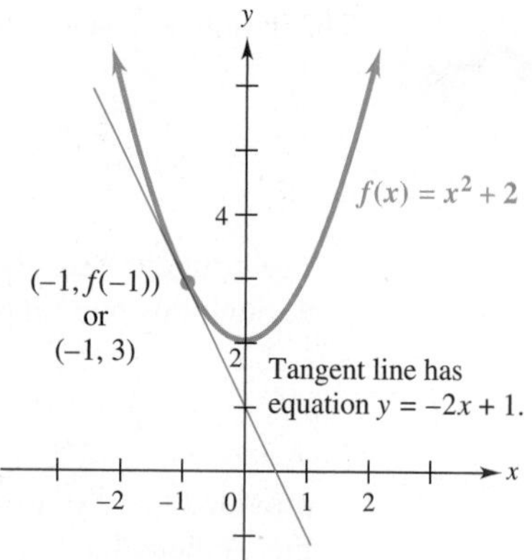

FIGURE 11.14

TECHNOLOGY TIP When finding the equation of a tangent line algebraically, you can confirm your answer with a graphing calculator by graphing both the function and the tangent line on the same screen to see if the tangent line appears to be correct. (2)

Once a function has been graphed, many graphing calculators can draw the tangent line at any specified point. Most also display the slope of the tangent line, and some actually display its equation. Look for TANGENT or TANLN in the MATH or DRAW or SKETCH menu.

In Figure 11.14, the tangent line $y = -2x + 1$ appears to coincide with the graph of $f(x) = x^2 + 2$ near the point $(-1, 3)$. This becomes more obvious when both $f(x)$ and the tangent line are graphed in a very small window on a graphing calculator (Figure 11.15). The graph and the tangent line now appear virtually identical near $(-1, 3)$. The table of values in Figure 11.16 (in which Y_1 is $f(x)$ and Y_2 the tangent line) confirms the fact that the tangent line is a good approximation of the function when x is very close to -1.

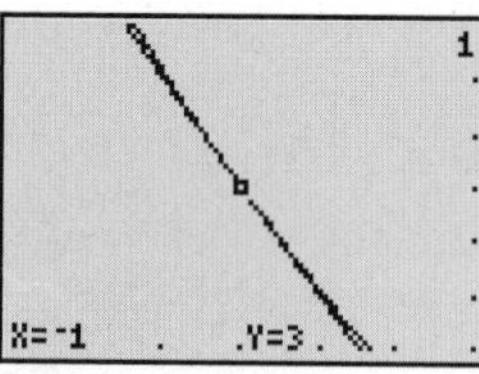

FIGURE 11.15

X	Y1	Y2
-1.1	3.21	3.2
-1.05	3.1025	3.1
-1.02	3.0404	3.04
-1	3	3
-.97	2.9409	2.94
-.94	2.8836	2.88
-.9	2.81	2.8

X=-.9

FIGURE 11.16

The same thing is true in the general case.

Tangent Line

If it exists, the tangent line to the graph of a function f at $x = a$ is a good approximation of the function f near $x = a$.

Suppose the graph of a function f is a straight line. The fact in the preceding box suggests that the tangent line to f at any point should be the graph of f itself (because it's certainly the best possible linear approximation of f). The next example shows that this is indeed the case.

EXAMPLE 2 Let a be any real number. Find the equation of the tangent line to the graph of $f(x) = 5x + 2$ at the point where $x = a$.

According to the definition, the slope of the tangent line is

$$\begin{aligned}\lim_{h\to 0}\frac{f(a+h)-f(a)}{h} &= \lim_{h\to 0}\frac{[5(a+h)+2]-[5a+2]}{h}\\ &= \lim_{h\to 0}\frac{[5a+5h+2]-5a-2}{h}\\ &= \lim_{h\to 0}\frac{5h}{h} = \lim_{h\to 0}5 = 5.\end{aligned}$$

Hence the equation of the tangent line at the point $(a, f(a))$ is

$$\begin{aligned}y - y_1 &= m(x - x_1)\\ y - f(a) &= 5(x - a)\\ y &= 5x - 5a + f(a)\\ y &= 5x - 5a + 5a + 2\\ y &= 5x + 2.\end{aligned}$$

Thus, the tangent line is precisely the graph of $f(x) = 5x + 2$. ◀

Secant lines and tangent lines (or more precisely, their slopes) are the geometric analogues of the average and instantaneous rates of change studied in the previous section, as summarized in the following chart.

Quantity	*Algebraic Interpretation*	*Geometric Interpretation*
$\frac{f(a+h)-f(a)}{h}$	Average rate of change of f from $x = a$ to $x = a + h$	Slope of the secant line through $(a, f(a))$ and $(a + h, f(a + h))$
$\lim_{h\to 0}\frac{f(a+h)-f(a)}{h}$	Instantaneous rate of change of f at $x = a$	Slope of the tangent line to the graph of f at $(a, f(a))$

The Derivative

If $y = f(x)$ is a function and a is a number in its domain, then we shall use the symbol $f'(a)$ to denote the special limit

$$\lim_{h \to 0} \frac{f(a+h) - f(a)}{h},$$

provided that it exists. In other words, to each number a, we can assign the number $f'(a)$ obtained by calculating this limit. This process defines an important new function.

Derivative

The **derivative** of the function f is the function denoted f' whose value at the number x is defined to be the number

$$f'(x) = \lim_{h \to 0} \frac{f(x+h) - f(x)}{h},$$

provided this limit exists.

The derivative function f' has as its domain all the points at which the specified limit exists, and the value of the derivative function at the number x is the number $f'(x)$. Using x instead of a here is similar to the way that $g(x) = 2x$ denotes the function that assigns to each number a the number $2a$.

If $y = f(x)$ is a function, then its derivative is denoted either by f' or by y'. If x is a number in the domain of $y = f(x)$ such that $y' = f'(x)$ is defined, then the function f is said to be **differentiable** at x. The process that produces the function f' from the function f is called **differentiation.**

The derivative function may be interpreted in many ways, two of which were discussed above.

1. The derivative function f' gives the *instantaneous rate of change* of $y = f(x)$ with respect to x. This instantaneous rate of change can be interpreted as marginal cost, marginal revenue, or marginal profit (if the original function represents cost, revenue, or profit) or as velocity (if the original function represents displacement along a line). From now on we will use "rate of change" to mean "instantaneous rate of change."
2. The derivative function f' gives the *slope* of the graph of f at any point. If the derivative is evaluated at $x = a$, then $f'(a)$ is the slope of the tangent line to the curve at the point $(a, f(a))$.

EXAMPLE 3 Use the graph of the function $f(x)$ in Figure 11.17 to answer the questions below.

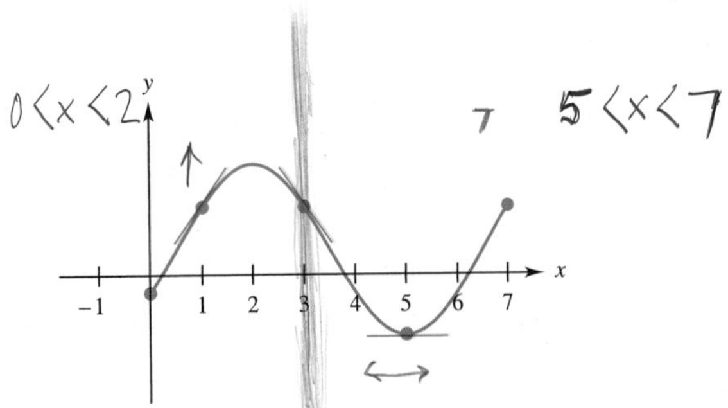

FIGURE 11.17

3

The graph of a function g is shown below. Determine whether the following numbers are *positive*, *negative*, or *zero*.

(a) $g'(0)$

(b) $g'(-1)$

(c) $g'(3)$

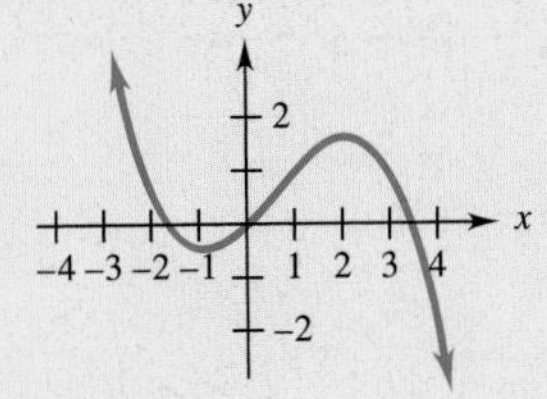

Answers:

(a) Positive

(b) 0

(c) Negative

(a) Is $f'(3)$ positive or negative?

We know that $f'(3)$ is the slope of the tangent line to the graph at the point where $x = 3$. Figure 11.17 shows that this tangent line slants downward from left to right, meaning that its slope is negative. Hence, $f'(3) < 0$.

(b) Which is larger, $f'(1)$ or $f'(5)$?

Figure 11.17 shows that the tangent line to the graph at the point where $x = 1$ slants upward from left to right, meaning that its slope, $f'(1)$, is a positive number. The tangent line at the point where $x = 5$ is horizontal, so it has slope 0, that is, $f'(5) = 0$. Therefore, $f'(1) > f'(5)$.

(c) For what values of x is $f'(x)$ positive?

Find the points on the graph, where the tangent line has positive slope (slants upward from left to right). At each such point, $f'(x) > 0$. Figure 11.17 shows that this occurs when $0 < x < 2$ and when $5 < x < 7$. ◀ 3

The rule of a derivative function can be found by using the definition of derivative and the following four-step procedure.

Finding $f'(x)$ from the Definition of the Derivative

Step 1 Find $f(x + h)$.

Step 2 Find $f(x + h) - f(x)$.

Step 3 Divide by h to get $\dfrac{f(x + h) - f(x)}{h}$.

Step 4 Let $h \to 0$; $f'(x) = \lim_{h \to 0} \dfrac{f(x + h) - f(x)}{h}$ if this limit exists.

EXAMPLE 4 Let $f(x) = x^3 - 4x$.

(a) Find the derivative $f'(x)$.
By definition

$$f'(x) = \lim_{h\to 0} \frac{f(x+h) - f(x)}{h}.$$

Step 1 Find $f(x+h)$.
Replace x with $x + h$ in the rule of $f(x)$.

$$f(x) = x^3 - 4x$$
$$f(x+h) = (x+h)^3 - 4(x+h)$$
$$= (x^3 + 3x^2h + 3xh^2 + h^3) - 4(x+h)$$
$$= x^3 + 3x^2h + 3xh^2 + h^3 - 4x - 4h.$$

Step 2 Find $f(x+h) - f(x)$.
Since $f(x) = x^3 - 4x$,

$$f(x+h) - f(x) = (x^3 + 3x^2h + 3xh^2 + h^3 - 4x - 4h) - (x^3 - 4x)$$
$$= x^3 + 3x^2h + 3xh^2 + h^3 - 4x - 4h - x^3 + 4x$$
$$= 3x^2h + 3xh^2 + h^3 - 4h.$$

Step 3 Form and simplify the quotient $\frac{f(x+h) - f(x)}{h}$.

$$\frac{f(x+h) - f(x)}{h} = \frac{3x^2h + 3xh^2 + h^3 - 4h}{h}$$
$$= \frac{h(3x^2 + 3xh + h^2 - 4)}{h}$$
$$= 3x^2 + 3xh + h^2 - 4.$$

Step 4 Find the limit as h approaches 0 of the result in Step 3.

$$f'(x) = \lim_{h\to 0} \frac{f(x+h) - f(x)}{h} = \lim_{h\to 0} (3x^2 + 3xh + h^2 - 4)$$
$$= 3x^2 - 4.$$

Therefore, the derivative of $f(x) = x^3 - 4x$ is $f'(x) = 3x^2 - 4$.

(b) Calculate and interpret $f'(1)$.

The procedure in part (a) works for *every* x and $f'(x) = 3x^2 - 4$. Hence, when $x = 1$,

$$f'(1) = 3 \cdot 1^2 - 4 = -1.$$

The number -1 is the slope of the tangent line to the graph of $f(x) = x^3 - 4x$ at the point where $x = 1$, that is, at $(1, f(1)) = (1, -3)$.

(c) Find the equation of the tangent line to the graph of $f(x) = x^3 - 4x$ at the point where $x = 1$.

By part (b), the point on the graph where $x = 1$ is $(1, -3)$ and the slope of the tangent line is $f'(1) = -1$. Therefore, the equation is

$$y - (-3) = (-1)(x - 1) \quad \text{Point-slope form}$$
$$y = -x - 2. \quad \text{Slope-intercept form}$$

Both $f(x)$ and the tangent line are shown in Figure 11.18. 

4

Let $f(x) = -2x^2 + 7$. Find the following.

(a) $f(x+h)$
(b) $f(x+h) - f(x)$
(c) $\frac{f(x+h) - f(x)}{h}$
(d) $f'(x)$
(e) $f'(4)$
(f) $f'(0)$

Answers:

(a) $-2x^2 - 4xh - 2h^2 + 7$
(b) $-4xh - 2h^2$
(c) $-4x - 2h$
(d) $-4x$
(e) -16
(f) 0

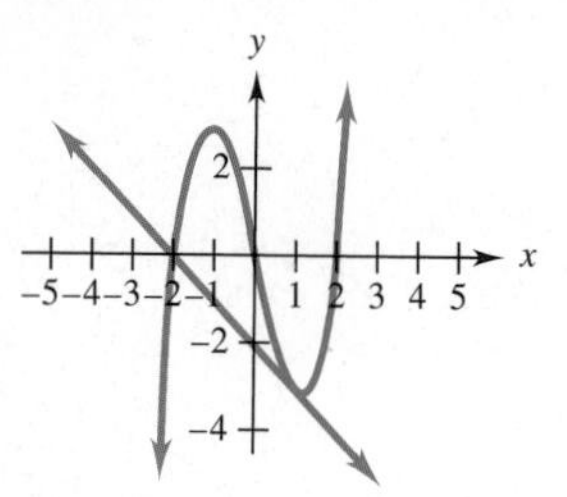

FIGURE 11.18

CAUTION

1. In Example 4(a) note that $f(x+h) \neq f(x) + h$ because by Step 1,

$$f(x+h) = x^3 + 3x^2h + 3xh^2 + h^3 - 4x - 4h,$$

but

$$f(x) + h = (x^3 - 4x) + h = x^3 - 4x + h.$$

2. In Example 4(b), do not confuse $f(1)$ and $f'(1)$. $f(1)$ is the value of the original function $f(x) = x^3 - 4x$ at $x = 1$, namely, -3, whereas $f'(1)$ is the value of the derivative function $f'(x) = 3x^2 - 4$ at $x = 1$, namely, -1.

EXAMPLE 5 Let $f(x) = 1/x$. Find $f'(x)$.

Step 1 $f(x+h) = \dfrac{1}{x+h}$

Step 2
$$f(x+h) - f(x) = \frac{1}{x+h} - \frac{1}{x}$$
$$= \frac{x - (x+h)}{x(x+h)} \quad \text{Find a common denominator.}$$
$$= \frac{x - x - h}{x(x+h)} \quad \text{Simplify the numerator.}$$
$$= \frac{-h}{x(x+h)}$$

Step 3
$$\frac{f(x+h) - f(x)}{h} = \frac{\frac{-h}{x(x+h)}}{h}$$
$$= \frac{-h}{x(x+h)} \cdot \frac{1}{h} \quad \text{Invert and multiply.}$$
$$= \frac{-1}{x(x+h)}$$

Step 4
$$f'(x) = \lim_{h \to 0} \frac{f(x+h) - f(x)}{h} = \lim_{h \to 0} \frac{-1}{x(x+h)}$$
$$= \frac{-1}{x(x+0)} = \frac{-1}{x(x)} = \frac{-1}{x^2}$$ 5

5

Let $f(x) = -5/x$. Find the following.

(a) $f(x+h)$

(b) $f(x+h) - f(x)$

(c) $\dfrac{f(x+h) - f(x)}{h}$

(d) $f'(x)$

(e) $f'(-1)$

Answers:

(a) $\dfrac{-5}{x+h}$

(b) $\dfrac{5h}{x(x+h)}$

(c) $\dfrac{5}{x(x+h)}$

(d) $\dfrac{5}{x^2}$

(e) 5

EXAMPLE 6 Let $g(x) = \sqrt{x}$. Find $g'(x)$.

Step 1 $g(x+h) = \sqrt{x+h}$

Step 2 $g(x+h) - g(x) = \sqrt{x+h} - \sqrt{x}$

Step 3 $\dfrac{g(x+h)-g(x)}{h}=\dfrac{\sqrt{x+h}-\sqrt{x}}{h}$

At this point, in order to be able to divide by h, multiply both numerator and denominator by $\sqrt{x+h}+\sqrt{x}$; that is, rationalize the *numerator*.

$$\begin{aligned}\frac{g(x+h)-g(x)}{h}&=\frac{\sqrt{x+h}-\sqrt{x}}{h}\cdot\frac{\sqrt{x+h}+\sqrt{x}}{\sqrt{x+h}+\sqrt{x}}\\&=\frac{(\sqrt{x+h})^2-(\sqrt{x})^2}{h(\sqrt{x+h}+\sqrt{x})}\\&=\frac{x+h-x}{h(\sqrt{x+h}+\sqrt{x})}=\frac{1}{\sqrt{x+h}+\sqrt{x}}\end{aligned}$$

Step 4 $g'(x)=\lim\limits_{h\to 0}\dfrac{1}{\sqrt{x+h}+\sqrt{x}}=\dfrac{1}{\sqrt{x}+\sqrt{x}}=\dfrac{1}{2\sqrt{x}}$ ◀

EXAMPLE 7 A sales representative for a textbook publishing company frequently makes a 4-hour drive from her home in a large city to a university in another city. If $s(t)$ represents her distance (in miles) from home t hours into the trip, then $s(t)$ is given by

$$s(t)=-5t^3+30t^2.$$

(a) How far from home will she be after 1 hour? After $1\frac{1}{2}$ hours?
Her distance from home after 1 hour is

$$s(1)=-5(1)^3+30(1)^2=25,$$

or 25 miles. After $1\frac{1}{2}$ (or 3/2) hours, it is

$$s\left(\frac{3}{2}\right)=-5\left(\frac{3}{2}\right)^3+30\left(\frac{3}{2}\right)^2=\frac{405}{8}=50.625,$$

or 50.625 miles.

Go through the four steps to find $s'(t)$, the velocity of the car at any time t.

Answer:

$s'(t)=-15t^2+60t$

(b) How far apart are the two cities?
Since the trip takes 4 hours and the distance is given by $s(t)$, the university city is $s(4)=160$ miles from her home.

(c) How fast is she driving 1 hour into the trip? $1\frac{1}{2}$ hours into the trip?
Velocity (or speed) is the instantaneous rate of change in position with respect to time. We need to find the value of the derivative $s'(t)$ at $t=1$ and $t=1\frac{1}{2}$. (6)

From problem 6 at the side, $s'(t)=-15t^2+60t$. At $t=1$, the velocity is

$$s'(1)=-15(1)^2+60(1)=45,$$

or 45 miles per hour. At $t=1\frac{1}{2}$, the velocity is

$$s'\left(\frac{3}{2}\right)=-15\left(\frac{3}{2}\right)^2+60\left(\frac{3}{2}\right)=56.25,$$

about 56 miles per hour.

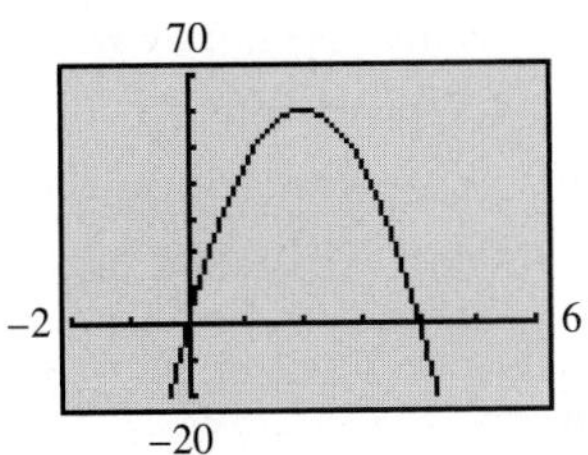

FIGURE 11.19

(d) Does she ever exceed the speed limit of 65 miles per hour on the trip?

To find the maximum velocity, notice that the graph of the velocity function $s'(t) = -15t^2 + 60t$ is a parabola opening downward (Figure 11.19). The maximum velocity will occur at the vertex. Use graphical or algebraic methods to verify that the vertex of the parabola is (2, 60). Thus, her maximum velocity during the trip is 60 miles per hour, so she never exceeds the speed limit.

Existence of the Derivative

The definition of the derivative includes the phrase "provided this limit exists." If the limit used to define $f'(x)$ does not exist, then of course the derivative does not exist at that x. For example, a derivative cannot exist at a point where the function itself is not defined. If there is no function value for a particular value of x, there can be no tangent line for that value. This was the case in Example 5—there was no tangent line (and no derivative) when $x = 0$.

Derivatives also do not exist at "corners" or "sharp points" on a graph. For example, the function graphed in Figure 11.20 is the *absolute value function*, defined by

$$f(x) = \begin{cases} x & \text{if } x \geq 0 \\ -x & \text{if } x < 0 \end{cases}$$

and written $f(x) = |x|$.

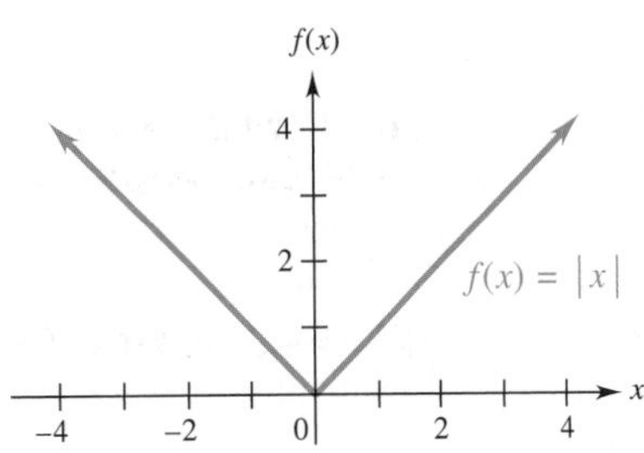

FIGURE 11.20

By the definition of derivative, the derivative at any value of x is given by

$$\lim_{h \to 0} \frac{f(x + h) - f(x)}{h},$$

provided this limit exists. To find the derivative at 0 for $f(x) = |x|$, replace x with 0 and $f(x)$ with $|0|$ to get

$$\lim_{h \to 0} \frac{|0 + h| - |0|}{h} = \lim_{h \to 0} \frac{|h|}{h}.$$

In Example 10 of Section 11.1 (with x in place of h) we showed that

$$\lim_{h \to 0} \frac{|h|}{h} \text{ does not exist.}$$

Therefore, there is no derivative at 0. However, the derivative of $f(x) = |x|$ *does* exist for all values of x other than 0.

Since a vertical line has an undefined slope, the derivative cannot exist at any point where the tangent line is vertical, as at x_5 in Figure 11.21. Figure 11.21 summarizes various ways that a derivative can fail to exist.

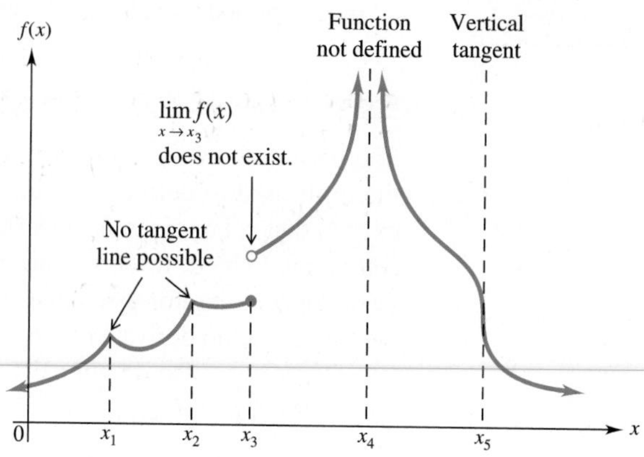

FIGURE 11.21

Derivatives and Technology

Many computer programs (such as Mathematica and Maple) and a few graphing calculators (such as TI-89 and Casio FX.2) can find symbolic formulas for the derivatives of most functions. Although other graphing calculators cannot find the rule of a derivative function, most of them can approximate the numerical value of the derivative function at any number where it is defined by using the **numerical derivative** feature. (See Exercise 46 at the end of this section for an explanation of how this feature works.)

7

If $f(x) = x^{3/4} - 7x^2 + 3^x$, use a graphing calculator to find the (approximate) value of

(a) $f'(1)$;

(b) $f'(4.5)$;

(c) $f'(8)$.

Answers:

(a) -9.9542

(b) 91.646

(c) 7096.4412

EXAMPLE 8 If $f(x) = x^3 + x^{3/2} + 6 \ln x$, use a graphing calculator to find the approximate value of $f'(4)$.

The numerical derivative feature is labeled Nderiv, or d/dx, or nDer, and is usually in the MATH menu or one of its submenus. Check your instruction manual for the correct syntax, but on many graphing calculators entering either

$$\text{NDeriv}(x^3 + x^{3/2} + 6 \ln x, x, 4) \quad \text{or} \quad \text{d/dx}(x^3 + x^{3/2} + 6 \ln x, 4)$$

produces the (approximate) value of $f'(4)$. Depending on the calculator, you are likely to get one of these answers.

$$52.4999982 \quad \text{or} \quad 52.5 \quad \text{or} \quad 52.50000102.$$

It can be shown that the exact value is $f'(4) = 52.5$. ◀ 7

TECHNOLOGY TIP On many graphing calculators, if you have the function $f(x)$ stored in the function memory as y_1, say, you can use y_1 with the NDeriv key, instead of typing in the rule of $f(x)$.

EXAMPLE 9 The cost (in thousands of dollars) to manufacture x thousand graphing calculators is given by

$$C(x) = .002x^4 + .1x^3 + .3x^2 - .1x + 50 \qquad (0 \le x \le 13).$$

Find the marginal cost (rate of change of cost) when 5000 and when 9000 calculators are made.

The marginal cost function is the derivative of the cost function and 5000 calculators corresponds to $x = 5$, so the marginal cost is $C'(5)$. In later sections you will learn how to find it exactly, but for now use the numerical derivative to approximate $C'(5)$. A typical calculator shows that $C'(5) \approx 11.4$, meaning that costs are changing at the rate of $11,400 per thousand calculators. When 9000 calculators are made, the marginal cost is

$$C'(9) \approx 35.432;\text{ that is, \$35,432 per thousand calculators.}$$

CAUTION Because of the approximation methods used, the NDeriv key may display an answer at numbers where the derivative is not defined. For instance, we saw above that the derivative of $f(x) = |x|$ is not defined when $x = 0$. But the NDeriv key on most calculators produces 1 or 0 or -1 as $f'(0)$.

11.3 Exercises

The derivatives of each of the following functions were found in Examples 4–7. Use them to find the equation of the tangent line to the graph of the function at the given point. (See Examples 1 and 2.)

1. $f(x) = x^3 - 4x$ at $x = 1$
2. $g(x) = \sqrt{x}$ at $x = 4$
3. $f(x) = 1/x$ at $x = -2$
4. $s(x) = -5x^3 + 30x^2$ at $x = 2$

For each of the following functions, (a) find the slope of the tangent line to the graph at the given point; (b) find the equation of the tangent line. (See Examples 1 and 2.)

5. $f(x) = x^2 - 2$ at $x = 3$
6. $g(x) = 6 - x^2$ at $x = -2$
7. $f(x) = \dfrac{5}{x}$ at $x = 2$
8. $g(x) = \dfrac{-2}{x}$ at $x = 4$
9. $f(x) = 4\sqrt{x}$ at $x = 9$
10. $g(x) = \sqrt{x+1}$ at $x = 15$ (*Hint:* In Step 3, multiply the numerator and denominator by $\sqrt{15+h} + \sqrt{15}$.)

Use the fact that f′(c) is the slope of the tangent line to the graph of f(x) at x = c to work these exercises. (See Example 3.)

11. In the graph of the function f, at which of the labeled x-values is
    (a) $f(x)$ the largest?
    (b) $f(x)$ the smallest?
    (c) $f'(x)$ the smallest?
    (d) $f'(x)$ the closest to 0?

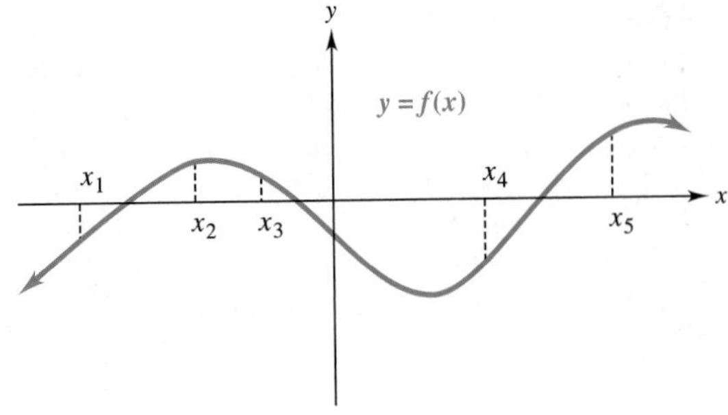

12. Sketch the graph of the *derivative* of the function g whose graph is shown on the next page. (*Hint:* Consider the

slope of the tangent line at each point along the graph of g. Are there any points where there is no tangent line?)

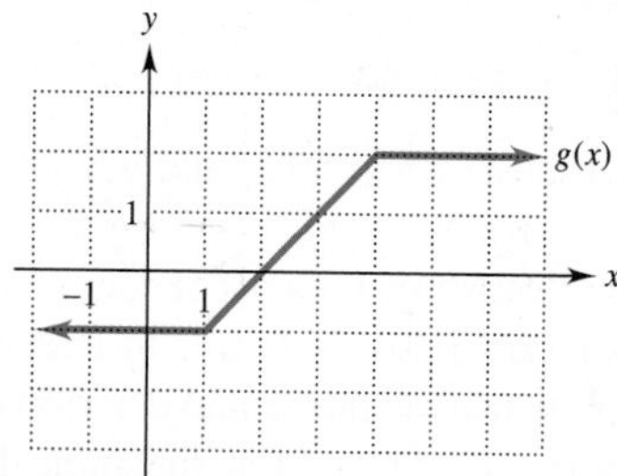

13. Sketch the graph of a function g with the property that $g'(x) > 0$ for $x < 0$ and $g'(x) < 0$ for $x > 0$. Many correct answers are possible.

14. **Physical Science** The graph shows the temperature in an oven during a self-cleaning cycle.* (The open circles on the graph are not points of discontinuity, but merely the times when the thermal door lock turns on and off.) The oven temperature is 100° when the cycle begins and 600° after half an hour. Let $T(x)$ be the temperature (in degrees Fahrenheit) after x hours.

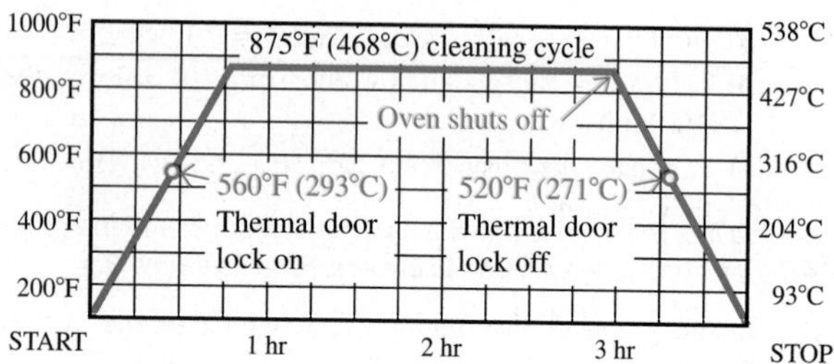

(a) Find the approximate values of x at which the derivative T' does not exist.

(b) Find and interpret $T'(.5)$.

(c) Find and interpret $T'(2)$.

(d) Find and interpret $T'(3.5)$.

15. **Physical Science** When a batter hits a baseball, the bat may not hit the center of the ball but might hit over or under the center by various amounts (measured in inches). The graph shows the trajectories of balls struck by a bat swung under the ball by given amounts.† Is the derivative for a bat swung under the ball by 1.5 inches positive or negative when the ball has traveled

(a) 100 ft? (b) 200 ft?

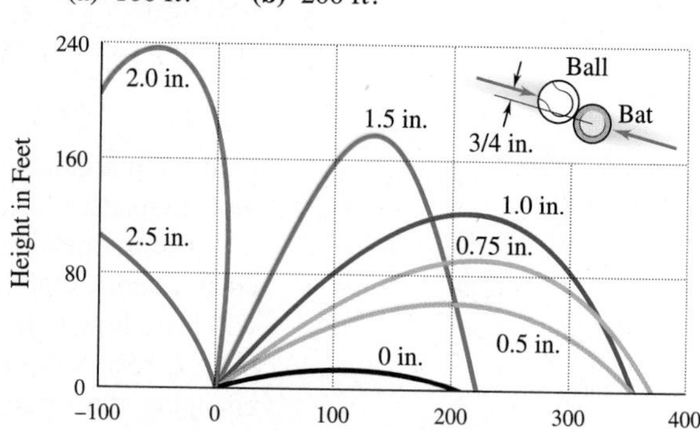

16. **Natural Science** The graph shows how the risk of coronary heart attack rises as blood cholesterol increases.*

(a) Approximate the average rate of change of the risk of coronary heart attack as blood cholesterol goes from 100 to 300 mg/dL.

(b) Is the rate of change when blood cholesterol is 100 mg/dL higher or lower than the average rate of change in part (a)? What feature of the graph shows this?

(c) Do part (b) when blood cholesterol is 300 mg/dL.

(d) Do part (b) when blood cholesterol is 200 mg/dL.

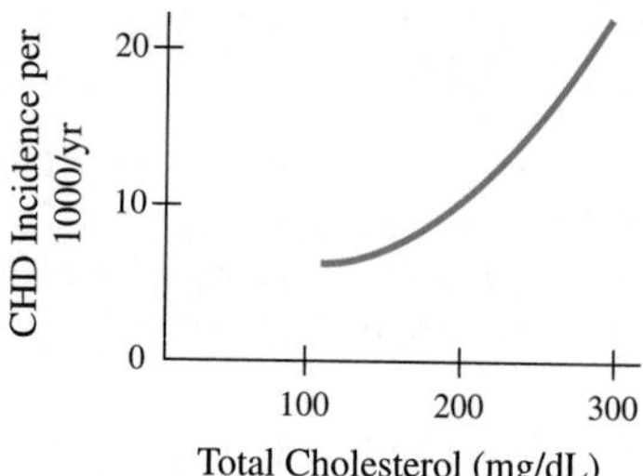

In Exercises 17–18, tell which graph, (a) or (b), represents velocity and which represents distance from a starting point. (Hint: Consider where the derivative is zero, positive, or negative.)

17. (a)

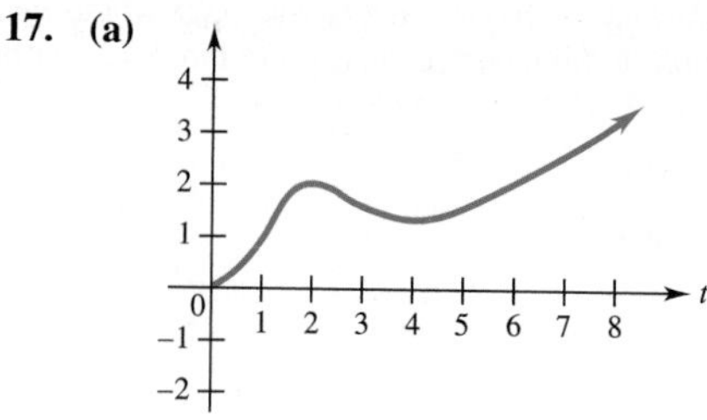

*Whirlpool Use and Care Guide, Self-Cleaning Electric Range.

†Adair, Robert K., *The Physics of Baseball*, Copyright © 1990 by HarperCollins, p. 83

*John C. LaRosa, et al., *The Cholesterol Facts: A Joint Statement by the American Heart Association and the National Heart, Lung, and Blood Institute*, from *Circulation* 81, no. 5 (May 1990): 1722.

(b)

18. (a)

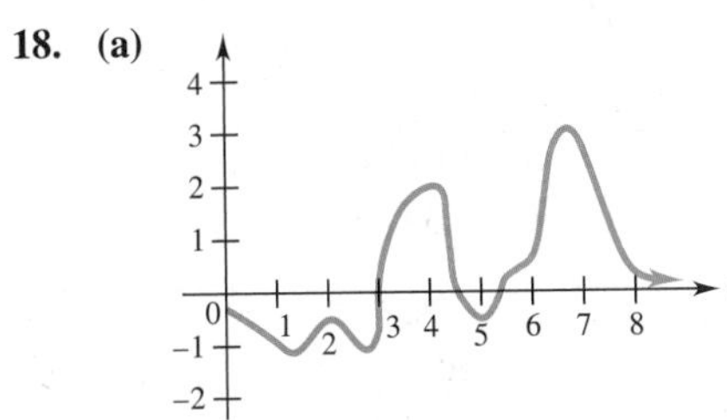

(b)

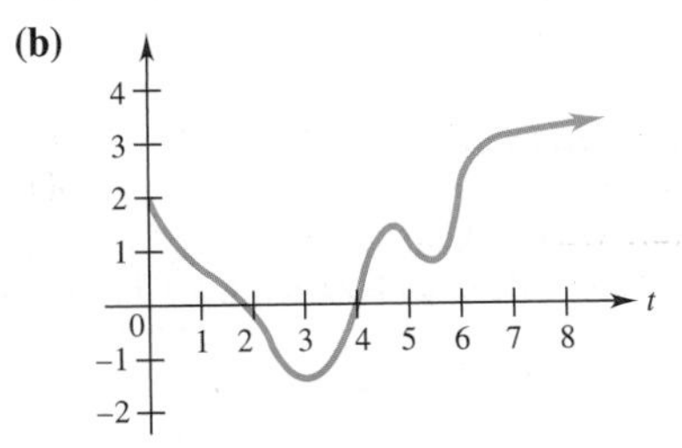

Find f′(x) for each function. Then find f′(2), f′(0), and f′(−3). (See Examples 4–6.)

19. $f(x) = -4x^2 + 11x$

20. $f(x) = 6x^2 - 4x$

21. $f(x) = 8x + 6$

22. $f(x) = x^3 + 3x$

23. $f(x) = -\dfrac{2}{x}$

24. $f(x) = \dfrac{6}{x}$

25. $f(x) = \dfrac{4}{x - 1}$

26. $f(x) = \sqrt{x}$

Find all points where the functions whose graphs are shown do not have derivatives. (See Figure 11.21 and the preceding discussion.)

27.

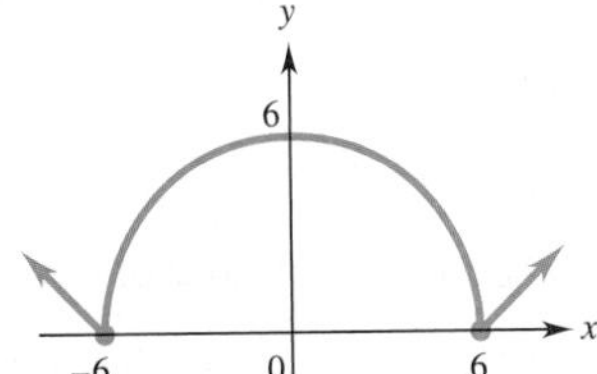

28.

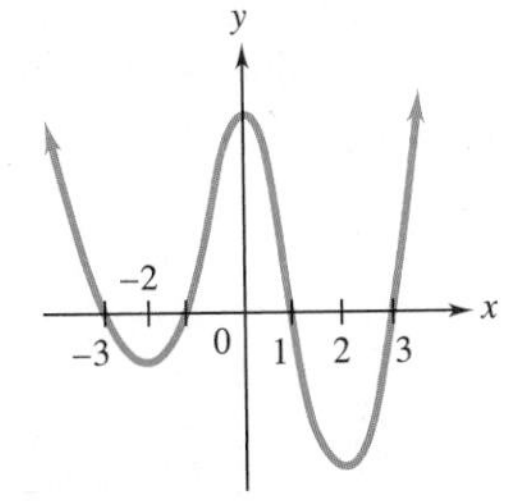

29.

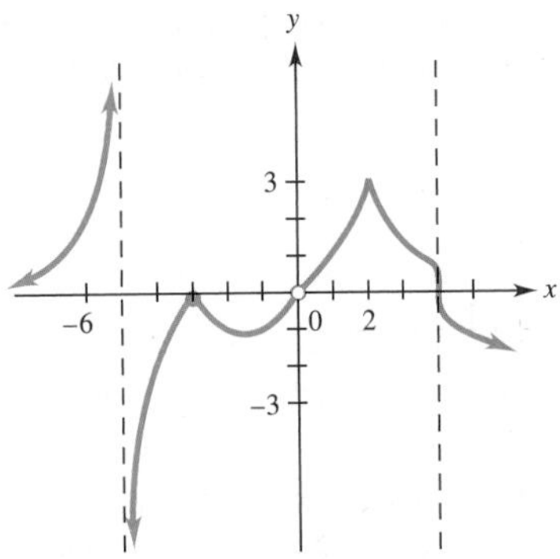

30. **(a)** Sketch the graph of $g(x) = \sqrt[3]{x}$ for $-1 \le x \le 1$.
(b) Explain why the derivative of $g(x)$ is not defined at $x = 0$. (*Hint:* What is the slope of the tangent line at $x = 0$?)

Work these exercises. (See Examples 4–7.)

31. Business The revenue generated from the sale of x picnic tables is given by

$$R(x) = 20x - \frac{x^2}{500}.$$

(a) Find the marginal revenue when $x = 1000$ units.
(b) Determine the actual revenue from the sale of the 1001st item.
(c) Compare the answers to parts (a) and (b). How are they related?

32. Business The cost of producing x tacos is

$$C(x) = 1000 + .24x^2, \qquad 0 \le x \le 30{,}000.$$

(a) Find the marginal cost, $C'(x)$.
(b) Find and interpret $C'(100)$.
(c) Find the exact cost to produce the 101st taco.
(d) Compare the answers to parts (b) and (c). How are they related?

33. Business Suppose the demand for a certain item is given by $D(p) = -2p^2 + 4p + 6$, where p represents the price of the item in dollars.
(a) Find the rate of change of demand with respect to price.
(b) Find and interpret the rate of change of demand when the price is \$10.

34. Business The profit (in dollars) from the expenditure of x thousand dollars on advertising is given by $P(x) = 1000 + 32x - 2x^2$. Find the marginal profit at the following expenditures. In each case, decide whether the firm should increase the expenditure.
(a) \$8000
(b) \$6000
(c) \$12,000
(d) \$20,000

 *Use technology to graph the (numerical) derivative of each function by graphing y = NDeriv(f(x), x).* See Example 8.)*

35. $f(x) = x^2 - 3x + 5$

36. $f(x) = .5x^3 + 3x^2 - x + 4$

37. $f(x) = \ln x$

38. $f(x) = e^x + 2$

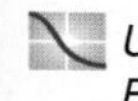 *Use numerical derivatives to work these exercises. (See Examples 8 and 9.)*

39. Natural Science In a research study, the population of shellfish in an area at time t (months) was approximated by the function $f(t) = 10{,}000(1 - e^{-.2t}) + 2000$.

(a) Estimate the value of the derivative function when $t = 2$, $t = 10$, $t = 30$, and $t = 60$.

(b) What does your answer to part (a) say about the rate at which the shellfish population is increasing?

40. Natural Science An inversion layer is formed when air at a higher altitude is warmer than air at sea level. For altitudes up to 10,000 feet, the function

$$T(h) = .014h^4 - .42h^3 + 3.72h^2 + 10h + 18$$

gives the temperature T (in degrees Celsius) at altitude h thousand feet during a recent inversion layer over Southern California. Estimate and interpret the value of the derivative at the following altitudes: 1000, 4000, and 8000 feet.

41. Business A small firm has begun a major cost cutting initiative. Their cost of operation in thousands of dollars (excluding salaries) in month t is given by $C(t) = 25e^{-.2t} + 45$.

(a) At what rate are costs decreasing at the beginning of the program, and at the end of the first, second, and third years?

(b) What do these numbers mean?

 *Use technology for Exercises 42–46.*

42. Finance The table gives the actual and projected year-end assets (in billions of dollars) in Social Security trust funds.

Year	*Assets*
1990	214
1996	550
1998	800
2000	1000
2010	2500
2020	3800
2030	250

(a) Let $x = 0$ correspond to 1990. Use quartic regression to find a fourth-degree polynomial function f that models this data. Round all coefficients to four decimal places. (Polynomial regression is discussed at the end of Section 3.6.)

(b) Use the model to estimate the assets in 2000, 2010, and 2020. Does it fit the data well?

(c) Use numerical derivatives to find the slope of the tangent line to the graph of f at $x = 10$, $x = 20$, $x = 30$, and $x = 35$.

(d) Use your results in part (c) to describe the graph of f and interpret the corresponding changes in Social Security assets.

43. (a) Graph the (numerical) derivative of

$$f(x) = .5x^5 - 2x^3 + x^2 - 3x + 2 \text{ for } -3 \le x \le 3.$$

(b) Graph $g(x) = 2.5x^4 - 6x^2 + 2x - 3$ on the same screen.

(c) How do the graphs of $f'(x)$ and $g(x)$ compare? What does this suggest that the derivative of $f(x)$ is?

44. Repeat Exercise 43 for $f(x) = (x^2 + x + 1)^{1/3}$ (with $-6 \le x \le 6$) and $g(x) = \dfrac{2x + 1}{3(x^2 + x + 1)^{2/3}}$.

45. By using a graphing calculator to compare graphs, as in Exercises 43 and 44, decide which of the following functions could *possibly* be the derivative of $y = \dfrac{4x^2 + x}{x^2 + 1}$.

(a) $f(x) = \dfrac{2x + 1}{2x}$

(b) $g(x) = \dfrac{x^2 + x}{2x}$

(c) $h(x) = \dfrac{2x + 1}{x^2 + 1}$

(d) $k(x) = \dfrac{-x^2 + 8x + 1}{(x^2 + 1)^2}$

46. If f is a function such that $f'(x)$ is defined, then it can be proved that

$$f'(x) = \lim_{h \to 0} \frac{f(x + h) - f(x - h)}{2h}.$$

Consequently, when h is very small, say $h = .001$, then

$$f'(x) \approx \frac{f(x + h) - f(x - h)}{2h} = \frac{f(x + .001) - f(x - .001)}{.002}.$$

(a) In Example 6 we saw that the derivative of $f(x) = \sqrt{x}$ is the function $f'(x) = \dfrac{1}{2\sqrt{x}}$. Make a table in which the first column lists $x = 1, 6, 11, 16, 21$; the second column lists the corresponding value of $f'(x)$, and the third column lists the corresponding value of

$$\frac{f(x + .001) - f(x - .001)}{.002}.$$

*On TI-83, use NDeriv($f(x), x, x$).

(b) How do the second and third columns of the table compare? (If you used the table feature on a graphing calculator, the entries in the table are rounded off, so move the cursor over each entry to see it fully displayed at the bottom of the screen.) Your answer to this question may explain why most graphing calculators use the method in the third column to compute numerical derivatives.

11.4 TECHNIQUES FOR FINDING DERIVATIVES

In the previous section, the derivative of a function was defined as a special limit. The mathematical process of finding this limit, called *differentiation,* resulted in a new function that was interpreted in several different ways. Using the definition to calculate the derivative of a function is a very involved process even for simple functions. In this section we develop rules that make the calculation of derivatives much easier. Keep in mind that even though the process of finding a derivative will be greatly simplified with these rules, *the interpretation of the derivative will not change.*

In addition to y' and $f'(x)$, there are several other commonly used notations for the derivative.

1

Use the results of some of Exercises 22–26 in the previous section to find each of the following.

(a) $\dfrac{d}{dx}(x^3 + 3x)$

(b) $\dfrac{d}{dx}\left(-\dfrac{2}{x}\right)$

(c) $D_x\left(\dfrac{4}{x-1}\right)$

(d) $D_x(\sqrt{x})$

Answers:

(a) $3x^2 + 3$

(b) $\dfrac{2}{x^2}$

(c) $\dfrac{-4}{(x-1)^2}$

(d) $\dfrac{1}{2\sqrt{x}}$

Notations for the Derivative

The derivative of the function $y = f(x)$ may be denoted in any of the following ways.

$$f'(x), \quad y', \quad \frac{dy}{dx}, \quad \frac{d}{dx}[f(x)], \quad D_x y, \quad D_x[f(x)]$$

The dy/dx notation for the derivative is sometimes referred to as *Leibniz notation,* named after one of the co-inventors of calculus, Gottfried Wilhelm Leibniz (1646–1716). (The other was Sir Isaac Newton (1642–1727).)

For example, the derivative of $y = x^3 - 4x$, which we found in Example 4 of the last section to be $y' = 3x^2 - 4$, can also be written

$$\frac{dy}{dx} = 3x^2 - 4$$

$$\frac{d}{dx}(x^3 - 4x) = 3x^2 - 4$$

$$D_x(x^3 - 4x) = 3x^2 - 4.$$ 1

A variable other than x may be used as the independent variable. For example, if $y = f(t)$ gives population growth as a function of time, then the derivative of y with respect to t could be written

$$f'(t), \quad \frac{dy}{dt}, \quad \frac{d}{dt}[f(t)], \quad \text{or} \quad D_t[f(t)].$$

The next rule shows how to find the derivative of the product of a constant and a function.

Constant Times a Function

Let k be a real number. If $g'(x)$ exists, then the derivative of $f(x) = k \cdot g(x)$ is

$$f'(x) = k \cdot g'(x).$$

(The derivative of a constant times a function is the constant times the derivative of the function.)

EXAMPLE 3 **(a)** If $y = 8x^4$, find y'.

Since the derivative of $g(x) = x^4$ is $g'(x) = 4x^3$ and $y = 8x^4 = 8g(x)$,

$$y' = 8g'(x) = 8(4x^3) = 32x^3.$$

(b) If $y = -\frac{3}{4}t^{12}$, find dy/dt.

$$\frac{dy}{dt} = -\frac{3}{4}\left[\frac{dy}{dt}(t^{12})\right] = -\frac{3}{4}(12t^{11}) = -9t^{11}.$$

(c) Find $D_x(15x)$.

$$D_x(15x) = 15 \cdot D_x(x) = 15(1) = 15.$$

(d) If $y = 6/x^2$, find y'.

Replace $\frac{6}{x^2}$ by $6 \cdot \frac{1}{x^2}$, or $6x^{-2}$. Then

$$y' = 6(-2x^{-3}) = -12x^{-3} = -\frac{12}{x^3}.$$

(e) Find $D_x(10x^{3/2})$ and use a graphing calculator to confirm your answer numerically and graphically.

$$D_x(10x^{3/2}) = 10\left(\frac{3}{2}x^{1/2}\right) = 15x^{1/2}.$$

To confirm this result numerically make a table of values for $y_1 = 15x^{1/2}$ and $y_2 =$ the numerical derivative of $10x^{3/2}$; check your instruction manual for the correct syntax for y_2, which is probably one of the following:

NDeriv($10x^{3/2}$,x) or NDeriv($10x^{3/2}$,x,x) or d/dx($10x^{3/2}$) or d/dx($10x^{3/2}$,x).

Figure 11.23(a) indicates that the corresponding values are identical to three decimal places. To confirm this result graphically, graph y_1 and y_2 on the same screen and verify that the graphs appear to be the same. (See Figure 11.23 (b)). ◀ 4

4

Find the derivatives of the following.

(a) $y = 12x^3$

(b) $f(t) = 30t^7$

(c) $y = -35t$

(d) $y = 5\sqrt{x}$

(e) $y = -10/t$

Answers:

(a) $36x^2$

(b) $210t^6$

(c) -35

(d) $(5/2)x^{-1/2}$ or $5/(2\sqrt{x})$

(e) $10t^{-2}$ or $10/t^2$

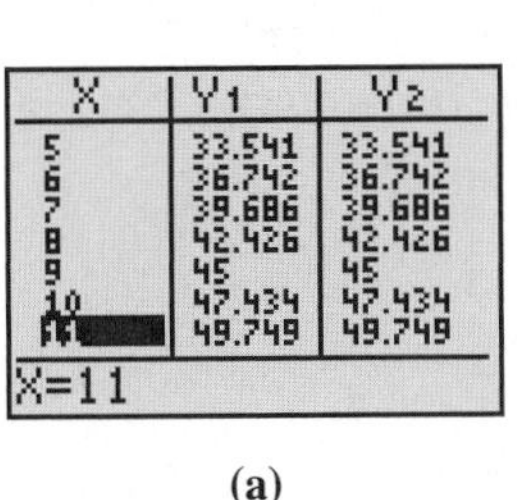

X	Y1	Y2
5	33.541	33.541
6	36.742	36.742
7	39.686	39.686
8	42.426	42.426
9	45	45
10	47.434	47.434
11	49.749	49.749

X=11

(a)

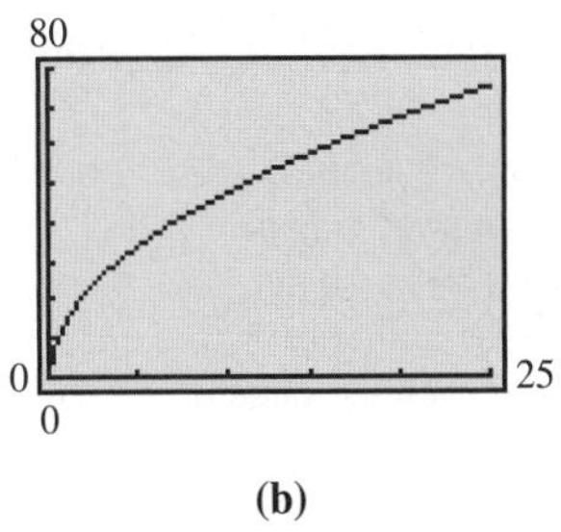

(b)

FIGURE 11.23

Confirming your calculations numerically or graphically, as in part (e) of Example 3, is a good way to detect algebraic errors. If you compute the rule of the derivative $f'(x)$, but its graph differs from the graph of the numerical derivative of $f(x)$, then you've made a mistake. If the two graphs appear to be identical, then you are *probably* correct. (The fact that two graphs appear identical on a calculator screen does not *prove* that they really are identical.)

The final rule in this section is for the derivative of a function that is a sum or difference of functions.

Sum or Difference Rule

If $f(x) = u(x) + v(x)$, and if $u'(x)$ and $v'(x)$ exist, then

$$\mathbf{f'(x) = u'(x) + v'(x).}$$

If $f(x) = u(x) - v(x)$, then

$$\mathbf{f'(x) = u'(x) - v'(x).}$$

(The derivative of a sum or difference of two functions is the sum or difference of the derivatives of the functions.)

For a proof of this rule, see Exercise 68 at the end of this section. This rule also works for sums and differences with more than two terms.

5

Find the derivatives of the following.

(a) $y = -4x^5 - 8x + 6$

(b) $y = 8t^{3/2} + 2t^{1/2}$

(c) $f(t) = -\sqrt{t} + 6/t$

Answers:

(a) $y' = -20x^4 - 8$

(b) $y' = 12t^{1/2} + t^{-1/2}$ or $12t^{1/2} + 1/t^{1/2}$

(c) $f'(t) = -1/(2\sqrt{t}) - 6/t^2$

EXAMPLE 4 Find the derivatives of the following functions.

(a) $y = 6x^3 + 15x^2$

Let $u(x) = 6x^3$ and $v(x) = 15x^2$; then $y = u(x) + v(x)$. Since $u'(x) = 18x^2$ and $v'(x) = 30x$,

$$\frac{dy}{dx} = 18x^2 + 30x.$$

(b) $p(t) = 8t^4 - 6\sqrt{t} + \dfrac{5}{t}$

Rewrite $p(t)$ as $p(t) = 8t^4 - 6t^{1/2} + 5t^{-1}$; then $p'(t) = 32t^3 - 3t^{-1/2} - 5t^{-2}$, which also may be written as $p'(t) = 32t^3 - \dfrac{3}{\sqrt{t}} - \dfrac{5}{t^2}$. (5)

Use a graphing calculator to confirm your answer to part (c) graphically by graphing Dx(f(x)) and the numerical derivative of $f(x)$ on the same screen.

Answer:

Both graphs look like this.

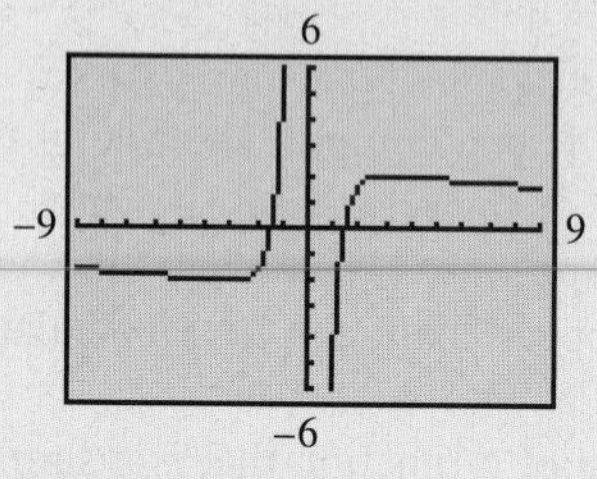

(c) $f(x) = 5\sqrt[3]{x^2} + 4x^{-2} + 7$

Rewrite $f(x)$ as $f(x) = 5x^{2/3} + 4x^{-2} + 7$. Then

$$D_x[f(x)] = \frac{10}{3}(x^{-1/3}) - 8x^{-3},$$

or

$$D_x[f(x)] = \frac{10}{3\sqrt[3]{x}} - \frac{8}{x^3}$$ ◀ 6

The rules developed in this section make it possible to find the derivative of a function more directly, so that applications of the derivative can be dealt with more effectively. The following examples illustrate some business applications.

Marginal Analysis

In business and economics the rates of change of such variables as cost, revenue, and profit are important considerations. Economists use the word *marginal* to refer to rates of change: for example, *marginal cost* refers to the rate of change of cost. Since the derivative of a function gives the rate of change of the function, a marginal cost (or revenue, or profit) function is found by taking the derivative of the cost (or revenue, or profit) function. Roughly speaking, the marginal cost at some level of production x is the cost to produce the $(x + 1)$st item, as we now show. (Similar statements could be made for revenue or profit.)

Look at Figure 11.24 where $C(x)$ represents the cost of producing x units of some item. Then the cost of producing $x + 1$ units is $C(x + 1)$. The cost of the $(x + 1)$st unit is, therefore, $C(x + 1) - C(x)$. This quantity is shown on the graph in Figure 11.24.

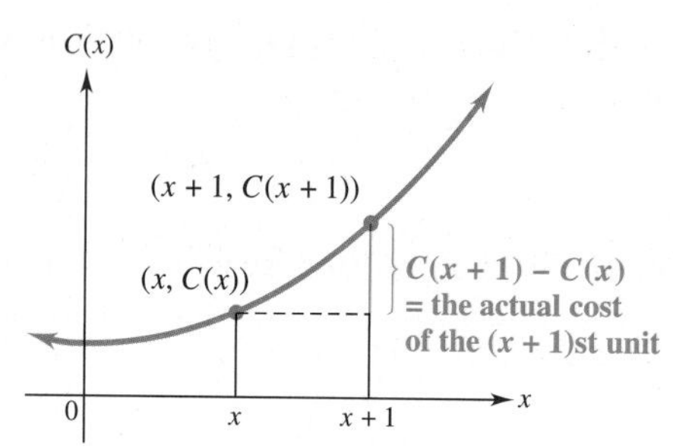

FIGURE 11.24

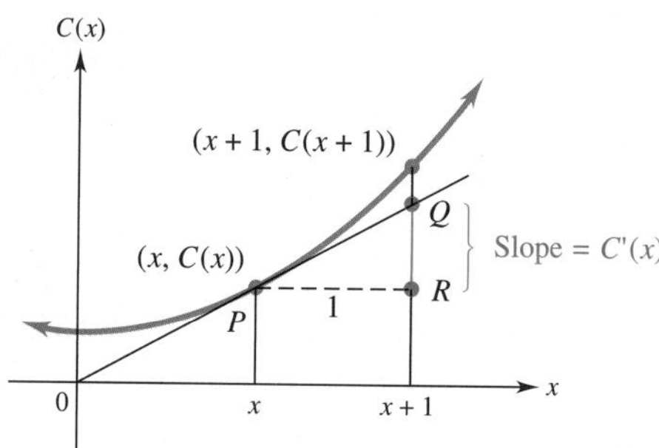

FIGURE 11.25

Now if $C(x)$ is the cost function, then the marginal cost $C'(x)$ represents the slope of the tangent line at any point $(x, C(x))$. The graph in Figure 11.25 shows the cost function $C(x)$ and the tangent line at point $P = (x, C(x))$. We know that the slope of the tangent line is $C'(x)$ and that the slope can be computed using the triangle PQR in Figure 11.25.

$$C'(x) = \text{slope} = \frac{QR}{PR} = \frac{QR}{1} = QR$$

So the length of the line segment QR is the number $C'(x)$.

Superimposing the graphs from Figures 11.24 and 11.25, as in Figure 11.26 below, shows that $C'(x)$ is indeed very close to $C(x + 1) - C(x)$. The two values are closest when $C'(x)$ is very large, so that 1 unit is relatively small.

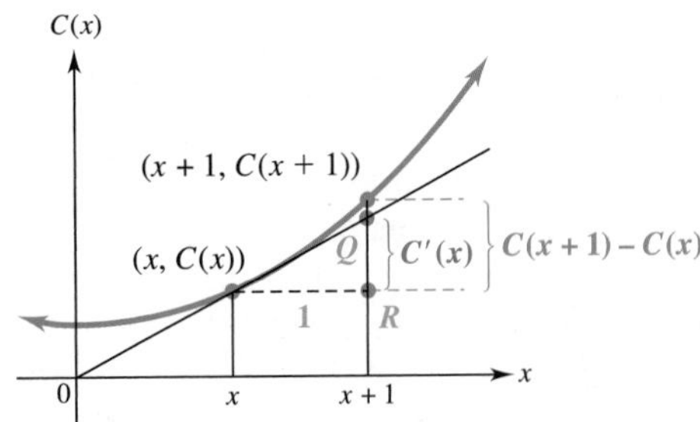

FIGURE 11.26

Therefore, we have the following conclusion.

Marginal Cost

If $C(x)$ is the cost function, then the marginal cost (rate of change of cost) is given by the derivative $C'(x)$.

$$C'(x) \approx \text{cost of making one more item after } x \text{ have been made.}$$

The marginal revenue $R'(x)$ and marginal profit $P'(x)$ are interpreted similarly.

EXAMPLE 5 Suppose that the total cost in hundreds of dollars to produce x thousand barrels of a beverage is given by

$$C(x) = 4x^2 + 100x + 500 \qquad (0 \le x \le 50).$$

Find the marginal cost for the following values of x.

(a) $x = 5$

To find the marginal cost, first find $C'(x)$, the derivative of the total cost function.

$$C'(x) = 8x + 100$$

When $x = 5$,

$$C'(5) = 8(5) + 100 = 140.$$

After 5 thousand barrels of the beverage have been produced, the cost to produce 1 thousand more barrels will be *approximately* 140 hundred dollars, or \$14,000.

The *actual* cost to produce 1 thousand more barrels is $C(6) - C(5)$.

$$\begin{aligned} C(6) - C(5) &= (4 \cdot 6^2 + 100 \cdot 6 + 500) - (4 \cdot 5^2 + 100 \cdot 5 + 500) \\ &= 1244 - 1100 \\ &= 144 \end{aligned}$$

The actual cost is 144 hundred dollars, or \$14,400.

7

The cost in dollars to produce x units of wheat is given by

$$C(x) = 5000 + 20x + 10\sqrt{x}.$$

Find the marginal cost when

(a) $x = 9$;

(b) $x = 16$;

(c) $x = 25$.

(d) As more wheat is produced, what happens to the marginal cost?

Answers:

(a) $65/3 \approx \$21.67$

(b) $85/4 = \$21.25$

(c) $21

(d) It decreases and approaches $20.

(b) $x = 30$

After 30 thousand barrels have been produced, the cost to produce 1 thousand more barrels will be approximately

$$C'(30) = 8(30) + 100 = 340,$$

or $34,000. Notice that the cost to produce an additional thousand barrels of beverage has increased by approximately $20,000 at a production level of 30 thousand barrels, compared with a production level of 5 thousand barrels. Management must be careful to keep track of marginal costs. If the marginal cost of producing an extra unit exceeds the revenue received from selling it, then the company will lose money on that unit. ◀ 7

Demand Functions

The **demand function,** defined by $p = f(x)$, relates the number of units x of an item that consumers are willing to purchase at the price p. (Demand functions were also discussed in Section 3.3.) The total revenue $R(x)$ is related to price per unit and the amount demanded (or sold) by the equation

$$R(x) = xp = x \cdot f(x).$$

EXAMPLE 6 The demand function for a certain product is given by

$$p = \frac{50{,}000 - x}{25{,}000}.$$

Find the marginal revenue when $x = 10{,}000$ units and p is in dollars.

From the given function for p, the revenue function is given by

$$\begin{aligned} R(x) &= xp \\ &= x\left(\frac{50{,}000 - x}{25{,}000}\right) \\ &= \frac{50{,}000x - x^2}{25{,}000} = 2x - \frac{1}{25{,}000}x^2. \end{aligned}$$

The marginal revenue is

$$R'(x) = 2 - \frac{2}{25{,}000}x.$$

When $x = 10{,}000$, the marginal revenue is

$$R'(10{,}000) = 2 - \frac{2}{25{,}000}(10{,}000) = 1.2,$$

or $1.20 per unit. Thus, the next unit sold (at sales of 10,000) will produce additional revenue of about $1.20. ◀ 8

Suppose the demand function for x units of an item is

$$p = 5 - \frac{x}{1000},$$

where x is the price in dollars. Find

(a) the marginal revenue;

(b) marginal revenue at $x = 500$;

(c) marginal revenue at $x = 1000$.

Answers:

(a) $R'(x) = 5 - \dfrac{x}{500}$

(b) $4

(c) $3

In economics, the demand function is written in the form $p = f(x)$, as shown above. From the perspective of a consumer, it is probably more reasonable to think of the quantity demanded as a function of price. Mathematically, these two viewpoints are equivalent. In Example 6, the demand function could have been written from the consumer's viewpoint as

$$x = 50{,}000 - 25{,}000p.$$

EXAMPLE 7 Suppose that the cost function for the product in Example 6 is given by

$$C(x) = 2100 + .25x, \quad \text{where } 0 \le x \le 30{,}000.$$

Find the marginal profit from the production of the following numbers of units.

(a) 15,000

From Example 6, the revenue from the sale of x units is

$$R(x) = 2x - \frac{1}{25{,}000}x^2.$$

Since profit, P, is given by $P = R - C$,

$$\begin{aligned} P(x) &= R(x) - C(x) \\ &= \left(2x - \frac{1}{25{,}000}x^2\right) - (2100 + .25x) \\ &= 2x - \frac{1}{25{,}000}x^2 - 2100 - .25x \\ &= 1.75x - \frac{1}{25{,}000}x^2 - 2100. \end{aligned}$$

The marginal profit from the sale of x units is

$$P'(x) = 1.75 - \frac{2}{25{,}000}x = 1.75 - \frac{1}{12{,}500}x.$$

At $x = 15{,}000$, the marginal profit is

$$P'(15{,}000) = 1.75 - \frac{1}{12{,}500}(15{,}000) = .55,$$

or \$.55 per unit.

(b) 21,875

When $x = 21{,}875$, the marginal profit is

$$P'(21{,}875) = 1.75 - \frac{1}{12{,}500}(21{,}875) = 0.$$

(c) 25,000

When $x = 25{,}000$, the marginal profit is

$$P'(25{,}000) = 1.75 - \frac{1}{12{,}500}(25{,}000) = -.25,$$

or $-$\$.25 per unit.

As shown by parts (b) and (c), if more than 21,875 units are sold, the marginal profit is negative. This indicates that increasing production beyond that level will *reduce* profit. 9

9

For a certain product, the cost is $C(x) = 1250 + .75x$, and the revenue is

$$R(x) = 5x - \frac{x^2}{10{,}000}$$

for x units.

(a) Find the profit $P(x)$.

(b) Find $P'(20{,}000)$.

(c) Find $P'(30{,}000)$.

(d) Interpret the results of parts (b) and (c).

Answers:

(a) $P(x) = 4.25x - x^2/10{,}000 - 1250$

(b) .25

(c) -1.75

(d) Profit is increasing by \$.25 per unit at 20,000 units in part (b) and decreasing by \$1.75 per unit at 30,000 units in part (c).

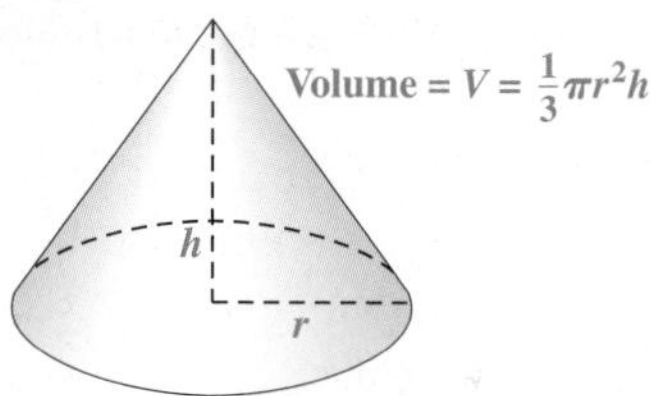

FIGURE 11.27

A balloon is spherical. The formula for the volume of a sphere is $V = (4/3)\pi r^3$, where r is the radius of the sphere. Find the following.

(a) dV/dr

(b) The rate of change of the volume when $r = 3$ inches

Answers:

(a) $4\pi r^2$

(b) 36π cubic inches per inch

The final example shows a medical application of the derivative as the rate of change of a function.

EXAMPLE 8 A tumor has the approximate shape of a cone. See Figure 11.27. The radius of the tumor is fixed by the bone structure at 2 centimeters, but the tumor is growing along the height of the cone. The formula for the volume of a cone is $V = \frac{1}{3}\pi r^2 h$, where r is the radius of the base and h is the height of the cone. Find the rate of change in the volume of the tumor with respect to the height.

To emphasize that the rate of change of the volume is found with respect to the height, we use the symbol dV/dh for the derivative. For this tumor, r is fixed at 2 cm. By substituting 2 for r,

$$V = \frac{1}{3}\pi r^2 h \quad \text{becomes} \quad V = \frac{1}{3}\pi \cdot 2^2 \cdot h \quad \text{or} \quad V = \frac{4}{3}\pi h.$$

Since $4\pi/3$ is constant,

$$\frac{dV}{dh} = \frac{4\pi}{3} \approx 4.2 \text{ cu cm per cm.}$$

For each additional centimeter that the tumor grows in height, its volume will increase approximately 4.2 cubic centimeters. (10)

11.4 Exercises

Find the derivatives of these functions. (See Examples 1–4.)

1. $f(x) = 4x^2 - 6x + 5$

2. $g(x) = 8x^2 + x - 12$

3. $y = 2x^3 + 3x^2 - 5x + 2$

4. $y = 4x^3 + 4x + 4$

5. $g(x) = x^4 + 3x^3 - 6x - 7$

6. $f(x) = 5x^6 - 3x^4 + x^3 - 3x + 9$

7. $f(x) = 6x^{1.5} - 4x^{.5}$

8. $f(x) = -2x^{2.5} + 8x^{.5}$

9. $y = -15x^{3/2} + 2x^{1.9}$

10. $y = 18x^{1.6} - 4x^{3.1}$

11. $y = 24t^{3/2} + 4t^{1/2}$

12. $y = -24t^{5/2} - 6t^{1/2}$

13. $y = 8\sqrt{x} + 6x^{3/4}$

14. $y = -100\sqrt{x} - 11x^{2/3}$

15. $g(x) = 6x^{-5} - x^{-1}$

16. $y = 4x^{-3} + x^{-1} + 5$

17. $y = 10x^{-2} + 3x^{-4} - 6x$

18. $y = x^{-5} - x^{-2} + 5x^{-1}$

19. $f(t) = \frac{6}{t} - \frac{8}{t^2}$

20. $f(t) = \frac{4}{t} + \frac{2}{t^3}$

21. $y = \frac{9 - 8x + 2x^3}{x^4}$

22. $y = \frac{3 + x - 7x^4}{x^6}$

23. $g(x) = 8x^{-1/2} - 5x^{1/2} + x$

24. $f(x) = -12x^{-1/2} + 12x^{1/2} - 12x$

25. $y = 4x^{-3/2} + 8x^{-1/2} + x^2$

26. $y = 2x^{1/2} + 5 + 2x^{-1/2} + x^{-3/2}$

27. $y = \frac{6}{\sqrt[4]{x}}$

28. $y = \frac{-2}{\sqrt[3]{x}}$

29. $y = \frac{-5t}{\sqrt[3]{t^2}}$

30. $g(t) = \frac{9t}{\sqrt{t^3}}$

Find each of the following.

31. $\frac{dy}{dx}$ if $y = 8x^{-5} - 9x^{-4}$

32. $\frac{dy}{dx}$ if $y = -3x^{-2} - 4x^{-5}$

33. $D_x\left(9x^{-1/2} + \frac{2}{x^{3/2}}\right)$

34. $D_x\left(\frac{8}{\sqrt[4]{x}} - \frac{3}{\sqrt{x^3}}\right)$

35. $f'(-2)$ if $f(x) = 6x^2 - 4x$

36. $f'(3)$ if $f(x) = 9x^3 - 8x^2$

37. $f'(4)$ if $f(t) = 2\sqrt{t} - \frac{3}{\sqrt{t}}$

38. $f'(8)$ if $f(t) = -5\sqrt[3]{t} + \frac{6}{\sqrt[3]{t}}$

39. If $f(x) = -\frac{(3x^2 + x)^2}{7}$, which of the following is *closest* to $f'(1)$?
(a) -12 **(b)** -9 **(c)** -6
(d) -3 **(e)** 0 **(f)** 3

40. If $g(x) = -3x^{3/2} + 4x^2 - 9x$, which of the following is *closest* to $g'(4)$?
(a) 3 **(b)** 6 **(c)** 9
(d) 12 **(e)** 15 **(f)** 18

Find the slope and the equation of the tangent line to the graph of each function at the given value of x.

41. $f(x) = x^4 - 2x^2 + 1; x = 1$

42. $g(x) = -x^5 + 4x^2 - 2x + 2; x = 2$

43. $y = 4x^{1/2} + 2x^{3/2} + 1; x = 4$

44. $y = -x^{-3} + 5x^{-1} + x; x = 2$

Work these exercises. (See Examples 5 and 7.)

45. Business The profit in dollars from the sale of x expensive cassette recorders is

$$P(x) = x^3 - 5x^2 + 7x + 10.$$

Find the marginal profit for the following values of x.
(a) $x = 4$ **(b)** $x = 8$ **(c)** $x = 10$ **(d)** $x = 12$

46. Business The total cost to produce x handcrafted weathervanes is

$$C(x) = 100 + 8x - x^2 + 4x^3.$$

Find the marginal cost for the following values of x.
(a) $x = 0$ **(b)** $x = 4$ **(c)** $x = 6$ **(d)** $x = 8$

47. Business Often sales of a new product grow rapidly at first and then level off with time. This is the case with the sales represented by the function

$$S(t) = 100 - 100t^{-1},$$

where t represents time in years. Find the rate of change of sales for the following values of t.
(a) $t = 1$ **(b)** $t = 10$

48. Business An analyst has found that a company's costs and revenues for one product are given by

$$C(x) = 2x \quad \text{and} \quad R(x) = 6x - \frac{x^2}{1000},$$

respectively, where x is the number of items produced.
(a) Find the marginal cost function.
(b) Find the marginal revenue function.
(c) Using the fact that profit is the difference between revenue and costs, find the marginal profit function.
(d) What value of x makes marginal profit equal 0?
(e) Find the profit when the marginal profit is 0. (As we shall see in the next chapter, this process is used to find *maximum* profit.)

49. Finance The total amount of money in circulation (in millions of dollars) for the years 1920–2000 can be approximated by

$$M(x) = .06x^4 - 10.84x^3 + 719.63x^2 - 19{,}637.5x + 189{,}011.12,$$

where x is the number of years since 1900.
(a) Find the derivative of $M(x)$.

Use $M'(x)$ to find the rate of change of money in circulation in the following years.
(b) 1920 **(c)** 1925 **(d)** 1930 **(e)** 1935
(f) 1960 **(g)** 1980 **(h)** 2000
(i) What do your answers in parts (b)–(h) tell you about the amount of money in circulation in those years?

50. Business U.S. postal rates have risen steadily since 1964 (when a first-class stamp cost 5¢). The cost (in cents) of a first-class stamp in year x can be approximated by

$$g(x) = -.0000042x^4 - .00018x^3 + .022x^2 + .485x + 4.81,$$

where $x = 0$ corresponds to 1964.
(a) Use the function g to find the estimated cost of a stamp in 1983, 1993, and 2003 (rounded to the nearest penny). How do these estimates compare with the actual costs (which were 20¢ in 1983 and 29¢ in 1993)?
(b) Find the rate of change of postage costs in 1983, 1993, and 2003 and interpret your results.
(c) Does this model have limitations? What are they?

Work these exercises. (See Example 6.)

51. Business Assume that a demand equation is given by $x = 5000 - 100p$. Find the marginal revenue for the following production levels (values of x). (*Hint:* Solve the demand equation for p and use $R(x) = xp$.)
(a) 1000 units **(b)** 2500 units **(c)** 3000 units

52. Business Suppose that for the situation in Exercise 51, the cost of producing x units is given by $C(x) = 3000 - 20x + .03x^2$. Find the marginal profit for each of the following production levels.
(a) 500 units **(b)** 815 units **(c)** 1000 units

Work the following exercises. (See Examples 5–8).

53. Natural Science A short length of blood vessel has a cylindrical shape. The volume of a cylinder is given by $V = \pi r^2 h$. Suppose an experimental device is set up to measure the volume of blood in a blood vessel of fixed length 80 mm as the radius changes.
(a) Find dV/dr.
Suppose a drug is administered which causes the blood vessel to expand. Evaluate dV/dr for the following values of r and interpret your answers.
(b) 4 mm **(c)** 6 mm **(d)** 8 mm

54. Social Science According to the U.S. Census Bureau, the number of Americans (in thousands) who are expected to be over 100 years old in year x is approximated by the function

$$f(x) = .4018x^2 + 2.039x + 50.071,$$

where $x = 0$ corresponds to 1994 and the formula is valid through 2004.
(a) Find a formula giving the rate of change in the number of Americans over 100 years old.
(b) What is the rate of change in the number of Americans expected to be over 100 years old in the year 2000?
(c) In 2000 is the number of Americans expected to be over 100 years old increasing or decreasing?

55. Social Science Living standards are defined by the total output of goods and services (gross domestic product) divided by the total population. In the United States during the 1990s, living standards were closely approximated by

$$f(x) = .002x^4 - .04x^3 + .33x^2 - .475x + 26.77,$$

where $x = 0$ corresponds to 1990 and $f(x)$ is in thousands of 1996 dollars.* Find the derivative of f and use it to find the rate of change in living standards in the following years.
(a) 1990 **(b)** 1993 **(c)** 1994 **(d)** 1997 **(e)** 1999
(f) What do your answers to parts (a)–(e) tell you about living standards in those years?

*Based on data from the U.S. Bureau of Economic Analysis.

56. Natural Science Insulation workers who were exposed to asbestos and employed before 1960 experienced an increased likelihood of lung cancer. If a group of insulation workers have a cumulative total of 100,000 years of work experience with their first date of employment t years ago, then the number of lung cancer cases occurring within the group can be modeled using the function

$$N(t) = .00437t^{3.2}.*$$

Find the rate of growth of the number of workers with lung cancer in the group when the first date of employment is:
(a) 5 years ago; **(b)** 10 years ago.

57. Natural Science The data in the table (from the Minneapolis *Star Tribune* of September 20, 1998) compares a dog's age to a human's. It can be modeled by a linear or by a quadratic function:

$$y_1 = 4.13x + 14.63 \quad \text{or}$$
$$y_2 = -.033x^2 + 4.647x + 13.347,$$

where x is the dog's actual age and y_1 or y_2 is the dog's age in human years.†

Dog Years	*Human Years*
1	16
2	24
3	28
5	36
7	44
9	52
11	60
13	68
15	76

(a) Find y_1 and y_2 when $x = 5$.
(b) Find $\frac{dy_1}{dx}$ and $\frac{dy_2}{dx}$ when $x = 5$ and interpret your answers.
(c) If the first three points are eliminated from the table, find the equation of a line that perfectly fits the reduced set of data. Interpret your findings.
(d) Of the three formulas, which do you prefer? Explain why.

58. Physical Science Researchers who have been studying the alarming rate at which the level of the Dead Sea has

*A. Walker, *Observation and Inference: An Introduction to the Methods of Epidemiology* (Epidemiology Resources Inc., 1991).

†Patrick Vennebush, "Media Clips: A Dog's Human Age," *Mathematics Teacher* 92 (1999):710–712.

been dropping have shown that the change in density $d(x)$ in g/cm^3 of the Dead Sea brine during evaporation can be estimated by the function

$$d(x) = 1.66 - .90x + .47x^2,$$

where x is the fraction of the remaining brine $(0 \le x \le 1)$.*

(a) Estimate the density of the brine when 50% of the brine remains.

(b) Find and interpret the instantaneous rate of change of the density when 50% of the brine remains.

59. Health To increase the velocity of the air flowing through the trachea when a human coughs, the body contracts the windpipe, producing a more effective cough. Tuchinsky determined that the velocity V of the air flowing through the trachea during a cough is given by

$$V = C(R_0 - R)R^2,$$

where C is a constant, based on individual body characteristics, R_0 is the radius of the windpipe before the cough, and R is the radius of the windpipe during the cough.†

(a) Think of V as a function of R. Multiply out the rule for V and find $\frac{dV}{dR}$. (*Hint:* C and R_0 are constants here; if you can't see what to do, it may help to consider the case when $C = 2$ and $R_0 = .15$.)

(b) It can be shown that the maximum velocity of the cough occurs when $\frac{dV}{dR} = 0$. Find the value of R that maximizes the velocity.‡ (Remember that R must be positive.)

60. Health The body mass index (BMI) is a number that can be calculated for any individual as follows: Multiply weight (lb) by 703 and divide by the person's height (in.) squared. That is,

$$BMI = \frac{703w}{h^2},$$

where w is in pounds and h is in inches. The National Heart, Lung, and Blood Institute uses the BMI to determine whether a person is "overweight" $(25 \le BMI < 30)$ or "obese" $(BMI \ge 30)$.

(a) Calculate the BMI for a male who weighs 220 lbs and is 6′2″ tall.

(b) How much weight would the person in part (a) have to lose until he reaches a BMI of 24.9 and is no longer "overweight"?

(c) For a 125 lb female, what is the rate of change of BMI with respect to height? $\left(\textit{Hint:}\text{ take the derivative of the function: } f(h) = \frac{703(125)}{h^2}.\right)$

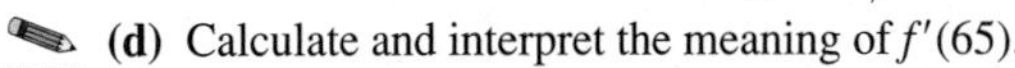

(d) Calculate and interpret the meaning of $f'(65)$.

(e) Use the table function on a graphing calculator to construct a table for BMI for various weights and heights.

Physical Science *We saw earlier that the velocity of a particle moving in a straight line is given by*

$$\lim_{h \to 0} \frac{s(t+h) - s(t)}{h},$$

where s(t) gives the position of the particle at time t. This limit is the derivative of s(t), so the velocity of a particle is given by s′(t). If v(t) represents velocity at time t, then v(t) = s′(t). For each of the following position functions, find (a) *v(t);* (b) *the velocity when t = 0, t = 5, and t = 10.*

61. $s(t) = 8t^2 + 3t + 1$

62. $s(t) = 10t^2 - 5t + 6$

63. $s(t) = 2t^3 + 6t^2$

64. $s(t) = -t^3 + 3t^2 + t - 1$

65. Physical Science If a rock is dropped from a 144-foot-high building, its position (in feet above the ground) is given by $s(t) = -16t^2 + 144$, where t is the time in seconds since it was dropped.

(a) What is its velocity 1 second after being dropped? 2 seconds after being dropped?

(b) When will it hit the ground?

(c) What is its velocity upon impact?

66. Physical Science A ball is thrown vertically upward from the ground at a velocity of 64 feet per second. Its distance from the ground at t seconds is given by $s(t) = -16t^2 + 64t$.

(a) How fast is the ball moving 2 seconds after being thrown? 3 seconds after being thrown?

(b) How long after the ball is thrown does it reach its maximum height?

(c) How high will it go?

67. Perform each step and give reasons for your results in the following proof that the derivative of $y = x^n$ is $y' = n \cdot x^{n-1}$. (We prove this result only for positive integer values of n, but it is valid for all values of n.)

(a) Recall the binomial theorem from algebra:

$$(p+q)^n = p^n + n \cdot p^{n-1}q + \frac{n(n-1)}{2}p^{n-2}q^2 + \cdots + q^n.$$

Evaluate $(x+h)^n$.

*Yoseph Yechieli, Ittai Gavrieli, Brian Berkowitz, and Daniel Ronen, "Will the Dead Sea Die?" *Geology* 26, no. 8 (August 1998); 755–758. These researchers have predicted that the Dead Sea will not die but will reach an equilibrium level.

†Philip Tuchinsky, "The Human Cough," *UMAP Module 211* (Lexington, MA: COMAP, Inc, 1979): 1–9

‡Interestingly, Tuchinsky also states that x-rays indicate that the body naturally contracts the windpipe to this radius during a cough.

(b) Find the quotient $\frac{(x+h)^n - x^n}{h}$.

(c) Use the definition of derivative to find y'.

68. Perform each step and give reasons for your result in the proof that the derivative of $y = f(x) + g(x)$ is

$$y' = f'(x) + g'(x).$$

(a) Let $s(x) = f(x) + g(x)$. Show that

$$s'(x) = \lim_{h\to 0} \frac{[f(x+h) + g(x+h)] - [f(x) + g(x)]}{h}.$$

(b) Show that

$$s'(x) = \lim_{h\to 0}\left[\frac{f(x+h) - f(x)}{h} + \frac{g(x+h) - g(x)}{h}\right].$$

(c) Finally, show that $s'(x) = f'(x) + g'(x)$.

Use a graphing calculator or computer to graph each function and its derivative on the same screen. Determine the values of x where the derivative is (a) positive, (b) zero, and (c) negative. (d) What is true of the graph of the function in each case?

69. $g(x) = 6 - 4x + 3x^2 - x^3$

70. $k(x) = 2x^4 - 3x^3 + x$

11.5 DERIVATIVES OF PRODUCTS AND QUOTIENTS

In the last section we saw that the derivative of the sum of two functions can be obtained by taking the sum of the derivatives. What about products? Is the derivative of a product of two functions equal to the product of their derivatives? For example, if

$$u(x) = 2x + 4 \quad \text{and} \quad v(x) = 3x^2,$$

then the product of u and v is

$$f(x) = (2x + 4)(3x^2) = 6x^3 + 12x^2.$$

Using the rules of the last section, we have

$$u'(x) = 2, \quad v'(x) = 6x, \quad \text{and} \quad f'(x) = 18x^2 + 24x,$$

so that

$$u'(x) \cdot v'(x) = 12x \quad \text{and} \quad f'(x) = 18x^2 + 24x.$$

Obviously, these two functions are *not* the same, which shows that the derivative of the product is *not* equal to the product of the derivatives. For graphical confirmation of this fact, work the problem in the margin. (1)

1

Let $u(x) = \ln x$ and $v(x) = \sqrt{x^2 + x}$. Using a graphing calculator and the viewing window with $0 \le x \le 5$ and $-2 \le y \le 4$, graph these two functions on the same screen:

NDeriv($u(x) \cdot v(x)$)
NDeriv($u(x)$) · NDeriv($v(x)$),

where NDeriv denotes the numerical derivative. What can you conclude from the graphs?

Answer:

Since the two graphs are not the same, the derivative y' of the product function $y = (\ln x)\sqrt{x^2 + x}$ is not equal to the product of the derivatives $u'(x) \cdot v'(x)$.

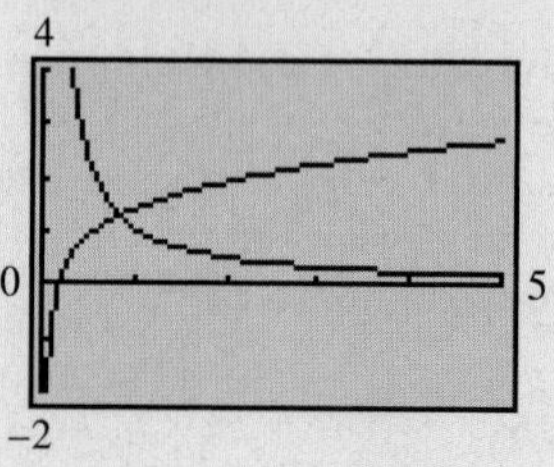

The correct rule for finding the derivative of a product is as follows.

Product Rule

If $f(x) = u(x) \cdot v(x)$, and if both $u'(x)$ and $v'(x)$ exist, then

$$\mathbf{f'(x) = u(x) \cdot v'(x) + v(x) \cdot u'(x).}$$

(The derivative of a product of two functions is the first function times the derivative of the second, plus the second function times the derivative of the first.)

To sketch the method used to prove the product rule, let

$$f(x) = u(x) \cdot v(x).$$

Then $f(x+h) = u(x+h)\cdot v(x+h)$, and, by definition, $f'(x)$ is given by

$$f'(x) = \lim_{h\to 0}\frac{f(x+h)-f(x)}{h}$$
$$= \lim_{h\to 0}\frac{u(x+h)\cdot v(x+h) - u(x)\cdot v(x)}{h}.$$

Now subtract and add $u(x+h)\cdot v(x)$ in the numerator, giving

$$f'(x) = \lim_{h\to 0}\frac{u(x+h)\cdot v(x+h) - u(x+h)\cdot v(x) + u(x+h)\cdot v(x) - u(x)\cdot v(x)}{h}$$

Factor $u(x+h)$ from the first two terms of the numerator and factor $v(x)$ from the last two terms. Then use the properties of limits (Section 11.1), as follows.

$$f'(x) = \lim_{h\to 0}\frac{u(x+h)[v(x+h)-v(x)] + v(x)[u(x+h)-u(x)]}{h}$$
$$= \lim_{h\to 0} u(x+h)\left[\frac{v(x+h)-v(x)}{h}\right] + \lim_{h\to 0} v(x)\left[\frac{u(x+h)-u(x)}{h}\right]$$
$$= \lim_{h\to 0} u(x+h)\cdot\lim_{h\to 0}\frac{v(x+h)-v(x)}{h} + \lim_{h\to 0} v(x)\cdot\lim_{h\to 0}\frac{u(x+h)-u(x)}{h}. \quad (*)$$

If u' and v' both exist, then

$$\lim_{h\to 0}\frac{u(x+h)-u(x)}{h} = u'(x) \quad \text{and} \quad \lim_{h\to 0}\frac{v(x+h)-v(x)}{h} = v'(x).$$

The fact that u' exists can be used to prove

$$\lim_{h\to 0} u(x+h) = u(x),$$

and since no h is involved in $v(x)$,

$$\lim_{h\to 0} v(x) = v(x).$$

Substituting these results into equation (*) gives

$$f'(x) = u(x)\cdot v'(x) + v(x)\cdot u'(x),$$

the desired result.

(2)

Use the product rule to find the derivatives of the following.

(a) $f(x) = (5x^2+6)(3x)$

(b) $g(x) = (8x)(4x^2+5x)$

Answers:

(a) $45x^2 + 18$

(b) $96x^2 + 80x$

EXAMPLE 1 Let $f(x) = (2x+4)(3x^2)$. Use the product rule to find $f'(x)$.

Here f is given as the product of $u(x) = 2x+4$ and $v(x) = 3x^2$. By the product rule and the fact that $u'(x) = 2$ and $v'(x) = 6x$,

$$f'(x) = u(x)\cdot v'(x) + v(x)\cdot u'(x)$$
$$= (2x+4)(6x) + (3x^2)(2)$$
$$= 12x^2 + 24x + 6x^2 = 18x^2 + 24x.$$

This result is the same as that found at the beginning of the section. ◀ (2)

Find the derivatives of the following.

(a) $f(x) = (x^2 - 3)(\sqrt{x} + 5)$

(b) $g(x) = (\sqrt{x} + 4)(5x^2 + x)$

Answers:

(a) $\frac{5}{2}x^{3/2} + 10x - \frac{3}{2}x^{-1/2}$

(b) $\frac{25}{2}x^{3/2} + 40x + \frac{3}{2}x^{1/2} + 4$

EXAMPLE 2 Find the derivative of $y = (\sqrt{x} + 3)(x^2 - 5x)$.

Let $u(x) = \sqrt{x} + 3 = x^{1/2} + 3$, and $v(x) = x^2 - 5x$. Then

$$\begin{aligned} y' &= u(x) \cdot v'(x) + v(x) \cdot u'(x) \\ &= (x^{1/2} + 3)(2x - 5) + (x^2 - 5x)\left(\frac{1}{2}x^{-1/2}\right) \\ &= 2x^{3/2} + 6x - 5x^{1/2} - 15 + \frac{1}{2}x^{3/2} - \frac{5}{2}x^{1/2} \\ &= \frac{5}{2}x^{3/2} + 6x - \frac{15}{2}x^{1/2} - 15. \end{aligned}$$

3

We could have found the derivatives above by multiplying out the original functions. The product rule then would not have been needed. In the next section, however, we shall see products of functions where the product rule is essential.

What about *quotients* of functions? To find the derivative of the quotient of two functions, use the next rule.

Quotient Rule

If $f(x) = \frac{u(x)}{v(x)}$, if all indicated derivatives exist, and if $v(x) \neq 0$, then

$$f'(x) = \frac{v(x) \cdot u'(x) - u(x) \cdot v'(x)}{[v(x)]^2}.$$

(The derivative of a quotient is the denominator times the derivative of the numerator, minus the numerator times the derivative of the denominator, all divided by the square of the denominator.)

The proof of the quotient rule is similar to that of the product rule and is omitted here.

CAUTION Just as the derivative of a product is *not* the product of the derivatives, the derivative of a quotient is *not* the quotient of the derivatives. If you are asked to take the derivative of a product or a quotient, it is essential that you recognize that the function contains a product or quotient and then use the appropriate rule.

EXAMPLE 3 Find $f'(x)$ if $f(x) = \frac{2x - 1}{4x + 3}$.

(4)

Find the derivatives of the following.

(a) $f(x) = \dfrac{3x+7}{5x+8}$

(b) $g(x) = \dfrac{2x+11}{5x-1}$

Answers:

(a) $\dfrac{-11}{(5x+8)^2}$

(b) $\dfrac{-57}{(5x-1)^2}$

Let $u(x) = 2x - 1$, with $u'(x) = 2$. Also, let $v(x) = 4x + 3$, with $v'(x) = 4$. Then, by the quotient rule,

$$f'(x) = \frac{v(x) \cdot u'(x) - u(x) \cdot v'(x)}{[v(x)]^2}$$
$$= \frac{(4x+3)(2) - (2x-1)(4)}{(4x+3)^2}$$
$$= \frac{8x+6-8x+4}{(4x+3)^2}$$
$$f'(x) = \frac{10}{(4x+3)^2}. \quad (4)$$

CAUTION In the second step of Example 3, we had the expression

$$\frac{(4x+3)(2) - (2x-1)(4)}{(4x+3)^2}.$$

Students often incorrectly "cancel" the $4x + 3$ in the numerator with one factor of the denominator. Because the numerator is a *difference* of two products, however, you must multiply and combine terms *before* looking for common factors in the numerator and denominator.

(5)

Find each derivative. Write answers with positive exponents.

(a) $D_x\left(\dfrac{x^{-2}-1}{x^{-1}+2}\right)$

(b) $D_x\left(\dfrac{2+x^{-1}}{x^3+1}\right)$

Answers:

(a) $\dfrac{-1-4x-x^2}{x^2+4x^3+4x^4}$

(b) $-\dfrac{6x^4+4x^3+1}{x^2(x^3+1)^2}$

EXAMPLE 4 Find $D_x\left(\dfrac{x-2x^2}{4x^2+1}\right)$.

Use the quotient rule.

$$D_x\left(\frac{x-2x^2}{4x^2+1}\right) = \frac{(4x^2+1)D_x(x-2x^2) - (x-2x^2)D_x(4x^2+1)}{(4x^2+1)^2}$$
$$= \frac{(4x^2+1)(1-4x) - (x-2x^2)(8x)}{(4x^2+1)^2}$$
$$= \frac{4x^2-16x^3+1-4x-8x^2+16x^3}{(4x^2+1)^2}$$
$$= \frac{-4x^2-4x+1}{(4x^2+1)^2} \quad (5)$$

EXAMPLE 5 Find $D_x\left(\dfrac{(3-4x)(5x+1)}{7x-9}\right)$.

This function has a product within a quotient. Instead of multiplying the factors in the numerator first (which is an option), we can use the quotient rule together with the product rule, as follows. Use the quotient rule first to get

$$D_x\left(\frac{(3-4x)(5x+1)}{7x-9}\right)$$
$$= \frac{(7x-9)[D_x(3-4x)(5x+1)] - [(3-4x)(5x+1)D_x(7x-9)]}{(7x-9)^2}.$$

Find each derivative.

(a) $D_x\left(\frac{(3x-1)(4x+2)}{2x}\right)$

(b) $D_x\left(\frac{5x^2}{(2x+1)(x-1)}\right)$

Answers:

(a) $\frac{6x^2+1}{x^2}$

(b) $\frac{-5x^2-10x}{(2x+1)^2(x-1)^2}$

Now use the product rule to find $D_x(3-4x)(5x+1)$ in the numerator.

$$= \frac{(7x-9)[(3-4x)5+(5x+1)(-4)]-(3+11x-20x^2)(7)}{(7x-9)^2}$$

$$= \frac{(7x-9)(15-20x-20x-4)-(21+77x-140x^2)}{(7x-9)^2}$$

$$= \frac{(7x-9)(11-40x)-21-77x+140x^2}{(7x-9)^2}$$

$$= \frac{-280x^2+437x-99-21-77x+140x^2}{(7x-9)^2}$$

$$= \frac{-140x^2+360x-120}{(7x-9)^2}$$ ◀ 6

Average Cost

Suppose $y = C(x)$ gives the total cost to manufacture x items. As mentioned earlier, the average cost per item is found by dividing the total cost by the number of items. The rate of change of average cost, called the *marginal average cost*, is the derivative of the average cost.

Average Cost

If the total cost to manufacture x items is given by $C(x)$, then the **average cost per item** is

$$\overline{C}(x) = \frac{C(x)}{x}.$$

The **marginal average cost** is the derivative of the average cost function, $\overline{C}'(x)$.

A company naturally would be interested in making the average cost as small as possible. We will see in the next chapter that this can be done by using the derivative of $C(x)/x$. The derivative often can be found with the quotient rule, as in the next example.

EXAMPLE 6 The total cost (in dollars) to manufacture x blank CDs is given by

$$C(x) = \frac{x^2+2x+2}{x+1} + 40{,}000.$$

(a) Find the average cost per CD.

The average cost is given by the total cost divided by the number of items:

$$\overline{C}(x) = \frac{C(x)}{x} = \frac{1}{x}C(x) = \frac{1}{x}\left(\frac{x^2+2x+2}{x+1} + 40{,}000\right) = \frac{x^2+2x+2}{x^2+x} + \frac{40{,}000}{x}.$$

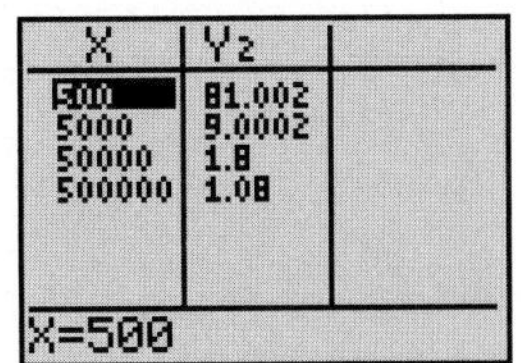

FIGURE 11.28

The total cost (in thousands of dollars) to make x dozen CB radios is given by

$$C(x) = 32x^2 + 7x + 80.$$

(a) Find the average cost.

(b) Find the marginal average cost.

Answers:

(a) $\dfrac{32x^2 + 7x + 80}{x}$

(b) $\dfrac{32x^2 - 80}{x^2}$

(b) Find the average cost per CD for each of these production levels:

$$500 \quad 5000 \quad 50{,}000 \quad 500{,}000.$$

Evaluate $\overline{C}(x)$ at each of these numbers, either by hand or by using technology, as in Figure 11.28. Note that the average cost per CD is \$81 when only 500 are produced but comes down to a more reasonable \$1.08 when 500,000 are made.

(c) Find the marginal average cost.

The marginal average cost is the derivative of the average cost function. Using the sum rule and the quotient rule,

$$\overline{C}'(x) = \frac{(x^2 + x)(2x + 2) - (x^2 + 2x + 2)(2x + 1)}{(x^2 + x)^2} - \frac{40{,}000}{x^2}$$

$$= \frac{(2x^3 + 4x^2 + 2x) - (2x^3 + 5x^2 + 6x + 2)}{(x^2 + x)^2} - \frac{40{,}000}{x^2}$$

$$= \frac{-x^2 - 4x - 2}{(x^2 + x)^2} - \frac{40{,}000}{x^2}.$$ ◀ 7

EXAMPLE 7 Suppose the cost in dollars of manufacturing x hundred items is given by

$$C(x) = 3x^2 + 7x + 12.$$

(a) Find the average cost.

The average cost is

$$\overline{C}(x) = \frac{C(x)}{x} = \frac{3x^2 + 7x + 12}{x} = 3x + 7 + \frac{12}{x}.$$

(b) Find the marginal average cost.

The marginal average cost is

$$\frac{d}{dx}(\overline{C}(x)) = \frac{d}{dx}\left(3x + 7 + \frac{12}{x}\right) = 3 - \frac{12}{x^2}.$$

(c) Find the marginal cost.

The marginal cost is

$$\frac{d}{dx}(C(x)) = \frac{d}{dx}(3x^2 + 7x + 12) = 6x + 7.$$

(d) Find the level of production at which the marginal average cost is zero.

Set the derivative $\overline{C}'(x) = 0$ and solve for x.

$$3 - \frac{12}{x^2} = 0$$

$$\frac{3x^2 - 12}{x^2} = 0$$

$$3x^2 - 12 = 0$$

$$x^2 = 4$$

$$x = \pm 2$$

You can't make a negative number of items, so $x = 2$. Since x is in hundreds, production of 200 items will produce a marginal average cost of zero dollars. ◀ 8

If the cost in Example 7 is given by

$$C(x) = x^2 + 10x + 16,$$

find the production level at which the marginal average cost is zero.

Answer:

400 items

11.5 Exercises

Use the product rule to find the derivatives of the following functions. (See Examples 1 and 2.) (Hint for Exercises 6–9: Write the quantity as a product.)

1. $y = (x^2 - 2)(3x + 1)$
2. $y = (2x^2 + 3)(4x + 5)$
3. $y = (6x^3 + 2)(5x - 3)$
4. $y = (2x^2 + 4x - 3)(5x^3 + x + 2)$
5. $y = (x^4 - 2x^3 + 2x)(4x^2 + x - 3)$
6. $y = (3x - 2)^2$
7. $y = (6x^2 + 4x)^2$
8. $y = (x^2 - 1)^2$
9. $y = (3x^3 + x^2)^2$

Use the quotient rule to find the derivatives of the following functions. (See Examples 3 and 4.)

10. $y = \dfrac{x+1}{2x-1}$
11. $y = \dfrac{3x-5}{x-4}$
12. $f(x) = \dfrac{7x+1}{3x+8}$
13. $f(t) = \dfrac{t^2-4t}{t+3}$
14. $y = \dfrac{4x+11}{x^2-3}$
15. $g(x) = \dfrac{3x^2+x}{2x^3-1}$
16. $k(x) = \dfrac{-x^2+6x}{4x^3+1}$
17. $y = \dfrac{x^2-4x+2}{x+3}$
18. $y = \dfrac{x^2+7x-2}{x-2}$
19. $r(t) = \dfrac{\sqrt{t}}{2t+3}$
20. $y = \dfrac{5x+6}{\sqrt{x}}$
21. $y = \dfrac{9x-8}{\sqrt{x}}$
22. $y = \dfrac{2}{3x-5}$
23. $y = \dfrac{9-7x}{1-x}$
24. $f(t) = \dfrac{t^2+t}{t-1}$

Find the derivative of each of the following. (See Example 5.)

25. $f(p) = \dfrac{(2p+3)(4p-1)}{3p+2}$
26. $g(t) = \dfrac{(5t-2)(2t+3)}{t-4}$
27. $g(x) = \dfrac{x^3+1}{(2x+1)(5x+2)}$
28. $f(x) = \dfrac{x^3-4}{(2x+1)(3x-2)}$

29. Find the error in the following work.

$$D_x\left(\frac{2x+5}{x^2-1}\right) = \frac{(2x+5)(2x)-(x^2-1)2}{(x^2-1)^2} = \frac{4x^2+10x-2x^2+2}{(x^2-1)^2} = \frac{2x^2+10x+2}{(x^2-1)^2}$$

30. Find the error in the following work.

$$D_x\left(\frac{x^2-4}{x^3}\right) = x^3(2x)-(x^2-4)(3x^2) = 2x^4-3x^4-12x^2 = -x^4+12x^2$$

Find the equation of the tangent line to the graph of f(x) at the given point.

31. $f(x) = \dfrac{x}{x-2}$ at $(3, 3)$
32. $f(x) = \dfrac{x}{x^2+1}$ at $(1, 1/2)$
33. $f(x) = \dfrac{(x-1)(2x+3)}{x-5}$ at $(9, 42)$
34. $f(x) = \dfrac{(x+3)^2}{x+2}$ at $(-4, -1/2)$

Work the following exercises. (See Examples 6 and 7.)

35. **Business** The total cost (in hundreds of dollars) to produce x units of perfume is

$$C(x) = \frac{3x+2}{x+4}.$$

Find the average cost for each of the following production levels.

(a) 10 units **(b)** 20 units **(c)** x units
(d) Find the marginal average cost function.

36. **Business** The total profit (in tens of dollars) from selling x self-help books is

$$P(x) = \frac{5x-6}{2x+3}.$$

Find the average profit from each of the following sales levels.

(a) 8 books **(b)** 15 books **(c)** x books
(d) Find the marginal average profit function.
(e) Is this a reasonable function for profit? Why?

37. Business A factory smokestack emits pollutants into the air. The cost (in dollars) of removing x percent of the pollutants is given by the cost-benefit function

$$C(x) = \frac{75{,}000x}{100 - x}.$$

(a) What is the cost of removing 70% of the pollutants? (Use $x = 70$, not .70)
(b) Find $C'(x)$.

Find the rate at which the cost of removing pollutants is changing when the following percentage of pollutants is removed.
(c) 75% (d) 90% (e) 95%

38. Business Suppose you are the manager of a trucking firm, and one of your drivers reports that, according to her calculations, her truck burns fuel at the rate of

$$G(x) = \frac{1}{200}\left(\frac{800}{x} + x\right)$$

gallons per mile when traveling at x miles per hour on a smooth, dry road.
(a) If the driver tells you that she wants to travel 20 miles per hour, what should you tell her? (*Hint:* Take the derivative of G and evaluate it for $x = 20$. Then interpret your results.)
(b) If the driver wants to go 40 miles per hour, what should you say? (*Hint:* Find $G'(40)$.)

39. Health During the course of an illlness, a patient's temperature (in degrees Fahrenheit) x hours after the start of the illness is given by

$$T(x) = \frac{10x}{x^2 + 5} + 98.6.$$

(a) Find dT/dx.
Evaluate dT/dx at the following times and interpret your answer.
(b) $x = 0$ (c) $x = 1$ (d) $x = 3$ (e) $x = 9$

40. Physical Science The relationship between the fixed focal length F of a camera, the distance x from the object being photographed to the lens, and the distance y from the lens to the film is given by $\frac{1}{F} = \frac{1}{x} + \frac{1}{y}$.

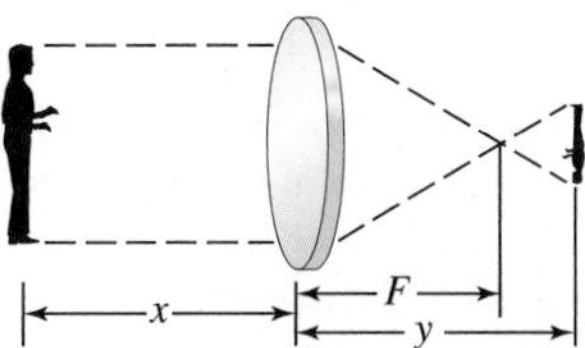

(a) Assume the focal length is 50 mm ($F = 50$). Express y as a function of x.
(b) Find dy/dx.
Suppose the person being photographed moves away from the lens, always being kept in focus. Find the rate at which the distance y from the film to the lens is changing when the person is the following distance from the lens.
(c) 3100 mm (about 10 ft)
(d) 4900 mm (about 16 ft)
(e) 7600 mm (about 25 ft)

41. Natural Science Murrell's formula for calculating the total amount of rest, in minutes, required after performing a particular type of work activity for 30 minutes is given by the formula

$$R(w) = 30\frac{w - 4}{w - 1.5},$$

where w is the work expended in kilocalories per min (kcal/min).*
(a) A value of 5 for w indicates light work, such as riding a bicycle on a flat surface at 10 miles per hour. Find $R(5)$.
(b) A value of 7 for w indicates moderate work, such as mowing grass with a push mower on level ground. Find $R(7)$.
(c) Find $R'(5)$ and $R'(7)$ and compare your answers. Explain whether or not these answers make sense.

42. Natural Science When a certain drug is introduced into a muscle, the muscle responds by contracting. The amount of contraction, s, in millimeters, is related to the concentration of the drug, x, in milliliters, by

$$s(x) = \frac{x}{m + nx},$$

where m and n are constants.
(a) Find $s'(x)$.
(b) Evaluate $s'(x)$ when $x = 50$, $m = 10$, and $n = 3$.
(c) Interpret your results in part (b).

*Mark Sanders and Ernest McCormick, *Human Factors in Engineering and Design,* Seventh Edition (New York: McGraw-Hill, 1993), pp. 243–246.

43. Business The average number of vehicles waiting in line to enter a parking lot can be modeled by the function

$$f(x) = \frac{x^2}{2(1-x)},$$

where x is a quantity between 0 and 1 known as the traffic intensity.* Find the rate of change of the number of vehicles waiting with respect to the traffic intensity for the following values of the intensity.

(a) $x = .1$ **(b)** $x = .6$

44. Natural Science Using data collected by zoologist Reto Zach, the work done by a crow to break open a whelk (large marine snail) can be estimated by the function.

$$W = \left(1 + \frac{20}{H - .93}\right)H,$$

where H is the height of the whelk (in meters) when it is dropped.†

(a) Find $\frac{dW}{dH}$.

(b) The amount of work is minimized when $\frac{dW}{dH} = 0$. Find the value of H that minimizes W.

(c) Interestingly, Zach observed the crows dropping the whelks from an average height of 5.23 meters. What does this imply?

45. Social Science After t hours of instruction, a typical typing student can type

$$N(t) = \frac{70t^2}{30 + t^2}$$

words per minute.

(a) Find $N'(t)$, the rate at which the student is improving after t hours.

(b) At what rate is the student improving after 3 hours? 5 hours? 7 hours? 10 hours? 15 hours?

(c) Describe the student's progress during the first 15 hours of instruction.

46. Physical Science The intensity I of a 100 watt light bulb (in watts per square meter) at distance d meters from the bulb is given by

$$I(d) = \frac{7.92}{d^2}.$$

(a) Find $I'(d)$.

Find the rate at which the intensity is changing when an object is the given distance from the bulb.

(b) 3 meters **(c)** 6 meters **(d)** 9 meters

*F. Mannering and W. Kilareski, *Principles of Highway Engineering and Traffic Control* Second Edition (John Wiley and Sons, 1997).

†Brian Kellar and Heather Thompson, "Whelk-come to Mathematics," *Mathematics Teacher* 92, no. 6 (September 1999): 475–481.

11.6 THE CHAIN RULE

Many of the most useful functions for applications are created by combining simpler functions. Viewing complex functions as combinations of simpler functions often makes them easier to understand and use.

Composition of Functions

Consider the function h whose rule is $h(x) = \sqrt{x^3}$. To compute $h(4)$, for example, you first find $4^3 = 64$ and then take the square root: $\sqrt{64} = 8$. So the rule of h may be rephrased as:

First apply the function $f(x) = x^3$,
then apply the function $g(x) = \sqrt{x}$ to the result.

The same idea can be expressed in functional notation like this:

	First apply f.	Then apply g to the result.
x →	$f(x)$ →	$g[f(x)]$
x	x^3	$\sqrt{x^3}$

Apply h.

So the rule of h may be written as $h(x) = g[f(x)]$, where $f(x) = x^3$ and $g(x) = \sqrt{x}$. We can think of the functions g and f as being "composed" to create the function h. Here is a formal definition of this idea.

Composite Function

Let f and g be functions. The **composite function,** or **composition,** of g and f is the function whose values are given by $g[f(x)]$ for all x in the domain of f such that $f(x)$ is in the domain of g.

EXAMPLE 1 Let $f(x) = 2x - 1$ and $g(x) = \sqrt{3x + 5}$. Find each of the following.

(a) $g[f(4)]$

Find $f(4)$ first.

$$f(4) = 2 \cdot 4 - 1 = 8 - 1 = 7$$

Then

$$g[f(4)] = g[7] = \sqrt{3 \cdot 7 + 5} = \sqrt{26}.$$

(b) $f[g(4)]$

Since $g(4) = \sqrt{3 \cdot 4 + 5} = \sqrt{17}$,

$$f[g(4)] = 2 \cdot \sqrt{17} - 1 = 2\sqrt{17} - 1.$$

(c) $f[g(-2)]$ does not exist since -2 is not in the domain of g. 1

1

Let $f(x) = 3x - 2$ and $g(x) = (x - 1)^5$. Find the following.

(a) $g[f(2)]$

(b) $f[g(2)]$

Answers:

(a) 243

(b) 1

EXAMPLE 2 Let $f(x) = 4x + 1$ and $g(x) = 2x^2 + 5x$. Find each of the following.

(a) $g[f(x)]$

Using the given functions,

$$\begin{aligned} g[f(x)] &= g[4x + 1] \\ &= 2(4x + 1)^2 + 5(4x + 1) \\ &= 2(16x^2 + 8x + 1) + 20x + 5 \\ &= 32x^2 + 16x + 2 + 20x + 5 \\ &= 32x^2 + 36x + 7. \end{aligned}$$

(b) $f[g(x)]$

By the definition above, with f and g interchanged,

$$\begin{aligned} f[g(x)] &= f[2x^2 + 5x] \\ &= 4(2x^2 + 5x) + 1 \\ &= 8x^2 + 20x + 1. \end{aligned}$$ 2

Let $f(x) = \sqrt{x + 4}$ and $g(x) = x^2 + 5x + 1$. Find $f[g(x)]$ and $g[f(x)]$.

Answers:

$f[g(x)] = \sqrt{x^2 + 5x + 5}$,

$g[f(x)] = (\sqrt{x + 4})^2 + 5\sqrt{x + 4} + 1 = x + 5 + 5\sqrt{x + 4}$

As Example 2 shows, $f[g(x)]$ usually is *not* equal to $g[f(x)]$. In fact, it is rare to find two functions f and g for which $f[g(x)] = g[f(x)]$.

The Chain Rule (Alternative Form)

If y is a function of u, say $y = f(u)$, and if u is a function of x, say $u = g(x)$, then $y = f(u) = f[g(x)]$, and

$$\frac{dy}{dx} = \frac{dy}{du} \cdot \frac{du}{dx}$$

provided dy/du and du/dx exist.

One way to remember the chain rule is to *pretend* that dy/du and du/dx are fractions, with du "canceling out."

EXAMPLE 5 Find dy/dx if $y = (3x^2 - 5x)^7$.

Let $y = u^7$, and $u = 3x^2 - 5x$. Then

$$\frac{dy}{dx} = \frac{dy}{du} \cdot \frac{du}{dx}$$

$$= 7u^6 \cdot (6x - 5).$$

Replacing u with $3x^2 - 5x$ gives

$$\frac{dy}{dx} = 7(3x^2 - 5x)^6(6x - 5).$$ 5

Use the chain rule to find dy/dx if $y = 10(2x^2 + 1)^4$.

Answer:

$160x(2x^2 + 1)^3$

Each of the functions in Examples 4 and 5 has the form $y = u^n$, with u a function of x:

$$y = (15x^2 + 1)^{1/2} \quad \text{and} \quad y = (3x^2 - 5x)^7.$$
$$y = u^{1/2} \qquad\qquad\qquad y = u^7$$

Their derivatives are

$$y' = \frac{1}{2}(15x^2 + 1)^{-1/2}(30x) \quad \text{and} \quad y' = 7(3x^2 - 5x)^6(6x - 5)$$
$$y' = \frac{1}{2}u^{-1/2}u' \qquad\qquad\qquad y' = 7u^6u'$$

Thus, these functions are examples of the following special case of the chain rule.

Generalized Power Rule

Let u be a function of x and let $y = u^n$, for any real number n. Then

$$y' = n \cdot u^{n-1} \cdot u',$$

provided that u' exists.

(The derivative of $y = u^n$ is found by decreasing the exponent on u by 1 and multiplying the result by the exponent n and by the derivative of u with respect to x.)

EXAMPLE 6 **(a)** Use the generalized power rule to find the derivative of $y = (3 + 5x)^2$.

Let $u = 3 + 5x$, and $n = 2$. Then $u' = 5$. By the generalized power rule,

$$y' = \frac{dy}{dx} = n \cdot u^{n-1} \cdot u'$$

$$\begin{aligned} &\quad n \quad\; u \quad n-1 \qquad\quad u' \\ &\quad\downarrow \quad \downarrow \quad \downarrow \qquad\quad \downarrow \\ &= 2 \cdot (3 + 5x)^{2-1} \cdot \overbrace{\frac{d}{dx}(3 + 5x)} \\ &= 2(3 + 5x)^{2-1} \cdot 5 \\ &= 10(3 + 5x) \\ &= 30 + 50x. \end{aligned}$$

(b) Find y' if $y = (3 + 5x)^{-3/4}$.

Use the generalized power rule with $n = -\frac{3}{4}$, $u = 3 + 5x$, and $u' = 5$.

$$\begin{aligned} y' &= -\frac{3}{4}(3 + 5x)^{(-3/4)-1}(5) \\ &= -\frac{15}{4}(3 + 5x)^{-7/4} \end{aligned}$$

⑥

⑥

Find dy/dx for the following.

(a) $y = (2x + 5)^6$

(b) $y = (4x^2 - 7)^3$

(c) $f(x) = \sqrt{3x^2 - x}$

(d) $g(x) = (2 - x^4)^{-3}$

Answers:

(a) $12(2x + 5)^5$

(b) $24x(4x^2 - 7)^2$

(c) $\dfrac{6x - 1}{2\sqrt{3x^2 - x}}$

(d) $\dfrac{12x^3}{(2 - x^4)^4}$

EXAMPLE 7 Find the derivative of the following.

(a) $y = 2(7x^2 + 5)^4$

Let $u = 7x^2 + 5$. Then $u' = 14x$, and

$$\begin{aligned} &\qquad\quad n \quad\; u \quad n-1 \qquad\quad u' \\ &\qquad\quad \downarrow \quad \downarrow \quad \downarrow \qquad\quad \downarrow \\ y' &= 2 \cdot 4(7x^2 + 5)^{4-1} \cdot \overbrace{\frac{d}{dx}(7x^2 + 5)} \\ &= 2 \cdot 4(7x^2 + 5)^3(14x) \\ &= 112x(7x^2 + 5)^3. \end{aligned}$$

(b) $y = \sqrt{9x + 2}$

Write $y = \sqrt{9x + 2}$ as $y = (9x + 2)^{1/2}$. Then

$$y' = \frac{1}{2}(9x + 2)^{-1/2}(9) = \frac{9}{2}(9x + 2)^{-1/2}.$$

The derivative also can be written as

$$y' = \frac{9}{2(9x + 2)^{1/2}} \quad \text{or} \quad y' = \frac{9}{2\sqrt{9x + 2}}.$$

⑦

⑦

Find dy/dx for the following.

(a) $y = 12(x^2 + 6)^5$

(b) $y = 8(4x^2 + 2)^{3/2}$

Answers:

(a) $120x(x^2 + 6)^4$

(b) $96x(4x^2 + 2)^{1/2}$

CAUTION **(a)** A common error is to forget to multiply by $g'(x)$ when using the generalized power rule. Remember, the generalized power rule is an example of the chain rule, and so the derivative must involve a "chain", or product, of derivatives.

(b) Another common mistake is to write the derivative as $n[g'(x)]^{n-1}$. Remember to leave $g(x)$ unchanged and then to multiply by $g'(x)$.

Sometimes both the generalized power rule and either the product or quotient rule are needed to find a derivative, as the next examples show.

EXAMPLE 8 Find the derivative of $y = 4x(3x + 5)^5$.

Write $4x(3x + 5)^5$ as the product

$$4x \cdot (3x + 5)^5.$$

To find the derivative of $(3x + 5)^5$, let $u = 3x + 5$ with $u' = 3$. Now use the product rule and the generalized power rule.

$$y' = 4x\underbrace{[5(3x + 5)^4 \cdot 3]}_{\text{Derivative of } (3x+5)^5} + (3x + 5)^5\underbrace{(4)}_{\text{Derivative of } 4x}$$

$$= 60x(3x + 5)^4 + 4(3x + 5)^5$$

$$= 4(3x + 5)^4[15x + (3x + 5)^1] \quad \text{Factor out the greatest common factor, } 4(3x + 5)^4.$$

$$= 4(3x + 5)^4(18x + 5).$$ ◀ 8

8

Find the derivatives of the following.

(a) $y = 6x(x + 2)^2$

(b) $y = -9x(2x^2 + 1)^3$

Answers:

(a) $6(x + 2)(3x + 2)$

(b) $-9(2x^2 + 1)^2(14x^2 + 1)$

EXAMPLE 9 Find the derivative of $y = \dfrac{(3x + 2)^7}{x - 1}$.

Use the quotient rule and the generalized power rule.

$$\frac{dy}{dx} = \frac{(x - 1)[7(3x + 2)^6 \cdot 3] - (3x + 2)^7(1)}{(x - 1)^2}$$

$$= \frac{21(x - 1)(3x + 2)^6 - (3x + 2)^7}{(x - 1)^2}$$

$$= \frac{(3x + 2)^6[21(x - 1) - (3x + 2)]}{(x - 1)^2} \quad \text{Factor out the greatest common factor, } (3x + 2)^6.$$

$$= \frac{(3x + 2)^6[21x - 21 - 3x - 2]}{(x - 1)^2} \quad \text{Simplify inside brackets.}$$

$$\frac{dy}{dx} = \frac{(3x + 2)^6(18x - 23)}{(x - 1)^2}$$ ◀ 9

9

Find the derivatives of the following.

(a) $y = \dfrac{(2x + 1)^3}{3x}$

(b) $y = \dfrac{(x - 6)^5}{3x - 5}$

Answers:

(a) $\dfrac{(2x + 1)^2(4x - 1)}{3x^2}$

(b) $\dfrac{(x - 6)^4(12x - 7)}{(3x - 5)^2}$

Applications

Some applications requiring the use of the chain rule or the generalized power rule are illustrated in the next examples.

Suppose the revenue in Example 10 is given by

$$R(n) = \frac{4500n}{n + 5}$$

and the work hours are decreasing at the rate of 4 per day. How fast is the revenue decreasing?

Answer:

About \$73.47 per day

EXAMPLE 10 The revenue realized by a small city from the collection of fines from parking tickets is given by

$$R(n) = \frac{8000n}{n + 2},$$

where n is the number of work hours each day that can be devoted to parking patrol. At the outbreak of a flu epidemic, 30 work hours are used daily in parking patrol, but during the epidemic that number is decreasing at the rate of 6 work hours per day. Thus, $dn/dt = -6$. How fast is revenue from parking fines decreasing during the epidemic?

We want to find dR/dt, the change in revenue with respect to time. By the chain rule,

$$\frac{dR}{dt} = \frac{dR}{dn} \cdot \frac{dn}{dt}.$$

First find dR/dn, as follows.

$$\frac{dR}{dn} = \frac{(n + 2)(8000) - 8000n(1)}{(n + 2)^2} = \frac{16{,}000}{(n + 2)^2}$$

Since $n = 30$, $dR/dn = 15.625$. Also, $dn/dt = -6$. Thus,

$$\frac{dR}{dt} = (15.625)(-6) = -93.75.$$

Revenue is being lost at the rate of \$93.75 per day. (10)

EXAMPLE 11 Suppose a sum of \$500 is deposited in an account with an interest rate of r percent per year compounded monthly. At the end of 10 years, the balance in the account is given by

$$A = 500\left(1 + \frac{r}{1200}\right)^{120}.$$

Find the rate of change of A with respect to r if $r = 5$, 4.2, or 3.

First find dA/dr using the generalized power rule.

$$\frac{dA}{dr} = (120)(500)\left(1 + \frac{r}{1200}\right)^{119}\left(\frac{1}{1200}\right) = 50\left(1 + \frac{r}{1200}\right)^{119}$$

If $r = 5$,

$$\frac{dA}{dr} = 50\left(1 + \frac{5}{1200}\right)^{119} \approx 82.01,$$

or \$82.01 per percentage point. If $r = 4.2$

$$\frac{dA}{dr} = 50\left(1 + \frac{4.2}{1200}\right)^{119} \approx 75.78,$$

or \$75.78 per percentage point. If $r = 3$,

$$\frac{dA}{dr} = 50\left(1 + \frac{3}{1200}\right)^{119} \approx 67.30,$$

or \$67.30 per percentage point.

The chain rule can be used to develop the formula for **marginal revenue product,** an economic concept that approximates the change in revenue when a manufacturer hires an additional employee. Start with $R = px$, where R is total revenue from the daily production of x units and p is the price per unit. The demand function is $p = f(x)$, as before. Also, x can be considered a function of the number of employees, n. Since $R = px$, and x and therefore p depend on n, R can also be considered a function of n. To find an expression for dR/dn, use the product rule for derivatives on the function $R = px$ to get

$$\frac{dR}{dn} = p \cdot \frac{dx}{dn} + x \cdot \frac{dp}{dn}. \tag{1}$$

By the chain rule,

$$\frac{dp}{dn} = \frac{dp}{dx} \cdot \frac{dx}{dn}.$$

Substituting for dp/dn in equation (1) gives

$$\begin{aligned} \frac{dR}{dn} &= p \cdot \frac{dx}{dn} + x\left(\frac{dp}{dx} \cdot \frac{dx}{dn}\right) \\ &= \left(p + x \cdot \frac{dp}{dx}\right)\frac{dx}{dn}. \qquad \text{Factor out } \frac{dx}{dn}. \end{aligned}$$

The expression for dR/dn gives the marginal revenue product.

EXAMPLE 12 Find the marginal revenue product dR/dn (in dollars) when $n = 20$ if the demand function is $p = 600/\sqrt{x}$ and $x = 5n$.

As shown above,

$$\frac{dR}{dn} = \left(p + x \cdot \frac{dp}{dx}\right)\frac{dx}{dn}.$$

Find dp/dx and dx/dn. From

$$p = \frac{600}{\sqrt{x}} = 600x^{-1/2},$$

we have the derivative

$$\frac{dp}{dx} = -300x^{-3/2}.$$

Also, from $x = 5n$,

$$\frac{dx}{dn} = 5.$$

Then, by substitution,

$$\frac{dR}{dn} = \left[\frac{600}{\sqrt{x}} + x(-300x^{-3/2})\right]5 = \frac{1500}{\sqrt{x}}.$$

Find marginal revenue product at $n = 10$ if the demand function is $p = 1000/x^2$ and $x = 8n$. Interpret your answer.

Answer:

$-\$1.25$; hiring an additional employee will produce a decrease in revenue of \$1.25.

If $n = 20$, then $x = 100$ and

$$\frac{dR}{dn} = \frac{1500}{\sqrt{100}} = 150.$$

This means that hiring an additional employee when production is at a level of 100 items will produce an increase in revenue of \$150. ◀ (11)

11.6 Exercises

Let $f(x) = 2x^2 + 3x$ and $g(x) = 4x - 1$. Find each of the following. (See Example 1.)

1. $f[g(3)]$ **2.** $f[g(-4)]$

3. $g[f(3)]$ **4.** $g[f(-4)]$

Find $f[g(x)]$ and $g[f(x)]$ in each of the following. (See Example 2.)

5. $f(x) = 8x + 12;\ g(x) = 3x - 1$

6. $f(x) = -6x + 9;\ g(x) = 5x + 7$

7. $f(x) = -x^3 + 2;\ g(x) = 4x$

8. $f(x) = 2x;\ g(x) = 6x^2 - x^3$

9. $f(x) = \frac{1}{x};\ g(x) = x^2$

10. $f(x) = \frac{2}{x^4};\ g(x) = 2 - x$

11. $f(x) = \sqrt{x + 2};\ g(x) = 8x^2 - 6$

12. $f(x) = 9x^2 - 11x;\ g(x) = 2\sqrt{x + 2}$

Write each function as a composition of two functions. (There may be more than one way to do this.) (See Example 3.)

13. $y = (4x + 3)^5$

14. $y = (x^2 + 2)^{1/3}$

15. $y = \sqrt{6 + 3x}$

16. $y = \sqrt{x + 3} - \sqrt[3]{x + 3}$

17. $y = \frac{\sqrt{x} + 3}{\sqrt{x} - 3}$

18. $y = \frac{2}{\sqrt{x} + 5}$

19. $y = (x^{1/2} - 3)^2 + (x^{1/2} - 3) + 5$

20. $y = (x^2 + 5x)^{1/3} - 2(x^2 + 5x)^{2/3} + 7$

Find the derivative of each of the following functions. (See Examples 4–7.)

21. $y = (3x + 4)^3$ **22.** $y = (6x - 1)^3$

23. $y = 6(3x + 2)^4$ **24.** $y = -5(2x - 1)^4$

25. $y = -2(8x^2 + 6)^4$ **26.** $y = -4(x^3 + 5x^2)^4$

27. $y = 12(2x + 5)^{3/2}$ **28.** $y = 45(3x - 8)^{3/2}$

Find the derivative of the function. (See Example 4.)

29. $y = -7(4x^2 + 9x)^{3/2}$ **30.** $y = 11(5x^2 + 6x)^{3/2}$

31. $y = 8\sqrt{4x + 7}$ **32.** $y = -3\sqrt{7x - 1}$

33. $y = -2\sqrt{x^2 + 4x}$ **34.** $y = 4\sqrt{2x^2 + 3}$

Use the product or quotient rule or the generalized power rule to find the derivative of each of the following functions. (See Examples 8 and 9.)

35. $y = (x + 1)(x - 3)^2$ **36.** $y = (2x + 1)^2(x - 5)$

37. $y = 5(x + 3)^2(2x - 1)^5$ **38.** $y = -9(x + 4)^2(2x - 3)^2$

39. $y = (3x + 1)^3\sqrt{x}$ **40.** $y = (3x + 5)^2\sqrt{x}$

41. $y = \frac{1}{(x - 4)^2}$ **42.** $y = \frac{-5}{(2x + 1)^2}$

43. $y = \frac{(4x + 3)^2}{2x - 1}$ **44.** $y = \frac{(x - 6)^2}{3x + 4}$

45. $y = \frac{x^2 + 4x}{(5x + 2)^3}$ **46.** $y = \frac{3x^2 - x}{(x - 1)^2}$

47. $y = (x^{1/2} + 1)(x^{1/2} - 1)^{1/2}$

48. $y = (3 - x^{2/3})(x^{2/3} + 2)^{1/2}$

Consider the following table of values of the functions f and g and their derivatives at various points:

x	1	2	3	4
$f(x)$	2	4	1	3
$f'(x)$	−6	−7	−8	−9
$g(x)$	2	3	4	1
$g'(x)$	2/7	3/7	4/7	5/7

Find each of the following.

49. **(a)** $D_x(f[g(x)])$ at $x = 1$
(b) $D_x(f[g(x)])$ at $x = 2$

50. **(a)** $D_x(g[f(x)])$ at $x = 1$
(b) $D_x(g[f(x)])$ at $x = 2$

51. If $f(x) = (2x^2 + 3x + 1)^{50}$, then which of the following is closest to $f'(0)$?
(a) 1 **(b)** 50 **(c)** 100 **(d)** 150 **(e)** 200 **(f)** 250

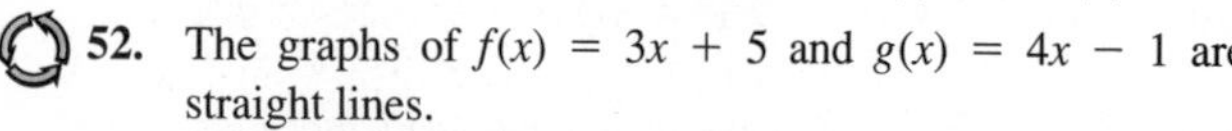

52. The graphs of $f(x) = 3x + 5$ and $g(x) = 4x - 1$ are straight lines.
(a) Show that the graph of $f[g(x)]$ is also a straight line.
(b) How are the slopes of the graphs of $f(x)$ and $g(x)$ related to the slope of the graph of $f[g(x)]$?

Work these exercises. (See Examples 10 and 11.)

53. Business The Acme Company's total cost for producing x widgets is given by

$$C(x) = 600 + \sqrt{50 + 15x^2} \qquad (0 \le x \le 200).$$

Find the marginal cost function.

54. Business A manufacturer's weekly profit from the sales of x souvenir cups is given by

$$P(x) = (x^3 + 12x + 120)^{1/3} - 200,$$

where $(0 \le x \le 2000)$.
(a) Use a calculator to find $P(50)$, $P(100)$, $P(200)$, and $P(1000)$.
(b) Explain why it is reasonable that some of the numbers found in part (a) are negative.
(c) Find the marginal profit function.

55. Business The revenue from the sale of x items is given by $R(x) = 10\sqrt{300x - 2x^2}$ $(0 \le x \le 150)$.
(a) Find the marginal revenue function.
(b) Evaluate the marginal revenue function at $x = 30, 60, 90$, and 120.
(c) Explain the significance of the answers found in part (b).

56. Finance A sum of \$1500 is deposited in an account with an interest rate of r percent per year, compounded daily. At the end of 5 years, the balance in the account is given by

$$A = 1500\left(1 + \frac{r}{365}\right)^{1825}.$$

Find the rate of change of A with respect to r (written as a decimal) for the following interest rates.
(a) 6% **(b)** 8% **(c)** 9%

Work these exercises. (See Example 12.)

57. Business Assume that the total revenue from the sale of r television sets is given by

$$R(x) = 1000\left(1 - \frac{x}{500}\right)^2$$

Find the marginal revenue when the following numbers of sets are sold.
(a) 400 **(b)** 500 **(c)** 600
(d) Find the average revenue from the sale of x sets.
(e) Find the marginal average revenue.

58. Business Find the marginal revenue product for a manufacturer with 8 workers if the demand function is $p = 300/x^{1/3}$ and if $x = 8n$.

59. Business Suppose the demand function for a product is $p = 200/x^{1/2}$. Find the marginal revenue product if there are 25 employees and if $x = 15n$.

Work these exercises. (See Examples 10–12.)

60. Natural Science An oil well off the Gulf Coast is leaking, with the leak spreading oil over the surface as a circle. At any time t, in minutes, after the beginning of the leak, the radius of the circular oil slick on the surface is $r(t) = t^2$ feet. Let $A(r) = \pi r^2$ represent the area of a circle of radius r. Find and interpret $A[r(t)]$.

61. Natural Science When there is a thermal inversion layer over a city (as happens often in Los Angeles), pollutants cannot rise vertically but are trapped below the layer and must disperse horizontally. Assume that a factory smokestack begins emitting a pollutant at 8 A.M. Assume that the pollutant disperses horizontally, forming a circle. If t represents the time, in hours, since the factory began emitting pollutants ($t = 0$ represents 8 A.M.), assume that the radius of the circle of pollution is $r(t) = 2t$ miles. Let $A(r) = \pi r^2$ represent the area of a circle of radius r. Find and interpret $A[r(t)]$.

62. Business Suppose a demand function is given by

$$x = 30\left(5 - \frac{p}{\sqrt{p^2 + 1}}\right),$$

where x is the demand for a product and p is the price per item in dollars. Find the rate of change in the demand for the product (i.e., find dx/dp).

63. Business Suppose the demand for a certain brand of vacuum cleaner is given by

$$D(p) = \frac{-p^2}{100} + 500,$$

where p is the price in dollars. If the price, in terms of the cost c, is expressed as

$$p(c) = 2c - 10,$$

find the demand in terms of the cost.

64. Business A certain truck depreciates according to the formula

$$V = \frac{6000}{1 + .3t + .1t^2},$$

where t is time measured in years and $t = 0$ represents the time of purchase. Find the rate at which the value of the truck is changing at the following times.

(a) 2 years **(b)** 4 years

65. Business Suppose the cost in dollars of manufacturing x items is given by

$$C = 2000x + 3500,$$

and the demand equation is given by

$$x = \sqrt{15{,}000 - 1.5p}.$$

In terms of the demand x,

(a) find an expression for the revenue R;

(b) find an expression for the profit P;

(c) find an expression for the marginal profit.

(d) determine the value of the marginal profit when the price is $25.

66. Natural Science The strength of a person's reaction to a certain drug is given by

$$R(Q) = Q\left(C - \frac{Q}{3}\right)^{1/2},$$

where Q represents the quantity of the drug given to the patient and C is a constant.

(a) The derivative $R'(Q)$ is called the *sensitivity* to the drug. Find $R'(Q)$.

(b) Find $R'(Q)$ if $Q = 87$ and $C = 59$.

67. Natural Science The volume and surface area of a "jawbreaker" candy for any radius is given by the formulas

$$V(r) = \frac{4}{3}\pi r^3 \quad \text{and} \quad S(r) = 4\pi r^2,$$

respectively. It is estimated that the radius of a jawbreaker while in a person's mouth is

$$r(t) = 6 - \frac{3}{17}t,$$

where $r(t)$ is in mm and t is in min.*

(a) What is the life expectancy of a "jawbreaker"?

(b) Find $\frac{dV}{dt}$ and $\frac{dS}{dt}$ when $t = 17$ and interpret your answer.

(c) Construct an analogous experiment using some other type of food or verify the results of this experiment.

68. Natural Science To test an individual's use of calcium, a researcher injects a small amount of radioactive calcium into the person's bloodstream. The calcium remaining in the bloodstream is measured each day for several days. Suppose the amount of the calcium remaining in the bloodstream in milligrams per cubic centimeter t days after the initial injection is approximated by

$$C(t) = \frac{1}{2}(2t + 1)^{-1/2}.$$

Find the rate of change of C with respect to time for each of the following.

(a) $t = 0$ **(b)** $t = 4$

(c) $t = 6$ **(d)** $t = 7.5$

69. Social Science Studies show that after t hours on the job, the number of items a supermarket cashier can ring up per minute is given by

$$F(t) = 60 - \frac{150}{\sqrt{8 + t^2}}.$$

(a) Find $F'(t)$, the rate at which the cashier's speed is increasing.

(b) At what rate is the cashier's speed increasing after 5 hours? 10 hours? 20 hours? 40 hours?

(c) Are your answers in part (b) increasing or decreasing with time? Is this reasonable? Explain.

Use a graphing calculator or computer to graph each function and its derivative on the same axes. Determine the values of x where the derivative is (a) *positive,* (b) *zero, and* (c) *negative.* (d) *What is true of the graph of the function in each case?*

70. $G(x) = \dfrac{2x}{(x - 1)^2}$ **71.** $K(x) = \sqrt[3]{(2x - 1)^2}$

*Roger Guffey, "The Life Expectancy of a Jawbreaker: An Application of the Composition of Functions," *Mathematics Teacher* 92, no. 2 (February 1999): 125–127.

2

Find each derivative.

(a) $y = 3e^{12x}$

(b) $y = -6e^{(-10x+1)}$

(c) $y = e^{-x^2}$

Answers:

(a) $y' = 36e^{12x}$

(b) $y' = 60e^{(-10x+1)}$

(c) $y' = -2xe^{-x^2}$

EXAMPLE 3 Find derivatives of the following functions.

(a) $y = e^{5x}$

Let $g(x) = 5x$, with $g'(x) = 5$. Then

$$y' = g'(x)\,e^{g(x)} = 5e^{5x}.$$

(b) $y = 3e^{-4x}$

Here $g(x) = -4x$ and $g'(x) = -4$, so that

$$y' = 3(-4e^{-4x}) = -12e^{-4x}$$

(c) $y = 10e^{3x^2}$

$$y' = 6x(10e^{3x^2}) = 60xe^{3x^2}$$ ◀ 2

Find each derivative.

(a) $y = \dfrac{e^x}{1 + x}$

(b) $y = \dfrac{10{,}000}{1 + 2e^x}$

Answers:

(a) $y' = \dfrac{xe^x}{(1 + x)^2}$

(b) $y' = \dfrac{-20{,}000e^x}{(1 + 2e^x)^2}$

EXAMPLE 4 Let $y = \dfrac{100{,}000}{1 + 100e^{-.3x}}$. Find y'.

Use the quotient rule.

$$y' = \frac{(1 + 100e^{-.3x})(0) - 100{,}000(-30e^{-.3x})}{(1 + 100e^{-.3x})^2}$$

$$= \frac{3{,}000{,}000e^{-.3x}}{(1 + 100e^{-.3x})^2}$$ ◀ 3

EXAMPLE 5 A 100 gram sample of a radioactive substance decays exponentially. The amount left after t years is given by $A(t) = 100e^{-.12t}$. Find the rate of change of the amount after 3 years.

The rate of change is given by the derivative dA/dt.

$$\frac{dA}{dt} = 100(e^{-.12t})(-.12) = -12e^{-.12t}$$

After 3 years ($t = 3$), the rate of change is

$$\frac{dA}{dt} = -12e^{-.12(3)} = -12e^{-.36} \approx -8.4$$

grams per year. ◀

Derivatives of Logarithmic Functions

To find the derivative of $g(x) = \ln x$, we use the definitions and properties of natural logarithms that were developed in Section 4.3:

$$g(x) = \ln x \text{ means } e^{g(x)} = x$$

and for all $x > 0$, $y > 0$ and every real number r,

$$\ln xy = \ln x + \ln y, \qquad \ln \frac{x}{y} = \ln x - \ln y, \qquad \ln x^r = r \ln x.$$

Note that $x > 0$, $y > 0$ because logarithms of negative numbers are not defined.

Differentiating with respect to x on each side of $e^{g(x)} = x$ shows that

$$D_x(e^{g(x)}) = D_x(x)$$
$$e^{g(x)} \cdot g'(x) = 1.$$

Because $e^{g(x)} = x$, this last equation becomes

$$x \cdot g'(x) = 1$$
$$g'(x) = \frac{1}{x}$$
$$\frac{d}{dx}(\ln x) = \frac{1}{x} \quad \text{for all } x > 0.$$

EXAMPLE 6 **(a)** Assume $x > 0$ and use properties of logarithms to find the derivative of $y = \ln 6x$.

$$y' = \frac{d}{dx}(\ln 6x)$$
$$= \frac{d}{dx}(\ln 6 + \ln x) \qquad \text{Product rule for logarithms}$$
$$= \frac{d}{dx}(\ln 6) + \frac{d}{dx}(\ln x) \qquad \text{Sum rule for derivatives}$$

Be careful here: ln 6 is a *constant* ($\ln 6 \approx 1.79$), so its derivative is 0 (*not* 1/6). Hence

$$y' = \frac{d}{dx}(\ln 6) + \frac{d}{dx}(\ln x) = 0 + \frac{1}{x} = \frac{1}{x}.$$

(b) Assume $x > 0$ and use the chain rule to find the derivative of $y = \ln 6x$.

Let $f(x) = \ln x$ and $g(x) = 6x$, so that $y = \ln 6x = \ln g(x) = f(g(x))$. Then by the chain rule,

$$y' = f'[g(x)] \cdot g'(x) = \frac{1}{g(x)} \cdot \frac{d}{dx}(6x) = \frac{1}{6x} \cdot 6 = \frac{1}{x}.$$

The argument used in Example 6(b) applies equally well in the general case. The derivative of $y = \ln g(x)$, where $g(x)$ is a function and $g(x) > 0$, can be found by letting $f(x) = \ln x$, so that $y = f[g(x)]$, and applying the chain rule:

$$y' = f'[g(x)] \cdot g'(x) = \frac{1}{g(x)} \cdot g'(x) = \frac{g'(x)}{g(x)}.$$

We summarize these results as follows.

Derivatives of ln x and ln $g(x)$

If $y = \ln x$, then $y' = \frac{1}{x}$ $\quad (x > 0)$.

If $y = \ln g(x)$, then $y' = \frac{g'(x)}{g(x)}$ $\quad (g(x) > 0)$.

EXAMPLE 7 Find the derivatives of the following functions.

(a) $y = \ln(3x^2 - 4x)$

Let $g(x) = 3x^2 - 4x$, so that $g'(x) = 6x - 4$. From the formula above,

$$y' = \frac{g'(x)}{g(x)} = \frac{6x - 4}{3x^2 - 4x}.$$

(b) $y = 3x \ln x^2$

Since $3x \ln x^2$ is the product of $3x$ and $\ln x^2$, use the product rule.

$$y' = (3x)\left(\frac{d}{dx} \ln x^2\right) + (\ln x^2)\left(\frac{d}{dx} 3x\right)$$

$$= 3x\left(\frac{2x}{x^2}\right) + (\ln x^2)(3) \qquad \text{Take derivatives.}$$

$$= 6 + 3 \ln x^2$$

$$= 6 + \ln(x^2)^3 \qquad \text{Property of logarithms}$$

$$y' = 6 + \ln x^6 \qquad \text{Property of exponents}$$

Find y' for the following.

(a) $y = \ln(7 + x)$

(b) $y = \ln(4x^2)$

(c) $y = \ln(8x^3 - 3x)$

(d) $y = x^2 \ln x$

Answers:

(a) $y' = \dfrac{1}{7 + x}$

(b) $y' = \dfrac{2}{x}$

(c) $y' = \dfrac{24x^2 - 3}{8x^3 - 3x}$

(d) $y' = x(1 + 2 \ln x)$

(c) $y = \ln[(x^2 + x + 1)(4x - 3)^5]$

Here we use the properties of logarithms before taking the derivative (the same thing could have been done in part (b), by writing $\ln x^2$ as $2 \ln x$).

$$y = \ln[(x^2 + x + 1)(4x - 3)^5]$$

$$= \ln(x^2 + x + 1) + \ln(4x - 3)^5 \qquad \text{Properties of logarithms}$$

$$= \ln(x^2 + x + 1) + 5 \ln(4x - 3) \qquad \text{Properties of logarithms}$$

$$y' = \frac{2x + 1}{x^2 + x + 1} + 5 \cdot \frac{4}{4x - 3} \qquad \text{Take derivatives.}$$

$$= \frac{2x + 1}{x^2 + x + 1} + \frac{20}{4x - 3}.$$ ◀ 4

The function $y = \ln(-x)$ is defined for all $x < 0$ (since $-x > 0$ when $x < 0$). Its derivative can be found by applying the derivative rule for $\ln g(x)$ with $g(x) = -x$.

$$y' = \frac{g'(x)}{g(x)}$$

$$= \frac{-1}{-x}$$

$$= \frac{1}{x}$$

This is the same as the derivative of $y = \ln x$, with $x > 0$. Since

$$|x| = \begin{cases} x & \text{if } x > 0 \\ -x & \text{if } x < 0, \end{cases}$$

we can combine two results into one, as follows.

5

Find each derivative.

(a) $y = e^{x^2} \ln |x|$

(b) $y = x^2/\ln |x|$

Answers:

(a) $y' = e^{x^2}\left(\frac{1}{x} + 2x \ln |x|\right)$

(b) $y' = \frac{2x \ln |x| - x}{(\ln |x|)^2}$

If $y = \ln |x|$, then $y' = \frac{1}{x}$ $(x \neq 0)$.

EXAMPLE 8 Let $y = e^x \cdot \ln |x|$. Find y'.

Use the product rule.

$$y' = e^x \cdot \frac{1}{x} + \ln |x| \cdot e^x = e^x\left(\frac{1}{x} + \ln |x|\right)$$ ◀ 5

Often a population, or the sales of a certain product, will start growing slowly, then grow more rapidly, and then gradually level off. Such growth can often be approximated by a *logistic function* of the form

$$f(x) = \frac{c}{1 + ae^{kx}}$$

for appropriate constants a, c, and k. (Logistic functions were introduced in Section 4.2.)

6

Suppose a deer population is given by

$$f(x) = \frac{10{,}000}{1 + 2e^x},$$

where x is time in years. (See side problem 3.) Find the rate of change of the population when

(a) $x = 0$;

(b) $x = 5$.

(c) Is the population increasing or decreasing?

Answers:

(a) About -2200

(b) About -33

(c) Decreasing

EXAMPLE 9 The number of people (in millions) who subscribe to basic cable TV is approximated by

$$S(x) = \frac{76.7}{1 + 16e^{-.1691x}},$$

where $x = 0$ corresponds to 1970.* At what rate is the number of subscribers changing in 2003?

Use the quotient rule to find the derivative of S.

$$S'(x) = \frac{(1 + 16e^{-.1691x})(0) - 76.7\,(16(-.1691)e^{-.1691x})}{(1 + 16e^{-.1691x})^2}$$

$$= \frac{207.51952e^{-.1691x}}{(1 + 16e^{-.1691x})^2}$$

The rate of change in 2003 ($x = 33$) is

$$S'(33) = \frac{207.51952e^{-.1691(33)}}{(1 + 16e^{-.1691(33)})^2} \approx .696102 \text{ million per year.}$$

This means that subscribers are increasing at the rate of approximately 696,102 per year. ◀ 6

*Based on data from *The Cable TV Financial Datebook* and *The Pay TV Newsletter.*

11.7 Exercises

Find the derivatives of the following functions. (See Examples 1–8.)

1. $y = e^{3x}$

2. $y = e^{-4x}$

3. $f(x) = 5e^{2x}$

4. $f(x) = 4e^{-3x}$

5. $g(x) = -4e^{-5x}$

6. $g(x) = 6e^{x/2}$

7. $y = e^{x^2}$

8. $y = e^{-x^2}$

9. $f(x) = e^{x^2/2}$

10. $y = 4e^{2x^2-4}$

11. $y = -3e^{3x^2+5}$

12. $y = xe^x$

13. $y = \ln(-8x^2 + 6x)$

14. $y = \ln\sqrt{x+5}$

15. $y = \ln\sqrt{2x+1}$

16. $y = \ln[(3x - 1)(5x + 2)]$

17. $f(x) = \ln[(2x - 3)(x^2 + 4)]$

18. $f(x) = \ln\left(\frac{4x+3}{5x-2}\right)$

19. $y = x^2e^{-2x}$

20. $y = (x - 3)^2e^{2x}$

21. $y = (3x^2 - 4x)e^{-3x}$

22. $y = \ln(3 - x)$

23. $y = \ln\left(\frac{6-x}{3x+5}\right)$

24. $y = \ln(x^4 + 5x^2)^{3/2}$

25. $y = \ln(5x^3 - 2x)^{3/2}$

26. $y = -3x\ln(x + 2)$

27. $y = x\ln(2 - x^2)$

28. $y = (2x^3 - 1)\ln|x|$

29. $y = \frac{\ln|x|}{x^3}$

30. $y = \frac{3\ln|x|}{3x+4}$

31. $y = \frac{-4\ln|x|}{5-2x}$

32. $y = \frac{3x^2}{\ln|x|}$

33. $y = \frac{x^3-1}{2\ln|x|}$

34. $y = [\ln(x + 1)]^4$

35. $y = \sqrt{\ln(x-3)}$

36. $y = \frac{e^x}{\ln|x|}$

37. $y = \frac{e^x-1}{\ln|x|}$

38. $y = \frac{e^x+e^{-x}}{x}$

39. $y = \frac{e^x-e^{-x}}{x}$

40. $y = e^{x^3}\ln|x|$

41. $f(x) = e^{3x+2}\ln(4x - 5)$

42. $f(x) = \frac{2400}{3+8e^{.2x}}$

43. $y = \frac{500}{7-10e^{.4x}}$

44. $y = \frac{10{,}000}{9+4e^{-.2x}}$

45. $y = \frac{500}{12+5e^{-.5x}}$

46. $y = \ln(\ln|x|)$

Find the equation of the tangent line to the graph of f at the given point.

47. $f(x) = \frac{e^x}{2x+1}$ at (0, 1)

48. $f(x) = \ln(1 + x^2)$ at (0, 0)

49. $f(x) = \frac{x^2}{e^x}$ when $x = 1$

50. $f(x) = \ln(2x^2 - 7x)$ when $x = -1$

51. If $f(x) = e^{2x}$, find $f'[\ln(1/4)]$.

52. If $g(x) = 3e\ln[\ln x]$, find $g'(e)$.

Work the following exercises.

53. Social Science The number of Internet users (in millions) in the United States is approximated by

$$f(x) = 54\,e^{.1513x},$$

where $x = 0$ corresponds to fall 1999.* Find the number of Internet users and the rate at which this number is changing in
(a) fall 2001; **(b)** fall 2003.

54. Social Science The Census Bureau projects the Latino-American population in the United States (in millions) to be given by

$$g(x) = 32.5\,e^{.02212x},$$

where $x = 0$ corresponds to 2000.†
(a) What is the Latino-American population expected to be in 2005?
(b) At what rate is this population changing in 2005?

55. Business Based on data from a car magazine, $P(x) = e^{-.02x}$ represents the proportion of cars manufactured by a given company that are still free of defects after x months of use. Find the proportion of cars free of defects after
(a) 1 month; **(b)** 10 months; **(c)** 100 months.
(d) Calculate and interpret $P'(100)$.

56. Natural Science The population of a bed of clams in the Great South Bay off Long Island is approximated by the logistic function

$$G(t) = \frac{52{,}000}{1+12e^{-.52t}},$$

where t is measured in years. Find the clam population and its rate of growth at the following times.
(a) 1 year **(b)** 4 years **(c)** 10 years
(d) What happens to the rate of growth over time?

57. Social Science According to work by the psychologist C. L. Hull, the strength of a habit is a function of the number of times the habit is repeated. If N is the number of repetitions and $H(N)$ is the strength of the habit, then

$$H(N) = 1000(1 - e^{-kN}),$$

*Based on projections by the U.S. Department of Commerce.

†*Statistical Abstracts of the United States* 2000.

where k is a constant. Find $H'(N)$ if $k = .1$ and
(a) $N = 10$; (b) $N = 100$; (c) $N = 1000$.
(d) Show that $H'(N)$ is always positive. What does this mean?

58. Health The percentage of people of any particular age group that will die in a given year may be approximated by

$$P(t) = .00239e^{.0957t},$$

where t is the age of the person in years.*
(a) Find $P(25)$, $P(50)$, and $P(75)$.
(b) Find $P'(25)$, $P'(50)$, and $P'(75)$.
(c) Interpret your answers for parts (a) and (b). What are the limitations of this formula?

59. Natural Science The age-weight relationship of female Arctic foxes, caught in Svalbard, Norway, can be estimated by the function

$$M(t) = 3102e^{-e^{-.022(t-56)}},$$

where t is the age of the fox in days and $M(t)$ is the weight of the fox in grams.†
(a) Estimate the weight of a female fox that is 200 days old.
(b) Estimate the rate of change in weight of an Arctic fox that is 200 days old.
(c) Use a graphing calculator to graph $M(t)$ and then describe the growth pattern.
(d) Use the table function on a graphing calculator or a spreadsheet to develop a chart that shows the estimated weight and growth rate of female foxes for days 50, 100, 150, 200, 250 and 300.

60. Social Science It has been observed that there has been an increase in the proportion of medical research papers that use the word *novel* in the title or abstract and that this proportion can be accurately modeled by the function

$$p(x) = .001131e^{.1268x},$$

where x is the number of years since 1970.‡
(a) Find $p(25)$.
(b) If this phenomenon continues, estimate the year in which every medical article will contain the word *novel* in its title or abstract.
(c) Estimate the rate of increase in the proportion of such medical papers in the year 2002.
(d) Explain some factors that may be contributing to researchers using this word.

61. Health The following function models the relationship between the size of a breast tumor and the amount of time it has been growing.

$$V(t) = \frac{1100}{(1 + 1023e^{-.02415t})^4},$$

where t is in months and $V(t)$ is in cubic centimeters.*
(a) Find the tumor volume at 240 months.
(b) Assuming that the shape of a tumor is spherical, find the radius of the tumor from part (a) (*Hint:* The volume of a sphere is given by the formula $v = \frac{4}{3}\pi r^3$.
(c) If a tumor of size .5 cm^3 is detected, according to the formula, how long has it been growing? What does this imply?
(d) Calculate the rate of change of tumor volume at 240 months and interpret.

62. Health One measure of whether a dialysis patient has been adequately dialyzed is by the urea reduction ratio (URR). It is generally agreed that a patient has been adequately dialyzed when URR exceeds a value of .65. The value of URR can be calculated for a particular patient using the following formula by Gotch:

$$URR = 1 - e^{-.0056t + .04} - \frac{8t(1 - e^{-.0056t+.04})}{126t + 900},$$

where t is measured in minutes.†
(a) Find the value of URR after this patient receives dialysis for 180 minutes. Has the patient received adequate dialysis?
(b) Find the value of URR after this patient receives dialysis for 240 minutes. Has the patient received adequate dialysis?
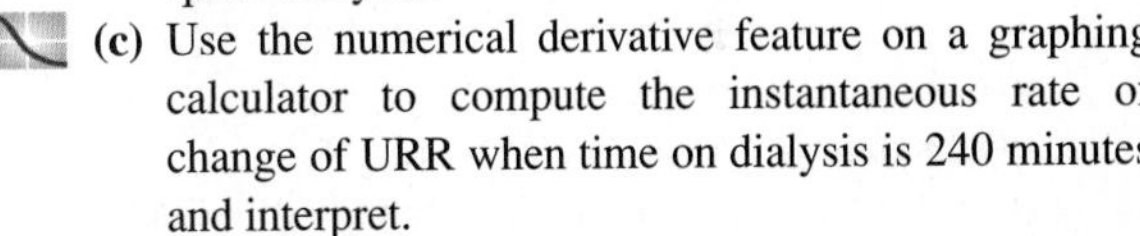
(c) Use the numerical derivative feature on a graphing calculator to compute the instantaneous rate of change of URR when time on dialysis is 240 minutes and interpret.

63. Business Suppose the demand function for x thousand of a certain item is

$$p = 100 + \frac{50}{\ln x}, \qquad x > 1,$$

where p is in dollars.
(a) Find the marginal revenue.
(b) Find the revenue from the next thousand items at a demand of 8000 ($x = 8$).

*Vital Statistics of the U.S. 1995.

†Pal Prestrud and Kjell Nilssen, "Growth, Size, and Sexual Dimorphism in Arctic Foxes," *Journal of Mammalogy* 76, no. 2 (May 1995): 522–530.

‡Simon H. Friedman and Jens O. Karlsson, "A Novel Paradigm," *Nature* 385, no. 6616 (February 6, 1997): 480.

*John Spratt et al., "Decelerating Growth and Human Breast Cancer," *Cancer* 71, no. 6 (1993): 2013–2019.

†Edward Kessler, Nathan Ritchey, et al., "Urea Reduction Ratio and Urea Kinetic Modeling: A Mathematical Analysis of Changing Dialysis Parameters," *American Journal of Nephrology* 18 (1998): 471–477.

64. Business The demand function for x units of a product is

$$p = 100 - 10 \ln x, \qquad 1 < x < 20{,}000,$$

where $x = 6n$ and n is the number of employees producing the product.

(a) Find the revenue function $R(x)$.

(b) Find the marginal revenue product function. (See Example 12 in Section 11.6.)

(c) Evaluate and interpret the marginal revenue product when $x = 20$.

65. Transportation A child is waiting at a street corner for a gap in traffic so that she can safely cross the street. A mathematical model for traffic shows that if the child's expected waiting time is to be at most one minute, then the maximum traffic flow (in cars per hour) is given by

$$f(x) = \frac{67{,}338 - 12{,}595 \ln x}{x},$$

where x is the width of the street in feet.* Find the maximum traffic flow and the rate of change of the maximum traffic flow with respect to street width when $x =$

(a) 30 feet; (b) 40 feet.

66. Social Science The number of visitors (in millions) to U.S. national parks since 1950 is approximated by

$$f(x) = -270.8 + 73.21 \ln x, \qquad (x \geq 50),$$

where $x = 50$ corresponds to 1950.†

(a) Estimate the number of visitors in 2001 and 2004.

(b) Find the rate of change of $f(x)$ when $x = 101$ and $x = 104$. Interpret your results.

67. Business The U.S. gross domestic product (GDP), in billions of dollars, is approximated by

$$f(x) = -6.168 + 5.069 \ln x, \qquad (x \geq 10),$$

where $x = 10$ corresponds to 1990.‡

(a) Estimate the GDP in 2000 and 2003.

(b) At what rate was the GDP changing in 2000 and in 2003?

*Edward A. Bender, *An Introduction to Mathematical Modeling*, Dover Publications, Inc., 2000.

†National Park Service.

‡U.S. Bureau of Economic Analysis.

68. Health The number of kidney transplants in year x is approximated by

$$g(x) = 951 + 4059.7 \ln x, \qquad (x \geq 5),$$

where $x = 5$ corresponds to 1985.*

(a) Estimate the number of transplants in 2001 and 2004.

(b) Find the rate of change in the number of transplants in 2001 and 2004.

69. Physical Science The net U.S. greenhouse gas emissions† from human activities for the years 1990–97 can be approximated by

$$E(t) = -15{,}790.44 + 3804.6 \ln(t),$$

where t is the year since 1900 and E is measured in millions of metric tons of carbon equivalent.‡ Assuming that the emissions continue to follow this formula, do the following.

(a) Calculate the net U.S. greenhouse gas emissions for the year 2002.

(b) Calculate $E'(102)$ and interpret your answer.

70. Physical Science The Richter scale provides a measure of the magnitude of an earthquake. In fact, the largest Richter number M ever recorded for an earthquake was 8.9 from the 1933 earthquake in Japan. The following formula shows a relationship between the amount of energy released E and the Richter number.

$$M = \frac{2 \ln E - 2 \ln .007}{3 \ln 10},$$

where E is measured in kilowatt-hours (kWh).§

(a) For the 1933 earthquake in Japan, what value of E gives a Richter number $M = 8.9$?

(b) If the average household uses 247 kilowatt-hours per month, how many months would the energy released by an earthquake of this magnitude power 10 million households?

(c) Find the rate of change of the Richter number M with respect to energy when $E = 70{,}000$ kWh.

(d) What happens to $\frac{dM}{dE}$ as E increases?

*U.S. Department of Health and Human Services and the United Network for Organ Sharing.

†The interpretation of "net" in this case is total emissions minus the carbon dioxide absorbed by forests and other means.

‡U.S. Environmental Protection Agency.

§Christopher Bradley, "Media Clips," *Mathematics Teacher* 93, no. 4 (April 2000): 300–303.

11.8 CONTINUITY AND DIFFERENTIABILITY

Intuitively speaking, a function is **continuous** at a point if you can draw the graph of the function near that point without lifting your pencil from the paper. Conversely, a function is **discontinuous** at a point if the pencil *must* be lifted from the paper in order to draw the graph on both sides of the point.

Looking at graphs having points of discontinuity will clarify the idea of continuity at a point. For example, the graph of the function in Figure 11.30(a) has an open circle at (2, 3), which indicates that there is a "hole" in the graph at that point. Therefore the function is discontinuous at $x = 2$ because to draw the graph from $x = 1$ to $x = 3$, you must lift the pencil for an instant as you pass through (2, 3).

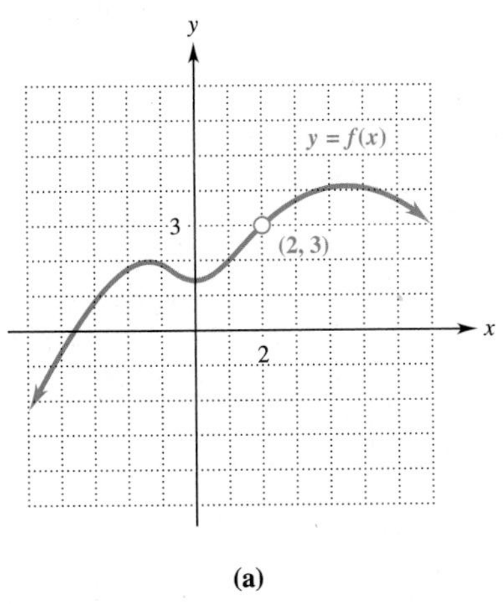

(a)

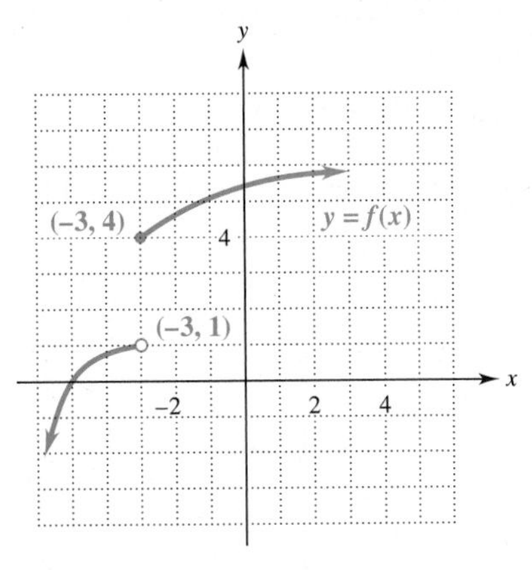

(b)

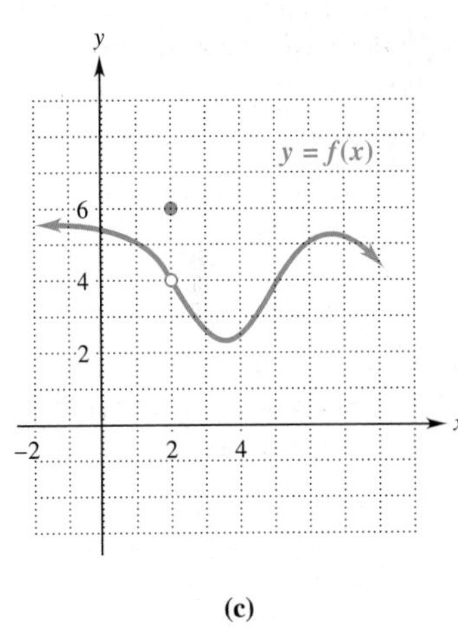

(c)

FIGURE 11.30

1

Find any points of discontinuity for the following functions.

(a)

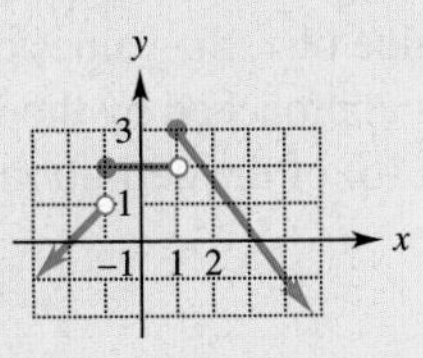

(b)

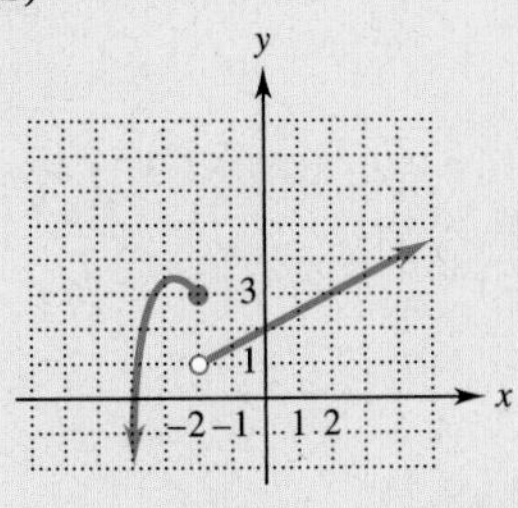

(continues on next page)

The function in Figure 11.30(b) is discontinuous at $x = -3$ because of the "jump" in the graph there (which necessitates lifting the pencil to draw the graph on both sides of $x = -3$). Although the function is not continuous at $x = -3$, it *is* continuous at $x = -1$ because you can draw the graph from $x = -2$ to $x = 0$, say, without lifting pencil from paper.

Finally, the function in Figure 11.30(c) is discontinuous at $x = 2$. When x is near (but not equal to) 2, all of the corresponding values of $f(x)$ are very near 4, so that on either side of $x = 2$, the graph is very near the point (2, 4) and can be drawn without lifting pencil from paper. But the instant $x = 2$, you must lift the pencil to the point $(2, f(2)) = (2, 6)$. Looked at from another point of view, as x gets closer and closer to 2, $f(x)$ gets closer and closer to 4, that is, $\lim_{x \to 2} f(x) = 4$. But $f(2) = 6$ and, hence,

$$\lim_{x \to 2} f(x) \neq f(2). \quad (1)$$

Now let's consider what it means for a function f to be continuous at $x = c$. If you *can* draw the graph of f around $x = c$ without lifting pencil from paper, then at the very least, $f(c)$ must be defined (otherwise there would be a hole in the graph). But the last example shows that this is not enough to guarantee continuity: as x gets very close to c, $f(x)$ must get very close to $f(c)$ (otherwise you have to lift the pencil at $x = c$). These considerations lead to this definition.

1

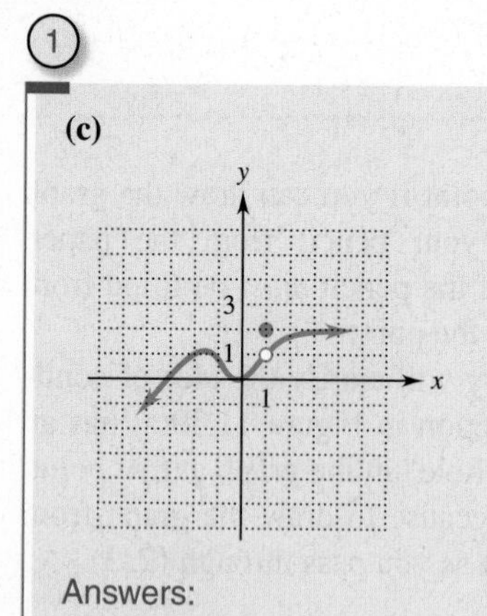

Answers:

(a) $x = -1, 1$

(b) $x = -2$

(c) $x = 1$

Definition of Continuity at a Point

A function f is **continuous** at $x = c$ if

(a) $f(c)$ is defined;

(b) $\lim_{x \to c} f(x)$ exists;

(c) $\lim_{x \to c} f(x) = f(c)$.

If f is not continuous at $x = c$, it is **discontinuous** there.

EXAMPLE 1 Tell why the following functions are discontinuous at the indicated points.

(a) $f(x)$ in Figure 11.31 at $x = 3$

The open circle on the graph of Figure 11.31 at the point where $x = 3$ means that $f(3)$ does not exist. Because of this, part (a) of the definition fails.

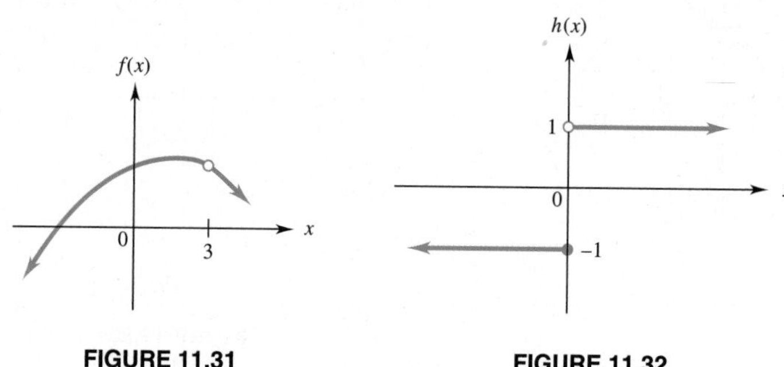

FIGURE 11.31

FIGURE 11.32

(b) $h(x)$ in Figure 11.32 at $x = 0$

The graph of Figure 11.32 shows that $h(0) = -1$. Also, as x approaches 0 from the left, $h(x)$ is -1. However, as x approaches 0 from the right, $h(x)$ is 1. As mentioned in Section 11.1, for a limit to exist at a particular value of x, the values of $h(x)$ must approach a single number. Since no single number is approached by the values of $h(x)$ as x approaches 0, $\lim_{x \to 0} h(x)$ does not exist, and part (b) of the definition fails.

(c) $g(x)$ at $x = 4$ in Figure 11.33

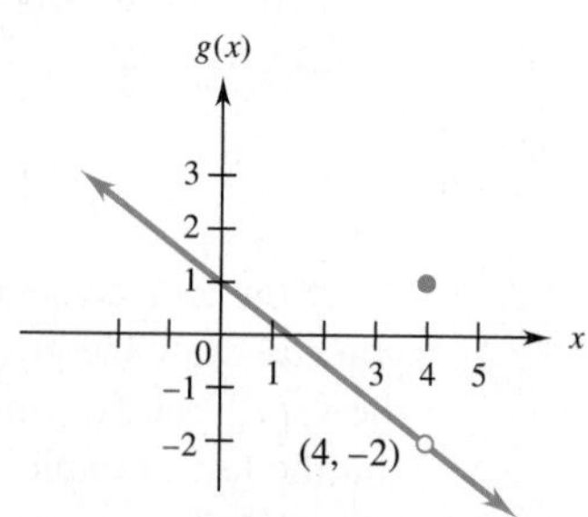

FIGURE 11.33

In Figure 11.33, the heavy dot above 4 shows that $g(4)$ is defined. In fact, $g(4) = 1$. However, the graph also shows that

$$\lim_{x \to 4} g(x) = -2,$$

so $\lim_{x \to 4} g(x) \neq g(4)$, and part (c) of the definition fails.

(d) $f(x)$ in Figure 11.34 at $x = -2$

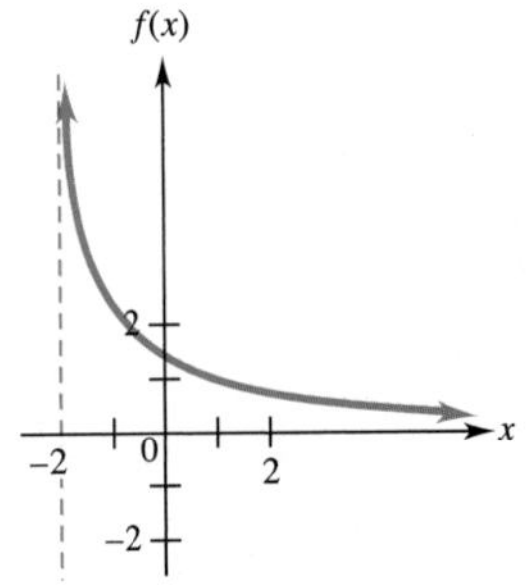

FIGURE 11.34

The function f graphed in Figure 11.34 is not defined at -2, and $\lim_{x \to -2} f(x)$ does not exist. Either of these reasons is sufficient to show that f is not continuous at -2. (Function f *is* continuous at any value of x greater than -2, however.) ◀ ②

When discussing the continuity of a function, it is often helpful to use interval notation, which was introduced in Chapter 1. The following chart should help you recall how it is used.

Interval	*Name*	*Description*	*Interval Notation*
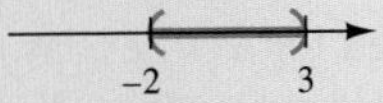	Open interval	$-2 < x < 3$	$(-2, 3)$
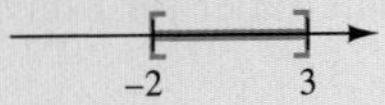	Closed interval	$-2 \leq x \leq 3$	$[-2, 3]$
	Open interval	$x < 3$	$(-\infty, 3)$
	Open interval	$x > -5$	$(-5, \infty)$

Remember, the symbol ∞ does not represent a number; ∞ is used for convenience in interval notation to indicate that the interval extends without bound in the positive direction. Also, $-\infty$ indicates no bound in the negative direction. ③

Continuity at a point was defined above; *continuity on an open interval* is defined as follows.

> If a function is continuous at each point of an open interval, it is said to be **continuous on the open interval.**

②

Tell why the following functions are discontinuous at the indicated points.

(a)

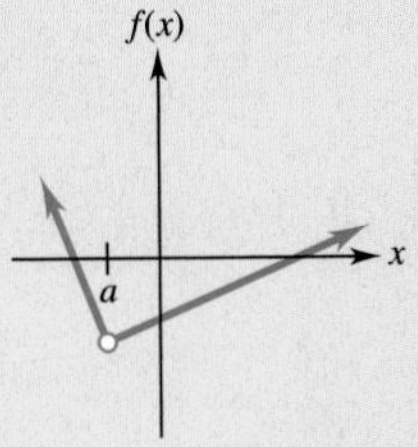

(b)

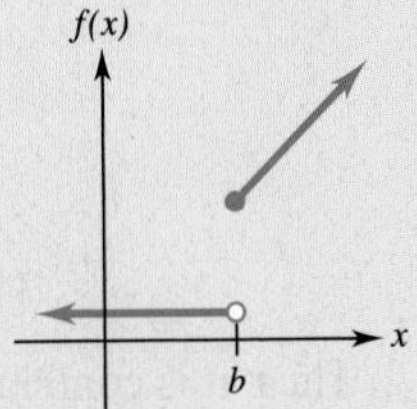

Answers:

(a) $f(a)$ does not exist

(b) $\lim_{x \to b} f(x)$ does not exist

③

Write each of the following using interval notation.

(a)

(b)

(c)

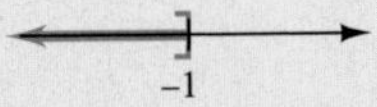

Answers:

(a) $(-5, 3)$

(b) $[4, 7]$

(c) $(-\infty, -1]$

4

Are the functions whose graphs are shown continuous on the indicated intervals?

(a) $(-4, -2)$; $(-3, 0)$

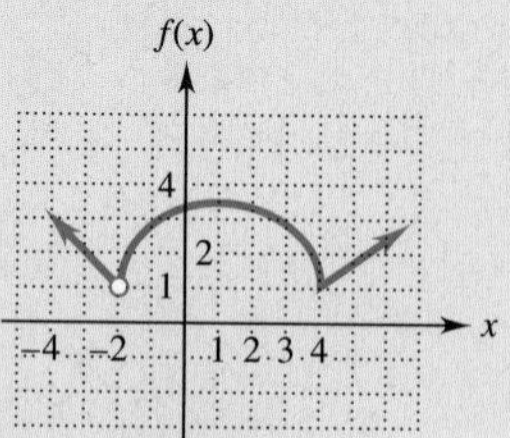

(b) $(-1, 1)$; $(0, 2)$

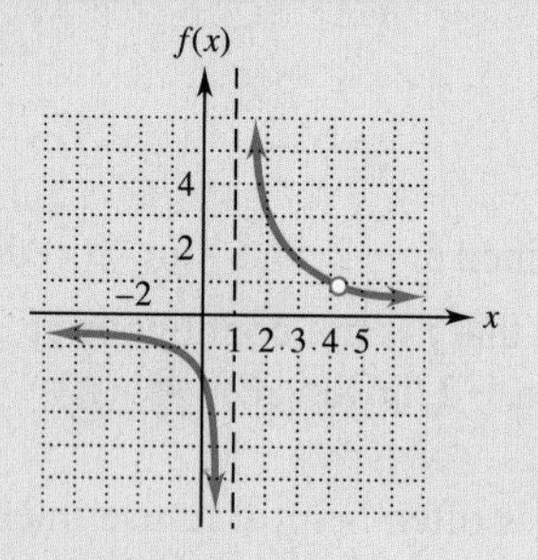

Answers:

(a) Yes; no

(b) Yes; no

Intuitively, the function f is continuous on the interval (a, b) if you can draw the graph between $x = a$ and $x = b$ without lifting your pencil from the paper.

EXAMPLE 2 Is the function of Figure 11.35 continuous on the x-intervals listed below?

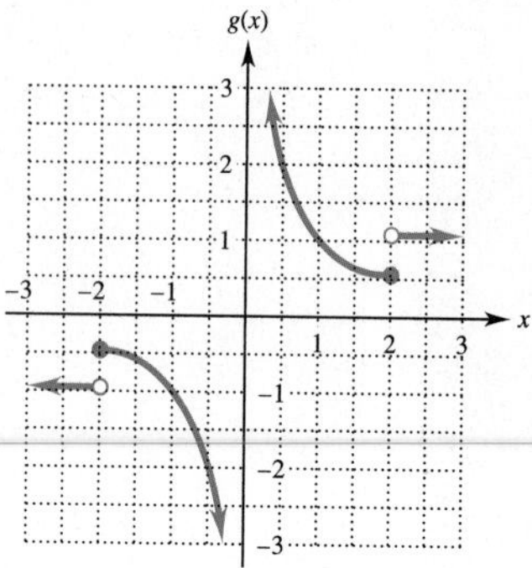

FIGURE 11.35

(a) $(-2, -1)$

The function is discontinuous only at $x = -2$, 0, and 2. Thus, it is continuous at every point of the open interval $(-2, -1)$ and, therefore, is continuous on the open interval.

(b) $(1, 3)$

Since this interval includes the point of discontinuity at $x = 2$, the function is not continuous on the open interval $(1, 3)$. 4

EXAMPLE 3 A trailer rental firm charges a flat \$4 to rent a hitch. The trailer itself is rented for \$11 per day or fraction of a day. Let $C(x)$ represent the cost of renting a hitch and trailer for x days.

(a) Graph C.

The charge for 1 day is \$4 for the hitch and \$11 for the trailer, or \$15. In fact, in the interval $(0, 1]$, $C(x) = 15$. To rent the trailer for more than 1 day, but not more than 2 days, the charge is $4 + 2 \cdot 11 = 26$ dollars. For any value of x in the interval $(1, 2]$, $C(x) = 26$. Also, in $(2, 3]$ $C(x) = 37$. These results lead to the graph of Figure 11.36.

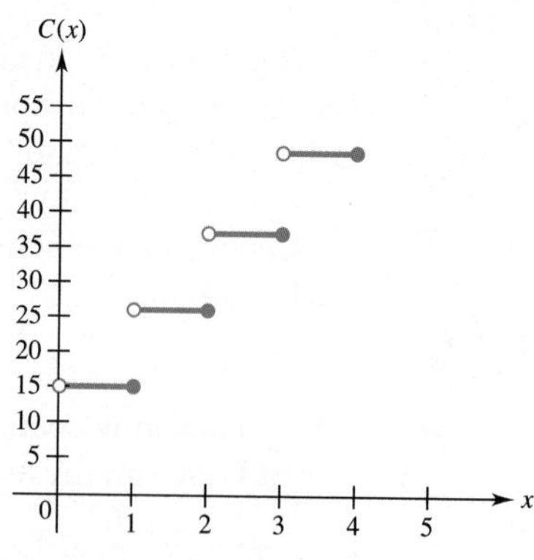

FIGURE 11.36

5

Suppose the cost is \$2.25 to mail a package weighing up to 1 pound plus \$.50 for each additional pound or fraction of a pound. Let $P(x)$ represent the cost of mailing a package weighing x pounds. Find any points of discontinuity for P.

Answers:

$x = 1, 2, 3, \ldots$

(b) Find any points of discontinuity for C.

As the graph suggests, C is discontinuous at 1, 2, 3, 4, and all other positive integers. ◀ 5

Continuity and Technology

Because of the way a graphing calculator or computer graphs functions, the graph of a continuous function may look like disconnected, closely spaced line segments (Figure 11.37). Furthermore, calculators often have trouble accurately portraying graphs at points of discontinuity. For instance, "jump discontinuities" (such as those at $x = 1, 2, 3$ in Figure 11.36), may appear as "stair steps" with vertical line segments connecting separate pieces of the graph (Figure 11.38). Similarly, a hole in the graph (such as the one at $x = 4$ in Figure 11.33) may not be visible on a calculator screen, depending on what viewing window is used. Nor can a calculator indicate whether or not an "endpoint" is included in the graph.

One reason these inaccuracies appear is due to the manner in which a calculator graphs: it plots points and connects them with line segments. In effect, the calculator assumes the function is continuous unless it actually computes a function value and determines it is undefined. In that case, it will skip a pixel and not connect points on either side of it.

The moral of this story is that a calculator or computer is only a tool. In order to use it correctly and effectively, you must understand the mathematics involved. When you do, it's usually easy to interpret screen images correctly.

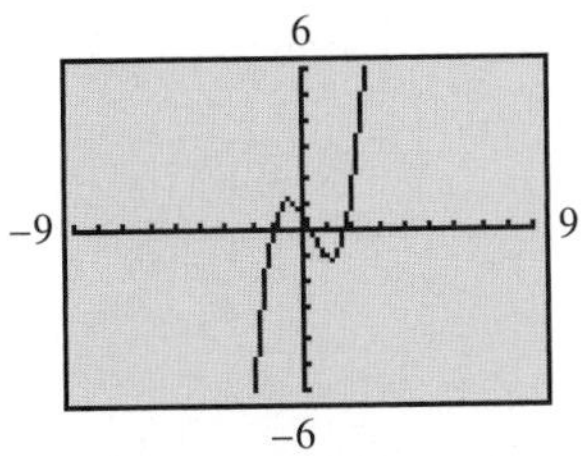

FIGURE 11.37

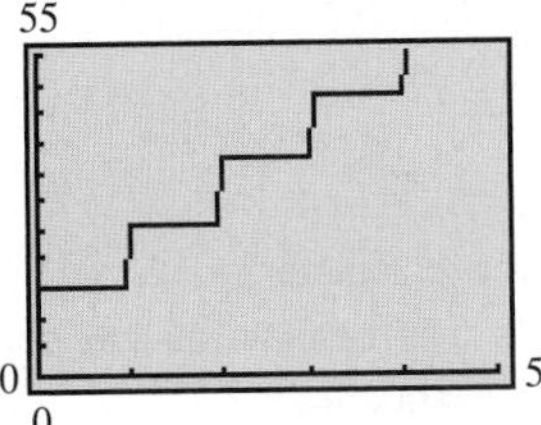

FIGURE 11.38

Continuity and Differentiability

As shown earlier in this chapter, a function fails to have a derivative at a point where the function is not defined, where the graph of the function has a "sharp point," or where the graph has a vertical tangent line. (See Figure 11.39.)

The function graphed in Figure 11.39 is continuous on the interval (x_1, x_2) and has a derivative at each point on this interval. On the other hand, the function is also continuous on the interval $(0, x_2)$ but does *not* have a derivative at each point on the interval (see x_1 on the graph).

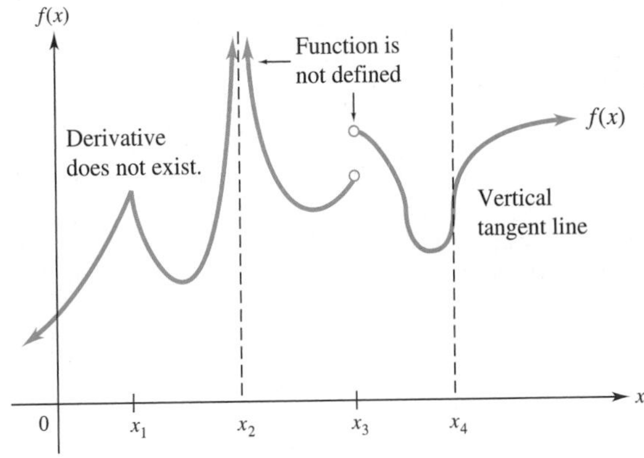

FIGURE 11.39

A similar situation holds in the general case.

If the derivative of a function exists at a point, then the function is continuous at that point. However, a function may be continuous at a point and not have a derivative there.

EXAMPLE 4 A nova is a star whose brightness suddenly increases and then gradually fades. The cause of the sudden increase in brightness is thought to be an explosion of some kind. The intensity of light emitted by a nova as a function of time is shown in Figure 11.40.* Notice that although the graph is a continuous curve, it is not differentiable at the point of the explosion.

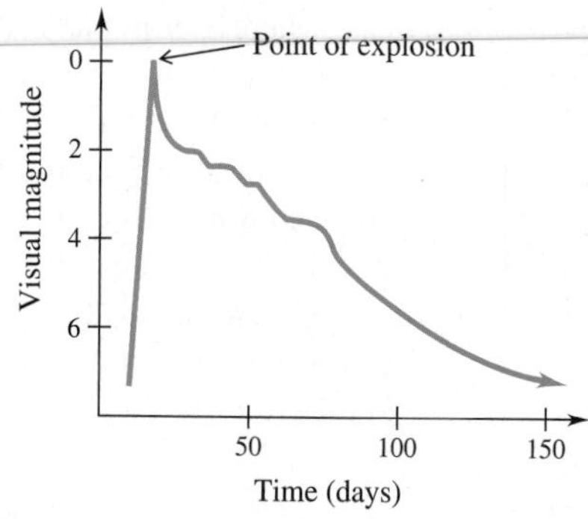

FIGURE 11.40

TECHNOLOGY TIP In some viewing windows a calculator graph may appear to have a sharp corner, when in fact the graph is differentiable at that point. When in doubt try a different window to see if the corner disappears.

*Reprinted with permission of Macmillan Publishing Company from *Astronomy: The Structure of the Universe* by William J. Kaufmann, III. Copyright © 1977 by William J. Kaufmann, III.

11.8 Exercises

Find all points of discontinuity for the functions whose graphs are shown below. (See Example 1.)

1.

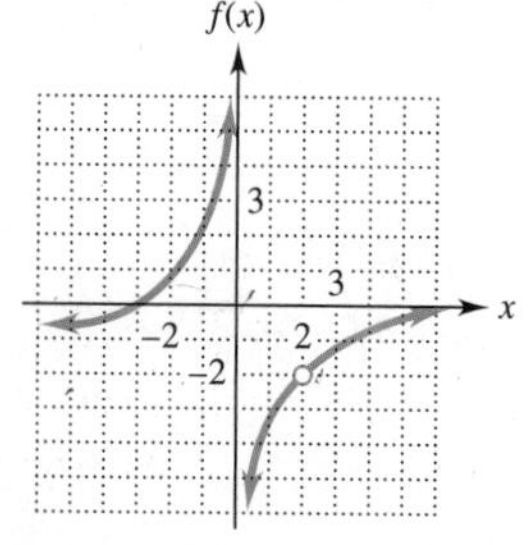

2.

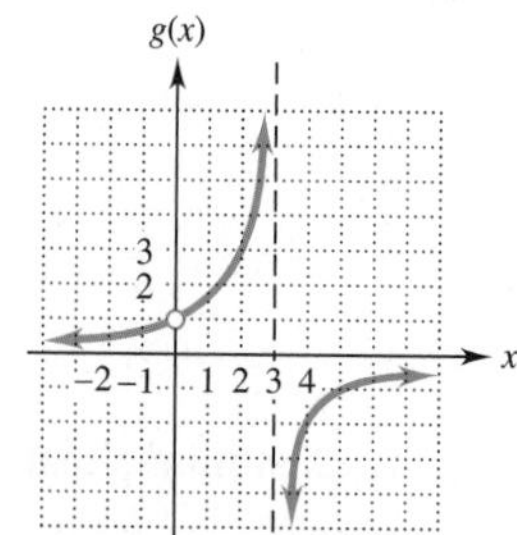

3.

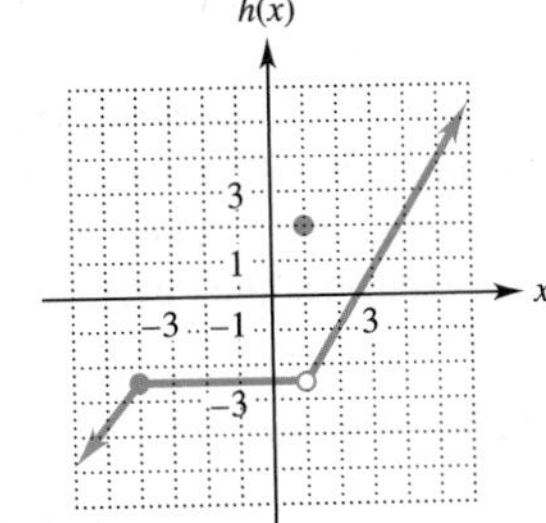

4.

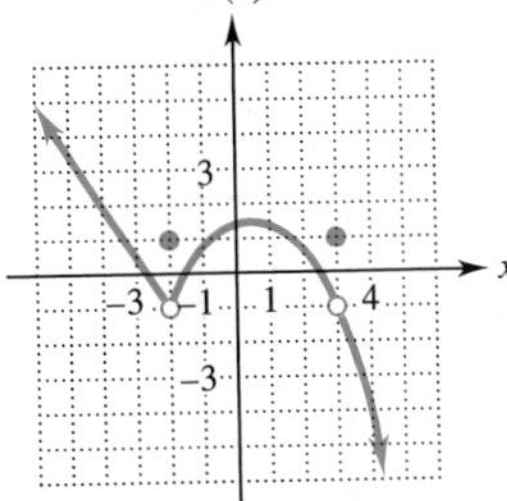

5.

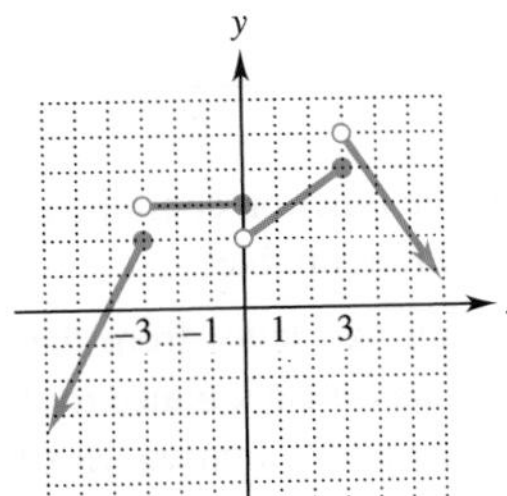

6.

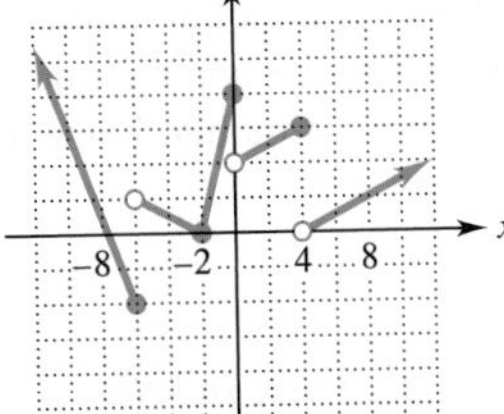

7.

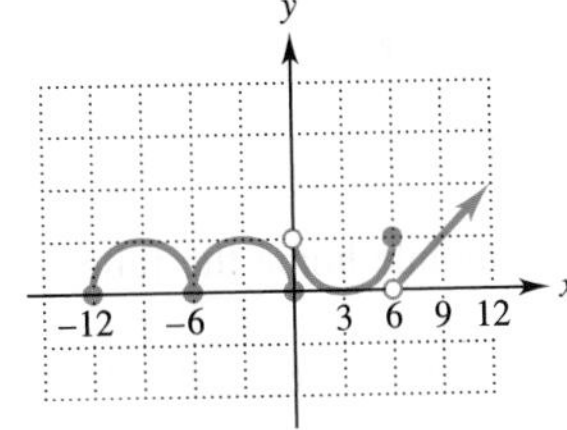

8.

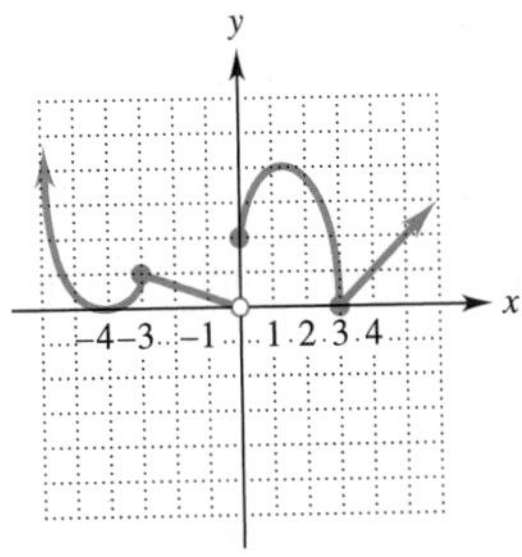

Are the following functions continuous at the given values of x?

9. $f(x) = \dfrac{4}{x - 2}$; $x = 0, x = 2$

10. $g(x) = \dfrac{5}{x + 5}$; $x = -5, x = 5$

11. $h(x) = \dfrac{1}{x(x - 3)}$; $x = 0, x = 3, x = 5$

12. $h(x) = \dfrac{-1}{(x - 2)(x + 3)}$; $x = 0, x = 2, x = 3$

13. $g(x) = \dfrac{x + 2}{x^2 - x - 2}$; $x = 1, x = 2, x = -2$

14. $h(x) = \dfrac{3x}{6x^2 + 15x + 6}$; $x = 0, x = -1/2, x = 3$

15. $g(x) = \dfrac{x^2 - 4}{x - 2}$; $x = 0, x = 2, x = -2$

16. $h(x) = \dfrac{x^2 - 25}{x + 5}$; $x = 0, x = 5, x = -5$

17. $p(x) = \dfrac{|x + 2|}{x + 2}$; $x = -2, x = 0, x = 2$

18. $r(x) = \dfrac{|5 - x|}{x - 5}$; $x = -5, x = 0, x = 5$

19. $f(x) = \begin{cases} x - 2 & \text{if } x \le 3 \\ 2 - x & \text{if } x > 3 \end{cases}$; $x = 2, x = 3$

20. $g(x) = \begin{cases} e^x & \text{if } x < 0 \\ x + 1 & \text{if } 0 \le x \le 3; \\ 2x - 3 & \text{if } x > 3 \end{cases}$ $x = 0, x = 3$

In Exercises 21 and 22, find the constant k that makes the given function continuous at x = 2.

21. $f(x) = \begin{cases} x + k & \text{if } x \le 2 \\ 5 - x & \text{if } x > 2 \end{cases}$

22. $g(x) = \begin{cases} x^k & \text{if } x \le 2 \\ 2x + 4 & \text{if } x > 2 \end{cases}$

Work the following problems. (See Example 4.)

23. Business Like many states, California suffered a large budget deficit in 1991. As part of the solution, officials raised the sales tax by 1.25 cents per dollar. The graph

3. $\lim_{x \to a} [f(x) \pm g(x)] = A \pm B = \lim_{x \to a} f(x) \pm \lim_{x \to a} g(x)$

4. $\lim_{x \to a} [f(x) \cdot g(x)] = A \cdot B = \lim_{x \to a} f(x) \cdot \lim_{x \to a} g(x)$

5. $\lim_{x \to a} = \dfrac{f(x)}{g(x)} = \dfrac{A}{B} = \dfrac{\lim_{x \to a} f(x)}{\lim_{x \to a} g(x)} \qquad (B \neq 0)$

6. For any real number r for which A^r exists,

$$\lim_{x \to a} [f(x)]^r = A^r = [\lim_{x \to a} f(x)]^r$$

Polynomial Limits

If f is a polynomial function, then $\lim_{x \to a} f(x) = f(a)$.

Limit Theorem

If f and g are functions that have limits as x approaches a, and $f(x) = g(x)$ for all $x \neq a$, then $\lim_{x \to a} f(x) = \lim_{x \to a} g(x)$.

The **instantaneous rate of change** of a function f when $x = a$ is

$$\lim_{h \to 0} \frac{f(a + h) - f(a)}{h},$$

provided this limit exists.

The **tangent line** to the graph of $y = f(x)$ at the point $(a, f(a))$ is the line through this point having slope $\lim_{h \to 0} \dfrac{f(a + h) - f(a)}{h}$, provided this limit exists.

The **derivative** of the function f is the function denoted f' whose value at the number x is $f'(x) = \lim_{h \to 0} \dfrac{f(x + h) - f(x)}{h}$, provided this limit exists.

Rules for Derivatives

(Assume all indicated derivatives exist.)

Constant Function

If $f(x) = k$, where k is any real number, then $\boldsymbol{f'(x) = 0}$.

Power Rule

If $f(x) = x^n$, for any real number n, then $\boldsymbol{f'(x) = n \cdot x^{n-1}}$.

Constant Times a Function

Let k be a real number. Then the derivative of $y = k \cdot f(x)$ is $\boldsymbol{y' = k \cdot f'(x)}$.

Sum or Difference Rule

If $f(x) = u(x) \pm v(x)$, then $\boldsymbol{f'(x) = u'(x) \pm v'(x)}$.

Product Rule

If $f(x) = u(x) \cdot v(x)$, then $\boldsymbol{f'(x) = u(x) \cdot v'(x) + v(x) \cdot u'(x)}$.

Quotient Rule

If $f(x) = \dfrac{u(x)}{v(x)}$, and $v(x) \neq 0$, then $\boldsymbol{f'(x) = \dfrac{v(x) \cdot u'(x) - u(x) \cdot v'(x)}{[v(x)]^2}}$.

Chain Rule

Let $y = f[g(x)]$. Then $\boldsymbol{y' = f'[g(x)] \cdot g'(x)}$.

Chain Rule (alternative form)

If y is a function of u, say $y = f(u)$, and if u is a function of x, say $u = g(x)$, then $y = f(u) = f[g(x)]$, and

$$\boldsymbol{\frac{dy}{dx} = \frac{dy}{du} \cdot \frac{du}{dx}}.$$

Generalized Power Rule

Let u be a function of x and let $y = u^n$ for any real number n. Then

$$\boldsymbol{y' = n \cdot u^{n-1} \cdot u'}.$$

Exponential Function If $y = e^{g(x)}$, then $\boldsymbol{y' = g'(x) \cdot e^{g(x)}}$.

Natural Logarithmic Function If $y = \ln|x|$, then $\boldsymbol{y' = \dfrac{1}{x}}$.

If $y = \ln[g(x)]$, then $\boldsymbol{y' = \dfrac{g'(x)}{g(x)}}$.

Continuity A function f is **continuous** at $x = c$ if $f(c)$ is defined, $\lim_{x \to c} f(x)$ exists, and $\lim_{x \to c} f(x) = f(c)$.

Chapter 11 Review Exercises

In Exercises 1–6, determine graphically or numerically if the limit exists. If a limit exists, find its (approximate) value.

1. $\lim_{x \to -3} f(x)$

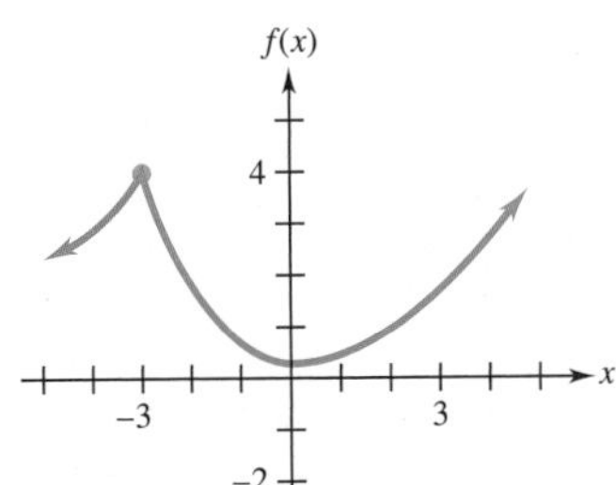

2. $\lim_{x \to -1} g(x)$

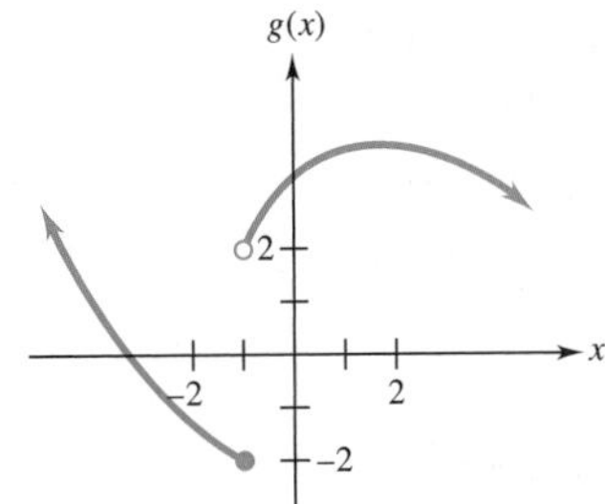

3. $\lim_{x \to 1} \dfrac{x^3 - 1.1x^2 - 2x + 2.1}{x - 1}$

4. $\lim_{x \to 2} \dfrac{x^4 + .5x^3 - 4.5x^2 - 2.5x + 3}{x - 2}$

5. $\lim_{x \to 0} \dfrac{\sqrt{2 - x} - \sqrt{2}}{x}$

6. $\lim_{x \to -1} \dfrac{10^x - .1}{x + 1}$

In Exercises 7–16, find the limit if it exists.

7. $\lim_{x \to 2} (x^2 - 3x + 1)$

8. $\lim_{x \to -1} (-2x^2 + x - 5)$

9. $\lim_{x \to 4} \dfrac{3x + 1}{x - 2}$

10. $\lim_{x \to 3} \dfrac{4x + 7}{x - 3}$

11. $\lim_{x \to 2} \dfrac{x^2 - 4}{x - 2}$

12. $\lim_{x \to -3} \dfrac{x^2 + 2x - 3}{x + 3}$

13. $\lim_{x \to -4} \dfrac{2x^2 + 3x - 20}{x + 4}$

14. $\lim_{x \to 3} \dfrac{3x^2 - 2x - 21}{x - 3}$

15. $\lim_{x \to 9} \dfrac{\sqrt{x} - 3}{x - 9}$

16. $\lim_{x \to 16} \dfrac{\sqrt{x} - 4}{x - 16}$

Use the graph to find the average rate of change of f on the following intervals.

17. $x = 0$ to $x = 4$

18. $x = 2$ to $x = 8$

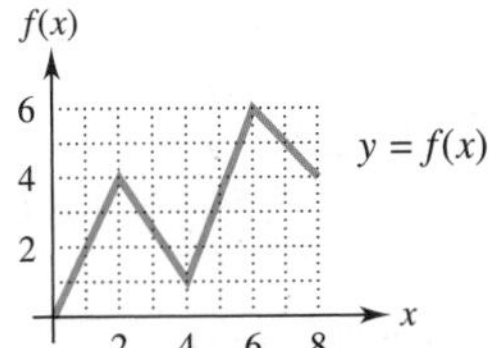

Find the average rate of change for each of the following functions.

19. $f(x) = 3x^2 - 5$, from $x = 1$ to $x = 6$

20. $g(x) = -x^3 + 2x^2 + 1$, from $x = -3$ to $x = 3$

21. $h(x) = \dfrac{6 - x}{2x + 3}$, from $x = 0$ to $x = 5$

22. $f(x) = e^{2x} + 5 \ln x$, from $x = 1$ to $x = 4$

Use the definition of the derivative to find the derivative of each of the following functions.

23. $y = 2x + 3$

24. $y = x^2 + 2x$

25. $y = 2x^2 - x - 1$

26. $y = x^3 + 5$

Find the slope of the tangent line to the given curve at the given value of x. Find the equation of each tangent line.

27. $y = x^2 - 6x$; at $x = 2$

28. $y = 8 - x^2$; at $x = 1$

29. $y = \dfrac{-2}{x + 5}$; at $x = -2$

30. $y = \sqrt{6x - 2}$; at $x = 3$

31. **Business** Suppose hardware store customers are willing to buy $T(p)$ boxes of nails at p dollars per box, where

$$T(p) = .06p^4 - 1.25p^3 + 6.5p^2 - 18p + 200 \qquad (0 < p \le 11).$$

(a) Find the average rate of change in demand for a change in price from \$5 to \$8.
(b) Find the instantaneous rate of change in demand when the price is \$5.
(c) Find the instantaneous rate of change in demand when the price is \$8.

32. Suppose the average rate of change of a function $f(x)$ from $x = 1$ to $x = 4$ is 0. Does this mean that f is constant between $x = 1$ and $x = 4$? Explain.

Find the derivative of each of the following.

33. $y = 5x^2 - 7x - 9$
34. $y = x^3 - 4x^2$
35. $y = 6x^{7/3}$
36. $y = -3x^{-2}$
37. $f(x) = x^{-3} + \sqrt{x}$
38. $f(x) = 6x^{-1} - 2\sqrt{x}$
39. $y = (3t^2 + 7)(t^3 - t)$
40. $y = (-5t + 4)(t^3 - 2t^2)$
41. $y = 8x^{3/4}(2x + 3)$
42. $y = 25x^{-3/5}(x^2 + 5)$
43. $f(x) = \dfrac{2x}{x^2 + 2}$
44. $g(x) = \dfrac{-4x^2}{3x + 4}$
45. $y = \dfrac{\sqrt{x} - 1}{x + 2}$
46. $y = \dfrac{\sqrt{x} + 6}{x - 3}$
47. $y = \dfrac{x^2 - x + 1}{x - 1}$
48. $y = \dfrac{2x^3 - 5x^2}{x + 2}$
49. $f(x) = (3x - 2)^4$
50. $k(x) = (5x - 1)^6$
51. $y = \sqrt{2t - 5}$
52. $y = -3\sqrt{8t - 1}$
53. $y = 2x(3x - 4)^3$
54. $y = 5x^2(2x + 3)^5$
55. $f(u) = \dfrac{3u^2 - 4u}{(2u + 3)^3}$
56. $g(t) = \dfrac{t^3 + t - 2}{(2t - 1)^5}$
57. $y = e^{-2x^3}$
58. $y = -4e^{x^2}$
59. $y = 5x \cdot e^{2x}$
60. $y = -7x^2 \cdot e^{-3x}$
61. $y = \ln(x^2 + 4x - 1)$
62. $y = \ln(4x^3 + 2x)$
63. $y = \dfrac{\ln 4x}{x^2 - 1}$
64. $y = \dfrac{\ln(3x + 5)}{x^2 + 5x}$
65. $y = \dfrac{x^2 + 3x - 10}{x - 3}$
66. $y = \dfrac{x^2 - x - 6}{x - 2}$
67. $y = -6e^{2x}$
68. $y = 8e^{.5x}$

Find each of the following.

69. $D_x\left(\dfrac{\sqrt{x} + 1}{\sqrt{x} - 1}\right)$
70. $D_x\left(\dfrac{2x + \sqrt{x}}{1 - x}\right)$
71. $\dfrac{dy}{dt}$ if $y = \sqrt{t^{1/2} + t}$
72. $\dfrac{dy}{dx}$ if $y = \dfrac{\sqrt{x - 1}}{x}$
73. $f'(1)$ if $f(x) = \dfrac{\sqrt{8 + x}}{x + 1}$
74. $f'(-2)$ if $f(t) = \dfrac{2 - 3t}{\sqrt{2 + t}}$

Find all points of discontinuity for the following.

75.

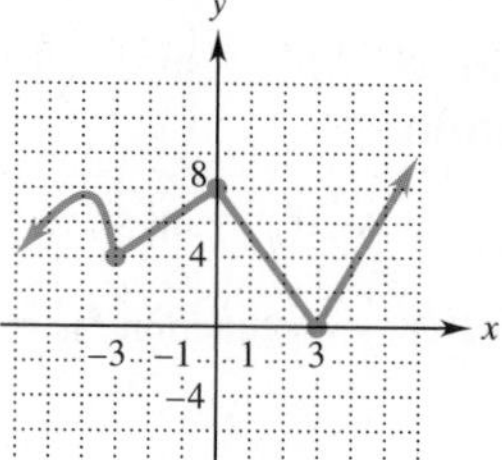

76.

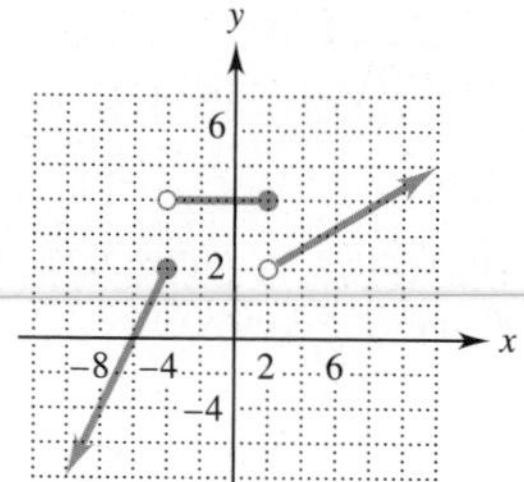

Are the following functions continuous at the given points?

77. $f(x) = \dfrac{2x - 3}{2x + 3}$; $x = -3/2, x = 0, x = 3/2$
78. $g(x) = \dfrac{2x - 1}{x^3 + x^2}$; $x = -1, x = 0, x = 1/2$
79. $h(x) = \dfrac{2 - 3x}{2 - x - x^2}$; $x = -2, x = 2/3, x = 1$
80. $f(x) = \dfrac{x^2 - 4}{x^2 - x - 6}$; $x = 2, x = 3, x = 4$
81. $f(x) = \dfrac{x - 6}{x + 5}$; $x = 6, x = -5, x = 0$
82. $f(x) = \dfrac{x^2 - 9}{x + 3}$; $x = 3, x = -3, x = 0$

Work the following problems.

83. **Business** As sales of DVD players have increased, prices have come down. The average price (in dollars) of a player in year x is approximated by

$$C(x) = -1.5x^3 + 26.5x^2 - 162x + 505.5 \qquad (0 \le x \le 6),$$

where $x = 0$ corresponds to 1997.*

(a) What was the average price in 1998? In 2002?
(b) What was the average rate of change in the price from 1997 to 2001? What does this mean?
(c) At what rate were prices decreasing in 1999? In 2001?

84. **Business** Sales of PCs (in millions of units) in Japan are approximated by

$$S(x) = -.0236x^4 + .295x^3 - .92x^2 + 1.32x + 6.1 \qquad (0 \le x \le 7).$$

*Videoscan, Consumer Electronics Association, VSDA/Vidtrack, and Showbiz data.

where $x = 0$ corresponds to 1995.*

(a) How many PCs were sold in 1996? In 2000?

(b) What was the average rate of change in the number of PCs sold from 1995 to 2001?

(c) At what rate were sales changing in 1997? In 2001?

85. Business The amount (in millions of dollars) spent by Intel on research and development each year is approximated by

$$f(x) = 1.3x^3 + 10.9x^2 + 91.4x + 524 \qquad (1 \le x \le 11).$$

where $x = 1$ corresponds to 1991.†

(a) How much was spent in 1994? In 2000?

(b) At what rate was spending changing in 1994? In 2000?

86. Business The book value of a share of Intel stock (in dollars, adjusted for stock splits) in year x is approximated by

$$g(x) = \frac{15.4}{1 + 30.3e^{-.29x}} \qquad (1 \le x \le 10),$$

where $x = 1$ corresponds to 1991.

(a) What was the book value of a share in 1995? In 2000?

(b) At what rate was the book value changing in 1995? In 1999?

87. Business India has become a major center for computer software and services. The revenue generated by this industry (in billions of dollars) in year x is approximated by

$$R(x) = 1.1\, e^{.41x} \qquad (0 \le x \le 6),$$

where $x = 0$ corresponds to 1995.‡

(a) Find the revenue in 1998 and 2000.

(b) Find the marginal revenue in 1998 and 2000.

88. Social Science The average daily prison population in Kansas is approximated by

$$P(x) = 2409 + 2606 \ln x \qquad (4 \le x \le 11).$$

where $x = 4$ corresponds to 1994.§

(a) How many prisoners were there in 1998? In 2000?

(b) At what rate was the prison population changing in 1996? In 2001?

89. Business Sales of DVDs are increasing rapidly. Total industry revenue (in billions of dollars) in year x is projected to be given by

$$g(x) = \frac{10.5}{1 + 11.6e^{-1.1x}} \qquad (0 \le x \le 6),$$

where $x = 0$ corresponds to 1999.||

(a) Find the revenue in 1999 and 2004.

(b) Find the marginal revenue in 2000 and 2003.

*Jeita.

†Intel Corporation Annual Report 2000.

‡NASSCOM.

§*2000 Corrections Yearbook,* Kansas Department of Corrections.

||Alexander and Associates.

90. Business The yearly amount spent by drug companies on direct-to-consumer advertising in the United States (in billions of dollars) is approximated by

$$S(x) = .8e^{.277x} \qquad (0 \le x \le 5),$$

where $x = 0$ corresponds to 1995.*

(a) How much was spent in 1996? In 2000?

(b) At what rate was spending increasing in 1997 and 2000?

91. Physical Science The graph shows how the velocities of the hands and baseball bat vary with the time of the swing.† Estimate and interpret the value of the derivative functions for the hands and for the bat at the time when the velocities of the two are equal. (*Note:* The rate of change of velocity is called *acceleration.*)

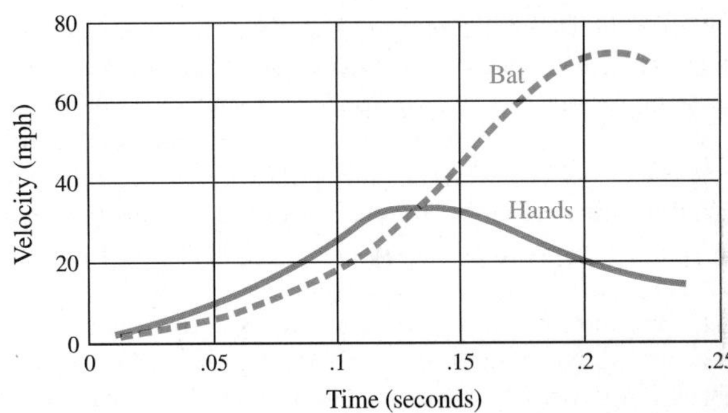

92. Natural Science Under certain conditions, the length of the monkeyface prickleback, a west coast game fish, can be approximated by

$$L = 71.5(1 - e^{-.1t})$$

and its weight by

$$W = .01289 \cdot L^{2.9},$$

where L is the length in cm, t is the age in years, and W is the weight in grams.‡

(a) Find the approximate length of a 5-year-old monkeyface.

(b) Find how fast the length of a 5-year-old monkeyface is increasing.

(c) Find the approximate weight of a 5-year-old monkeyface. (*Hint:* Use your answer from part (a).)

(d) Find the rate of change of the weight with respect to length for a 5-year-old monkeyface.

(e) Using the chain rule and your answers to parts (b) and (d), find how fast the weight of a 5-year-old monkeyface is increasing.

*IMS Health, Competitive Media Reporting.

†Adair, Robert K., *The Physics of Baseball,* Second Revised Edition, HarperCollins, 1994.

‡William H. Marshall and Tina Wyllie Echeverria, "Characteristics of the Monkeyface Prickleback," California Fish and Game 78, no. 2 (spring 1992). For more details, see Case 3 on page 229.

Case 10

Price Elasticity of Demand

Any retailer who sells a product or a service is concerned with how a change in price affects demand for the article. The sensitivity of demand to price changes varies with different items. For smaller items such as soft drinks, food staples, and lightbulbs, small percentage changes in price will not affect the demand for the item much. However, sometimes a small change in price on big ticket items such as cars, homes, and furniture can have significant effects on demand.

One way to measure the sensitivity of changes in price to demand is by the ratio of percent change in demand to percent change in price. If q represents the quantity demanded and p the price of the item, this ratio can be written as

$$\frac{\Delta q/q}{\Delta p/p},$$

where Δq represents the change in q and Δp represents the change in p. The ratio is always negative, because q and p are positive, while Δq and Δp have opposite signs. (An *increase* in price causes a *decrease* in demand.) If the absolute value of this quantity is large, it shows that a small increase in price can cause a relatively large decrease in demand.

Applying some algebra, this ratio can be rewritten as

$$\frac{\Delta q/q}{\Delta p/p} = \frac{\Delta q}{q} \cdot \frac{p}{\Delta p} = \frac{p}{q} \cdot \frac{\Delta q}{\Delta p}.$$

Suppose $q = f(p)$. (Note that this is the inverse of the way our demand functions have been expressed so far; previously we had $p = D(q)$.) Then $\Delta q = f(p + \Delta p) - f(p)$. It follows that

$$\frac{\Delta q}{\Delta p} = \frac{f(p + \Delta p) - f(p)}{\Delta p}.$$

As $\Delta p \to 0$, this quotient becomes

$$\lim_{\Delta p \to 0} \frac{\Delta q}{\Delta p} = \lim_{\Delta p \to 0} \frac{f(p + \Delta p) - f(p)}{\Delta p} = \frac{dq}{dp}$$

and

$$\lim_{\Delta p \to 0} \frac{p}{q} \cdot \frac{\Delta q}{\Delta p} = \frac{p}{q} \cdot \frac{dq}{dp}.$$

The quantity

$$E = -\frac{p}{q} \cdot \frac{dq}{dp}$$

is positive because dq/dp is negative. E is called **elasticity of demand** and measures the instantaneous responsiveness of demand to price. For example, E may be .2 for medical expenses (even though these expenses have considerable price increases each year, they still have a high demand) but may be 1.2 for stereo equipment (high cost and nice to have items). These numbers indicate that the demand for medical services is much less responsive to price changes than the demand for stereo equipment.

If $E < 1$, the relative change in demand is less than the relative change in price, and the demand is called **inelastic.** If $E > 1$, the relative change in demand is greater than the relative change in price, and the demand is called **elastic.** When $E = 1$, the percentage changes in price and demand are relatively equal and the demand is said to have **unit elasticity.**

Sometimes elasticity is counterintuitive. The addiction to illicit drugs is an excellent example. The quantity of the drug demanded by addicts, if anything, increases, no matter what the cost. Thus illegal drugs are an inelastic commodity.

EXAMPLE 1 The VCR market provides a good example of price elasticity of consumer electronics. The demand for VCRs can be expressed by the equation $q = -.025p + 20.45$ where q is the annual demand in millions of VCRs and p is the price of the product.*

(a) Calculate and interpret the elasticity of demand when $p = \$200$; when $p = \$500$.

Since $q = -.025p + 20.45$, we have $dq/dp = -.025$, so that

$$\begin{aligned} E &= -\frac{p}{q} \cdot \frac{dq}{dp} \\ &= -\frac{p}{-.025p + 20.45} \cdot (-.025) \\ &= \frac{.025p}{-.025p + 20.45}. \end{aligned}$$

*Adapted from Todd Thibodeaux, *Pricing Plots Products' Destinies,* Consumer Electronics Vision, July/August 1998.

Let $p = 200$ to get

$$E = \frac{.025(200)}{-.025(200) + 20.45} \approx .324.$$

Since $.324 < 1$, the demand is inelastic, and a percentage change in price will result in a smaller percentage change in demand. For example, a 10% increase in price will cause a 3.24% decrease in demand.

If $p = 500$, then

$$E = \frac{.025(500)}{-.025(500) + 20.45} \approx 1.57.$$

Since $1.57 > 1$, the price is elastic. At this point a percentage increase in price will result in a greater percentage decrease in demand. A 10% increase in price will result in a 15.7% decrease in demand.

(b) Determine the price at which demand will have unit elasticity ($E = 1$). What is the significance of this price?

Demand will have unit elasticity at the price p that makes $E = 1$, so we must solve the equation

$$E = \frac{.025p}{-.025p + 20.45} = 1$$

$$.025p = -.025p + 20.45$$

$$.05p = 20.45$$

$$p = 409.$$

Demand will have unit elasticity at a price of \$409 per VCR. Unit elasticity indicates that the changes in price and demand are about the same.

The definitions from the above discussion can be expressed as follows.

Elasticity of Demand

Let $q = f(p)$, where q is the demand at a price p. The **elasticity of demand** is as follows.

Demand is **inelastic** if $E < 1$.

Demand is **elastic** if $E > 1$.

Demand has **unit elasticity** if $E = 1$.

Exercises

1. The monthly demand for beef in a given region can be expressed by the equation $q = -3.003p + 675.23$, where q is the monthly demand in tons and p is the price in dollars per 100 pounds. Determine the elasticity of demand when the price is \$70.*

2. Acme Stationery sells designer brand pens. The demand equation for annual sales of these pens is $q = -1000p + 70{,}000$, where p is the price per pen. Normally, these pens sell for \$30 each. They are very popular, and Acme has been thinking of raising the price by 1/3.†

(a) Find the elasticity of demand if $p = \$30$.

(b) Find the elasticity of demand if the price is raised by 1/3. Is this a good idea?

3. The monthly demand for lodging in a certain city is given by $q = -2481.52p + 472{,}191.2$ where p is the nightly rate.‡

(a) Find E when $p = \$100$ and $p = \$75$.

(b) At what price is there unit elasticity?

4. Although there are other contributing factors, increasing the price of cigarettes between 1991 and 2000 shows decreasing consumption. Given that $q = -2.35p + 28.26$ represents the annual demand in billions of packs and p is the price per pack, find and interpret the demand elasticity when $p = \$3.00$.§

5. What must be true about the demand if $E = 0$ everywhere?

*Adapted from *How Demand and Supply Determine Price*, Agricultural Marketing Manual, Alberta, Canada, February 1999.

†Taken from R. Horn, *Economics 331: Warm-up Problems: Supply, Demand, Elasticity*, http://cob.jmu.edu/hornrn/prob1.html.

‡Extracted from Bjorn Hanson, *Price Elasticity of Lodging Demand* (PricewaterhouseCoopers, 2000)

§Obtained from Frank J. Chaloupka, *Policy Levers for the Control of Tobacco Consumption* (Impact Teen, 2000).

(1) For what values of x is the function whose graph is shown increasing? decreasing?

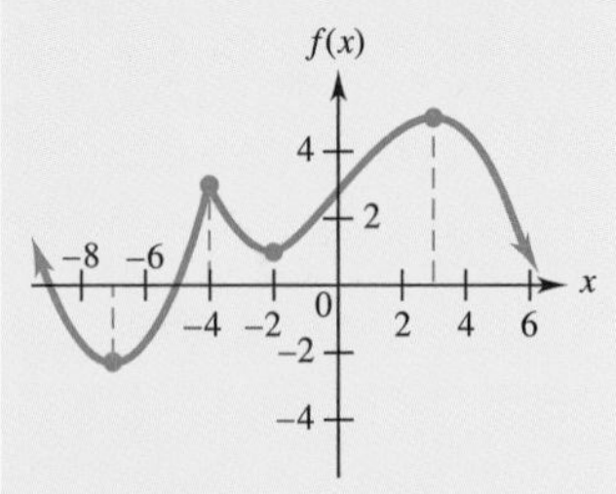

Answer:

Increasing on $(-7, -4)$ and $(-2, 3)$; decreasing on $(-\infty, -7)$, $(-4, -2)$, and $(3, \infty)$

Moving from left to right, the function is increasing up to -4, then decreasing from -4 to 0, constant (neither increasing nor decreasing) from 0 to 4, increasing from 4 to 6, and finally, decreasing from 6 on. In interval notation, the function is increasing on $(-\infty, -4)$ and $(4, 6)$, decreasing on $(-4, 0)$ and $(6, \infty)$, and constant on $(0, 4)$. (1)

In order to examine the connection between the graph of a function f and the derivative of f, it is sometimes helpful to think of the graph of f as a roller coaster track, with a roller coaster car moving from left to right along the graph, as shown in Figure 12.2. At any point along the graph, the floor of the car (a straight line segment) represents the tangent line to the graph at that point.

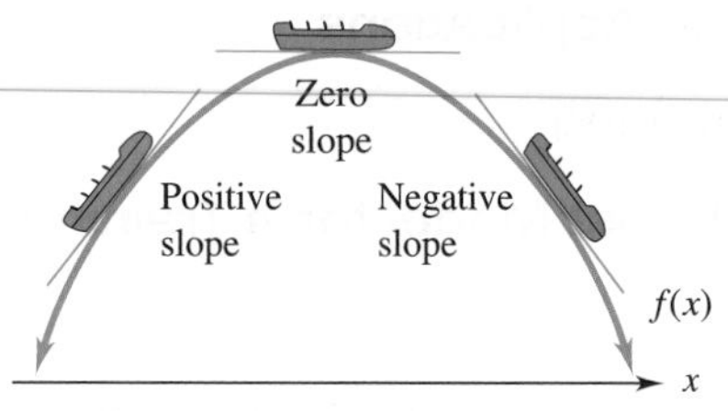

FIGURE 12.2

The slope of the tangent line is positive when the car travels uphill (the function is *increasing*) and the slope of the tangent line is negative when the car travels downhill (the function is *decreasing*). Since the slope of the tangent line at the point $(x, f(x))$ is given by the derivative $f'(x)$, we have these useful facts.

Increasing and Decreasing Functions

Suppose a function f has a derivative at each point in an open interval:

1. If $f'(x) > 0$ for each x in the interval, f is *increasing* on the interval.
2. If $f'(x) < 0$ for each x in the interval, f is *decreasing* on the interval.
3. If $f'(x) = 0$ for each x in the interval, f is *constant* on the interval.

EXAMPLE 2 Find the intervals on which the function

$$f(x) = x^3 + 3x^2 - 9x + 4,$$

whose graph is shown in Figure 12.3, is increasing, decreasing, or constant.

The graph in Figure 12.3 does not clearly show the intervals on which f is increasing or decreasing and suggests that f is constant near $x = -3$. To get more precise information, we first find the derivative:

$$f'(x) = 3x^2 + 6x - 9.$$

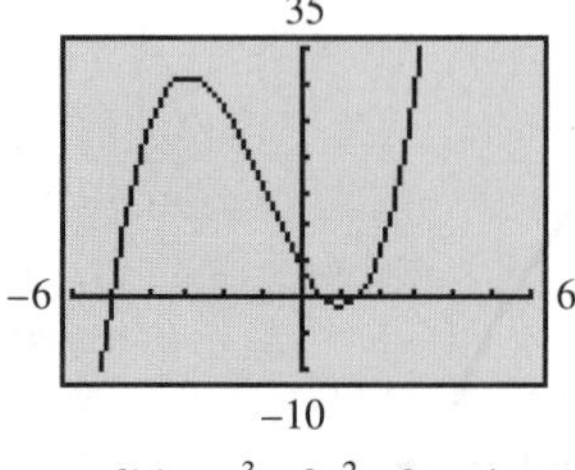

$f(x) = x^3 + 3x^2 - 9x + 4$

FIGURE 12.3

According to the box preceding the example, the function f is increasing on the intervals where $f'(x) > 0$ and decreasing on the intervals where $f'(x) < 0$. So we must solve these inequalities:

$$f'(x) > 0 \qquad \text{and} \qquad f'(x) < 0$$
$$3x^2 + 6x - 9 > 0 \qquad\qquad 3x^2 + 6x - 9 < 0$$

As shown in Section 2.5, the first step is to find the x-intercepts of the graph of $f'(x)$ by solving the equation $f'(x) = 0$.

$$\begin{aligned} 3x^2 + 6x - 9 &= 0 \\ 3(x^2 + 2x - 3) &= 0 \\ 3(x + 3)(x - 1) &= 0 \\ x = -3 \quad \text{or} \quad x &= 1. \end{aligned}$$

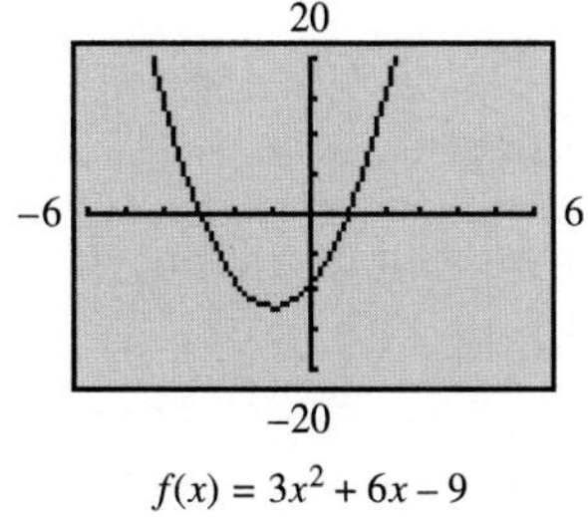

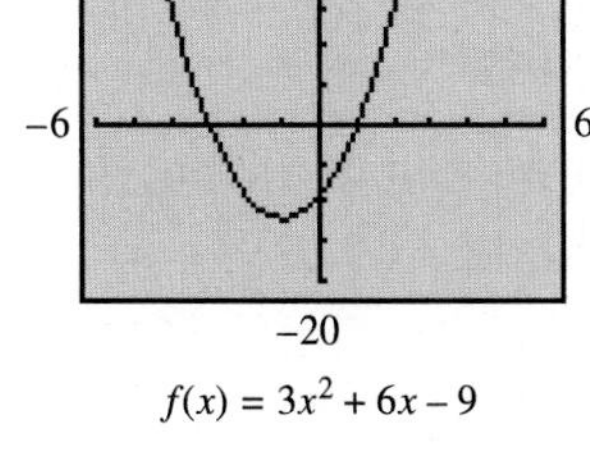

FIGURE 12.4

Hence, the x-intercepts are $x = -3$ and $x = 1$. Now use either graphical or algebraic methods to solve the inequalities.

Graphical The graph of the derivative $f'(x) = 3x^2 + 6x - 9$ in Figure 12.4 lies above the x-axis on the intervals $(-\infty, -3)$ and $(1, \infty)$, meaning that $f'(x) > 0$ on these intervals. Similarly, $f'(x) < 0$ on $(-3, 1)$ because the graph of $f'(x)$ is below the x-axis on $(-3, 1)$.

Algebraic The x-intercepts divide the x-axis into three intervals: $(-\infty, -3)$, $(-3, 1)$ and $(1, \infty)$. Determine the sign of $f'(x)$ on each interval by testing a number in that interval, as summarized below.

Interval	$(-\infty, -3)$	$(-3, 1)$	$(1, \infty)$
Test Value in Interval	-4	0	2
Value of $f'(x) = 3x^2 + 6x - 9$	15	-9	15
Graph of $f'(x)$	Above x-axis	Below x-axis	Above x-axis
Conclusion	$f'(x) > 0$	$f'(x) < 0$	$f'(x) > 0$

2

Find all intervals on which $f(x) = 4x^3 + 3x^2 - 18x + 1$ is increasing or decreasing.

Answer:

Increasing on $(-\infty, -3/2)$ and $(1, \infty)$; decreasing on $(-3/2, 1)$

Regardless of method, the conclusion is the same: the original function f is increasing on $(-\infty, -3)$ and $(1, \infty)$, the intervals where $f'(x) > 0$, and decreasing on $(-3, 1)$, where $f'(x) < 0$. In particular, f is *not* constant near $x = -3$; the "flat" portion of the graph in Figure 12.3 is not actually horizontal (as you can confirm by using the trace feature of a graphing calculator). ◀ 2

Critical Numbers

In Example 2, the numbers for which $f'(x) = 0$ were essential for determining exactly where the function was increasing and decreasing. The situation is a bit different with the absolute value function $f(x) = |x|$, whose graph is shown in Figure 12.5. Clearly, f is decreasing on the left of $x = 0$ and increasing on the right of $x = 0$. But as we saw on page 631, the derivative of $f(x) = |x|$ does not exist at $x = 0$. These examples suggest that the points where the derivative is 0 or undefined play an important role.

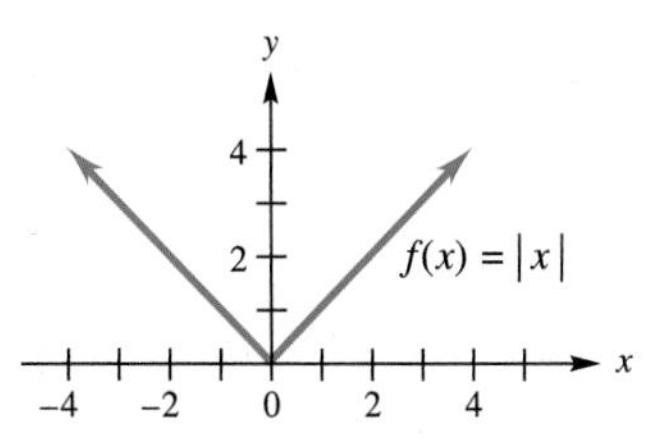

FIGURE 12.5

In view of the preceding discussion, we have this definition.

Critical Numbers

If f is a function, then a number c for which $f(c)$ is defined and

$$\text{either } f'(c) = 0 \quad \text{or} \quad f'(c) \text{ does not exist}$$

is called a **critical number** of f. The corresponding point $(c, f(c))$ on the graph of f is called a **critical point.**

The procedure used in Example 2, which applies to all functions treated in this book, can now be summarized as follows.

Increasing/Decreasing Test

To find the intervals on which a function f is increasing or decreasing, do the following:

Step 1 Compute the derivative f'.

Step 2 Find the critical numbers of f.

Step 3 Solve the inequalities $f'(x) > 0$ and $f'(x) < 0$ graphically or algebraically (by testing a number in each of the intervals determined by the critical numbers).

The solutions of $f'(x) > 0$ are intervals on which f is increasing and the solutions of $f'(x) < 0$ are intervals on which f is decreasing.

Maxima and Minima

We have seen that the graph of a typical function may have "peaks" or "valleys," as illustrated in Figure 12.6.

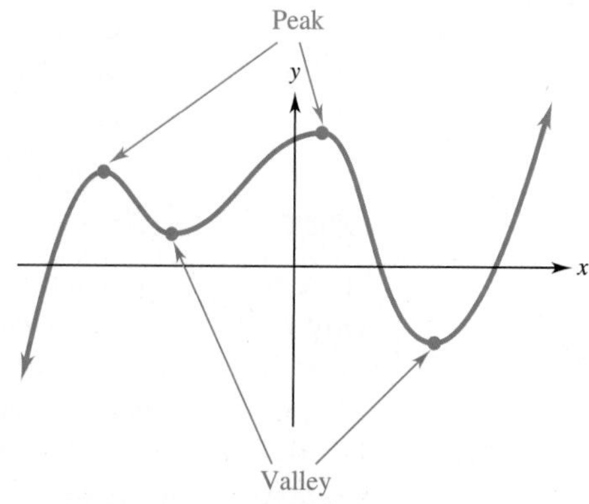

FIGURE 12.6

A peak is the highest point in its neighborhood but not necessarily the highest point on the graph. Similarly, a valley is the lowest point in its neighborhood but not necessarily the lowest point on the graph. Consequently, a peak is called a *local maximum* and a valley a *local minimum*. More precisely, we have the following definitions.

Local Extrema

Let c be a number in the domain of a function f.

1. f has a **local maximum** at c if $f(x) \leq f(c)$ for all x near c.

2. f has a **local minimum** at c if $f(x) \geq f(c)$ for all x near c.

The function f is said to have a **local extremum** at c if it has a local maximum or minimum there.

NOTE The plurals of maximum, minimum, and extremum, respectively, are maxima, minima, and extrema.

EXAMPLE 3 Identify the local extrema of the function whose graph is shown in Figure 12.7.

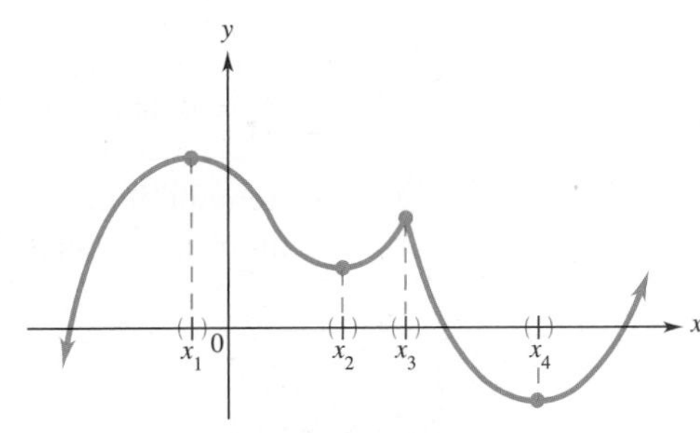

FIGURE 12.7

The function has local maxima at x_1 and x_3 and local minima at x_2 and x_4. 

(3)

Identify the x-values of all points where these graphs have local maxima or local minima.

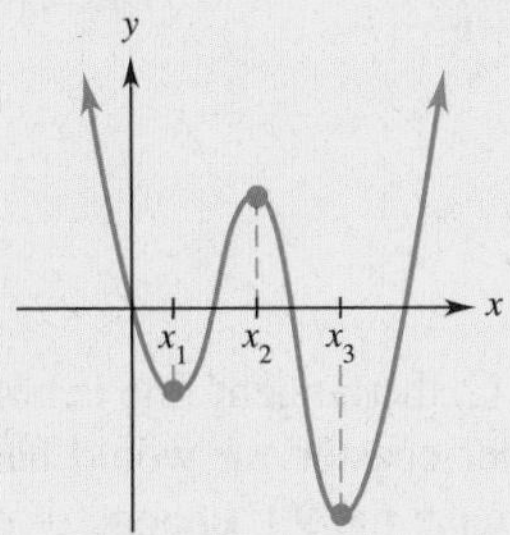

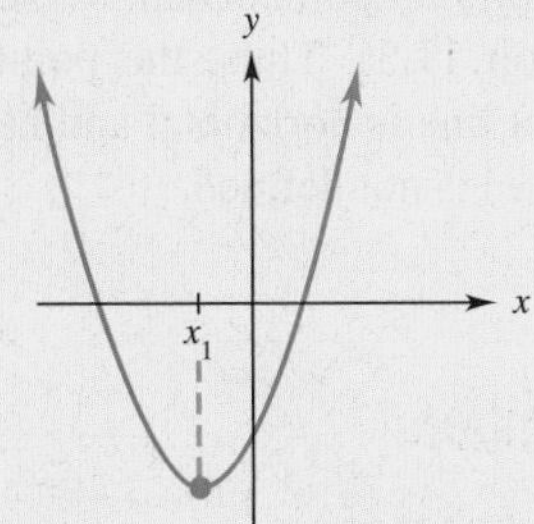

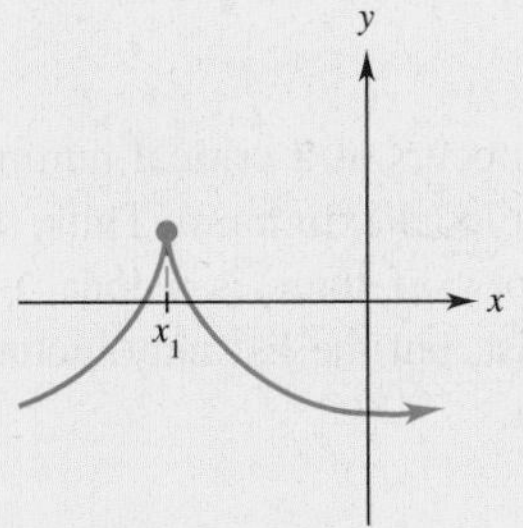

Answers:

(a) Local maximum at x_2; local minima at x_1 and x_3

(b) No local maximum; local minimum at x_1

(c) Local maximum at x_1; no local minimum

TECHNOLOGY TIP Most graphing calculators have a maximum/minimum finder that can approximate local extrema to a high degree of accuracy. Check your instruction manual.

The *exact* location of a local extremum (rather than a calculator's approximation) can normally be found by using derivatives. To see why this is so, let f be a function and, once again, think of the graph of f as a roller coaster track, as shown in Figure 12.8 on the next page.

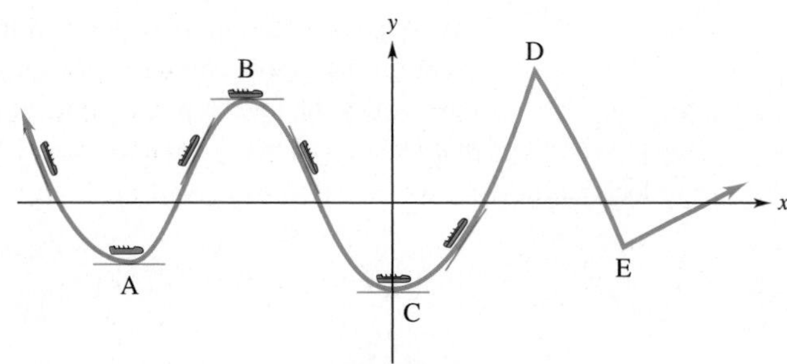

FIGURE 12.8

As the car passes through the local extrema at A, B, and C, the tangent line is horizontal and has slope 0. At D and E, however, a real roller coaster car would have trouble. It would fly off the track at D and be unable to make the 90° change of direction at E. Notice that the graph does not have tangent lines at points D and E (see the discussion of the existence of derivatives in Section 11.3). Thus, the points where local extrema occur have this property: the tangent line is horizontal and has slope 0 *or* there is no tangent line, that is, $f'(c) = 0$ or $f'(c)$ is not defined.

Local Extrema and Critical Numbers

If f has a local extremum at $x = c$, then c is a critical number of f.

CAUTION This result says that every local extremum occurs at a critical number but *not* that every critical number produces a local extremum. Thus, the critical numbers provide a list of *possibilities*: if there is a local extremum, it must occur at a number on the list, but the list may include numbers at which there is no local extremum.

EXAMPLE 4 Find the critical numbers of the following functions.

(a) $f(x) = 2x^3 - 3x^2 - 72x + 15$

We have $f'(x) = 6x^2 - 6x - 72$, so $f'(x)$ exists for every x. Setting $f'(x) = 0$ shows that

$$6x^2 - 6x - 72 = 0$$
$$6(x^2 - x - 12) = 0$$
$$x^2 - x - 12 = 0$$
$$(x + 3)(x - 4) = 0$$
$$x + 3 = 0 \quad \text{or} \quad x - 4 = 0$$
$$x = -3 \quad \text{or} \quad x = 4.$$

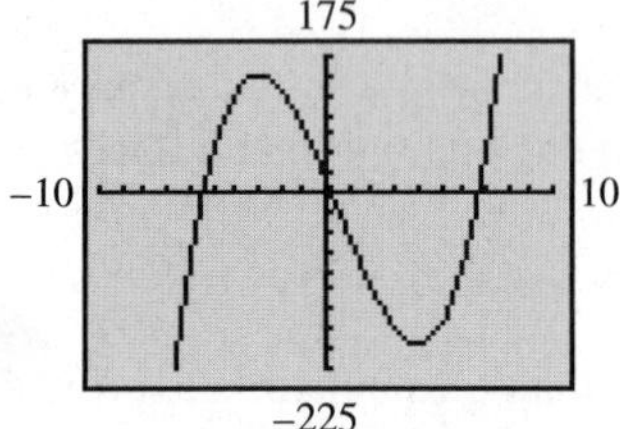

$f(x) = 2x^3 - 3x^2 - 72x + 15$

FIGURE 12.9

Therefore, -3 and 4 are the critical numbers of f; these are the only places where local extrema could possibly occur. The graph of f in Figure 12.9 shows that there is a local maximum at $x = -3$ and a local minimum at $x = 4$.

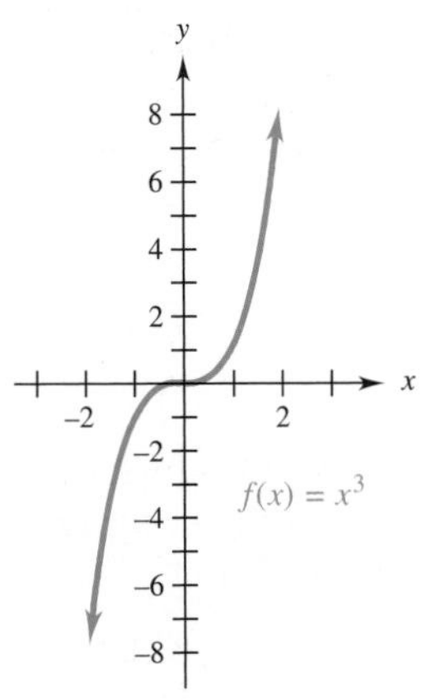

FIGURE 12.10

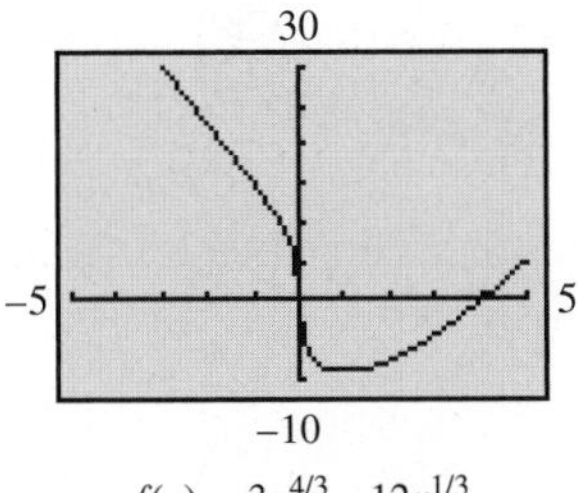

$f(x) = 3x^{4/3} - 12x^{1/3}$

FIGURE 12.11

(b) $f(x) = x^3$

The derivative $f'(x) = 3x^2$ is 0 exactly when $x = 0$. So $x = 0$ is the only critical number of f and the only possible location for a local maximum or minimum. In this case, however, we know what the graph of f looks like (see Figure 12.10). The graph shows that there is no local maximum or minimum at $x = 0$.

(c) $f(x) = 3x^{4/3} - 12x^{1/3}$.

We first compute the derivative.

$$f'(x) = 3 \cdot \frac{4}{3}x^{1/3} - 12 \cdot \frac{1}{3}x^{-2/3}$$

$$= 4x^{1/3} - \frac{4}{x^{2/3}}$$

$$= \frac{4x^{1/3}x^{2/3}}{x^{2/3}} - \frac{4}{x^{2/3}}$$

$$= \frac{4x - 4}{x^{2/3}}$$

The derivative fails to exist when $x = 0$. Since the original function f is defined when $x = 0$, 0 is a critical number of f. If $x \neq 0$, then $f'(x)$ is 0 only when the numerator $4x - 4 = 0$, that is, when $x = 1$. So the critical numbers of f are 0 and 1. These numbers are the *possible* locations for local extrema. However, the graph of f in Figure 12.11 suggests that there is a local minimum at $x = 1$, but no local extremum at $x = 0$. ◀ 4

Find the critical numbers for each of these functions.

(a) $\frac{1}{3}x^3 - x^2 - 15x + 6$

(b) $6x^{2/3} - 4x$

Answers:

(a) $-3, 5$

(b) $0, 1$

The First Derivative Test

When all the critical numbers of a function f have been found, you must then determine which ones lead to local extrema. Sometimes this can be done graphically, as in Example 4. It can also be done algebraically by using the following observation:

> At a local maximum f changes from increasing to decreasing and at a local minimum f changes from decreasing to increasing, as illustrated in Figure 12.12.

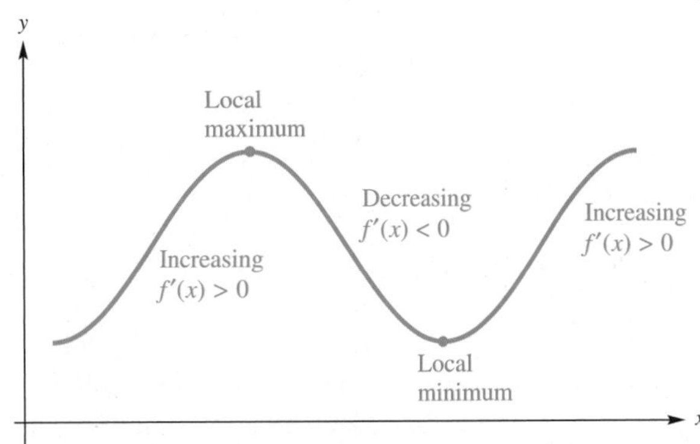

FIGURE 12.12

When f changes from increasing to decreasing, its derivative f' changes from positive to negative. Similarly, when f changes from decreasing to increasing, f' changes from negative to positive. These facts lead to the following test for local extrema, whose formal proof will be omitted.

First Derivative Test

Assume that $a < c < b$ and that c is the only critical number for a function f in the interval $[a, b]$. Assume that f is differentiable for all x in $[a, b]$, except possibly at $x = c$.

1. If $f'(a) > 0$ and $f'(b) < 0$, then there is a local maximum at c.
2. If $f'(a) < 0$ and $f'(b) > 0$, then there is a local minimum at c.
3. If $f'(a)$ and $f'(b)$ are both positive, or both negative, then there is no local extremum at c.

The sketches in the following table show how the first derivative test works. Assume the same conditions on a, b, c as those stated in the box.

f(x) has:	*Sign of $f'(a)$*	*Sign of $f'(b)$*	*Sketches*
Local Maximum	+	−	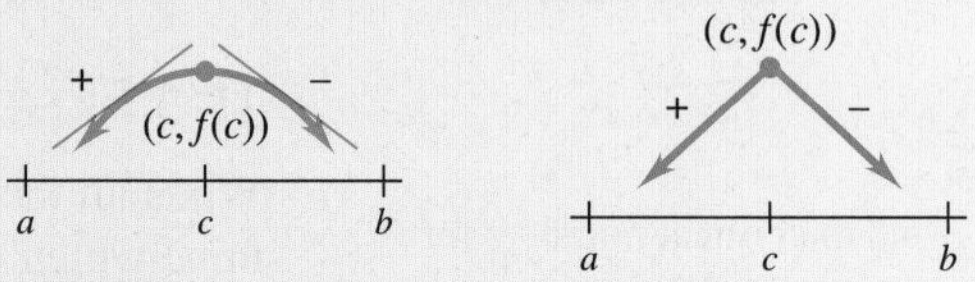 + $(c, f(c))$ − ; a c b — + $(c, f(c))$ − ; a c b
Local Minimum	−	+	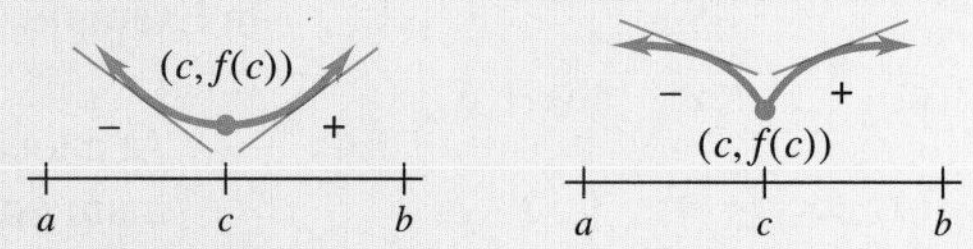 $(c, f(c))$ − + ; a c b — − + $(c, f(c))$; a c b
No Local Extrema	+	+	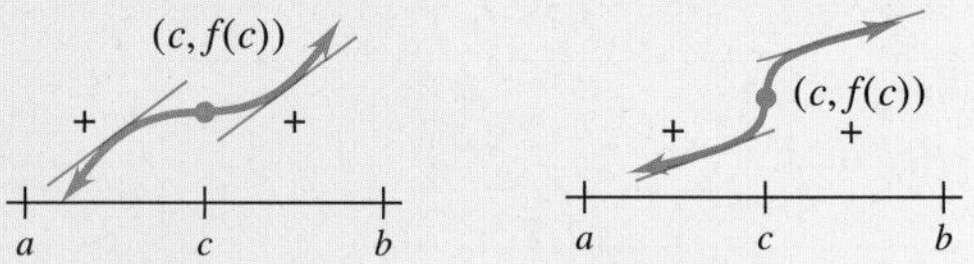 $(c, f(c))$ + + ; a c b — $(c, f(c))$ + + ; a c b
No Local Extrema	−	−	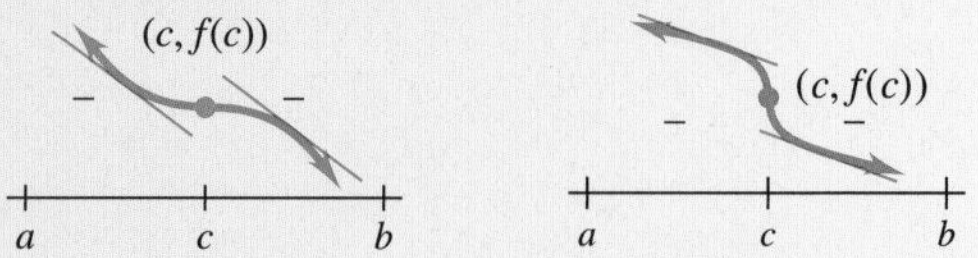 $(c, f(c))$ − − ; a c b — $(c, f(c))$ − − ; a c b

The local extrema of the functions in Example 4(a) and 4(c) were determined by using a graphing calculator (once the critical numbers had been found algebraically). Examples 5 and 6 show how the first derivative test can be used in place of graphing.

EXAMPLE 5 Use the first derivative test to find the local extrema of

$$f(x) = 2x^3 - 3x^2 - 72x + 15.$$

Example 4(a) shows that the critical numbers of f are -3 and 4. To use the first derivative test on -3, we must choose an interval $[a, b]$ containing -3 but not containing the other critical number. We shall use $a = -4$ and $b = 0$. Many other choices of a and b are possible, but we try to select ones that will make the computations easy. Since

$$f'(x) = 6x^2 - 6x - 72 = 6(x^2 - x - 12) = 6(x + 3)(x - 4),$$

we see that

$$f'(-4) = 6(-4 + 3)(-4 - 4) = 6(-1)(-8) > 0;$$

and

$$f'(0) = 6(0 + 3)(0 - 4) = 6(3)(-4) < 0.$$

(Note that it's not necessary to finish calculating the exact value of $f'(x)$ in order to determine its sign.) Thus, the value of the derivative is positive to the left of -3 and negative to the right of -3, as shown in Figure 12.13. By part 1 of the first derivative test, there is a local maximum at $x = -3$, which is $f(-3) = 150$.

Similarly, we can use $a = 0$ and $b = 5$ to test the critical number $c = 4$. We just saw that $f'(0) < 0$; and

$$f'(5) = 6(5 + 3)(5 - 4) = 6(8)(1) > 0.$$

Hence, by part 2 of the first derivative test, there is a local minimum at $x = 4$, where $f(4) = -193$. ◀ 5

Find the location of all local extrema of these functions.

(a) $f(x) = 2x^2 - 8x + 1$

(b) $g(x) = x^3 - 9x^2 - 48x + 195$

Answers:

(a) Local minimum at 2

(b) Local maximum at -2; local minimum at 8

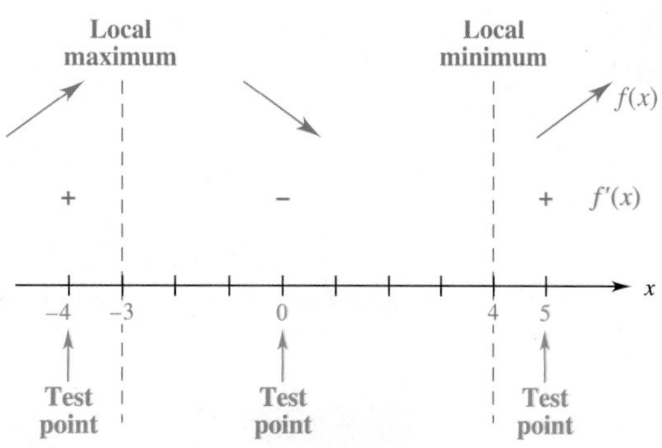

FIGURE 12.13

EXAMPLE 6 Use the first derivative test to find the local extrema of

$$f(x) = 3x^{4/3} - 12x^{1/3}.$$

Example 4(c) shows that the critical numbers of f are 0 and 1. We can use the intervals $[-1, 1/2]$ and $[1/2, 2]$ to test these critical numbers.

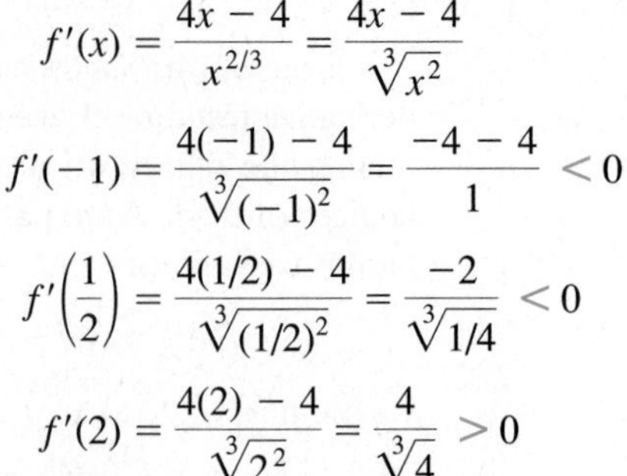

$$f'(x) = \frac{4x - 4}{x^{2/3}} = \frac{4x - 4}{\sqrt[3]{x^2}}$$

$$f'(-1) = \frac{4(-1) - 4}{\sqrt[3]{(-1)^2}} = \frac{-4 - 4}{1} < 0$$

$$f'\left(\frac{1}{2}\right) = \frac{4(1/2) - 4}{\sqrt[3]{(1/2)^2}} = \frac{-2}{\sqrt[3]{1/4}} < 0$$

$$f'(2) = \frac{4(2) - 4}{\sqrt[3]{2^2}} = \frac{4}{\sqrt[3]{4}} > 0$$

Since $f'(x)$ is negative at both -1 and $1/2$, part 3 of the first derivative test shows that there is no local extremum at $x = 0$. Since $f'(1/2) < 0$ and $f'(2) > 0$, there is a local minimum at $x = 1$, by part 2 of the first derivative test. These results are shown schematically in Figure 12.14. ◀ 6

6

Find any local extrema for $f(x) = x^3 + 4x^2 - 3x + 5$.

Answer:

Local maximum at -3 of $f(-3) = 23$; local minimum at $1/3$ of $f(1/3) = 121/27$

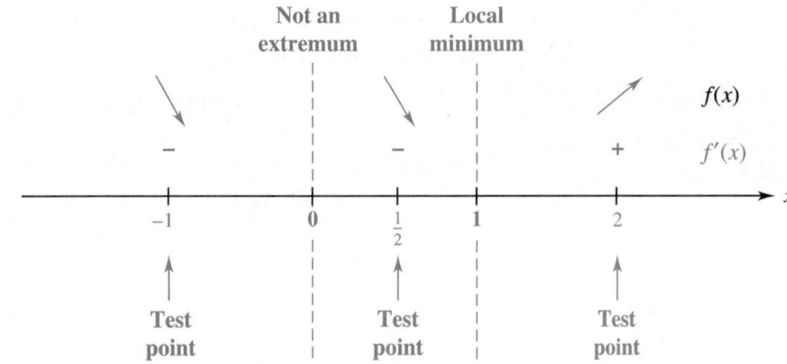

FIGURE 12.14

Although graphing technology sometimes eliminates the need to use the first derivative test, this is not always the case, as the next example illustrates.

EXAMPLE 7 Find all local extrema of $f(x) = x^2e^{5x}$.

Although the graph of f in Figure 12.15 suggests that there are no local extrema, algebraic analysis shows otherwise. Using the product rule for derivatives, we have

$$f'(x) = x^2(5e^{5x}) + 2x(e^{5x}) = 5x^2e^{5x} + 2xe^{5x}.$$

So the critical numbers are the solutions of

$$5x^2e^{5x} + 2xe^{5x} = 0$$
$$e^{5x}(5x^2 + 2x) = 0$$
$$e^{5x}(5x + 2)x = 0$$

Find a viewing window on a graphing calculator that clearly shows the local maximum and minimum of the function f in Example 7.

Answer:

There are many possibilities, including

$-2 \le x \le 2$ and
$-.1 \le y \le .1$.

Find all local extrema of

$g(x) = x^3e^x$.

Answer:

Local minimum at $x = -3$; no local extremum at $x = 0$

Since e^{5x} is never 0, this derivative can equal 0 only when

$$5x + 2 = 0 \qquad \text{or} \quad x = 0$$
$$x = -2/5 = -.4$$

We choose the interval $[-1, -.2]$ to test $-.4$ and the interval $[-.2, 1]$ to test 0. Using a calculator to compute values of $y_1 = f'(x)$, we obtain Figure 12.16.

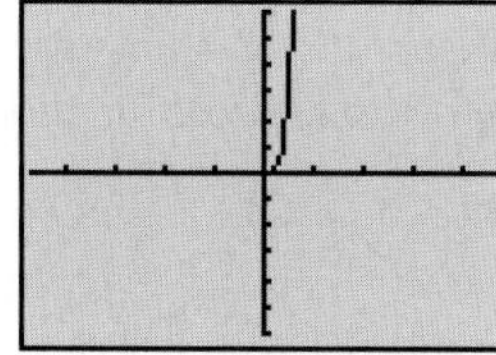

FIGURE 12.15

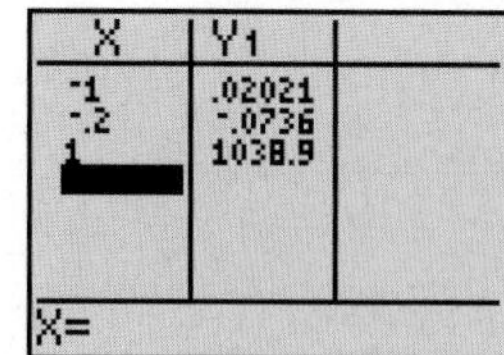

FIGURE 12.16

By the first derivative test, there is a local maximum at $x = -.4$ and there is a local minimum at $x = 0$, as indicated schematically in Figure 12.17. ◀ (7) (8)

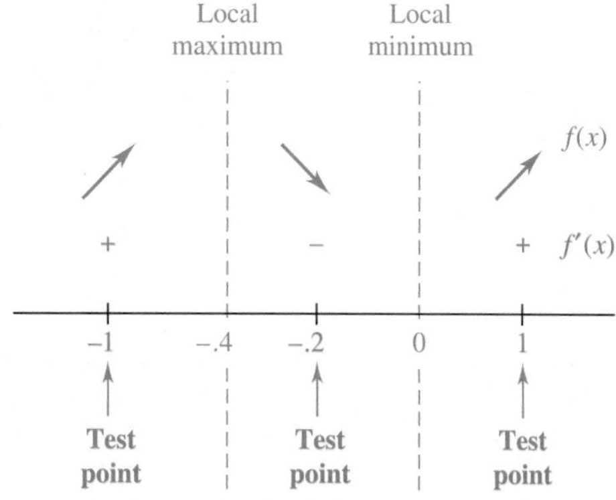

FIGURE 12.17

EXAMPLE 8 A company that makes digital alarm clocks has the following cost and revenue functions.

$$R(x) = 36x - .0015x^2 \qquad 0 \le x \le 12{,}000$$
$$C(x) = .00000034x^3 - .005x^2 + 27x + 25{,}000 \qquad 0 \le x \le 12{,}000$$

Determine where the profit function is increasing and find the maximum possible profit.

The profit function is $P(x) = R(x) - C(x)$, so

$$\begin{aligned} P(x) &= (36x - .0015x^2) - (.00000034x^3 - .005x^2 + 27x + 25{,}000) \\ &= -.00000034x^3 + .0035x^2 + 9x - 25{,}000. \end{aligned}$$

To determine the critical numbers we find $P'(x)$ and use the quadratic formula to solve $P'(x) = 0$.

$$P'(x) = -.00000102x^2 + .007x + 9 = 0$$

$$x = \frac{-.007 \pm \sqrt{(.007)^2 - 4(-.00000102)(9)}}{2(-.00000102)} \approx \begin{cases} 7969.86 \\ -1107.11 \end{cases}$$

Find the intervals where profit is increasing and $P(x) > 0$ if the profit function is defined as $P(x) = 1000 + 90x - x^2$.

Answer:

(0, 45)

Since $x \geq 0$ here, the only relevant critical number is 7969.86. Applying the first derivative test with $a = 0$ and $b = 10{,}000$ shows that

$$P'(0) = 9 > 0 \quad \text{and} \quad P'(10{,}000) \approx -23 < 0.$$

Therefore, $P(x)$ is increasing on the (approximate) interval (0, 7969.86), decreasing on (7969.86, 12000), and has a local maximum at approximately 7969.86. Hence maximum profit occurs when about 7970 clocks are manufactured and sold. As the graphs in Figure 12.18 show, profit increases as long as the revenue function increases faster than the cost function. ◀ 9

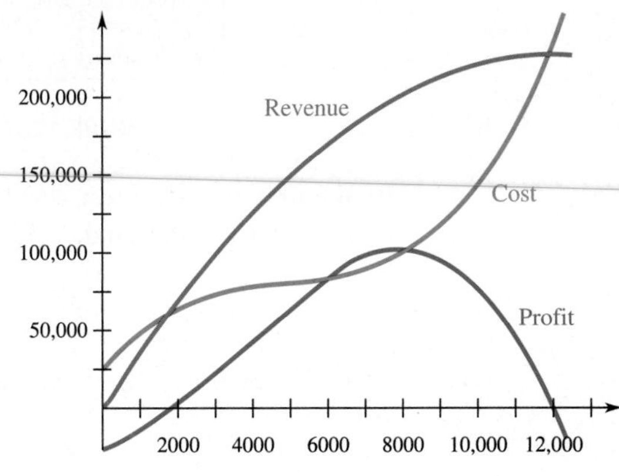

FIGURE 12.18

12.1 Exercises

For each function, list the intervals where the function is increasing, the intervals where it is decreasing, and the location of all local extrema. (See Examples 1 and 3.)

1.

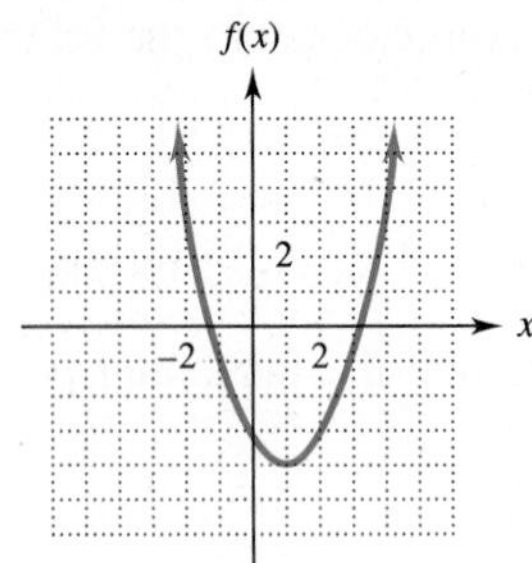

2.

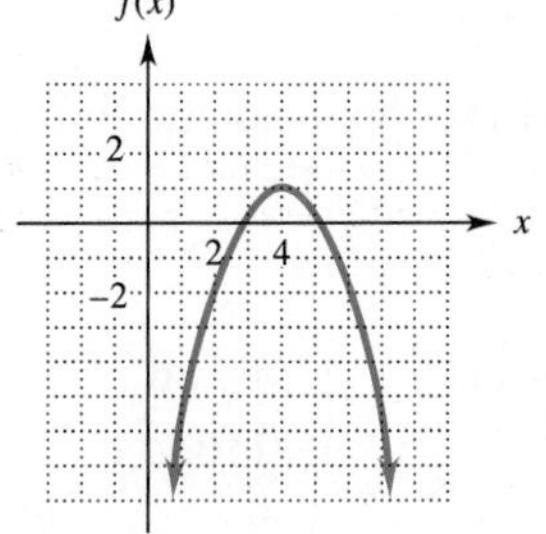

3.

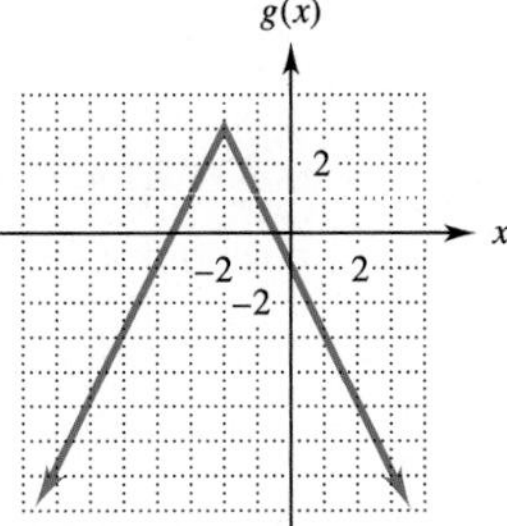

4.

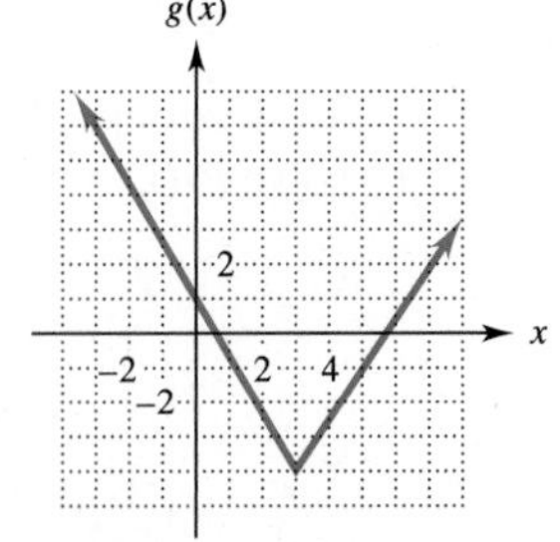

5. 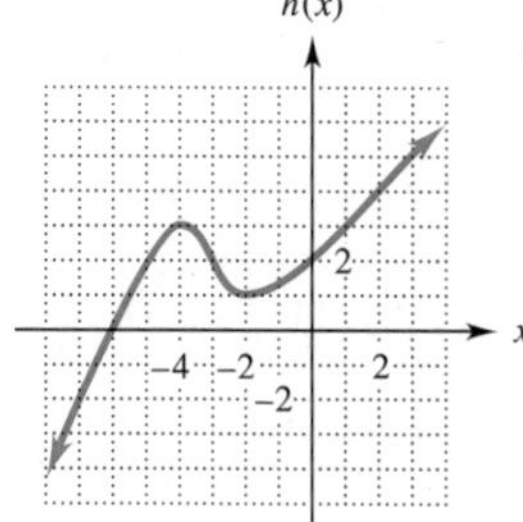

6. 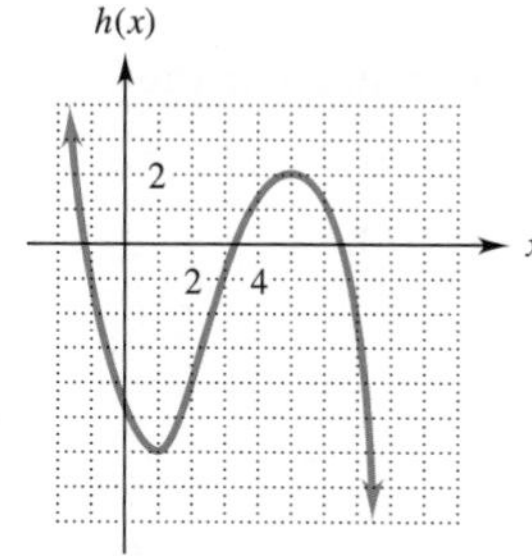

7. 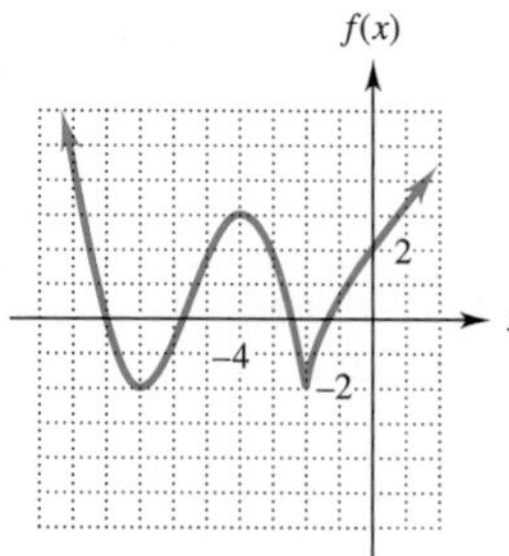

8. 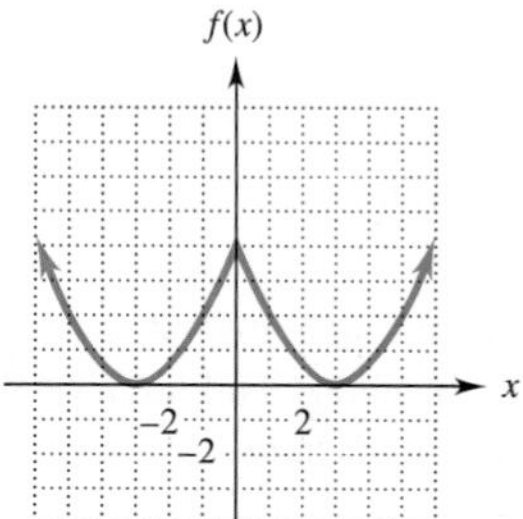

Find the intervals where each function is increasing or decreasing. (See Example 2.)

9. $f(x) = 2x^3 - 5x^2 - 4x + 2$
10. $f(x) = 4x^3 - 9x^2 - 12x + 7$
11. $f(x) = \dfrac{x + 1}{x + 4}$
12. $f(x) = \dfrac{x^2 + 1}{x}$
13. $f(x) = \sqrt{5 - x}$
14. $f(x) = \sqrt{x^2 + 1}$
15. $f(x) = 2x^3 - 3x^2 - 12x + 2$
16. $f(x) = 4x^3 - 15x^2 - 72x + 5$

The graph of the derivative function f' is given; list the critical numbers of the function f.

17. 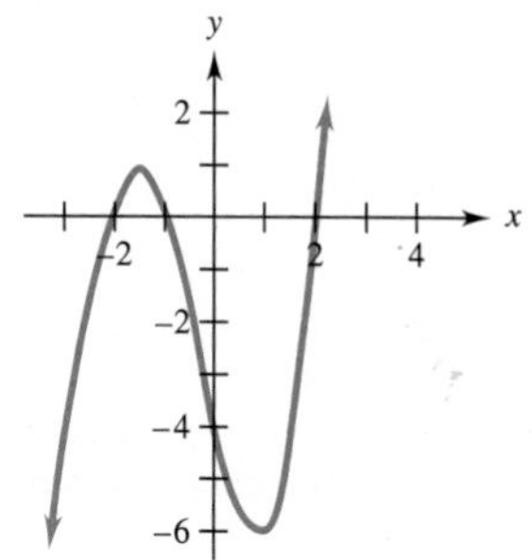

18. 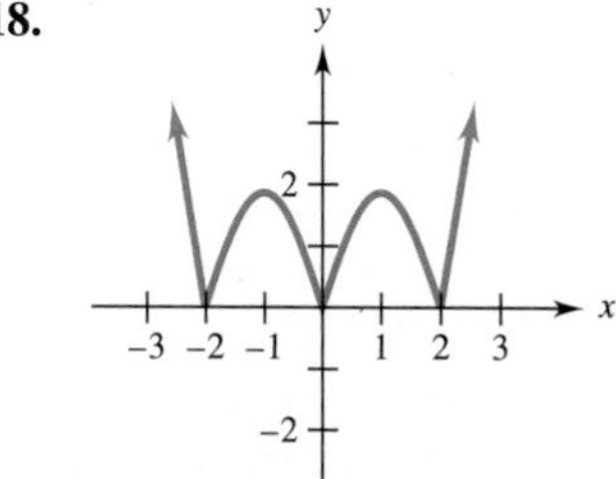

Determine the location of each local extremum of the function. (See Examples 4–7.)

19. $f(x) = x^3 - 3x^2 + 1$
20. $f(x) = x^3 - x^2 - 5x + 1$
21. $f(x) = x^3 + 6x^2 + 9x + 2$
22. $f(x) = x^3 + 3x^2 - 24x + 2$
23. $f(x) = -\dfrac{4}{3}x^3 - \dfrac{21}{2}x^2 - 5x + 8$
24. $f(x) = -\dfrac{2}{3}x^3 - \dfrac{1}{2}x^2 - 3x - 4$
25. $f(x) = \dfrac{2}{3}x^3 - x^2 - 12x + 2$
26. $f(x) = \dfrac{4}{3}x^3 - 10x^2 + 24x - 1$
27. $f(x) = x^5 - 20x^2 + 3$
28. $f(x) = 3x^3 - 18.5x^2 - 4.5x - 45$

In Exercises 29–42, use the first derivative test to determine the location of each local extremum and the value of the function at this extremum. (See Examples 5–7.)

29. $f(x) = x^{11/5} - x^{6/5} + 1$
30. $f(x) = (7 - 2x)^{2/3} - 2$

31. $f(x) = -(3 - 4x)^{2/5} + 4$

32. $f(x) = x^2 + \frac{1}{x}$

33. $f(x) = \frac{x^2}{x^2 + 1}$

34. $f(x) = \frac{x^2 - 2x + 1}{x - 3}$

35. $f(x) = -xe^x$

36. $f(x) = xe^{-x}$

37. $f(x) = x \cdot \ln|x|$

38. $f(x) = x - \ln|x|$

39. $f(x) = xe^{3x} - 2$

40. $f(x) = x^3e^{4x} + 1$

41. $f(x) = e^x + e^{-x}$

42. $f(x) = -x^2e^x$

Use the maximum/minimum finder on a graphing calculator to determine the approximate location of all local extrema of these functions.

43. $f(x) = .1x^4 - x^3 - 12x^2 + 99x - 10$

44. $f(x) = x^5 - 12x^4 - x^3 + 232x^2 + 260x - 600$

45. $f(x) = .01x^5 + x^4 - x^3 - 6x^2 + 5x + 4$

46. $f(x) = .1x^5 + 3x^4 - 4x^3 - 11x^2 + 3x + 2$

Work the following exercises. (See Example 8.)

47. **Natural Science** The graph shows the amount of air pollution removed by trees in the Chicago urban region for each month of the year.* From the graph we see that the ozone level increases up to June and then abruptly decreases.
    (a) At what points are the derivatives of the functions whose graphs are shown undefined?
    (b) Look at the graph for particulates. Where is the function increasing? decreasing? constant?
    (c) On what intervals do all four lower graphs indicate the corresponding functions are constant? Why do you think the functions are constant on those intervals?

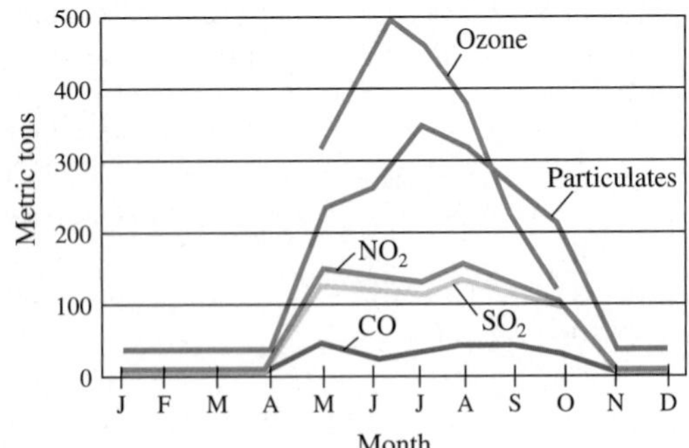

*National Arbor Day Foundation, 100 Arbor Ave., Nebraska City, NE 68410, ad in *Chicago Tribune*, Sunday 2/4/96, sec. 2, p. 11.

48. **Physical Science** The following graph shows the horsepower and torque as a function of the engine speed for a 1991 Porsche 928 GT.
    (a) On what intervals is the horsepower increasing with engine speed?
    (b) On what intervals is the horsepower decreasing with engine speed?
    (c) On what intervals is the torque increasing with engine speed?
    (d) On what intervals is the torque decreasing with engine speed?

928 GT Performance and Torque

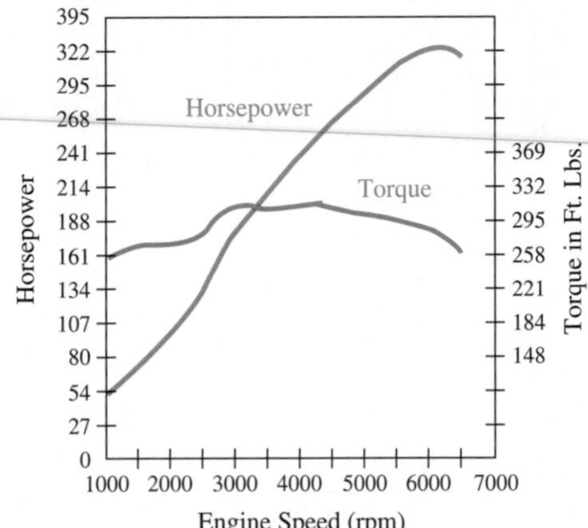

49. **Business** The total profit $P(x)$ (in thousands of dollars) from the sale of x thousand units of a new prescription drug is given by

$$P(x) = -x^3 + 3x^2 + 72x \quad (0 \le x \le 10).$$

    (a) Find the number of units that should be sold in order to maximize the total profit.
    (b) What is the maximum profit?

50. **Social Science** For 1975 through 1990, the number of master's degrees awarded in library and archival science in the United States can be modeled by

$$M(x) = 670\sqrt{147 - 21.6x + x^2},$$

where x is the number of years since 1975.* Find the years during which the number of degrees was increasing and decreasing.

51. **Social Science** The standard normal probability function is used to describe many different populations. Its graph is the well-known normal curve. This function is defined by

$$f(x) = \frac{1}{\sqrt{2\pi}}e^{-x^2/2}.$$

Give the intervals where the function is increasing and decreasing.

Mathematics Teacher 89, no. 3 (March 1996): 216.

52. Business A county realty group estimates that the number of housing starts per year over the next 3 years will be

$$H(r) = \frac{300}{1 + .03r^2},$$

where r is the mortgage rate (in percent).

(a) Where is $H(r)$ increasing?
(b) Where is $H(r)$ decreasing?

53. Business A manufacturer sells cutlery with the following cost and revenue functions, where x is the number of sets sold.

$$C(x) = 4000 - 4x, \quad 0 \le x \le 17{,}000$$
$$R(x) = 20x - x^2/1000, \quad 0 \le x \le 17{,}000$$

Determine the intervals on which the profit function is increasing.

54. Natural Science During a four-week long flu epidemic, the number of people $P(t)$ infected t days after the epidemic begins is approximated by

$$P(t) = t^3 - 60t^2 + 900t + 20 \qquad (0 \le t \le 28).$$

When will the number of people infected start to decline?

55. Natural Science The function

$$A(x) = -.15x^3 + 1.058x$$

approximates the alcohol concentration (in tenths of a percent) in an average person's bloodstream x hours after drinking 8 ounces of 100-proof whiskey. The function applies only for the interval [0, 8].

(a) On what time intervals is the alcohol concentration increasing?
(b) On what intervals is it decreasing?

56. Natural Science The percent of concentration of a drug in the bloodstream x hours after the drug is administered is given by

$$K(x) = \frac{4x}{3x^2 + 27}.$$

(a) On what time intervals is the concentration of the drug increasing?
(b) On what intervals is it decreasing?

57. Business The graph at the top of the next column shows the amount spent by Mexico (in millions of dollars) on imports of diesel fuel.* List the intervals in which the function is

(a) increasing;
(b) decreasing.

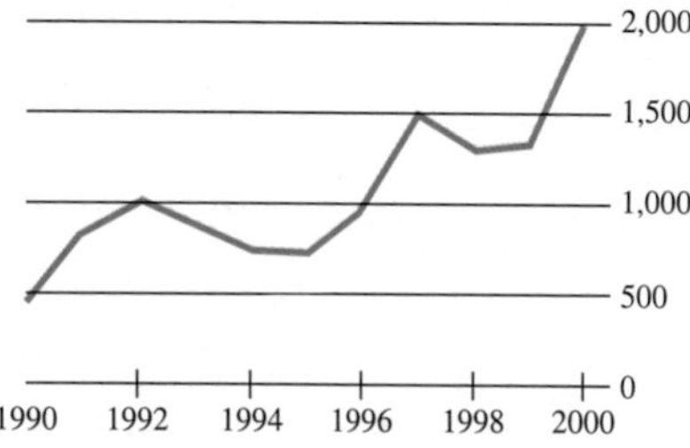

(c) List the years when there was a local maximum or minimum of this function.

58. Social Science The graph below shows the stockpile of nuclear weapons held by the United States and by the Soviet Union and its successor states from 1945 to 2000.*

(a) On what intervals was the stockpile of U.S. weapons increasing?
(b) On what intervals was the stockpile of Soviet weapons increasing?
(c) In what years was the U.S. stockpile of weapons at a local maximum?

Warhead Stockpiles—A New Look

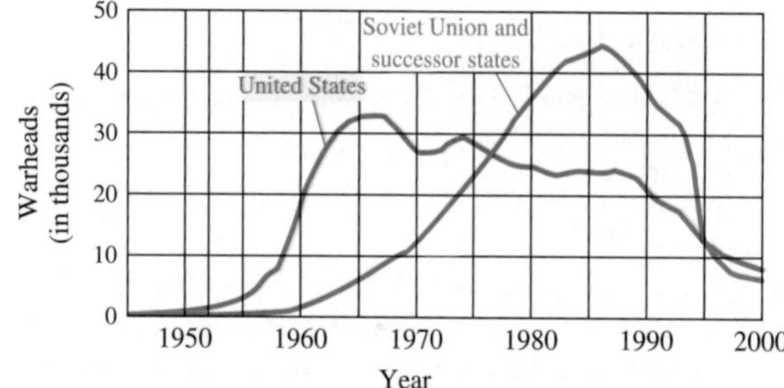

59. Social Science The minimum wage rose from \$1.25 an hour in 1965 to \$5.15 in 2000. But the actual purchasing power of the minimum wage has varied greatly. The graph on the next page shows the value of the minimum wage in constant 2000 dollars adjusted for inflation, from 1965 to 2000.†

(a) What was the approximate value of the minimum wage in these years: 1969, 1982, 1989, 1998, 2000?
(b) Between 1965 and 2000, how many local maxima are there?
(c) List the years in which local minima occurred between 1980 and 2000.

*Economist, August 25, 2001.

*Chicago Tribune and Bulletin of the Atomic Scientists.

†U.S. Department of Labor, Bureau of Labor Statistics.

(d) On what time intervals between 1980 and 2000 was the actual value of the minimum wage decreasing?

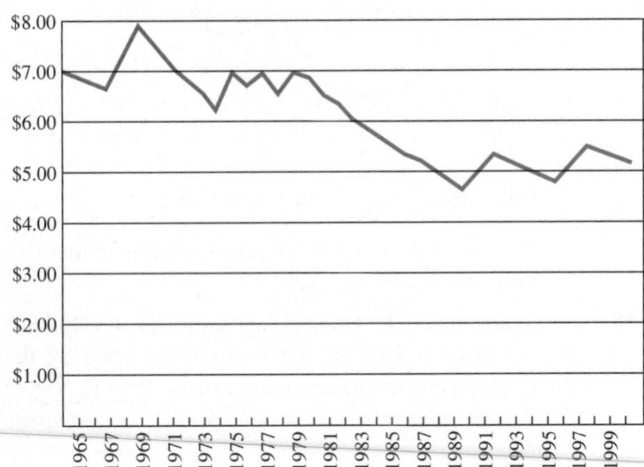

60. Physical Science A Boston Red Sox pitcher stands on top of the 37-foot-high left field wall (the "Green Monster") in Fenway Park and fires a fast ball straight up. The position function that gives the height of the ball (in feet) at time t seconds is given by $s(t) = -16t^2 + 140t + 37$.* Find
(a) the maximum height of the ball;
(b) the time when the ball hits the ground;
(c) the velocity of the ball when it hits the ground.

61. Social Science The percentage of new Ph.D.s in mathematics who are unemployed can be roughly approximated by the function

$$f(x) = .044x^3 - .92x^2 + 5x + 1.4 \qquad (1 \le x \le 11),$$

where $x = 1$ corresponds to 1991.† In what years was the percentage a local maximum or minimum? (Round your answers to the nearest year.)

62. Natural Science A marshy region used for agricultural drainage has become contaminated with selenium. It has been determined that flushing the area with clean water will reduce the selenium for a while, but it will then begin to build up again. A biologist has found that the percent of selenium in the soil x months after the flushing begins is given by

$$f(x) = \frac{x^2 + 36}{2x}, \qquad 1 \le x \le 12.$$

When will the selenium be reduced to a minimum? What is the minimum percent?

63. Finance The interest rate on 10-year U.S. Treasury bonds can be approximated by

$$g(x) = .0005x^3 - .037x^2 + .66x + 6 \qquad (0 \le x \le 31),$$

where $x = 0$ corresponds to 1970. Find the local extrema of this function (within the given range of x-values). Interpret your answer.

64. Natural Science The number of salmon swimming upstream to spawn is approximated by

$$S(x) = -x^3 + 3x^2 + 400x + 5000, \qquad 6 \le x \le 20,$$

where x represents the temperature of the water in degrees Celsius. Find the water temperature that produces the maximum number of salmon swimming upstream.

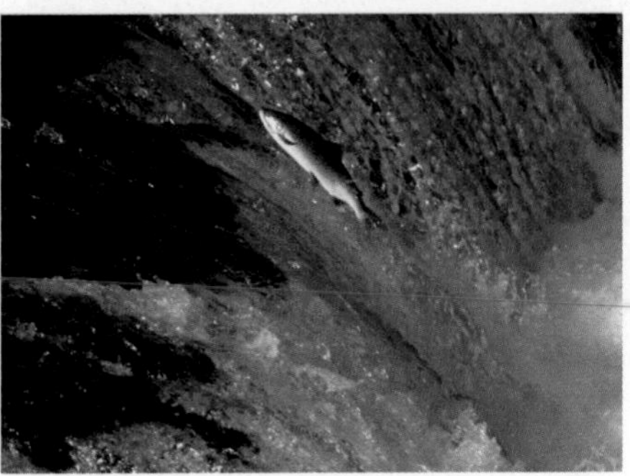

65. Social Science Far more American troops have died in training accidents than in combat since the Vietnam War. Recently the military has been redoubling its efforts to increase safety standards and prevent violence in the ranks. The number of casualties (in hundreds) each year is approximated by

$$f(x) = .198x^3 - 1.56x^2 + 2.03x + 14.8,$$

where x is the number of years since 1990.* The function is accurate only on the finite domain $\{0, 1, 2, 3, 4, 5\}$. Find the location and value of any extrema for this function.

66. Natural Science In the summer the activity level of a certain type of lizard varies according to the time of day. A biologist has determined that the activity level is given by the function

$$a(t) = .008t^3 - .27t^2 + 2.02t + 7,$$

where t is the number of hours after 12 noon and $0 \le t \le 24$. When is the activity level highest? When is it lowest?

67. Social Science Social psychologists have found that as the discrepancy between the views of a speaker and those of an audience increases, the attitude change in the audience also increases to a point but decreases when the discrepancy becomes too large, particularly if the communicator is viewed by the audience as having low credi-

*Exercise provided by Frederick Russell of Charles County Community College.

†Notices of the American Mathematical Society, August 2001.

*New York Times, December 25, 1995, p. 8.

bility.* Suppose that the degree of change can be approximated by the function

$$D(x) = -x^4 + 8x^3 + 80x^2, \qquad 0 \le x \le 13,$$

where x is the discrepancy between the views of the speaker and those of the audience, as measured by scores on a questionnaire. Find the amount of discrepancy the speaker should aim for to maximize the attitude change in the audience.

68. Natural Science The microbe concentration, $B(x)$, in appropriate units, of Lake Tom depends approximately on the oxygen concentration, x, again in appropriate units, according to the function

$$B(x) = x^3 - 7x^2 - 160x + 1800, \qquad 0 \le x \le 20.$$

(a) Find the oxygen concentration that will lead to the minimum microbe concentration.

(b) What is the minimum concentration?

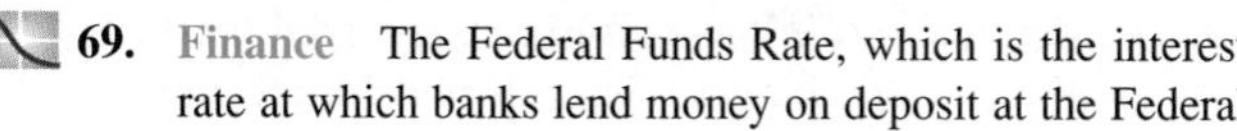

69. Finance The Federal Funds Rate, which is the interest rate at which banks lend money on deposit at the Federal Reserve to other banks overnight, can be approximated by

$$f(x) = -.038x^4 + .6x^3 - 3.1x^2 + 5.8x + 2.25 \qquad (0 \le x \le 7).$$

where $x = 0$ corresponds to 1994.* Use a graphing calculator to determine when this function has local maxima.

70. Consider the function†

$$g(x) = \frac{1}{x^{12}} - 2\left(\frac{1000}{x}\right)^6.$$

(a) Using a graphing calculator, try to find any local minima or tell why finding a local minimum is difficult for this function.

(b) Find any local minima using the techniques of calculus.

(c) Based on your results in parts (a) and (b), describe circumstances under which relative extrema are easier to find using the techniques of calculus than using a graphing calculator.

*See A. H. Eagly and K. Telaak, "Width of the Latitude of Acceptance as a Determinant of Attitude Change," *Journal of Personality and Social Psychology* 23 (1972): 388–397.

*The Federal Reserve Board.

†From Ed Dubinsky, "Is Calculus Obsolete?", *Mathematics Teacher*, 88, no. 2 (February 1995): 146–148.

12.2 THE SECOND DERIVATIVE

The first derivative of a function indicates when the function is increasing or decreasing. In many applications, however, this isn't enough. For example, Figure 12.19 shows the prices of two different stocks over a period of months. Both stocks are worth \$5 per share at the beginning and are continually increasing in price.

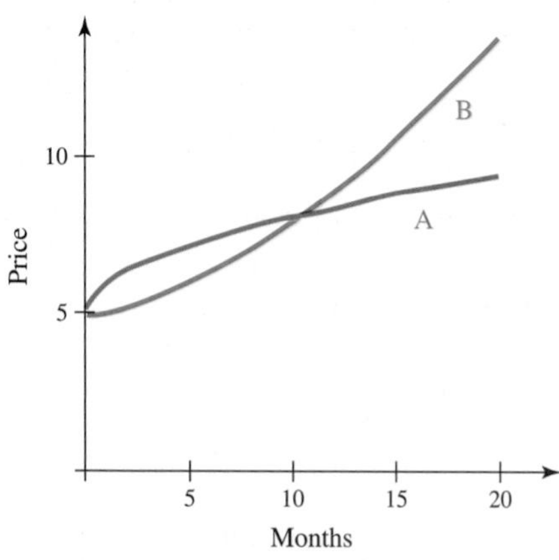

FIGURE 12.19

It's easy to see from the graph that stock B is the better long-term investment. In real life, however, you see only *part* of the graph (the past performance of a stock). How can that help you to predict future performance? To answer this question, look at the graphs again. Both are always increasing, but after the first ten months, stock B increases at a much faster rate than stock A. So the *rate* of increase plays a crucial role here.

In mathematical terms, if the price of the stock in month x is $f(x)$, then the derivative $f'(x)$ tells you when the price is increasing (or decreasing). The *rate* of increase (or decrease) of the derivative function $f'(x)$ is given by *its* derivative. In other words, the rate at which the stock price is increasing (or decreasing) is given by the derivative of the derivative of the price function. In order to deal with such situations, we need some additional terminology and notation.

1

Let $f(x) = 4x^3 - 12x^2 + x - 1$. Find

(a) $f''(0)$;

(b) $f''(4)$;

(c) $f''(-2)$.

Answers:

(a) -24

(b) 72

Higher Derivatives

If a function f has a derivative f', then the derivative of f', if it exists, is the **second derivative** of f, written $f''(x)$. The derivative of $f''(x)$, if it exists, is called the **third derivative** of f, and so on. By continuing this process, we can find **fourth derivatives** and other higher derivatives. For example, if $f(x) = x^4 + 2x^3 + 3x^2 - 5x + 7$, then

$$f'(x) = 4x^3 + 6x^2 + 6x - 5, \quad \text{First derivative of } f$$
$$f''(x) = 12x^2 + 12x + 6, \quad \text{Second derivative of } f$$
$$f'''(x) = 24x + 12, \quad \text{Third derivative of } f$$
$$f^{(4)}(x) = 24. \quad \text{Fourth derivative of } f$$ 1

The second derivative of $y = f(x)$ can be written with any of the following notations:

$$f''(x), \quad y'', \quad \frac{d^2y}{dx^2}, \quad \text{or} \quad D_x^2[f(x)].$$

The third derivative can be written in a similar way. For $n \geq 4$, the nth derivative is written $f^{(n)}(x)$.

EXAMPLE 1 Find the second derivative of the following functions.

(a) $f(x) = 8x^3 - 9x^2 + 6x + 4$

Here $f'(x) = 24x^2 - 18x + 6$. The second derivative is the derivative of $f'(x)$, or

$$f''(x) = 48x - 18.$$

(b) $y = \dfrac{4x + 2}{3x - 1}$

Use the quotient rule to find y'.

$$y' = \frac{(3x - 1)(4) - (4x + 2)(3)}{(3x - 1)^2} = \frac{12x - 4 - 12x - 6}{(3x - 1)^2} = \frac{-10}{(3x - 1)^2}$$

Find the second derivatives of the following.

(a) $y = -9x^3 + 8x^2 + 11x - 6$

(b) $y = -2x^4 + 6x^2$

(c) $y = \dfrac{x + 2}{5x - 1}$

(d) $y = e^x + \ln x$

Answers:

(a) $y'' = -54x + 16$

(b) $y'' = -24x^2 + 12$

(c) $y'' = \dfrac{110}{(5x - 1)^3}$

(d) $y'' = e^x - \dfrac{1}{x^2}$

Use the quotient rule again to find y''.

$$y'' = \frac{(3x - 1)^2(0) - (-10)(2)(3x - 1)(3)}{[(3x - 1)^2]^2}$$

$$= \frac{60(3x - 1)}{(3x - 1)^4} = \frac{60}{(3x - 1)^3}$$

(c) $y = xe^x$

Using the product rule gives

$$\frac{dy}{dx} = x \cdot e^x + e^x \cdot 1 = xe^x + e^x.$$

Differentiate this result to get $\dfrac{d^2y}{dx^2}$.

$$\frac{d^2y}{dx^2} = (xe^x + e^x) + e^x = xe^x + 2e^x = (x + 2)e^x$$ ◀ 2

Applications

In the previous chapter we saw that the first derivative of a function represents the rate of change of the function. The second derivative, then, represents the rate of change of the first derivative. This fact has a variety of applications. For one such application, we take another look at stocks A and B from the beginning of this section.

EXAMPLE 2 Suppose that over a ten-month period, the price of stock A is given by $f(x) = x^{1/2} + 5$ and that the price of stock B is given by $g(x) = .1x^{3/2} + 5$.

(a) When are stocks A and B increasing in price?

The first derivatives of the price functions are

$$f'(x) = \frac{1}{2}x^{-1/2} = \frac{1}{2\sqrt{x}} \quad \text{and} \quad g'(x) = .1\left(\frac{3}{2}x^{1/2}\right) = \frac{.3}{2}\sqrt{x}.$$

Both derivatives are always positive (because $\sqrt{x}$ is positive for $x > 0$). So both price functions are increasing for all $x > 0$.

(b) At what rate are these stock prices increasing in the tenth month? What does this suggest about their future performance?

The rate at which $f'(x) = \frac{1}{2}x^{-1/2}$ and $g'(x) = \frac{.3}{2}x^{1/2}$ are increasing is given by *their* derivatives (the second derivatives of f and g):

$$f''(x) = \frac{1}{2} \cdot \frac{-1}{2}x^{-3/2} = \frac{-1}{4\sqrt{x^3}} \quad \text{and} \quad g''(x) = \frac{.3}{2} \cdot \frac{1}{2}x^{-1/2} = \frac{3}{40\sqrt{x}}.$$

When $x = 10$,

$$f''(10) = \frac{-1}{4\sqrt{10^3}} \approx -.0079 \quad \text{and} \quad g''(10) = \frac{3}{40\sqrt{10}} \approx .0237.$$

The rate of increase for stock A is negative, meaning that its price is increasing at a decreasing rate. The rate of increase for stock B is positive, meaning that its price is increasing at an increasing rate. In other words, the price of stock A is increasing more and more slowly, while the price of stock B is increasing faster and faster. This suggests that stock B is a probably a better investment for the future.

The preceding discussion assumes that present trends continue (which is not guaranteed in the stock market). If they do continue for another ten months, then Figure 12.19 shows what will happen to the stocks during this period. ◀

Although Example 2 is much simpler than real life, second derivatives are actually used by some investors. According to an article in the *Wall Street Journal* on August 1, 2001 (when the market was in a downturn), "many investors hope it will help them get into the market ahead of any big rallies. In essence, the second-derivative approach involves looking at such things as profit warnings, analysts' earning estimates and equipment orders (including cancellations) for signs that the downward momentum is slowing."

The second derivative also plays a role in the physics of a moving particle. If a function describes the position of a moving object (along a straight line) at time t, then the first derivative gives the velocity of the object, That is, if $y = s(t)$ describes the position (along a straight line) of the object at time t, then $v(t) = s'(t)$ gives the velocity at time t.

The rate of change of velocity is called **acceleration.** Since the second derivative gives the rate of change of the first derivative, the acceleration is the derivative of the velocity. Thus, if $a(t)$ represents the acceleration at time t, then

$$\mathbf{a(t) = \frac{d}{dt} v(t) = s''(t).}$$

▶ EXAMPLE 3 Suppose that an object is moving along a straight line, with its position in feet at time t in seconds given by

$$s(t) = t^3 - 2t^2 - 7t + 9.$$

Find the following.

(a) The velocity at any time t

The velocity is given by

$$v(t) = s'(t) = 3t^2 - 4t - 7.$$

(b) The acceleration at any time t

Acceleration is given by

$$a(t) = v'(t) = s''(t) = 6t - 4.$$

(c) The object stops when velocity is zero. For $t \geq 0$, when does that occur?

Set $v(t) = 0$.

$$3t^2 - 4t - 7 = 0$$
$$(3t - 7)(t + 1) = 0$$
$$3t - 7 = 0 \quad \text{or} \quad t + 1 = 0$$
$$t = \frac{7}{3} \qquad\qquad t = -1$$

Since we want $t \geq 0$, only $t = 7/3$ is acceptable here. The object will stop at 7/3 seconds. ◀ 3

Rework Example 3 if $s(t) = t^4 - t^3 + 10$.

Answers:

(a) $v(t) = 4t^3 - 3t^2$

(b) $a(t) = 12t^2 - 6t$

(c) At 0 and 3/4 second

Concavity

We shall now see how the second derivative provides information about how the graph of the function "bends," which is often hard to see on a calculator or computer screen. A graph is **concave upward** on an interval if it bends upward over the interval and **concave downward** if it bends downward, as shown in Figure 12.20. The graph is concave downward on the interval (a, b) and concave upward on the interval (b, c).* A point on the graph where the concavity changes (such as the point where $x = b$ in Figure 12.20) is called a **point of inflection.**

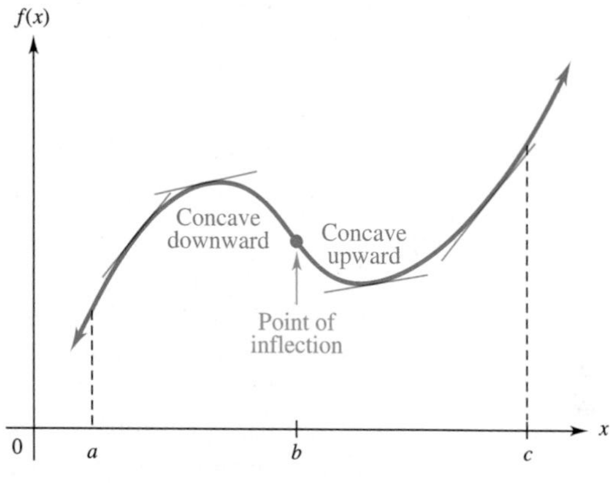

FIGURE 12.20

A function that is increasing on an interval may have either kind of concavity; the same is true for a decreasing function. Some of the possibilities are illustrated in Figure 12.21.

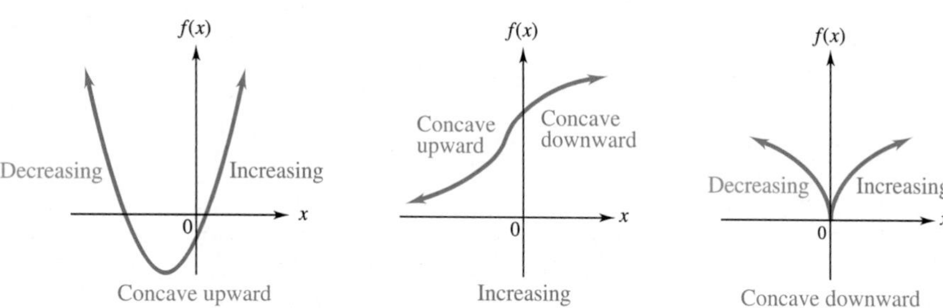

FIGURE 12.21

Next we examine the relationship between the second derivative of a function f and the concavity of the graph of f. We have seen that when the derivative of any function is positive, then that function is increasing. Consequently, if the *second* derivative of f (the derivative of the first derivative) is positive, then the *first* derivative of f is increasing. Since the first derivative gives the slope of the tangent line to the graph of f at each point, the fact that the first derivative is increasing means that the

*Figure 12.20 also illustrates the formal definition of concavity: a function is *concave downward* on an interval if its graph lies below the tangent line at each point in the interval and *concave upward* if its graph is above the tangent line at each point in the interval.

tangent line slopes are increasing as you move from left to right along the graph of f, as illustrated in Figure 12.22.

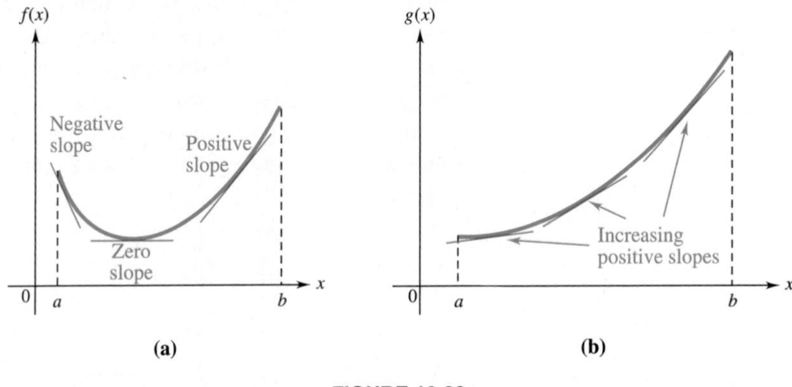

FIGURE 12.22

In Figure 12.22(a), the slopes of the tangent lines increase from negative at the left, to 0 in the center, to positive at the right. In Figure 12.22(b), the slopes are all positive but are increasing as the tangent lines get steeper. Note that both graphs in Figure 12.22 are *concave upward.*

Similarly, when the second derivative is negative, then the first derivative (slope of the tangent line) is decreasing, as illustrated in Figure 12.23. In Figure 12.23(a), the tangent line slopes decrease from positive, to 0, to negative. In Figure 12.23(b), the slopes get more and more negative as the tangent lines drop downward more steeply. Note that both graphs are *concave downward.*

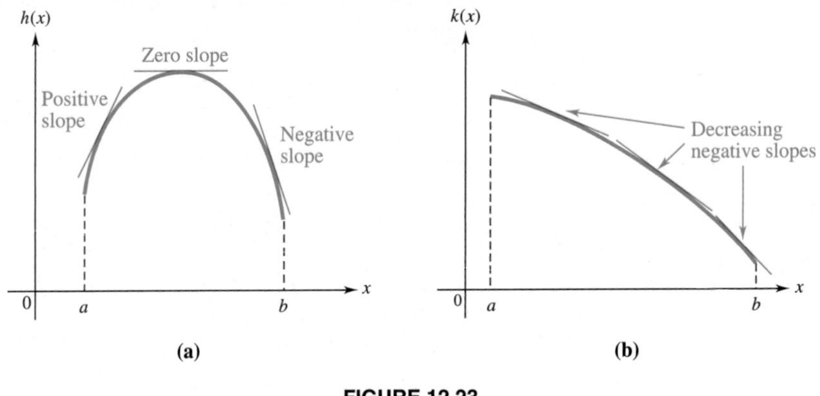

FIGURE 12.23

The preceding discussion suggests the following result.

Concavity Test

Let f be a function whose first and second derivatives exist at all points in the interval (a, b).

1. If $f''(x) > 0$ for all x in (a, b), then f is concave upward on (a, b).

2. If $f''(x) < 0$ for all x in (a, b), then f is concave downward on (a, b).

EXAMPLE 4 Find the intervals over which the function is concave upward or downward and find any points of inflection.

(a) $f(x) = .5x^3 - 2x^2 + 5x + 1$

The graph of f (Figure 12.24) suggests that f is concave upward to the right of the y-axis but the concavity near the y-axis isn't clear. Even in a window where the concavity is visible, the location of the point of inflection is not clear (try it!). So we use the test in the box above. The first derivative is $f'(x) = 1.5x^2 - 4x + 5$ and the second derivative is $f''(x) = 3x - 4$. The function is concave upward when $f''(x) > 0$, that is, when

$$3x - 4 > 0$$
$$3x > 4$$
$$x > 4/3$$

Similarly, f is concave downward if $3x - 4 < 0$, that is, when $x < 4/3$. Therefore, f is concave downward on $(-\infty, 4/3)$ and concave upward on $(4/3, \infty)$, with a point of inflection when $x = 4/3$.

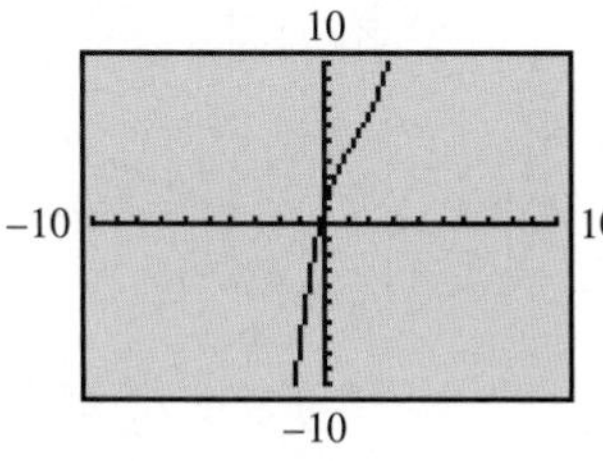

FIGURE 12.24

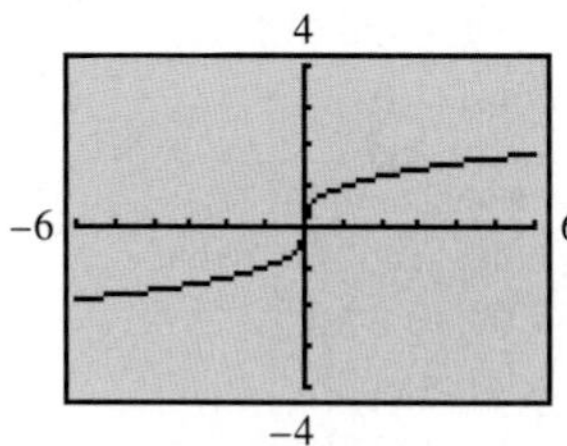

FIGURE 12.25

(b) $g(x) = x^{1/3}$

The graph of g in Figure 12.25 suggests that g is concave upward when $x < 0$ and concave downward when $x > 0$. We can confirm this conclusion algebraically by noting that $g'(x) = \frac{1}{3}x^{-2/3}$ and

$$g''(x) = -\frac{2}{9}x^{-5/3} = \frac{-2}{9\sqrt[3]{x^5}}.$$

When $x < 0$, then the denominator of $g''(x)$ is negative (why?), so $g''(x)$ is positive and g is concave upward on $(-\infty, 0)$. When $x > 0$, then the denominator of $g''(x)$ is positive, so $g''(x)$ is negative and g is concave downward on $(0, \infty)$. The graph changes concavity at $(0, 0)$, which is a point of inflection.

In Example 4(a) the point of inflection on the graph of f occurs at $x = 4/3$, the number at which the second derivative $f''(x) = 3x - 4$ is 0. In Example 4(b), the point of inflection occurs when $x = 0$, the number at which the second derivative $g''(x) = \frac{-2}{9\sqrt[3]{x^5}}$ is not defined. These facts suggest the following result.

If a function f has a point of inflection at $x = c$, then $f''(c) = 0$ or $f''(c)$ does not exist.

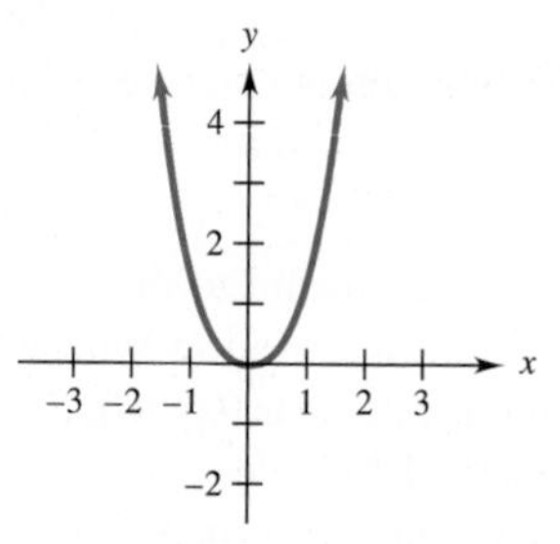

FIGURE 12.26

CAUTION The reverse of this fact is not always true. The second derivative may be 0 at a point that is not a point of inflection. For example, if $f(x) = x^4$, then $f'(x) = 4x^3$ and $f''(x) = 12x^2$. Hence $f''(x) = 0$ when $x = 0$. However, the graph of $f(x)$ in Figure 12.26 is always concave upward, so it has no point of inflection at $x = 0$ (or anywhere else). (4)

Find the intervals where the following are concave upward. Identify any inflection points.

(a) $f(x) = 6x^3 - 24x^2 + 9x - 3$

(b) $f(x) = 2x^2 - 4x + 8$

Answers:

(a) Concave upward on $(4/3, \infty)$; point of inflection is $(4/3, -175/9)$

(b) $f''(x) = 4$, which is always positive; function is always concave upward, no inflection point

The **law of diminishing returns** in economics is related to the idea of concavity. The graph of the function f in Figure 12.27 shows the output y from a given input x. For instance, the input might be advertising costs and the output the corresponding revenue from sales.

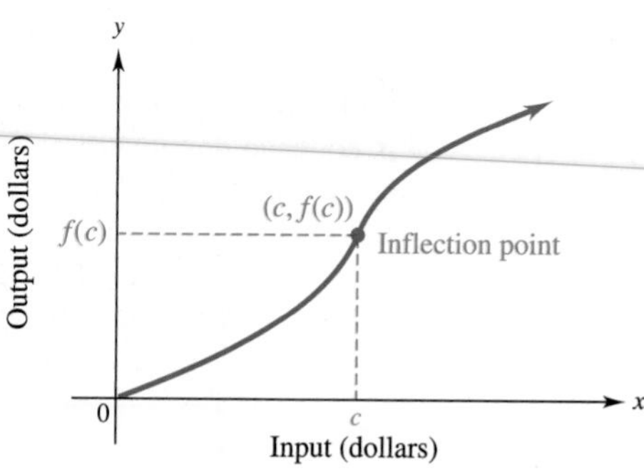

FIGURE 12.27

The graph in Figure 12.27 shows an inflection point at $(c, f(c))$. For $x < c$, the graph is concave upward, so the rate of change of the slope is increasing. This indicates that the output y is increasing at a faster rate with each additional dollar spent. When $x > c$, however, the graph is concave downward, the rate of change of the slope is decreasing, and the increase in y is smaller with each additional dollar spent. Thus, further input beyond c dollars produces diminishing returns. The point of inflection at $(c, f(c))$ is called the **point of diminishing returns.** Any investment beyond the value c is not considered a good use of capital.

EXAMPLE 5 The revenue $R(x)$ generated from sales of a certain product is related to the amount x spent on advertising by

$$R(x) = \frac{1}{150{,}000}(600x^2 - x^3), \qquad 0 \le x \le 600,$$

where x and $R(x)$ are in thousands of dollars. Is there a point of diminishing returns for this function? If so, what is it?

Since a point of diminishing returns occurs at an inflection point, look for an x-value that makes $R''(x) = 0$. Write the function as

$$R(x) = \frac{600}{150{,}000}x^2 - \frac{1}{150{,}000}x^3 = \frac{1}{250}x^2 - \frac{1}{150{,}000}x^3.$$

Now find $R'(x)$ and then $R''(x)$.

$$R'(x) = \frac{2x}{250} - \frac{3x^2}{150{,}000} = \frac{1}{125}x - \frac{1}{50{,}000}x^2$$

$$R''(x) = \frac{1}{125} - \frac{1}{25{,}000}x$$

Set $R''(x)$ equal to 0 and solve for x.

$$\frac{1}{125} - \frac{1}{25{,}000}x = 0$$

$$-\frac{1}{25{,}000}x = -\frac{1}{125}$$

$$x = \frac{25{,}000}{125} = 200$$

Test a number in the interval (0, 200) to see that $R''(x)$ is positive there. Then test a number in the interval (200, 600) to find $R''(x)$ negative in that interval. Since the sign of $R''(x)$ changes from positive to negative at $x = 200$, the graph changes from concave upward to concave downward at that point, and there is a point of diminishing returns at the inflection point $(200, 106\frac{2}{3})$. Any investment in advertising beyond \$200,000 would not pay off. ◀ 5

5

In Example 5,

$$R(x) = \frac{1}{50{,}000}(200x^3 - x^4)$$

for another product. What is the point of diminishing returns?

Answer:

(100, 2000)

Maxima and Minima

If a function f has a local maximum at c and $f'(c)$ is defined, then the graph of f is necessarily concave downward near $x = c$. Similarly, if f has a local minimum at d and $f'(d)$ exists, then the graph is concave upward near $x = d$, as shown in Figure 12.28.

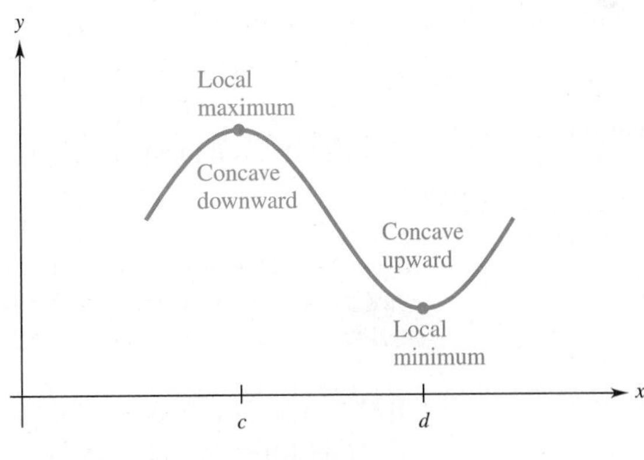

FIGURE 12.28

These facts should make the following result plausible.

The Second Derivative Test

Let c be a critical number of the function f such that $f'(c) = 0$ and $f''(x)$ exists for all x in some open interval containing c.

1. If $f''(c) > 0$, then f has a local minimum at c.
2. If $f''(c) < 0$, then f has a local maximum at c.
3. If $f''(c) = 0$, then this test gives no information; use the first derivative test.

EXAMPLE 6 The number of passenger cars imported into the United States (in millions) in year x can be approximated by

$$f(x) = -.0376x^3 + .37x^2 - .77x + 5.4 \qquad (0 \le x \le 6),$$

where $x = 0$ corresponds to 1990.* Find all local extrema of this function and interpret your answers.

First, find the critical numbers. The derivative is

$$f'(x) = 3(-.0376)x^2 + 2(.37)x - .77 = -.1128x^2 + .74x - .77.$$

Use the quadratic formula and a calculator to solve the equation $f'(x) = 0$.

$$-.1128x^2 + .74x - .77 = 0$$

$$x = \frac{-.74 \pm \sqrt{.74^2 - 4(-.1128)(-.77)}}{2(-.1128)} = \frac{-.74 \pm \sqrt{.200176}}{-.2256} \approx \begin{cases} 1.3 \\ 5.3 \end{cases}$$

Now use the second derivative test on each critical number. The second derivative is.

$$f''(x) = 2(-.1128)x + .74 = -.2256x + .74.$$

Then

$$f''(1.3) = -.2256(1.3) + .74 = .44672 > 0,$$

so that there is a local minimum at $x = 1.3$. Also,

$$f''(5.3) = -.2256(5.3) + .74 = -.45568 < 0.$$

Hence, there is a local maximum at $x = 5.3$. Thus, passenger car imports were at a low point in early 1991 ($x = 1.3$) and at a high point in early 1995 ($x = 5.3$). ◀ 6

6

Find all local maxima and local minima for the following functions. Use the second derivative test.

(a) $f(x) = 6x^2 + 12x + 1$

(b) $f(x) = x^3 - 3x^2 - 9x + 8$

Answers:

(a) Local minimum of -5 at $x = -1$

(b) Local maximum of 13 at $x = -1$; local minimum of -19 at $x = 3$

NOTE The second derivative test works only for those critical points c that make $f'(c) = 0$. This test does not work for those critical points c for which $f'(c)$ does not exist (since $f''(c)$ would not exist either). Also, the second derivative test does not work for critical points c that make $f''(c) = 0$. In both of these cases, use the first derivative test.

*U.S. Department of Commerce.

12.2 Exercises

For each of these functions, find $f''(x)$, $f''(0)$, $f''(2)$, and $f''(-3)$. (See Examples 1 and 2.)

1. $f(x) = x^3 - 5x^2 + 1$

2. $f(x) = 2x^4 + x^3 - 3x^2 + 2$

3. $f(x) = (x + 2)^4$

4. $f(x) = \dfrac{2x + 5}{x - 3}$

5. $f(x) = \dfrac{x^2}{1 + x}$

6. $f(x) = \dfrac{-x}{1 - x^2}$

7. $f(x) = \sqrt{x + 4}$

8. $f(x) = \sqrt{2x + 9}$

9. $f(x) = 5x^{3/5}$

10. $f(x) = -2x^{2/3}$

11. $f(x) = 2e^x$

12. $f(x) = \ln(2x - 3)$

13. $f(x) = 5e^{2x}$

14. $f(x) = 2 + e^{-x}$

15. $f(x) = \ln|x|$

16. $f(x) = \dfrac{1}{x}$

17. $f(x) = x \ln|x|$

18. $f(x) = \dfrac{\ln|x|}{x}$

Business *In Exercises 19 and 20, $P(t)$ is the price of a certain stock at time t during a particular day. (see Example 2.)*

19. If the price of the stock is falling faster and faster, are $P'(t)$ and $P''(t)$ positive or negative? Explain your answer.

20. When the stock reaches its highest price during the day, are $P'(t)$ and $P''(t)$ positive or negative? Explain your answer.

Physical Science *Each of the functions in Exercises 21–24 gives the distance from a starting point at time t of a particle moving along a line. Find the velocity and acceleration functions. Then find the velocity and acceleration at $t = 0$ and $t = 4$. Assume that time is measured in seconds and distance is measured in centimeters. Velocity will be in centimeters per second (cm/sec) and acceleration in centimeters per second per second (cm/sec2). (See Example 3.)*

21. $s(t) = 6t^2 + 2t$

22. $s(t) = 4t^3 - 6t^2 + 3t - 4$

23. $s(t) = 3t^3 - 4t^2 + 8t - 9$

24. $s(t) = \dfrac{-2}{3t + 4}$

Find the largest open intervals on which each function is concave upward or concave downward and find the location of any points of inflection. (See Example 4.)

25. $f(x) = x^2 + 3x - 5$

26. $f(x) = -x^2 + 8x - 7$

27. $f(x) = x^3 + 4x^2 - 6x + 3$

28. $f(x) = 5x^3 + 12x^2 - 32x - 14$

29. $f(x) = \dfrac{2}{x - 3}$

30. $f(x) = \dfrac{-2}{x + 1}$

31. $f(x) = x^4 + 8x^3 - 30x^2 + 24x - 3$

32. $f(x) = x^4 + 8x^3 + 18x^2 + 12x - 84$

Business *In Exercises 33 and 34, find the point of diminishing returns for the given functions, where $R(x)$ represents revenue in thousands of dollars and x represents the amount spent on advertising in thousands of dollars. (See Example 5.)*

33. $R(x) = 10{,}000 - x^3 + 42x^2 + 800x;\ 0 \le x \le 20$

34. $R(x) = \dfrac{4}{27}(-x^3 + 66x^2 + 1050x - 400);\ 0 \le x \le 25$

Find all critical numbers of the functions in Exercises 35–46. Then use the second derivative test on each critical number to determine whether it leads to a local maximum or minimum. (See Example 6.)

35. $f(x) = -2x^3 - 3x^2 - 72x + 1$

36. $f(x) = \dfrac{2}{3}x^3 + \dfrac{1}{2}x^2 - x - \dfrac{1}{4}$

37. $f(x) = x^3 + \dfrac{3}{2}x^2 - 60x + 100$

38. $f(x) = (x - 2)^5$

39. $f(x) = x^4 - 8x^2$

40. $f(x) = x^4 - 32x^2 + 7$

41. $f(x) = x + \frac{3}{x}$

42. $f(x) = x - \frac{1}{x}$

43. $f(x) = \frac{x^2 + 9}{2x}$

44. $f(x) = \frac{x^2 + 16}{2x}$

45. $f(x) = \frac{2 - x}{2 + x}$

46. $f(x) = \frac{x + 2}{x - 1}$

In Exercises 47–50, the rule of the derivative of a function f is given (but not the rule of f itself). Find the location of all local extrema and points of inflection of the function f.

47. $f'(x) = (x - 1)(x - 2)(x - 4)$

48. $f'(x) = (x^2 - 1)(x - 2)$

49. $f'(x) = (x - 2)^2(x - 1)$

50. $f'(x) = (x - 1)^2(x - 3)$

51. In each part, list the points (A–E) on the graph of f that satisfy the given conditions.
    (a) $f'(x) > 0$ and $f''(x) > 0$
    (b) $f'(x) < 0$ and $f''(x) > 0$
    (c) $f'(x) = 0$ and $f''(x) < 0$
    (d) $f'(x) = 0$ and $f''(x) > 0$
    (e) $f'(x) < 0$ and $f''(x) = 0$

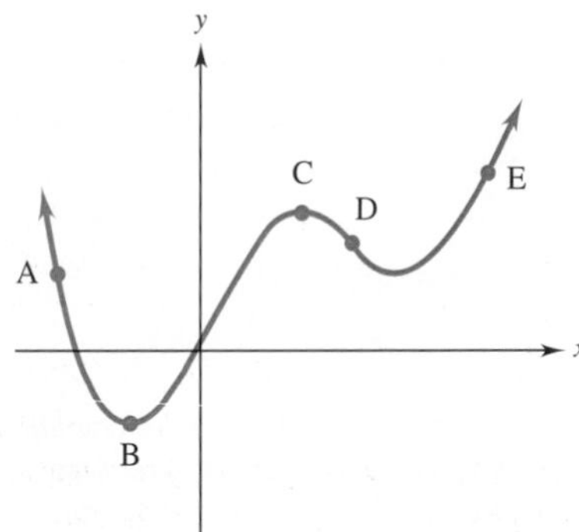

52. An abstract for an article states, "We tentatively conclude that Olympic weightlifting ability in trained subjects undergoes a nonlinear decline with age in which the second derivative of the performance versus age curve repeatedly changes sign."*
    (a) What does this quote tell you about the first derivative of the performance versus age curve?
    (b) Describe what you know about the performance versus age curve based upon the information in the quote.

Work these problems. (See Example 3.)

53. **Physical Science** When an object is dropped straight down, the distance in feet that it falls in t seconds is given by
$$s(t) = -16t^2,$$
where negative distance (or velocity) indicates downward motion. Find the velocity at each of the following times.
    (a) After 3 seconds
    (b) After 5 seconds
    (c) After 8 seconds
    (d) Find the acceleration. (The answer here is a constant, the acceleration due to the influence of gravity alone.)

54. **Physical Science** If an object is thrown directly upward with a velocity of 256 ft/sec, its height above the ground after t seconds is given by $s(t) = 256t - 16t^2$. Find the velocity and the acceleration after t seconds. What is the maximum height the object reaches? When does it hit the ground?

Work these problems. (See Example 5.)

55. **Business** A national chain has found that advertising produces sales, but that too much advertising for a product tends to make consumers "turn off" so that sales are reduced. Based on past experience the chain expects that the number $N(x)$ of cameras sold during a week is related to the amount spent on advertising by the function
$$N(x) = -3x^3 + 135x^2 + 3600x + 12{,}000,$$
where x (with $0 \le x \le 40$) is the amount spent on advertising in thousands of dollars. What is the point of diminishing returns?

56. **Business** Because of raw material shortages, it is increasingly expensive to produce fine cigars. In fact, the profit in thousands of dollars from producing x hundred thousand cigars is approximated by
$$P(x) = -x^3 + 28x^2 + 20x - 60,$$
where $0 \le x \le 20$. Find the point of diminishing returns.

57. **Business** In the United States, the average annual expenditures for entertainment can be approximated by the function
$$f(x) = -1.9x^3 + 21x^2 - 4.3x + 1431 \qquad (1 \le x \le 10),$$
where $x = 1$ corresponds to 1991.* Find and interpret the point of inflection.

58. **Business** The U.S. gross domestic product (GDP) in billions of dollars can be modeled by
$$g(x) = -6.168 + 5.069 \ln x \qquad (x \ge 10),$$
where $x = 10$ corresponds to 1990.†

*David E. Meltzer, "Age Dependence of Olympic Weightlifting," *Medicine and Science in Sports and Exercise*, vol. 26, no. 8, (August 1994): 1053.

*U.S. Bureau of Labor Statistics, Consumer Expenditure Survey.

†U.S. Bureau of Economic Analysis.

(a) When is this function increasing?

(b) What does the second derivative tell you about the growth rate of the GDP?

Work these problems. In several of them, you will need to use the quadratic formula to find the critical numbers. (See Example 6.)

59. Business U.S. national defense expenditures (in billions of constant 1996 dollars) are approximately by

$$f(x) = .1x^3 - 3.75x^2 + 36.3x + 242 \quad (0 \le x \le 20),$$

where $x = 0$ corresponds to 1980.*

(a) Find the critical numbers of this function. (*Hint:* Use the quadratic formula to solve $f'(x) = 0$).

(b) In what years was defense spending at a local minimum or maximum?

(c) Find the point of inflection. What does it indicate?

60. Social Science The population of Memphis, Tennessee (in thousands), is approximated by

$$g(x) = -.0027x^3 + .38x^2 - 6.2x + 102 \quad (0 \le x \le 98),$$

where $x = 0$ corresponds to 1900.† According to this model, when was the population of Memphis largest? What was the population at that time?

61. Business A small company must hire expensive temporary help to supplement its full-time staff. It estimates that the weekly costs $C(x)$ of salaries and benefits are related to the number x of full-time employees by the function

$$C(x) = 250x + \frac{16{,}000}{x} + 1000 \qquad (1 \le x \le 30).$$

How many full-time employees should the company have on its staff to minimize these costs?

62. Natural Science The percent of concentration of a certain drug in the bloodstream x hours after the drug is administered is given by

$$K(x) = \frac{3x}{x^2 + 4}.$$

For example, after 1 hour the concentration is given by $K(1) = 3(1)/(1^2 + 4) = (3/5)\% = .6\% = .006$.

(a) Find the time at which concentration is a maximum.

(b) Find the maximum concentration.

63. Social Science The number of burglaries in the United States per 100,000 population is approximated by

$$g(x) = .28x^3 - 12.5x^2 + 126.6x + 950 \quad (4 \le x \le 20),$$

where $x = 4$ corresponds to 1984.‡

(a) According to this model, when do burglaries peak after 1984?

(b) Assuming this model remains accurate until 2007, when are burglaries at their lowest point?

64. Business U.S. exports of coal (measured in quadrillion BTUs) are approximated by

$$y = -.000115x^3 + .0043x^2 + .025x + 1.1 \qquad (0 \le x \le 40),$$

where $x = 0$ corresponds to 1960.* When were coal exports at their highest?

65. Health A new communicable disease is running rampant in Gambier, Ohio. Epidemiologists estimate that t days after the disease was first observed in the community, the percent of the population infected by the disease is approximated by

$$p(t) = \frac{20t^3 - t^4}{1000} \qquad (0 \le t \le 20).$$

(a) After how many days is the percent of the population infected a maximum?

(b) What is the maximum percent of the population infected?

66. Social Science The number of U.S. civilian workers, 16 or older, who are unemployed in year x is approximated by

$$f(x) = 5x^3 - 124x^2 + 581x + 7045 \qquad (0 \le x \le 10),$$

where $x = 0$ corresponds to 1990.† When during this period was unemployment the highest?

67. Social Science The number of children under 18 years old who are living below the poverty level is approximated by

$$g(x) = -12.9x^3 + 400x^2 - 3443x + 20{,}509 \qquad (3 \le x \le 18),$$

where $x = 3$ corresponds to 1983.‡ In what year between 1983 and 1998 was the number of children in poverty at its lowest? At its highest?

68. Business When a company has to pay large amounts of overtime, or build a larger factory, its profits may go down even though sales are going up. The Wizard Widget Company expects that its profit (in hundreds of dollars) during the next six months will be given by

$$P(x) = -x + 200\sqrt{x} - 2000 \qquad (0 \le x \le 35{,}000),$$

where x is the number of units sold. Find the number of units that produces maximum profit.

*U.S. Office of Management and Budget.

†U.S. Census Bureau.

‡FBI Uniform Crime Reports.

*U.S. Department of Energy.

†U.S. Bureau of Labor Statistics.

‡U.S. Census Bureau.

12.3 OPTIMIZATION APPLICATIONS

In most applications, the domains of the functions involved are restricted to numbers in a particular interval. For example, a factory that can produce a maximum of 40 units (because of market conditions, availability of labor, etc.) might have this cost function:

$$C(x) = -3x^3 + 135x^2 + 3600x + 12{,}000 \qquad (0 \le x \le 40).$$

Even though the rule of C is defined for all numbers x, only the numbers in the interval [0, 40] are relevant because the factory can't produce a negative number of units or more than 40 units. In such applications, we often want to find a smallest or largest quantity—for instance, the minimum cost or the maximum profit—when x is restricted to the relevant interval. So we begin with the mathematical description of such a situation.

Let f be a function that is defined for all x in the closed interval $[a, b]$. Let c be a number in the interval. We say that f has an **absolute maximum on the interval** at c if

$$f(x) \le f(c) \quad \text{for all } x \text{ with } a \le x \le b,$$

that is, if $(c, f(c))$ is the highest point on the graph of f over the interval $[a, b]$. Similarly, f has an **absolute minimum on the interval** at c if

$$f(x) \ge f(c) \quad \text{for all } x \text{ with } a \le x \le b,$$

that is, if $(c, f(c))$ is the lowest point on the graph of f over the interval $[a, b]$.

1

Find the location of the absolute maximum and absolute minimum of the function f in Figure 12.29 on the interval [−2, 1].

Answer:

Absolute maximum at 0; absolute minimum at −2

EXAMPLE 1 Figure 12.29 shows the graph of a function f. Consider the function f on the interval [−2, 6]. Since we are interested only in the interval [−2, 6], the values of the function outside this interval are irrelevant. On the interval [−2, 6], f has an absolute minimum at 3 (which is also a local minimum) and an absolute maximum at 6 (which is not a local maximum of the entire function). ◀ 1

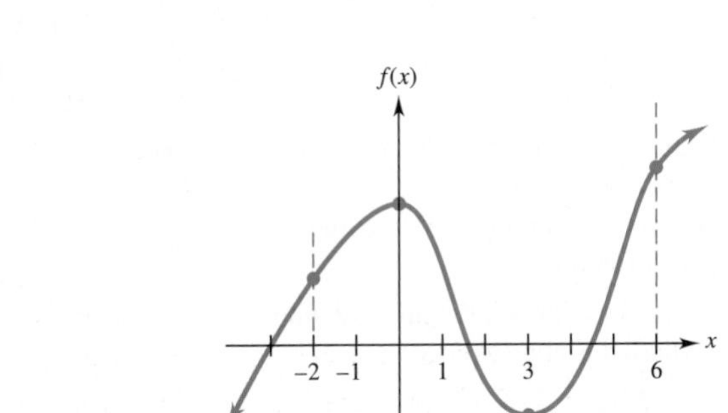

FIGURE 12.29

The absolute maximum in Example 1 occurred at $x = 6$, which is an endpoint of the interval, and the absolute minimum occurred at $x = 3$, which is a critical number of f (because f has a local minimum there). Similarly, in problem 1 in the margin, the absolute maximum occurred at a critical number and the absolute minimum at an endpoint. These examples illustrate the following result, whose proof is omitted.

Extreme Value Theorem

If a function f is continuous on a closed interval $[a, b]$, then f has both an absolute maximum and an absolute minimum on the interval. Each of these occurs either at an endpoint of the interval or at a critical number of f.

CAUTION The Extreme Value Theorem may not hold on intervals that are not closed (that is, intervals that do not include one or both endpoints). For example, $f(x) = 1/x$ does not have an absolute maximum on the interval $(0, 1)$; the values of $f(x)$ get larger and larger as x approaches 0, as you can easily verify with a calculator.

EXAMPLE 2 Use the Extreme Value Theorem to find the absolute extrema of $f(x) = 4x + \dfrac{36}{x}$ on the interval $[1, 6]$.

According to the Extreme Value Theorem we need only consider the critical numbers of f and the endpoints, 1 and 6, of the interval. Begin by finding the derivative and determining the critical numbers.

$$f'(x) = 4 - \frac{36}{x^2} = 0$$

$$\frac{4x^2 - 36}{x^2} = 0$$

Since we are looking for critical numbers in $[1, 6]$, $x \neq 0$. Since $x \neq 0$, then $f'(x) = 0$ when

$$4x^2 - 36 = 0$$
$$4x^2 = 36$$
$$x^2 = 9$$
$$x = -3 \quad \text{or} \quad x = 3.$$

2

Find the absolute maximum and absolute minimum values.

$f(x) = -x^2 + 4x - 8$ on $[-4, 4]$.

Answer:

Absolute maximum of -4 at $x = 2$; absolute minimum of -40 at $x = -4$

Since -3 is not in the interval $[1, 6]$, disregard it; 3 is the only critical number of interest. By the Extreme Value Theorem, the absolute maximum and minimum must occur at critical numbers or endpoints, that is, at 1, 3, or 6. Evaluate f at these three numbers to see which ones give the largest and smallest values.

x-Value	*Value of Function*	
1	40	← Absolute maximum
3	24	← Absolute minimum
6	30	

◀ 2

A graphing calculator or other graphing technology sometimes makes it easy to find absolute extrema without taking the derivative. Technology, however, has its limits, so you should confirm your results algebraically whenever feasible.

EXAMPLE 3 Find the absolute extrema of $g(x) = -.02x^3 + 600x - 20{,}000$ on [60, 135].

The graph of g on [60, 135] in Figure 12.30 shows that the absolute maximum occurs at the local maximum, which is close to $x = 100$. The absolute minimum occurs at one of the endpoints, but you can't determine which one from the graph.

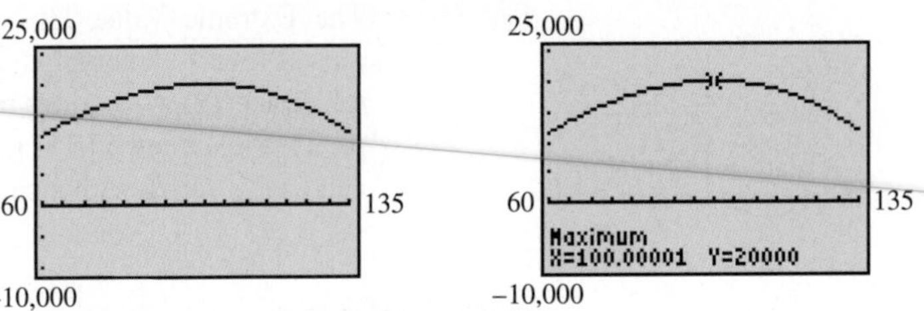

FIGURE 12.30

We first determine the critical number exactly. Since $g'(x) = -.06x^2 + 600$, the critical numbers are the solutions of

$$-.06x^2 + 600 = 0$$
$$-.06x^2 = -600$$
$$x^2 = 10{,}000$$
$$x = -100 \quad \text{or} \quad x = 100$$

We disregard $x = -100$ because it is not in [60, 135] and evaluate g at the endpoints and the critical number 100, as shown below. The absolute maximum of 20,000 occurs at $x = 100$ (as we suspected from the graph) and the absolute minimum of 11,680 occurs at $x = 60$.

x-Value	Value of Function	
60	11,680	← Absolute minimum
100	20,000	← Absolute maximum
135	11,792.5	

 3

3

Find the absolute extrema of $g(x) = x^3 - 15x^2 + 48x + 50$ on [1, 6].

Answer:

Absolute maximum of 94 at $x = 2$; absolute minimum of 14 at $x = 6$.

Applications

When solving applied problems that involve maximum and minimum values, follow these guidelines:

Solving Applied Problems

Step 1 Read the problem carefully. Make sure you understand what is given and what is asked for.

Step 2 If possible, sketch a diagram and label the various parts.

Step 3 Decide which variable is to be maximized or minimized. Express that variable as a function of *one* other variable. Be sure to determine the domain of this function.

Step 4 Find the critical numbers for the function in Step 3.

Step 5 If the domain is a closed interval, evaluate the function at the endpoints and at each critical number to see which yields the maximum or minimum. If the domain is an open interval, test each critical number either graphically or by using the first or second derivative test to see which yields a maximum or a minimum.

CAUTION Do not skip Step 5 in the box above. If you are looking for a maximum and you find a critical number in Step 4, do not automatically assume that the maximum occurs there. It may occur at an endpoint or may not exist at all.

EXAMPLE 4 The cost (in thousands of dollars) of manufacturing x thousand rewritable CDs is given by

$$C(x) = .00009x^3 - .016x^2 + x + 31.$$

(a) Find the average cost function.

As we saw in Section 11.5, the average cost $\overline{C}(x)$ is given by

$$\overline{C}(x) = \frac{C(x)}{x} = \frac{.00009x^3 - .016x^2 + x + 31}{x}$$

$$= .00009x^2 - .016x + 1 + \frac{31}{x}.$$ (4)

4

Find the average cost per CD for the following production levels: 1000, 50,000, and 150,000.

Answers:

$31.98; $1.05; $0.83

(b) How many CDs should be made to minimize the average cost per CD? What is the minimum average cost?

The average cost function $\overline{C}(x) = .00009x^2 - .016x + 1 + 31x^{-1}$ is defined for all $x > 0$, so endpoints play no role here. Its derivative is

$$\overline{C}'(x) = .00018x - .016 - 31x^{-2},$$

which is defined for all $x > 0$. Hence, the critical numbers are the solutions of

$$.00018x - .016 - \frac{31}{x^2} = 0$$

$$.00018x^3 - .016x^2 - 31 = 0 \quad \text{Multiply both sides by } x^2.$$

Technology is needed to solve this equation. Using graphical methods, we see that the only real solution is $x \approx 104.62279$ (Figure 12.31). We use the second derivative test to determine whether this critical number is a minimum.

$$\overline{C}'(x) = .00018x - .016 - 31x^{-2}$$

$$\overline{C}''(x) = .00018 - 31(-2)x^{-3} = .00018 + \frac{62}{x^3}$$

The second derivative is positive for all $x > 0$, so $\overline{C}(x)$ has a local minimum at $x \approx 104.623$, that is, when 104,623 CDs are produced. At this production level the average cost is just below 61¢, namely, $\overline{C}(104.623) \approx \$0.6075$ per CD.

Alternate Solution Once we realize that technology and approximations are necessary here, we can also graph $\overline{C}(x) = .00009x^2 - .016x + 1 + 31x^{-1}$ and use a minimum finder to determine the local minimum, as in Figure 12.32. Both methods agree to three decimal places.

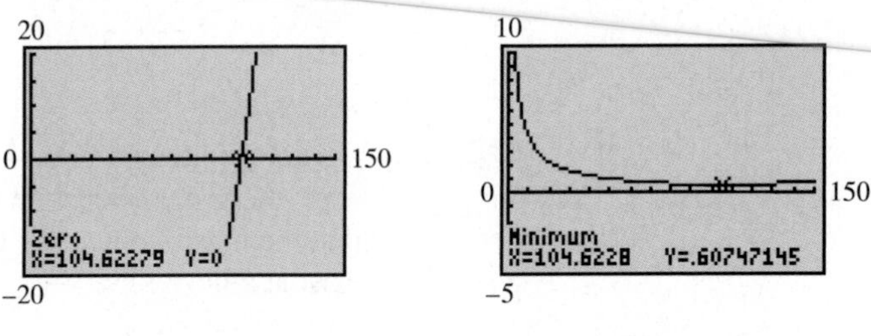

FIGURE 12.31

FIGURE 12.32

EXAMPLE 5 An open box is to be made by cutting a square from each corner of a 12-inch by 12-inch piece of metal and then folding up the sides. The finished box must be at least 1.5 inches deep but not deeper than 3 inches. What size square should be cut from each corner in order to produce a box of maximum volume?

Let x represent the length of a side of the square that is cut from each corner, as shown in Figure 12.33(a). The width of the box is $12 - 2x$, while the length is also $12 - 2x$. As shown in Figure 12.33(b), the depth of the box will be x inches.

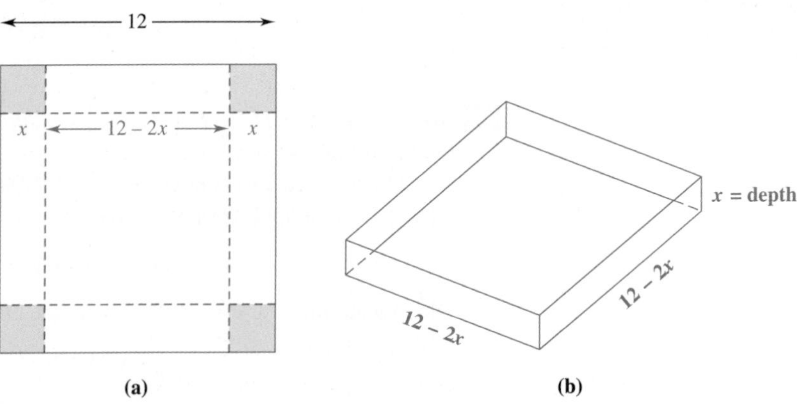

FIGURE 12.33

We must maximize the volume $V(x)$ of the box, which is given by

$$\text{Volume} = \text{length} \cdot \text{width} \cdot \text{height}$$
$$V(x) = (12 - 2x) \cdot (12 - 2x) \cdot x = 144x - 48x^2 + 4x^3.$$

Since the height x must be between 1.5 and 3 inches, the domain of this volume function is the closed interval [1.5, 3]. First find the critical numbers by setting the derivative equal to 0:

$$V'(x) = 12x^2 - 96x + 144 = 0$$
$$12(x^2 - 8x + 12) = 0$$
$$12(x - 2)(x - 6) = 0$$
$$x - 2 = 0 \quad \text{or} \quad x - 6 = 0$$
$$x = 2 \quad \text{or} \quad x = 6.$$

Since 6 is not in the domain, the only critical number of interest here is $x = 2$. The maximum volume must occur at $x = 2$ or at the endpoints $x = 1.5$ or $x = 3$.

x	$V(x)$	
1.5	121.5	
2	128	← Maximum
3	108	

The table shows that the box has maximum volume when $x = 2$ and that this maximum volume is 128 cubic inches. ◀ 5

5

An open box is to be made by cutting squares from each corner of a 20 cm by 32 cm piece of metal and folding up the sides. Let x represent the length of the side of the square to be cut out. Find

(a) an expression for the volume of the box, $V(x)$;

(b) $V'(x)$;

(c) the value of x that leads to maximum volume; (*Hint:* The solutions of the equation $V'(x) = 0$ are 4 and 40/3.);

(d) the maximum volume.

Answers:

(a) $V(x) = 640x - 104x^2 + 4x^3$

(b) $V'(x) = 640 - 208x + 12x^2$

(c) $x = 4$

(d) $V(4) = 1152$ cubic centimeters

EXAMPLE 6 The U.S. Postal Service requires that boxes to be mailed have a length plus girth of no more than 108 inches, as shown in Figure 12.34. Find the dimensions of the box with largest volume that can be mailed, assuming its width and height are equal.

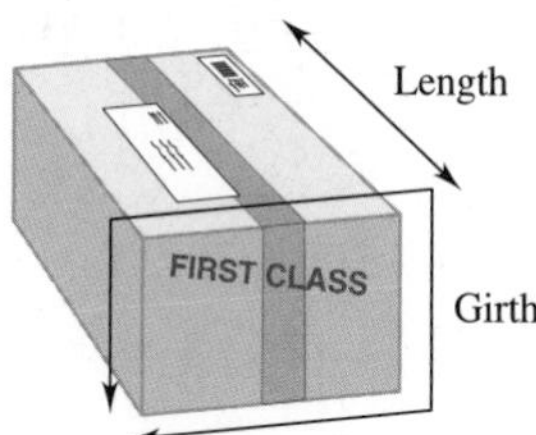

FIGURE 12.34

Let x be the width and y the length of the box. Since width and height are the same, the volume of the box is

$$V = y \cdot x \cdot x = yx^2.$$

Now express V in terms of just *one* variable. Use the facts that the girth is $x + x + x + x = 4x$ and that the length plus girth is 108, so that

$$y + 4x = 108, \quad \text{or equivalently,} \quad y = 108 - 4x.$$

Substitute for y in the expression for V to get

$$V = yx^2 = (108 - 4x)x^2 = 108x^2 - 4x^3.$$

Since x and y are dimensions, we must have $x > 0$ and $y > 0$. Now $y = 108 - 4x > 0$ implies that

$$4x < 108, \quad \text{or equivalently,} \quad x < 27.$$

Therefore, the domain of the volume function V (the values of x that make sense in the situation) is the open interval (0, 27). Since the domain is an open interval, we need not test the endpoints. Find the critical numbers for V by setting its derivative equal to 0 and solving the equation.

$$V' = 216x - 12x^2 = 0$$
$$x(216 - 12x) = 0$$
$$x = 0 \quad \text{or} \quad 12x = 216$$
$$x = 18$$

Use the second derivative test to check $x = 18$, the only critical number in the domain of V. Since $V'(x) = 216x - 12x^2$,

$$V''(x) = 216 - 24x.$$

Hence, $V''(18) = 216 - 24 \cdot 18 = -216$ and V is maximized when $x = 18$. In this case, $y = 108 - 4 \cdot 18 = 36$. Therefore, a box with dimensions 18 by 18 by 36 inches satisfies postal regulations and yields the maximum volume of $18^2 \cdot 36 = 11{,}664$ cubic inches. ◀

EXAMPLE 7 A landscape gardener wants to build an 8000 square foot rectangular display garden along the side of a river. The garden will be fenced on three sides (no fence is necessary along the river). She plans to use ornamental steel fencing at \$12 per foot on the side opposite the river and chain link fencing at \$3 per foot on the other two sides. What dimensions for the garden will minimize her costs?

First draw a sketch of the situation and label the sides of the gardens, as in Figure 12.35.

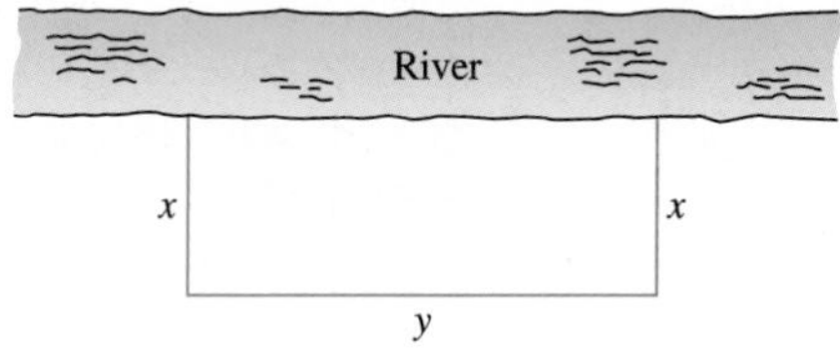

FIGURE 12.35

$2x$ feet of fencing (at \$3 per foot) are needed for the parallel sides and y feet (at \$12 per foot) for the side opposite the river. So the cost is

$$C = 2x(3) + y(12) = 6x + 12y.$$

6

Show that the function C in Example 7 has a minimum at $x = \sqrt{16{,}000} \approx 126.5$ in either of the following ways.

(a) Find $C''(x)$ and $C''(\sqrt{16{,}000})$. Then use the second derivative test.

(b) Graph C on a graphing calculator and use the minimum finder.

Answer:

(a) $C''(x) = 192{,}000x^{-3}$ and $C''(\sqrt{16{,}000}) \approx .095 > 0$, so there is a minimum at $x = \sqrt{16{,}000}$ by the second derivative test.

(b)

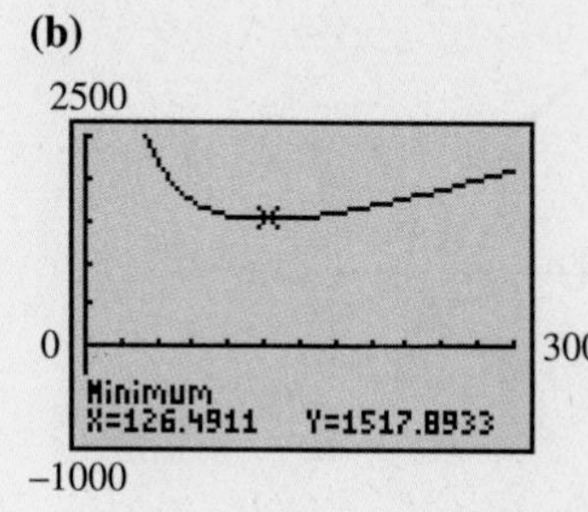

We must find the values of x and y that make C as small as possible. First, however, we must express C in terms of a single variable. Since the garden is to have an area of 8000 square feet and the area of a rectangle is its length times its width, we have

$$xy = \text{area} = 8000$$
$$y = \frac{8000}{x}.$$

Substituting this expression for y in the cost function, we have

$$C = 6x + 12y = 6x + 12\left(\frac{8000}{x}\right) = 6x + \frac{96{,}000}{x}.$$

Since x is a length, it is positive. Hence, the domain of the cost function is the open interval $(0, \infty)$ and we need only look at the critical numbers. The derivative is

$$C'(x) = 6 - 96{,}000x^{-2},$$

so that the critical numbers are the solutions of

$$6 - \frac{96{,}000}{x^2} = 0$$
$$6x^2 - 96{,}000 = 0$$
$$6x^2 = 96{,}000$$
$$x^2 = 16{,}000$$
$$x = \pm\sqrt{16{,}000} \approx \pm 126.5.$$

The only relevant critical number is $x = \sqrt{16{,}000}$ (why?). Do problem 6 in the margin to show that this number makes C a minimum. (6)

Therefore, the garden with the least expensive fence has dimensions

$$x = 126.5 \text{ ft} \quad \text{and} \quad y = \frac{8000}{x} \approx \frac{8000}{126.5} = 63.24 \text{ ft.}$$

The preceding examples illustrate some of the factors that may affect applications in the real world. First, you must be able to find a function that models the situation. The rule of this function may be defined for values of x that do not make sense in the context of the application, so the domain must be restricted to the relevant values of x.

The techniques of calculus apply to functions that are defined and continuous at every real number in some interval, so the maximum or minimum for the mathematical model (function) may not be feasible in the setting of the problem. For instance, if $C(x)$ has a minimum at $x = 80\sqrt{3}$ (≈ 138.564), where $C(x)$ is the cost of hiring x employees, then the real-life minimum occurs at either 138 or 139, whichever one leads to lower cost.

Economic Lot Size

Suppose that a company manufactures a constant number of units of a product per year and that the product can be manufactured in several batches of equal size during the year. If the company were to manufacture the item only once per year, it would minimize setup costs but incur high warehouse costs. On the other hand, if it were to make many small batches, this would increase setup costs. Calculus can be used to find the number of batches per year that should be manufactured in order to minimize total cost. This number is called the **economic lot size.**

Figure 12.36 shows several of the possibilities for a product having an annual demand of 12,000 units. The top graph shows the results if only one batch of the product is made annually; in this case, an average of 6000 items will be held in a warehouse. If four batches (of 3000 each) are made at equal time intervals during a year, the average number of units in the warehouse falls to only 1500. If twelve batches are made, an average of 500 items will be in the warehouse.

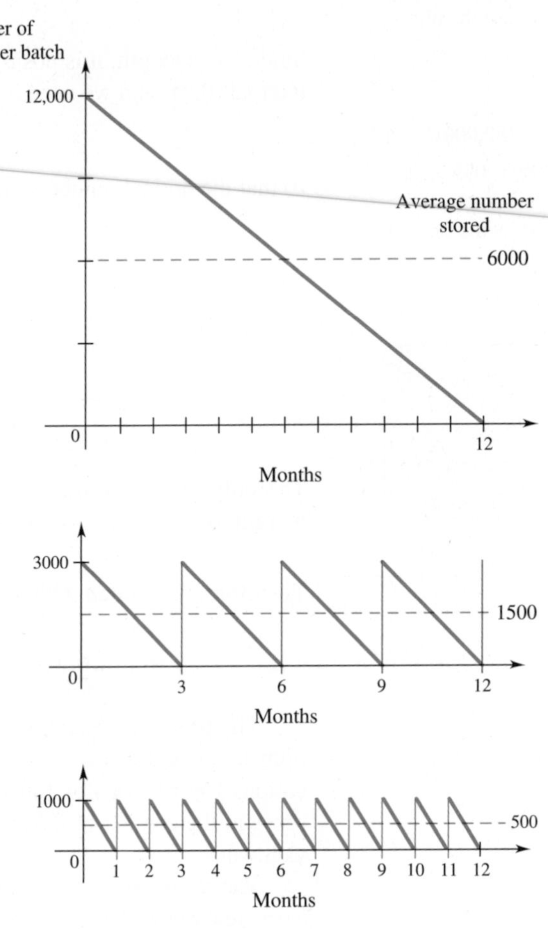

FIGURE 12.36

The following variables will be used in our discussion of economic lot size.

x = number of batches to be manufactured annually

k = cost of storing 1 unit of the product for 1 year

a = fixed setup cost to manufacture the product

b = variable cost of manufacturing a single unit of the product

M = total number of units produced annually

The company has two types of costs associated with the production of its product: a cost associated with manufacturing the item and a cost associated with storing the finished product.

During a year the company will produce x batches of the product, with M/x units of the product produced per batch. Each batch has a fixed cost a and a variable cost b per unit, so that the manufacturing cost per batch is

$$a + b\left(\frac{M}{x}\right).$$

There are x batches per year, so the total annual manufacturing cost is

$$\left[a + b\left(\frac{M}{x}\right)\right]x. \tag{1}$$

Each batch consists of M/x units and demand is constant; therefore, it is common to assume an average inventory of

$$\frac{1}{2}\left(\frac{M}{x}\right) = \frac{M}{2x}$$

units per year. The cost to store 1 unit of the product for a year is k, so the total storage cost is

$$k\left(\frac{M}{2x}\right) = \frac{kM}{2x}. \tag{2}$$

The total production cost is the sum of the manufacturing and storage costs, or the sum of expressions (1) and (2). If $T(x)$ is the total cost of producing x batches,

$$T(x) = \left[a + b\left(\frac{M}{x}\right)\right]x + \frac{kM}{2x} = ax + bM + \left(\frac{kM}{2}\right)x^{-1}.$$

Now find the value of x that will minimize $T(x)$. (Remember that a, b, k, and M are constants.) Find $T'(x)$.

$$T'(x) = a - \frac{kM}{2}x^{-2}$$

Set this derivative equal to 0 and solve for x (remember that $x > 0$).

$$a - \frac{kM}{2}x^{-2} = 0$$

$$a = \frac{kM}{2x^2}$$

$$2ax^2 = kM$$

$$x^2 = \frac{kM}{2a}$$

$$x = \sqrt{\frac{kM}{2a}} \tag{3}$$

The second derivative test can be used to show that $\sqrt{kM/(2a)}$ is the annual number of batches that gives minimum total production cost.

A manufacturer of business forms has an annual demand for 30,720 units of form letters to people delinquent in their payments of installment debt. It costs \$5 per year to store 1 unit of the letters and \$1200 to set up the machines to produce them. Find the number of batches that should be made annually to minimize total cost.

Answer:
8 batches

EXAMPLE 8 A paint company has a steady annual demand for 24,500 cans of automobile primer. The cost accountant for the company says that it costs \$2 to store 1 can of paint for 1 year and \$500 to set up the plant for the production of the primer. Find the number of batches of primer that should be produced for the minimum total production cost.

Use equation (3) above.

$$x = \sqrt{\frac{kM}{2a}}$$

$$x = \sqrt{\frac{2(24{,}500)}{2(500)}} \qquad \text{Let } k = 2,\ M = 24{,}500,\ a = 500.$$

$$x = \sqrt{49} = 7$$

Seven batches of primer per year will lead to minimum production costs. 

12.3 Exercises

Find the location of the absolute maximum and absolute minimum of the function on the given interval. (See Example 1.)

1. [0, 4]

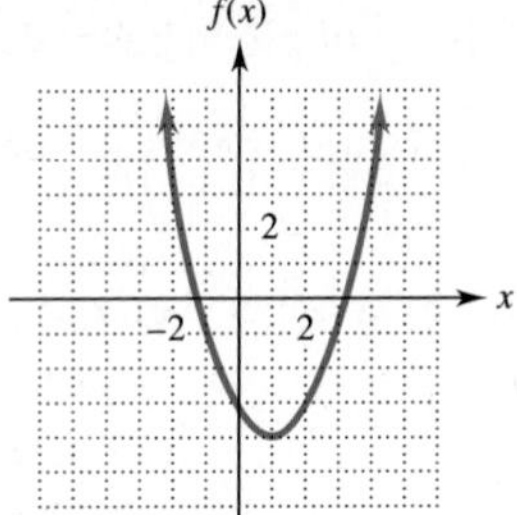

2. [2, 5]

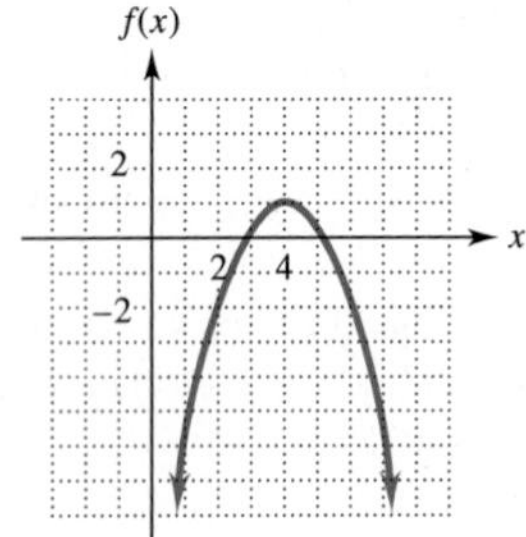

3. [−4, 2]

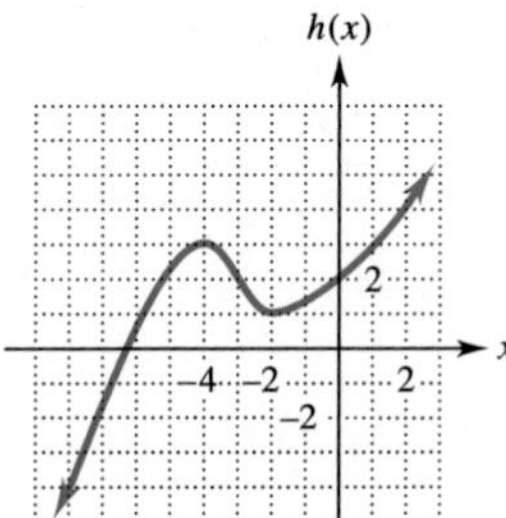

4. [−1, 2]

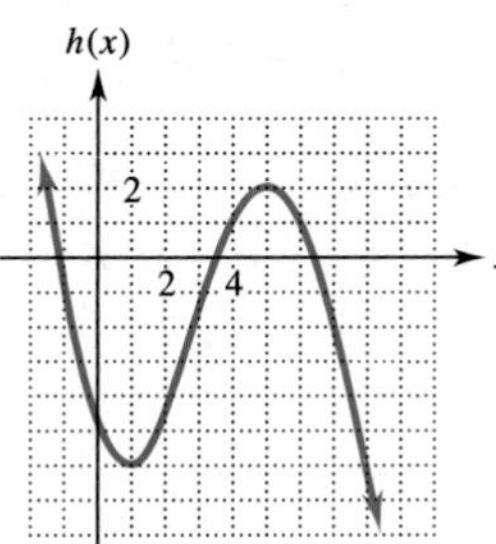

5. [−8, 0]

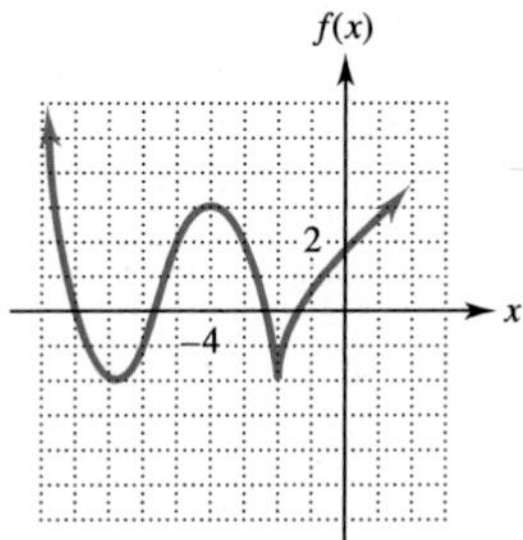

6. $[-4, 4]$

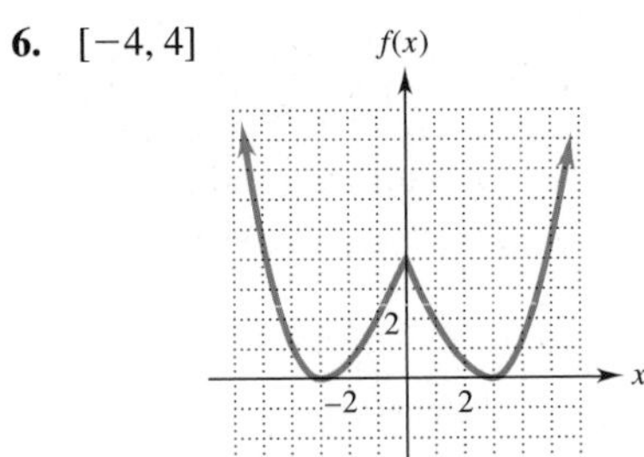

Find the absolute extrema of each function on the given interval. (See Examples 2 and 3.)

7. $f(x) = x^4 - 32x^2 - 7$; $[-5, 6]$
8. $f(x) = x^4 - 18x^2 + 1$; $[-4, 4]$
9. $f(x) = \dfrac{8 + x}{8 - x}$; $[4, 6]$
10. $f(x) = \dfrac{1 - x}{3 + x}$; $[0, 3]$
11. $f(x) = \dfrac{x}{x^2 + 2}$; $[0, 4]$
12. $f(x) = \dfrac{x - 1}{x^2 + 1}$; $[1, 5]$
13. $f(x) = (x^2 + 18)^{2/3}$; $[-3, 3]$
14. $f(x) = (x^2 + 4)^{1/3}$; $[-2, 2]$
15. $f(x) = \dfrac{1}{\sqrt{x^2 + 1}}$; $[-1, 1]$
16. $f(x) = \dfrac{3}{\sqrt{x^2 + 4}}$; $[-2, 2]$

Work these problems. (See Example 4.)

17. Business The daily cost of producing x flat computer monitors is given by $C(x) = .13x^3 - 70x^2 + 10{,}000x$, and no more than 300 monitors can be produced each day. What production level will give the lowest average cost per monitor? What is the minimum average cost?

18. Business The number of unfilled pharmacist positions in year x is approximated by

$$g(x) = -583x^3 + 2068x^2 + 323x + 2670 \qquad (0 \le x \le 3),$$

where $x = 0$ corresponds to 1998.* In what year was the shortage of pharmacists most severe?

19. Business The number of PCs sold worldwide (in millions) in year x is approximated by

$$f(x) = -1.23x^3 + 8x^2 + 1.5x + 72 \qquad (0 \le x \le 5),$$

where $x = 0$ corresponds to 1996. In what year was the largest number sold?

20. Natural Science A lake polluted by bacteria is treated with an antibacterial chemical. After t days, the number N of bacteria per ml of water is approximated by

$$N(t) = 20\left(\frac{t}{12} - \ln\left(\frac{t}{12}\right)\right) + 30 \qquad (1 \le t \le 15).$$

(a) When during this period will the number of bacteria be a minimum?
(b) What is this minimum number of bacteria?
(c) When during this period will the number of bacteria be a maximum?
(d) What is this maximum number of bacteria?

21. Business An auto parts manufacturer makes radiators that sell for \$350 each. The total cost of producing x radiators is approximated by the function $C(x) = 600{,}000 - 25x + .01x^2$.
(a) Write the rule of the revenue function in this situation.
(b) Write the rule of the profit function.
(c) What number of radiators should be made to guarantee a maximum profit? What will that profit be?

22. Business Name tags can be sold for \$29 per thousand. The total cost of manufacturing x thousand tags is $.001x^3 + .06x^2 - 1.5x$ dollars. Assume that all tags manufactured are sold.
(a) What is the revenue function in this situation? What is the profit function?
(b) What number of tags will produce the largest possible profit?

23. Health A disease has hit College Station, Texas. The percent of the population infected t days after the disease arrives is approximated by $p(t) = 10t\,e^{-t/8}$ for $0 \le t \le 40$.
(a) After how many days is the percent of infected people a maximum?
(b) What is the maximum percent of the population infected?

*National Association of Chain Drugstores.

24. Social Science Suppose dots and dashes are transmitted over a telegraph line so that dots occur a fraction p of the time (where $0 \le p \le 1$) and dashes occur a fraction $1 - p$ of the time. The information content of the telegraph line is given by $I(p)$, where

$$I(p) = -p \ln p - (1 - p) \ln(1 - p).$$

(a) Show that $I'(p) = -\ln p + \ln(1 - p)$.
(b) Let $I'(p) = 0$ and find the value of p that maximizes the information content.

Work these problems. (See Examples 5 and 6.)

25. Geometry An open box is to be made by cutting a square from each corner of a 3 foot by 8 foot piece of cardboard and then folding up the sides. What size square should be cut from each corner in order to produce a box of maximum volume?

26. Business A television manufacturing firm needs to design an open-topped box with a square base. The box must hold 32 cubic inches. Find the dimensions of the box that can be built with the minimum amount of materials.

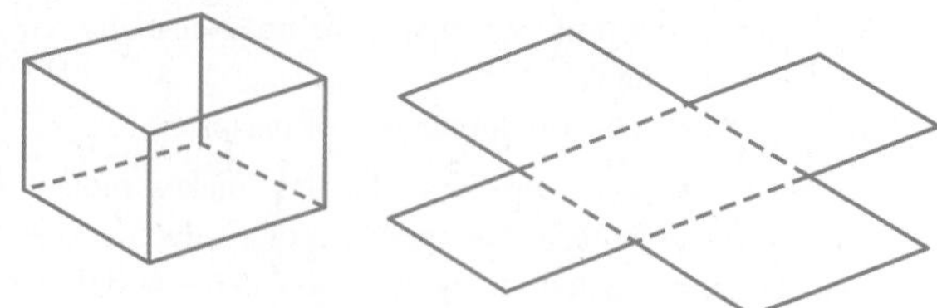

27. Business An artist makes a closed box with a square base which is to have a volume of 16,000 cubic centimeters. The material for the top and bottom of the box costs 3 cents per square centimeter, while the material for the sides costs 1.5 cents per square centimeter. Find the dimensions of the box that will lead to minimum total cost. What is the minimum total cost?

28. Business A cylindrical box will be tied up with ribbon as shown in the figure. The longest piece of ribbon available is 130 cm long, and 10 cm of that are required for the bow. Find the radius and height of the box with the largest possible volume.

29. Business A company wishes to manufacture a rectangular box with a volume of 36 cubic feet that is open on top and that is twice as long as it is wide. Find the dimensions of the box produced from the minimum amount of material.

30. Business A cylindrical can of volume 58 cubic inches (approximately one quart) is to be designed. For convenient handling, it must be at least one inch high and two inches in diameter. What dimensions (radius of top, height of can) will use the least amount of material?

Business *Work these problems. (See Example 7.)*

31. A farmer has 1200 m of fencing. He wants to enclose a rectangular field bordering a river, with no fencing needed along the river. Let x represent the width of the field.
(a) Write an expression for the length of the field.
(b) Find the area of the field.
(c) Find the value of x leading to the maximum area.
(d) Find the maximum area.

32. A rectangular field is to be enclosed with a fence. One side of the field is against an existing fence, so that no fence is needed on that side. If material for the fence costs \$2 per foot for the two ends and \$4 per foot for the side parallel to the existing fence, find the dimensions of the field of largest area that can be enclosed for \$1000.

33. A rectangular field is to be enclosed on all four sides with a fence. Fencing material costs \$3 per foot for two opposite sides and \$6 per foot for the other two sides. Find the maximum area that can be enclosed for \$2400.

34. A fence must be built to enclose a rectangular area of 20,000 ft^2. Fencing material costs \$3 per foot for the two sides facing north and south and \$6 per foot for the other two sides. Find the cost of the least expensive fence.

35. A fence must be built in a large field to enclose a rectangular area of 15,625 m^2. One side of the area is bounded by an existing fence; no fence is needed there. Material for the fence costs \$2 per meter for the two ends and \$4 per meter for the side opposite the existing fence. Find the cost of the least expensive fence.

36. Business A mathematics book is to contain 36 square inches of printed matter per page, with margins of 1 inch along the sides, and $1\frac{1}{2}$ inches along the top and bottom. Find the dimensions of the page that will lead to the minimum amount of paper being used for a page.

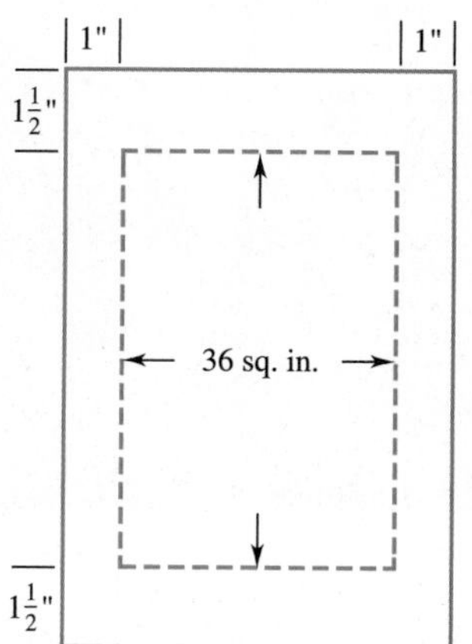

Work these problems. (See Examples 4–7.)

37. Business If the price charged for a candy bar is $p(x)$ cents, then x thousand candy bars will be sold in a certain city, where

$$p(x) = 100 - \frac{x}{10}.$$

(a) Find an expression for the total revenue from the sale of x thousand candy bars. (*Hint:* Find the product of $p(x)$, x, and 1000.)
(b) Find the value of x that leads to maximum revenue.
(c) Find the maximum revenue.

38. Business The sale of cassette tapes of "lesser" performers is very sensitive to price. If a tape manufacturer charges $p(x)$ dollars per tape, where

$$p(x) = 6 - \frac{x}{8},$$

then x thousand tapes will be sold.
(a) Find an expression for the total revenue from the sale of x thousand tapes. (*Hint:* Find the product of $p(x)$, x, and 1000.)
(b) Find the value of x that leads to maximum revenue.
(c) Find the maximum revenue.

39. Social Science The population of China (in millions) in year x is projected to be given by

$$P(x) = -.00096x^3 - .1x^2 + 11.3x + 1274 \qquad (0 \le x \le 50),$$

where $x = 0$ corresponds to 2000.*
(a) When will China's population be at its highest? (Round your answer to the nearest year.)
(b) What will the population be in that year?

40. Business A company makes novelty bookmarks that sell for \$142 per hundred. The total cost (in dollars) of making x hundred bookmarks is $x^3 - 8x^2 + 20x + 40$. Assume that the company can sell all the bookmarks it makes.
(a) Because of other projects a maximum of 600 bookmarks per day can be manufactured. How many should the company make per day to maximize its profits?
(b) As a result of change in other orders, as many as 1600 bookmarks can now be made each day. How many should be made in order to maximize profits?

41. Business A truck burns fuel at the rate of $G(x)$ gallons per mile, where

$$G(x) = \frac{1}{32}\left(\frac{64}{x} + \frac{x}{50}\right),$$

while traveling x miles per hour.
(a) If fuel costs \$1.60 per gallon, find the speed that will produce minimum total cost for a 400-mile trip.
(b) Find the minimum total cost.

*United Nations.

42. Business A rock-and-roll band travels from engagement to engagement in a large bus. This bus burns fuel at the rate of $G(x)$ gallons per mile, where

$$G(x) = \frac{1}{50}\left(\frac{200}{x} + \frac{x}{15}\right),$$

while traveling x miles per hour.
(a) If fuel costs \$2 per gallon, find the speed that will produce minimum total cost for a 250-mile trip.
(b) Find the minimum total cost.

43. Business A company wishes to run a utility cable from point A on the shore to an installation at point B on the island. The island is 6 miles from the shore (at point C) and point A is 9 miles from point C. It costs \$400 per mile to run the cable on land and \$500 per mile underwater. Assume that the cable starts at A and runs along the shoreline, then angles and runs underwater to the island. Find the point at which the line should begin to angle in order to yield the minimum total cost. (*Hint:* The length of the line underwater is $\sqrt{x^2 + 36}$.)

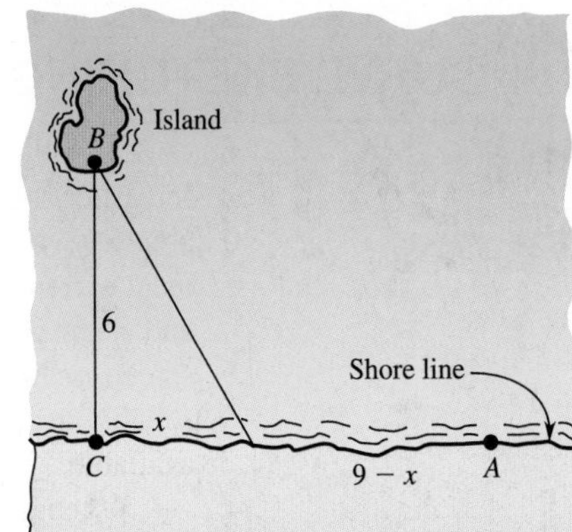

44. Natural Science Homing pigeons avoid flying over large bodies of water, preferring to fly around them instead. (One possible explanation is the fact that extra energy is required to fly over water because air pressure drops over water in the daytime.) Assume that a pigeon released from a boat 1 mile from the shore of a lake (point B in the figure) flies first to point P on the shore and then along the straight edge of the lake to reach its home at L. Assume that L is 2 miles from point A, the point on the shore closest to the boat, and that a pigeon needs 4/3 as much energy to fly over water as over land. Find the location of point P if the pigeon uses the least possible amount of energy.

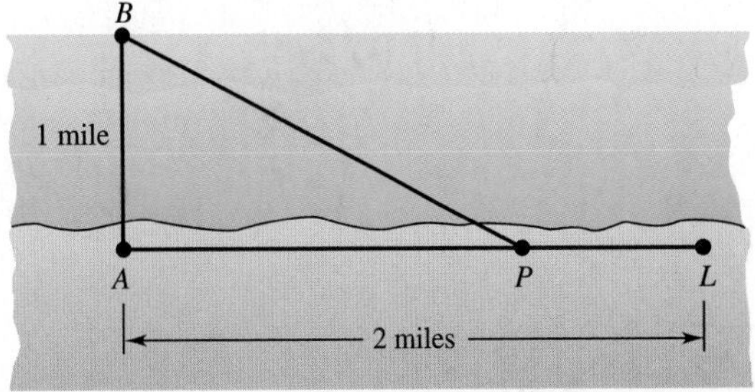

Business *The following exercises refer to economic lot size. (See Example 8.)*

45. Find the approximate number of batches that should be produced annually if 100,000 units are to be manufactured. It costs \$1 to store a unit for 1 year, and it costs \$500 to set up the factory to produce each batch.

46. How many units per batch will be manufactured in Exercise 45?

47. A market has a steady annual demand for 16,800 cases of sugar. It costs \$3 to store 1 case for 1 year. The market pays \$7 for each order that is placed. Find the number of orders for sugar that should be placed each year. (*Hint:* Use the formula for economic lot size, with ordering cost in place of setup cost.)

48. Find the number of cases per order in Exercise 47.

49. A bookstore has an annual demand for 100,000 copies of a best-selling book. It costs \$.50 to store one copy for one year, and it costs \$60 to place an order. Find the optimum number of copies per order.

50. A restaurant has an annual demand for 900 bottles of a California wine. It costs \$1 to store 1 bottle for 1 year, and it costs \$5 to place a reorder. Find the number of orders that should be placed annually.

51. Choose the correct answer:* The economic order quantity formula assumes that

(a) Purchase costs per unit differ due to quantity discounts.

(b) Costs of placing an order vary with quantity ordered.

(c) Periodic demand for the goods is known.

(d) Erratic usage rates are cushioned by safety stocks.

*Question from the Uniform CPA Examination of the American Institute of Certified Public Accountants, May, 1991. Reprinted by permission of the Institute of Certified Public Accountants.

12.4 CURVE SKETCHING

In earlier sections we saw that the first and second derivatives of a function provide a variety of information about the graph of the function, such as the location of its local extrema, the concavity of the graph, and the intervals on which it is increasing and decreasing. This information can be very helpful for interpreting misleading screen images on a graphing calculator or computer. It also enables us to make reasonably accurate graphs of many functions by hand, if graphing technology is not available.

When graphing functions by hand, you should use the following guidelines. It may not always be feasible to carry out all the steps, but you should do as many as necessary, in any convenient order, to obtain a reasonable graph.

To sketch the graph of a function $y = f(x)$:

1. Find the y-intercept (if it exists) by letting $x = 0$ and computing $y = f(0)$.
2. Find the x-intercepts (if any) by letting $y = 0$ and solving the equation $f(x) = 0$, if this is not too difficult.
3. If f is a rational function, find any vertical asymptotes by finding the numbers for which the denominator is 0 but the numerator is nonzero. Find any horizontal asymptotes by using the techniques of Section 3.7, as summarized in the box on page 182.
4. Find $f'(x)$ and $f''(x)$.
5. Locate any critical numbers by solving the equation $f'(x) = 0$ and determining where $f'(x)$ does not exist, but $f(x)$ does. Find the local extrema by using the first or second derivative test. Find the intervals where f is increasing or decreasing by solving the inequalities $f'(x) > 0$ and $f'(x) < 0$.

(continued)

6. Locate potential points of inflection by solving the equation $f''(x) = 0$ and determining where $f''(x)$ does not exist, but $f(x)$ does. Find the intervals where f is concave upward or downward by solving the inequalities $f''(x) > 0$ and $f''(x) < 0$. Use this information to determine the points of inflection.
7. Use the preceding results and any other information that may be available to determine the general shape of the graph.
8. Plot the intercepts, critical points, points of inflection, and other points as needed.

EXAMPLE 1 Graph $f(x) = 2x^3 - 3x^2 - 12x + 1$.

Step 1 The y-intercept is $f(0) = 2 \cdot 0^3 - 3 \cdot 0^2 - 12 \cdot 0 + 1 = 1$.

Step 2 To find the x-intercepts, we must solve the equation

$$2x^3 - 3x^2 - 12x + 1 = 0.$$

There is no easy way to do this by hand, so skip this step. Since $f(x)$ is a polynomial function, the graph has no asymptotes, so we can also skip Step 3.

Step 4 The first derivative is $f'(x) = 6x^2 - 6x - 12$ and the second derivative is $f''(x) = 12x - 6$.

Step 5 The first derivative is defined for all x, so the only critical numbers are the solutions of $f'(x) = 0$.

$$6x^2 - 6x - 12 = 0$$

$$x^2 - x - 2 = 0 \qquad \text{Divide both sides by 6.}$$

$$(x + 1)(x - 2) = 0 \qquad \text{Factor.}$$

$$x = -1 \quad \text{or} \quad x = 2$$

Using the second derivative test on the critical number $x = -1$, we have

$$f''(-1) = 12(-1) - 6 = -18 < 0.$$

Hence there is a local maximum when $x = -1$, that is, at the point $(-1, f(-1)) = (-1, 8)$. Similarly,

$$f''(2) = 12(2) - 6 = 18 > 0,$$

so that there is a local minimum when $x = 2$ (at the point $(2, f(2)) = (2, -19)$).

Next, we determine the intervals on which f is increasing or decreasing by solving the inequalities

$$f'(x) > 0 \qquad \text{and} \qquad f'(x) < 0$$

$$6x^2 - 6x - 12 > 0 \qquad\qquad 6x^2 - 6x - 12 < 0.$$

The critical numbers divide the x-axis into 3 regions. Testing a number from each region, as indicated in Figure 12.37, we conclude that f is increasing on the intervals $(-\infty, -1)$ and $(2, \infty)$ and decreasing on $(-1, 2)$.

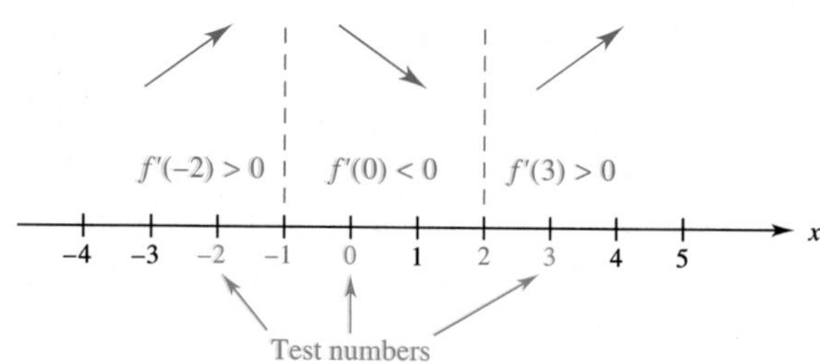

FIGURE 12.37

Step 6 The possible points of inflection are determined by the solutions of $f''(x) = 0$.

$$12x - 6 = 0$$
$$x = 1/2$$

Determine the concavity of the graph by solving

$$f''(x) > 0 \qquad \text{and} \qquad f''(x) < 0$$
$$12x - 6 > 0 \qquad\qquad 12x - 6 < 0$$
$$x > 1/2 \qquad\qquad x < 1/2$$

Therefore, f is concave upward on the interval $(1/2, \infty)$ and concave downward on $(-\infty, 1/2)$. Consequently, the only point of inflection is $(1/2, f(1/2)) = (1/2, -5.5)$.

Step 7 Since f is a third-degree polynomial function, we know from Section 3.6 that when x is very large in absolute value, its graph must resemble the graph of its highest degree term $2x^3$; that is, the graph must rise sharply on the right side and fall sharply on the left. Combining this fact with the information obtained in the preceding steps, we see that the graph of f must have the general shape shown in Figure 12.38.

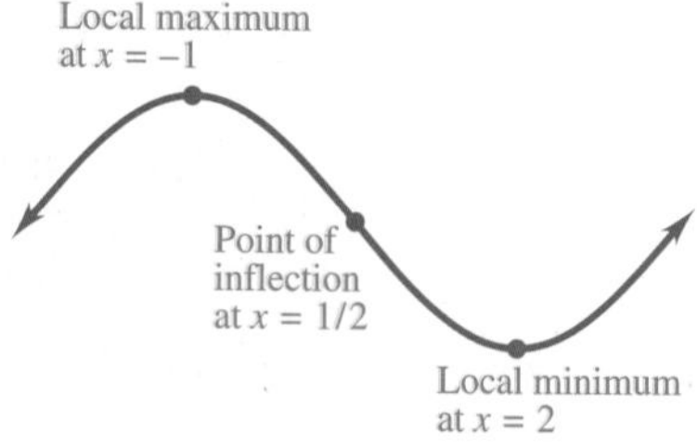

FIGURE 12.38

Step 8 Now we plot the points determined in Steps 1, 5, and 6, together with a few additional points, to obtain the graph in Figure 12.39. ◀ (1)

(1)

Sketch the graph of $f(x) = x^3 - 3x^2$.

Answer:

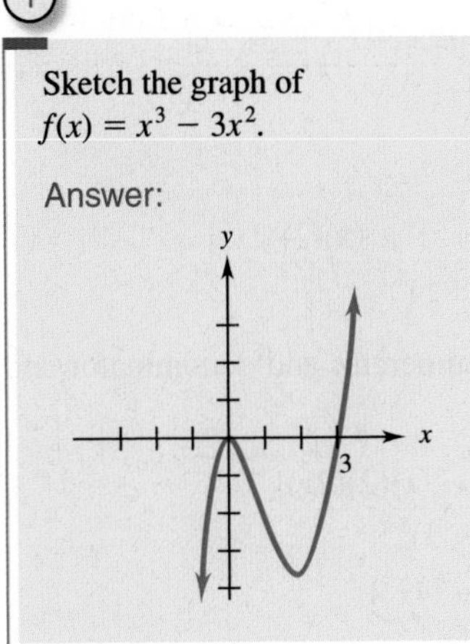

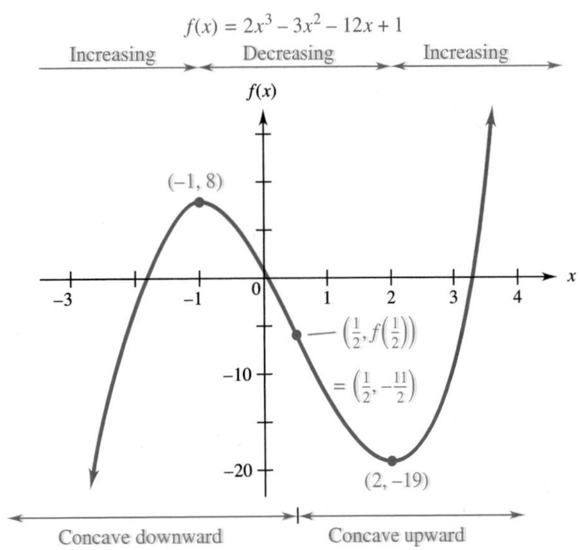

FIGURE 12.39

EXAMPLE 2 Graph $f(x) = \dfrac{3x^2}{x^2 + 5}$.

Step 1 The y-intercept is $f(0) = 0/5 = 0$.

Step 2 To find the x-intercepts, note that $f(x)$ is always defined because $x^2 + 5$ is always positive. Hence, $f(x) = 0$ when the numerator $3x^2 = 0$; this occurs when $x = 0$. So the point $(0, f(0)) = (0, 0)$ is both the x- and y-intercept.

Step 3 This is a rational function, but its denominator is always nonzero (why?), so there are no vertical asymptotes. Using the techniques presented in Section 3.7, we see that

$$f(x) = \frac{3x^2}{x^2 + 5} = \frac{\dfrac{3x^2}{x^2}}{\dfrac{x^2}{x^2} + \dfrac{5}{x^2}} = \frac{3}{1 + \dfrac{5}{x^2}}.$$

When x is very large in absolute value, so is x^2, so that $5/x^2$ is very close to 0, and hence $f(x)$ is very close to $3/(1 + 0) = 3$. Consequently, the horizontal line $y = 3$ is a horizontal asymptote.

Step 4 The first derivative is

$$f'(x) = \frac{(x^2 + 5)(6x) - (3x^2)(2x)}{(x^2 + 5)^2} = \frac{30x}{(x^2 + 5)^2}.$$

The second derivative is

$$f''(x) = \frac{(x^2 + 5)^2(30) - (30x)(2)(x^2 + 5)(2x)}{(x^2 + 5)^4}.$$

Factor $30(x^2 + 5)$ out of the numerator.

$$f''(x) = \frac{30(x^2 + 5)[(x^2 + 5) - (x)(2)(2x)]}{(x^2 + 5)^4}$$

Divide a factor of $(x^2 + 5)$ out of the numerator and denominator and simplify the numerator.

$$\begin{aligned} f''(x) &= \frac{30[(x^2 + 5) - (x)(2)(2x)]}{(x^2 + 5)^3} \\ &= \frac{30[(x^2 + 5) - 4x^2]}{(x^2 + 5)^3} \\ &= \frac{30(5 - 3x^2)}{(x^2 + 5)^3} \end{aligned}$$

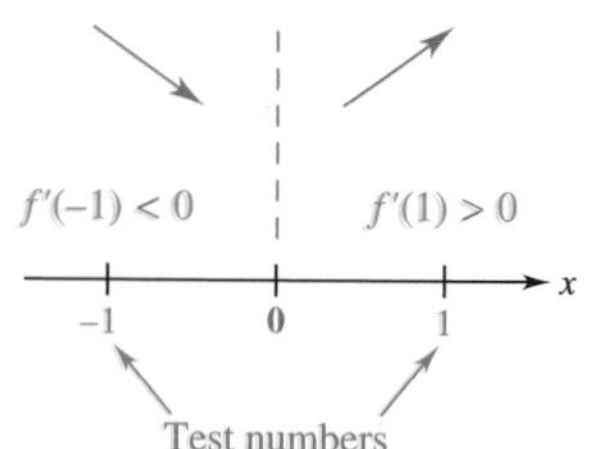

FIGURE 12.40

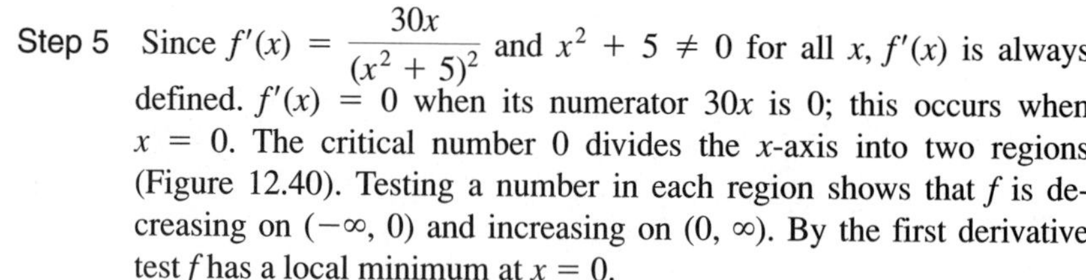

Step 5 Since $f'(x) = \dfrac{30x}{(x^2 + 5)^2}$ and $x^2 + 5 \neq 0$ for all x, $f'(x)$ is always defined. $f'(x) = 0$ when its numerator $30x$ is 0; this occurs when $x = 0$. The critical number 0 divides the x-axis into two regions (Figure 12.40). Testing a number in each region shows that f is decreasing on $(-\infty, 0)$ and increasing on $(0, \infty)$. By the first derivative test f has a local minimum at $x = 0$.

Step 6 The numerator of $f''(x) = \dfrac{30(5 - 3x^2)}{(x^2 + 5)^3}$ is 0 when

$$\begin{aligned} 30(5 - 3x^2) &= 0 \\ 3x^2 &= 5 \\ x &= \pm\sqrt{5/3} \approx \pm 1.29. \end{aligned}$$

Testing a point in each of the three intervals defined by these points shows that f is concave downward on $(-\infty, -1.29)$ and $(1.29, \infty)$ and concave upward on $(-1.29, 1.29)$. The graph has inflection points at $(\pm\sqrt{5/3}, f(\pm\sqrt{5/3})) \approx (\pm 1.29, \pm .75)$.

Step 7 The information about the shape of the graph obtained in Steps 4, 5, and 6 is summarized in this chart.

Interval	$(-\infty, -1.29)$	$(-1.29, 0)$	$(0, 1.29)$	$(1.29, \infty)$
Sign of f'	−	−	+	+
Sign of f''	−	+	+	−
f Increasing or Decreasing	Decreasing	Decreasing	Increasing	Increasing
Concavity of f	Downward	Upward	Upward	Downward
Shape of Graph				

Step 8 Plot some points (several are needed near the origin), including the intercept at the origin, and use the fact that $y = 3$ is a horizontal asymptote to obtain the graph in Figure 12.41. ◀

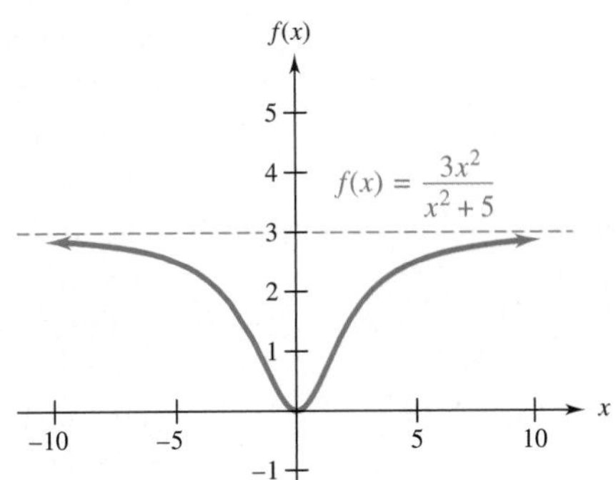

FIGURE 12.41

EXAMPLE 3 Graph $f(x) = x + 1/x$.

Step 1 Since $x = 0$ is not in the domain of the function (why?), there is no y-intercept.

Step 2 To find the x-intercepts, solve $f(x) = 0$.

$$x + \frac{1}{x} = 0$$

$$x = -\frac{1}{x}$$

$$x^2 = -1$$

Since x^2 is always positive, there is also no x-intercept.

Step 3 Note that the rule of f can be written as

$$f(x) = x + \frac{1}{x} = \frac{x^2+1}{x}.$$

When $x = 0$, the denominator is 0, but the numerator is nonzero, so there is a vertical asymptote at $x = 0$. Since the numerator of $f(x)$ has higher degree than the denominator, there is no horizontal asymptote.

Step 4 Since $f(x) = x + 1/x = x + x^{-1}$, we have

$$f'(x) = 1 - x^{-2} = 1 - 1/x^2,$$

so that

$$f''(x) = 2x^{-3} = 2/x^3.$$

Step 5 $f'(x) = 0$ when

$$\frac{1}{x^2} = 1$$

$$x^2 = 1$$

$$x = 1 \quad \text{or} \quad x = -1.$$

Hence $x = -1$ and $x = 1$ are critical numbers. The derivative does not exist when $x = 0$, but the function is not defined there either, so $x = 0$ is not a critical number. Evaluating $f'(x)$ in each of the regions determined by the critical numbers and the asymptote shows that f is increasing on $(-\infty, -1)$ and $(1, \infty)$ and decreasing on $(-1, 0)$ and $(0, 1)$, as summarized in the chart in Step 7. By the first derivative test, f has a relative maximum of $y = f(-1) = -2$, when $x = -1$, and a relative minimum of $y = f(1) = 2$ when $x = 1$.

Step 6 The second derivative $f''(x) = 2/x^3$ is never equal to 0 and does not exist when $x = 0$. (The function itself also does not exist at 0.) Because of this, there may be a change of concavity, but not an inflection point, when $x = 0$. The second derivative is negative when x is negative, making f concave downward on $(-\infty, 0)$. Also, $f''(x) > 0$ when $x > 0$, making f concave upward on $(0, \infty)$, as indicated in the chart in Step 7.

Step 7 The preceding information is summarized in the following chart.

Interval	$(-\infty, -1)$	$(-1, 0)$	$(0, 1)$	$(1, \infty)$
Sign of f'	+	−	−	+
Sign of f''	−	−	+	+
f Increasing or Decreasing	Increasing	Decreasing	Decreasing	Increasing
Concavity of f	Downward	Downward	Upward	Upward
Shape of Graph				

We can determine the shape of the graph when x is very large in absolute value by noting that as x gets very large, the second term of its rule, $1/x$, gets very small, so that $f(x) = x + 1/x \approx x$. Hence the graph gets closer and closer to the straight line $y = x$ as x becomes larger and larger. The line $y = x$ is called an **oblique** or **slant asymptote.**

Step 8 Plot several points and use the information above to obtain the graph of $f(x)$ in Figure 12.42.

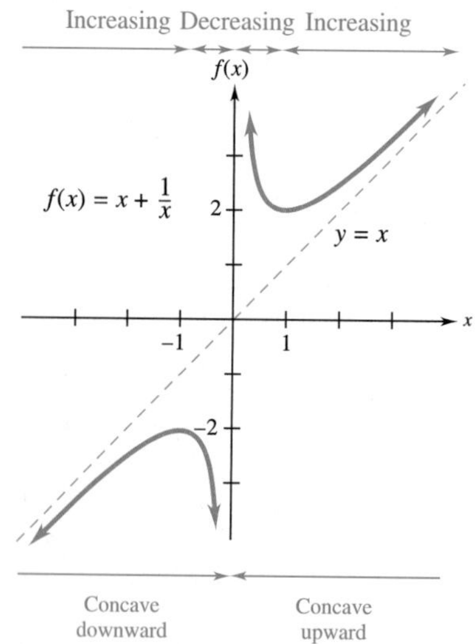

FIGURE 12.42

12.4 Exercises

Sketch the graph of the function. Identify any local extrema and points of inflection. (See Examples 1–3.)

1. $f(x) = -x^2 - 10x - 25$ **2.** $f(x) = x^2 - 12x + 36$

3. $f(x) = 3x^3 - 3x^2 + 1$ **4.** $f(x) = 2x^3 - 4x^2 + 2$

5. $f(x) = -2x^3 - 9x^2 + 108x - 10$

6. $f(x) = -2x^3 - 9x^2 + 60x - 8$

7. $f(x) = 2x^3 + \frac{7}{2}x^2 - 5x + 3$

8. $f(x) = x^3 - \frac{15}{2}x^2 - 18x - 1$

9. $f(x) = (x + 3)^4$ **10.** $f(x) = x^3$

11. $f(x) = x^4 - 18x^2 + 5$ **12.** $f(x) = x^4 - 8x^2$

13. $f(x) = x - \frac{1}{x}$ **14.** $f(x) = 2x + \frac{8}{x}$

15. $f(x) = \frac{x^2 + 25}{x}$ **16.** $f(x) = \frac{x^2 + 4}{x}$

17. $f(x) = \frac{x - 1}{x + 1}$ **18.** $f(x) = \frac{x}{1 + x}$

In Exercises 19–24, sketch the graph of a function f that has all of the properties listed. There are many correct answers and your graph need not be given by an algebraic formula.

19. **(a)** The domain of f is $[0, 10]$.
(b) $f'(x) > 0$ and $f''(x) > 0$ for all x in the domain of f.

20. **(a)** The domain of f is $[0, 10]$.
(b) $f'(x) > 0$ and $f''(x) < 0$ for all x in the domain of f.

21. **(a)** Continuous and differentiable for all real numbers
(b) Increasing on $(-\infty, -3)$ and $(1, 4)$
(c) Decreasing on $(-3, 1)$ and $(4, \infty)$
(d) Concave downward on $(-\infty, -1)$ and $(2, \infty)$
(e) Concave upward on $(-1, 2)$
(f) $f'(-3) = f'(4) = 0$
(g) Inflection points at $(-1, 3)$ and $(2, 4)$

22. **(a)** Continuous for all real numbers
(b) Increasing on $(-\infty, -2)$ and $(0, 3)$
(c) Decreasing on $(-2, 0)$ and $(3, \infty)$
(d) Concave downward on $(-\infty, 0)$ and $(0, 5)$
(e) Concave upward on $(5, \infty)$
(f) $f'(-2) = f'(3) = 0$

23. **(a)** Continuous for all real numbers
(b) Decreasing on $(-\infty, -6)$ and $(1, 3)$
(c) Increasing on $(-6, 1)$ and $(3, \infty)$
(d) Concave upward on $(-\infty, -6)$ and $(3, \infty)$
(e) Concave downward on $(-6, 3)$
(f) A y-intercept at $(0, 2)$

24. **(a)** Continuous and differentiable everywhere except at $x = 1$, where it has a vertical asymptote
(b) Decreasing everywhere it is defined
(c) Concave downward on $(-\infty, 1)$ and $(2, 4)$
(d) Concave upward on $(1, 2)$ and $(4, \infty)$

25. Natural Science The figure shows how the risk of chromosomal abnormality in a child increases with the age of the mother.*

(a) What is the sign of the first derivative on the interval $(20, 50)$? Why?
(b) What is the sign of the second derivative on this interval? What does this tell you about the rate of risk?

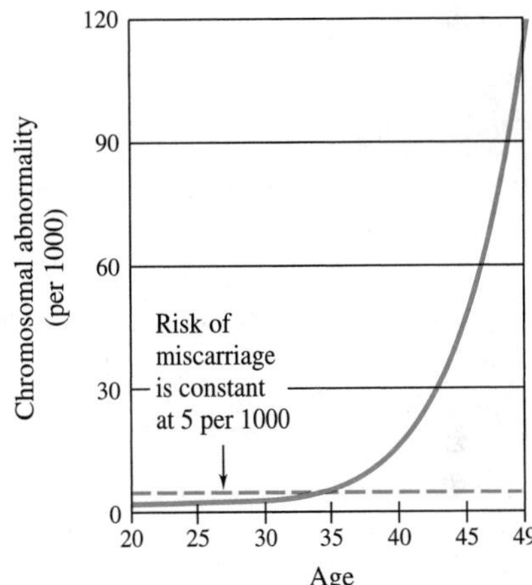

Source: American College of Obstetricians and Gynocologists.

26. Business The figure on the next page shows the *product life cycle* graph, with typical products marked on it. It illustrates the fact that a new product is often purchased at a faster and faster rate as people become familiar with it. In time, saturation is reached and the purchase rate stays constant until the product is made obsolete by newer products, after which it is purchased less and less.†

(a) Which products on the left side of the graph are closest to the left-hand point of inflection? What does the point of inflection mean here?

**New York Times*, February 5, 1994, p. 24.

†Based on "The Product Life Cycle: A Key to Strategic Marketing Planning" in *MSU Business Topics* (Winter 1973), p. 30. Reprinted by permission of the publisher. Graduate School of Business Administration, Michigan State University.

(b) Which product on the right side of the graph is closest to the right-hand point of inflection? What does the point of inflection mean here?

(c) Discuss where home computers, fax machines, and other new technologies should be placed on the graph.

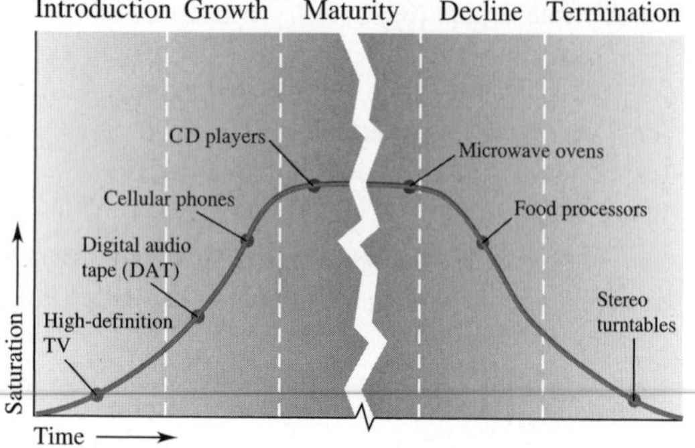

Graph used with permission from Whirlpool Corporation.

Use calculus and a graphing calculator or computer to find the approximate location of all relative extrema and points of inflection of these functions. Several viewing windows may be needed to see some of the graphs clearly. Be on the lookout for "hidden behavior," such as extrema that may not be obvious at first glance.

27. $f(x) = .1x^3 - .1x^2 - .005x + 1$

28. $f(x) = 2x^3 - .33x^2 - .006x + 5$

29. $f(x) = .01x^5 + x^4 - x^3 - 6x^2 + 5x + 4$

30. $f(x) = .1x^5 + 3x^4 - 4x^3 - 11x^2 + 3x + 2$

CHAPTER 12 SUMMARY

Key Terms and Symbols

12.1 increasing function on an interval
decreasing function on an interval
critical number
critical point
local maximum (maxima)
local minimum (minima)
local extremum (extrema)

12.2 $f''(x)$ or y'' or $\dfrac{d^2y}{dx^2}$ or $D_x^2[f(x)]$ second derivative of f
$f'''(x)$ third derivative of f
$f^{(n)}(x)$ nth derivative of f
acceleration
concave upward
concave downward
point of inflection
point of diminishing returns

12.3 absolute maximum on an interval
absolute minimum on an interval
Extreme Value Theorem
economic lot size

12.4 curve sketching
oblique asymptote

Key Concepts

If $f'(x) > 0$ for each x in an interval, then f is **increasing** on the interval; if $f'(x) < 0$ for each x in the interval, then f is **decreasing** on the interval; if $f'(x) = 0$ for each x in the interval, then f is **constant** on the interval.

Local Extrema

Let c be a number in the domain of a function f. Then f has a **local maximum** at c if $f(x) \leq f(c)$ for all x near c, and f has a **local minimum** at c if $f(x) \geq f(c)$ for all x near c. If f has a local extremum at c, then $f'(c) = 0$ or $f'(c)$ does not exist.

First Derivative Test Let f be a differentiable function for all x in $[a, b]$, except possibly at $x = c$. Assume $a < c < b$ and that c is the only critical number for f in $[a, b]$. If $f'(a) > 0$ and $f'(b) < 0$, then there is a local maximum at c. If $f'(a) < 0$ and $f'(b) > 0$, then there is a local minimum at c.

Concavity Let f have derivatives f' and f'' for all x in (a, b). f is **concave upward** on (a, b) if $f''(x) > 0$ for all x in (a, b). f is **concave downward** on (a, b) if $f''(x) < 0$ for all x in (a, b). f has a **point of inflection** at $x = c$ if $f''(x)$ changes sign at $x = c$.

Second Derivative Test Let c be a critical number of f such that $f'(c) = 0$ and $f''(x)$ exists for all x in some open interval containing c. If $f''(c) > 0$, then there is a local minimum at c. If $f''(c) < 0$, then there is a local maximum at c. If $f''(c) = 0$, then the test gives no information.

Absolute Extrema Let c be in an interval $[a, b]$ where f is defined. Then f has an **absolute maximum** on the interval at c if $f(x) \leq f(c)$ for all x in $[a, b]$ and f has an **absolute minimum** on the interval at c if $f(x) \geq f(c)$ for all x in $[a, b]$.

Chapter 12 Review Exercises

1. When the rule of a function is given, how can you determine where it is increasing and where it is decreasing?
2. When the rule of a function is given, how can you determine where the local extrema are located? State two algebraic ways to test whether a local extremum is a maximum or a minimum.
3. What is the difference between a local extremum and an absolute extremum? Can a local extremum be an absolute extremum? Is a local extremum necessarily an absolute extremum?
4. What information about a graph can be found from the first derivative? From the second derivative?

Find the largest open intervals on which the following functions are increasing or decreasing.

5. $f(x) = x^2 + 7x - 9$
6. $f(x) = -3x^2 - 2x + 11$
7. $g(x) = 2x^3 - x^2 - 4x + 7$
8. $g(x) = -4x^3 - 5x^2 + 8x + 1$
9. $f(x) = \dfrac{4}{x - 3}$
10. $f(x) = \dfrac{6}{3x + 2}$

Find the locations and the values of all local maxima and minima for the following functions.

11. $f(x) = 2x^3 + 3x^2 - 36x + 20$
12. $f(x) = 2x^3 + 3x^2 - 12x + 5$
13. $f(x) = x^4 + \dfrac{8}{3}x^3 - 6x^2 + 1$
14. $f(x) = x \cdot e^x$
15. $f(x) = 3x \cdot e^{-x}$
16. $f(x) = \dfrac{e^x}{x - 1}$

Find the second derivatives of the following functions; then find $f''(1)$ and $f''(-2)$.

17. $f(x) = 2x^5 - 4x^3 + 2x - 1$
18. $f(x) = \dfrac{3 - 2x}{x + 2}$
19. $f(x) = -5e^{4x}$
20. $f(x) = \ln|5x + 2|$

Sketch the graph of each of these functions. List the location of each local extremum and point of inflection, the intervals on which the function is increasing and decreasing, and the intervals on which it is concave upward and concave downward.

21. $f(x) = -2x^3 - \dfrac{1}{2}x^2 - x - 3$
22. $f(x) = -\dfrac{4}{3}x^3 + x^2 + 30x - 7$
23. $f(x) = x^4 - \dfrac{4}{3}x^3 - 4x^2 + 1$
24. $f(x) = -\dfrac{2}{3}x^3 + \dfrac{9}{2}x^2 + 5x + 1$
25. $f(x) = \dfrac{x - 1}{2x + 1}$
26. $f(x) = \dfrac{2x - 5}{x + 3}$
27. $f(x) = -4x^3 - x^2 + 4x + 5$

28. $f(x) = x^3 + \frac{5}{2}x^2 - 2x - 3$

29. $f(x) = x^4 + 2x^2$

30. $f(x) = 6x^3 - x^4$

31. $f(x) = \frac{x^2 + 4}{x}$

32. $f(x) = x + \frac{8}{x}$

Find the locations and values of all absolute maxima and absolute minima for the following functions on the given intervals.

33. $f(x) = -x^2 + 5x + 1$; $[1, 4]$

34. $f(x) = 4x^2 - 8x - 3$; $[-1, 2]$

35. $f(x) = x^3 + 2x^2 - 15x + 3$; $[-4, 2]$

36. $f(x) = -2x^3 - x^2 + 4x - 1$; $[-3, 1]$

Work the following exercises. You may need the quadratic formula to find some of the critical numbers.

37. **Natural Science** The number of bacteria in a culture is approximated by

$$S(x) = -x^3 + 3x^2 + 360x + 5000 \qquad (6 \le x \le 20),$$

where x represents the temperature of the culture in degrees Celsius. Find the temperature that produces the maximum number of bacteria.

38. **Social Science** The unemployment rate in the United States is approximated by

$$g(x) = .055x^3 - .22x^2 - .24x + 5.2 \qquad (0 \le x < 5),$$

where $x = 0$ corresponds to 1997.* When was the unemployment rate lowest?

39. **Business** Mexico's production of natural gas (in billion cubic feet per day) is approximated by

$$f(x) = -6.94x^3 + 113x^2 - 327x + 3638 \qquad (0 \le x \le 10),$$

where $x = 0$ corresponds to 1990.† When was production at a maximum and when was it at a minimum?

40. **Business** The amount (in millions of dollars) spent by Intel on capital additions to property, plant, and equipment on the interval [5, 10] is approximated by

$$g(x) = 166.9x^4 - 4868.2x^3 + 52{,}277.4x^2 - 244{,}448.4x + 423{,}084,$$

where $x = 5$ corresponds to 1995.‡ When were these expenditures the largest? The smallest?

41. **Social Science** The number of families on welfare (in millions) is approximated by

$$f(x) = -.0037x^3 + .1027x^2 - .73x + 5 \qquad (2 \le x < 21),$$

where $x = 2$ corresponds to 1982.* When was this number largest?

42. **Business** The packaging department of a corporation is designing a box with a square base and top. The volume is to be 27 cubic meters. To reduce cost, the box is to have minimum surface area. What dimensions (height, length, and width) should the box have?

43. **Social Science** The city park department is planning an enclosed play area in a new park. One side of the area will be against an existing building, with no fence needed there. Find the dimensions of the rectangular space of maximum area that can be enclosed with 900 meters of fence.

44. **Business** A company plans to package its product in a cylinder that is open at one end. The cylinder is to have a volume of 27π cubic inches. What radius should the circular bottom of the cylinder have to minimize the cost of the material? (*Hint:* The volume of a circular cylinder is $\pi r^2 h$, where r is the radius of the circular base and h is the height; the surface area of an open cylinder is $2\pi rh + r^2$.)

45. **Business** In 1 year, a health food manufacturer produces and sells 240,000 cases of vitamins. It costs \$2 to store a case for 1 year and \$15 to produce each batch. Find the number of batches that should be produced annually.

46. **Business** A company produces 128,000 cases of a soft drink annually. It costs \$1 to store a case for 1 year and \$10 to produce one lot. Find the number of lots that should be produced annually.

47. **Social Science** If the play area referred to in Exercise 43 needs fencing on all four sides, find the dimensions of the maximum rectangular area that can be made with 900 meters of fence.

*Bureau of Labor Statistics.

†Pemex.

‡Intel Corporation Annual Report 2000.

*Congressional Research Service.

Case 11

A Total Cost Model for a Training Program*

In this application, we set up a mathematical model for determining the total costs in setting up a training program. Then we use calculus to find the time interval between training programs that produces the minimum total cost. The model assumes that the demand for trainees is constant and that the fixed cost of training a batch of trainees is known. Also, it is assumed that people who are trained, but for whom no job is readily available, will be paid a fixed amount per month while waiting for a job to open up.

The model uses the following variables.

D = demand for trainees per month

N = number of trainees per batch

C_1 = fixed cost of training a batch of trainees

C_2 = variable cost of training per trainee per month

C_3 = salary paid monthly to a trainee who has not yet been given a job after training

m = time interval in months between successive batches of trainees

t = length of training program in months

$Z(m)$ = total monthly cost of program

The total cost of training a batch of trainees is given by $C_1 + NtC_2$. However, $N = mD$, so that the total cost per batch is $C_1 + mDtC_2$.

After training, personnel are given jobs at the rate of D per month. Thus, $N - D$ of the trainees will not get a job the first month, $N - 2D$ will not get a job the second month, and so on. The $N - D$ trainees who do not get a job the first month produce total costs of $(N - D)C_3$, those not getting jobs during the second month produce costs of $(N - 2D)C_3$, and so on. Since $N = mD$, the costs during the first month can be written as

$$(N - D)C_3 = (mD - D)C_3 = (m - 1)DC_3,$$

while the costs during the second month are $(m - 2)DC_3$, and so on. The total cost for keeping the trainees without a job is thus

$$(m - 1)DC_3 + (m - 2)DC_3 + (m - 3)DC_3 + \cdots + 2DC_3 + DC_3,$$

which can be factored to give

$$DC_3[(m - 1) + (m - 2) + (m - 3) + \cdots + 2 + 1].$$

The expression in brackets is the sum of the terms of an arithmetic sequence discussed in most algebra texts. Using formulas for arithmetic sequences, the expression in brackets can be shown to equal $m(m - 1)/2$, so that we have

$$DC_3\left[\frac{m(m - 1)}{2}\right] \qquad (1)$$

as the total cost for keeping jobless trainees.

The total cost per batch is the sum of the training cost per batch, $C_1 + mDtC_2$, and the cost of keeping trainees without a proper job, given by equation (1). Since we assume that a batch of trainees is trained every m months, the total cost per month, $Z(m)$, is given by

$$Z(m) = \frac{C_1 + mDtC_2}{m} + \frac{DC_3\left[\frac{m(m - 1)}{2}\right]}{m}$$

$$= \frac{C_1}{m} + DtC_2 + DC_3\left(\frac{m - 1}{2}\right).$$

Exercises

1. Find $Z'(m)$.
2. Solve the equation $Z'(m) = 0$.

As a practical matter, it is usually required that m be a whole number. If m does not come out to be a whole number in Exercise 2, then m^+ and m^-, the two whole numbers closest to m, must be chosen. Calculate both $Z(m^+)$

*Based on "A Total Cost Model for a Training Program" by P. L. Goyal and S. K. Goyal, Faculty of Commerce and Administration, Concordia University. Used with permission

and $Z(m^-)$; the smaller of the two provides the optimum value of Z.

3. Suppose a company finds that its demand for trainees is 3 per month, that a training program requires 12 months, that the fixed cost of training a batch of trainees is \$15,000, that the marginal cost per trainee per month is \$100, and that trainees are paid \$900 per month after training but before going to work. Use your result from Exercise 2 and find m.
4. Since m is not a whole number, find m^+ and m^-.
5. Calculate $Z(m^+)$ and $Z(m^-)$.
6. What is the optimum time interval between successive batches of trainees? How many trainees should be in a batch?

Chapter 13

Integral Calculus

The derivative and its applications, which were studied in Chapters 11 and 12, are part of what is called *differential calculus*. This chapter is devoted to the other main branch of calculus, *integral calculus*. Integrals are used to find areas, to determine the length of curved paths, to solve complicated probability problems, to calculate the location of an object (such as the height of a space shuttle) from its velocity, and in a variety of other ways. Such topics may seem quite different from those studied earlier, but the so-called Fundamental Theorem of Calculus in Section 13.4 will reveal a surprisingly close connection between differential and integral calculus.

13.1 ANTIDERIVATIVES

Functions used in applications in previous chapters have provided information about a *total amount* of a quantity, such as cost, revenue, profit, temperature, gallons of oil, or distance. Derivatives of these functions provided information about the rate of change of these quantities and allowed us to answer important questions about the extrema of the functions. It is not always possible to find ready-made functions that provide information about the total amount of a quantity, but it is often possible to collect enough data to come up with a function that gives the *rate of change* of a quantity. We know that derivatives give the rate of change when the total amount is known. In this section we shall see that this process can be reversed: when the rate of change is known, a function that gives the total amount can be obtained by a process called *antidifferentiation*. Here is the basic definition.

Antiderivatives

If $F(x)$ and $f(x)$ are functions such that $F'(x) = f(x)$, then $F(x)$ is said to be an **antiderivative** of $f(x)$.

1

Find an antiderivative for each of the following.

(a) $3x^2$

(b) $5x$

(c) $8x^7$

Answers:

Only one possible antiderivative is given for each.

(a) x^3

(b) $\frac{5}{2}x^2$

(c) x^8

EXAMPLE 1 **(a)** If $F(x) = 10x$, then $F'(x) = 10$, so $F(x) = 10x$ is an antiderivative of $f(x) = 10$.

(b) For $F(x) = x^5$, $F'(x) = 5x^4$, which means that $F(x) = x^5$ is an antiderivative of $f(x) = 5x^4$. ◀

EXAMPLE 2 Find an antiderivative of $f(x) = 2x$.

By remembering formulas for derivatives, it's easy to see that $F(x) = x^2$ is an antiderivative of $f(x)$ because $F'(x) = 2x = f(x)$. Note that $G(x) = x^2 + 2$ and $H(x) = x^2 - 4$ are also antiderivatives of $f(x)$ because

$$G'(x) = 2x + 0 = f(x) \quad \text{and} \quad H'(x) = 2x - 0 = f(x).$$ ◀ 1

Any two of the antiderivatives of $f(x) = 2x$ that were found in Example 2 differ by a constant. For instance, $G(x) - F(x) = 2$ and $H(x) - G(x) = -6$. The same thing is true in the general case.

If $F(x)$ and $G(x)$ are both antiderivatives of $f(x)$, then there is a constant C such that

$$F(x) - G(x) = C.$$

(Two antiderivatives of a function can differ only by a constant.)

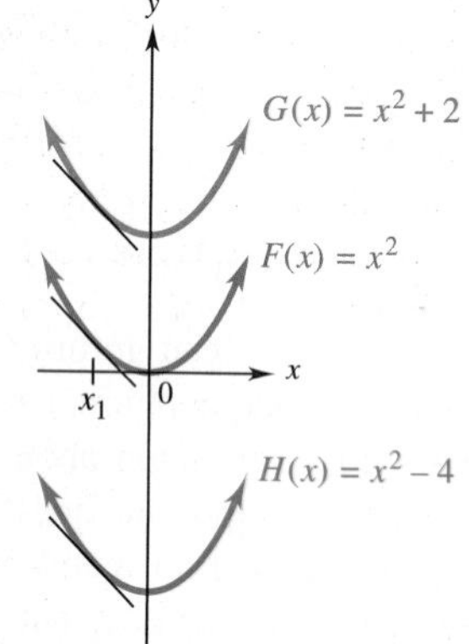

Slopes of the tangent lines at $x = x_1$ are the same.

FIGURE 13.1

The statement in the box reflects a geometric fact about derivatives, namely, that the derivative of a function gives the slope of the tangent line at any number x. For example, if you graph the three antiderivatives of $f(x) = 2x$ found in Example 2, you will see that the graphs all have the same shape because at any x-value, their tangent lines all have the same slope, as shown in Figure 13.1.

The family of all antiderivatives of the function f is indicated by

$$\int f(x)\,dx.$$

The symbol $\int$ is the **integral sign,** $f(x)$ is the **integrand,** and $\int f(x)\,dx$ is called an **indefinite integral.** Since any two antiderivatives of $f(x)$ differ by a constant (which means that one is a constant plus the other), we can describe the indefinite integral as follows.

Indefinite Integral

If $F'(x) = f(x)$, then

$$\int f(x)\,dx = F(x) + C,$$

for any real number C.

For example, using this notation,

$$\int 2x\,dx = x^2 + C.$$

NOTE The dx in the indefinite integral $\int f(x)\,dx$ indicates that x is the variable of the function whose antiderivative is to be found, in the same way that dy/dx denotes the derivative when y is a function of the variable x. For example, in the indefinite integral $\int 2ax\,dx$, the variable of the function is x, whereas in the indefinite integral $\int 2ax\,da$, the variable is a.

The symbol $\int f(x)\,dx$ was created by G. W. Leibniz (1646–1716) in the latter part of the seventeenth century. The $\int$ is an elongated S from *summa,* the Latin word for *sum.* The word *integral* as a term in the calculus was coined by Jakob Bernoulli (1654–1705), a Swiss mathematician who corresponded frequently with Leibniz. The relationship between sums and integrals will be clarified in Section 13.3.

Because finding an antiderivative is the inverse of finding a derivative, each formula for derivatives leads to a rule for antiderivatives. For instance, the power rule for derivatives tells us that

$$\text{if } F(x) = x^4, \quad \text{then } F'(x) = 4x^3.$$

Consequently,

$$\text{if } F(x) = \frac{1}{4}x^4, \quad \text{then } F'(x) = \frac{1}{4}(4x^3) = x^3.$$

In other words, an antiderivative of $f(x) = x^3$ is $F(x) = \frac{1}{4}x^4$. Similarly, an antiderivative of $g(x) = x^7$ is $G(x) = \frac{1}{8}x^8$ because $G'(x) = \frac{1}{8}(8x^7) = x^7$. The same pattern holds in the general case: to find the antiderivative of x^n, increase the exponent by 1 and divide by that same number.

Power Rule for Antiderivatives

For any real number $n \neq -1$,

$$\int x^n\,dx = \frac{1}{n+1}x^{n+1} + C.$$

This result can be verified by differentiating the expression on the right above.

$$\frac{d}{dx}\left(\frac{1}{n+1}x^{n+1} + C\right) = \frac{n+1}{n+1}x^{(n+1)-1} + 0 = x^n$$

(If $n = -1$, the expression in the denominator is 0, and the rule above cannot be used. We will see later how to find an antiderivative in this case.)

EXAMPLE 3 Find each antiderivative.

(a) $\int x^3\,dx$

Use the power rule with $n = 3$.

$$\int x^3\,dx = \frac{1}{3+1}x^{3+1} + C = \frac{1}{4}x^4 + C$$

(b) $\int \frac{1}{t^2}\,dt$

First, write $1/t^2$ as t^{-2}. Then

$$\int \frac{1}{t^2}\,dt = \int t^{-2}\,dt = \frac{1}{-2+1}t^{-2+1} = \frac{t^{-1}}{-1} + C = \frac{-1}{t} + C.$$

(c) $\int \sqrt{u}\,du$

Since $\sqrt{u} = u^{1/2}$,

$$\int \sqrt{u}\,du = \int u^{1/2}\,du = \frac{1}{1/2+1}u^{1/2+1} + C = \frac{1}{3/2}u^{3/2} + C = \frac{2}{3}u^{3/2} + C.$$

To check this, differentiate $(2/3)u^{3/2} + C$; the derivative is $u^{1/2}$, the original function.

(d) $\int dx$

Writing dx as $1 \cdot dx$, and using the fact that $x^0 = 1$ for any nonzero number x,

$$\int dx = \int 1\,dx = \int x^0\,dx = \frac{1}{1}x^1 + C = x + C.$$ ◀ 2

2

Find each of the following.

(a) $\int x^5\,dx$

(b) $\int \sqrt[3]{x}\,dx$

(c) $\int 5\,dx$

Answers:

(a) $\frac{1}{6}x^6 + C$

(b) $\frac{3}{4}x^{4/3} + C$

(c) $5x + C$

As shown in Chapter 11, the derivative of the product of a constant and a function is the product of the constant and the derivative of the function. A similar rule applies to antiderivatives. Also, since derivatives of sums or differences are found term by term, antiderivatives can also be found term by term.

Properties of Antiderivatives

Let $f(x)$ and $g(x)$ be functions that have antiderivatives and let k be any real number.

Constant Multiple Rule: $\int k \cdot f(x)\, dx = k \int f(x)\, dx.$

Sum or Difference Rule: $\int [f(x) \pm g(x)]\, dx = \int f(x)\, dx \pm \int g(x)\, dx.$

CAUTION The constant multiple rule requires that k be a *number*. The rule does not apply to a *variable*. For example,

$$\int x\sqrt{x-1}\, dx \neq x \int \sqrt{x-1}\, dx.$$

EXAMPLE 4 Find each of the following.

(a) $\int 2x^3\, dx$

By the constant multiple rule and the power rule,

$$\int 2x^3\, dx = 2\int x^3\, dx = 2\left(\frac{1}{4}x^4\right) + C = \frac{1}{2}x^4 + C.$$

Since C represents any real number, it is not necessary to multiply it by 2 in the next-to-last step.

(b) $\int \frac{12}{z^5}\, dz$

Convert to negative exponents and use the power rule.

$$\int \frac{12}{z^5}\, dz = \int 12z^{-5}\, dz$$

$$= 12\int z^{-5}\, dz \qquad \text{Constant multiple rule}$$

$$= 12\left(\frac{z^{-4}}{-4}\right) + C \qquad \text{Power rule}$$

$$= -3z^{-4} + C$$

$$= \frac{-3}{z^4} + C$$

3

Find each of the following.

(a) $\int (-6x^4)\,dx$

(b) $\int 9x^{2/3}\,dx$

(c) $\int \frac{8}{x^3}\,dx$

(d) $\int (5x^4 - 3x^2 + 6)\,dx$

(e) $\int \left(3\sqrt{x} + \frac{2}{x^2}\right)dx$

Answers:

(a) $-\frac{6}{5}x^5 + C$

(b) $\frac{27}{5}x^{5/3} + C$

(c) $-4x^{-2} + C$ or $-\frac{4}{x^2} + C$

(d) $x^5 - x^3 + 6x + C$

(e) $2x^{3/2} - \frac{2}{x} + C$

(c) $\int (3z^2 - 4z + 5)\,dz$

By extending the sum or difference property given above to more than two terms,

$$\begin{aligned}\int (3z^2 - 4z + 5)\,dz &= 3\int z^2\,dz - 4\int z\,dz + 5\int dz\\ &= 3\left(\frac{1}{3}z^3\right) - 4\left(\frac{1}{2}z^2\right) + 5z + C\\ &= z^3 - 2z^2 + 5z + C.\end{aligned}$$

Only one constant C is needed in the answer: the three constants from the term by term antiderivatives are combined. ◀ 3

The nice thing about working with antiderivatives is that you can always check your work by taking the derivative of the result. For instance, in Example 4(c) check that $z^3 - 2z^2 + 5z + C$ is the required antiderivative by taking the derivative:

$$\frac{d}{dz}(z^3 - 2z^2 + 5z + C) = 3z^2 - 4z + 5.$$

The result is the original function to be integrated, so the work checks.

EXAMPLE 5 Find each of the following.

(a) $\int \frac{x^2 + 1}{\sqrt{x}}\,dx$

First rewrite the integrand as follows.

$$\begin{aligned}\int \frac{x^2+1}{\sqrt{x}}\,dx &= \int \left(\frac{x^2}{\sqrt{x}} + \frac{1}{\sqrt{x}}\right)dx\\ &= \int \left(\frac{x^2}{x^{1/2}} + \frac{1}{x^{1/2}}\right)dx\\ &= \int (x^{3/2} + x^{-1/2})\,dx \qquad \text{Quotient rule for exponents}\end{aligned}$$

Now find the antiderivative.

$$\begin{aligned}\int (x^{3/2} + x^{-1/2})\,dx &= \frac{x^{5/2}}{5/2} + \frac{x^{1/2}}{1/2} + C\\ &= \frac{2}{5}x^{5/2} + 2x^{1/2} + C\end{aligned}$$

(b) $\int (x^2 - 1)^2\,dx$

Square the binomial first and then find the antiderivative.

$$\begin{aligned}\int (x^2 - 1)^2\,dx &= \int (x^4 - 2x^2 + 1)\,dx\\ &= \frac{x^5}{5} - \frac{2x^3}{3} + x + C\end{aligned}$$ ◀ 4

4

Find each of the following.

(a) $\int \frac{\sqrt{x} + 1}{x^2}\,dx$

(b) $\int (\sqrt{x} + 2)^2\,dx$

Answers:

(a) $-\frac{2}{\sqrt{x}} - \frac{1}{x} + C$

(b) $\frac{x^2}{2} + \frac{8}{3}x^{3/2} + 4x + C$

As shown in Chapter 11, the derivative of $f(x) = e^x$ is $f'(x) = e^x$. Also, the derivative of $f(x) = e^{kx}$ is $f'(x) = k \cdot e^{kx}$. These results lead to the following formulas for antiderivatives of exponential functions.

Antiderivatives of Exponential Functions

If k is a real number, $k \neq 0$, then

$$\int e^x\,dx = e^x + C;$$

$$\int e^{kx}\,dx = \frac{1}{k} \cdot e^{kx} + C.$$

5

Find each of the following.

(a) $\int (-4e^x)\,dx$

(b) $\int e^{3x}\,dx$

(c) $\int (e^{2x} - 2e^x)\,dx$

(d) $\int (-11e^{-x})\,dx$

Answers:

(a) $-4e^x + C$

(b) $\frac{1}{3}e^{3x} + C$

(c) $\frac{1}{2}e^{2x} - 2e^x + C$

(d) $11e^{-x} + C$

EXAMPLE 6 Here are some antiderivatives of exponential functions.

(a) $\int 9e^x\,dx = 9\int e^x\,dx = 9e^x + C$

(b) $\int e^{9t}\,dt = \frac{1}{9}e^{9t} + C$

(c) $\int 3e^{(5/4)u}\,du = 3\left(\frac{1}{5/4}e^{(5/4)u}\right) + C = 3\left(\frac{4}{5}\right)e^{(5/4)u} + C$

$$= \frac{12}{5}e^{(5/4)u} + C$$ 5

The antiderivative formula $\int x^n\,dx = \frac{x^{n+1}}{n+1} + C$ does not hold when $n = -1$ because the denominator is 0 in that case. Nevertheless, the function $g(x) = x^{-1}$ does have an antiderivative. Recall that the derivative of $f(x) = \ln|x|$ is $f'(x) = 1/x = x^{-1}$.

Antiderivative of x^{-1}

$$\int x^{-1}\,dx = \int \frac{1}{x}\,dx = \ln|x| + C, \qquad \text{where } x \neq 0.$$

6

Find each of the following.

(a) $\int (-9/x)\,dx$

(b) $\int (8e^{4x} - 3x^{-1})\,dx$

Answers:

(a) $-9 \cdot \ln|x| + C$

(b) $2e^{4x} - 3 \cdot \ln|x| + C$

CAUTION The domain of the logarithmic function is the set of positive real numbers. However, $y = x^{-1} = 1/x$ has as domain the set of all nonzero real numbers, so the absolute value of x *must* be used in the antiderivative.

EXAMPLE 7 Here are some antiderivatives of logarithmic functions.

(a) $\int \frac{4}{x}\,dx = 4\int \frac{1}{x}\,dx = 4 \cdot \ln|x| + C$

(b) $\int \left(-\frac{5}{x} + e^{-2x}\right)dx = -5 \cdot \ln|x| - \frac{1}{2}e^{-2x} + C$ 6

Applications

In the preceding examples, a family of antiderivative functions was found. In many applications, however, the given information allows us to determine the value of the constant C. The next examples illustrate this idea.

EXAMPLE 8 According to data from the Cellular Telecommunications Industry Association, the rate of increase in the number of cell phone subscribers (in millions) since 1990 is approximated by

$$S'(x) = 1.7x + 1.32,$$

where $x = 0$ corresponds to 1990. There were 86 million subscribers in 1999. Find a function $S(x)$ that gives the number of subscribers (in millions) in year x.

Since $S'(x)$ gives the rate of change of subscribers, the number of subscribers is given by

$$\begin{aligned} S(x) &= \int (1.7x + 1.32)\,dx \\ &= 1.7\frac{x^2}{2} + 1.32x + C \\ &= .85x^2 + 1.32x + C. \end{aligned}$$

To find the value of C, use the fact that in 1999 ($x = 9$) there were 86 million subscribers, which says that $S(9) = 86$. This gives an equation that can be solved for C.

$$\begin{aligned} S(9) &= .85(9^2) + 1.32(9) + C \\ 86 &= .85(9^2) + 1.32(9) + C \\ 86 &= 80.73 + C \\ C &= 5.27 \end{aligned}$$

Thus, the number of subscribers in year x is

$$S(x) = .85x^2 + 1.32x + 5.27.$$

EXAMPLE 9 Suppose the marginal revenue from selling x units of a product is given by $40/e^{.05x} + 10$.

(a) Find the revenue function.

The marginal revenue is the derivative of the revenue function.

$$\begin{aligned} \frac{dR}{dx} &= \frac{40}{e^{.05x}} + 10 \\ R &= \int \left(\frac{40}{e^{.05x}} + 10\right) dx = \int (40e^{-.05x} + 10)\,dx \\ &= 40\left(\frac{-1}{.05}\right)e^{-.05x} + 10x + k = -800e^{-.05x} + 10x + k, \end{aligned}$$

where k is a constant. If $x = 0$, then $R = 0$ (no items sold means no revenue), and

$$\begin{aligned} 0 &= -800e^0 + 10 \cdot 0 + k \\ 800 &= k. \end{aligned}$$

The marginal cost at a level of production of x items is

$$C'(x) = 2x^3 + 6x - 5.$$

The fixed cost is \$800. Find the cost function $C(x)$.

Answer:

$$C(x) = \frac{1}{2}x^4 + 3x^2 - 5x + 800$$

Thus,

$$R = -800e^{-.05x} + 10x + 800$$

gives the revenue function.

(b) Find the demand function for this product.

Recall that $R = xp$, where p is the demand function. Hence,

$$-800e^{-.05x} + 10x + 800 = xp$$

$$\frac{-800e^{-.05x} + 10x + 800}{x} = p.$$

The demand function is $p = \dfrac{-800e^{-.05x} + 10x + 800}{x}$. ◀ 7

EXAMPLE 10 Mexico's population was 68.7 million in 1980. Its growth rate (in millions per year) is approximated by

$$f(t) = 1.1679\, e^{.017t} \qquad (0 \le t \ge 30),$$

where $t = 0$ corresponds to 1980.*

(a) Find the rule of the population function $F(t)$ that gives the population (in millions) in year t.

The derivative of the population function $F(t)$ is the rate at which the population is growing, that is, $F'(t) = 1.1679e^{.017t}$. Therefore,

$$F(t) = \int 1.1679e^{.017t}\, dt = 1.1679 \cdot \frac{1}{.017}e^{.017t} + C = 68.7e^{.017t} + C.$$

Since the population was 68.7 million in 1980 (that is, when $t = 0$), we have

$$68.7 = F(0) = 68.7e^{.017(0)} + C = 68.7e^0 + C = 68.7 + C,$$

so that $C = 0$. Therefore, the population function is $F(t) = 68.7e^{.017t}$.

(b) What is the population in the year 2003?

Since 2003 corresponds to $t = 23$, the population is

$$F(23) = 68.7e^{.017(23)} = 68.7e^{.391} \approx 101.57 \text{ million.}$$ ◀

*Based on data and projections from the U.S. Census Bureau.

13.1 Exercises

1. What must be true of $F(x)$ and $G(x)$ if both are antiderivatives of $f(x)$?

2. How is the antiderivative of a function related to the function?

3. In your own words, describe what is meant by an integrand.

4. Explain why the restriction $n \neq -1$ is necessary in the rule $\int x^n\, dx = \frac{1}{n+1}x^{n+1} + C$.

Find each of the following. (See Examples 3–7.)

5. $\int 10x\,dx$

6. $\int 25r\,dr$

7. $\int 8p^2\,dp$

8. $\int 5t^3\,dt$

9. $\int 100\,dx$

10. $\int 35\,dt$

11. $\int (5z - 1)\,dz$

12. $\int (2m + 3)\,dm$

13. $\int (z^2 - 4z + 2)\,dz$

14. $\int (2y^2 + 4y + 7)\,dy$

15. $\int (x^3 - 14x^2 + 20x + 3)\,dx$

16. $\int (x^3 + 5x^2 - 10x - 4)\,dx$

17. $\int 6\sqrt{y}\,dy$

18. $\int 8z^{1/2}\,dz$

19. $\int (6t\sqrt{t} + 3\sqrt{t})\,dt$

20. $\int (12\sqrt{x} - x\sqrt{x})\,dx$

21. $\int (56t^{1/2} + 18t^{7/2})\,dt$

22. $\int (10u^{3/2} - 14u^{5/2})\,du$

23. $\int \frac{24}{x^3}\,dx$

24. $\int \frac{-20}{x^2}\,dx$

25. $\int \left(\frac{1}{y^2} - \frac{2}{\sqrt{y}}\right)dy$

26. $\int \left(\frac{3 + 2u}{\sqrt{u}}\right)du$

27. $\int (6x^{-3} + 4x^{-1})\,dx$

28. $\int (3x^{-1} - 10x^{-2})\,dx$

29. $\int 4e^{3u}\,du$

30. $\int -e^{-4x}\,dx$

31. $\int 3e^{-.2x}\,dx$

32. $\int -4e^{.2v}\,dv$

33. $\int \left(\frac{3}{x} + 4e^{-.5x}\right)dx$

34. $\int \left(\frac{9}{x} - 3e^{-.4x}\right)dx$

35. $\int \frac{1 + 2t^3}{t}\,dt$

36. $\int \frac{2y^{1/2} - 3y^2}{y}\,dy$

37. $\int \left(e^{2u} + \frac{u}{4}\right)du$

38. $\int \left(\frac{2}{v} - e^{3v}\right)dv$

39. $\int (x + 1)^2\,dx$

40. $\int (2y - 1)^2\,dy$

41. $\int \frac{\sqrt{x} + 1}{\sqrt[3]{x}}\,dx$

42. $\int \frac{1 - 2\sqrt[3]{z}}{\sqrt[3]{z}}\,dz$

43. The slope of the tangent line to a curve is given by

$$f'(x) = 6x^2 - 4x - 3.$$

If the point (0, 1) is on the curve, find the equation of the curve.

44. Find the equation of the curve whose tangent line has a slope of

$$f'(x) = x^{2/3},$$

if the point (1, 3/5) is on the curve.

Work the following problems. (See Example 8.)

45. **Social Science** Imports (in billions of dollars) from Canada to the United States have changed at a rate given by $g(x) = 4.8x^2 - 16.4x + 20.6$, where x is the number of years since 1995.* The United States imported $144 billion worth of goods in 1995.
 (a) Find a function giving the imports in year x.
 (b) What was the value of imports from Canada in 1999?

46. **Social Science** There were 17.5 million golfers in 1985. The rate at which this number was changing between 1985 and 1998 is approximated by

$$f(x) = -.01976x^3 + .5112x^2 - 3.92x + 8.2 \qquad (0 \le x \le 13),$$

where $x = 0$ corresponds to 1985.†

*U.S. Census Bureau.

†National Golf Foundation.

(a) Find a function that gives the number of golfers (in millions) in year x.
(b) Estimate the number of golfers in 1998.

Work the following problems. (See Example 9.)

47. Business Revenue from the nationwide sales of recreational vehicles in 1980 was \$1952 million. The approximate marginal revenue function for sales in year x is

$$r(x) = -1.72x^3 + 42x^2 - 262x + 917 \qquad (0 \le x \le 17),$$

where $x = 0$ corresponds to 1980.*
(a) Find the revenue function.
(b) What was the revenue in 1996?

48. Business The marginal revenue from a product is given by $r(x) = 50 - 3x - x^2$. Find the demand function for the product.

49. Business The marginal profit from the sale of x hundred items of a product is $P'(x) = 4 - 6x + 3x^2$, and the "profit" when no items are sold is $-\$40$. Find the profit function.

Business *Find the cost function for each of the following marginal cost functions.*

50. $C'(x) = x^{1/2}$; 16 units cost \$60.

51. $C'(x) = x^{2/3} + 2$; 8 units cost \$58.

52. $C'(x) = x^2 - 2x + 3$; 3 units cost \$15.

53. $C'(x) = .2x^2 + .4x + .2$; 6 units cost \$29.60.

54. $C'(x) = .0015x^3 + .033x^2 + .044x + .25$; 10 units cost \$25.

*Recreational Vehicle Industry Association.

55. $C'(x) = -\dfrac{40}{e^{.05x}} + 100$; 5 units cost \$1200.

56. $C'(x) = 1.2e^{.02x}$; 2 units cost \$95

Work the following problems. (See Example 10.)

57. Social Science The world population was 4.52 billion in 1980. The approximate growth rate of world population is given by $f(t) = .064184\, e^{.0142t}$, where $t = 0$ corresponds to 1980.*
(a) Find a function that gives the population of the world (in billions) in year t.
(b) Estimate the world population in 2005.

58. Natural Science If the rate of excretion of a biochemical compound is given by

$$f'(t) = .01e^{-.01t},$$

the total amount excreted by time t (in minutes) is $f(t)$.
(a) Find an expression for $f(t)$.
(b) If 0 units are excreted at time $t = 0$, how many units are excreted in 10 minutes?

59. Social Science There were about 2.08 million U.S. military personnel on active duty in 1990. This number has been changing at a rate that is approximated by

$$g(x) = -\frac{.3}{x} \qquad (1 \le x \le 8),$$

where $x = 1$ corresponds to 1990.†
(a) Find a function that gives the approximate number of military personnel on active duty in year x.
(b) How many were on active duty in 1997?

*U.S. Census Bureau.

†U.S Department of Defense.

13.2 INTEGRATION BY SUBSTITUTION

In Section 13.1 we saw how to find the antiderivatives of a few simple functions. The antiderivatives of more complicated functions can sometimes be found by a process called *integration by substitution.* The technique depends on the following concept. If $u = f(x)$, then the **differential** of u, written du, is defined as

$$du = f'(x)\, dx.$$

For example, if $u = 6x^4$, then $du = 24x^3\, dx$. (1)

Find du for the following.

(a) $u = 9x$

(b) $u = 5x^3 + 2x^2$

(c) $u = e^{-2x}$

Answers:

(a) $du = 9\, dx$

(b) $du = (15x^2 + 4x)\, dx$

(c) $du = -2e^{-2x}\, dx$

Differentials have many useful interpretations that are studied in more advanced courses. We shall use them only as a convenient notational device when finding an antiderivative such as

$$\int (3x^2 + 4)^4\, 6x\, dx.$$

The function $(3x^2 + 4)^4 6x$ is reminiscent of the chain rule and so we shall try to use differentials and the chain rule in *reverse* to find the antiderivative. Let $u = 3x^2 + 4$; then $du = 6x\,dx$. Now substitute u for $3x^2 + 4$ and du for $6x\,dx$ in the indefinite integral above.

$$\int (3x^2 + 4)^4 6x\,dx = \int \overbrace{(3x^2 + 4)}^{u}{}^4\overbrace{(6x\,dx)}^{du}$$
$$= \int u^4\,du$$

This last integral can now be found by the power rule.

$$\int u^4\,du = \frac{u^5}{5} + C$$

Finally, substitute $3x^2 + 4$ for u and $6x\,dx$ for du.

$$\int (3x^2 + 4)^4 6x\,dx = \frac{u^5}{5} + C = \frac{(3x^2 + 4)^5}{5} + C$$

We can check the accuracy of this result by using the chain rule to take the derivative.

$$\frac{d}{dx}\left[\frac{(3x^2 + 4)^5}{5} + C\right] = \frac{1}{5} \cdot 5(3x^2 + 4)^4(6x) - 0$$
$$= (3x^2 + 4)^4 6x,$$

which is the original function.

This method of integration is called **integration by substitution.** As shown above, it is simply the chain rule for derivatives in reverse. The results can always be verified by differentiation.

2

Find the following.

(a) $\int 8x(4x^2 - 1)^5\,dx$

(b) $\int (3x - 8)^5\,dx$

(c) $\int 18x^2(x^3 - 5)^{3/2}\,dx$

Answers:

(a) $\frac{(4x^2 - 1)^6}{6} + C$

(b) $\frac{(3x - 8)^6}{18} + C$

(c) $\frac{12(x^3 - 5)^{5/2}}{5} + C$

EXAMPLE 1 Find $\int (4x + 5)^9\,dx$.

We choose $4x + 5$ as u. Then $du = 4\,dx$. We are missing the constant 4. We can rewrite the integral by using the fact that $4(1/4) = 1$, as follows.

$$\int (4x + 5)^9\,dx = \frac{1}{4} \cdot 4 \int (4x + 5)^9\,dx$$
$$= \frac{1}{4}\int (4x + 5)^9(4\,dx) \qquad k\int f(x)\,dx = \int kf(x)\,dx$$
$$= \frac{1}{4}\int u^9\,du \qquad \text{Substitute.}$$
$$= \frac{1}{4} \cdot \frac{u^{10}}{10} + C = \frac{u^{10}}{40} + C \qquad \text{Find antiderivative}$$
$$= \frac{(4x + 5)^{10}}{40} + C \qquad \text{Substitute.}$$

2

CAUTION When changing the x-problem to the u-problem, make sure that the change is complete, that is, that no x's are left in the u-problem.

EXAMPLE 2 Find $\int x^2\sqrt{x^3+1}\,dx$.

Rewrite the function as $x^2(x^3+1)^{1/2}$. An expression raised to a power is usually a good choice for u, so we let $u = x^3 + 1$; then $du = 3x^2\,dx$. The integrand does not contain the constant 3, which is part of du. One way to take care of this is to solve the differential $du = 3x^2\,dx$ for $x^2\,dx$.

$$du = 3x^2\,dx$$

$$\frac{1}{3}du = x^2\,dx$$

Substitute (1/3) du for $x^2\,dx$.

$$\int x^2\sqrt{x^3+1}\,dx = \int \sqrt{x^3+1}(x^2\,dx) = \int \sqrt{u}\cdot\frac{1}{3}\,du$$

Now use the constant multiple rule to bring the 1/3 outside the integral sign.

$$\int x^2\sqrt{x^3+1}\,dx = \int \sqrt{u}\cdot\frac{1}{3}\,du = \frac{1}{3}\int u^{1/2}\,du$$

$$= \frac{1}{3}\cdot\frac{u^{3/2}}{3/2} + C = \frac{2}{9}u^{3/2} + C$$

Since $u = x^3 + 1$,

$$\int x^2\sqrt{x^3+1}\,dx = \frac{2}{9}(x^3+1)^{3/2} + C.$$ ◀ 3

3

Find the following.

(a) $\int x(5x^2+6)^2\,dx$

(b) $\int x\sqrt{x^2+16}\,dx$

Answers:

(a) $\frac{1}{30}(5x^2+6)^3 + C$

(b) $\frac{1}{3}(x^2+16)^{3/2} + C$

CAUTION The substitution method given in the examples above *will not always work*. For example, we might try to find

$$\int x^3\sqrt{x^3+1}\,dx$$

by substituting $u = x^3 + 1$, so that $du = 3x^2\,dx$. However, there is no *constant* that can be inserted inside the integral sign to give $3x^2$. This integral, and a great many others, cannot be evaluated by substitution.

With practice, choosing u will become easy if you keep two principles in mind. First, u should equal some expression in the integral that, when replaced with u, tends to make the integral simpler. Second and more important, u must be an expression whose derivative is also present in the integral. The substitution should include as much of the integral as possible, so long as its derivative is still present. In Example 2, we could have chosen $u = x^3$, but $u = x^3 + 1$ is better, because it has the same derivative as x^3 and captures more of the original integral. If we carry this reasoning further, we might try $u = \sqrt{x^3+1} = (x^3+1)^{1/2}$, but this is a poor choice because $du = (1/2)(x^3+1)^{-1/2}(3x^2)\,dx$, an expression not present in the original integral.

Find the following.

(a) $\int z(z^2 + 1)^2\, dz$

(b) $\int \frac{x^2 + 3}{\sqrt{x^3 + 9x}}\, dx$

Answers:

(a) $\frac{(z^2 + 1)^3}{6} + C$

(b) $\frac{2}{3}\sqrt{x^3 + 9x} + C$

EXAMPLE 3 Find $\int \frac{x + 3}{(x^2 + 6x)^2}\, dx$.

Let $u = x^2 + 6x$, so that $du = (2x + 6)\, dx = 2(x + 3)\, dx$. The integral is missing the 2, so multiply by 2/2, putting 2 inside the integral sign and 1/2 outside.

$$\begin{aligned}\int \frac{x + 3}{(x^2 + 6x)^2}\, dx &= \frac{1}{2}\int \frac{2(x + 3)}{(x^2 + 6x)^2}\, dx \\ &= \frac{1}{2}\int \frac{du}{u^2} = \frac{1}{2}\int u^{-2}\, du \\ &= \frac{1}{2} \cdot \frac{u^{-1}}{-1} + C = \frac{-1}{2u} + C\end{aligned}$$

Substituting $x^2 + 6x$ for u gives

$$\int \frac{x + 3}{(x^2 + 6x)^2}\, dx = \frac{-1}{2(x^2 + 6x)} + C.$$ 4

Recall that if $f(x)$ is a function, then by the chain rule, the derivative of the exponential function $y = e^{f(x)}$ is

$$\frac{dy}{dx} = e^{f(x)} \cdot f'(x).$$

Both the function $f(x)$ and its derivative appear on the right. This suggests that the antiderivative of a function of the form $e^{f(x)}$ can be found by letting u be the exponent.

EXAMPLE 4 Find $\int 2e^{x^2} x\, dx$.

Let $u = x^2$, so that $du = 2x\, dx$. Then

$$\begin{aligned}\int 2e^{x^2} x\, dx &= \int e^{x^2}\, 2x\, dx \\ &= \int e^u\, du \\ &= e^u + C = e^{x^2} + C.\end{aligned}$$

Check this answer by using the chain rule to take the derivative

$$\frac{d}{dx}\left(e^{x^2} + C\right) = e^{x^2}(2x) + 0 = 2e^{x^2} x.$$

The techniques used in the preceding examples can be summarized as follows.

Substitution Method

Let $u(x)$ be some function of x.

Form of the Integral	*Form of the Antiderivative*
1. $\int [u(x)]^n \cdot u'(x)\,dx,\ n \neq -1$	$\frac{[u(x)]^{n+1}}{n+1} + C$
2. $\int e^{u(x)} \cdot u'(x)\,dx$	$e^{u(x)} + C$
3. $\int \frac{u'(x)\,dx}{u(x)}$	$\ln\lvert u(x)\rvert + C$

EXAMPLE 5 Find the following.

(a) $\int e^{-11x}\,dx$

Choose $u = -11x$, so $du = -11\,dx$. Multiply the integral by $(-1/11)(-11)$, and use the rule for $\int e^u\,du$.

$$\begin{aligned}\int e^{-11x}\,dx &= -\frac{1}{11} \cdot -11 \int e^{-11x}\,dx\\ &= -\frac{1}{11}\int e^{-11x}(-11\,dx)\\ &= -\frac{1}{11}\int e^u\,du\\ &= -\frac{1}{11}e^u + C\\ &= -\frac{1}{11}e^{-11x} + C\end{aligned}$$

(b) $\int x^2 \cdot e^{x^3}\,dx$

Let $u = x^3$, the exponent on e. Then $du = 3x^2\,dx$, and $(1/3)\,du = x^2\,dx$,

$$\begin{aligned}\int x^2 \cdot e^{x^3}\,dx &= \int e^{x^3}(x^2\,dx)\\ &= \int e^u\left(\frac{1}{3}\,du\right) && \text{Substitute.}\\ &= \frac{1}{3}\int e^u\,du && \text{Constant multiple rule}\\ &= \frac{1}{3}e^u + C && \text{Integrate.}\\ &= \frac{1}{3}e^{x^3} + C. && \text{Substitute.}\end{aligned}$$

◀ 5

5

Find the following.

(a) $\int e^{5x}\,dx$

(b) $\int 8xe^{3x^2}\,dx$

(c) $\int 2x^3e^{x^4-1}\,dx$

Answers:

(a) $\frac{1}{5}e^{5x} + C$

(b) $\frac{4}{3}e^{3x^2} + C$

(c) $\frac{1}{2}e^{x^4-1} + C$

Recall that the antiderivative of $f(x) = 1/x$ is $\ln|x|$. The next example uses $\int x^{-1}\,dx = \ln|x| + C$, and the method of substitution.

EXAMPLE 6 Find the following.

(a) $\int \frac{dx}{9x+6}$

Choose $u = 9x + 6$, so $du = 9\,dx$. Multiply by (1/9)(9).

$$\int \frac{dx}{9x+6} = \frac{1}{9}\cdot 9\int \frac{dx}{9x+6} = \frac{1}{9}\int \frac{1}{9x+6}(9\,dx)$$

$$= \frac{1}{9}\int \frac{1}{u}\,du = \frac{1}{9}\ln|u| + C = \frac{1}{9}\ln|9x+6| + C$$

(b) $\int \frac{(2x-3)\,dx}{x^2-3x}$

Let $u = x^2 - 3x$, so that $du = (2x - 3)\,dx$. Then

$$\int \frac{(2x-3)\,dx}{x^2-3x} = \int \frac{du}{u} = \ln|u| + C = \ln|x^2 - 3x| + C.$$ 6

6

Find the following.

(a) $\int \frac{4\,dx}{x-3}$

(b) $\int \frac{(3x^2+8)\,dx}{x^3+8x+5}$

Answers:

(a) $4\ln|x-3| + C$

(b) $\ln|x^3+8x+5| + C$

A slightly more complicated substitution technique is needed in some cases.

EXAMPLE 7 Find $\int x\sqrt{1-x}\,dx$.

Let $u = 1 - x$. Then $x = 1 - u$ and $dx = -du$. Now substitute:

$$\int x\sqrt{1-x}\,dx = \int (1-u)\sqrt{u}\,(-du) = \int (u-1)u^{1/2}\,du$$

$$= \int (u^{3/2} - u^{1/2})\,du = \frac{2}{5}u^{5/2} - \frac{2}{3}u^{3/2} + C$$

$$= \frac{2}{5}(1-x)^{5/2} - \frac{2}{3}(1-x)^{3/2} + C.$$ 7

7

Find

$$\int x(x+1)^{2/3}\,dx.$$

Answer:

$\frac{3}{8}(x+1)^{8/3} - \frac{3}{5}(x+1)^{5/3} + C$

Applications

Integration by substitution enables you to handle a wider variety of real-life applications.

EXAMPLE 8 During the late 1990s the approximate marginal revenue function (in billions of dollars per year) for the computer and office equipment industry was

$$R'(x) = \frac{-13{,}333(2x-29)}{(x^2-29x+271)^2} \qquad (5 \le x \le 10),$$

where $x = 5$ corresponds to 1995.* The actual revenue in 1995 was \$91.343 billion. Find the revenue function and estimate the revenue in 2000.

*U.S. Census Bureau, Current Industrial Reports.

The revenue function is

$$R(x) = \int R'(x)\,dx = \int \frac{-13{,}333(2x - 29)}{(x^2 - 29x + 271)^2}\,dx.$$

Let $u = x^2 - 29x + 271$. Then $du = (2x - 29)\,dx$ and

$$\begin{aligned} R(x) &= -13{,}333 \int (x^2 - 29x + 271)^{-2}(2x - 29)\,dx \\ &= -13{,}333 \int u^{-2}\,du \\ &= -13{,}333 \frac{u^{-2+1}}{-2+1} + C \\ &= \frac{13{,}333}{u} + C \\ R(x) &= \frac{13{,}333}{x^2 - 29x + 271} + C. \end{aligned}$$

Find the value of C by using the fact that revenue in 1995 was \$91.343 billion, that is, $R(5) = 91.343$:

$$\begin{aligned} 91.343 &= \frac{13{,}333}{5^2 - 29(5) + 271} + C \\ 91.343 &= 88.298 + C \\ C &= 3.045. \end{aligned}$$

8 Sales of a new company, in thousands, are changing at a rate of

$$S'(t) = 27e^{-3t},$$

where t is time in months. Ten units were sold when $t = 0$. Find the sales function.

Answer:

$S(t) = 19 - 9e^{-3t}$

Therefore, the revenue function is

$$R(x) = \frac{13{,}333}{x^2 - 29x + 271} + 3.045.$$

The revenue in 2000 ($x = 10$) is

$$R(10) = \frac{13{,}333}{10^2 - 29(10) + 271} + 3.045 \approx \$167.650 \text{ billion.}$$ 8

EXAMPLE 9 To determine the top 100 popular songs of each year since 1956, Jim Quirin and Barry Cohen developed a function that represents the rate of change on the charts of *Billboard* magazine required for a song to earn a "star" on the *Billboard* "Hot 100" survey.* They developed the function

$$f(x) = \frac{A}{B + x},$$

where $f(x)$ represents the rate of change in position on the charts, x is the position on the "Hot 100" survey, and A and B are appropriate constants. The function

$$F(x) = \int f(x)\,dx$$

is defined as the "Popularity Index." Find $F(x)$.

*Formula for the "Popularity Index" from *Chartmasters' Rock 100*, Fourth Edition, by Jim Quirin and Barry Cohen. Copyright © 1987 by Chartmasters. Reprinted by permission.

Integrating $f(x)$ gives

$$F(x) = \int f(x)\,dx$$
$$= \int \frac{A}{B + x}\,dx$$
$$= A\int \frac{1}{B + x}\,dx. \qquad \text{Constant multiple rule}$$

Let $u = B + x$, so that $du = dx$. Then

$$F(x) = A\int \frac{1}{u}\,du = A \ln u + C$$
$$= A \ln(B + x) + C.$$

(Absolute value is not necessary, since $B + x$ is always positive here.)

13.2 Exercises

1. Integration by substitution is related to what differentiation method? What type of integrand suggests using integration by substitution?

2. For each of the following integrals, decide what factor should be u. Then find du.

   (a) $\int (3x^2 - 5)^4\, 2x\,dx$ (b) $\int \sqrt{1 - x}\,dx$

   (c) $\int \frac{x^2}{2x^3 + 1}\,dx$ (d) $\int (8x - 8)(4x^2 - 8x)\,dx$

Use substitution to find the following indefinite integrals. (See Examples 1–6 for Exercises 3–36 and Example 7 for Exercises 37–40.)

3. $\int 3(12x - 1)^2\,dx$

4. $\int 5(4 - 2t)^3\,dt$

5. $\int \frac{2}{(3t + 1)^2}\,dt$

6. $\int \frac{4}{\sqrt{5u - 1}}\,du$

7. $\int \frac{x + 1}{(x^2 + 2x - 4)^{3/2}}\,dx$

8. $\int \frac{3x^2 - 2}{(2x^3 - 4x)^{5/2}}\,dx$

9. $\int r^2\sqrt{r^3 + 3}\,dr$

10. $\int y^3\sqrt{y^4 - 6}\,dy$

11. $\int (-3e^{5k})\,dk$

12. $\int (-2e^{-3z})\,dz$

13. $\int 4w^2e^{2w^3}\,dw$

14. $\int 5ze^{-z^2}\,dz$

15. $\int (2 - t)e^{4t - t^2}\,dt$

16. $\int (3 - x^2)e^{9x - x^3}\,dx$

17. $\int \frac{e^{\sqrt{y}}}{\sqrt{y}}\,dy$

18. $\int \frac{e^{1/z^2}}{z^3}\,dz$

19. $\int \frac{-4}{2 + 5x}\,dx$

20. $\int \frac{7}{3 - 4x}\,dx$

21. $\int \frac{e^{2t}}{e^{2t} + 1}\,dt$

22. $\int \frac{e^{w+1}}{2 - e^{w+1}}\,dw$

23. $\int \frac{x + 2}{(2x^2 + 8x)^3}\,dx$

24. $\int \frac{4y - 2}{(y^2 - y)^4}\,dy$

25. $\int \left(\frac{1}{r} + r\right)\left(1 - \frac{1}{r^2}\right)dr$

26. $\int \left(\frac{2}{a} - a\right)\left(\frac{-2}{a^2} - 1\right)da$

27. $\int \frac{x^2 + 1}{(x^3 + 3x)^{2/3}}\,dx$

28. $\int \frac{B^3 - 1}{(2B^4 - 8B)^{3/2}}\,dB$

29. $\int \frac{x + 2}{3x^2 + 12x + 8}\,dx$

30. $\int \frac{x^2}{x^3 + 3}\,dx$

31. $\int 2x(x^2 + 1)^3\,dx$

32. $\int y^2(y^3 - 4)^3\,dy$

33. $\int \left(\sqrt{x^2 + 12x}\right)(x + 6)\,dx$

34. $\int \left(\sqrt{x^2 - 6x}\right)(x - 3)\,dx$

35. $\int \frac{(1 + \ln x)^2}{x} dx$

36. $\int \frac{1}{x(\ln x)} dx$

37. $\int \frac{u}{\sqrt{u - 1}} du$

38. $\int \frac{2x}{(x + 5)^6} dx$

39. $\int t\sqrt{5t - 1}\, dt$

40. $\int 4r\sqrt{8 - r}\, dr$

Work these problems. (See Examples 8 and 9.)

41. Business A company has found that the marginal cost of a new production line (in thousands of dollars) is

$$C(x) = \frac{60x}{5x^2 + 1},$$

where x is the number of years the line is in use.

(a) If the fixed costs are \$10,000, find the total cost function for the production line.

(b) The company will add a new production line if the total cost of this one for the first five years stays below \$40,000. Should they add the new line?

42. Business The rate of expenditure for maintenance of a particular machine is given by

$$M'(x) = \sqrt{x^2 + 12x}\,(2x + 12),$$

where x is time measured in years. Total maintenance costs through the fourth year are \$612.

(a) Find the total maintenance function.

(b) How many years must pass before the total maintenance costs reach \$2000?

43. Transportation In 1980, there were 22.5 deaths per 100,000 population from motor vehicle accidents. In the past two decades, the rate at which this number has been changing is given by

$$g(x) = \frac{-2.41}{x + 1},$$

where $x = 0$ corresponds to 1980.*

(a) Find a function that gives the number of deaths per 100,000 population in year x.

(b) Estimate the number of deaths in 1998.

44. Business In 1992, sales of pharmaceutical preparations by U.S. drug companies were \$50.5 billion. The rate at which these sales have been changing since 1991 is given by

$$f(x) = \frac{3x^2 - 1}{x^2},$$

where $x = 1$ corresponds to 1991.*

(a) Find a function that gives the sales of pharmaceutical preparations (in billions of dollars) in year x.

(b) Estimate the sales in 2000.

45. Business Revenues from tourism in Spain were \$18.6 billion in 1990. Since then, revenues have been growing at a rate given by

$$g(x) = 1.11e^{.06x},$$

where $x = 0$ corresponds to 1990.†

(a) Find a function that gives Spanish tourism revenues in year x.

(b) Estimate the tourism revenues in 2001.

46. Business Total revenues from all local phone companies in the United States were \$73 billion in 1985. The growth rate of revenues is approximated by

$$f(x) = 2.16e^{.03x} \qquad (0 \le x \le 13),$$

where $x = 0$ corresponds to 1985.‡

(a) Find a function that gives the phone companies revenues in year x.

(b) Did revenues exceed \$110 billion by 1998?

*National Highway Traffic Safety Administration.

*U.S. Department of Commerce.

†World Tourism Organization, Madrid, Spain.

‡U.S. Federal Communications Commission.

13.3 AREA AND THE DEFINITE INTEGRAL

Suppose a car travels along a straight road at a constant speed of 50 mph. The speed of the car at time t is given by the constant function $v(t) = 50$ whose graph is a horizontal straight line, as shown in Figure 13.2(a).

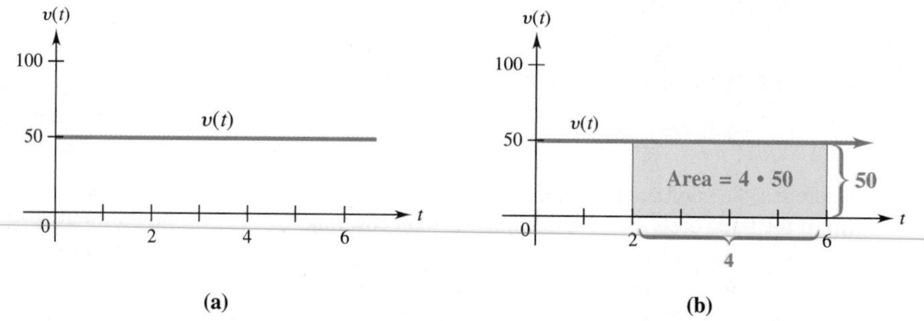

FIGURE 13.2

How far does the car travel from time $t = 2$ to $t = 6$? Since this is a 4-hour period, the answer, of course, is $4 \cdot 50 = 200$ miles. Note that 200 is precisely the *area* under the graph of the speed function $v(t)$ and above the t-axis from $t = 2$ to $t = 6$, as shown in Figure 13.2(b) above.

As we saw in Chapter 12, the speed function $v(t)$ is the rate of change of distance with respect to time, that is, the rate of change of the distance function $s(t)$ (which gives the distance traveled by the car at time t). Now the distance traveled from time $t = 2$ to $t = 6$ is the amount that the distance has changed from $t = 2$ to $t = 6$. In other words, the *total change* in distance from $t = 2$ to $t = 6$ is the area under the graph of the speed (rate of change) function from $t = 2$ to $t = 6$.

A more complicated argument (that is omitted here) shows that a similar situation holds in the general case.

Total Change in $F(x)$

Let f be a function such that f is continuous on the interval $[a, b]$ and $f(x) \geq 0$ for all x in $[a, b]$. If $f(x)$ is the rate of change of a function $F(x)$, then the **total change in $F(x)$** as x goes from a to b is the area between the graph of $f(x)$ and the x-axis from $x = a$ to $x = b$.

EXAMPLE 1 Figure 13.3 shows the graph of the function that gives the rate of change of the annual maintenance charges for a certain machine. The rate function is increasing because maintenance tends to cost more as the machine gets older. Estimate the total maintenance charges over the 10-year life of the machine.

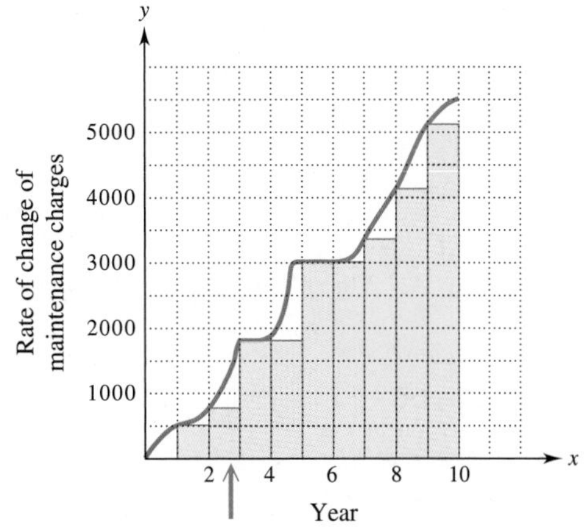

FIGURE 13.3

This is the situation described in the preceding box, with $F(x)$ being the maintenance cost function and $f(x)$, whose graph is given, the rate of change function. The total maintenance charges are the total change in $F(x)$ from $x = 0$ to $x = 10$, that is, the area between the graph of the rate function and the x-axis from $x = 0$ to $x = 10$. We can approximate this area by using the shaded rectangles in Figure 13.3. For instance, the rectangle marked with an arrow has base 1 (from year 2 to year 3) and height 750 (the rate of change at $x = 2$), so its area is $1 \times 750 = 750$. Similarly, each of the other rectangles has base 1 and height determined by the rate of change at the beginning of the year. Consequently, we estimate the area to be the sum

$$1 \cdot 0 + 1 \cdot 500 + 1 \cdot 750 + 1 \cdot 1800 + 1 \cdot 1800 + 1 \cdot 3000 + 1 \cdot 3000 + 1 \cdot 3400 + 1 \cdot 4200 + 1 \cdot 5200 = 23{,}650.$$

Hence, the total maintenance charges over the 10 years are at least \$23,650 (the unshaded areas under the rate graph have not been accounted for in this estimate). 

Use Figure 13.3 to estimate the maintenance charge during

(a) the first 6 years of the machine's life;

(b) the first 8 years.

Answers:

(a) \$7850

(b) \$14,250

Area

The preceding examples show that the area between a graph and the x-axis has useful interpretations. In this section and the next, we develop a means of measuring such areas precisely when the function is given by an algebraic formula. The underlying idea is the same as in Example 1: Use rectangles to approximate the area under the graph.

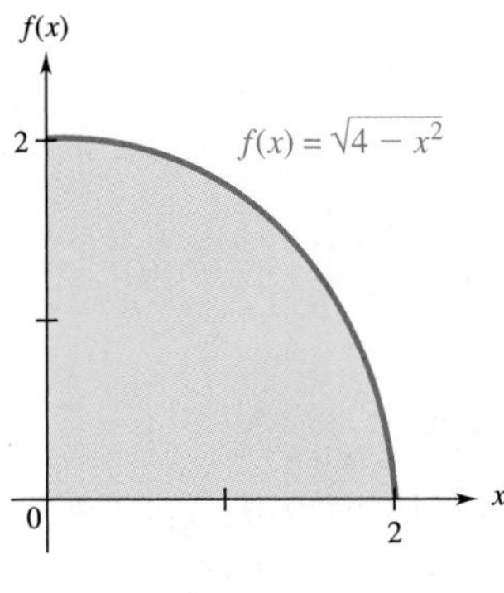

FIGURE 13.4

EXAMPLE 2 Find the area under the graph of $f(x) = \sqrt{4 - x^2}$ from $x = 0$ to $x = 2$, shown in Figure 13.4.

A very rough approximation of the area of this region can be found by using two rectangles, as in Figure 13.5 on the next page. The height of the rectangle on the left is $f(0) = 2$ and the height of the rectangle on the right is $f(1) = \sqrt{3}$. The width of each rectangle is 1, making the total area of the two rectangles

$$1 \cdot f(0) + 1 \cdot f(1) = 2 + \sqrt{3} \approx 3.7321 \text{ square units.}$$

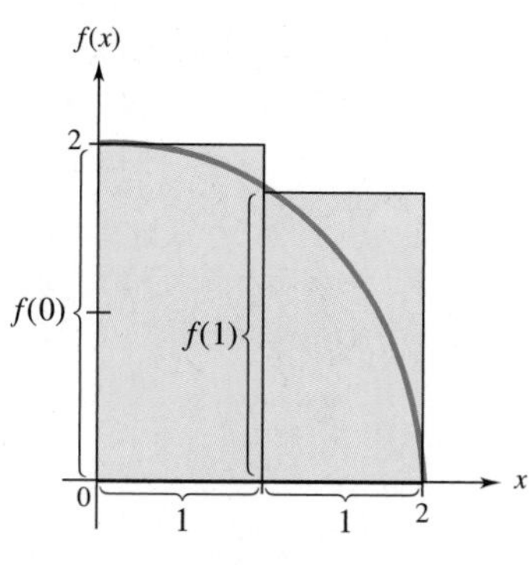

FIGURE 13.5

2

Calculate the sum

$$\frac{1}{4}\cdot f(0)+\frac{1}{4}\cdot f\left(\frac{1}{4}\right)+\cdots+\frac{1}{4}\cdot f\left(\frac{7}{4}\right),$$

using the following information.

x	$f(x)$
0	2
1/4	1.98431
1/2	1.93649
3/4	1.85405
1	1.73205
5/4	1.56125
3/2	1.32288
7/4	.96825

Answer:

3.33982 square units

As Figure 13.5 suggests, this approximation is greater than the actual area. To improve the accuracy of the approximation, we could divide the interval from $x = 0$ to $x = 2$ into four equal parts, each of width 1/2, as shown in Figure 13.6. As before, the height of each rectangle is given by the value of f at the left-hand side of the rectangle, and its area is the width, 1/2, multiplied by the height. The total area of the four rectangles is

$$\frac{1}{2}\cdot f(0)+\frac{1}{2}\cdot f\left(\frac{1}{2}\right)+\frac{1}{2}\cdot f(1)+\frac{1}{2}\cdot f\left(\frac{3}{2}\right)$$
$$=\frac{1}{2}(2)+\frac{1}{2}\left(\frac{\sqrt{15}}{2}\right)+\frac{1}{2}(\sqrt{3})+\frac{1}{2}\left(\frac{\sqrt{7}}{2}\right)$$
$$=1+\frac{\sqrt{15}}{4}+\frac{\sqrt{3}}{2}+\frac{\sqrt{7}}{4}\approx 3.4957 \text{ square units.}$$

This approximation looks better, but it is still greater than the actual area desired. To improve the approximation, divide the interval from $x = 0$ to $x = 2$ into eight parts with equal widths of 1/4. (See Figure 13.7.) The total area of all these rectangles is

$$\frac{1}{4}\cdot f(0)+\frac{1}{4}\cdot f\left(\frac{1}{4}\right)+\frac{1}{4}\cdot f\left(\frac{1}{2}\right)+\frac{1}{4}\cdot f\left(\frac{3}{4}\right)+\frac{1}{4}\cdot f(1)+\frac{1}{4}\cdot f\left(\frac{5}{4}\right)$$
$$+\frac{1}{4}\cdot f\left(\frac{3}{2}\right)+\frac{1}{4}\cdot f\left(\frac{7}{4}\right).$$ 2

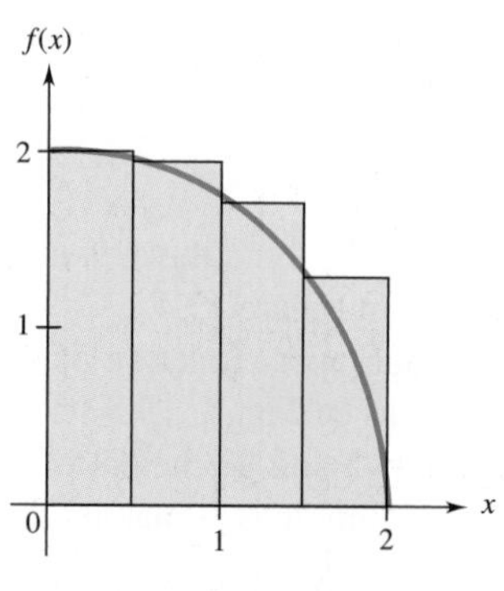

FIGURE 13.6

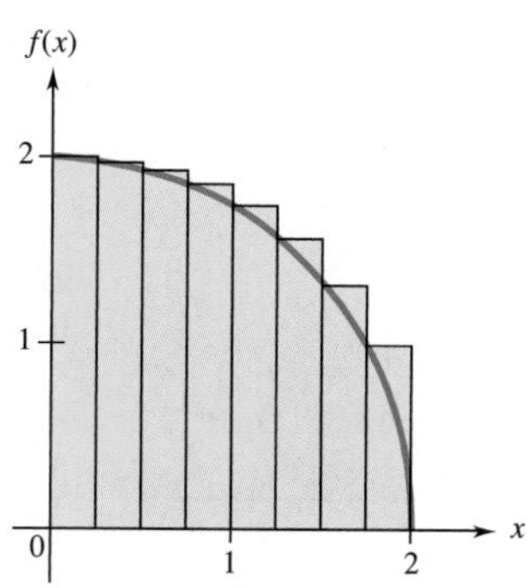

FIGURE 13.7

This process of approximating the area under the curve by using more and more rectangles to get a better and better approximation can be extended. To do this, divide the interval from $x = 0$ to $x = 2$ into n equal parts. Each of these n intervals has width

$$\frac{2-0}{n} = \frac{2}{n},$$

n	*Area*
125	3.15675
2000	3.14257
8000	3.14184
32,000	3.14165
128,000	3.14160
512,000	3.14159

so each rectangle has width $2/n$ and height determined by the value of the function $f(x) = \sqrt{4 - x^2}$ at the left side of the rectangle. A computer was used to find approximations of the area for several values of n given in the table at the side.

As the number n of rectangles gets larger and larger, the sum of their areas gets closer and closer to the actual area of the region. In other words, the actual area is the *limit* of these sums as n gets larger and larger without bound, which can be written

$$\text{area} = \lim_{n\to\infty} (\text{sum of areas of } n \text{ rectangles}).$$

The table suggests that this limit is a number whose decimal expansion begins 3.14159 . . . , which is the same as the beginning of the decimal approximation of π. Therefore, it seems plausible that

$$\text{area} = \lim_{n\to\infty} (\text{sum of areas of } n \text{ rectangles}) = \pi.$$

It can be shown that the region whose area was found in Example 2 is one-fourth of the interior of a circle of radius 2 with center at the origin (see Figure 13.4). Hence, its area is

$$\frac{1}{4}(\pi r^2) = \frac{1}{4}(\pi \cdot 2^2) = \pi,$$

which agrees with our answer in Example 2.

The method used in Example 2 can be generalized to find the area bounded by the curve $y = f(x)$, the x-axis, and the vertical lines $x = a$ and $x = b$, as shown in Figure 13.8. To approximate this area, we could divide the region under the curve first into ten rectangles (Figure 13.8(a)) and then into twenty rectangles (Figure 13.8(b)). In each case, the sum of the areas of the rectangles gives an approximation of the area under the curve.

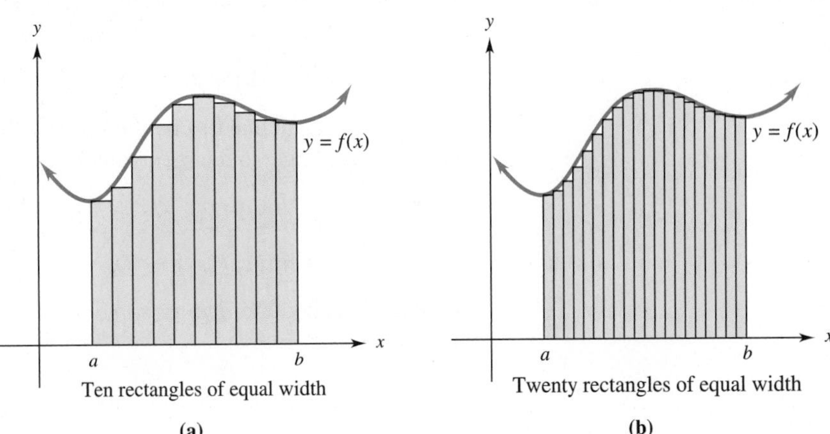

FIGURE 13.8

We can get better and better approximations by increasing the number n of rectangles. Here is a description of the general procedure. Let n be a positive integer. Divide the interval from a to b into n pieces of equal length. The symbol Δx is traditionally used to denote the length of each piece. Since the length of the entire interval is $b - a$, each of the n pieces has length

$$\Delta x = \frac{b-a}{n}.$$

Use each of these pieces as the base of a rectangle, as shown in Figure 13.9, where the endpoints of the n intervals are labeled $x_1, x_2, x_3, \ldots, x_{n+1}$. A typical rectangle, the one whose lower left corner is at x_i, has red shading. The base of this rectangle is Δx and its height is the height of the graph over x_i, namely, $f(x_i)$, so

$$\text{Area of } i\text{th rectangle} = f(x_i) \cdot \Delta x.$$

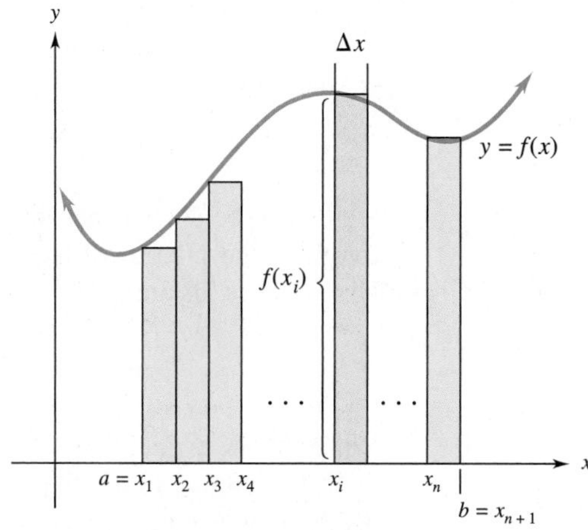

FIGURE 13.9

The total area under the curve is approximated by the sum of the areas of all n of the rectangles, namely,

$$f(x_1) \cdot \Delta x + f(x_2) \cdot \Delta x + f(x_3) \cdot \Delta x + \cdots + f(x_n) \cdot \Delta x.$$

The exact area is defined to be the limit of this sum (if it exists) as the number of rectangles gets larger and larger, without bound. Hence,

$$\begin{aligned}\text{Exact area} &= \lim_{n\to\infty} (f(x_1) \cdot \Delta x + f(x_2) \cdot \Delta x + f(x_3) \cdot \Delta x + \cdots + f(x_n) \cdot \Delta x) \\ &= \lim_{n\to\infty} ([f(x_1) + f(x_2) + f(x_3) + \cdots + f(x_n)] \cdot \Delta x).\end{aligned}$$

This limit is called the *definite integral* of $f(x)$ from a to b and is denoted by the symbol

$$\int_a^b f(x)\, dx.$$

The preceding discussion can be summarized as follows.

The Definite Integral

If f is a continuous function on the interval $[a, b]$, then the **definite integral** of f from a to b is the number

$$\int_a^b f(x)\,dx = \lim_{n\to\infty} [(f(x_1) + f(x_2) + f(x_3) + \cdots + f(x_n))\,\Delta x],$$

where $\Delta x = (b - a)/n$ and x_i is the left-hand endpoint of the ith interval.

For instance, the area of the region in Example 2 could be written as the definite integral

$$\int_0^2 \sqrt{4 - x^2}\,dx = \pi.$$

Although the definition in the box is obviously motivated by the problem of finding areas, it is applicable to many other situations, as we shall see in later sections. In particular, *the definition of the definite integral is valid even when $f(x)$ takes negative values* (that is, when the graph goes below the x-axis). In that case, however, the resulting number is not the area between the graph and the x-axis. In this section we shall only deal with the area interpretation of the definite integral.

The elongated "S" in the notation for the definite integral stands for the word "Sum," which plays a crucial role in the definition. This notation looks very similar to that used for antiderivatives (also called *indefinite* integrals) in earlier sections. The connection between the definite integral and antiderivatives, which is the reason for the similar terminology and notation, will be explained in the next section.

CAUTION Children learning to read sometimes confuse b and d. Both consist of a half-circle and a vertical line segment, but the location of the line segment makes all the difference. Similarly, the symbols $\int f(x)\,dx$ and $\int_a^b f(x)\,dx$ have totally different meanings—the a and b make all the difference. The indefinite integral $\int f(x)\,dx$ denotes a set of *functions* (the antiderivatives of $f(x)$), whereas the definite integral $\int_a^b f(x)\,dx$ represents a *number* (that can be interpreted as the area under the graph when $f(x) \geq 0$).

EXAMPLE 3 Approximate $\int_1^4 3x\,dx$, the area under the graph of $f(x) = 3x$, above the x-axis, and between $x = 1$ and $x = 4$, by using six rectangles of equal width, whose heights are the values of the function at the left endpoint of each rectangle.

3

Divide the region of Figure 13.10 into 12 rectangles of equal width whose heights are the values of the function at the left endpoint of each rectangle.

(a) Complete this table.

i	x_i	$f(x_i)$
1	1	
2	1.25	
3	1.5	
4	1.75	
5	2	
6	2.25	
7		
8		
9		
10		
11		
12		

(b) Use the results from the table to approximate $\int_1^4 3x\,dx$.

Answers:

(a)

i	x_i	$f(x_i)$
1	1	3
2	1.25	3.75
3	1.5	4.5
4	1.75	5.25
5	2	6
6	2.25	6.75
7	2.5	7.5
8	2.75	8.25
9	3	9
10	3.25	9.75
11	3.5	10.5
12	3.75	11.25

(b) 21.375

We want to find the area of the shaded region in Figure 13.10.

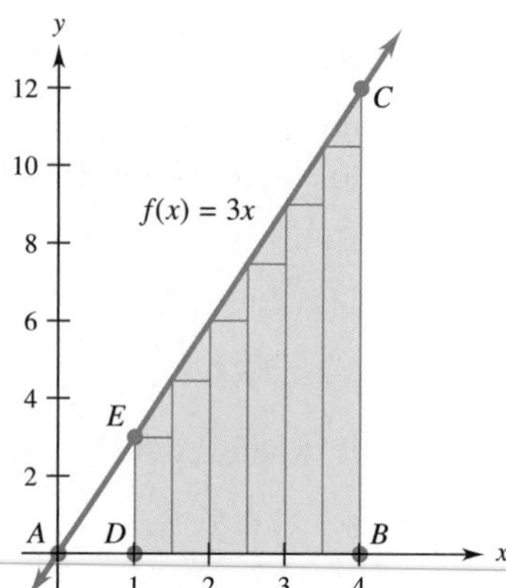

FIGURE 13.10

The heights of the six rectangles given by $f(x_i)$ are as follows.

i	x_i	$f(x_i)$
1	$x_1 = 1$	$f(1) = 3$
2	$x_2 = 1.5$	$f(1.5) = 4.5$
3	$x_3 = 2$	$f(2) = 6$
4	$x_4 = 2.5$	$f(2.5) = 7.5$
5	$x_5 = 3$	$f(3) = 9$
6	$x_6 = 3.5$	$f(3.5) = 10.5$

The width of each rectangle is $\Delta x = \dfrac{4-1}{6} = \dfrac{1}{2} = .5$. The sum of the areas of the six rectangles is

$$\begin{aligned}&f(x_1)\Delta x + f(x_2)\Delta x + f(x_3)\Delta x + f(x_4)\Delta x + f(x_5)\Delta x + f(x_6)\Delta x\\&= f(1)\Delta x + f(1.5)\Delta x + f(2)\Delta x + f(2.5)\Delta x + f(3)\Delta x + f(3.5)\Delta x\\&= (3)(.5) + (4.5)(.5) + (6)(.5) + (7.5)(.5) + (9)(.5) + (10.5)(.5)\\&= 20.25.\end{aligned}$$

We can check the accuracy of this approximation by noting that the area of the shaded region in Figure 13.10 is the difference of the areas of triangle ABC and triangle ADE. Triangle ABC has base 4 and height 12 and triangle ADE has base 1 and height 3. Using the formula for the area of a triangle, $A = (1/2)bh$, we see that the area of the shaded region is

$$\text{Area } ABC - \text{Area } ADE = (1/2)(4)(12) - (1/2)(1)(3) = 24 - 1.5 = 22.5.$$

So our approximation is a bit low. Using more rectangles will produce a better approximation of the area. ◀ 3

Technology and Definite Integrals

The preceding examples illustrate the difficulty of evaluating definite integrals by direct computation. In the next section an algebraic evaluation technique that works in many cases will be developed. In all cases, technology can be used to evaluate definite integrals. Computer programs and calculators that have symbolic capabilities, such as Maple, Mathematica, the TI-89, or the Casio FX.2, will generally produce exact results. Other graphing calculators amd spreadsheets usually provide highly accurate approximations.

TECHNOLOGY TIP There are two ways to approximate a definite integral with a graphing calculator, each of which is illustrated here for $\int_{-2}^{1} (x^3 - 5x + 6)\, dx$.

1. Approximate the limit in the definition by computing the sum for a large value of n, say, $n = 100$. Then $\Delta x = (b - a)/n = (1 - (-2))/100 = 3/100 = .03$ and the sum is

$$[f(x_1) + f(x_2) + f(x_3) + \cdots + f(x_n)] \cdot \Delta x$$
$$= [f(x_1) + f(x_2) + f(x_3) + \cdots + f(x_{100})](.03).$$

   To compute this sum automatically on most TI calculators, look in the submenus of the LIST menu and enter

   sum(seq($x^3 - 5x + 6$, x, -2, $1 - .03$, $.03$)) × (.03).

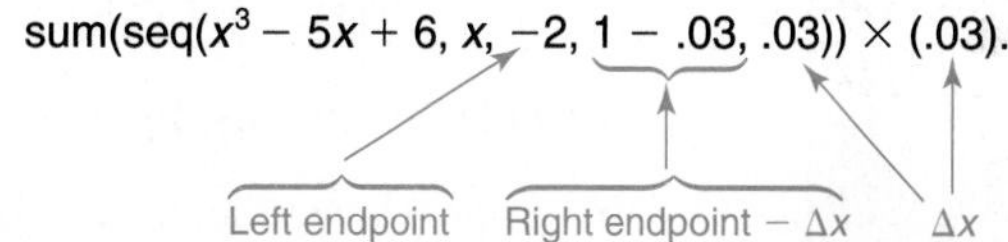

   On HP-39, look in the LIST submenu of the MATH menu and enter

   ΣLIST(MAKELIST($x^3 - 5x + 6$, x, -2, $1 - .03$, $.03$)) × (.03).

   On other calculators, use the RECTANGLE program in Appendix A. In each case, the calculator produces the approximation 21.8393.

2. Use **numerical integration**. This feature approximates the definite integral by using more complicated summing techniques than used above (which typically involve trapezoids or regions with parabolic boundaries). On most TI calculators, look in the MATH or CALC menu and enter

   fnInt($x^3 - 5x + 6$, x, -2, 1).

   On Casio, use the $\int dx$ key on the keyboard and enter

$$\int (x^3 - 5x + 6, x, -2, 1).$$

   On HP-39, look for $\int$ in the CALCULUS submenu of the MATH menu and enter

$$\int (-2, 1, x^3 - 5x + 6, x).$$

   In each case, the calculator approximates the integral as 21.75, which actually is its exact value, as we shall see in the next section.

EXAMPLE 4 Figure 13.11(a) on the next page shows the graph of

$$f(x) = x^3 - 8x^2 + 18x + 5.$$

The area under the graph from $x = 1$ to $x = 4.5$, which is shown in Figure 13.11(b), is given by

$$\int_1^{4.5} (x^3 - 8x^2 + 18x + 5)\, dx.$$

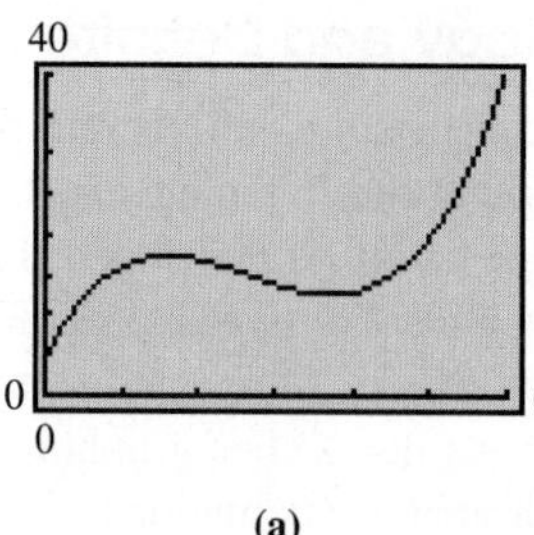

(a)

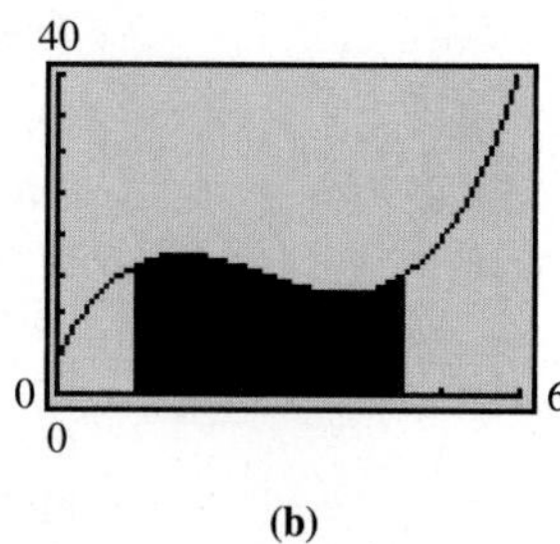

(b)

FIGURE 13.11

If $f(x) = x^3 - 8x^2 + 18x + 5$, use the first part of the Technology Tip to compute the sum

$$[f(1) + f(1.035) + f(1.07) + \cdots + f(4.465)](.035).$$

Answer:

52.6977

(a) Use a calculator to approximate this integral by using the definition, with 100 rectangles.

For $n = 100$ rectangles, $\Delta x = (b - a)/n = (4.5 - 1)/100 = .035$. The left-hand endpoints of the bases of the rectangles are 1, 1.035, 1.07, 1.105, 1.14, . . . , 4.43, and 4.465 (the right-hand endpoint of the base of the last rectangle is 4.5). We must compute this sum:

$$[f(1) + f(1.035) + f(1.07) + \cdots + f(4.43) + f(4.465)](.035).$$

Compute this sum by doing problem 4 in the margin. (4)

FIGURE 13.12

(b) Approximate the integral by using numerical integration on a calculator.

The numerical integration program on a typical calculator (see part 2 of the Technology Tip above) produces the answer in Figure 13.12. It differs slightly from the answer in part (a) and is accurate to seven decimal places, as the techniques in the next section will show. ◀

At the beginning of this section we saw that if $f(x)$ (with $f(x) \geq 0$) is the rate of change of a function $F(x)$, then the total change in $F(x)$ as x goes from a to b is the area between the graph of $f(x)$ and the x-axis from $x = a$ to $x = b$. In other words.

$$\text{Total Change in } F(x) = \int_a^b f(x)\,dx.$$

FIGURE 13.13

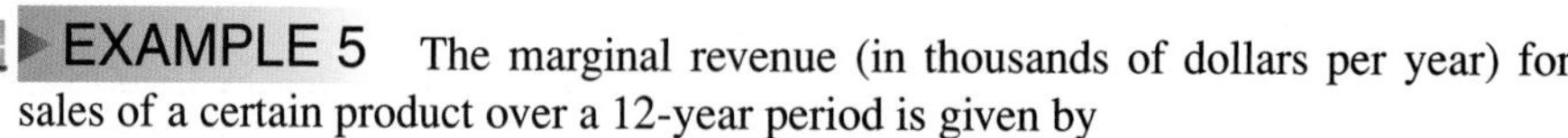

EXAMPLE 5 The marginal revenue (in thousands of dollars per year) for sales of a certain product over a 12-year period is given by

$$MR(x) = .12x^3 - 2.5x^2 + 15x \qquad (0 \leq x \leq 12).$$

Find the total sales revenue for the 12-year period.

Total revenue is the change in revenue from $x = 0$ to $x = 12$. Since the marginal revenue is the derivative of the revenue function,

$$\text{Total Revenue} = \int_0^{12} (.12x^3 - 2.5x^2 + 15x)\,dx.$$

Numerical integration (Figure 13.13) shows that the total revenue is 262.08 thousand dollars, that is, \$262,080. (This is the exact value of the integral, as we shall see in the next section.) ◀ (5)

If the marginal revenue (in hundreds of dollars per year) is given by

$$MR(x) = .3x^3 - .5x^2 + x,$$

find the total revenue for the first six years (that is, from $x = 0$ to $x = 6$).

Answer:

Approximately \$7920

13.3 Exercises

In the following exercises, estimate the required areas by using rectangles, whose height is given by the value of the function at the left side of each rectangle. (See Example 1.)

1. Business The graph shows the rate of use of electrical energy (in kilowatt hours) in a certain city on a very hot day. Estimate the total usage of electricity on that day. Let the width of each rectangle be 2 hours.

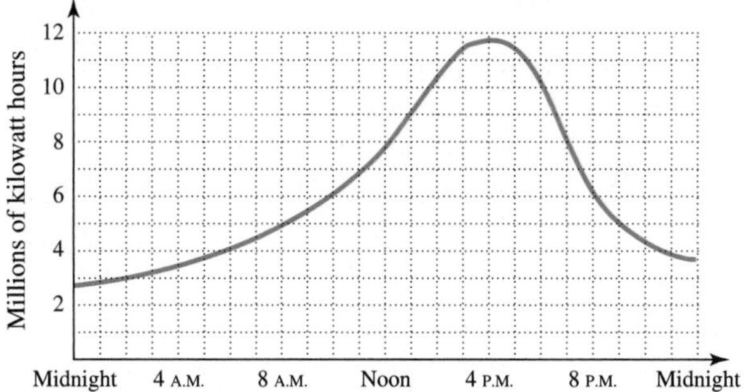

2. Business The graphs shows U.S. oil production and consumption rates in millions of barrels per day. The rates for the years beyond 1991 are projected using the policy in place in 1990 (labeled "Current policy base" in the graph) and using a new policy proposed by the Bush Administration in 1991 (labeled "With strategy" in the graph).* Estimate the amount of oil produced between 1990 and 2010 using the policy in place in 1990. Use rectangles with widths of 5 years. (*Hint:* There are 365 days in a year.)

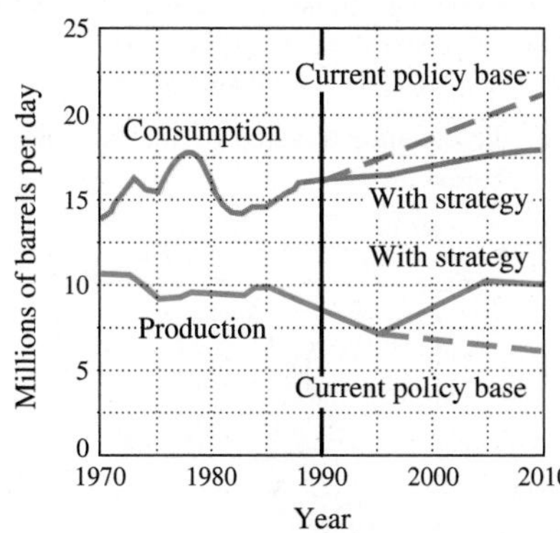

3. Health The graph at the top of the next column shows the approximate concentration of alcohol in a person's bloodstream t hours after drinking 2 ounces of alcohol. Estimate the total amount of alcohol in the bloodstream by estimating the area under the curve. Use rectangles with widths of 1 hour.

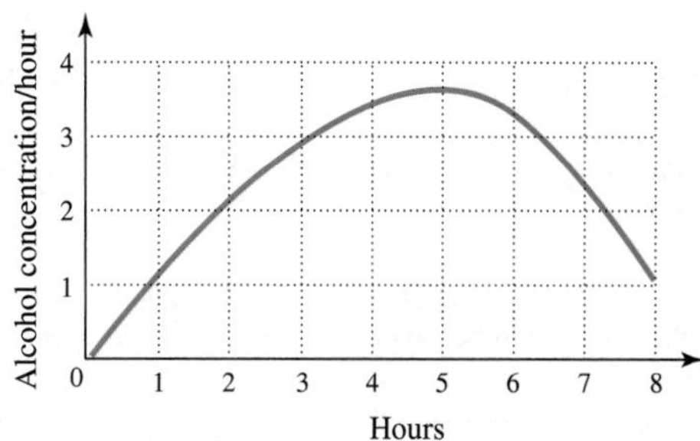

4. Natural Science The graph below shows the rate of inhalation of oxygen by a person riding a bicycle very rapidly for 10 minutes. Estimate the total volume of oxygen inhaled in the first 20 minutes after the beginning of the ride. Use rectangles of width 1 minute.

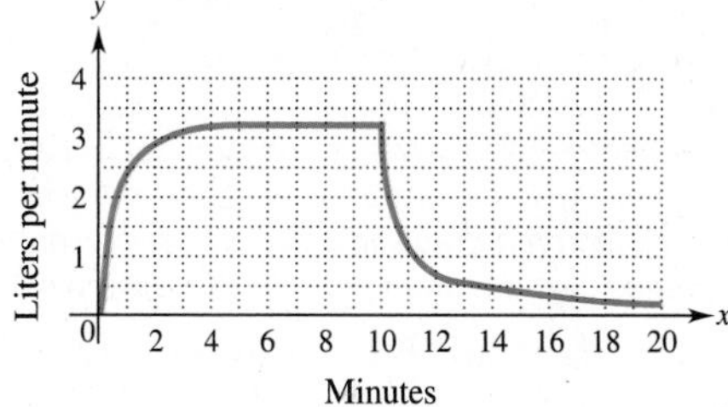

5. Explain the difference between an indefinite integral and a definite integral.

6. Complete the following statement.

$$\int_0^3 (x^2 + 2)\, dx = \lim_{n \to \infty} \_\_\_\_\_, \text{ where } \Delta x = \_\_\_\_\_.$$

Approximate the area under each given curve and above the x-axis on the given interval, using rectangles whose height is given by the value of the function at the left side of the rectangle.

(a) Use two rectangles. (See Examples 2 and 3.)
(b) Use four rectangles.
(c) Use a graphing calculator (or other technology) and 40 rectangles. (See Example 4.)

7. $f(x) = 2x + 5$; $[0, 4]$

8. $f(x) = 4 - x$; $[0, 4]$

9. $f(x) = 4 - x^2$; $[-2, 2]$

10. $f(x) = x^2 + 1$; $[-2, 2]$

11. $f(x) = e^x - 1$; $[0, 4]$

12. $f(x) = \frac{2}{x}$; $[1, 9]$

*Graph, "Uncle Sam's Energy Strategy" from National Energy Strategy, February 1991. United States Department of Energy, Washington, D.C.

Work the following exercises. (See Examples 2 and 3.)

13. Consider the region below $f(x) = x/2$, above the x-axis, between $x = 0$ and $x = 4$. Let x_i be the left endpoint of the ith subinterval.
(a) Approximate the area of the region using four rectangles.
(b) Approximate the area of the region using eight rectangles.
(c) Find $\int_0^4 f(x)\,dx$ by using the formula for the area of a triangle.

14. Find $\int_0^5 (5 - x)\,dx$ by using the formula for the area of a triangle.

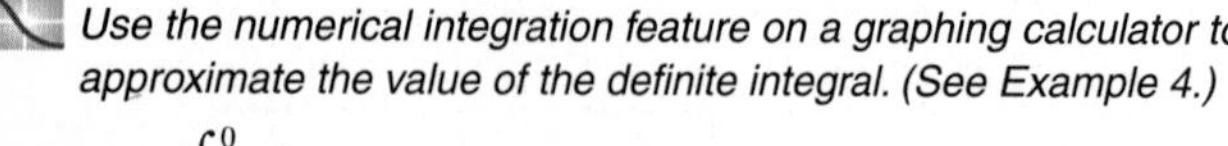
Use the numerical integration feature on a graphing calculator to approximate the value of the definite integral. (See Example 4.)

15. $\displaystyle\int_{-5}^{0} (x^3 + 2x^2 - 15x + 2)\,dx$

16. $\displaystyle\int_{-1}^{3} (-x^4 + 3x^3 - x^2 + 9)\,dx$

17. $\displaystyle\int_{2}^{7} 5\ln(x^2 + 1)\,dx$ **18.** $\displaystyle\int_{1}^{5} x \ln x\,dx$

19. $\displaystyle\int_{-1}^{3} x^2 e^{-x}\,dx$ **20.** $\displaystyle\int_{1}^{5} \frac{\ln x}{x}\,dx$

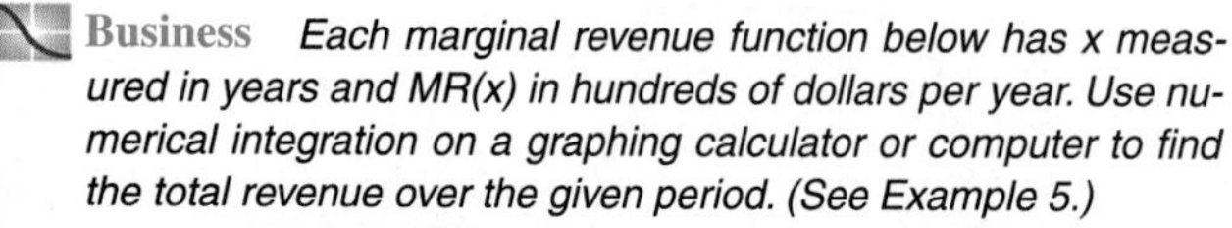
Business *Each marginal revenue function below has x measured in years and MR(x) in hundreds of dollars per year. Use numerical integration on a graphing calculator or computer to find the total revenue over the given period. (See Example 5.)*

21. $MR(x) = .04x^3 - .6x^2 + 3x \quad (0 \le x \le 12)$

22. $MR(x) = .04x^4 - .5x^3 + 2x^2 - x \quad (0 \le x \le 6)$

23. $MR(x) = .26x^4 - 6.25x^3 + 30.2x^2 + 87.5x \quad (0 \le x \le 5)$

24. $MR(x) = \dfrac{210}{1 + 5e^{-.8x}} - 35 \quad (0 \le x \le 8)$

Work the following exercises.

25. **Social Science** The hourly wage of an average U.S. production worker in year x is approximated by

$$g(x) = .36x + 10 \qquad (0 \le x \le 10),$$

where $x = 0$ corresponds to 1990.*
(a) Let f be the function that gives the *yearly* wage rate of this worker. Assume the person works 2000 hours per year. Find the rule of f.
(b) Graph the function f on the interval $0 \le x \le 10$.
(c) Estimate the total wages of the worker from January 1, 1990, through December 31, 1999, by using rectangles of width 2 years. Is this estimate greater or less than the actual total wages?
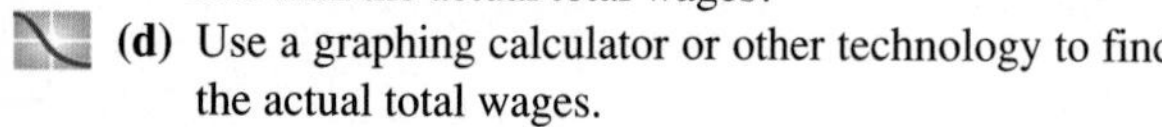
(d) Use a graphing calculator or other technology to find the actual total wages.

*Bureau of Labor Statistics.

26. **Social Science** The table shows the federal minimum hourly wage rate over an 11-year period.

Year	*Hourly Wage*
1990	$3.80
1991–95	$4.25
1996	$4.75
1997–2000	$5.15

(a) Let f be the function that gives the *yearly* wage rate of a minimum wage worker, with $x = 0$ corresponding to 1990. Assume the person works 2000 hours per year. Graph the function f on the interval $0 \le x \le 11$.
(b) Find the total earnings of the worker from January 1, 1990, through December 31, 2000.
(c) What was the worker's average yearly income during this period? For comparison purposes, the poverty level for a family of three was $10,419 in 1990 and $13,003 in 1998.

Physical Science *The graphs for Exercises 27 and 28 are from* Road & Track *magazine.* *The curve shows the velocity at time t, in seconds, when the car accelerates from a dead stop. To find the total distance traveled by the car in reaching 100 miles per hour, we must estimate the definite integral*

$$\int_0^T v(t)\,dt,$$

where T represents the number of seconds it takes for the car to reach 100 mph.

Use the graphs to estimate this distance by adding the areas of rectangles with widths of 5 seconds. The last rectangle has a width of 3. To adjust your answer to miles per hour, divide by 3600 (the number of seconds in an hour). You then have the number of miles that the car traveled in reaching 100 mph. Finally, multiply by 5280 feet per mile to convert the answers to feet.

27. Estimate the distance traveled by the Porsche 928, using the graph below.

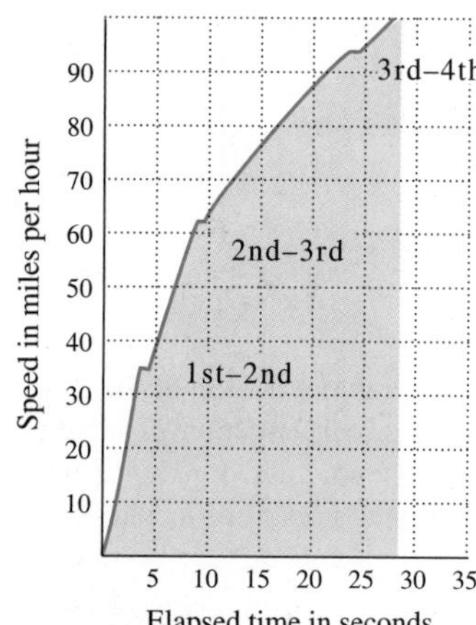

*From *Road & Track*, April and May, 1978. Reprinted with permission of *Road & Track*.

28. Estimate the distance traveled by the BMW 733i, using the graph below.

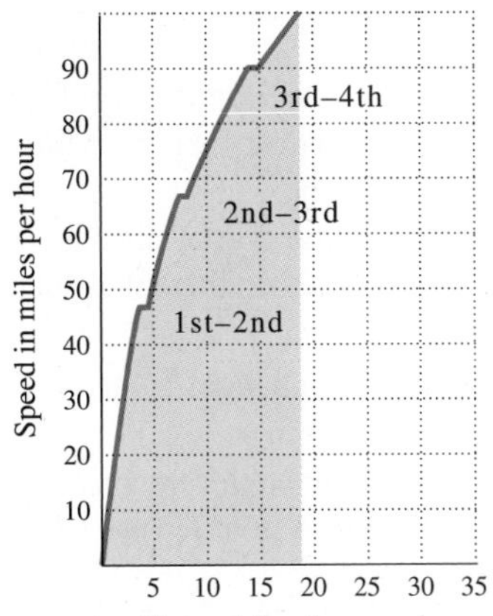

29. Two cars start from rest at a traffic light and accelerate for several minutes. The graph below shows their velocities (in feet per second) as a function of time (in second). Car A is the one that initially has greater velocity.*

(a) How far has car A traveled after 2 seconds? (*Hint:* Use formulas from geometry.)

(b) When is car A farthest ahead of car B?

(c) Estimate the farthest that car A gets ahead of car B. For car A, use formulas from geometry. For car B, use $n = 4$ and the value of the function at the midpoint of each interval.

(d) Give a rough estimate of when car B catches up with car A.

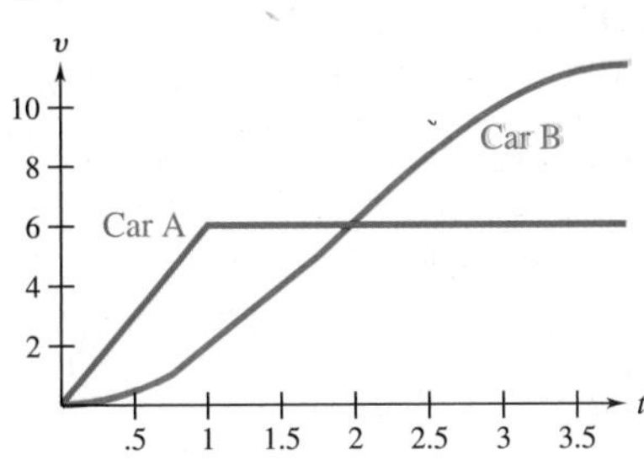

*Based on an example given by Steve Monk of the University of Washington.

30. The booklet *All About Lawns* published by Ortho Books gives the following instructions for measuring the area of an irregularly shaped region.

> **Irregular Shapes** (within 5% accuracy)
> Measure a long (L) axis of the area. Every 10 feet along the length line, measure the width at right angles to the length line. Total widths and multiply by 10.
>
> Area = $(A_1A_2 + B_1B_2 + C_1C_2$ etc.$) \times 10$
> A = $(40' + 60' + 32') \times 10$
> A = $132' \times 10'$
> A = 1320 square feet

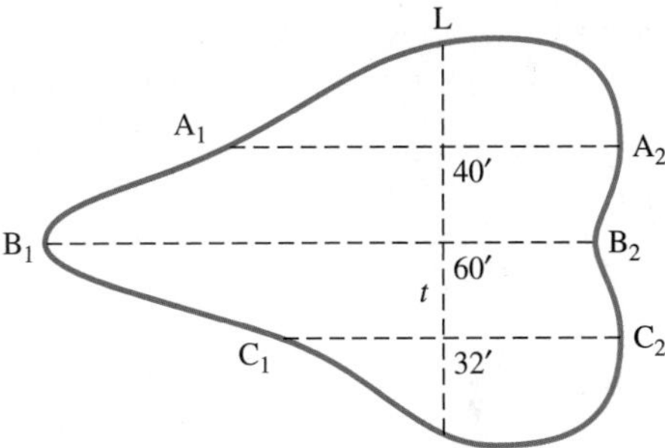

How does this method relate to the discussion in this section?

13.4 THE FUNDAMENTAL THEOREM OF CALCULUS

We now develop the connection between definite integrals and antiderivatives, which will explain the similar notation for these two concepts. More important, it provides a way to calculate definite integrals exactly.

In the last section we saw that when $f(x)$ is the rate of change of the function $F(x)$ and $f(x) \geq 0$ on $[a, b]$, then the definite integral has this interpretation:

$$\int_a^b f(x)\,dx = \text{Total Change in } F(x) \text{ as } x \text{ changes from } a \text{ to } b. \qquad (*)$$

The total change in $F(x)$ as x changes from a to b is the difference between the value of F at the end and the value of F at the beginning, that is, $F(b) - F(a)$. So statement $(*)$ can be stated as

$$\int_a^b f(x)\,dx = F(b) - F(a).$$

Now $f(x)$ is the rate of change of $F(x)$, which means that $f(x)$ is the derivative of $F(x)$, or equivalently, $F(x)$ is an antiderivative of $f(x)$. This relationship is an example of the following result.

Fundamental Theorem of Calculus

Suppose f is continuous on the interval $[a, b]$ and F is *any* antiderivative of f. Then

$$\int_a^b f(x)\,dx = F(b) - F(a).$$

The proof of the Fundamental Theorem is discussed at the end of this section. For now, we shall concentrate on applying it. Note these facts:

1. The Fundamental Theorem applies to every continuous function $f(x)$. It does not require $f(x) > 0$.
2. If the antiderivative $F(x)$ is replaced by $F(x) + C$ for any constant C, the conclusion of the Fundamental Theorem is the same because C is eliminated in the final answer.

$$\begin{aligned}\int_a^b f(x)\,dx &= (F(b) + C) - (F(a) + C)\\ &= F(b) - F(a).\end{aligned}$$

3. The variable used in the integrand does not matter. Each of the following integrals represents the number $F(b) - F(a)$.

$$\int_a^b f(x)\,dx = \int_a^b f(t)\,dt = \int_a^b f(u)\,du$$

4. The definition of $\int_a^b f(x)\,dx$ assumed that $a < b$, that is, that the lower limit of integration is the smaller number. When $a > b$, we make this definition:

$$\int_a^b f(x)\,dx = -\int_b^a f(x)\,dx.$$

For example, $\int_3^1 x^2\,dx = -\int_1^3 x^2\,dx$. Similarly, when the limits of integration are the same, we define $\int_a^a f(x)\,dx$ to be 0. The Fundamental Theorem is valid for such integrals.

The Fundamental Theorem of Calculus certainly deserves its name since it is the key connection between differential calculus and integral calculus, which were originally developed separately without knowledge of this connection between them. Most important for our purposes, it provides an algebraic means to evaluate definite integrals exactly.

1

Let $C(x) = x^3 + 4x^2 - x + 3$. Find the following.

(a) $C(x)\Big|_1^5$

(b) $C(x)\Big|_3^4$

Answers:

(a) 216

(b) 64

Applying the Fundamental Theorem

When evaluating definite integrals, the number $F(b) - F(a)$ is denoted by the symbol

$$F(x)\Big|_a^b$$

For example, if $F(x) = x^3$, then

$$x^3\Big|_1^2 \quad \text{means } F(2) - F(1) = 2^3 - 1^3 = 7.$$ (1)

2

Find each of the following.

(a) $\int_4^6 5z\,dz$

(b) $\int_2^5 8t^3\,dt$

(c) $\int_1^9 \sqrt{z}\,dz$

Answers:

(a) 50

(b) 1218

(c) 52/3

EXAMPLE 1 Evaluate each of these integrals.

(a) $\int_1^2 4t^3\,dt$.

We know that t^4 is an antiderivative of $4t^3$. By the Fundamental Theorem

$$\int_1^2 4t^3\,dt = t^4\Big|_1^2 = 2^4 - 1^4 = 15.$$

(b) $\int_0^5 e^{2x}\,dx$

By the antiderivative rules in Section 13.1, we know that an antiderivative of e^{2x} is $\frac{1}{2}e^{2x}$. Therefore,

$$\int_0^5 e^{2x}\,dx = \frac{1}{2}e^{2x}\Big|_0^5 = \frac{1}{2}e^{10} - \frac{1}{2}e^0 = \frac{1}{2}e^{10} - \frac{1}{2}.$$ (2)

To evaluate more complicated integrals, we need the following properties.

Properties of Definite Integrals

For any real numbers a and b for which the definite integrals exist,

1. $\int_a^b k \cdot f(x)\,dx = k \cdot \int_a^b f(x)\,dx$, for any real constant k (constant multiple of a function);
2. $\int_a^b [f(x) \pm g(x)]\,dx = \int_a^b f(x)\,dx \pm \int_a^b g(x)\,dx$ (sum or difference of functions)
3. $\int_a^b f(x)\,dx = \int_a^c f(x)\,dx + \int_c^b f(x)\,dx$, for any real number c.

Properties 1 and 2 follow directly from the similar properties of antiderivatives. Property 3 won't be proved here, but Figure 13.14 illustrates it when $f(x) > 0$. In this case,

$$\begin{aligned}\int_a^b f(x)\,dx &= \text{area of the entire shaded region}\\ &= \text{area of blue region} + \text{area of red region}\\ &= \int_a^c f(x)\,dx + \int_c^b f(x)\,dx\end{aligned}$$

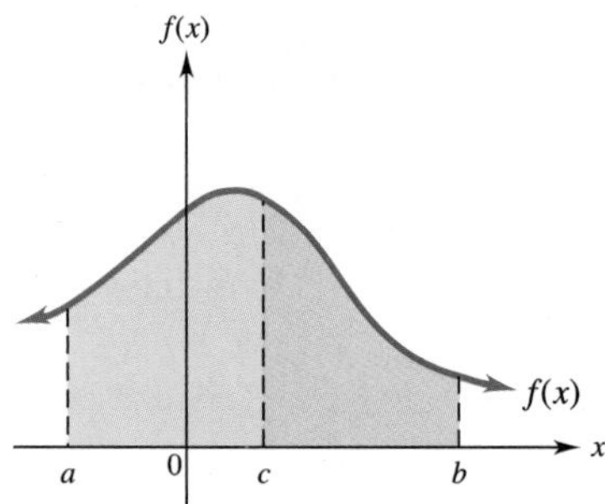

FIGURE 13.14

EXAMPLE 2 Evaluate $\int_2^5 (6x^2 - 3x + 5)\,dx$.

Use the properties above, and the Fundamental Theorem, along with the power rule from Section 13.1.

Evaluate each definite integral.

(a) $\int_1^3 (x + 3x^2)\, dx$

(b) $\int_2^4 (6k^2 - 2k + 1)\, dk$

Answers:

(a) 30

(b) 102

$$\int_2^5 (6x^2 - 3x + 5)\, dx = 6\int_2^5 x^2\, dx - 3\int_2^5 x\, dx + 5\int_2^5 dx \qquad \text{Constant multiple; sum or difference}$$

$$= 2x^3\Big|_2^5 - \frac{3}{2}x^2\Big|_2^5 + 5x\Big|_2^5 \qquad \text{Integrate.}$$

$$= 2(5^3 - 2^3) - \frac{3}{2}(5^2 - 2^2) + 5(5 - 2) \qquad \text{Evaluate the limits.}$$

$$= 2(125 - 8) - \frac{3}{2}(25 - 4) + 5(3)$$

$$= 234 - \frac{63}{2} + 15 = \frac{435}{2}$$ ③

④

Evaluate the following.

(a) $\int_0^4 e^x\, dx$

(b) $\int_3^5 \frac{dx}{x}$

(c) $\int_2^8 \frac{4}{x}\, dx$

Answers:

(a) 53.59815

(b) .51083

(c) 5.54518

EXAMPLE 3 Find $\int_1^2 \frac{dy}{y}$.

Using a result from Section 13.1,

$$\int_1^2 \frac{dy}{y} = \ln|y|\Big|_1^2 = \ln|2| - \ln|1| \approx .6931 - 0 = .6931.$$ ④

EXAMPLE 4 Evaluate $\int_0^5 x\sqrt{25 - x^2}\, dx$.

Use substitution to find the antiderivative. Let $u = 25 - x^2$, so that $du = -2x\, dx$. Now find the antiderivative and *express it in terms of x.*

$$\int x\sqrt{25 - x^2}\, dx = -\frac{1}{2}\int \sqrt{25 - x^2}\,(-2x\, dx)$$

$$= -\frac{1}{2}\int \sqrt{u}\, du \qquad \text{Substitute.}$$

$$= -\frac{1}{2}\int u^{1/2}\, du \qquad \text{Fractional exponent}$$

$$= -\frac{1}{2}\cdot\frac{u^{3/2}}{3/2} \qquad \text{Use antiderivative formula.}$$

$$= -\frac{1}{2}\cdot\frac{2}{3}u^{3/2} = -\frac{1}{3}u^{3/2} \qquad \text{Simplify}$$

$$= -\frac{1}{3}(25 - x^2)^{3/2} \qquad \text{Express in terms of } x.$$

Find $\int_0^2 \frac{x}{x^2 + 1}\, dx$.

Answer:

$\frac{1}{2}\ln 5$

Now evaluate the definite integral.

$$\int_0^5 x\sqrt{25 - x^2}\, dx = -\frac{1}{3}(25 - x^2)^{3/2}\Big|_0^5$$

$$= -\frac{1}{3}(25 - 5^2)^{3/2} - \left[-\frac{1}{3}(25 - 0^2)^{3/2}\right]$$

$$= 0 + \frac{1}{3}\cdot 25^{3/2} = \frac{125}{3}$$ ⑤

CAUTION When using substitution, be sure to express the antiderivative in terms of x before evaluating it. In Example 4, for instance, if you evaluate the antiderivative $-\frac{1}{3}u^{3/2}$ at 5 and 0, you get a wrong answer.

Area

When the graph of $f(x)$ lies above the x-axis between a and b, then the definite integral $\int_a^b f(x)\,dx$ gives the area between the graph of $f(x)$ and the x-axis from a to b. Now consider a function f whose graph lies below the x-axis from a to b. The shaded area in Figure 13.15 can be approximated by sums of areas of rectangles, that is, by sums of the form

$$f(x_1)\cdot\Delta x + f(x_2)\cdot\Delta x + f(x_3)\cdot\Delta x + \cdots + f(x_n)\cdot\Delta x,$$

with one difference: since $f(x)$ is negative, the sum represents the negative of the sum of the areas of the rectangles. Consequently, the definite integral $\int_a^b f(x)\,dx$, which is the limit of such sums as n gets larger and larger, is the *negative* of the shaded area in Figure 13.15.

6

Find each area.

(a)

(b)

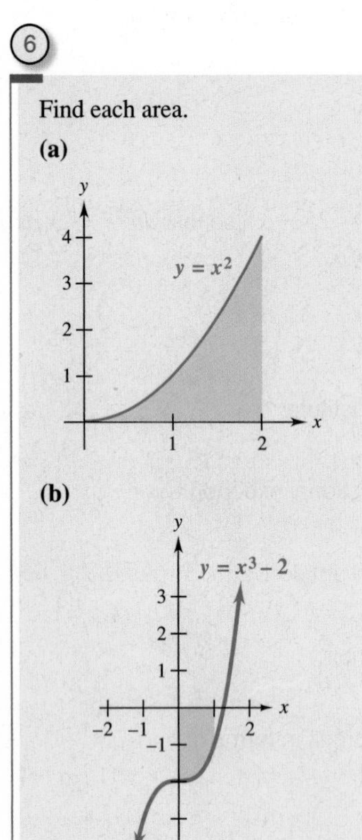

Answers:

(a) 8/3

(b) 7/4

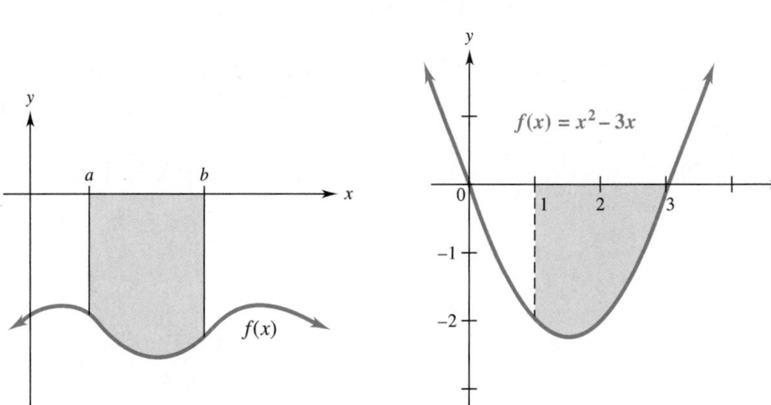

FIGURE 13.15

FIGURE 13.16

EXAMPLE 5 Find the area between the x-axis and the graph of $f(x) = x^2 - 3x$ from $x = 1$ to $x = 3$.

The region, which is shaded in Figure 13.16, lies below the x-axis and the definite integral gives the negative of its area:

$$\int_1^3 (x^2 - 3x)\,dx = \left(\frac{x^3}{3} - \frac{3x^2}{2}\right)\Bigg|_1^3 = \left(\frac{27}{3} - \frac{27}{2}\right) - \left(\frac{1}{3} - \frac{3}{2}\right) = -\frac{10}{3}.$$

Therefore, the area of the region is 10/3. ◀ 6

EXAMPLE 6 Find the area between the graph of $f(x) = 6x^2 - 7x - 3$ and the x-axis from $x = 0$ to $x = 2$, which is the shaded region in Figure 13.17.

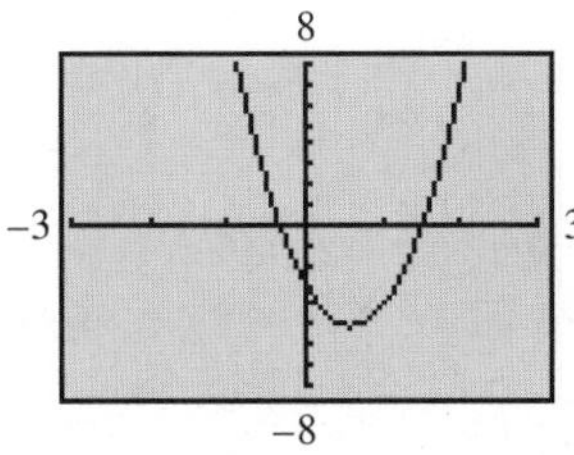

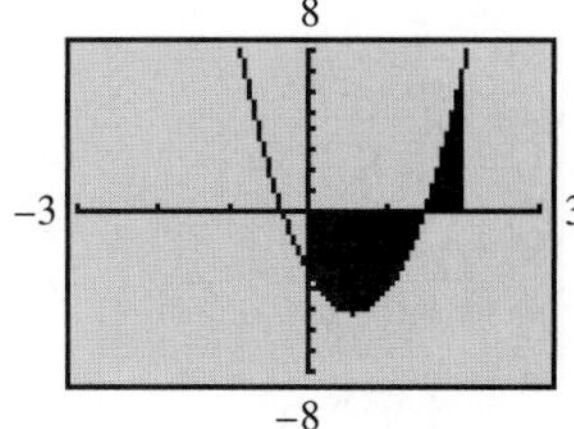

$f(x) = 6x^2 - 7x - 3$

FIGURE 13.17

To find the total area, compute the area under the x-axis and the area over the x-axis separately. Start by finding the x-intercepts of the graph by solving

$$6x^2 - 7x - 3 = 0$$
$$(2x - 3)(3x + 1) = 0$$
$$x = 3/2 \quad \text{or} \quad x = -1/3.$$

Since we are concerned only with the graph between 0 and 2, the only relevant x-intercept is 3/2. The area below the x-axis is the negative of

$$\int_0^{3/2} (6x^2 - 7x - 3)\, dx = \left(2x^3 - \frac{7}{2}x^2 - 3x\right)\Bigg|_0^{3/2}$$
$$= \left(2\left(\frac{3}{2}\right)^3 - \frac{7}{2}\left(\frac{3}{2}\right)^2 - 3\left(\frac{3}{2}\right)\right) - (0 - 0 - 0) = -\frac{45}{8}.$$

So the area of the lower region is 45/8. The area above the x-axis is

$$\int_{3/2}^{2} (6x^2 - 7x - 3)\, dx = \left(2x^3 - \frac{7}{2}x^2 - 3x\right)\Bigg|_{3/2}^{2}$$
$$= \left(2(2^3) - \frac{7}{2}(2^2) - 3(2)\right) - \left(2\left(\frac{3}{2}\right)^3 - \frac{7}{2}\left(\frac{3}{2}\right)^2 - 3\left(\frac{3}{2}\right)\right)$$
$$= \frac{13}{8}.$$

The area of the upper region is 13/8 and the total area between the graph and the x-axis is

$$\frac{45}{8} + \frac{13}{8} = \frac{58}{8} = 7.25 \text{ square units.}$$

Evaluate

$$\int_0^2 (6x^2 - 7x - 3)\, dx.$$

Answer:

-4

CAUTION If you attempt to find the area in Example 6 by using a single integral over the entire interval [0, 2], you will get a wrong answer, as shown in problem 7 at the side. It is essential to compute the upper and lower areas separately, using the two integrals over [0, 3/2] and [3/2, 2]. (7)

Although it may be helpful, it is not necessary to graph a function in order to find the entire area between its graph and the x-axis. As Example 6 shows, the procedure depends only on finding the x-intercepts of the graph that are the limits of integration on the various integrals. Here is a summary of the procedure used there.

Finding Area

To find the area bounded by $y = f(x)$, the vertical lines $x = a$ and $x = b$, and the x-axis, use the following steps.

Step 1 Sketch a graph, if convenient.

Step 2 Find any x-intercepts in $[a, b]$. These divide the total region into subregions.

Step 3 The definite integral will be *positive* for subregions above the x-axis and *negative* for subregions below the x-axis. Use separate integrals to find the areas of the subregions.

Step 4 The total area is the sum of the areas of all of the subregions.

We have seen a variety of interpretations of certain definite integrals in terms of area. We now present an interpretation that applies to all definite integrals. Consider, for example, $\int_0^2 (6x^2 - 7x - 3)\,dx$. By property 3 of definite integrals,

$$\int_0^2 (6x^2 - 7x - 3)\,dx = \int_0^{3/2} (6x^2 - 7x - 3)\,dx + \int_{3/2}^2 (6x^2 - 7x - 3)\,dx.$$

Each of the integrals on the right was evaluated in Example 6. Using those results we see that

$$\int_0^2 (6x^2 - 7x - 3)\,dx = -\frac{45}{8} + \frac{13}{8} = \frac{13}{8} - \frac{45}{8}.$$

As Example 6 shows, this last difference is precisely

$$(\text{area above the } x\text{-axis}) - (\text{area below the } x\text{-axis}).$$

The same result holds in the general case.

If f is a continuous function on $[a, b]$, then

$$\int_a^b f(x)\,dx = \begin{pmatrix}\text{area between the}\\ \text{graph and the } x\text{-axis}\\ \textit{above} \text{ the axis}\end{pmatrix} - \begin{pmatrix}\text{area between the}\\ \text{graph and the } x\text{-axis}\\ \textit{below} \text{ the axis}\end{pmatrix}$$

Applications

In the last section we saw that the area under the graph of a rate of change function $f'(x)$ from $x = a$ to $x = b$ gives the total change in $f(x)$ from a to b. We can now use the Fundamental Theorem to compute this change.

EXAMPLE 7 The United States Energy Information Administration projects that the yearly rate of consumption of natural gas (in quadrillion BTUs) in the next two decades will be given by

$$C'(t) = 23\, e^{.0177t},$$

where t is time in years and $t = 0$ corresponds to 2000. At this consumption rate, what is the total amount of natural gas that will be used in the 10-year period from 2000 to 2010?

The amount used over the 10 years is the total change in consumption from year 0 to year 10, so it is given by the definite integral.

$$\int_0^{10} 23\, e^{.0177t}\, dt = 23 \cdot \frac{e^{.0177t}}{.0177}\Bigg|_0^{10} \approx 1299.435 e^{.0177t}\Bigg|_0^{10}$$

$$= 1299.435\, e^{.177} - 1299.435\, e^{0}$$

$$\approx 251.6$$

Therefore, a total of about 251.6 quadrillion BTUs of natural gas will be used during the first decade of this century, if the projected consumption rate remains valid. ◄ 8

8

In Example 7, suppose a conservation campaign and higher prices cause the rate of consumption to be given by $c'(x) = 22.4\, e^{.01347t}$. Find the total amount of gas used from 2000 to 2010.

Answer:

About 239.8 quadrillion BTUs.

Proof of the Fundamental Theorem

Although we cannot give a rigorous proof of the Fundamental Theorem of Calculus here, we can indicate why it is true in an important case. Suppose $f(x) > 0$ when $a \le x \le b$, so that the definite integral represents the area under the curve and above the x-axis.

Define a new function $A(x)$ by this rule:

$$A(x) = \text{area between the graph of } f(x) \text{ and the } x\text{-axis from } a \text{ to } x,$$

as shown in Figure 13.18. For instance, if the area under the graph from a to 4 is 35, then $A(4) = 35$.

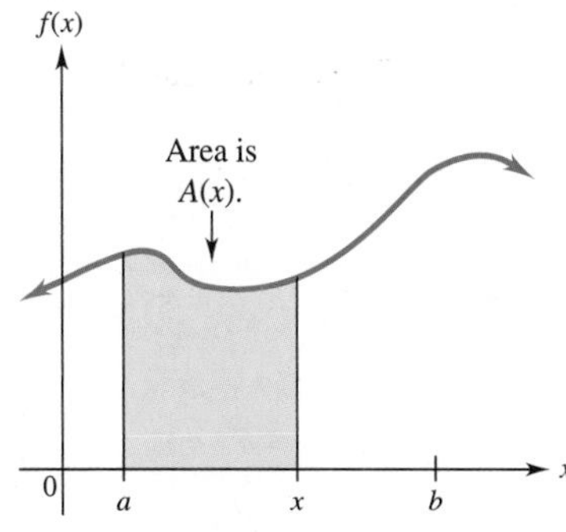

FIGURE 13.18

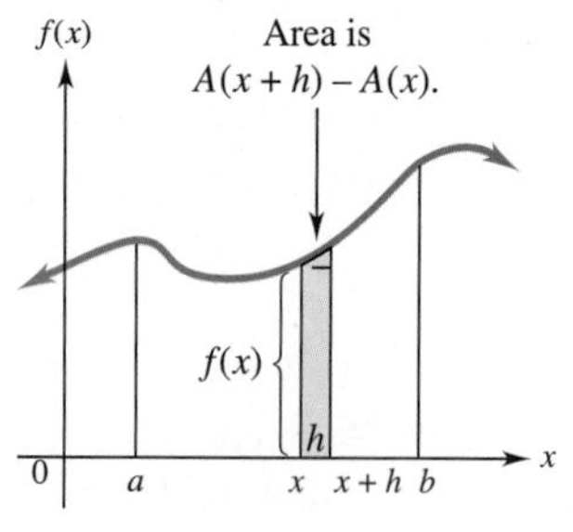

FIGURE 13.19

We now show that $A(x)$ is an antiderivative of $f(x)$. We must show that

$$A'(x) = \lim_{h\to 0} \frac{A(x+h) - A(x)}{h}$$

is actually $f(x)$. Look at the situation geometrically. When h is a small positive number, then $A(x+h)$ is the area under the graph of f from a to $x+h$ and $A(x)$ is the area from a to x. Consequently, $A(x+h) - A(x)$ is the area of the shaded strip in Figure 13.19. This area can be approximated with a rectangle having base h and height $f(x)$. The area of the rectangle is $h \cdot f(x)$ and

$$A(x+h) - A(x) \approx h \cdot f(x).$$

Dividing both sides by h gives

$$\frac{A(x+h) - A(x)}{h} \approx f(x).$$

This approximation improves as h gets smaller and smaller. Take the limit on the left as h approaches 0.

$$\lim_{h\to 0} \frac{A(x+h) - A(x)}{h} = f(x)$$

This limit is simply $A'(x)$, so

$$A'(x) = f(x).$$

This result means that A is an antiderivative of f, as we set out to show.

Since $A(x)$ is the area under the graph of f from a to x, we see that $A(a) = 0$ and $A(b)$ is the area under the graph from a to b. But this last area is the definite integral $\int_a^b f(x)\,dx$. Putting these facts together we have

$$\int_a^b f(x)\,dx = A(b) = A(b) - 0 = A(b) - A(a).$$

This argument suggests that the Fundamental Theorem is true when $f(x) > 0$ and the area function $A(x)$ is used as the antiderivative. Other arguments, which are omitted here, handle the case when $f(x)$ may not be positive and any antiderivative $F(x)$ is used.

13.4 Exercises

Evaluate each of the following definite integrals. (See Examples 1–4.)

1. $\int_{-1}^{3} (6x^2 - 4x + 3)\,dx$
2. $\int_{0}^{2} (-3x^2 + 2x + 5)\,dx$
3. $\int_{0}^{2} 3\sqrt{4u + 1}\,du$
4. $\int_{3}^{9} \sqrt{2r - 2}\,dr$
5. $\int_{0}^{1} 2(t^{1/2} - t)\,dt$
6. $\int_{0}^{4} -(3x^{3/2} - x^{1/2})\,dx$
7. $\int_{1}^{4} (5y\sqrt{y} + 3\sqrt{y})\,dy$
8. $\int_{4}^{9} (4\sqrt{r} - 3r\sqrt{r})\,dr$
9. $\int_{4}^{6} \frac{2}{(x-3)^2}\,dx$
10. $\int_{1}^{4} \frac{-3}{(2p+1)^2}\,dp$
11. $\int_{1}^{5} (5n^{-1} + n^{-3})\,dn$
12. $\int_{2}^{3} (3x^{-1} - x^{-4})\,dx$
13. $\int_{2}^{3} \left(2e^{-.1A} + \frac{3}{A}\right) dA$
14. $\int_{1}^{2} \left(\frac{-1}{B} + 3e^{.2B}\right) dB$

15. $\int_1^2 \left(e^{5u} - \frac{1}{u^2}\right) du$

16. $\int_{.5}^1 (p^3 - e^{4p})\, dp$

17. $\int_{-1}^0 y(2y^2 - 3)^5\, dy$

18. $\int_0^3 m^2(4m^3 - 2)^3\, dm$

19. $\int_1^{64} \frac{\sqrt{z} - 2}{\sqrt[3]{z}}\, dz$

20. $\int_1^8 \frac{3 - y^{1/3}}{y^{2/3}}\, dy$

21. $\int_1^2 \frac{\ln x}{x}\, dx$

22. $\int_1^3 \frac{\sqrt{\ln x}}{x}\, dx$

23. $\int_0^8 x^{1/3}\sqrt{x^{4/3} + 9}\, dx$

24. $\int_1^2 \frac{3}{x(1 + \ln x)}\, dx$

25. $\int_0^1 \frac{e^t}{(3 + e^t)^2}\, dt$

26. $\int_0^1 \frac{e^{2z}}{\sqrt{1 + e^{2z}}}\, dz$

27. $\int_1^{49} \frac{(1 + \sqrt{x})^{4/3}}{\sqrt{x}}\, dx$

28. $\int_1^8 \frac{(1 + x^{1/3})^6}{x^{2/3}}\, dx$

29. Suppose the function in Example 6, $f(x) = 6x^2 - 7x - 3$ from $x = 0$ to $x = 2$, represented the annual rate of profit of a company over a two-year period. What might the negative integral for the first year and a half indicate? What integral would represent the overall profit for the two-year period?

 30. In your own words describe how the Fundamental Theorem relates definite and indefinite integrals.

Use the definite integral to find the area between the x-axis and f(x) over the indicated interval. Check first to see if the graph crosses the x-axis in the given interval. (See Examples 5 and 6.)

31. $f(x) = 4 - x^2$; $[0, 3]$

32. $f(x) = x^2 - 2x - 3$; $[0, 4]$

33. $f(x) = x^3 - 1$; $[-2, 2]$

34. $f(x) = x^3 - 2x$; $[-2, 4]$

35. $f(x) = e^x - 1$; $[-1, 2]$

36. $f(x) = 1 - e^{-x}$; $[-1, 2]$

37. $f(x) = \frac{1}{x}$; $[1, e]$

38. $f(x) = \frac{1}{x}$; $[e, e^2]$

Find the area of each shaded region.

39.

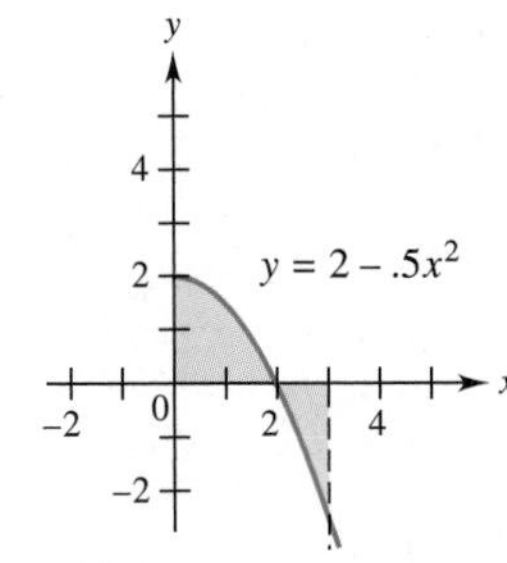

40.

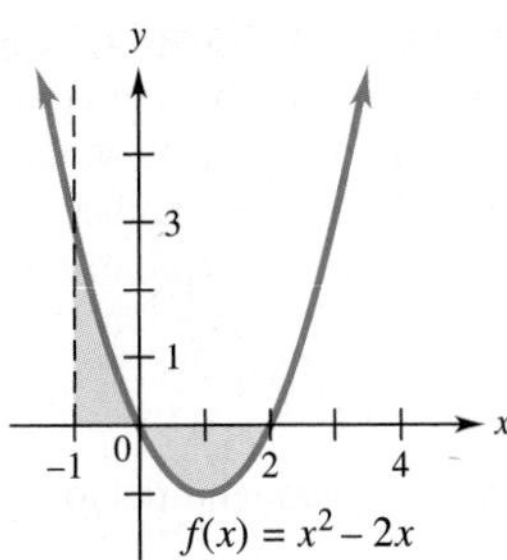

41.

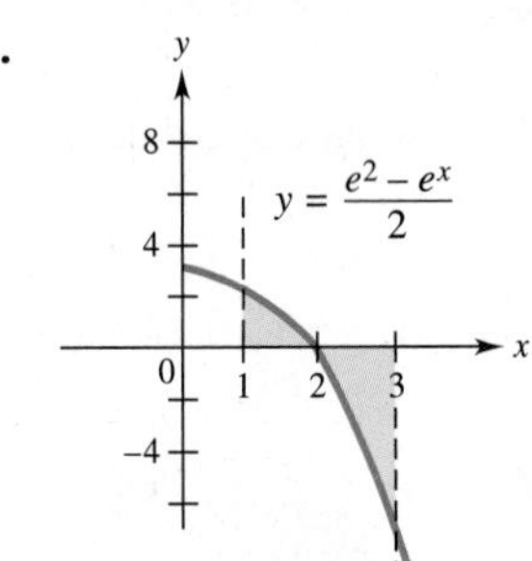

42.

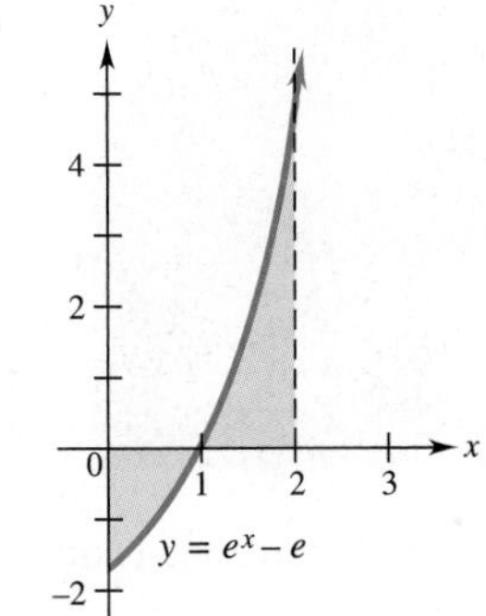

Use Property 4 of definite integrals to find $\int_1^4 f(x)\, dx$ for the following functions.

43. $f(x) = \begin{cases} 2x + 3 & \text{if } x \le 2 \\ -.5x + 8 & \text{if } x > 2 \end{cases}$

44. $f(x) = \begin{cases} x^2 - 2 & \text{if } x \le 3 \\ -x^2 + 16 & \text{if } x > 3 \end{cases}$

Work these problems. (See Example 7.)

45. **Business** De Win Enterprises has found that its expenditure rate per day (in hundreds of dollars) on a certain type of job is given by

$$E(x) = 4x + 2,$$

where x is the number of days since the start of the job.

(a) Find the total expenditure if the job takes 10 days.

(b) How much will be spent on the job from the 10th to the 20th day?

(c) If the company wants to spend no more than \$50,000 on the job, in how many days must they complete it?

46. Business A worker new to a job will improve his efficiency with time so that it takes him fewer hours to produce an item with each day on the job, up to a certain point. Suppose the rate of change of the number of hours it takes a worker in a certain factory to produce the xth item is given by

$$H'(x) = 20 - 2x.$$

(a) What is the total number of hours required to produce the first 5 items?

(b) What is the total number of hours required to produce the first 10 items?

47. Business The sales of CDs (in millions of dollars) in year x is approximated by

$$f(x) = 15.6x^3 - 271.8x^2 + 2367.4x + 1927 \qquad (1 \leq x \leq 10),$$

where $x = 1$ corresponds to 1991.* What were total CD sales from 1992 to 2000?

48. Business The rate of depreciation for a certain truck is

$$f(t) = \frac{6000(.3 + .2t)}{(1 + .3t + .1t^2)^2},$$

where t is in years and $t = 0$ is the year of purchase.

(a) Find the total depreciation at the end of 3 years.

(b) In what year will the total depreciation be at least 3000?

49. Business World energy consumption (in quadrillion BTUs) in year x is projected to be given by

$$g(x) = -.0028x^3 + .168x^2 + 7.8x + 404 \qquad (0 \leq x \leq 20),$$

where $x = 0$ corresponds to 2000.† Find the projected total consumption from 2004 to 2020.

50. Business Nuclear power consumption in the United States (in quadrillion BTUs) over the next two decades is projected to be given by

$$f(x) = -.0000295x^4 + .0017x^3 - .035x^2 + .11x + 7.5,$$

where $x = 0$ corresponds to 2000.† How much nuclear power will be consumed from 2000 to 2010?

*Recording Industry Association of America.

†U.S. Energy Information Administration.

51. Natural Science An oil tanker is leaking oil at a rate given in barrels per hour by

$$L'(t) = \frac{80 \ln(t + 1)}{t + 1},$$

where t is the time in hours after the tanker hits a hidden rock (when $t = 0$).

(a) Find the total number of barrels that the ship will leak on the first day.

(b) Find the total number of barrels that the ship will leak on the second day.

(c) What is happening over the long run to the amount of oil leaked per day?

52. Business Americans are eating more chicken. Production of chickens (in millions) is approximated by

$$C(x) = 351.495\, e^{.0225x},$$

where $x = 0$ corresponds to 1990.* How many chickens were produced from 1995 to 2002?

53. Social Science The 2000 U.S. Census gives an age distribution which is approximated by

$$f(x) = -.77x^2 + 2.5x + 40 \qquad (0 \leq x \leq 9),$$

where x is in decades, with $x = 0$ corresponding to age 0, $x = 1$ to age 10, and so on, and $f(x)$ is in millions.† The population of an age group can be found by integrating this function over the interval for that age group.

(a) Find $\int_0^9 f(x)\, dx$. What does this integral represent?

(b) Baby Boomers are those born between 1945 and 1965, that is, those in the range of 3.5 to 5.5 decades. Find the number of Baby Boomers.

54. Business Suppose that the rate of consumption of a natural resource is $c'(t)$, where

$$c'(t) = ke^{rt}.$$

Here t is time in years, r is a constant, and k is the consumption in the year when $t = 0$. In 1990, an oil company sold 1.2 billion barrels of oil. Assume that $r = .04$.

(a) Set up a definite integral for the amount of oil that the company will sell in the next 10 years.

(b) Evaluate the definite integral of part (a).

(c) The company has about 20 billion barrels of oil in reserve. To find the number of years that this amount will last, solve this equation for T:

$$\int_0^T 1.2e^{.04t}\, dt = 20.$$

(d) Rework part (c), assuming that $r = .02$.

*U.S. Department of Agriculture.

†Exercise suggested by Ralph DeMarr, University of New Mexico.

13.5 APPLICATIONS OF INTEGRALS

If a function gives the rate of change of a quantity, then the area between the graph of the rate function and the x-axis is the total amount of the quantity. This fact has numerous applications, some of which are explored here.

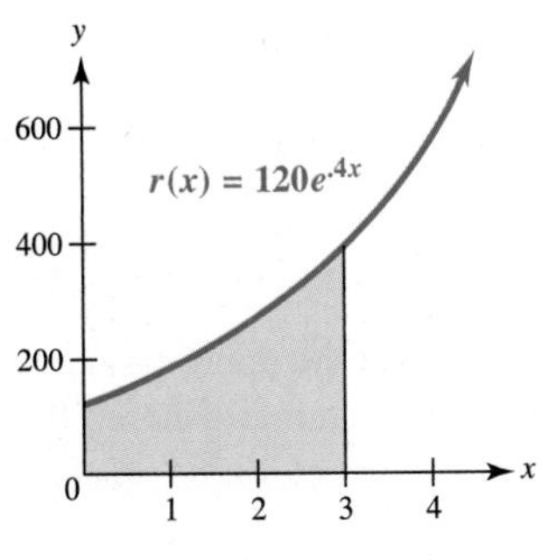

FIGURE 13.20

EXAMPLE 1 When you buy a used car, you may be offered the opportunity to buy a warranty that covers repair costs for a certain period of time. A company that provides such warranties uses current data and past experience to determine that the annual rate of repair costs for a particular model is given in year x by

$$r(x) = 120\,e^{.4x}.$$

What will the total repair costs be for one year and for three years?

The total repair costs are represented by the area under the rate curve over the appropriate interval. For the first three years, the area is shown in Figure 13.20 and is given by the integral

$$\int_0^3 120e^{.4x}\,dx = \frac{120}{.4}e^{.4x}\Big|_0^3 = 300e^{.4x}\Big|_0^3$$
$$= 300\,e^{.4(3)} - 300\,e^{.4(0)}$$
$$= 300\,e^{1.2} - 300 \approx \$696.04.$$

Similarly, the total repair costs for one year are

$$\int_0^1 120e^{.4x}\,dx = 300\,e^{.4x}\Big|_0^1 = 300\,e^{.4} - 300 \approx \$147.55.$$

In order to make a profit, the company must charge more than \$147.55 for a one-year warranty and more than \$696.04 for a three-year warranty. ◀ 1

In Example 1, find the total repair costs for

(a) Two years

(b) Four years

Answers:

(a) \$367.66

(b) \$1185.91

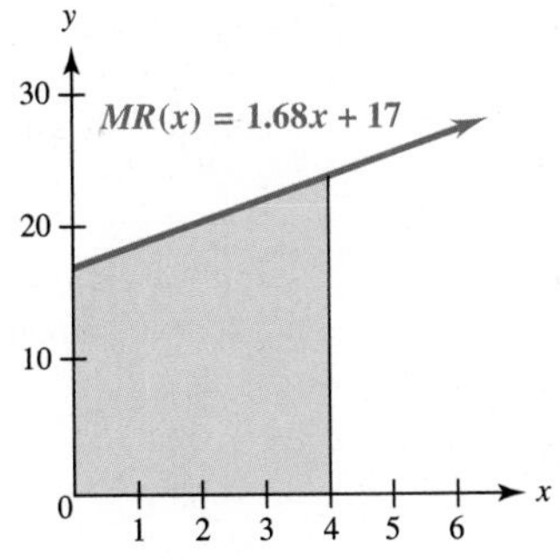

EXAMPLE 2 According to data from the Toy Manufacturers of America, Inc., the marginal revenue from video game sales (in billions of dollars per year) is approximately given by

$$MR(x) = 1.68x + 17,$$

where $x = 0$ corresponds to 1996.

(a) What was the total video game revenue from 1996 to 2000?

Marginal revenue is the derivative of the revenue function. So the total revenue from 1996 to 2000 is the area under the marginal revenue curve and above the x-axis from $x = 0$ to $x = 4$, as shown in Figure 13.21. This area is given by

$$\int_0^4 (1.68x + 17)\,dx = 1.68\frac{x^2}{2} + 17x\Big|_0^4 = .84x^2 + 17x\Big|_0^4$$
$$= .84(4^2) + 17(4) - 0$$
$$= \$81.44 \text{ billion}.$$

FIGURE 13.21

(b) At what point in time did total revenue reach \$50 billion?
Total revenue was \$50 billion at time t such that

$$\int_0^t 1.68x + 17\,dx = 50. \qquad (*)$$

In Example 2, find:

(a) the total revenue from 1997 to 1999;

(b) the time when total revenue reached \$100 billion

Answers:

(a) About \$40.72 billion

(b) About September of 2000 ($x = 4.76$)

Evaluating the definite integral on the left side, we have

$$.84x^2 + 17x \Big|_0^t = .84t^2 + 17t - [.84(0^2) + 17(0)] = .84t^2 + 17t.$$

Therefore, equation (*) becomes

$$.84t^2 + 17t = 50$$
$$.84t^2 + 17t - 50 = 0.$$

Use the quadratic formula to solve this equation and verify that the approximate solutions are 2.64 and −25.61. Only the positive solution is meaningful here. Total revenue reached \$50 billion when $t = 2.64$, that is, about August of 1998. ◀ 2

Area between Two Curves

In some applications, it is necessary to find the area between two curves. For example, the area between the graphs of $f(x)$ and $g(x)$ from $x = a$ to $x = b$ in Figure 13.22 is shaded. This area is the area under the graph of $f(x)$ *minus* the area under the graph of $g(x)$, that is,

$$\int_a^b f(x)\,dx - \int_a^b g(x)\,dx,$$

which can be written as

$$\int_a^b [f(x) - g(x)]dx.$$

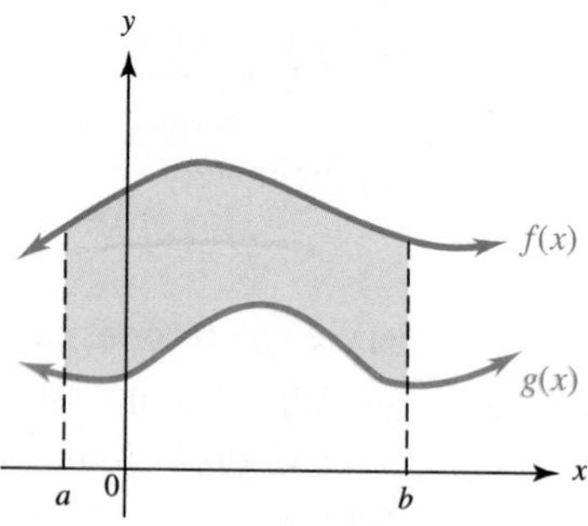

FIGURE 13.22

Similar arguments in other cases (even when the graphs are not above the x-axis) produce the following important result.

Area Between Two Curves

If f and g are continuous functions and $f(x) \geq g(x)$ on the interval $[a, b]$, then the area between the graphs of $f(x)$ and $g(x)$ from $x = a$ to $x = b$ is given by

$$\int_a^b [f(x) - g(x)]dx.$$

EXAMPLE 3 The revenue of major investor-owned electric utilities (in billions of dollars) during the period 1993–98 is approximated by $f(x) = 5.18x + 159$, where $x = 3$ corresponds to 1993. Their costs during the same period are approximated by $g(x) = -.16x^3 + 3.36x^2 - 16.62x + 170$.* Find the profit earned from 1993 to 1998.

The cost and revenue functions are graphed in Figure 13.23. Total revenue during this period is the area under the revenue curve and total cost is the area under the cost curve. So the profit is the area between the two curves from $x = 3$ to $x = 8$, as shown in Figure 13.24.

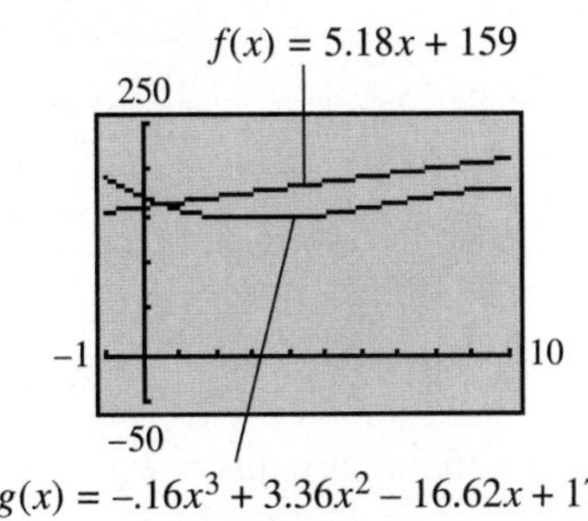

FIGURE 13.23

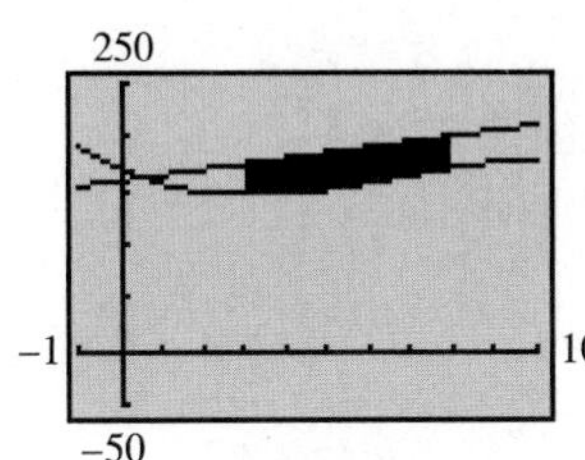

FIGURE 13.24

This area is

$$\int_3^8 [f(x) - g(x)]dx = \int_3^8 [(5.18x + 159) - (-.16x^3 + 3.36x^2 - 16.62x + 170)]\, dx$$

$$= \int_3^8 (.16x^3 - 3.36x^2 + 21.8x - 11)]\, dx$$

$$= .16\frac{x^4}{4} - 3.36\frac{x^3}{3} + 21.8\frac{x^2}{2} - 11x\Big|_3^8$$

$$= .04x^4 - 1.12x^3 + 10.9x^2 - 11x\Big|_3^8$$

$$= [.04(8^4) - 1.12(8^3) + 10.9(8^2) - 11(8)] - [.04(3^4) - 1.12(3^3) + 10.9(3^2) - 11(3)]$$

$$= 161.9.$$

The profit was $161,900,000,000. ◀ 3

Find the area between the graphs of $f(x) = 8 - x^2$ and $g(x) = -x + 1$ from $x = -1$ to $x = 2$.

Answer:

19.5

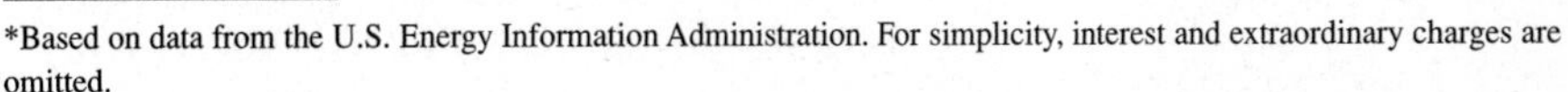

*Based on data from the U.S. Energy Information Administration. For simplicity, interest and extraordinary charges are omitted.

EXAMPLE 4 A company is considering a new manufacturing process in one of its plants. The new process provides substantial initial savings, with the savings declining with time x according to the rate-of-savings function

$$S(x) = 100 - x^2,$$

where $S(x)$ is in thousands of dollars. At the same time, the cost of operating the new process increases with time x, according to the rate-of-cost function (in thousands of dollars)

$$C(x) = x^2 + \frac{14}{3}x.$$

(a) For how many years will the company realize savings?

Figure 13.25 shows the graphs of the rate-of-savings and the rate-of-cost functions. The rate-of-cost (marginal cost) is increasing, while the rate-of-savings (marginal savings) is decreasing.

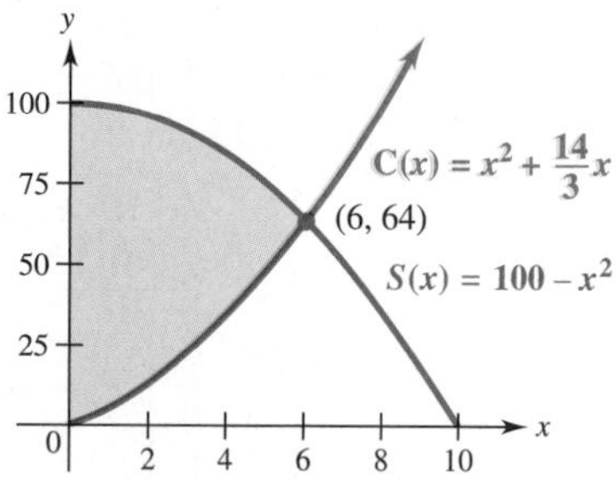

FIGURE 13.25

The company should use this new process until the difference between these quantities is zero—that is, until the time at which these graphs intersect. The graphs intersect when

$$S(x) = C(x),$$

or

$$100 - x^2 = x^2 + \frac{14}{3}x.$$

Solve this equation as follows.

$$0 = 2x^2 + \frac{14}{3}x - 100$$

$$0 = 3x^2 + 7x - 150 \quad \text{Multiply by 3/2.}$$

$$0 = (x - 6)(3x + 25) \quad \text{Factor.}$$

Set each factor equal to 0 and solve both equations to get

$$x = 6 \quad \text{or} \quad x = -\frac{25}{3}.$$

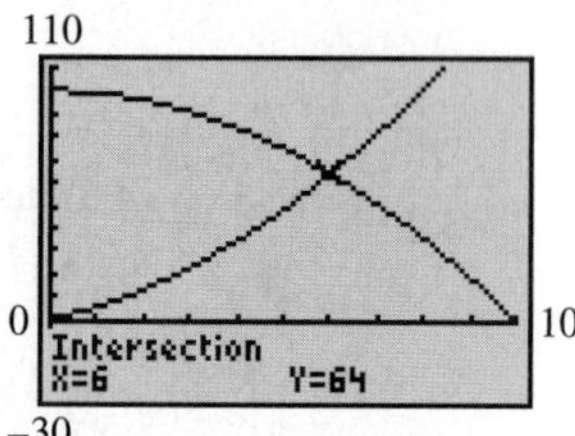

FIGURE 13.26

Only 6 is a meaningful solution here. Alternatively, the intersection point can be found by using the intersection finder on a calculator, as in Figure 13.26. The company should use the new process for 6 years.

In Example 4, find the total net savings if pollution control regulations permit the new process for only 4 years.

Answer:

$320,000

5

If a company's marginal savings and marginal cost functions are

$$S(x) = 150 - 2x^2$$

and

$$C(x) = x^2 + 15x,$$

where $S(x)$ and $C(x)$ give amounts in thousands of dollars, find the following.

(a) Number of years until savings equal costs

(b) Total savings

Answers:

(a) 5

(b) $437,500

(b) Taking the additional costs into account, how much will the company actually save during this period?

The actual savings (or net savings) are the difference between the savings and the additional costs. Total savings are given by the area under the rate-of-savings curve and total additional costs by the area under the rate-of-costs curve from $x = 0$ to $x = 6$. So total net savings are given by the difference beween these two areas, that is, by the shaded area in Figure 13.25. This area is given by the integral

$$\begin{aligned}
\text{Net savings} &= \int_0^6 [S(x) - C(x)]\,dx \\
&= \int_0^6 \left[(100 - x^2) - \left(x^2 + \frac{14}{3}x\right)\right] dx \\
&= \int_0^6 \left(100 - \frac{14}{3}x - 2x^2\right) dx && \text{Combine terms.} \\
&= 100x - \frac{7}{3}x^2 - \frac{2}{3}x^3 \Big|_0^6 && \text{Integrate.} \\
&= 100(6) - \frac{7}{3}(36) - \frac{2}{3}(216) = 372.
\end{aligned}$$

The company will save a total of $372,000 over the 6-year period. 

EXAMPLE 5 A farmer has been using a new fertilizer that gives him a better yield, but because it exhausts other nutrients in the soil, he must use other fertilizers in greater and greater amounts, so that his costs increase each year. The new fertilizer produces a rate of increase in revenue (in hundreds of dollars) given by

$$R(t) = -.4t^2 + 8t + 10,$$

where t is measured in years. The rate of increase in yearly costs (also in hundreds of dollars) due to use of the fertilizer is given by

$$C(t) = 2t + 5.$$

How long can the farmer profitably use the fertilizer? What will be his net increase in revenue over this period?

The farmer should use the new fertilizer until the marginal costs equal the marginal revenue. Find this point by solving the equation $R(t) = C(t)$ as follows.

$$\begin{aligned}
-.4t^2 + 8t + 10 &= 2t + 5 \\
-4t^2 + 80t + 100 &= 20t + 50 && \text{Multiply by 10.} \\
-4t^2 + 60t + 50 &= 0
\end{aligned}$$

Find the positive solution by using a graphing calculator (Figure 13.27) or the quadratic formula:

$$t = \frac{-60 - \sqrt{(-60)^2 - 4(-4)(50)}}{2(-4)} \approx 15.8.$$

The new fertilizer will be profitable for about 15.8 years.

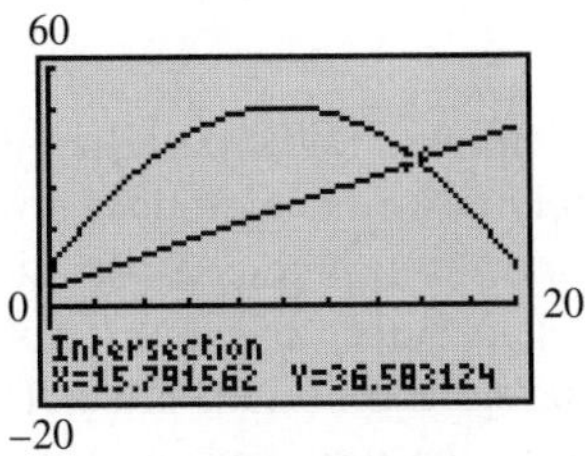

FIGURE 13.27

To find the net amount of additional revenue over the 15.8-year period, find the area between the graphs of the rate of revenue and the rate of cost functions, shown in Figure 13.28.

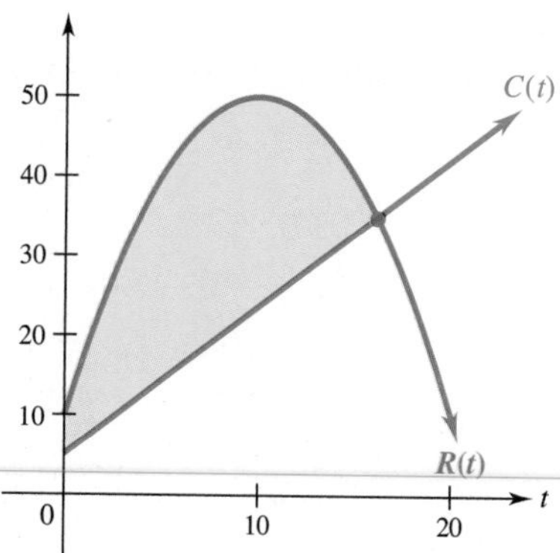

FIGURE 13.28

The area (net revenue) is

$$\begin{aligned}\text{Net revenue} &= \int_0^{15.8} [R(t) - C(t)]\,dt \\ &= \int_0^{15.8} [(-.4t^2 + 8t + 10) - (2t + 5)]\,dt \\ &= \int_0^{15.8} (-.4t^2 + 6t + 5)\,dt && \text{Combine terms.} \\ &= \left(\frac{-.4t^3}{3} + \frac{6t^2}{2} + 5t\right)\Bigg|_0^{15.8} && \text{Integrate.} \\ &\approx 302.01.\end{aligned}$$

The net additional revenue will amount to about \$30,201 over the 15.8-year period.

It is not realistic to say that the farmer will need to use the new process for 15.8 years—he will probably have to use it for 15 years or for 16 years. In this case, when the mathematical result is not in the domain of the function, it will be necessary to find the net revenue after 15 years and after 16 years and then select the best result. ◀

Consumers' and Producers' Surplus

The market determines the price at which a product is sold. As indicated earlier, the point of intersection of the demand curve and the supply curve for a product gives the equilibrium price. At the equilibrium price, consumers will purchase the same amount of the product that the manufacturers want to sell. Some consumers, however, would be willing to spend more for an item than the equilibrium price. The total of the differences between the equilibrium price of the item and the higher prices all those individuals would be willing to pay is thought of as savings realized by those individuals and is called the **consumers' surplus.**

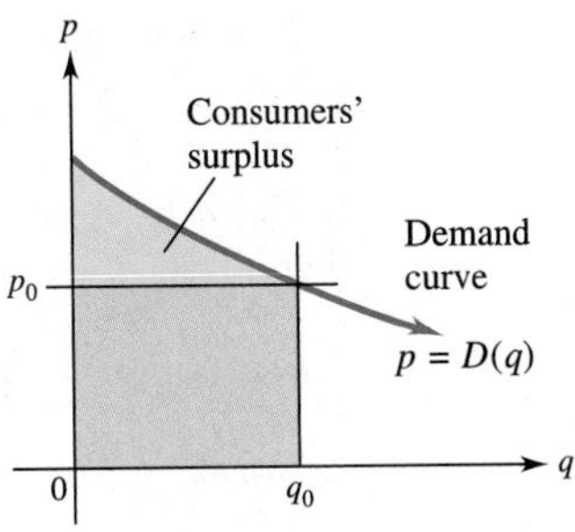

FIGURE 13.29

In Figure 13.29 the colored area under the demand curve is the total amount consumers are willing to spend for q_0 items. The red area under the line $y = p_0$ shows the total amount consumers actually will spend at the equilibrium price of p_0. The blue area represents the consumers' surplus. As the figure suggests, the consumers' surplus is given by an area between the two curves $p = D(q)$ and $p = p_0$, so its value can be found with a definite integral as follows.

Consumers' Surplus

If $D(q)$ is a demand function with equilibrium price p_0 and equilibrium demand q_0 then

$$\textbf{Consumers' surplus} = \int_0^{q_0} [D(q) - p_0]\, dq.$$

Similarly, if some manufacturers would be willing to supply a product at a price *lower* than the equilibrium price p_0, the total of the differences between the equilibrium price and the lower prices at which the manufacturers would sell the product is considered added income for the manufacturers and is called the **producers' surplus.** Figure 13.30 shows the (red) total area under the supply curve from $q = 0$ to $q = q_0$, which is the minimum total amount the manufacturers are willing to realize from the sale of q_0 items. The total area under the line $p = p_0$ is the amount actually realized. The difference between these two areas, the producers' surplus (blue), is also given by a definite integral.

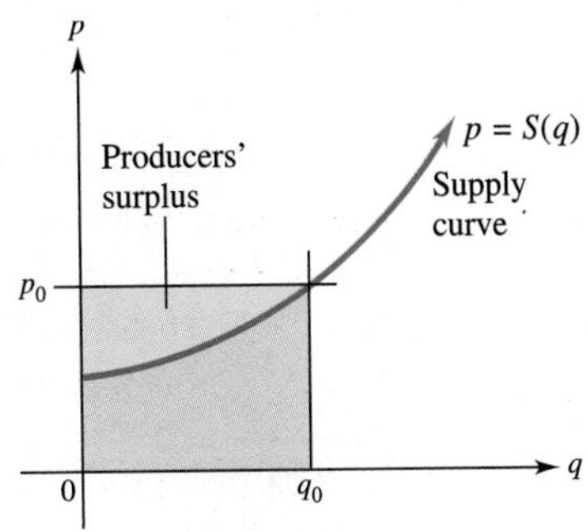

FIGURE 13.30

Producers' Surplus

If $S(q)$ is a supply function with equilibrium price p_0 and equilibrium supply q_0 then

$$\textbf{Producers' surplus} = \int_0^{q_0} [p_0 - S(q)]\, dq.$$

EXAMPLE 6 Suppose the price (in dollars per ton) for oat bran is

$$D(q) = 900 - 20q - q^2,$$

when the demand for the product is q tons. Also, suppose the function

$$S(q) = q^2 + 10q$$

gives the price (in dollars per ton) when the supply is q tons. Find the consumers' surplus and the producers' surplus.

We begin by finding the equilibrium quantity by setting the two equations equal.

$$\begin{aligned} 900 - 20q - q^2 &= q^2 + 10q \\ 0 &= 2q^2 + 30q - 900 \\ 0 &= q^2 + 15q - 450 \end{aligned}$$

Use the quadratic formula or factor to see that the only positive solution of this equation is $q = 15$. At the equilibrium point where the supply and demand are both 15 tons, the price is

$$S(15) = 15^2 + 10(15) = 375,$$

or \$375. Verify that the same answer is found by computing $D(15)$. The consumers' surplus, represented by the blue area shown in Figure 13.31, is

$$\int_0^{15} [(900 - 20q - q^2) - 375]\, dq = \int_0^{15} (525 - 20q - q^2)\, dq.$$

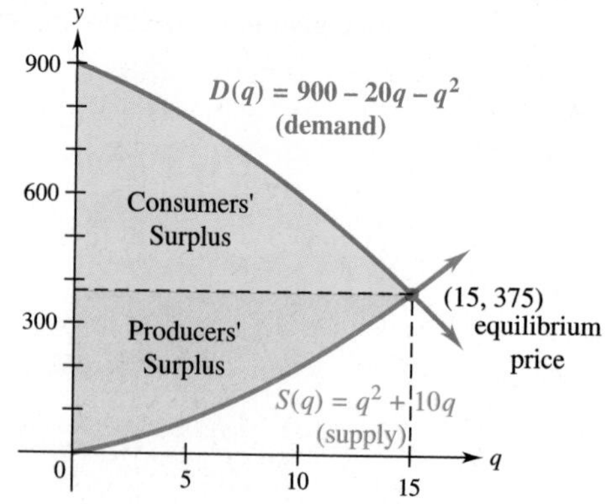

FIGURE 13.31

Evaluating this definite integral gives

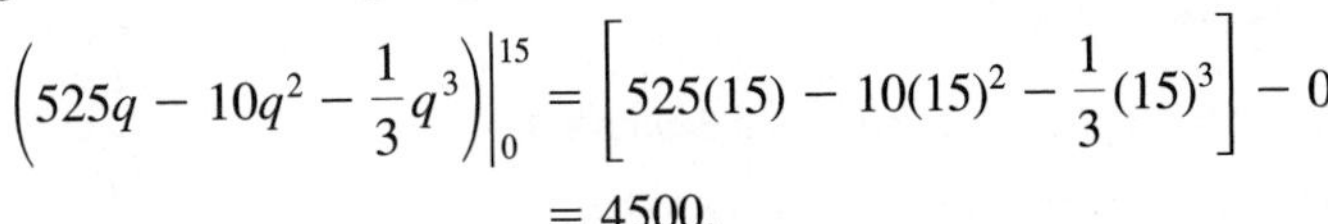

$$\left(525q - 10q^2 - \frac{1}{3}q^3\right)\Big|_0^{15} = \left[525(15) - 10(15)^2 - \frac{1}{3}(15)^3\right] - 0$$
$$= 4500.$$

Here, the consumers' surplus is $4500. The producers' surplus, also shown in Figure 13.31, is given by

$$\int_0^{15} [375 - (q^2 + 10q)]\, dq = \int_0^{15} (375 - q^2 - 10q)\, dq$$
$$= 375q - \frac{1}{3}q^3 - 5q^2 \Big|_0^{15}$$
$$= \left[375(15) - \frac{1}{3}(15)^3 - 5(15)^2\right] - 0$$
$$= 3375.$$

The producers' surplus is $3375. ◀ 6

Given the demand function $D(q) = 12 - .07q$ and the supply function $S(q) = .05q$, where $D(q)$ and $S(q)$ are in dollars, find

(a) the equilibrium point;

(b) the consumers' surplus;

(c) the producers' surplus.

Answers:

(a) $x = 100$

(b) $350

(c) $250

13.5 Exercises

Management *Work the following exercises. (See Examples 1 and 2.)*

1. A car-leasing firm must decide how much to charge for maintenance on the cars it leases. After careful study, the firm decides that the rate of maintenance, $M(x)$, on a new car will approximate $M(x) = 60(1 + x^2)$, where x is the number of years the car has been in use. What total maintenance cost can the company expect for a 2-year lease? What minimum amount should be added to the monthly lease payments to pay for maintenance?

2. Using the function of Exercise 1, find the maintenance cost the company can expect during the third year. Find the total charge during the first 3 years. What minimum monthly charge should be added to cover a 3-year lease?

3. According to data from the National Automobile Dealers Association, the marginal revenue of franchised new car dealerships (in billions of dollars per year) is approximated by

$$MR(x) = 31.64x + 300,$$

where $x = 0$ corresponds to 1990.

(a) What was total revenue from 1995 to 2000?

(b) When did the total revenue since 1990 reach $3000 billion?

4. The marginal revenue of all hardware stores in the United States (in billions of dollars per year) can be approximated by

$$MR(x) = 12.351e^{.023x},$$

where $x = 0$ corresponds to 1990.*

(a) What was total revenue from 1990 to 2000?

(b) In what year did total revenue since 1990 reach $100 billion? (*Hint:* Use the methods of Section 4.4 or technology.)

5. A company is considering a new manufacturing process. It knows that the rate of savings from the process, $S(t)$, will be about $S(t) = 1000(t + 2)$, where t is the number of years the process has been in use. Find the total savings during the first year. Find the total savings during the first 6 years.

6. Assume that the new process in Exercise 5 costs $16,000. About when will it pay for itself?

7. A company is introducing a new product. Production is expected to grow slowly because of difficulties in the start-up process. It is expected that the rate of production, $P(x)$, will be approximated by $P(x) = 1000e^{.2x}$, where x is the number of years since the introduction of the product. Will the company be able to supply 20,000 units during the first 4 years?

*U.S. Census Bureau.

8. About when will the company of Exercise 7 be able to supply its 15,000th unit?

Find the area between the two curves. (See Example 3.)

9. $y = 2x$ and $y = x^2 - 3$ from $x = -1$ to $x = 1$

10. $y = x^2 - 30$ and $y = 10 - 3x$ from $x = 0$ to $x = 5$

11. $y = x^2$ and $y = x^3$ from $x = 0$ to $x = 1$

12. $y = e^x$ and $y = 3 - e^x$ from $x = -2$ to $x = 0$

Use a graphing calculator to approximate the area between the graphs of each pair of functions on the given interval.

13. $y = \ln x$ and $y = xe^x$; $[1, 4]$

14. $y = \ln x$ and $y = 4 - x^2$; $[2, 4]$

15. $y = \sqrt{9 - x^2}$ and $y = \sqrt{x + 1}$; $[-1, 2]$

16. $y = \sqrt{4 - 4x^2}$ and $y = \sqrt{\frac{9 - x^2}{3}}$; $[-1, 1]$

Work these problems. (See Examples 4 and 5.)

17. **Business** Suppose a company wants to introduce a new machine that will produce a rate of annual savings in dollars given by

$$S(x) = 150 - x^2,$$

where x is the number of years of operation of the machine, while producing a rate of annual costs in dollars of

$$C(x) = x^2 + \frac{11}{4}x.$$

(a) For how many years will it be profitable to use this new machine?
(b) What are the net total savings during the first year of use of the machine?
(c) What are the net total savings over the entire period of use of the machine?

18. **Natural Science** A new smog-control device will reduce the output of sulfur oxides from automobile exhausts. It is estimated that the rate of savings to the community from the use of this device will be approximated by

$$S(x) = -x^2 + 4x + 8,$$

where $S(x)$ is the rate of savings (in millions of dollars) after x years of use of the device. The new device cuts down on the production of sulfur oxides, but it causes an increase in the production of nitrous oxides. The rate of additional costs (in millions) to the community after x years is approximated by

$$C(x) = \frac{3}{25}x^2.$$

(a) For how many years will it pay to use the new device?
(b) What will be the net savings over this period of time?

19. **Business** Lahr Enterprises had an expenditure rate of $E(x) = e^{.1x}$ dollars per day and an income rate of $I(x) = 98.8 - e^{.1x}$ dollars per day on a particular job, where x was the number of days from the start of the job. The company's profit on that job will equal total income less total expenditures. Profit will be maximized if the job ends at the optimum time, which is the point where the two curves meet. Find the following.
(a) The optimum number of days for the job to last
(b) The total income for the optimum number of days
(c) The total expenditures for the optimum number of days
(d) The maximum profit for the job

20. **Business** A factory at Harold Levinson Industries has installed a new process that will produce an increased rate of revenue (in thousands of dollars) of

$$R(t) = 104 - .4e^{t/2},$$

where t is time measured in years. The new process produces additional costs (in thousands of dollars) at the rate of

$$C(t) = .3e^{t/2}.$$

(a) When will it no longer be profitable to use this new process?
(b) Find the total net savings.

21. **Business** After t years, a mine is producing at the rate of

$$P(t) = \frac{15}{t + 1}$$

tons per year. At the same time, the ore produced is consumed at a rate of $C(t) = .1t + 2$ tons per year.
(a) In how many years will the rate of consumption equal the rate of production?
(b) What is the total excess production before consumption and production are equal?
(c) Consumption equals production when $t = 0$. Why is 0 not the correct answer in part (a)?

22. **Business** The rate of expenditure (in dollars) for maintenance of a certain machine is given by

$$M(x) = x^2 + 6x,$$

where x is time measured in years. The machine produces a rate of savings (in dollars) given by

$$S(x) = 360 - 2x^2.$$

(a) In how many years will the maintenance rate equal the savings rate?
(b) What will be the total net savings?

23. **Natural Science** Pollution from a factory is entering a lake. The rate of concentration of the pollutant at time t (in years) is $P'(t) = 140t^{.4}$. A cleaning substance is introduced into the lake that cleans it at a rate given by $C'(t) = 1.6t^{2.5}$. How long will it be before the total net effect is 0?

Business *Work the following supply and demand exercises, where the price is given in dollars. (See Example 5.)*

24. Find the consumers' surplus and the producers' surplus for an item having supply function

$$S(q) = 3q^2$$

and demand function

$$D(q) = 144 - \frac{q^2}{6}.$$

25. Suppose the supply function of a certain item is given by

$$S(q) = \frac{7}{5}q$$

and the demand function is given by

$$D(q) = -\frac{3}{5}q + 10.$$

(a) Graph the supply and demand curves.
(b) Find the point at which supply and demand are in equilibrium.
(c) Find the consumers' surplus.
(d) Find the producers' surplus.

26. Find the producers' surplus if the supply function for pork bellies is given by

$$S(q) = q^{5/2} + 2q^{3/2} + 50.$$

Assume supply and demand are in equilibrium at $q = 16$.

27. Suppose the supply function for concrete is given by

$$S(q) = 100 + 3q^{3/2} + q^{5/2}$$

and that supply and demand are in equilibrium at $q = 9$. Find the producers' surplus.

28. Find the consumers' surplus if the demand function for grass seed is given by

$$D(q) = \frac{100}{(3q + 1)^2},$$

assuming supply and demand are in equilibrium at $q = 3$.

29. Find the consumers' surplus if the demand function for extra virgin olive oil is given by

$$D(q) = \frac{16{,}000}{(2q + 8)^3}$$

and if supply and demand are in equilibrium at $q = 6$.

30. Suppose the supply function of a certain item is given by

$$S(q) = e^{q/2} - 1$$

and the demand function is given by

$$D(q) = 400 - e^{q/2}.$$

(a) Graph the supply and demand curves.
(b) Find the point at which supply and demand are in equilibrium.
(c) Find the consumers' surplus.
(d) Find the producers' surplus.

31. Repeat the four steps in Exercise 30 for the supply function

$$S(q) = q^2 + \frac{11}{4}q$$

and the demand function

$$D(q) = 150 - q^2.$$

Work the following exercises.

32. Business In an article in the December 1994 *Scientific American* magazine, the authors estimated future gasoline use.* Without a change in U.S. policy, auto fuel use is forecasted to rise along the projection shown at the right in the figure. The shaded band predicts gas use if the technologies for increased fuel economy are phased in by the year 2010. The moderate estimate (center curve) corresponds to an average of 46 miles per gallon for all cars on the road. Discuss the interpretation of the shaded area and other regions of the graph that pertain to the topic in this section.

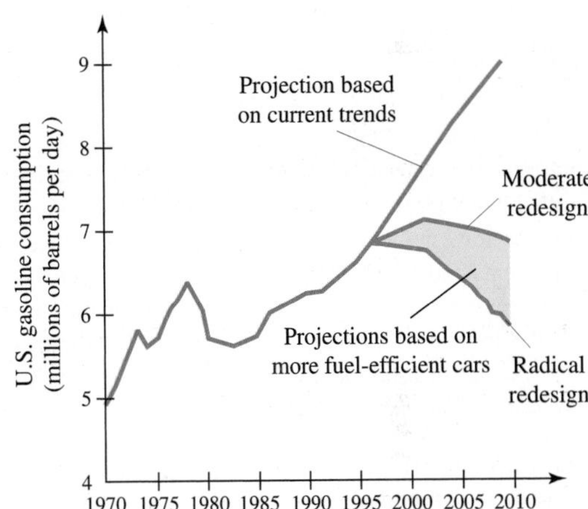

33. Physical Science A worker sketches the curves $y = \sqrt{x}$ and $y = x/2$ on a sheet of metal and cuts out the region between the curves to form a metal plate. Find the area of the plate.

34. Business If a large truck has been driven x thousand miles, the rate of repair costs in dollars per mile is given by $R(x)$, where

$$R(x) = .05x^{3/2}.$$

Find the total repair costs if the truck is driven
(a) 100,000 miles;
(b) 400,000 miles.

*Adapted from "Improving Automotive Efficiency," by John DeCicco and Marc Ross (Dec. 1994). Copyright © 1994 by Scientific American, Inc. All rights reserved.

35. Business In a recent inflationary period, costs of a certain industrial process, in millions of dollars, were increasing according to the function

$$i(t) = .45t^{3/2},$$

where $i(t)$ is the rate of increase in costs at time t measured in years. Find the total increase in costs during the first 4 years.

36. Natural Science From 1905 to 1920, most of the predators of the Kaibab Plateau of Arizona were killed by hunters. This allowed the deer population there to grow rapidly until they had depleted their food sources, which caused a rapid decline in population. The rate of change of this deer population during that time span is approximated by the function

$$D(t) = \frac{25}{2}t^3 - \frac{5}{8}t^4,$$

where t is the time in years ($0 \le t \le 25$).

(a) Find the function for the deer population if there were 4000 deer in 1905 ($t = 0$).

(b) What was the population in 1920?

(c) When was the population at a maximum?

(d) What was the maximum population?

37. Social Science Suppose that all the people in a country are ranked according to their incomes, starting at the bottom. Let x represent the fraction of the community making the lowest income ($0 \le x \le 1$); $x = .4$, therefore, represents the lower 40% of all income producers. Let $I(x)$ represent the proportion of the total income earned by the lowest x of all the people. Thus, $I(.4)$ represents the fraction of total income earned by the lowest 40% of the population. Suppose

$$I(x) = .9x^2 + .1x.$$

Find and interpret the following.

(a) $I(.1)$ **(b)** $I(.5)$ **(c)** $I(.9)$

If income were distributed uniformly, we would have $I(x) = x$. The area under this line of complete equality is 1/2. As $I(x)$ dips farther below $y = x$, there is less equality of income distribution. This inequality can be quantified by the ratio of the area between $I(x)$ and $y = x$ to 1/2. This ratio is called the *coefficient of inequality* and equals $2\int_0^1 (x - I(x))\,dx$.

(d) Graph $I(x) = x$ and $I(x) = .9x^2 + .1x$ for $0 \le x \le 1$ on the same axes.

(e) Find the area between the curves. What does this area represent?

38. Natural Science For a certain drug, the rate of reaction in appropriate units is given by

$$R(t) = \frac{5}{t} + \frac{2}{t^2},$$

where t is measured in hours after the drug is administered. Find the total reaction to the drug

(a) from $t = 1$ to $t = 12$;

(b) from $t = 12$ to $t = 24$.

39. Management A worker new to a job will improve his efficiency with time so that it takes him fewer hours to produce an item with each day on the job up to a certain point. Suppose the rate of change of the number of hours it takes a worker in a certain factory to produce the xth item is given by

$$H(x) = 20 - 2x.$$

The production rate per item is a maximum when $\int_0^T H(x)\,dx$ is a maximum.

(a) How many items must be made to achieve the maximum production rate? Assume 0 items are made in 0 hours.

(b) What is the maximum production rate per item?

13.6 TABLES OF INTEGRALS (OPTIONAL)

Although the Fundamental Theorem of Calculus is a powerful tool, it cannot be used unless you can find the antiderivative of the function to be integrated. Sometimes this is difficult or even impossible. For instance,

$$\int_0^1 e^{x^2}\,dx$$

cannot be evaluated exactly. In such cases, the only option is to use various approximation methods (such as those used by graphing calculators).

Even when a function has an antiderivative, the methods presented in earlier sections may not be adequate for finding it. Computer algebra programs and some graphing calculators, such as TI-89 and Casio FX2, can find the antiderivatives of a wide variety of functions. If you do not have access to such technology, there are tables of integrals available. One such table is Table 3 in Appendix B at the back of this book. The examples below illustrate its use.

(1)

Find the following.

(a) $\displaystyle\int \frac{4}{\sqrt{x^2+100}}\,dx$

(b) $\displaystyle\int \frac{-9}{\sqrt{x^2-4}}\,dx$

Answers:

(a)
$4\ln\left|x+\sqrt{x^2+100}\right|+C$

(b)
$-9\ln\left|x+\sqrt{x^2-4}\right|+C$

EXAMPLE 1 Find $\displaystyle\int \frac{1}{\sqrt{x^2+16}}\,dx.$

By inspecting the table, we see that if $a = 4$, this antiderivative is the same as entry 5 of the table. Entry 5 of the table is

$$\int \frac{1}{\sqrt{x^2+a^2}}\,dx = \ln\left|x+\sqrt{x^2+a^2}\right|+C.$$

Substituting 4 for a in this entry, we get

$$\int \frac{1}{\sqrt{x^2+16}}\,dx = \ln\left|x+\sqrt{x^2+16}\right|+C.$$

This last result could be verified by taking the derivative of the right-hand side of this last equation. ◀ (1)

(2)

Find the following.

(a) $\displaystyle\int \frac{1}{x^2-4}\,dx$

(b) $\displaystyle\int \frac{-6}{x\sqrt{25-x^2}}\,dx$

Answers:

(a) $\displaystyle\frac{1}{4}\ln\left|\frac{x-2}{x+2}\right|+C$

(b)
$\displaystyle\frac{6}{5}\ln\left|\frac{5+\sqrt{25-x^2}}{x}\right|+C$

EXAMPLE 2 Find $\displaystyle\int \frac{8}{16-x^2}\,dx.$

Convert this antiderivative into the one given in entry 7 of the table by writing the 8 in front of the integral sign (permissible only with constants) and by letting $a = 4$. Doing this gives

$$8\int \frac{1}{16-x^2}\,dx = 8\left[\frac{1}{2\cdot 4}\ln\left|\frac{4+x}{4-x}\right|\right]+C$$
$$= \ln\left|\frac{4+x}{4-x}\right|+C.$$ ◀ (2)

EXAMPLE 3 Find $\int \sqrt{9x^2 + 1}\,dx$.

This antiderivative seems most similar to entry 15 of the table. However, entry 15 requires that the coefficient of the x^2 term be 1. We can satisfy that requirement here by factoring out the 9.

$$\int \sqrt{9x^2+1}\,dx = \int \sqrt{9\left(x^2+\frac{1}{9}\right)}\,dx$$
$$= \int 3\sqrt{x^2+\frac{1}{9}}\,dx$$
$$= 3\int \sqrt{x^2+\frac{1}{9}}\,dx$$

Now, use entry 15 with $a = 1/3$.

$$\int \sqrt{9x^2+1}\,dx = 3\left[\frac{x}{2}\sqrt{x^2+\frac{1}{9}} + \frac{\left(\frac{1}{3}\right)^2}{2}\cdot \ln\left|x+\sqrt{x^2+\frac{1}{9}}\right|\right] + C$$
$$= \frac{3x}{2}\sqrt{x^2+\frac{1}{9}} + \frac{1}{6}\ln\left|x+\sqrt{x^2+\frac{1}{9}}\right| + C$$ ◀ 3

3

Find the following.

(a) $\int \frac{3}{16x^2-1}\,dx$

(b) $\int \frac{-1}{100x^2-1}\,dx$

Answers:

(a) $\frac{3}{8}\ln\left|\frac{x-\frac{1}{4}}{x+\frac{1}{4}}\right| + C$

(b) $-\frac{1}{20}\ln\left|\frac{x-\frac{1}{10}}{x+\frac{1}{10}}\right| + C$

13.6 Exercises

Use the table of integrals to find each antiderivative. (See Examples 1–3.)

1. $\int \frac{-4}{\sqrt{x^2+36}}\,dx$
2. $\int \frac{9}{\sqrt{x^2+9}}\,dx$
3. $\int \frac{6}{x^2-9}\,dx$
4. $\int \frac{-12}{x^2-16}\,dx$
5. $\int \frac{-4}{x\sqrt{9-x^2}}\,dx$
6. $\int \frac{3}{x\sqrt{121-x^2}}\,dx$
7. $\int \frac{-2x}{3x+1}\,dx$
8. $\int \frac{6x}{4x-5}\,dx$
9. $\int \frac{2}{3x(3x-5)}\,dx$
10. $\int \frac{-4}{3x(2x+7)}\,dx$
11. $\int \frac{4}{4x^2-1}\,dx$
12. $\int \frac{-6}{9x^2-1}\,dx$
13. $\int \frac{3}{x\sqrt{1-9x^2}}\,dx$
14. $\int \frac{-2}{x\sqrt{1-16x^2}}\,dx$
15. $\int \frac{4x}{2x+3}\,dx$
16. $\int \frac{4x}{6-x}\,dx$
17. $\int \frac{-x}{(5x-1)^2}\,dx$
18. $\int \frac{-3}{x(4x+3)^2}\,dx$
19. $\int x^4 \ln|x|\,dx$
20. $\int 4x^2 \ln|x|\,dx$
21. $\int \frac{\ln|x|}{x^2}\,dx$
22. $\int \frac{-2\ln|x|}{x^3}\,dx$
23. $\int xe^{-2x}\,dx$
24. $\int xe^{3x}\,dx$

Use Table 3 in Appendix B to solve the following problems.

25. **Business** The rate of change of revenue in dollars from the sale of x units of small desk calculators is

$$R'(x) = \frac{1000}{\sqrt{x^2+25}}.$$

Find the total revenue from the sale of the first 20 calculators.

26. Health The rate of reaction to a drug is given by

$$r'(x) = 2x^2e^{-x},$$

where x is the number of hours since the drug was administered. Find the total reaction to the drug from $x = 1$ to $x = 6$.

27. Natural Science The rate of growth of a microbe population is given by

$$m'(x) = 30xe^{2x},$$

where x is time in days. What is the total accumulated growth after 3 days?

28. Social Science The rate (in hours per item) at which a worker in a certain job produces the xth item is

$$h'(x) = \sqrt{x^2 + 16}.$$

What is the total number of hours it will take this worker to produce the first 7 items?

13.7 DIFFERENTIAL EQUATIONS

Suppose that an economist wants to develop an equation that will forecast interest rates. By studying data on previous changes in interest rates, she hopes to find a relationship between the level of interest rates and their rate of change. A function giving the rate of change of interest rates would be the derivative of the function describing the level of interest rates. A **differential equation** is an equation that involves an unknown function, $y = f(x)$, and a finite number of its derivatives. Solving the differential equation for y would give the unknown function to be used for forecasting interest rates.

Differential equations have been important in the physical sciences and engineering for several centuries. More recently, they have been used in the social sciences, life sciences, and economics to solve a variety of problems, including population growth and biological balance, as well as interest rates.

Up to now we have considered only equations whose solutions are *numbers*. The solutions of differential equations, however, are *functions*. For example, the solutions of the differential equation

$$\frac{dy}{dx} = 3x^2 - 2x \tag{1}$$

consist of all functions y that satisfy the equation. Since the left side of the equation is the derivative of y with respect to x, we can solve the equation for y by finding an antiderivative on each side. On the left, the antiderivative is $y + C_1$. On the right side,

$$\int (3x^2 - 2x)\,dx = x^3 - x^2 + C_2.$$

The solutions of equation (1) are given by

$$y + C_1 = x^3 - x^2 + C_2$$

or

$$y = x^3 - x^2 + C_2 - C_1.$$

Replacing the constant $C_2 - C_1$ with the single constant C gives

$$y = x^3 - x^2 + C. \tag{2}$$

(From now on we will add just one constant, with the understanding that it represents the difference between the two constants obtained in the two integrations.)

Find the general solution of

(a) $dy/dx = 4x$;

(b) $dy/dx = -x^3$;

(c) $dy/dx = 2x^2 - 5x$.

Answers:

(a) $y = 2x^2 + C$

(b) $y = -\frac{1}{4}x^4 + C$

(c) $y = \frac{2}{3}x^3 - \frac{5}{2}x^2 + C$

Each different value of C in equation (2) leads to a different solution of equation (1), showing that a differential equation can have an infinite number of solutions. Equation (2) is the **general solution** of the differential equation (1). Some of the solutions of equation (1) are graphed in Figure 13.32. (1)

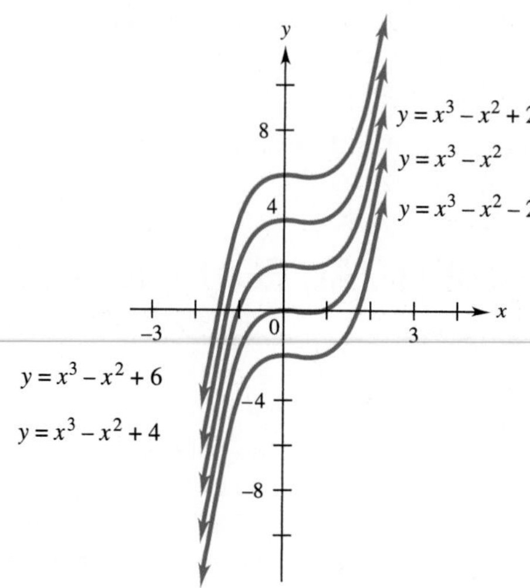

FIGURE 13.32

The simplest kind of differential equation has the form

$$\frac{dy}{dx} = f(x).$$

Equation (1) has this form, so the solution of equation (1) suggests the following generalization.

General Solution of $dy/dx = f(x)$

The general solution of the differential equation $dy/dx = f(x)$ is

$$y = \int f(x)\,dx.$$

EXAMPLE 1 The population P of a flock of birds is growing exponentially so that

$$\frac{dP}{dx} = 20e^{.05x},$$

where x is time in years. Find P in terms of x if there were 20 birds in the flock initially.

Solve the differential equation.

$$P = \int 20e^{.05x}\,dx = \frac{20}{.05}e^{.05x} + C = 400e^{.05x} + C$$

Since P is 20 when x is 0,

$$20 = 400e^{0} + C$$
$$-380 = C,$$

and

$$P = 400e^{.05x} - 380.$$

Find the particular solution in Example 1 if there were 100 birds in the flock after 2 years.

Answer:

$P = 400e^{.05x} - 342$

In Example 1, the given information was used to produce a solution with a specific value of C. Such a solution is called a **particular solution** of the given differential equation. The given information, $P = 20$ when $x = 0$, is called an **initial condition.** (2)

Sometimes a differential equation must be rewritten in the form

$$\frac{dy}{dx} = f(x)$$

before it can be solved.

EXAMPLE 2 Find the particular solution of

$$\frac{dy}{dx} - 2x = 5,$$

given that $y = 2$ when $x = -1$.

Add $2x$ to both sides of the equation to get

$$\frac{dy}{dx} = 2x + 5.$$

The general solution is

$$y = \frac{2x^2}{2} + 5x + C = x^2 + 5x + C.$$

Substituting 2 for y and -1 for x gives

$$2 = (-1)^2 + 5(-1) + C$$
$$C = 6.$$

The particular solution is $y = x^2 + 5x + 6$. (3)

Find the particular solution of

$$2\frac{dy}{dx} - 4 = 6x^2,$$

given that $y = 4$ when $x = 1$.

Answer:

$y = x^3 + 2x + 1$

The solution method used so far in this section is essentially the same as that in Sections 13.1 and 13.2, where we studied antiderivatives. But not all differential equations can be solved so easily. Suppose, for example, that y is a function of x such that

$$\frac{dy}{dx} = \frac{x^2}{y}.$$

We cannot simply integrate both sides since the right side involves both variables. Nevertheless, there is a solution method that we now develop.

Consider a differential equation of the form

$$\frac{dy}{dx} = \frac{f(x)}{g(y)}.$$

(The example in the preceding paragraph is the case when $f(x) = x^2$ and $g(y) = y$.) Up to this point we have used the symbol dy/dx to denote the derivative function of the function y. In advanced courses it is shown that dy/dx can also be interpreted as a quotient of two differentials, dy divided by dx. (You may recall that differentials were used in integration by substitution.) Then we can multiply both sides of the equation above by $g(y)\,dx$ to obtain

$$g(y)\,dy = f(x)\,dx.$$

In this form all terms involving y (including dy) are on one side of the equation, and all terms involving x (and dx) are on the other side. A differential equation in this form is said to be **separable,** since the variables x and y can be separated. A separable differential equation may be solved by integrating each side after separating the variables.

EXAMPLE 3 Find the general solution of $\dfrac{dy}{dx} = -\dfrac{6x}{y}$.

Separate the variables by multiplying both sides by $y\,dx$:

$$y\,dy = -6x\,dx.$$

To solve this equation take antiderivatives on each side.

$$\int y\,dy = \int -6x\,dx$$

$$\frac{y^2}{2} = -3x^2 + C$$

$$3x^2 + \frac{y^2}{2} = C.$$

Find the general solution of

$$\frac{dy}{dx} = \frac{x^2}{y}.$$

Answer:

$$\frac{y^2}{2} - \frac{x^3}{3} = C$$

Since powers of y are involved, it is better to leave the solution in this form rather than trying to solve for y. For each positive constant C, the graph of the solution is an ellipse, as shown in Figure 13.33. ◀ 4

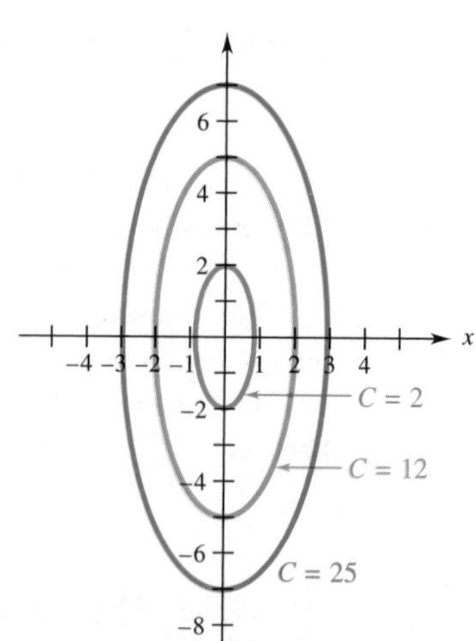

FIGURE 13.33

EXAMPLE 4 Find the general solution of $\frac{dy}{dx} = 5y$.

Separating variables leads to

$$\frac{1}{y}\,dy = 5\,dx.$$

To solve this equation, take antiderivatives on each side.

$$\int \frac{1}{y}\,dy = \int 5\,dx$$

$$\ln|y| = 5x + C$$

$$|y| = e^{5x+C} \quad \text{Definition of logarithm}$$

$$|y| = e^{5x}e^{C} \quad \text{Property of exponents}$$

$$y = e^{5x}e^{C} \quad \text{or} \quad y = -e^{5x}e^{C} \quad \text{Definition of absolute value}$$

Since e^C and $-e^C$ are constants, replace them with the constant M, which may have any nonzero real-number value, to get the single equation

$$y = Me^{5x}.$$

The equation $y = Me^{5x}$ defines an exponential growth function, such as those considered in Section 4.2. ◀

We can now explain how the exponential growth and decay functions presented in Section 4.2 were obtained. In the absence of inhibiting conditions, a population y (which might be human, animal, bacterial, etc.) grows in such a way that the rate of change of population is proportional to the population at time x; that is, there is a constant k such that

$$\frac{dy}{dx} = ky.$$

The constant k is called the **growth rate constant.** Example 4 is the case when $k = 5$. The same argument used there (with k in place of 5) shows that the population y at time x is given by

$$y = Me^{kx},$$

where M is the population at time $x = 0$. A positive value of k indicates growth, while a negative value of k indicates decay. (5)

(5)

Find the particular solution of $\frac{dy}{dx} = .05y$ if y is 2000 when x is 0.

Answer

$y = 2000e^{.05x}$

As a model of population growth, the equation $y = Me^{kx}$ is not realistic over the long run for most populations. As shown by graphs of functions of the form $y = Me^{kx}$, with both M and k positive, growth would be unbounded. Additional factors, such as space restrictions or a limited amount of food, tend to inhibit growth of populations as time goes on. In an alternative model that assumes a maximum population of size N, the rate of growth of a population is proportional to how close the population is to that maximum. These assumptions lead to the differential equation

$$\frac{dy}{dx} = k(N - y),$$

the limited growth function mentioned in Chapter 4. Graphs of limited growth functions look like the graph in Figure 13.34, where y_0 is the initial population.

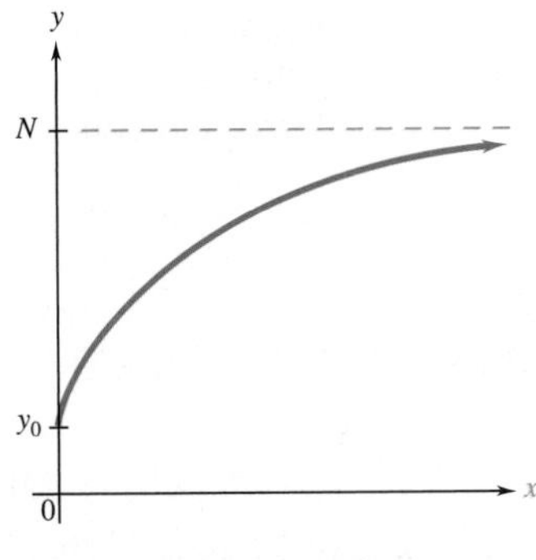

FIGURE 13.34

EXAMPLE 5 A certain area can support no more than 4000 mountain goats. There are 1000 goats in the area at present, with a growth constant of .20.

(a) Write a differential equation for the rate of growth of this population.

Let $N = 4000$ and $k = .20$. The rate of growth of the population is given by

$$\frac{dy}{dx} = .20(4000 - y).$$

To solve for y, first separate the variables.

$$\frac{dy}{4000 - y} = .2\,dx$$

$$\int \frac{dy}{4000 - y} = \int .2\,dx$$

$$-\ln(4000 - y) = .2x + C$$

$$\ln(4000 - y) = -.2x - C$$

$$4000 - y = e^{-.2x-C} = (e^{-.2x})(e^{-C})$$

The absolute value bars are not needed for $\ln(4000 - y)$ because y must be less than 4000 for this population so that $4000 - y$ is always nonnegative. Let $e^{-C} = B$. Then

$$4000 - y = Be^{-.2x}$$

$$y = 4000 - Be^{-.2x}.$$

Find B by using the fact that $y = 1000$ when $x = 0$.

$$1000 = 4000 - B$$

$$B = 3000$$

Notice that the value of B is the difference between the maximum population and the initial population. Substituting 3000 for B in the equation for y gives

$$y = 4000 - 3000e^{-.2x}.$$

(b) What will the goat population be in 5 years?

In 5 years, the population will be

$$y = 4000 - 3000e^{(-.2)(5)} = 4000 - 3000e^{-1}$$

$$= 4000 - 1103.6 = 2896.4,$$

or about 2900 goats. ◀ 6

6

An animal population is growing at a constant rate of 4%. The habitat will support no more than 10,000 animals. There are 3000 animals present now. Write an equation giving the population y in x years.

Answer:

$y = 10{,}000 - 7000e^{-.04x}$

Marginal productivity is the rate at which production changes (increases or decreases) for a unit change in investment. Thus, marginal productivity can be expressed as the first derivative of the function that gives production in terms of investment.

EXAMPLE 6 Suppose the marginal productivity of a manufacturing process is given by

$$P'(x) = 3x^2 - 10, \qquad (3)$$

where x is the amount of the investment in hundreds of thousands of dollars. If the process produces 100 units per month with the present investment of \$300,000 (that

7

In Example 6, if marginal productivity is changed to

$$P'(x) = 3x^2 + 2x,$$

with the same initial conditions,

(a) find an equation for production;

(b) find the increase in production if investment increases to $500,000.

Answers:

(a) $P(x) = x^3 + x^2 + 64$

(b) Production goes from 100 units to 214 units, an increase of 114 units.

is, $x = 3$), by how much would production increase if the investment is increased to $500,000?

To obtain an equation for production, we can take antiderivatives on both sides of equation (3) to get

$$P(x) = x^3 - 10x + C.$$

To find C, use the given initial values: $P(x) = 100$ when $x = 3$.

$$100 = 3^3 - 10(3) + C$$
$$C = 103$$

Production is thus given by

$$P(x) = x^3 - 10x + 103,$$

and if investment is increased to $500,000, production becomes

$$P(5) = 5^3 - 10(5) + 103 = 178.$$

An increase to $500,000 in investment will increase production from 100 units to 178 units. ◀ 7

13.7 Exercises

Find general solutions for the following differential equations. (See Examples 1–4.)

1. $\frac{dy}{dx} = -2x + 3x^2$

2. $\frac{dy}{dx} = 3e^{-2x}$

3. $3x^3 - 2\frac{dy}{dx} = 0$

4. $3x^2 - 3\frac{dy}{dx} = 2$

5. $y\frac{dy}{dx} = x$

6. $y\frac{dy}{dx} = x^2 - 1$

7. $\frac{dy}{dx} = 2xy$

8. $\frac{dy}{dx} = x^2y$

9. $\frac{dy}{dx} = 3x^2y - 2xy$

10. $(y^2 - y)\frac{dy}{dx} = x$

11. $\frac{dy}{dx} = \frac{y}{x}, x > 0$

12. $\frac{dy}{dx} = \frac{y}{x^2}$

13. $\frac{dy}{dx} = y - 5$

14. $\frac{dy}{dx} = 3 - y$

15. $\frac{dy}{dx} = y^2e^x$

16. $\frac{dy}{dx} = \frac{e^x}{e^y}$

Find particular solutions for the following equations. (See Examples 1–4.)

17. $\frac{dy}{dx} + 2x = 3x^2$; $y = 2$ when $x = 0$

18. $\frac{dy}{dx} = 4x^3 - 3x^2 + x$; $y = 0$ when $x = 1$

19. $\frac{dy}{dx}(x^3 + 28) = \frac{x^2}{y}$; $y^2 = 6$ when $x = -3$

20. $\frac{y}{x-3}\frac{dy}{dx} = \sqrt{x^2 - 6x}$; $y^2 = 44$ when $x = 8$

21. $\frac{dy}{dx} = \frac{x^2}{y}$; $y = 3$ when $x = 0$

22. $x^2\frac{dy}{dx} = y$; $y = -1$ when $x = 1$

23. $(2x + 3)y = \frac{dy}{dx}$; $y = 1$ when $x = 0$

24. $x\frac{dy}{dx} - y\sqrt{x} = 0$; $y = 1$ when $x = 0$

25. $\frac{dy}{dx} = \frac{2x + 1}{y - 3}$; $y = 4$ when $x = 0$

26. $\frac{dy}{dx} = \frac{x^2 + 5}{2y - 1}$; $y = 11$ when $x = 0$

27. What is the difference between a general solution and a particular solution of a differential equation?

28. What is meant by a separable differential equation?

Work the following problems. (See Examples 4–6.)

29. Business The marginal productivity of a process is given by

$$\frac{dy}{dx} = \frac{100}{32 - 4x},$$

where x represents the investment (in thousands of dollars). Find the productivity for each of the following investments if productivity is 100 units when the investment is $1000.

(a) $3000 **(b)** $5000

(c) Can investments ever reach $8000 according to this model? Why?

30. Business Sales (in thousands) of a certain product are declining at a rate proportional to the amount of sales, with a decay constant of 25% per year.

(a) Write a differential equation to express the rate of sales decline.

(b) Find a general solution to the equation in part (a).

(c) How much time will pass before sales become 30% of their original value?

31. Natural Science The time dating of dairy products depends on the solution of a differential equation. The rate of growth of bacteria in such products increases with time. If y is the number of bacteria (in thousands) present at a time t (in days), then the rate of growth of bacteria can be expressed as dy/dt and we have

$$\frac{dy}{dt} = kt,$$

where k is an appropriate constant. For a certain product, $k = 10$ and $y = 50$ (in thousands) when $t = 0$.

(a) Solve the differential equation for y.

(b) Suppose the maximum allowable value for y is 550 (thousand). How should the product be dated?

32. Social Sciences A recent report by the U.S. Census Bureau predicts that the Latino-American population will increase from 26.7 million in 1995 to 96.5 million in 2050.* Assuming the unlimited growth model $dy/dt = ky$ fits this population growth, express the population y as a function of the year t. Let 1995 correspond to $t = 0$.

33. Social Sciences (Refer to Exercise 32.) The report also predicted that the African-American population of the United States would increase from 31.4 million in 1995 to 53.6 million in 2050.* Repeat Exercise 32 using this data.

34. Business In a certain area, 1500 small business firms are threatened by bankruptcy. Assume the rate of change in the number of bankruptcies is proportional to the number of small firms that are not yet bankrupt. If the growth constant is 6% and if 100 firms are bankrupt initially, how many will be bankrupt in 2 years?

*"Population Projections of the U.S. by Age, Racc, and Hispanic Origin: 1995 to 2050," U.S. Census Bureau.

35. Natural Science The rate at which the number of bacteria in a culture is changing after the introduction of a bactericide is given by

$$\frac{dy}{dx} = 50 - y,$$

where y is the number of bacteria (in thousands) present at time x. Find the number of bacteria present at each of the following times if there were 1000 thousand bacteria present at time $x = 0$.

(a) $x = 2$ **(b)** $x = 5$ **(c)** $x = 10$

36. Natural Science The amount of a tracer dye injected into the bloodstream decreases exponentially, with a decay constant of 3% per minute. If 6 cc are present initially, how many cc are present after 10 minutes? (Here k will be negative.)

Business *Elasticity of demand was discussed in Case 10 at the end of Chapter 11, where it was defined as*

$$E = -\frac{p}{q} \cdot \frac{dq}{dp}$$

for demand q and price p. Find the general demand equation $q = f(p)$ for each of the following elasticity functions. (Hint: *Set each elasticity function equal to $-\frac{p}{q} \cdot \frac{dq}{dp}$ and then solve for q. Write the constant of integration as ln C.)*

37. $E = 2$

38. $E = \dfrac{4p^2}{q^2}$

39. Social Science Suppose the rate at which a rumor spreads—that is, the number of people who have heard the rumor over a period of time—increases with the number of people who have heard it. If y is the number of people who have heard the rumor, then

$$\frac{dy}{dt} = ky,$$

where t is the time in days and k is a constant.

(a) If y is 1 when $t = 0$, and y is 5 when $t = 2$, find k.

Using the value of k from part (a), find y for each of the following times.

(b) $t = 3$ **(c)** $t = 5$ **(d)** $t = 10$

40. Social Science A company has found that the rate at which a person new to the assembly line produces items is

$$\frac{dy}{dx} = 7.5e^{-.3y},$$

where x is the number of days the person has worked on the line. How many items can a new worker be expected to produce on the eighth day if he produces none when $x = 0$?

Physical Science *Newton's law of cooling states that the rate of change of temperature of an object is proportional to the difference in temperature between the object and the surrounding medium. Thus, if T is the temperature of the object after t hours and C is the (constant) temperature of the surrounding medium, then*

$$\frac{dT}{dt} = -k(T - C),$$

where k is a constant. When a dead body is discovered within 48 hours of the death and the temperature of the medium (air or water, for example) has been fairly constant, Newton's law of cooling can be used to determine the time of death. (The medical examiner does not actually solve the equation for each case but uses a table that is based on the formula.) Use Newton's law of cooling to work the following problems.*

41. Assume the temperature of a body at death is 98.6°F, the temperature of the surrounding air is 68°F, and at the end of one hour the body temperature is 90°F.

(a) Find an equation that gives the body temperature T after t hours.

(b) What was the temperature of the body after two hours?

(c) When will the temperature of the body be 75°F?

(d) Approximately when will the temperature of the body be within .01° of the surrounding air?

42. Repeat Exercise 41 under these conditions: The temperature of the surrounding air is 38°F and after one hour the body temperature is 81°.

*Dennis Callas and David J. Hildreth, "Snapshots of Applications in Mathematics," *College Mathematics Journal* 26, No. 2 (March 1995).

CHAPTER 13 SUMMARY

Key Terms and Symbols

13.1 $\int f(x)\,dx$ indefinite integral of f
antiderivative
integral sign
integrand
power rule
constant multiple rule
sum or difference rule

13.2 differential
integration by substitution

13.3 $\int_a^b f(x)\,dx$ definite integral of f
total change in $F(x)$

13.4 $F(x)\Big|_a^b = F(b) - F(a)$

13.5 consumers' surplus
producers' surplus

13.6 tables of integrals

13.7 differential equation
general solution
particular solution
initial condition
separable differential equation
growth rate constant
marginal productivity

Key Concepts

Indefinite Integral

$F(x)$ is an antiderivative of $f(x)$ if $F'(x) = f(x)$.

If $F'(x) = f(x)$, then $\int f(x)\,dx = F(x) + C$, for any real number C.

Properties of Integrals

$\int k \cdot f(x)\,dx = k \cdot \int f(x)\,dx$, for any real number k.

$\int [f(x) \pm g(x)]\,dx = \int f(x)\,dx \pm \int g(x)\,dx.$

Rules for Integrals

For $u = f(x)$ and $du = f'(x)\,dx$,

$$\int u^n\,du = \frac{u^{n+1}}{n+1} + C; \qquad \int e^u\,du = e^u + C; \qquad \int u^{-1}\,du = \int \frac{du}{u} = \ln|u| + C.$$

The Definite Integral

If f is continuous on $[a, b]$, the definite integral of f from a to b is

$$\int_a^b f(x)\,dx = \lim_{n\to\infty} ([f(x_1) + f(x_2) + f(x_3) + \cdots + f(x_n)] \cdot \Delta x),$$

provided, that this limit exists, where $\Delta x = \dfrac{b-a}{n}$ and x_i is the left endpoint of the ith interval.

Total Change in $F(x)$

Let f be continuous on $[a, b]$ and $f(x) \geq 0$ for all x in $[a, b]$. If $f(x)$ is the rate of change of $F(x)$, then the total change in $F(x)$ as x goes from a to b is given by

$$\int_a^b f(x)\,dx.$$

Fundamental Theorem of Calculus

Let f be continuous on $[a, b]$ and let F be any antiderivative of f. Then

$$\int_a^b f(x)\,dx = F(x)\Big|_a^b = F(b) - F(a).$$

General Solution of $\dfrac{dy}{dx} = f(x)$

The general solution of the differential equation $dy/dx = f(x)$ is

$$y = \int f(x)\,dx.$$

General Solution of $\dfrac{dy}{dx} = ky$

The general solution of the differential equation $dy/dx = ky$ is

$$y = Me^{kx}.$$

Chapter 13 Review Exercises

Find each indefinite integral.

1. $\int (x^2 - 3x - 2)\,dx$
2. $\int (6 - x^2)\,dx$
3. $\int 3\sqrt{x}\,dx$
4. $\int \frac{\sqrt{x}}{2}\,dx$
5. $\int (x^{1/2} + 3x^{-2/3})\,dx$
6. $\int (2x^{4/3} + x^{-1/2})\,dx$
7. $\int \frac{-4}{x^3}\,dx$
8. $\int \frac{5}{x^4}\,dx$
9. $\int -3e^{2x}\,dx$
10. $\int 5e^{-x}\,dx$
11. $\int \frac{2}{x-1}\,dx$
12. $\int \frac{-4}{x+2}\,dx$
13. $\int xe^{3x^2}\,dx$
14. $\int 2xe^{x^2}\,dx$
15. $\int \frac{3x}{x^2-1}\,dx$
16. $\int \frac{-x}{2-x^2}\,dx$
17. $\int \frac{x^2\,dx}{(x^3+5)^4}$
18. $\int (x^2 - 5x)^4(2x - 5)\,dx$
19. $\int \frac{4x-5}{2x^2-5x}\,dx$
20. $\int \frac{12(2x+9)}{x^2+9x+1}\,dx$
21. $\int \frac{x^3}{e^{3x^4}}\,dx$
22. $\int e^{3x^2+4}\,x\,dx$
23. $\int -2e^{-5x}\,dx$
24. $\int e^{-4x}\,dx$
25. Explain how rectangles are used to approximate the area under a curve.
26. Use a graphing calculator for this exercise.
    (a) Use 5 rectangles to approximate the area between the graph of $f(x) = 16x^2 - x^4 + 2$ and the x-axis from $x = -2$ to $x = 3$.
    (b) Use numerical integration to approximate this area.
27. Repeat Exercise 26 for the function $g(x) = -x^4 + 12x^2 + x + 5$ from $x = -3$ to $x = 3$.

28. Approximate the area under the graph of $f(x) = 2x + 3$ and above the x-axis from $x = 0$ to $x = 4$ using four rectangles. Let the height of each rectangle be the function value on the left side.

29. Find $\int_0^4 (2x + 3)\, dx$ by using the formula for the area of a trapezoid: $A = \frac{1}{2}(B + b)h$, where B and b are the lengths of the parallel sides and h is the distance between them. Compare with Exercise 28. If the answers are different, explain why.

30. Explain under what circumstances substitution is useful in integration.

Find each definite integral.

31. $\int_0^1 (x^3 - x^2)\, dx$

32. $\int_0^1 e^{3t}\, dt$

33. $\int_1^5 (3x^{-2} + x^{-3})\, dx$

34. $\int_2^3 (5x^{-2} + x^{-4})\, dx$

35. $\int_1^3 2x^{-1}\, dx$

36. $\int_1^6 8x^{-1}\, dx$

37. $\int_0^4 2e^x\, dx$

38. $\int_1^6 \frac{5}{2}e^{4x}\, dx$

39. $\int_{\sqrt{5}}^5 2x\sqrt{x^2 - 3}\, dx$

40. $\int_0^1 x\sqrt{5x^2 + 4}\, dx$

Find the area between the x-axis and f(x) over each of the given intervals.

41. $f(x) = e^x$; $[0, 2]$

42. $f(x) = 1 + e^{-x}$; $[0, 4]$

Management *Find the cost function for each of the marginal cost functions in Exercises 43–46.*

43. $C'(x) = 10 - 2x$; fixed cost is \$4.

44. $C'(x) = 2x + 3x^2$; 2 units cost \$12.

45. $C'(x) = 3\sqrt{2x - 1}$; 13 units cost \$270.

46. $C'(x) = \frac{1}{x + 1}$; fixed cost is \$18.

Work the following exercises.

47. **Business** The rate of change of sales of a new brand of tomato soup, in thousands, is given by

$$S(x) = \sqrt{x} + 2,$$

where x is the time in months that the new product has been on the market. Find the total sales after 9 months.

48. **Business** The curve shown at the top of the next column gives the rate that an investment accumulates income (in dollars per year). Use rectangles of width 2 units and height determined by the function value at the left endpoint to find the total income accumulated over 10 years.

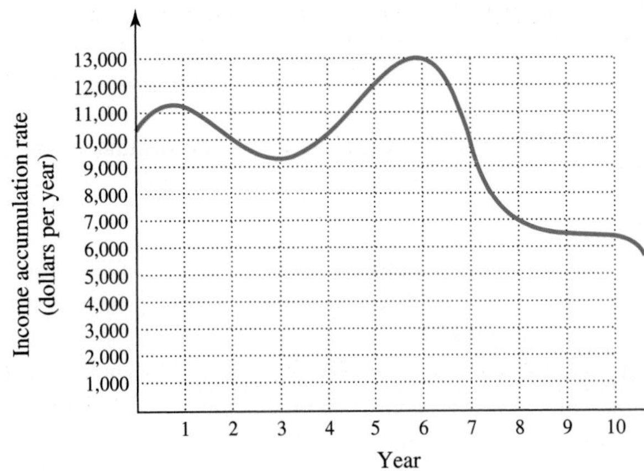

49. **Business** A manufacturer of electronic equipment requires a certain rare metal. He has a reserve supply of 4,000,000 units that he will not be able to replace. If the rate at which the metal is used is given by

$$f(t) = 100{,}000e^{.03t},$$

where t is the time in years, how long will it be before he uses up the supply? (*Hint:* Find an expression for the total amount used in t years and set it equal to the known reserve supply.)

50. **Natural Science** The rate of change of the population of a rare species of Australian spider is given by

$$f(t) = 100 - \sqrt{2.4t + 1},$$

where $f(t)$ is the number of spiders present at time t, measured in months. Find the total number of additional spiders in the first 10 months.

51. **Social Science** Based on data from the New York City Police Department, the homicide rate in New York City between 1988 and 1999 can be approximated by

$$f(t) = 5.45x^3 - 105x^2 + 391x + 1798,$$

where t is the number of years since 1988.* Find the total number of homicides during the 12-year period from the beginning of 1988 to the end of 1999.

**New York Times*, December 19, 1999, p. 55.

52. **Business** The rate at which life insurance proceeds (in billions of dollars per year) are paid out is approximated by $f(t) = 24.5e^{.063t}$, where $t = 0$ corresponds to 1990.* Find the total paid out from 1990 to 1998.

53. **Social Science** Sunday newspaper circulation (in millions of papers per year) is approximated by

$$g(x) = 550(x + 1)^{1/7},$$

where $x = 0$ corresponds to 1970.† Did total circulation from 1970 to 2000 reach 23 billion papers? What was this total circulation?

54. **Business** A company has installed new machinery that will produce a savings rate (in thousands of dollars) of

$$S'(x) = 225 - x^2,$$

where x is the number of years the machinery is to be used. The rate of additional costs (in thousands of dollars) to the company due to the new machinery is expected to be

$$C'(x) = x^2 + 25x + 150.$$

For how many years should the company use the new machinery? Find the net savings (in thousands of dollars) over this period.

55. Explain what consumer's surplus and producers' surplus are.

56. **Business** Suppose that the supply function of some commodity is

$$S(q) = q^2 + 5q + 100$$

and the demand function for the commodity is

$$D(q) = 350 - q^2.$$

(a) Find the producers' surplus.
(b) Find the consumers' surplus.

57. **Natural Science** The rate of infection of a disease (in people per month) is given by the function

$$I'(t) = \frac{100t}{t^2 + 1},$$

where t is the time in months since the disease broke out. Find the total number of infected people over the first four months of the disease.

Use technology or a table of integrals to find the following.

58. $\displaystyle\int \frac{1}{\sqrt{x^2 - 64}}\,dx$

59. $\displaystyle\int \frac{5}{x\sqrt{25 + x^2}}\,dx$

60. $\displaystyle\int \frac{12}{x^2 - 9}\,dx$

61. $\displaystyle\int \frac{15x}{2x - 5}\,dx$

62. What is a differential equation? What is it used for?

Find general solutions for the following differential equations.

63. $\dfrac{dy}{dx} = 2x^3 + 6x$

64. $\dfrac{dy}{dx} = x^2 + 5x^4$

65. $\dfrac{dy}{dx} = \dfrac{3x + 1}{y}$

66. $\dfrac{dy}{dx} = \dfrac{e^x + x}{y - 1}$

Find particular solutions for the following differential equations.

67. $\dfrac{dy}{dx} = 5(e^{-x} - 1)$; $y = 17$ when $x = 0$

68. $\dfrac{dy}{dx} = \dfrac{x}{x^2 - 3}$; $y = 52$ when $x = 2$

69. $(5 - 2x)y = \dfrac{dy}{dx}$; $y = 2$ when $x = 0$

70. $\sqrt{x}\dfrac{dy}{dx} = xy$; $y = 4$ when $x = 1$

71. **Business** A deposit of $10,000 is made to a savings account at 5% interest compounded continuously. Assume that continuous *withdrawals* of $1000 per year are made.
(a) Write a differential equation to describe the situation.
(b) How much will be left in the account after 1 year?

72. **Business** The marginal sales (in hundreds of dollars) of a computer software company are given by

$$\frac{dy}{dx} = 5e^{.2x},$$

where x is the number of months the company has been in business. Assume that sales were 0 initially.
(a) Find the sales after 6 months.
(b) Find the sales after 12 months.

73. **Business** The rate at which a new worker in a certain factory produces items is given by

$$\frac{dy}{dx} = .2(125 - y),$$

where y is the number of items produced by the worker per day, x is the number of days worked, and the maximum production per day is 125 items. Assume the worker produced 20 items the first day on the job ($x = 0$).
(a) Find the number of items the new worker will produce in 10 days.
(b) According to the function that is the solution of the differential equation, can the worker ever produce 125 items in a day?

74. **Physical Science** A roast at a temperature of 40° is put in a 300° oven. After 1 hour the roast has reached a temperature of 150°. Newton's law of cooling states that

$$\frac{dT}{dt} = k(T - T_F),$$

where T is the temperature of an object, the surrounding medium has temperature T_F at time t, and k is a constant.

*American Council of Life Insurance.

†Editor and Publisher Co.

(a) Use Newton's law to find the temperature of the roast after 2 hours.

(b) How long does it take for the roast to reach a temperature of 250°?

75. Natural Science Find an equation relating x to y given the following equations, which describe the interaction of two competing species and their growth rates.

$$\frac{dx}{dt} = .2x - .5xy$$

$$\frac{dy}{dt} = -.3y + .4xy$$

Find the values of x and y for which both growth rates are 0. (*Hint:* Solve the system of equations found by setting each growth rate equal to 0.)

Case 12

Bounded Population Growth

The population growth model $\frac{dP}{dt} = kP$, whose solution is the exponential growth function $P = P_0e^{kt}$, can work well in a situation where a population is changing rapidly over a short period of time but may not be useful in the long term. Exponential functions grow very large very quickly and will eventually exceed any values for the population under consideration. For this reason, the exponential function is said to be *unbounded.* A more realistic growth model would take into account any natural bonds on the population size.

For example, consider a population of cattle or sheep that lives and grazes on a piece of farmland. Sheep, in particular, obtain the majority of their food intake from foraging in their enviroment. The sheep happily graze on the grass, plants, and weeds without exhausting the supply. This allows time for the plant life to grow back, and the eating-growing cycle continues in this way. If the farmer populates the farmland with too many sheep, the plants may be eaten too quickly, failing to grow enough before some of the sheep develop health problems and possibly die. The maximum number of animals the farmland can comfortably sustain over a long period of time is called the *carrying capacity* of the land.

One model used for bounded population growth is the *logistic differential equation*

$$\frac{dP}{dt} = kP(C - P),$$

where k is a positive proportionality constant and C is the carrying capacity of the environment under consideration. When the population satisfies

$$0 < P < C$$

the expression $kP(C - P)$ will be positive, and a positive derivative indicates the population is increasing. If the population P were to exceed the carrying capacity C, the factor $C - P$ becomes negative and

$$\frac{dP}{dt} < 0,$$

indicating that the population is decreasing, which must be the case when the carrying capacity is exceeded. The model agrees with the expected behavior of a bounded population.

For example, given an initial population of 10 sheep on land with carrying capacity $C = 200$ and a proportionality factor of $k = .005$, the logistic differential equation becomes

$$\frac{dP}{dt} = .005P(200 - P).$$

Separating variables and integrating gives

$$\int \frac{1}{P(200-P)}\,dP = \int .005dt. \tag{1}$$

The right side of equation (1) is easily integrated.

$$\int .005dt = .005t + K \tag{2}$$

The left side of equation (1) does not match the standard integration formulas but can be integrated by using the following identity, which you are asked to verify in Exercise 1.

$$\frac{1}{P(C-P)} = \frac{1/C}{P} + \frac{1/C}{C-P} \tag{3}$$

Applying this identity with $C = 200$ allows the left-hand side of equation (1) to be easily integrated.

$$\begin{aligned}\int \frac{1}{P(200-P)}\,dP &= \int \frac{1/200}{P}\,dP + \int \frac{1/200}{200-P}\,dP \\ &= \frac{1}{200}\ln(P) - \frac{1}{200}\ln(200-P)\end{aligned} \tag{4}$$

Substituting (2) and (4) in equation (1) shows that

$$\frac{1}{200}\ln(P) - \frac{1}{200}\ln(200-P) = .005t + K. \tag{5}$$

In Exercise 2, you are asked to solve equation (5) for $P(t)$ and use the fact that $P(0) = 10$ to obtain the population function

$$P(t) = \frac{200e^t}{e^t + 19}.$$

After one year the population would consist of $P(1) \approx 25$ sheep.

The graph of the population function in the Figure illustratres the typical S-shaped curve for a population modeled by the logistic differential equation. The population increases rapidly at first, and then the rate of increase decreases as the population approaches the carrying capacity of its environment.

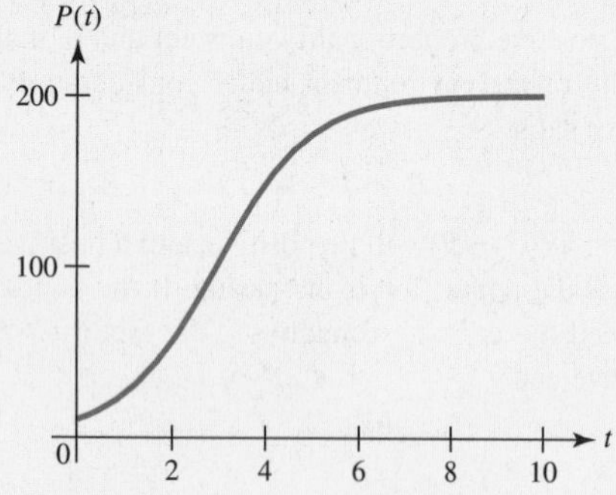

Before the logistic model can be applied, an estimate of the carrying capacity is needed. In the case of cattle and sheep, veterinary and agricultural science along with hundreds of years of recorded experience tell a farmer how much livestock can survive comfortably on a given plot of land. A much more difficult problem is to determine the carrying capacity of a population where many factors contribute to the carrying capacity.

The carrying capacity of Earth, with respect to its human population, is one such problem. Factors such as energy supply, existing food sources, available farmland, climate, and quality of life all enter into an estimate. One estimate, assuming a relatively high standard of living for each inhabitant, puts the carrying capacity between 1.5 and 2 billion. Given the current population of approximately 6 billion, this estimate states we have already exceeded the carrying capacity. More optimistic estimates take into account technologies that are the focus of active research. For example, some say by assuming alternative energy sources, such as solar energy, can be harnessed fully, the carrying capacity can be increased to 23 billion. Further consideration of alternative food sources and methods provided by aquaculture and hydroponics raises that estimate to 50 billion. As you might expect, estimates of Earth's carrying capacity are the subject of great debate.

Exercises

1. Verify equation (3).
2. Use the techniques of Section 13.7 and the fact that $P(0) = 10$ to solve equation (5).
3. Assuming a carrying capacity of 20 billion people and a current population of 6 billion, determine when the world population will reach 18 billion. Use $k = .0011$, a value based on observed world population growth during the 1990s.
4. Under the assumptions of Exercise 3, will Earth's population ever reach 20 billion? Explain.

Chapter 14

Multivariate Calculus

Many of the ideas developed for functions of one variable also apply to functions of more than one variable. In particular, the fundamental idea of derivative generalizes in a very natural way to functions of more than one variable.

14.1 FUNCTIONS OF SEVERAL VARIABLES

If a company produces x items at a cost of \$10 per item, then the total cost $C(x)$ of producing the items is given by

$$C(x) = 10x.$$

The cost is a function of one independent variable, the number of items produced. If the company produces two products, with x of one product at a cost of \$10 each and y of another product at a cost of \$15 each, then the total cost to the firm is a function of *two* independent variables, x and y. By generalizing $f(x)$ notation, the total cost can be written as $C(x, y)$, where

$$C(x, y) = 10x + 15y.$$

When $x = 5$ and $y = 12$, the total cost is written $C(5, 12)$, with

$$C(5, 12) = 10 \cdot 5 + 15 \cdot 12 = 230.$$

Here is a general definition.

> $z = f(x, y)$ is a **function of two independent variables** if a unique value of z is obtained from each ordered pair of real numbers (x, y). The variables x and y are **independent variables;** z is the **dependent variable.** The set of all ordered pairs of real numbers (x, y) such that $f(x, y)$ is a real number is the **domain** of f; the set of all values of $f(x, y)$ is the **range.**

Let $f(x, y) = x^3 - 4x^2 + xy$. Find

(a) $f(2, 4)$;

(b) $f(-2, 3)$.

Answers:

(a) 0

(b) −30

EXAMPLE 1 Let $f(x, y) = 4x^2 + 2xy + 3/y$ and find each of the following.

(a) $f(-1, 3)$.

Replace x with -1 and y with 3.

$$f(-1, 3) = 4(-1)^2 + 2(-1)(3) + \frac{3}{3} = 4 - 6 + 1 = -1$$

(b) $f(2, 0)$

Because of the quotient $3/y$, it is not possible to replace y with 0, so $f(2, 0)$ is undefined. By inspection we see that the domain of the function f consists of all ordered pairs (x, y) such that $y \neq 0$. ◀ ①

EXAMPLE 2 Let x represent the number of milliliters (ml) of carbon dioxide released by the lungs in 1 minute. Let y be the change in the carbon dioxide content of the blood as it leaves the lungs (y is measured in ml of carbon dioxide per 100 ml of blood). The total output of blood from the heart in one minute (measured in ml) is given by C, where C is a function of x and y such that

$$C(x, y) = \frac{100x}{y}.$$

Find $C(320, 6)$.

Replace x with 320 and y with 6 to get

$$C(320, 6) = \frac{100(320)}{6}$$

$$\approx 5333 \text{ ml of blood per minute.}$$ ◀

The definition given before Example 1 was for a function of two independent variables, but similar definitions could be given for functions of three, four, or more independent variables. Functions of more than one independent variable are called **multivariate functions.**

Graphing Equations in Three Variables

Equations in two variables, such as $3x + 2y = 5$, are graphed by using an x-axis and a y-axis to locate points in a plane. The plane determined by the x- and y-axes is called the ***xy*-plane.** In order to graph an equation in three variables, such as $2x + y + z = 6$, a third axis is needed—the z-axis, which goes through the origin in the xy-plane and is perpendicular to both the x-axis and the y-axis.

Figure 14.1 shows one possible way to draw the three axes. In Figure 14.1, the yz-plane is in the plane of the page, with the x-axis perpendicular to the plane of the page.

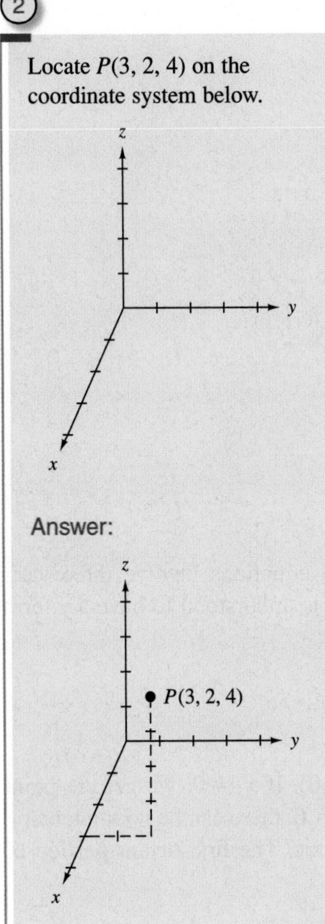

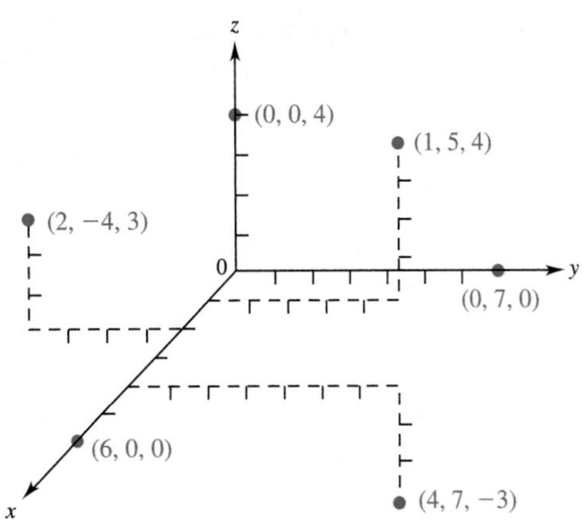

FIGURE 14.1

Just as we graphed ordered pairs earlier, we can now graph **ordered triples** of the form (x, y, z). For example, to locate the point corresponding to the ordered triple $(2, -4, 3)$, start at the origin and go 2 units along the positive x-axis. Then go 4 units in a negative direction (to the left), parallel to the y-axis. Finally, go up 3 units, parallel to the z-axis. The point representing $(2, -4, 3)$ is shown in Figure 14.1, along with several other points. The region of three-dimensional space where all coordinates are positive is called the **first octant.** (2)

Some simple equations in three variables can be graphed by hand. In Chapter 2 we saw that the graph of $ax + by = c$ (where a, b, and c are constants and not both a and b are zero) is a straight line. This result generalizes to three dimensions.

Planes

If a, b, c, and d are real numbers, with a, b, and c not all zero, then the graph of

$$\mathbf{ax + by + cz = d}$$

is a plane.

EXAMPLE 3 Graph $2x + y + z = 6$.

By the result above, the graph of this equation is a plane. Earlier, we graphed straight lines by finding x- and y-intercepts. A similar idea helps graph a plane. To find the x-intercept, the point where the graph crosses the x-axis, let $y = 0$ and $z = 0$.

$$2x + 0 + 0 = 6$$
$$x = 3$$

The point $(3, 0, 0)$ is on the graph. Letting $x = 0$ and $z = 0$ gives the point $(0, 6, 0)$, while $x = 0$ and $y = 0$ lead to $(0, 0, 6)$. The plane through these three points includes the triangular surface shown in Figure 14.2. This region is the first-octant part of the plane that is the graph of $2x + y + z = 6$.

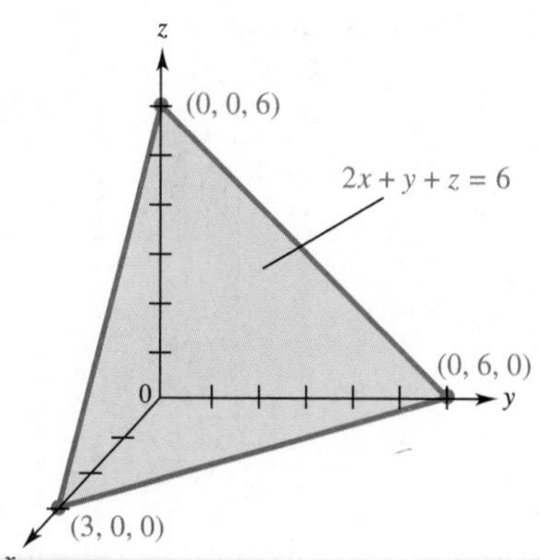

FIGURE 14.2

Throughout this discussion we assume that all equations involve three variables. Consequently, an equation such as $x + z = 6$ is understood to have a y-term with zero coefficient: $x + 0y + z = 6$.

EXAMPLE 4 Graph $x + z = 6$.

To find the x-intercept, let $z = 0$, giving (6, 0, 0). If $x = 0$, we get the point (0, 0, 6). Because there is no y in the equation $x + z = 6$, there can be no y-intercept. A plane that has no y-intercept is parallel to the y-axis. The first-octant portion of the graph of $x + z = 6$ is shown in Figure 14.3.

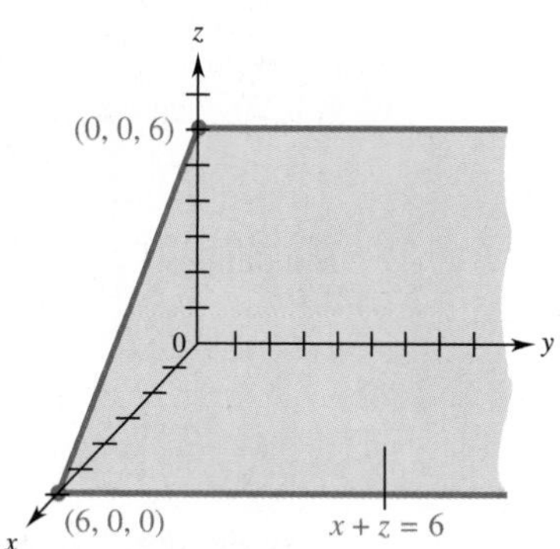

FIGURE 14.3

EXAMPLE 5 Graph each of the following functions in two independent variables.

(a) $x = 3$

This graph, which goes through (3, 0, 0), can have no y-intercept and no z-intercept. It is, therefore, a plane parallel to the y-axis and the z-axis and, therefore, to the yz-plane. The first-octant portion of the graph is shown in Figure 14.4.

FIGURE 14.4

(b) $y = 4$

This graph goes through (0, 4, 0) and is parallel to the xz-plane. The first-octant portion of the graph is shown in Figure 14.5.

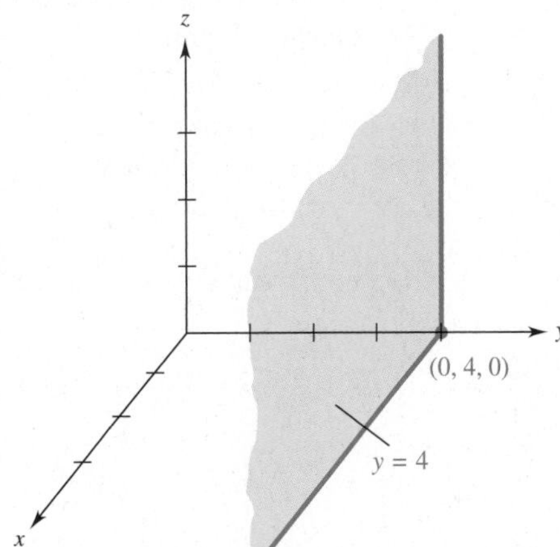
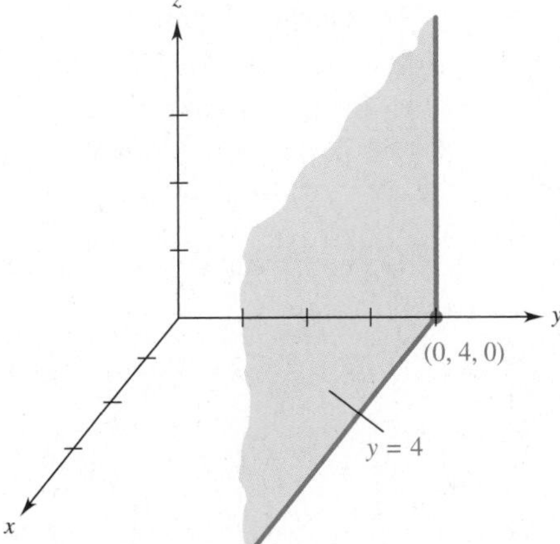

FIGURE 14.5

3

Describe each graph and give any intercepts.

(a) $2x + 3y - z = 4$

(b) $x + y = 3$

(c) $z = 2$

Answers:

(a) A plane; (2, 0, 0), (0, 4/3, 0), (0, 0, −4)

(b) A plane parallel to the z-axis; (3, 0, 0), (0, 3, 0)

(c) A plane parallel to the x-axis and the y-axis; (0, 0, 2)

(c) Graph $z = 1$.

The graph is a plane parallel to the xy-plane, passing through (0, 0, 1). Its first-octant portion is shown in Figure 14.6. ◀ 3

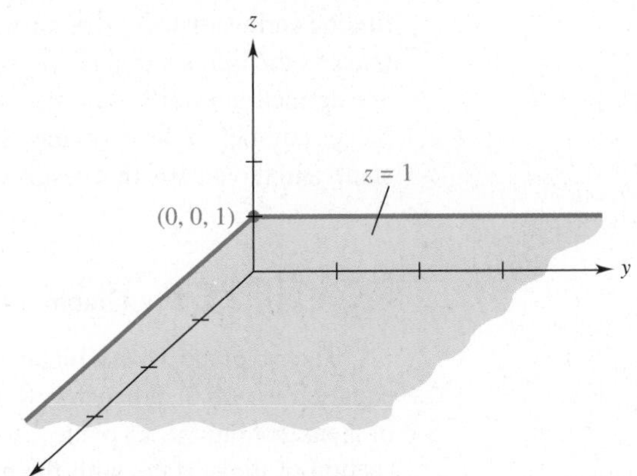

FIGURE 14.6

Graphing Functions of Two Independent Variables

The graph of a function of one variable, $y = f(x)$, is a curve in the plane. If x_0 is in the domain of f, the point $(x_0, f(x_0))$ on the graph lies directly above, or below (possibly on) the number x_0 on the x-axis, as shown in Figure 14.7.

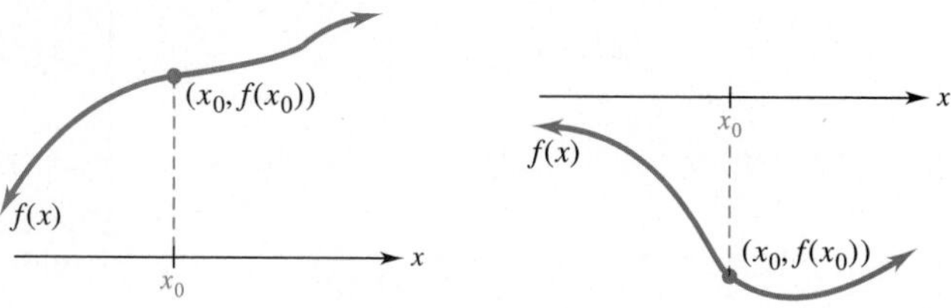

FIGURE 14.7

The graph of a function of two variables, $z = f(x, y)$, is a **surface** in three-dimensional space. If (x_0, y_0) is in the domain of f, the point $(x_0, y_0, f(x_0, y_0))$ lies directly above, or below (possibly on) the point (x_0, y_0) in the xy-plane, as shown in Figure 14.8.

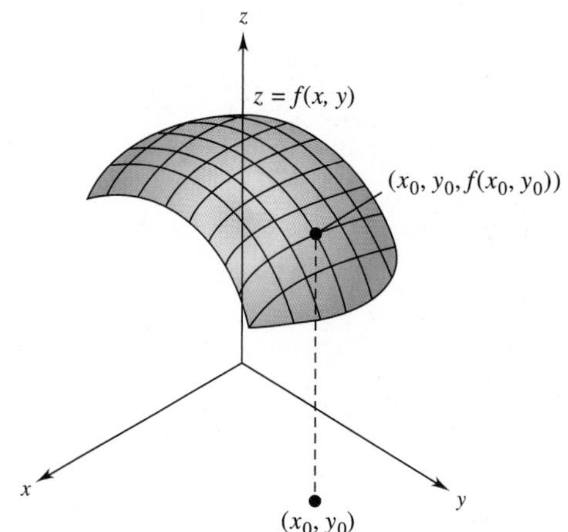

FIGURE 14.8

Computer software is available for drawing the graphs of functions of two independent variables, but you can sometimes get a good picture of a graph without it by finding various **traces**—the curves that result when a surface is cut by a plane. The ***xy*-trace** is the intersection of the surface with the xy-plane. The ***yz*-trace** and ***xz*-trace** are defined similarly. You can also determine the intersection of the surface with planes parallel to the xy-plane. Such planes are of the form $z = k$, where k is a constant, and the curves that result when they cut the surface are called **level curves.**

EXAMPLE 6 Graph $z = x^2 + y^2$.

The yz-plane is the plane in which every point has first coordinate 0, so its equation is $x = 0$. When $x = 0$, the equation becomes $z = y^2$, which is the equation of a parabola in the yz-plane, as shown in Figure 14.9(a). Similarly, to find the intersection of the surface with the xz-plane (whose equation is $y = 0$), let $y = 0$ in the equation. It then becomes $z = x^2$, which is the equation of a parabola in the xz-plane (shown in Figure 14.9(a)). The xy-trace (the intersection of the surface with the plane $z = 0$) is the single point (0, 0, 0) because $x^2 + y^2$ is never negative and is equal to 0 only when $x = 0$ and $y = 0$.

Next, we find the level curves by intersecting the surface with the planes $z = 1$, $z = 2$, $z = 3$, etc. (all of which are parallel to the xy-plane). In each case, the result is a circle

$$x^2 + y^2 = 1, \qquad x^2 + y^2 = 2, \qquad x^2 + y^2 = 3,$$

and so on, as shown in Figure 14.9(b). Drawing the traces and level curves on the same set of axes suggests that the graph of $z = x^2 + y^2$ is the bowl-shaped figure, called a **paraboloid,** that is shown in Figure 14.9(c). ◀

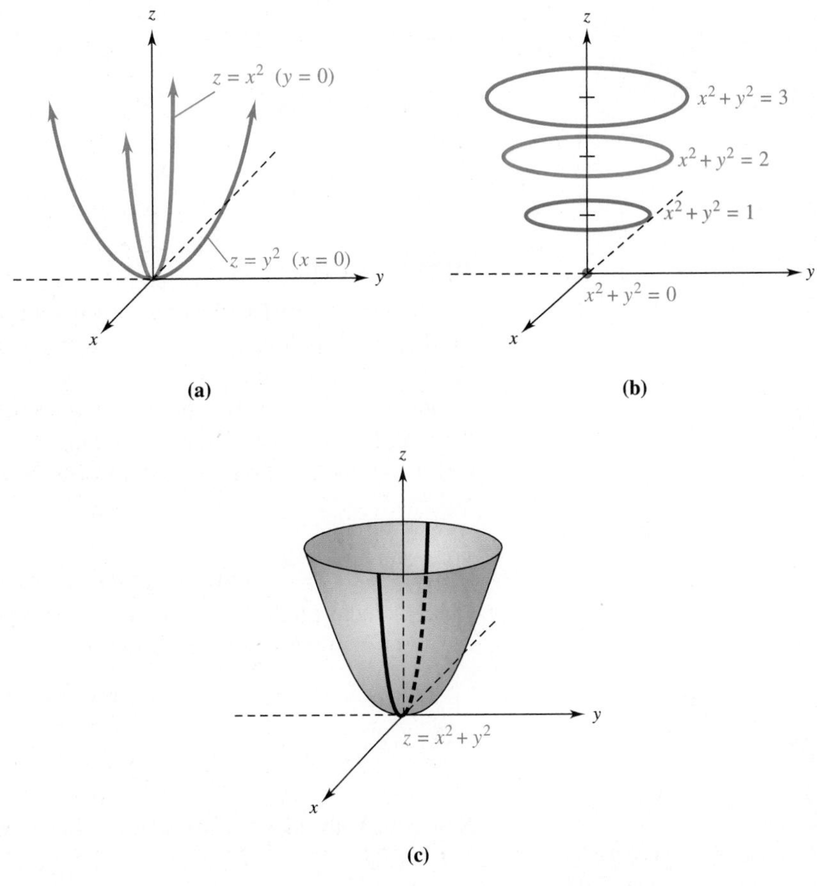

FIGURE 14.9

Figure 14.10 on the next page shows the level curves from Example 6 plotted in the xy-plane. The picture can be thought of as a topographical map that describes the surface generated by $z = x^2 + y^2$, just as the topographical map in Figure 14.11 describes the surface of the land in a part of New York State.

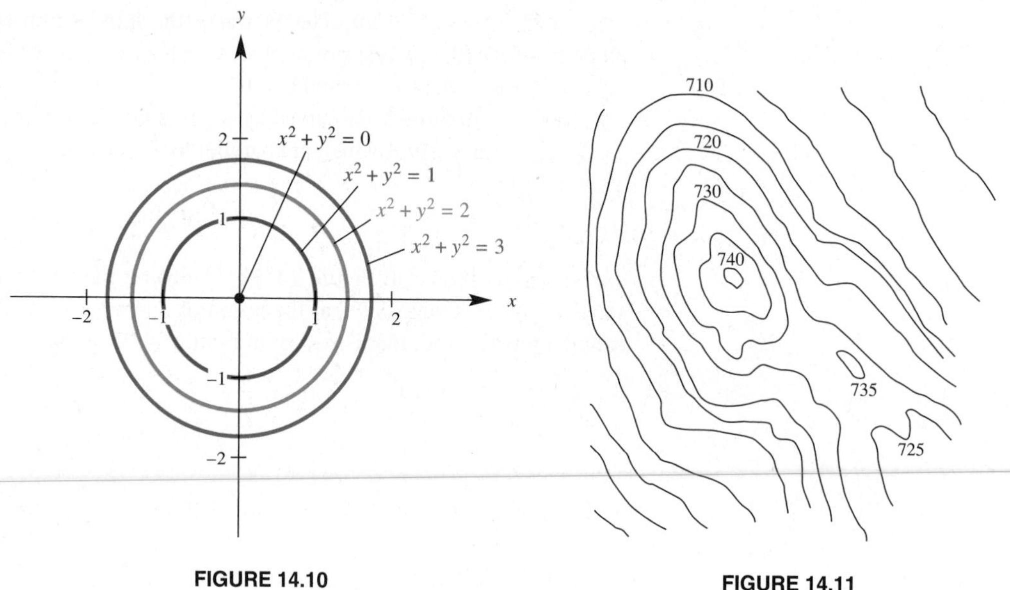

FIGURE 14.10

FIGURE 14.11

Applications

One application of level curves in economics occurs with production functions. A **production function** $z = f(x, y)$ is a function that gives the quantity z of an item produced as a function of x and y, where x is the amount of labor and y is the amount of capital (in appropriate units) needed to produce z units. If the production function has the special form $z = P(x, y) = Ax^a y^{1-a}$, where A is a constant and $0 < a < 1$, the function is called a **Cobb-Douglas production function.**

EXAMPLE 7 Find the level curve at a production of 100 items for the Cobb-Douglas production function $z = x^{2/3}y^{1/3}$.

Let $z = 100$ and solve for y to get

$$100 = x^{2/3}y^{1/3}$$

$$\frac{100}{x^{2/3}} = y^{1/3}.$$

Now cube both sides to express y as a function of x.

$$y = \frac{100^3}{x^2}$$

$$y = \frac{1{,}000{,}000}{x^2}$$

The level curve of height 100 is shown graphed in three dimensions in Figure 14.12(a) and on the familiar xy-plane in Figure 14.12(b). The points of the graph correspond to those values of x and y that lead to production of 100 items. ◀ 4

Find the equation of the level curve for production of 100 items if the production function is $z = 5x^{1/4}y^{3/4}$.

Answer:

$$y = \frac{20^{4/3}}{x^{1/3}}$$

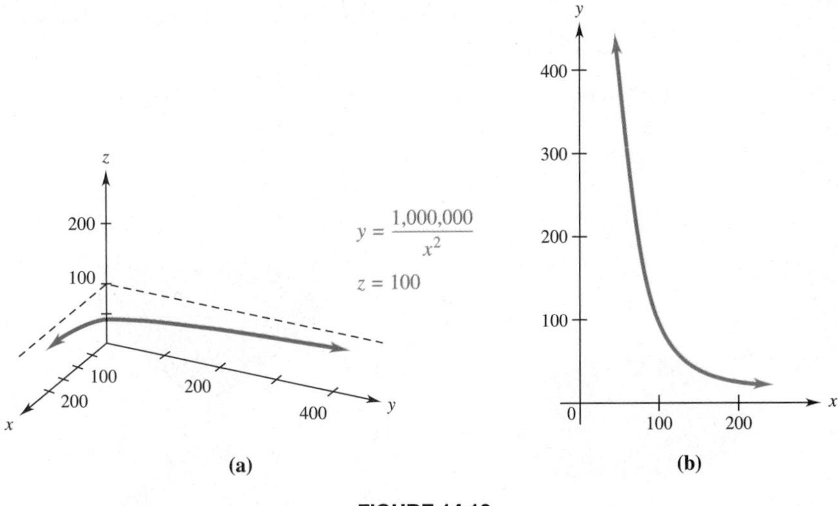

FIGURE 14.12

The curve in Figure 14.12 is called an *isoquant,* for *iso* (equal) and *quant* (amount). In Example 7, the "amounts" all "equal" 100.

Other Three-Dimensional Graphs

The graphing techniques discussed above may not always be effective. So technology is normally used to graph equations in three variables. Figure 14.13 shows the graph of $z = x^2 + y^2$, which was discussed in Example 6, on a TI-92 graphing calculator (most graphing calculators cannot do three-dimensional graphing). Figure 14.14 shows the same graph drawn by the computer program Maple™. The computer-generated graph has more detail than a graphing calculator can produce.

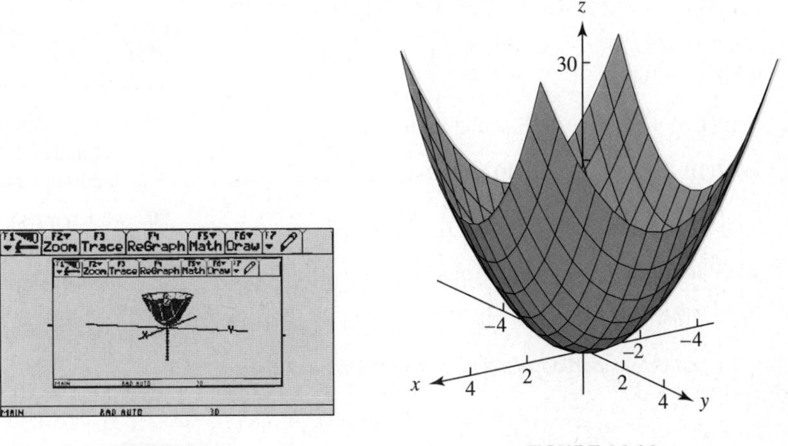

FIGURE 14.13

FIGURE 14.14

Figure 14.15 shows some other computer-generated graphs.

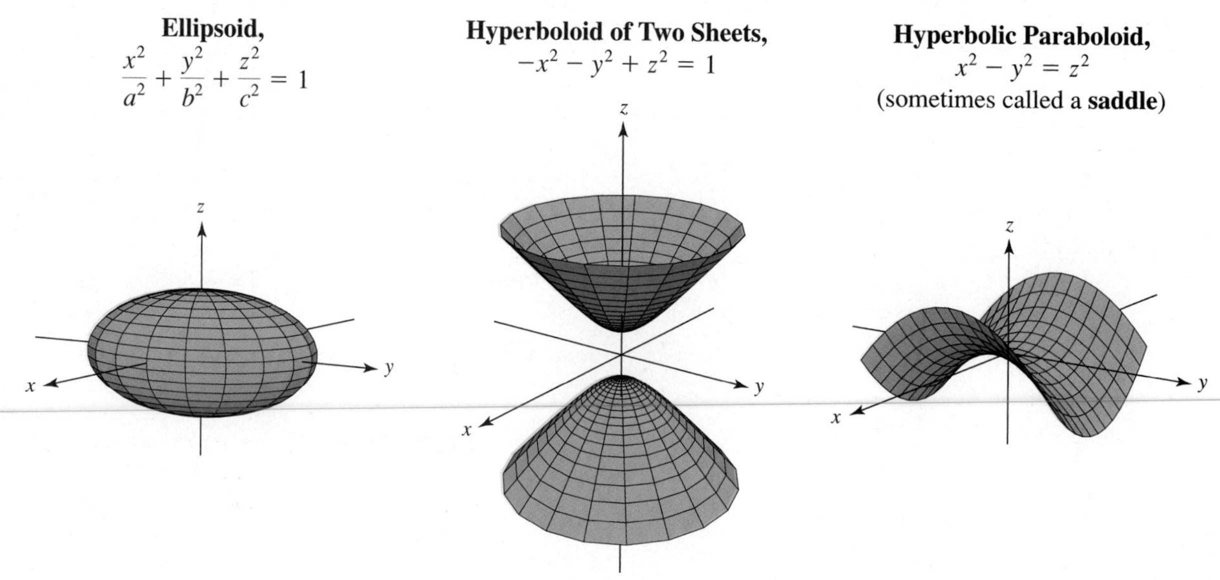

FIGURE 14.15

14.1 Exercises

For each of the following functions, find f(2, −1), f(−4, 1), f(−2, −3), and f(0, 8). (See Example 1.)

1. $f(x, y) = 5x + 2y - 4$ **2.** $f(x, y) = 2x^2 - xy + y^2$

3. $f(x, y) = \sqrt{y^2 + 2x^2}$ **4.** $f(x, y) = \frac{3x + 4y}{\ln|x|}$

5. Let $g(x, y) = -x^2 - 4xy + y^3$. Find the following.

(a) $g(-2, 4)$ **(b)** $g(-1, -2)$
(c) $g(-2, 3)$ **(d)** $g(5, 1)$

6. Let $f(x, y) = \frac{\sqrt{9x + 5y}}{\log x}$. Find the following.

(a) $f(10, 2)$ **(b)** $f(100, 1)$
(c) $f(1000, 0)$ **(d)** $f\left(\frac{1}{10}, 5\right)$

7. What are xy-, xz-, and yz-traces of a graph?

8. What is a level curve?

Graph the first-octant portion of each of the following planes. (See Examples 3–5.)

9. $3x + 2y + z = 12$ **10.** $2x + 3y + 3z = 18$

11. $x + y = 5$ **12.** $y + z = 3$

13. $z = 4$ **14.** $y = 3$

Graph the level curves in the first octant at heights of z = 0, z = 2, and z = 4 for the following equations. (See Example 6.)

15. $3x + 2y + z = 18$ **16.** $x + 3y + 2z = 8$

17. $y^2 - x = -z$ **18.** $2y - \frac{x^2}{3} = z$

Business *Find the level curve at a production of 500 for each of the production functions in Exercises 19 and 20. Graph each function on the xy-plane. (See Example 7.)*

19. The production function z for the United States was once estimated as $z = x^{.7}y^{.3}$, where x stands for the amount of labor and y stands for the amount of capital.

20. If x represents the amount of labor and y the amount of capital, a production function for Canada is approximately $z = x^{.4}y^{.6}$.

Business *The multiplier function*

$$M = \frac{(1 + i)^n(1 - t) + t}{[1 + (1 - t)\, i]^n}$$

compares the growth of an Individual Retirement Account (IRA) with the growth of the same deposit in a regular savings account. The function M depends on the three variables n, i, and t, where n represents the number of years an amount is left at interest, i represents the interest rate in both types of accounts, and t represents the income tax rate. Values of M > 1 indicate that the IRA grows faster than the savings account. Let M = f(n, i, t) and work Exercises 21 and 22.

21. Find the multiplier when funds are left for 25 years at 5% interest and the income tax rate is 28%. Which account grows faster?

22. What is the multiplier when money is invested for 25 years at 6% interest and the income tax rate is 33%? Which account grows faster?

23. Natural Science The surface area of a human (in square meters) is approximated by

$$A = .202W^{.425}H^{.725},$$

where W is the weight of the person in kilograms and H is the height in meters.* Find A for the following data.

(a) Weight = 72 kg; height = 1.78 m
(b) Weight = 65 kg; height = 1.40 m
(c) Weight = 70 kg; height = 1.60 m
(d) Using your weight and height, find your own surface area.

24. Natural Science The oxygen consumption of a well-insulated mammal that is not sweating is approximated by

$$m = \frac{2.5(T - F)}{w^{.67}},$$

where T is the internal body temperature of the animal (in °C), F is the temperature of the outside of the animal's fur (in °C), and w is the animal's weight in kilograms.* Find m for the following data.

(a) Internal body temperature = 38°C; outside temperature = 6°C; weight = 32 kg
(b) Internal body temperature = 40°C; outside temperature = 20°C; weight = 43 kg

25. Natural Science An article entitled "How Dinosaurs Ran" explains that the locomotion of different sized animals can be compared when they have the same Froude number, defined as

$$F = \frac{v^2}{gl},$$

where v is the velocity, g is the acceleration of gravity (9.81 m/sec²), and l is the leg length.†

(a) One result described in the article is that different animals change from a trot to a gallop at the same Froude number, roughly 2.56. Find the velocity at which this change occurs for a ferret, with a leg length of .09 m, and a rhinoceros, with a leg length of 1.2 m.

(b) Ancient footprints in Texas of a sauropod, a large herbivorous dinosaur, are roughly 1m in diameter, corresponding to a leg length of roughly 4m. By comparing the stride divided by the leg length with that of various modern creatures, it can be determined that the Froude number for these dinosaurs is roughly .025. How fast were the sauropods traveling?

26. Natural Science Using data collected by the U.S. Forest Service, the annual number of deer-vehicle accidents for any given county in Ohio can be estimated by the function

$$A(L, T, U, C) = 53.02 + .383L + .0015T + .0028U - .0003C,$$

where A is the estimated number of accidents, L is the road length in kilometers, T is the total county land area in hundred acres (Ha), U is the urban land area in hundred acres, and C is the number of hundred acres of crop land.*

(a) Use this formula to estimate the number of deer-vehicle accidents for Mahoning County, where $L = 266$ km, $T = 107{,}484$ Ha, $U = 31{,}697$ Ha, and $C = 24{,}870$ Ha. The actual value was 396.

(b) Given the magnitude and nature of the input numbers, which of the variables have the greatest potential to influence the number of deer-vehicle accidents? Explain your answer.

27. Business Extra postage is charged for parcels sent by U.S. mail that are more than 108 inches in length and girth combined. (Girth is the distance around the parcel perpendicular to its length. See the figure.) Express the combined length and girth as a function of L, W, and H.

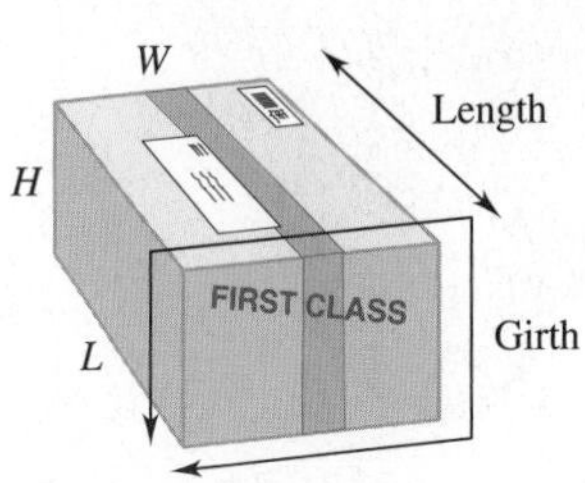

*Exercises 23 and 24 from Clow, Duane J., and N. Scott Urquhart, *Mathematics in Biology*. Ardsley House, 1984.

†R, MacNeill Alexander, "How Dinosaurs Ran," *Scientific American* 264 (April 1991): 4.

*Aaron Iverson and Louis Iverson, "Spatial and Temporal Trends of Deer Harvest and Deer-Vehicle Accidents in Ohio," *Ohio Journal of Science* 99 (1999): 84–94.

28. Natural Science Using data collected by the U.S. Forest Service, the annual number of deer that are harvested for any given county in Ohio can be estimated by the function

$$N(R, C) = 329.32 + .0377R - .0171C,$$

where N is the estimated number of harvested deer, R is the rural land area in hundred acres, and C is the number of hundred acres of crop land.* Use this formula to estimate the number of harvested deer for Tuscarawas County, where $R = 141{,}319$ Ha and $C = 37{,}960$ Ha. The actual value in 1995 was 4925 deer harvested.

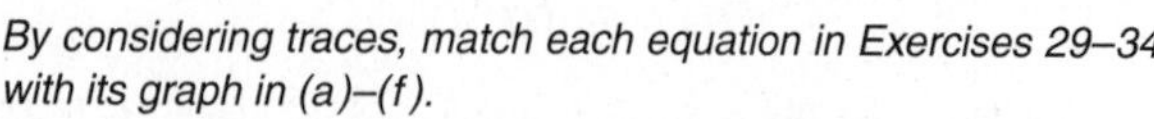

By considering traces, match each equation in Exercises 29–34 with its graph in (a)–(f).

29. $z = x^2 + y^2$

30. $z^2 - y^2 - x^2 = 1$

31. $x^2 - y^2 = z$

32. $z = y^2 - x^2$

33. $\dfrac{x^2}{16} + \dfrac{y^2}{25} + \dfrac{z^2}{4} = 1$

34. $z = 5(x^2 + y^2)^{-1/2}$

(a)

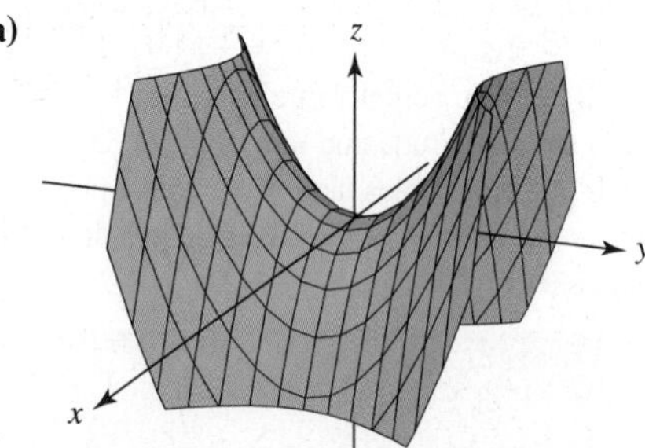

(b)

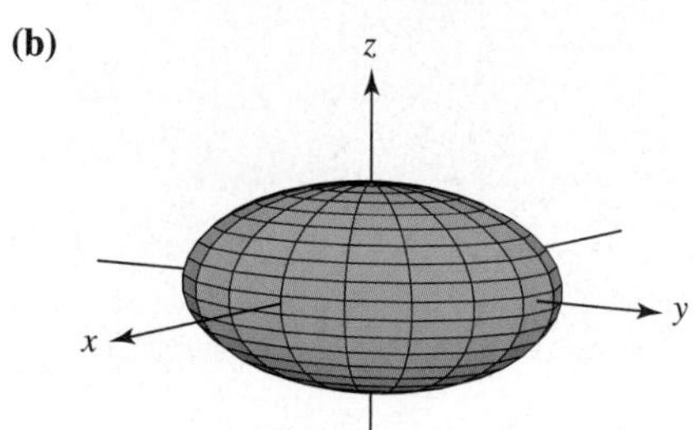

(c)

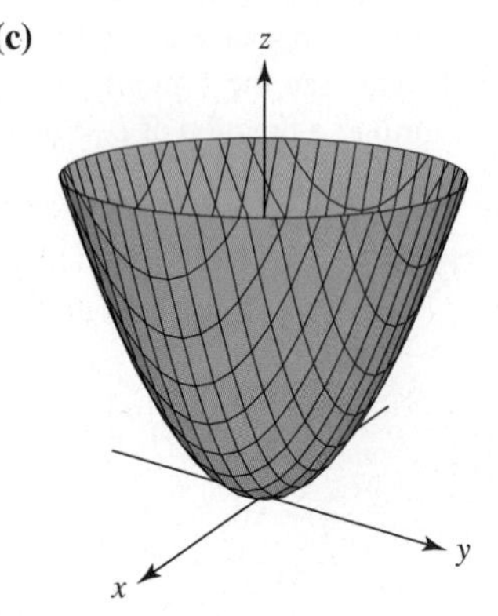

(d)

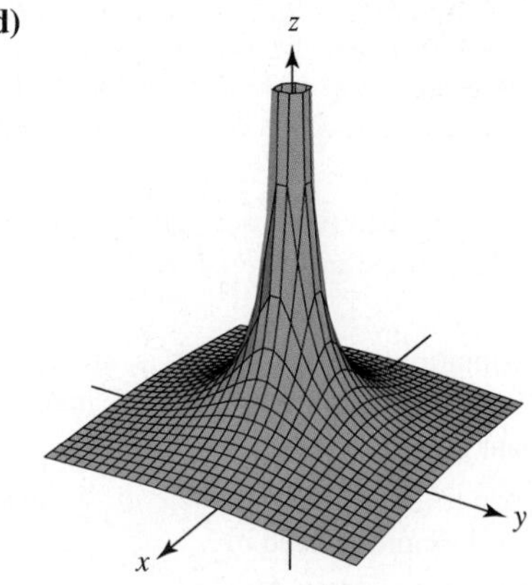

(e)

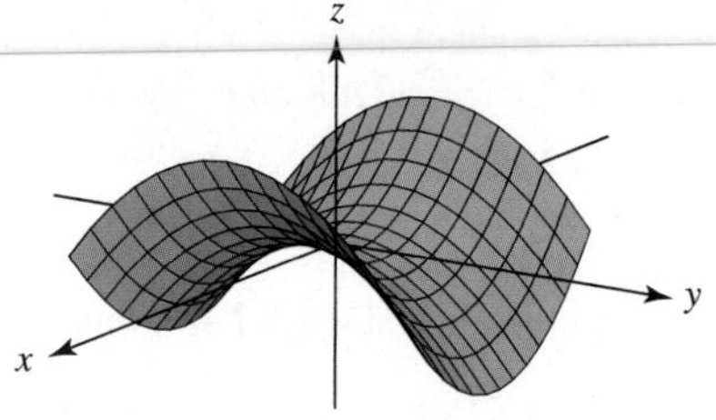

(f)

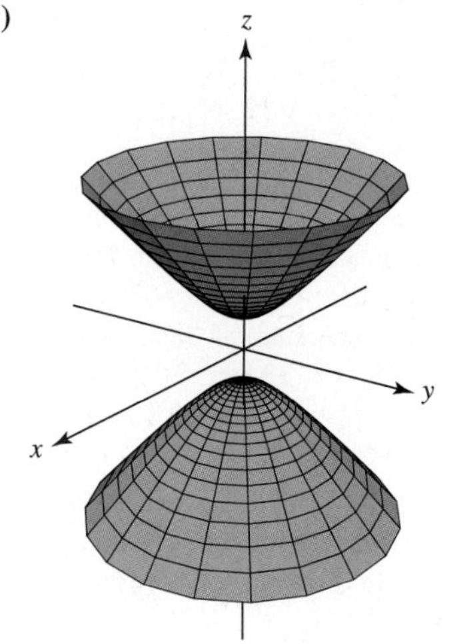

*Aaron Iverson and Louis Iverson, "Spatial and Temporal Trends of Deer Harvest and Deer-Vehicle Accidents in Ohio," *Ohio Journal of Science* 99 (1999): 84–94.

14.2 PARTIAL DERIVATIVES

A small firm makes only two products, radios and audiocassette recorders. The profits of the firm are given by

$$P(x, y) = 40x^2 - 10xy + 5y^2 - 80,$$

where x is the number of units of radios sold and y is the number of units of recorders sold. How will a change in x or y affect P?

Suppose that sales of radios have been steady at 10 units; only the sales of recorders vary. The management would like to find the marginal profit with respect to y, the number of recorders sold. Recall that marginal profit is given by the derivative of the profit function. Here, x is fixed at 10. Using this information, we begin by finding a new function, $f(y) = P(10, y)$. Let $x = 10$ to get

$$\begin{aligned} f(y) = P(10, y) &= 40(10)^2 - 10(10)y + 5y^2 - 80 \\ &= 3920 - 100y + 5y^2. \end{aligned}$$

The function $f(y)$ shows the profit from the sale of y recorders, assuming that x is fixed at 10 units. Find the derivative df/dy to get the marginal profit with respect to y.

$$\frac{df}{dy} = -100 + 10y$$

In this example, the derivative of the function $f(y)$ was taken with respect to y only; we assumed that x was fixed. To generalize, let $z = f(x, y)$. An intuitive definition of the *partial derivatives* of f with respect to x and y follows.

Partial Derivatives (Informal Definition)

The **partial derivative of f with respect to x** is the derivative of f obtained by treating x as a variable and y as a constant.

The **partial derivative of f with respect to y** is the derivative of f obtained by treating y as a variable and x as a constant.

The symbols $f_x(x, y)$ (no prime used), $\partial z/\partial x$, and $\partial f/\partial x$ are used to represent the partial derivative of $z = f(x, y)$ with respect to x, with similar symbols used for the partial derivative with respect to y. The symbol $f_x(x, y)$ is often abbreviated as just f_x, with $f_y(x, y)$ abbreviated f_y.

Generalizing from the definition of derivative given earlier, partial derivatives of a function $z = f(x, y)$ are formally defined as follows.

Partial Derivatives (Formal Definition)

Let $z = f(x, y)$ be a function of two variables. Then the **partial derivative of f with respect to x** is

$$f_x(x, y) = \lim_{h \to 0} \frac{f(x + h, y) - f(x, y)}{h};$$

the **partial derivative of f with respect to y** is

$$f_y(x, y) = \lim_{h \to 0} \frac{f(x, y + h) - f(x, y)}{h};$$

provided these limits exist.

Similar definitions could be given for functions of more than two independent variables.

1

Find f_x and f_y.

(a) $f(x, y) = -x^2y + 3xy + 2xy^2$

(b) $f(x, y) = x^3 + 2x^2y + xy$

Answers:

(a) $f_x = -2xy + 3y + 2y^2$; $f_y = -x^2 + 3x + 4xy$

(b) $f_x = 3x^2 + 4xy + y$; $f_y = 2x^2 + x$

EXAMPLE 1 Let $f(x, y) = 4x^2 - 9xy + 6y^3$. Find f_x and f_y.

To find f_x, treat y as a constant and x as a variable. The derivative of the first term, $4x^2$, is $8x$. In the second term, $-9xy$, the constant coefficient of x is $-9y$, so the derivative with x as the variable is $-9y$. The derivative of $6y^3$ is zero, since we are treating y as a constant. Thus,

$$f_x = 8x - 9y.$$

Now, to find f_y, treat y as a variable and x as a constant. Since x is a constant, the derivative of $4x^2$ is zero. In the second term, the coefficient of y is $-9x$ and the derivative of $-9xy$ is $-9x$. The derivative of the third term is $18y^2$. Thus,

$$f_y = -9x + 18y^2.$$ ◀ 1

Find f_x and f_y.

(a) $f(x, y) = \ln(2x + 3y)$

(b) $f(x, y) = e^{xy}$

Answers:

(a) $f_x = \frac{2}{2x + 3y}$; $f_y = \frac{3}{2x + 3y}$

(b) $f_x = ye^{xy}$; $f_y = xe^{xy}$

EXAMPLE 2 Let $f(x, y) = \ln(x^2 + y)$. Find f_x and f_y.

Recall the formula for the derivative of a natural logarithmic function. If $y = \ln(g(x))$, then $y' = g'(x)/g(x)$. Using this formula,

$$f_x = \frac{D_x(x^2 + y)}{x^2 + y} = \frac{2x}{x^2 + y},$$

and

$$f_y = \frac{D_y(x^2 + y)}{x^2 + y} = \frac{1}{x^2 + y}.$$ ◀ 2

The notation

$$f_x(a, b) \quad \text{or} \quad \frac{\partial f}{\partial x}(a, b)$$

represents the value of a partial derivative when $x = a$ and $y = b$, as shown in the next example.

EXAMPLE 3 Let $f(x, y) = 2x^2 + 3xy^3 + 2y + 5$. Find the following.

3

Let $f(x, y) = x^2 + xy^2 + 5y - 10$. Find the following.

(a) $f_x(2, 1)$

(b) $\frac{\partial f}{\partial y}(-1, 0)$

Answers:

(a) 5

(b) 5

(a) $f_x(-1, 2)$

First, find f_x by holding y constant.

$$f_x = 4x + 3y^3$$

Now let $x = -1$ and $y = 2$.

$$f_x(-1, 2) = 4(-1) + 3(2)^3 = -4 + 24 = 20$$

(b) $\frac{\partial f}{\partial y}(-4, -3)$

Since $\partial f/\partial y = 9xy^2 + 2$,

$$\frac{\partial f}{\partial y}(-4, -3) = 9(-4)(-3)^2 + 2 = 9(-36) + 2 = -322.$$ 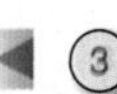

The derivative of a function of one variable can be interpreted as the slope of the tangent line to the graph at that point. With some modification, the same is true of partial derivatives of functions of two variables. At a point on the graph of a function of two variables, $z = f(x, y)$, there may be many tangent lines, all of which lie in the same tangent plane, as shown in Figure 14.16.

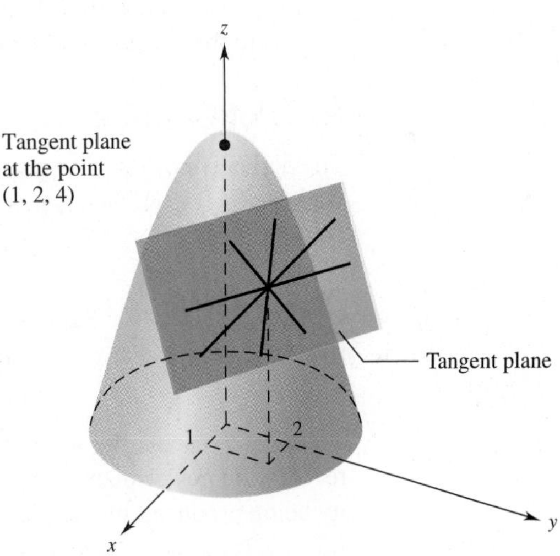

FIGURE 14.16

In any particular direction, however, there will be only one tangent line. We use partial derivatives to find the slope of the tangent lines in the x- and y-directions as follows.

Figure 14.17 shows a surface $z = f(x, y)$ and a plane that is parallel to the xz-plane. The equation of the plane is $y = a$. (This corresponds to holding y fixed.)

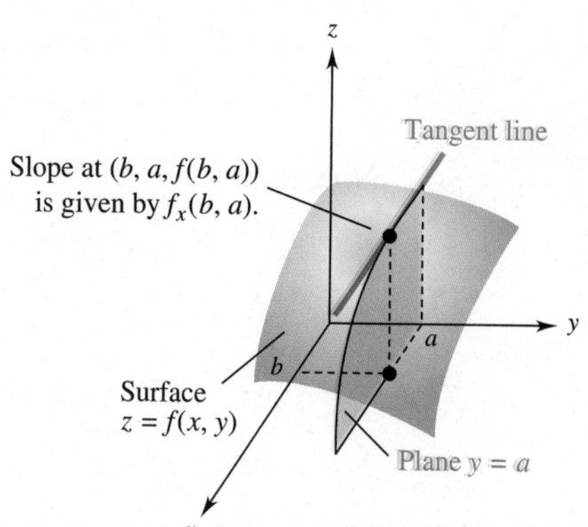

FIGURE 14.17

Because $y = a$ for points on the plane, any point on the curve that represents the intersection of the plane and the surface must have the form $(x, a, f(x, a))$. Thus, this curve can be described as $z = f(x, a)$. Since a is constant, $z = f(x, a)$ is a function of one variable. When the derivative of $z = f(x, a)$ is evaluated at $x = b$, it gives the slope of the line tangent to this curve at the point $(b, a, f(b, a))$, as shown in Figure 14.17. Thus, the partial derivative of f with respect to x, $f_x(b, a)$, gives the rate of change of the surface $z = f(x, y)$ in the x-direction at the point $(b, a, f(b, a))$. In the same way, the partial derivative with respect to y will give the slope of the line tangent to the surface in the y-direction at the point $(b, a, f(b, a))$.

Rate of Change

The derivative of $y = f(x)$ gives the rate of change of y with respect to x. In the same way, if $z = f(x, y)$, then f_x gives the rate of change of z with respect to x, if y is held constant.

EXAMPLE 4 Suppose that the temperature of the water at the point on a river where a nuclear power plant discharges its hot wastewater is approximated by

$$T(x, y) = 2x + 5y + xy - 40.$$

Here, x represents the temperature of the river water in degrees Celsius before it reaches the power plant and y is the number of megawatts (in hundreds) of electricity being produced by the plant.

(a) Find and interpret $T_x(9, 5)$.
First, find the partial derivative T_x.

$$T_x = 2 + y$$

This partial derivative gives the rate of change of T with respect to x. Replacing x with 9 and y with 5 gives

$$T_x(9, 5) = 2 + 5 = 7.$$

Just as marginal cost is the approximate cost of one more item, this result, 7, is the approximate change in temperature of the output water if input water temperature changes by 1 degree, from $x = 9$ to $x = 9 + 1 = 10$, while y remains constant at 5 (500 megawatts of electricity produced).

4

Use the function of Example 4 to find and interpret the following.

(a) $T_x(5, 4)$

(b) $T_y(8, 3)$

Answers:

(a) $T_x(5, 4) = 6$; the approximate increase in temperature if the input temperature increases from 5 to 6 degrees

(b) $T_y(8, 3) = 13$; the approximate increase in temperature if the production of electricity increases from 300 to 400 megawatts

(b) Find and interpret $T_y(9, 5)$.
The partial derivative T_y is

$$T_y = 5 + x.$$

This partial derivative gives the rate of change of T with respect to y, with

$$T_y(9, 5) = 5 + 9 = 14.$$

This result, 14, is the approximate change in temperature resulting from a 1-unit increase in production of electricity from $y = 5$ to $y = 5 + 1 = 6$ (from 500 to 600 megawatts), while the input water temperature x remains constant at 9°C. ◀ 4

As mentioned in the previous section, if $P(x, y)$ gives the output P produced by x units of labor and y units of capital, $P(x, y)$ is a production function. The partial derivatives of this production function have practical implications. For example, $\partial P/\partial x$ gives the marginal productivity of labor. This represents the rate at which the output is changing with respect to changes in labor for a fixed capital investment. That is, if the capital investment is held constant and labor is increased by 1 work hour, $\partial P/\partial x$ will yield the approximate change in the production level. Likewise, $\partial P/\partial y$ gives the marginal productivity of capital, which represents the rate at which the output is changing with respect to changes in capital for a fixed labor value. So if the labor force is held constant and the capital investment is increased by 1 unit, $\partial P/\partial y$ will approximate the corresponding change in the production level.

EXAMPLE 5 A company that manufactures computers has determined that its production function is given by

$$P(x, y) = 500x + 800y + 3x^2y - x^3 - \frac{y^4}{4},$$

where x is the size of the labor force (in work hours per week) and y is the amount of capital (in units of \$1000) invested. Find the marginal productivity of labor and the marginal productivity of capital when $x = 50$ and $y = 20$ and interpret the results.

The marginal productivity of labor is found by taking the derivative of P with respect to x.

$$\frac{\partial P}{\partial x} = 500 + 6xy - 3x^2$$

$$\frac{\partial P}{\partial x}(50, 20) = 500 + 6(50)(20) - 3(50)^2$$

$$= -1000$$

Thus, if the capital investment is held constant at \$20,000 and labor is increased from 50 to 51 work hours per week, production will decrease by about 1000 units. In the same way, the marginal productivity of capital is $\partial P/\partial y$.

$$\frac{\partial P}{\partial y} = 800 + 3x^2 - y^3$$

$$\frac{\partial P}{\partial y}(50, 20) = 800 + 3(50)^2 - 20^3$$

$$= 300$$

Suppose a production function is given by $P(x, y) = 10x^2y + 100x + 400y - 5xy^2$, where x and y are defined as in Example 5. Find the marginal productivity of labor and capital when $x = 30$ and $y = 50$.

Answer:

Marginal productivity of labor is 17,600. Marginal productivity of capital is -5600.

If work hours are held constant at 50 hours per week and the capital investment is increased from \$20,000 to \$21,000, production will increase by about 300 units. ◀ 5

Second-Order Partial Derivatives

The second derivative of a function of one variable is very useful in determining local maxima and minima. **Second-order partial derivatives** (partial derivatives of a partial derivative) are used in a similar way for functions of two or more variables. The situation is somewhat more complicated, however, with more independent variables. For example, $f(x, y) = 4x + x^2y + 2y$ has two first-order partial derivatives,

$$f_x = 4 + 2xy \quad \text{and} \quad f_y = x^2 + 2.$$

Because each of these has two partial derivatives, one with respect to y and one with respect to x, there are *four* second-order partial derivatives of function f. The notations for these four second-order partial derivatives are given below.

Second-Order Partial Derivatives

For a function $z = f(x, y)$, if all indicated partial derivatives exist, then

$$\frac{\partial}{\partial x}\left(\frac{\partial z}{\partial x}\right) = \frac{\partial^2 z}{\partial x^2} = f_{xx} \qquad \frac{\partial}{\partial y}\left(\frac{\partial z}{\partial y}\right) = \frac{\partial^2 z}{\partial y^2} = f_{yy}$$

$$\frac{\partial}{\partial y}\left(\frac{\partial z}{\partial x}\right) = \frac{\partial^2 z}{\partial y \partial x} = f_{xy} \qquad \frac{\partial}{\partial x}\left(\frac{\partial z}{\partial y}\right) = \frac{\partial^2 z}{\partial x \partial y} = f_{yx}.$$

As seen above, f_{xx} is used as an abbreviation for $f_{xx}(x, y)$, with f_{yy}, f_{xy}, and f_{yx} used in a similar way. The symbol f_{xx} is read "the partial derivative of f_x with respect to x," and f_{xy} is read "the partial derivative of f_x with respect to y." Also, the symbol $\partial^2 z/\partial y^2$ is read "the partial derivative of $\partial z/\partial y$ with respect to y."

NOTE For most functions found in applications and all the functions in this book, the second-order partial derivatives f_{xy} and f_{yx} are equal. Therefore, it is not necessary for us to be particular about the order in which these derivatives are found.

EXAMPLE 6 Find all second-order partial derivatives for

$$f(x, y) = -4x^3 - 3x^2y^3 + 2y^2.$$

First find f_x and f_y.

$$f_x = -12x^2 - 6xy^3 \quad \text{and} \quad f_y = -9x^2y^2 + 4y$$

To find f_{xx}, take the partial derivative of f_x with respect to x.

$$f_{xx} = -24x - 6y^3$$

Take the partial derivative of f_y with respect to y; this gives f_{yy}.

$$f_{yy} = -18x^2y + 4$$

6

Let $f(x, y) = 4x^2y^2 - 9xy + 8x^2 - 3y^4$. Find all second-order partial derivatives.

Answer:

$f_{xx} = 8y^2 + 16$
$f_{yy} = 8x^2 - 36y^2$
$f_{xy} = 16xy - 9$
$f_{yx} = 16xy - 9$

Find f_{xy} by starting with f_x, then taking the partial derivative of f_x with respect to y.

$$f_{xy} = -18xy^2$$

Finally, find f_{yx} by starting with f_y; take its partial derivative with respect to x.

$$f_{yx} = -18xy^2$$ ◀ 6

EXAMPLE 7 Let $f(x, y) = 2e^x - 8x^3y^2$. Find all second-order partial derivatives.

Here

$$f_x = 2e^x - 24x^2y^2 \quad \text{and} \quad f_y = -16x^3y.$$

(Recall: if $g(x) = e^x$, then $g'(x) = e^x$.) Now find the second-order partial derivatives.

$$f_{xx} = 2e^x - 48xy^2 \qquad f_{xy} = -48x^2y$$
$$f_{yy} = -16x^3 \qquad f_{yx} = -48x^2y$$ ◀ 7

7

Let $f(x, y) = 4e^{x+y} + 2x^3y$. Find all second-order partial derivatives.

Answer:

$f_{xx} = 4e^{x+y} + 12xy$
$f_{yy} = 4e^{x+y}$
$f_{xy} = 4e^{x+y} + 6x^2$
$f_{yx} = 4e^{x+y} + 6x^2$

Partial derivatives of multivariate functions with more than two independent variables are found in a way similar to that for functions with two independent variables. For example, to find f_x for $w = f(x, y, z)$ treat y and z as constants and differentiate with respect to x.

EXAMPLE 8 Let $f(x, y, z) = xy^2z + 2x^2y - 4xz^2$. Find f_x, f_y, f_z, f_{xy}, and f_{yz}.

$$f_x = y^2z + 4xy - 4z^2$$
$$f_y = 2xyz + 2x^2$$
$$f_z = xy^2 - 8xz$$

To find f_{xy}, differentiate f_x with respect to y.

$$f_{xy} = 2yz + 4x$$

In the same way, differentiate f_y with respect to z to get

$$f_{yz} = 2xy.$$ ◀ 8

8

Let $f(x, y, z) = xyz + x^2yz + xy^2z^3$. Find f_x, f_y, f_z, and f_{xz}.

Answer:

$f_x = yz + 2xyz + y^2z^3$
$f_y = xz + x^2z + 2xyz^3$
$f_z = xy + x^2y + 3xy^2z^2$
$f_{xz} = y + 2xy + 3y^2z^2$

14.2 Exercises

For each of these functions, find

(a) $\dfrac{\partial z}{\partial x}$ (b) $\dfrac{\partial z}{\partial y}$ (c) $f_x(2, 3)$ (d) $f_y(1, -2)$.

1. $z = f(x, y) = 8x^3 - 4x^2y + 9y^2$
2. $z = f(x, y) = -3x^2 - 2xy^2 + 5y^3$

In Exercises 3–14, find f_x and f_y. Then find $f_x(2, -1)$ and $f_y(-4, 3)$. Leave the answers in terms of e in Exercises 5–8 and 13 and 14. (See Examples 1–3.)

3. $f(x, y) = -x^2y + 3x^4 - 8$
4. $f(x, y) = 5y^2 - 6xy^2 + 7$

5. $f(x, y) = e^{2x+y}$

6. $f(x, y) = -4e^{x-y}$

7. $f(x, y) = \dfrac{-2}{e^{x+2y}}$

8. $f(x, y) = \dfrac{6}{e^{4x-y}}$

9. $f(x, y) = \dfrac{x + 3y^2}{x^2 + y^3}$

10. $f(x, y) = \dfrac{8x^2y}{x^3 - y}$

11. $f(x, y) = \ln|2x - x^2y|$

12. $f(x, y) = \ln|4xy^2 + 3y|$

13. $f(x, y) = x^2e^{2xy}$

14. $f(x, y) = ye^{5x+2y}$

Find all second-order partial derivatives. (See Examples 6 and 7.)

15. $f(x, y) = 10x^2y^3 - 5x^3 - 3y$

16. $g(x, y) = 8x^3y + 2x^4 + 6y^3$

17. $h(x, y) = -3y^2 - 4x^2y^2 + 7xy^2$

18. $P(x, y) = -16x^3 + 3xy^2 - 12x^4y^2$

19. $R(x, y) = \dfrac{3y}{2x + y}$

20. $C(x, y) = \dfrac{8x}{x - 4y}$

21. $z = 4xe^y$

22. $z = -3ye^x$

23. $r = \ln(x + y)$

24. $k = \ln(5x - 7y)$

25. $z = x\ln(xy)$

26. $z = (y + 1)\ln(x^3y)$

In Exercises 27 and 28, evaluate $f_{xy}(2, 1)$ and $f_{yy}(1, 2)$.

27. $f(x, y) = x\ln(xy)$

28. $f(x, y) = (y + 1)\ln(x^3y)$

Find values of x and y such that both $f_x(x, y) = 0$ and $f_y(x, y) = 0$.

29. $f(x, y) = 6x^2 + 6y^2 + 6xy + 36x - 5$

30. $f(x, y) = 50 + 4x - 5y + x^2 + y^2 + xy$

31. $f(x, y) = 9xy - x^3 - y^3 - 6$

32. $f(x, y) = 2200 + 27x^3 + 72xy + 8y^2$

Find f_x, f_y, f_z, and f_{yz} for these functions. In Exercises 33 and 34, also find $f_y(2, -1, 3)$ and $f_{yz}(-1, 1, 0)$. (See Example 8.)

33. $f(x, y, z) = x^2 + yz + z^4$

34. $f(x, y, z) = 3x^5 - x^2 + y^5$

35. $f(x, y, z) = \dfrac{6x - 5y}{4z + 5}$

36. $f(x, y, z) = \dfrac{2x^2 + xy}{yz - 2}$

37. $f(x, y, z) = \ln(x^2 - 5xz^2 + y^4)$

38. $f(x, y, z) = \ln(8xy + 5yz - x^3)$

39. How many partial derivatives does a function with three independent variables have? How many second-order partial derivatives? Explain why.

40. Suppose $z = f(x, y)$ describes the cost to build a certain structure, where x represents the labor costs and y represents the cost of materials. Describe what f_x and f_y represent.

41. **Business** A car dealership estimates that the total weekly sales of its most popular model are a function of the car's list price, p, and the interest rate in percent, i, offered by the manufacturer. The approximate weekly sales are given by

$$f(p, i) = 132p - 2pi - .01p^2.$$

(a) Find the weekly sales if the average list price is \$9400 and the manufacturer is offering an 8% interest rate.
(b) Find and interpret f_p and f_i.
(c) What would be the effect on weekly sales if the price is \$9400 and interest rates rise from 8% to 9%?

42. **Business** Suppose the production function of a company is given by

$$P(x, y) = 100\sqrt{x^2 + y^2},$$

where x represents units of labor and y represents units of capital. (See Example 5.) Find the following when $x = 4$ and $y = 3$.
(a) The marginal productivity of labor
(b) The marginal productivity of capital

43. **Business** A manufacturer estimates that production (in hundreds of units) is a function of the amounts x and y of labor and capital used, as follows.

$$f(x, y) = \left[\frac{1}{3}x^{-1/3} + \frac{2}{3}y^{-1/3}\right]^{-3}$$

(a) Find the number of units produced when 27 units of labor and 64 units of capital are utilized.
(b) Find and interpret $f_x(27, 64)$ and $f_y(27, 64)$.
(c) What would be the approximate effect on production of increasing labor by 1 unit?

44. **Business** A manufacturer of automobile batteries estimates that his total production in thousands of units is given by

$$f(x, y) = 3x^{1/3}y^{2/3},$$

where x is the number of units of labor and y is the number of units of capital utilized.
(a) Find and interpret $f_x(64, 125)$ and $f_y(64, 125)$ if the current level of production uses 64 units of labor and 125 units of capital.
(b) What would be the approximate effect on production of increasing labor to 65 units while holding capital at the current level?
(c) Suppose that sales have been good and management wants to increase either capital or labor by 1 unit. Which option would result in a larger increase in production?

45. **Business** The production function z for the United States was once estimated as

$$z = x^{.7}y^{.3},$$

where x stands for the amount of labor and y stands for the amount of capital. Find the marginal productivity of labor (find $\partial z/\partial x$) and of capital. (See Example 5.)

46. **Business** A similar production function for Canada is

$$z = x^{.4}y^{.6},$$

with x, y, and z as in Exercise 45. Find the marginal productivity of labor and of capital. (See Example 5.)

47. **Natural Science** In one method of computing the quantity of blood pumped through the lungs in 1 minute, a researcher first finds each of the following (in milliliters).

b = quantity of oxygen used by body in 1 minute
a = quantity of oxygen per liter of blood that has just gone through the lungs
v = quantity of oxygen per liter of blood that is about to enter the lungs

In 1 minute,

Amount of oxygen used
= amount of oxygen per liter
× number of liters of blood pumped.

If C is the number of liters pumped through the blood in 1 minute, then

$$b = (a - v) \cdot C \quad \text{or} \quad C = \frac{b}{a - v}.$$

(a) Find C if $a = 160$, $b = 200$, and $v = 125$.
(b) Find C if $a = 180$, $b = 260$, and $v = 142$.
Find the following partial derivatives.
(c) $\partial C/\partial b$ **(d)** $\partial C/\partial v$

48. **Health** A weight-loss counselor has prepared a program of diet and exercise for a client. If the client sticks to the program, the weight loss that can be expected (in pounds per week) is given by

$$\text{Weight loss} = f(n, c) = \frac{1}{8}n^2 - \frac{1}{5}c + \frac{1937}{8},$$

where c is the average daily calorie intake for the week and n is the number of 40-minute aerobic workouts per week.

(a) How many pounds can the client expect to lose by eating an average of 1200 calories per day and participating in four 40-minute workouts in a week?
(b) Find and interpret $\partial f/\partial n$.
(c) The client currently averages 1100 calories per day and does three 40-minute workouts each week. What would be the approximate impact on weekly weight loss of adding a fourth workout per week?

49. **Physical Science** The gravitational attraction F on a body a distance r from the center of the earth, where r is greater than the radius of the earth, is a function of its mass m and the distance r as follows:

$$F = \frac{mgR^2}{r^2},$$

where R is the radius of the earth and g is the force of gravity—about 32 feet per second per second (ft/sec^2).

(a) Find and interpret F_m and F_r.
(b) Show that $F_m > 0$ and $F_r < 0$. Why is this reasonable?

50. **Natural Science** In 1908, J. Haldane constructed diving tables that provide a relationship between the water pressure on body tissues for various water depths and dive times. The tables were successfully used by divers to virtually eliminate decompression sickness. The pressure in atmospheres for a no-stop dive is given by the following formula:*

$$p(l, t) = 1 + \frac{l}{33}(1 - e^{-.1386t}),$$

where t is in minutes, l is in feet, and p is in atmospheres (atm).†

(a) Find the pressure at 33 feet for a 10-minute dive.
(b) Find $p_l(33, 10)$ and $p_t(33, 10)$.
(c) Haldane estimated that decompression sickness for no-stop dives could be avoided if the diver's tissue pressure did not exceed 2.15 atmospheres. Find the maximum amount of time that a diver could stay down (time includes going down and coming back up) if he or she wants to dive to a depth of 66 feet.

51. **Health** The body mass index (BMI) is a number that can be calculated for any individual as follows: Multiply a person's weight by 703 and divide by the person's height squared. That is,

$$B = \frac{703w}{h^2},$$

where w is in pounds and h is in inches.‡ The National Heart, Lung and Blood Institute uses the BMI to determine whether a person is "overweight" ($25 \le B < 30$) or "obese" ($B \ge 30$).

(a) Calculate the BMI for a person who weighs 220 pounds and is 74 inches tall.
(b) Calculate $\dfrac{\partial B}{\partial w}$ and $\dfrac{\partial B}{\partial h}$.

52. **Natural Science** In 1941, explorers Paul Siple and Charles Passel discovered that the amount of heat lost when an object is exposed to cold air depends on both the

*These estimates are conservative. Please consult modern dive tables before making a dive.

†David Westbrook, "The Mathematics of Scuba Diving," *UMAP Journal* 18, no. 2 (1997): 2–19.

‡The National Institutes of Health.

temperature of the air and the velocity of the wind. They developed the Wind-Chill Index as a way to measure the danger of frostbite while performing outdoor activities. Based on new findings and additional data, the Wind-Chill Index was recalculated in 2001. The new index is given by

$$W(V, T) = 35.74 + .6215T - 35.75V^{.16} + (.4275\,T)V^{.16},$$

where V is the wind speed in miles per hour and T is the temperature in Fahrenheit for wind speeds between 4 and 45 mph.*

*Joint Action Group of the National Weather Service and the Meteorological Services of Canada, Press release, August 2001.

(a) Find the wind chill for a wind speed of 20 mph and 10°F.

(b) If a weather report indicates that the wind chill is −25°F and the actual outdoor temperature is 5°F, use a graphing calculator to find the corresponding wind speed to the nearest mile per hour.

(c) Find $W_V(20, 10)$ and $W_T(20, 10)$ and interpret.

(d) Using the table command on a graphing calculator or a spreadsheet, develop a wind-chill chart for various wind speeds of 5, 10, 15, and 20 mph and temperatures of 0°, 5°, 15° and 25°.

14.3 EXTREMA OF FUNCTIONS OF SEVERAL VARIABLES

One of the most important applications of calculus is in finding maxima and minima for functions. Earlier, we studied this idea extensively for functions of a single independent variable; now we shall see that extrema can be found for functions of two variables. In particular, an extension of the second derivative test can be derived and used to identify maxima or minima. We begin with the definitions of local maxima and minima.

Local Maxima and Minima

Let (a, b) be in the domain of a function f.

1. f has a **local maximum** at (a, b) if there is a circular region in the xy-plane with (a, b) in its interior such that
$$f(a, b) \geq f(x, y)$$
for all points (x, y) in the circular region.
2. f has a **local minimum** at (a, b) if there is a circular region in the xy-plane with (a, b) in its interior such that
$$f(a, b) \leq f(x, y)$$
for all points (x, y) in the circular region.

As before, the term *local extremum* is used for either a local maximum or a local minimum. Examples of a local maximum and a local minimum are given in Figures 14.18 and 14.19.

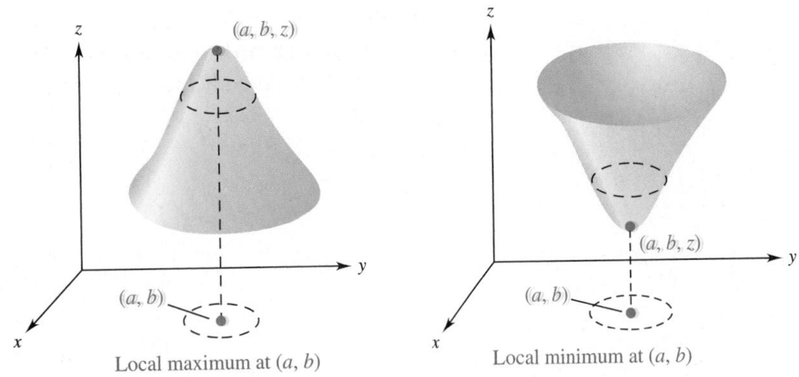

FIGURE 14.18 **FIGURE 14.19**

NOTE With functions of a single variable, we made a distinction between local extrema and absolute extrema. The methods for finding absolute extrema are quite involved for functions of two independent variables, so we will discuss only local extrema.

As suggested by Figure 14.20, at a local maximum, the tangent line parallel to the x-axis has a slope of 0, as does the tangent line parallel to the y-axis. (Notice the similarity to functions of one variable.) That is, if the function $z = f(x, y)$ has a local extremum at (a, b), then $f_x(a, b) = 0$ and $f_y(a, b) = 0$, as stated in the theorem below.

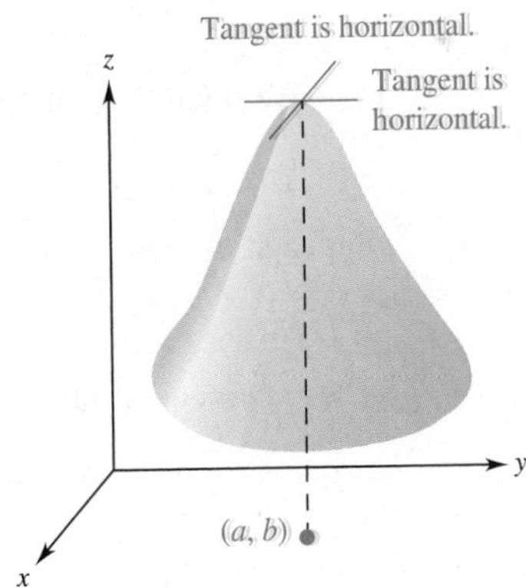

FIGURE 14.20

Location of Extrema

If a function $z = f(x, y)$ has a local maximum or local minimum at the point (a, b), and $f_x(a, b)$ and $f_y(a, b)$ both exist, then

$$\mathbf{f_x(a, b) = 0} \quad \text{and} \quad \mathbf{f_y(a, b) = 0.}$$

Just as with functions of one variable, the fact that the slopes of the tangent lines are 0 is no guarantee that a local extremum has been located. For example, Figure 14.21 shows the graph of $z = f(x, y) = x^2 - y^2$. Both $f_x(0, 0) = 0$ and $f_y(0, 0) = 0$, and yet (0, 0) leads to neither a local maximum nor a local minimum for the function. The point (0, 0, 0) on the graph of this function is called a **saddle point;** it is a minimum when approached from one direction but a maximum when approached from another direction. A saddle point is neither a maximum nor a minimum.

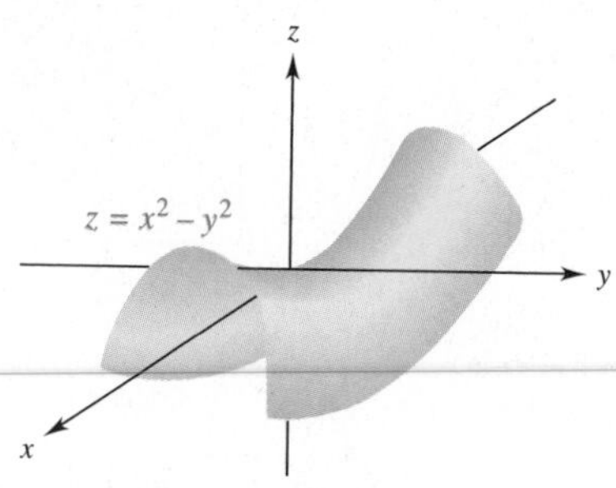

FIGURE 14.21

The theorem on location of extrema suggests a useful strategy for finding extrema. First, locate all points (a, b) where $f_x(a, b) = 0$ and $f_y(a, b) = 0$. Then test each of these points separately, using the test given after the next example. For a function $f(x, y)$, the points (a, b) such that $f_x(a, b) = 0$ and $f_y(a, b) = 0$ (or such that $f_x(a, b)$ or $f_y(a, b)$ does not exist) are called **critical points.**

EXAMPLE 1 Find all critical points for

$$f(x, y) = 6x^2 + 6y^2 + 6xy + 36x - 54y - 5.$$

We must find all points (a, b) such that $f_x(a, b) = 0$ and $f_y(a, b) = 0$. Here

$$f_x = 12x + 6y + 36 \quad \text{and} \quad f_y = 12y + 6x - 54.$$

Set each of these two partial derivatives equal to 0.

$$12x + 6y + 36 = 0 \quad \text{and} \quad 12y + 6x - 54 = 0$$

These two equations form a system of linear equations that we can rewrite as

$$\begin{aligned} 12x + 6y &= -36 \\ 6x + 12y &= 54. \end{aligned}$$

To solve this system by elimination, multiply the first equation by -2 and add.

$$\begin{array}{rcr} -24x - 12y &=& 72 \\ 6x + 12y &=& 54 \\ \hline -18x \qquad &=& 126 \\ x &=& -7 \end{array}$$

Substituting $x = -7$ in the first equation of the system, we have

$$\begin{aligned} 12(-7) + 6y &= -36 \\ 6y &= 48 \\ y &= 8. \end{aligned}$$

1

Find all critical points for the following.

(a) $f(x, y) = 4x^2 + 3xy + 2y^2 + 7x - 6y - 6$

(b) $f(x, y) = e^{-2x} + xy$

Answers:

(a) $(-2, 3)$

(b) $(0, 2)$

Therefore, $(-7, 8)$ is the solution of the system. Since this is the only solution, $(-7, 8)$ is the only critical point for the given function. By the theorem above, if the function has a local extremum, it must occur at $(-7, 8)$. ◀ 1

The results of the next theorem can be used to decide whether $(-7, 8)$ in Example 1 leads to a local maximum, a local minimum, or neither. The proof of this theorem is beyond the scope of this course.

Test for Local Extrema

Suppose $f(x, y)$ is a function such that f_{xx}, f_{yy}, f_{xy} all exist. Let (a, b) be a critical point for which

$$f_x(a, b) = 0 \text{ and } f_y(a, b) = 0.$$

Let M be the number defined by

$$M = f_{xx}(a, b) \cdot f_{yy}(a, b) - [f_{xy}(a, b)]^2.$$

1. If $M > 0$ and $f_{xx}(a, b) < 0$, then f has a **local maximum** at (a, b).
2. If $M > 0$ and $f_{xx}(a, b) > 0$, then f has a **local minimum** at (a, b).
3. If $M < 0$, then f has a **saddle point** at (a, b).
4. If $M = 0$, the test gives **no information.**

The chart below summarizes the conclusions of the theorem.

	$f_{xx}(a, b) < 0$	$f_{xx}(a, b) > 0$
$M > 0$	Local maximum	Local minimum
$M = 0$	No information	
$M < 0$	Saddle point	

EXAMPLE 2 Example 1 showed that the only critical point for the function

$$f(x, y) = 6x^2 + 6y^2 + 6xy + 36x - 54y - 5$$

is $(-7, 8)$. Does $(-7, 8)$ lead to a local maximum, a local minimum, or neither?

We can find out by using the test above. From Example 1,

$$f_x(-7, 8) = 0 \text{ and } f_y(-7, 8) = 0.$$

Now find the various second-order partial derivatives used in finding M. From $f_x = 12x + 6y + 36$ and $f_y = 12y + 6x - 54$,

$$f_{xx} = 12, \quad f_{yy} = 12, \quad \text{and} \quad f_{xy} = 6.$$

(If these second-order partial derivatives had not all been constants, we would have had to evaluate them at the point $(-7, 8)$.) Now

$$M = f_{xx}(-7, 8) \cdot f_{yy}(-7, 8) - [f_{xy}(-7, 8)]^2 = 12 \cdot 12 - 6^2 = 108.$$

Find any local maxima or minima for the functions defined in side problem 1.

Answers:

(a) Local minimum at $(-2, 3)$

(b) Neither a local minimum nor a local maximum at $(0, 2)$

Since $M > 0$ and $f_{xx}(-7, 8) = 12 > 0$, Part 2 of the theorem applies, showing that $f(x, y) = 6x^2 + 6y^2 + 6xy + 36x - 54y - 5$ has a local minimum at $(-7, 8)$. This local minimum value is $f(-7, 8) = -347$. ◀ (2)

EXAMPLE 3 Find all points where the function

$$f(x, y) = 9xy - x^3 - y^3 - 6$$

has any local maxima or local minima.

First find any critical points. Here

$$f_x = 9y - 3x^2 \quad \text{and} \quad f_y = 9x - 3y^2.$$

Set each of these partial derivatives equal to 0.

$$\begin{aligned} f_x &= 0 & f_y &= 0 \\ 9y - 3x^2 &= 0 & 9x - 3y^2 &= 0 \\ 9y &= 3x^2 & 9x &= 3y^2 \\ 3y &= x^2 & 3x &= y^2 \end{aligned}$$

In the first equation ($3y = x^2$), notice that since $x^2 \geq 0$, $y \geq 0$. Also, in the second equation ($3x = y^2$), $y^2 \geq 0$, so $x \geq 0$.

The substitution method can be used to solve the system of equations

$$\begin{aligned} 3y &= x^2 \\ 3x &= y^2. \end{aligned}$$

The first equation, $3y = x^2$, can be rewritten as $y = x^2/3$. Substitute this into the second equation to get

$$3x = y^2 = \left(\frac{x^2}{3}\right)^2$$

$$3x = \frac{x^4}{9}.$$

Solve this equation as follows.

$$\begin{aligned} 27x &= x^4 && \text{Multiply both sides by 9.} \\ x^4 - 27x &= 0 \\ x(x^3 - 27) &= 0 && \text{Factor.} \\ x = 0 \quad \text{or} \quad x^3 - 27 &= 0 && \text{Set each factor equal to 0.} \\ x^3 &= 27 \\ x &= 3 && \text{Take the cube root on each side.} \end{aligned}$$

Use these values of x, along with the equation $3x = y^2$, to find y.

$$\begin{aligned} &\text{If } x = 0, & &\text{If } x = 3, \\ &3x = y^2 & &3x = y^2 \\ &3(0) = y^2 & &3(3) = y^2 \\ &0 = y^2 & &9 = y^2 \\ &0 = y. & &3 = y \quad \text{or} \quad -3 = y. \end{aligned}$$

The points (0, 0), (3, 3), and (3, −3) appear to be critical points; however, (3, −3) does not have $y \geq 0$. The only possible local extrema for $f(x, y) = 9xy - x^3 - y^3 - 6$ occur at the critical points (0, 0) or (3, 3). To identify any extrema, use the test. Here

$$f_{xx} = -6x, \quad f_{yy} = -6y, \quad \text{and } f_{xy} = 9.$$

Test each of the critical points.

For (0, 0):

$$f_{xx}(0, 0) = -6(0) = 0, \; f_{yy}(0, 0) = -6(0) = 0, \quad f_{xy}(0, 0) = 9,$$

so that $M = 0 \cdot 0 - 9^2 = -81$. Since $M < 0$, there is a saddle point at (0, 0).

For (3, 3):

$$f_{xx}(3, 3) = -6(3) = -18, \quad f_{yy}(3, 3) = -6(3) = -18, \quad f_{xy}(3, 3) = 9,$$

so that $M = (-18)(-18) - 9^2 = 243$. Since $M > 0$, and $f_{xx}(3, 3) = -18 < 0$, there is a local maximum at (3, 3). If you examine Figure 14.22 carefully, you will see both this local maximum and the saddle point at (0, 0) ◀ 3

3

Find any local extrema for

$$f(x, y) = \frac{2\sqrt{2}}{3}x^3 - xy + \frac{1}{3}y^3 - 10.$$

Answer:

Local minimum at $\left(\frac{1}{2}, \frac{\sqrt{2}}{2}\right)$; saddle point at (0, 0)

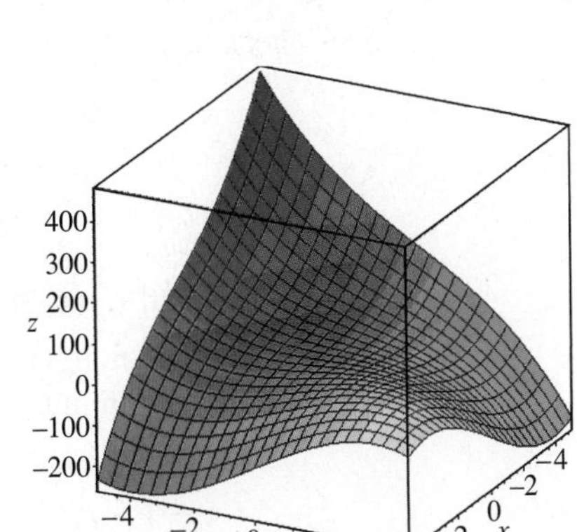

FIGURE 14.22

EXAMPLE 4 A company is developing a new soft drink. The cost in dollars to produce a batch of the drink is approximated by

$$C(x, y) = 2200 + 27x^3 - 72xy + 8y^2,$$

where x is the number of kilograms of sugar per batch and y is the number of grams of flavoring per batch.

(a) Find the amounts of sugar and flavoring that result in minimum cost for a batch.

Start with the following partial derivatives.

$$C_x = 81x^2 - 72y \quad \text{and} \quad C_y = -72x + 16y$$

Set each of these equal to 0 and solve for y.

$$81x^2 - 72y = 0 \qquad\qquad -72x + 16y = 0$$
$$-72y = -81x^2 \qquad\qquad 16y = 72x$$
$$y = \frac{9}{8}x^2 \qquad\qquad y = \frac{9}{2}x$$

From the equation on the left, $y \geq 0$. Since $(9/8)x^2$ and $(9/2)x$ both are equal to y, they are equal to each other. Set $(9/8)x^2$ and $(9/2)x$ equal and solve the resulting equation for x.

$$\frac{9}{8}x^2 = \frac{9}{2}x$$
$$9x^2 = 36x$$
$$9x^2 - 36x = 0$$
$$9x(x - 4) = 0$$
$$9x = 0 \quad \text{or} \quad x - 4 = 0$$
$$x = 0 \qquad\qquad x = 4$$

The answer $x = 0$ is not a useful answer for our problem. Substitute $x = 4$ into $y = (9/2)x$ to find y.

$$y = \frac{9}{2}x = \frac{9}{2}(4) = 18$$

Now check to see if the critical point (4, 18) leads to a local minimum. For (4, 18),

$$C_{xx} = 162x = 162(4) = 648, \quad C_{yy} = 16, \quad \text{and } C_{xy} = -72.$$

Also,

$$M = (648)(16) - (-72)^2 = 5184.$$

Since $M > 0$ and $C_{xx}(4, 18) > 0$, the cost at (4, 18) is a minimum.

(b) What is the minimum cost?
To find the minimum cost, go back to the cost function and evaluate $C(4, 18)$.

$$C(x, y) = 2200 + 27x^3 - 72xy + 8y^2$$
$$C(4, 18) = 2200 + 27(4)^3 - 72(4)(18) + 8(18)^2 = 1336$$

The minimum cost for a batch is \$1336. ◀

TECHNOLOGY TIP The most widely used spreadsheet programs have built-in solvers that are able to optimize complicated functions of several variables. Check your instruction manual for details. Typically, the solver requires you to enter an initial guess. Unfortunately, the solver will not find a maximum or minimum for a function if the initial guess is the exact place at which a saddle point occurs. So it is a good idea to run the solver for two different initial values and compare the solutions.

14.3 Exercises

1. Compare and contrast the way critical points are found for functions with one independent variable and functions with more than one independent variable.
2. Compare and contrast the second derivative test for $y = f(x)$ and the test for local extrema for $z = f(x, y)$.

Find all points where these functions have any local extrema. Give the values of any local extrema. Identify any saddle points. (See Examples 1–3.)

3. $f(x, y) = 2x^2 + 4xy + 6y^2 - 8x - 10$
4. $f(x, y) = x^2 + xy + y^2 - 6x - 3$
5. $f(x, y) = x^2 - xy + 2y^2 + 2x + 6y + 8$
6. $f(x, y) = 3x^2 + 6xy + y^2 - 6x - 3y$
7. $f(x, y) = 2x^2 + 4xy + y^2 - 4x + 4y$
8. $f(x, y) = 4xy - 4x^2 - 2y^2 + 4x - 8y - 7$
9. $f(x, y) = 4xy - 10x^2 - 4y^2 + 8x + 8y + 9$
10. $f(x, y) = x^2 + xy + 3x + 2y - 6$
11. $f(x, y) = x^2 + xy - 2x - 2y + 2$
12. $f(x, y) = x^2 + xy + y^2 - 3x - 5$

13. $f(x, y) = x^2 - y^2 - 2x + 4y - 7$

14. $f(x, y) = 4x + 2y - x^2 + xy - y^2 + 3$

15. $f(x, y) = 2x^3 + 2y^2 - 12xy + 15$

16. $f(x, y) = 2x^2 + 4y^3 - 24xy + 18$

17. $f(x, y) = 3x^2 + 6y^3 - 36xy + 27$

18. $f(x, y) = 2y^3 + 5x^2 + 60xy + 25$

19. $f(x, y) = e^{xy}$ **20.** $f(x, y) = x^2 + e^y$

Figures (a)–(f) show the graphs of the functions defined in Exercises 21–26. Find all local extrema for each function and then match the equation to its graph.

21. $z = -3xy + x^3 - y^3 + \dfrac{1}{8}$

22. $z = \dfrac{3}{2}y - \dfrac{1}{2}y^3 - x^2y + \dfrac{1}{16}$

23. $z = y^4 - 2y^2 + x^2 - \dfrac{17}{16}$

24. $z = -2x^3 - 3y^4 + 6xy^2 + \dfrac{1}{16}$

25. $z = -x^4 + y^4 + 2x^2 - 2y^2 + \dfrac{1}{16}$

26. $z = -y^4 + 4xy - 2x^2 + \dfrac{1}{16}$

(a)

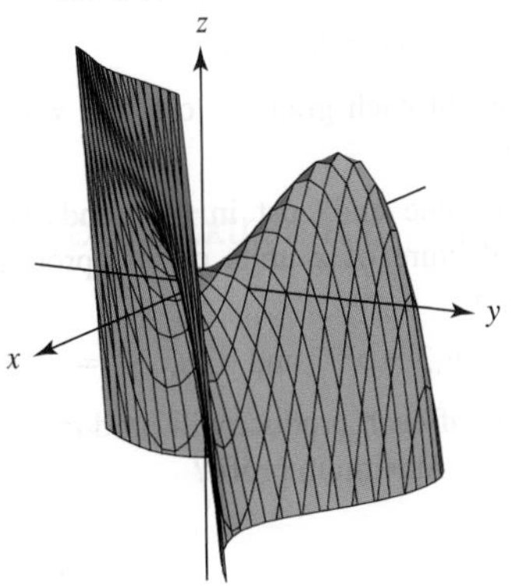

(b)

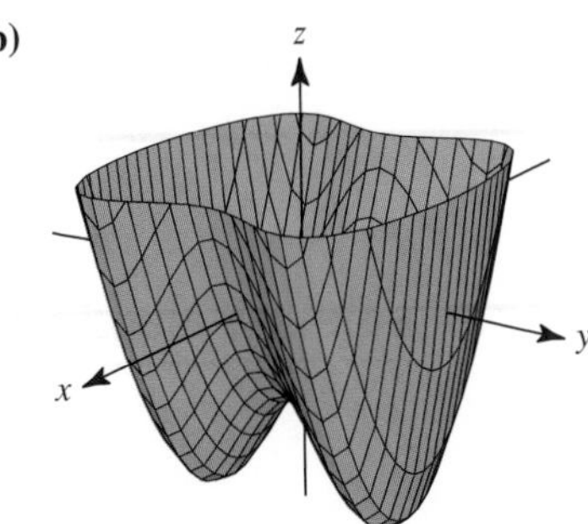

(c)

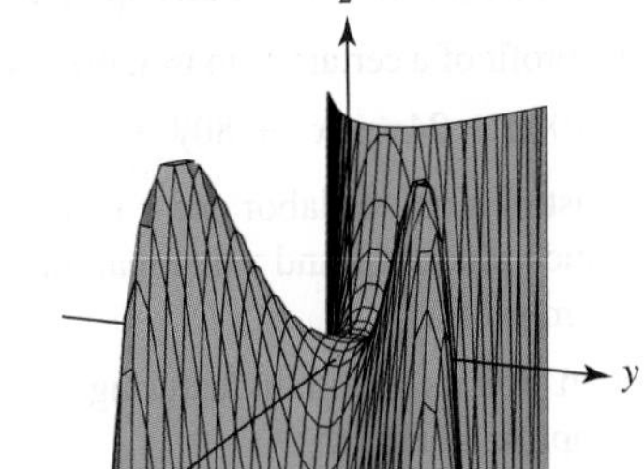

(d)

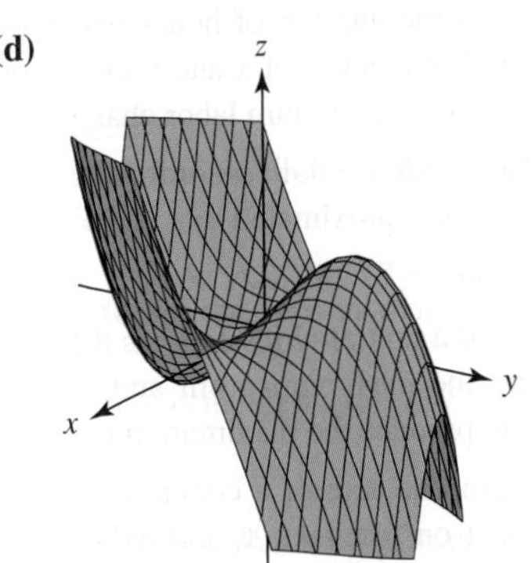

(e)

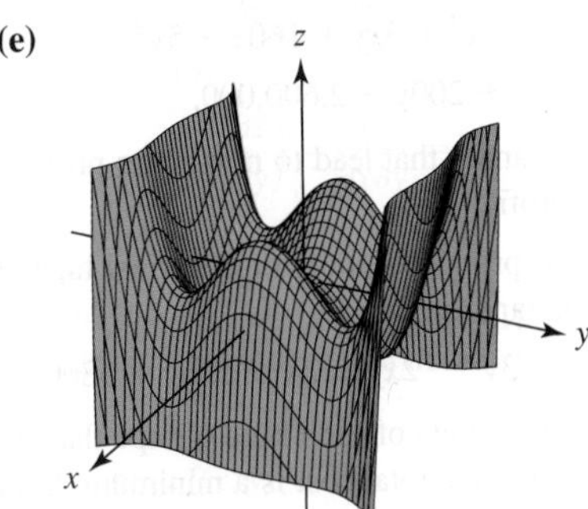

(f)

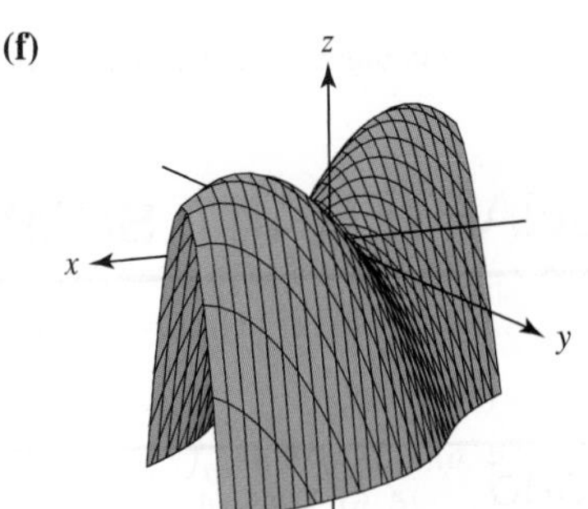

Chapter 14 Review Exercises

Find f(−1, 2) and f(6, −3) for each of the following.

1. $f(x, y) = 6y^2 - 5xy + 2x$
2. $f(x, y) = -3x + 2x^2y^2 + 5y$
3. $f(x, y) = \dfrac{2x - 4}{x + 3y}$
4. $f(x, y) = x\sqrt{x^2 + y^2}$
5. Describe the graph of $2x + y + 4z = 12$.
6. Describe the graph of $y = 2$ on a three-dimensional grid.

Graph the first-octant portion of each plane.

7. $x + 2y + 4z = 4$
8. $3x + 2y = 6$
9. $4x + 5y = 20$
10. $x = 6$
11. Let $z = f(x, y) = -2x^2 + 5xy + y^2$. Find the following.

    (a) $\dfrac{\partial z}{\partial x}$ **(b)** $\dfrac{\partial z}{\partial y}(-1, 4)$ **(c)** $f_{xy}(2, -1)$

12. Let $z = f(x, y) = \dfrac{2y + x^2}{3y - x}$. Find the following.

    (a) $\dfrac{\partial z}{\partial y}$ **(b)** $\dfrac{\partial z}{\partial x}(0, 2)$ **(c)** $f_{yy}(-1, 0)$

13. Explain the difference between $\dfrac{\partial z}{\partial x}$ and $\dfrac{\partial z}{\partial y}$.

Find f_x and f_y.

14. $f(x, y) = 3y - 7x^2y^3$
15. $f(x, y) = 4x^3y + 10xy^4$
16. $f(x, y) = \sqrt{3x^2 + 2y^2}$
17. $f(x, y) = \dfrac{3x - 2y^2}{x^2 + 4y}$
18. $f(x, y) = x^3e^{3y}$
19. $f(x, y) = (y + 1)^2e^{2x+y}$
20. $f(x, y) = \ln|x^2 - 4y^3|$
21. $f(x, y) = \ln|1 + x^3y^2|$
22. Explain the difference between f_{xx} and f_{xy}.

Find f_{xx} and f_{xy}.

23. $f(x, y) = 4x^3y^2 - 8xy$
24. $f(x, y) = -6xy^4 + x^2y$
25. $f(x, y) = \dfrac{2x}{x - 2y}$
26. $f(x, y) = \dfrac{3x + y}{x - 1}$
27. $f(x, y) = x^2e^y$
28. $f(x, y) = ye^{x^2}$
29. $f(x, y) = \ln(2 - x^2y)$
30. $f(x, y) = \ln(1 + 3xy^2)$

Find all points where these functions have any local extrema. Find any saddle points.

31. $z = x^2 + 2y^2 - 4y$
32. $z = x^2 + y^2 + 9x - 8y + 1$
33. $f(x, y) = x^2 + 5xy - 10x + 3y^2 - 12y$
34. $z = x^3 - 8y^2 + 6xy + 4$
35. $z = x^3 + y^2 + 2xy - 4x - 3y - 2$
36. $f(x, y) = 7x^2 + y^2 - 3x + 6y - 5xy$
37. **Geometry** Write a function in terms of L, W, and H that gives the total material required to build the closed box shown in the figure.

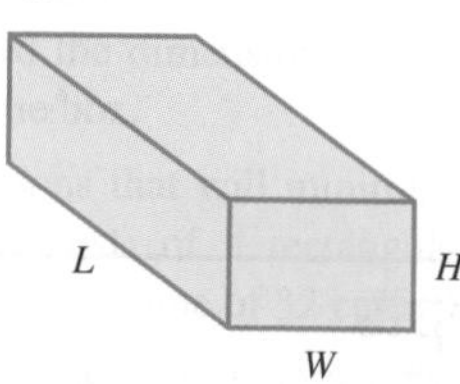

38. **Business** The manufacturing cost in dollars for a medium-sized business computer is given by

$$c(x, y) = 2x + y^2 + 4xy + 25,$$

where x is the memory capacity (RAM) of the computer in megabytes and y is the number of hours of labor required. Find each of the following.

(a) $\dfrac{\partial c}{\partial x}(64, 6)$ **(b)** $\dfrac{\partial c}{\partial y}(128, 12)$

39. **Business** The total cost in dollars to manufacture x solar cells and y solar collectors is

$$c(x, y) = x^2 + 5y^2 + 4xy - 70x - 164y + 1800.$$

(a) Find values of x and y that produce minimum total cost.

(b) Find the minimum total cost.

40. **Business** The total profit from 1 acre of a certain crop depends on the amount spent on fertilizer, x, and on hybrid seed, y, according to the model

$$P(x, y) = .01(-x^2 + 3xy + 160x - 5y^2 + 200y + 2600).$$

The budget for fertilizer and seed is limited to \$280.

(a) Use the budget constraint to express one variable in terms of the other. Then substitute into the profit function to get a function with one independent variable. Use the method shown in Chapter 12 to find the amounts spent on fertilizer and seed that will maximize profit. What is the maximum profit per acre?

(*Hint:* Throughout this problem you may ignore the coefficient of .01 until you need to find the maximum profit.)

(b) Find the amounts spent on fertilizer and seed that will maximize profit using the method shown in this chapter. (*Hint:* You will not need to use the budget constraint.)

(c) Discuss any relationships between these methods.

Case 13

Global Warming and the Method of Least Squares

The least squares regression line for modeling data that is approximately linear was introduced in Section 2.3. The discussion there explained why least squares are used to measure how well a line fits the data. In this case, we show how the theory of extrema of functions of several variables is used to construct the least squares regression line.

Global warming, the idea that the atmospheric temperature of Earth is slowly increasing over time, is an issue under great debate. Some claim global warming is a serious problem for which action must be taken; others acknowledge a warming trend but do not feel the trend indicates a serious problem; while still others believe there is no verifiable warming trend. In any case, it is clear that a large amount of data must be collected and analyzed before conclusions may be drawn.

The National Oceanic and Atmospheric Administration (NOAA) has compiled a great deal of historical global temperature data. Data from recent years is compiled by satellite, but a variety of methods, including the analysis of temperature-sensitive phenomenon such as tree ring density, allow for temperature estimates from centuries past. The data is often presented in terms of temperature anomalies, that is, as deviations from some norm.

For example the NOAA reports the following temperature anomalies. The baseline temperature against which the anomalies were determined is the average global temperature over land from the years 1880 to 2000, an average computed to be 8.5°C. Adding 8.5 to each anomaly gives the actual temperature.

GLOBAL LAND SURFACE AVERAGE TEMPERATURE

Year	*Temperature Anomaly (°C)*	*Temperature (°C)*
1900	−0.03	8.47
1950	−0.17	8.33
2000	0.59	9.09

Figure 1 on the next page shows a plot of the adjusted temperature data.

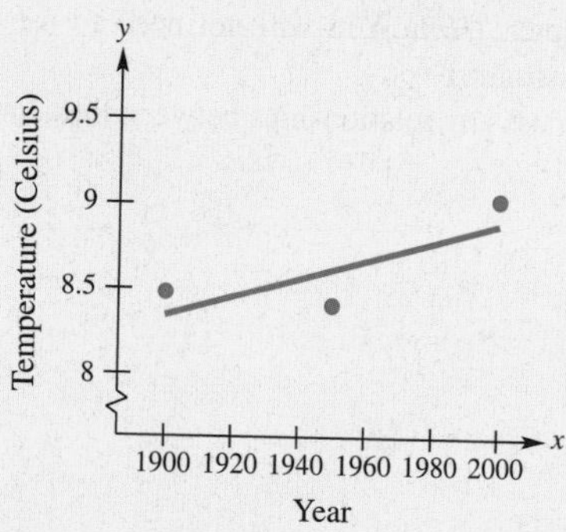

FIGURE 1 Global Land Surface Temperature

If the three points lie on a straight line, the rate at which the temperature is changing could be determined by the slope of that line. The points clearly are not on the same line but can be approximated by a straight line as in the figure. The purpose of the least squares method is to find the best such approximation.

Figure 2 shows the distance from each point to the approximating line.

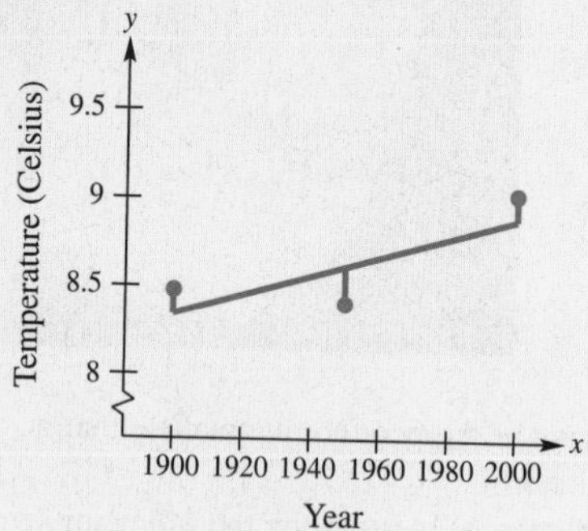

FIGURE 2 Deviations from Line

If the first point is (x_1, y_1) and the equation of the line is $y_1 = m\,x_1 + b$, then the first vertical distance is $y_1 - (mx_1 + b) = y_1 - mx_1 - b$. In the least squares method, the values of m and b are determined so that the sum of the squares of these distances is minimized. For three points (x_1, y_1), (x_2, y_2), and (x_3, y_3), the function of two variables

$$f(m, b) = (y_1 - mx_1 - b)^2 + (y_2 - mx_2 - b)^2 + (y_3 - mx_3 - b)^2 \qquad (1)$$

must be minimized. Using the points

$$(1900, 8.47), \quad (1950, 8.33), \quad \text{and} \quad (2000, 9.09),$$

equation (1) becomes

$$f(m, b) = 223.76 - 101{,}033m - 51.78b + 11{,}412{,}500m^2 + 11{,}700mb + 3b^2.$$

The partial derivatives of f with respect to m and b are

$$\frac{\partial f}{\partial m} = -101{,}033 + 22{,}825{,}000m + 11{,}700b$$

$$\frac{\partial f}{\partial b} = -51.78 + 11{,}700m + 6b$$

and the resulting critical point is $m = .0062$ and $b = -3.46$. The line that best fits this data is $y = .0062x - 3.46$ and is the line drawn in Figure 1.

According to this analysis, the global land temperature increases at a rate of approximately 0.0062°C/year. It is not fair to say this is evidence of global warming as only three pieces of data were used. The data set was kept small to illustrate the least squares method. This method can be generalized to use any amount of data and also to fit more complicated functions such as exponentials and sine waves for a more detailed analysis.

Exercises

1. Verify that the critical point $m = .0062$ and $b = -3.46$ corresponds to a minimum of the function $f(m,b)$.
2. By differentiating equation (1) with respect to m and b, without substituting the specific values for points (x_1, y_1), (x_2, y_2), and (x_3, y_3), the formulas

$$m = \frac{3(x_1y_1 + x_2y_2 + x_3y_3) - (x_1 + x_2 + x_3)(y_1 + y_2 + y_3)}{3(x_1^2 + x_2^2 + x_3^2) - (x_1 + x_2 + x_3)^2}$$

$$b = \frac{y_1 + y_2 + y_3}{3} - m\left(\frac{x_1 + x_2 + x_3}{3}\right)$$

for m and b can be derived. Show that these formulas give the same values for m and b given earlier.

3. The formulas in Exercise 2 generalize to any number of data points. For example, given four data points, the formulas for m and b become

$$m = \frac{4(x_1y_1 + x_2y_2 + x_3y_3 + x_4y_4)}{4(x_1^2 + x_2^2 + x_3^2 + x_4^2) - (x_1 + x_2 + x_3 + x_4)^2} - \frac{(x_1 + x_2 + x_3 + x_4)(y_1 + y_2 + y_3 + y_4)}{4(x_1^2 + x_2^2 + x_3^2 + x_4^2) - (x_1 + x_2 + x_3 + x_4)^2}$$

$$b = \frac{y_1 + y_2 + y_3 + y_4}{4} - m\left(\frac{x_1 + x_2 + x_3 + x_4}{4}\right)$$

Using a similar formula for five data points, add the two temperature data points (1925, 8.46) and (1975, 8.55) to the three above and find the least squares straight line fit. Plot the points and the line and discuss the value of m.

APPENDIX A

Graphing Calculators

A summary of the contents of this appendix appears below. The entire appendix is available at our Web site: www.aw.com/MWA8. You can print as much or as little of it as you want.

Part 1: Using Your Calculator

This is a brief introduction to the features of a graphing calculator that are relevant to the topics in this text, including the following.

Basics: keyboard, edit/replay, scientific notation
Function Graphing: trace, zoom, maximum/minimum finder, intersection finder
Function Evaluation: tables, functional notation, eval key
Equation Solving: root finder, polynomial solver, equation solver
Linear Regression and Other Regression Procedures
Matrices: matrix editor, matrix arithmetic, row operations

Part 2: Calculator Programs

The programs listed below are available for TI, Sharp 9600, and Casio 9850 (they work on some other Casio models as well). TI users may download them directly, if they have the appropriate link hardware and software. All programs may also be printed and then entered into your calculator by hand.

1. Fraction Conversion for Casio and Sharp 9600
2. Quadratic Formula for TI-82/83 and Sharp 9600
3. Table Maker for TI-85
4. Present and Future Value of an Annuity
5. Loan Payment
6. Loan Balance after n Payments
7. Amortization Table
8. RREF Program for TI-82 and Casio 9850
9. Simplex Method
10. Two-Stage Method
11. Rectangle Approximation of $\int_a^b f(x)\,dx$ (using left endpoints)

Programs 1–3 and 8 provide the specified calculators with programs that are built-in on most other models. Programs 7 and 9–11 are not built-in on any calculators. Programs 4–6 are included in the TI-83 Financial menu (and in similar menus that can be downloaded from TI for TI-86 and TI-89), though some students may find the versions here easier to use.

Graphing Calculators

APPENDIX B

Tables

Table 1 Formulas from Geometry

CIRCLE
Area: $A = \pi r^2$
Circumference: $C = 2\pi r$

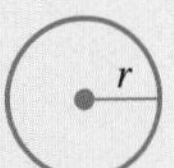

RECTANGLE
Area: $A = lw$
Perimeter: $P = 2l + 2w$

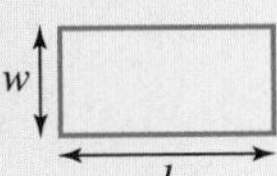

PARALLELOGRAM
Area: $A = bh$
Perimeter: $P = 2a + 2b$

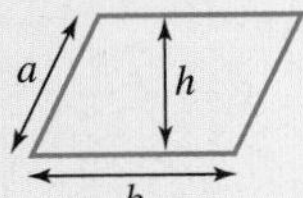

TRIANGLE
Area: $A = \frac{1}{2}bh$

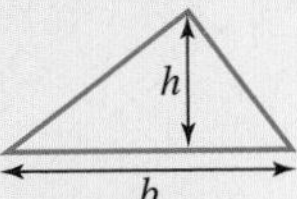

SPHERE
Volume: $V = \frac{4}{3}\pi r^3$
Surface area: $A = 4\pi r^2$

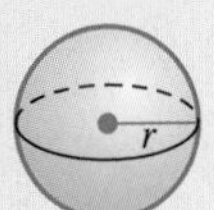

RECTANGULAR BOX
Volume: $V = lwh$
Surface area: $A = 2lh + 2wh + 2lw$

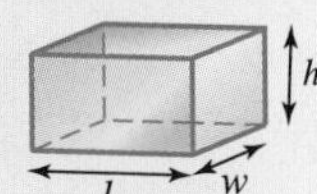

CIRCULAR CYLINDER
Volume: $V = \pi r^2 h$
Surface area: $A = 2\pi r^2 + 2\pi rh$

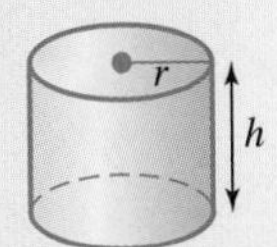

TRIANGULAR CYLINDER
Volume: $V = \frac{1}{2}bhl$

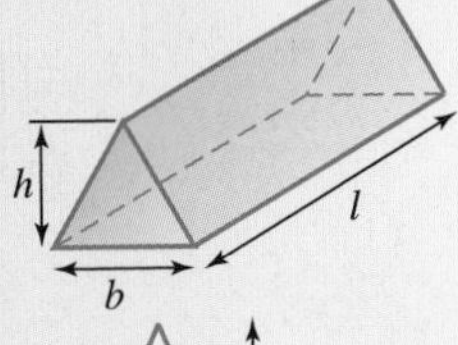

CONE
Volume: $V = \frac{1}{3}\pi r^2 h$

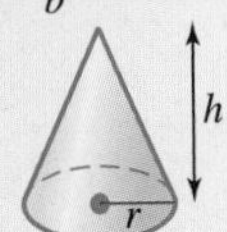

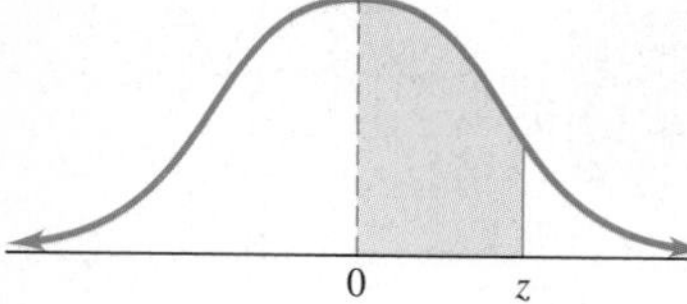

Table 2 Areas under the Normal Curve
The column under A gives the proportion of the area under the entire curve that is between $z = 0$ and a positive value of z.

z	A	z	A	z	A	z	A
.00	.0000	.48	.1844	.96	.3315	1.44	.4251
.01	.0040	.49	.1879	.97	.3340	1.45	.4265
.02	.0080	.50	.1915	.98	.3365	1.46	.4279
.03	.0120	.51	.1950	.99	.3389	1.47	.4292
.04	.0160	.52	.1985	1.00	.3413	1.48	.4306
.05	.0199	.53	.2019	1.01	.3438	1.49	.4319
.06	.0239	.54	.2054	1.02	.3461	1.50	.4332
.07	.0279	.55	.2088	1.03	.3485	1.51	.4345
.08	.0319	.56	.2123	1.04	.3508	1.52	.4357
.09	.0359	.57	.2157	1.05	.3531	1.53	.4370
.10	.0398	.58	.2190	1.06	.3554	1.54	.4382
.11	.0438	.59	.2224	1.07	.3577	1.55	.4394
.12	.0478	.60	.2258	1.08	.3599	1.56	.4406
.13	.0517	.61	.2291	1.09	.3621	1.57	.4418
.14	.0557	.62	.2324	1.10	.3643	1.58	.4430
.15	.0596	.63	.2357	1.11	.3665	1.59	.4441
.16	.0636	.64	.2389	1.12	.3686	1.60	.4452
.17	.0675	.65	.2422	1.13	.3708	1.61	.4463
.18	.0714	.66	.2454	1.14	.3729	1.62	.4474
.19	.0754	.67	.2486	1.15	.3749	1.63	.4485
.20	.0793	.68	.2518	1.16	.3770	1.64	.4495
.21	.0832	.69	.2549	1.17	.3790	1.65	.4505
.22	.0871	.70	.2580	1.18	.3810	1.66	.4515
.23	.0910	.71	.2612	1.19	.3830	1.67	.4525
.24	.0948	.72	.2642	1.20	.3849	1.68	.4535
.25	.0987	.73	.2673	1.21	.3869	1.69	.4545
.26	.1026	.74	.2704	1.22	.3888	1.70	.4554
.27	.1064	.75	.2734	1.23	.3907	1.71	.4564
.28	.1103	.76	.2764	1.24	.3925	1.72	.4573
.29	.1141	.77	.2794	1.25	.3944	1.73	.4582
.30	.1179	.78	.2823	1.26	.3962	1.74	.4591
.31	.1217	.79	.2852	1.27	.3980	1.75	.4599
.32	.1255	.80	.2881	1.28	.3997	1.76	.4608
.33	.1293	.81	.2910	1.29	.4015	1.77	.4616
.34	.1331	.82	.2939	1.30	.4032	1.78	.4625
.35	.1368	.83	.2967	1.31	.4049	1.79	.4633
.36	.1406	.84	.2996	1.32	.4066	1.80	.4641
.37	.1443	.85	.3023	1.33	.4082	1.81	.4649
.38	.1480	.86	.3051	1.34	.4099	1.82	.4656
.39	.1517	.87	.3079	1.35	.4115	1.83	.4664
.40	.1554	.88	.3106	1.36	.4131	1.84	.4671
.41	.1591	.89	.3133	1.37	.4147	1.85	.4678
.42	.1628	.90	.3159	1.38	.4162	1.86	.4686
.43	.1664	.91	.3186	1.39	.4177	1.87	.4693
.44	.1700	.92	.3212	1.40	.4192	1.88	.4700
.45	.1736	.93	.3238	1.41	.4207	1.89	.4706
.46	.1772	.94	.3264	1.42	.4222	1.90	.4713
.47	.1808	.95	.3289	1.43	.4236	1.91	.4719

Table 2 *(continued)*

z	A	z	A	z	A	z	A
1.92	.4726	2.42	.4922	2.92	.4983	3.42	.4997
1.93	.4732	2.43	.4925	2.93	.4983	3.43	.4997
1.94	.4738	2.44	.4927	2.94	.4984	3.44	.4997
1.95	.4744	2.45	.4929	2.95	.4984	3.45	.4997
1.96	.4750	2.46	.4931	2.96	.4985	3.46	.4997
1.97	.4756	2.47	.4932	2.97	.4985	3.47	.4997
1.98	.4762	2.48	.4934	2.98	.4986	3.48	.4998
1.99	.4767	2.49	.4936	2.99	.4986	3.49	.4998
2.00	.4773	2.50	.4938	3.00	.4987	3.50	.4998
2.01	.4778	2.51	.4940	3.01	.4987	3.51	.4998
2.02	.4783	2.52	.4941	3.02	.4987	3.52	.4998
2.03	.4788	2.53	.4943	3.03	.4988	3.53	.4998
2.04	.4793	2.54	.4945	3.04	.4988	3.54	.4998
2.05	.4798	2.55	.4946	3.05	.4989	3.55	.4998
2.06	.4803	2.56	.4948	3.06	.4989	3.56	.4998
2.07	.4808	2.57	.4949	3.07	.4989	3.57	.4998
2.08	.4812	2.58	.4951	3.08	.4990	3.58	.4998
2.09	.4817	2.59	.4952	3.09	.4990	3.59	.4998
2.10	.4821	2.60	.4953	3.10	.4990	3.60	.4998
2.11	.4826	2.61	.4955	3.11	.4991	3.61	.4999
2.12	.4830	2.62	.4956	3.12	.4991	3.62	.4999
2.13	.4834	2.63	.4957	3.13	.4991	3.63	.4999
2.14	.4838	2.64	.4959	3.14	.4992	3.64	.4999
2.15	.4842	2.65	.4960	3.15	.4992	3.65	.4999
2.16	.4846	2.66	.4961	3.16	.4992	3.66	.4999
2.17	.4850	2.67	.4962	3.17	.4992	3.67	.4999
2.18	.4854	2.68	.4963	3.18	.4993	3.68	.4999
2.19	.4857	2.69	.4964	3.19	.4993	3.69	.4999
2.20	.4861	2.70	.4965	3.20	.4993	3.70	.4999
2.21	.4865	2.71	.4966	3.21	.4993	3.71	.4999
2.22	.4868	2.72	.4967	3.22	.4994	3.72	.4999
2.23	.4871	2.73	.4968	3.23	.4994	3.73	.4999
2.24	.4875	2.74	.4969	3.24	.4994	3.74	.4999
2.25	.4878	2.75	.4970	3.25	.4994	3.75	.4999
2.26	.4881	2.76	.4971	3.26	.4994	3.76	.4999
2.27	.4884	2.77	.4972	3.27	.4995	3.77	.4999
2.28	.4887	2.78	.4973	3.28	.4995	3.78	.4999
2.29	.4890	2.79	.4974	3.29	.4995	3.79	.4999
2.30	.4893	2.80	.4974	3.30	.4995	3.80	.4999
2.31	.4896	2.81	.4975	3.31	.4995	3.81	.4999
2.32	.4898	2.82	.4976	3.32	.4996	3.82	.4999
2.33	.4901	2.83	.4977	3.33	.4996	3.83	.4999
2.34	.4904	2.84	.4977	3.34	.4996	3.84	.4999
2.35	.4906	2.85	.4978	3.35	.4996	3.85	.4999
2.36	.4909	2.86	.4979	3.36	.4996	3.86	.4999
2.37	.4911	2.87	.4980	3.37	.4996	3.87	.5000
2.38	.4913	2.88	.4980	3.38	.4996	3.88	.5000
2.39	.4916	2.89	.4981	3.39	.4997	3.89	.5000
2.40	.4918	2.90	.4981	3.40	.4997		
2.41	.4920	2.91	.4982	3.41	.4997		

Table 3 Integrals

(C is an arbitrary constant.)

1. $\int x^n \, dx = \frac{1}{n+1} x^{n+1} + C \quad (n \neq -1)$

2. $\int e^{kx} \, dx = \frac{1}{k} e^{kx} + C$

3. $\int \frac{a}{x} \, dx = a \ln |x| + C$

4. $\int \ln |ax| \, dx = x(\ln |ax| - 1) + C$

5. $\int \frac{1}{\sqrt{x^2 + a^2}} \, dx = \ln |x + \sqrt{x^2 + a^2}| + C$

6. $\int \frac{1}{\sqrt{x^2 - a^2}} \, dx = \ln |x + \sqrt{x^2 - a^2}| + C$

7. $\int \frac{1}{a^2 - x^2} \, dx = \frac{1}{2a} \cdot \ln \left| \frac{a + x}{a - x} \right| + C \quad (a \neq 0)$

8. $\int \frac{1}{x^2 - a^2} \, dx = \frac{1}{2a} \cdot \ln \left| \frac{x - a}{x + a} \right| + C \quad (a \neq 0)$

9. $\int \frac{1}{x\sqrt{a^2 - x^2}} \, dx = -\frac{1}{a} \cdot \ln \left| \frac{a + \sqrt{a^2 - x^2}}{x} \right| + C \quad (a \neq 0)$

10. $\int \frac{1}{x\sqrt{a^2 + x^2}} \, dx = -\frac{1}{a} \cdot \ln \left| \frac{a + \sqrt{a^2 + x^2}}{x} \right| + C \quad (a \neq 0)$

11. $\int \frac{x}{ax + b} \, dx = \frac{x}{a} - \frac{b}{a^2} \cdot \ln |ax + b| + C \quad (a \neq 0)$

12. $\int \frac{x}{(ax + b)^2} \, dx = \frac{b}{a^2(ax + b)} + \frac{1}{a^2} \cdot \ln |ax + b| + C \quad (a \neq 0)$

13. $\int \frac{1}{x(ax + b)} \, dx = \frac{1}{b} \cdot \ln \left| \frac{x}{ax + b} \right| + C \quad (b \neq 0)$

14. $\int \frac{1}{x(ax + b)^2} \, dx = \frac{1}{b(ax + b)} + \frac{1}{b^2} \cdot \ln \left| \frac{x}{ax + b} \right| \quad (b \neq 0)$

15. $\int \sqrt{x^2 + a^2} \, dx = \frac{x}{2}\sqrt{x^2 + a^2} + \frac{a^2}{2} \cdot \ln |x + \sqrt{x^2 + a^2}| + C$

16. $\int x^n \cdot \ln |x| \, dx = x^{n+1} \left[\frac{\ln|x|}{n + 1} - \frac{1}{(n + 1)^2} \right] + C \quad (n \neq -1)$

17. $\int x^n e^{ax} \, dx = \frac{x^n e^{ax}}{a} - \frac{n}{a} \cdot \int x^{n-1} e^{ax} \, dx + C \quad (a \neq 0)$

Answers to Selected Exercises

CHAPTER 1

Section 1.1 (Page 9)

1. true **3.** identity property of addition **5.** commutative property of addition **7.** associative property of multiplication

11. -54 **13.** $-\frac{1}{2}$ **15.** 18% **17.** .75% **19.** 4

21. -4 **23.** -1 **25.** irrational; 9.4248

27. irrational; 1.7321 **29.** $5 < 7$ **31.** $y \le 8.3$ **33.** $t > 0$

35. -8 -1

37. -2 2

39. -4 -2 0 2 4

41. 3

43. -2

45. 4 **47.** 6 **49.** 0 **51.** **(a)** 17.6 **(b)** no

53. **(a)** 29.3 **(b)** no **55.** 47° **57.** 22° **59.** 4

61. -19 **63.** = **65.** = **67.** = **69.** =

71. $-(a - 7) = 7 - a$

77. Let x represent the number of dollars contributed annually by foreign students to the California economy. Then $x > 1{,}000{,}000{,}000$.

79. Let x represent the percent of foreign students now in the United States from Middle Eastern countries. Then $0 \le x < 7.5$.

81. Let x represent the percent of foreign students in the United States that are in California. Then $x > 13$.

Section 1.2 (Page 16)

1. 37,974,983.3583 **3.** 289.099133864 **7.** 2^7

9. $(-5)^7 = -5^7$ **11.** $(2z)^{11}$ **13.** $-x^3 + x^2 + 3x$

15. $-6y^2 + 3y + 10$ **17.** $-10x^2 + 4x - 2$

19. $-18m^3 - 27m^2 + 9m$ **21.** $12z^3 + 14z^2 - 7z + 5$

23. $12k^2 - 20k + 3$ **25.** $6y^2 + 7y - 5$

27. $18k^2 - 7kq - q^2$ **29.** $4.34m^2 + 5.68m - 4.42$

31. $-k + 3$ **33.** $R = 5000x$; $C = 150{,}000 + 2250x$; $P = 2750x - 150{,}000$ **35.** $R = 7500x$; $C = -3x^2 + 3480x + 299{,}675$; $P = 3x^2 + 4020x - 299{,}675$

37. **(a)** 9,400,000 **(b)** 8,900,000

39. **(a)** 32,700,000 **(b)** 32,036,000

41. 43,844,000 **43.** 49,076,000

45. **(a)** approximately 60,501,067 cu ft **(b)** The shape becomes a rectangular box with a square base, with volume b^2h. **(c)** yes

47. **(a)** 0, 1, 2, 3, or no degree (if one is the negative of the other) **(b)** 0, 1, 2, 3, or no degree (if they are equal) **(c)** 6

Section 1.3 (Page 22)

1. $12x(x - 2)$ **3.** $r(r^2 - 5r + 1)$ **5.** $6z(z^2 - 2z + 3)$

7. $2(2y - 1)^2(5y - 1)$ **9.** $(x + 5)^4(x^2 + 10x + 28)$

11. $(x + 1)(x + 4)$ **13.** $(x + 3)(x + 4)$ **15.** $(z + 4)(z + 6)$

17. $(2x - 1)(x - 4)$ **19.** $(3p - 4)(5p - 1)$

21. $(2z - 5)(2z - 3)$ **23.** $(2x + 1)(3x - 4)$

25. $(5y - 2)(2y + 5)$ **27.** $(2x - 1)(3x + 4)$

29. $(2a + 5)(a - 1)$ **31.** $(x + 8)(x - 8)$ **33.** $(3p - 4)^2$

35. $(r + 2t)(r - 5t)$ **37.** $(m - 3n)^2$ **39.** $(2p + 3)(2p - 3)$

41. $3(x - 4z)^2$ **43.** cannot be factored

45. $(-x + 4)(x - 3)$ or $(x - 4)(-x + 3)$ **47.** $(3a + 5)(a - 6)$

49. $(7m + 2n)(3m + n)$ **51.** $(4y - x)(5y + 11x)$

53. $(y - 7z)(y + 3z)$ **55.** $(11x + 8)(11x - 8)$

57. $m^3(m - 1)^2(m^2 + m + 1)^2(5 + 3m^2 - 3m^5)$

59. $(a - 6)(a^2 + 6a + 36)$ **61.** $(2r - 3s)(4r^2 + 6rs + 9s^2)$

63. $(4m + 5)(16m^2 - 20m + 25)$

65. $(10y - z)(100y^2 + 10yz + z^2)$

Section 1.4 (Page 27)

1. $\frac{x}{5}$ **3.** $\frac{4}{7p}$ **5.** $\frac{5}{4}$ **7.** $\frac{4}{w + 3}$ **9.** $\frac{y - 4}{3y^2}$ **11.** $\frac{m - 2}{m + 3}$

13. $\frac{x + 4}{x + 1}$ **15.** $\frac{2p}{7}$ **17.** $\frac{9a^3}{4}$ **19.** $\frac{15}{8c}$ **21.** $\frac{3}{4}$ **23.** $\frac{3}{10}$

25. $\frac{2(a + 4)}{a - 3}$ **27.** $\frac{k + 2}{k + 3}$ **31.** $-\frac{1}{15z}$ **33.** $\frac{4}{3}$ **35.** $\frac{12 + x}{3x}$

37. $\frac{3m - 2}{m(m - 1)}$ **39.** $\frac{14}{3(a - 1)}$ **41.** $\frac{23}{20(k - 2)}$

43. $\frac{7x + 9}{(x - 3)(x + 1)(x + 2)}$ **45.** $\frac{y^2}{(y + 4)(y + 3)(y + 2)}$

47. $\frac{k^2 - 13k}{(2k - 1)(k + 2)(k - 3)}$ **49.** $\frac{x + 1}{x - 1}$ **51.** $\frac{-1}{x(x + h)}$

53. **(a)** $\frac{x - 5}{x^2 - 10x}$ **(b)** 32.7 sec; 8 sec; 4.5 sec

Section 1.5 (Page 37)

1. 1 **3.** $\frac{1}{6}$ **5.** $\frac{1}{32}$ **7.** $-\frac{1}{2401}$ **9.** 216 **11.** $\frac{9}{16}$

13. 7 **15.** 1.7 **17.** 9 **19.** -16 **21.** $\frac{81}{16}$ **23.** $\frac{1}{4^5}$

25. 4^3 **27.** 2 **29.** 8^5 **31.** $\frac{1}{9^{32/15}}$ **33.** z^3 **35.** $\frac{p}{4}$

37. $\frac{q^5}{r^2}$ **39.** $\frac{8}{25p^7}$ **41.** $2^{5/6}p^{3/2}$ **43.** $2p + 5p^{5/3}$ **45.** (f)

47. (h) **49.** (g) **51.** (c) **53.** 4 **55.** 5 **57.** -2

59. 9 **61.** $\sqrt{77}$ **63.** 7 **65.** 3 **67.** $\frac{\sqrt{5} - 1}{2}$

69. $-1-\sqrt{3}$ **71.** $\dfrac{1}{7+4\sqrt{3}}$ **73. (a)** 14 **(b)** 85 **(c)** 58
75. (a) $-63.4°$ **(b)** $-46.7°$ **(c)** Approximation is very good.
77. 166.46 **79.** 175.36 **81.** about 9440 **83.** about 12,533
85. about 14,216 **87. (a)** about 2,069,800; about 3,159,800; about 4,579,000 **(b)** about 2021; no

Section 1.6 (Page 48)

1. 5 **3.** 7 **5.** $-\dfrac{9}{8}$ **7.** 2 **9.** $\dfrac{40}{7}$ **11.** $\dfrac{26}{3}$
13. $-\dfrac{12}{5}$ **15.** $-\dfrac{59}{6}$ **17.** $-\dfrac{9}{4}$ **19.** $x = .72$
21. $r \approx -13.26$ **23.** $x = \dfrac{5a-b}{6}$ **25.** $x = \dfrac{3b}{a+5}$
27. $V = \dfrac{k}{P}$ **29.** $g = \dfrac{V-V_0}{t}$ **31.** $B = \dfrac{2A}{h} - b$ or $B = \dfrac{2A-bh}{h}$
33. 2, -3 **35.** -2, 8 **37.** $\dfrac{5}{2}, \dfrac{7}{2}$ **39.** $-7, -\dfrac{3}{7}$
41. 13% **43.** \$4000 **45.** 1975 **47.** 1995 **49.** 1996
51. 2001 **53.** \$205.41 **55. (a)** .0352 **(b)** about .015, or 1.5% **(c)** approximately one case **57.** \$8000 **59.** \$5000
61. about 838 mi **63.** $\dfrac{400}{3}$ L **65.** 142 mi **67.** 70 mph
69. 10.025 gal premium; 5.475 gal regular **71.** 6 cm **73.** 4 ft

Section 1.7 (Page 58)

1. -3, 12 **3.** -5, 0 **5.** 0, 2 **7.** -7, -8 **9.** $\dfrac{3}{2}$, 1
11. $-\dfrac{1}{2}, \dfrac{1}{3}$ **13.** $\dfrac{5}{2}$, 4 **15.** 5, 2 **17.** $\dfrac{4}{3}, -\dfrac{4}{3}$ **19.** 0, 1
21. $2 \pm \sqrt{7}$ **23.** $\dfrac{1 \pm 2\sqrt{5}}{4}$ **25.** $\dfrac{-5 \pm \sqrt{17}}{4}$; $-.219, -2.281$
27. $\dfrac{-1 \pm \sqrt{5}}{4}$; .309, $-.809$ **29.** $\dfrac{-3 \pm \sqrt{19}}{5}$; .272, -1.472
31. no real number solutions **33.** $\dfrac{5}{2}$, 1 **35.** no real number solutions **37.** -5, 3/2 **39.** 1 **41.** 2
43. $x \approx .4701$ or 1.8240 **45.** $x \approx -1.0376$ or .6720
47. (a) 1.5 sec **(b)** about 3.35 sec **49. (a)** about 32 and 62 **(b)** about 17 and 77 **51. (a)** 1985 **(b)** 2001 **53. (a)** $x + 20$ **(b)** northbound: $5x$; eastbound: $5(x + 20)$ or $5x + 100$ **(c)** $(5x)^2 + (5x + 100)^2 = 300^2$ **(d)** about 31.23 mph and 51.23 mph
55. (a) $150 - x$ **(b)** $x(150 - x) = 5000$ **(c)** length 100 m; width 50 m **57.** 9 ft by 12 ft **59. (a)** 2 sec **(b)** $\dfrac{1}{2}$ sec or $\dfrac{7}{2}$ sec **(c)** It reaches the given height twice—once on the way up and once on the way down. **61.** $t = \pm\dfrac{\sqrt{2Sg}}{g}$ **63.** $h = \dfrac{\pm d^2\sqrt{kL}}{L}$
65. $R = \dfrac{-2Pr + E^2 \pm E\sqrt{E^2 - 4Pr}}{2P}$ **67. (a)** $x^2 - 2x = 15$ **(b)** $x = 5$ or $x = -3$ **(c)** $z = \pm\sqrt{5}$ **69.** $\pm\dfrac{\sqrt{6}}{2}$
71. $\pm\sqrt{\dfrac{3 + \sqrt{13}}{2}}$

Chapter 1 Review (Page 61)

1. 0, 6 **3.** $-12, -6, -\dfrac{9}{10}, -\sqrt{4}, 0, \dfrac{1}{8}, 6$
5. commutative property of multiplication **7.** distributive property
9. $x \geq 6$ **11.** $-7, -3, -2, 0, \pi, 8$
13. $-|3 - (-2)|, -|-2|, |6 - 4|, |8 + 1|$ **15.** -3 **17.** -1
19.

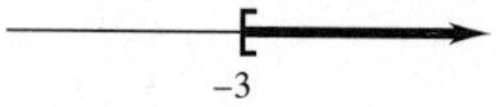

21.

23. -8 **25.** $-\dfrac{3}{4}$
27. $4x^4 - 4x^2 + 13x$ **29.** $2q^4 + 14q^3 - 4q^2$ **31.** $15z^2 - 4z - 4$
33. $16k^2 - 9h^2$ **35.** $36x^2 + 36xy + 9y^2$ **37.** $k(2h^2 - 4h + 5)$
39. $a^2(3a + 1)(a + 4)$ **41.** $(2y - 1)(5y - 3)$ **43.** $(2a - 5)^2$
45. $(12p + 13q)(12p - 13q)$ **47.** $(2y - 1)(4y^2 + 2y + 1)$
49. $\dfrac{7x^2}{3}$ **51.** 4 **53.** $\dfrac{(y-5)(4y-5)}{6y}$ **55.** $\dfrac{(m-1)^2}{3(m+1)}$
57. $\dfrac{1}{6z}$ **59.** $\dfrac{64}{35q}$ **61.** $\dfrac{1}{5^3}$ or $\dfrac{1}{125}$ **63.** -1 **65.** $\dfrac{5^2}{6^2}$ or $\dfrac{25}{36}$
67. 4^3 **69.** $\dfrac{1}{9}$ **71.** 9^3 **73.** $\dfrac{3}{4}$ **75.** 25 **77.** $\dfrac{1}{3^5}$
79. $\dfrac{1}{5^{2/3}}$ **81.** $3^{7/2}a^{5/2}$ **83.** 3 **85.** $2\sqrt{6}$ **87.** $3pq\sqrt[3]{2q^2}$
89. $\dfrac{n\sqrt{30m}}{6m}$ **91.** $-8\sqrt{3}$ **93.** 4 **95.** $\dfrac{-\sqrt{2} + \sqrt{6}}{2}$
97. \$64,000,000 **99.** \$10,081,152 **101.** 3 **103.** -2
105. no solution **107.** $x = \dfrac{1}{5a - 1}$ **109.** $x = \dfrac{c - 3}{3a - ac - 2}$
111. 12, -6 **113.** -38, 42 **115. (a)** 1999 **(b)** 1996
117. (a) 1993 **(b)** 2000 **119.** \$60,000 at 8%; \$40,000 at 5%
121. 2 **122.** 2 **123.** 1 **125.** 0 **127.** $\dfrac{-1 \pm \sqrt{7}}{2}$
129. $3, -\dfrac{5}{2}$ **131.** $-2 \pm \sqrt{5}$ **133.** $\dfrac{-1 \pm \sqrt{3}}{3}$ **135.** $\dfrac{1}{2}, \dfrac{1}{6}$
137. $-\dfrac{8}{3}$, 2 **139.** $\dfrac{\pm\sqrt{3}}{3}$ **141.** $\dfrac{\pm\sqrt{3}}{3}$
143. $r = \dfrac{-Rp \pm E\sqrt{Rp}}{p}$ **145.** $s = \dfrac{a \pm \sqrt{a^2 + 4K}}{2}$
147. (a) 2000 **(b)** 2006 **149.** 50 m by 225 m or 112.5 m by 100 m

Case 1 (Page 65)

1. $c = 700 + 85x$ **3.** The \$700 refrigerator costs \$300 more over 10 years.

CHAPTER 2

Section 2.1 (Page 73)

1. yes **3.** no
5.

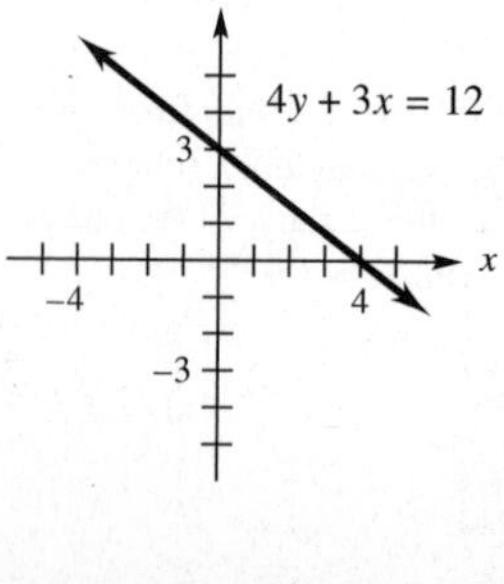

7.

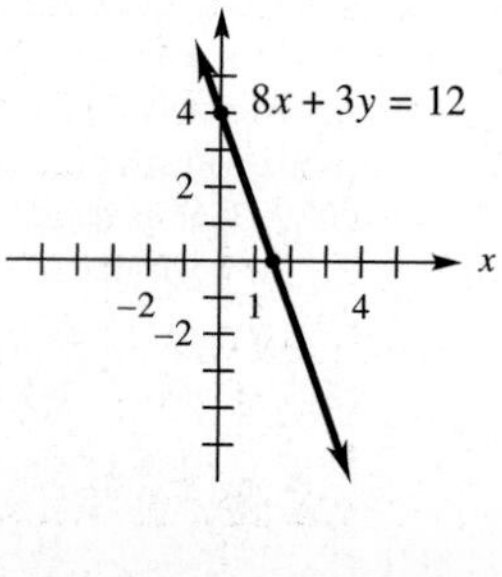

9.

11. x-intercepts -1, 3.5; y-intercept 1 **13.** x-intercepts -2, 0, 2; y-intercept 0 **15.** x-intercept 8; y-intercept 6 **17.** x-intercept 3; y-intercept -2 **19.** x-intercepts 3 and -3; y-intercept -9

21.

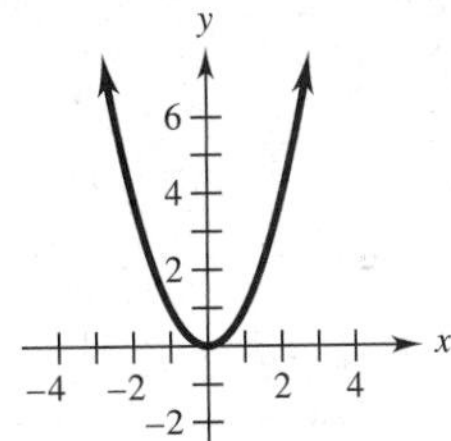

23.

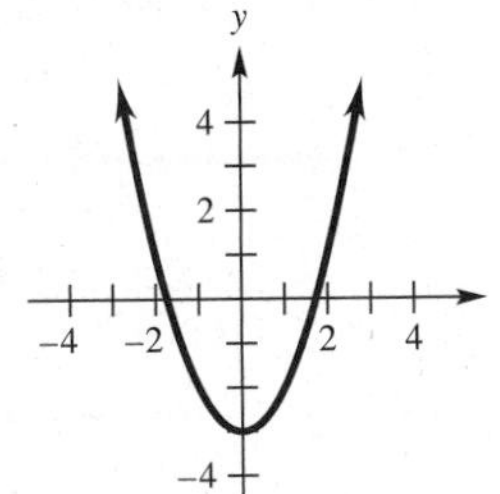

25.

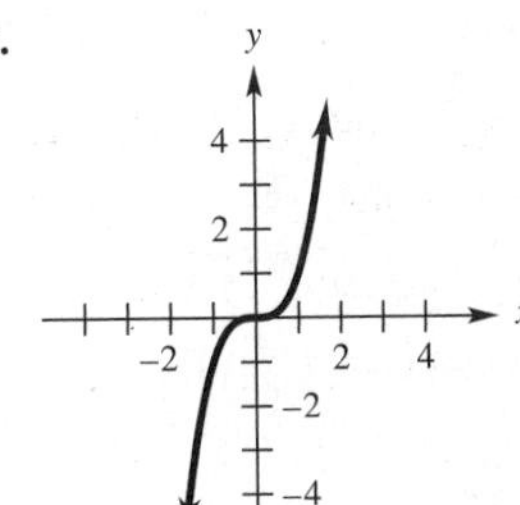

27.

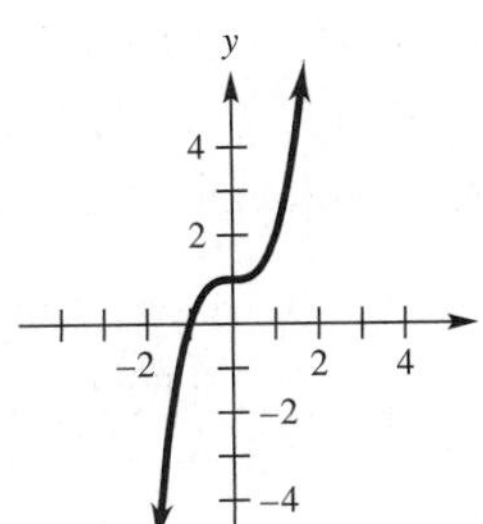

29.

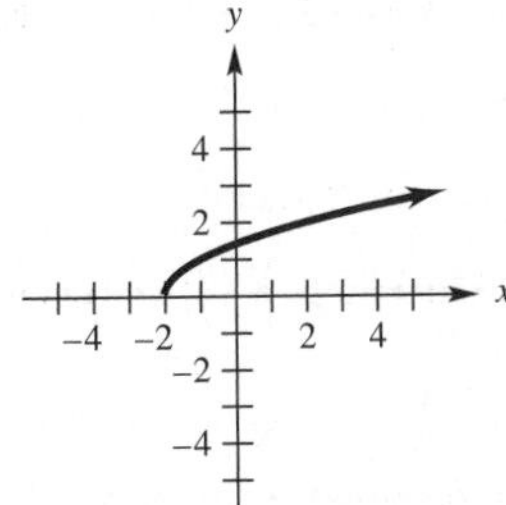

31.

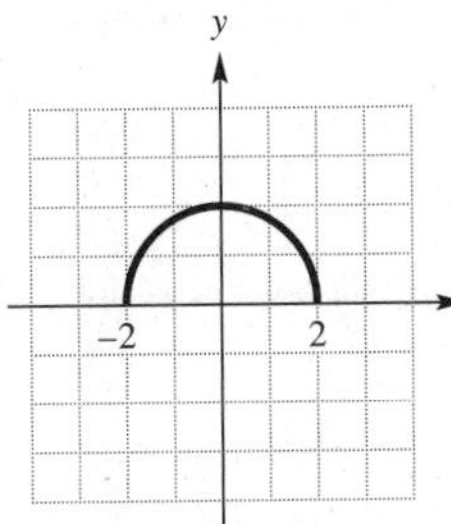

33.

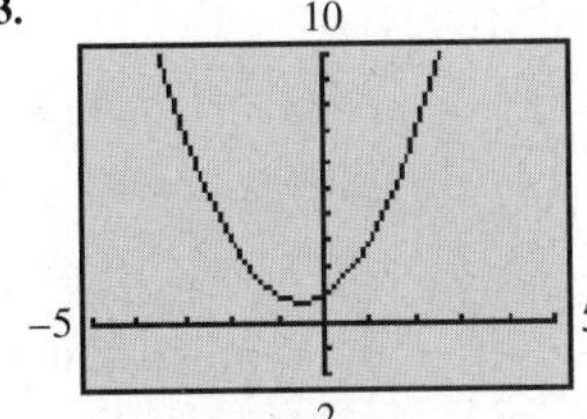

35.

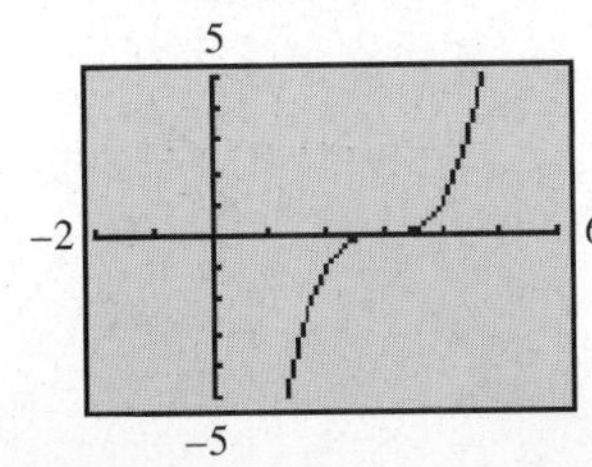

37.

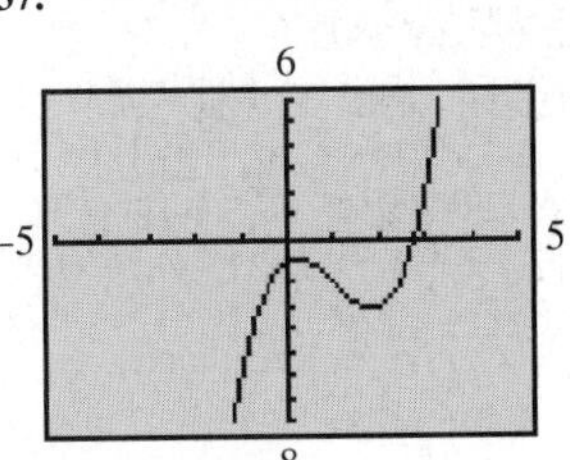

39.

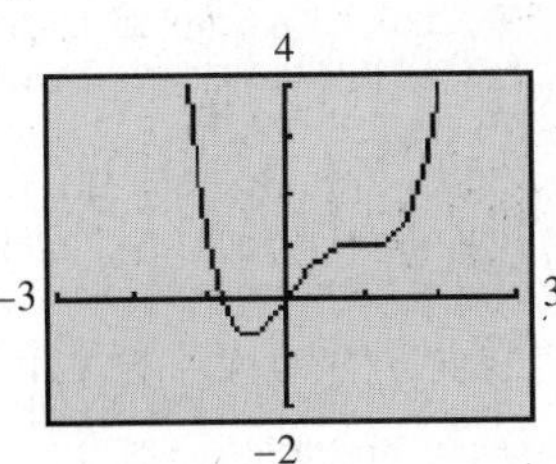

41. $x \approx -1.1038$ **43.** $x \approx 2.1017$ **45.** $x \approx -1.7521$ **47.** $r \approx 4.6580$ in. **49.** 1989 **51.** In mid-1985 ($x \approx 10.46$) about 3,349,000 boys were involved in sports. **53.** Fargo, about 2:00 P.M.; Seattle, about 5 P.M. **55.** from 11 A.M. until 6 P.M. **57.** **(a)** about $1,250,000 **(b)** about $1,750,000 **(c)** about $4,250,000 **59.** **(a)** about $500,000 **(b)** about $1,000,000 **(c)** about $1,500,000 **61.** approximately 12 sec **63.** **(a)** about $750, $600, and $300. **(b)** during the 22nd year (that is, between $t = 21$ and $t = 22$) **65.** from 1990 to the middle of 1993 **67.** It rose from about 117 at the beginning of 1997 to a high of about 138 in mid-1998, then dropped sharply to about 119, and then started up again, finishing the year around 127. **69.** from June 1996 to approximately April 1997, briefly in fall 1998, and in the first two months of 2001

Section 2.2 (Page 86)

1. $-\frac{1}{2}$ **3.** -1 **5.** $-\frac{3}{2}$ **7.** undefined **9.** $y = 3x + 5$ **11.** $y = -2.3x + 1.5$ **13.** $y = -\frac{3x}{4} + 4$ **15.** $m = 2; b = -7$ **17.** $m = 3; b = -2$ **19.** $m = \frac{2}{3}; b = -\frac{14}{9}$ **21.** $m = \frac{2}{3}; b = 0$ **23.** $m = 1; b = 5$ **25.** **(a)** C **(b)** B **(c)** B **(d)** D

27.

29.

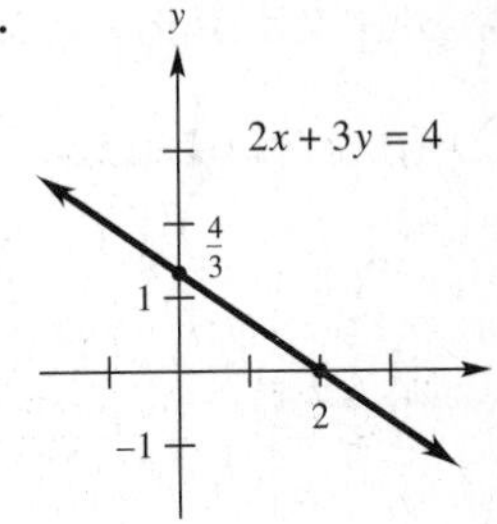

31.

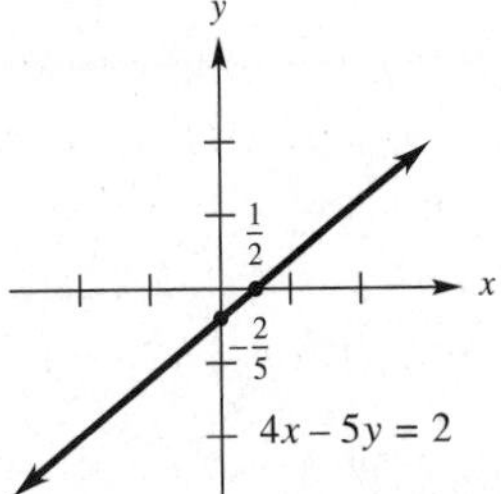

33. perpendicular **35.** parallel **37.** neither **39.** **(a)** $\frac{2}{5}; \frac{9}{8}; -\frac{5}{2}$ **(b)** yes **41.** $3y = -2x + 4$ or $y = -\frac{2}{3}x + \frac{4}{3}$ **43.** $y = 2x + 2$ **45.** $y = 2$ **47.** $x = 6$ **49.** $3y = 4x + 7$ **51.** $2y = 5x - 1$ **53.** $y = 7x$ **55.** $x = 5$ **57.** $y = 2x - 2$ **59.** $y = x - 5$ **61.** $y = -x + 2$ **63.** $1248.20

65. \$5178.57 **67.** **(a)** 4,130,000 **(b)** 4,610,000 **(c)** 4,850,000 **(d)** 2004 **69.** **(a)** \$6,928,230,000 **(b)** 2005 **71.** **(a)** (0, 54), (8, 63) **(b)** $y = 1.125x + 54$ **(c)** 56,250,000 **(d)** 2005 **73.** **(a)** $y = .19x + 1.1$ **(b)** about 7.4 million **75.** **(a)** $T = .03t + 15$ **(b)** 2103 approximately **77.** **(a)** The slope of $-.0221$ indicates that on the average, from 1912 to 1992, the 5000-meter run is being run .0221 second faster every year. It is negative because the times are generally decreasing as time progresses. **(b)** World War II was going on in 1940 and 1944, and no Olympic Games were held.

Section 2.3 (Page 96)

1. **(a)** $y = \frac{5}{9}(x - 32)$ **(b)** 10°C; 23.89°C **3.** 462.2°C **5.** $y = 4x + 130.7$; 150.7; 182.7 **7.** $y = 9x + 1$; \$109 billion **9.** $y = -.3125x + 14.25$; 42,500 sq mi **11.** **(a)** For $y = .5x + 1.5$: 0, $-.5$, 0, $-.5$, 1; sum 0. For $y = x$: 1, 0, 0, -1, 0; sum 0. **(b)** 1.5 for $y = .5x + 1.5$ and 2 for $y = x$ **(c)** $y = .5x + 1.5$ **13.** A straight line is a poor model ($r \approx .1$). **15.** **(a)** two-point model: $y = -4.4615x + 522$; regression line model: $y = -4.8647x + 530.3210$ **(b)** two-point model: about 321; regression line model: about 311 **17.** **(a)** two-point model: $y = 15x + 2750$; regression line model: $y = 14.9x + 2822$ **(b)** two-point model: 5000, 6950, 9050; regression line model: 5055, 6992, 9078 **(c)** two-point model: 6275, 6500; regression line model: 6321.5, 6500 **19.** **(a)** (1, 4.8), (2, 5.8), (3, 8.8), (4, 11.5), (5, 16.2), (6, 20.8), (7, 25.1), (8, 26.3), (9, 29.4), (10, 33.7) **(b)** $y = 3.383x - .367$ **(c)** \$43.6 billion; \$50.4 billion **21.** **(a)** $y = 1.5607x + 31.914$ **(b)** 2008 **23.** **(a)** $y = 56.4889x + 597.0404$ **(b)** Yes; coefficient of correlation is $r \approx .995$. **(c)** about \$1444.4 billion **(d)** 2015 **25.** **(a)** $y = 50.9335x + 401.5133$ **(b)** Yes; coefficient of correlation is $r \approx .986$ **(c)** \$1114.6 billion **(d)** 2011

Section 2.4 (Page 106)

3. $[-4, \infty)$ **5.** $(-\infty, 0)$ **7.** $\left(-\infty, \frac{8}{3}\right]$ **9.** $(-\infty, -7]$ **11.** $(-\infty, 3)$ **13.** $(-1, \infty)$

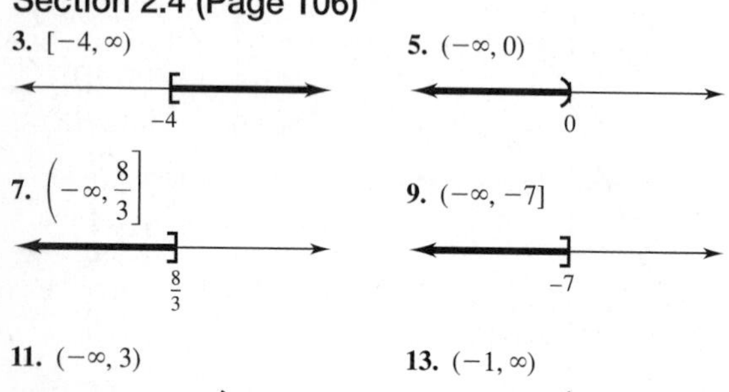

15. $(-\infty, 1]$ **17.** $\left(\frac{1}{5}, \infty\right)$

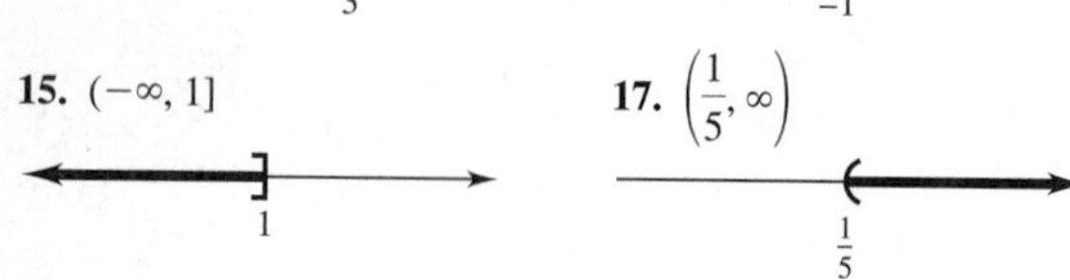

19. $(-5, 6)$ **21.** $\left[\frac{7}{3}, 4\right]$

23. $\left[-\frac{11}{2}, \frac{7}{2}\right]$ **25.** $\left[-\frac{17}{7}, \infty\right)$

27. $x \geq 2$ **29.** $-2 < x \leq 5$ **31.** **(a)** Let x represent the number of mg per L of lead in the water. **(b)** $.038 \leq x \leq .042$ **(c)** yes **33.** **(a)** $0 \leq x \leq 6000$; $6000 < x \leq 27{,}950$; $27{,}950 < x \leq 67{,}700$; $67{,}700 < x \leq 141{,}250$; $141{,}250 < x \leq 307{,}050$; $x > 307{,}050$ **(b)** $0 \leq T \leq 600$; $600 < T \leq 3892.50$; $3892.50 < T \leq 14{,}625$; $14{,}625 < T \leq 36{,}690$; $36{,}690 < T \leq 94{,}720$; $x > 94{,}720$ **35.** $(-1, 1)$ **37.** no solution

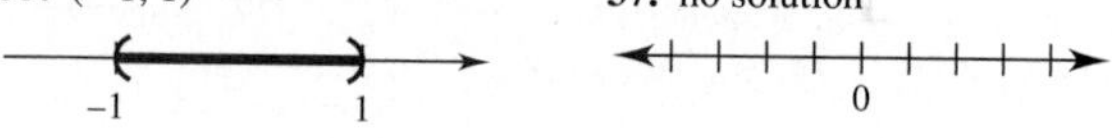

39. $(-4, -1)$

41. $\left(-\infty, -\frac{8}{3}\right]$ or $[2, \infty)$ **43.** $\left(-\frac{3}{2}, \frac{13}{10}\right)$

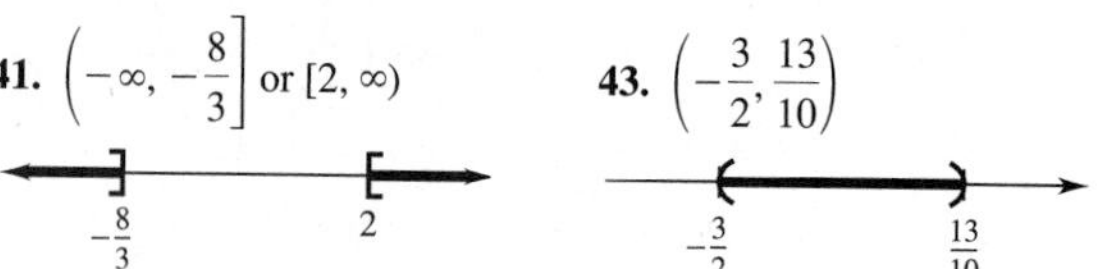

45. $29 \leq T \leq 69$ **47.** $43 \leq T \leq 81$ **49.** **(a)** $25.33 \leq R_L \leq 28.17$; $36.58 \leq R_E \leq 40.92$ **(b)** $5699.25 \leq T_L \leq 6338.25$ $8230.5 \leq T_E \leq 9207$ **51.** 1999 through 2003 **53.** 20 through 35 **55.** between 2 and 6 mi **57.** $[500, \infty)$ **59.** $[45, \infty)$ **61.** impossible to break even **63.** $|x - 2| \leq 4$ **65.** $|z - 12| \geq 2$

Section 2.5 (Page 113)

1. $\left[-5, \frac{3}{2}\right]$ **3.** $(-\infty, -3)$ or $(-1, \infty)$ **5.** $\left[-2, \frac{1}{4}\right]$ **7.** $(-\infty, -1)$ or $\left(\frac{1}{4}, \infty\right)$ **9.** $[-5, 5]$ **11.** $(-\infty, 0)$ or $(16, \infty)$

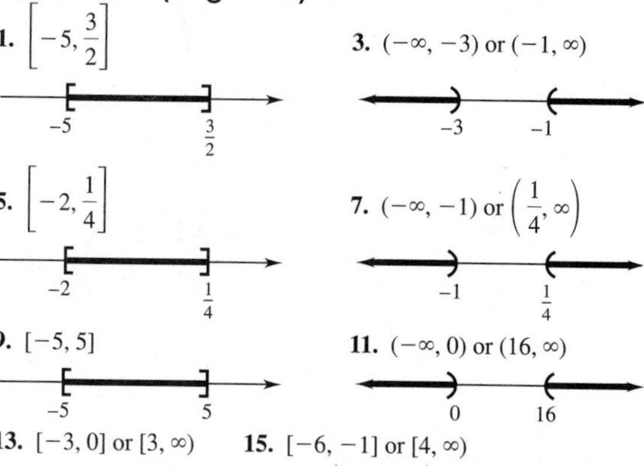

13. $[-3, 0]$ or $[3, \infty)$ **15.** $[-6, -1]$ or $[4, \infty)$ **17.** $(-\infty, -4)$ or $(-1, 3)$ **19.** $\left(-\infty, -\frac{1}{2}\right)$ or $\left(0, \frac{4}{3}\right)$ **21.** no **23.** $(-.0806, 2.4806)$ **25.** $[-2.2635, .7556]$ or $[3.5079, \infty)$ **27.** $(.5, .8393)$ **29.** $(-\infty, 1)$ or $[3, \infty)$ **31.** $\left(\frac{7}{2}, 5\right)$ **33.** $(-\infty, 2)$ or $(5, \infty)$ **35.** $(-\infty, -1)$ **37.** $(-\infty, -2)$ or $(0, 3)$ **39.** $[-1, .5]$ **41.** $\left(0, \frac{5}{3}\right)$ or $(10, \infty)$ **43.** $(100, 150)$ **45.** $[4, 9.75]$

Chapter 2 Review (Page 115)

1. $(-2, 3), (0, -5), (3, -2), (4, 3)$

3.

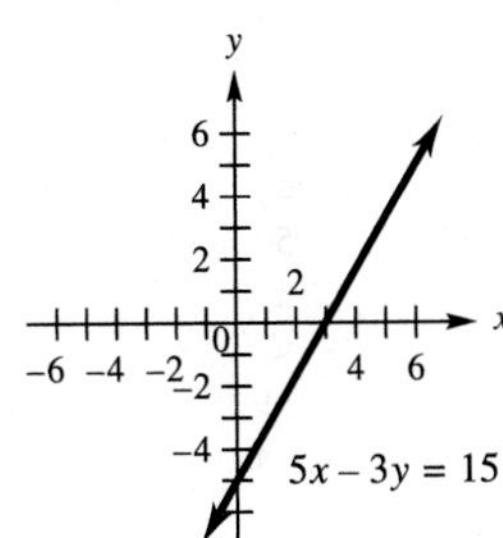

5.

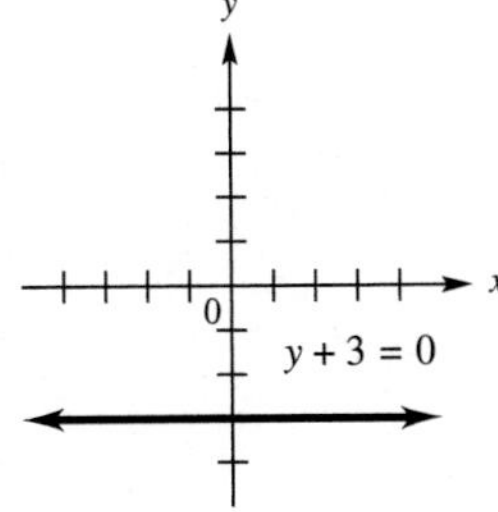

7.

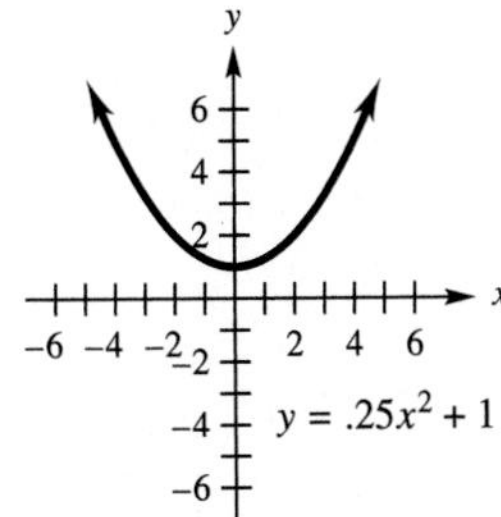

9. (a) about 11:30 A.M. to about 7:30 P.M. **(b)** from midnight until about 5 P.M. and after about 10:30 P.M. **13.** $-\frac{5}{6}$ **15.** $\frac{1}{4}$

17. 4 **19.** 0 **21.** -3

23.

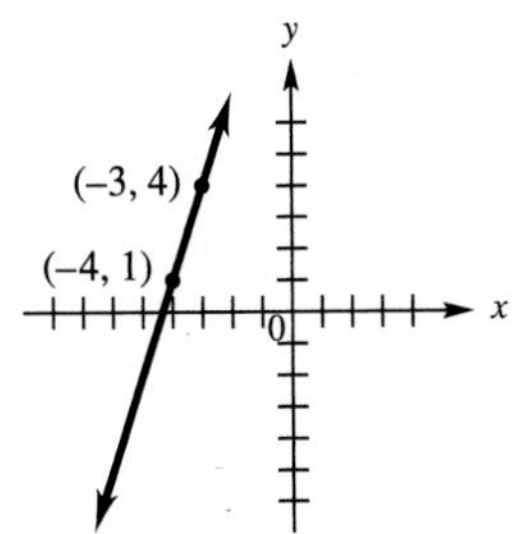

25. $3y = 2x - 13$ **27.** $4y = -5x + 17$ **29.** $x = -1$
31. $3y = 5x + 15$ **33. (a)** $y = -.1625x + 25.3$ **(b)** Negative; smoking is decreasing. **(c)** 22.7% **35. (a)** $y = .365x + 9.955$ **(b)** $y = .3645x + 9.7975$ **(c)** two-point model: \$12.15; regression model: \$11.98; models high by \$0.34 and \$0.17, respectively **(d)** two-point model: \$15.07; regression model: \$14.90
37. (a) $y = 695.88x + 14831.15$ **(b)** $r \approx .9878$; yes **(c)** about \$24,573 **39.** $\left(\frac{3}{8}, \infty\right)$ **41.** $\left(-\infty, \frac{1}{4}\right]$

43. $\left[-\frac{1}{2}, 2\right]$ **45.** $[-8, 8]$ **47.** $(-\infty, 2]$ or $[5, \infty)$

49. $\left[-\frac{9}{5}, 1\right]$ **51.** $5.5 \le y \le 9.5$ **53. (a)** $y = 23x + 415$ **(b)** until 1998 **(c)** after 2003 **55.** $(-3, 2)$

57. $(-\infty, -5]$ or $\left[\frac{3}{2}, \infty\right)$ **59.** $(-\infty, -5]$ or $[-2, 3]$ **61.** $[-2, 0)$

63. $\left(-1, \frac{3}{2}\right)$ **65.** $[-19, -5)$ or $(2, \infty)$ **67.** 1950–1992

Case 2 (Page 118)

3. The poor prediction isn't surprising, since we were extrapolating far beyond the range of the original data. **5.** It's not clear that any simple smooth function will fit this data—there seems to be a break in the pattern between 1970 and 1980. This will make it difficult to predict the life expectancy for females born in 2010. **7.** They used a regression equation of some kind to predict this value!

CHAPTER 3

Section 3.1 (Page 126)

1. function **3.** function **5.** not a function **7.** function **9.** $(-\infty,\infty)$ **11.** $(-\infty,\infty)$ **13.** $(-\infty, 0]$ **15.** all real numbers except 1 **17.** $(-\infty,\infty)$ **19. (a)** 6 **(b)** 6 **(c)** 6 **(d)** 6
21. (a) 48 **(b)** 6 **(c)** 25.38 **(d)** 28.42
23. (a) $\sqrt{7} \approx 2.6458$ **(b)** 0 **(c)** $\sqrt{5.7} \approx 2.3875$ **(d)** not defined **25. (a)** -18.2 **(b)** -97.3 **(c)** -14.365 **(d)** -289.561 **27. (a)** 12 **(b)** 23 **(c)** 12.91 **(d)** 49.41
29. (a) $\frac{\sqrt{3}}{15} \approx .1155$ **(b)** not defined **(c)** $\frac{\sqrt{1.7}}{6.29} \approx .2073$ **(d)** not defined

31.

X	Y_1
3.5	414.31
3.9	642.51
4.3	954.73
4.7	1369.5
5.1	1907.1
5.5	2589.8
5.9	3441.6

Y_1=414.3125

33. (a) $5 - p$ **(b)** $5 + r$ **(c)** $2 - m$
35. (a) $\sqrt{4 - p}\ (p \le 4)$ **(b)** $\sqrt{4 + r}\ (r \ge -4)$ **(c)** $\sqrt{1 - m}\ (m \le 1)$ **37. (a)** $p^3 + 1$ **(b)** $-r^3 + 1$ **(c)** $m^3 + 9m^2 + 27m + 28$

39. (a) $\frac{3}{p - 1}\ (p \ne 1)$ **(b)** $\frac{3}{-r - 1}\ (r \ne -1)$ **(c)** $\frac{3}{m + 2}\ (m \ne -2)$ **41.** 2 **43.** $2x + h$ **45. (a)** about 89% **(b)** about 47% **47. (a)** about 50,000; about 55,700 **(b)** about 157,300

49. (a)

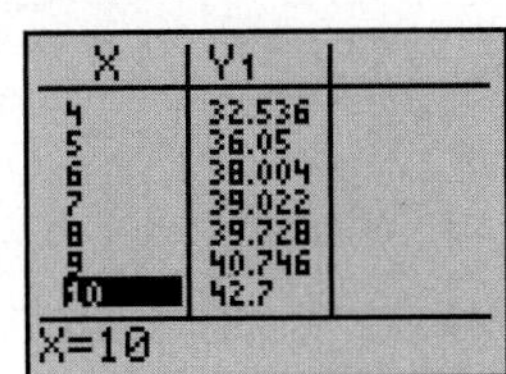

X	Y_1
4	32.536
5	36.05
6	38.004
7	39.022
8	39.728
9	40.746
10	42.7

X=10

51. $f(t) = 200 - 475t$ **53. (a)** $C(x) = 125x + 36,000$ **(b)** \$185

Section 3.2 (Page 137)

1.

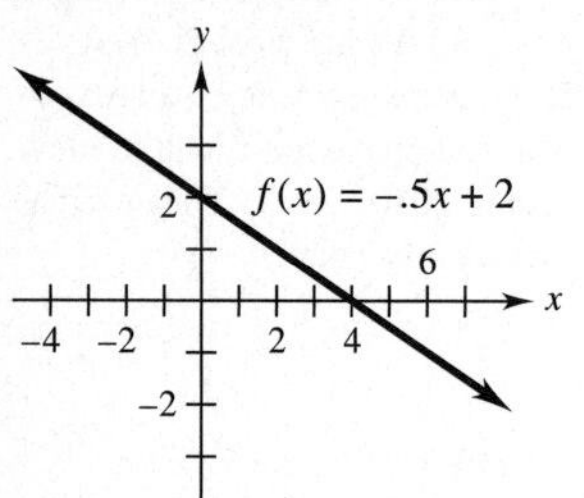

3.

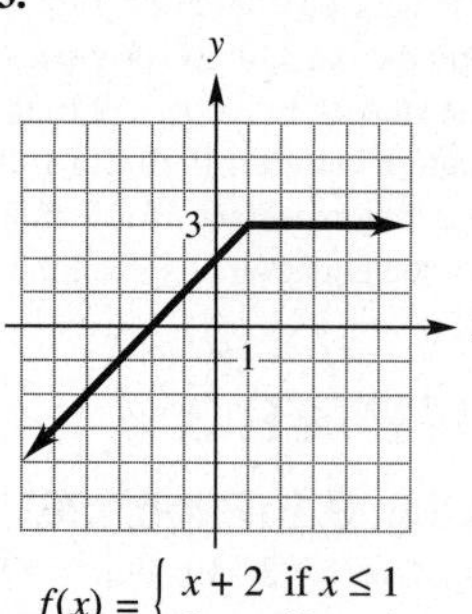

$$f(x) = \begin{cases} x + 2 & \text{if } x \le 1 \\ 3 & \text{if } x > 1 \end{cases}$$

5.

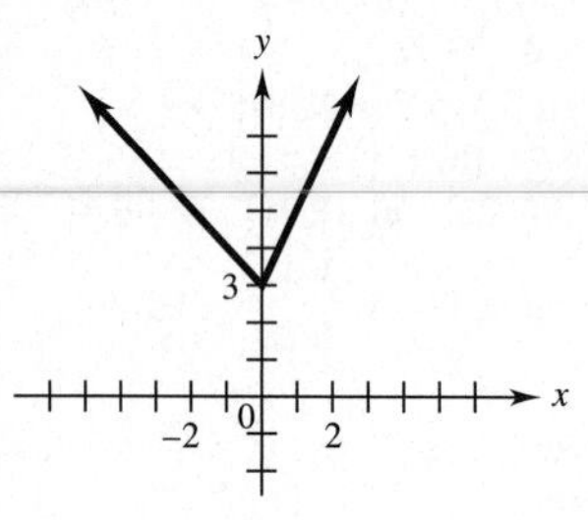

$$y = \begin{cases} 3 - x & \text{if } x \le 0 \\ 2x + 3 & \text{if } x > 0 \end{cases}$$

7.

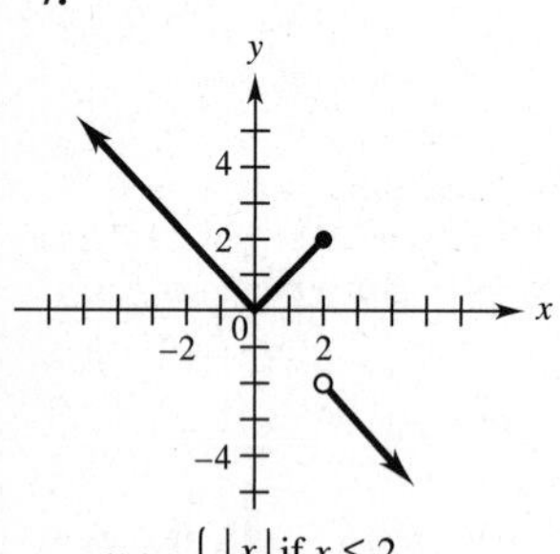

$$f(x) = \begin{cases} |x| & \text{if } x \le 2 \\ -x & \text{if } x > 2 \end{cases}$$

9.

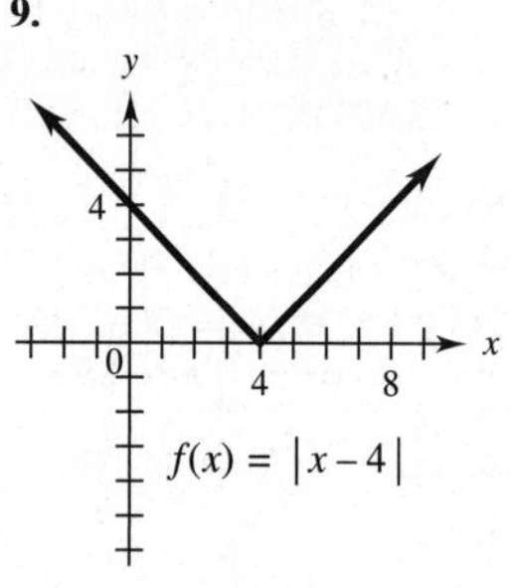

11.

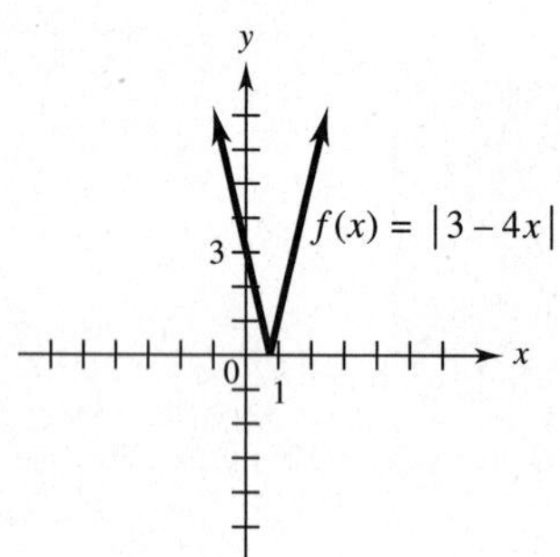

13.

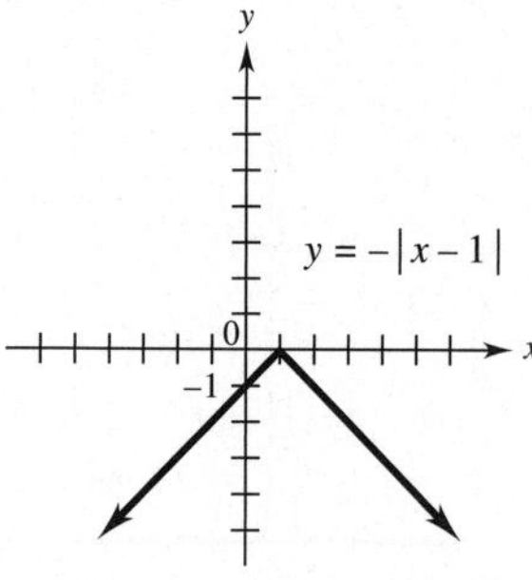

15.

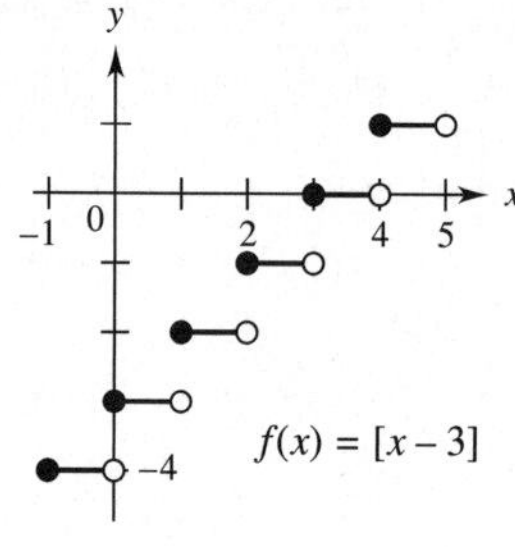

17.

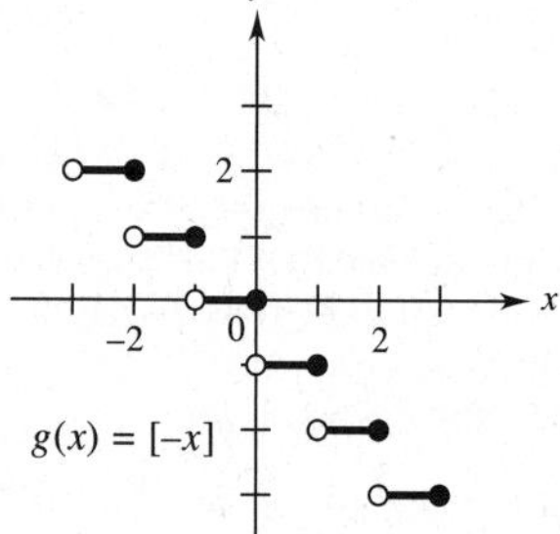

19.

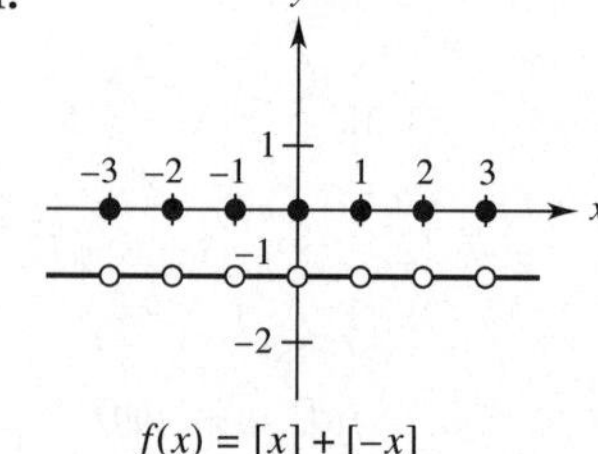

21.

y

1

$-3 \quad -2 \quad -1 \qquad 1 \quad 2 \quad 3$

x

-1

-2

$f(x) = [x] + [-x]$

23.

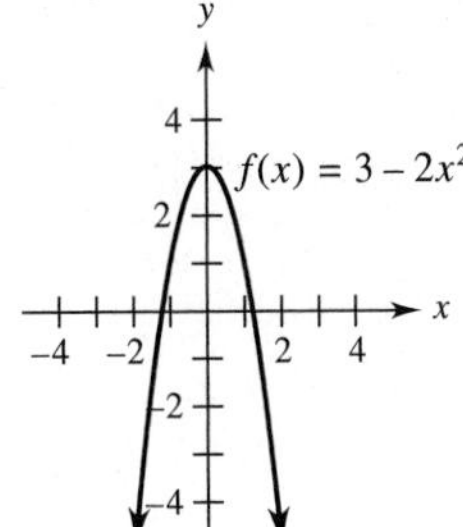

25.

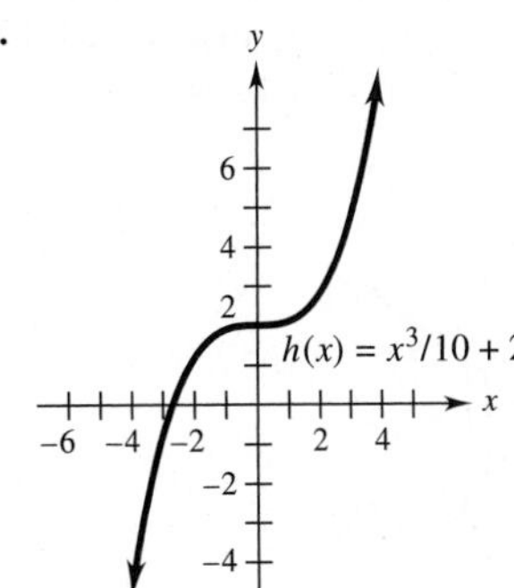

27.

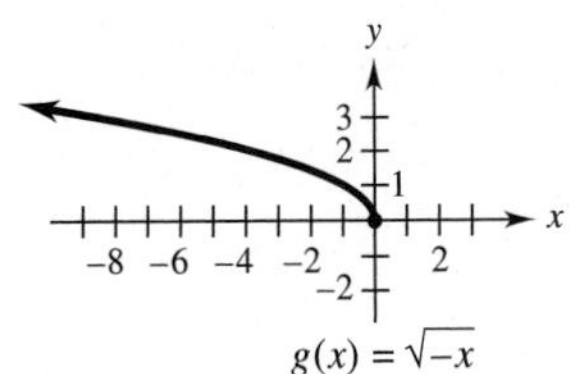

29.

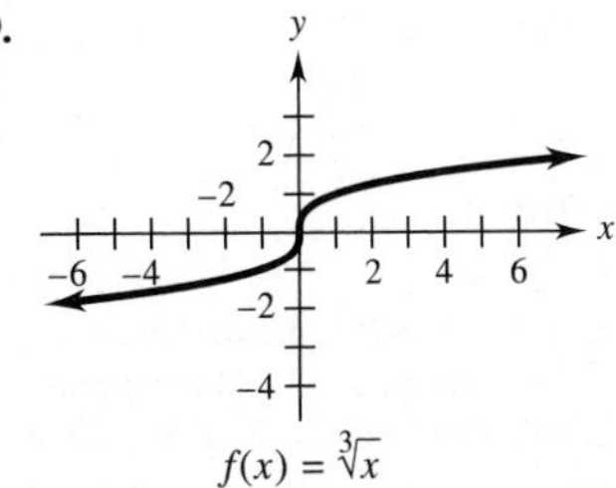

31. function **33.** not a function **35.** function

37. (a)

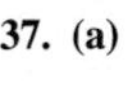

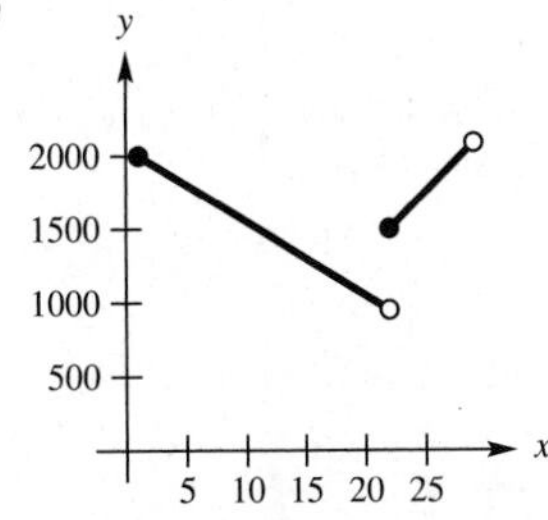

(b) When adjusted for inflation, the maximum yearly IRA contribution fell from \$2000 to \$1000 during this period.

39. (a)

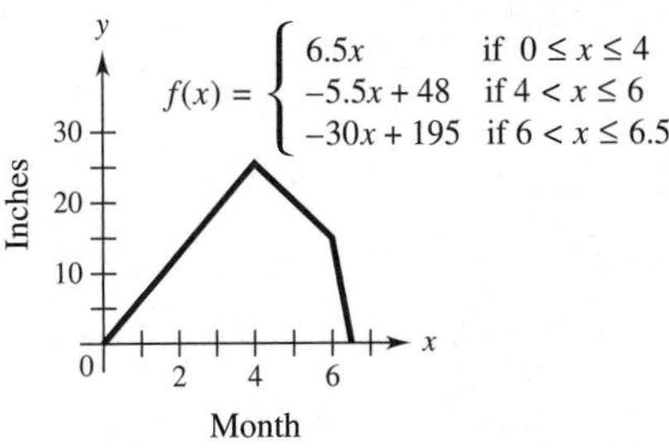

(b) at the beginning of February; 26 in. **(c)** begins in early October; ends in mid-April

41. (a)

$$f(x) = \begin{cases} 1.3x + 15.1 & \text{if } 0 \le x \le 25 \\ 8.5x - 164.1 & \text{if } 25 < x \le 50 \end{cases}$$

(b)

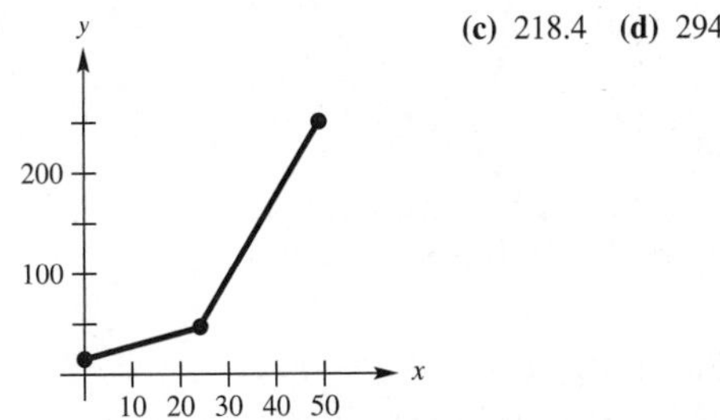

(c) 218.4 **(d)** 294.9

43. (a) No. The graph of the CPI for all items is always above the x-axis, so the percent of change is always positive; this means that the CPI is always increasing. **(b)** Increasing 1980–1982, briefly in 1983–1984, 1987– late 1990, 1995–1997 and at the end of 1998. The percent of change was positive or zero during these years, which means the CPI was increasing. **(c)** Decreasing 1982– late 1983, mid 1984–1987, late 1990–1995, 1997–1999. The percent of change was negative during these years, which means the CPI was decreasing.

45. (a)

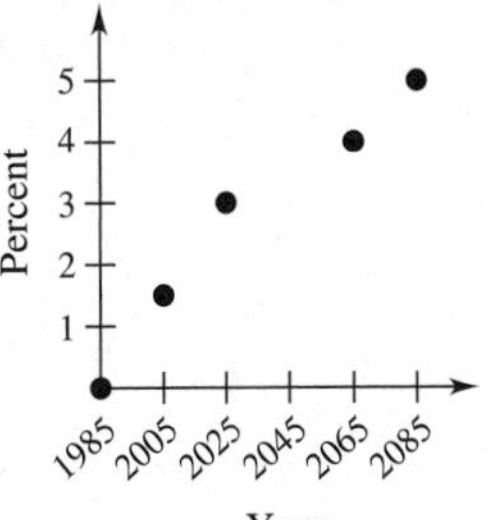

(b) $y = .044x - 86.7$ **(c)** $f(x) = .044x - 86.7$ **(d)** 4.16; yes; yes

47. (a) 25; 33 **(b)** The figure has vertical line segments, which can't be part of the graph of a function (why?). To make the figure into the graph of f, delete the vertical line segments; then for each horizontal segment of the graph, put a closed dot on the left end and an open circle dot on the right end (as in Figure 3.9).

49. (a) \$29 **(b)** \$29 **(c)** \$33 **(d)** \$33

(e)

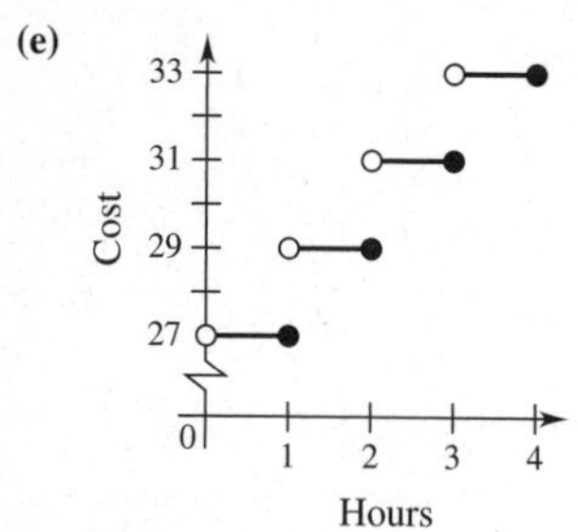

51. **(a)** There are many correct answers, including

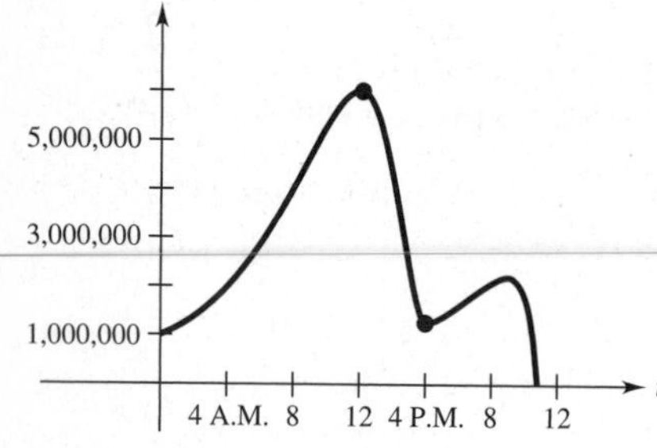

Section 3.3 (Page 148)

1. Let $C(x)$ be the cost of renting a saw for x for hours; $C(x) = 12 + x$. **3.** Let $P(x)$ be the cost (in cents) of parking for x half hours; $P(x) = 30x + 35$. **5.** $C(x) = 30x + 100$ **7.** $C(x) = 120x + 3800$ **9.** \$48; \$15.60; \$13.80 **11.** \$55.50; \$11.40; \$8.46 **13.** **(a)** \$80,000 **(b)** \$42.50 **(c)** \$122,500; \$1,440,000 **(d)** \$122.50; \$45 **15.** **(a)** $C(x) = .097x + 1.32$ **(b)** \$98.32 **(c)** \$98.417 **(d)** \$.097, or 9.7¢ **(e)** \$.097, or 9.7¢ **17.** **(a)** $C(x) = 10x + 500$ **(b)** $R(x) = 35x$ **(c)** $P(x) = 25x - 500$ **(d)** \$2000 **19.** **(a)** $C(x) = 18x + 250$ **(b)** $R(x) = 28x$ **(c)** $P(x) = 10x - 250$ **(d)** \$750 **21.** **(a)** $C(x) = 12.5x + 18{,}000$ **(b)** $R(x) = 25x$ **(c)** $P(x) = 12.5x - 18{,}000$ **(d)** $-\$16{,}750$ (loss)

23. $(3, -1)$ **25.** $\left(-\frac{11}{4}, -\frac{61}{4}\right)$

27. **(a)** 200,000 policies ($x = 200$)
(b)

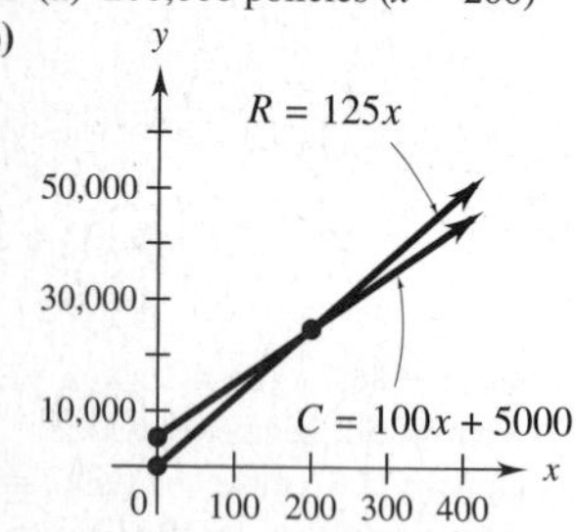

(c) revenue: \$12,500; cost: \$15,000 **29.** **(a)** $c = .126x + 1.5$ **(b)** \$2.382 million **(c)** about 17.857 units **31.** Break-even point is about 467 units; do not produce the item. **33.** Break-even point is about 1037 units; produce the item. **35.** about (1988, \$14.68) **37.** **(a)** Ireland: $f(x) = 1.44x + 6.88$; Italy: $g(x) = .50x + 13.12$

(b)

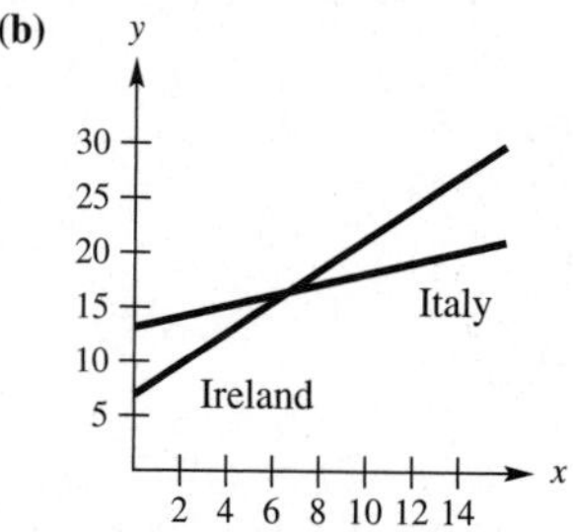

(c) About (6.64, 16.44), investment in both countries was about \$16.44 billion in late 1996. **39.** \$140 **41.** 10 items; 10 items **43.** **(a)** \$16 **(b)** \$11 **(c)** \$6 **(d)** 8 units **(e)** 4 units **(f)** 0 units **(g)**

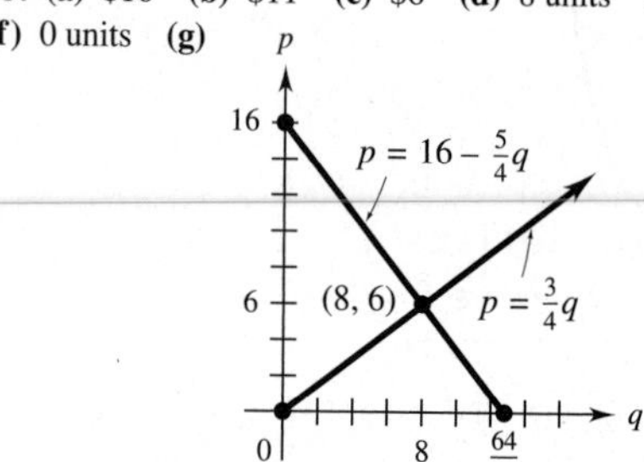

(h) 0 units **(i)** $\frac{40}{3}$ units **(j)** $\frac{80}{3}$ units **(k)** See part (g). **(l)** 8 units **(m)** \$6
45. **(a)**

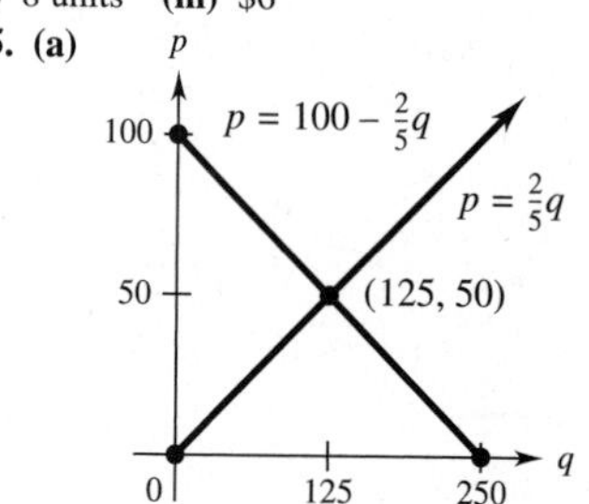

(b) 125 units **(c)** 50¢ **(d)** [0, 125) **47.** Total cost increases when more items are made (because it includes the cost of all previously made items), so the graph cannot move downward. No; the average cost can decrease as more items are made, so its graph may move downward.

Section 3.4 (Page 158)

1. (5, 2); upward **3.** $(-6, -35)$; upward **5.** $(-1, -1)$; upward **7.** x-intercepts 1, 3; y-intercept 9 **9.** x-intercepts $-1, -3$; y-intercept 6 **11.** $f(x) = \frac{1}{4}(x - 1)^2 + 2$ **13.** $f(x) = (x + 1)^2 - 2$ **15.** $(-2, 0), x = -2$ **17.** $(1, -3), x = 1$

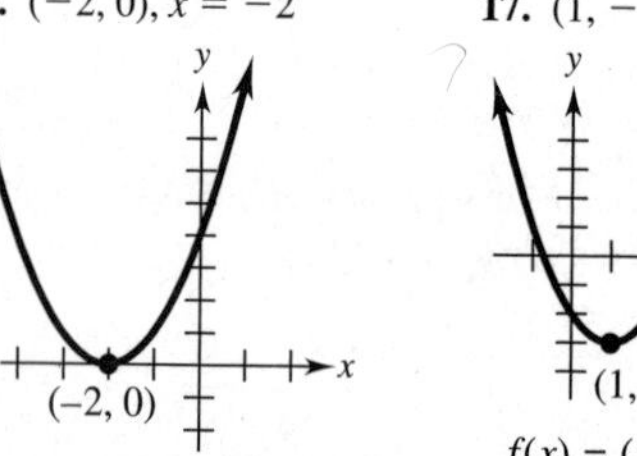

19. (2, 2), $x = 2$

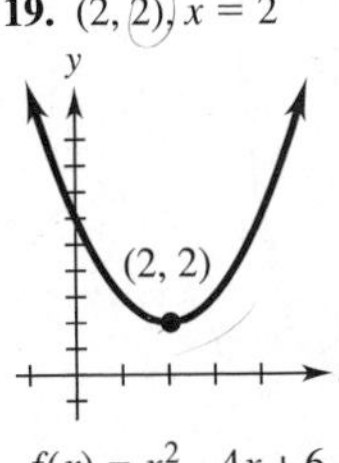

$f(x) = x^2 - 4x + 6$
$f(x) = (x - 2)^2 + 2$

21. (1, 3), $x = 1$

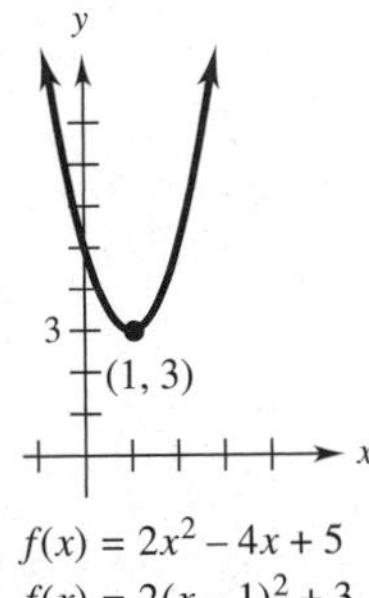

$f(x) = 2x^2 - 4x + 5$
$f(x) = 2(x - 1)^2 + 3$

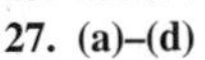

23. about 47 **25.** 1995

27. **(a)–(d)**

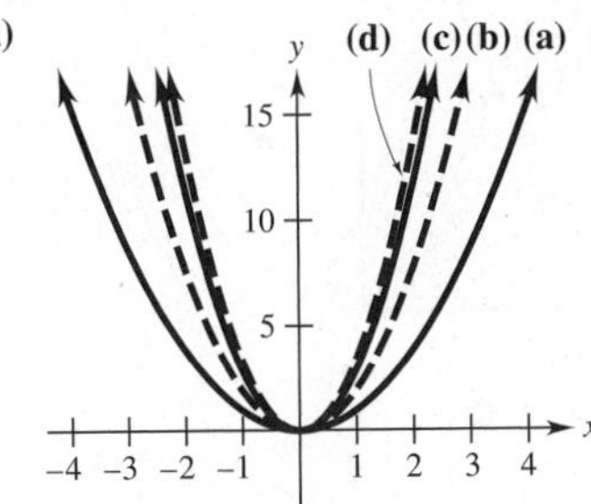

(e) The graph of $f(x) = ax^2$ is the graph of $k(x) = x^2$ stretched away from the x-axis by a factor of a.

29. **(a)–(d)**

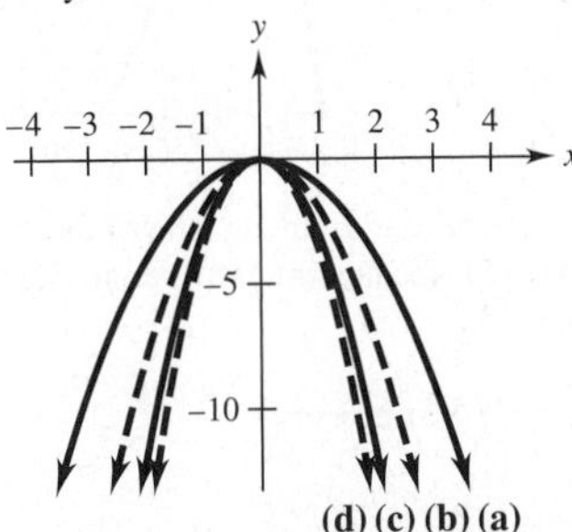

(e) Changing the sign of a reflects the graph in the x-axis.

31. **(a)–(d)**

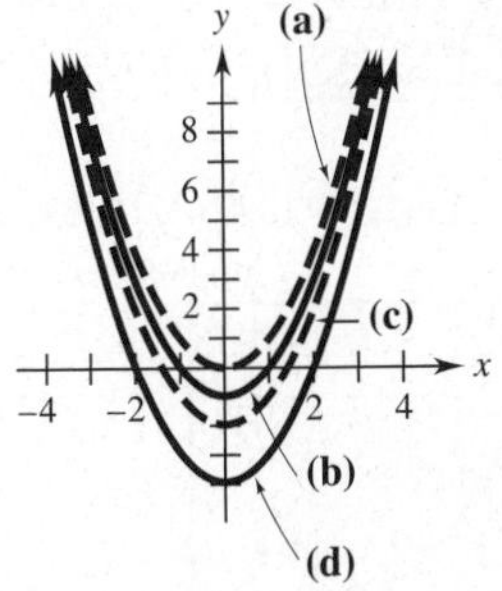

(e) The graph of $f(x) = x^2 - c$ is the graph of $k(x) = x^2$ shifted c units downward. **33.** $f(x) = 3x^2$

Section 3.5 (Page 165)

1. **(a)** \$30; \$9; \$105 **(b)**

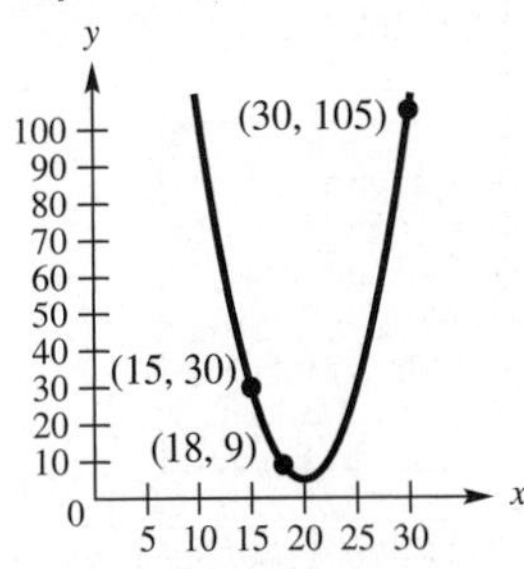

(c) (20, 5) **(d)** 20 boxes per day; \$5 per box
3. **(a)** 10 milliseconds **(b)** 40 responses per millisecond
5. **(a)** 27 cases **(c)** 15 cases **7.** **(a)** about 12 books **(b)** 10 books **(c)** about 7 books **(d)** 0 books **(e)** 5 books **(f)** about 7 books **(g)** 10 books **(h)** about 12 books
(i)

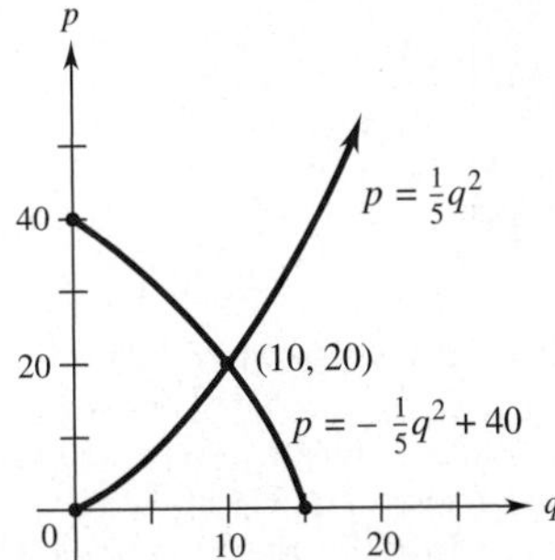

9. **(a)** \$640 **(b)** \$515 **(c)** \$140
(d)

p
1000
800
600
400
200
$p = 640 - 5q^2$
$p = 5q^2$
2 4 6 8 10
q

(e) 800 units **(f)** \$320 **11.** equilibrium quantity 80; equilibrium price \$3600 **13.** equilibrium quantity 30; equilibrium price \$1500
15. 20 **17.** 10
19. **(a)** $R(x) = (100 - x)(200 + 4x)$
$= 20{,}000 + 200x - 4x^2$

(b)

y
(25, 22,500)
22,500
20,000
15,000
10,000
5000
$R(x) = (100 - x)(200 + 4x)$
0 25 50 75 100
x

(c) 25 seats **(d)** \$22,500 **21.** 13 weeks; \$96.10/hog

23. (a) $f(x) = 1574.4023(x - 1)^2 + 159$
(b) $g(x) = 1738.7098x^2 - 4323.7917x - 4313.0848$

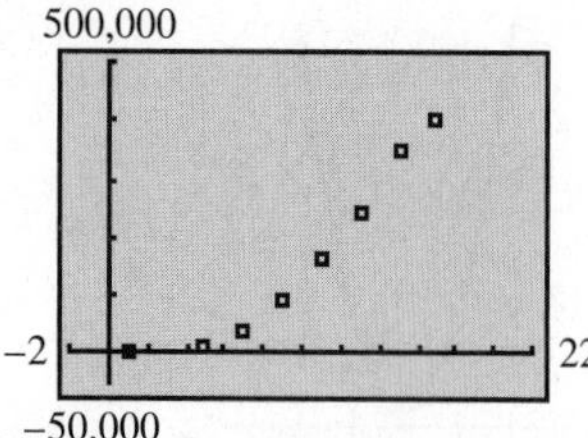

(c) $f(12) = 190{,}662$; $f(16) = 354{,}400$; $f(20) = 568{,}518$; $g(12) = 194{,}176$; $g(16) = 371{,}616$; $g(20) = 604{,}695$
25. (a) $f(x) = .08677(x - 5)^2 + 109$
(b)

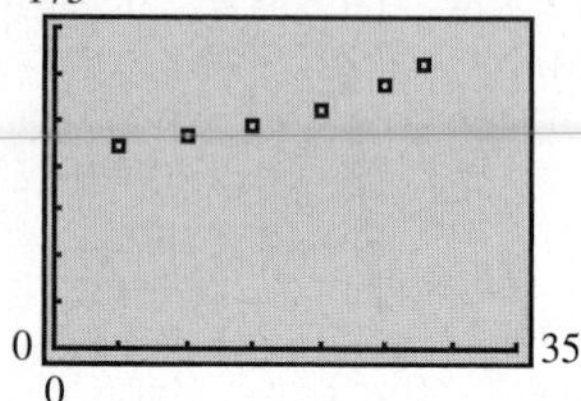

$g(x) \approx .060305x^2 - .0585x + 108.4373$; a very good model since $r^2 \approx .998$, which is very close to 1
(c) $f(23) = 137.11$; $f(27) = 151$; $g(23) = 138.99$; $g(27) = 150.82$
(d) for $f(x)$, = 163.23 and 177.03; for $g(x)$, 160.96 and 172.18
27. 80 ft by 160 ft **29. (a)** 11.3 and 88.7 **(b)** 50 **(c)** \$3000
(d) $x < 11.3$ or $x > 88.7$ **(e)** $11.3 < x < 88.7$

Section 3.6 (Page 175)

1.

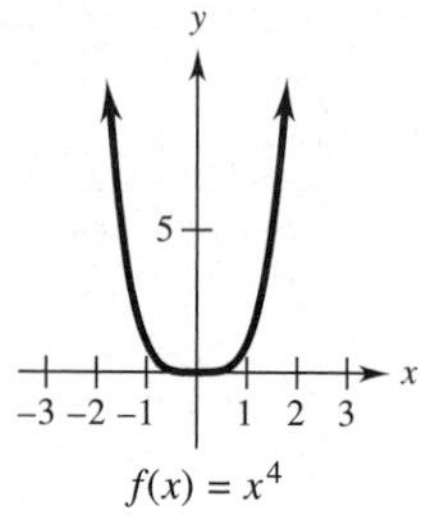

3.

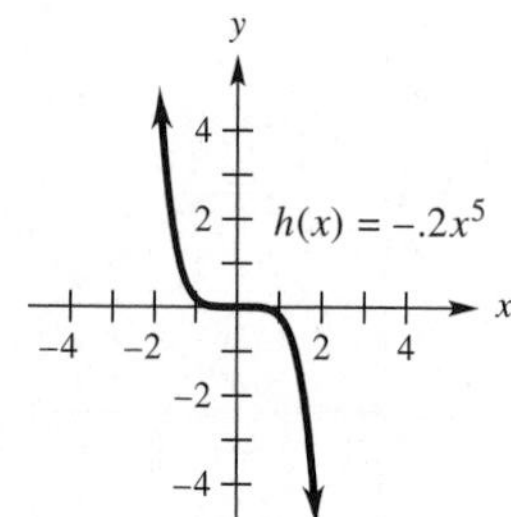

5. (a) yes **(b)** no **(c)** no **(d)** yes **7. (a)** yes **(b)** no
(c) yes **(d)** no **9.** D **11.** B **13.** E

15.

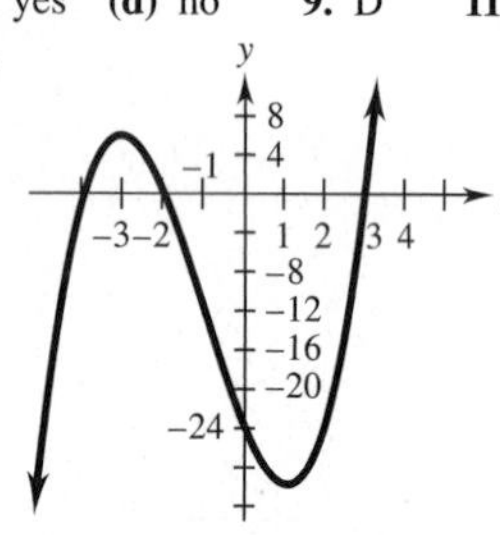

17.

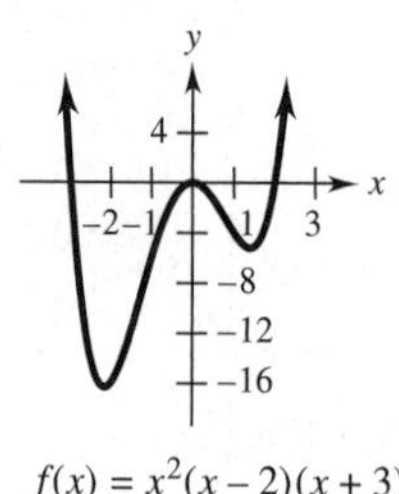

19.

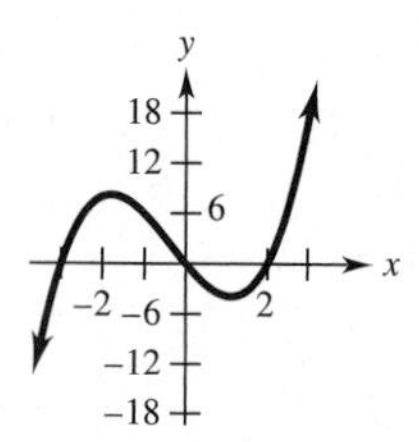

21.

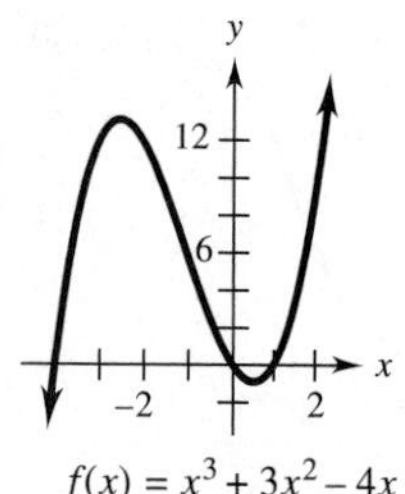

23. $-3 \le x \le 5$ and $-20 \le y \le 5$ **25.** $-3 \le x \le 4$ and $-35 \le y \le 20$ **27. (a)** \$933.33 billion **(b)** \$1200 billion
(c) \$1145.8 billion **(d)** \$787.5 billion
(e)

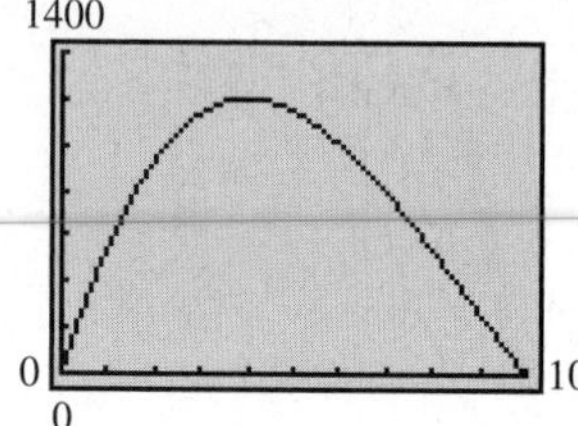

29. (a) 0; 108; 28; 10 **(b)**

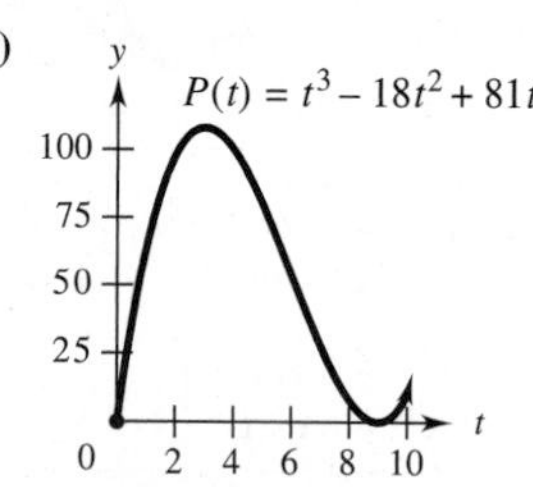

(c) increasing for years 0 to 3 and from the 9th year on; decreasing for the years 3 to 9 **31. (a)** about 5880 **(b)** about 17,742
(c) about 47,134 **(d)** 2001
33. (a)

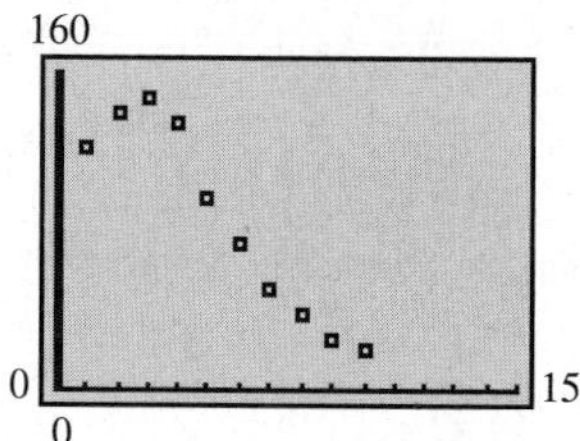

(b) $f(x) = -.037005x^4 + 1.51962x^3 - 18.4164x^2 + 63.5282x + 78.8951$ **(c)**

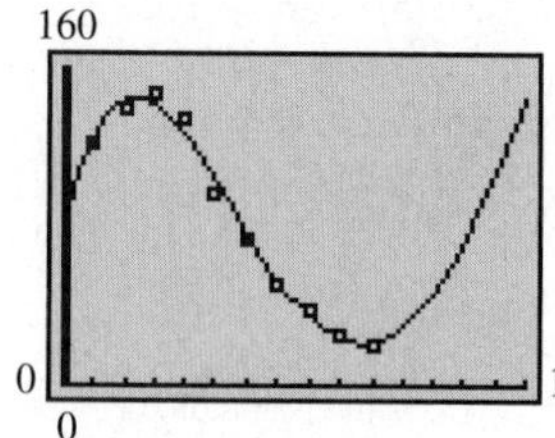

(d) a sharp increase, which seems unlikely

Section 3.7 (Page 185)

1. $x = -5, y = 0$

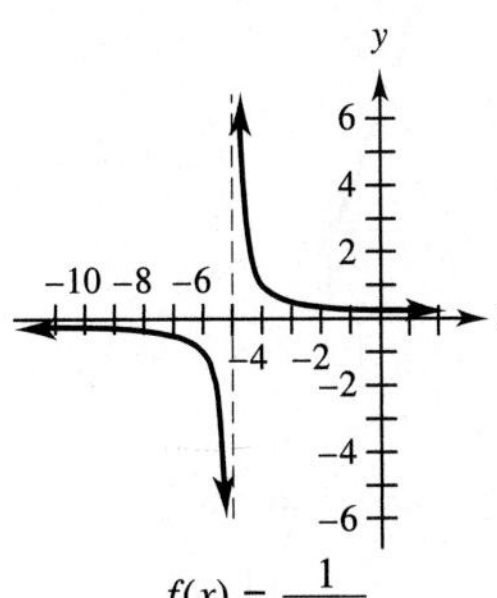

3. $x = -\frac{5}{2}, y = 0$

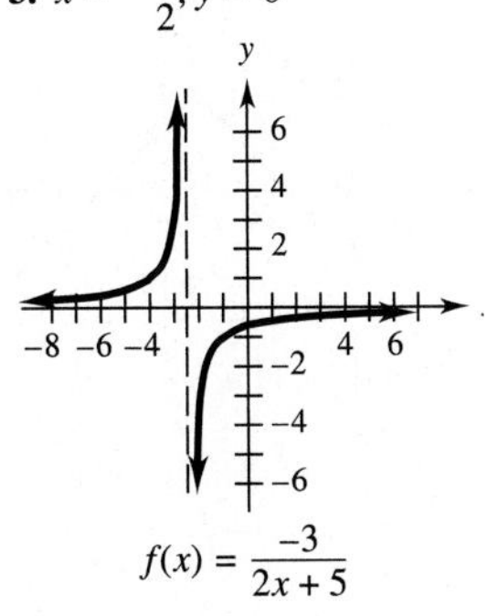
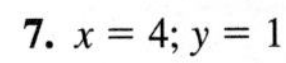

5. $x = 1, y = 3$

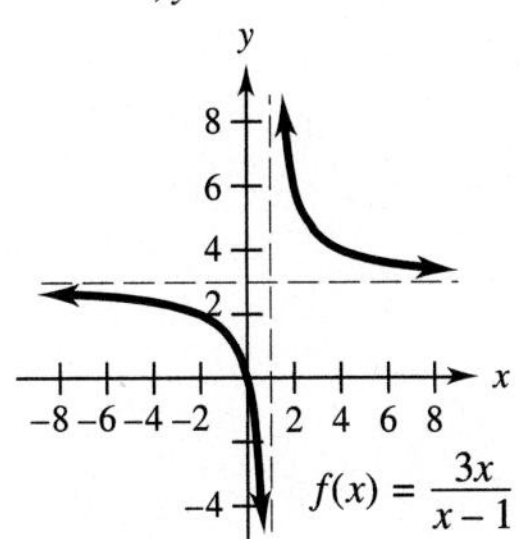

7. $x = 4; y = 1$

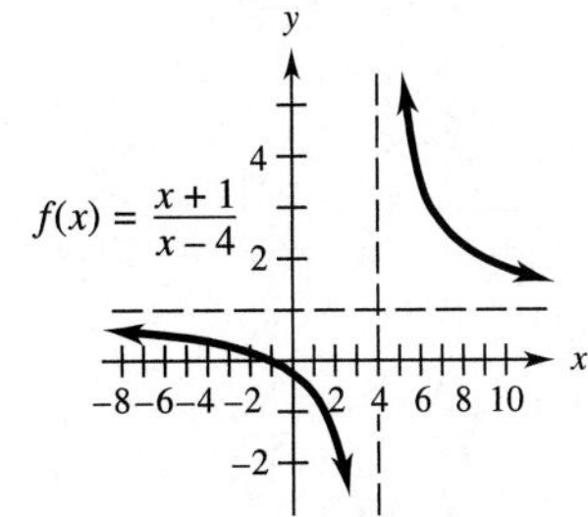

9. $x = 3; y = -1$

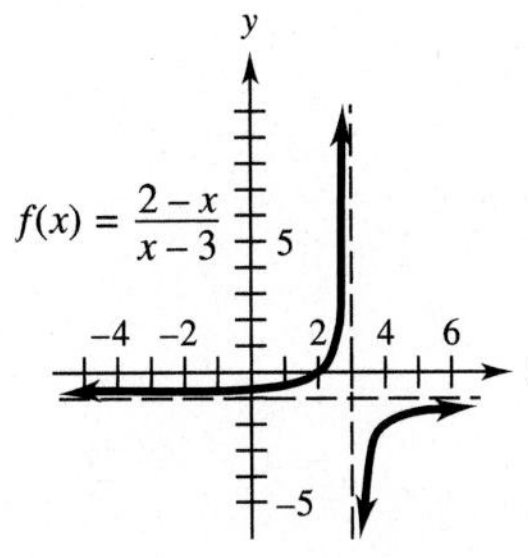

11. $x = -\frac{1}{2}; y = \frac{1}{2}$

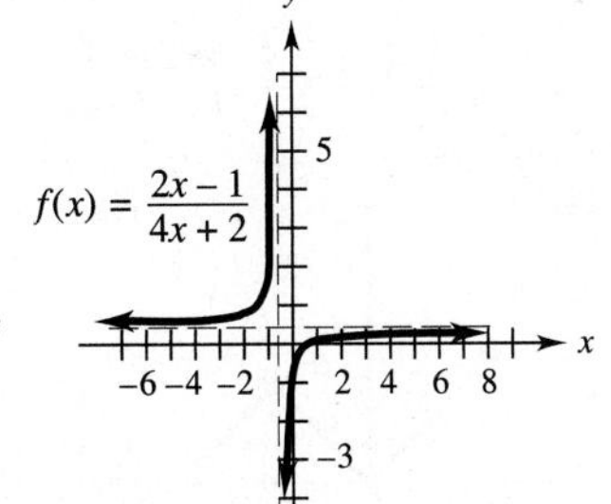

13. $x = -4, x = 1, y = 0$

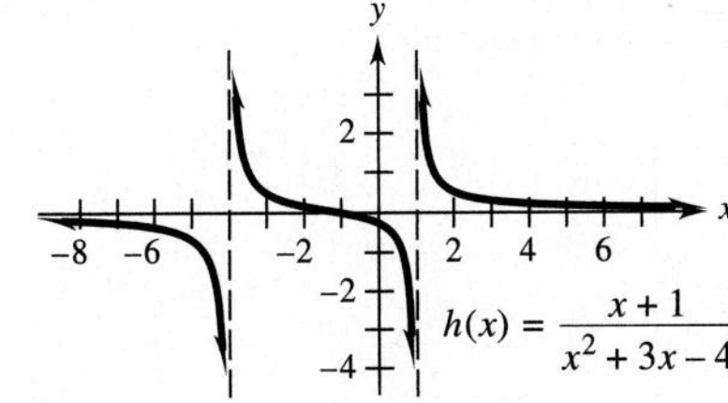

15. $x = -1, x = 1, y = 1$

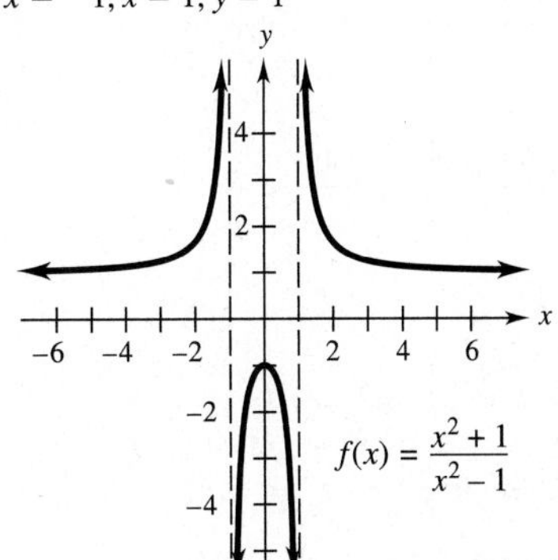

17. $x = -2; x = 1$ **19.** $x = -1; x = 5$ **21. (a)** \$4300 **(b)** \$10,033.33 **(c)** \$17,200 **(d)** \$38,700 **(e)** \$81,700 **(f)** \$210,700 **(g)** \$425,700 **(h)** no
(i)

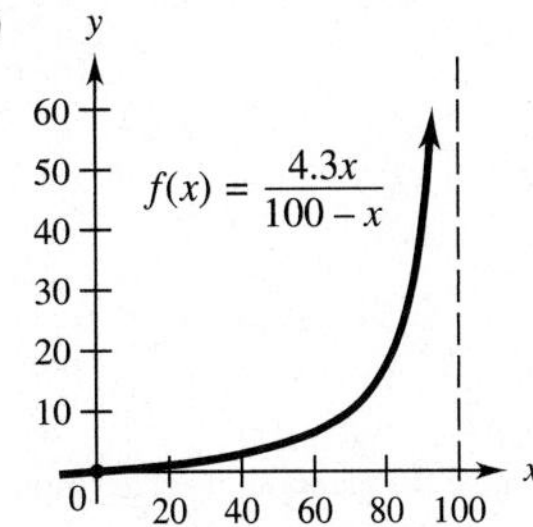

23. (a) $[0, \infty)$

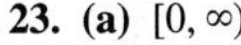

(b)

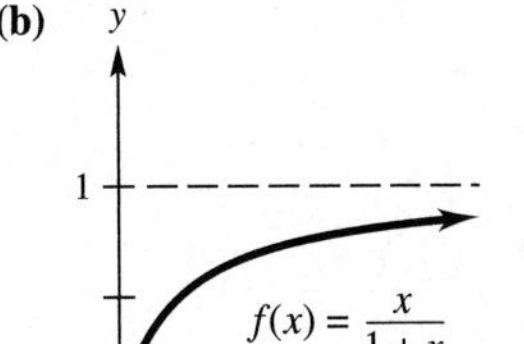

(c)

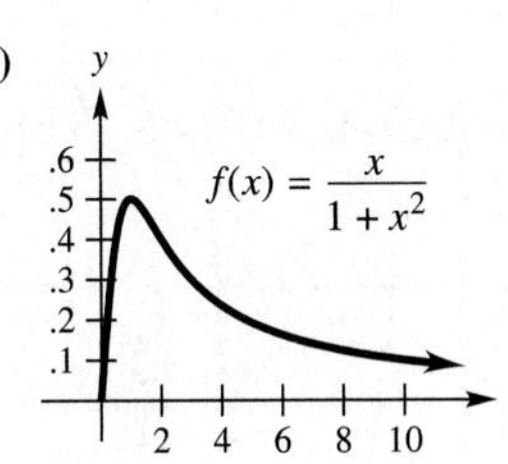

(d) Increasing b makes the next generation smaller when this generation is larger. **25. (a)** 6 min **(b)** 1.5 min **(c)** .6 min **(d)** $A = 0$ **(e)**

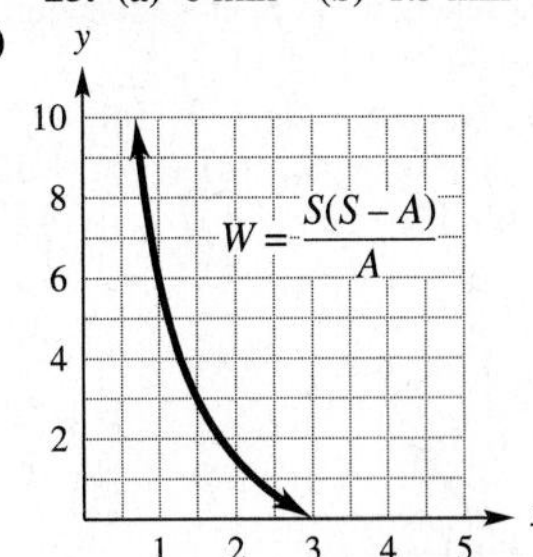

(f) W becomes negative. The waiting time approaches 0 as A approaches 3. The formula does not apply for $A > 3$ because there will be no waiting if people arrive more than 3 min apart.

27. **(a)** 30,000 reds; 10,000 blues

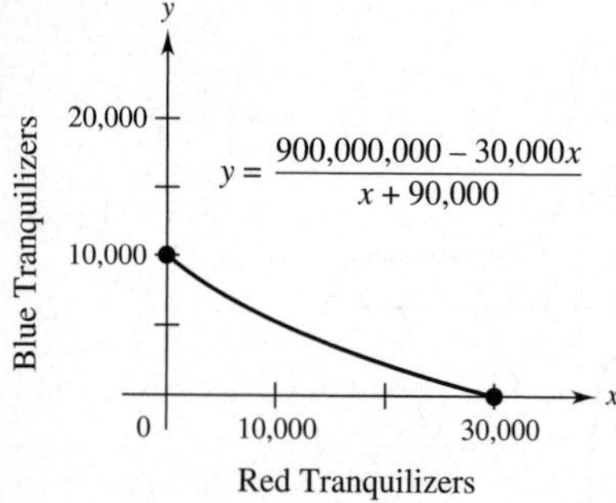

29. **(a)** $C(x) = 2.6x + 40{,}000$

(b) $\overline{C}(x) = \dfrac{2.6x + 40{,}000}{x} = 2.6 + \dfrac{40{,}000}{x}$

(c) $y = 2.6$; the average cost may get close to, but will never equal, \$2.60. **31.** about 73.9

33. **(a)**

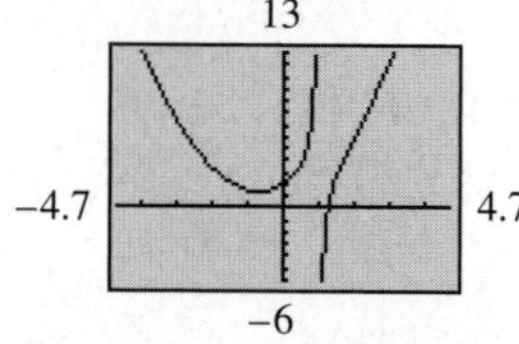

(b) They appear almost identical because the parabola is an asymptote of the graph.

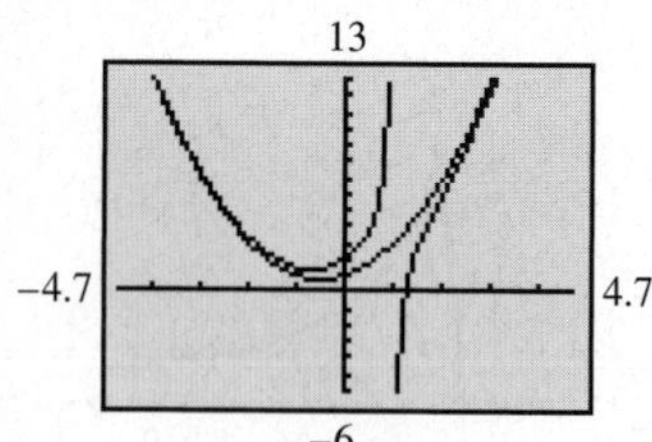

Chapter 3 Review (Page 188)

1. not a function **3.** function **5.** not a function
7. **(a)** 23 **(b)** -9 **(c)** $4p - 1$ **(d)** $4r + 3$
9. **(a)** -28 **(b)** -12 **(c)** $-p^2 + 2p - 4$ **(d)** $-r^2 - 3$
11. **(a)** -13 **(b)** 3 **(c)** $-k^2 - 4k$ **(d)** $-9m^2 + 12m$
(e) $-k^2 + 14k - 45$ **(f)** $12 - 5p$

13.

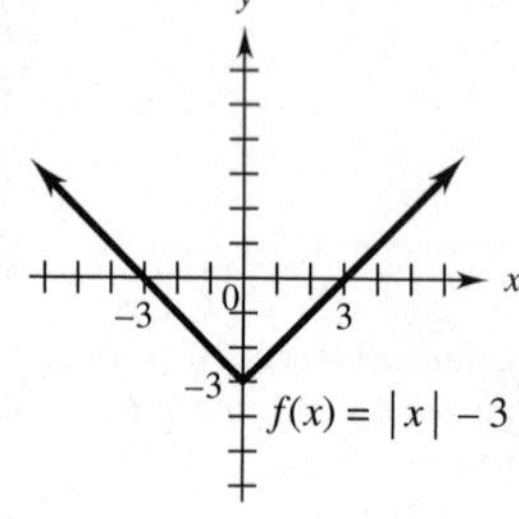

15.

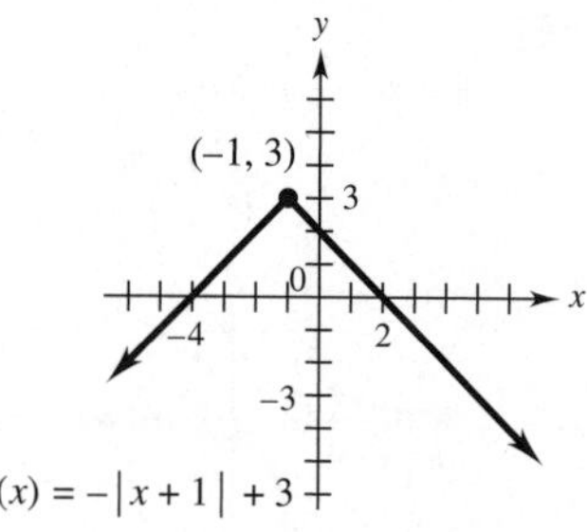

17.

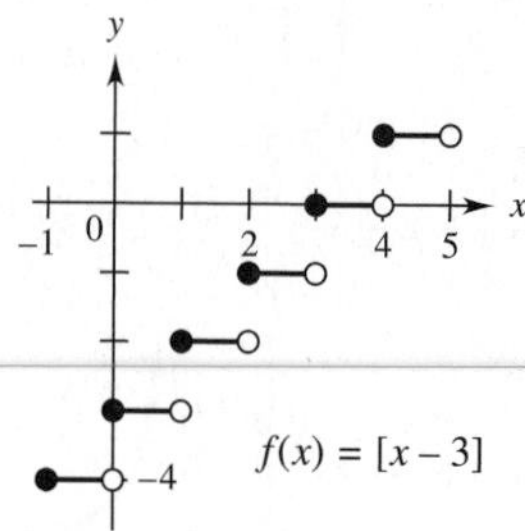

19.

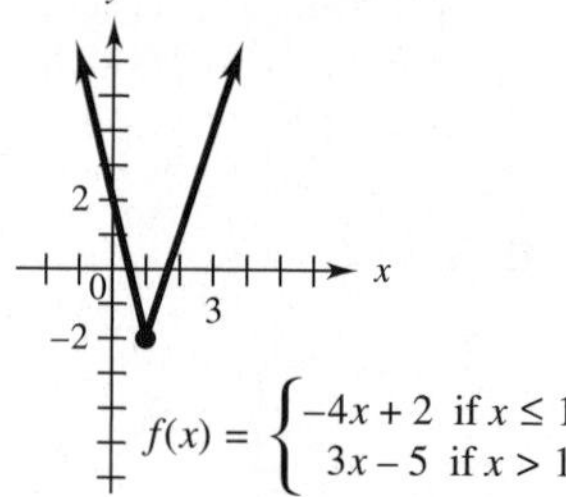

21.

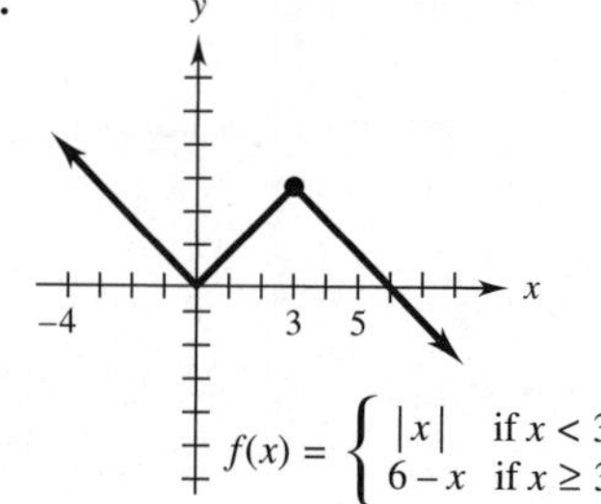

23.

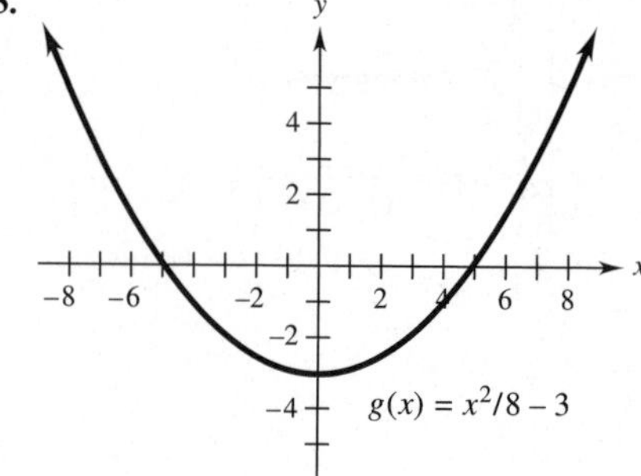

25. **(a)**

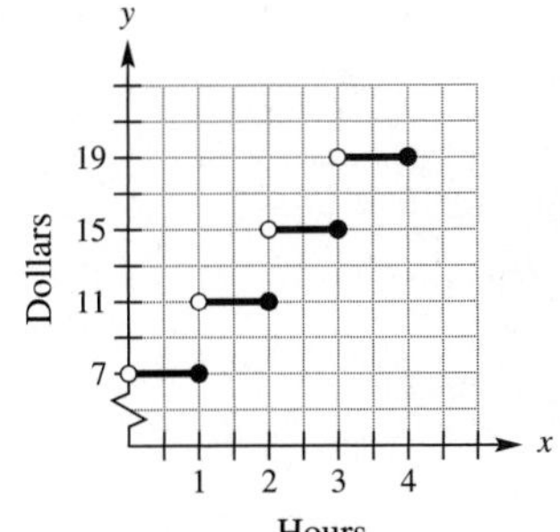

(b) domain: $(0, \infty)$; range: $\{7, 11, 15, 19, \ldots\}$ **(c)** 3 days

27. These births appear to be leveling off.

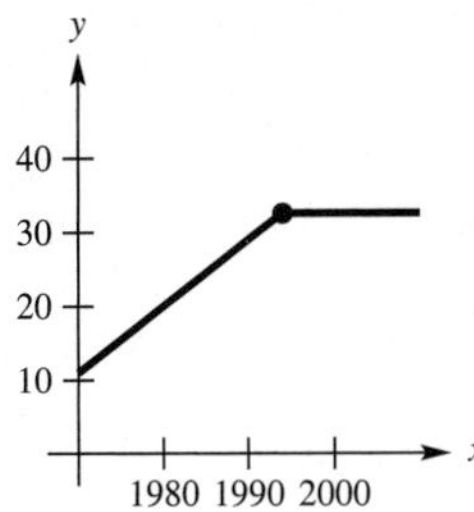

29. **(a)** $C(x) = 30x + 60$ **(b)** \$30 **(c)** \$30.60 **31.** **(a)** $C(x) = 30x + 85$ **(b)** \$30 **(c)** \$30.85 **33.** **(a)** \$18,000 **(b)** $R(x) = 28x$ **(c)** 4500 cartridges **(d)** \$126,000 **35.** Equilibrium quantity is 36 million subscribers at a price of \$12.95 per month.

37. upward; $(2, 6)$ **39.** downward; $(-1, 8)$

41.

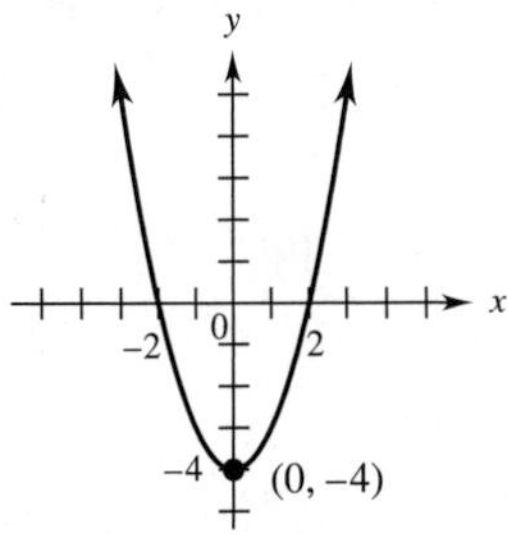

$f(x) = x^2 - 4$

43.

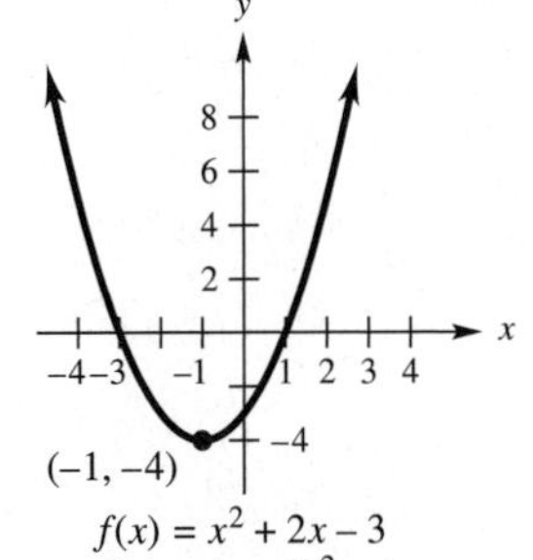

$f(x) = x^2 + 2x - 3$
$f(x) = (x + 1)^2 - 4$

45.

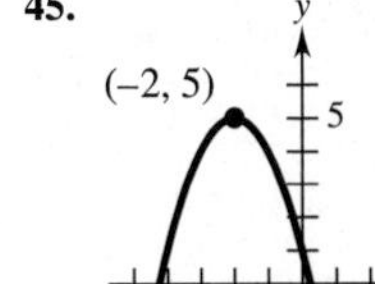
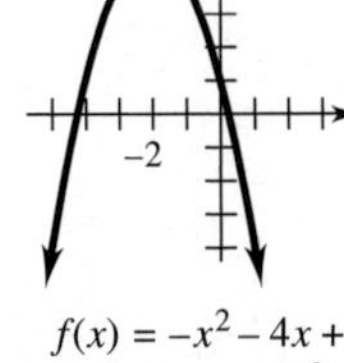

$f(x) = -x^2 - 4x + 1$
$f(x) = -(x + 2)^2 + 5$

47.

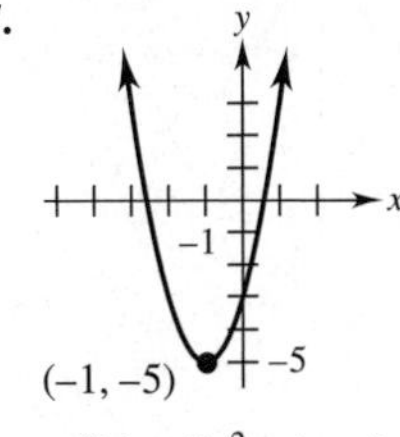

$f(x) = 2x^2 + 4x - 3$
$f(x) = 2(x + 1)^2 - 5$

49. minimum value; -11 **51.** maximum value; 7

53. 4 months after she began **55.** 125 units

57. **(a)** $f(x) = \frac{11}{36}(x - 5)^2 + 39$ **(b)** \$54 billion

59.

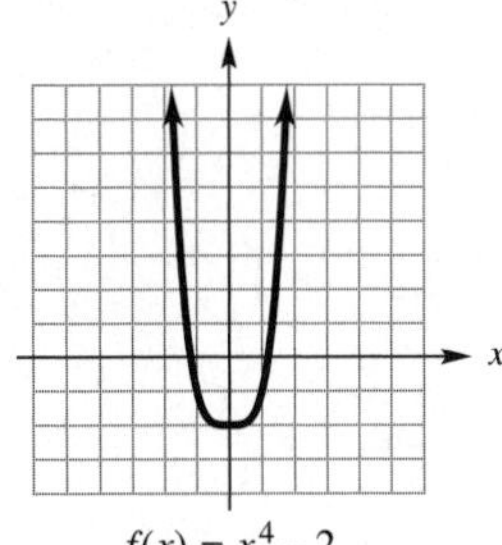

$f(x) = x^4 - 2$

61.

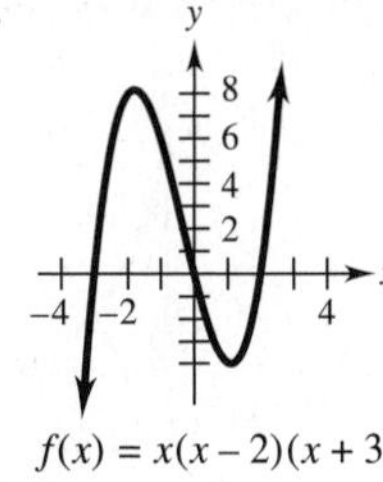

$f(x) = x(x - 2)(x + 3)$

63.

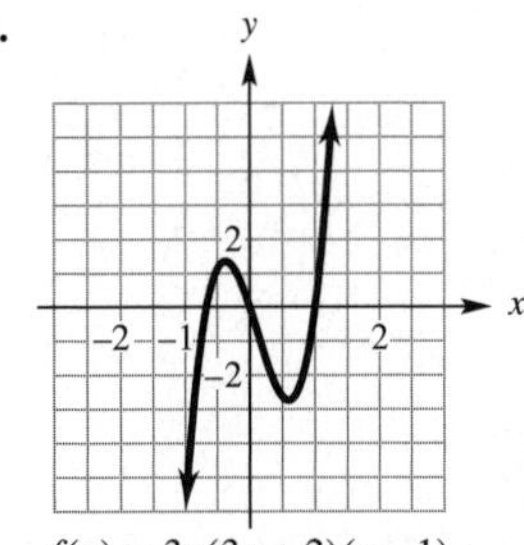

$f(x) = 3x(3x + 2)(x - 1)$

65.

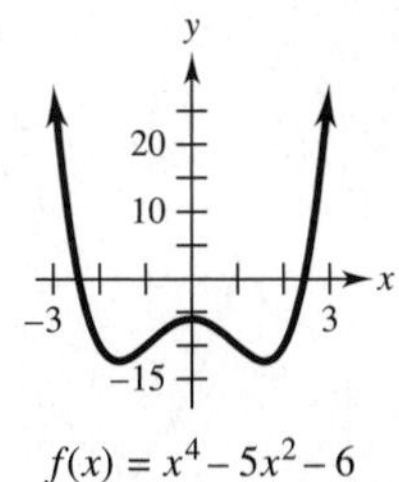

$f(x) = x^4 - 5x^2 - 6$

67. about 313,152; about \$690.72 per thousand

69. **(a)** $R(x) = 23x$; $P(x) = .000006x^3 - .07x^2 + 21x - 1200$ **(b)** about 76.54, which means that 77 must be made to earn a profit (you can't make .54 of a rack) **(c)** 230 **(d)** 153; about \$395.86

71. $x = 3, y = 0$

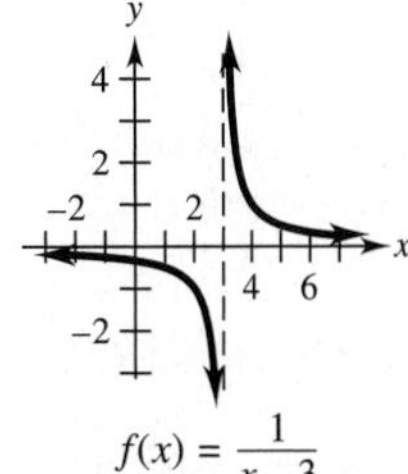

$f(x) = \frac{1}{x - 3}$

73. $x = 2, y = 0$

$f(x) = \frac{-3}{2x - 4}$

75. $x = -\frac{1}{2}, x = \frac{3}{2}, y = 0$

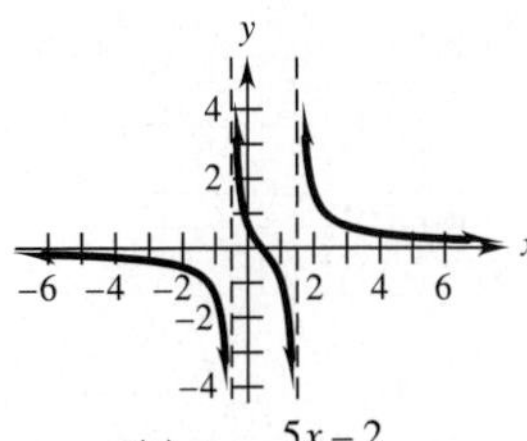

$g(x) = \frac{5x - 2}{4x^2 - 4x - 3}$

77. **(a)** about \$10.83 **(b)** about \$4.64 **(c)** about \$3.61 **(d)** about \$2.71 **(e)**

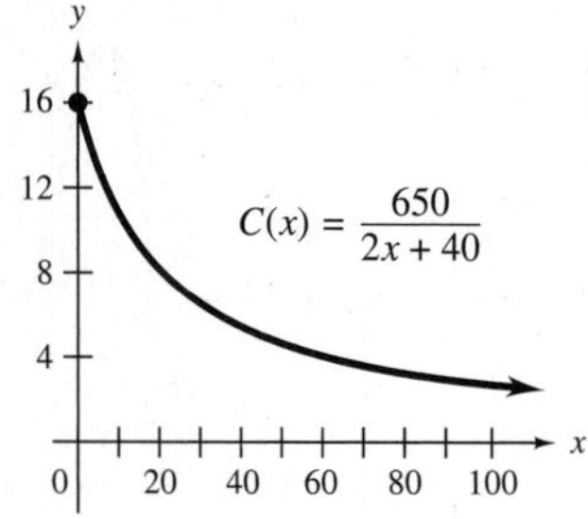

79. **(a)** (10, 50)

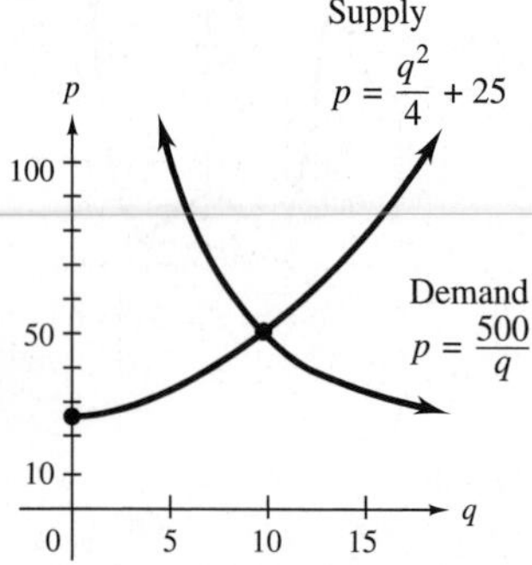

(b) $(10, \infty)$ **(c)** (0, 10)

CHAPTER 4

Section 4.1 (Page 197)

1. quadratic **3.** exponential **5.** **(a)** The graph lies entirely above the x-axis and falls from left to right. It falls relatively steeply until it reaches the y-intercept 1 and then falls slowly, with the positive x-axis as a horizontal asymptote. **(b)** (0, 1), (1, .8)

7. **(a)** The graph lies entirely above the x-axis and rises from left to right. The negative x-axis is a horizontal asymptote. The graph rises slowly until it reaches the y-intercept 1 and then rises quite steeply. **(b)** $(0, 1), (1, 5^{.4})$

9.

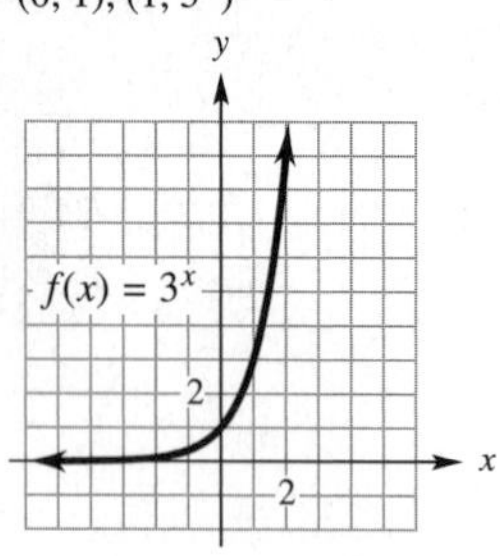

11.

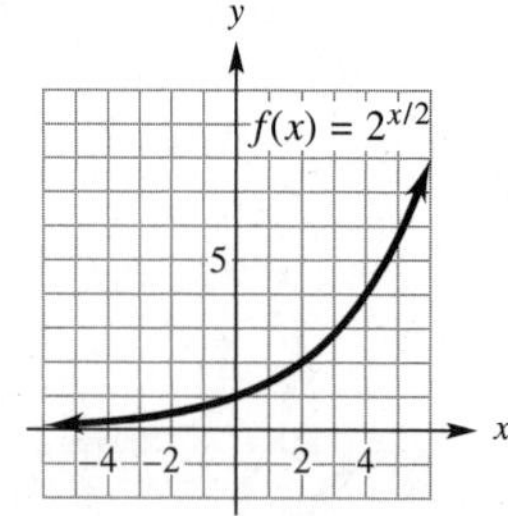

13.

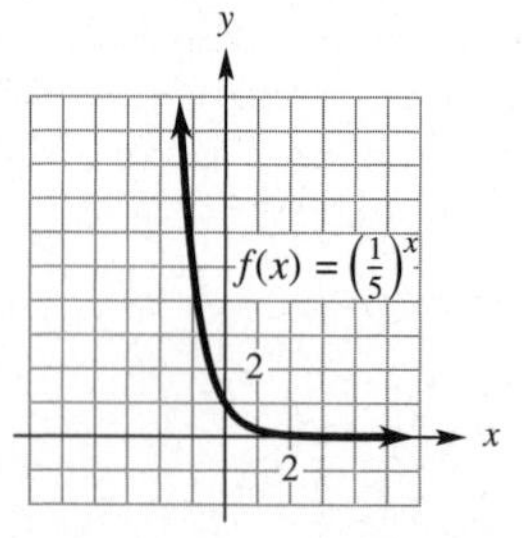

15. **(a)–(c)**

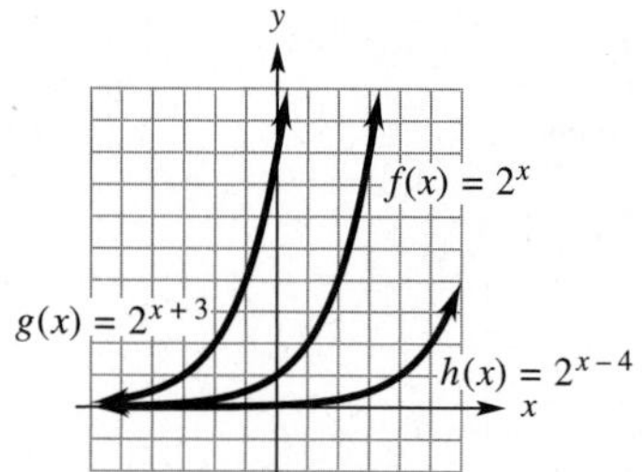

17. 2.3 **19.** .75 **21.** .31 **23.** **(a)** $a > 1$
(b) domain: $(-\infty, \infty)$; range; $(0, \infty)$
(c)

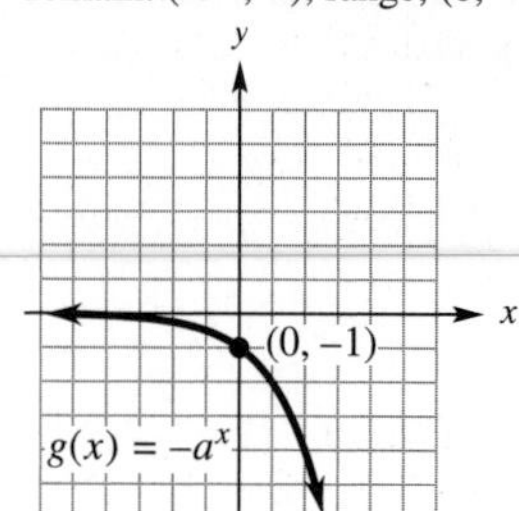

(d) domain: $(-\infty, \infty)$; range; $(-\infty, 0)$
(e)

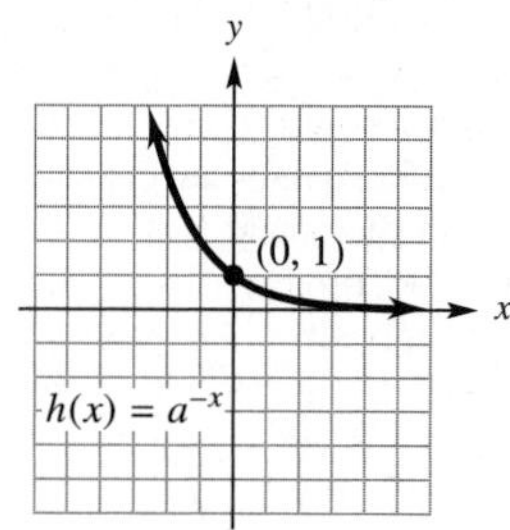

(f) domain: $(-\infty, \infty)$; range; $(0, \infty)$ **25.** **(a)** 3 **(b)** $\frac{1}{3}$ **(c)** 9 **(d)** 1

27.

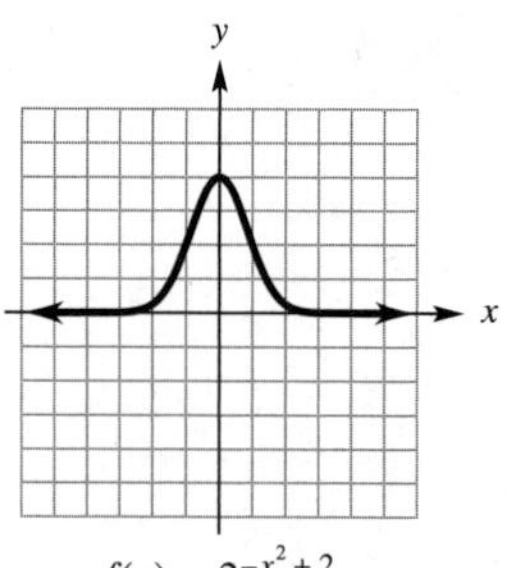

$f(x) = 2^{-x^2+2}$

29.

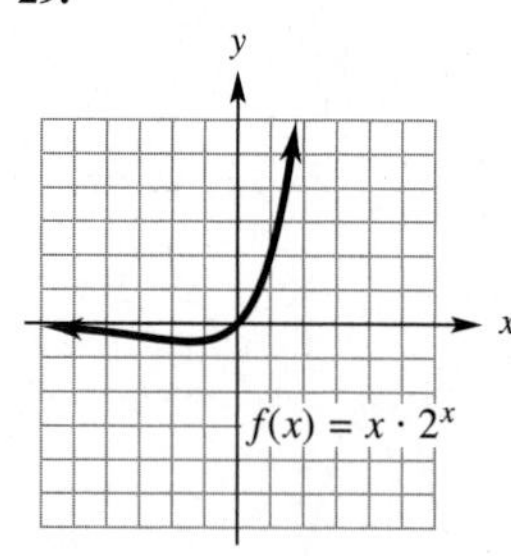

31. **(a)**

t	0	1	2	3	4	5	6	7	8	9	10
y	1	1.06	1.12	1.19	1.26	1.34	1.42	1.50	1.59	1.69	1.79

(b)

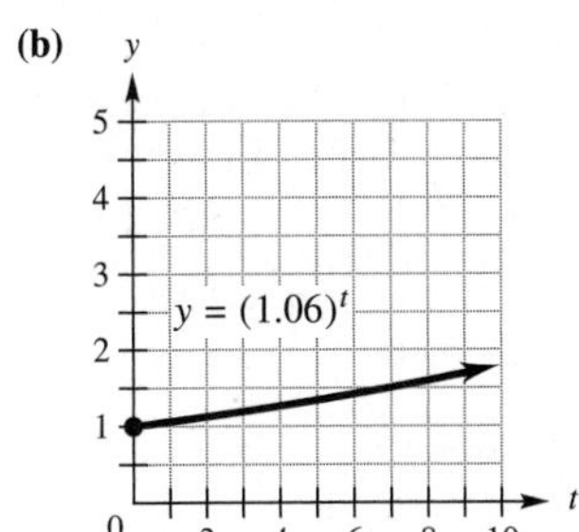

33. (a) about $141,892 **(b)** about $64.10 **35. (a)** about 113,210,000 **(b)** about 145,150,000 **(c)** No; the model predicts over 500 million cell phones in 2006, when the U.S. population is expected to be less than 300 million. **37. (a)** about .97 kg **(b)** about .75 kg **(c)** about .65 kg **(d)** 24,360 yr **39. (a)** about 6.2 billion **(b)** about 6.5 billion **(c)** about 8 billion **41.** $19,420.26

Section 4.2 (Page 203)

1. (a) 100 g **(b)** about 70.9 g **(c)** about 50.3 g **(d)** about 11.6 g **3. (a)** about $53.5 billion **(b)** about $132.9 billion **(c)** about $289.9 billion **5. (a)** $f(t) = 32.5(1.022362)^t$ **(b)** 40.5 million; 70.5 million **(c)** about 2023 **7. (a)** $f(t) = 800(1.01)^t$ **(b)** $956.92; $1015.79 **9. (a)** two-point: $f(t) = 1.003(.9684)^t$; regression: $g(t) = .9887(.96728)^t$ **(b)** $f(18) = .563$; $f(20) = .528$; $f(22) = .495$ and $g(18) = .543$; $g(20) = .508$; $g(22) = .476$ **(c)** in 2010 (regression model) or 2011 (two-point model) **11. (a)** two-point: $f(t) = 253.6(.9757)^t$; regression: $g(t) = 255(.9755)^t$ **(b)** $f(15) = 175.3$ and $f(30) = 121.2$; $g(15) = 175.8$ and $g(30) = 121.2$ **(c)** in 2007 **13. (a)** about 6 items **(b)** about 23 items **(c)** 25 items **15.** 6.25°C **17. (a)** .13 **(b)** .23 **(c)** about 2 weeks **19. (a)** about 49.7 million; about 73.5 million **(b)**

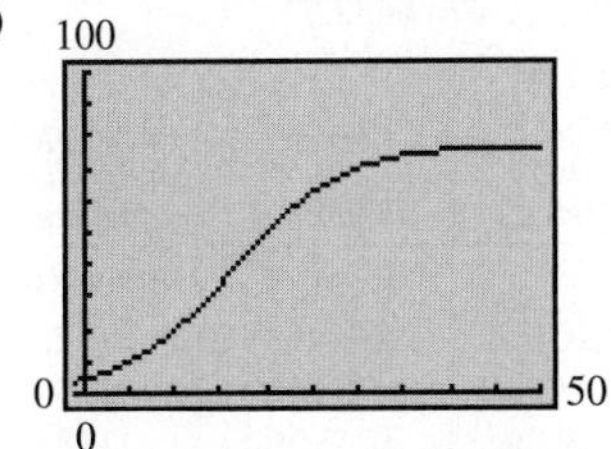

(c) in 2000 **(d)** No; the graph suggests that the number of subscribers will level off below the 80 million level. **21. (a)** about $4695.1 billion; about $5795.7 billion; about $6453.4 billion **(b)**

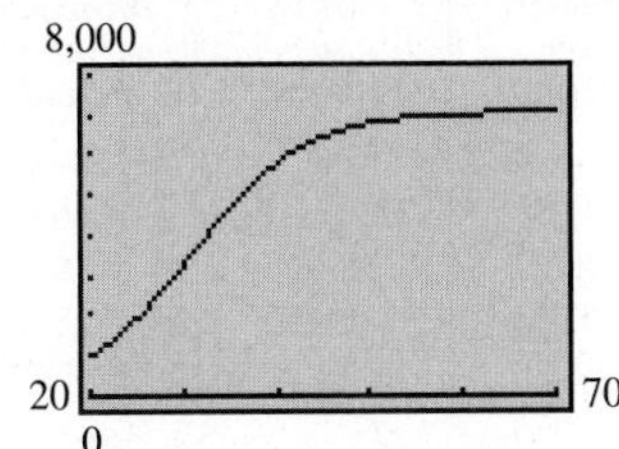

(c) in 2010 **(d)** yes, around 2020

Section 4.3 (Page 213)

1. a^y **3.** It is missing the value that equals b^y. If that value is x, the expression should read $y = \log_b x$.

5. $10^5 = 100{,}000$ **7.** $3^4 = 81$ **9.** $\log 75 = 1.8751$ **11.** $\log_3\left(\frac{1}{9}\right) = -2$ **13.** 3 **15.** 2 **17.** 3 **19.** -2 **21.** $\frac{1}{2}$ **23.** 3.78 **25.** 1.672 **27.** -1.047 **31.** log 16 **33.** ln 5 **35.** $\log\left(\frac{u^2w^3}{v^6}\right)$ **37.** $\ln\left(\frac{(x+1)^2}{x+2}\right)$ **39.** $\frac{1}{2}\ln 6 + 2\ln m + \ln n$ **41.** $\frac{1}{2}\log x - \frac{5}{2}\log z$ **43.** $2u + 5v$ **45.** $3u - 2v$ **47.** 3.5145 **49.** 2.4851 **51.** There are many correct answers, including $b = 1$, $c = 2$. **53.** **55.**

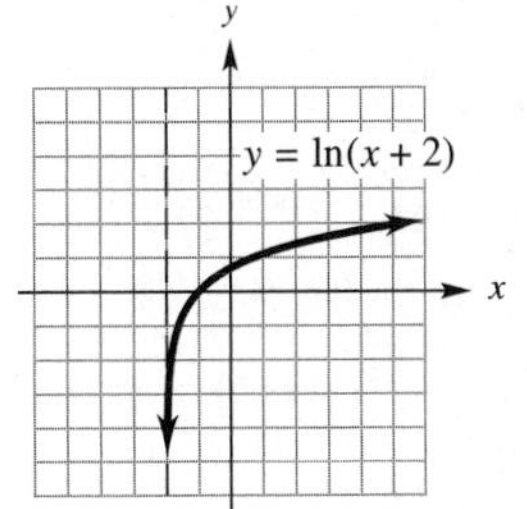

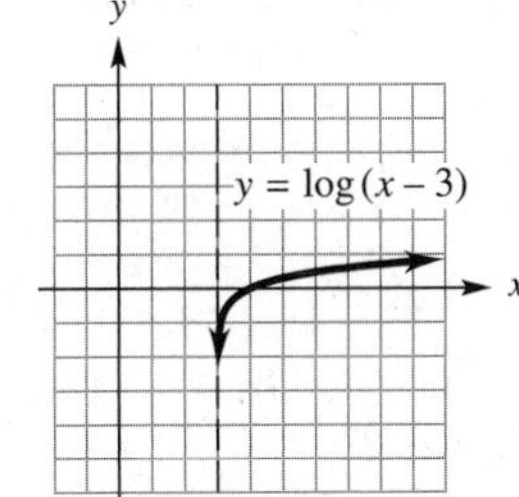

59. $\ln 2.75 = 1.0116009$; $e^{1.0116009} = 2.75$ **61. (a)** 17.67 yr **(b)** 9.01 yr **(c)** 4.19 yr **(d)** 2.25 yr **(e)** $\frac{72}{4} = 18$; $\frac{72}{8} = 9$; $\frac{72}{18} = 4$; $\frac{72}{36} = 2$. So it takes about $\frac{72}{k}$ years for money to double at k% interest. **63. (a)** The percent of change in rents increased rapidly from 1992 to 1993, leveled off to a slow increase from 1993 to 1994, and then increased rapidly again from 1994 to 1996. From 1996 to 1999 (estimated) the percent of increase slowed and became constant at about 7%. **(b)** $f(92) \approx -3.4$; $f(99) \approx 7.1$; $f(92) \approx 3.4$ agrees closely with the y-value on the graph of about -3; $f(99) \approx 7.1$ is also reasonably close to the y-value on the graph of about 7.5.

Section 4.4 (Page 222)

1. 5 **3.** no solution **5.** 11 **7.** $\frac{8}{5}$ **9.** $\frac{1}{3}$ **11.** 10 **13.** 5.2378 **15.** 5 **17.** $\frac{4+b}{4}$ **19.** $\frac{10^{2-b} - 5}{6}$ **23.** 3 **25.** $-\frac{5}{4}$ **27.** -2 **29.** -2 **31.** 1.465 **33.** 2.710 **35.** -1.825 **37.** .805 **39.** $-.123$ **41.** $\frac{\log d + 3}{4}$ **43.** $\frac{\ln b + 1}{2}$ **45.** 3 **47.** no solution **49.** $-2, 2$ **51.** 1, 10 **53.** 1 **55.** 4, -4 **57.** ± 2.141 **59.** 1.386 **63. (a)** about 3390; about 8692 **(b)** in 2005 **65.** 14.2 hr **67. (a)** 25 g **(b)** about 4.95 yr **69.** about 3689 yr old **71. (a)** approximately $3{,}981{,}072i_0$ **(b)** approximately $3{,}162{,}278i_0$ **(c)** about 1.26 times stronger. **73. (a)** 21 **(b)** 100 **(c)** 105 **(d)** 120 **(e)** 140 **75. (a)** 27.5% **(b)** $130.14 **77. (a)**

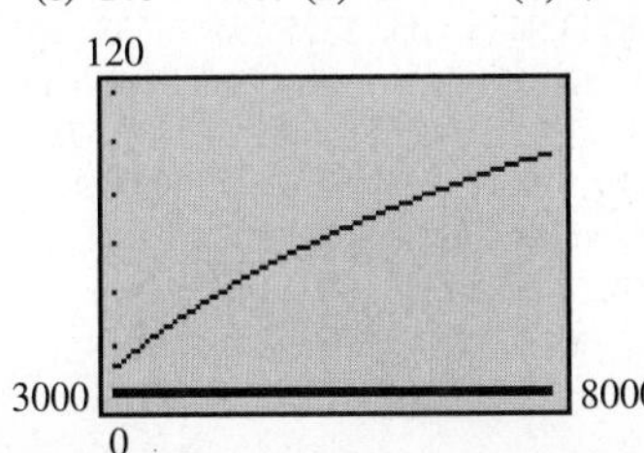

(b) about 4717 ft **79.** **(a)** $f(x) = 951.1369 + 4059.6835 \ln x$ **(b)** 13,500 **(c)** 2008

Chapter 4 Review (Page 226)

1. (c) **3.** (d) **5.** $0 < a < 1$ **7.** $(0, \infty)$

9.

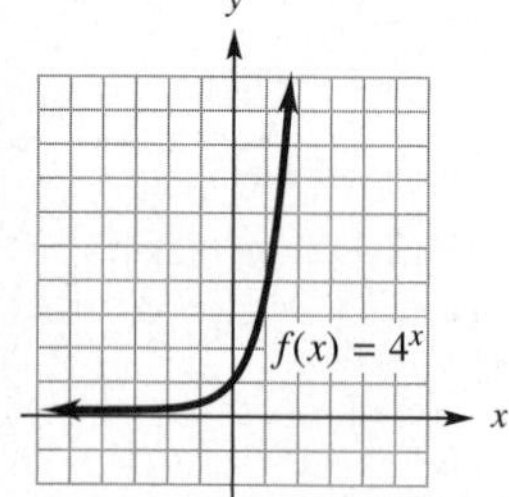

11.

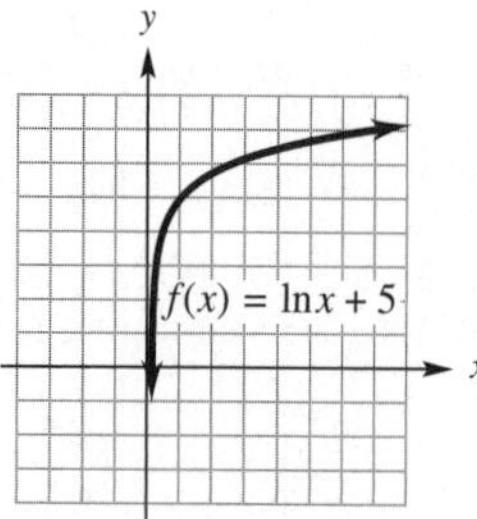

13. **(a)** about 2542.1 million **(b)** 2002 **15.** $\log 47 = 1.6721$ **17.** $\ln 39 = 3.6636$ **19.** $10^3 = 1000$ **21.** $e^{4.5581} = 95.4$ **23.** 3 **25.** 7.4 **27.** $\frac{4}{3}$ **29.** $\log 20k^4$ **31.** $\log\left(\frac{b^2}{c^3}\right)$ **33.** 3 **35.** 8 **37.** 6 **39.** 2 **41.** -1 **43.** 2 **45.** 1.416 **47.** -2.807 **49.** -3.305 **51.** .747 **53.** 28.463 **55.** **(a)** $15 million **(b)** $15.6 million **(c)** $16.4 million **57.** **(a)** 10 g **(b)** about 140 days **(c)** about 243 days **59.** 81.25° Celsius **61.** **(a)** $f(x) = 345.4(.8142^x)$ **(b)** $g(x) = 336.8267(.818^x)$ **(c)** about 66.7 or 67.5 million, depending on the model **(d)** in 2003 **63.** **(a)** $f(x) = \frac{73.27}{1 + 44e^{-.285x}}$ **(b)** in 2004

Case 3 (Page 230)

1. 23.6; 47.7; 58.4; the estimates are a bit low.

CHAPTER 5

Section 5.1 (Page 236)

1. time and interest rate **3.** $231.00 **5.** $286.75 **7.** $119.15 **9.** $241.56 **11.** $304.38 **13.** $3654.10 **17.** $46,265.06 **19.** $28,026.37 **21.** $8898.75 **23.** $48,140.65 **25.** 10.7% **27.** 11.8% **29.** $27,894.30 **31.** $3056.25 **33.** 5.0% **35.** 11.4% **37.** $1732.90 **39.** $5778.50 **41.** $3773; 13.58% **43.** $7278.68; no **45.** **(a)** $372.77 **(b)** $26.94 **(c)** No, he will still owe $166.45.

Section 5.2 (Page 246)

1. r is the interest rate per year, while i is the interest rate per compounding period. t is number of years, while n is the number of periods. **3.** the interest rate and number of compounding **7.** $1593.85 **9.** $1515.80 **11.** $13,213.16 **13.** $10,466.35 **15.** $2307.95 **17.** $1968.48 **19.** $26,545.91 **21.** $45,552.97 **23.** $83,002.92 **25.** 4.04% **27.** 8.16% **29.** 12.36% **31.** $8954.58 **33.** $3255.55 **35.** $11,572.58 **37.** $1000 now **39.** $30,611.30 **41.** about $1.946 million **43.** 7.11% **45.** 4.91, 5.20, 5.34, 5.56, 5.63 **47.** $7522.50 **49.** **(a)** $16,659.95 **(b)** $21,472.67 **51.** $1000 now **53.** about 14 yr **55.** about 35 yr **57.** 10.00% **59.** **(a)** $16,288.95 **(b)** $16,436.19 **(c)** $16,470.09 **(d)** $16,486.65 **61.** **(a)**

Section 5.3 (Page 257)

3. **(a)** $a_1 = 1169$; $r = .916$ **(b)** $a_{10} = 531$; $a_{20} = 221$; This means that a person who is 10 years from retirement should have savings of 531% of his or her annual salary; a person 20 years from retirement should have savings of 221% of his or her annual salary. **5.** 135 **7.** 6 **9.** 2315.25 **11.** 15 **13.** 6.24 **15.** 594.048 **17.** 15.91713 **19.** 21.82453 **21.** 22.01900 **23.** $159,663.02 **25.** $17,435.70 **27.** $72,482.38 **29.** $748.19 **31.** $2038.01 **35.** $7382.54 **37.** $1463.16 **39.** $258.73 **41.** $6603.39 **43.** $158,260.36 **45.** $26,671.23 **47.** $4421.89 **49.** $144,872.14 **51.** $4168.30 **53.** **(a)** $149,850.69 **(b)** $137,895.79 **(c)** $11,954.90 **55.** $130,159.72 **57.** $284,527.35 **59.** **(a)** $226.11 **(b)** $245.77 **61.** $522.85 **63.** $7.397% **65.** **(a)** $1200 **(b)** $3511.58

(c)

Payment Number	*Amount of Deposit*	*Interest Earned*	*Total*
1	$3511.58	$0.00	$3511.58
2	3511.58	105.35	7128.51
3	3511.58	213.86	10,853.95
4	3511.58	325.62	14,691.15
5	3511.58	440.73	18,643.46
6	3511.58	559.30	22,714.34

Section 5.4 (Page 265)

1. (c) **3.** 9.71225 **5.** 12.65930 **7.** 14.71787 **11.** $8045.29 **13.** $1,367,773.96 **15.** $205,724.40 **17.** $103,796.58 **21.** $5309.69 **23.** $11,727.32 **25.** $260.19 **27.** $1267.07 **29.** $847.91 **31.** $86.24 **33.** $13.02 **35.** $48,677.34 **37.** **(a)** $158 **(b)** $1584 **39.** **(a)** $623,110.52 **(b)** $456,427.28 **(c)** $563,757.78 **(d)** $392,903.18 **41.** $280.46; $32,310.74 **43.** **(a)** $2349.51; $197,911.80 **(b)** $2097.30; $278,352 **(c)** $1965.82; $364,746 **(d)** after 170 payments **45.** **(a)** $4025.90 **(b)** $2981.93 **47.** $432.53

49.

Payment Number	*Amount of Payment*	*Interest for Period*	*Portion to Principal*	*Principal at End of Period*
0	—	—	—	$4000.00
1	$1207.68	$320.00	$887.68	3112.32
2	1207.68	248.99	958.69	2153.63
3	1207.68	172.29	1035.39	1118.24
4	1207.70	89.46	1118.24	0

51.

Payment Number	*Amount of Payment*	*Interest for Period*	*Portion to Principal*	*Principal at End of Period*
0	—	—	—	$7184.00
1	$189.18	$71.84	117.34	7066.66
2	189.18	70.67	118.51	6948.15
3	189.18	69.48	119.70	6828.45
4	189.18	68.28	120.90	6707.55

Chapter 5 Review (Page 269)

1. $426.88 **3.** $62.91 **7.** $78,717.19 **9.** $739.09 **13.** $70,538.38; $12,729.04 **15.** $5282.19; $604.96 **17.** $12,857.07 **19.** $1923.09 **21.** $4, 2, 1, \frac{1}{2}$ **23.** -32

25. 5500 **29.** \$137,925.91 **31.** \$25,396.38
35. \$2619.29 **37.** \$916.12 **39.** \$31,921.91
41. \$14,222.42 **43.** \$3628.00 **45.** \$546.93 **47.** \$896.06
49. \$2696.12 **51.** \$10,550.54 **53.** \$8369.51
55. (a) about 37.44% **(b)** about 24.74% **57.** 3.6%
59. \$24,818.76; \$2418.76 **61.** \$32.49 **63.** \$5596.62
65. \$3560.61 **67.** (d) **69.** \$5927.56

Case 4 (Page 274)

1.

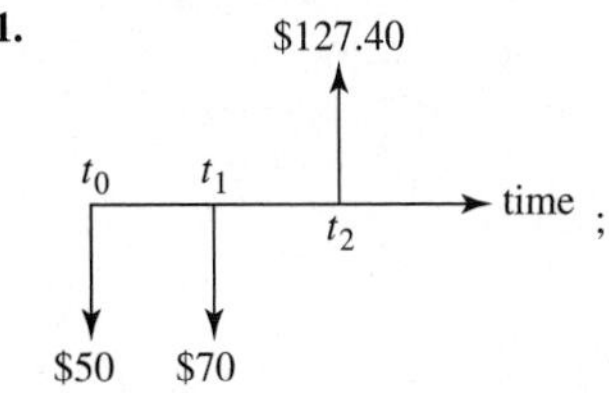

$50(1 + i)^2 + 70(1 + i) = 127.40$; 4.3%

3. (a)

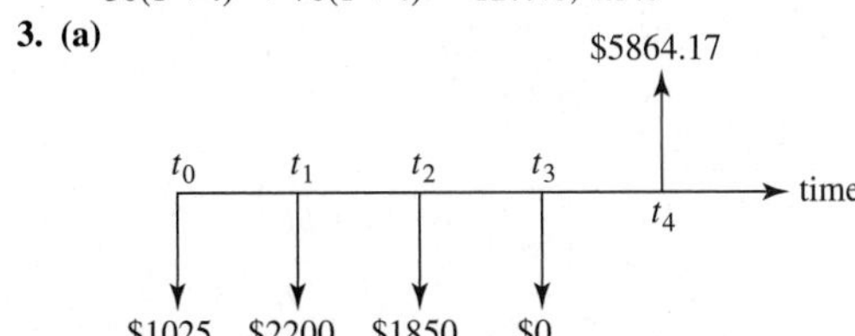

(b) 5.2%

5. (a)

\$90
t_0 t_1 t_2 time
\$50 \$50

$50(1 + i)^2 + 50(1 + i) - 90 = 0$; $-.068, -2.932$
(b) $-.068$ is reasonable, -2.932 is not.

CHAPTER 6

Section 6.1 (Page 287)

1. yes **3.** $\left(\frac{11}{5}, -\frac{7}{5}\right)$ **5.** (28, 22) **7.** (2, −1)

9. no solution **11.** $(4y + 1, y)$ for any real number y
13. no solution **15.** (a) **17.** (5, 10)
19. (a) $r = 175{,}000$ and $b = 375{,}000$

21. $\left[\begin{array}{cccc|c} 3 & 4 & -2 & -3 & 0 \\ 1 & -3 & 7 & 4 & 9 \\ 2 & 0 & 5 & -6 & 0 \end{array}\right]$

23. $3x + 2y + 6z = 18$
$2x - 2y + 5z = 7$
$x \qquad + 5z = 20$

25. $\left[\begin{array}{cccc|c} 1 & 5 & 2 & 0 & -1 \\ 8 & 5 & 4 & 6 & 6 \\ 0 & -15 & 1 & 1 & -1 \end{array}\right]$ **27.** $\left[\begin{array}{ccc|c} 2 & 5 & 1 & -1 \\ -4 & 0 & 4 & 6 \\ \frac{3}{2} & 0 & 2 & -1 \end{array}\right]$

29. dependent **31.** inconsistent **33.** independent
35. (2, 1, 4) **37.** (−1, 4, 2) **39.** no solution **41.** (2, 0, 1)

43. $\left(\frac{(-7 - 2z)}{5}, \frac{(11z + 21)}{5}, z\right)$ for any real number z

45. (1, 2) **47. (a)** $y = x + 1$ **(b)** $y = 3x + 4$ **(c)** $\left(-\frac{3}{2}, -\frac{1}{2}\right)$

51. $W_1 = \frac{300}{1 + \sqrt{3}} \approx 110$ lb
$W_2 = \frac{300 - W_1}{\sqrt{2}} \approx 134$ lb

53. 260 skirts; 270 blouses
55. (a) using the first equation, approximately 245 lb; using the second equation, approximately 253 lb **(b)** for the first, 7.46 lb; and for the second, 7.93 lb **(c)** at approximately 66 in. and 118 lb
57. .75 lb cashews; 2.25 lb peanuts
59. (a) $b + c$ **(b)** .35, .45, .2; A is tumorous, B is bone, C is healthy.
61. (a) $x_2 + x_3 = 700, x_3 + x_4 = 600$
(b) $(1000 - x_4, 100 + x_4, 600 - x_4, x_4)$ **(c)** 0, 600
(d) 400, 1000 for x_1; 100, 700 for x_2; 0, 600 for x_3 **(e)** If you know the number of cars entering or leaving three of the intersections, then the number entering or leaving the fourth is automatically determined, because the number leaving must equal the number entering.

Section 6.2 (Page 299)

1. (3/2, 5, −2, 0) **3.** $(2 - w, -3 - 2w, 4, w)$ for any real number w
5. (−3, 4, 0) **7.** (1, 0, −1) **9.** $\left(\frac{-9z + 5}{23}, \frac{10z - 3}{23}, z\right)$ for any real number z **11.** (−1, 23, 16) **13.** no solution
15. (−7, 5) **17.** $(-3, z - 17, z)$ for any real number z
19. (−1, 1, −3, −2) **21.** no solution
23. Garcia 20 hr, Wong 15 hr **25.** 40 lb pretzels, 20 lb dried fruit, 80 lb nuts **27. (a)** .056057, 1.06657 **(b)** 228 ft
29. \$15,000 in the mutual fund, \$30,000 in bonds, and \$25,000 in the food franchise **31.** 6 cups of Roasted Chicken Rotini, 9 cups of Hearty Chicken, and 2 cups of Chunky Chicken Noodle; serving size 1.7 cups **33. (a)** \$12,000 at 6%, \$7000 at 7%, and \$6000 at 10%
(b) \$10,000 at 6%, \$15,000 at 7%, and \$5000 at 10%
(c) \$20,000 at 6%, \$10,000 at 7%, and \$10,000 at 10%
35. (a) $y = .01x^2 - .3x + 4.24$ **(b)** 15 platters; \$1.99
37. (a) $C = .015t^2 + .7t + 315$ **(b)** 2081
39. (a) $L = 92.517t^2 - 1181.32t + 7053$ **(b)** 3296
(c) $L = -20.758t^3 + 362.38t^2 - 1928.6t + 7053$

Section 6.3 (Page 308)

1. 2 × 3; $\begin{bmatrix} -7 & 8 & -4 \\ 0 & -13 & -9 \end{bmatrix}$

3. 3 × 3; square matrix; $\begin{bmatrix} 3 & 0 & -11 \\ -1 & -\frac{1}{4} & 7 \\ -5 & 3 & -9 \end{bmatrix}$

5. 2 × 1; column matrix; $\begin{bmatrix} -7 \\ -11 \end{bmatrix}$ **7.** B is a 5 × 3 zero matrix.

9. $\begin{bmatrix} 9 & 12 & 0 & 2 \\ 1 & -1 & 2 & -4 \end{bmatrix}$ **11.** $\begin{bmatrix} 5 & 13 & 0 \\ 3 & 1 & 8 \end{bmatrix}$ **13.** $\begin{bmatrix} 3 & -7 & 7 \\ 4 & 4 & -8 \end{bmatrix}$

15. $\begin{bmatrix} -4 & 8 \\ 0 & 6 \end{bmatrix}$ **17.** $\begin{bmatrix} 24 & -8 \\ -16 & 0 \end{bmatrix}$ **19.** $\begin{bmatrix} -22 & -6 \\ 20 & -12 \end{bmatrix}$

21. $\begin{bmatrix} 4 & -\frac{7}{2} \\ 4 & \frac{21}{2} \end{bmatrix}$

23. $X + T = \begin{bmatrix} x & y \\ z & w \end{bmatrix} + \begin{bmatrix} r & s \\ t & u \end{bmatrix} = \begin{bmatrix} x+r & y+s \\ z+t & w+u \end{bmatrix}$ a 2 × 2 matrix

25. $X + (T + P) = \begin{bmatrix} x + (r+m) & y + (s+n) \\ z + (t+p) & w + (u+q) \end{bmatrix}$

$= \begin{bmatrix} (x+r)+m & (y+s)+n \\ (z+t)+p & (w+u)+q \end{bmatrix}$

$= (X + T) + P$

27. $P + O = \begin{bmatrix} m+0 & n+0 \\ p+0 & q+0 \end{bmatrix} = \begin{bmatrix} m & n \\ p & q \end{bmatrix} = P$

29. Several possible correct answers, including:

	basketball	hockey	football	baseball
percent of no shows	16	16	20	18
lost revenue per fan ($)	18.20	18.25	19	15.40
lost annual revenue (millions $)	22.7	35.8	51.9	96.3

31.

	1990	1998	2000
heart	1788	4121	4143
lung	308	3171	3614
liver	1237	12,070	15,359
kidney	17,883	38,270	45,273

; other answers are possible

33. (a) $A = \begin{bmatrix} 80 & 926 \\ 43 & 122 \\ 7 & 41 \end{bmatrix}$

(b) $B = \begin{bmatrix} 744 & 879 \\ 48 & 70 \\ 2 & 33 \end{bmatrix}$

(c) $A + B = \begin{bmatrix} 824 & 1805 \\ 91 & 192 \\ 9 & 74 \end{bmatrix}$; the matrix gives the total number of women's and men's teams in the three sports in 1981–82 and 1998–99.

(d) $C = \begin{bmatrix} 1019 \\ 134 \\ 45 \end{bmatrix}$

35. (a) $A = \begin{bmatrix} 2.61 & 4.39 & 6.29 & 9.08 \\ 1.63 & 2.77 & 4.61 & 6.92 \\ .92 & .75 & .62 & .54 \end{bmatrix}$

(b) $B = \begin{bmatrix} 1.38 & 1.72 & 1.94 & 3.31 \\ 1.26 & 1.48 & 2.82 & 2.28 \\ .41 & .33 & .27 & .40 \end{bmatrix}$

(c) $\begin{bmatrix} 1.23 & 2.67 & 4.35 & 5.77 \\ .37 & 1.29 & 1.79 & 4.64 \\ .51 & .42 & .35 & .14 \end{bmatrix}$

Section 6.4 (Page 321)

1. 2 × 2; 2 × 2 **3.** 3 × 2; *BA* does not exist

5. *AB* does not exist; 3 × 2 **7.** columns; rows **9.** $\begin{bmatrix} 13 \\ 25 \end{bmatrix}$

11. $\begin{bmatrix} -2 & 10 \\ 0 & 8 \end{bmatrix}$ **13.** $\begin{bmatrix} -4 & 1 \\ 2 & -3 \end{bmatrix}$ **15.** $\begin{bmatrix} 3 & -5 & 7 \\ -2 & 1 & 6 \\ 0 & -3 & 4 \end{bmatrix}$

17. $\begin{bmatrix} 16 & 11 \\ 37 & 32 \\ 58 & 53 \end{bmatrix}$ **19.** $AB = \begin{bmatrix} -30 & -45 \\ 20 & 30 \end{bmatrix}$, but $BA = \begin{bmatrix} 0 & 0 \\ 0 & 0 \end{bmatrix}$.

21. $(A + B)(A - B) = \begin{bmatrix} -7 & -24 \\ -28 & -33 \end{bmatrix}$, but $A^2 - B^2 = \begin{bmatrix} -37 & -69 \\ -8 & -3 \end{bmatrix}$.

23. $P(XT) =$
$\begin{bmatrix} (mx+nz)r + (my+nw)t & (mx+nz)s + (my+nw)u \\ (px+qz)r + (py+qw)t & (px+qz)s + (py+qw)u \end{bmatrix}$
$P(XT)$ is the same so $(PX)T = P(XT)$.

25. $k(X + T) = k\begin{bmatrix} x+r & y+s \\ z+t & w+u \end{bmatrix}$

$= \begin{bmatrix} k(x+r) & k(y+s) \\ k(z+t) & k(w+u) \end{bmatrix}$

$= \begin{bmatrix} kx+kr & ky+ks \\ kz+kt & kw+ku \end{bmatrix}$

$= \begin{bmatrix} kx & ky \\ kz & kw \end{bmatrix} + \begin{bmatrix} kr & ks \\ kt & ku \end{bmatrix} = kX + kT$

27. no **29.** yes **31.** no **33.** $\begin{bmatrix} 2 & -3 \\ -1 & 2 \end{bmatrix}$

35. no inverse **37.** $\begin{bmatrix} 2 & -3 \\ -\frac{1}{2} & 1 \end{bmatrix}$ **39.** $\begin{bmatrix} 3 & 3 & -1 \\ -2 & -2 & 1 \\ -4 & -5 & 2 \end{bmatrix}$

41. $\begin{bmatrix} 2 & 1 & -1 \\ 8 & 2 & -5 \\ -11 & -3 & 7 \end{bmatrix}$ **43.** no inverse

45. $\begin{bmatrix} \frac{7}{4} & \frac{5}{2} & 3 \\ -\frac{1}{4} & -\frac{1}{2} & 0 \\ -\frac{1}{4} & -\frac{1}{2} & -1 \end{bmatrix}$ **47.** $\begin{bmatrix} \frac{1}{2} & \frac{1}{2} & -\frac{1}{4} & \frac{1}{2} \\ -1 & 4 & -\frac{1}{2} & -2 \\ -\frac{1}{2} & \frac{5}{2} & -\frac{1}{4} & -\frac{3}{2} \\ \frac{1}{2} & -\frac{1}{2} & \frac{1}{4} & \frac{1}{2} \end{bmatrix}$

49. (a) $R = \begin{bmatrix} .024 & .008 \\ .025 & .007 \\ .015 & .009 \\ .011 & .011 \end{bmatrix}$

(b) $P = \begin{bmatrix} 1996 & 286 & 226 & 460 \\ 2440 & 365 & 252 & 484 \\ 2906 & 455 & 277 & 499 \\ 3683 & 519 & 310 & 729 \\ 4723 & 697 & 364 & 702 \end{bmatrix}$

(c) $PR = \begin{bmatrix} 63.504 & 25.064 \\ 76.789 & 29.667 \\ 90.763 & 34.415 \\ 114.036 & 43.906 \\ 143.959 & 53.661 \end{bmatrix}$

(d) Rows represent the years 1970, 1980, 1990, 2000, 2025. Column one gives the total births in those years; column two the total deaths. **(e)** 114,036,000; 53,661,000

51. **(a)** $C = \begin{bmatrix} 394 & 2753 \\ 595 & 3093 \\ 601 & 3085 \\ 610 & 3119 \end{bmatrix}$

(b) $D = \begin{bmatrix} 15.1 & 14.2 & 13.7 & 14.1 \\ 6.3 & 6.0 & 6.0 & 6.0 \end{bmatrix}$

(c) $DC = \begin{bmatrix} 31{,}233.1 & 171{,}733.3 \\ 13{,}318.2 & 73{,}125.9 \end{bmatrix}$

(d) row 1, column 1 **(e)** row 2, column 2

53. **(a)**

	A	B
Dept. 1	57	70
Dept. 2	41	54
Dept. 3	27	40
Dept. 4	39	40

(b) supplier A: \$164; supplier B: \$204

Section 6.5 (Page 332)

1. $\begin{bmatrix} 8 \\ 6 \end{bmatrix}$ **3.** $\begin{bmatrix} \frac{1}{2} & 1 \\ \frac{3}{2} & 1 \end{bmatrix}$ **5.** $\begin{bmatrix} 11 \\ -3 \\ 5 \end{bmatrix}$ **7.** $(-31, 24, -4)$

9. $(-31, -131, 181)$ **11.** $(15, -5, -1)$ **13.** $(-7, -34, -19, 7)$

15. $\begin{bmatrix} -6 \\ -14 \end{bmatrix}$ **17.** $\begin{bmatrix} \frac{32}{3} \\ \frac{25}{3} \end{bmatrix}$

19. 848 units of agriculture, 516 units of manufacturing, and 2970 units of households **21.** 195.5 million pounds of agriculture, 25.9 million pounds of manufacturing, and 13.6 million pounds of energy **23.** either 10 buffets, 5 chairs, no tables, or 11 buffets, 1 chair, 1 table **25.** 2340 of the first species, 10,128 of the second species, 224 of the third species **27.** jeans \$34.50; jacket \$72; sweater \$44; shirt \$21.75 **29.** about 1073 metric tons of wheat, about 1431 metric tons of oil **31.** gas \$98 million, electric \$123 million

33. **(a)** $\frac{7}{4}$ bushels of yams, $\frac{15}{8} \approx 2$ pigs **(b)** 167.5 bushels of yams, $153.75 \approx 154$ pigs

35. **(a)** $\begin{bmatrix} 1.67 & .56 & .56 \\ .19 & 1.17 & .06 \\ 3.15 & 3.27 & 4.38 \end{bmatrix}$

(b) These multipliers imply that if the demand for one community's output increases by \$1 then the output in the other community will increase by the amount in the row and column of that matrix. For example, if the demand for Hermitage's output increases by \$1 then output from Sharon will increase by \$.56, from Farrell by \$.06, and from Hermitage by \$4.38.

37. $\begin{bmatrix} 23 \\ 51 \end{bmatrix}, \begin{bmatrix} 13 \\ 30 \end{bmatrix}, \begin{bmatrix} 45 \\ 96 \end{bmatrix}, \begin{bmatrix} 69 \\ 156 \end{bmatrix}, \begin{bmatrix} 87 \\ 194 \end{bmatrix}, \begin{bmatrix} 23 \\ 51 \end{bmatrix}, \begin{bmatrix} 51 \\ 110 \end{bmatrix}, \begin{bmatrix} 45 \\ 102 \end{bmatrix}, \begin{bmatrix} 69 \\ 157 \end{bmatrix}$

39. **(a)** 3 **(b)** 3 **(c)** 5 **(d)** 3

41. **(a)** $B = \begin{bmatrix} 0 & 2 & 3 \\ 2 & 0 & 4 \\ 3 & 4 & 0 \end{bmatrix}$ **(b)** $B^2 = \begin{bmatrix} 13 & 12 & 8 \\ 12 & 20 & 6 \\ 8 & 6 & 25 \end{bmatrix}$ **(c)** 12 **(d)** 14

43. **(a)**

$C =$

	Dogs	Rats	Cats	Mice
Dogs	0	1	1	1
Rats	0	0	0	1
Cats	0	1	0	1
Mice	0	0	0	0

(b) $C^2 = \begin{bmatrix} 0 & 1 & 0 & 2 \\ 0 & 0 & 0 & 0 \\ 0 & 0 & 0 & 1 \\ 0 & 0 & 0 & 0 \end{bmatrix}$; C^2 gives the number of food sources once removed from the feeder. Thus, since dogs eat rats and rats eat mice, mice are an indirect as well as direct food source for dogs.

Chapter 6 Review (Page 337)

1. $(-5, 7)$ **3.** $(0, -2)$ **5.** no solution; inconsistent system. **7.** $(-35, 140, 22)$ **9.** 8000 standard clips, 6000 extra large clips **11.** \$7000 in the first fund, \$11,000 in the second fund **13.** 50 from source I, 150 from source II, 100 from source III **15.** $(0, 3, 3)$ **17.** no solution **19.** $(-79, 99, -8)$ **21.** 2×2; square **23.** 1×4; row **25.** 2×3

27. $\begin{bmatrix} 8 & 8 & 8 \\ 10 & 5 & 9 \\ 7 & 10 & 7 \\ 8 & 9 & 7 \end{bmatrix}$ **29.** $\begin{bmatrix} -2 & -3 & 2 \\ -2 & -4 & 0 \\ 0 & -1 & -2 \end{bmatrix}$ **31.** $\begin{bmatrix} 2 & 30 \\ -4 & -15 \\ 10 & 13 \end{bmatrix}$

33. not defined

35. Next day $\begin{bmatrix} 2310 & -\frac{1}{4} \\ 1258 & -\frac{1}{4} \\ 5061 & \frac{1}{2} \\ 1812 & \frac{1}{2} \end{bmatrix}$; Two-day total $\begin{bmatrix} 4842 & -\frac{1}{2} \\ 2722 & -\frac{1}{8} \\ 10{,}035 & -1 \\ 3566 & 1 \end{bmatrix}$

37. $\begin{bmatrix} 18 & 80 \\ -7 & -28 \\ 21 & 84 \end{bmatrix}$ **39.** $\begin{bmatrix} 13 & 43 \\ 17 & 46 \end{bmatrix}$ **41.** $\begin{bmatrix} 222 & 632 \\ -77 & -224 \\ 231 & 672 \end{bmatrix}$

43. **(a)** $\begin{bmatrix} \frac{1}{4} & \frac{1}{2} \\ \frac{1}{3} & \frac{1}{3} \end{bmatrix}$ **(b)** cutting 34 hr; shaping 46 hr

45. There are many correct answers, including $\begin{bmatrix} 1 & 2 \\ 3 & 4 \end{bmatrix}$.

47. $\begin{bmatrix} -\frac{1}{4} & \frac{1}{6} \\ 0 & \frac{1}{3} \end{bmatrix}$ **49.** no inverse **51.** $\begin{bmatrix} \frac{1}{4} & \frac{1}{2} & \frac{1}{2} \\ \frac{1}{4} & -\frac{1}{2} & \frac{1}{2} \\ \frac{1}{8} & -\frac{1}{4} & -\frac{1}{4} \end{bmatrix}$

53. no inverse

55. $\begin{bmatrix} -\frac{2}{3} & -\frac{17}{3} & -\frac{14}{3} & -3 \\ \frac{1}{3} & \frac{1}{3} & \frac{1}{3} & 0 \\ -\frac{1}{3} & -\frac{10}{3} & -\frac{7}{3} & -2 \\ 0 & 2 & 1 & 1 \end{bmatrix}$ **57.** $\begin{bmatrix} -\frac{7}{19} & \frac{4}{19} \\ \frac{3}{19} & \frac{1}{19} \end{bmatrix}$

59. $\begin{bmatrix} 1 & 1 \\ -2 & -3 \end{bmatrix}$ **61.** no inverse **63.** $\begin{bmatrix} 18 \\ -7 \end{bmatrix}$ **65.** $\begin{bmatrix} -22 \\ -18 \\ 15 \end{bmatrix}$

67. $(2, 2)$ **69.** $(2, 1)$ **71.** $(-1, 0, 2)$ **73.** no inverse; no solution for the system **75.** 16 liters of the 9%, 24 liters of the 14% **77.** 30 liters of 40% solution, 10 liters of 60% solution

79. 80 bowls, 120 plates **81.** \$12,750 at 8%, \$27,250 at $8\frac{1}{2}$%, \$10,000 at 11%

83. **(a)** $\begin{bmatrix} 1 & -\frac{1}{4} \\ -\frac{1}{2} & 1 \end{bmatrix}$ **(b)** $\begin{bmatrix} \frac{8}{7} & \frac{2}{7} \\ \frac{4}{7} & \frac{8}{7} \end{bmatrix}$ **(c)** $\begin{bmatrix} 2800 \\ 2800 \end{bmatrix}$

85. (a) $\begin{bmatrix} 1.30 & .045 & .567 & .012 & .068 & .020 \\ .204 & 1.03 & .183 & .004 & .022 & .006 \\ .155 & .038 & 1.12 & .020 & .114 & .034 \\ .018 & .021 & .028 & 1.08 & .016 & .033 \\ .537 & .525 & .483 & .279 & 1.73 & .419 \\ .573 & .346 & .497 & .536 & .087 & 1.94 \end{bmatrix}$; every dollar of increased demand for livestock will result in an increase of production demand of \$.204 in crops. **(b)** In millions of dollars make \$3855 of livestock, \$1476 in crops, \$2726 in food products, \$1338 in mining and manufacturing, \$8439 in households, and \$10,256 in other business sectors. (Solutions from nonrounded part (a) answer.)
87. agriculture \$140,909, manufacturing \$95,455
89. (a) $\begin{bmatrix} 54 \\ 32 \end{bmatrix}, \begin{bmatrix} 134 \\ 89 \end{bmatrix}, \begin{bmatrix} 172 \\ 113 \end{bmatrix}, \begin{bmatrix} 118 \\ 74 \end{bmatrix}, \begin{bmatrix} 208 \\ 131 \end{bmatrix}$ **(b)** $\begin{bmatrix} 2 & -3 \\ -\frac{1}{2} & 1 \end{bmatrix}$

Case 5 (Page 342)

1. Boston, Hyannis, Martha's Vineyard, Nantucket, New Bedford, and Providence may be reached by a two-flight sequence from New Bedford; all Cape Air cities may be reached by a three-flight sequence. **3.** The connection between Provincetown and Providence and the connection between Provincetown and New Bedford each take three flights. **5.** All Big Sky cities may be reached by a three-flight sequence from Helena. At least three flights must be used to get from Helena to Havre, Glendive, and Bismarck.

CHAPTER 7

Section 7.1 (Page 349)

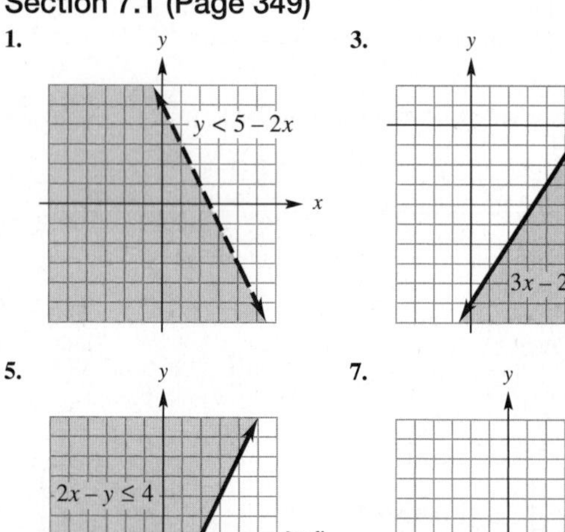

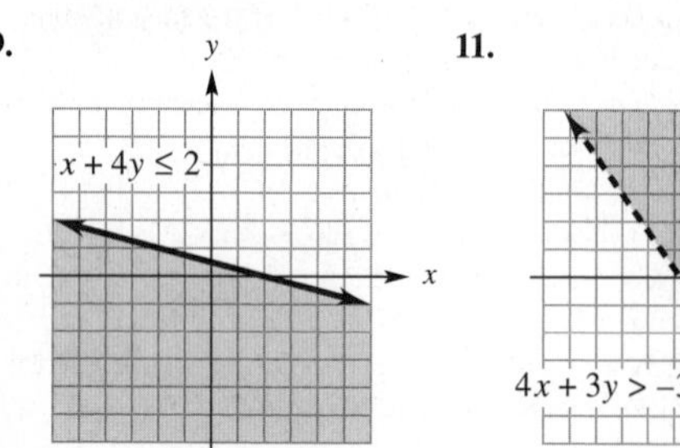

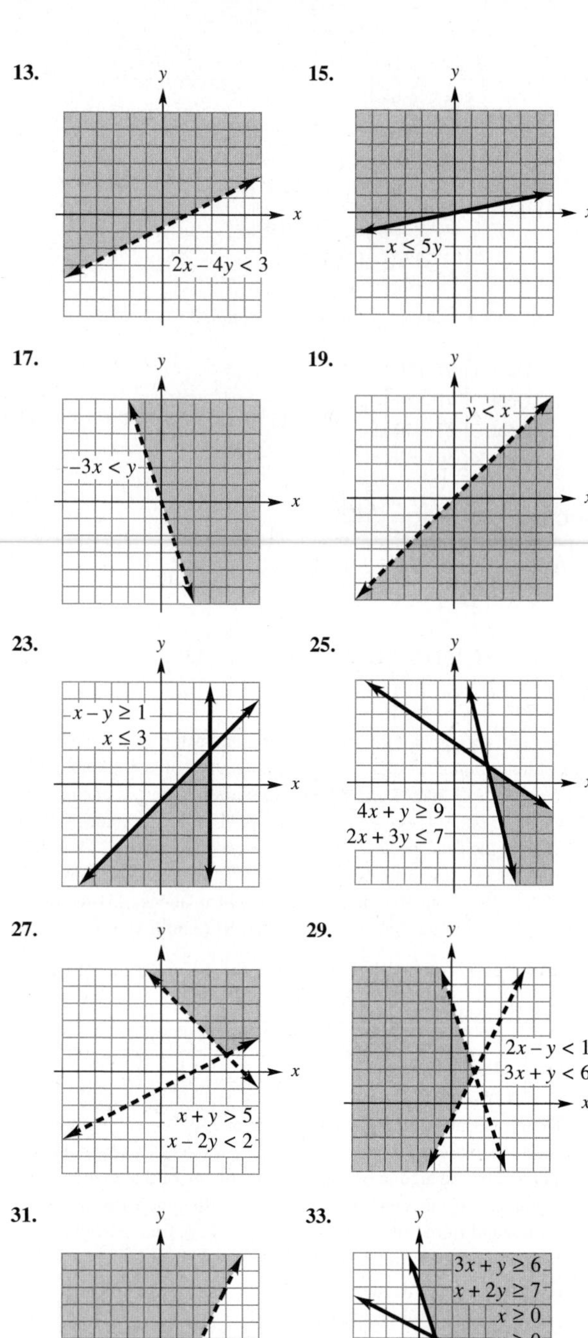

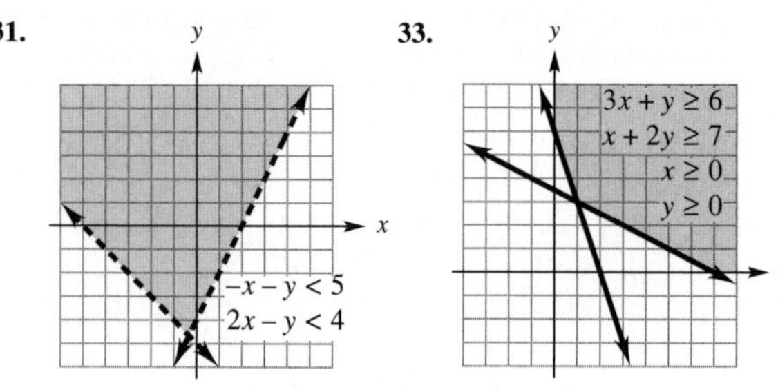

35.

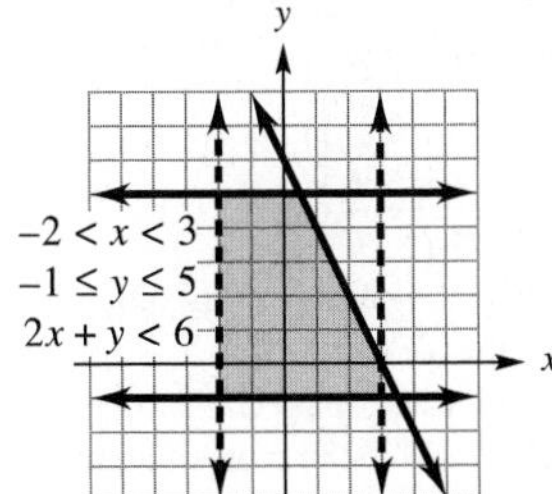

37.

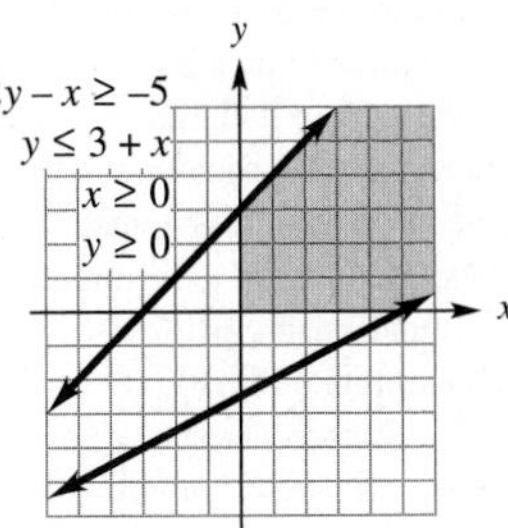

39.

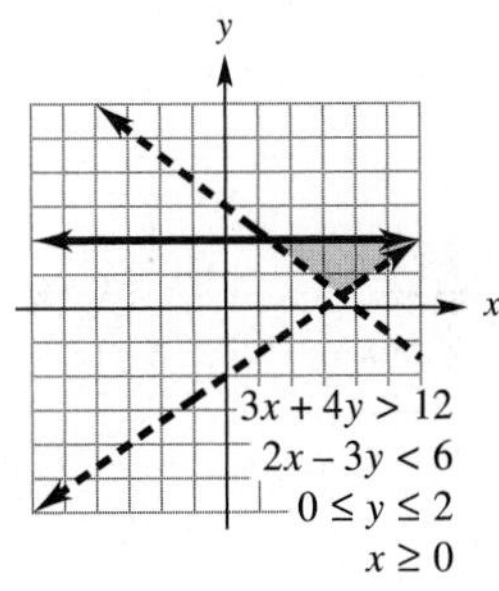

41. $2 < x < 7$
$-1 < y < 3$

43. (a)

Number	*Hours Spinning*	*Hours Dyeing*	*Hours Weaving*
Shawls x	1	1	1
Afghans y	2	1	4
Maximum Number of Hours Available	8	6	14

(b) $x + 2y \le 8$; $x + y \le 6$; $x + 4y \le 14$; $x \ge 0$; $y \ge 0$

(c)

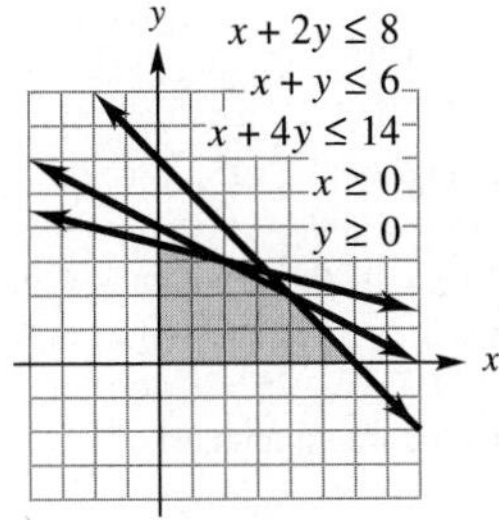

45. $x \ge 3000$; $y \ge 5000$; $x + y \le 10{,}000$

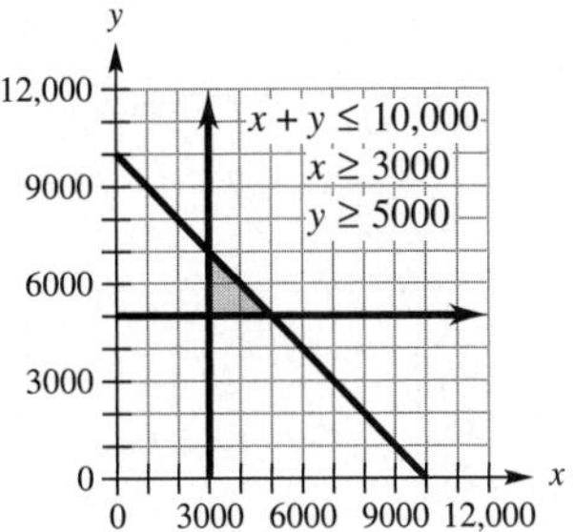

47. $x + y \ge 3.2$; $.16x + .20y \le .8$; $.5x + .3y \le 1.8$; $x \ge 0$; $y \ge 0$

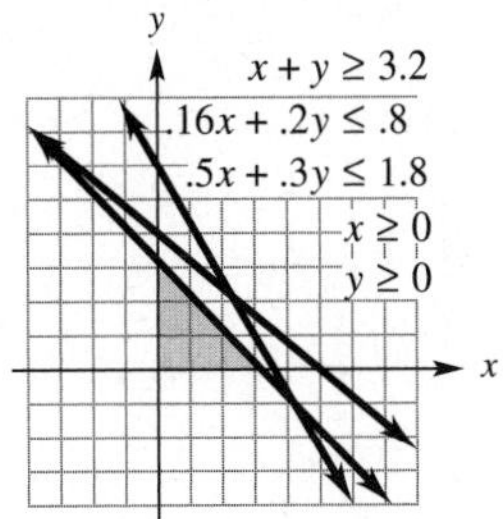

Section 7.2 (Page 357)

1. maximum of 65 at (5, 10); minimum of 8 at (1, 1)
3. maximum of 9 at (0, 12); minimum of 0 at (0, 0)
5. (a) no maximum; minimum of 16 at (0, 8)
(b) no maximum; minimum of 18 at (3, 4)
(c) no maximum; minimum of 21 at $\left(\frac{13}{2}, 2\right)$
(d) no maximum; minimum of 12 at (12, 0)
7. maximum of $\frac{42}{5}$ at $x = \frac{6}{5}$, $y = \frac{6}{5}$ or $\left(\frac{6}{5}, \frac{6}{5}\right)$
9. minimum of 13 at $x = 5$, $y = 3$ or (5, 3)
11. maximum of $\frac{235}{4}$ at $x = \frac{105}{8}$, $y = \frac{25}{8}$ or $\left(\frac{105}{8}, \frac{25}{8}\right)$
13. no maximum; minimum of 9 at $\left(1, \frac{3}{2}\right)$
15. maximum of 22 at (2, 4); no minimum
17. (a) (18, 2) **(b)** $\left(\frac{12}{5}, \frac{39}{5}\right)$ **(c)** no maximum

Section 7.3 (Page 363)

1. $6x + 4y \le 90$, where x = number of canoes, y = number of rowboats **3.** $150x + 750y \le 8500$, where x = number of radio spots, y = number of TV ads **5.** 12 chain saws; no chippers **7.** 10 lb deluxe; 40 lb regular **9. (a)** 6 units of policy *A* and 16 units of policy *B* for a minimum cost of \$940 **(b)** 30 units of policy *A* and 0 units of policy *B* for a minimum cost of \$750 **11.** $3\frac{3}{4}$ servings of A and $1\frac{7}{8}$ servings of B for a minimum cost of \$1.69 **13.** 800 type 1 and 1600 type 2 for a maximum revenue of \$272 **15.** 0 plants and 18 animals, for a minimum of 270 hr **17.** From warehouse I ship 60 boxes to San Jose and 300 boxes to Memphis. From warehouse II ship 290 boxes to San Jose and none to Memphis, for a minimum cost of \$147.70. **19.** \$20 million in bonds and \$20 million in mutual funds for a maximum annual interest of \$4 million **21.** 5 humanities, 12 science **23.** (b) **25.** (c)

Section 7.4 (Page 378)

1. **(a)** 3
(b) x_3, x_4, x_5
(c)
$$\begin{aligned} 4x_1 + 2x_2 + x_3 \qquad\qquad &= 20 \\ 5x_1 + x_2 \qquad + x_4 \qquad &= 50 \\ 2x_1 + 3x_2 \qquad\qquad + x_5 &= 25 \end{aligned}$$

3. **(a)** 3
(b) x_4, x_5, x_6
(c)
$$\begin{aligned} 3x_1 - x_2 + 4x_3 + x_4 \qquad\qquad &= 95 \\ 7x_1 + 6x_2 + 8x_3 \qquad + x_5 \qquad &= 118 \\ 4x_1 + 5x_2 + 10x_3 \qquad\qquad + x_6 &= 220 \end{aligned}$$

5.
$$\begin{array}{c} \begin{array}{cccccc} x_1 & x_2 & x_3 & x_4 & x_5 & z \end{array} \\ \left[\begin{array}{cccccc|c} 2 & 3 & 1 & 0 & 0 & 0 & 6 \\ 4 & 1 & 0 & 1 & 0 & 0 & 6 \\ 5 & 2 & 0 & 0 & 1 & 0 & 15 \\ \hline -5 & -1 & 0 & 0 & 0 & 1 & 0 \end{array}\right] \end{array}$$

7.
$$\begin{array}{c} \begin{array}{ccccccc} x_1 & x_2 & x_3 & x_4 & x_5 & x_6 & z \end{array} \\ \left[\begin{array}{ccccccc|c} 1 & 2 & 3 & 1 & 0 & 0 & 0 & 10 \\ 2 & 1 & 1 & 0 & 1 & 0 & 0 & 8 \\ 3 & 0 & 2 & 0 & 0 & 1 & 0 & 6 \\ \hline -1 & -5 & -10 & 0 & 0 & 0 & 1 & 0 \end{array}\right] \end{array}$$

9. 3 in row 2, column 1 **11.** 6 in row 1, column 1

13.
$$\begin{array}{c} \begin{array}{cccccc} x_1 & x_2 & x_3 & x_4 & x_5 & z \end{array} \\ \left[\begin{array}{cccccc|c} -1 & 0 & 3 & 1 & -1 & 0 & 16 \\ 1 & 1 & \frac{1}{2} & 0 & \frac{1}{2} & 0 & 20 \\ \hline 2 & 0 & -\frac{1}{2} & 0 & \frac{3}{2} & 1 & 60 \end{array}\right] \end{array}$$

15.
$$\begin{array}{c} \begin{array}{ccccccc} x_1 & x_2 & x_3 & x_4 & x_5 & x_6 & z \end{array} \\ \left[\begin{array}{ccccccc|c} -\frac{1}{2} & \frac{1}{2} & 0 & 1 & -\frac{1}{2} & 0 & 0 & 10 \\ \frac{3}{2} & \frac{1}{2} & 1 & 0 & \frac{1}{2} & 0 & 0 & 50 \\ -\frac{7}{2} & \frac{1}{2} & 0 & 0 & -\frac{3}{2} & 1 & 0 & 50 \\ \hline 2 & 0 & 0 & 0 & 1 & 0 & 1 & 100 \end{array}\right] \end{array}$$

17. **(a)** basic: x_3, x_5; nonbasic: x_1, x_2, x_4 **(b)** $x_1 = 0, x_2 = 0, x_3 = 16, x_4 = 0, x_5 = 29, z = 11$ **(c)** not maximum
19. **(a)** basic: x_1, x_2, x_5; nonbasic: x_3, x_4, x_6 **(b)** $x_1 = 6, x_2 = 13, x_3 = 0, x_4 = 0, x_5 = 21, x_6 = 0, z = 18$ **(c)** maximum
21. Maximum is 30 when $x_1 = 0, x_2 = 10, x_3 = 0, x_4 = 0, x_5 = 16$.
23. Maximum is 8 when $x_1 = 4, x_2 = 0, x_3 = 8, x_4 = 2, x_5 = 0$.
25. no maximum **27.** no maximum **29.** Maximum is 34 when $x_1 = 17, x_2 = 0, x_3 = 0, x_4 = 0, x_5 = 14$ or when $x_1 = 0, x_2 = 17, x_3 = 0, x_4 = 0, x_5 = 14$. **31.** Maximum is 26,000 when $x_1 = 60, x_2 = 40, x_3 = 0, x_4 = 0, x_5 = 80, x_6 = 0$. **33.** Maximum is 64 when $x_1 = 28, x_2 = 16, x_3 = 0, x_4 = 0, x_5 = 28, x_6 = 0$.
35. Maximum is 250 when $x_1 = 0, x_2 = 0, x_3 = 0, x_4 = 50, x_5 = 0, x_6 = 50$. **37.** **(a)** Maximum is 24 when $x_1 = 12, x_2 = 0, x_3 = 0, x_4 = 0, x_5 = 6$. **(b)** Maximum is 24 when $x_1 = 0, x_2 = 12, x_3 = 0, x_4 = 0, x_5 = 18$. **(c)** The unique maximum value of z is 24, but this occurs at two different basic feasible solutions.

Section 7.5 (Page 384)

1.
$$\begin{array}{c} \begin{array}{ccccc} x_1 & x_2 & x_3 & x_4 & x_5 \end{array} \\ \left[\begin{array}{ccccc|c} 2 & 1 & 1 & 0 & 0 & 90 \\ 1 & 2 & 0 & 1 & 0 & 80 \\ 1 & 1 & 0 & 0 & 1 & 50 \\ \hline -12 & -10 & 0 & 0 & 0 & 0 \end{array}\right] \end{array}$$
where x_1 is the number of Siamese cats and x_2 is the number of Persian cats

3.
$$\begin{array}{c} \begin{array}{ccccccc} x_1 & x_2 & x_3 & x_4 & x_5 & x_6 & x_7 \end{array} \\ \left[\begin{array}{ccccccc|c} 0 & 0 & .375 & .625 & 1 & 0 & 0 & 500 \\ 0 & .75 & .5 & .375 & 0 & 1 & 0 & 600 \\ 1 & .25 & .125 & 0 & 0 & 0 & 1 & 300 \\ \hline -90 & -70 & -60 & -50 & 0 & 0 & 0 & 0 \end{array}\right] \end{array}$$
where x_1 is the number of kg of P, x_2 is the number of kg of Q, x_3 is the number of kg of R, and x_4 is the number of kg of S
5. **(a)** Make no 1-speed or 3-speed bicycles; make 2295 10-speed bicycles; maximum profit is $55,080. **(b)** 5280 units of aluminum are unused; all the steel is used. **7.** **(a)** 12 min to the senator, 9 min to the congresswoman, and 6 min to the governor for a maximum of 1,320,000 viewers **(b)** $x_4 = 0$ means that all of the 27 available min are used; $x_5 = 0$ means that the senator had *exactly* twice as much time as the governor; $x_6 = 0$ means that the senator and governor had a total time *exactly* twice the time of the congresswoman.
9. 4 radio ads, 6 TV ads, and no newspaper ads for a maximum exposure of 64,800 people **11.** **(a)** 22 fund-raising parties, no mailings, and 3 dinner parties for a maximum of $6,200,000
13. **(a)** (3) **(b)** (4) **(c)** (3) **15.** The breeder should raise 40 Siamese and 10 Persian cats for a maximum gross income of $580.
17. 163.6 kg of food P, none of Q, 1090.9 kg of R, 145.5 kg of S; maximum is 87,454.5.

Section 7.6 (Page 394)

1. $\begin{bmatrix} 3 & 1 & 0 \\ -4 & 10 & 3 \\ 5 & 7 & 6 \end{bmatrix}$ **3.** $\begin{bmatrix} 3 & 4 \\ 0 & 17 \\ 14 & 8 \\ -5 & -6 \\ 3 & 1 \end{bmatrix}$

5. Maximize $z = 4x_1 + 6x_2$
subject to: $3x_1 - x_2 \le 3$
$x_1 + 2x_2 \le 5$
$x_1 \ge 0, x_2 \ge 0.$

7. Maximize $z = 18x_1 + 15x_2 + 20x_3$
subject to: $x_1 + 4x_2 + 5x_3 \le 2$
$7x_1 + x_2 + 3x_3 \le 8$
$x_1 \ge 0, x_2 \ge 0, x_3 \ge 0.$

9. Maximize $z = 8x_1 + 12x_2$
subject to: $3x_1 + x_2 \le 1$
$4x_1 + 5x_2 \le 2$
$6x_1 + 2x_2 \le 6$
$x_1 \ge 0, x_2 \ge 0.$

11. Maximize $z = 5x_1 + 4x_2 + 15x_3$
subject to: $x_1 + x_2 + 2x_3 \le 8$
$x_1 + x_2 + x_3 \le 9$
$x_1 + 3x_3 \le 3$
$x_1 \ge 0, x_3 \ge 0, x_3 \ge 0.$

13. $y_1 = 0, y_2 = 100, y_3 = 0$; minimum is 100. **15.** $y_1 = 0, y_2 = 12, y_3 = 0$; minimum is 12. **17.** $y_1 = 0, y_2 = 0, y_3 = 2$; minimum is 4. **19.** $y_1 = 10, y_2 = 10, y_3 = 0$; minimum is 320.
21. $y_1 = 0, y_2 = 0, y_3 = 4$; minimum is 4. **23.** 4 servings of A, 2 servings of B for a minimum cost of $1.76 **25.** 28 units of regular beer and 14 units of light beer for a minimum cost of $1,680,000 **27.** (a)
29. **(a)** Minimize $w = 200y_1 + 600y_2 + 90y_3$
subject to: $y_1 + 4y_2 \ge 1$
$2y_1 + 3y_2 + y_3 \ge 1.5$
$y_1 \ge 0, y_2 \ge 0, y_3 \ge 0.$
(b) $y_1 = .6, y_2 = .1, y_3 = 0, w = 180$ **(c)** $186 **(d)** $179

Section 7.7 (Page 406)

1. (a) Maximize $z = 5x_1 + 2x_2 - x_3$
subject to:

$$\begin{aligned} 2x_1 + 3x_2 + 5x_3 - x_4 \quad &= 8 \\ 4x_1 - x_2 + 3x_3 + \quad x_5 &= 7 \end{aligned}$$

$x_1 \geq 0, x_2 \geq 0, x_3 \geq 0, x_4 \geq 0, x_5 \geq 0.$

(b)

$$\begin{array}{ccccc|c} x_1 & x_2 & x_3 & x_4 & x_5 & \\ \hline 2 & 3 & 5 & -1 & 0 & 8 \\ 4 & -1 & 3 & 0 & 1 & 7 \\ \hline -5 & -2 & 1 & 0 & 0 & 0 \end{array}$$

3. (a) Maximize $z = 2x_1 - 3x_2 + 4x_3$
subject to:

$$\begin{aligned} x_1 + x_2 + x_3 + x_4 \quad &= 100 \\ x_1 + x_2 + x_3 \quad - x_5 &= 75 \\ x_1 + x_2 \quad - x_6 &= 27 \end{aligned}$$

$x_1 \geq 0, x_2 \geq 0, x_3 \geq 0, x_4 \geq 0, x_5 \geq 0, x_6 \geq 0.$

(b)

$$\begin{array}{cccccc|c} x_1 & x_2 & x_3 & x_4 & x_5 & x_6 & \\ \hline 1 & 1 & 1 & 1 & 0 & 0 & 100 \\ 1 & 1 & 1 & 0 & -1 & 0 & 75 \\ 1 & 1 & 0 & 0 & 0 & -1 & 27 \\ \hline -2 & 3 & -4 & 0 & 0 & 0 & 0 \end{array}$$

5. Maximize $z = -2y_1 - 5y_2 + 3y_3$
subject to:

$$\begin{aligned} y_1 + 2y_2 + 3y_3 &\geq 115 \\ 2y_1 + y_2 + y_3 &\leq 200 \\ y_1 \quad + y_3 &\geq 50 \end{aligned}$$

$y_1 \geq 0, y_2 \geq 0, y_3 \geq 0.$

$$\begin{array}{cccccc|c} y_1 & y_2 & y_3 & y_4 & y_5 & y_6 & \\ \hline 1 & 2 & 3 & -1 & 0 & 0 & 115 \\ 2 & 1 & 1 & 0 & 1 & 0 & 200 \\ 1 & 0 & 1 & 0 & 0 & -1 & 50 \\ \hline 2 & 5 & -3 & 0 & 0 & 0 & 0 \end{array}$$

7. Maximize $z = -y_1 + 4y_2 - 2y_3$
subject to:

$$\begin{aligned} 7y_1 - 6y_2 + 8y_3 &\geq 18 \\ 4y_1 + 5y_2 + 10y_3 &\geq 20 \end{aligned}$$

$y_1 \geq 0, y_2 \geq 0, y_3 \geq 0$

$$\begin{array}{ccccc|c} y_1 & y_2 & y_3 & y_4 & y_5 & \\ \hline 7 & -6 & 8 & -1 & 0 & 18 \\ 4 & 5 & 10 & 0 & -1 & 20 \\ \hline 1 & -4 & 2 & 0 & 0 & 0 \end{array}$$

9. Maximum is 480 when $x_1 = 40, x_2 = 0$. **11.** Maximum is 114 when $x_1 = 38, x_2 = 0, x_3 = 0$. **13.** Maximum is 90 when $x_1 = 12, x_2 = 3$ or when $x_1 = 0, x_2 = 9$. **15.** Minimum is 40 when $y_1 = 0, y_2 = 20$. **17.** Maximum is $133\frac{1}{3}$ when $x_1 = 33\frac{1}{3}, x_2 = 16\frac{2}{3}$.
19. Minimum is 512 when $y_1 = 6, y_2 = 8$.

21.

$$\begin{array}{ccccccc|c} y_1 & y_2 & y_3 & y_4 & y_5 & y_6 & y_7 & \\ \hline 1 & 1 & 1 & -1 & 0 & 0 & 0 & 10 \\ 1 & 1 & 1 & 0 & 1 & 0 & 0 & 15 \\ 1 & -\frac{1}{4} & 0 & 0 & 0 & -1 & 0 & 0 \\ -1 & 0 & 1 & 0 & 0 & 0 & -1 & 0 \\ \hline .30 & .09 & .27 & 0 & 0 & 0 & 0 & 0 \end{array}$$

23.

$$\begin{array}{cccccccccc|c} y_1 & y_2 & y_3 & y_4 & y_5 & y_6 & y_7 & y_8 & y_9 & y_{10} & \\ \hline 1 & 1 & 0 & 0 & -1 & 0 & 0 & 0 & 0 & 0 & 32 \\ 1 & 1 & 0 & 0 & 0 & 1 & 0 & 0 & 0 & 0 & 32 \\ 0 & 0 & 1 & 1 & 0 & 0 & -1 & 0 & 0 & 0 & 20 \\ 0 & 0 & 1 & 1 & 0 & 0 & 0 & 1 & 0 & 0 & 20 \\ 1 & 0 & 1 & 0 & 0 & 0 & 0 & 0 & 1 & 0 & 25 \\ 0 & 1 & 0 & 1 & 0 & 0 & 0 & 0 & 0 & 1 & 30 \\ \hline 14 & 12 & 22 & 10 & 0 & 0 & 0 & 0 & 0 & 0 & 0 \end{array}$$

25. Ship 200 barrels of oil from supplier S_1 to distributor D_1; ship 2800 barrels of oil from supplier S_2 to distributor D_1; ship 2800 barrels of oil from supplier S_1 to distributor D_2; ship 2200 barrels of oil from supplier S_2 to distributor D_2. Minimum cost is $180,400.

27. Use 1000 lb of bluegrass, 2400 lb of rye, and 1600 lb of Bermuda for a minimum cost of $560.

29. Allot $3,000,000 in commercial loans and $22,000,000 in home loans for a maximum return of $2,940,000.

31. Make 32 units of regular beer and 10 units of light beer for a minimum cost of $1,632,000.

33. $1\frac{2}{3}$ ounces of ingredient I, $6\frac{2}{3}$ ounces of ingredient II, and $1\frac{2}{3}$ ounces of ingredient III produce a minimum cost of $1.55 per barrel.

35. 22 from W_1 to D_1, 10 from W_2 to D_1, none from W_1 to D_2, and 20 from W_2 to D_2 for a minimum cost of $628.

Chapter 7 Review (Page 410)

1.

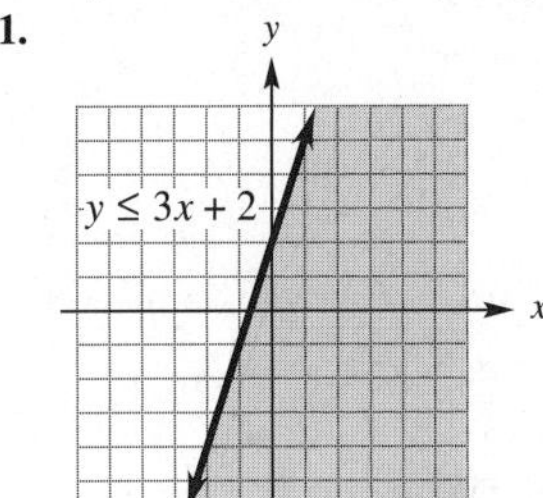

3.

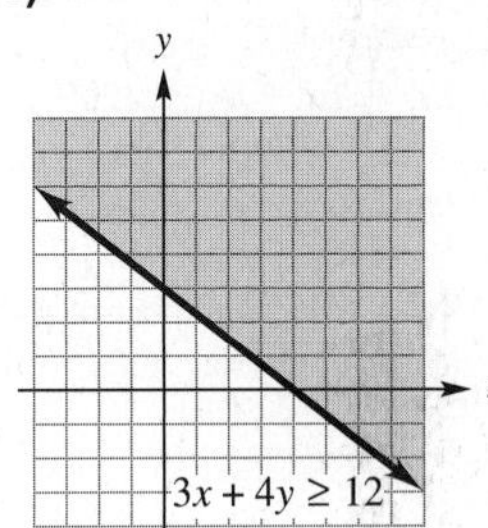

5.

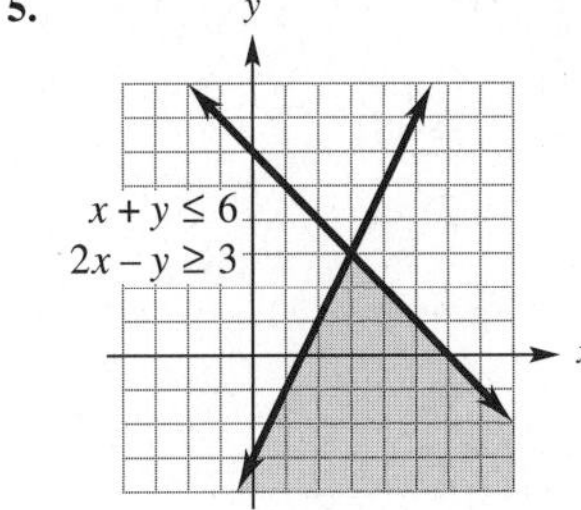

7.

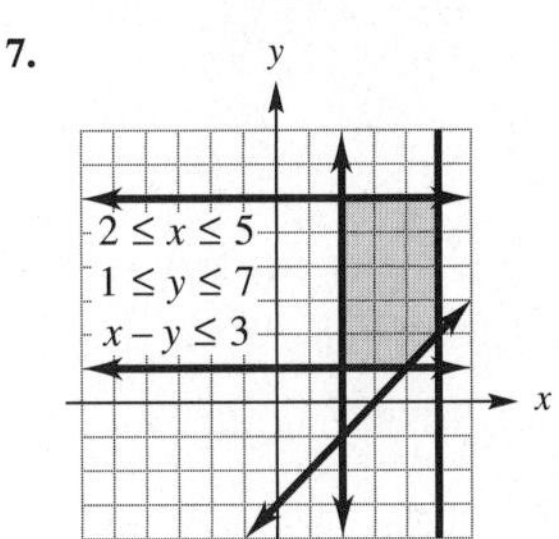

9. Let x represent the number of batches of cakes and let y represent the number of batches of cookies. Then

$$2x + \left(\frac{3}{2}\right)y \le 15$$
$$3x + \left(\frac{2}{3}\right)y \le 13$$
$$x \ge 0$$
$$y \ge 0.$$

(0, 10) $2x + \frac{3y}{2} \le 15$ $3x + \frac{2y}{3} \le 13$ (3, 6) $x \ge 0$ $y \ge 0$ $\left(\frac{13}{3}, 0\right)$

11. minimum of 8 at (2, 1); maximum of 40 at (6, 7) **13.** Maximum is 15 when $x = 5, y = 0$. **15.** Minimum is 60 when $x = 0, y = 20$. **17.** Make 3 batches of cakes and 6 batches of cookies for a maximum profit of $210. **19. (a)** Let x_1 = number of item A; x_2 = number of item B; x_3 = number of item C.
(b) $z = 4x_1 + 3x_2 + 3x_3$
(c) $2x_1 + 3x_2 + 6x_3 \le 1200$
$x_1 + 2x_2 + 2x_3 \le 800$
$2x_1 + 2x_2 + 4x_3 \le 500$
$x_1 \ge 0, x_2 \ge 0, x_3 \ge 0$
21. (a) Let x_1 = number of gallons of Fruity wine and x_2 = number of gallons of Crystal wine. **(b)** $z = 12x_1 + 15x_2$
(c) $2x_1 + x_2 \le 110$
$2x_1 + 3x_2 \le 125$
$2x_1 + x_2 \le 90$
$x_1 \ge 0, x_2 \ge 0$
23. when there are more than 2 variables
25. any standard minimization problem
27. (a)

$$\begin{aligned} 3x_1 + 5x_2 + x_3 \qquad\qquad &= 47 \\ x_1 + x_2 \qquad + x_4 \qquad &= 25 \\ 5x_1 + 2x_2 \qquad\quad + x_5 \quad &= 35 \\ 2x_1 + x_2 \qquad\qquad + x_6 &= 30 \end{aligned}$$

(b)

$$\begin{array}{cccccc|c} x_1 & x_2 & x_3 & x_4 & x_5 & x_6 & \\ \hline 3 & 5 & 1 & 0 & 0 & 0 & 47 \\ 1 & 1 & 0 & 1 & 0 & 0 & 25 \\ 5 & 2 & 0 & 0 & 1 & 0 & 35 \\ 2 & 1 & 0 & 0 & 0 & 1 & 30 \\ \hline -2 & -7 & 0 & 0 & 0 & 0 & 0 \end{array}$$

29. (a)

$$\begin{aligned} x_1 + x_2 + x_3 + x_4 \qquad\qquad &= 100 \\ 2x_1 + 3x_2 \qquad\quad + x_5 \quad &= 500 \\ x_1 \quad + 2x_3 \qquad\qquad + x_6 &= 350 \end{aligned}$$

(b)

$$\begin{array}{cccccc|c} x_1 & x_2 & x_3 & x_4 & x_5 & x_6 & \\ \hline 1 & 1 & 1 & 1 & 0 & 0 & 100 \\ 2 & 3 & 0 & 0 & 1 & 0 & 500 \\ 1 & 0 & 2 & 0 & 0 & 1 & 350 \\ \hline -4 & -6 & -3 & 0 & 0 & 0 & 0 \end{array}$$

31. Maximum is 80 when $x_1 = 16, x_2 = 0, x_3 = 0, x_4 = 12, x_5 = 0$. **33.** Maximum is 35 when $x_1 = 5, x_2 = 0, x_3 = 5, x_4 = 35, x_5 = 0, x_6 = 0$. **35.** Maximize $z = -18y_1 - 10y_2$ with the same constraints. **37.** Maximize $z = -6y_1 + 3y_2 - 4y_3$ with the same constraints. **39.** Minimum is 40 when $y_1 = 10$ and $y_2 = 0$. **41.** minimum of 1957 at (47, 68, 0, 92, 35, 0, 0) **43.** (9, 5, 8, 0, 0, 0); minimum is 62. **45.** Get 250 of A, none of B or C for a maximum profit of $1000. **47.** Make 17.5 gal of Crystal and 36.25 gal of Fruity for a maximum profit of $697.50. **49.** Produce 660 cases of corn, no beans, and 340 cases of carrots for a minimum cost of $15,100. **51.** Use 1,060,000 kg for whole tomatoes and 80,000 kg for sauce for a minimum cost of $4,500,000.

Case 6 (Page 414)

1. The answer in 100-gram units is 0.243037 units feta cheese, 2.35749 units lettuce, 0.3125 units salad dressing, and 1.08698 units tomato. Converting into kitchen units gives approximately 1/6 cup feta cheese, $4\frac{1}{4}$ cups lettuce, 1/8 cup salad dressing, and 7/8 of a tomato.

CHAPTER 8

Section 8.1 (Page 422)

1. false **3.** true **5.** true **7.** true **9.** false **13.** $\subseteq$ **15.** $\not\subseteq$ **17.** $\subseteq$ **19.** $\subseteq$ **21.** 8 **23.** 64 **25.** $\{x \mid x$ is an integer less than or equal to 0 or greater than or equal to 7.$\}$ **29.** $\cap$ **31.** $\cap$ **33.** $\cup$ **35.** $\cup$ **37.** $\{3, 5\}$ **39.** $\{7, 9\}$ **41.** $\emptyset$ **43.** U or $\{2, 3, 4, 5, 7, 9\}$ **45.** all students in this school not taking this course **47.** all students in this school taking accounting and zoology **49.** C and D, A and E, C and E, D and E **51.** C' is the set of all stocks with a zero or negative price change; $C' = \{\text{GenElec}\}$. **53.** $(A \cup B)'$ is the set of all stocks with a high price less than or equal to $100 and a last price either less than or equal to $20 or greater than or equal to $40; $(A \cup B)' = \{\text{GenElec, Lucent}\}$. **55.** $M \cap E$ is the set of all male, employed applicants. **57.** $M' \cup S'$ is the set of all female or married applicants. **59.** {Magazines, Yellow Pages, Internet} **61.** F = {The Disney Channel, HBO/Cinemax} **63.** H = {Showtime/The Movie Channel} **65.** $H \cup F$ = {The Disney Channel, HBO/Cinemax, Showtime/The Movie Channel} **67.** $\{s, d, c\}$ **69.** $\{g\}$

Section 8.2 (Page 431)

1.

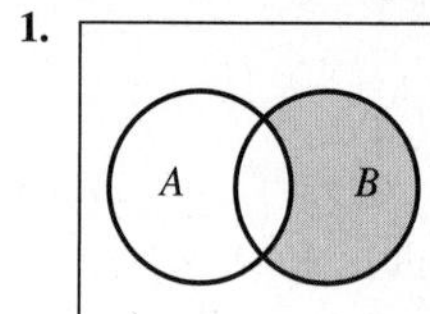

3.

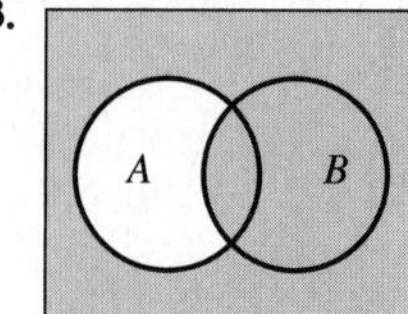

5.

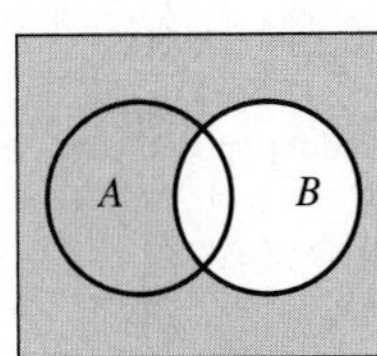

7. ∅ **9.** 8

11.

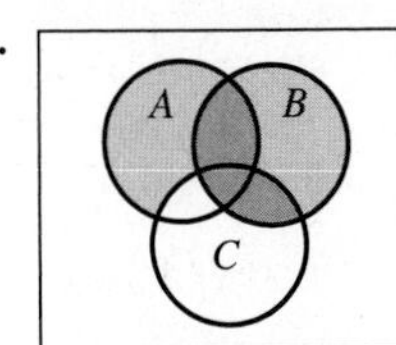

13.

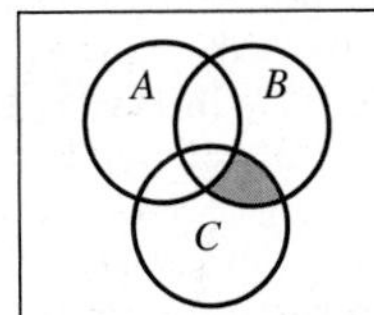

15.

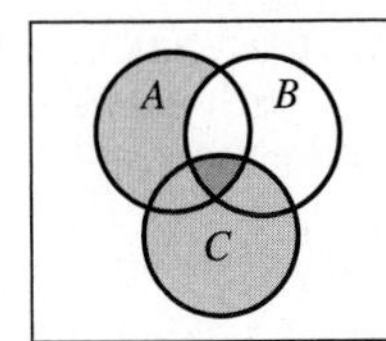

17. 17.3%

19. **(a)** 342 **(b)** 192 **(c)** 72 **(d)** 86 **21.** **(a)** 50 **(b)** 2 **(c)** 14 **23.** **(a)** 54 **(b)** 17 **(c)** 10 **(d)** 7 **(e)** 15 **(f)** 3 **(g)** 12 **(h)** 1 **25.** **(a)** 34 **(b)** 16 **(c)** 431 **(d)** 481 **27.** **(a)** 120.1 million **(b)** 88.5 million **(c)** 63.7 million **(d)** 39.5 million **(e)** 239.5 million **(f)** 36.0 million **31.** 9 **33.** 27

35.

A 20, 6, B 4, 13

37.

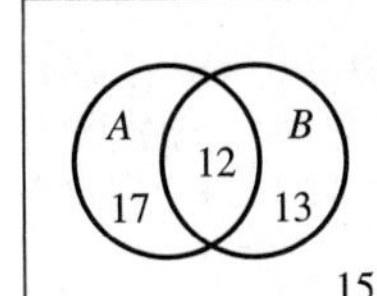

39.

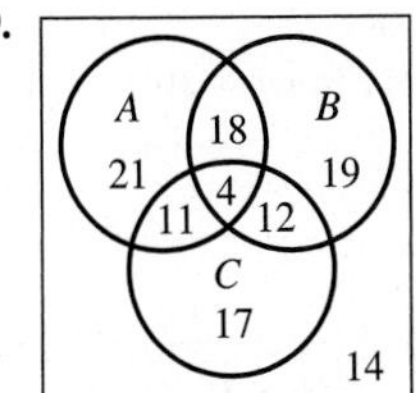

Section 8.3 (Page 439)

3. {January, February, March, . . ., December}
5. {0, 1, 2, . . . , 80} **7.** {go ahead, cancel}
11. $\{y_1, y_2, y_3, w_1, w_2, w_3, w_4, b_1, b_2, b_3, b_4, b_5, b_6, b_7, b_8\}$
(a) $\{y_1, y_2, y_3\}$ **(b)** $\{b_1, b_2, b_3, b_4, b_5, b_6, b_7, b_8\}$
(c) $\{w_1, w_2, w_3, w_4\}$ **(d)** ∅
13. {ccc, ccw, cwc, cww, wcc, wcw, wwc, www}
(a) {www} **(b)** {ccw, cwc, wcc} **(c)** {cww}
15. {(1, 1), (1, 2), (1, 3), (1, 4), (1, 5), (2, 1), (2, 2), (2, 3), (2, 4), (2, 5), (3, 1), (3, 2), (3, 3), (3, 4), (3, 5), (4, 1), (4, 2), (4, 3), (4, 4), (4, 5)}
(a) {(2, 1), (2, 2), (2, 3), (2, 4), (2, 5), (4, 1), (4, 2), (4, 3), (4, 4), (4, 5)} **(b)** {(1, 2), (1, 4), (2, 2), (2, 4), (3, 2), (3, 4), (4, 2), (4, 4)}
(c) {(1, 4), (2, 3), (3, 2), (4, 1)} **(d)** ∅

17. $\frac{1}{6}$ **19.** $\frac{1}{3}$ **21.** $\frac{1}{3}$ **23.** $\frac{1}{13}$ **25.** $\frac{1}{26}$ **27.** $\frac{1}{52}$

29. $\frac{1}{13}$ **31.** $\frac{1}{5}$ **33.** $\frac{9}{25}$ **35.** $\frac{21}{25}$ **37.** $\frac{9}{25}$
39. **(a)** .0644 **(b)** .1869 **(c)** .7331 **41.** **(a)** The person smokes or has a family history of heart disease. **(b)** The person does not smoke and has a family history of heart disease. **(c)** The person does not have a family history of heart disease or is not overweight.

Section 8.4 (Page 447)

3. no **5.** yes **7.** no **9.** **(a)** $\frac{5}{36}$ **(b)** $\frac{1}{9}$ **(c)** $\frac{1}{12}$ **(d)** 0

11. **(a)** $\frac{5}{18}$ **(b)** $\frac{5}{12}$ **(c)** $\frac{1}{3}$ **13.** $\frac{2}{9}$

15. **(a)** $\frac{2}{13}$ **(b)** $\frac{4}{13}$ **(c)** $\frac{7}{13}$

17. **(a)** $\frac{3}{13}$ **(b)** $\frac{4}{13}$ **(c)** $\frac{7}{13}$ **(d)** $\frac{4}{13}$ **(e)** $\frac{8}{13}$

19. **(a)** $\frac{1}{2}$ **(b)** $\frac{7}{10}$ **(c)** $\frac{1}{2}$ **21.** **(a)** $\frac{1}{10}$ **(b)** $\frac{9}{10}$ **(c)** $\frac{7}{20}$

23. **(a)** .39 **(b)** .84 **(c)** .74 **(d)** .26 **25.** 1 to 5 **27.** 2 to 1 **29.** **(a)** 1 to 4 **(b)** 8 to 7 **(c)** 4 to 11 **31.** 2 to 7 **35.** not empirical **37.** empirical **39.** empirical **41.** not empirical **45.** possible **47.** Not possible; the sum of the probabilities is greater than 1. **49.** Not possible; a probability cannot be negative.
In Exercises 51 and 53, theoretical answers are given; actual answers will vary.
51. **(a)** .2778 **(b)** .4167 **53.** **(a)** .15625 **(b)** .3125 **55.** **(a)** .51 **(b)** .67 **(c)** .39 **(d)** .12
57. **(a)** .89 **(b)** .38 **(c)** .18 **59.** **(a)** .56 **(b)** .20 **(c)** .31

61. **(a)** $\frac{3}{4}$ **(b)** $\frac{1}{4}$ **63.** **(a)** .90 **(b)** .23

65. **(a)** .06 **(b)** .44 **(c)** .74 **(d)** .18
67. **(a)**

	A	***B***	***C***	***D***
O	.052	.060	.039	.004
E	.306	.270	.213	.045
M	.003	.003	.003	0

(b) about .31 **(c)** about .04 **(d)** about .7 **(e)** about .88

Section 8.5 (Page 462)

1. $\frac{1}{3}$ **3.** 1 **5.** $\frac{1}{6}$ **7.** $\frac{4}{17}$ **9.** $\frac{11}{51}$ **11.** .012

13. .245 **19.** No; yes **21.** .43 **23.** **(a)** .815 **(b)** .79 **25.** **(a)** .46 **(b)** .29 **(c)** .23 **(d)** .61 **27.** .031 **29.** .834 **31.** The probability that a person in the group is employed in a managerial/professional capacity, given that the person is male, is about .505. **33.** The probability that a person in the group is employed in a technical/sales/administrative support capacity, given that the person is female, is .638. **35.** .527 **37.** .042

39. $\frac{6}{7}$, or about .857 **41.** yes **43.** .05 **45.** .25

47. .01254 **49.** .791 **51.** .236 **53.** .288 **55.** .0457 **57.** .999985 **59.** **(a)** 3 backups **61.** no

Section 8.6 (Page 469)

1. $\frac{1}{3}$ **3.** $\frac{2}{41} \approx .0488$ **5.** $\frac{21}{41} \approx .5122$ **7.** $\frac{8}{17} \approx .4706$

9. .0135 **11.** .342 **13.** .824 **15.** .006 **17.** .706

19. (c) **21.** .481 **23.** (a) .333 **25.** (a) .155 (b) .990 (c) about 59 **27.** (a) .53 (b) .1623 **29.** .524 **31.** .162 **33.** .086 **35.** .178

Chapter 8 Review (Page 474)

1. false **3.** false **5.** true **7.** true **9.** {New Year's Day, Martin Luther King's Birthday, Presidents' Day, Memorial Day, Independence Day, Labor Day, Columbus Day, Veterans' Day, Thanksgiving, Christmas} **11.** {1, 2, 3, 4} **13.** $\{B_1, B_2, B_3, B_6, B_{12}\}$ **15.** $\{A, C, E\}$ **17.** $\{A, B_3, B_6, B_{12}, C, D, E\}$, **19.** The set of all students in the class who are redheaded males **21.** The set of all students in the class who are male or younger than 21 **23.** The set of all students in the class who are not A students and do not have red hair

25.

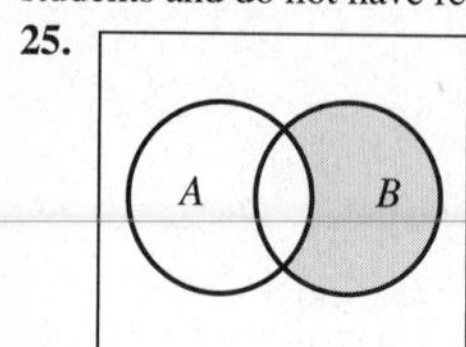

27.

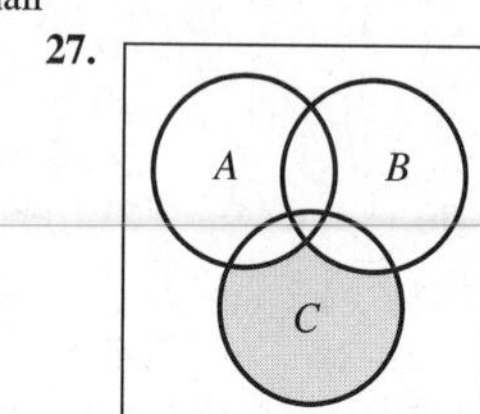

29. 5 **31.** {1, 2, 3, 4, 5, 6} **33.** {(red, 10), (red, 20) (red, 30), (blue, 10), (blue, 20), (blue, 30), (green, 10), (green, 20), (green, 30)} **35.** {(2, blue), (4, blue), (6, blue), (8, blue), (10, blue)} **37.** no

39. $E \cup F$ **45.** $\frac{3}{13}$ **47.** $\frac{1}{2}$ **49.** 1 **51.** 1 to 25

53. .86

55.

	N_2	T_2
N_1	N_1N_2	N_1T_2
T_1	T_1N_2	T_1T_2

57. $\frac{1}{2}$ **59.** $\frac{5}{36} \approx .139$ **61.** $\frac{5}{18} \approx .278$ **63.** $\frac{1}{3}$ **65.** .66

67. .71 **69.** (a) .7 (b) $\frac{2}{15} \approx .133$ **73.** $\frac{3}{22}$ **75.** $\frac{2}{3}$

77. (a) .271 (b) .379 (c) .625 (d) .342 (e) .690 (f) .514

Additional Probability Review (Page 476)

1. (a) .88 (b) .30 (c) .70 **3.** (a) .187 (b) .540 (c) .769 **5.** (a) 0 (b) 1/3 (c) 1 **7.** (a) .102 (b) .153 (c) .120 (d) .135 **9.** .091 **11.** no **13.** $\frac{1}{6}$ **15.** 82.4%

17. .187 **19.** .333 **21.** .286 **23.** $\frac{1}{5}$

25. (a) $\frac{2}{5}$ (b) $\frac{3}{5}$ (c) $\frac{1}{10}$ (d) $\frac{7}{10}$

Case 7 (Page 479)

1. .076 **3.** .051

CHAPTER 9

Section 9.1 (Page 492)

1. 12 **3.** 56 **5.** 8 **7.** 24 **9.** 792 **11.** 156 **13.** 6,375,600 **15.** 240,240 **17.** 352,716 **19.** 2,042,975 **21.** Undefined **23.** (a) 8 (b) 64 **25.** 840

27. $10^9 = 1{,}000{,}000{,}000$; yes **29.** $10^9 = 1{,}000{,}000{,}000$ **31.** (a) 25,200 (b) 15,120 **33.** (a) 160; 8,000,000 (b) Some numbers, such as 911, 800, and 900, are reserved for special purposes. **35.** 1600 **39.** 479,001,600 **41.** 1440 **43.** 1,256,640 **45.** 2730 **47.** (a) 220 (b) 220 **49.** (a) 0 (b) 792 (c) 658,008 (d) 652,080 (e) 79,092 **53.** (a) 9 (b) 6 (c) 3; yes, from both **55.** (a) 84 (b) 10 (c) 40 (d) 74 **57.** (a) 961 (b) 29,791 **59.** (a) 1,120,529,256 (b) 806,781,064,300 **61.** 81 **63.** 1 **65.** 55,440 **67.** (a) 7,059,052 (b) $3,529,526 (c) no **69.** 5.524×10^{26} **71.** (a) 840 (b) 180 (c) 420 **73.** (a) 362,880 (b) 6 (c) 1260

Section 9.2 (Page 501)

1. .424 **3.** $\frac{7}{9}$ **5.** $\frac{5}{12}$ **7.** .147 **9.** .020 **11.** .314

13. 1326 **15.** .851 **17.** .765 **19.** .941

23. $\dfrac{\binom{4}{2}\binom{4}{3}\binom{44}{8}}{\binom{52}{13}}$ **25.** (a) 8.9×10^{-10} (b) 1.2×10^{-12}

27. 3.3×10^{-16} **29.** (a) .218 (b) .272 (c) .728

31. (a) .003 (b) .005 (c) .057 (d) .252 **33.** $1 - \frac{_{365}P_{100}}{(365)^{100}}$

35. .350 **37.** There were 10 balls with 9 of them blue. **39.** Answers will vary. The theoretical answers are: (a) .0399 (b) .5191 (c) .0226

Section 9.3 (Page 508)

1. .067 **3.** .066 **5.** .934 **7.** .0000000005 **9.** .269

11. .875 **13.** $\frac{1}{32}$ **15.** $\frac{13}{16}$ **19.** .00655 **21.** .9932

23. .1285 **25.** .7941 **27.** .8159 **29.** .4913 **31.** .0025 **33.** (a) .1493 (b) .9116 **35.** (a) .0060 (b) .99998 **37.** .000079 **39.** .2162 **41.** (a) .9995 **43.** (a) 1 chance in 1024 (b) about 1 chance in 1.1×10^{12} (c) about 1 chance in 2.6×10^6

Section 9.4 (Page 518)

1.

Number of Heads	0	1	2	3	4
Probability	1/16	1/4	3/8	1/4	1/16

3.

Number of Aces	0	1	2	3
Probability	.7826	.2042	.0130	.0002

5.

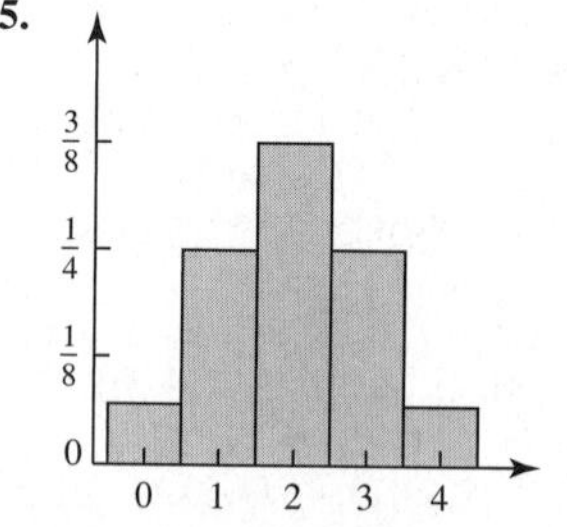

7.

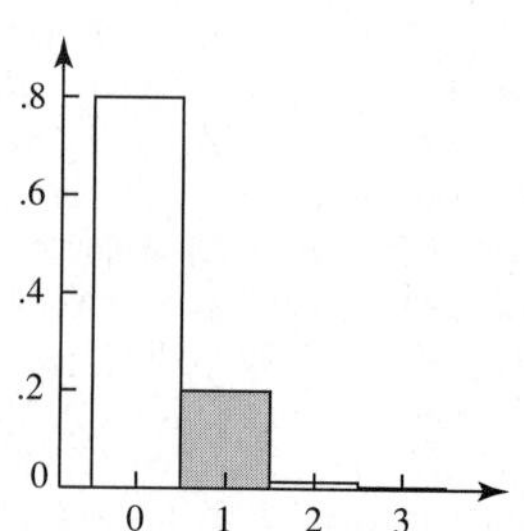

9. 3.6 **11.** 14.64 **13.** 2.7 **15.** 18 **17.** Yes, the probability of a match is still $\frac{1}{2}$.

19. (a)

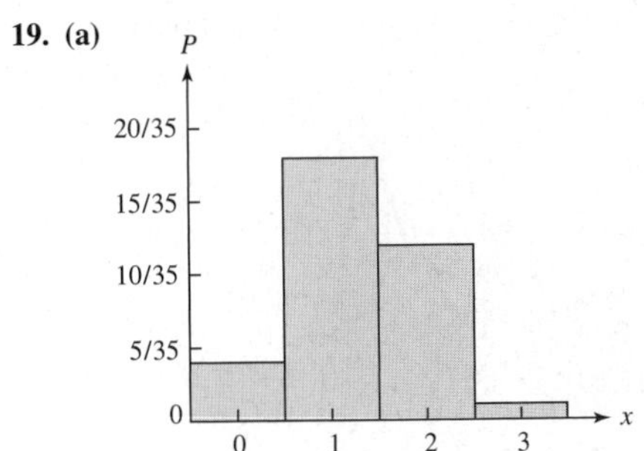

21. (a) $5/3 \approx 1.667$ **(b)** $4/3 \approx 1.333$ **23.** 1 **25.** No, the expected value is about $-21¢$. **29.** $-32¢$; no **31.** $-66¢$ **33.** $-2.7¢$ **35.** $-20¢$ **37.** $-87.8¢$ **39.** 3.51

41.

Account	*Expected Value*	*Total*	*Class*
3	\$2000	\$22,000	C
4	\$1000	\$51,000	B
5	\$25,000	\$30,000	C
6	\$60,000	\$60,000	A
7	\$16,000	\$46,000	B

43. (a) amoxicillin \$68.51; cefaclor \$72.84 **(b)** amoxicillin
45. (a) \$50,000 **(b)** \$65,000
47. (a)

Number	0	1	2	3	4	5
Probability	.0000	.0000	.0000	.0003	.0024	.0141

Number	6	7	8	9	10
Probability	.0573	.1600	.2929	.3178	.1552

(b) .0027 **(d)** 8.3

Section 9.5 (Page 531)

1. yes **3.** yes **5.** no **7.** no
9. yes **11.** no

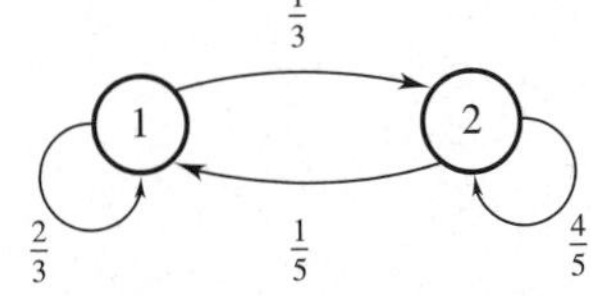

13. not a transition diagram

15.
$$\begin{array}{c} \\ A \\ B \\ C \end{array}\begin{array}{c} \begin{array}{ccc} A & B & C \end{array} \\ \begin{bmatrix} .6 & .2 & .2 \\ .9 & .02 & .08 \\ .4 & 0 & .6 \end{bmatrix} \end{array}$$

17. yes **19.** yes **21.** no

23. $\begin{bmatrix} \frac{1}{3} & \frac{2}{3} \end{bmatrix}$ **25.** $\begin{bmatrix} \frac{3}{11} & \frac{8}{11} \end{bmatrix}$ **27.** [.4633 .1683 .3684]

29. $\begin{bmatrix} \frac{1}{4} & \frac{1}{4} & \frac{1}{2} \end{bmatrix}$

31. The first power is the given transition matrix;

$$\begin{bmatrix} .23 & .21 & .24 & .17 & .15 \\ .26 & .18 & .26 & .16 & .14 \\ .23 & .18 & .24 & .19 & .16 \\ .19 & .19 & .27 & .18 & .17 \\ .17 & .20 & .26 & .19 & .18 \end{bmatrix}; \begin{bmatrix} .226 & .192 & .249 & .177 & .156 \\ .222 & .196 & .252 & .174 & .156 \\ .219 & .189 & .256 & .177 & .159 \\ .213 & .192 & .252 & .181 & .162 \\ .213 & .189 & .252 & .183 & .163 \end{bmatrix};$$

$$\begin{bmatrix} .2205 & .1916 & .2523 & .1774 & .1582 \\ .2206 & .1922 & .2512 & .1778 & .1582 \\ .2182 & .1920 & .2525 & .1781 & .1592 \\ .2183 & .1909 & .2526 & .1787 & .1595 \\ .2176 & .1906 & .2533 & .1787 & .1598 \end{bmatrix};$$

$$\begin{bmatrix} .21932 & .19167 & .25227 & .17795 & .15879 \\ .21956 & .19152 & .25226 & .17794 & .15872 \\ .21905 & .19152 & .25227 & .17818 & .15898 \\ .21880 & .19144 & .25251 & .17817 & .15908 \\ .2185 & .19148 & .25253 & .17824 & .15918 \end{bmatrix}; .17794$$

33. 24.02% will be 30-year fixed loans, 69.72% will be 15-year fixed loans and 6.26% will be adjustable loans. **35.** $\frac{17}{18}$

37.
$$\begin{array}{c} \\ R \\ P \\ W \end{array}\begin{array}{c} \begin{array}{ccc} R & P & W \end{array} \\ \begin{bmatrix} \frac{1}{2} & \frac{1}{2} & 0 \\ \frac{1}{4} & \frac{1}{2} & \frac{1}{4} \\ 0 & \frac{1}{2} & \frac{1}{2} \end{bmatrix} \end{array}; \begin{bmatrix} \frac{1}{4} & \frac{1}{2} & \frac{1}{4} \end{bmatrix}$$

39. homeowners: .769; renters: .231; homeless: 0

41. (a) 42,500 in G_0; 5000 in G_1; 2500 in G_2 **(b)** 37,125 in G_0; 8750 in G_1; 4125 in G_2 **(c)** 33,281 in G_0; 11,513 in G_1; 5206 in G_2 **(d)** $\begin{bmatrix} \frac{28}{59} & \frac{22}{59} & \frac{9}{59} \end{bmatrix}$ or [.475 .373 .152]; in the long run, the probabilities of no accidents, one accident, and more than one accident are .475, .373, and .152, respectively.

43. (a) $\begin{bmatrix} .348 & .379 & .273 \\ .320 & .378 & .302 \\ .316 & .360 & .323 \end{bmatrix}$ **(b)** .348 **(c)** .320

45. (b) $\begin{bmatrix} \frac{5}{12} & \frac{13}{36} & \frac{2}{9} \\ \frac{5}{12} & \frac{13}{36} & \frac{2}{9} \\ \frac{1}{4} & \frac{5}{12} & \frac{1}{3} \end{bmatrix}$ **(c)** $\frac{2}{9}$

47. [0 0 .102273 .897727]

Section 9.6 (Page 536)

1. (a) choose the coast **(b)** choose the highway **(c)** choose the highway; \$38,000 **(d)** choose the coast **3. (a)** make no repairs **(b)** make repairs **(c)** make no repairs; \$10,750

5. (a)
$$\begin{array}{c} \\ \text{Overhaul} \\ \text{Don't} \end{array}\begin{array}{c} \begin{array}{cc} \text{Fails} & \text{Doesn't} \end{array} \\ \begin{bmatrix} -8600 & -2600 \\ -6000 & 0 \end{bmatrix} \end{array}$$

(b) do not overhaul before shipping **7. (a)** no campaign **(b)** campaign for all **(c)** campaign for all **9.** emphasize the environment; 14.25

Chapter 9 Review (Page 539)

1. 120 **3.** 220 **5.** 120 **7.** **(a)** 72 **(b)** 72 **11.** .245

13. .362 **15.** .091 **17.** .455 **19.** **(a)** $\binom{n}{0}$, or 1; $\binom{n}{1}$, or n; $\binom{n}{2}$; $\binom{n}{n}$, or 1 **(b)** $\binom{n}{0} + \binom{n}{1} + \binom{n}{2} + \cdots + \binom{n}{n}$ **21.** .99996

23. .00004 **25.** 1.0000 **27.** $8500

29. **(a)**

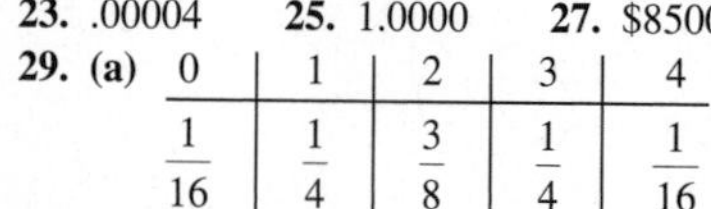

0	1	2	3	4
$\frac{1}{16}$	$\frac{1}{4}$	$\frac{3}{8}$	$\frac{1}{4}$	$\frac{1}{16}$

(b)

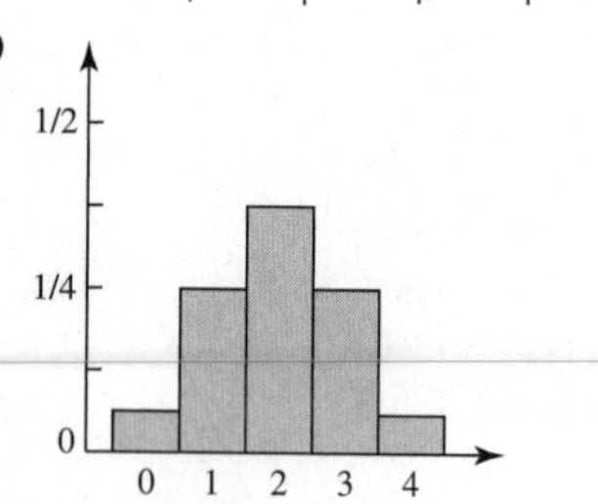

(c) 2

31. **(a)**

Number	0	1	2
Probability	10/28	15/28	3/28

(b)

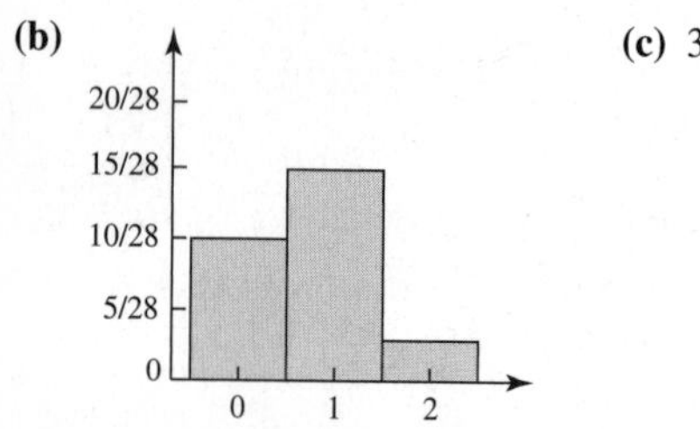

(c) 3/4

33. $1.29 **35.** −28¢ **37.** **(a)** $\frac{3}{13}$, or .23 **(b)** $\frac{3}{4}$, or .75

39. no **41.** yes **43.** **(a)** [.54 .46] **(b)** [.6464 .3536]

(c) $\frac{2}{3}$ **45.** **(a)** oppose **(b)** oppose **(c)** oppose; 2700

(d) oppose; 1400 **47.** (c)

Case 8 (Page 543)

1. **(a)** $69.01 **(b)** $79.64 **(c)** $58.96 **(d)** $82.54 **(e)** $62.88 **(f)** $64.00 **3.** The events of needing parts 1, 2, and 3 are not the only events in the sample space.

CHAPTER 10

Section 10.1 (Page 557)

1. **(a)–(b)**

Interval	*Frequency*
0–24	4
25–49	3
50–74	6
75–99	3
100–124	5
125–149	9

(c)–(d)

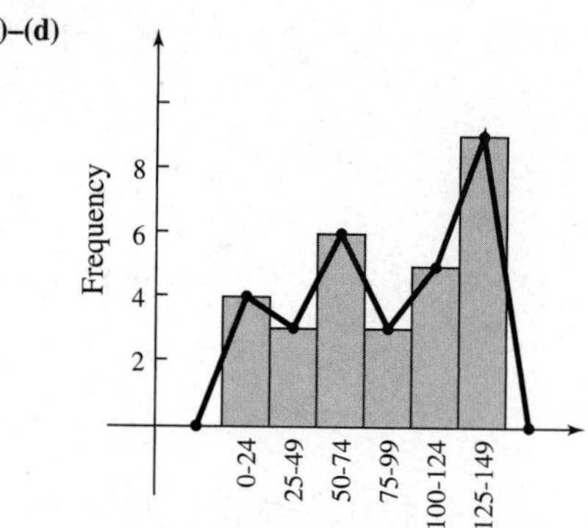

3. **(a)–(b)**

Interval	*Frequency*
70–74	2
75–79	1
80–84	3
85–89	2
90–94	6
95–99	5
100–104	6
105–109	4
110–114	2

(c)–(d)

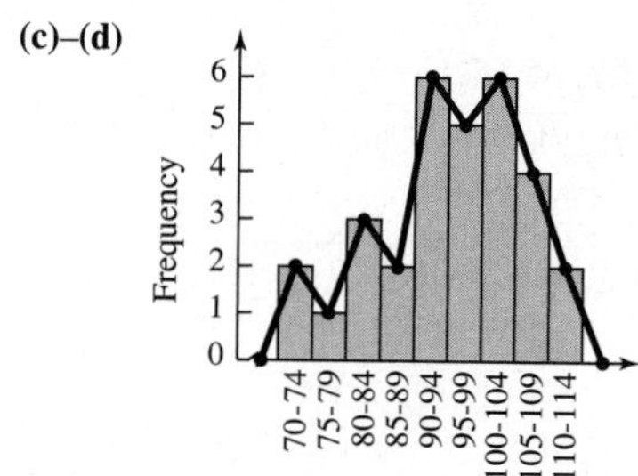

5.

Stem	*Leaves*
07	249
08	23448
09	02224467799
10	0122446678
11	01

9. 27,955 **11.** 7.7 **13.** 6.7 **15.** 17.2 **17.** 51 **19.** 130 **21.** 29.1 **23.** 9 **25.** 68 and 74 **27.** 6.3 **31.** 94.9; 90–94 and 100–104 **35.** **(a)** $90,599,700 **(b)** $75,694,500 **(c)** There is one very large salary that affects the value of the mean. **(d)** $49,326,000 **37.** 7.38; 7; There are three modes: 7, 5, and 4. **39.** 55.5°F; 50.5°F **41.** $3.42; $3.32 **43.** $29,952.28 **45.** **(a)** 14 **(b)** 7 **(c)** 30–39 and 40–49

Section 10.2 (Page 567)

3. 6; 2.2 **5.** 12; 4.7 **7.** 53; 21.8 **9.** 46; 16.1 **11.** 45.2 **13.** $\bar{x} = 5.0876$; $s = .1087$ **15.** at least $\frac{3}{4}$ **17.** at least $\frac{5}{9}$ **19.** at least 93.75% **21.** no more than 11.1%

23. (a) Britelite: $\overline{x} = 26.2, s = 4.4$; Brand X: $\overline{x} = 25.5, s = 7.2$ (b) Britelite (c) Britelite **25.** (a) 7.635; 1995 (b) 1.122 (c) 7 (d) 10 **27.** (a) 14.8; 3.8 (b) 10 **29.** (a) 127.71 days; 30.16 days (b) all of them **31.** (a) $\frac{1}{3}$; 2; $-\frac{1}{3}$; 0; $\frac{5}{3}, \frac{7}{3}$; 1; $\frac{4}{3}, \frac{7}{3}, \frac{2}{3}$ (b) 2.1; 2.6; 1.5; 2.6; 2.5; .6; 1.0; 2.1; .6; 1.2 (c) 1.13 (d) 1.68 (e) 4.41; −2.15; the process is out of control. **33.** mean = 80.17; standard deviation = 12.2

Section 10.3 (Page 581)

1. The mean **5.** 45.99% **7.** 16.64% **9.** 7.7% **11.** 47.35% **13.** 92.42% **15.** 1.64 or 1.65 **17.** −1.04 **19.** .5; .5 **21.** .889; .997 **23.** .3821 **25.** .5762 **27.** .8887 **29.** .0021 **31.** 5000 **33.** 642 **35.** 9918 **37.** 372; 628 **39.** 35.2 mph **41.** 24.17% **45.** 625 units **47.** 189 units **49.** 15.87% **51.** .5000 **53.** .9050 **55.** .8413 **57.** .5987

Section 10.4 (Page 588)

1. The number of trials and the probability of success on each trial **3.** (a) .1964 (b) .1974 **5.** (a) .0106 (b) .0122 **7.** .0240 **9.** .9032 **11.** .0956 **13.** .6443 **15.** .1841 **17.** .0222 **19.** .2578 **21.** .0301 **23.** (a) .0472 (b) .9875 (c) .8814 **25.** (a) .0000 (b) .0018 **27.** .5977 **29.** .9898 **31.** (a) 1.2139×10^{-7} (b) essentially 0 (c) .7939

Chapter 10 Review (Page 591)

3. (a)

Interval	*Frequency*
450–474	5
475–499	6
500–524	5
525–549	2
550–574	2

(b)–(c)

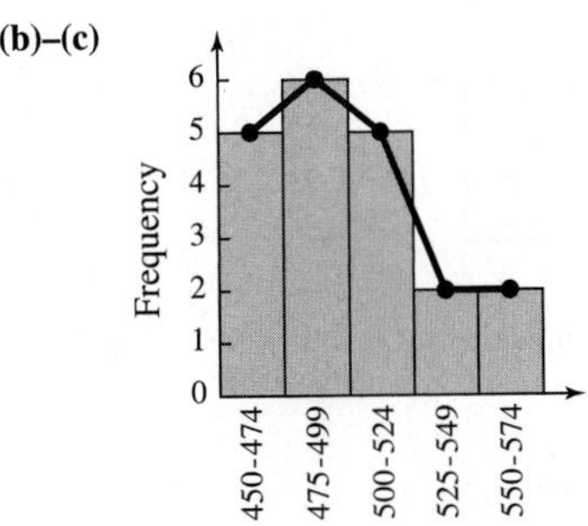

5. 73 **7.** 34.9 **11.** 38; 36 and 38 **13.** 55–59 **17.** 67; 23.9 **19.** 6.2 **23.** .6591 **25.** .0606 **29.** (a) .0317 (b) 1.482; 1.056 **31.** (a) Diet A: (mean = 2.7) (b) Diet B: (s = .95) **33.** (a) $\overline{x} = 19.45\%$; $s = 2.39\%$ (b) 1.1; 1.1 **35.** 15.87% **37.** (a) .0305 (b) .9938 (c) 1.000 (d) .1049; the binomial distribution gives .1326. **39.** (a) .1190 (b) .3936 (d) .9987 (e) .1336

Case 9 (Page 595)

1. $z \approx 29.1$ **3.** (a) 174 (b) 6.03 (c) −12.3 (d) The probability is less than .004. (Actually, it is less than 10^{-30}, which is the smallest probability in the table)

CHAPTER 11

Section 11.1 (Page 606)

1. (a) 3 (b) 0 **3.** (a) does not exist (b) $-\frac{1}{2}$ **5.** (a) 2 (b) does not exist **7.** (a) 1 (b) −1 **11.** 1 **13.** 2 **15.** 0 **17.** 8 **19.** does not exist **21.** 8 **23.** 2 **25.** 4 **27.** $\frac{3}{2}$

29. (a)

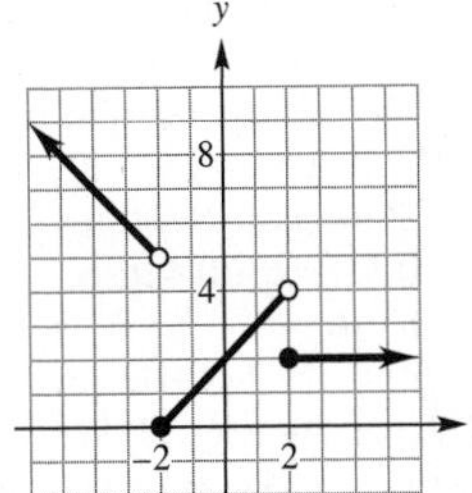

(b) does not exist (c) 3 (d) does not exist **31.** 41 **33.** $\frac{9}{7}$ **35.** 6 **37.** −5 **39.** $\frac{1}{2}$ **41.** does not exist **43.** $\sqrt{5}$ **45.** does not exit **47.** $-\frac{1}{9}$ **49.** $\frac{1}{10}$ **51.** $\frac{1}{2\sqrt{5}}$ or $\frac{\sqrt{5}}{10}$

53. (a) approx. 12.86 (b) approx. 58.24 (c) approx. 58.24 **55.** (a) 25 (b) 5.2 (c) 7 **57.** (a) 1.5 (b) 1.2 (c) 0 (d) 3

59. (a) $P(x) = \begin{cases} .99 & \text{if } 0 \le x \le 20 \\ 1.06 & \text{if } 20 < x \le 21 \\ 1.13 & \text{if } 21 < x \le 22 \\ 1.20 & \text{if } 22 < x \le 23 \end{cases}$

(b)

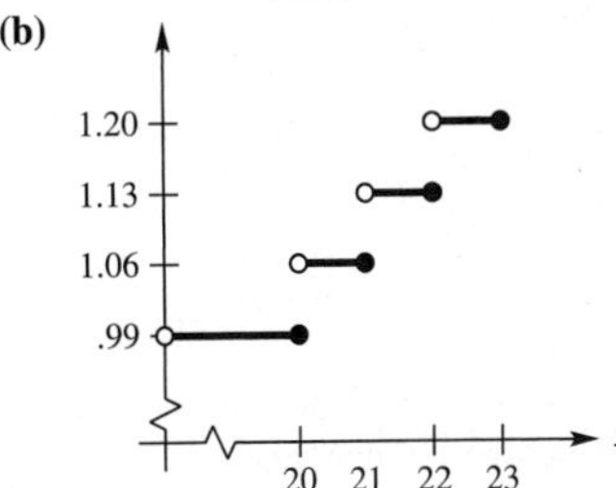

(c) .99 (d) does not exist (e) 1.20

Section 11.2 (Page 617)

1. 7 **3.** 8 **5.** $\frac{1}{3}$ **7.** $-\frac{1}{3}$

9. (a) 1; from catalog distributions of 10,000 to 20,000, sales will have an average increase of $1000 for each additional 1000 catalogs distributed. (b) $\frac{3}{5}$; from catalog distributions of 20,000 to 30,000, sales will have an average increase of $600 for each additional 1000 catalogs distributed. (c) $\frac{2}{5}$; from catalog distributions of 30,000 to 40,000, sales will have an average increase of $400 for each additional 1000 catalogs distributed. (d) As more catalogs are distributed, sales increase at a smaller and smaller rate.

11. **(a)** approximately $-\$6.75$ billion **(b)** approximately \$14.58 billion **(c)** approximately \$0 billion **13.** **(a)** 261.4 restaurants per year **(b)** 362 restaurants per year **(c)** 439.6 restaurants per year **(d)** 287.2 restaurants per year **(e)** 344.7 restaurants per year **(f)** After increasing for a number of years, it is now decreasing. **15.** **(a)** about $-.233$ billion per year, which means that during this period, use of mass transportation was decreasing at a rate of about 233,000,000 trips per year **(b)** .32 billion per year, which means that during this period, use of mass transportation was increasing at a rate of 320,000,000 trips per year **(c)** The average rate of change during this period is 0. This does not accurately reflect the fact that transportation use decreased significantly and then increased again during this period. **19.** 20 ft per sec **21.** 60 ft per sec **23.** 7 ft per sec **25.** **(a)** 5 ft/sec **(b)** 2 ft/sec **(c)** 3 ft/sec **(d)** 5 ft/sec **(e)** (i) 3 ft/sec; (ii) 2.5 ft/sec **(f)** (i) 5 ft/sec; (ii) 4 ft/sec **27.** **(a)** $a^2 + 2ah + h^2 - a - h - 1$ **(b)** $2a + h - 1$ **(c)** 9 **29.** **(a)** $a^3 + 3a^2h + 3ah^2 + h^3$ **(b)** $3a^2 + 3ah + h^2$ **(c)** 75 **31.** **(a)** \$5998 per unit **(b)** \$6000 per unit **(c)** \$5998 **33.** **(a)** 6% per day **(b)** 7% per day **35.** **(a)** increasing at a rate of about 2,959,100 subscribers per year **(b)** increasing at a rate of about 597,920 subscribers per year

Section 11.3 (Page 633)

1. $y = -x - 2$ **3.** $y = -\frac{1}{4}x - 1$

5. **(a)** 6 **(b)** $y = 6x - 11$ **7.** **(a)** -1.25 **(b)** $y = -1.25x + 5$

9. **(a)** 2/3 **(b)** $y = \frac{2}{3}x + 6$ **11.** **(a)** x_5 **(b)** x_4 **(c)** x_3 **(d)** x_2

13. One of many possible graphs is:

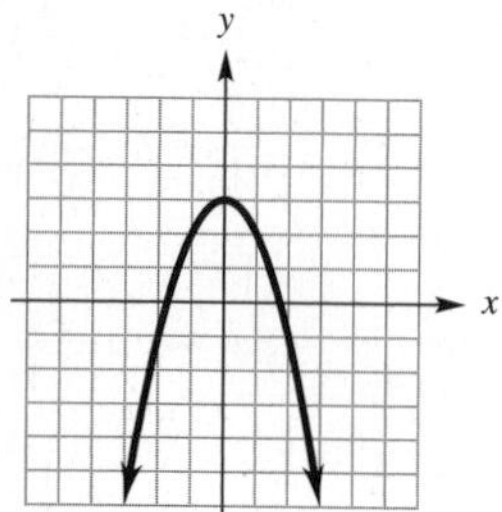

15. **(a)** The derivative is positive because the tangent line is rising from left to right at $x = 100$ and thus has positive slope.
(b) The derivative is negative because the tangent line is falling from left to right at $x = 200$ and thus has negative slope.
17. **(a)** distance **(b)** velocity **19.** $f'(x) = -8x + 11; -5; 11; 35$

21. $f'(x) = 8; 8; 8; 8$ **23.** $f'(x) = \frac{2}{x^2}; \frac{1}{2};$ does not exist; $\frac{2}{9}$

25. $f'(x) = \frac{-4}{(x-1)^2}; -4; -4; -\frac{1}{4}$ **27.** $-6; 6$

29. $-5; -3; 0; 2; 4$ **31.** **(a)** \$16 per table **(b)** \$15.998 or \$16 **(c)** The marginal revenue found in part (a) approximates the actual revenue from the sale of the 1001st item found in part (b).
33. **(a)** $-4p + 4$ **(b)** -36; demand is decreasing at the rate of about 36 items for each increase in price of \$1.

35.

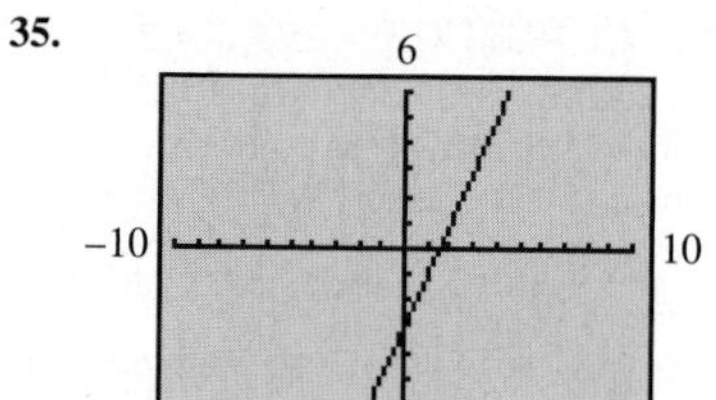

37.

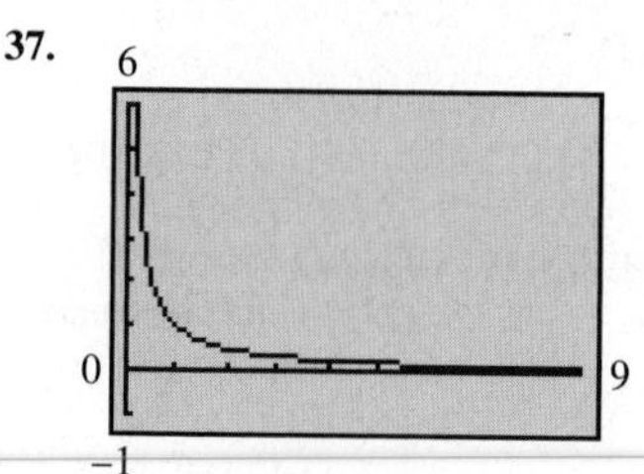

39. **(a)** 1340.640; 270.671; 4.958; .012 **(b)** The rate of increase is getting smaller and smaller as time goes on, which means that the population is almost stable after 60 months.
41. **(a)** about \$5000 per month; about \$453.59 per month; about \$41.15 per month; about \$3.73 per month
43. **(a)**

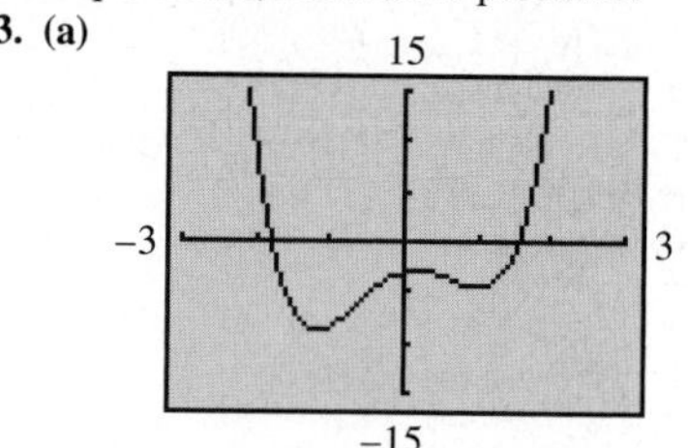

(b)

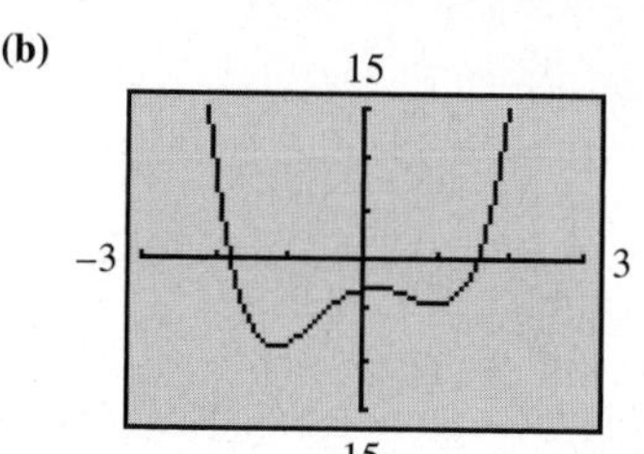

(c) The graphs appear identical, which suggests that $f'(x) = g(x)$.
45. (d)

Section 11.4 (Page 646)

1. $f'(x) = 8x - 6$ **3.** $y' = 6x^2 + 6x - 5$

5. $g'(x) = 4x^3 + 9x^2 - 6$ **7.** $f'(x) = 9x^{.5} - 2x^{-.5}$ or $9x^{.5} - \frac{2}{x^{.5}}$

9. $y' = -22.5x^{1/2} + 3.8x^{.9}$ **11.** $y' = 36t^{1/2} + 2t^{-1/2}$ or $36t^{1/2} + \frac{2}{t^{1/2}}$

13. $y' = 4x^{-1/2} + \left(\frac{9}{2}\right)x^{-1/4}$ or $\frac{4}{x^{1/2}} + \frac{9}{2x^{1/4}}$

15. $g'(x) = -30x^{-6} + x^{-2}$ or $-\frac{30}{x^6} + \frac{1}{x^2}$

17. $y' = -20x^{-3} - 12x^{-5} - 6$ or $-\frac{20}{x^3} - \frac{12}{x^5} - 6$

19. $f'(t) = -6t^{-2} + 16t^{-3}$ or $-\frac{6}{t^2} + \frac{16}{t^3}$

21. $y' = -36x^{-5} + 24x^{-4} - 2x^{-2}$ or $-\frac{36}{x^5} + \frac{24}{x^4} - \frac{2}{x^2}$

23. $g'(x) = -4x^{-3/2} - \frac{5}{2}x^{-1/2} + 1$ or $-\frac{4}{x^{3/2}} - \frac{5}{2x^{1/2}} + 1$

25. $y' = -6x^{-5/2} - 4x^{-3/2} + 2x$ or $-\frac{6}{x^{5/2}} - \frac{4}{x^{3/2}} + 2x$

27. $y' = -\frac{3}{2}x^{-5/4}$ or $-\frac{3}{2x^{5/4}}$ **29.** $y' = -\frac{5}{3}t^{-2/3}$ or $-\frac{5}{3t^{2/3}}$

31. $\frac{dy}{dx} = -40x^{-6} + 36x^{-5}$ or $\frac{-40}{x^6} + \frac{36}{x^5}$

33. $-\frac{9}{2}x^{-3/2} - 3x^{-5/2}$ or $-\frac{9}{2x^{3/2}} - \frac{3}{x^{5/2}}$ **35.** -28 **37.** $\frac{11}{16}$

39. (b) **41.** 0; $y = 0$ **43.** 7; $y = 7x - 3$
45. (a) \$15 **(b)** \$119 **(c)** \$207 **(d)** \$319
47. (a) 100 **(b)** 1
49. (a) $M'(x) = .24x^3 - 32.52x^2 + 1439.26x - 19{,}637.5$ **(b)** -1940.3 **(c)** -231 **(d)** 752.3 **(e)** 1189.6 **(f)** 1486.1 **(g)** 10,255.3 **(h)** 39,088.5 **(i)** The amount of money in circulation was decreasing at a rate of \$1940.3 million per year in 1920, it was decreasing more slowly in 1925 (\$231 million per year), and by 1930 it was increasing at a rate of \$752.3 million per year. In 1960 it was increasing at a slightly larger rate. By 1980 it was increasing at a much larger rate and by 2000 at a rate of \$39,088.5 million per year (that is, at more than \$39 billion per year). **51. (a)** 30 **(b)** 0 **(c)** -10

53. (a) $\frac{dV}{dr} = 160\pi r$ **(b)** 640π; the volume of blood is increasing 640π cu mm per mm of change in the radius. **(c)** 960π; the volume of blood is increasing 960π cu mm per mm of change in the radius. **(d)** 1280π; the volume of blood is increasing 1280π cu mm per mm of change in the radius. **55.** $f'(x) = .008x^3 - .12x^2 + .66x - .475$ **(a)** $-.475$ **(b)** .641 **(c)** .757 **(d)** 1.009 **(e)** 1.577 **(f)** Living standards were decreasing in 1990 but then began to increase. The rate of increase steadily rose from 1991–99, with the rate in 1999 almost double the rate in 1994.
57. (a) $y_1 = 35$; $y_2 = 36$ **(b)** 4.13; 4.32; these values are fairly close, and they represent the rate of change of approximately four years for a dog for one year of a human. **(c)** $y = 4x + 16$
59. (a) $V = CR_0R^2 - CR^3$, so that $dV/dR = 2\,CR_0R - 3CR^2$ **(b)** $R = \frac{2}{3}R_0$ **61. (a)** $v(t) = 16t + 3$ **(b)** 3; 83; 163
63. (a) $v(t) = 6t^2 + 12t$ **(b)** 0; 210; 720
65. (a) -32 ft per sec; -64 ft per sec **(b)** in 3 sec **(c)** -96 ft per sec **69. (a)** none **(b)** none **(c)** $(-\infty, \infty)$ **(d)** The derivative is always negative, so the graph of $g(x)$ is always decreasing.

Section 11.5 (Page 656)

1. $y' = 9x^2 + 2x - 6$ **3.** $y' = 120x^3 - 54x^2 + 10$
5. $y' = 24x^5 - 35x^4 - 20x^3 + 42x^2 + 4x - 6$
7. $y' = 144x^3 + 144x^2 + 32x$ **9.** $y' = 54x^5 + 30x^4 + 4x^3$
11. $y' = \frac{-7}{(x-4)^2}$ **13.** $f'(t) = \frac{t^2 + 6t - 12}{(t+3)^2}$
15. $g'(x) = \frac{-6x^4 - 4x^3 - 6x - 1}{(2x^3 - 1)^2}$ **17.** $y' = \frac{x^2 + 6x - 14}{(x+3)^2}$

19. $r'(t) = \frac{-\sqrt{t} + \frac{3}{2\sqrt{t}}}{(2t+3)^2}$ or $\frac{(3-2t)}{2\sqrt{t}(2t+3)^2}$ **21.** $y' = \frac{\frac{9\sqrt{x}}{2} + \frac{4}{\sqrt{x}}}{x}$ or $\frac{9x + 8}{2x\sqrt{x}}$ **23.** $y' = \frac{2}{(1-x)^2}$ **25.** $f'(p) = \frac{24p^2 + 32p + 29}{(3p+2)^2}$
27. $g'(x) = \frac{10x^4 + 18x^3 + 6x^2 - 20x - 9}{(2x+1)^2(5x+2)^2}$
29. In the first step, the numerator should be $(x^2 - 1)(2) - (2x + 5)(2x)$.
31. $y = -2x + 9$ **33.** $y = -1.25x + 53.25$
35. (a) \$22.86 per unit **(b)** \$12.92 per unit **(c)** $\frac{3x + 2}{x^2 + 4x}$ hundred dollars per unit **(d)** $\overline{C}'(x) = \frac{-3x^2 - 4x - 8}{(x^2 + 4x)^2}$
37. (a) \$175,000 **(b)** $C'(x) = \frac{7{,}500{,}000}{(100 - x)^2}$ **(c)** increasing at the rate of \$12,000 for each 1% removed **(d)** increasing at the rate of \$75,000 for each 1% removed **(e)** increasing at the rate of \$300,000 for each 1% removed
39. (a) $T'(x) = \frac{-10x^2 + 50}{(x^2 + 5)^2}$ **(b)** 2; temperature increasing at 2 degrees per hour **(c)** about 1.1111; temperature increasing at about 1.1111 degrees per hour **(d)** about $-.2041$; temperature decreasing at about .2041 degrees per hour **(e)** about $-.1028$; temperature decreasing at about .1028 degrees per hour
41. (a) 8.57 min **(b)** 16.36 min **(c)** 6.12 min/(kcal/min); 2.48 min/(kcal/min) **43. (a)** .1173 **(b)** 2.625
45. (a) $N'(t) = \frac{4200t}{(30 + t^2)^2}$ **(b)** 8.28 words per min; 6.94 wpm; 4.71 wpm; 2.49 wpm; .97 wpm

Section 11.6 (Page 667)

1. 275 **3.** 107 **5.** $24x + 4$; $24x + 35$
7. $-64x^3 + 2$; $-4x^3 + 8$ **9.** $\frac{1}{x^2}$; $\frac{1}{x^2}$
11. $\sqrt{8x^2 - 4}$ or $2\sqrt{2x^2 - 1}$; $8x + 10$
13. If $f(x) = x^5$ and $g(x) = 4x + 3$, then $y = f[g(x)]$.
15. If $f(x) = \sqrt{x}$ and $g(x) = 6 + 3x$, then $y = f[g(x)]$.
17. If $f(x) = \frac{x+3}{x-3}$ and $g(x) = \sqrt{x}$, then $y = f[g(x)]$.
19. If $f(x) = x^2 + x + 5$ and $g(x) = x^{1/2} - 3$, then $y = f[g(x)]$.
21. $y' = 9(3x + 4)^2$ **23.** $y' = 72(3x + 2)^3$
25. $y' = -128x(8x^2 + 6)^3$ **27.** $y' = 36(2x + 5)^{1/2}$
29. $y' = -\frac{21}{2}(8x + 9)(4x^2 + 9x)^{1/2}$ **31.** $y' = \frac{16}{\sqrt{4x + 7}}$
33. $y' = \frac{-2x - 4}{\sqrt{x^2 + 4x}}$ **35.** $y' = 2(x + 1)(x - 3) + (x - 3)^2$
37. $y' = 70(x + 3)(2x - 1)^4(x + 2)$
39. $y' = \frac{(3x + 1)^2(21x + 1)}{2\sqrt{x}}$ **41.** $y' = -2(x - 4)^{-3}$ or $\frac{-2}{(x-4)^3}$
43. $y' = \frac{2(4x + 3)(4x - 7)}{(2x - 1)^2}$ **45.** $y' = \frac{-5x^2 - 36x + 8}{(5x + 2)^4}$
47. $y' = \frac{3x^{1/2} - 1}{4x^{1/2}(x^{1/2} - 1)^{1/2}}$ **49. (a)** -2 **(b)** $-\frac{24}{7}$
51. (d) **53.** $C'(x) = \frac{15x}{\sqrt{50 + 15x^2}}$
55. (a) $R'(x) = \frac{5(300 - 4x)}{\sqrt{300x - 2x^2}}$ **(b)** \$10.61; \$2.89; $-\$2.89$; $-\$10.61$

57. **(a)** $-.8$ **(b)** 0 **(c)** .8 **(d)** $\overline{R}(x) = \frac{1000}{x}\left(1 - \frac{x}{500}\right)^2$
(e) $\frac{d\overline{R}(x)}{dx} = \left(1 - \frac{x}{500}\right)\left(-\frac{2}{x} - \frac{1000}{x^2}\right)$
59. \$77.46 per additional employee **61.** $A[r(t)] = A(2t) = 4\pi t^2$; this function gives the area of the pollution in terms of the time since the pollutants were first emitted.
63. $D(c) = \frac{-c^2 + 10c - 25}{25} + 500$
65. **(a)** $R(x) = \frac{30{,}000x - 2x^3}{3}$ **(b)** $P(x) = 8000x - \frac{2x^3}{3} - 3500$
(c) $\frac{dP}{dx} = 8000 - 2x^2$ **(d)** $-\$21{,}925$
67. **(a)** 34 min **(b)** $-\frac{108}{17}\pi$ mm³/min; $-\frac{72}{17}\pi$ mm²/min
69. **(a)** $F'(t) = \frac{150t}{(8 + t^2)^{3/2}}$ **(b)** about 3.96; 1.34; .36; .09
71. **(a)** $(.5, \infty)$ **(b)** none **(c)** $(-\infty, .5)$ **(d)** The derivative does not exist at $x = .5$, which corresponds to a sharp point on the graph of $K(x)$. The derivative is positive when $K(x)$ is increasing and negative when $K(x)$ is decreasing.

Section 11.7 (Page 676)

1. $y' = 3e^{3x}$ **3.** $f'(x) = 10e^{2x}$ **5.** $g'(x) = 20e^{-5x}$
7. $y' = 2xe^{x^2}$ **9.** $f'(x) = xe^{x^2/2}$ **11.** $y' = -18xe^{3x^2+5}$
13. $y' = \frac{-8x + 3}{-4x^2 + 3x}$ **15.** $y' = \frac{1}{2x + 1}$
17. $y' = \frac{6x^2 - 6x + 8}{(2x - 3)(x^2 + 4)}$ **19.** $y' = 2x(1 - x)e^{-2x}$
21. $y' = (-9x^2 + 18x - 4)e^{-3x}$ **23.** $y' = \frac{-23}{(6 - x)(3x + 5)}$
25. $y' = \frac{3(15x^2 - 2)}{2x(5x^2 - 2)}$ **27.** $y' = \frac{-2x^2}{2 - x^2} + \ln(2 - x^2)$
29. $y' = \frac{1 - 3\ln|x|}{x^4}$ **31.** $y' = \frac{\frac{-20}{x} + 8 - 8\ln|x|}{(5 - 2x)^2}$
33. $y' = \frac{3x^3\ln|x| - (x^3 - 1)}{2x(\ln|x|)^2}$ **35.** $y' = \frac{1}{2(x - 3)\sqrt{\ln(x - 3)}}$
37. $y' = \frac{xe^x\ln|x| - e^x + 1}{x(\ln|x|)^2}$
39. $y' = \frac{xe^x + xe^{-x} - e^x + e^{-x}}{x^2}$ or $\frac{e^x(x - 1) + e^{-x}(x + 1)}{x^2}$
41. $f'(x) = \frac{4e^{3x+2}}{4x - 5} + 3e^{3x+2}\ln(4x - 5)$ **43.** $y' = \frac{2000e^{.4x}}{(7 - 10e^{.4x})^2}$
45. $y' = \frac{1250e^{-.5x}}{(12 + 5e^{-.5x})^2}$ **47.** $y = -x + 1$ **49.** $y = \frac{1}{e}x$
51. 1/8 **53.** **(a)** about 73 million; increasing at approximately 11.06 million people per year **(b)** about 99 million; increasing at approximately 14.96 million people per year
55. **(a)** .98 **(b)** .82 **(c)** .14 **(d)** $-.0027$; it is the rate of change in the proportion of defect-free cars when $x = 100$ months.
57. **(a)** 36.8 **(b)** .00454 **(c)** $3.72 \times 10^{-42} \approx 0$ **(d)** $100e^{-.1N}$ is always positive since powers of e are never negative. This means that repetition always makes a habit stronger.
59. **(a)** 2974.15 g **(b)** 2.75 g/day
(c)

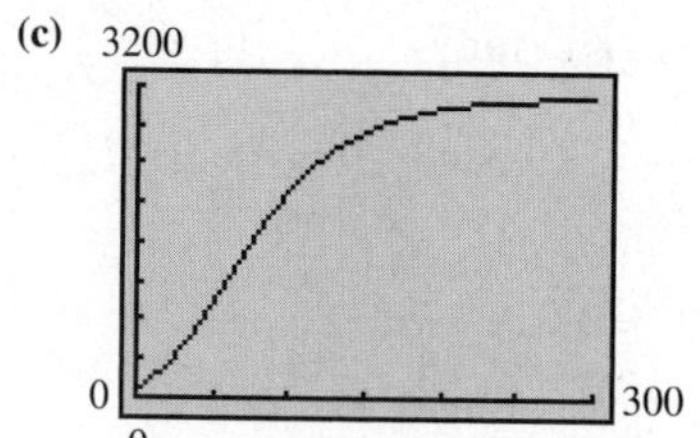

(d)

Weight	Day	Rate
990.9797	50	24.87793
2121.673	100	17.72981
2733.571	150	7.603823
2974.153	200	2.753855
3058.845	250	0.942782
3087.568	300	0.316772

61. **(a)** 3.857 cm³ **(b)** .973 cm **(c)** about 214 mo **(d)** The tumor is 240 months old; it is increasing in volume at the instantaneous rate of .282 cm³/mo.
63. **(a)** $\frac{dR}{dx} = 100 + \frac{50(\ln x - 1)}{(\ln x)^2}$ **(b)** \$112.48
65. **(a)** about 817 cars/hr; -41.2 cars/hr per ft **(b)** about 522 cars/hr; -20.9 cars/hr per ft
67. **(a)** about \$9 billion; about \$9.7 billion **(b)** increasing at approximately \$253.45 million per year; increasing at approximately \$220.39 million per year.
69. **(a)** 1805.73 million metric tons **(b)** The greenhouse gas emissions were increasing at the rate of 37.3 million metric tons per year at the end of 2002.

Section 11.8 (Page 684)

1. $x = 0, 2$ **3.** $x = 1$ **5.** $x = -3, 0, 3$ **7.** $x = 0, 6$
9. yes, no **11.** no, no, yes **13.** yes, no, yes
15. yes, no, yes **17.** no, yes, yes **19.** yes, no **21.** 1
23. **(a)** 6¢ **(b)** does not exist **(c)** any three of the years 1935, 1943, 1949, 1967, 1973, 1989, 1991, 2001
25. discontinuous at $t = m$; differentiable everywhere except $t = m$ and the endpoints
27. **(a)**

(b) continuous at every number except $x = 1, 2, 3, 4, 5, 6, 7$
29. **(a)** \$120 **(b)** \$150 **(c)** \$150 **(d)** \$150 **(e)** \$180 **(f)** \$150 **(g)** \$150 **31.** $[-9, 15]$ **33.** $(10, \infty)$
35. $(-6, 0), (0, 3), (4, 8)$

Chapter 11 Review (Page 689)

1. 4 **3.** -1.2 **5.** $-.35$ **7.** -1 **9.** $\frac{13}{2}$ **11.** 4

13. -13 **15.** $\frac{1}{6}$ **17.** $\frac{1}{4}$ **19.** 21 **21.** about $-.3846$

23. $y' = 2$ **25.** $y' = 4x - 1$ **27.** -2; $y + 2x = -4$

29. $\frac{2}{9}$; $2x - 9y = 2$ **31. (a)** -25.33 per dollar **(b)** -16.75 per dollar **(c)** -31.12 per dollar **33.** $y' = 10x - 7$

35. $y' = 14x^{4/3}$ **37.** $f'(x) = -3x^{-4} + \left(\frac{1}{2}\right)x^{-1/2}$ or $-\frac{3}{x^4} + \frac{1}{2x^{1/2}}$

39. $y' = 15t^4 + 12t^2 - 7$ **41.** $y' = 16x^{3/4} + 6x^{-1/4}(2x + 3)$

43. $f'(x) = \frac{-2x^2 + 4}{(x^2 + 2)^2}$ **45.** $y' = \frac{2 - x + 2\sqrt{x}}{x\sqrt{x}(x + 2)^2}$

47. $y' = \frac{x^2 - 2x}{(x - 1)^2}$ **49.** $f'(x) = 12(3x - 2)^3$ **51.** $y' = \frac{1}{\sqrt{2t - 5}}$

53. $y' = 18x(3x - 4)^2 + 2(3x - 4)^3$

55. $f'(u) = \frac{(2u + 3)(6u - 4) - 6(3u^2 - 4u)}{(2u + 3)^4}$ **57.** $y' = -6x^2e^{-2x^3}$

59. $y' = 10xe^{2x} + 5e^{2x}$ or $5e^{2x}(2x + 1)$ **61.** $y' = \frac{2x + 4}{x^2 + 4x - 1}$

63. $y' = \frac{x - \frac{1}{x} - 2x \ln 4x}{(x^2 - 1)^2}$ **65.** $y' = \frac{x^2 - 6x + 1}{(x - 3)^2}$

67. $y' = -12e^{2x}$ **69.** $y' = \frac{-1}{x^{1/2}(x^{1/2} - 1)^2}$

71. $\frac{dy}{dt} = \frac{1 + 2t^{1/2}}{4t^{1/2}(t^{1/2} + t)^{1/2}}$ **73.** $-\frac{2}{3}$ **75.** none

77. no, yes, yes **79.** no, yes, no **81.** yes, no, yes

83. (a) \$368.50; \$170.50 **(b)** $-$\$80; prices were decreasing at an average rate of \$80 per year. **(c)** \$74 per year; \$22 per year

85. (a) about \$1147.2 million; about \$3828 million **(b)** increasing about \$241 million per year; increasing about \$699.4 million per year

87. (a) about \$3.76 billion; about \$8.54 billion **(b)** about \$1.5 billion; about \$3.5 billion

89. (a) about \$833 million; about \$10 billion **(b)** about \$1.887 billion; about \$1.26 billion **91.** 0 mph per sec for the hands and approximately 640 mph per sec for the bat. This represents the acceleration of the hands and the bat at the moment when their velocities are equal.

Case 10 (Page 693)

1. about .452 **3. (a)** about 1.1; about .65 **(b)** \$95.14

5. Demand is constant, no matter what the price.

CHAPTER 12

Section 12.1 (Page 706)

1. increasing on $(1, \infty)$; decreasing on $(-\infty, 1)$; local minimum of -4 at $x = 1$ **3.** increasing on $(-\infty, -2)$; decreasing on $(-2, \infty)$; local maximum of 3 at $x = -2$ **5.** increasing on $(-\infty, -4)$ and $(-2, \infty)$; decreasing on $(-4, -2)$; local maximum of 3 at $x = -4$; local minimum of 1 at $x = -2$ **7.** increasing on $(-7, -4)$ and $(-2, \infty)$; decreasing on $(-\infty, -7)$ and $(-4, -2)$; local maximum of 3 at $x = -4$; local minimum of -2 at $x = -7$ and $x = -2$

9. increasing on $\left(-\infty, -\frac{1}{3}\right)$ and $(2, \infty)$; decreasing on $\left(-\frac{1}{3}, 2\right)$

11. increasing on $(-\infty, -4)$ and $(-4, \infty)$

13. decreasing on $(-\infty, 5)$ **15.** increasing on $(-\infty, -1)$ and $(2, \infty)$; decreasing on $(-1, 2)$ **17.** $-2, -1, 2$

19. local maximum at 0; local minimum at 2

21. local maximum at -3; local minimum at -1

23. local maximum at $-\frac{1}{4}$; local minimum at -5

25. local maximum at -2; local minimum at 3

27. local maximum at 0; local minimum at 2

29. local maximum at 0; $f(0) = 1$; local minimum at $\frac{6}{11}$; $f\left(\frac{6}{11}\right) \approx .7804$

31. local maximum at $\frac{3}{4}$; $f\left(\frac{3}{4}\right) = 4$

33. local minimum at 0; $f(0) = 0$

35. local maximum at -1; $f(-1) = \frac{1}{e}$

37. local maximum at $-\frac{1}{e}$; $f\left(-\frac{1}{e}\right) = \frac{1}{e}$; local minimum at $\frac{1}{e}$; $f\left(\frac{1}{e}\right) = -\frac{1}{e}$ **39.** local minimum at $-\frac{1}{3}$; $f\left(-\frac{1}{3}\right) \approx -2.1226$

41. local minimum at $x = 0$; $f(0) = 2$ **43.** approximate local maximum: (3.35, 161.98); approximate local minima: $(-6.77, -709.87)$ and $(10.92, -240.08)$ **45.** approximate local maxima: $(-80.7064, 8{,}671{,}701.6)$ and (.3982, 5.0017); approximate local minima: $(-1.6166, -8.8191)$ and $(1.9249, -1.7461)$

47. (a) The derivative is undefined at any point where the graph shows a "peak"—that is, a sharp change from increasing to decreasing. **(b)** increasing from April to July; decreasing from July to November; constant from January to April and from November to December **(c)** from January to April and for November and December. Air pollution is greatly reduced when the temperature is low, as is the case during these months. **49. (a)** 6 units **(b)** \$324

51. increasing on $(-\infty, 0)$; decreasing on $(0, \infty)$ **53.** (0, 12,000)

55. (a) (0, 1.5) **(b)** (1.5, 8) **57. (a)** (1990, 1992), (1995, 1997), (1998, 2000) **(b)** (1992, 1995), (1997, 1998) **(c)** local maxima in 1992 and 1997; local minima in 1995 and 1998

59. (a) about \$8.00 in 1969, \$6.00 in 1982, \$4.65 in 1989, \$5.44 in 1998; exactly \$5.15 in 2000 **(b)** 6 **(c)** around 1989 and 1995 **(d)** approximately (1980, 1989), (1992, 1995), (1997, 2000)

61. local maximum in 1994; local minimum in 2000

63. local maximum at 11.68, which means that interest rates peaked around 1981–82 **65.** maximum of about 1550 casualties in 1991; minimum of about 1060 casualties in 1994 **67.** 10

69. local maxima at $x \approx 1.54$ and $x \approx 6.34$, which correspond to mid-1995 and spring 2000, respectively

Section 12.2 (Page 721)

1. $f''(x) = 6x - 10$; -10; 2; -28

3. $f''(x) = 12(x + 2)^2$; 48, 192; 12 **5.** $f''(x) = \frac{2}{(1 + x)^3}$; 2; $\frac{2}{27}$; $-\frac{1}{4}$

7. $f''(x) = -\frac{(x + 4)^{-3/2}}{4}$ or $-\frac{1}{4(x + 4)^{3/2}}$; $-\frac{1}{32}$; $-\frac{1}{4 \cdot 6^{3/2}}$; $-\frac{1}{4}$

9. $f''(x) = \left(-\frac{6}{5}\right)x^{-7/5}$ or $-\frac{6}{5x^{7/5}}$; $f''(0)$ does not exist; $-\frac{6}{5 \cdot 2^{7/5}}$; $-\frac{6}{5(-3)^{7/5}}$ **11.** $f''(x) = 2e^x$; 2; $2e^2$; $2e^{-3}$ or $\frac{2}{e^3}$

13. $f''(x) = 20e^{2x}$; 20; $20e^4$; $20e^{-6}$

15. $f''(x) = -\frac{1}{x^2}$; not defined; $-\frac{1}{4}$; $-\frac{1}{9}$

17. $f''(x) = \frac{1}{x}$; not defined; $\frac{1}{2}$; $-\frac{1}{3}$ **21.** $v(t) = 12t + 2$; $a(t) = 12$; $v(0) = 2$ cm/sec; $v(4) = 50$ cm/sec; $a(0) = a(4) = 2$ cm/sec^2 **23.** $v(t) = 9t^2 - 8t + 8$; $a(t) = 18t - 8$; $v(0) = 8$ cm/sec; $v(4) = 120$ cm/sec; $a(0) = -8$ cm/sec^2; $a(4) = 64$ cm/sec^2 **25.** concave upward on $(-\infty, \infty)$; no points of inflection **27.** concave upward on $\left(-\frac{4}{3}, \infty\right)$; concave downward on $\left(-\infty, -\frac{4}{3}\right)$; point of inflection at $\left(-\frac{4}{3}, \frac{425}{27}\right)$ **29.** concave upward on $(3, \infty)$; concave downward on $(-\infty, 3)$; no points of inflection **31.** concave upward on $(-\infty, -5)$ and $(1, \infty)$; concave downward on $(-5, 1)$; points of inflection at $(-5, -1248)$ and $(1, 0)$ **33.** (14, 26,688) **35.** no critical numbers; no local extrema **37.** local maximum at $x = -5$; local minimum at $x = 4$ **39.** local maximum at $x = 0$; local minima at $x = 2$ and $x = -2$ **41.** local maximum at $x = -\sqrt{3}$; local minimum at $x = \sqrt{3}$ **43.** local maximum at $x = -3$; local minimum at $x = 3$ **45.** no critical numbers; no local extrema **47.** local maximum at $x = 2$; local minima at $x = 1$ and $x = 4$; points of inflection at $x = \frac{7 + \sqrt{7}}{3}$ and $x = \frac{7 - \sqrt{7}}{3}$ **49.** local minimum at $x = 1$; points of inflection at $x = \frac{4}{3}$ and $x = 2$ **51.** (a) E (b) A (c) C (d) B (e) D **53.** (a) -96 ft/sec (b) -160 ft/sec (c) -256 ft/sec (d) -32 ft/sec^2 **55.** (15, 86,250) **57.** Point of inflection when $x = 3.68$. In late 1993, the rate of spending begins to slow down (although it is still increasing). **59.** (a) $x = 6.56$; $x \approx 18.44$ (b) local maximum in mid-1986 ($x = 6.56$); local minimum in mid-1998 ($x = 18.44$) (c) $x = 12.5$; the rate at which expenditures were decreasing began to slow down in mid-1992. **61.** 8 **63.** (a) in mid-1986 ($x = 6.47$) (b) in early 2003 ($x = 23.29$) **65.** (a) 15 days (b) 16.875% **67.** lowest in early 1986 ($x = 6.1$) and highest in mid-1994 ($x = 14.56$)

Section 12.3 (Page 734)

1. absolute maximum at $x = 4$; absolute minimum at $x = 1$ **3.** absolute maximum at $x = 2$; absolute minimum at $x = -2$ **5.** absolute maximum at $x = -4$; absolute minimum at $x = -7$ and $x = -2$ **7.** absolute maximum at $x = 6$; absolute minimum at $x = -4$ and $x = 4$ **9.** absolute maximum at $x = 6$; absolute minimum at $x = 4$ **11.** absolute maximum at $x = \sqrt{2}$; absolute minimum at $x = 0$ **13.** absolute maximum at $x = -3$ and $x = 3$; absolute minimum at $x = 0$ **15.** absolute maximum at $x = 0$; absolute minimum at $x = -1$ and $x = 1$ **17.** about 269 monitors; about \$577 per monitor **19.** in 2000 ($x = 4.4278$) **21.** (a) $R(x) = 350x$ (b) $P(x) = R(x) - C(x) = 350x - (600{,}000 - 25x + .01x^2) = -.01x^2 + 375x - 600{,}000$ (c) 18,750 radiators; \$2,915,625 **23.** (a) 8 (b) 29.43% **25.** 8 in. by 8 in. **27.** 20 cm by 20 cm by 40 cm; \$72 **29.** 3 ft by 6 ft by 2 ft **31.** (a) $1200 - 2x$ (b) $A(x) = 1200x - 2x^2$ (c) 300 m (d) 180,000 m^2 **33.** 20,000 ft^2 **35.** \$1000 **37.** (a) $R(x) = 100{,}000x - 100x^2$ (b) 500 (c) 25,000,000¢, or \$250,000 **39.** (a) 2037 (b) about 1,507,000,000 **41.** (a) 56.6 mph (b) \$45.25 **43.** 1 mile from point *A* **45.** 10 **47.** 60 **49.** 5000 **51.** (c)

Section 12.4 (Page 745)

1. $f(x) = -x^2 - 10x - 25$
no points of inflection

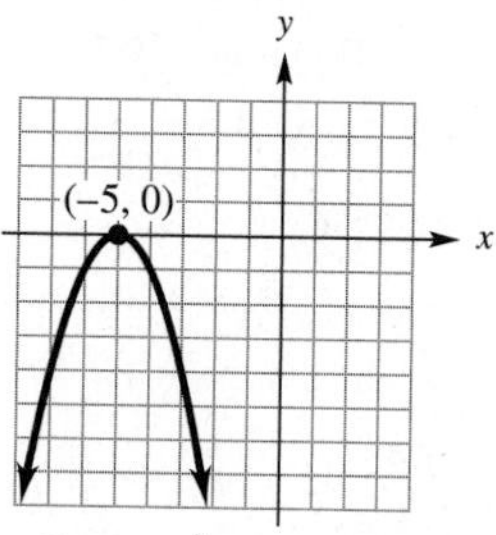

$f(x) = -x^2 - 10x - 25$

3. $f(x) = 3x^3 - 3x^2 + 1$
point of inflection at $x = \frac{1}{3}$

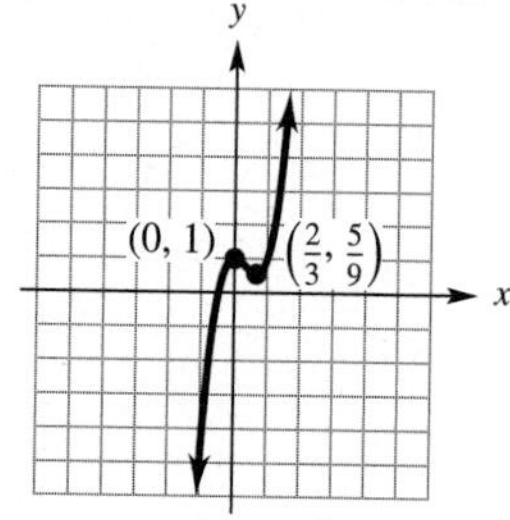

$f(x) = 3x^3 - 3x^2 + 1$

5. $f(x) = -2x^3 - 9x^2 + 108x - 10$
point of inflection at $x = -\frac{3}{2}$

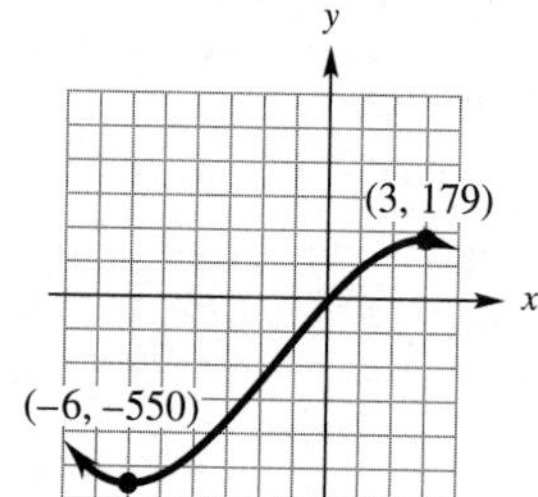

$f(x) = -2x^3 - 9x^2 + 108x - 10$

7. $f(x) = 2x^3 + \frac{7}{2}x^2 - 5x + 3$
point of inflection at $x = -\frac{7}{12}$

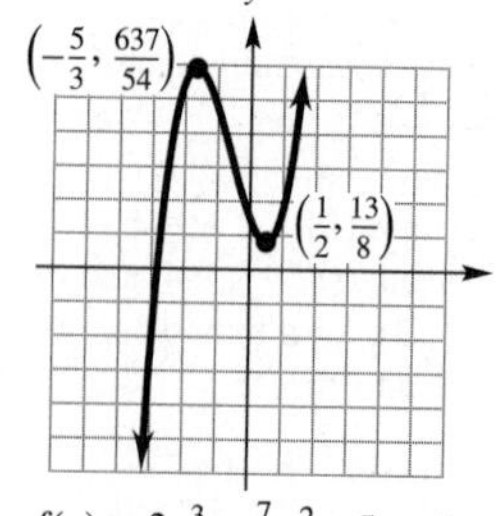

$f(x) = 2x^3 + \frac{7}{2}x^2 - 5x + 3$

9. $f(x) = (x + 3)^4$
no points of inflection

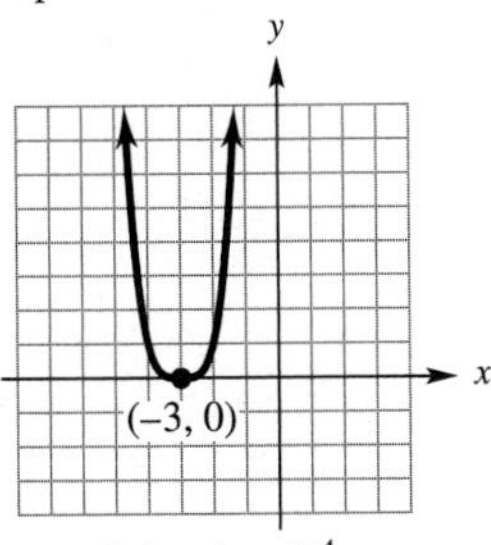

$f(x) = (x + 3)^4$

11. $f(x) = x^4 - 18x^2 + 5$
points of inflection at $x = -\sqrt{3}$ and $x = \sqrt{3}$

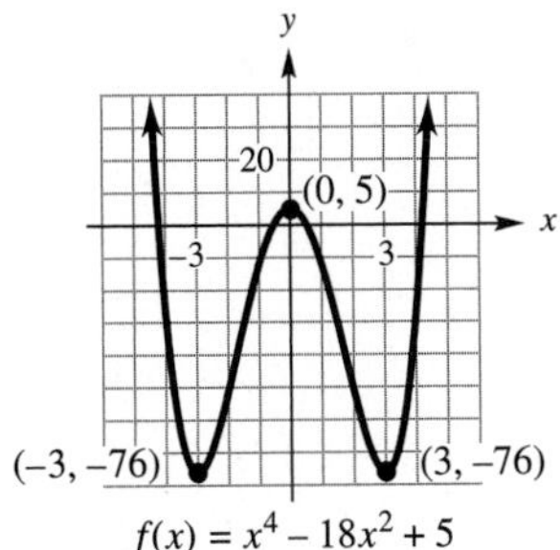

$f(x) = x^4 - 18x^2 + 5$

13. $f(x) = x - \dfrac{1}{x}$
no points of inflection

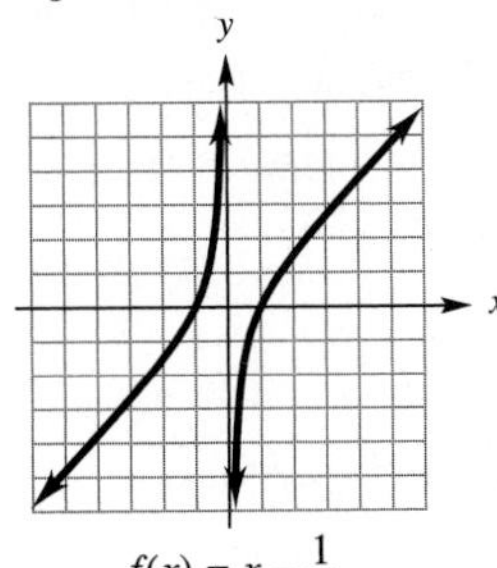

$f(x) = x - \frac{1}{x}$

15. $f(x) = \dfrac{x^2 + 25}{x}$
no points of inflection

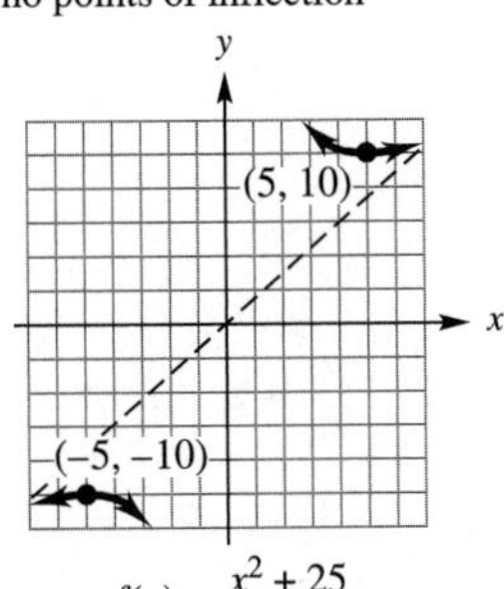

$f(x) = \dfrac{x^2 + 25}{x}$

17. $f(x) = \dfrac{x - 1}{x + 1}$
no points of inflection

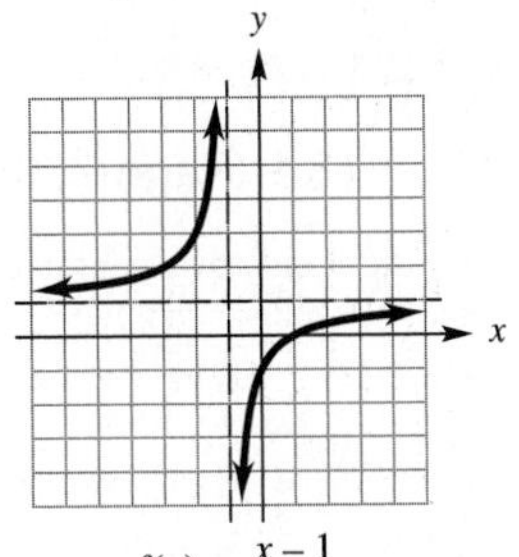

$f(x) = \dfrac{x - 1}{x + 1}$

For problems 19–23, many correct answers are possible, including the following:

19.

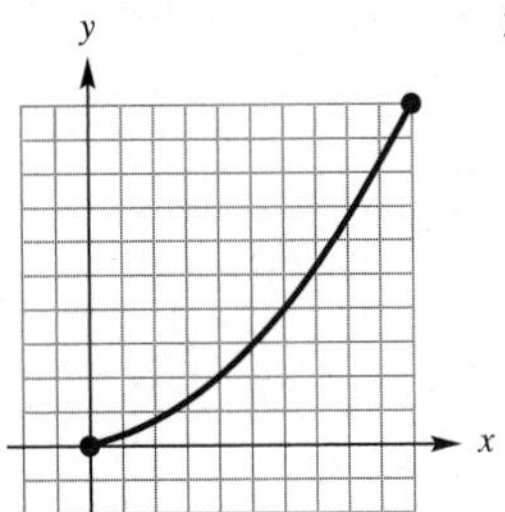

21.

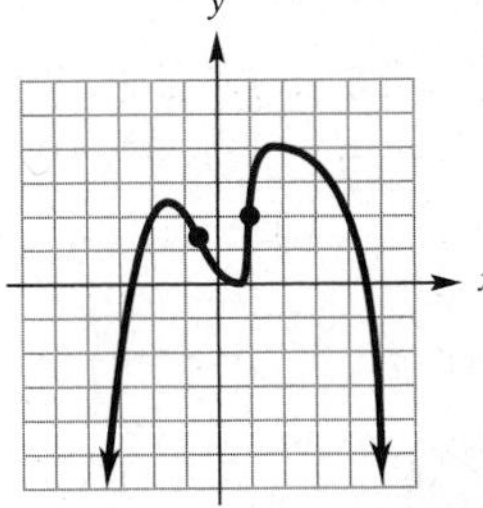

23.

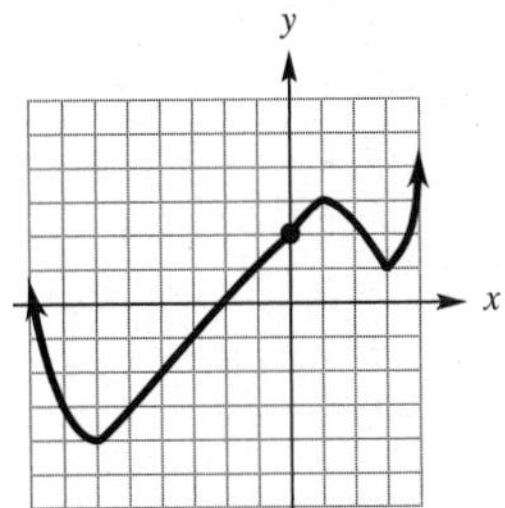

25. **(a)** The first derivative is always positive because the function is increasing. **(b)** The second derivative is positive because the graph is concave upward. This means that the risk is increasing at a faster and faster rate. **27.** local maximum at $(-.0241, 1.0001)$; local minimum at $(.6908, .9818)$; point of inflection at $\left(\frac{1}{3}, \frac{5351}{5400}\right)$

29. local maxima at $(-80.7064, 8{,}671{,}701.6)$ and $(.3982, 5.0017)$; local minima at $(-1.6166, -8.8191)$ and $(1.9249, -1.7461)$; points of inflection at $(-60.4796, 5{,}486{,}563.4)$, $(-.7847, -2.7584)$, and $(1.2643, 1.2972)$

Chapter 12 Review (Page 747)

5. increasing on $\left(-\frac{7}{2}, \infty\right)$; decreasing on $\left(-\infty, -\frac{7}{2}\right)$

7. increasing on $\left(-\infty, -\frac{2}{3}\right)$ and $(1, \infty)$; decreasing on $\left(-\frac{2}{3}, 1\right)$

9. decreasing on $(-\infty, 3)$ and $(3, \infty)$ **11.** local maximum of 101 at $x = -3$; local minimum of -24 at $x = 2$ **13.** local maximum of 1 at $x = 0$; local minima of -44 at $x = -3$ and $-\frac{4}{3}$ at $x = 1$

15. local maximum of $\frac{3}{e}$ at $x = 1$ **17.** $f''(x) = 40x^3 - 24x$; 16; -272

19. $f''(x) = -80e^{4x}$; $-80e^4$; $-80e^{-8}$ **21.** no local extrema; point of inflection at $x = -\frac{1}{12}$; decreasing on $(-\infty, \infty)$; concave upward on $\left(-\infty, -\frac{1}{12}\right)$; concave downward on $\left(-\frac{1}{12}, \infty\right)$

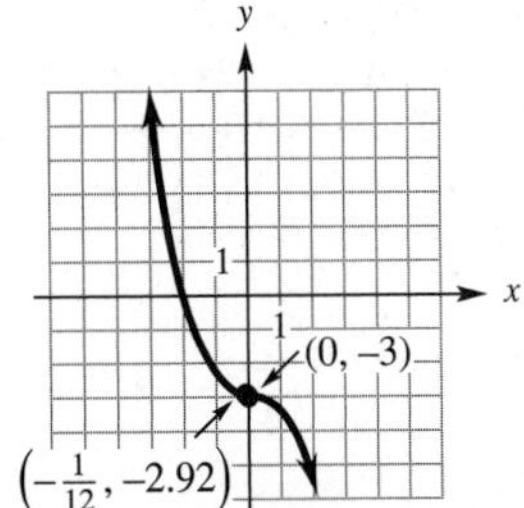

23. local maximum at $x = 0$; local minima at $x = -1$ and $x = 2$; points of inflection at $x = \frac{1-\sqrt{7}}{3}$ and $x = \frac{1+\sqrt{7}}{3}$; increasing on $(-1, 0)$ and $(2, \infty)$; decreasing on $(-\infty, -1)$ and $(0, 2)$; concave upward on $\left(-\infty, \frac{1-\sqrt{7}}{3}\right)$ and $\left(\frac{1+\sqrt{7}}{3}, \infty\right)$; concave downward on $\left(\frac{1-\sqrt{7}}{3}, \frac{1+\sqrt{7}}{3}\right)$

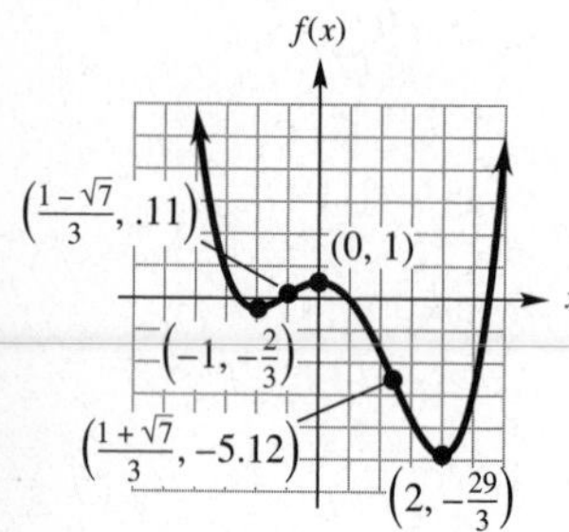

25. no local extrema or points of inflection; increasing on $\left(-\infty, -\frac{1}{2}\right)$ and $\left(-\frac{1}{2}, \infty\right)$; concave upward on $\left(-\infty, -\frac{1}{2}\right)$; concave downward on $\left(-\frac{1}{2}, \infty\right)$

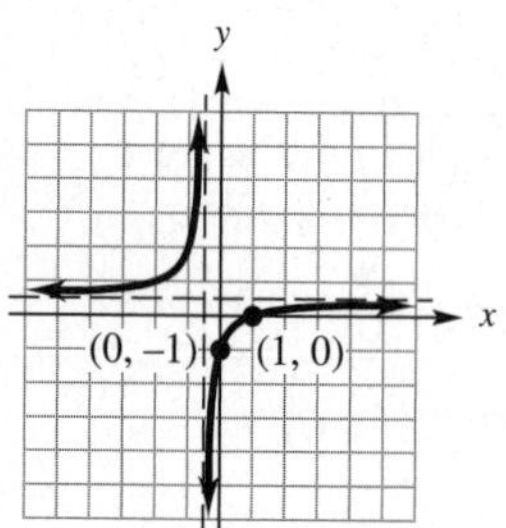

27. local maximum at $x = \frac{1}{2}$; local minimum at $x = -\frac{2}{3}$; point of inflection at $x = -\frac{1}{12}$; increasing on $\left(-\frac{2}{3}, \frac{1}{2}\right)$; decreasing on $\left(-\infty, -\frac{2}{3}\right)$ and $\left(\frac{1}{2}, \infty\right)$; concave upward on $\left(-\infty, -\frac{1}{12}\right)$; concave downward on $\left(-\frac{1}{12}, \infty\right)$

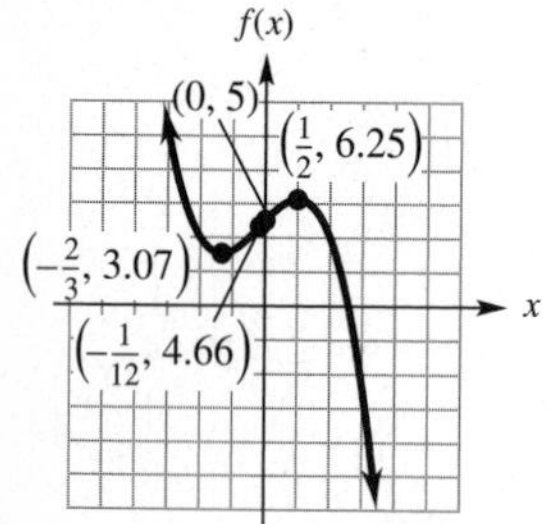

29. local minimum at $x = 0$; concave upward on $(-\infty, \infty)$. In the graph, the horizontal lines are 10 units apart.

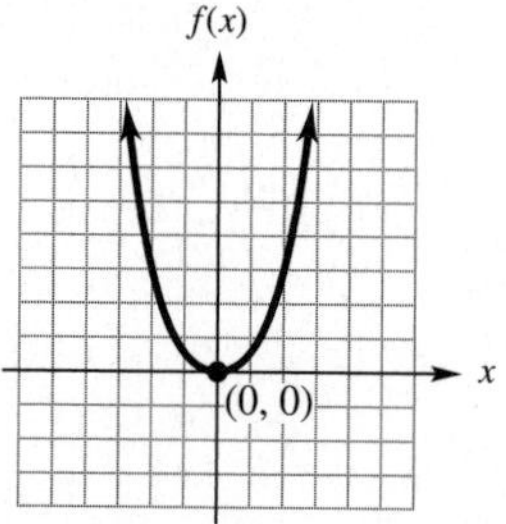

31. local maximum at $x = -2$; local minimum at $x = 2$; no point of inflection; increasing on $(-\infty, -2)$ and $(2, \infty)$; decreasing on $(-2, 0)$ and $(0, 2)$; concave upward on $(0, \infty)$; concave downward on $(-\infty, 0)$. In the graph, the horizontal lines are 2 units apart.

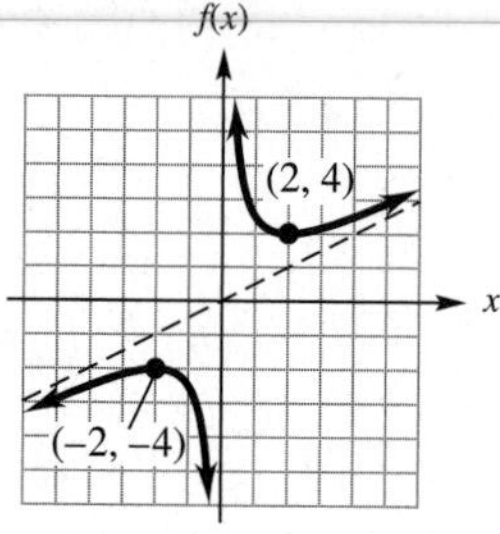

33. absolute maximum of $\frac{29}{4}$ at $x = \frac{5}{2}$; absolute minimum of 5 at $x = 1$ and $x = 4$ **35.** absolute maximum of 39 at $x = -3$; absolute minimum of $-\frac{319}{27}$ at $x = \frac{5}{3}$ **37.** 12° **39.** minimum in late 1991 ($x = 1.719$); maximum in early 1999 ($x = 9.136$) **41.** in late 1993 ($x = 13.71$) **43.** 225 m by 450 m **45.** 126 **47.** 225 m by 225 m

Case 11 (Page 749)

1. $Z'(m) = -\frac{C_1}{m^2} + \frac{DC_3}{2}$ **3.** $D = 3, t = 12, C_1 = 15{,}000, C_3 = 900 \quad m = \frac{10}{3}$

5. $Z(m^+) = Z(4) = \$11{,}400$
$Z(m^-) = Z(3) = \$11{,}300$

CHAPTER 13

Section 13.1 (Page 759)

1. They differ only by a constant. **5.** $5x^2 + C$ **7.** $\frac{8p^3}{3} + C$ **9.** $100x + C$ **11.** $\frac{5z^2}{2} - z + C$ **13.** $\frac{z^3}{3} - 2z^2 + 2z + C$ **15.** $\frac{x^4}{4} - \frac{14x^3}{3} + 10x^2 + 3x + C$ **17.** $4y^{3/2} + C$ **19.** $\frac{12t^{5/2}}{5} + 2t^{3/2} + C$ **21.** $\frac{112t^{3/2}}{3} + 4t^{9/2} + C$ **23.** $-\frac{12}{x^2} + C$ **25.** $-\frac{1}{y} - 4\sqrt{y} + C$ **27.** $-3x^{-2} + 4\ln|x| + C$ **29.** $\frac{4e^{3u}}{3} + C$ **31.** $-15e^{-.2x} + C$ **33.** $3\ln|x| - 8e^{-.5x} + C$

35. $\ln|t| + \frac{2t^3}{3} + C$ **37.** $\frac{e^{2u}}{2} + \frac{u^2}{8} + C$ **39.** $\frac{x^3}{3} + x^2 + x + C$

41. $\frac{6x^{7/6}}{7} + \frac{3x^{2/3}}{2} + C$ **43.** $f(x) = 2x^3 - 2x^2 + 3x + 1$
45. **(a)** $G(x) = 1.6x^3 - 8.2x^2 + 20.6x + 144$ **(b)** \$ 197.6 billion
47. **(a)** $R(x) = -.43x^4 + 14x^3 - 131x^2 + 917x + 1952$
(b) about \$12,252 million **49.** $P(x) = x^3 - 3x^2 + 4x - 40$
51. $C(x) = \frac{3x^{5/3}}{5} + 2x + \frac{114}{5}$ **53.** $C(x) = \frac{x^3}{15} + \frac{x^2}{5} + .2x + 6.8$
55. $C(x) = \frac{800}{e^{.05x}} + 100x + 76.96$ **57.** **(a)** $F(t) = 4.52\,e^{.0142t}$
(b) about 6.45 billion **59.** **(a)** $G(x) = 2.08 - .3 \ln x$ **(b)** about 1.46 million

Section 13.2 (Page 768)

3. $\frac{(12x-1)^3}{12} + C$ **5.** $-\frac{2}{3(3t+1)} + C$
7. $-\frac{1}{\sqrt{x^2+2x-4}} + C$ **9.** $\frac{2(r^3+3)^{3/2}}{9} + C$
11. $-\frac{3e^{5k}}{5} + C$ **13.** $\frac{2e^{2w^3}}{3} + C$ **15.** $\frac{e^{4t-t^2}}{2} + C$
17. $2e^{\sqrt{y}} + C$ **19.** $-\frac{4\ln|2+5x|}{5} + C$ **21.** $\frac{1}{2}\ln|e^{2t}+1| + C$
23. $-\frac{1}{8(2x^2+8x)^2} + C$ **25.** $\frac{\left(\frac{1}{r}+r\right)^2}{2} + C$
27. $(x^3+3x)^{1/3} + C$ **29.** $\frac{1}{6}\ln|3x^2+12x+8| + C$
31. $\frac{(x^2+1)^4}{4} + C$ **33.** $\frac{(x^2+12x)^{3/2}}{3} + C$
35. $\frac{(1+\ln x)^3}{3} + C$ **37.** $\frac{2(u-1)^{3/2}}{3} + 2(u-1)^{1/2} + C$
39. $\frac{2(5t-1)^{5/2}}{125} + \frac{2(5t-1)^{3/2}}{75} + C$
41. **(a)** $C(x) = 6\ln(5x^2+1) + 10$ **(b)** yes
43. **(a)** $G(x) = 22.5 - 2.41\ln(x+1)$ **(b)** about 15.4 deaths per 100,000 population **45.** **(a)** $G(x) = 18.5e^{.06x} + .1$
(b) about \$35.9 billion

Section 13.3 (Page 779)

1. about 151 million kWh **3.** a concentration of about 19 units
7. **(a)** 28 **(b)** 32 **(c)** 35.6 **9.** **(a)** 8 **(b)** 10 **(c)** 10.66
11. **(a)** 12.8 **(b)** 27.2 **(c)** 46.9629 **13.** **(a)** 3 **(b)** 3.5 **(c)** 4
15. 124.5833 **17.** 74.0439 **19.** 1.8719 **21.** \$7776
23. \$153,802 **25.** **(a)** $f(x) = 2000(.36x + 10) = 720x + 20,000$
(b)

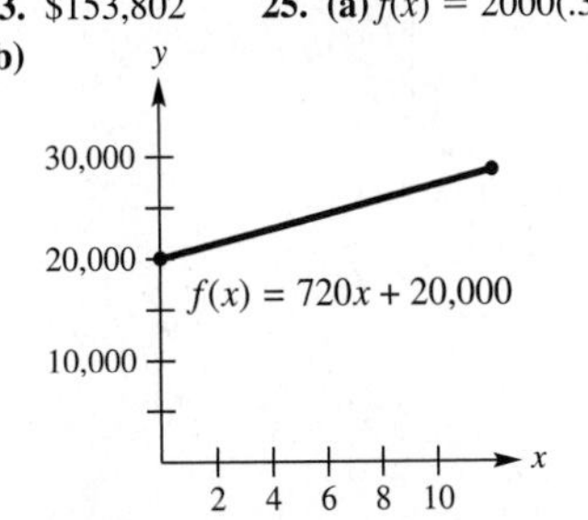

(c) \$228,800; less than the actual total **(d)** \$236,000
27. about 2400 ft **29.** **(a)** 9 ft **(b)** 2 sec **(c)** about 4.4 ft
(d) between 3 and 3.5 seconds

Section 13.4 (Page 790)

1. 52 **3.** 13 **5.** $\frac{1}{3}$ **7.** 76 **9.** $\frac{4}{3}$
11. $5\ln 5 + \frac{12}{25} \approx 8.527$
13. $20e^{-.2} - 20e^{-.3} + 3\ln 3 - 3\ln 2 \approx 2.775$
15. $\frac{e^{10}}{5} - \frac{e^5}{5} - \frac{1}{2} \approx 4375.1$ **17.** $\frac{91}{3} \approx 30.333$
19. $\frac{447}{7} \approx 63.857$ **21.** $\frac{(\ln 2)^2}{2} \approx .24023$ **23.** 49
25. $\frac{1}{4} - \frac{1}{3+e} \approx .075122$ **27.** $\frac{6}{7}(128 - 2^{7/3}) \approx 105.3946$
29. a loss for the first year and a half; $\int_0^2 (6x^2 - 7x - 3)\,dx$
31. $\frac{23}{3} \approx 7.666$ **33.** 9.5 **35.** $e^2 + e^{-1} - 3 \approx 4.757$
37. 1 **39.** $\frac{23}{6} \approx 3.833$ **41.** $\frac{e + e^3 - 2e^2}{2} \approx 4.01$ **43.** 19
45. **(a)** \$22,000 **(b)** \$62,000 **(c)** about 4.5 days
47. about \$78,113,600,000 **49.** about 8294.1952 quadrillion BTUs **51.** **(a)** about 414 barrels **(b)** about 191 barrels
(c) decreasing to 0 **53.** **(a)** about 274.14 million; the total number of people from age 0 to age 90 **(b)** about 70.8 million

Section 13.5 (Page 801)

1. \$280; \$11.67 **3.** **(a)** \$2686.5 billion **(b)** about March of 1997 ($x = 7.24$) **5.** \$2500; \$30,000 **7.** no **9.** 16/3
11. 1/12 **13.** 161.2 **15.** 4.999
17. **(a)** 8 yr **(b)** about \$148 **(c)** about \$771
19. **(a)** 39 days **(b)** \$3369.18 **(c)** \$484.02 **(d)** \$2885.16
21. **(a)** 5 yr **(b)** about 15.6 tons **23.** 13 yr
25. **(a)**

p
8
6
4
2
(5, 7)
$D(q) = -\frac{3}{5}q + 10$
$S(q) = \frac{7}{5}q$
2 4 6 8
q

(b) (5, 7) **(c)** \$7.50 **(d)** \$17.50 **27.** \$1999.54 **29.** \$40.50
31. **(a)**

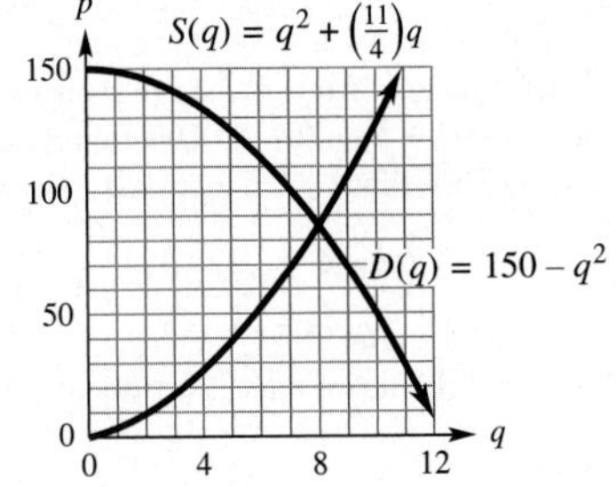

(b) (8, 86) **(c)** \$341.33 **(d)** \$429.33 **33.** 4/3

35. \$5.76 million **37. (a)** .019; the lower 10% of the income producers earn 1.9% of the total income of the population. **(b)** .275; the lower 50% of the income producers earn 27.5% of the total income of the population. **(c)** .819; the lower 90% of the income producers earn 81.9% of the total income of the population.
(d)

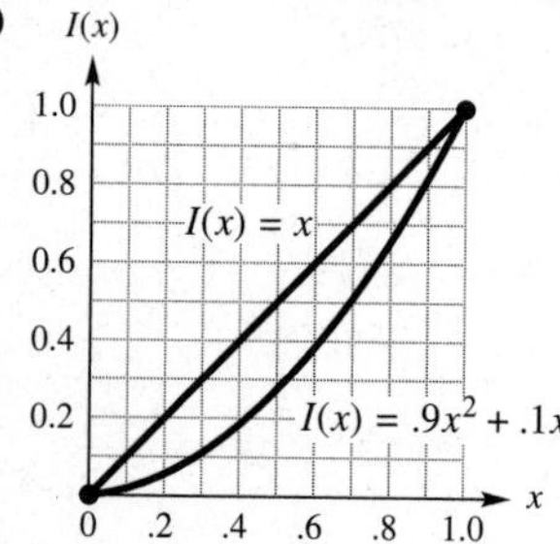

(e) .15; the amount of inequality of income distribution
39. (a) 10 items **(b)** 10 hr

Section 13.6 (Page 806)

1. $-4 \ln|x + \sqrt{x^2 + 36}| + C$ **3.** $\ln\left|\frac{x-3}{x+3}\right| + C, (x^2 > 9)$
5. $\frac{4}{3}\ln\left[\frac{3 + \sqrt{9 - x^2}}{x}\right] + C, (0 < x < 3)$
7. $-\frac{2x}{3} + 2\frac{\ln|3x + 1|}{9} + C$ **9.** $-\frac{2}{15}\ln\left|\frac{x}{3x - 5}\right| + C$
11. $\ln\left|\frac{2x - 1}{2x + 1}\right| + C, \left(x^2 > \frac{1}{4}\right)$ **13.** $-3\ln\left|\frac{1 + \sqrt{1 - 9x^2}}{3x}\right| + C,$
$\left(0 < x < \frac{1}{3}\right)$ **15.** $2x - 3\ln|2x + 3| + C$
17. $\frac{1}{25(5x - 1)} - \frac{\ln|5x - 1|}{25} + C$ **19.** $x^5\left[\frac{\ln|x|}{5} - \frac{1}{25}\right] + C$
21. $\frac{1}{x}(-\ln|x| - 1) + C$ **23.** $-\frac{xe^{-2x}}{2} - \frac{e^{-2x}}{4} + C$
25. about \$2094.71 **27.** about 15,100 microbes

Section 13.7 (Page 813)

1. $y = -x^2 + x^3 + C$ **3.** $y = \frac{3x^4}{8} + C$ **5.** $y^2 = x^2 + C$
7. $y = Me^{x^2}$ **9.** $y = Me^{(x^3 - x^2)}$ **11.** $y = Mx$
13. $y = Me^x + 5$ **15.** $-\frac{1}{e^x + C}$ **17.** $y = x^3 - x^2 + 2$
19. $y^2 = \frac{2\ln|x^3 + 28|}{3} + 6$ **21.** $y^2 = \frac{2x^3}{3} + 9$
23. $y = e^{x^2 + 3x}$ **25.** $\frac{y^2}{2} - 3y = x^2 + x - 4$
29. (a) 108.4 **(b)** 121.2 **(c)** no; if $x = 8$, the denominator becomes 0. **31. (a)** $y = 5t^2 + 50$ **(b)** 10 days from time $t = 0$
33. $y = 31.4e^{.009723t}$ **35. (a)** about 178.6 thousand **(b)** about 56.4 thousand **(c)** about 50.0 thousand **37.** $q = \frac{C}{p^2}$
39. (a) $k \approx .8$ **(b)** 11 **(c)** 55 **(d)** 2981
41. (a) $T = 68 + 30.6e^{-.33t}$ **(b)** about 83.8° **(c)** about 4.5 hr **(d)** about 24.3 hr

Chapter 13 Review (Page 816)

1. $\frac{x^3}{3} - \frac{3x^2}{2} + 2x + C$ **3.** $2x^{3/2} + C$ **5.** $\frac{2x^{3/2}}{3} + 9x^{1/3} + C$
7. $2x^{-2} + C$ **9.** $-\frac{3e^{2x}}{2} + C$ **11.** $2\ln|x - 1| + C$
13. $\frac{e^{3x^2}}{6} + C$ **15.** $\frac{3\ln|x^2 - 1|}{2} + C$ **17.** $\frac{-(x^3 + 5)^{-3}}{9} + C$
19. $\ln|2x^2 - 5x| + C$ **21.** $\frac{-e^{-3x^4}}{12} + C$ **23.** $\frac{2e^{-5x}}{5} + C$
27. (a) 140 **(b)** 148.8 **29.** 28 **31.** $-1/12$
33. $\frac{72}{25} \approx 2.88$ **35.** $2\ln 3$ or $\ln 9 \approx 2.1972$
37. $2e^4 - 2 \approx 107.1963$ **39.** $\frac{2}{3}(22^{3/2} - 2^{3/2}) \approx 66.907$
41. $e^2 - 1 \approx 6.3891$ **43.** $C(x) = 10x - x^2 + 4$
45. $C(x) = (2x - 1)^{3/2} + 145$ **47.** 36,000 **49.** about 26.3 yr
51. about 17,500 **53.** yes; about 23,884,742,540
57. about 142 **59.** $-\ln\left|\frac{5 + \sqrt{25 + x^2}}{x}\right| + C$
61. $15\left[\frac{x}{2} + \frac{5\ln|2x - 5|}{4}\right] + C$ **63.** $y = \frac{x^4}{2} + 3x^2 + C$
65. $y^2 = 3x^2 + 2x + C$ **67.** $y = -5e^{-x} - 5x + 22$
69. $y = 2e^{5x - x^2}$ **71. (a)** $\frac{dA}{dt} = .05A - 1000$ **(b)** \$9487.29
73. (a) about 111 items **(b)** not exactly, but for practical purposes, yes **75.** $.2\ln y - .5y = -.3\ln x + .4x + C$; $x = \frac{3}{4}$ unit; $y = \frac{2}{5}$ unit

Case 12 (Page 820)

1. Using a common denominator of $P(C - P)$, $\frac{1/C}{P} + \frac{1/C}{C - P} =$
$\frac{(1/C)(C - P)}{P(C - P)} + \frac{(1/C)P}{P(C - P)} = \frac{(1/C)C - (1/C)P + (1/C)P}{P(C - P)}$
$= \frac{1}{P(C - P)}$
3. $P(t) = \frac{60e^{.022t}}{7 + 3e^{.022t}} = 18$ in $t \approx 138$ years

CHAPTER 14

Section 14.1 (Page 830)

1. 4; −22; −20; 12 **3.** 3; $\sqrt{33}$; $\sqrt{17}$; 8 **5. (a)** 92 **(b)** −17 **(c)** 47 **(d)** −44
9.

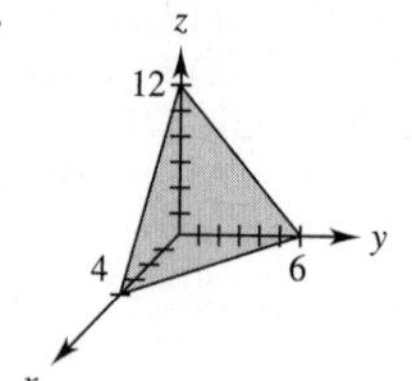

11.

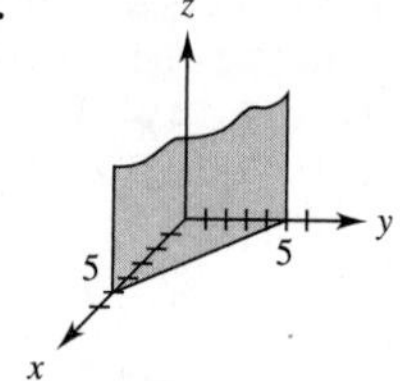

13.

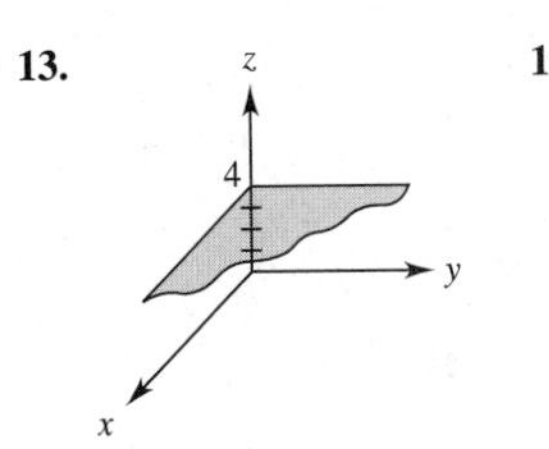

15. $3x + 2y + z = 18$

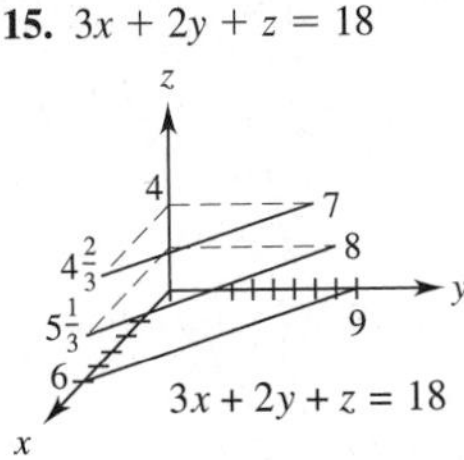

17. $y^2 - x = -z$

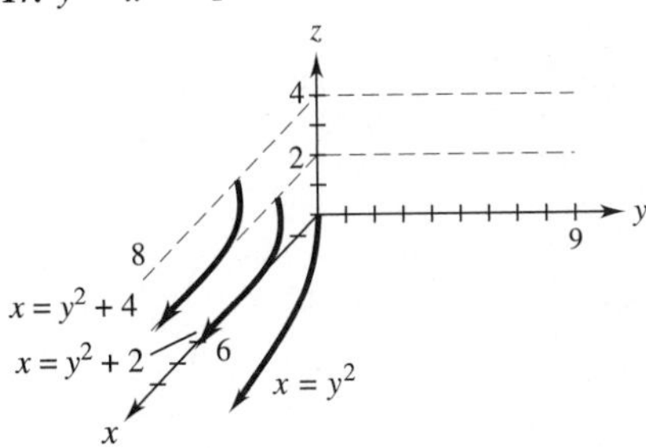

19. $y = 500^{10/3}x^{-7/3}$

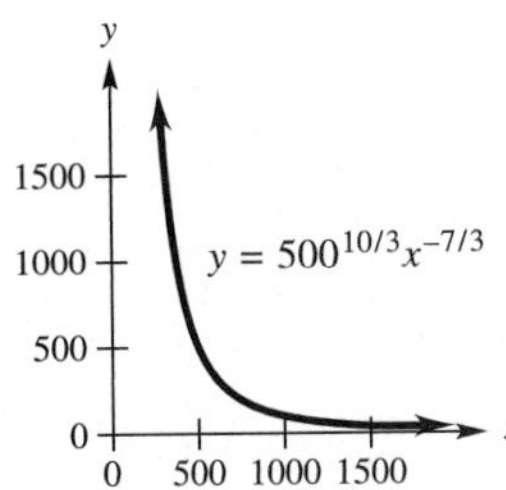

21. 1.123; the IRA account grows faster. **23. (a)** 1.8892 **(b)** 1.5198 **(c)** 1.7278 **25. (a)** 1.5 m/sec; 5.5 m/sec **(b)** 1 m/sec **27.** $f(L,W,H) = L + 2H + 2W$ **29. (c)** **31. (e)** **33. (b)**

Section 14.2 (Page 839)

1. (a) $24x^2 - 8xy$ **(b)** $-4x^2 + 18y$ **(c)** 48 **(d)** -40 **3.** $f_x = -2xy + 12x^3$; $f_y = -x^2$; 100; -16 **5.** $f_x = 2e^{2x+y}$; $f_y = e^{2x+y}$; $2e^3$; e^{-5} **7.** $f_x = \frac{2}{e^{x+2y}}$; $f_y = \frac{4}{e^{x+2y}}$; 2; $\frac{4}{e^2}$

9. $f_x = \frac{y^3 - x^2 - 6xy^2}{(x^2 + y^3)^2}$; $f_y = \frac{6x^2y - 3xy^2 - 3y^4}{(x^2 + y^3)^2}$; $-\frac{17}{9}$; $\frac{153}{1849} \approx .083$

11. $f_x = \frac{2 - 2xy}{2x - x^2y}$; $f_y = -\frac{x}{2 - xy}$; $\frac{3}{4}$; $\frac{2}{7}$

13. $f_x = 2x^2ye^{2xy} + 2xe^{2xy}$; $f_y = 2x^3e^{2xy}$; $-4e^{-4}$; $-128e^{-24}$

15. $f_{xx} = 20y^3 - 30x$; $f_{xy} = f_{yx} = 60xy^2$; $f_{yy} = 60x^2y$

17. $h_{xx} = -8y^2$; $h_{xy} = h_{yx} = -16xy + 14y$; $h_{yy} = -6 - 8x^2 + 14x$

19. $R_{xx} = \frac{24y}{(2x + y)^3}$; $R_{xy} = R_{yx} = \frac{-12x + 6y}{(2x + y)^3}$; $R_{yy} = \frac{-12x}{(2x + y)^3}$

21. $z_{xx} = 0$; $z_{yy} = 4xe^y$; $z_{xy} = z_{yx} = 4e^y$

23. $r_{xx} = -\frac{1}{(x + y)^2}$; $r_{yy} = -\frac{1}{(x + y)^2}$; $r_{xy} = r_{yx} = -\frac{1}{(x + y)^2}$

25. $z_{xx} = \frac{1}{x}$; $z_{yy} = -\frac{x}{y^2}$; $z_{xy} = z_{yx} = \frac{1}{y}$ **27.** 1; $-\frac{1}{4}$

29. $x = -4, y = 2$ **31.** $x = 0, y = 0$; or $x = 3, y = 3$

33. $f_x = 2x$; $f_y = z$; $f_z = y + 4z^3$; $f_{yz} = 1$; 3; 1

35. $f_x = \frac{6}{4z + 5}$; $f_y = -\frac{5}{4z + 5}$; $f_z = -\frac{4(6x - 5y)}{(4z + 5)^2}$; $f_{yz} = \frac{20}{(4z + 5)^2}$

37. $f_x = \frac{2x - 5z^2}{x^2 - 5xz^2 + y^4}$; $f_y = \frac{4y^3}{x^2 - 5xz^2 + y^4}$; $f_z = -\frac{10xz}{x^2 - 5xz^2 + y^4}$; $f_{yz} = \frac{40xy^3z}{(x^2 - 5xz^2 + y^4)^2}$ **41. (a)** \$206,800 **(b)** $f_p = 132 - 2i - .02p$; $f_i = -2p$; the rate at which sales revenue is changing per unit of change in price (f_p) or interest rate (f_i) **(c)** a sales revenue drop of \$18,800 **43. (a)** 46.656 **(b)** $f_x(27, 64) = .6912$ is the rate at which production is changing when labor changes by 1 unit from 27 to 28 and capital remains constant; $f_y(27, 64) = .4374$ and is the rate at which production is changing when capital changes by 1 unit from 64 to 65 and labor remains constant. **(c)** Production would be increased by about 69 units when labor is increased by 1 unit.

45. $.7x^{-.3}y^{.3}$, or $\frac{.7y^{.3}}{x^{.3}}$; $.3x^{.7}y^{-.7}$, or $\frac{.3x^{.7}}{y^{.7}}$ **47. (a)** about 5.71 **(b)** about 6.84 **(c)** $\frac{1}{a - v}$ **(d)** $\frac{b}{(a - v)^2}$

49. (a) $F_m = \frac{gR^2}{r^2}$; the rate of change in force per unit change in mass; $F_r = \frac{-2mgR^2}{r^3}$; the rate of change in force per unit change in distance

51. (a) 28.24 **(b)** $\frac{\partial B}{\partial w} = \frac{703}{h^2}$; $\frac{\partial B}{\partial h} = \frac{-1406w}{h^3}$

Section 14.3 (Page 848)

3. local minimum of -22 at $(3, -1)$ **5.** local minimum of 0 at $(-2, -2)$ **7.** saddle point at $(-3, 4)$ **9.** local maximum of 17 at $\left(\frac{2}{3}, \frac{4}{3}\right)$ **11.** saddle point at $(2, -2)$ **13.** saddle point at $(1, 2)$ **15.** saddle point at $(0, 0)$; local minimum of -201 at $(6, 18)$ **17.** saddle point at $(0, 0)$; local minimum of -5157 at $(72, 12)$ **19.** saddle point at $(0, 0)$ **21.** local maximum of $\frac{9}{8}$ at $(-1, 1)$; saddle point at $(0, 0)$; (a) **23.** local minima of $-\frac{33}{16}$ at $(0, 1)$ and $(0, -1)$; saddle point at $(0, 0)$; (b) **25.** local maxima of $\frac{17}{16}$ at $(1, 0)$ and $(-1, 0)$; local minima of $-\frac{15}{16}$ at $(0, 1)$ and $(0, -1)$; saddle points at $(0, 0)$, $(-1, 1)$, $(1, -1)$, $(1, 1)$, and $(-1, -1)$; (e) **27.** $P(12, 40) = 2744$ **29.** $P(12, 72) = \$2,528,000$ **31.** $C(12, 25) = 2237$ **33.** 3 m by 3 m by 3 m **35.** 36 in. (length) by 18 in. by 18 in. **37.** 1000 tons of grade A ore and 2000 tons of grade B ore for maximum profit of \$1400

Chapter 14 Review (Page 852)

1. 32; 156 **3.** $-\frac{6}{5}$, $-\frac{8}{3}$ **7.**

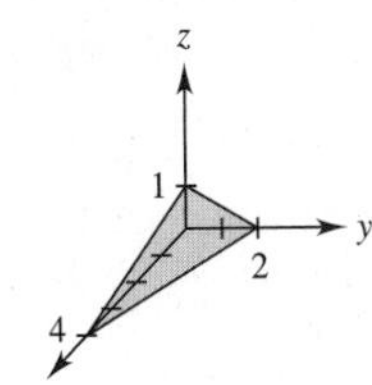

9.

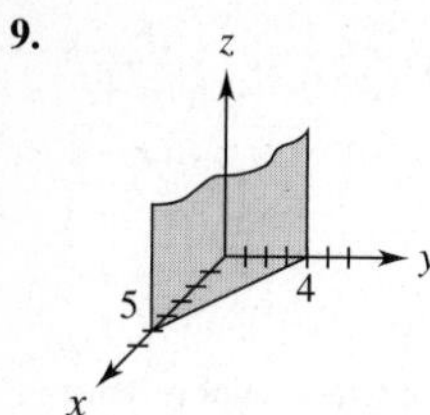

11. (a) $-4x + 5y$ (b) 3 (c) 5
15. $f_x = 12x^2y + 10y^4$; $f_y = 4x^3 + 40xy^3$
17. $f_x = \dfrac{12y - 3x^2 + 4xy^2}{(x^2 + 4y)^2}$; $f_y = \dfrac{-4x^2y - 8y^2 - 12x}{(x^2 + 4y)^2}$
19. $f_x = 2(y + 1)^2e^{2x+y}$; $f_y = (y + 1)(y + 3)e^{2x+y}$
21. $f_x = \dfrac{3x^2y^2}{1 + x^3y^2}$; $f_y = \dfrac{2x^3y}{1 + x^3y^2}$ 23. $f_{xx} = 24xy^2$; $f_{xy} = 24x^2y - 8$
25. $f_{xx} = \dfrac{8y}{(x - 2y)^3}$; $f_{xy} = \dfrac{-4x - 8y}{(x - 2y)^3}$ 27. $f_{xx} = 2e^y$; $f_{xy} = 2xe^y$
29. $f_{xx} = \dfrac{-2x^2y^2 - 4y}{(2 - x^2y)^2}$; $f_{xy} = -\dfrac{4x}{(2 - x^2y)^2}$
31. local minimum at (0, 1) 33. saddle point at (0, 2)
35. local minimum at $\left(1, \dfrac{1}{2}\right)$; saddle point at $\left(-\dfrac{1}{3}, \dfrac{11}{6}\right)$
37. $F(L, W, H) = 2LW + 2WH + 2LH$ 39. (a) local minimum at (11, 12) (b) $431

Case 13 (Page 854)

1. At $m = .0062$, $b = -3.46$, we have $f_m = f_b = 0$, $f_{mm} = 22{,}825{,}000 > 0$, $f_{mm}f_{bb} - [f_{mb}]^2 = 60{,}000 > 0$, so f has a minimum. 3. $m = .00532$, $b = -1.794$

Index of Applications

Business

Finance

Geometry

Health

Natural Science

Physical Science

Social Science

Transportation

Index

C

D

E

F

N

The **derivative** of the function f is the function denoted f' whose value at the number x is defined to be the number

$$f'(x) = \lim_{h \to 0} \frac{f(x+h) - f(x)}{h},$$

provided this limit exists.

RULES FOR DERIVATIVES

Assume all indicated derivatives exist.

Constant Function If $f(x) = k$, where k is any real number, then

$$f'(x) = 0.$$

Power Rule If $f(x) = x^n$, for any real number n, then

$$f'(x) = n \cdot x^{n-1}.$$

Constant Times a Function Let k be a real number. Then the derivative of $y = k \cdot f(x)$ is

$$y' = k \cdot f'(x).$$

Sum or Difference Rule If $y = f(x) \pm g(x)$, then

$$y' = f'(x) \pm g'(x).$$

Product Rule If $f(x) = g(x) \cdot k(x)$, then

$$f'(x) = g(x) \cdot k'(x) + k(x) \cdot g'(x).$$

Quotient Rule If $f(x) = \frac{g(x)}{k(x)}$, and $k(x) \neq 0$, then

$$f'(x) = \frac{k(x) \cdot g'(x) - g(x) \cdot k'(x)}{[k(x)]^2}.$$

Chain Rule Let $y = f[g(x)]$. Then

$$y' = f'[g(x)] \cdot g'(x).$$

Chain Rule (Alternative Form) If y is a function of u, say $y = f(u)$, and if u is a function of x, say $u = g(x)$, then $y = f[g(x)]$, and

$$\frac{dy}{dx} = \frac{dy}{du} \cdot \frac{du}{dx}.$$

Generalized Power Rule Let u be a function of x, and let $y = u^n$ for any real number n. Then

$$y' = n \cdot u^{n-1} \cdot u'.$$

Exponential Function If $y = e^{g(x)}$, then

$$y' = g'(x) \cdot e^{g(x)}.$$

Natural Logarithmic Function If $y = \ln |g(x)|$, then

$$y' = \frac{g'(x)}{g(x)}.$$